THE CLASSIC TEXT REISSUED

BASIC PHYSICS

A RESOURCE FOR PHYSICS TEACHERS

THE CLASSIC TEXT REISSUED

BASIC PHYSICS

A RESOURCE FOR PHYSICS TEACHERS

KENNETH W. FORD

World Scientific

NEW JERSEY • LONDON • SINGAPORE • BEIJING • SHANGHAI • HONG KONG • TAIPEI • CHENNAI • TOKYO

Published by

World Scientific Publishing Co. Pte. Ltd.
5 Toh Tuck Link, Singapore 596224
USA office: 27 Warren Street, Suite 401-402, Hackensack, NJ 07601
UK office: 57 Shelton Street, Covent Garden, London WC2H 9HE

British Library Cataloguing-in-Publication Data
A catalogue record for this book is available from the British Library.

Design and layout by Adam B. Ford. The body copy is set in Noto Serif. The text accompanying the figures and the new introduction text are set in Noto Sans. The Noto family of fonts was designed by Goole, Inc..
Chapter heading type is Eurostile T by URW Software.

Cover photo courtesy of NASA. This photograph of the eclipse of the sun was taken with a 16mm motion picture camera from the Apollo 12 spacecraft during its trans-Earth journey home from the moon.

BASIC PHYSICS

This Work was originally published by Blaisdell Publishing Company,
a Division of Ginn and Company, in 1968.

This 2016 edition is published by World Scientific Publishing Company Pte. Ltd. by arrangement with Kenneth W. Ford.

ISBN 978-981-3208-00-1
ISBN 978-981-3208-01-8 (pbk)

Desk Editor: Christopher Teo

Printed in Singapore

TO *Star, Jason, Adam, Caroline, Nina, Sarah, and Paul*

INTRODUCTION

This is not a new or revised or updated edition of *Basic Physics.* What's included—although redesigned and with a few words changed here and there—is the unvarnished original, dating from 1968. This means that you will encounter cgs units and an embarrassment of sexist language. I actually chose cgs units—at a time when many textbooks were transitioning to SI units—for pedagogical reasons. Expressed in these units, the equations of electricity and magnetism are simpler in appearance and more obviously symmetric. As to the sexist language, I ask the reader to remember that in 1968, "he" meant "he or she" and "man" meant "person" or "humankind." At the time, one's ear did not react as it does now.

Why reissue a book that is nearly fifty years old? I am offering it again (with the help of World Scientific) because it contains numerous discussions that remain relevant to 21st-century physics teaching—discussions that I like to think can help teachers gain new insights and deeper understanding of many topics in both classical and modern physics. Over the years, some teachers have told me that they valued *Basic Physics* as a resource, and a few suggested that it be made available again. I set about re-reading it, and surprised myself to find how little of it is really dated. True, neutrinos are said to be massless, and quarks and gluons are not mentioned, but throughout the book, from Kepler to Feynman, almost all of the discussions link to the way physics is still taught.

If you, as a physics teacher, want to use this book for reference, you will very likely be more interested in discussions of laws and principles—more interested in my efforts to get at underlying meaning—than in specific derivations. With this thought in mind, I have identified 174 passages that I call "Features." These range in length from less than a page to several pages. The Features are set forth in a separate table of contents at the beginning of the book and are indexed in a separate index at the back.

I do include among the Features a few derivations that deviate from standard textbook presentations. One of these is my treatment of the Bohr atom, which follows Bohr's original path, not the later common path of quantizing angular momentum for electrons in circular orbits. I also echo Einstein's original derivation of $E = mc^2$.

Needless to say, I will be pleased if some teachers find it rewarding to browse beyond the Features.

Bringing this book back to life involved scanning and digitizing an old copy. This gave us a chance to correct many typos in the original and may well have introduced as many or more new typos. Adam Ford has worked diligently to redesign the book and make it as "clean" as possible. I have assisted him in this effort. I am indebted to Adam and also to Chad Hollingsworth of World Scientific, who saw potential in the project and shepherded it through the publication process.

Ken Ford

An Answer Manual for *Basic Physics* is available.
For free access and download, please visit

hbarpress.com/answers

FEATURES CONTENTS

PART 4 THERMODYNAMICS

PART 5 ELECTROMAGNETISM

PART 6 RELATIVITY

EPILOGUE

WHAT FOLLOWS IS FROM THE ORIGINAL BOOK

BASIC PHYSICS

KENNETH W. FORD

Original Preface

This is a textbook for a one-year introductory course in physics. It is intended especially for the many students whose mode of reasoning is more often verbal than mathematical. I have tried to use equations only where they are essential, and words where they suffice. Despite this limitation, the book is not intended to be superficial or simple. I believe that mathematically adept students may also find it challenging, and I hope that all students who use it will find it, at least part of the time, exciting.

The book is organized in the main about the great theories of physics: mechanics, thermodynamics, electromagnetism, relativity, and quantum mechanics. The organization, the choice of individual topics, and the various recurring themes have all been governed by one idea: to help the student come to understand physics as a human activity dedicated to organizing a certain part of experience in the simplest, most economical, most general, and most satisfying way. I have tried to keep the reader's attention focused on fundamental questions. Whenever possible, examples are chosen which illustrate the fundamentals in a direct way. I believe that the book can be read at several levels. It contains subtleties enough for the best students, but these should not stand in the way of basic understanding for the student who wants to skip the subtleties.

Much of my book, *The World of Elementary Particles,* has found its way into this book (especially in Chapters Two, Three, Four, Twenty-Three, Twenty-Six, and Twenty-Seven). I do not apologize for this cannibalism, since that was the intended fate of the elementary particle book when it was first written. Instructors who prefer less emphasis on the submicroscopic world may treat Part 1 lightly. However, I would recommend that students be asked to read through this part, even if it does not receive much class time. Particles are used from time to time throughout the book in illustrative examples. Conservation laws, treated first in Chapter Four, also recur often. Although applications are often modern, classical physics is not neglected. It occupies more than half of the book.

I have avoided the division of exercises into "questions" and "problems." Those requiring calculation or algebraic manipulation are intermixed with those which do not. Often calculation and commentary are mixed in one exercise. Extremely straightforward exercises, requiring only substitution in an equation or repetition of material in the chapter, are not very numerous. Instructors who feel that their students need more drill may easily add such exercises. Exercises that are not too difficult or special are arranged at the end of a chapter in the same order as the material in the chapter. After these are placed exercises that are more general, or more challenging, or somewhat peripheral, or require reference to other books. Occasionally some new material is introduced in exercises, such as the Lorentz transformation for energy and momentum in Chapter Twenty-One. Some exercises refer to other exercises, but always without pyramiding, so that the second can be answered without the first. In the exercises as in the text, I have tried to avoid difficulties of a purely mathematical character.

Although logic more than history dictates the sequence of topics in this book, certain ideas are presented in a historical context, particularly when the history helps to illuminate the ideas or seems likely to make them more memorable. Although the historical material is mostly from secondary sources, and certainly does not represent the results of serious historical research on my part, I hope that I have succeeded in avoiding much of the imaginary or folklore history that finds its way into some textbooks. My treatment of the Bohr atom, although not as "neat" as the usual angular momentum quantization in circular orbits, is closer to Bohr's actual treatment and is perhaps pedagogically superior, for none of Bohr's basic ideas as used in this approach need to be discarded later.

The system of units employed is the cgs system, supplemented by some practical (mks) units of electricity in discussions of capacitance, inductance, and simple circuits. There are two principal reasons for this choice. First, for an elementary presentation of the fundamentals of electromagnetic theory, the cgs system has certain pedagogic advantages. It preserves the symmetry between electricity and magnetism that is inherent in nature, and it restricts the number of new proportionality constants to one, the speed of light. Second, for students using this book, further work in chemistry or biology (where the cgs-plus-practical system remains nearly universal), is more likely than further work in physics, where the mks system is predominant.

Some instructors will find this book wordy. It is long, and the ratio of words to equations is high. However, the density of new ideas is correspondingly less than in more concise books. A reading assignment of 40 pages need not be more onerous than one of 20 pages, if the longer assignment holds the student's interest. Only experience in class use will reveal to what extent I have succeeded in making the book readable.

I owe a debt of gratitude to the many interested students at Brandeis University who inspired me to start this book and to the many more at Irvine who inspired me to finish it. Numerous colleagues also contributed in a variety of ways. I am indebted above all to Neal D. Newby, Jr., who read the manuscript and fashioned many of the exercises, and to Brenton Stearns, whose careful criticism has always been helpful. Many useful comments came from Alfred M. Bork and Richard L. Jacob, who read much of the manuscript. Other important help was provided by Richard J. Drachman, Gerald Holton, J. Carson Mark, Thomas E. Stark, John A. Wheeler, and Victor F. Weisskopf. Mrs. Barbara MacDonald and Mrs. June B. Marks rendered superb service as typists, and the staff of Blaisdell Publishing Company proved unwaveringly competent and agreeable. For encouragement and patience, both in large measure, I am grateful to my wife Joanne.

Kenneth W. Ford

Original Contents

PART 3 MECHANICS

PART 5 ELECTROMAGNETISM

Original Acknowledgments

Grateful acknowledgment is made to the following sources for the quotations and photographs appearing on the title page and part title pages.

Title page and cover photograph of elementary particle production and decay in a bubble chamber. Courtesy of the Lawrence Radiation Laboratory, University of California, Berkeley.

Prologue: quotation from Richard P. Feynman, "The Value of Science," in *Frontiers in Science—A Survey,* Edward Hutchings, Jr., ed. (New York: Basic Books, Inc., 1958). Photograph courtesy of the American Institute of Physics.

Part 1: quotation from Enrico Fermi, "Attempt at a Theory of Beta-Radiation," *Zeitschrift für Physik,* 88, 161-171 (1934). (Original in German.) Photograph courtesy of the American Institute of Physics.

Part 2: quotation from Karl Friedrich Gauss, in Eric Temple Bell, *Mathematics, Queen and Servant of Science* (New York: McGraw-Hill Book Company, 1951). Photograph courtesy of The Bettman Archive, Inc.

Part 3: quotation from Isaac Newton, *Mathematical Principles of Natural Philosophy* (1687); translated from the Latin by Andrew Motte (1729); revised translation by Florian Cajori (Berkeley, California: University of California Press, 1960). Photograph courtesy of the American Institute of Physics.

Part 4: quotation from James P. Joule, "On Matter, Living Force, and Heat," lectures printed in the *Manchester Courier,* May 5 and 12, 1847; reprinted in F. R. Moulton and J. J. Schifferes, *The Autobiography of Science* (Garden City, New York: Doubleday and Co., Inc., 1960). Photograph courtesy of The Bettman Archive, Inc.

Part 5: quotation from James Clerk Maxwell, "A Dynamical Theory of the Electromagnetic Field," *Philosophical Transactions, 155,* 459 (1865). Photograph courtesy of the American Institute of Physics.

Part 6: quotations from Albert Einstein, "On the Electrodynamics of Moving Bodies," *Annalen der Physik, 17,* 891 (1905); English translation in Morris H. Shamos, *Great Experiments in Physics* (New York: Holt, Rinehart & Winston, Inc., 1959); and Albert Einstein and Leopold Infeld, *The Evolution of Physics* (New York: Simon and Schuster, Inc., 1961). Photograph courtesy of the American Institute of Physics.

Part 7: quotations from Niels Bohr, "Unity of Knowledge," in *Atomic Physics and Human Knowledge* (New York: John Wiley & Sons, Inc., 1958); and Werner Heisenberg, *Physics and Philosophy* (New York: Harper & Bros., Publishers, 1958). Photographs courtesy of the American Institute of Physics.

Epilogue: quotation from Bertrand Russell, *The ABC of Relativity* (New York: The New American Library of World Literature, Inc., 1959). Photograph courtesy of Robert Karplus and Science Curriculum Improvement Study.

Original Notes on the text

The number of chapters in this book is 28, about the same as the number of weeks of instruction in a typical course. However, the chapters vary greatly in length and difficulty. Some will require more than a week of class time, some less. Some particularly full chapters are Eight, Twelve, Thirteen, Eighteen, Twenty-Four, and Twenty-Five. Some that can be covered most quickly are One, Three, Five, Nineteen, and Twenty-Eight. It should be possible to cover most of the material in the book in a one-year course, although certain sections or even chapters can be omitted without difficulty.

Classical mechanics, despite its traditional place at the beginning of a physics course, is not an easy subject. It is approached gradually in this book in a series of introductory chapters. An instructor may move more or less rapidly through the first seven chapters, depending on his taste. It is to be noted that these early chapters do contain certain material on mechanics: basic concepts, conservation laws, vectors, and some kinematics. The early chapters also establish themes that recur throughout the book.

Prologue. It is more meaningful to discuss science after it is studied than before, but students bring enough knowledge and prejudice to the study of physics to make a general introductory chapter appropriate. The student might profitably reread this chapter at the end of the course to see if its meaning or impact is any different the second time.

Part 1. Chapter Two, a survey of the elementary particles, gives the student an immediate view of a research frontier and also establishes familiarity with simple entities useful for many later illustrations and examples. Concepts employed in this chapter, such as baryon number, although less familiar than force or mass to older physicists, are in essence simpler than most mechanical concepts. Sections 2.6 and 2.7 can be omitted or postponed.

A number of basic concepts are introduced and discussed in semiquantitative terms in Chapter Three. Most are redefined more fully in later chapters. In the meantime the student can understand and use such ideas as energy, charge, and spin at a certain level. Section 3.9 is optional.

Chapter Four deals with seven absolute conservation laws at an early level of sophistication, with deeper levels to follow. Sections 4.8 and 4.9 provide qualitative insight into the link between conservation laws and symmetry principles.

Part 2. Chapter Five is concerned with the nature of mathematics and the difference between mathematics and science, while Chapters Six and Seven are concerned with more practical goals—the development of certain mathematical tools. Uniform motion in a circle is treated in Chapter Six; one-dimensional kinematics in Chapter Seven. Section 5.5 on the history of mathematics is optional but easy reading. Questions of dimensions can sometimes be difficult without being very rewarding. Section 7.3 contains dimensional subtleties which can be treated as optional.

Part 3. Newton's first and second laws appear in Chapter Eight, where more than average attention is given to the meaning of the basic concepts and to the content of the first law. Newton's third law appears in Chapter Nine, where it is linked to the conservation of momentum. The question of the homogeneity of space, discussed in Chapter Four, recurs in Section 9.9.

I have taken the view that angular momentum is basically a simpler concept than that of energy, and have arranged Chapters Ten and Eleven accordingly. The law of areas appears in Chapter Ten as a manifestation of angular momentum conservation and recurs in Chapter Twelve as Kepler's second law. Levers and pulleys find a place in Section 11.6. Section

11.8 is a rather difficult discussion of mass and energy which could be treated as optional or postponed until Chapter Twenty-One.

Chapter Twelve, although lengthy, is no meatier than the shorter preceding chapters. Much of the first five sections is historical. Gravitational potential energy in the large is introduced first in Section 12.9.

Part 4. Chapter Thirteen is a long chapter which includes the zeroth and the first laws of thermodynamics; heat and temperature; the ideal gas law; and kinetic theory in an elementary form. The viewpoint is generally microscopic, but the links between microscopic and macroscopic descriptions are also emphasized. Heat and internal energy are carefully distinguished.

Chapter Fourteen approaches entropy and the second law of thermodynamics from the microscopic side, with enough detail on probability to make this approach meaningful. Instructors who prefer to emphasize the macroscopic approach might find the material in this chapter somewhat unbalanced. A final section on the arrow of time ties in with the discussion of time-reversal invariance in Chapter Twenty-Seven.

Part 5. Without calculus and without many weeks, it is impossible to develop electromagnetic theory to the same quantitative level as can be reached in mechanics. I have included in this part most of the basic laws and ideas of electromagnetism, but only a circumscribed set of them are developed mathematically. The rest are qualitative. I hope this will make it easier for the student to comprehend the whole subject without bogging down in details. As in earlier parts, the microscopic viewpoint is stressed.

In Chapter Fifteen, magnetism is introduced in an old-fashioned way, with bar magnets and magnetic poles. This makes it possible to emphasize the difference between electric and magnetic fields before their connection is established. Section 15.4, on charge in nature, is a detour off the main track of the chapter.

In Chapter Sixteen, all of the fundamental laws of electromagnetism are surveyed in the first section, and then are dealt with one at a time in later sections. Potential is postponed to Chapter Seventeen, because it is not essential to any of the fundamentals of the two preceding chapters, but is indispensable in treating circuits. The mks units of resistance, capacitance, inductance, potential, and current find a place in Chapter Seventeen. Otherwise only cgs units are used.

Chapter Eighteen could have been divided equally well into two chapters. Sections 18.1 to 18.4 are concerned with electromagnetic radiation. The remaining sections (except the last) deal with wave phenomena in general. Geometrical optics is somewhat curtailed, being restricted to Section 18.10. The last section, on line spectra, provides a link to Part 7.

Part 6. These chapters are more concise than most of the rest. This brevity, coupled with the difficulty of the new ideas, means that the rate of progress in pages per week will be considerably less than in earlier parts.

Chapter Nineteen is introductory. For pedagogical reasons, I have chosen to give prominence to the Michelson-Morley experiment, despite evidence that it did not have much influence on Einstein's thinking. The Lorentz transformation and the properties of space and time are found in Chapter Twenty, momentum and energy in Chapter Twenty-One. Dynamics does not go beyond collisions and decay processes in one dimension.

The equivalence principle is used as a basis for understanding important ideas of general relativity in Chapter Twenty-Two. This chapter could be considered optional.

Part 7. Because it is impossible to organize a mathematical and deductive treatment of quantum mechanics at this level, I have chosen a framework of key ideas. They are discussed and illustrated in Chapter Twenty-Three, then elaborated further and applied to atoms, nuclei, and elementary particles in the remaining chapters of this part. These remaining chapters contain, of course, many additional facts about the structure of matter, but no effort is made to pursue the subject of chemistry.

Remarks about the Bohr atom treatment in Chapter Twenty-Four appear in the preface. It differs from the conventional textbook treatment* in several ways, notably by using the correspondence principle in an essential way (as Bohr did), and in setting as the goal the derivation of a formula for the Rydberg constant, not the derivation of the whole Balmer formula (again following Bohr). In a final section on lasers and stimulated emission, Chapter Twenty-Four circles back to photons and radiation, where it began.

In Chapter Twenty-Five, radioactive decay chains and the complicated early history of nuclear physics are postponed to Section 25.8. Section 25.3 on pions provides a link with the next two chapters. The subjects of vital practical concern, fission and fusion, are discussed in some detail in Sections 25.9 and 25.10.

Chapters Twenty-Six and Twenty-Seven are optional, so far as the development of the key ideas of quantum mechanics is concerned, but they bring the subject up to its contemporary frontier, and round out the topic opened up in Chapter Two. It did not seem practical to include material on the other quantum frontier, the many-body frontier of condensed matter and low temperature.

Epilogue. This brief chapter calls on the student to think about the significance and the future of the subject he or she has been studying.

* Not all textbooks are conventional, of course. See, for instance, Arnold B. Arons, Development of Concepts of Physics (Reading, Massachusetts: Addison-Wesley Publishing Co., Inc., 1965).

Prologue

RICHARD PHILLIPS FEYNMAN
(1918 – 1988)

"I would like not to underestimate the value of the world view which is the result of scientific effort. We have been led to imagine all sorts of things infinitely more marvelous than the imaginings of poets and dreamers of the past."

ONE

The scientist's view of the world

Four hundred years ago, most men believed that they lived on a stationary earth at the center of the universe. The world beyond the solar system was a mystery. The submicroscopic domain of atoms and molecules was a complete unknown. Even the immediate environment impinging on man's senses was largely not understood, or else misunderstood. Except for the simplest facts about the balancing of static forces, not a single law of nature governing man's own world was accurately formulated. The Copernican theory of the solar system, which places the sun at the center, had been published, but it had few adherents and many powerful opponents. There was no science-based technology. There was scarcely any activity that we would today call science. Mathematics was in its infancy.

Now, four hundred years later—the mere blink of an eye in the lifetime of the human race—man, surveying the panorama of nature from elementary particles to galaxies, has reason to stand in awe of his own achievements. In terms of a remarkably few fundamental theories of nature, his understanding spans a vast scale of sizes in the universe. He sees, woven into the rich complexity of the world of his senses, simple patterns. In domains far beyond the range of his sense perceptions, he has discovered surprisingly different but equally simple patterns. As a human activity, science has become a large enterprise. The technology spawned by science touches every facet of life. Standing on the shoulders of physical science, the biological and medical sciences have grown to powerful stature. In parallel with science, mathematics has matured.

In all of human history, there has been no more stunning triumph of the intellect than the creation over the past few centuries of a scientific structure encompassing a large part of the physical world. Yet the insights into the workings of nature afforded by this development are not so widely appreciated as they could be and should be. The aim of this book is to present the modern scientist's view of the physical world. Everywhere the emphasis will be on the concepts used to describe nature and the pictures of nature to which physical theories have led. But pictorial description alone is not sufficient for understanding. Science is an area of human activity in which precision is vital. This is a book *of* science, not merely a book *about* science. Definitions will be exact; the essential content of most important physical theories will be presented; and at key points where it is necessary, mathematics will be used.

Two themes run through the following chapters. The first is a theme of simplicity. The abstractions of science have led from the complexity of everyday experience to the simplicity of the underlying laws of nature. The second is a theme of activity. Science and our picture of the physical world are creations of the human mind. Science has provided insight, not absolute truth. The creative activity of science is continually altering as well as enlarging our view of the physical world.

1.1 *The faith in simplicity*

The commonly known and accepted miracle of science is the enormous power it has given to man to change the world around him. There is another miracle of science, less often appreciated. It is the miracle of the simplicity, generality, and beauty of fundamental physical theories. The straitjacket of experimental confirmation, far from enclosing the imagination of the scientist in a cheerless and airless room, has spurred on his imagination, and the successful theories of physical science that have been created possess a simplicity and inner harmony as satisfying to the mind of man as any creation of the free and unfettered imagination. Expressed differently, we should be grateful to nature for revealing its secrets to us in such a rewarding way.

Scientists like to say that *nature* is simple. What they mean is that it has been found possible to describe parts of nature—the parts we understand—in a simple way. Without the underlying faith in simplicity and the rewards of success, man would lack the stamina to overcome the obstacles to understanding that line the route to the discovery of simplicity. As every student of science is well aware, the simplicity of nature is not synonymous with ease of comprehension. The theory of relativity, viewed as a formal structure, is exceedingly simple. But to understand it and to use it requires a heroic effort of the mind, for the concepts employed, and the ways of thinking required, are largely foreign to our everyday experience. Simplicity to the scientist means economy and compactness—of assumptions, of fundamental concepts, of mathematical equations. The fewer the basic elements of a theory and the greater the range of phenomena described by the theory, the simpler does he declare nature to be.

Throughout human history, the faith in simplicity has been a primary motivating force in science. In essence it is the faith in the possibility of science at all, that nature, or parts of nature, follow an orderly and predictable pattern governed by fixed laws. But more than a mere faith in the existence of laws of nature, the faith in simplicity adds a conviction that these laws are sufficiently simple that man dare hope to find and comprehend them. So powerful was the faith in simplicity among some Greek philosophers that simplicity itself came to be regarded as a sufficient test of the truth of a theory. Aristotle accepted circular motion as the rule in the heavens not because careful measurements showed the stars and planets to move in circles, but because the circle is the simplest (or most "harmonious" or most "perfect") of plane figures. In describing motion on earth, he advanced a theory of utmost economy. It required but four elements (earth, water, air, and fire) and but two kinds of natural motion, vertically up and vertically down. Surely the simplicity of Aristotelian physics must have contributed to its durability.

The birth of modern science in the sixteenth and seventeenth centuries brought with it no change in the ancient faith in the simplicity of nature. Rather a new element was added: the reliance on accurate observations (usually in carefully controlled experiments) to test the acceptability of a theory. The pyramiding successes of science in the past few centuries have richly rewarded the scientist's faith in simplicity coupled with his insistence on experimental confirmation.

Because the simplicity of basic science is so little appreciated, it is proper to emphasize it, even to extol it. At the same time, we must keep in mind that, at least in part, science is fundamentally simple because man has made it so. From science man gets simple answers because he asks easy questions. But what exactly is an easy question? Compare these two, the first scientific, the second nonscientific:

1. What is the electric dipole transition probability from the metastable 2*s* state in hydrogen?

2. What are the advantages of foreign travel?

The average reader without scientific training will regard the first question as incomprehensible and therefore obviously difficult. But anyone should be willing to venture an answer to the second, which appears easy enough. Upon a little reflection, of course, you should convince yourself that it is after all the first question that is easy, the second that is difficult. Assuming the first question is meaningful (it is) and that science has progressed to the point that an answer exists (it does), then the answer is just a simple number upon which the thousands who can give an answer amicably agree, and which the millions of others as amicably accept. The second question, on the other hand, deals with what is good for man, and there are no harder questions that man has posed to himself than questions of good and evil.

1.2 Science as creative activity

There are two outstanding misconceptions about science. The first is that it is a cold, dull, 2
emotionless cataloging of old facts coupled with the plodding relentless discovery and ordering of new facts according to something called the "scientific method." The second is that science is devoted to the invention and development of machines and gadgets, that the highly developed technology of the modern world *is* science. Each of these misconceptions contains a germ of truth; but as a characterization of science, each falls very wide of the mark. No such thing as the scientific method can be discerned in the creative human activity that has produced most of the advances in science. And technology is a poor substitute for the magnificent structure of facts and theories that are the body of science.

Modern science got its start around 1600, when men began to ask answerable questions about nature—to seek relationships rather than final causes, and to test ideas against experiment rather than just against logic. Francis Bacon was one of the first to see how man could and should proceed to learn about nature. Yet Bacon, and after him Descartes, went too far in their speculation and foresaw an all-embracing power of a scientific method which could act as a kind of recipe for scientific discovery. Their vision never came to fruition. It is true that there have been important elements of *method* in science—the experimental method for example, and the mathematical method. Yet at the key points of important new scientific discoveries, chance, intuition, insight, and trial and error have been more evident than any well-defined scientific method. The trouble is that if one does not know where one is going, one does not know how to get there. Scientific progress is the advance into the unknown, and so far no reliable recipes for the best way to make that advance have been found. Probing into the unknown, the scientist is as much on his own as is a composer before a blank page or an artist facing an empty canvas. Science, like man's other achievements, has emerged from the triumph of individual genius over human frailty; the approaches of successful scientists have varied as much as human personalities vary.

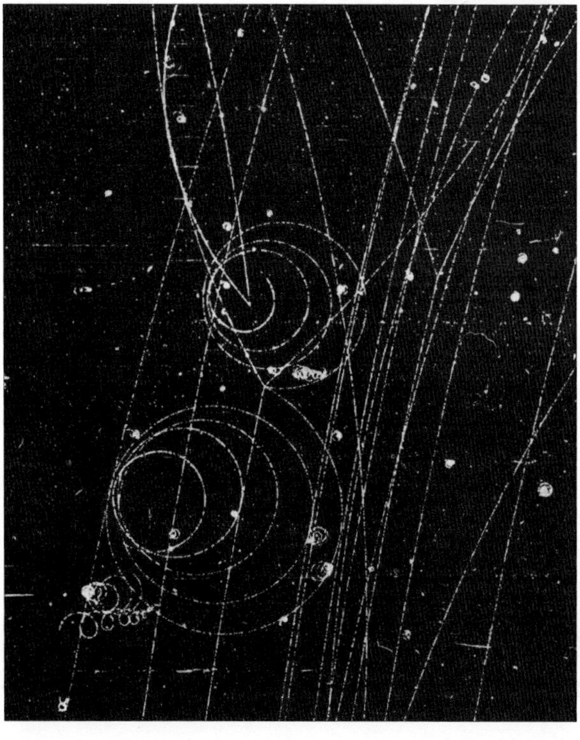

FIGURE 1.1 The frontier of the very small. Tracks of elementary particles in a bubble chamber. At the point in the chamber from which the two spirals seem to emanate, two particles were destroyed and four others created in the submicroscopic violence of a collision extending over 10^{-13} centimeter in space and 10^{-23} second in time. (Photograph courtesy of Lawrence Radiation Laboratory, University of California, Berkeley.)

1.3 The structures of the world

Much of the history of science can be characterized as a probing upward and downward away from the world of man's immediate sense experience. The frontiers of physical science have left the human-sized world which we call the macroscopic world and moved away to the infinitesimal submicroscopic world (Figure 1.1) and the enormous cosmological world (Figure 1.2). Between these distant limits is arrayed that part of the physical world that is understood, if only imperfectly.

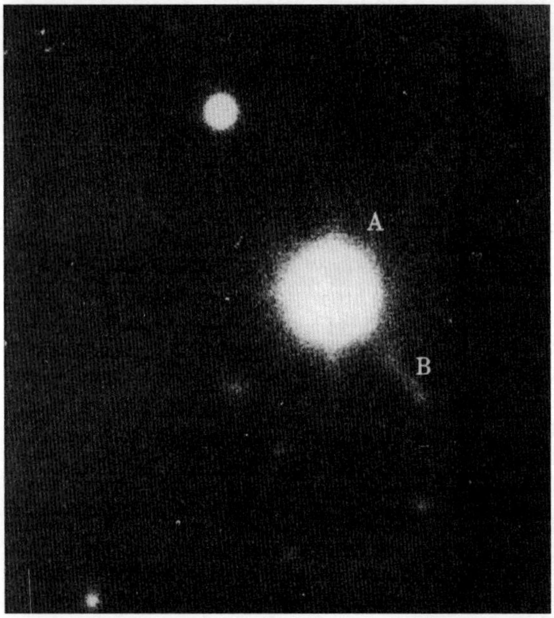

FIGURE 1.2 The frontier of the very large. A quasi-stellar object (named 3C 273) moving away from the earth at a speed of more than 27,000 miles/sec. Its light may have started toward the earth about 1.5 billion years ago. The central object (A) is a copious source of both light and radio waves. It is overexposed in this picture in order to render visible a mysterious "jet" (B) which is a separate strong source of radio waves. (Photograph courtesy of Mount Wilson and Palomar Observatories.)

TABLE 1.1 The Structures of the World

OBJECT	SIZE*	SPECIAL ASSOCIATED BRANCH OF SCIENCE
Elementary particle	10^{-13} cm or less	Particle physics
Atomic nucleus	10^{-12} cm	Nuclear physics
Atom	10^{-8} cm	Atomic physics
Molecule	10^{-7} cm	Chemistry
Giant molecule	10^{-5} cm	Biochemistry
Solids		Solid-state physics
Liquids		Hydrodynamics
Gases		Aerodynamics
Plants and animals	10^{-5} cm to 10^{4} cm	Biology
The planet Earth	10^{9} cm	Geology
Star	10^{9} cm to 10^{14} cm	Astrophysics
Galaxy	10^{22} cm	Astronomy
Galactic cluster	10^{25} cm	
The known part of the universe	10^{28} cm	Cosmology

* Lengths in this table are expressed in centimeters, usually abbreviated cm. There are 2.54 cm in one inch, and 30.5 cm in one foot.

From the subatomic elementary particles up to the collection of galactic clusters which is optimistically called the universe, man is now familiar with a hierarchy of objects joined together in structures of ever increasing size. Incomplete though the picture may be, it is a grand panorama spanning a factor of 10^{41} in dimensions. The names of some of the structures and of some of the special branches of science concerned with particular levels of the hierarchy are given in Table 1.1.

As a matter of convenience when dealing with large and small numbers, scientists use what is called the exponential notation, a practice followed in Table 1.1 and in the paragraph above. (Students who already have facility with the exponential notation should skip the next few paragraphs.) The number one hundred, or ten times ten, is written 10^2, and spoken "ten squared" or "ten to the two." One thousand, or 1,000, or ten times ten times ten, is written 10^3. One million, which may be written as a one followed by six zeros, 1,000,000, is more compactly written 10^6. One billion is 10^9. This is spoken "ten to the nine," which is short for "ten raised to the ninth power," that is, ten multiplied by itself nine times. When numbers become exceedingly large, the value of the exponential notation becomes obvious. The number 10^{41} in the paragraph above, which is the ratio of the largest distance man knows anything about to the smallest distance he has been able to study, would be written out as 100,000,000,000,000,000,000,000,000,000,000,000,000,000. Even expressed in words, as one hundred thousand million million million million million million, it is unwieldy.

The exponential notation for small numbers follows a similar pattern. One tenth is written 10^{-1} ("ten to the minus one"), one hundredth is written 10^{-2} ("ten to the minus two"), one thousandth is 10^{-3}, one millionth is 10^{-6}, and one billionth is 10^{-9}. Note that 10^{-3} is the same as the number one divided by 10^3, 10^{-6} is 1 divided by 10^6, and so on. In decimal notation, 10^{-6} is 0.000001. The number of zeros to the right of the decimal point is *not* six; it is five. The rule for transforming the exponential notation to the decimal notation is the following. Start with the number one with the decimal point to the right (1.). Then let the exponent be an instruction for moving the decimal point—to the right for a positive exponent, to the left for a negative exponent. Thus, for 10^{-1}, the decimal point is moved one place to the left from 1. to give 0.1. For 10^{-6}, it is moved six places to the left from 1. to give 0.000001. For 10^3, it is moved three places to the right to give 1,000.

The rule for multiplying exponential numbers can easily be developed from a few examples. Ten times one hundred is one thousand, or $10^1 \times 10^2 = 10^3$. One hundred times one thousand is one hundred thousand, or $10^2 \times 10^3 = 10^5$. The rule is simply this: To multiply powers of ten, add the exponents. The same rule holds for negative exponents. For example, $10^{-3} \times 10^{-3} = 10^{-6}$, one thousandth of one thousandth is one millionth. (The sum of −3 and −3 is −6.) For mixed positive and negative exponents, the same rule continues to hold true. Since the sum of −2 and +3 is +1, the rule gives the result $10^{-2} \times 10^3 = 10^1$, that is, one hundredth of one thousand is ten. Notice that 10^1 is the same as 10 itself. What then is 10^0? According to the rule for shifting the decimal point away from 1, the number 10^0 must be exactly 1. There is no shift of the decimal point. The same conclusion can be reached from the rule for multiplication. For example, $10^{-2} \times 10^2 = 10^0$, one hundredth of one hundred is one.

Let us return now to an examination of Table 1.1. It is worth expending some effort trying to visualize the scale of this physical picture of the world. Ten miles or 1,000 miles or 3 inches conveys to us an immediate sense of distance, but 10^{-12} cm, the size of a nucleus, without some thought and extrapolation and analogy, is almost meaningless. Since 10^{-6} is one millionth, 10^{-12} is one millionth of one millionth. If we could place one million nuclei in a line, this line would be only one millionth of one cm long. If we diligently lined up nuclei, adding one nucleus each second (night and day), we would have a line of nuclei one cm long after 30,000 years. If our one-centimeter line were expanded to stretch from New York to San Francisco, how big would a nucleus become? It would be blown up to a tiny speck which would still require a microscope to see.

Oddly enough, the structures of the world appear to grow simpler as we depart in either direction from the size of man. There is no organization of constituents either in the world of the very small or in the world of the very large that begins to approach the complexity and degree of organization found in living creatures. It might be argued that there exist larger and more complicated degrees of organization in the universe which man's limited intellect is incapable of grasping, but this is an argument outside the scope of science. So far there is no evidence for any such organization and much evidence against it. Indeed the simple theories of the submicroscopic world have succeeded in dealing with the structure of the stars and to some extent with the structure of galaxies. There have even been some hints that the properties of the universe at large may be linked intimately with the laws governing the submicroscopic world. To discover if this is true remains one of the most challenging problems for the future of science.

1.4 *The structure of science*

As indicated in Table 1.1, special branches of science have come to be associated with particular sizes and levels of arrangement of the matter in the world. But at a deeper level science has evolved in quite a different way than according to the particular object studied. The structure of physics, the science of concern to us here, is built not around objects or the physical structure of the world, or even about particular phenomena in the world. The basic framework of physics is rather a set of general theories (a very small set, as we shall see), each of which describes a wide range of phenomena and of objects. Mechanics, for example, is one such general theory, accounting for the behavior of matter over almost the whole explored range of sizes from the submicroscopic to the cosmological. Electromagnetism is another, whose area of application extends from the emission of gamma rays by elementary particles up to the transmission of starlight throughout the universe. Some of the branches of study listed in Table 1.1 are merely special areas of application of some general theory. Atomic physics, for example, is the application of the theory of quantum mechanics to the properties of atoms. Most of the special branches, however, draw upon more than one general theory. Hydrodynamics utilizes the theories of mechanics and of thermodynamics, and astrophysics draws upon every general theory of physics in its effort to account for the world of the very large.

The arrangement of this book is in the main according to the broad theories that form the most natural and most beautiful structure of physical science. There will be some exceptions, however. Indeed, Chapter Two on elementary particles is an exception, for in this fascinating field of modern exploration, there exist as yet no satisfactory general theories. Particles come first in our survey of physics because they are the basic building blocks of the universe.

1.5 *Theory and experiment in science*

Some areas of science, especially in the fields of biology and psychology, are purely experimental. Facts are being gathered and knowledge increased, but there does not exist the body of ideas, concepts, and relationships that are collectively called theory to tie together the facts into a coherent whole, that is, to "explain" the facts. In other areas of science, for example mechanics, the equations of a well-tested theory have been so elaborately developed that the branch of science seems to be almost a part of pure mathematics. But every area of science, regardless of its state of mathematical development, differs in a very fundamental way from pure mathematics and from nearly every other area of human activity. Science is essentially empirical. No idea in science survives because it is esthetically pleasing, or mathematically elegant, or magnificently general, although many ideas in science are all of these things. The idea must weather the test of experiment, and not just one experiment.

It will be attacked from all sides by every device that the experimenter can muster. A scientist checking a theory is like a test-pilot wringing out a new airplane. He tries his best to break it apart while hoping against hope that it will hang together. And with the theory as with the airplane, one flaw is sufficient to bring about its destruction. (The parachute of the scientist is his sense of detachment and caution. Occasionally scientists have thrown away their parachutes by falling in love with an idea. The disintegration of the idea can result in the destruction of the scientist.)

Experiment is the final arbiter in science, but a science with only experiment would be a dull thing indeed. In attempting to understand nature, man has sought much more than mere empirical facts. It is the theories tying facts together that provide the challenge and the reward of science. The new way of looking at nature, the unexpected relation between different facts, the single equation governing a vast range of phenomena—these are the things that give to science its stature and nobility.

Although the hard evidence of experiment can destroy a theory, no amount of experimental verification can "prove" a theory. Every theory has to remain tentative, for two reasons. First, the theory is likely to be capable of making an infinite number of different predictions, but man's finite capabilities limit his ability to test the predictions. The test pilot, no matter how long he flies the airplane, can never put it through every conceivable maneuver under all possible conditions. He must test only what he deems most important and recognize that a subtle hidden flaw may go undetected. So it is with theories. Newton's theory of mechanics survived two centuries of exhaustive tests, but finally the flaw appeared.

Second, no theory is unique. The possibility must always remain open that a theory is supplanted, not because experimental evidence forces its rejection, but just because an alternative theory is found which, although neither better nor worse experimentally, is in some way more satisfying to man. It may be conceptually simpler, possess a more economical mathematical framework, or appear in some way to be deeper, more profound, and therefore more pleasing esthetically. This human judgment of theory is as important as experiment itself in shaping the structure and the progress of science. The scientist's faith in simplicity may indeed cause him to reject a complicated and cumbersome theory even if no better alternative is at hand, just because of his conviction that a simpler description of nature must exist.

An idealized version of scientific progress goes something like this: 3

Experimental facts,

Laws tying the facts together,

Hypothesis,

Test of the hypothesis against past facts,

Prediction of new facts and further tests,

Theory,

Elaboration and application.

In fact, no such set pattern has been realized in the evolution of any theory in physical science, but elements of the pattern can be discerned throughout the history of science. Near the end of the sixteenth century, for example, Tycho de Brahe made accurate observations of the positions of the planets in the sky. These proved to be vital experimental facts in the evolution of the theory of mechanics. Brahe's assistant, Johannes Kepler, discovered the laws of planetary motion which in capsule form neatly summarized the myriad of individual observations of his master, without in any way "explaining" those observations. But Kepler's laws tied the facts together and made further progress possible. Some decades later Newton drew together contributions to mechanics by Galileo and by Hooke, coupled with his own inspired hypothesis of universal gravitation, and created the theory of mechanics.

The theory at once accounted for the past observations summarized by Kepler and led to the prediction of new observations. It passed its most crucial test in 1846 when astronomers pointed their telescopes at a certain point in the sky and discovered the new planet Neptune where it was predicted to be. Elaborated by mathematicians and applied by astronomers and by practical men, mechanics evolved and still stands today as a comprehensive theory embracing the subject of motion of material objects over a wide range (but, as we now know, less than the infinite range once imagined for it).

More often, theory and experiment have developed side by side through mutual cross-fertilization. The experimenter without ideas can discover an endless sequence of useless facts. The theorist unbridled by the limitations of experiment can produce a stream of fanciful ideas that have nothing to do with nature.

1.6 Mathematics and machines

Theory and experiment are the two essential and inseparable units of science, but there are in addition two auxiliary services nurtured by science and nurturing science—mathematics and machines. Machines—that is, modern technology—are the outgrowth of science and provide the essential tools for further experimental research in science. In a somewhat similar way, much mathematical discovery has been stimulated by science, and mathematics becomes itself the tool of theoretical research and the vehicle of expression for theoretical results. Mathematics, much more than technology, has a life of its own independent of science. That mathematics can exist as a kind of scientific theory divorced from scientific fact has been realized for less than two hundred years. Nevertheless, much of mathematics can be used for the description of nature—indeed it provides the most elegant description—and part of the scientist's faith in simplicity is a faith in the possibility of expressing nature's laws in mathematical form.

Mathematics and machines also form a part of the bridge between pure science and applied science. Although the motivations of the search for new knowledge and the application of already acquired knowledge for practical purposes are entirely different, both efforts employ mathematics, indeed often nearly identical mathematics, and both make use of similar mechanical devices. Through discoveries in mathematics and the development of machines, pure science and applied science have enriched each other.

FIGURE 1.3 Science serves technology. The transistor, a by-product of fundamental research in solid-state physics, makes possible compact portable radios as well as a host of other modern electronic devices. (Photograph courtesy of Jurgen H. Stehnike.)

Because the technological by-products of science are more readily comprehensible than fundamental science, and because they have a greater direct impact on our lives, technology is often confused with science (Figure 1.3). Science is the discovery of the facts of nature and the unification of these facts by means of structures of ideas and equations which are collectively known as theories. Technology is the application of known facts of nature for practical purposes. Michael Faraday, after discovering the induction of electricity by magnetism in 1831, foresaw accurately the revolutionary consequences of this scientific discovery for the technology of electric power. But he resisted every temptation to follow up the technological development himself and returned to pure science and the search for new knowledge. Technology today sometimes requires scientific training and mathematical skill of a high order. Moreover, some technology has acted back upon science, as a tool for research (Figure 1.4). For these reasons, the lines between science and technology are blurred, but the fundamentally different motivations of the search for knowledge and the goal of practical application remain clear. The modern laboratory of pure research is in fact a marvel of technological achievement. An electron accelerator, for example, is a technological achievement of a high order made possible by the fundamental theories of electromagnetism and relativity. The accelerator in turn makes possible exploration of the subatomic world, enriching contemporary science.

FIGURE 1.4 Technology serves science. Experimental research in elementary-particle physics harnesses products of technology, in great number and great complexity. Shown here is an experimental area at the Cosmotron accelerator in Brookhaven. New York. (Photograph courtesy of Brookhaven National Laboratory.)

Our modern technology illustrates one of the most characteristic aspects of science: its power. Besides the obvious example of nuclear bombs, every airplane, every automobile, every household gadget—indeed, almost all of the physical aspects of modern industrialized society attest to the power of science. This power springs essentially from the quantitative and predictive character of science. If science merely accounted for what had been observed and experienced in the past, this power and control would not be one of its dominant features. It is the fact that simultaneously with the explanation of a previously known phenomenon comes the ability to predict new phenomena which gives to science its power to modify the world around us. Occasionally technological inventions have appeared without real scientific understanding—the steam engine is a good example—but more frequently technology has been the daughter of science. In many cases, in fact, the invention could not even have been visualized before the creation of the scientific theory upon which it rests. One can scarcely imagine the vacuum tube used in radios having been invented before Thomson's discovery of the electron, or the atomic bomb before the theory of relativity, the knowledge of atomic structure, and the discovery of the neutron.

1.7 The unity of science

4 Striking as the power of science may be, other aspects of science have probably had a more profound influence on man's view of nature and of himself, and it is these subtler aspects that underlie the presentation of the fundamentals of physics in this book. It is the unity of science and the simplicity of nature which gives to the physical world a beauty that is both the motivating power for scientific research and the reward of scientific understanding.

The overall unity of physical science could scarcely be suspected by considering the multitude of different kinds of natural phenomena even within reach of our senses, quite aside from the greater multitude that escape our notice. Nor is this unity even obvious in a superficial survey of the fundamentals of physics. We speak of the theory of mechanics, the theory of electromagnetism, the theory of relativity, and a few others, apparently as distinct theories describing different parts of nature. Within each theory an enormous unification has already been achieved. The trajectory of a baseball, the flight of a rocket, and the majestic sweep of the planets around the sun are all described by the same equation of mechanics. A hi-fi system, the light of the sun, and the Van Allen radiation belt merely illustrate different aspects of the same laws of electromagnetism.

Yet the unity of physical science goes further than the immense generality encompassed by each theory separately, for through all the branches of physical science run certain common concepts linking the theories together. The most important of these is energy. The conservation of energy is a principle which applies not to any branch of physical science, but to all of nature. Energy of one kind may disappear, but only if an equal amount of energy of another kind appears.

One theory in particular, relativity, introduces a special kind of unity into physical science, since it is in fact a kind of supertheory, or theory of theories. The theory of relativity imposes on lesser theories—those known as well as, very likely, those yet to be discovered—a requirement of form and structure. It is common for scientists to reject as unsatisfactory any hypothesis that does not conform to this relativistic requirement, in no matter what field of science it occurs. Finally there is a kind of hidden unity in physical science arising from the fact that there are not as many different theories as there appear to be. Ordinary mechanics, or "classical" mechanics, is now known to be contained within quantum mechanics. Thermodynamics, the theory of heat and temperature, is known to be derivable from ordinary mechanics. Nevertheless we go on talking of them as three different theories for purely practical reasons. In a region where mechanics is valid to a high degree of precision, for example, the solar system, the mathematical techniques are far simpler than those of quantum mechanics. It would be possible in principle to derive the planetary motions directly from quantum mechanics, but to do so would be a very formidable and quite unnecessary task.

Although physics possesses remarkable unifying features, its unity is incomplete. The number of different theories, different particles, and different interactions that are now required to describe all that we know of the physical world are so tantalizingly few that most scientists very naturally suspect that the future will reveal further connections and lead to a truly unified view of the physical world.

1.8 The great theories of physics

In the list on page 9, a law has been placed at a lower level in the development of science than a theory. This sounds surprising, for we are accustomed to thinking of a law as something definite and well established, and a theory as something hypothetical and unsure. Although these terms are frequently used confusingly in science (and sometimes even interchangeably!), a scientific law does retain a kind of definiteness not possessed by a scientific theory. Nevertheless, it has to be recognized that an uncertain theory can be much more profound than a certain law. Someone surveying the stock market from 1930 to 1940 might discover a law of its behavior—for instance, a periodic fluctuation superimposed upon a steady rise—that could be expressed mathematically. It is quite definite; it happened. But a theory that gives a reasonable account of why it happened in that way would be more interesting. If the theory then made possible a successful prediction of the market behavior from 1940 to 1950 it would be a much more noteworthy achievement than the discovery of the law—even if later the theory had to give way under the pressure of new evidence while the law stood fast.

In this book we will be concerned with a few theories that have progressed well beyond 5
the stage of hypothesis, theories that deserve to be called the great theories of physics. Each is a structure of ideas and equations thoroughly tested by experiment, and in addition satisfying for its internal simplicity and its generality of application.

The great theories of physics are just five in number:

1. Mechanics (sometimes called Newtonian mechanics or classical mechanics). The theory of the motion of material objects.
2. Thermodynamics. The theory of heat, temperature, and the behavior of large arrays of particles.
3. Electromagnetism. The theory of electricity, magnetism, and electromagnetic radiation.
4. Relativity. The theory of invariance in nature, and the theory of high-speed motion.
5. Quantum mechanics. The theory of the mechanical behavior of the submicroscopic world.

As mentioned in the preceding section, even these five are closely interrelated. Every one of the myriad of phenomena in the physical world that is understood is explained in terms of one or more of these few theories. The behavior of single atoms, for example, is governed by quantum mechanics, relativity, and electromagnetism. A collection of many atoms requires also the theories of mechanics and thermodynamics for its description.

About these five theories we can say with some confidence that none will ever be completely overthrown. If history is a reliable guide, we can say with equal confidence that none will prove to be entirely correct. Consider mechanics, already known to be "incorrect." Relativity "overthrew" mechanics when it showed Newton's laws of mechanics to be incorrect for describing ultra-high-speed motion. Then quantum mechanics showed classical mechanics to be incorrect for describing the internal motions within atoms. Each of these twentieth-century developments proved mechanics to be wrong. Yet we doggedly list mechanics as one of the great theories of physical science. Why? Because over what is still a vast domain of sizes and speeds, mechanics is so extremely accurate that it is for all practical purposes completely correct. It is the best and simplest tool for describing nature in a certain domain. It is better to say that relativity and quantum mechanics have chipped away at the boundaries of mechanics, reducing it from an infinite to a finite domain, than to say that either theory has overthrown mechanics. The situation is analogous to the "overthrow" of the automobile by the airplane. Using quantum mechanics to describe a phenomenon, such as the trajectory of a satellite, that is accurately described by classical mechanics is like taxiing an airplane on the ground a few miles from home to office in order to avoid using an old-fashioned vehicle, the automobile.

Electromagnetism, relativity, and quantum mechanics are theories with no known flaws. No scientist believes this situation will persist indefinitely. Undoubtedly new discoveries will show that these theories too have a limited domain of validity. But when that happens—when the theories are "overthrown"—they will very likely also refuse to disappear and will remain in the scientist's bag of tools as the most satisfactory description of that part of nature where they have already been well tested.

There are a good many branches of physical science which will occur to the reader that do not appear in the list of great theories. Yet we assert that these theories encompass all of physical science. Where, for example, is the enormous subject of chemistry? Where are astronomy, geology, and aerodynamics? They are omitted because, being less fundamental, they derive their sustenance from these theories. This is not to say that they are less interesting, or less complicated (they are indeed more complicated), but that they are further removed from the fundamental underlying laws of nature. Throughout this book we are concerned primarily with the basic concepts, laws, and theories of physics, and only secondarily with applications and extensions to special systems (for example, complex molecules or air masses or the interior of the earth).

Most of the applied areas of physical science have in common that they deal with such complicated systems that direct application of the fundamental theories is impractical. We have every reason to believe, for instance, that the precise and detailed behavior of the air blanket surrounding the earth is in principle exactly predictable from the already known theories of physics. This belief is of no value whatever to the meteorologist, because the practical task of carrying out this prediction is far beyond the capabilities of all the men and all the computing machines now on the earth. The meteorologist must make approximations and seek simplifying features. In this process he introduces new concepts and new methods so that meteorology becomes almost transformed into an independent science, in spite of its ultimate dependence on the basic theories. In a similar way, the chemist, in studying the interactions among atoms and groups of atoms, is forced to introduce new concepts, new vocabulary, and new ways of thinking about nature. Although working within the framework of the fundamental theories of physics, he must break new ground and create an independent branch of science.

1.9 The nature of scientific theory

6 What exactly is a theory in physical science? This question does not have a simple answer. All of the great theories are basically mathematical descriptions of nature, yet they are more than bare mathematics. What we call Maxwell's equations of electromagnetism are central to the theory of electromagnetism, but one cannot point to the equations and say they *are* the theory. This would be equivalent to pointing to the numbers in one's bank book and saying, that is my wealth. Rather the numbers represent one's wealth, and give accurate information about it only because of a number of understandings and assumptions which seem usually to be true. The assumption is made that money is conserved—if left alone, it will neither increase nor diminish; that money transactions follow the laws of arithmetic

(indeed, arithmetic was invented to make this true); that bankers are honest men.

As the bank balance represents wealth, so the equations of a physical theory represent nature. But the connection between representation and reality requires a set of ideas, agreements, and assumptions which the equations themselves do not provide. These must stand beside the equations to give them life. Even then the equations are made to correspond only with experimental results; their correspondence with a deeper underlying reality in nature requires more assumptions which are tempting to make but are probably inessential to science. In the same way one's bank balance represents only the tangible (experimental) thing called money; that the money in turn represents "real" wealth is a further assumption which it is satisfying to make but which is of no help in balancing the books.

Altogether a physical theory (a more or less completed theory) is a set of concepts; assumptions about the mathematical representation of these concepts; mathematical relationships among the concepts; prescriptions for relating the mathematical structure to actual measurements; the accumulated evidence in favor of the assumptions and prescriptions; and a way of looking at nature provoked by the ideas and their success. All of these things taken together constitute the understanding of past facts, the predictions of new facts, and the view of nature that comprise a full theory in physical science. Equally important to the theory are its skeleton of equations and its flesh and blood of ideas, interpretation, and visualizable pictures.

The word "concept" used above requires some discussion because it is a vitally important word in every theory of physics, naming the link joining the mathematical, the verbal, and the experimental. In everyday use it normally refers to an idea that is not directly visualizable, either because it is abstract (bravery) or because it is beyond reach of our senses (an atom). It is sometimes used in the latter sense in science, but we shall most often mean by a concept a thing or a quantity that is measurable. This may be called a quantitative concept. Roughly speaking, a quantitative concept is anything that can be given both a name and a number. Here number means a numerical value, not a serial number—soldiers, sailors, and prisoners do not qualify as quantitative concepts. But money is such a concept. It has a name—which means there can be a common basis of agreement about what it is—and it can be assigned a numerical value.

Some typical physical concepts that should be already familiar are force, energy, temperature, and velocity. In the course of this book we shall encounter many more, some of which, such as angular momentum and entropy, may be new. But even the familiar ones will require a new look, for precision of definition is required to pin down the exact meaning of these concepts as used in science.

In comprehending a theory there are two major tasks. The first is to learn what the concepts the theory works with mean—how they are defined and how measured. The second is to learn the relationships that exist among the concepts—usually quantitative relationships expressed through equations. Accomplishing the first of these tasks without the second would tell you what the theory is all about, but not what it has to say about nature. Accomplishing the second without the first would be an exercise in pure mathematics, since the equations would not be interpretable. It is the equations coupled with the physical interpretation of the symbols that together comprise the essence of a theory.

There are, of course, many half-born, uncertain trial theories in science, some with ideas but no equations, some (less often) with equations with uncertain interpretation. But our concern will be principally with those already completed great structures of theory about whose validity over some wide domain there can be little doubt.

1.10 Explanation in science

The untrained novice in science may be impressed by the scope and power of science. The beginner who has learned a little of science may on the contrary be disappointed by the modesty and circumscribed boundary of science. The goals of nature is an idea foreign to science, and although the scientist in unguarded moments frequently enough uses the word "cause," he will if pressed say that he does not seek causes of effects, but only relationships among phenomena. Explanation in science is reduced to relationships; one thing is "explained" by being related to something else. A good theory is not one that gives an ultimate cause of events, but merely one that relates many different events through a few simple ideas and equations. An equation is itself, of course, a particular kind of mathematical relationship.

Yet there can be little doubt that explanation in science is probing gradually to deeper and more fundamental levels. The practicing scientist uses an intuitive guide to acceptable explanation. If one thing is explained only in terms of something else of equal or greater complexity, the scientist will naturally tend to reject the explanation as not interesting or fruitful. But if the phenomenon is explained in terms of something simpler and more general, he will accept it as really an "explanation," a tentative answer to the question why. Newton "explained" planetary motion by postulating a universal law of gravitational force of a particular simple mathematical form. The explanation works so beautifully that we willingly discard cautious phraseology and say that planets move as they do *because* they are attracted to the sun and to each other by a gravitational force. It would sound rather silly, but it would in fact be logically as valid to say that there exists a gravitational force because the planets move as they do. To be quite accurate, we can only say that the assumption of such a force can be related to the motion of the planets. Yet because it is a relationship between something complicated (the motion of many planets) and something much simpler (a single law of force), we very naturally call the relationship an explanation. We cannot help feeling that the relationship has carried us one step closer to the underlying reality of nature.

This progression of explanation to successively deeper levels can be illustrated by backing up one level before Newton. The primary observations of the planets show them to move in a confusingly erratic way through the sky overhead. If we should draw on a piece of paper the position in the sky of a single planet for a long period of time, a strange and convoluted curve would be traced out. Kepler's giant achievement was to demonstrate that the complex motion apparent to the eye could be "explained" by supposing that the planets move in simple elliptical paths about the sun. Still there was a separate ellipse for each planet. Newton explained all of the ellipses in terms of one law of force. But why the particular law of force known as the inverse square law, according to which the force varies as the square of the inverse of the distance between objects? In symbols,

$$F \sim \frac{1}{d^2}\,.$$

(The sign ~ means "proportional to.") The deeper "explanation" for this force law came only a few decades ago. Forces arise from the rapid and incessant exchange back and forth of tiny particles. The gravitational particles have been christened gravitons, and have no mass. If they possessed some mass, they would give rise to a different force law. Newton's inverse square law exists "because" the tiny (so far unobserved) graviton has no mass.

It is quite natural to push on relentlessly and ask, why does the graviton have no mass? That is, does there exist any relationship between the zero mass of the graviton and some still deeper, simpler, more all-embracing fact of nature? A sure answer to this question does not yet exist, although there is reason to believe that the massless graviton may be related to some general principles of invariance in nature. Thus, through the chain of successive relationships, constancy—absolute unchanging aspects of nature—may be related to the erratic and complex motions of the planets through the sky. We would be less than human if we did not regard such a relationship as explanation, and beautifully elegant explanation at that. The seeker of ultimate truth may not yet find satisfaction in such explanation, but he must be reminded that had not some men several centuries ago abandoned the search for ultimate truth in favor of the search for quantitative relationships, even this approach toward ultimate understanding would never have been achieved.

1.11 *The beauty of science*

Power, unity, simplicity of structure, the pleasurable surprise of unforeseen connections. 7
These attributes of science add up to what can be called the beauty of science, as surely as these same attributes define the beauty of a Picasso painting. A goal of this book is to enable the reader, while learning the fundamentals of physics and while acquiring something of the scientist's view of the world, to glimpse this beauty of science which can be comprehended only when the whole of physical science is viewed as a coherent entity created by the mind of man.

But physical science is not a completed reproduction of nature to be hung on the wall and admired. In spite of the enormous strides made in the past four centuries, there are as many unanswered questions now as then. They are deeper but no less numerous, and the uncertainties and perplexities will deserve our attention as much as the elegantly completed parts of science.

EXERCISES

1.1. Carry out the following arithmetic operations: **(a)** $10^{13} \times 10^{-12}$; **(b)** $3 \times 10^8/6 \times 10^8$; **(c)** $21{,}000 \times 3{,}000{,}000 \times 0.001$; **(d)** $3 \times 10^{-7} + 3 \times 10^{-8}$.

1.2. **(1)** Using the fact that light travels about 10^{18} cm in one year, calculate from Table 1.1 the time required for light to cross **(a)** a galaxy; **(b)** a galactic cluster; **(c)** the known part of the universe. **(2)** If a fast electron covers 10^{10} cm in one second, calculate how long it would spend in traversing **(a)** an atomic nucleus; **(b)** an atom; **(c)** a giant molecule.

1.3. There are about 10^{59} elementary particles in a typical star, about 10^{11} stars in a galaxy, about 10^3 galaxies in a cluster, and about 10^9 clusters in the known part of the universe. Approximately how many elementary particles comprise the known part of the universe? If all of this matter were so compressed that each particle occupied only 10^{-40} cm^3, what would be the approximate radius of this highly condensed universe? With what presently known structure does this dimension compare?

1.4. If the known part of the universe were shrunk to the size of the earth, how large would the earth be on this scale? How large would an atom be?

1.5. If an atom were enlarged to the size of a man—two meters in diameter—how large would its nucleus be? If all atoms were so enlarged, how tall would a man be?

1.6. Which of the following are quantitative concepts: **(a)** time; **(b)** beauty; **(c)** temperature; **(d)** progress; **(e)** galaxy; **(f)** atomic energy; **(g)** disorder?

1.7. Which of the following statements constitute explanation in science? **(1)** Grass is green because it contains chlorophyll. **(2)** The earth rotates on its axis because living creatures need an alternation of day and night. **(3)** The burning of coal releases stored chemical energy. **(4)** Tides are caused by the moon. **(5)** Each molecule of water contains one oxygen atom.

1.8. Why is an atom so small? (HINT: Consider why, on the atomic scale of size, a man is so large.)

1.9. In an experiment, natural events are deliberately manipulated. Discuss what you think has probably been the significance of experiment in science, as contrasted with mere observation. Have you ever learned through experiment in your everyday life?

1.10. Give several aspects of your daily life in which you rely on the simplicity of nature, or at least on its orderliness and predictability.

1.11. In Section 1.1 it is stated that the simplicity of nature is not synonymous with ease of comprehension. Discuss one example of something from outside of science that has a kind of simplicity, yet is not easily understood.

1.12. A law of nature summarizes in compact form a sequence of observations. Study the following observational data with care and try to find a single "law of rocket flight." Predict the height of each rocket after 25 seconds. Suggest a searching experiment that would test your law.

Time Since Launch	Height Above Ground of Rocket A	Height Above Ground of Rocket B	Height Above Ground of Rocket C
5 sec	125 ft	0.1 mile	91 ft
10 sec	500 ft	0.4 mile	364 ft
15 sec	1125 ft	0.9 mile	819 ft
20 sec	2000 ft	1.6 mile	1456 ft

1.13. Alfred North Whitehead wrote: "It is a profoundly erroneous truism, repeated by all copy books and by eminent people when they are making speeches, that we should cultivate the habit of thinking of what we are doing. The precise opposite is the case. Civilization advances by extending the number of important operations which we can perform without thinking about them." Illustrate Whitehead's point with some examples of your own "thoughtless" use of mathematics and machines.

1.14. Do you agree that unity and simplicity of structure are among the criteria of beauty (Section 1.11)? Illustrate your agreement or disagreement with two examples, at least one of them from outside of science.

1.15. Go to another book to learn the meaning of an "operational definition." **(1)** Give an operational definition of any scientific concept. **(2)** Attempt a brief operational definition of science itself.

1.16. Read Edgar Allan Poe's sonnet, "To Science." Write a paragraph contrasting his view of science with the view presented in this chapter. (How serious is Poe? Where do you stand?)

1.17 **(1)** In the *Saturday Evening Post* of February 20, 1960, Alexander Eliot wrote: "The best poets are those who speak comprehensibly of incomprehensible things. Scientists, on the other hand, speak in riddles about things that soon will be understood by all." Interpret and comment on this statement, relating it to remarks made in the text on the nature of science. **(2)** In the same article Professor Eliot wrote: "The scientist's heavy load of learning may oppress and darken his mind." What do you understand Professor Eliot to mean by this remark? Do you agree with him?

Part 1
The Submicroscopic World

ENRICO FERMI
(1901 – 1954)

"The total number of electrons, as well as neutrinos, is not necessarily constant. Electrons (or neutrinos) can be created and annihilated."

TWO

Elementary particles

Today is the particle era in physics. The downward probing of the past few centuries has led finally in the last few decades to the world of elementary particles. Actually no scientist believes that all the particles we now know are truly elementary, but for lack of a better word, and more especially for lack of any deeper knowledge of a lower substratum of matter, they bear the name elementary. It is the effort to understand the particles, to tie them together somehow through deeper and simpler concepts, that occupies the attention of many scientists throughout the world.

2.1 The submicroscopic frontier

8

Why do we begin a study of physics by examining the modern microscopic frontier represented by the elementary particles? In part it is to give a view of one of the furthest points of advance of physical science in order to provide a reference point as the theories of physics are developed, approximately in historical order, in the subsequent parts of the book. There is about the particles also the paradoxical fact that, although their discovery rested upon all the earlier developments of science, they are themselves rather easily visualizable and some of their main properties can be understood with no background of science at all. We can think of a golf ball or a marble, and extrapolating downward many orders of magnitude, picture an elementary particle as a tiny speck of matter, a basic building block of the universe. Of course, some of the properties of the particles, especially the way in which they interact with one another, require for their understanding a knowledge of some or all of the main theories of physics. These aspects of the particles are saved for appropriate places in later chapters.

The modern scientist, whether biologist, chemist, or physicist, is decidedly microscopically oriented. He pictures the large-scale world as put together from smaller and smaller units, down to the elementary particles and someday soon, perhaps, to a still deeper unit. And he pictures events and phenomena as arising from laws which ultimately have their simplest expression in the world of the very small. If Aristotelian physics was a science of final causes, modern physics is a science of microscopic causes. “Explanation” in physical science is, as often as not, the description of something larger in terms of something smaller. In looking first at the elementary particles, we are examining the basis of physical science as well as its frontier.

The elementary particles are useful objects for illustrating many aspects of physics. Their very simplicity means they are better adapted for illustrating basic principles of science than are the unwieldy and more complicated objects of our everyday life. We shall see in Chapter Four how the elementary particles illustrate the all-important conservation laws. Throughout the remainder of the book, we shall find the particles convenient for demonstrating aspects of each of the great theories of physics.

About the particles we now know a great deal, but not nearly enough. They remain very puzzling. Looking first at the particles will emphasize, as eighteenth-and nineteenth-century science could not do nearly so well, the exploratory and tentative nature of science. With the particles, the known and the unknown appear equally exciting.

To the scientist, the elementary particles provide the greatest challenge in modern physics. To the student they provide important insight into the scientist's view of the world. It is well to recall that the most familiar points of impact of physical science on man—our satellite communications, our detergents, our bombs, our household gadgets—form only a sideline off the mainstream of scientific advance. The true frontiers of science lie far more remote from everyday life, and one of these is the submicroscopic frontier inhabited by the particles. Being pursued by the most refined experimental and mathematical techniques, the particles may soon give way and allow the frontier to be extended to a still smaller domain.

2.2 *The early particles*

Man and the familiar objects of his world are constructed from atoms and molecules. At the turn of this century, atoms were known to exist but, as with today's elementary particles, the structure of the atom and the relation of one atom to another were mysteries. It was known that the atom was the smallest unit of an element such as hydrogen or oxygen or sodium or uranium; that there were somewhat over eighty different kinds of atoms (today we know just over a hundred); and that atoms were all of about the same size, such that one hundred million of them side by side would stretch into a line less than one inch long. It was also known that groups of different kinds of atoms could join together to form tiny structures called molecules, which, in turn, became the basic building blocks of the vast and wonderful variety of substances we encounter in the world.

In the first decade of the twentieth century discoveries came thick and fast, leading to a giant step downward toward the subatomic world of particles. The electron, the first known particle, had been discovered in 1897 by J. J. Thomson in England. He showed that the "cathode rays" produced by high voltage applied within an evacuated vessel could be deflected both electrically and magnetically. The rays were most simply interpreted as a beam of high-speed identical particles carrying negative electric charge (the same sign as the charge on an electrified rubber balloon). That the mass of each of these particles was considerably less than the mass of a single atom was verified by the ease of deflecting them. That each also had a size much less than the size of an atom was shown by their power to penetrate through a gas. Thomson realized, as those who had studied cathode rays before him had not, that he was dealing with truly subatomic objects, even though he was still far from isolating a single one of them.

The electron, as soon as it was identified, was strongly suspected to be a particle contained within atoms. As a common constituent of atoms, it provided the first definite link between distinct atoms. Another important link was forged when, in 1902, it was discovered that a radioactive atom (one capable of spontaneously emitting powerful radiation) could transmute itself into an entirely different kind of atom. This transmutation strongly suggested that atoms must be not independent, indivisible entities, but structures built up of some common, more elementary building blocks.

The alpha particles ejected at high speed from radioactive atoms were used shortly thereafter as the first projectiles for bombarding atoms. (These atomic "bullets," conveniently provided free by nature, are not energetic enough for modern purposes; today they are replaced by particles artificially pushed to still higher speeds in giant accelerators.) A result of the early bombardments was the revelation that the interior of atoms is largely empty space. By 1911 the experimenter Ernest Rutherford had discovered that the atom contained a massive, positively charged core—the nucleus—at least ten thousand times smaller than the atom as a whole, and that the remaining space was occupied by a few light-weight, negatively charged electrons. Two years later the theorist Niels Bohr provided a successful mathematical description of the motion of the electrons in the atom. Despite later modifications of detail, this description remains, in its essentials, our picture of atomic structure up to the present. The electrons whirl rapidly about the nucleus, providing a sort of atomic skin just as a whirling propeller seems to form a disk. Like a bullet that passed safely between the propeller blades of a World War I fighter plane, a high-speed particle can readily penetrate the electron cloud to reach deep within the atom. Another atom, however, drifting up with a relatively low speed, is turned back at the periphery, much as a stone tossed slowly at a spinning propeller might be batted back.

The nucleus of the lightest atom, hydrogen, was christened the proton and joined the electron to bring the list of elementary particles of fifty years ago up to two. Heavier nuclei were believed to be formed of a number of protons and electrons packed closely together. The details of this picture of nuclear structure were never clear; indeed, it was abandoned in 1932. However, it was a most attractive idea to have just two fundamental particles—the negatively charged electron and the positively charged proton—from which all of the matter in the universe was constructed. The only puzzling feature (a puzzle not yet resolved) was why the proton should be much heavier than the electron. In any case, this idyllic two-particle situation was not to last.

Several things happened in the early 1930s to begin the disturbing increase in the number of known elementary particles which has continued until the present time. A new particle, the neutron, was discovered; it was about as massive as the proton but carried no electric charge. The neutron was quite welcome, for it was just the particle needed to join the proton to form atomic nuclei. The picture of the nucleus immediately adopted and still

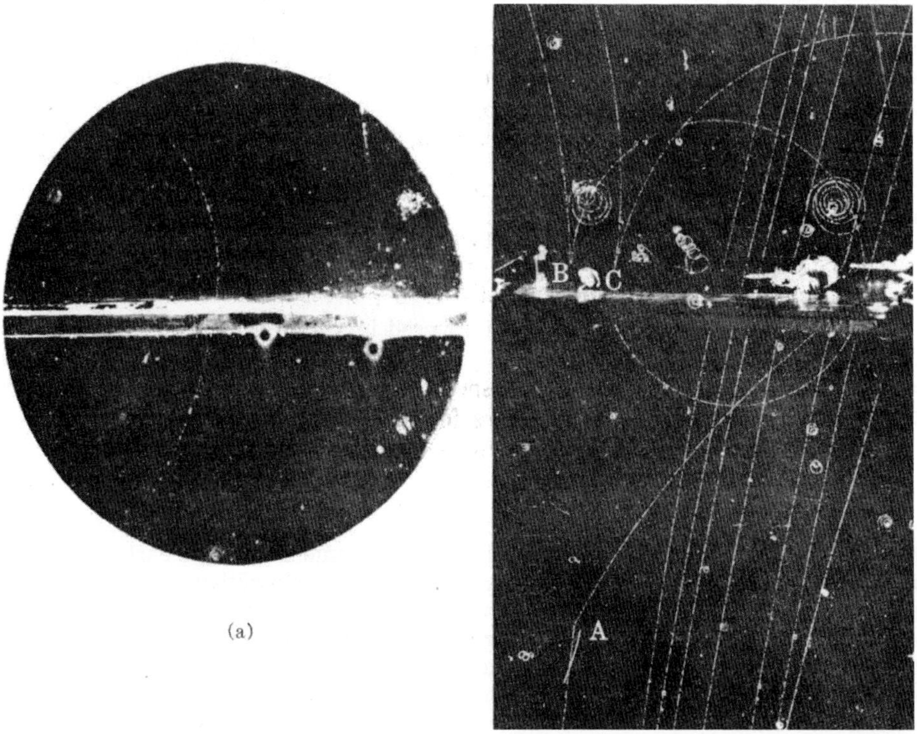

FIGURE 2.1 Positron tracks. **(a)** Carl Anderson's photograph first served to identify the positron in 1932. Entering the cloud chamber at the bottom, the positron is deflected into a curved path by a magnetic force. After being slowed down in the central metal plate, it is more strongly deflected. An electron trajectory would have curved in the opposite direction. **(b)** This more recent bubble chamber photograph shows tracks of both positrons and electrons (as well as other particles). High-energy photons (gamma rays) emanating from point A create electron-positron pairs at points B and C. The positrons arc upward to the left. The electron tracks bend away to the right. [(a) Photograph courtesy of Carl D. Anderson, California Institute of Technology. (b) Photograph courtesy of Lawrence Radiation Laboratory, University of California, Berkeley.]

accepted was that of a collection of protons and neutrons glued tightly together by a new strong force, simply called the nuclear force. For example, U^{235}, the most famous isotope of uranium, has a nucleus consisting of 92 protons and 143 neutrons. The far simpler nucleus of helium (the same thing as an alpha particle*) contains two neutrons and two protons.

At nearly the same time, a fourth particle was discovered by the track it left in a cloud chamber exposed to cosmic radiation in Pasadena, California. This new particle, the positron, was as light as the electron but carried a positive instead of a negative charge. Some positron tracks are shown in Figure 2.1. Like the neutron, the positron arrived at an opportune time. A few years earlier, in 1928, Paul Dirac had constructed a new theory of the electron that was brilliantly successful in accounting for the fine details of atomic structure. But Dirac's theory seemed to have one flaw. It predicted a sister particle for the electron, alike in all respects except the sign of its electric charge. A slot in the structure of theoretical physics was ready and waiting for the positron when Carl Anderson discovered it in 1932. (Dirac's theory also predicted a negatively charged sister for the proton, called the antiproton, but many years had to go by before it was seen. The construction of the six-billion-volt Bevatron in Berkeley, California, made possible the production of antiprotons, and they were first observed in Berkeley in 1955.)

The advance of physical theory in the late 1920s was responsible for the "rediscovery" of an old particle, the photon. Back in 1905, the same year his first important paper on relativity was published, Albert Einstein had shown that a phenomenon called the photoelectric effect (which is discussed in Section 24.1) could best be understood with the assumption that light waves are absorbed only in bundles of a definite energy. These energy packets, now called photons, behaved in some ways like particles, yet were quite different from ordinary material particles. Although they carried energy, they had no mass. They could be neither speeded up nor slowed down, but traveled always at the same invariable (and enormous) speed. They could be born and die (that is, be emitted and absorbed), whereas ordinary particles—or so it was then believed—remained in existence forever. And, unlike material particles, the photon could never be isolated at a particular point except during the moment of its birth or death; otherwise it spread in a diffuse manner through space. For all of these reasons, the photon was not associated with the electron and the proton as a true elementary particle.

The theory of quantum mechanics, discovered in 1925 and developed over the next decade, changed this view of the photon. It showed that, from a fundamental point of view, the difference between a photon and a material particle was not so great. The particle happened to have mass; the photon did not. All of their other dissimilarities could simply be understood as arising from this single difference. In particular, the quantum theory suggested that it should be possible for material particles to be created and annihilated. The photon appeared not to be so distinctive after all and it joined the list of elementary particles.

9 Very shortly a theory developed by Enrico Fermi showed that man had, in fact, been witnessing the creation of material particles for some time. Ever since early studies of radioactivity at the beginning of this century, scientists had known that some radioactive atoms shoot out high-speed electrons which, in this particular manifestation, were called beta particles. The radioactive transformation giving rise to these electrons was known as beta decay, but its revolutionary import was not suspected for many years. Since atoms were known to contain electrons, it did not seem surprising that electrons should sometimes be ejected from atoms.

Even after the discovery of the atomic nucleus in 1911, when it became clear that the beta electrons must emerge from the nucleus, the significance of beta decay was not appreciated. Electrons were simply assumed to exist within the nucleus as well as in the space surrounding the nucleus. But when the discovery of the neutron in 1932, together with vari-

* For historical reasons, a helium nucleus is called an alpha "particle," even though we now know it to be a composite of other particles.

ous theoretical difficulties, finally banished electrons from the nucleus, beta decay became perplexing. Fermi in 1934 suggested that, at the instant of the radioactive transformation, an electron came suddenly into existence in the nucleus and swiftly departed, to be recorded as the beta particle. In short, beta decay, already known for many years, represented the creation of material particles. Fermi's suggestion was, of course, more than a verbal hypothesis. Couched in mathematical language within the framework of the quantum theory, it gave a satisfying explanation of beta decay. Among other things, it predicted the possibility that the beta transformation might, for some atoms, result in positrons rather than electrons; this was at once verified with artificially produced radioactive materials. Only slightly modified, Fermi's theory still gives an adequate explanation of all beta-decay phenomena.

As so often happens with successful theories in physical science, there is an unexpected bonus. The theory does what is expected of it and then a bit more. Dirac's theory of the electron magnificently explained details of atomic structure and then surprisingly predicted the positron as well. Fermi's theory, in a similar way, accounted for beta decay and as a bonus predicted a strange new particle, the neutrino. (The first prediction of the neutrino was actually made by Wolfgang Pauli several years before Fermi's work. More accurately stated, Fermi provided a definite mathematical framework to accommodate Pauli's speculative suggestion.) According to the theory, a particle with no electric charge and little or no mass (it is now believed to be, like the photon, precisely massless) is also created at the moment of beta decay and leaves the nucleus along with the electron. This most elusive of all particles leaves no trail in a cloud chamber and travels through miles of solid material as if nothing were there. Nevertheless, theory demanded it as a new member of the elementary-particle family, and physicists would have been most upset if the neutrino had not finally been detected. The story of its observation in 1956 and the discovery of a second kind of neutrino in 1962 is told in Chapter Twenty-Six.

By the middle 1930s the electron and the proton had been joined by the photon, the neutron, and the positron (or antielectron). The neutrino, although not directly observed, was a theoretical necessity, and was added with considerable confidence to the list of particles. In addition, there was every reason to believe that antiprotons and antineutrons existed. The graviton had also been hypothesized, but little hope was held out then (or now) for its observation.

2.3 *Yukawa and the pion*

Before most scientists all over the world left their fundamental research to devote their talents to the technology of warfare, two more particles were predicted and one more particle was discovered (which was neither of the predicted particles). In a brilliant piece of theoretical work in Japan in 1935, Hideki Yukawa predicted the particle we now call the pi meson or pion. For this he was awarded a Nobel prize—after his prediction proved to be correct. Yukawa initiated a line of reasoning about the nature of forces which has become one of the key steps in the transition from our "everyday" way of looking at the world to the new way of looking at the submicroscopic world.

Force is a familiar idea in our daily life. It is a push or a pull, usually seeming to be associated with physical contact. The seat of one's chair exerts a force to hold one up. A bump into another pedestrian on the sidewalk feels like a force pushing one aside. A road exerts a force on the tires of a car to push the car forward, and if the contact between tire and road is not firm enough (as on an icy pavement), not enough force can be exerted. But we also know of forces that act without direct contact. A comb run through the hair on a dry day can attract a bit of paper without touching it. Two magnets brought close together can push or pull each other without contact. Moreover, if we think on an atomic scale, the idea of contact—of touching—becomes ill defined. In the collision between two pedestrians, the atoms in the coat sleeve of one run up against the atoms in the coat sleeve of the other. But an atom has no well-defined boundary; it is a fuzzy-edged object. It is impossible to say whether two atoms are touching or not touching. One can say only that two atoms close together exert forces on each other, and two atoms far apart feel little or no force.

By the middle of the nineteenth century, scientists had given up the naive idea of force through contact and had replaced it by "action at a distance" (Figure 2.2). The force of gravity acts over vast stretches of empty space. The electric force of comb on paper and the magnetic force between magnets require no physical contact. Even the normal contact forces—between a boxer's glove and his opponent's chin, for instance—can be understood as forces acting over a distance (true, an uncomfortably short distance) between different collections of atoms. A new entity, the "field," was invented to explain *why* forces can act through empty space, but we shall save our discussion of fields for Chapters Fifteen and Twenty-Seven.

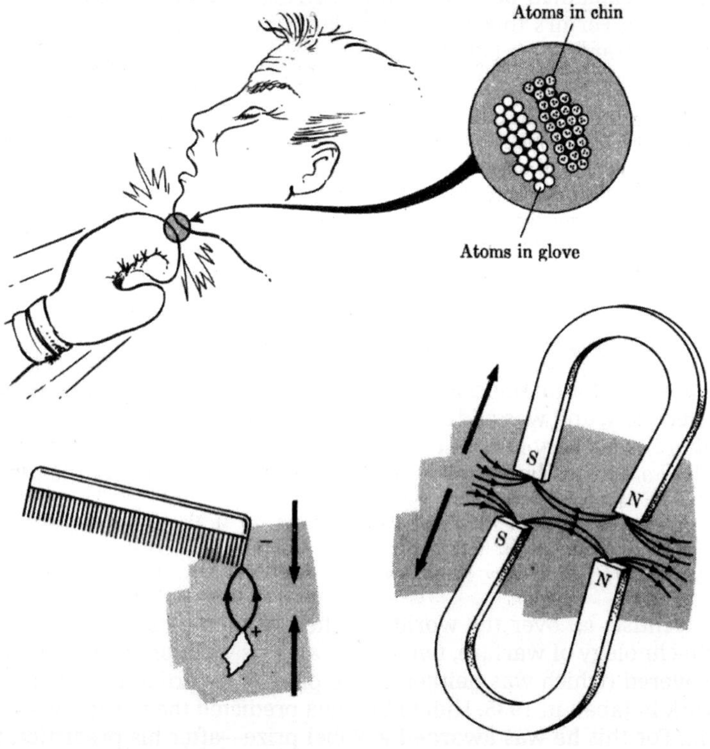

FIGURE 2.2 The nature of forces. The apparent contact force between boxer's glove and chin is really "action at a distance." Different groups of atoms exert forces on each other over an intervening space just as an electrified comb attracts a bit of paper or two magnets push on each other without direct contact.

A few years after the discovery of quantum mechanics, not long before Yukawa's work, a quantum theory of electric and magnetic forces was developed. This theory, which was responsible for the new view of the photon as an elementary particle, assigned to the photon the role of carrier of the force. A proton and an electron, for example, which attract each other electrically, continually exchange photons. Each electrically charged particle is continually emitting and absorbing photons, and it is this ceaseless exchange that gives rise to the force. Just how the exchange produces a force cannot really be visualized, but a crude analogy might help. A solitary charged particle sends out and then recaptures photons like a boy batting out a ball attached to a rubber cord which draws the ball back to him after each blow. If he is approached by another boy engaged in the same sport, each might "absorb" the ball of the other, that is, reach out and grab the other's ball as it approaches. The rubber cords would then tend to pull the boys together; in other words, the exchange would produce an attractive force. This is about as close as one can come to visualizing the process of exchanging photons; but it is not very close. In truth, photons are not attached to anything, and their exchange may give rise either to a repulsive or to an attractive force. In any case, the quantum theory of photon exchange initiated a revolutionary new understanding of the nature of electric force, and of the closely related magnetic force as well.

Yukawa gave thought to the powerful new nuclear force that acted between the nuclear particles, protons and neutrons, to hold them together in the tiny nuclei within atoms. He asked the question: What if this force also arises from an exchange between the nuclear particles? The thing exchanged, he demonstrated, could not be a photon, nor could it be any other known particle. (Yukawa's reasoning, which rests on basic ideas in the quantum theory, is explained in Chapter Twenty-Five.) The new particle should be 200 to 300 times more massive than an electron (but still six to nine times lighter than a proton). A dozen years elapsed before the Yukawa particle was finally discovered, first in cosmic radiation in 1947, and the following year in the debris from nuclear collisions in an accelerator in Berkeley, California. This particle, now called the pion, is today a very well known member of the elementary-particle family. It is 274 times more massive than the electron, and there is little doubt that it is the intermediate particle primarily responsible for nuclear forces.

2.4 The modern particles

Not long after the pion and the graviton were predicted, the muon was discovered. This discovery roughly marks the beginning of the chaotic period in elementary-particle physics which extends to the present—the period during which unexpected particles have kept appearing on the scene. The muon's trail was first identified beyond question in cloud-chamber photographs of cosmic radiation in 1936; its properties came slowly into focus over the next decade. With a mass about 200 times that of the electron, at first the muon looked like the Yukawa particle. Gradually, however, it became clear that the muon could not play the role that Yukawa had written into the drama of science. His hypothesized particle, being responsible for nuclear forces, should interact strongly with nuclei. The muon, on the contrary, was indifferent to nuclei, responding electrically to the proton's charge, but otherwise penetrating nuclei as if they were not there. Today we know the muon's properties with remarkable accuracy, but its place in the scheme of nature remains a mystery.

The collection of known particles expanded rapidly between 1947 and 1954 with the discovery of four more groups whose members are known collectively as "strange" particles (the physicist's straightforward way of confessing his mystification about these unexpected particles). The kaons (or K particles) constitute the lightest of the new groups, each kaon being about half as heavy as a proton; each of the other strange particles, consisting of the groups of lambda (Λ), sigma (Σ), and xi (Ξ) particles, is somewhat heavier than a proton. These particles were all seen first in cosmic radiation. Studies of their properties are now in progress at the sites of high-energy accelerators in the United States, western Europe, and the Soviet Union. One of these accelerators, in Stanford, California, is pictured in Figure 2.3.

Before the construction of these accelerators, physicists had to rely exclusively on cosmic radiation as a source of particles. Luckily, the earth is continually bombarded by particles from outer space, enough to provide the physicist with particles to study, but not enough to be a health hazard to the world's population sheltered beneath the blanket of air. Most of the particles are protons, some of which have exceedingly great energy. When the protons strike the air, nuclear collisions occur, giving rise to a shower of various particles, including the short-lived muons, pions, and strange particles as well as photons, electrons, and positrons. It was through cloud chambers exposed to cosmic radiation that some of the known particles were discovered. But the physicist carrying out particle experiments with the random cosmic radiation is about like an aeronautical engineer mounting a wing to be tested in an open field and hoping a high wind will spring up. As the engineer turns to the controlled conditions of a wind tunnel, the physicist has turned to the controlled conditions of the beams of particles accelerated in high-energy machines. Only for studying processes at energies beyond any yet reached by machines does he now return to the cosmic radiation (Figure 2.4).

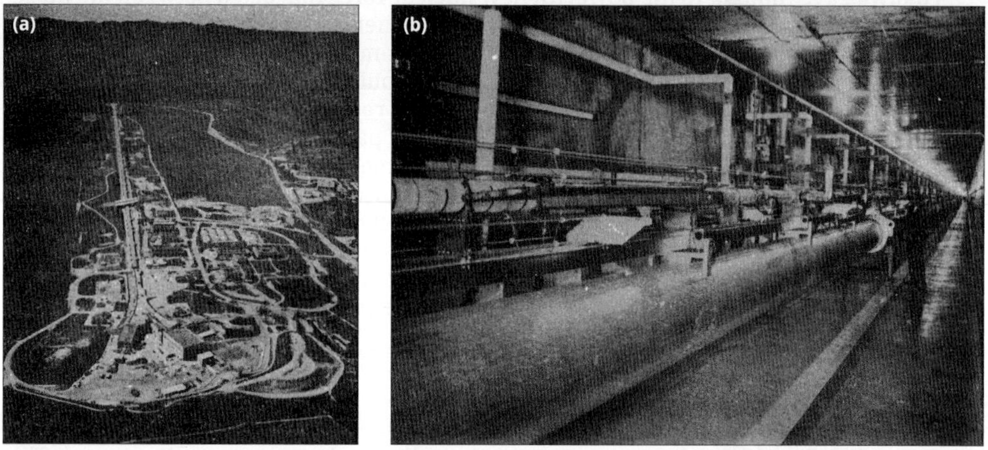

FIGURE 2.3 The Stanford Linear Accelerator. **(a)** Air view of the two-mile-long accelerator, with the "switchyard" in the foreground where energetic particles are deflected into experimental research buildings on the left or the right. **(b)** Interior view of the accelerator. The large tube at the bottom provides support and alignment. Electrons fly through the smaller beam tube mounted atop a girder. [(a) Photograph courtesy of Roland Quintero; (b) photograph courtesy of Stanford Linear Accelerator Center.]

The need for machines the size of a football field to study the tiniest things in nature is a paradox related to two remarkable connecting links discovered in this century: the connection between mass and energy, and the connection between waves and particles. These subtle connections, which have so strongly altered man's view of the small-scale world, are discussed in later chapters.

For the most part, physicists seeking new particles have been like trappers in an unfamiliar forest. They set out their bubble chambers and other detectors and wait in suspense to see what might wander into them. But in 1962 a second kind of neutrino was deliberately sought and found, proving the worth of the big new accelerators. Neutrinos cannot be created alone; they are born only in conjunction with other particles. The neutrino of Pauli and Fermi is a partner of the electron; the muon also has its neutrino partner. Once the Brookhaven accelerator was in operation, it became possible to answer the question: Are those two neutrinos the same? The answer was no, and one more member of the elementary-particle family was captured and classified.

Although the controlled conditions of the modern accelerator make possible searches for specific particles such as the muon's neutrino, the suspense and uncertainty are far from removed. There have been unexpected surprises too; recent experiments with the new machines have revealed a whole new class of super-short-lived particles. These particles come into existence and vanish again so quickly that they cannot move a noticeable distance, and they leave no direct record of their presence. Most of the "old-fashioned" particles live long enough to move at least a few centimeters—an enormous distance on the submicroscopic scale—and to leave a track in a cloud chamber or a bubble chamber. The new particles are detected only by more indirect means, and lead to the semi-philosophical question: When is a particle a particle? Some people prefer to call these new ephemeral particles "resonances." Whatever they are called, it seems clear that they are elementary structures closely related to the structures generally called particles which happen to be dignified by longer lives. Nevertheless, in the interest of avoiding complication, we shall say relatively little about these newly discovered objects.

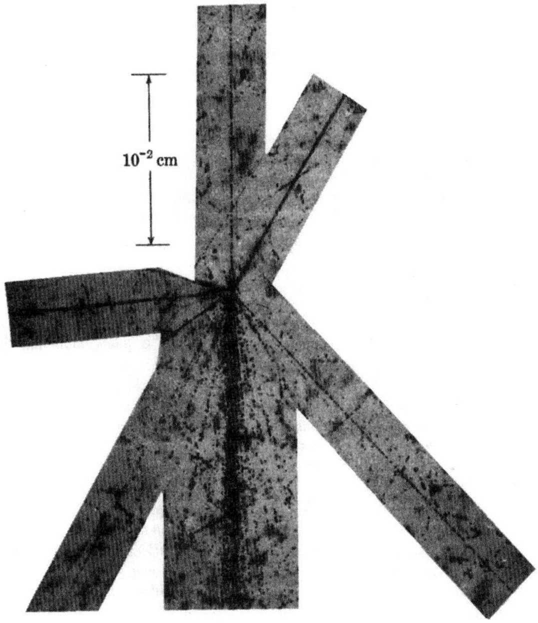

FIGURE 2.4 Cosmic ray event at very high energy. An alpha particle with an energy of 3,000 GeV (3×10^{12} eV), a component of the cosmic radiation, made the vertical track at the top. It struck a nucleus in the photographic emulsion, causing a spray of several dozen particles, most of them projected downward. Many of the tracks were made by pions and other unstable particles created at the instant of the collision. This figure is a composite of many separate photographs made through a microscope. The scale from top to bottom is a few hundredths of a centimeter. [From C. F. Powell, P. H. Fowler, and D. H. Perkins, *The Study of Elementary Particles by the Photographic Method* (New York: Pergamon Press, Inc., 1959), p. 627. Reprinted by permission of the authors and publisher.]

2.5 *The properties of the particles*

Although the number of elementary particles is disturbingly large to the theorist seeking a simple explanation of their structure and their interconnections, it is still a manageably small number—smaller, for example, than the number of different atoms known in 1900. Table 2.1 lists the names and the identifying characteristics of every known species in the elementary-particle family, excluding the resonances. (In Appendix 1 is a Greek alphabet as a helpful guide to the symbols that have come into common use in designating the particles.)

In Table 2.1, the number of different kinds of particles known with certainty is thirteen or, if we include the graviton, whose existence is so strongly suspected, fourteen. Certain very closely related particles are lumped together. The electron and its antiparticle, the positron, for example, are of the same kind. The neutron and proton are also united under one heading, the nucleon (so named because protons and neutrons are the constituents of nuclei), along with their opposites, the antiproton and the antineutron. The number of distinct individual particles, not counting the graviton, is thirty-five.

Perhaps the most important identifying characteristic of a particle is its mass. The photon, the graviton, and the neutrinos are massless. Having no mass is the same as having no inertia, that is, having no resistance to being speeded up; consequently, the massless particles always move as fast as it is possible to move, at nature's speed limit, the invariable speed of light (which could therefore just as well be called the speed of gravity or the speed of neutrinos). The lightest particle with mass is the electron (and the equally massive positron); its mass therefore provides a convenient unit for weighing in the other particles. Next comes the muon, more than two hundred times heavier than the electron. The pion is somewhat heavier than the muon (but close enough to explain why the muon, when first discovered, was mistaken for the pion). The list continues on to the heaviest known "elementary" particle, the omega, with a mass 3,276 times greater than that of the electron. There are, of course, heavier known particles— nuclei, atoms, molecules, or a speck of dirt in the eye. But all of these are understood as composites of two or more of those listed in Table 2.1. The omega particle is the heaviest particle that might be elementary, since it has not yet been explained as a composite of any of the lighter particles. That *all* of these particles are built up from some more primordial material remains, of course, a strong possibility.

TABLE 2.1 Some of the More Important Elementary Particles*

Family Name	Particle Name	Symbol	Mass	Spin	Electric Charge	Antiparticle	No. of Distinct Particles	Average Lifetime (seconds)	Typical Mode of Decay
	photon	γ	0	1	neutral	same particle	1	infinite	—
	graviton	—	0	2	neutral	same particle	1	infinite	—
Electron family	electron's neutrino	ν_e	0	½	neutral	$\overline{\nu_e}$	2	infinite	—
	electron	e^-	1	½	negative	e^+ (positron)	2	infinite	—
Muon family	muon's neutrino	ν_μ	0 (?)	½	neutral	$\overline{\nu_\mu}$	2	infinite	—
	muon	μ^-	206.77	½	negative	μ^+	2	2.20×10^{-6}	$\mu^- \to e^- + \overline{\nu_e} + \nu_\mu$
Mesons	pion	π^+	273.1	0	positive	π^- (same as the particles)		2.61×10^{-8}	$\pi^+ \to \mu^+ + \nu_\mu$
		π^-	273.1	0	negative	π^+ (same as the particles)	3	2.61×10^{-8}	$\pi^- \to \mu^- + \overline{\nu_\mu}$
		π^0	264.1	0	neutral	π^0 (same as the particles)		0.9×10^{-16}	$\pi^0 \to \gamma + \gamma$
	kaon	K^+	966.4	0	positive	$\overline{K^+}$ (negative)	4	1.23×10^{-8}	$K^+ \to \pi^+ + \pi^0$
		K^0	974.3	0	neutral	$\overline{K^0}$		0.87×10^{-10} and† 5.7×10^{-8}	$K^0 \to \pi^+ + \pi^-$
	eta	η	1074	0	neutral	same particle	1	more than 10^{-22}	$\eta \to \gamma + \gamma$
Baryons	nucleon	p (proton)	1836.10	½	positive	$\overline{p}$ (negative)	4	infinite	—
		n (neutron)	1838.63	½	neutral	$\overline{n}$		882	$n \to p + e^- + \overline{\nu_e}$
	lambda	Λ^0	2183.1	½	neutral	$\overline{\Lambda^0}$	2	2.51×10^{-10}	$\Lambda^0 \to p + \pi^-$
	sigma	Σ^+	2327.7	½	positive	$\overline{\Sigma^+}$ (negative)		8.1×10^{-11}	$\Sigma^+ \to n + \pi^+$
		Σ^-	2343.3	½	negative	$\overline{\Sigma^-}$ (positive)	6	1.6×10^{-10}	$\Sigma^- \to n + \pi^-$
		Σ^0	2333.7	½	neutral	$\overline{\Sigma^0}$		about 10^{-20}	$\Sigma^0 \to \Lambda^0 + \gamma$
	xi	Ξ^-	2585.5	½	negative	$\overline{\Xi^-}$ (positive)	4	1.75×10^{-10}	$\Xi^- \to \Lambda^0 + \pi^-$
		Ξ^0	2573	½	neutral	$\overline{\Xi^0}$		3.0×10^{-10}	$\Xi^0 \to \Lambda^0 + \pi^0$
	omega	Ω^-	3276	3/2	negative	$\overline{\Omega^-}$ (positive)	2	1.5×10^{-10}	$\Omega^- \to \Xi^0 + \pi^-$
							36		

* This table includes all known particles that do not decay by means of the strong interactions.

† The K^0 meson has two different lifetimes. All other particles have only one.

The Dirac theory of the electron first predicted that a particle should be accompanied in nature by a sister particle, identical in mass but opposite in electric charge and in some other intrinsic properties. This sister particle is usually called an antiparticle (although it is itself a perfectly good particle), and it appears that most particles in nature have distinct antiparticles. (The antiparticle of the antiparticle is the original particle.) The photon, the graviton, the neutral pion, and the eta are special; for each of them, the particle and the antiparticle are identical. For all the others, particle and antiparticle differ. The antineutron, for example, is distinguishable from the neutron, even though both are neutral. The proton and antiproton are even more easily distinguishable, since one is positively charged, the other negatively charged. A horizontal bar over a particle symbol is used to designate the antiparticle.

The fact that particle-antiparticle pairs exist in nature was a startling and unexpected consequence of merging the theory of relativity and the theory of quantum mechanics. It is scarcely possible to describe in words how that came about. Relativity is a kind of legalistic theory which imposes its rules on all other theories. "Lesser" theories (for example, quantum mechanics) must satisfy the rule that experimenters carrying out the same measurements in different frames of reference must get the same results. It was found impossible to construct a theory of single particles that conformed to this invariance requirement of relativity. Only if antiparticles were admitted as well could the laws of relativity be satisfied. It is rather as if one were given a set of identical colored strips like those pictured in Figure 2.5, red on one side, green on the other, and were asked to construct from them any designs whatever, subject only to the invariance requirement that the completed design should appear the same to a man who could see only red as to a man who could see only green. It would not take long for one to discover the impossibility of this task unless one were also furnished with some "antistrips" like those in Figure 2.6. In a similar way, particles alone proved to be insufficient to construct a relativistic theory; antiparticles were needed as well. There is now ample evidence for antiparticles, and their existence gives support to the theories of both relativity and quantum mechanics. Figure 2.7 is an interesting photograph showing an antilambda and two antiprotons.

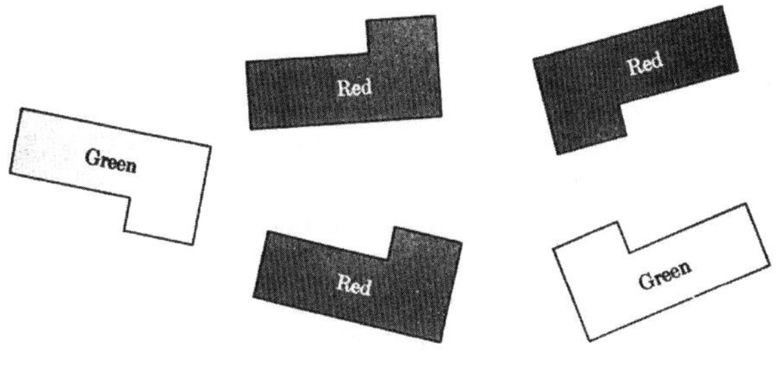

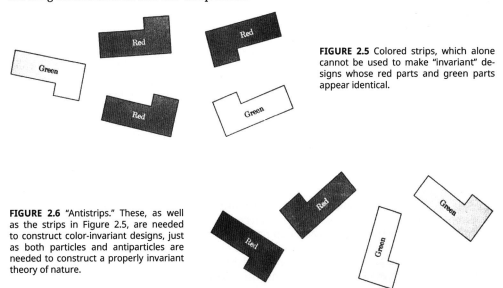

FIGURE 2.5 Colored strips, which alone cannot be used to make "invariant" designs whose red parts and green parts appear identical.

FIGURE 2.6 "Antistrips." These, as well as the strips in Figure 2.5, are needed to construct color-invariant designs, just as both particles and antiparticles are needed to construct a properly invariant theory of nature.

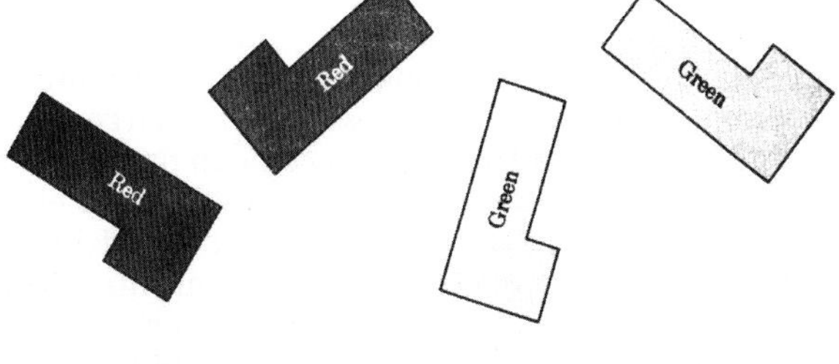

A fascinating property possessed by most of the particles is spin. This means that each particle is spinning about an axis like a top. There is actually no way to tell how fast a particle is spinning. The "amount" of its spin cannot be measured in terms of revolutions per minute, but must be measured in units of angular momentum. Angular momentum, the subject of Chapter Ten, is a somewhat more complicated quantity, depending not only on speed of rotation but also on the mass and on the size of the rotating object. For the present discussion, we need note only that more speed, or more mass, or greater size, or any combination of increases of these quantities, produces greater angular momentum. A spinning electron has a certain angular momentum; a spinning top has much more because it is vastly bigger and heavier; a spinning merry-go-round has still more, even though its rate of turning is lower. Angular momentum can be thought of as the strength of rotation, and it is related to the effort required to start or stop the rotational motion.

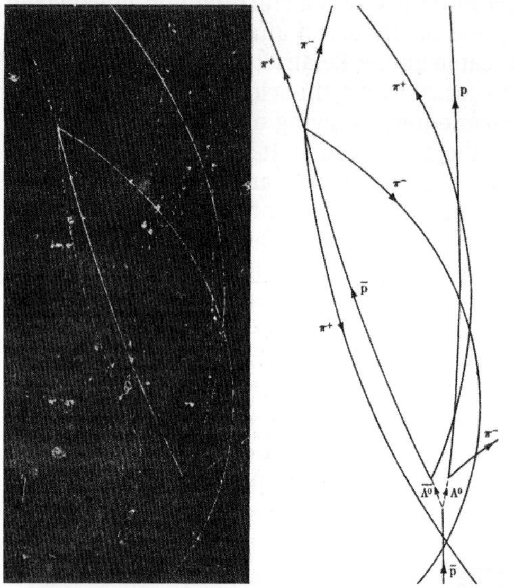

FIGURE 2.7 Particles and antiparticles. An antiproton entering the bubble chamber at the bottom vanishes in collision with a proton in the chamber, giving rise, in a rare event, to a lambda-antilambda pair. These neutral particles leave no tracks, but decay into charged particles which do. One of the products of the antilambda decay is another antiproton whose annihilation in the upper left part of the picture produces a spray of pions. Because the chamber is in a magnetic field, positive particles are deflected in one direction, negative particles in the other direction. (Photograph courtesy of Lawrence Radiation Laboratory, University of California, Berkeley.)

Spin in the submicroscopic world is conveniently measured in units of the photon spin, given in the table simply as 1. The electron has a spin of ½; that is, it has half as much rotational angular momentum as a photon. A few particles, such as pions, have no spin at all.

According to the quantum theory, some physical quantities can take on only distinct, separated values ("discrete values," in the language of mathematics), and angular momentum is one of these. If the same type of law applied to speed in the everyday world, one might have a car that could travel only at 10 miles per hour, or at 20, or at 30, but not at any intermediate speed. This would make traveling rather jerky. Indeed the quantum theory gives to the submicroscopic world just such a jerkiness; physical systems remain in one state of motion for awhile, then change suddenly to another state of motion. The spins of the elementary particles, however, although pegged only at certain allowed values, remain fixed for each particle as long as that particle continues to exist.

Some of the particles possess electric charge, some are neutral. Charge, like spin, is one of those quantities in nature that come in packages of a certain size. There is no such thing as half an electron charge or three and a third electron charges. The amount of electric charge on the electron is nature's unit of electricity. In fact, every other elementary particle in Table 2.1 is either neutral or has the same magnitude of electric charge as the electron, either positive or negative. The reason for this is not known. Some of the resonances do have two units of charge, but none of the longer-lived objects we have chosen to call "bona-fide" particles has more charge than the electron.

A given species of particle may exist in different "charge states." An electron has a single charge state; it is negative. The neutral lambda also has only one charge state. Each of these has an antiparticle, the positively charged positron, and the neutral antilambda, respectively. The sigma particle, a diversified species, exists in three charge states, positive, negative, and neutral, and to each of these corresponds an antisigma, so that altogether there are six different sigma particles. The pion species is a special case, for the antiparticle of the positive pion is the negative pion, itself one of the pion particles.

Roughly, to say that a particle exists in different charge states means that several par- 10
ticles having different charge are so nearly alike in all other respects that it is only logical to regard them as, in some way, different manifestations of the "same" particle. The proton and neutron, for example, the building blocks of atomic nuclei, differ greatly in their electric and magnetic properties, but are nearly identical in mass and seem to experience identical nuclear forces. They are, therefore, now regarded as different charge states of the same basic particle, the nucleon. It is roughly like issuing the same basic automobile chassis and engine in two quite different body styles.

The mass of a particle, its charge, and its spin are among its most important identifying features, but these by no means exhaust the properties a particle may have. To know a particle fully one must know how it interacts with each other particle. The interaction properties are not yet fully known, and are the subject of the most intensive investigation.

The most dramatic consequence of particle interaction is the transformation of one kind of particle into two or more other lighter particles. A charged pion, for example, left to itself, will live about two hundredths of one millionth of a second, and then spontaneously vanish, to be replaced most probably by a muon and a neutrino, but possibly by an electron and a neutrino (Figure 2.8). The pion is said to "decay" into the two lighter particles. Most of the known particles are unstable and decay in a variety of ways after a very short time. Table 2.1 shows the characteristic lifetimes for the particles and the typical modes of decay for those that are unstable. Only the proton, the electron, and the massless particles are stable; that is, each has—as far as we know—an infinite lifetime.

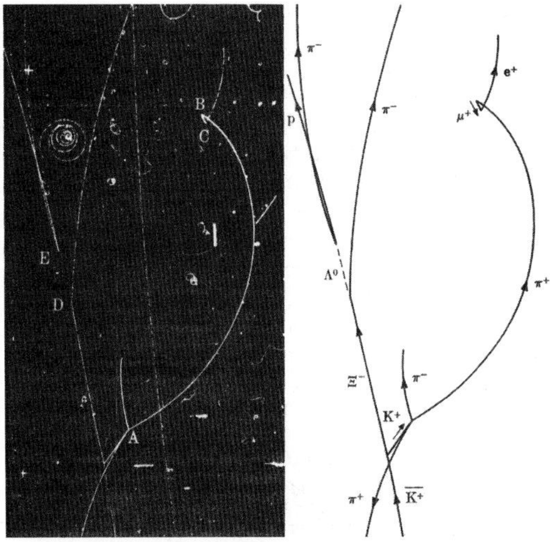

FIGURE 2.8 Decay of unstable particles. This unusual bubble-chamber photograph shows the decay of five different elementary particles. At point A, a positive kaon decays into three pions. At B, one of these pions decays into a muon and an unseen neutrino. At C, the muon decays into a positron (plus a neutrino and an antineutrino). At point D, a xi particle decays into a lambda particle and a pion. The invisible neutral lambda decays into a proton and a pion at point E. (Photograph courtesy of Lawrence Radiation Laboratory, University of California, Berkeley.)

Without the stability of the electron and the proton, man would not be on hand to study the elementary particles. The stability of the massless particles is of no value for constructing matter, for the photon, graviton, and neutrinos cannot be corralled and held in one place for use as bricks in the structure of the universe. They fly forever onward at the speed of light. This leaves only the electron and proton as suitable building blocks of the world. In addition, by a stroke of good luck, the neutron can be stabilized when joined with protons. A solitary neutron decays into a proton, an electron, and an antineutrino after an average lifetime of about 16 minutes. But when the neutron joins a proton, the attractive force holding them together causes a lowering of the energy of the neutron which, in turn, prohibits its decay. Exactly how this effect, which is dependent on the mass-energy equivalence, comes about is discussed in Chapter Twenty-Five. Without this stabilizing effect the world would contain only hydrogen, for the hydrogen atom is the only atom built exclusively of electrons and protons (one of each). The existence of every other substance in the world depends upon the fact that the nuclear forces are powerful enough to stabilize the normally unstable neutron and make it also available as a universal building block.

Table 2.1 contains one more piece of information, the classification of particles into family groups: the baryons (heavy particles), the mesons (intermediate particles), and the muon and electron families. The muon, the electron, and their respective neutrinos, which comprise the last two families, are also known as leptons (light particles). The pion, kaon, and eta are mesons; the proton and all heavier particles—neutron through omega—are the baryons. These words of Greek origin do much more than label particles according to their mass. Note in Table 2.1, for example, that the mesons have spin 0, while the leptons and baryons do not. More important, the groups deserve to be separated and given family names because there are laws of conservation for three of the four families. Whenever one baryon disappears, another one appears in its place. It is the law of baryon conservation that stabilizes the proton. The proton cannot decay into any lighter particles because it is itself the lightest baryon, and such a decay would require the disappearance of a baryon with no replacement. Similar laws apply to the muon family and to the electron family, but not to the mesons, which, like photons, can be created and annihilated in arbitrary numbers. The law of baryon conservation has been tested experimentally to phenomenally high precision (so that there is no chance that our world will collapse from the instability of protons within the next few hundred billion years), and the two lepton-conservation laws are also reasonably certain. Nevertheless, we have no theoretical understanding of the basis of these laws.

2.6 Experimental tools

Modern science got its start with the study of objects visible to the human eye—the sun, the planets in the night sky, wooden balls rolling down inclined surfaces. The most marked characteristic of contemporary science is that it deals with domains of nature outside the range of human perception—the interior of stars, the structure of molecules, the interactions of elementary particles. Man's picture of nature must now rest on the indirect evidence supplied by instruments of his own design rather than on the direct evidence of his senses.

Actually, the fact that we cannot directly perceive the parts of nature we now study is a much less revolutionary aspect of contemporary science than might appear at first. The true revolution of science was wrought in the sixteenth and seventeenth centuries, when man supplemented the subjectivity of perception with the objectivity of careful quantitative measurements. Our dependence now on instruments and machines of remarkable complexity and refinement to reveal phenomena beyond the range of human perception is only the natural evolutionary result of turning from qualitative observation to quantitative measurement. There has been a revolution of twentieth-century physics, but it is a revolution in our view of nature and our outlook on the form of natural laws, not a true revolution in how we study nature. The great surprise and great insight afforded by science in this century is the fact that the laws of nature governing domains beyond the range of human perception violate common sense. Nature at its most fundamental behaves quite differently from nature in the macroscopic world. Extrapolation of observational techniques to new domains has been successful, but extrapolation of ideas consistent with human perception has not worked. This is the true revolution of recent scientific advance, and it will be a dominant theme in Parts 6 and 7 of this book.

Man's perceptions are limited in scope. More important for the scientist, they are severely limited in accuracy. It is by no means necessary to leave the macroscopic world in order to go beyond the range of human perception. A butcher who weighs a piece of meat on a scale is turning to a technical device to extend the range and precision of his senses. That we cannot "see" an atom is fundamentally not a more serious problem than the fact that we cannot weigh a piece of meat by hand. In either case we must supplement human perception by mechanical devices and must form a judgment about something in the world outside ourselves based on information supplied by these devices. Even the simple measuring tape of a housewife is a technical aid to observation. Throughout the whole history of modern science man has leaned heavily on machines, devices, and equipment whose sole purpose was to extend the range and the precision of his sense perceptions. In the sixteenth century, Tycho de Brahe needed what we would now call a giant protractor to measure angles of inclination of stars and planets in the sky. Later, Galileo, in studying motion, needed an accurate way to measure time intervals and invented a new clock for the purpose. Only because of the telescope could Olaus Römer in the seventeenth century use observations of the moons of Jupiter to determine that light has a finite speed. With every step forward in fundamental understanding, man learned how to base new experimental equipment on his new knowledge. With every new device he could probe to new domains, or measure more accurately in old domains, laying the experimental basis for still deeper understanding. Now in studying elementary particles and seeking a pattern of nature in the subatomic domain, man uses a remarkable array of technological aids to observation. These devices, the most recent and the most complicated in a long history of mechanical links between man and nature, rest on every earlier fundamental theory of nature.

The devices used in elementary particle research are basically of two kinds, those used to create particles and those used to detect particles. The creation of material particles requires energy, and that energy must be highly concentrated within the dimensions of an atomic nucleus or less. This means that the energy must itself be carried by a particle. Very energetic particles bombard the earth continually from outer space (about one each second in each square centimeter), and these supplied the concentrated energy needed in early studies of the particles in the 1930s and 1940s. In most contemporary research, however, beams of "man-made" particles are used. An accelerator, or high-energy machine as it is frequently called, by repeated application of electric forces, pushes charged particles to high energy. The energetic particles, concentrated in a narrow beam, strike atomic nuclei in a target and there undergo interactions or create other particles which fly on to interact in another target or in a detector.

The earliest accelerators were called cyclotrons, Cockcroft-Walton accelerators, and Van de Graaf accelerators. The principle of operation of the cyclotron is discussed in Chapter Seventeen. Such accelerators are still in operation, but they do not achieve energies as great as more recent machines called linear accelerators and synchrotrons. So far no one has succeeded in accelerating neutral particles or unstable particles—the first because charge is the only convenient "handle" for pushing a particle, the second because even the longest lived unstable particle with charge, the muon, lives but two millionths of a second. Projectiles in accelerators have therefore been limited to electrons, protons, and atomic nuclei.

In the Alternating Gradient Synchrotron (AGS) in Brookhaven, New York (Figure 2.9), the acceleration of protons to 33 GeV (billion electron volts) is an achievement resting upon the theories of mechanics and electromagnetism, the more recent theory of relativity, and a host of engineering and technical developments of this century. Electromagnetism enters in several vital ways. Bombardment of hydrogen atoms by low-energy electrons strips away the orbital electrons from the atoms, leaving the central proton alone. In a linear accelerator electric forces pull the freed protons down a straight tube until they are "injected" at modest energy (50 MeV) into the slender evacuated doughnutshaped tube of the synchrotron—about four inches across and half a mile around its periphery. Here a weak magnetic field causes the protons to be deflected in such a way that they follow around the curve of the beam tube without striking its walls. At intervals the protons receive impulses of electric force to accelerate them forward. At the same time the strength of the magnetic field is gradually increased, for as the protons gain energy, a stronger magnetic field is required to bend their trajectories along the same curved path. After a time less than one second, the protons have covered 150,000 miles in 300,000 trips around the synchrotron. They strike a target and the cycle can begin again. Throughout their acceleration, precise timing of the accelerating impulses and of the increasing magnetic field has been controlled by electronic circuitry attached to the machine. When they strike the target at full energy, the protons are traveling at 99.96 per cent of the speed of light. Since relativity is that theory describing the motion of high-speed objects, it is clear that the successful acceleration of the protons to such enormous speed while guiding them within a 4-inch channel over a distance of 150,000 miles would have been out of the question without an understanding of relativity.

FIGURE 2.9 The Alternating Gradient Synchrotron at Brookhaven, New York. **(a)** An air view of the AGS, visible only as a doughnutshaped mound. **(b)** The injection area of the accelerator. Protons of 50 MeV energy coming from a linear accelerator in the left background may proceed straight ahead to be injected into the circular accelerator (only a small portion of which is visible here), or they may be deflected down the tube in the left foreground for separate analysis. The AGS accelerates about 7×10^{12} protons per minute to an energy greater than 30 GeV. (Photographs courtesy of Brookhaven National Laboratory.)

J. J. Thomson detected electrons by observing a fluorescent glow where the beam of electrons struck the glass wall of his cathode-ray tube. At about the same time (near 1900) Ernest Rutherford detected alpha particles by noticing that they electrified the air in passing through it and thereby caused an electrified object (an electroscope) to lose its charge to the air more rapidly than otherwise. Very early in this century methods were discovered to detect single particles rather than the cumulative effect of many particles. Rutherford observed through a microscope the tiny flashes of light at points on a zinc sulphide screen where alpha particles struck. In 1908, his assistant Hans Geiger invented what we now call the Geiger counter. Within the Geiger counter a highly charged central wire is almost but not quite able to cause a spark to jump to or from the walls of the counter tube. The electrification of the gas within the tube caused by the passage of a single energetic charged particle through the counter is enough to trigger the spark and record the particle. Other counters have been invented since then. Especially popular detectors of more recent invention allow the experimenter not merely to detect the passage of a single particle by means of a single pulse of energy or light, but actually to see (or to photograph) the path of the particle as it passes through the detector. Such detectors include the cloud chamber, the spark chamber, and the bubble chamber. A cloud-chamber photograph appears in Figure 2.1, bubble-chamber photographs in Figures 2.7 and 2.8.

Almost every method of particle detection rests upon the fact that the particle to be detected has an energy larger by far than the average energy of the atoms and molecules of the matter through which the particle is passing. The constituent atoms and molecules of ordinary matter have a ceaseless random motion. Like a restless flock of sheep, they mill about colliding frequently with each other. Into their midst charges the more energetic particle like a wolf, literally tearing some atoms apart, and leaving behind a swath of destruction. (The difference is that the atoms quickly repair the damage and resume their normal behavior.) To make this comparison quantitative, we note that atoms at normal temperature move about with an energy of about one twenty-fifth of an electron volt,* whereas the particles to be detected might have energies of millions or even billions of electron volts. The energy required to strip an electron away from an atom is about ten electron volts, several hundred times greater than the energy of random motion of the atoms. Thus in the jostling of one atom against another, not enough energy is involved in the collision to set free an electron, and the atoms remain neutral. But an energetic particle flying through the material can strip off one electron after another as it collides with successive atoms. It leaves behind a trail of freed electrons and positively charged atoms called ions. This process, called ionization, is what is responsible for the deleterious effects of high-energy radiation in living cells. It is also what makes it relatively easy to detect a high-energy charged particle.

In the Geiger counter (Figure 2.10), the trail of ions and electrons left behind the energetic particle are set into motion by the electric field acting between the central wire and the outer wall of the counter. They gain enough energy to do some more ionizing themselves, and a cascade of electric current rapidly builds up—that is, a spark jumps the gap. A similar principle operates in the more modern spark chamber. The wire and the wall of the Geiger counter are replaced by two parallel metal plates, also charged in such a way that a spark is almost but not quite able to jump the gap between them. Then many more plates are added parallel to the first pair, charged alternately positive and negative to prepare the possibility of a spark jumping between any plate and either of its neighbors. In research applications, the plates are charged for only a few millionths of a second just after the passage of one or more particles through the chamber. A sufficiently energetic particle can penetrate the entire stack of parallel plates. Each time it crosses the gas-filled gap between two plates, it leaves a trail of ions which serves as a localized trigger of sparking. Looking in from the side one sees a series of sparks that provide a visual record of the path followed by the particle. Figure 2.11 shows a spark-chamber photograph.

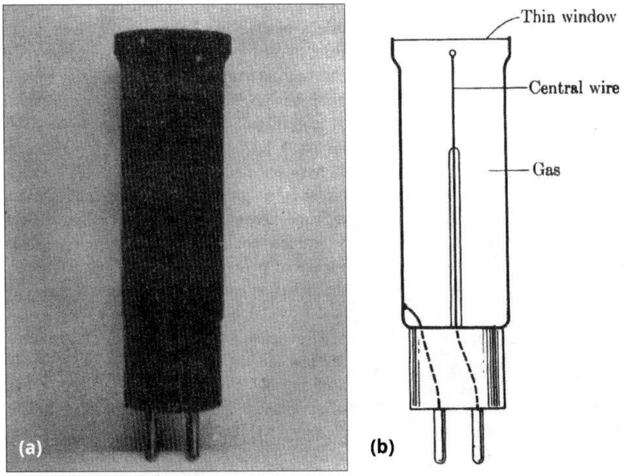

FIGURE 2.10 The Geiger counter. **(a)** A typical modern Geiger counter. **(b)** Its essential ingredients: a central wire maintained at high voltage relative to the wall of the tube, a low-density gas able to conduct a short burst of current, and a thin window to allow particles of relatively low energy to enter the tube.

* The electron volt is a common unit of energy in the world of particles. It is defined in Section 3.6.

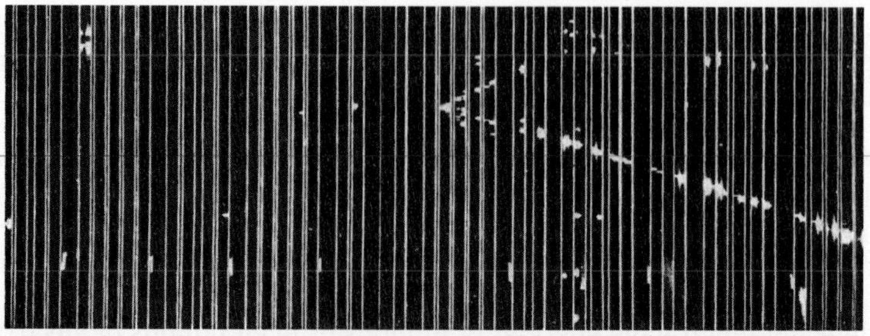

FIGURE 2.11 Spark-chamber photograph. An unseen neutrino entering the chamber from the left interacted with a neutron within an atomic nucleus in one of the plates to produce a muon and a proton. The longer line of sparks traces the path of the muon; the shorter path is that of the proton. (Photograph courtesy of European Organization for Nuclear Research [CERN].)

The process of ionization is also crucial to the operation of the cloud chamber and the bubble chamber. In the cloud chamber, ions left in the wake of the fast charged particles form centers of condensation for vapor in the gas-filled chamber. Tiny liquid droplets form in a line along the track of the particle, and are photographed before they have had a chance to drift away. (Ordinary water droplets in the air are also frequently centered about ions, formed by random cosmic rays or by ultraviolet light from the sun.) The particle trails seen in a bubble chamber are gas bubbles in a liquid rather than liquid droplets in a gas. In the bubble chamber, ions act as centers of boiling. With timing even more critical than in a cloud chamber, the thin line of boiling liquid along the trail of the particle is photographed before the boiling spreads throughout the liquid. It should be added that both the cloud chamber and the bubble chamber must be prepared to be in a particular critical condition just before the photograph is taken. The vapor in the cloud chamber must be "supersaturated." This means roughly that it is more than ready to condense into liquid, and will do so readily on any convenient center of condensation, such as an ion. In a somewhat similar way, the bubble chamber liquid must be "superheated," that is, more than ready to boil. Supersaturation is achieved by sudden cooling of the cloud chamber. Superheating is

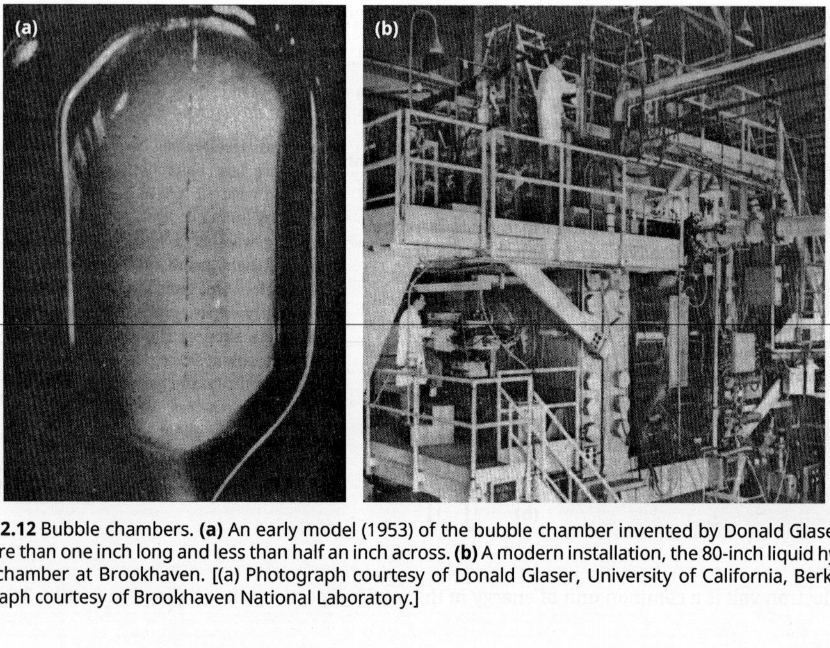

FIGURE 2.12 Bubble chambers. **(a)** An early model (1953) of the bubble chamber invented by Donald Glaser. It is a little more than one inch long and less than half an inch across. **(b)** A modern installation, the 80-inch liquid hydrogen bubble chamber at Brookhaven. [(a) Photograph courtesy of Donald Glaser, University of California, Berkeley; (b) photograph courtesy of Brookhaven National Laboratory.]

achieved by a sudden release of pressure in the bubble chamber. In Figure 2.12, one of the first bubble chambers is compared with a more recent installation.

Neutral particles in passing through matter cause a great deal less disturbance than do charged particles. They leave very few ions in their wake, and as a result they leave no tracks in cloud chambers or bubble chambers. In these detectors the paths of neutral particles must be inferred indirectly (such inference can be made in the photographs of Figures 2.7 and 2.8). In a Geiger counter, on the other hand, the neutral photon, or gamma ray, can be detected because even very little ionization can trigger a spark if the counter is suitably adjusted to be only a hairline away from sparking. The neutron, even less obtrusive than a gamma ray, requires other kinds of detectors. It can, for example, be allowed to strike protons. These charged particles recoil to produce ionization, and so by secondary means the neutron can be detected. Or the neutron may initiate a nuclear reaction which gives rise to gamma rays, these in turn being observable.

Only the neutrino now offers a serious challenge to physicists wanting to record particles. The story of the first detection of the neutrino is told in Chapter Twenty-Six.

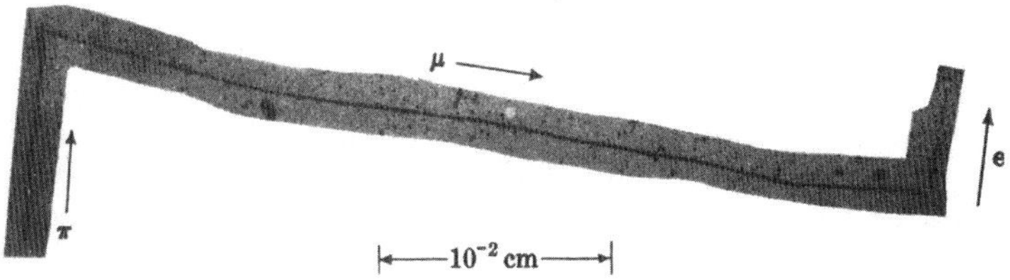

FIGURE 2.13 Early example of pion-to-muon decay. After being brought to rest in a photographic emulsion, the pion decays into a muon and an unseen neutrino. The muon, as it slows, leaves an increasingly heavy track. After being stopped, the muon also decays, as is evidenced by the track of a high-speed electron on the right. [From C. F. Powell, P. H. Fowler, and D. H. Perkins, *The Study of Elementary Particles by the Photographic Method* (New York: Pergamon Press, Inc., 1959), p. 245. Reprinted by permission of the authors and publisher.]

2.7 *How do we know?*

The acts of accelerating particles, creating new particles, and detecting single particles are now commonplace in the laboratory. Questioning, as some philosophers do, whether particles "really" exist, since they are at best indirectly observed by man, is by now an idle pastime. It makes little more sense than questioning whether a field is "really" one acre, since it has been surveyed by transit and light beam rather than by pacing it off; or whether a roast of beef really weighs three pounds since that is only the reading of a scale, not something directly experienced by the customer. A massive accumulation of interdependent data support the concept of particle. Although a theoretical possibility, it is, practically speaking, impossible to conceive that the world of the very small might successfully and simply be described without molecules, atoms, nuclei, and particles. Nevertheless, an important question to ask is: How do we know what we claim to know about the particles? We pose this question here not because it can be answered in a few paragraphs, or even in a chapter, but only because its importance needs to be emphasized. In fact, there is not one answer, but many different answers. The charge of the electron has been measured in one way, the lifetime of the muon in another, the size of the proton in another, the spin of the lambda in another, and so on. The ingenuity and technical skill of scientists has been applied in a thousand ways to measure the properties of the particles, as well as the quantities of interest in other fields of physics. So numerous are the approaches to experimental problems, and so complicated the techniques frequently employed, that it would be out of place (as well as impossible) in a book about the concepts and ideas of physics to try to buttress every fact with a description of how that fact was determined. Instead the reader will have to accept most of the facts without the evidence. However, in the remaining chapters, certain key experiments are discussed. The selection is based on two criteria—that the experiment illustrate some fundamental facet of a physical theory, or that it be a measurement of a quantity that seems especially important. Fortunately, many of the most crucial experiments in physics can be understood and appreciated without advanced technical or mathematical skill.

To give an example, we may consider the discovery of the pion in 1947. The identification of the pion as a new and different particle made use of still another detector not mentioned so far, the photographic emulsion. Actually the photographic emulsion is nearly the oldest of particle detectors, having figured in the discovery of radioactivity by Becquerel in 1896. At that time the emulsion reacted to particles only by a general fogging or blackening of the plate. Later, individual tracks of alpha particles and protons were observed, and in the 1940s came technical developments that converted the emulsion into a precision research tool. Especially in the emulsions produced by Ilford, Ltd., specific features of a track, such as its grain density and its deviation from a straight line, could be used to reveal detailed information about the mass, speed, and charge of the particle that made the track. While studying the tracks left in emulsions by cosmic ray particles, Cecil F. Powell and his co-workers learned in 1947 that there were two different particles intermediate in mass between the electron and the proton (we now know more), not one as had been supposed on the basis of cloud-chamber studies. The heavier of these intermediate particles (or mesons), now called the pion, decayed into the lighter of the pair, now called the muon. One of the earliest examples of a pion-to-muon decay recorded in a photographic emulsion is shown in Figure 2.13. Shortly thereafter the charged kaon was discovered by Powell in a similar way.*

2.8 *The significance of the particles*

In view of the incredibly short lifetimes of most of the particles, one may be tempted to ask why they seem to be so important. A proton is obviously important; all matter, animate and inanimate, contains protons. But why a lambda particle, which is not a constituent of anything we know? In a sufficiently violent nuclear collision brought about with the help of a man-made accelerator, a lambda particle may be created. It travels a few centimeters in less than one billionth of a second, then decays into a nucleon and a pion. The pion, after a slightly longer time, decays into a muon and a neutrino. The muon decays shortly into an electron, a neutrino, and an antineutrino. In a total time of about a millionth of a second, and all within a few feet of the point of the initial collision, a sequence of transient particles has been born and died, with no net effect beyond the addition to the universe of a few more neutrinos.

There are two reasons why physicists believe that the short-lived unstable particles are fully as important and interesting as the few stable particles that compose our world. In the first place, the unstable particles may have a vitally important effect on the properties of the stable particles. To give the most important example, the force that holds nuclei together (and therefore makes possible the existence of all atoms heavier than hydrogen) arises from the exchange of unstable pions between the nuclear particles. The second, perhaps deeper, reason is that it appears to be entirely a matter of "chance" which particles are stable and which are unstable. The muon and the electron, for example, appear to be nearly identical in all respects except that the muon happens to be heavier than the electron. It can therefore release its extra mass energy and decay spontaneously into an electron (and neutrinos). The muon lives two millionths of a second; the electron apparently lives forever. Yet this difference is less striking to the physicist than are the many points of similarity between

* For his development of the technique of particle detection by photographic emulsions and for his discoveries of mesons, Powell was awarded the Nobel prize in physics in 1950.

muon and electron. It seems very unlikely that the "true" nature of the electron will ever be understood unless the closely related muon is understood at the same time.

All the elementary particles seem to belong to one big family. No one of them is independent of all the others. The "normal" thing is for a particle to undergo decay and transmute itself into other lighter particles. For reasons that are not yet fully understood, there are two "abnormal" particles, the proton and the electron, which are prohibited from decaying. According to this larger view of the particles, there are certain rules of nature, described in Chapter Four, that happen to prevent the decay of these two particles. Because of this chance, the construction of a material world is possible.

Of course, since there is only one universe, and one set of natural laws, it does not make much sense to say that a particular state of affairs in the world exists by chance. But this view of the multiplicity of particles continues the process, begun by Copernicus, of making man feel more and more humble when facing the design of nature. We and our world exist by the grace of certain conservation laws which stabilize a few particles and permit an orderly structure to be built upon the normal chaos of the submicroscopic world.*

Is the exploring physicist nearing the end in the discovery of new particles? It would be very rash to say so. Since 1960 the muon's neutrino has been identified, many new short-lived resonances have been found and the omega particle has been discovered. The list of "elementary" objects will probably increase before it decreases again, and will include a spectrum of particles whose lifetimes vary from the incredibly short to the infinitely long.

In Chapter Four we shall be coming to grips with some ideas about nature that the particles have generated or have illuminated. Before doing so, it is essential to gain some understanding of various quantities such as charge, mass, and energy, and to visualize the scale of the particle world. These are the goals of Chapter Three.

EXERCISES

2.1. An alpha particle is the same thing as a nucleus of a helium atom (He^4). Is an alpha particle an elementary particle? Is a beta particle an elementary particle? Explain.

2.2. **(1)** A charged particle that moves 10 microns in a photographic emulsion leaves a clearly discernible track. If the particle traveled at close to the speed of light (3×10^{10} cm/sec), how long must it live in order to move 10 microns? (One micron is 10^{-6} meters or 10^{-4} cm.) Compare this time with average lifetimes of charged particles in Table 2.1. **(2)** Moving at 10^9 cm/sec, how far does a neutral sigma progress in its average lifetime? Is this distance greater or less than the diameter of a single atom?

2.3. **(1)** Moving at 10^{10} cm/sec, which of the particles in Table 2.1 move too short a distance to leave a discernible track? **(2)** At the same speed, which particle moves an average distance greater than the diameter of the earth before decaying?

2.4. Study Table 2.1 in order to answer the following questions. **(1)** What are the final end products of a Ξ^0 particle after the successive decays have run their course? **(2)** As one moves from nucleon to omega through successively heavier baryons, what happens to the average electric charge of each particle type?

2.5. **(1)** Examine with care the five pion tracks in Figure 2.8. **(a)** What rule relates the sign of the charge to the direction of curvature? **(b)** What is the correlation between intensity of track and speed of particle? (A slower particle is more strongly deflected). **(2)** A small spiral track in the upper left part of the picture was made by an electron. Was this electron positive or negative? How can you deduce its direction of motion?

* This theme is elaborated in Section 27.8.

2.6. Name several objects in the realm of your own experience that you "explain" by describing their composition in terms of smaller units.

2.7. **(1)** Mention several ways in which you use technological aids to observation, to increase either the accuracy or the scope of your sense perception. **(2)** Give any example you may have learned about of a device whose invention rested on basic science and which itself became a tool of basic science. (Do not repeat examples given in the text.)

2.8. Do you think it would be possible to make an "atom" out of a proton and a negative muon? a proton and a negative pion? a positive electron and a negative electron? a proton and a positron?

2.9. Comment in a short paragraph on this statement: "It is not reasonable that all the particles we know are truly elementary."

2.10. When a physicist calls a particle elementary, what does he mean? When Sherlock Holmes says, "Elementary, my dear Watson," what does he mean?

THREE

The large and the small

It is easy to talk about the "incredibly short" lifetime of an elementary particle or about the "fantastically small" size of an atomic nucleus. It is not so easy to visualize these things. On the submicroscopic frontier of science—as well as on the cosmological frontier—man has proceeded so far away from the familiar scale of the world encompassed by his senses, that he must make a real effort of the imagination to relate these new frontiers to the ordinary world. The reward of being able to think pictorially over the whole panorama, from infinitesimal to enormous, adequately repays the effort.

3.1 Measurements

In order to describe nature, the scientist needs a number of concepts which are so well defined that they are not merely descriptive ideas but are measurable quantities. The simplest of these are the ideas of size (length measurement) and duration (time measurement). Each property of an elementary particle—its mass, electric charge, energy, spin, angular momentum—is such a quantitative concept.

The quantities to be discussed here are the very core—the basic vocabulary— of physics. Although we choose to introduce them in connection with particles, they will be the recurring concepts in every part of this book. Unfortunately it is not possible to set down precise definitions of all, or even most, of the key concepts in physics at one time, and then build the theories of physics upon these, a technique that might seem appealing from a logical point of view. This comes about simply because the concepts of physics cannot be divorced from the laws of physics. Electric charge can be defined with quantitative precision only in conjunction with the law of electric force. Mass can be defined only in terms of certain mechanical measurements whose success hinges on Newton's laws of motion. Energy, the most ubiquitous concept in physics other than space and time, can be defined in its manifold forms only with the help of some knowledge of every great theory of physics. In spite of these limitations, it is important early in the study of physics to gain what the physicist would call a "feeling" for the important concepts, in order to be able to think meaningfully about the natural world. Most of the definitions in the remainder of this chapter will require refining at a later point in the book, but the discussions of the "meaning" of the concepts will require no alteration.

For each of the concepts used to describe nature, a unit of measurement is introduced to allow comparisons of measurements in different places or at different times. This is a long way of saying something we all know from daily life. Our height is expressed in terms of feet and inches, our weight in pounds, our age in years. Each quantity needs a unit in terms of which it can be expressed. Unfortunately there is no international agreement about units (although the situation is much less chaotic within science than in the everyday world), but this circumstance of a multiplicity of units serves all the better to demonstrate the need for units. An American, asked his weight, might answer, "I weigh 154." An Englishman of the same size weighs 11. A Frenchman of similar build might claim to weigh 70. Their weights are all the same, of course, but the American is reckoning in pounds, the Englishman in stone, and the Frenchman in kilograms. A statement of the number without the unit is quite meaningless, although in a sufficiently provincial gathering the unit may, by common agreement, be understood and therefore not stated.

There are in science some "pure" numbers, also called dimensionless numbers. These numbers in fact hold a special fascination for scientists just because they are independent of any set of units. If we say, "Table 2.1 lists fourteen kinds of particles," the "fourteen" is a pure number, the result of counting, and refers to no particular unit. But in the main the quantities needed to describe nature do require units.

The normal scientific units, such as the centimeter of length (frequently abbreviated cm) and the second (sec) of time, are "man-sized" units adopted for convenience in our macroscopic world. The centimeter, about half an inch, is roughly the thickness of a man's finger; in one second a man can blink a few times, say "one thousand and one," or stroll about 100 cm. These units are handy and easy to visualize. But in the worlds of the large and the small they become ridiculously inappropriate. The distance from earth to sun is an enormous number of centimeters ; the size of a hydrogen atom, a tiny fraction of one centimeter. The age of the earth is a vast number of seconds, the lifetime of a pion an imperceptible part of a second. Journalists are fond of writing out numbers from the cosmological or submicroscopic worlds in their full glory, with a string of zeros before or after the decimal point: The pion lives 0.000000026 second; the number of hydrogen atoms in a quart of water is 60,000,000,000,000,000,000,000,000. These numbers are impressive, but a bit confusing and not very instructive.

TABLE 3.1 Table of Measurements

Physical Quantity	Common Unit in the Large-Scale World	Scale of the Submicroscopic World
Length	Centimeter (about one-half inch)	Size of atom about 10^{-8} cm = 1 angstrom Size of particle about 10^{-13} cm = 1 fermi
Speed	Centimeter per second (speed of a snail)	Speed of light = 3×10^{10} cm/sec
Time	Second (the swing of a pendulum)	Natural time unit of particle: about 10^{-23} sec Typical lifetime of "long-lived" particle about 10^{-10} sec
Mass	Gram (mass of cubic centimeter of water)	Mass of electron = 9×10^{-28} gm
Energy	Erg (energy of a lazy bug) Food calorie (40 billion ergs)	1 eV (electron volt) = 1.6×10^{-12} ergs Air molecule has about 0.04 eV Proton in largest accelerator; about 30 billion eV
Charge	Coulomb (lights a lamp for one second)	Electron charge = 1.6×10^{-19} coulomb
Spin	Gm × cm × cm/sec (grasshopper turning around)	Spin of photon = $\hbar = 10^{-27}$ gm × cm × cm/sec

Scientists have done two things about this situation. First, and most simply, they have replaced the lengthy newspaper notation for large and small numbers by the exponential notation, which was discussed in Section 1.3. In this notation the two numbers in the paragraph above are much more compactly written as 2.6×10^{-8} and 6×10^{25}. Section 1.3 also contains the rules of arithmetic for numbers written in the exponential notation.

The second approach to dealing with large and small quantities is to invent new units more appropriate to the domain being considered. Thus, for cosmological purposes, the light-year (which is 9×10^{17} cm) is a convenient unit of length. For dealing with atoms, the angstrom unit, equal to 10^{-8} cm, is frequently used; for nuclear and particle phenomena the fermi (10^{-13} cm), one hundred thousand times smaller, is a more suitable unit. In Table 2.1

we followed this approach, adopting the spin of the photon as the unit spin, the charge of the proton as the unit charge, and the mass of the electron as the unit mass. It is, of course, still necessary to know how to convert these units back to the conventional units, just as it is necessary to know how to convert among centimeters, inches, feet, yards, miles, and light-years.

Table 3.1 summarizes some of the important concepts of physics, indicating typical magnitudes encountered in the macroscopic world and in the submicroscopic world. Some of these magnitudes are included among the important constants of nature listed in Appendix 1. Useful conversions of units appear in Appendix 2.

3.2 Length

The concept of distance, or spatial separation, is probably as primitive as any idea in science other than counting. Primitive or not, length has proved to be a tricky concept, as the later chapters on relativity will show. According to our modern insight, the length of an object depends on whether or not it is in motion with respect to the measurer. This fact need not concern us now. Geometry, the first branch of mathematics that was brought to a sophisticated level, is based simply on the idea of length—or more accurately stated, of spatial relationships. The definition of length rests on the size of some standard object. For everyday purposes, that standard object might be a yardstick or meter stick. For many years, the scientific standard of length was a metal bar kept in a temperature-controlled vault in Paris, its length—by definition—one meter. Now the standard "object" is a light wave, a very particular wave of red-orange light emitted by atoms of the isotope krypton 86. Its wavelength—again by definition—is $6.057802105 \times 10^{-5}$ cm. (This odd value was chosen in order to keep the new meter as nearly as possible the same as the length of the previous standard.) This is a far more satisfactory standard than a single piece of metal in an inaccessible spot. Krypton atoms are available in any laboratory. Moreover, we have much more confidence in the immutability and identity of krypton atoms than we have in the eternal constancy of a large piece of metal.

One of the best ways to try to visualize large and small distances is by analogy, as in Section 1.3. To picture the nucleus, whose size is about 10^{-4} to 10^{-5} of the size of an atom, one may imagine the atom expanded to, for example, 10,000 feet (10^4 feet), or nearly two miles. This is about the length of a runway at a large air terminal such as New York's John F. Kennedy Airport. A fraction 10^{-4} of this is one foot, or about the diameter of a basketball. A fraction 10^{-5} is ten times smaller, or about the diameter of a golf ball. A golf ball in the middle of Kennedy International Airport is about as lonely as the proton at the center of a hydrogen atom. The basketball would-correspond to a heavy nucleus such as uranium. On this scale, one centimeter would be expanded to 2×10^8 (200 million) miles, or about twice the distance from earth to sun.

The size of a proton is about 10^{-13} cm; this distance has been given the name fermi, in honor of Enrico Fermi, who pioneered studies of the nuclear particles in the 1930s. The smallest distance probed in any experiment so far conducted is about a tenth of a fermi, or 10^{-14} cm.

It is interesting to compare the astronomical scale of size with the submicroscopic. The known part of the universe extends out to about 10^{10} light-years, or 10^{28} cm. Man, a creature about 10^2 cm high, is thus smaller than the universe by a factor of 10^{26}, but larger than a proton by a factor of "only" 10^{15}. The universe is about as much larger than the whole solar system as man is larger than the proton. From the smallest known structure (10^{-13} cm) to the largest (10^{28} cm), man has spanned a factor of 10^{41} in size. The number 10^{41} is so enormous that not even analogy is of much help in comprehending it. Suppose that we let the population explosion run its course until there were 10^{41} people. The earth itself can accommodate only a bit over 10^{15} people standing shoulder to shoulder. A million earths similarly packed could handle 10^{21}. To reach our goal of 10^{41} people, we would have to find and people to the limit some 10^{26} earths, a substantial number. There are only about 10^{23} stars in the universe. If every star had ten planets, and every planet was packed with people like sardines in a can, the universe would still not hold as many as 10^{41} people.

3.3 Speed

A snail in a hurry can travel at a speed of about 1 centimeter per second (or, in briefer scientific notation, 1 cm/sec). A man strolls at about 10^2 cm/sec, drives a car at 3×10^3 cm/sec, and rides in a jet plane at near the speed of sound, which is 3×10^4 cm/sec (about 700 miles per hour [miles/hr]).

In distance, man has encountered no limit, large or small. But in speed, nature seems to have established a very definite limit, the speed of light, 3×10^{10} cm/sec. This is a million times the speed of sound in air, an easy ratio to remember. Even an astronaut falls short of the speed of light by a factor of forty thousand. He needs an hour and a half to get once around the earth, while a photon (if it could be caused to travel in a curved path) could complete the trip in a tenth of a second. Still, man is not so far removed from nature's top speed as he is from the frontiers in space and time.

Atoms and molecules, executing their continual restless motion in solids, liquids, and gases on earth, move sluggishly about at only one to ten times the speed of sound in air, a factor 10^5 to 10^6 short of the speed of light. But for elementary particles, speeds near the speed of light are common. The photon and neutrinos have no choice, of course, and travel at exactly the speed of light, as does also the graviton. In all of the larger modern accelerators, particles with mass—usually electrons or protons—are pushed very near to the speed of light, and the unstable particles formed in nuclear collisions also frequently emerge with speeds near the speed of light.

Astronauts in science fiction frequently shift into "superdrive" and scoot about the galaxy above the speed of light. Is there any chance that this will become reality? It is extremely unlikely, and for a very simple reason. The lighter an object is, the more easily it can be accelerated. Freight trains lumber slowly up to speed, automobiles more quickly, and protons in a cyclotron still more quickly. A particle with no mass whatever should be the easiest to accelerate; indeed, the massless photon jumps instantaneously to the speed of light when it is created. But not beyond. If anything at all were able to go faster than light, then light itself, being composed of massless photons, should go faster.

3.4 Time

11 The idea of time is essentially an idea of periodic repetition. That time is a "visualizable" concept, a built-in part of every man's view of the world, stems from two facts: that man has a memory, and that man is constantly subjected to regularly repeated stimuli. We are conscious of the alternation of the seasons, the steady pattern of day and night, our own recurring hunger or fatigue. A clock amounts to nothing more than any device that counts off the number of repetitions of some recurring motion. An old-fashioned pendulum clock is an obvious example. The earth itself is a clock, because of its daily rotation and its annual trip around the sun. One second of time was defined originally as 1/86,400 of a day. Then the international standard time unit, the ephemeris second, was defined as a certain fraction of the period of rotation of the earth about the sun (for the year 1900). Many atoms and molecules are clocks—and very accurate ones— because of internal oscillatory motion. For all precision laboratory work, a particular atomic time standard is now used. Cesium atoms emit and absorb radiation whose frequency, 9,192,631,770 vibrations per second, provides the present definition of the second. How can we be sure that the time intervals between

successive cycles of a clock's repeating motion are the same? We cannot. The uniformity of time scale can be based only upon self-consistency (that clocks of different design continue to agree) and upon agreement with simple laws of nature (that the periodic motion is theoretically regular). Time, like every other concept, cannot escape to a status independent of the laws in which it is employed.

In dealing with the time scale of the submicroscopic world, it is essential first to get rid of preconceived notions about what is a "short" time or a "long" time. A millionth of a second certainly seems to qualify as a short time; yet for an elementary particle it is an exceedingly long time. On the other hand, a million years is the mere blink of an eye for the stately cosmological march of events.

If an automobile had its last bolt tightened at the end of the assembly line, then was driven a hundred feet away, where it promptly collapsed in a heap, we should say that it had had a short lifetime. If it covered some 20 billion miles before collapsing, we should say that it had had an amazingly long lifetime, indeed, that it was the most extraordinarily long-lived car ever built. Let us translate these distances to the elementary-particle world. The size of a particle is about 10^{-13} cm; it travels typically at about 10^{10} cm/sec. Thus, to cover a distance ten times its own size (comparable to the car moving 100 feet), the particle needs only 10^{-22} sec. The duration, 10^{-23} sec, in which it can move a distance about equal to its own size, is a kind of natural time unit for a particle. Yet most of the particles in Table 2.1 live at least 10^{-10} sec, an enormously long time compared to 10^{-23} sec. In 10^{-10} sec the particle can cover a whole centimeter, more than a million million times its own size. A particle moving one centimeter is comparable to a car going 20 billion miles. Any particle that can move one centimeter away from its birthplace before dying deserves to be called long-lived. The pion and muon, with lifetimes greater than 10^{-8} sec and 10^{-6} sec, respectively, can move much farther even than 1 cm. The neutron is a strange special case. Its average lifetime of 15 minutes is practically infinite for the elementary-particle world.

The new particles, or "resonances," now being discovered have lifetimes of 10^{-20} sec or less. They are indeed short-lived, and perhaps they do not even deserve to be called particles. They are like the car that collapses before it gets out of the factory gate. The manufacturer might be tempted to say, "That was no car; it was just an unstable phenomenon with a transitory existence" (for which the physicist uses the word "resonance").

Since the smallest object probed experimentally is about 10^{-13} cm in size, it is fair to say that the shortest time interval studied is 3×10^{-24} sec (although direct time measurements are still very far from reaching this short an interval). The longest known time is the "lifetime of the universe," that is, the apparent duration of the expansion of the universe, which is also a few times the age of the earth. This amounts to 3×10^{17} sec (10 billion years). The ratio of these is 10^{41}, the same enormous number as the ratio of the largest and smallest distances. This is not a coincidence. The outermost reaches of the universe are moving away from us at a speed near the speed of light and the particles used to probe the submicroscopic world are moving at speeds near the speed of light. On both the cosmological and submicroscopic frontiers, the speed of light appears to be the natural link between distance and time measurements.

3.5 Mass

To begin with, we may think of the mass of an object as the amount of material in the object. This is actually a very unscientific definition, but it provides a way of thinking about mass. The idea of mass is confused by the circumstance that the pull of gravity on an object is proportional to the mass in the object. A parcel of large mass on a postal scale is pulled down more strongly than a parcel of small mass, and consequently it causes the scale to show a higher reading. We say that the more massive object "weighs" more; that is, it is pulled more strongly by the earth, and in fact in exact proportion to its mass.

To understand mass a bit better, imagine some astronauts floating freely inside the cabin of their space ship in a weightless condition (Figure 3.1). If two of them join hands, then push and let go, they will float apart—in a particular way. The larger man will move away a bit more slowly than the smaller man; we attribute this to his greater mass. The important fact about mass is that it is a measure of resistance to a change of motion, a property usually called inertia. If a single astronaut floating in mid-cabin tosses a baseball, he will recoil and drift backward slowly as the ball moves swiftly off in the opposite direction. If the ball leaves the point of separation five hundred times faster than the astronaut, it is because its mass is only one five-hundredth as great as the astronaut's, and it has, therefore, five hundred times less resistance to being set into motion. Anyone who has fired a gun knows about recoil. If a hunter had no more mass than his bullet, he and the bullet would be accelerated to equal speed. Because he has a great deal more mass than the bullet, he has more resistance to being set into motion and is moved much more slowly than the bullet.

This inertial property of mass may be used as a technical definition of mass. The astronauts in their space ship, for example, might *define* the mass of their baseball arbitrarily to be 100 grams (gm). (Actually the gram is equal to the mass of one cubic centimeter of water at a temperature of 4 degrees Centigrade.) Then each astronaut could determine his mass by throwing the standard baseball and having his fellow astronauts measure his speed of recoil and the speed of the baseball. The astronaut who recoiled at one five-hundredth the speed of the baseball would have a mass five hundred times greater than the baseball, that is 50,000 gm, or 50 kilograms (kg) (about 110 pounds). A heavier astronaut might recoil at one seven-hundredth the speed of the baseball. His mass would be 70 kg. A still heavier astronaut, recoiling at only one-thousandth the speed of the baseball, would have a mass of 100 kg. That this way of defining mass is *permissible* cannot be denied, since it is arbitrary. But is it also useful? That depends on the answer to two other questions. Is the definition self-consistent? Does the mass so defined appear in a simple way in the laws of nature? The

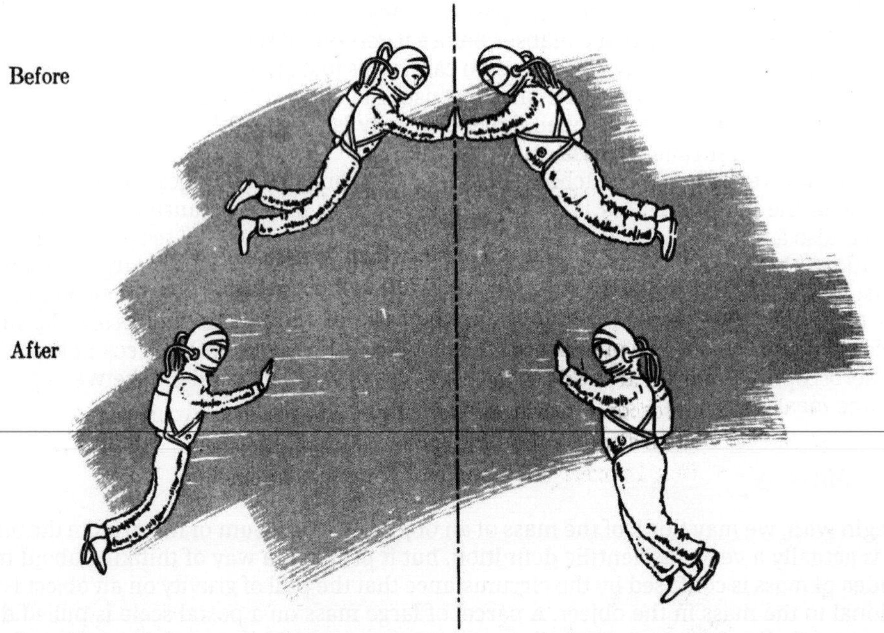

FIGURE 3.1 A way to define mass. The ratio of the masses of the astronauts is the inverse of the ratio of their recoil speeds if they push apart when initially at rest. Before: a thin astronaut and a fat astronaut start to push apart. After: they are moving apart, the thin astronaut faster than the fat astronaut.

answer to both of these questions is yes. The self-consistency can be checked in a number of ways. For example, the 50-kg astronaut and the 100-kg astronaut might push each other and float apart. The heavier man would move away at only half the speed of the lighter man. Their relative mass would be two to one, the same as deduced from their separate experiments with the baseball. Or these two astronauts might link arms and together perform the baseball experiment. Together they would recoil from the thrown baseball at only one fifteen-hundredth of the speed of the baseball. Their combined mass would be the sum of their separate masses. The question of the appearance of mass in the laws of nature is postponed to Chapter Eight.

If the astronauts came back to earth, frictional forces and the pull of gravity would put practical obstacles in the way of their mass-measuring experiments. But gravity, the very thing that prevents their free-floating experiments, can be harnessed to measure mass in another way. Since the pull of gravity on an object is found to be proportional to its mass, mass on the earth can simply be weighed. The 50-kg astronaut on one arm of a balance would be in equilibrium with 500 identical baseballs loaded onto the other arm of the balance. To balance gravity's pull on the 100-kg astronaut would require 1,000 baseballs.

The electron is the lightest particle with nonvanishing mass, and it is therefore usual to adopt the mass of the electron as a convenient unit in the submicroscopic world, as we did in Table 2.1. The heaviest elementary particle, the omega, is more than 3,000 times heavier than the electron, yet a million million omega particles would not be able to tip the world's most sensitive balance.

The scientific unit of mass is the gram, about the mass of an average vitamin pill. A quart of water has a mass of about 900 gm. A liter of water, somewhat larger, is exactly 1,000 gm. A European "pound" is half the mass of a liter of water, or 500 gm. The English and American pound (lb) is half the mass of a quart of water, or 454 gm. The electron's mass is 9×10^{-28} gm, that of the "heavy" omega particle about 3×10^{-24} gm.

To end the discussion of mass on a cosmic note we ask, What is the mass of the universe? This is certainly not well known, but a rough estimate can be made. There are about 10^{23} stars in the universe (this number is roughly the same as the number of molecules in a gram of water). An average star weighs about 10^{35} gm, making the total mass approximately 10^{58} gm. Each gram of matter contains about 10^{24} protons, so that what we know of the universe contains (very roughly) 10^{82} protons. The known part of the universe presumably contains a like number of electrons, about 10^{82}. It contains fewer neutrinos, perhaps about 10^{79}, but uncountably many photons and gravitons. The unstable particles, on the other hand, are far less numerous than protons.

3.6 *Energy*

The most remarkable fact about energy is its diversity. Like a clever actor who can assume many guises, energy appears in a variety of forms, and can shift from one role to another. Because of this richness of form, energy appears in nearly every part of the description of nature and can make a good claim to be the most important single concept in science.

One common form of energy is energy of motion, kinetic energy, which is a measure of how much force over how great a distance is required to set an object into motion or to bring it to rest. The faster a particle moves, the more kinetic energy it possesses. For speeds that are not too large (not too close to the speed of light) or masses that are not too small, the formula for kinetic energy is

$$K.E. = \tfrac{1}{2}mv^2 , \tag{3.1}$$

one half of the mass m of an object multiplied by the square of its velocity v. As accident investigators are well aware, an automobile moving at 60 miles/hr has four times as much kinetic energy as an automobile moving at 30 miles/hr. (This formula is no longer true for particles moving near or at the speed of light. The corrected version, supplied by the theory of relativity, is dealt with in Chapter Twenty-One.)

The great significance of energy springs in part from the variety of its manifestations, in part from the fact that it is conserved—that the total amount of energy remains always the same, with the loss of one kind of energy being compensated by the gain of another kind of energy. Trace, for example, the flow of energy from sun to earth to man; this illustrates both the variety and the conservation of energy. When protons in the sun unite to form helium nuclei, nuclear energy is released. This energy may go first to the kinetic energy of motion of nuclei. Some of the energy is then carried away from the sun by photons, the particles which are bundles of electromagnetic energy. The energy content of the photons may be transformed, by the complicated and not yet fully understood process of photosynthesis, into stored chemical energy in plants. Either by eating plants or by eating animals that ate plants, man acquires this solar energy, which is then made available to power his brain and muscles and to keep him warm.

That mass is one of the forms of energy was first realized at the beginning of this century. The energy of mass and the energy of motion are the two forms of energy that dominate the elementary-particle world. Mass energy can be thought of as the "energy of being," matter possessing energy just by virtue of existing. A material particle is nothing more than a highly concentrated and localized bundle of energy. The amount of concentrated energy for a motionless particle is proportional to its mass. If the particle is moving it has still more energy, its kinetic energy. A massless particle such as a photon has only energy of motion (kinetic energy) and no energy of being (mass).

12 Einstein's most famous equation,

$$E = mc^2 , \qquad (3.2)$$

provides the relation between the mass, m, of a particle and its intrinsic energy, or energy of being, E. The quantity c in this formula is the speed of light. The important statement Einstein's equation makes is that energy is proportional to mass. Twice as much mass means twice as much intrinsic energy; no mass means no intrinsic energy. The factor c^2 is a constant of proportionality; it does the job of converting from the units in which mass is expressed to the units in which energy is expressed. By analogy, consider the equation giving the cost of filling a car with gasoline,

$$C = GP.$$

The cost, C, is equal to the number of gallons, G, multiplied by the price per gallon, P. The cost is proportional to the number of gallons, and P is the constant of proportionality which converts the number of gallons to a total cost. In a similar way, c^2 is a price. It is energy per unit mass, the price that must be paid in energy to create a unit of mass.

In an isolated nuclear collision, or particle reaction, the total energy remains unchanged. In fact, practically every event in the submicroscopic world *is* isolated, for the distance over which the particles interact with one another is generally exceedingly small compared with the distance between neighboring atoms, about 10^{-8} cm. The individual event occurs with the particles effectively unaware of anything else in the universe. In a collision, or a reaction, or a decay process, there are two ways in which energy may be supplied. A particle may be slowed down, thus giving up some of its kinetic energy; or a particle with mass may be destroyed, thus giving up its intrinsic energy. Analogously, energy may be taken up in two ways. A particle may be speeded up or a new particle may be created. The rule of energy conservation can be stated as follows: The total energy supplied must be equal to the total energy taken up; that is, the energy loss must equal the energy gain.

Consider, for example, the decay of a pion at rest. Since it is motionless, its only energy is the intrinsic energy associated with its mass. It decays spontaneously into a muon and a neutrino. The vanishing of the pion makes its mass energy available. Some of this is used to create the mass of the lighter muon. The rest is supplied as energy of motion, and the muon and neutrino fly off at high speed with just enough kinetic energy to make up the difference. This example makes clear why a particle can decay spontaneously only into other particles lighter than itself. In a high-energy collision, on the other hand, such as those that occur near a particle accelerator, the projectile particle may be slowed down and some of its kinetic energy made available to create the mass of new particles. This is the way in which antiprotons and the various unstable particles are created in the modern high-energy laboratory.

The average citizen who pays an electric bill and watches his weight is probably aware of at least two of the energy units in common use, the kilowatt-hour and the food calorie. Ten 100-watt light bulbs require a kilowatt; if they all remain lit for one hour, one kilowatt-hour of energy is used up. Reasonably enough, it is electrical energy for which one pays. An overweight person might remark that it is also calories for which one pays. One food calorie is the energy released by a good-sized pinch of sugar when it is oxidized. A thousand or more calories of energy are needed each day to keep the human machine running efficiently.

A more usual scientific unit of energy is called the erg. It is actually quite small, being the energy of motion of a 2-gm bug crawling at 1 cm/sec. A food calorie is some 40 billion (4×10^{10}) ergs, and a kilowatt-hour is nearly a thousand times larger still, being 3.6×10^{13} ergs. Nevertheless, the erg is considerably larger than the energies usually encountered in the submicroscopic world, where another energy unit, the electron volt (eV), has been introduced (the last energy unit we shall mention!). The electron volt is about one millionth of one millionth of an erg (1.6×10^{-12} erg, to be exact.) The electron volt has nothing in particular to do with the electron. It is the energy given by an electric potential of one volt to *any* particle carrying one quantum unit of charge.

The average kinetic energy of the incessant random motion of molecules and atoms provides a kind of a baseline for energy comparisons in the world of the very small. A single molecule at normal temperatures moves about with an average kinetic energy of about one twenty-fifth of an electron volt. On the much hotter surface of the sun this thermal motion averages about half an electron volt. But in accelerators man can readily push the energies of particles to much-higher values. Early cyclotrons accelerated protons to over one million electron volts (1 MeV). Other accelerators completed after World War II pushed the energy of protons toward one billion electron volts (1 GeV). The Bevatron in Berkeley, California accelerates protons to 6 GeV. The largest accelerator, operational in 1968, is in Serpukhov, U.S.S.R. Its proton projectiles reach an energy of 70 GeV. Some of the cosmic-ray particles from outer space arrive with energies even much greater than this. How these particles are accelerated to such enormous energies remains uncertain.

Since mass and energy are equivalent, how much energy is needed to create the mass of an elementary particle? Apart from the massless particles, the electron and positron are easiest to manufacture, each requiring 500,000 eV. The older accelerators were capable of supplying sufficient energy to create these light particles. But the proton mass is equivalent to nearly 1 GeV. The production of antiprotons had to await the construction of the 6-GeV Bevatron. The newer machines can create all of the known particles with energy to spare; new and still heavier particles are being discovered with the help of these accelerators.

Mass is a very potent and highly concentrated form of energy. The power of the atomic bomb, in which less than a tenth of one per cent of the mass is converted to energy, serves as an eloquent reminder of this fact. To understand this in terms of the energy equivalents given above, recall the typical energies of thermal motion. Even on the white-hot surface of the sun, a proton has a kinetic energy of less than 1 eV, a mere nothing compared with the billion eV locked up in its mass. The mass of a single proton converted to energy on the earth is sufficient to heat up a billion atoms to a temperature higher than that of the sun's surface.

3.7 *Charge*

13 Electric charge is best described as being like French perfume. It is that certain something worn by particles which makes them attractive—specifically, attractive to the opposite kind of particle. Particles that do not have it are called neutral and have no influence (at least no electric influence) on other particles. Charge can lead to pairing of particles; the hydrogen atom, for example, consists of a proton and electron held together by electric attraction. More energetic particles are not held together by the electric force; it merely causes them to deviate slightly from a straight course.

The charge carried by a particle may be positive or negative. Two charges of the same kind repel each other, two of opposite kind attract. The protons within a nucleus, for example, repel each other, but the overpowering nuclear force nevertheless holds the nucleus together. Eventually, however, for very heavy nuclei, the electric repulsion becomes more than the nuclear forces can counteract, and the nucleus flies apart. It is for this reason that no nuclei heavier than uranium exist in nature.

Which charge is called positive and which negative is entirely arbitrary, and is the result of historical accident. The definition that led to electrons being negative and protons positive probably stems from a guess made by Benjamin Franklin about the middle of the eighteenth century. His choice of nomenclature was based on the erroneous supposition that it is positive electricity that flows most readily from one object to another. We now know that it is the negative electrons that are mobile and account for the flow of electric current in metals.

Charge is still very mysterious to the physicist. He understands it little better than he understands the workings of French perfume. If we think of a particle as a small structure spread out over a tiny region of space, it is logical to think of its charge as being spread out too. But if this is so, why do the various bits of charge making up the particle not repel each other and cause the whole particle to disintegrate and fly apart? No one knows a satisfactory answer to this question. It is also perplexing that almost all particles have exactly the same magnitude of charge. If the charge of the electron is called $-e$ (it is negative) then the charge of every other "long-lived" particle is either $-e$ or $+e$ or zero. No other possibilities are realized in nature. We have no understanding of this fact, and also we have no clue as to why the electron has the charge it does have, rather than some other value. The true nature of charge and the reason it comes only in lumps of a certain size are among the most important problems in elementary-particle physics.

The fact that we do not understand charge at a fundamental level has been no hindrance to making extensive use of charge for practical purposes. Electrons can be detached from atoms rather readily—at least in certain metals called conductors—and by means of electrical forces they can be pushed and pulled through wires or sent flying through empty space, as within a vacuum tube or a television tube. Almost all of the fine control and all of the communications in the world are effected by electrons in electronic circuits. A great part of the world's heavy labor is also done by electrons turning motors or supplying heat.

The fact that there is a voltage across the two holes in a household electrical outlet means that an electric force is standing ready to do some work. If an electric light is plugged in, the electrons are sent scurrying from one prong of the plug, through the light (where they expend some energy, which appears as light and heat), then back to the other prong. The number of electrons involved in such a flow is enormous. In a typical household light bulb, about 10^{19} electrons flow through the filament each second. In heavy machinery or in the high-tension lines connecting cities, the number is far greater. Even through the tiniest and most delicate electronic circuit, many billions of electrons flow each second.

If a comb is passed through dry hair, perhaps a million million electrons leave the hair and stick on the comb. Nevertheless, the comb is almost neutral. For every extra electron it has acquired, the comb has a million million neutral atoms. It is fortunate for us that the objects in our macroscopic world remain always almost neutral. If the comb acquired anything close to an electrification of one extra electron per atom, the consequence would be dire. Either there would be a powerful and deadly bolt of lightning from comb to man as the charge was neutralized, or the enormous force of electric attraction would draw the comb back so violently that it would be a dangerous weapon.

Intrinsically, the electric force in nature is a great deal stronger than the gravitational force, and in the submicroscopic world the gravitational force is usually ignored altogether. The matter in our macroscopic world, however, exists in such a fine state of electric balance that gravity has a chance to make itself felt. In every object in our world, the number of positive charges is almost precisely equal to the number of negative charges. The effects nearly cancel, and what we regard as marked electric effects arise from an exceedingly tiny imbalance in positive and negative charge. If a big imbalance were ever realized (there is no chance of this) the disastrous result would make the force of gravity appear to be truly inconsequential.

One of the common scientific units of charge is the coulomb, named for the French scientist, Charles A. Coulomb, who discovered the exact law of electric force in 1785. One coulomb is roughly the amount of charge that moves through a 100-watt light bulb in a second, or through an electric iron in a fifth of a second. (The coulomb is not commonly encountered in the house, but a very closely related unit, the ampere, is. On a household fuse one might see stamped "15 amp." The ampere is one coulomb per second. If more than 15 coulombs of charge flow through this fuse each second, a wire inside it will melt and stop the flow.) The basic unit of charge in the world of particles is the electron charge, 1.6×10^{-19} coulombs. This is less than one billionth of one billionth of one coulomb.

3.8 *Spin*

Rotational motion seems to be a characteristic of most of the structures in the universe, from neutrinos to galaxies. Our earth rotates on its axis once a day, and rotates once around the sun in a year. The sun itself turns on its axis once in 26 days, and, along with "neighboring" stars of our galaxy, travels once around the galaxy in 230 million years. It is not known whether larger structures, such as clusters of galaxies, have an overall rotational motion, but it would be surprising if they do not.

Going down the scale, the atoms that compose molecules can rotate about each other, and do, although at a rate that varies from time to time as the molecule is disturbed by its neighbors. Within the atom, electrons rotate about the nucleus at speeds of from 1% up to more than 10% of the speed of light, thereby giving a kind of solidity to the sphere of empty space in which they move. The remarkable fact that the electron also spins like a top about its own axis was first discovered in 1925, and now we know that many particles have this property of intrinsic spin.

The spin of an elementary particle, unlike the rotation of a molecule, is an invariable property of that particle, always fixed at a definite value. The electron cannot be stopped from rotating, nor can its rotation be speeded up. The spin of an electron is such an essential feature of the electron that it can be changed only at the expense of destroying the electron altogether. Actually, this is a rather subtle point. It is probably more accurate to say that if the electron is caused to spin more rapidly, this so drastically changes the properties of the electron that it is more convenient to think of the resulting structure as a completely new and distinct particle. To what extent the different particles are really independent and to what extent they are only different states of motion of some common underlying structure is, of course, the great, unsolved problem of particle physics, and it would be idle to speculate further about this point here. It can only be re-emphasized that physicists retain their faith that there exists a simpler structure underlying the particles.

Spin is measured in terms of the quantity called angular momentum (introduced in

Chapter Two), which is a combined measure of the mass, the size, and the speed of the rotating system. The rate of rotation of an electron cannot actually be measured but it has to be so great that the charge within the electron is moving at nearly the speed of light. In spite of this frantic rate of rotation, the electron is not able to generate much angular momentum because of its small size and small mass. A man swiveling slowly in his chair to watch a tennis match has at least 10^{33} times more angular momentum than a single electron.

As the electron carries nature's smallest unit of electric charge, it also carries the indivisible unit of spin, which, for historical reasons, is denoted by $\frac{1}{2}\hbar$. At the turn of this century, Max Planck discovered the existence of a constant in nature which relates the frequency and the energy of photons (this relation is discussed in Chapter Twenty-Four). We now call it Planck's constant and write it h. About ten years later Niels Bohr discovered that this constant also has something to do with the rotation of electrons about the nucleus in an atom. In their orbits about the nucleus, the electrons have an angular momentum that is always equal to h divided by 2π ($h/2\pi$) or two times $h/2\pi$ or three times $h/2\pi$ and so on, never any value between. Since it is a nuisance to write out the extra factor of 2π whenever it occurs, the notation $\hbar$ (pronounced "h bar") has come into common use for $h/2\pi$. Finally, when Samuel Goudsmit and George Uhlenbeck discovered in 1925 that the electron spins, they found that its spin angular momentum is not $\hbar$, which had been thought to be the indivisible unit, but only $\frac{1}{2}\hbar$. The quantity $\hbar$ has been adopted as the unit of spin and angular momentum in the submicroscopic world, even though the smallest indivisible unit is only half so great. Most of the leptons and baryons have spin $\frac{1}{2}$, the photon has spin 1, and the graviton spin 2, in this unit.

The principle of spin quantization applies in the macroscopic world as well as in the microscopic, but it is too subtle an effect ever to be measured for a large object. When the spectator at a tennis tournament turns to follow the ball, his angular momentum, in units of $\hbar$, might be 10^{33} or $10^{33} + 1$ or $10^{33} + 2$, but never $10^{33} + \frac{1}{3}$. The increment $\hbar$ between allowed values is so infinitesimal that it is entirely beyond hope ever to notice this discreteness in the macroscopic world. A change of one penny in the gross national product of the United States is a disturbance more than a billion billion times greater than a change by one unit of the spectator's angular momentum. It is no wonder that man did not discover the quantum theory of angular momentum until he was able to study in detail the structure of a system as small as an atom.

3.9 Natural units and dimensionless physics

14 The units of measurement man normally uses, even in scientific work, have been defined in an arbitrary way, chosen merely for convenience. They have nothing to do with "natural units" and are not in particular harmony with the basic structure of the world. Yet in this century we have learned of the existence of two natural units which are now commonly employed in studying the elementary particles. It seems quite likely that a deeper understanding of the particles will be accompanied by the discovery of a third natural unit.

The meter was originally defined as one ten-millionth of the distance from the earth's pole to the equator. (It is not quite that, because the meter was standardized in the nineteenth century but the knowledge of the size of the earth has been improved since then.) The centimeter, in turn, is one-hundredth of a meter. The gram was defined initially as the mass of a cube of water one centimeter on a side. Both the centimeter and the gram thus depend on the size of the earth, and there is no reason to believe that there is anything very special about the size of the earth. The third basic unit, the second, also depends on a property of our earth, its rate of rotation—again, nothing very special. For no better reasons than that the Egyptians divided the day and the night into twelfths and the Sumerians liked to count in sixties, the hour is one twenty-fourth of a day, the minute a sixtieth of an hour, and the second a sixtieth of a minute.

Within the first five years of this century two natural units were discovered which appeared to be obvious choices for a basis of measurement in the submicroscopic world. These were the speed of light c and Planck's constant h. Neither is directly a mass, a length, or a time, but they are simple combinations of these three. If they were joined by a third natural unit, they would form a basis of measurement as complete as, and much more satisfying than, the gram, the centimeter, and the second. (The thoughtful reader might propose the charge of the electron as an obvious candidate for the third natural unit. Unfortunately, it will not serve, for it is not independent of c and h—just as speed is not independent of time and distance.)

The fact that light travels at a fixed speed c has been known for several centuries, but the central significance of this speed in nature could be appreciated only when the theory of relativity was developed. Relativity revealed first of all that c is a natural speed limit, attainable not only by light but by any massless particle. The theory also showed that this constant appeared in various surprising places that had nothing to do with speed, for instance, in the mass-energy relation, $E = mc^2$. Planck's constant was brand new in 1900, but its significance also mounted over the next few decades as it came to be recognized as the fundamental constant of the quantum theory, governing not only the allowed values of spin, but every other quantized quantity.

It has to be recognized that every measurement is really the statement of a ratio. If you say you weigh 151 pounds you are, in effect, saying your weight is 151 times greater than the weight of a standard object (a pint of water), which is arbitrarily called one pound. A 50-minute class is 50 times longer than the arbitrarily defined time unit, the minute. When one uses natural units, the ratio is taken with respect to some physically significant unit rather than an arbitrary one. On the natural scale, a jet-plane speed of $10^{-6}c$ is very slow, a particle speed of $0.99c$ is very fast. An angular momentum of $10{,}000\hbar$ is large, an angular momentum of $\frac{1}{2}\hbar$ is small.

The difficult point to recognize here is that once the speed of light has been adopted as the unit of speed, it no longer makes any sense to ask how fast light travels. The only answer is: Light goes as fast as it goes. Since every measurement is really a comparison, there must always be at least one standard that cannot be compared with anything but itself. This leads to the idea of a "dimensionless physics." Having agreed on a standard of speed, we can say a jet plane travels at a speed of 10^{-6}, that is, at one one-millionth of the speed of light. The 10^{-6} is a pure number to which no unit need be attached; it is the ratio of the speed of the plane to the speed of light. To make possible a dimensionless physics we need one more independent natural unit, and this has not yet emerged. This unit, if it is found, may be a length, and there is much speculation that such a unit will be connected with a whole new view of the nature of space (and of time) in the world of the very small.

It should be added that a dimensionless physics is not so profound as it may sound, nor would it necessarily be a terminus of man's downward probing. Its lack of profundity springs from the fact that it, too, would rest on an arbitrary agreement among men about units. The hope is, however, that all scientists will be led naturally and uniquely to agree that there is only one sensible set of natural units, in contrast with the present situation, where the fact about which we all agree is that there is nothing special whatever about the centimeter, the gram, and the second.

EXERCISES

3.1. Fill in the blank. Man is a rather small creature, much closer in size to an atomic nucleus than to the universe at large. If a shrinkage occurred which compressed a man to the size of a nucleus, the known part of the universe would condense down to only about ______ miles.

3.2. The highest speed reached by an astronaut thus far is about 5 miles per second. **(1)** Express this speed in cm/sec. **(2)** If such a speed were attained in air, what would be its Mach number (the ratio of the speed to the speed of sound) ?

3.3. At the speed of an orbital astronaut (5 miles/sec), how long would it take to reach the nearest star, Alpha Centauri, about 4 light years distant? Express the answer in years.

3.4. Is 1500 meters (sometimes known as a "metric mile") greater or less than one mile? What should be the time for a four-minute miler to cover 1500 meters? *Optional:* Look up the current world records for the 1500 meter run and the mile. Compare the ratio of these two times with the ratio of the distances.

3.5. Calculate and record for possible later use the following conversion factors: **(a)** the number of inches in one meter; **(b)** the number of astronomical units, or A.U., in one light-year (the A.U. is the mean distance from earth to sun); and **(c)** the number of ft/sec in one mile/hr.

3.6. **(1)** How many electrons are needed to make a mass of one gram? **(2)** Make a rough estimate of the number of electrons in the earth, remembering that for every electron in the earth, there is one proton and about one neutron.

3.7. How many hydrogen atoms are there in a mass equal to that of an average star?

3.8. Explain in simple terms why the height of a man is "only" about 10^{10} times greater than the diameter of an atom, whereas his mass is about 10^{28} times greater than the mass of a typical atom.

3.9. **(1)** Calculate in ergs the kinetic energy of an electron traveling at 3×10^9 cm/sec (fast enough to circumnavigate the earth in one second). **(2)** How fast would you have to walk in order to have the same kinetic energy? At this speed, how long would it take you to cover 100 cm?

3.10. From the kinetic energy formula, $K.E. = \frac{1}{2}mv^2$, deduce the connection between the cgs unit of energy, the erg, and the cgs units of mass, length, and time (gm, cm, and sec).

3.11. An astronaut in a low earth orbit has a speed of 7.5×10^5 cm/sec. **(1)** What is the ratio of his kinetic energy in orbit to his kinetic energy as he motored to the launching site at 60 miles/hr? **(2)** What is the ratio of his kinetic energy in orbit to his intrinsic energy, mc^2? (HINT: Neither answer requires a knowledge of the mass of the astronaut.)

3.12. A supersonic jet is flying at a speed of 6×10^4 cm/sec (about Mach 2). What is the ratio of its intrinsic mass energy to its kinetic energy? Give an answer first in algebraic form, then in numerical form.

3.13. **(1)** How many electrons are there in one coulomb of electric charge? **(2)** If 10 coulombs of electric charge pass through a 15-amp fuse in ½ sec, will the fuse "blow"?

3.14. The cgs unit of charge is the e.s.u. (electrostatic unit), also known sometimes as the statcoulomb. It is related to the coulomb by

$$1 \text{ coulomb} = 3 \times 10^9 \text{ e.s.u.}$$

(1) Calculate the charge-to-mass ratio of the electron in e.s.u./gm. **(2)** If a one-gram sample of hydrogen carries a net positive charge of 1 e.s.u., what fraction of the hydrogen atoms have lost electrons?

3.15. Consider an astronaut in three positions: standing on the earth; in orbit around the earth; and standing on the moon. **(1)** Would his mass be different in these three locations? If so, where would it be greatest and where least? **(2)** Would his weight be different in these three locations? If so, where would it be greatest and where least?

3.16. Give an account of the various energy transformations that culminate in an increase of the kinetic energy of an automobile when the accelerator is depressed.

3.17. When the ammonia-beam maser oscillator was first operated it was proclaimed the most accurate clock ever built. With it, irregularities in the rate of rotation of the earth on its axis were detected. How could the accuracy of such a clock be tested, if no other kind of clock could match it in accuracy?

3.18. How, if at all, do you think that time might be defined if there were in nature no regularly recurring phenomena whatever? Defend either one of these two positions: **(a)** the concept of time would be meaningless, and it would be impossible to design a clock; or **(b)** it would still be possible to define time and design a clock.

FOUR

Conservation laws

In a slow and subtle, yet inexorable, way conservation laws have moved in the past few centuries from the role of an interesting sidelight in physics to that of the most central position. What little we now understand about the interactions and transformations of particles comes in large part through certain conservation laws which govern elementary-particle behavior.

4.1 Laws of permission and laws of prohibition

15 A conservation law is a statement of constancy in nature. If there is a room full of people, for example, at a party, and no one comes in or leaves, we can say that there is a law of conservation of the number of people; that number is a constant. This would be a rather uninteresting law. But suppose the conservation law remained valid as guests came and went. This would be more interesting, for it would imply that the rate of arrival of guests was exactly equal to the rate of departure. During a process of change, something is remaining constant. The significant conservation laws in nature are of this type: laws of constancy during change. It is not surprising that scientists, in their search for simplicity, fasten on conservation laws with particular enthusiasm, for what could be simpler than a quantity that remains absolutely constant during complicated process of change? To cite an example from the world of particles, the total electric charge remains precisely constant in every collision, regardless of how many particles may be created or annihilated in the process.

The classical laws of physics are expressed primarily as laws of change, rather than as laws of constancy. Newton's law of motion describes how the motion of an object responds to forces that act upon it. Maxwell's equations of electromagnetism connect the rate of change of electric and magnetic fields in space and time. The early emphasis in fundamental science was rather naturally on discovering those laws that successfully describe the changes actually occurring in nature. Briefly, the "classical" philosophy concerning nature's laws is that man can imagine countless possible laws, indeed infinitely many, that might describe a particular phenomenon. Of these, nature has chosen only one simple law, and the job of science is to find it. Having successfully found laws of change, man may derive from them certain conservation laws, such as the conservation of energy in mechanics. These appear as particularly interesting and useful consequences of the theory, but are not themselves taken as fundamental statements of the theory.

Gradually conservation laws have percolated to the top in the hierarchy of natural laws. This importance is not merely because of their simplicity, although that has been an important factor. It comes about also for two other reasons. One is the connection between conservation laws and principles of invariance and symmetry in nature—surely, one of the most beautiful aspects of modern science. The meaning of this connection is discussed near the end of this chapter. The other reason, which we wish to discuss here, might best be described simply as a new view of the world, in which conservation laws appear naturally as the most fundamental statements of natural law. This new view is a view of order upon chaos—the order of conservation laws imposed upon the chaos of continual annihilation and creation taking place in the submicroscopic world. The strong hint emerging from recent studies of elementary particles is that the only inhibition imposed upon the chaotic flux of events in the world of the very small is that imposed by the conservation laws. Everything that *can* happen without violating a conservation law *does* happen.

This new view of democracy in nature—freedom under law—represents a revolutionary change in man's view of natural law. The older view of a fundamental law of nature was that it must be a law of *permission.* It defined what *can* (and must) happen in natural phenomena. According to the new view, the more fundamental law is a law of *prohibition.* It defines what *cannot* happen. A conservation law is, in effect, a law of prohibition. It prohibits any phenomenon that would change the conserved quantity, but otherwise allows any events. Consider, for example, the production of pions in a proton-proton collision,

$$\mathrm{p} + \mathrm{p} \rightarrow \mathrm{p} + \mathrm{p} + \pi + \pi + \pi + \cdots .$$

If a law of permission were operative, one might expect that, for protons colliding in a particular way, the law would specify the number and the type of pions produced. A conservation law is less restrictive. The conservation of energy limits the number of pions that can be produced, because the mass of each one uses up some of the available energy. It might say, for example, that not more than six pions can be produced. In the actual collision there might be none, or one, or any number up to six. The law of charge conservation says that the total charge of the pions must be zero, but places no restriction on the charge of any particular pion; this could be positive, negative, or neutral.

To make more clear the distinction between laws of permission and laws of prohibition, let us return to the party. A law of change, which is a law of permission, might describe the rate of arrival and the rate of departure of guests as functions of time. In simplest form, it might say that three guests per minute arrive at 6:00, two guests per minute at 6:15, and so on. Or it might say, without changing its essential character as a law of permission, that the rate of arrival of guests is given by the formula:

$$R = \frac{A}{\pi D} \frac{1}{1 + \left(T - 5 - \frac{A}{D}\right)^2} ,$$

where R is the number of guests arriving per minute, A is the annual income of the host in thousands of dollars, D is the distance in miles from the nearest metropolitan center, and T is the number of hours after noon. This law resembles, in spirit, a classical law of physics. It covers many situations, but for any particular situation it predicts exactly what will happen.

A conservation law is simpler and less restrictive. Suppose it is observed that between 7 and 10 o'clock the number of guests is conserved at all parties. This is a grand general statement, appealing for its breadth of application and its simplicity. It would, were it true, be regarded as a deep truth, a very profound law of human behavior. But it gives much less detailed information than the formula for R above. The conservation law allows the guests to arrive at any rate whatever, so long as guests depart at the same rate. To push the analogy with natural law a bit further, we should say that according to the old view, since going to parties is a fundamental aspect of human behavior, we seek and expect to find simple explicit laws governing the flow of guests. According to the new view, we expect to find the flux of arriving and departing guests limited only by certain conservation principles. Any behavior not prohibited by the conservation laws will, sooner or later, at some party, actually occur.

This view of nature is closely related to an insight afforded by the theory of quantum mechanics: The fundamental laws of nature are laws of probability, not laws of certainty. The role of probability in nature is dealt with in Chapter Twenty-Three. Here we wish only to emphasize how probability has helped to elevate conservation laws to a more dominant role in physics. If a conservation law does not prohibit various possible results of an experiment, as in the proton-proton collision cited above, then these various possibilities will occur, each with some definite probability. A law of permission is a law of certainty. A law of prohibition, by leaving open more possibilities for the results of some experiment, goes naturally hand in hand with laws of probability, which leave open the possibility for each of the allowed results of the experiment to occur some of the time. As the endless changes in the particle world take place within the limitations imposed by conservation laws, man can know at best the probability, never the certainty, of any chain of events.

We have so far emphasized that a conservation law is less restrictive than an explicit law of change, or law of permission. However, there are a number of different conservation laws and, taken all together, they may be very strongly restrictive, far more so than any one taken alone. In the ideal case, they may leave open only one possibility. The laws of prohibition, all taken together, then imply a unique law of permission. The most beautiful example of this kind of power of conservation laws concerns the nature of the photon. From conservation principles alone, it has been possible to show that the photon must be a massless particle of unit spin and no charge, emitted and absorbed by charged particles in a particular characteristic way. This truly amazing result has been expressed vividly by J. J. Sakurai, who wrote, "The Creator was supremely imaginative when he declared, 'Let there be light.' "* In the world of human law, a man so hemmed in by restrictions that there is only one course of action open to him is not very happy. In the world of natural law it is remarkable and satisfying to learn that a few simple statements about constant properties in nature can have locked within them such latent power that they determine uniquely the nature of light and its interaction with matter.

4.2 *Absolute conservation laws*

16 There are conservation laws and conservation laws. That is, some things in nature are constant, but others are even more constant. To convert this jargon into sense, some quantities in nature seem to be absolutely conserved, remaining unchanged in all events whatever; other quantities seem to be conserved in some kinds of processes and not in others. The rules governing the latter are still called conservation laws, but nature is permitted to violate them under certain circumstances. We shall postpone the discussion of these not-quite-conservation laws to Chapter Twenty-Seven, and consider here only seven of the recognized absolute conservation laws.

We begin by listing by name the seven quantities that are conserved:

1. Energy (including mass)
2. Momentum
3. Angular momentum, including spin
4. Charge
5. Electron-family number
6. Muon-family number
7. Baryon-family number.

There are two different kinds of quantities here, which can be called (a) properties of motion and (b) intrinsic properties, but the two are not clearly separated. The intrinsic particle properties that enter into the conservation laws are mass, spin, charge, and the several "family numbers." The properties of motion are kinetic energy, momentum, and angular momentum, the last frequently being called orbital angular momentum to avoid possible confusion with intrinsic spin, which is a form of angular momentum. In the laws of energy conservation and angular-momentum conservation, the intrinsic properties and properties of motion become mixed.

The interactions and transformations of the elementary particles serve admirably to illustrate the conservation laws, and we shall focus attention on the particles for illustrative

* *Annals of Physics, 11* (1960), 5.

purposes. It is through studies of the particles that all of these conservation laws have been verified, although the first four were already known in the macroscopic world. The particles provide the best possible testing ground for conservation laws, for any law satisfied by small numbers of particles is necessarily satisfied for all larger collections of particles, including the macroscopic objects of our everyday world. Whether the extrapolation of the submicroscopic conservation laws on into the cosmological domain is justified is uncertain, since gravity, whose effects in the particle world appear to be entirely negligible, becomes of dominant importance in the astronomical realm.

Various intrinsic properties of the particles were discussed in Chapter Two, and we shall examine first the conservation laws that have to do with the intrinsic properties.

4.3 Charge conservation

Every particle in Table 2.1 carries the same electric charge as the electron (defined to be negative), or the equal and opposite charge of the proton (positive), or is neutral. The charge is a measure of the strength of electric force which the particle can exert and, correspondingly, a measure of the strength of electric force which the particle experiences. A neutral particle, of course, neither exerts nor responds to an electric force. A charged particle does both.

The law of charge conservation requires that the total charge remain unchanged during every process of interaction or transformation. For any event involving particles, then, the total charge before the event must add up to the same value as the total charge after it. In the decay of a lambda into a neutron and a pion,

$$\Lambda^0 \rightarrow \mathrm{n} + \pi^0,$$

the charge is zero both before and after. In the positive pion decay,

$$\pi^+ \rightarrow \mu^+ + \nu_\mu,$$

the products are a positive muon and a neutral neutrino. A possible high-energy nuclear collision might proceed as follows:

$$\mathrm{p} + \mathrm{p} \rightarrow \mathrm{n} + \Lambda^0 + \mathrm{K}^+ + \pi^+.$$

Neither positively charged proton survives the collision, but the net charge +2 appears on the particles created.

Note that the law of charge conservation provides a partial explanation for the fact that particle charges come in only one size. If the charge on a pion were 0.73 electron charge, it would be quite difficult to balance the books in transformation processes and maintain charge conservation. Actually, according to the present picture of elementary processes, the charge is conserved not only from "before" to "after," but at every intermediate stage of the process. One can visualize a single charge as an indivisible unit which, like a baton in a relay race, can be handed off from one particle to another, but never dropped or divided.

Perhaps the most salutary effect of the law of charge conservation in human affairs is the stabilization of the electron. The electron is the lightest charged particle and, for this reason alone, it cannot decay. The only lighter particles, the photon and neutrinos (and graviton) are neutral, and a decay of the electron would therefore necessarily violate the law of charge conservation. The stability of the electron is one of the simplest, yet one of the most stringent tests of the law of charge conservation. Nothing else prevents electron decay. If the law were almost, but not quite, valid, the electron should have a finite lifetime. A recent experiment places the electron lifetime beyond 10^{21} years; this means that charge conservation must be regarded as at least a very good approximation to an absolute law.

4.4 *Family-number conservation laws*

Unlike the other four laws, which were already known in the macroscopic world, the laws of family-number conservation were discovered through studies of particle transformation. We can best explain their meaning through examples. Recall that the proton and all heavier particles are called baryons, that is, they belong to the baryon family. In the decay of the unstable Λ particle,

$$\Lambda^0 \rightarrow p + \pi^-,$$

one baryon, the Λ, disappears, but another, the proton, appears. Similarly, in the decay of the Σ^0,

$$\Sigma^0 \rightarrow \Lambda^0 + \gamma,$$

the number of baryons is conserved. Note that, in one of these examples, a pion is created; in the other, a photon. Pions and photons belong to none of the special family groups and can come and go in any number. In a typical proton-proton collision the number of baryons (2) remains unchanged, as in the example,

$$p + p \rightarrow p + \Sigma^+ + K^0.$$

These and numerous other examples have made it appear that the number of baryons remains forever constant—in every single event, and therefore, of course, in any larger structure.

Each of the Ω, Ξ, Σ, and Λ particles, and the neutron, undergoes spontaneous decay into a lighter baryon. But the lightest baryon, the proton, has nowhere to go. The law of baryon conservation stabilizes the proton and makes possible the structure of nuclei and atoms and, therefore, of our world. From the particle physicist's point of view, this is a truly miraculous phenomenon, for the proton stands perched at a mass nearly 2,000 times the electron mass, having an intrinsic energy of about one billion electron volts, while beneath it lie several unstable particles. Only the law of baryon conservation holds this enormous energy locked within the proton and makes it a suitable building block for the universe. The proton appears to be absolutely stable. If it is unstable it has, according to a recent experimental result, a half life greater than 2×10^{28} years, or more than a billion billion times the age of the earth.

Our statement of the law of baryon conservation needs some amplification, for we have not yet taken into account antibaryons. A typical antiproton-production event at the Berkeley Bevatron might go as follows:

$$p + p \rightarrow p + p + p + \bar{p}.$$

(The bar over the letter designates the antiparticle. Since the antiproton has negative charge, the total charge of plus 2 is conserved.) It appears that we have transformed two baryons into four. Similarly, in the antiproton annihilation event,

$$p + \bar{p} \rightarrow \pi^+ + \pi^- + \pi^0,$$

two baryons have apparently vanished. The obvious way to patch up the law of baryon conservation is to assign to the antiparticles baryon number –1, and to the particles baryon number +1. Then the law would read: In every event the total number of baryons *minus* the total number of antibaryons is conserved; or, equivalently, the total baryon number remains unchanged.

The cynic might say that with so many arbitrary definitions—which particles should be called baryons and which not, and the use of negative baryon numbers—it is no wonder that a conservation law can be constructed. To this objection, two excellent answers can be given. The first is that it is not so easy to find an absolute conservation law. To find any quantity absolutely conserved in nature is so important that it easily justifies a few arbitrary definitions. The arbitrariness at this stage of history only reflects our lack of any deep understanding of the reason for baryon conservation, but it does not detract from the obvious significance of baryon conservation as a law of nature. The other answer, based on the mathematics of the quantum theory, is that the use of negative baryon number for antiparticles is perfectly natural, in fact, is demanded by the theory. This comes about because the description of the appearance of an antiparticle is "equivalent" (in a mathematical sense we cannot delve into*) to the description of the disappearance of a particle; and conversely antiparticle annihilation is "equivalent" to particle creation.

The "electron family" contains only the electron and its neutrino; the "muon family" only the muon and its neutrino. For each of these small groups, there is a conservation of family members exactly like the conservation of baryons. The antiparticles must be considered negative members of the families, the particles positive members. These light-particle conservation laws are not nearly as well tested as the other absolute conservation laws because of the difficulties of studying neutrinos, but there are no known exceptions to them.

The beta decay of the neutron,

$$n \rightarrow p + e^- + \overline{\nu}_e\,, \tag{4.1}$$

illustrates nicely the conservation laws we have discussed. Initially, the single neutron has charge zero, baryon number 1, and electron family number zero. The oppositely charged proton and electron preserve zero charge; the single proton preserves the baryon number; and the electron with its antineutrino $\overline{\nu}_e$) together preserve zero electron-family number. In the pion decay processes,

$$\pi^+ \rightarrow \mu^+ + \nu_\mu \quad \text{and} \quad \pi^- \rightarrow \mu^- + \overline{\nu}_\mu\,,$$

muon-family conservation demands that a neutrino accompany the μ^+ antimuon, and an antineutrino accompany the μ^- muon. The muon, in turn, decays into three particles, for example,

$$\mu^- \rightarrow e^- + \nu_\mu + \overline{\nu}_e\,,$$

which conserves the members of the muon family and of the electron family.

The general rule enunciated earlier in this chapter was that whatever *can* happen without violating a conservation law *does* happen. Until 1962, there was a notable exception to this rule; its resolution has beautifully strengthened the idea that conservation laws play a central role in the world of elementary particles. The decay of a muon into an electron and a photon,

$$\mu^- \rightarrow e^- + \gamma,$$

has never been seen, a circumstance that had come to be known as the μ-e-γ puzzle. Before the discovery of the muon's neutrino it was believed that electron, muon, and one neutrino formed a single family (called the lepton family) with a single family-conservation law. If this were the case, no conservation law prohibited the decay of muon into electron and photon, since the lost muon was replaced with an electron, and charge and all other quantities were conserved as well. According to the classical view of physical law, the absence of this process should have caused no concern. There was, after all, no law of permission which said that it should occur. There was only the double negative: No conservation law was known to prohibit the decay.

* In Chapter Twenty-Seven antiparticles are described as "particles moving backward in time."

However, the view of the fundamental role of conservation laws in nature as the only inhibition on physical processes had become so ingrained in the thinking of physicists that the absence of this particular decay mode of the muon was regarded as a significant mystery. It was largely this mystery that stimulated the search for a second neutrino belonging exclusively to the muon. The discovery of the muon's neutrino established as a near certainty that the electron and muon belong to two different small families which are separately conserved. With the electron and muon governed by two separate laws of conservation, the prohibition of the μ-e-γ decay became immediately explicable, and the faith that what can happen does happen was further bolstered.

We turn now to the conservation laws that involve properties of motion (the first three in the list in Section 4.2).

4.5 Energy conservation

17 In the world of particles there are only two kinds of energy: energy of motion, or kinetic energy, and energy of being, which is equivalent to mass. Whenever particles are created or annihilated (except the massless particles), energy is transformed from one form to the other, but the total energy in every process always remains conserved. The simplest consequence of energy conservation for the spontaneous decay of unstable particles is that the total mass of the products must be less than the mass of the parent. For each of the following decay processes the masses on the right add up to less than the mass on the left:

$$K^+ \rightarrow \pi^+ + \pi^+ + \pi^- ,$$

$$\Xi^- \rightarrow \Lambda^0 + \pi^- ,$$

$$\mu^+ \rightarrow e^+ + \nu_e + \bar{\nu}_\mu ,$$

These are allowed processes, "downhill" in mass; "uphill" decays are forbidden. An unstable particle at rest has only its energy of being, no energy of motion. The difference between this parent mass and the mass of the product particles is transformed into kinetic energy which the product particles carry away as they rapidly leave the scene.

One might suppose that if the parent particle is moving when it decays it has some energy of motion of its own that might be transformed to mass. The conservation of momentum prohibits this. The extra energy of motion is in fact "unavailable" for conversion into mass. If a particle loses energy, it also loses momentum. Momentum conservation therefore prohibits the conversion of all of the energy into mass. It turns out that momentum and energy conservation taken together forbid uphill decays into heavier particles no matter how fast the initial particle might be moving.

If two particles collide, on the other hand, some—but not all—of their energy of motion is available to create mass. It is in this way that the various unstable particles are manufactured in the laboratory. In an actual typical collision in the vicinity of an accelerator, one of the two particles, the projectile, is moving rapidly, and the other, the target, is at rest. Under these conditions, the requirement that the final particles should have as much momentum as the initial projectile severely restricts the amount of energy that can be converted into mass. This is too bad, for the projectile has been given a great energy at a great expense. To make a proton-antiproton pair, for example, by the projectile-hitting-fixed-target method, the projectile must have a kinetic energy of 6 GeV (billion electron volts), of which only 2 GeV goes into making the mass. The 6 GeV Berkeley Bevatron was designed with this fact in mind, in order to be able to make antiprotons and antineutrons. Typical processes for protons striking protons are:

$$p + p \rightarrow p + p + p + \bar{p} ,$$

$$p + p \rightarrow p + p + n + \bar{n}\,.$$

The unfortunate waste of 4 GeV in these processes could be avoided if the target proton were not quiescent, but flew at the projectile with equal and opposite speed. It is hard enough to produce one high-energy beam, and far more difficult to produce two at once. Nevertheless, the gain in available energy makes it worth the trouble, and a technique for producing "clashing beams" is now employed at Stanford University, where oppositely directed beams of electrons collide. The device is sometimes called by physicists the synchroclash.

4.6 *Momentum conservation*

Momentum is purely a property of motion—that is, if there is no motion, there is no momentum. In classical mechanics, momentum is defined as the product of mass and velocity,

$$\mathbf{p} = m\mathbf{v} \tag{4.2}$$

($\mathbf{p}$ is the usual symbol for momentum). In relativistic mechanics (where even massless particles have momentum), its definition is slightly more complicated. In one respect, momentum is trickier than energy, for momentum is what is called a vector quantity. It has direction as well as magnitude. Vectors are actually familiar in everyday life, whether or not we know them by that name. The velocity of an automobile is a vector, with a magnitude (50 mph, for example) and a direction (northbound, for example). Force is a vector, a push or pull of some strength in some direction. Mass, on the other hand, is not a vector. It points in no particular direction. Energy also has no direction. The momentum of a rolling freight car, however, is directed along the tracks, and the momentum of an elementary particle is directed along its course through space.

In order to appreciate the law of momentum conservation, one must know how to add vectors. Two men pushing on a stalled car are engaged in adding vectors. If they push with equal strength *and* in the same direction, the total force exerted is twice the force each one exerts and, of course, in the direction they are pushing [Figure 4.1(a)], If they push with equal strength but at opposite ends of the car, their effort comes to naught, for the sum of two vector quantities that are equal in strength but opposite in direction is zero [Figure 4.1(b)], If they get on opposite sides of the car and push partly inward, partly forward, the net force exerted will be forward, but less than twice the force of each [Figure 4.1(c)]. Depending on their degree of cooperation, the two men may achieve a strength of force from zero up to twice the force each can exert. This is a general characteristic of the sum of two vectors. It may have a wide range of values depending on the orientation of the two vectors. (The topic of vector addition is pursued more fully in Chapter Six.)

Consider the law of momentum conservation applied to the decay of a kaon into muon and neutrino,

$$K^+ \rightarrow \mu^+ + \nu_\mu\,.$$

Before the decay, suppose the kaon is at rest [Figure 4.2(a)], After the decay, momentum conservation requires that muon and neutrino fly off with equal magnitudes of momenta *and* that the momenta be oppositely directed [Figure 4.2(b)]. Only in this way can the vector sum of the two final momenta be equal to the original momentum, namely, zero. This type of decay, called a two-body decay, is rather common, and is always characterized by particles emerging in exactly opposite directions.

In a three-body decay, the emerging particles have more freedom. Look ahead to Figure 9.8 to see the decay of a kaon at rest into three pions, with the tracks pointing in three different directions. Recalling the analogy between momentum and force, one can visualize a situation in which three different forces are acting and producing no net effect—two fighters and a referee all pushing in different directions in a clinch. Similarly, the momentum vectors must adjust themselves to produce no net effect; that is, they must add up to give zero. Momentum conservation on a grander scale is shown in Figure 4.3, where eight particles emerge from a single event.

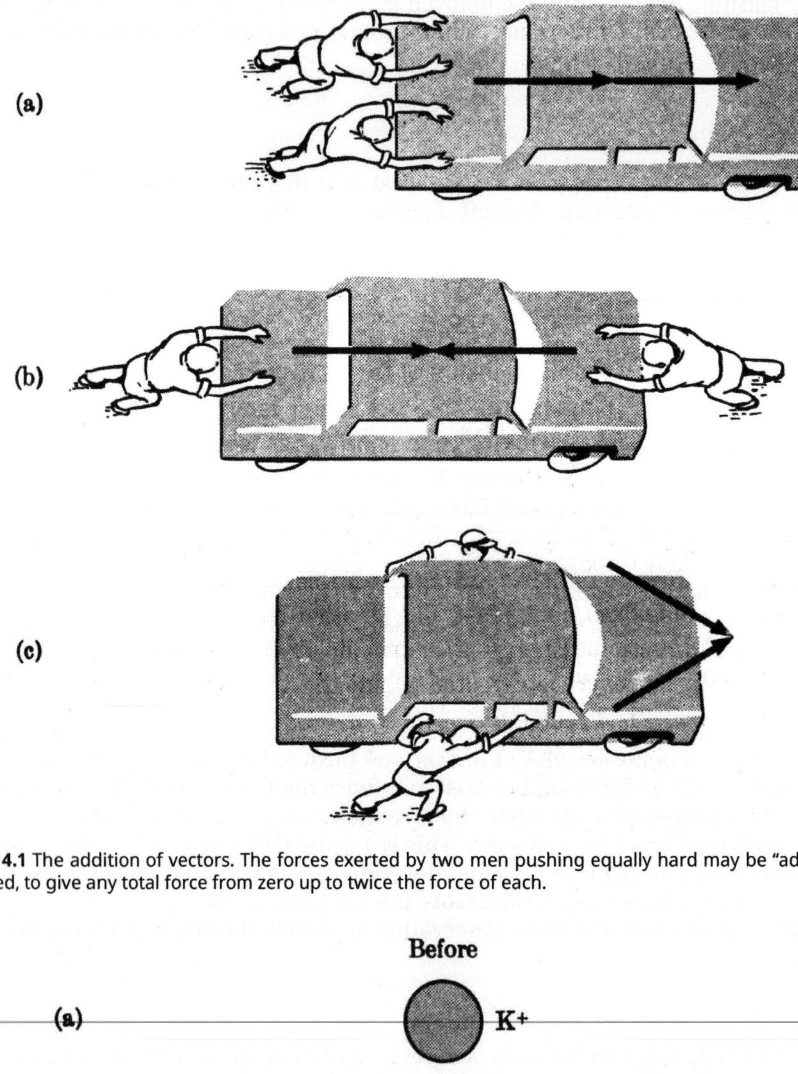

FIGURE 4.1 The addition of vectors. The forces exerted by two men pushing equally hard may be "added," that is, combined, to give any total force from zero up to twice the force of each.

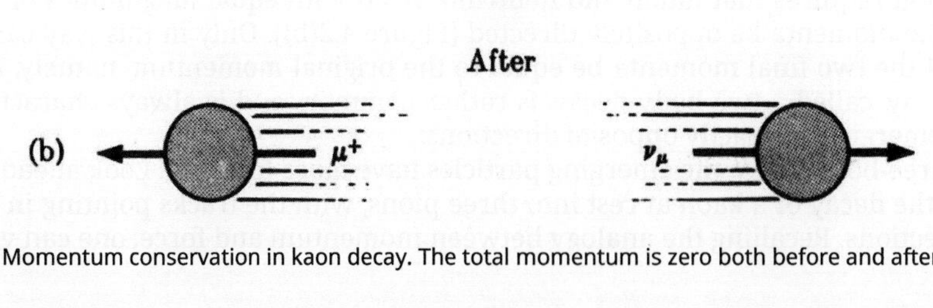

FIGURE 4.2 Momentum conservation in kaon decay. The total momentum is zero both before and after the decay.

One vital prohibition of the law of momentum conservation is that against one-body decays. Consider this possibility,

$$K^+ \rightarrow \pi^+ ,$$

the transformation of kaon to pion. It satisfies the laws of charge and family-number conservation. It is consistent with energy conservation, for it is downhill in mass, and it also satisfies spin conservation. But the kaon-pion mass difference must get converted to energy of motion, so that if the kaon was at rest, the pion will fly away. In whatever direction it moves, it has some momentum and therefore violates momentum conservation, since the kaon had none. On the other hand, if we enforce the law of momentum conservation, and keep the pion at rest, we shall have violated energy conservation, for in this case the extra energy arising from the mass difference will be unaccounted for.

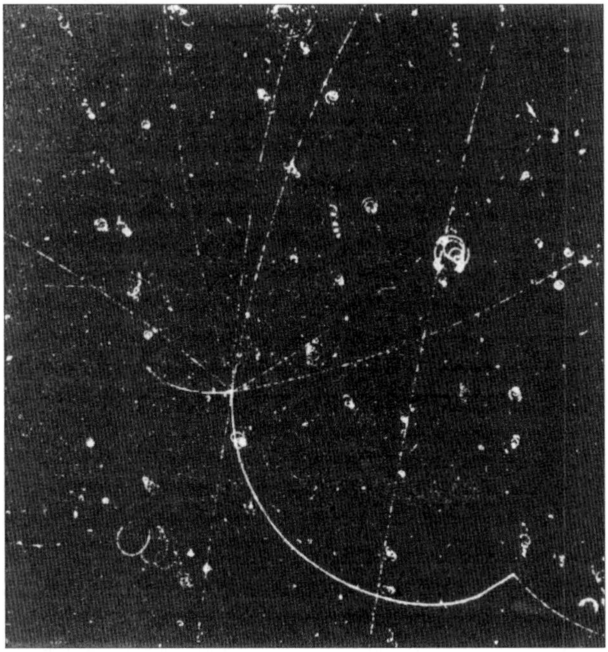

FIGURE 4.3 Momentum conservation in an antiproton annihilation event. An antiproton entering from the bottom collides with a proton in the bubble chamber. Eight pions, four negative and four positive, spray off from the annihilation event. The momentum of each can be measured from the curvature of the track; the eight momenta added together as vectors are just equal to the momentum of the single incoming antiproton. The kink in the track at the lower right is a pion decay, $\pi^+ \rightarrow \mu^+ + \nu_\mu$. (In what general direction did the unseen neutrino fly off?) (Photograph courtesy of Lawrence Radiation Laboratory, University of California, Berkeley.)

4.7 *Angular-momentum conservation*

Angular momentum, a measure of the strength of rotational motion, has been a key concept in physics since the time of Kepler. Actually, Kepler did not recognize it as such, but the second of his three laws of planetary motion—the so-called law of areas—is equivalent to a law of conservation of angular momentum. According to this law, an imaginary straight line drawn from the earth to the sun sweeps out area in space at a constant rate. During a single day this line sweeps across a thin triangular region with apex at the sun and base along the earth's orbit. The area of this triangle is the same for every day of the year. So, when the earth is closer to the sun, it must move faster in order to define a triangle with the same area. It speeds up just enough, in fact, to maintain a constant value of its angular momentum, and the law of areas can be derived as a simple consequence of the law of conservation of angular momentum. Angular momentum is precisely defined in Chapter Ten.

18 The earth also serves to illustrate approximately the two kinds of angular momentum that enter into the conservation law—orbital and spin. The earth possesses angular momentum because of its orbital motion around the sun and because of its daily (spin) rotation about its own axis.

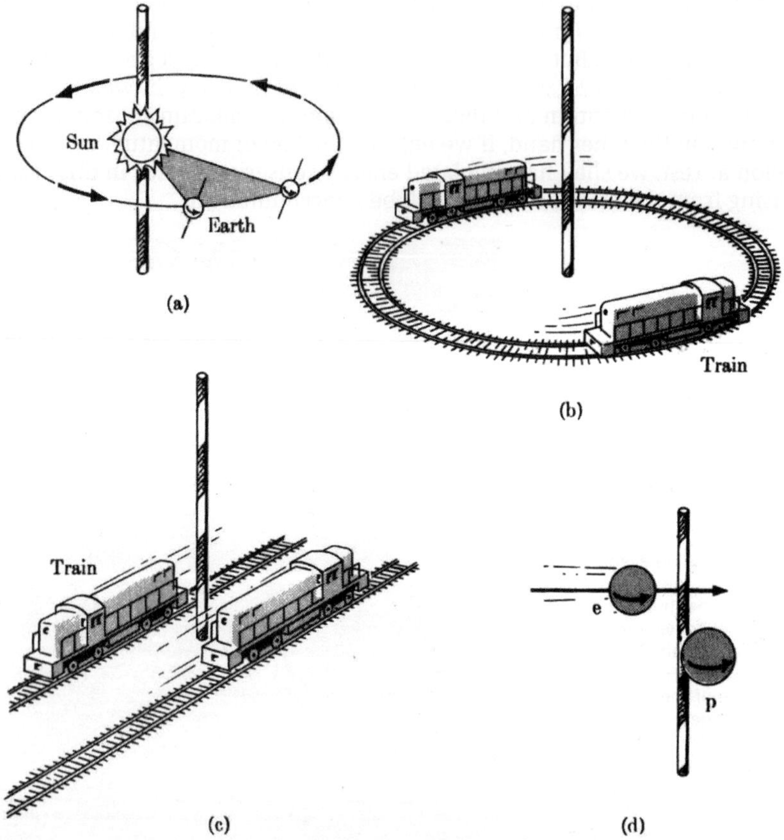

FIGURE 4.4 Examples of motion with angular momentum. **(a)** The earth possesses spin angular momentum about its axis as well as orbital angular momentum about an axis designated by the giant barber pole. **(b)** Trains on a circular track possess angular momentum about a vertical axis. **(c)** Even on straight tracks, angular momentum is associated with a similar relative motion of trains. **(d)** An electron flies past a proton. Both particles possess spin angular momentum and, because they are not on a collision course, they also have orbital angular momentum.

If a photographer in space took a time exposure of the earth and sun, his photograph would contain a short blur for the sun and a longer blur for the earth (for both are actually in motion). He would notice that the blurs were not directed toward each other, and from this fact alone he could conclude that earth and sun possess relative angular momentum. He would not need to know whether the earth swings around the sun or whether it proceeds into interstellar space. The key fact defining orbital angular momentum is some transverse motion of two objects. Any two moving objects, not aimed directly at each other, possess relative angular momentum. Two trains passing on the Great Plains have relative angular momentum, even though each is proceeding straight as an arrow. But if, through some mischance, both were on the same track on a collision course, they would have zero angular momentum. In particle collisions and decays, orbital angular momentum is usually of this trains-in-the-plains type, not involving actual orbiting of one particle round another. Figure 4.4 illustrates several examples of motion with angular momentum.

Angular momentum is a vector quantity. Its direction is taken to be the axis of rotation. The axis is well defined for spin, but what about orbital motion? For the passing trains, imagine again a blurred photograph indicating their direction of motion. Then ask: What would the axis be if the trains rotated about each other, instead of proceeding onward? The answer is a vertical axis; the angular momentum is directed upward. Of course an axis has two directions, not one. Why upward and not downward? This is an entirely arbitrary matter. By convention physicists have adopted a "right-hand rule" to define the direction of the angular momentum vector. If the curved fingers of the right hand follow the direction of rotational motion, the right thumb indicates the direction of angular momentum along the axis of rotation. This convention is easy to remember and is equally suitable for spin motion or for orbital motion. One more fact about orbital angular momentum needs to be known. Unlike spin, which comes in units of $\frac{1}{2}\hbar$, it comes only in units of $\hbar$.

The spinless pion decays into muon and neutrino, each with spin ½. In Figure 4.5 we use artistic license and represent the particles by little spheres with arrows to indicate their direction of spin. Muon and neutrino spin oppositely in order to preserve the total zero angular momentum. In this case, no orbital angular momentum is involved.

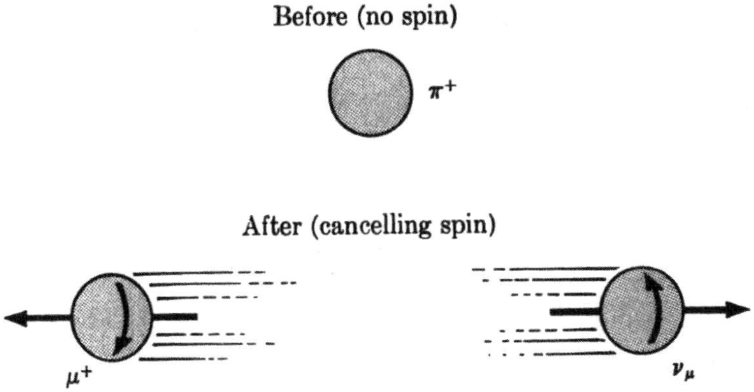

FIGURE 4.5 Angular-momentum conservation in pion decay. The total angular momentum is zero before and after the decay.

Another two-body decay, that of the Λ, illustrates the coupling of spin and orbital motion. The Λ, supposed initially at rest [Figure 4.6(a)], has spin ½. One of its possible decay modes is

$$\Lambda^0 \rightarrow p + \pi^- .$$

This may proceed in two ways. The proton and pion may move apart with no orbital angular momentum, the proton spin directed upward to match the initial Λ spin [Figure 4.6(b)]; or the proton spin may be flipped to point downward while proton and pion separate with one unit of orbital angular momentum, directed upward [Figure 4.6(c)]. In the first case,

$$\text{original spin } \tfrac{1}{2} \text{ (up)} \rightarrow \text{final spin } \tfrac{1}{2} \text{ (up)} .$$

In the second case,

$$\text{original spin } \tfrac{1}{2} \text{ (up)} \rightarrow \text{final spin } \tfrac{1}{2} \text{ (down)} + \text{orbital angular momentum 1 (up)} .$$

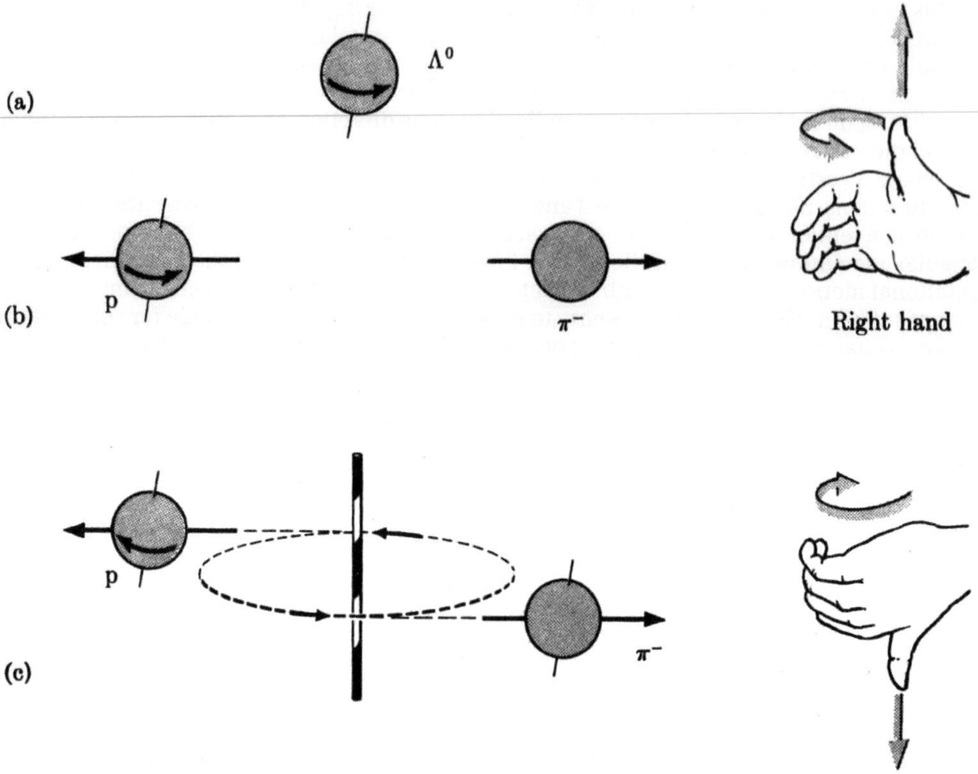

FIGURE 4.6 Angular-momentum conservation in lambda decay. The direction of angular momentum is defined by the right-hand rule. If the curved fingers of the right hand point in the direction of rotational motion, the right thumb defines the direction assigned to the angular momentum. Thus the particle spin is up in diagrams **(a)** and **(b)** and down in diagram **(c)**; the orbital angular momentum is up in diagram **(c)**.

4.8 Conservation laws and symmetry principles

19 The aspect of conservation laws that makes them appear to the theorist and the philosopher to be the most beautiful and profound statements of natural law is their connection with principles of symmetry in nature. Otherwise stated, energy, momentum, and angular momentum are all conserved because space and time are isotropic (the same in every direction) and homogeneous (the same at every place). This is a breath-taking statement when one reflects upon it, for it says that three of the seven absolute conservation laws arise solely because empty space has no distinguishing characteristics, and is everywhere equally empty and equally undistinguished. (Because of the relativistic link between space and time, we really mean space-time.) It seems, in the truest sense, that we are getting something from nothing.

Yet there can be no doubt about the connection between the properties of empty space and the fundamental conservation laws that govern elementary-particle behavior. This connection raises philosophical questions which we will mention but not pursue at any length. On the one hand, it may be interpreted to mean that conservation laws, being based on the most elementary and intuitive ideas, are the most profound statements of natural law. On the other hand, one may argue, as Bertrand Russell* has done, that it only demonstrates the

* Bertrand Russell, *The ABC of Relativity* (New York: New American Library, 1959).

hollowness of conservation laws ("truisms," according to Russell), energy, momentum, and angular momentum all being defined in just such a way that they must be conserved. Now, in fact, it is not inconsistent to hold both views at once. If the aim of science is the self-consistent description of natural phenomena based upon the simplest set of basic assumptions, what could be more satisfying than to have basic assumptions so completely elementary and self-evident (the uniformity of space-time) that the laws derived from them can be called truisms? Since the scientist generally is inclined to call most profound that which is most simple and most general, he is not above calling a truism profound. Speaking more pragmatically, we must recognize the discovery of *anything* that is absolutely conserved as something of an achievement, regardless of the arbitrariness of definition involved. Looking at those conservation laws whose basis we do not understand (the three family-number-conservation laws) also brings home the fact that it is easier to call a conservation law a truism after it is understood than before. It seems quite likely that we shall gain a deeper understanding of nature and of natural laws before the conservation of baryon number appears to anyone to be a self-evident truth.

Before trying to clarify through simple examples the connection between conservation laws and the uniformity of space, we shall consider the question, "What is symmetry?" In most general terms, symmetry means that in spite of some particular change in one thing, something else remains unchanged (see Figure 4.7). A symmetrical face is one whose appearance would remain the same if its two sides were interchanged. If a square figure is rotated through 90 deg, its appearance is not changed. Among plane figures, the circle is the most symmetrical, for if it is rotated about its center through any angle whatever, it remains indistinguishable from the original circle—or, in the language of modern physics, its form remains invariant. In the language of ancient Greece, the circle is the most perfect and most beautiful of plane figures.

Aristotle regarded the motion of the celestial bodies as necessarily circular because of the perfection (the symmetry) of the circle. Now, from a still deeper symmetry of space-time, we can derive the ellipses of Kepler. Modern science, which could begin only after breaking loose from the centuries-old hold of Aristotelian physics, now finds itself with an unexpected Aristotelian flavor, coming both from the increasingly dominant role of symmetry principles and from the increasingly geometrical basis of physics.

We are accustomed to think of symmetry in spatial terms. The symmetry of the circle, the square, and the face are associated with rotations or inversions in space. Symmetry in time is an obvious extension of spatial symmetry; the fact that nature's laws appear to remain unchanged as time passes is a fundamental symmetry of nature. However, there exist some subtler symmetries, and it is reasonable to guess that the understanding of baryon conservation, for example, will come through the discovery of new symmetries not directly connected with space and time.

In the symmetry of interest to the scientist, the unchanging thing—the invariant element—is the form of natural laws. The thing changed may be orientation in space, or position in space or time, or some more abstract change (not necessarily realizable in practice) such as the interchange of two particles. The inversion of space and the reversal of the direction of flow of time are other examples of changes not realizable in practice, but nonetheless of interest for the symmetries of natural law. These latter two phenomena are discussed in Chapter Twenty-Seven.

If scientists in Chicago, New York, and Geneva perform the same experiment and get the same answer (within experimental error), they are demonstrating one of the symmetries of nature, the homogeneity of space. If the experiment is repeated later with the same result, no one is surprised, for we have come to accept the homogeneity of time. The laws of nature are the same, so far as we know, at all points in space, and for all times. This invariance is important and is related to the laws of conservation of energy and momentum, but ordinary experience conditions us to expect such invariance so that it seems at first to

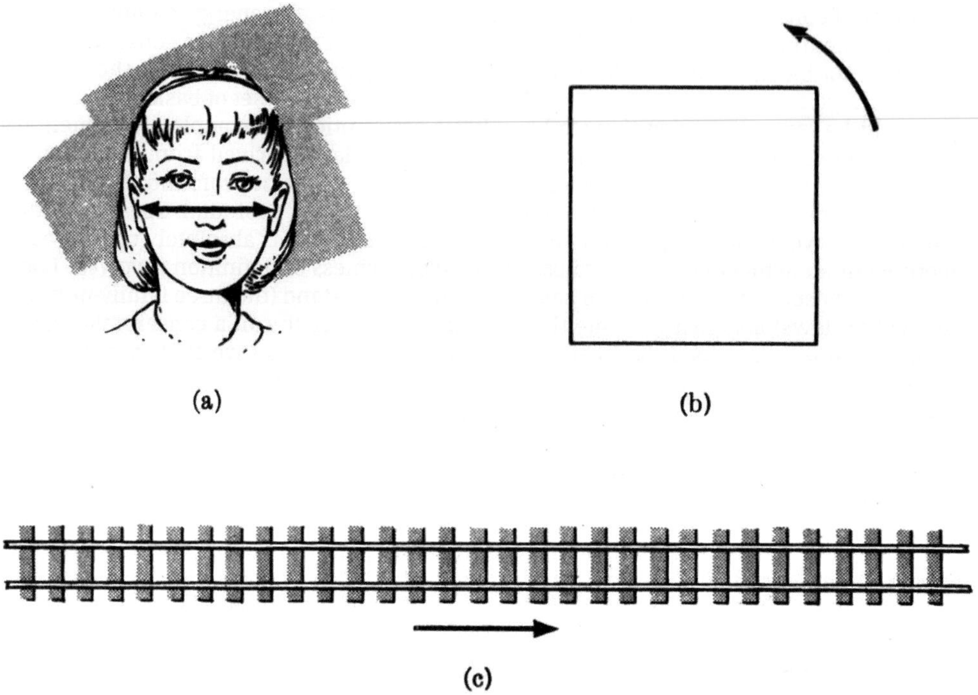

FIGURE 4.7 Three kinds of symmetry. **(a)** Reflection symmetry. A symmetrical face is unchanged if left and right are interchanged (reflection through a vertical plane). **(b)** Rotation symmetry. The appearance of a square is unchanged if it is rotated through 90 deg or any integral multiple of 90 deg. **(c)** Translation symmetry. A straight railroad line, hypothetically infinite in extent, is unchanged if it is shifted by any integral number of ties.

be trivial or self-evident. It might seem hard to visualize any science at all if natural law changed from place to place and time to time, but, in fact, quantitative science would be perfectly possible without the homogeneity of space-time. Imagine yourself, for example, on a merry-go-round that speeded up and slowed down according to a regular schedule. If you carried out experiments to deduce the laws of mechanics and had no way of knowing that you were on a rotating system, you would conclude that falling balls were governed by laws that varied with time and with position (distance from the central axis), but you would be quite able to work out the laws in detail and predict accurately the results of future experiments, provided you knew where and when the experiment was to be carried out. Thanks to the actual homogeneity of space and time, the results of future experiments can in fact be predicted without any knowledge of the where or when.

A slightly less obvious kind of invariance, although one also familiar from ordinary experience, is the invariance of the laws of nature for systems in uniform motion. Passengers on an ideally smooth train or in an ideally smooth elevator are unaware of motion. If the laws of mechanics were significantly altered, the riders would be aware of it through unusual bodily sensations. Such a qualitative guide is, of course, not entirely reliable, but careful experiments performed inside the ideal uniformly moving train would reveal the same laws of nature revealed by corresponding experiments conducted in a stationary laboratory. This particular invariance underlies the theory of relativity, and is a manifestation of the isotropy of four-dimensional space-time. A discussion of space-time, the subject of Section 20.7, is out of place here. However, a brief statement about the invariance of uniform motion, even if scarcely comprehensible, may serve to illustrate the kind of thinking engendered by the theory of relativity. What, to our limited three-dimensional vision, appears to be uniform motion is, to a more enlightened brain capable of encompassing four dimensions, merely a rotation. Instead of turning, say, from north to east, the experimenter who climbs aboard the train is, from the more general view, turning from space partly toward the time direction. According to relativity, which joins space and time together in a four-dimensional space-time, the laws of nature should no more be changed by "turning" experimental apparatus toward the time direction (that is, loading it aboard the train) than by turning it through 90 deg in the laboratory.

The chain of connection we have been discussing is: Symmetry → invariance → conservation. The symmetry of space and time, or possibly some subtler symmetry of nature, implies the invariance of physical laws under certain changes associated with the symmetry. In the simplest case, for example, the symmetry of space which we call its homogeneity implies the invariance of experimental results when the apparatus is moved from one place to another. This invariance, in turn, implies the existence of certain conservation laws. The relation between conservation laws and symmetry principles is what we now wish to illuminate through two examples. Unfortunately, an adequate discussion of this important connection requires the use of mathematics beyond the scope of this book.

4.9 *The uniformity of space*

Suppose we imagine a single isolated hydrogen atom alone and at rest in empty space. If we 20
could draw up a chair and observe it without influencing it, what should we expect to see? (For this discussion, we ignore quantum mechanics and the wave nature of particles, pretending that electron and proton may be localized at points in space and be uninfluenced by the observer. The reader will have to accept the fact that these false assumptions are permissible and irrelevant for the present discussion.) We should see an electron in rapid motion circling about a proton, and the proton itself moving more slowly in a smaller circle. Were we to back off until the whole atom could be discerned only as a single spot, that spot, if initially motionless, would remain at rest forever. We now must ask whether this circumstance is significant or insignificant, important or dull. It certainly does not seem surprising. Why should the atom move? we may ask. It is isolated from the rest of the universe; no forces act upon it from outside; therefore there is nothing to set it into motion. If we leave a book on a table and come back later, we expect to find it there. Everyday experience conditions us to expect that an object on which no external forces act will not spontaneously set itself into motion. There is no more reason for the atom to begin to move than for the book to migrate across the table and fly into a corner. The trouble with this argument is that it makes use of the common sense of ordinary experience, without offering any explanation for the ordinary experience.

If we put aside "common sense" and ask what the atom might do, it is by no means obvious that it should remain at rest. In spite of the fact that no external forces are acting, strong internal forces are at work. The proton exerts a force on the electron which constantly alters its motion; the electron, in turn, exerts a force on the proton. Both atomic constituents are experiencing force. Why should these forces not combine to set the atom as a whole into motion? Having put the question in this way, we may consider the book on the table again. It consists of countless billions of atoms, each one exerting forces on its neighboring atoms. Through what miracle do these forces so precisely cancel out that no net force acts upon the book as a whole and it remains quiescent on the table?

The classical approach to this problem is to look for a positive, or permissive, law, a law that tells what *does* happen. Newton first enunciated this law which (except for some modification made necessary by the theory of relativity) has withstood the test of time to the present day. It is called Newton's third law, and it says that all forces in nature occur in equal and opposite balanced pairs. The proton's force on the electron is exactly equal and opposite to the electron's force on the proton. The sum of these two forces (the *vector* sum) is zero, so that there is no tendency for the structure as a whole to move in any direction. The balancing of forces, moreover, can be related to a balancing of momenta. By making use of Newton's second law,* which relates the motion to the force, one can discover that, in a hydrogen atom initially at rest, the balanced forces will cause the momenta of electron and proton to be equal and opposite. At a given instant, the two particles are moving in opposite directions. The heavier proton moves more slowly, but has the same momentum as the electron. As the electron swings to a new direction and a new speed in its track, the proton swings too in just such a way that its momentum remains equal and opposite to that of the electron. In spite of the continuously changing momenta of the two particles, the total momentum of the atom remains zero; the atom does not move. In this way— by "discovering" and applying two laws, Newton's second and third laws of motion—one derives the law of momentum conservation and finds an explanation of the fact that an isolated atom does not move.

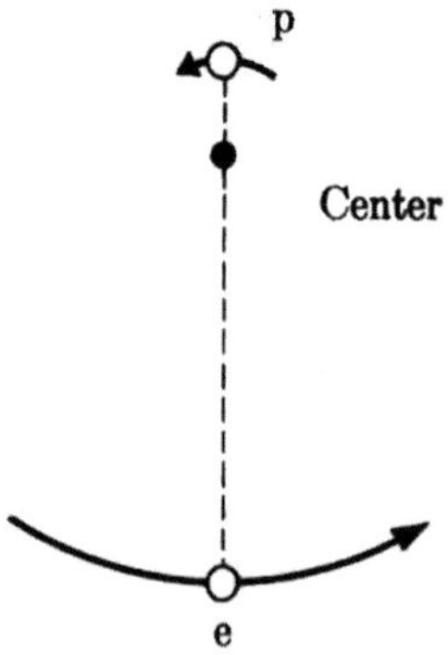

FIGURE 4.8 Schematic view of "motionless" hydrogen atom. In order for the center of mass of the atom to remain at rest, proton and electron must continually move in such a way that their momenta remain equal and opposite. Otherwise spontaneous acceleration of the atom as a whole would show space to be nonhomogeneous. (For clarity, the diagram exaggerates the motion of the proton and shows the center of mass to be farther from the proton than it actually is.)

Without difficulty, the same arguments may be applied to the book on the table. Since all forces come in equal and opposite pairs, the forces between every pair of atoms cancel, so that the total force is zero, no matter how many billions of billions of atoms and individual forces there might be.

It is worth reviewing the steps in the argument above. Two laws of permission were discovered, telling what does happen. One law relates the motion to the force; the other says that the forces between pairs of particles are always equal and opposite. From these laws, the conservation of momentum was derived as an interesting consequence, and this conservation law in turn explained the fact that an isolated atom at rest remains at rest.

The modern approach to the problem starts in quite a different way, by seeking a law of prohibition, a principle explaining why the atom does *not* move. This principle is the invariance of laws of nature to a change of position. Recall the chain of key ideas referred to on page 73: Symmetry → invariance → conservation. In the example of the isolated hydrogen atom, the symmetry of interest is the homogeneity of space. Founded upon this symmetry is the invariance principle just cited. Finally, the conservation law resting on this invariance principle is the conservation of momentum.

In order to clarify, through the example of the hydrogen atom, the connecting links between the assumed homogeneity of space and the conservation of momentum, we must begin with an exact statement of the invariance principle as applied to our isolated atom. The principle is this: No aspect of the motion of an isolated atom depends upon the location

* Newton's second law, usually written $\mathbf{F} = m\mathbf{a}$, says that the acceleration $\mathbf{a}$ experienced by a particle multiplied by its mass m is equal to the force $\mathbf{F}$ acting upon it. The law may also be stated in this way: The rate at which the momentum of a particle is changing is equal to the force applied. These ideas are expanded in Chapters Eight and Nine.

of the center of mass of the atom. The center of mass of any object is the average position of all of the mass in the object. In a hydrogen atom, the center of mass is a point in space between the electron and the proton, close to the more massive proton.

Let us visualize our hydrogen atom isolated in empty space with its center of mass at rest (Figure 4.8). Suppose now that its center of mass starts to move. In which direction should it move? We confront at once the question of the homogeneity of space. Investing our atom with human qualities for a moment, v/e can say that it has no basis upon which to "decide" how to move. To the atom surveying the possibilities, every direction is precisely as good or bad as every other direction. It is therefore frustrated in its "desire" to move and simply remains at rest.

This anthropomorphic description of the situation can be replaced by sound mathematics. What the mathematics shows is that an acceleration of the center of mass—for example, changing from a state of rest to a state of motion—is not consistent with the assumption that the laws of motion of the atom are independent of the location of the center of mass. If the center of mass of the atom is initially at rest at point A and it then begins to move, it will later pass through another point B. At point A, the center of mass had no velocity. At point B it does have a velocity. Therefore, the state of motion of the atom depends on the location of the center of mass, contrary to the invariance principle. Only if the center of mass remains at rest can the atom satisfy the invariance principle. If the center of mass of the atom had been moving initially, the invariance principle requires that it continue moving with constant velocity. The immobility of the center of mass requires, in turn, that the two particles composing the atom have equal and opposite momenta. A continual balancing of the two momenta means that their sum, the total momentum, is a constant.

The argument thus proceeds directly from the symmetry principle to the conservation law without making use of Newton's laws of motion. That this is a deeper approach to conservation laws, as well as a more esthetically pleasing one, has been verified by history. Although Newton's laws of motion have been altered by relativity and by quantum mechanics, the direct connection between the symmetry of space and the conservation of momentum has been unaffected—or even strengthened—by these modern theories, and momentum conservation remains one of the pillars of modern physics. We must recognize that a violation of the law of momentum conservation would imply an inhomogeneity of space; this is not an impossibility, but it would have far-reaching consequences for our view of the universe.

Through similar examples it is possible to relate the law of conservation of angular momentum to the isotropy of space. A compass needle that is held pointing east and is then released will swing toward the north because of the action of the earth's magnetic field upon it. But if the same compass needle is taken to the depths of empty space, far removed from all external influences, and set to point in some direction, it will remain pointing in that direction. A swing in one direction or the other would imply a nonuniformity* of space. If the uniformity of space is adopted as a fundamental symmetry principle, it can be concluded that the total angular momentum of all the atomic constituents of the needle must be a constant. Otherwise, the internal motions within the needle could set the whole needle into spontaneous rotation, and its motion would violate the symmetry principle.

Energy conservation, in a way that is not so easy to see, is related to the homogeneity of time. Thus all three conservation laws—of energy, momentum, and angular momentum—are "understood" in terms of the symmetry of space-time; and indeed the theory of relativity has shown that these three laws are all parts of a single general conservation law in the four-dimensional world.

---

* Strictly, momentum conservation rests on the *homogeneity* of space (uniformity of place), and angular momentum conservation rests on the *isotropy* of space (uniformity of direction). The distinction is not important for our purposes, and it is satisfactory to think of space simply as everywhere the same, homogeneity and isotropy being summarized by the word uniformity.

EXERCISES

4.1. If you walk across a rug on a dry day, your body acquires an electric charge. Why does this *change* of charge not violate the principle of charge conservation?

4.2. Is it possible in principle to convert one gram of matter entirely into energy? Is it possible in practice to do so? Explain.

4.3. A shell fired vertically upward explodes, breaking into two equal fragments, just at the moment it reaches its greatest height (at which moment it is at rest). In how many directions may its fragments fly apart? Explain why the two fragments must start apart with exactly equal kinetic energy.

4.4. An astronaut floating in space fires a gun with a rifled barrel, which gives the bullet a spin. Discuss the motion of the astronaut after the bullet is fired, assuming that he was motionless before.

4.5. Two automobiles approaching an intersection from opposite directions both turn right, as shown in the figure. What is the direction of their relative angular momentum **(a)** before the turn; **(b)** during the turn; and **(c)** after the turn? Justify each part of your answer.

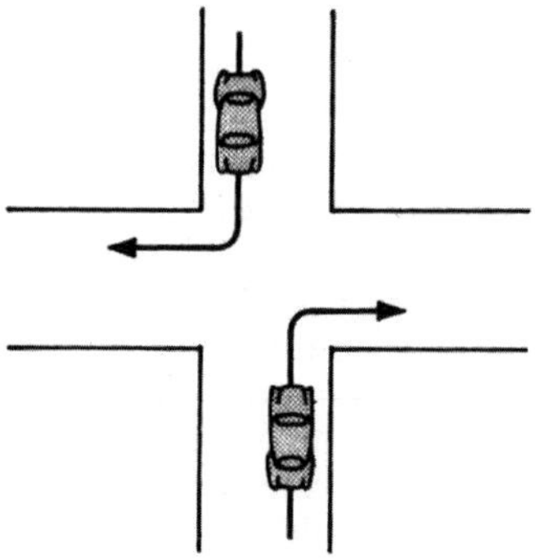

4.6. Make a drawing analogous to Figure 4.5 showing "before" and "after" diagrams for neutron decay, a three-body decay. Pay special attention to momentum conservation and angular momentum conservation.

4.7. A neutral pion (π^0) at rest has a mass-energy of 135 MeV. It decays into two photons. **(1)** Give **(a)** the energy of each photon in MeV and in ergs, **(b)** the momentum of each photon in gm cm/sec (for a photon, $E = pc$), and **(c)** the relative direction of the photons. **(2)** What conservation laws are relevant in the decay? **(3)** If the pion had decayed in flight instead of at rest, how would the results have differed qualitatively?

4.8. Write a reaction formula that expresses the decay of μ^+. Show that it satisfies the relevant conservation laws.

4.9. **(1)** Explain explicitly how the beta decay of the neutron conforms to the conservation laws of energy, charge, electron-family number, and baryon number. **(2)** If a neutron at rest decays, the initial paths of the three product particles must be in a plane. What conservation law imposes this condition? Explain how it does so.

4.10. Table 2.1 shows that the Σ^+ particle can decay into a neutron and a positive pion. Give one other way in which it should be able to decay. Justify your answer by pointing out how each relevant conservation law is satisfied.

4.11. Each of the following particle decays except one violates some conservation law. Which one is all right and which law or laws is violated by each of the others?

(a) $\mu^+ \rightarrow e^+ + e^-$
(b) $\pi^- \rightarrow K^- + \nu_\mu$
(c) $K^- \rightarrow \mu^- + \bar{\nu}_\mu$
(d) $n \rightarrow \pi^+ + \pi^-$
(e) $e^- \rightarrow \nu_e + \gamma$
(f) $\Lambda^0 \rightarrow p + e^-$

4.12. Two of the following reaction processes do occur and two do not. Select the two allowed reactions and explain how the relevant conservation laws are satisfied. (Assume that sufficient energy is available in the collision to provide for any necessary mass increase.)

(a) $n + p \rightarrow \Lambda^0 + n + K^+ + \pi^0$
(b) $p + p \rightarrow p + \Lambda^0 + \Sigma^+$
(c) $e^+ + e^- \rightarrow \mu^+ + \pi^-$
(d) $e^+ + e^- \rightarrow \gamma + \gamma$

4.13. Pick any four of the fundamental conservation laws and for each, name one decay or transformation process which it forbids. For at least two of the four, pick a process forbidden *only* by this one conservation law.

4.14. Write down several possible results of a proton-proton collision from which four particles emerge. (Assume that enough energy is at hand to create the new particles.) For each, explain why you consider the result possible.

4.15. Mention a few simple ways in which you rely daily on the homogeneity of space and of time.

4.16. From the "party equation" in Section 4.1, plot R vs. T for two different combinations of A and D, chosen to be more or less reasonable. Under what limiting circumstance must this theoretical formula fail?

4.17. Some conservation laws do not require a quantity to have a fixed value but do require it to have some other fixed property, such as oddness or evenness. Consider the guests at a reception who mix and from time to time shake hands. At any particular instant, a certain number of them will have shaken hands an odd number of times and the rest will have shaken hands not at all or else an even number of times. Fix attention on the number who have shaken hands an odd number of times. **(a)** Explain why this number need not be constant. **(b)** What property does this number have that is conserved? Explain the basis of your answer.

Part 2
Mathematics

KARL FRIEDRICH GAUSS
(1777 – 1855)

"Mathematics is the Queen of the Sciences and Arithmetic the Queen of Mathematics. She often condescends to render service to astronomy and other natural sciences, but under all circumstances the first place is her due."

FIVE

Mathematics in science

A grocer adding a column of figures is doing mathematics. So is an algebraist manipulating symbols that represent abstract concepts unrelated to anything known in the physical world. So is a child wrestling with the problem of how to get a fox and a hare and a cabbage safely across a river in a boat that carries only one of them at a time. From the limit of pure application to the limits of pure abstraction and of pure amusement, mathematics spans a remarkably wide range of human experience, perhaps a wider range than any other single discipline of thought. The all-too-common "nonmathematical" person would probably be startled if confronted with a list of occasions when he regularly uses mathematics or mathematical reasoning—in balancing his bank book (application), in trying to apply logical analysis to some human problem (abstraction), in playing a game of cards (amusement). In most general terms, mathematics is simply logical analysis. Some form of logic has no doubt been a part of the human scene for as long as man has existed. As a recognized distinct field of study, mathematics has existed at least since the earliest civilizations five thousand or more years ago.

5.1 The breadth of mathematics

Everyone knows that mathematics and science are closely related. Less well known are the characteristic features of mathematics and science that set them apart as independent fields of study. In Chapter One we considered the nature of science. Here we wish to consider the nature of mathematics, and most particularly two key aspects of mathematics—how it is an essential ingredient of science, yet how it has a life of its own independent of science. Unless the nature of mathematics as a separate human enterprise is understood, its role in science cannot be fully appreciated.

There are probably as many popular misconceptions about the nature of mathematics as about the nature of science. One important misconception concerns the scope of mathematics. All too often, mathematics is regarded as merely the tool that enables man to perform necessary but mundane calculations. Intellectuals frequently display an ignorance of mathematics and an ignorance of plumbing with the same degree of pride. This attitude toward mathematics is very nearly as wide of the mark as the view that language is "merely" the tool that enables man to communicate. It is scarcely possible to imagine a race of creatures capable of thought without language. No more easily can one visualize a society devoid of any mathematical heritage, yet capable of logical reasoning.

To the Greeks, the most valuable training for philosophy lay in the study of mathematics, music, and astronomy. At least since that time, mathematics and philosophy have been frequent partners not only in societies, but in individuals. A leading mathematician of the seventeenth century, Gottfried Leibniz, is now more widely known as a philosopher. In our own century, two leading philosophers, Bertrand Russell and Alfred North Whitehead, have made fundamental contributions to mathematics. Mathematics, too often regarded as a mere tool of the engineer and the banker, with about as much intellectual content as a screwdriver, deserves rather to be considered one of the noblest pursuits of the human mind, because of the challenges it offers to the intellect and because of its large historical role in the development of human thought. In our modern era the rise of technology has caused confusion about the nature of mathematics, and of science as well. By calling attention to practical application, technology tends to obscure the intellectual significance of the disciplines of thought upon which it is based.

5.2 Physical mathematics and nonphysical mathematics

So far as the role of mathematics in science is concerned, there exists a popular misunderstanding about mathematics even more significant than the common misunderstanding about its scope and philosophical content. This has to do with the nature of mathematical truth. A mathematical fact, so runs the common view, is absolutely true. It is not a matter of opinion, it is just a plain fact, and there is no room for discussion about it. This view of mathematics as a collection of incontrovertible facts (perhaps "cold, dull, emotionless" facts) no doubt accounts for the aversion to mathematics expressed by many people who find the warm, sharp, emotional problems of human relations more challenging. Of course everyone knows that a mathematical statement may be true or false. "Two plus two equals five" is false. "Two plus two equals four" is true. A mathematician would be quite willing to endorse these statements of fact and fallacy, as well as the view that his subject is a collection of facts. The misconception we are trying to get at is not so obvious as a completely wrong view of what is true and what is false in mathematics. Rather it is concerned with the subtler question of the exact nature of mathematical truth.

If a student is asked *why* two plus two equals four, he is likely to respond that it is simply a self-evident truth. If given more time to reflect on the question, he would perhaps recognize that he accepts the statement as true for two different reasons. The first is a purely logical or mathematical reason. The numerical concepts "two" and "four" and the operational concepts "plus" and "equal" have all. been defined in such a way that the statement is true. We may say that two plus two equals four by definition. Actually it is not true directly by definition, but by certain definitions plus rules of logical deduction from these definitions. This is the truth of the mathematician. We might call it the "inner truth" of the mathematical statement.

There is quite another kind of truth in the statement "two plus two equals four" that we might call the "external truth." This is the more convincing to the nonmathematically trained person, and the reason he calls the truth of the statement self-evident. We all know from our own experience that if we have two objects and add to them two more objects, we then have four objects. "Two plus two equals four" is not just a logical statement about a relationship among mathematical symbols. It is a statement about the "real" world, about nature, therefore a scientific truth. In human terms, the inner truth of mathematics is a purely intellectual truth, a truth of agreement among men. The external truth of mathematics is a truth about the physical world, the world of man's sense perceptions.

Now in terms of this example, we may express the popular misconception about mathematical truth. It is commonly believed that a mathematical fact, if it is true at all, possesses both inner truth and external truth. This belief is hardly surprising, for all of the elementary mathematics usually taught in schools does indeed possess both kinds of truth. Yet, from the point of view of pure mathematics, the external truth is a bonus, by no means necessary. The mathematician requires of his facts only that they are logically consistent with other facts and are deducible from his basic definitions and assumptions (called axioms). If by chance they happen to conform to something known in the physical world, and thereby become useful tools of the scientist or the practical man, it makes the facts neither more nor less acceptable to the mathematician. From the scientist's point of view, it should be regarded as the miracle of nature that mathematics—and often very simple mathematics—can be used successfully to describe natural phenomena in quantitative detail.

Inner truth and external truth have not always been distinguished in mathematics. Their clear separation came very late indeed in the history of mathematics, in the period from the mid-nineteenth century to the early twentieth century. Perhaps the most important single discovery ever made in mathematics was the discovery that mathematics is a logical structure independent of the physical world. This "discovery," made neither by one man nor at one time, came over a period of decades as "nonphysical" mathematics was gradually accepted as a significant part of mathematics, and as "physical" mathematics was placed on a solid logical basis independent of experiment. Nonphysical mathematics had been knocking at the door to be let in for many centuries. In some ancient civilizations, negative numbers were known but were regarded as outside mathematics, because negative length or negative quantity seemed to have no physical meaning. (Negative numbers were, in fact, a bit suspect as late as the seventeenth century.) The Greeks discovered "irrational" lengths, but rejected irrational numbers* because they involved the nonphysical concept of infinity. Imaginary numbers,† first studied carefully in the sixteenth century, had to wait until the nineteenth century to become completely acceptable members of the arithmetic family.

The particular developments of the nineteenth century that led to the explosive expansion of mathematics into nonphysical domains were the study of non-Euclidean geometry and the study of spaces of more than three dimensions. The geometry of Euclid was the branch of mathematics whose "external truth" was the spatial relationships of our three-dimensional world. It was at one and the same time a branch of mathematics (a logically self-consistent structure of ideas) and a branch of physics (a quantitative description of distances and shapes in our physical world). Mathematicians first broke free of the requirement of external truth when they courageously altered one of the basic axioms of Euclidean geometry and built a new self-consistent theoretical structure upon the new axioms. Then followed geometries of more than three dimensions, and algebras that violated the laws of arithmetic. Finally the old familiar branches of "physical" mathematics were recast in a purely logical form, as sets of basic axioms and relationships derived from those axioms, so that it no longer mattered whether they conformed to observations in the physical world.

From the modern point of view, mathematics can be defined roughly in this way: It is the study of logical relationships among mathematical "objects" based upon a few basic assumptions (axioms) about the properties of these objects. The "objects" of mathematical study are abstract concepts whose basic nature is determined exclusively by the axioms that specify their behavior. The ordinary numbers of mathematics—integers, positive numbers, negative numbers, real numbers, rational numbers, imaginary numbers—are examples of such mathematical objects. Most numbers do not seem formidably abstract to us, only because we happen to know that the rules governing the behavior of numbers are the same as the rules governing combinations of real physical objects in the external world. One of the axioms of the arithmetic of numbers, for example, is called the commutative law of addition. It states:

$$a + b = b + a\,;$$

the result of adding two numbers is the same regardless of the order of the addition. For millennia this was regarded not as an assumption about the behavior of abstract concepts, but as a self-evident truth about perfectly concrete things called numbers. You know that if you deposit ten dollars in your bank account today, and twenty dollars tomorrow, your balance will increase by exactly the same amount as if you had deposited first twenty and then ten. It is "obviously" so. The great discovery of the nineteenth century was the realization that the commutative law could be regarded as an entirely arbitrary assumption. Being arbitrary, it could be discarded or changed at will and a whole new arithmetic could be

* An irrational number is one that cannot be expressed as one integer divided by another integer. The number 0.75, three divided by four, is rational. The number π, the ratio of the circumference to the diameter of a circle, is an example of an irrational number. We still keep the Greek designation, "irrational," although such numbers are perfectly sensible.

† An imaginary number is one which, multiplied by itself, gives a negative number.

developed upon the basis of new axioms. In what is called noncommutative algebra, $a + b$ is *not* the same as $b + a$. In this nonphysical branch of mathematics,* the symbols a and b represent objects that are no longer visualizable, since we know of no physical objects obeying a noncommutative law of addition. From a purely mathematical point of view, however, the noncommuting objects are no more abstract or unreal than the commuting objects. It just happens that the commuting objects, which we call numbers, bear an exact relationship to things we perceive in the physical world, and the noncommuting objects do not. This fact affords an important insight into the nature of the physical world, but it in no way affects the structure of mathematics.

A similar situation prevails in geometry. The objects of Euclidean geometry— points, lines, areas, angles, volumes—correspond to the reality of our physical world, and can be visualized in concrete terms. The mathematically no-more-abstract objects of non-Euclidean geometry and multidimensional geometry defy visualization, and accordingly seem to the practical man less real. To the mathematician, the physical and nonphysical branches of mathematics seem equally real and equally challenging. He need not be concerned with which is which. Indeed, the lines are not clear. The big scientific surprise of the twentieth century has been that some "nonphysical" branches of mathematics turned out ta be physical after all. We shall return to this point below.

5.3 Mathematics—a tool and a toy

21 Pure mathematics as an independent structure of thought can best be looked at as a game, for it is precisely that. Its axioms define the game, which is played according to rules of logic. The mathematician's symbols are the pieces to be moved about. The wealth of relationships that can be derived from the basic axioms correspond to the rich variety of situations that can arise in a game with comparatively simple rules. In some particularly simple area of mathematics, such as the study of a group of only three objects, all possible consequences of the axioms can be derived, just as every possibility in tic-tac-toe can be spelled out. In more complex fields of mathematics, the seemingly limitless wealth of consequences that flow from the axioms continue to challenge the minds of generations of mathematicians, just as the inexhaustible possibilities in bridge and chess fascinate generations of players. To pursue this deep-going analogy a bit further, consider the game of chess. To make it correspond to a branch of mathematics, we must regard the board as the writing tablet of the mathematician, and the chessmen as the symbols representing the abstract objects of study. One set of chessmen may be of ivory, another of wood, another of plastic; but every knight, regardless of his size or shape or material, represents the same essence. There is but one single abstract object called "knight." The piece on the board is no more that basic concept than is a pencil stroke on a piece of paper the abstract object of mathematical study. The statements of the way in which each chessman moves are the axioms of the game. A minor change of a single axiom produces a drastic change in the game, just as a single alteration of the axioms of arithmetic produces a whole new kind of mathematics. Finally, there is the question of what constitutes complete knowledge of a concept or of a chessman. When the way in which a knight moves, the way he takes other pieces, and the way he can be taken are specified, that constitutes a complete definition of the abstract concept of a knight. There is nothing else to know. Questions such as how would a knight move on a triangular board are meaningless, for they are outside the game of chess. There is still much that can be learned about the usefulness of a knight and his relation to other chessmen, but nothing more that can be said to define what he is. In the same way, a mathematical concept is completely defined in terms of a few axioms. Its very abstraction rests on the fact that so few of its properties are specified. The beginning student

* Some noncommutative algebra has found a place in modern physics. It is nonphysical so far as the description of the everyday world is concerned.

of mathematics might say, "Yes, I understand how vectors behave, but what *is* a vector?" The only answer is that a vector is an idea defined by a few rules of its behavior. A scientist, more interested in making the idea visualizable and useful than in sticking to logical rigor, might give the answer, "Well, force, for example, is a vector." What he means by this is that there exists a physical concept called force that behaves like the mathematical concept called vector. The mathematical concept has been found to correspond to a physical concept and thereby to have acquired external truth as well as inner truth. His answer is exactly the same as saying about a particular carved piece of ivory, "This is a knight." More precisely, the piece is a physical object that behaves like the abstract concept of knight. The essence of knight is an idea quite independent of whether the piece is black or white, large or small, ivory or plastic. In the same way, the mathematical ideas of number, vector, length, or angle are abstractions quite independent of any correspondence to things in the "real" world.

Mathematics, then, can be regarded as a creation of the human mind independent of science and independent of nature. As such, it is not merely *like* a game; it *is* a game. Its rules are arbitrary, and its only criterion of truth is the inner truth of self-consistency. In terms of the nature of truth, mathematics and science can be clearly separated. Mathematics, no matter how numerous its applications to the real world, finds its ultimate criterion of validity only in the harmony of inner truth. Science, no matter how abstract its concepts or how theoretical its reasoning, is ultimately justified only by the external truth of experimental confirmation. In human terms, the dichotomy can be expressed this way. Science rests basically upon man's awareness of an orderly world outside himself. Mathematics rests basically upon man's awareness of an orderly world within himself.

To the scientist, of course, mathematics is more than a game. It is half a tool, and half a toy. Science without mathematics is unthinkable, for it is mathematics that gives to science its quantitative character and its predictive power. But taking the modern point of view about the nature of mathematics, it must be regarded as a miraculous chance that mathematics has found useful application in the description of the physical world.

Or is it a chance? Of course, it is no surprise that arithmetic and algebra and Euclidean geometry have something to do with the real world. They were invented and developed for practical ends, in order to describe reality. Application came first, abstraction later. But the question goes deeper than that. Basically it is: Is man capable of conceiving the nonphysical? Since man is himself a part of the physical world, does it make any sense to distinguish his inner world and his external world? Must any mathematics the human mind is capable of inventing have some connection with the physical world, whether or not that connection has yet been discovered? In other words, is the popular "misconception" about the external truth of mathematics not a misconception after all? Whatever the future holds, the fact is that there now exist branches of mathematics with no known external truth; that part of mathematics remains, so far, an intellectual exercise.

At the same time, much of the "nonphysical" mathematics of the nineteenth century has become the physical mathematics of the twentieth century. Our three-dimensional space has been generalized to a four-dimensional space-time. The general theory of relativity has shown that space is not really Euclidean, although it seems so in any small region. The mathematical formulation of the quantum theory has required the use of imaginary numbers, of quantities that do not commute with each other, and of vectors in a space of infinitely many dimensions. Impressed by this great expansion in the mathematical basis of physics, Paul Dirac* wrote in 1931: "Non-Euclidean geometry and

* P. A. M. Dirac, an outstanding theoretical physicist in England, was himself responsible for the introduction into physics of some previously nonphysical mathematics. In 1928 he discovered that the electron must be described not only by its position in ordinary space and time, but by its position in a peculiar abstract "spinor space." This new mathematics led to an improved theory of the electron and to the prediction of the positron. For this work, Dirac received the Nobel Prize in physics in 1933. (The quotation above appeared in the *Proceedings of the Royal Society, A133* (1931), 60.)

non-commutative algebra, which were at one time considered to be purely fictions of the mind and pastimes for logical thinkers, have now been found to be very necessary for the description of general facts of the physical world. It seems likely that this process of increasing abstraction will continue in the future and that advance in physics is to be associated with a continual modification and generalization of the axioms at the base of the mathematics rather than with a logical development of any one mathematical scheme on a fixed foundation." Dirac does not say that *all* mathematics will eventually turn out to be useful in the scientific description of nature, but he expresses his belief that ever wider ranges of mathematics will prove to be physical mathematics, that is, to have an external truth as well as an inner truth. Even if it were true that any self-consistent mathematical scheme that man can invent necessarily bears some correspondence to the physical world (a proposition that seems unlikely to this author), it still seems impossible that man could prove this to himself. In the race for truth, the mathematicians would remain always a lap ahead of the scientists. At all times there would exist some branches of mathematics that had as yet found no physical application and would remain, at least for a time, "purely fictions of the mind and pastimes for logical thinkers."

5.4 The branches of mathematics

There is no escaping the fact that the mathematical foundations of physics have in this century become more abstract and more difficult of comprehension. Correspondingly, the physical concepts have themselves become further removed from everyday experience and more difficult to visualize. This means that the contemporary research physicist must undergo a longer and more rigorous training in physics and in advanced mathematics than did physicists of earlier generations. Does this mean that physics is moving into an esoteric realm beyond the bounds of a liberal education? It is perhaps moving in that direction, although the modern student, faced with a technically complex world and with greater abstraction in every field of study, be it science or philosophy or economics, is very likely gaining in sophistication of thought at a corresponding rate. In any case, it is still possible to understand most of the concepts and theories of physics—and to understand them in much more than a superficial way—with a minimal background of mathematical and technical knowledge. Indeed it is a principal aim of this book to demonstrate this fact. The ideas of modern physics stretch the imagination, but do not necessarily require a high degree of mathematical skill. Aided by description, by analogy, by history, and by elementary mathematics at key points, the student without any professional orientation toward physics can still gain a rather profound insight into the meaning of the great theories of physics and into the modern ways of thinking about the physical world.

In the chapters that follow we shall employ as mathematical tools arithmetic, elementary algebra, and some ideas of geometry, assumed to be known already. A few other necessary topics—vectors, graphical representations, and trigonometric functions—will not be assumed to be already familiar; they will be clarified in the next two chapters. This means that we shall be basing twentieth-century physics on seventeenth-century mathematics. Although this situation is not ideal, it will suffice. Fortunately, many of the most important results throughout all of physics can be expressed in terms of simple algebraic formulas. The discrepancy between the mathematical and the physical levels in this book will perhaps seem less shocking if it is pointed out that the contemporary research physicist himself is about half a century behind the mathematicians.

Although modern mathematics has diversified into numerous specialties, most of its branches have retained the flavor of one or the other of the two ancient mainstreams of mathematics, arithmetic and geometry. In broadest terms, arithmetic and its derivatives may be called studies of quantity, geometry and its derivatives, studies of shape and size and form. Even these two major branches are hard to distinguish. Connections between arithmetic and geometry have been known since ancient times, and they have been very closely linked since the analytic geometry of the late seventeenth century. Indeed every branch of mathematics has connecting links to every other branch. The discovery of such links, often startling and beautiful because they join seemingly distinct ideas, can be one of the chief joys of the mathematician. To give a simple example, the quantity π, the ratio of the circumference to the diameter of a circle, crops up at numerous places in mathematics that apparently have nothing whatever to do with circles. Similarly, the prime numbers* appear quite mysteriously in areas of mathematics seemingly unrelated to these special integers.

The branch of modern mathematics most directly descended from early arithmetic is called number theory, the study of the positive integers. Far from being finished, this branch remains very much alive, and is often regarded as the queen of mathematics.† Another main descendant of arithmetic is algebra. In its earliest form (the form most often employed in this book), algebra is simply symbolic arithmetic. The evolution of algebra has paralleled the continual development and generalization of the idea of quantity, first as the scope of numbers expanded, then as mathematical objects more general and more abstract than numbers were introduced. Numbers began with the positive integers. To these were added fractions (rational numbers), zero, negative numbers, irrational numbers, and imaginary numbers. Then came vectors, tensors, matrices, and other "objects" that could be manipulated algebraically, but that no longer obeyed the same arithmetic laws as the numbers. Several of the abstract objects of modern algebra have found a place in physics. Among these we shall consider only vectors in this book.

A third major offshoot of arithmetic is called analysis, which includes calculus. The bulk of the mathematics used in science and technology comes under the general heading of analysis. It is impossible in a few words to give more than a vague idea of the nature of this vast field of mathematics. Basically, analysis rests on two key ideas, the idea of "rate of change" and the idea of "dependence." Since great parts of the study of nature are the study of change and the study of the dependence of one quantity upon another, it is not surprising that analysis is a chief tool of the scientist. The reason that the study of rate of change requires a distinct branch of mathematics can best be explained by example. The speed of an automobile could be determined by measuring off a known distance and dividing this distance (an arithmetic process) by the time required for the automobile to cover the distance. But unless the automobile were moving at constant speed, this process would yield only the average speed during the time interval, not the speed at any particular instant. In order to approach the idea of instantaneous speed, it would be necessary to divide an ever smaller distance by an ever smaller time interval. Algebra was unable to cope with the concept of the limit in which the automobile moved no distance in no time, yet had some speed at that particular moment. A new branch of mathematics, called calculus, had to be invented. Calculus makes it possible to describe rate of change quantitatively, even if that rate of change is not steady. The other key idea of analysis, "dependence," leads to the concept of a function. A function is a quantity (usually, but not necessarily, a number) whose value depends on the value of another quantity. If a change in a quantity x is always accompanied by a change in a quantity y, then y is said to be a function of x. Obviously the ideas of function and dependence are related to the idea of change. Obviously, too, the concept of function must enter into the description of nature, where the behavior of one quantity is linked to that of another.

As the evolution of algebra has paralleled the generalization of the concept of quantity, the evolution of geometry has paralleled the generalization of the concept of space. The

* A prime number is any integer that cannot be factored into the product of any other integers except unity times itself.

† At a meeting in Berlin in 1886, Leopold Kronecker, an outstanding German mathematician, said, "God made the integers; everything else is the work of man."

space of the earliest geometry was the flat plane (specifically, the Euclidean plane, in which the sum of the angles of every triangle add up to 180 deg, and parallel lines never meet). Then came three-dimensional Euclidean geometry; geometry on the surface of a sphere; and the algebraic description of geometry, known as analytic geometry. Trigonometry is likewise an analytic description of geometry. Modern geometry includes the study of spaces with any number of dimensions, and non-Euclidean spaces with a bumpy structure quite unlike the featureless smooth space of ordinary experience. A major derivative branch of geometry is called topology, the study of abstract geometrical relationships without reference to particular sizes or shapes or coordinate systems. The geometry of major concern to physics is the Euclidean geometry of two and three dimensions. In the latter part of this book, we shall have something to say about the four-dimensional geometry of space-time.

The branches of mathematics named so far—arithmetic, number theory, algebra, calculus, analysis, geometry, and topology—by no means exhaust the list. Some other branches are not so easily connected with the mainstreams of arithmetic and geometry. Probability theory, for instance, with its origin in the practical needs of insurance companies and gamblers, is an invention of the modern era that has flowered into a well-defined branch of mathematics with numerous applications in science. An even younger branch that has found recent application in physics is group theory, the study of sets of abstract objects of a special kind (which we shall not attempt to define here). The manipulation of these abstractions is a part of modern algebra, but group theory has more than algebraic significance, for the members of many groups have been given a geometrical interpretation. Group theory provides a link between modern algebra and modern geometry not unlike the link between older and less abstract forms of algebra and geometry that was provided by analytic geometry and trigonometry. Group theory has shown its power in modern physics by relating algebraic quantities such as energy and momentum to the geometrical properties of space and time.

5.5 *Some history**

Mathematical abstraction has a most interesting history, for it got its foot in the door of mathematics in ancient times, and then needed over two thousand years to get itself all the way in. The foot in the door was the axiomatic foundation of geometry discovered by the Greeks and brought to a beautifully complete form by Euclid in his geometry textbooks written about 300 B.C. The Greek thinking about geometry included such abstractions as the perfect point with no extent, the perfect line with no width, the perfect circle unrealizable in nature, and so on. It included also the concept of infinity, as in the idea of parallel lines that never meet. So "modern" is Euclid's geometry, that contemporary instruction in geometry still closely follows the presentation of Euclid.

The fact that mathematics, in its long history of development in the interim, was chained to the idea of physical reality had several main consequences. One of these was that a rigorous axiomatic basis for most branches of mathematics was not sought. Although the axioms of Euclidean geometry had been formulated in Biblical times, arithmetic was not put on a solid axiomatic basis until late in the nineteenth century. Arithmetic was so "obviously" true that no one looked for the arbitrary assumptions on which it could be founded.

Another and a very natural consequence of the physical orientation of mathematics until very recent times was that its development closely paralleled the development of science—particularly physics. Not only did mathematical discovery enrich physics and make possible its more rapid advance, but mathematicians themselves devoted their talents to solving physical problems. Many scientists, in turn, made important mathematical discov-

* Most of the facts in this section came from Dirk J. Struik, *A Concise History of Mathematics* (New York: Dover Publications, Inc., 1948).

eries. Both Kepler and Galileo, at the beginning of the seventeenth century, made discoveries in geometry and had ideas about infinitesimal quantities, ideas that pointed toward, but did not achieve, calculus. Newton, the most towering scientific figure of the seventeenth century, invented calculus in order to facilitate his calculations on gravitational force and motion. (Newton's contemporary Leibniz independently discovered calculus.) Throughout the eighteenth century, problems in physics provided the chief stimulus to mathematical discovery. Such leading mathematicians of the time as Leonhard Euler, Joseph Lagrange, Joseph Fourier, Jean D'Alembert, and Pierre Laplace have become part of the vocabulary of every physicist; we refer to Eulerian and Lagrangian coordinates in the study of fluids, to Fourier transforms in the study of wave motion, to the D'Alembertian and Laplacian operators in the theory of electromagnetism. And these are but a few. Although eighteenth-century experimenters such as Benjamin Franklin did not participate in the stream of mathematical advance, the theoretically oriented scientist and the mathematician were closely allied, and often were one and the same man.

Finally, a very important consequence of the physical orientation of mathematics throughout most of its history was that new abstract concepts to which mathematicians were naturally led had a very difficult time of it if their physical interpretation was not immediately evident. The fact that we still refer to imaginary numbers and irrational numbers serves as a reminder that these were once regarded as unacceptable concepts. Even the negative numbers had a hard time establishing their legitimacy. As late as 1703, Newton was considered bold to use negative coordinates. In order to understand how "illegitimate" concepts make their way into "legitimate" mathematics, consider the simple example of the addition and subtraction of positive integers. The sum of any two positive integers is another positive integer. The difference of two positive integers is sometimes a positive integer, but not always. Four subtracted from three is something new, a negative integer. The mathematician must either admit the new kind of number or he must declare that the subtraction of a larger number from a smaller number is meaningless. Early mathematicians preferred the latter alternative, since negative quantity seemed to be physically meaningless. In a similar way, irrational numbers and imaginary numbers intruded themselves into mathematics without being invited, and for a long time remained known but not wanted.

One thread of mathematical history, whose significance is easily overlooked, deserves some attention. This is the role of notation. Cumbersome notation probably did more to inhibit arithmetic development than did limitations of human imagination. Conversely, every advance in notational convenience gave birth to a rapid series of mathematical discoveries. Undoubtedly the most significant single advance in arithmetic notation was the decimal-positional method of writing numbers, which originated in India around A.D. 600,* and made its way westward to Europe via the Arabs some centuries later. The positional notation means that the value of a digit depends upon its position in a written number. For example the symbol 2 alone means two; the same symbol 2 in a different position, for example in 214, means two hundred. The decimal system by itself is of far less significance than the positional notation. Roman numerals, for example, are basically a decimal system, there being separate symbols for ten, one hundred, one thousand, besides the intermediate symbols for five, fifty, and five hundred. But one need only think of the relative difficulty of multiplying 35 by 48 and multiplying XXXV by XLVIII in order to appreciate the advantages of the positional notation. This advantage is not merely a matter of familiarity. In fourteenth-century Italy, bankers and merchants introduced the Hindu-Arabic number system despite the greater familiarity of Roman numerals, and despite official objections to the new numbers (a Florentine law of 1299 forbade the use of Arabic numerals). The extension of the positional notation to fractions, for example 0.75 replacing ¾, occurred in the late

* The Babylonians had a form of positional notation much earlier, based on the sexagesimal rather than the decimal system. However, our modern notation stems from the Hindus.

sixteenth century, and made fractional calculations and integer calculations equally easy. The only notational extension of any importance in scientific calculation since that time has been the introduction of the exponential notation for large and small numbers.

The decimal-positional notation is only one of numerous examples of the significance of mathematical notation. The sixteenth-century introduction of letters for unknowns in equations facilitated algebraic discovery. The graphical representation of functions in the seventeenth century did much to bring about the union of algebra and geometry. Because his calculus notation was simpler than Newton's, Leibniz had a more important influence on later mathematicians than did Newton. The Greeks, impeded by a cumbersome notation for numbers, made but little progress in arithmetic and algebra while they were taking giant strides in geometry.

The power of good notation has been as evident in science as in mathematics. The faith in simplicity that is such a powerful stimulus to scientific investigation is itself related to notation, for whether or not a given law can be written in a brief and elegant form may influence the scientist's view of the importance of that law. Einstein's general theory of relativity, for example, although in practical application extremely difficult to work with, is regarded as one of the most profound theories in nature, in part at least because its basic equations can be written in a remarkably simple and symmetric form using the modern notation of the theory of tensors. In most general terms, a good notation is one that gives simple concrete expression to abstract ideas, making them easy to visualize and easy to work with. The models and pictures that the scientist uses to help him think about the parts of nature out of reach of his sense perceptions may themselves be regarded as "notational" achievements. As the ideas of science follow the ideas of mathematics into new realms of abstraction, it becomes ever more important for the scientist to devise simple concrete ways of thinking about his abstractions.

By the end of the eighteenth century, what we might call classical mathematics—the mathematics based firmly upon external reality—had reached a crest in its development, an apex that led many mathematicians to take a pessimistic view of the future of mathematics. Exactly a century later, when the great theories of mechanics, electromagnetism, and thermodynamics seemed to be in beautiful and final form, a similar sense of pessimism was to be found among physicists. To the older generation of mathematicians around the turn of the nineteenth century, and to the older generation of physicists around the turn of the twentieth, it seemed that the rich lode of their specialty had been thoroughly mined, leaving only the dregs of tedious detail to be worked over by future generations. In both cases, the experts seriously underestimated the power of the human imagination (and the genius of their younger colleagues). The nineteenth century witnessed the revolutionary enlargement of the scope of mathematics, as new branches were created and old branches expanded into nonphysical domains. No less revolutionary was the impact on physics of the twentieth-century theories of relativity and quantum mechanics. So far in this century neither mathematicians nor physicists have found any reasons for renewed pessimism about the future potentialities of their fields of endeavor.

Specifically, the situation in mathematics at the end of the eighteenth century was something like this. Trigonometry, analytic geometry, and a comprehensive understanding of two and three dimensions had rounded out the ancient geometry of Euclid and linked it to algebra. The arithmetic of integers had been formalized into the well-developed number theory. The arithmetic of real numbers (rational and irrational) had led to a well-developed algebra and to the extremely rich field of calculus. From calculus flowed differential equations and the theory of functions. The theory of imaginary numbers was in a primitive state, but it did not seem to be needed to describe nature. Mathematics was tied to theoretical science, and theoretical science at the time meant principally only one thing, Newtonian mechanics. Over a century old, mechanics had served to stimulate a vast amount of mathematical research, and mathematicians in turn had applied mechanics to all manner of systems, including planets (celestial mechanics), fluids (hydrodynamics), and atoms (kinetic theory). On the one hand, the science of mechanics seemed to be fully exploited. On the other hand, the ancient mainstreams of mathematics, arithmetic and geometry, seemed to have been fully worked out. Of course no one saw an end to these subjects; but their future advance was foreseen to lie within fixed boundaries rather than to probe across unknown frontiers. Insofar as these subjects were concerned, the pessimistic view was not wrong. It was wrong only in not seeing the possibility of wholly new and unexpected areas of research.

Despite the eighteenth-century tie of mathematics to science, it was not new science that led to new mathematics, for nothing in nineteenth-century science required anything more than eighteenth-century mathematics. Rather the rebirth of mathematics must be attributed to the independent spirit of mathematicians, particularly of those few who had the courage in their search for new horizons to break the tie with the "real" world. The very facts of mathematical development that led to pessimism were the facts that led a new generation of mathematicians to seek wholly new fields to challenge their imagination. But the revolution of modern mathematics was not swift. Unlike an advance in science, where experimental confirmation can do much to make radical ideas palatable, the advance in mathematics, which amounted to a new way of looking at the basic nature of mathematics, had to evolve slowly. Some of the important enlargements of the boundaries of mathematics in the nineteenth century were the extension of analysis to the domain of complex numbers,* the extension of geometry to more than three dimensions and to non-Euclidean space, and the extension of the concepts of algebra from numbers to vectors and thence to tensors and more abstract quantities. The development of complex analysis by Augustin Cauchy and Karl Friedrich Gauss dates from about 1814. In 1816 Gauss studied non-Euclidean geometry but was unwilling to publish his findings. The first published work on the subject was by Nikolai Lobachevsky in 1829 and Janos Bolyai in 1832. Little noticed by mathematicians, the subject was revived in 1854 by Bernhard Riemann, then a young assistant at Göttingen, who perhaps more than anyone else of the time, caused mathematicians to begin to see their subject as an exercise in the logical manipulation of pure abstractions. The geometry of many dimensions was studied first by Hermann Grassmann in 1844 and was further developed by Riemann. Higher algebra also dates from the 1840s; its early development is attributable to William Rowan Hamilton and other mathematicians, mainly in the British Isles.

More important perhaps than specific achievements was the new attitude of mathematicians toward their own subject. Indicative of the new approach is this statement by Carl Jacobi (1804–1851), about Joseph Fourier, a distinguished mathematician of the preceding generation: "It is true that Mr. Fourier believed that the main end of mathematics was public usefulness and the explanation of natural phenomena, but such a scholar as he was should have known that the only end of science is the honor of the human mind, and that accordingly a question concerning number is as important as a question concerning the structure of the world." Gauss, despite his original reluctance to publish his discoveries in non-Euclidean geometry, came to realize that the nature of the geometry of the physical world was a question to be settled experimentally. That is, he recognized that Euclid's axioms were arbitrary assumptions, not self-evident truths about nature. Accordingly, he actually set out to measure whether or not space is Euclidean. Using surveying instruments on hilltops near Göttingen, he measured and summed the angles of triangles, assuming that beams of light defined straight lines. He found that Euclid had not been wrong. The space near Göttingen was Euclidean. More than half a century elapsed before the non-Euclidean character of space was demonstrated, using methods of measurement far more sensitive than those available to Gauss. But to Gauss goes the credit for being the first to see the possibility that the established geometry did not necessarily have to describe the real world.

* The full field of numbers—positive and negative, rational and irrational, real and imaginary—are called complex numbers.

These historical remarks may help to give perspective on the two themes of this chapter: the essential role of mathematics in science on the one hand, the fundamental difference between mathematics and science on the other hand. So closely linked have been these subjects throughout most of history that their distinction was unclear. Only in the nineteenth century did mathematics and physics diverge as clearly separate intellectual disciplines. Now with our modern insight, we are forced to marvel that so much of nature can be quantitatively described with such simple mathematics. Far from being obvious that it should be so, it seems a miracle. At the same time, we witness more and more of the modern mathematics of the last 150 years finding a place in the latest theories of nature, and must wonder how much of the mathematics conceivable by man has already been conceived and used by nature.

EXERCISES

5.1. The following questions provide a review of exponential arithmetic:

1. The speed of light is 3×10^{10} cm/sec. How long does it take light to travel across an atom whose diameter is 10^{-8} cm?
2. If it takes light 10 minutes to reach earth from Mars at a certain time of year, how far away is Mars at that time (a) in cm? (b) in miles?
3. Express the speed of light in ft/sec, miles/sec, miles/hr, and cm/microsec (a microsecond is 10^{-6} sec).
4. The product of wavelength and frequency for electromagnetic radiation is equal to the speed of light. What is the frequency of radar whose wavelength is 1 cm? What is the wavelength radiated at a radio station whose frequency is 1,340 kilocycles (1.34×10^6 cycles)? What is the frequency of green light whose wavelength is 5,000 angstroms?

5.2. When Li'l Abner needed about 10^6 dollars, he was advised to get a dollar and then double his money n times. How big is n? Express this $n$fold doubling process by means of a simple equation.

5.3. If an automobile traveling at 30 miles/hr doubles its speed, it moves at 60 miles/hr. Does this statement possess the inner truth of pure mathematics, the external truth of physical verification, or both? Explain your answer.

5.4. State any mathematical result you know that shows a connection between arithmetic and geometry.

5.5. Name one discipline other than geometry which can be considered to be a branch of both mathematics and physics.

5.6. **(1)** Name any pair of *physical* operations that are *not* commutative (the result of carrying out the operations in one order differs from the result of carrying out the operations in the other order). (2) Name one pair of physical operations that *are* commutative (the result of carrying out the operations in either order is the same).

5.7. Define carefully the meaning of subtraction in terms of addition, and the meaning of division in terms of multiplication. Explain how the desire to give meaning to these operations for *all* pairs of positive integers leads one to introduce new kinds of numbers. What are these new kinds of numbers?

5.8. **(1)** Construct the complete multiplication table for the four mathematical objects +1; −1; $\sqrt{-1}$; $-\sqrt{-1}$. **(2)** Evaluate $\frac{1}{2}(1 + \sqrt{-1})^2$.

5.9. An interesting algebra can be developed using "numbers" of the form $(a + b\epsilon)$. Here a and b are positive or negative integers and ϵ is a special quantity with the property, $\epsilon^2 = 0$ (this is part of the *definition* of ϵ). Numbers of the above type can be manipulated following the ordinary rules of algebra with the additional point that ϵ^2 when it occurs is to be replaced by zero. Evaluate

(a) $(2 + 3\epsilon) + (5 + \epsilon)$

(b) $(4 + 5\epsilon) - (6 + 3\epsilon)$

(c) $(7 + 8\epsilon) \times (-3 + 2\epsilon)$

(d) $(1 + \epsilon) \times \epsilon$

(e) $\dfrac{1}{1 + \epsilon}$

(HINT: try to find a quantity that, when multiplied by $(1 + \epsilon)$, yields 1.)

5.10. Find a way to get a fox, a hare, and a cabbage across a stream in a boat that carries one of them at a time, in such a way that fox and hare are never left together on one bank, and hare and cabbage are never left together on one bank. What do you think is the least number of trips that can accomplish this mission? Give a reason for your answer. Is this science? Is it mathematics? Why?

5.11. Consider a standard 8 × 8 chessboard and a set of 31 rectangular dominoes, each of which can cover two squares of the board. Prove that no possible arrangement of the dominoes can be found which will cover all the squares except two diagonally opposite corner squares. Is this puzzle mathematical? Why? Do you consider your proof "elegant"? Why?

5.12. A race of ants live on part of the surface of a large smooth sphere. They believe their world to be a flat plane. How could they put this belief to experimental test without circumnavigating the sphere?

5.13. In the middle of the nineteenth century, according to Dirk Struik (in the book referred to on page 88), "geometry of more than three dimensions was received with distrust and incredulity." How does it strike you personally? Why do scientists no longer find the idea distasteful?

5.14. Write a paragraph either defending or attacking the proposition that for any nation, culture, or group, abstract thought in philosophy or religion goes hand in hand with richness of mathematical development.

SIX

Vectors

A vector is a mathematical object with both numerical and geometrical properties. The physicist has been able to make good use of it in his description of nature. As background for the study of vectors, it is very useful to be able to think graphically about ordinary numbers. Accordingly, we begin this chapter on vectors with a geometrical view of arithmetic.

6.1 *The real axis: geometrical arithmetic*

Let us proceed as follows. First, draw a straight horizontal line. Pick any point on the line and label it 0 (zero). This point is called the origin. Label the point one unit to the right of the origin 1. (Any unit will do—a centimeter, an inch, or whatever is convenient.) Label the point two units to the right 2, the point three units to the right 3, and so on. The points one, two, and three units to the left of the origin should be labeled –1, –2, and –3. This line, with its labeled points, as shown in Figure 6.1, is called the real axis. It must be imagined to extend to infinity in both directions. The connection between numbers and the geometry of the real axis is this: To every real number, there corresponds a point on the line; to every point on the line there corresponds a real number. This is what mathematicians call a one-to-one correspondence. The number +2, for instance, corresponds to the labeled point 2. The number 2.45 corresponds to a point 45% of the way from the labeled point 2 to the labeled point 3. Which points are labeled in the diagram is an entirely arbitrary matter. The integers are merely chosen for convenience. It is best to think of every point as "labeled" with whatever numerical value corresponds to that point. We imagine a point a bit more than 41% of the way from 1 to 2, for example, labeled $\sqrt{2}$ (this irrational number is approximately equal to 1.414). If you could touch the line with an infinitely sharp pencil point, then no matter what point on the line is touched, there corresponds to that point a real number. The magnitude of the number is equal to the distance of the point from the origin (in whatever units have been chosen). If the point is to the right of the origin, the number is positive; if it is to the left of the origin, the number is negative. Like so many birds perched on a telephone wire, the real numbers are arranged on the real axis, each point "supporting" a different number.

Having assigned numbers to points on a line, we may carry out geometrical arithmetic (Figure 6.2). For the sum, $2 + 3 = 5$, the geometrical instructions are: Begin at the origin. Move two units to the right. Then move three more units to the right. The final point reached corresponds to the sum. Another way to say this is that the distance from the starting point, the origin, to the final point is the sum. Now it is easy to guess how to extend these rules. To add together a positive number and a negative number, move right for the positive number and left for the negative number. For example, to perform geometrically the sum, $2 + (-3) = -1$, again begin at the origin, move two units to the right, then three units to the left. The final

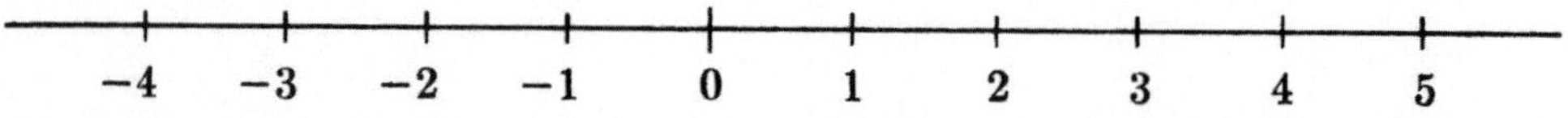

FIGURE 6.1 The real axis. There is a one-to-one correspondence between points on the line and real numbers.

point is one unit to the left of the origin. Therefore the sum is –1. For subtraction, reverse the directions for addition. To subtract a positive number, move to the left. To subtract a negative number, move to the right. These geometrical directions are obviously equivalent to the arithmetic rule, "To subtract, change the sign and add." In summary, for positive numbers, add to the right, subtract to the left; for negative numbers, add to the left, subtract to the right. Several more examples may help to clarify these left-right rules. To perform the subtraction, 7 – 3 = 4, geometrically, move seven units to the right, then three to the left, to reach a point four units to the right of the starting point. To add two negative numbers, (–5) + (–4) = –9, move five units to the left, then four more units to the left. The geometrical path five units to the right, then three units to the right, corresponds either to the sum, 5 + 3 = 8, or to the difference, 5 – (–3) = 8.

Geometrical arithmetic applies equally well to all real numbers. The choice of integers in the example above was only for simplicity. Moving 2.36 units to the left or right is, of course, as easy to visualize as moving exactly two units. It is also clear that geometrical arithmetic need not be restricted to two numbers. The operation

$$2.0 + 3.5 - 4.9 - (-3.0) + (-2.5) = 1.1$$

may be performed geometrically by moving 2.0 units to the right, then 3.5 units to the right, then 4.9 units to the left, then 3.0 units to the right, then 2.5 units to the left. One ends up 1.1 unit to the right of the starting point. Finally, geometrical arithmetic illustrates clearly an important formal property of addition called the commutative property. This means that the order of terms in a sum may be changed without changing the sum. In algebraic notation, $a + b = b + a$. It is evident that whether one moves first two units to the left and then three to the right or first three to the right and then two to the left, the end point is the same.

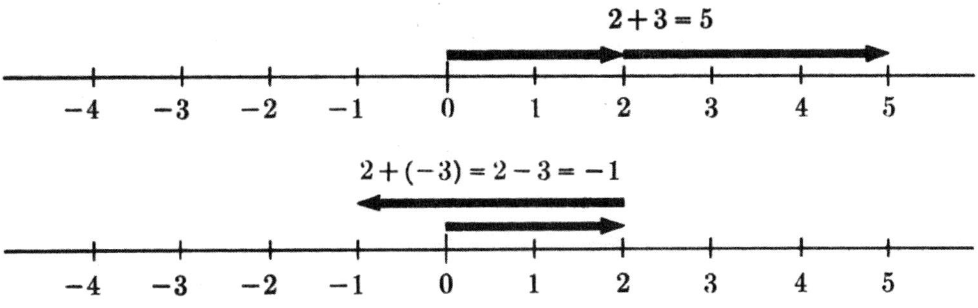

FIGURE 6.2 Geometrical arithmetic. Move right to add a positive number or subtract a negative number; move left to subtract a positive number or add a negative number.

Since multiplication is repeated addition, it may also be carried out geometrically. The product, 3 × 4 = 12, corresponds to moving four units to the right three times in succession to reach the point twelve units to the right. The product, 2.5 × 6 = 15, may be performed geometrically by moving six units to the right, then six more, then half of six more, that is 6 + 6 + 3 = 15. However, there is another and perhaps more instructive way to look at multiplication geometrically. That is in terms of stretching and shrinking. To perform the operation, 3 × 4 = 12, imagine a string stretched from the origin to the point labeled 4. Then stretch the string further to treble its length, so that it extends from the origin to the point labeled 12. To carry out the multiplication, 0.35 × 10 = 3.5, imagine a string first extended from the origin ten units to the right, then contracted to 35% of its original length, so that it reaches only 3.5 units from the origin. Thus, multiplication by a number larger than one corresponds to stretching, multiplication by a number smaller than one corresponds to shrinking.

Another formal property of numbers, called the distributive property, has to do with a combination of addition and multiplication. Algebraically, the distributive law may be written

$$c(a + b) = ca + cb. \tag{6.1}$$

It says that the result of adding two numbers together and multiplying the sum by a third number is the same as the result of multiplying each of the first two numbers separately by the third and adding the two products. Our everyday experience with numbers tells us that the distributive law is a "self-evident truth." To the mathematician, it is rather an arbitrary assumption, one of the defining axioms of numerical arithmetic. Because addition and multiplication of numbers can be represented geometrically, the distributive law can be illustrated geometrically. Adding together the length of two strings, for example, and then stretching the total length by a certain factor, produces the same result as first stretching each string by the given factor, and then adding their new lengths together.

Because of the exact correspondence between numbers and points on a line, it is easy and convenient to begin to think of the points as numbers. We might, for example, say that moving three units to the right and then four to the right brings us "to the number 7," rather than the more cautiously phrased "to the point on the line that corresponds to the number 7." There is nothing harmful in this locution, any more than there is harm in pointing to a picture of Mrs. Smith and saying, "That is Mrs. Smith." With Mrs. Smith, it is easy to distinguish the representation from the thing represented. With numbers, it is somewhat less easy, but it is equally essential. Actually, the physicist thinks of numbers in three different ways. Although he (and this author) may sometimes mix these different ways verbally, it is important that they be clearly separated mentally. First is the mathematician's number, the abstract concept "four," for instance, a mathematical object whose reality and existence are defined by the axioms of arithmetic. Second is the geometrical representation, a point on a line or a distance from the origin to the point. The geometrical representation makes the abstract concept pictorial, and is useful because rules of combining lengths along a straight line can be made to correspond exactly to the rules of combining abstract numbers. Third is the physical number, a measure of size or quantity of some measurable concept. Usually when a number is assigned to a physical concept, it means more than that the concept is quantitative (measurable). It means as well that the quantity in question obeys the laws of arithmetic. If we say that a piece of meat weighs four pounds, it is a useful statement only because it is an experimental fact that weights combine like numbers. A four-pound piece of meat and a three-pound piece of meat together weigh seven pounds. If this were not so—and there is no self-evident reason that it must be so—then numbers might not usefully be attached to weight. Instead the physical concept, weight, might have to be associated with some mathematical concept more abstract than numbers. This has indeed happened with some of the concepts of modern physics. Nevertheless, a great many physical concepts are measured in terms of numbers. Whenever this is the case, it is not because there is any law requiring that it be true, rather it is because, experimentally, numbers have been found to suffice. The assignment of a number to a physical concept is provisional in exactly the same sense that a law of nature is provisional; it is accepted so long as it is successful and free from contradiction.

6.2 *Vectors*

We have learned in the preceding section that addition and subtraction can be performed by marching back and forth along the real axis. The result of a calculation is always given by the distance from the starting point to the end point, with the sign of the answer determined by whether the end point is to the right or left of the starting point. This geometrical view of arithmetic suggests an interesting possibility for generalization. Suppose that we abandon the real axis and allow the possibility of moving off in an arbitrary direction. We could, for example, "add" together an eastbound line segment and a northbound line segment, each two units long (Figure 6.3). As before, we begin at an arbitrarily selected origin. From the origin we move two units to the east, then two units to the north. The "sum" is represented by a line segment drawn from the starting point to the end point. It is 2.828 units long and points northeastward. In our new version of arithmetic, both direction and magnitude have meaning, so it is important to specify the direction as well as the length of the sum line. The line segments being added, and their sum, can conveniently be designated by arrows, as in Figure 6.3. This new directional arithmetic is called vector arithmetic, or, if it is performed symbolically, vector algebra.

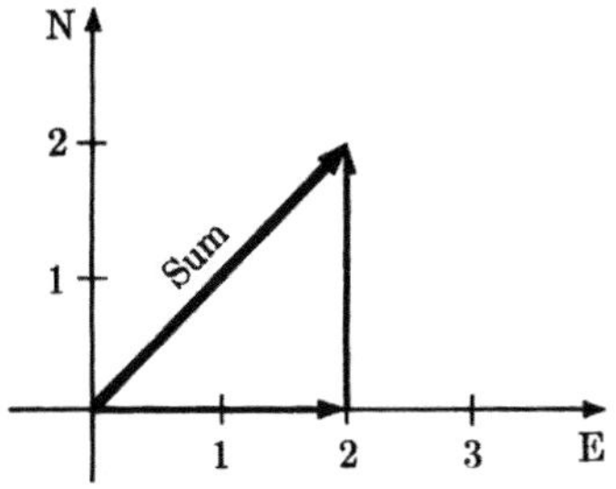

FIGURE 6.3 Vector arithmetic. Two units east plus two units north equal 2.828 units northeast.

A natural question at this point is, "Why?" What is the reason for introducing vector arithmetic, and what does it mean? There are two good reasons, one mathematical, one physical. To the mathematician, any generalization of an existing facet of mathematics that can be carried through in a rigorous yet simple and self-contained way is naturally fascinating to investigate. Vector arithmetic is just an interesting generalization of numerical arithmetic. It can be founded upon an axiomatic basis and takes its place beside numerical arithmetic as an abstract branch of mathematics. Distances measured along the real axis (right or left) *represent* numbers (positive or negative). In exactly the same way, line segments, or arrows, pointing in any direction whatever *represent* vectors. The vector itself is an abstract mathematical object no more the same as the arrow representing it than is the fundamental concept of number the same as a distance measured on the real axis. But just as the real axis provides a good pictorial way of visualizing numbers and numerical arithmetic, arrows in space provide a good pictorial way of visualizing vectors and vector arithmetic. Between the arrow and the abstract concept of vector there is a one-to-one correspondence. The geometry of "adding" arrows in various directions has an exact correspondence to the formal arithmetic of adding vectors.

To the mathematician, the vector is an interesting generalization of the number concept. 22
To the physicist it is an essential tool for describing nature. There exist many physical quantities that are not adequately represented by numbers alone. It is found instead that their mathematical behavior is identical to that of vectors. We shall encounter many quantities of this type, such as velocity, force, momentum, and angular momentum. Such quantities are themselves sometimes called vectors. However, it is preferable to reserve the word vector, like the word number, to refer to the mathematical entity. The physical concept (force, for instance) that "behaves like" a vector we shall call a vector quantity. The statement that force behaves like a vector means simply that physical forces combine, or add, in exactly the same way as mathematical vectors. Physical concepts that behave like numbers rather than like vectors are called scalar quantities. Mass is an example of a scalar quantity. A mass of 2 gm added to a mass of 2 gm produces a mass of exactly 4 gm. On the other hand, a force of two dynes added to a force of two dynes could produce any total force from zero up to a maximum of four dynes, depending on the relative orientation of the forces being added.

With vectors, as with numbers, it is important to distinguish three separate ideas—the abstract mathematical object, the concrete geometrical representation, and the physical quantity. With numbers, the geometrical representation is interesting but hardly necessary. With vectors, the geometrical representation is almost a necessity. Since vectors are somewhat more abstract than numbers, they defy easy visualization. Only with the help of arrow diagrams can we think easily about vectors and vector arithmetic. A vector quantity, as defined above, is a physical concept that behaves like a vector, in an exact mathematical sense. Sometimes a vector quantity is defined roughly as a quantity with both magnitude and direction. This is not quite an adequate definition, for to be a vector quantity it must not only have magnitude and direction. It must also follow precisely the laws of vector arithmetic when it enters into a physical law. Velocity, for instance, is a quantity with both magnitude and direction. So long as its magnitude is small compared with the speed of light, it behaves like a vector. But at enormous speeds its behavior deviates from that of a vector. Although it has magnitude and direction, it is no longer a vector quantity. This is an exceptional case, however. The student who assumes that any quantity with magnitude and direction is a vector quantity will almost always be correct. Even the exception, velocity, will be treated as a vector quantity throughout Parts 3 and 4 of this book, for it is so regarded throughout classical physics. The fact that a quantity once believed to be a vector quantity turns out not to be a vector quantity emphasizes a very important point about all physical concepts. The association of any physical quantity with a certain mathematical quantity— be it number, vector, or something else—is based purely on experiment and is always subject to later change.

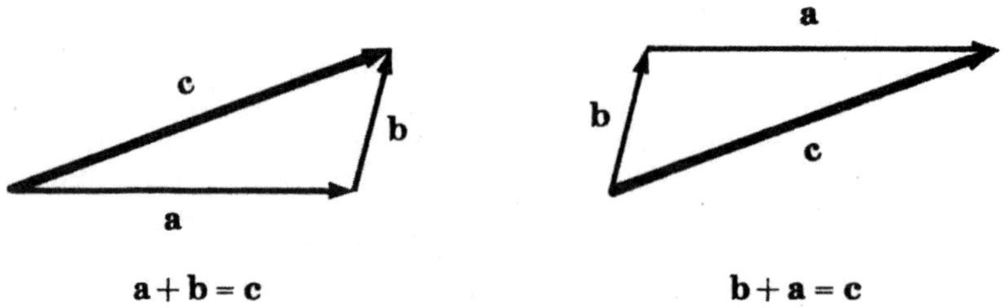

FIGURE 6.4 Vector addition. The diagrams show the commutative law of vector addition: **a** + **b** = **b** + **a**.

6.3 Vector arithmetic

Only three operations with vectors will be needed for the physical applications in this book: addition of vectors, subtraction of vectors, and multiplication of a vector by a scalar. There are two different ways to give meaning to the multiplication of one vector by another, but we shall not need these operations. Addition of vectors has been illustrated in Figure 6.3. To add two vectors geometrically, measure off a distance equal to the magnitude of the first vector and in the direction of the first vector, then measure off a distance equal to the magnitude of the second vector and in the direction of the second vector. These two displacements may conveniently be designated by arrows. Their sum is represented by an arrow drawn from the starting point to the end point. Figure 6.4 shows another example of vector addition, which may be written symbolically

$$\mathbf{a} + \mathbf{b} = \mathbf{c}\,.$$

We shall follow the practice of designating vector quantities by **boldface** type. Figure 6.4 also illustrates the fact that vectors, like numbers, obey the commutative law of addition. In symbols,

$$\mathbf{a} + \mathbf{b} = \mathbf{b} + \mathbf{a}\,. \tag{6.2}$$

Whether the arrow representing **a** is drawn first or second, the end point is the same, and the arrow representing the sum is the same.

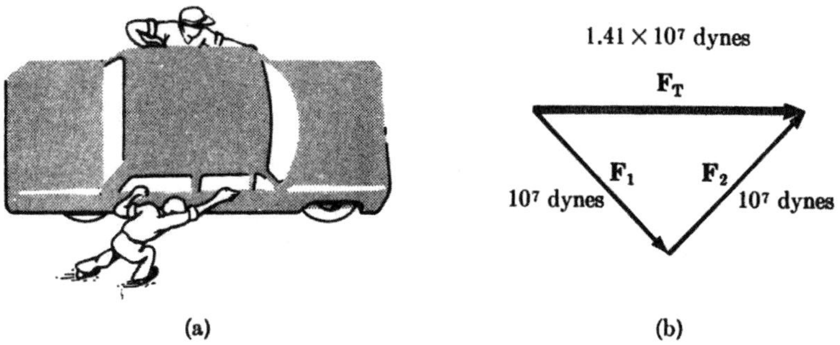

FIGURE 6.5 Vector addition of forces: $\mathbf{F}_1 + \mathbf{F}_2 = \mathbf{F}_T$.

Notice that in case two vectors are parallel, adding them is almost equivalent to adding numbers, for then the arrow diagram lies all along a single straight line. We say "almost," for even in the special case of parallel vectors, the vectors retain an individuality and are not the same as numbers. Most important, their direction still has meaning. The sum of an eastbound vector of magnitude 4 and an eastbound vector of magnitude 3 is an eastbound vector of magnitude 7. The magnitudes add like numbers, $4 + 3 = 7$, but the adjective "eastbound" has a meaning only for a vector, not for a number. The real axis could be laid out in an east-west direction or in any other direction. It would not matter. However, in adding parallel vectors, the orientation does matter; the direction as well as the magnitude of the sum is important.

In order to pin down vector addition, let us consider some practical examples. First, suppose that two men on opposite sides of an automobile are pushing partly inward, partly forward, such that each is exerting a force of 10^7 dynes (this unit of force will be defined later) at an angle of 45 deg to the long axis of the car [Figure 6.5(a)], What is the total force they exert on the car? Since force is a vector quantity, we may proceed to sum the two forces with the aid of an arrow diagram, as in Figure 6.5(b). According to the commutative law, the order in which the two arrows are drawn is irrelevant. The sum is represented by an arrow parallel to the axis of the car, of length about 1.41 times the length of each of the arrows being summed. Therefore the total force is forward and has magnitude 1.41×10^7 dynes. This result could be obtained by drawing the arrow diagram with great care and measuring the length of the sum arrow. Or it could be obtained (much more easily) by trigonometry,* as illustrated Figure 6.6. Half of the length of the sum arrow divided by the length of one of the

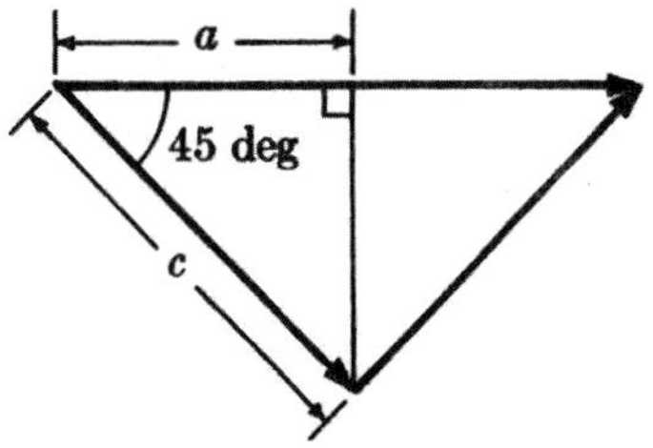

FIGURE 6.6 Use of trigonometry to sum vectors. The length c represents the magnitude of vector $\mathbf{F}_1$. The length a is, in this example, half the magnitude of the total force $\mathbf{F}_T$. The lengths a and c are connected by the equation, $a = c \cos(45 \text{ deg})$.

* Some elements of trigonometry are reviewed in Section 6.10.

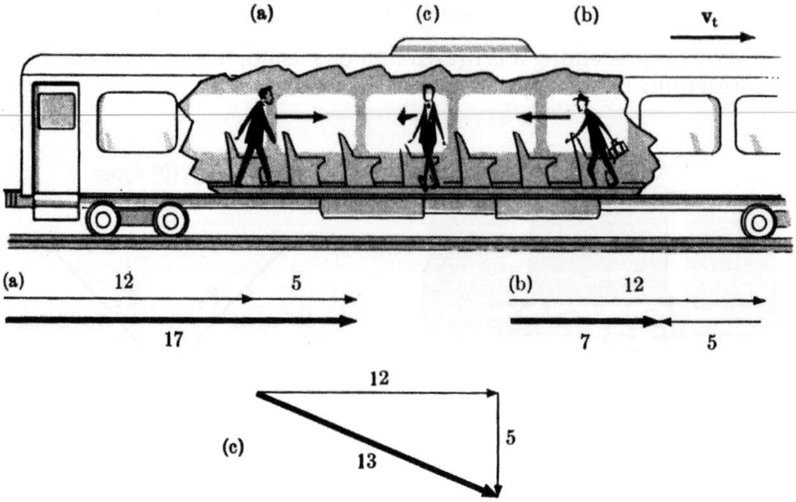

FIGURE 6.7 Vector diagrams for examples of relative motion. The magnitude of the train's velocity $\mathbf{v}_t$ is 12 ft/sec. The magnitude of the passenger's velocity $\mathbf{v}_p$ relative to the train is 5 ft/sec. The sum, $\mathbf{v}_t + \mathbf{v}_p$, is the passenger's velocity relative to the ground.

original arrows is equal to the cosine of 45 deg, which is 0.707. Therefore half of the total force is 0.707×10^7 dynes; the total force is about 1.41×10^7 dynes.

As a second example, consider a problem of relative motion. If a passenger walks at 5 ft/sec within a train which is moving forward at 12 ft/sec, what is the velocity of the passenger with respect to the ground? This is again a problem of vector addition. To the passenger's velocity within the train (call it v_p) the train adds its own velocity ($\mathbf{v}_t$) to produce the total velocity ($\mathbf{v}$) of the passenger with respect to the ground:

$$\mathbf{v} = \mathbf{v}_p + \mathbf{v}_t \, .$$

If the passenger walks forward in the train, the velocities $\mathbf{v}_p$ and $\mathbf{v}_t$ are parallel; the arrow diagram lies in a straight line [Figure 6.7 (a)]. The total velocity $\mathbf{v}$ then points forward along the tracks and has magnitude 17 ft/sec. If the passenger walks to the rear, the vector diagram is again simple [Figure 6.7(b)]. The forward-pointing total velocity has magnitude 7 ft/sec. A more complicated vector sum occurs when the passenger walks transversely across the car [Figure 6.7(c)]. The arrows representing the vectors $\mathbf{v}_p$ and $\mathbf{v}_t$ are at right angles. With the third arrow representing their sum, $\mathbf{v}$, they form a right triangle. It is convenient to let the lightface symbol v designate the numerical magnitude of the vector quantity $\mathbf{v}$, and so on for the other vectors.* In this notation, for the example of transverse motion (and *only* for this example), the Pythagorean theorem for right triangles may be applied:

$$v^2 = v_p{}^2 + v_t{}^2 \, .$$

With $v_p = 5$ ft/sec and $v_t = 12$ ft/sec, the magnitude of the total velocity is 13 ft/sec. This velocity makes an angle with the track whose sine is 5/13, approximately 23 deg. This kind of relative-motion problem is of practical importance to pilots. The velocity $\mathbf{v}$ of an airplane with respect to the ground is the vector sum of its velocity with respect to the air ($\mathbf{v}_p$) and

* The use of the same symbol for a vector quantity and for its numerical magnitude should not be a source of confusion. All equations in lightface type will be ordinary numerical equations. Boldface will serve as a flag of warning: Caution! Vector equation! Numerical rules do not apply!

the velocity of the air, that is, the wind velocity ($\mathbf{v}_W$):

$$\mathbf{v} = \mathbf{v}_p + \mathbf{v}_W .$$

The rule for subtracting vectors is very simple: reverse the direction of the arrow representing the vector to be subtracted, then add. This is obviously similar to the rule for subtracting a number: Reverse its sign and add. Suppose that an observer on the ground measures a certain wind velocity $\mathbf{v}_W$ and also measures the velocity $\mathbf{v}$ of a passing airplane with respect to the ground. He wants to know the velocity $\mathbf{v}_p$ of the airplane with respect to the air. From the observed airplane velocity he must subtract the wind velocity,

$$\mathbf{v}_p = \mathbf{v} - \mathbf{v}_W .$$

Such an example is illustrated diagrammatically in Figure 6.8.

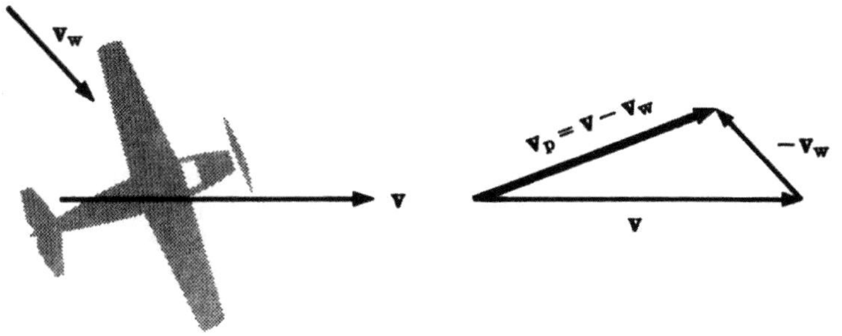

FIGURE 6.8 An airplane moves with velocity **v** relative to the ground. The wind velocity is $\mathbf{v}_W$. The velocity $\mathbf{v}_p$ of the airplane relative to the air is the difference $\mathbf{v} - \mathbf{v}_W$.

Any number of vectors may be successively added and subtracted graphically. The result of such a vector calculation is always represented by an arrow drawn from the starting point to the end point. An example of the addition of four vectors is shown in Figure 6.9. An important practical application of the addition of numerous vectors occurs in the branch of mechanics called statics. If an object subjected to many forces is at rest and remains at rest, the sum of all the forces acting on it must be zero. The arrow diagram for any combination of forces producing static equilibrium always begins and ends at the same point, so that the sum vector is zero.

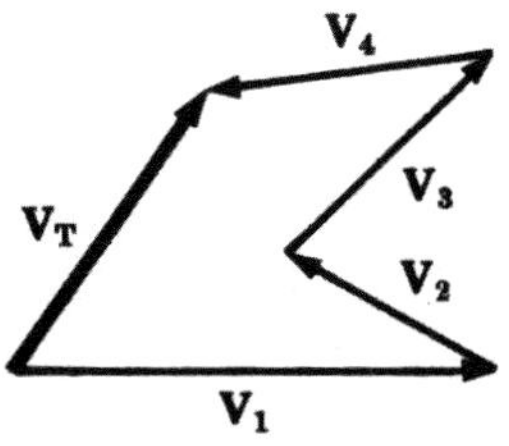

FIGURE 6.9 The addition of four vectors: $\mathbf{V}_T = \mathbf{V}_1 + \mathbf{V}_2 + \mathbf{V}_3 + \mathbf{V}_4$.

The third vector operation of interest is the multiplication of a vector by a scalar (here a scalar means the same thing as a number). This operation can be viewed geometrically as a stretching or shrinking of a vector without changing its direction. For example, the vector equation

$$\mathbf{b} = 3\mathbf{a}$$

means that the vector **b** has three times the magnitude of **a** and has the same direction as **a** (Figure 6.10). The arrow representing **b** is parallel to the arrow representing **a**, but is three times as long. If another pair of vectors are related by

$$\mathbf{c} = 0.25\mathbf{d}\,,$$

this means that the vector **d**, if shrunk to one-quarter its magnitude without change of direction, yields the vector **c**. A simple physical example of this kind of multiplication occurs in the definition of the momentum of a particle. The defining equation is

$$\mathbf{p} = m\mathbf{v}\,,$$

where **v** is velocity, a vector quantity; m is mass, a scalar quantity; and **p** is the momentum, again a vector quantity. The momentum of a particle is represented by an arrow parallel to its velocity whose length is the mass of the particle multiplied by its speed. (The magnitude of velocity is usually called speed.)

The operations of addition and multiplication by a scalar may be combined to express the distributive law of vector arithmetic,

$$n(\mathbf{a} + \mathbf{b}) = n\mathbf{a} + n\mathbf{b}\,. \tag{6.3}$$

The law states that summing two vectors and then multiplying the sum by a scalar gives the same result as separately multiplying the two vectors by the scalar before adding them. The distributive law may be regarded as a fundamental axiom of vector arithmetic, but its truth follows from the geometrical rules of vector manipulation that we have already laid down. The arrow diagram in Figure 6.11 illustrates the distributive law in a simple way. The fact that vectors as well as numbers obey a commutative law and a distributive law means that many familiar rules of the algebra of numbers hold true also for the algebra of vectors. Note, however, that we have not defined the multiplication of one vector by another, nor the division of one vector by another.* This means that the range of vector algebraic manipulations open to us is quite limited, although adequate for a large number of physical applications.

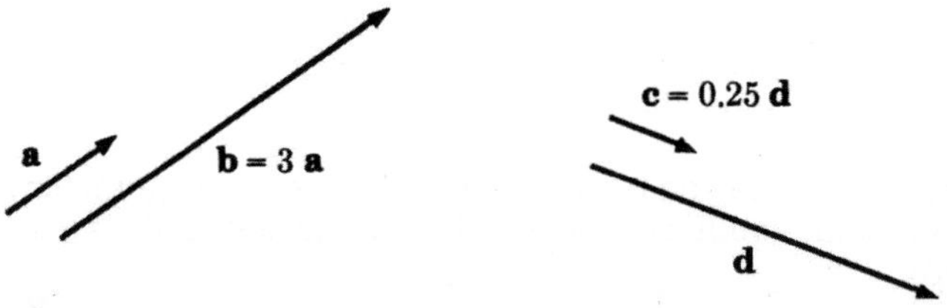

FIGURE 6.10 Multiplication of a vector by a scalar is equivalent to stretching or shrinking the vector without changing its direction. (Its unit of measurement may also be changed if the scalar is not dimensionless.)

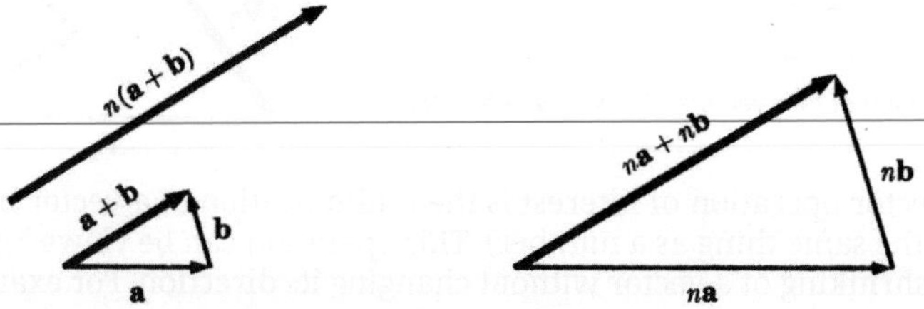

FIGURE 6.11 The distributive law of vector arithmetic. See Equation 6.3.

* The division of one vector by another is undefined (except for parallel vectors), even in the complete theory of vector algebra.

6.4 *The geometry of vectors*

Two points of difference between vectors and numbers, already implicit in the discussion above, are worth reviewing here. The first difference is that vectors are *essentially* geometric, numbers are not. Numbers may be represented geometrically, either by points on a line, or by distances measured along a line, but they need not be. And even if they are, the line is arbitrary; its direction in physical space is without significance. A vector, on the other hand, has meaning only when its direction in space is specified. For the mathematician, this "space" of the vector may be itself an abstraction, such as a space of many dimensions, but it is still basically geometrical. For the physicist, the space of most vectors is the ordinary three-dimensional space of classical physics, the space of human perception. In the twentieth century, he has learned also to define and manipulate vectors in four-dimensional space-time.

The second difference between vectors and numbers worthy of special notice is the fact that the vector lacks one quality possessed by numbers: a sign. To call a vector positive or negative has no meaning. A number has magnitude and sign, a vector has magnitude and direction. Direction may be thought of as the generalization of the idea of sign. Instead of two possibilities (plus and minus), there are infinitely many possibilities for direction. Of course a vector may be added or subtracted, and it may appear in an equation preceded by a plus sign or a minus sign, but the vector itself has no intrinsic sign. Every vector, whether pointing north, south, east, west, up, or down, is said to have a positive magnitude. The idea of direction replaces the idea of sign.

This essentially geometric nature of a vector is manifested only in the direction of the vector, not in its magnitude. It is the replacement of sign by direction that converts the nongeometric number to the geometric vector. But the magnitude of the vector remains, like the magnitude of a number, usually without any geometric significance. Vector quantities such as force or momentum, for instance, are expressed in units that are not spatial. The nongeometric character of vector magnitude is reflected also in the arrows used to represent vectors. A force of one dyne might be represented by an arrow one inch long, or one millimeter, or any other arbitrarily chosen unit. The length of the representative arrow is adjustable to suit our convenience. However, its direction in space, an essential part of the definition of the vector, is not free to be varied at will.

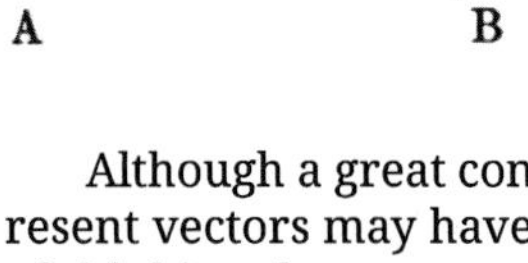

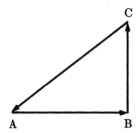

FIGURE 6.12 Balanced forces on an electron. Despite the spatial extent of the diagram, the forces all act at a single point.

Although a great convenience for visualization and calculation, the arrows used to represent vectors may have an undesirable side effect. They make it easy to fall into the error of thinking that a vector extends from one point to another point. It is very important to be able to abstract one's thinking away from the arrow on the page with its finite length to the concept of a vector existing at a single point in space. Consider, for example, a single electron at rest at some point in space. Suppose that the electron at this point is subjected to three different forces whose sum is zero, so that it remains at rest. The arrow diagram representing these forces forms a triangle (Figure 6.12), one arrow extending on the page from A to B, another from B to C, and the third from C to A. Although the forces themselves all act at one single point in space, the diagram required to represent the forces has to span some region of space (or region of page). The irrelevance of this spatial extension of the diagram is already suggested by the arbitrariness of the length scale used for the arrows. We could, if we like, imagine the diagram shrunk down to subatomic size, hardly a practical procedure but a good way to remind ourselves that the physical vectors exist at a single point. As another example, consider the same electron moving along some path through space (Figure 6.13). Its velocity is a vector quantity and may be represented at any instant by an arrow. In Figure 6.13, the velocity vectors are indicated at three different points. Although the arrows must extend over a finite distance in the diagram, it is important again to realize that they apply each to one single point. The vector velocity may be different at each point along the track of the electron.

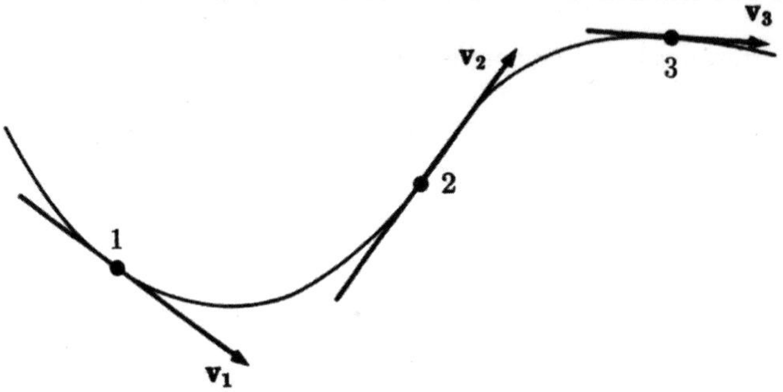

FIGURE 6.13 An electron track in space, and representation of velocity vectors at three points along the track.

6.5 The position vector

Among physical vector quantities, one of the most important is one of the simplest (in principle) and at the same time one of the most troublesome (in practice). This quantity is *position.* The location in space of a particle (or for that matter of any point, whether occupied by a particle or not) is a vector quantity. Its magnitude is the distance of the point from an arbitrarily chosen origin, its direction is the direction of the point from the origin. The position vector is special first in that it is an exception to the rule that the magnitude of a vector is without geometric significance. The magnitude of the position vector does have geometric significance. It is a length. The position vector is special also in depending for its definition on the choice of an origin. An automobile moving down a highway has a velocity and a momentum and is experiencing forces, all vector quantities independent of a choice of spatial origin. But its position can be specified only with respect to some reference point, which we may call the origin.

Suppose that a square classroom is laid out with its walls parallel to east-west and north-south lines. In the room are three students, Alice, Barbara, and Charlotte. Now we choose to call the southwest comer of the room the origin. It is clear that the location of each girl can be uniquely specified by giving her *distance* from the origin, and her *direction* from the origin. These quantities determine the position vector of each girl. These vectors, which we may call **A**, **B**, and **C**, are represented by arrows in Figure 6.14. Now comes the conceptually troublesome point. It is very natural to think of the position vector **A** extending from the origin corner to Alice. This is wrong. Even though the magnitude of **A** is Alice's distance from the origin, her position vector must be thought of as existing only at the point where she is standing. It is more difficult to visualize, but the position vector is exactly like all other vectors in this respect. To facilitate right thinking, the representative arrows in Figure 6.14

have been drawn deliberately much smaller than the dimensions of the room. The position vector is a quantity that specifies the location in space of a particular point, and is defined at that point.

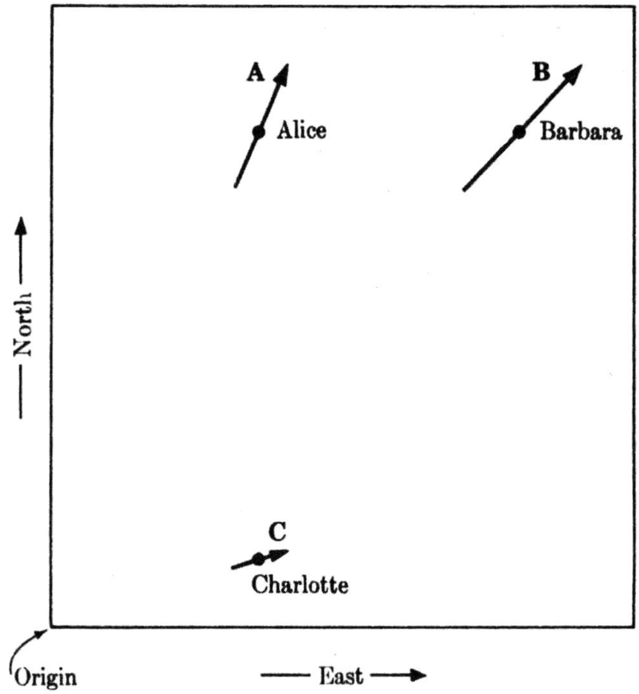

FIGURE 6.14 Position vectors. The three arrows represent the positions of Alice, Barbara, and Charlotte in a frame of reference whose origin is the southwest corner of the room.

To clarify this idea of the "local" position vector, consider another example on a grander scale. Suppose that Eureka, Kansas, has been chosen as an origin for specifying positions in the United States. As an aid to lost pilots, city fathers across the country order that there shall be painted atop the tallest building in each city an arrow whose length in inches is equal to the distance of that city from Eureka in miles, and whose direction is the direction of that city from Eureka. Each of these arrows represents the position vector of a particular city. The lost pilot, examining an arrow with care, could determine his exact position relative to Eureka. The position vector of Chicago is defined at Chicago and is represented by an arrow at Chicago. It would be quite unnecessary and very wasteful of paint to lay down an arrow extending from Eureka to Chicago. Moreover, it would be both useless and wrong. It would be useless because the pilot over Chicago would gain no information seeing only the tip of the arrow. It would be wrong because the arrow would exist partly on some farm in Iowa whose position vector is represented not by this arrow but by a different one. The arrow of position must exist only at one point, for otherwise it would exist also at some points whose position it does not correctly represent.

To return to the girls in the classroom, let us consider the difference in the position vectors **A, B**, and **C**. The differences **B** – **A** and **C** – **A**, illustrated diagrammatically in Figure 6.15, represent the position of Barbara relative to Alice and of Charlotte relative to Alice. These differences, as we can guess from their physical meaning, do not depend upon our original choice of origin. For example, had we chosen the southeast corner of the room as the origin, the position vectors **A**, **B**, and **C** would be different, but the differences **B** – **A** and **C** – **A**, representing relative positions, would be the same. This is fortunate, for it means that in all physical problems, the choice of origin is irrelevant. In studying the solar system, we are interested in the position of the earth relative to the sun, or of the moon relative to the earth, but not in the position of any one of these objects by itself. If the origin were moved, or if the whole solar system were moved bodily to another part of space, the relative motions within the solar system would be unaffected. The fact that the laws of nature do not depend upon the choice of origin has been elevated by the theory of relativity from the role of useful fact to the role of fundamental postulate. Its significance will be discussed in later chapters. For the moment, we remark only that its truth is not surprising. We would scarcely expect the behavior of Alice, Barbara, and Charlotte to be affected by whether some outside observer chooses to call the southwest corner or the southeast corner of their room the origin.

FIGURE 6.15 Relative position vectors. The position of Barbara relative to Alice is indicated by the vector **B** – **A**, and of Charlotte relative to Alice by **C** – **A**. Relative position vectors are uninfluenced by a change of origin.

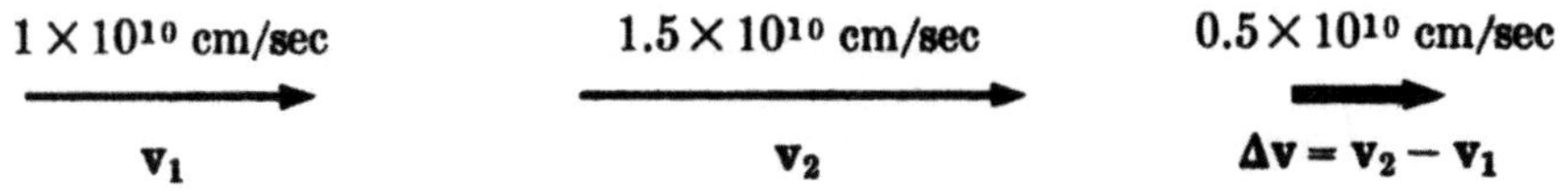

FIGURE 6.16 Vector changing in magnitude but not in direction.

6.6 *Vectors changing in time*

Much of physics is concerned with motion and change. Therefore, we must examine the important idea of the rate of change of a vector. If a baby's mass changes in one month from 11 lb to 12 lb, the *amount* of change is 1 lb, and the average *rate* of change is 1 lb per month. This is an example of the rate of change of a scalar quantity. A scalar changes simply by increasing or decreasing its magnitude. Because a vector is a more complicated concept than a number, the ways in which it can change are also more complicated. A vector has both magnitude and direction; a change in either one of these properties (or in both) constitutes a change of the vector. The simplest change, analogous to the change of a scalar, is a change in magnitude only. An electron in a linear accelerator might increase its speed from 1.0×10^{10} cm/sec to 1.5×10^{10} cm/sec without change of direction. The *change* in its velocity is a vector in the same direction as the velocity, with magnitude 0.5×10^{10} cm/sec (Figure 6.16). On the other hand, a proton in a circular accelerator might change its velocity from 1.0×10^{10} cm/sec east-bound to 1.0×10^{10} cm/sec northbound (Figure 6.17), a change of direction without change of magnitude. In that case the change of velocity, $\Delta\mathbf{v}$, has a magnitude 1.41×10^{10} cm/sec and a direction northwestbound. In the laws of nature, changes of direction are as important as changes of magnitude. Note, in the second example just given, that even when the magnitude of a vector undergoes no change, the vector itself, through change of direction, may experience a change which has a large magnitude.

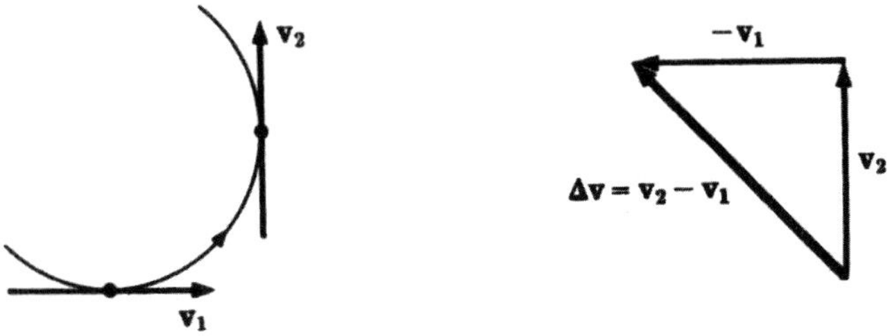

FIGURE 6.17 Vector changing in direction but not in magnitude.

The change of any vector in some interval of time may be found by subtracting the vector at the earlier time from the vector at the later time. In symbols,

$$\Delta \mathbf{V} = \mathbf{V}_2 - \mathbf{V}_1 ,$$

where $\mathbf{V}_2$ is the vector at the later time, $\mathbf{V}_1$ is the vector at the earlier time, and $\Delta\mathbf{V}$ is the change of the vector $\mathbf{V}$. (The symbol Δ is frequently used to mean "change of.") For example, if Alice leaves her position **A** and walks over to Barbara's position **B**, the change of Alice's position is **B** – **A**, that is,

$$\Delta \mathbf{A} = \mathbf{B} - \mathbf{A} .$$

The direction of the change, $\Delta\mathbf{A}$, may of course be very different from the directions of **A** or of **B**. The average *rate* of change of a vector is calculated by dividing the change of the vector by the interval of time, Δt. This average rate of change, written $\Delta\mathbf{V}/\Delta t$, may be expressed

$$\frac{\Delta \mathbf{V}}{\Delta t} = \frac{\mathbf{V}_2 - \mathbf{V}_1}{\Delta t} = \left(\frac{1}{\Delta t}\right)(\mathbf{V}_2 - \mathbf{V}_1) . \tag{6.4}$$

The rightmost expression makes clear that division of a vector by the scalar quantity Δt is equivalent to multiplying the vector by $(1/\Delta t)$, an operation that has been defined. For example, if Barbara was standing 6 feet from Alice, and Alice covered the distance in 2 sec, her average velocity is the vector **B** – **A**, divided by the scalar, 2 sec. This is a vector of magnitude 3 ft/sec having the same direction as the change $\Delta\mathbf{A}$, the direction from Alice's original position to Barbara's position. Note that in calculating a rate of change, the only real difficulty is in finding the change of the vector, a process of vector subtraction. Once this has been done, finding the rate of change only requires dividing the magnitude of the change vector, $\Delta\mathbf{V}$, by the number Δt, a purely numerical process.

It was in part the problem of developing an exact mathematical description of rate of change that led Newton and Leibniz in the late seventeenth century to develop calculus. The nature of the problem was discussed in the last chapter. The following example may illuminate it further. Suppose that you drive from Boston to New York, a distance of 200 miles, in 5 hours. Then your average speed for the trip is 40 miles/hr. But this is of course not your speed all along the way. Sometimes you drive faster, sometimes slower than 40 miles/hr. In order to get a better figure for your speed at some place along the way, you might measure the time required to move from one telephone pole to the next, and divide the spatial separation of the poles by the time interval. In symbols,

$$v = \frac{\Delta x}{\Delta t} , \tag{6.5}$$

where Δx designates change of position, and v designates speed. From a mathematical point of view, this is scarcely any more satisfactory than dividing the full distance, 200 miles, by the total time, 5 hours. The calculation still yields only an average value. Between the two telephone poles the speed of the car might have varied. Yet at one single instant of time and point in space, the car was certainly moving. It had a speed at that point. It is this instantaneous speed, not the average speed over any interval, that is needed for the exact description of the motion. To imagine the measurement of the instantaneous speed, we must hypothesize a series of measurements of ever smaller space intervals divided by ever smaller time intervals. Calculus is concerned with the precise description of the limit in which nothing divided by nothing (no motion in no time) yields something (a finite speed). Vector calculus differs from numerical calculus only in that it treats rates of change of vectors rather than rates of change of numbers. As a concept, rate of change is not hard to grasp—we are all familiar with velocity, the rate of change of position, in everyday life. To treat rate of change with precision, however, requires the use of calculus, beyond the mathematical level of this book. Therefore in specific examples we shall usually deal with average rates of change over finite intervals.

6.7 *Uniform motion in a circle*

A physical example of considerable practical importance that rather thoroughly tests one's understanding of vectors is the uniform motion of a particle in a circle. Circular, or nearly circular, motion is encountered at many places in nature—a stone on the end of a rope, a proton in an accelerator, the earth in its orbit round the sun. The description of such motion belongs to what is called kinematics. Kinematics is the study of motion without particular reference to the question of why the motion is what it is. It is simply an effort to give a quantitative account of motion without explaining the motion. For circular motion, our goal is to understand the behavior of three vector quantities—the position; the velocity, which is the rate of change of position; and, taking one further step, the acceleration, which is the rate of change of velocity.

A convenient origin for describing the position is the center of the circle. Then the position vector has a constant magnitude, since the rotating particle is always the same distance from the center. The position is represented by an arrow of fixed length pointing upward when the particle is at 12 o'clock, outward to the right when it is at 3 o'clock, downward when it is at 6 o'clock, and so on (Figure 6.18). Now we inquire about the velocity. Our postulated uniform motion around the circle means that the speed (the magnitude of the velocity) is constant. The representative arrow is of constant length. If the particle moves clockwise, this velocity arrow points to the right when the particle is at 12 o'clock, downward when the particle is at 3 o'clock, and so on. This seems simple enough, because velocity is already a familiar concept. However, let us pretend less familiarity with velocity, and derive the behavior of the velocity vector only from its fundamental definition, rate of change of position. How is the position vector changing as the particle moves around the circle? Figure 6.19 shows the change of the position vector from the time the particle is at the 1-o'clock position until it is at the 2-o'clock position. Call the position vector at the 1-o'clock position $\mathbf{r}_1$ and the position vector at the 2-o'clock position $\mathbf{r}_2$. The change of position is

$$\Delta\mathbf{r} = \mathbf{r}_2 - \mathbf{r}_1 ,$$

a vector directed approximately tangent to the circle in the direction of motion. In the limit of a very small interval, $\Delta\mathbf{r}$ becomes precisely tangent to the circle. That is the direction of the velocity. A simple way to get this result without the aid of diagrams is the following: Place a pencil, representing the position vector, on a desk or table, pointing upward for example to the 12-o'clock position. Now holding the eraser fixed, swing the pencil in a small arc. The motion of the point of the pencil gives the change of the position vector. As the pencil starts to move, its point moves in a direction perpendicular to its length. Thus the velocity vector is perpendicular to the position vector.

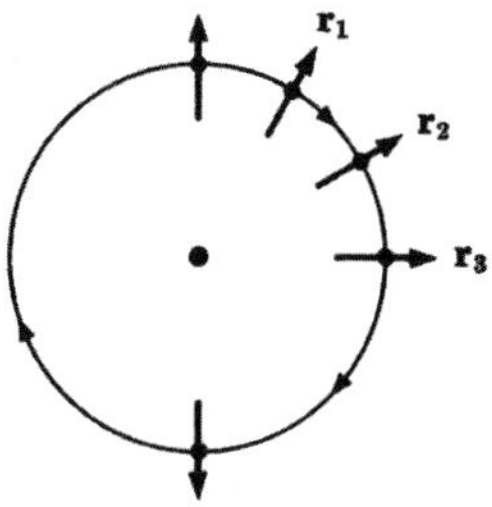

FIGURE 6.18 Uniform motion in a circle. The arrows designate position vectors of a particle at various points along its trajectory if the origin is at the center of the circle.

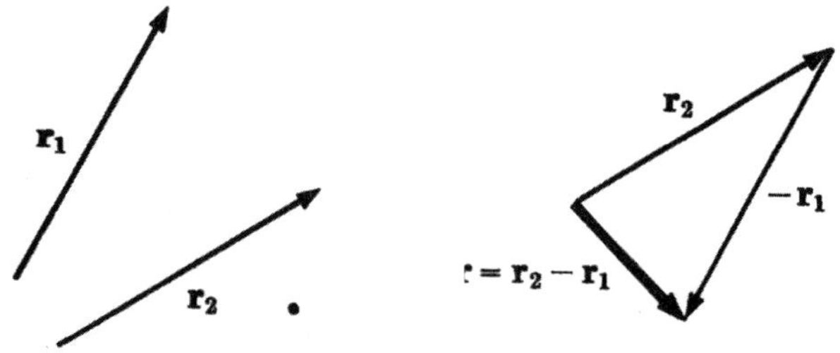

FIGURE 6.19 The change of position of the circling particle from its 1-o'clock position to its 2-o'clock position is $\Delta\mathbf{r} = \mathbf{r}_2 - \mathbf{r}_1$, a vector directed 45 deg below the horizontal. This is the average direction of the particle's velocity vector in this part of its motion.

Acceleration, a vector quantity, is defined as the rate of change of velocity. It bears exactly the same relation to velocity that velocity bears to position. But being one stage further removed than velocity from the basic human perception of position, it is somewhat more difficult to think about. Normally we think about acceleration as change of speed. An automobile gaining speed is accelerating; losing speed, it is decelerating. The physicist uses the single word, acceleration, to describe both gain and loss of speed, assigning only a different direction to the acceleration. An automobile gaining speed is accelerating in the direction of its motion. Losing speed, it is accelerating in the direction opposite to its motion. Since velocity is a vector quantity, a change in its direction produces acceleration as surely as a change in its magnitude. The acceleration associated with change of direction of velocity, although less familiar as an everyday concept, is just as important in physics as the acceleration associated with change of speed. In the example being considered, uniform motion in a circle, all of the acceleration comes from change of direction of velocity. As the particle passes the 12-o'clock position, its velocity vector is directed horizontally to the right. A moment later, the velocity has swung to an angle slightly below the horizontal. The *change* of velocity is a vector directed downward, toward the center of the circle, and this is the direction of the acceleration (Figure 6.20). The pencil on the desk may again be used for illustration. This time the pencil is the arrow representing velocity. As the particle swings to a new direction without change of speed, the pencil (of constant length) swings to a corresponding new direction, and the motion of its point gives the direction of acceleration. The particle is continually accelerated toward the center of the circle. A satellite in orbit, for instance, is continually "falling" toward the center of the earth as it continues to circle around the earth.

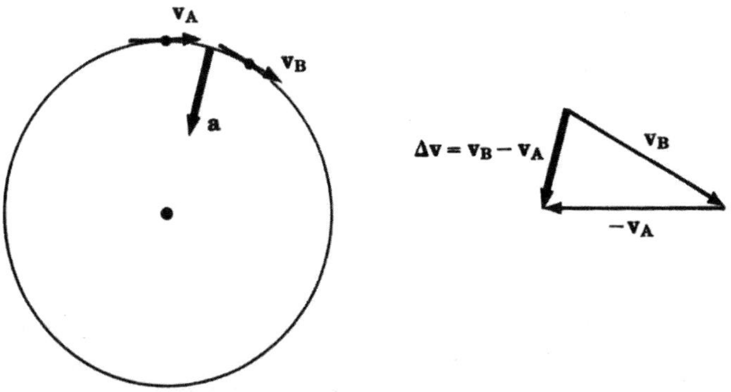

FIGURE 6.20 Vector diagram showing that the acceleration of the uniformly rotating particle is directed toward the center of the circle. The acceleration a has the same direction as the small change of velocity, $\Delta\mathbf{v}$.

In using the rotating pencil to find the direction of acceleration, one might ask: Since not only the velocity, but also the position, of the particle changes, should not the pencil be displaced as well as rotated? This wrong way to rotate the arrow of velocity is compared in Figure 6.21 with the right way. In order to understand why it is correct to keep the eraser end of the pencil fixed, consider the following simpler problem. An automobile in Denver and an automobile in Chicago are both traveling due east at 30 miles/hr. What is the difference in their velocities? Obviously, you respond, there is no difference. They have the same velocity. This is indeed correct. Each is moving in the same direction at the same speed. Each velocity is represented by an arrow of the same length and the same direction. But if we thought of one of these arrows located in Denver, the other located in Chicago, and tried to describe the difference in the velocities in terms of the difference in their locations, we would be led to quite a wrong conclusion. We must instead think of picking up one of the arrows, laying it down for comparison next to the other, and noting whether they coincide. In the same way, to compare the velocity of the particle at one point in its circle with the velocity at another point, we must pay attention only to the velocity vectors themselves, not to their location.

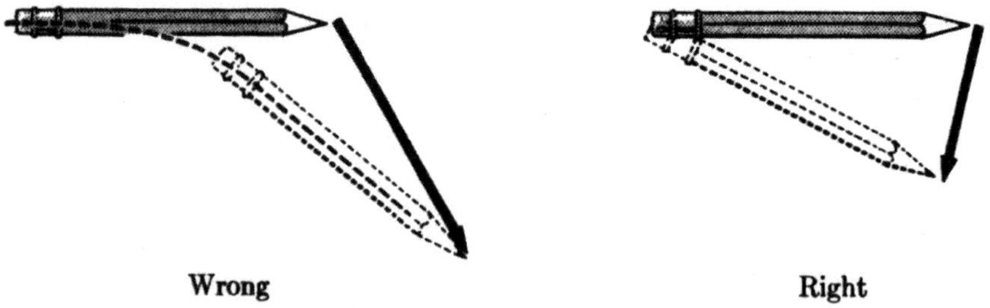

Figure 6.21 The wrong way and the right way to find the difference of two vectors. The pencils represent velocity vectors. The arrows represent the change of velocity, proportional to acceleration.

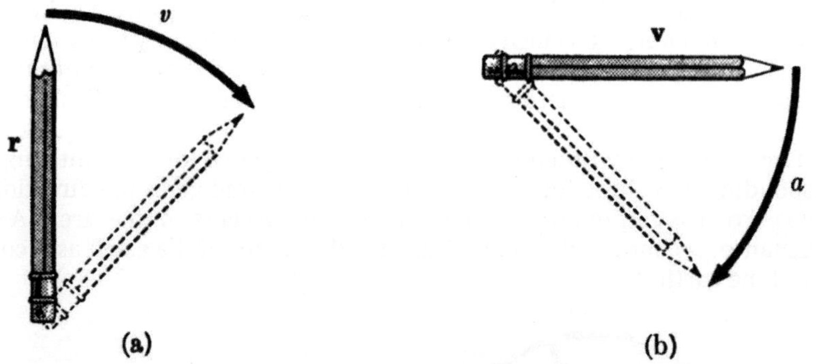

FIGURE 6.22 (a) If the pencil represents the position vector, its point moves at speed v. **(b)** If the pencil represents the velocity vector, the rate of progression of its point is equal to the acceleration a. Because both **r** and **v** rotate at the same rate, the ratio v/r is equal to the ratio a/v.

Besides testing one's comprehension of vectors, the example of circular motion illustrates particularly clearly the value of the vector concept in describing nature. One unified concept, acceleration, describes velocity changes of two quite different kinds—changes in magnitude and changes in direction. Both kinds of change are vitally important in mechanics. If we had restricted the term acceleration to mean only change of speed, uniform motion in a circle would have been motion without acceleration. Then some different concept would have been necessary to describe the change of direction of the motion.

The most important formula associated with uniform circular motion gives the magnitude of the inward acceleration, a, in terms of the speed, v, and the radius of the circle, r. It is

$$a = \frac{v^2}{r} \, . \tag{6.6}$$

This formula may be proved with the aid of the pencil on the desk and a little reasoning (Figure 6.22). If, first, the pencil represents the position vector, its length is the radius r and the speed of its point as it rotates uniformly is v. If, second, the pencil represents the velocity vector, its length is v, and the speed of its point is a. The key point is that the position arrow and the velocity arrow rotate at exactly the same rate. Each goes through a complete turn each time the particle completes a full circle. Therefore the speed bears exactly the same mathematical relation to the radius that the acceleration bears to the speed. Now the speed and the radius may easily be related. Speed is distance divided by time:

$$v = \frac{2\pi r}{t} \, ,$$

where t is the time to go once around the circumference, whose length is $2\pi r$. Dividing both sides by r gives

$$\frac{v}{r} = \frac{2\pi}{t} \, .$$

According to the argument above, a bears the same relation to v as v to r. Therefore if v/r is equal to $2\pi/t$, a/v is also equal to $2\pi r/t$, and v/r and a/v are equal to each other:

$$\frac{a}{v} = \frac{v}{r} \, .$$

Multiplication of both sides by v gives the fundamental Equation 6.6 for acceleration in uniform circular motion. It would be possible to go further and study the rate of change of acceleration. We do not do so because the study of nature does not seem to require it. The laws of motion involve no kinematic concepts beyond acceleration.

6.8 Components

In order to define the position vector, it is necessary to select an arbitrary origin, which may be located at any convenient point. Although not necessary, it is often very convenient in studying vectors also to select an arbitrary set of reference directions. Lines radiating from the origin, called axes, specify these directions. For example, in the square classroom with origin chosen to be in the southwest comer of the floor, the lines running east, north, and up from this point could be called the x-axis, the y-axis, and the z-axis. Axes are almost always chosen to be mutually perpendicular. For studying vectors in a plane, two axes suffice. In three dimensions, three axes are necessary. On a map of some small portion of the earth, east-west and north-south lines play the role of axes, serving as convenient reference directions.

Suppose that Barbara, in Figure 6.14, is located 14.2 feet from the origin corner, and that her position vector makes an angle of 45 deg with each wall. These two numbers, giving the magnitude and direction of the position vector, specify her location precisely. Another way to describe her location would be to say that she is 10 feet east of the origin (10.04 feet, to be more exact), and 10 feet north of the origin. Her position vector may be written as the sum of two vectors, one eastbound and one northbound, as illustrated graphically in Figure 6.23. In symbols,

$$\mathbf{B} = \mathbf{B}_x + \mathbf{B}_y \, . \tag{6.7}$$

Here **B** is Barbara's position vector (magnitude, about 14 feet, direction northeast); $\mathbf{B}_x$ is a vector parallel to the *x*-axis with magnitude 10 feet, direction east; and $\mathbf{B}_y$ is a vector parallel to the *y*-axis, magnitude 10 feet, direction north. The magnitude of $\mathbf{B}_x$ (written B_x) is called the *x*-component of **B**; the magnitude of $\mathbf{B}_y$ (written B_y) is called the *y*-component of **B**. Expressing the vector **B** as the sum of other vectors parallel to the axes is called "resolving" **B** into its components.

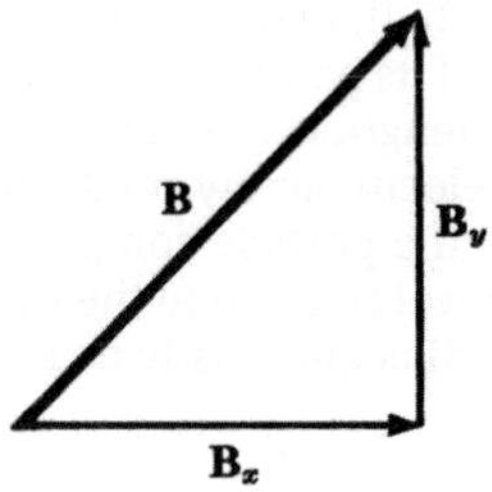

FIGURE 6.23 Components. Barbara's position vector is expressed as the sum of an eastbound vector $\mathbf{B}_x$ and a northbound vector $\mathbf{B}_y$.

The resolution of a vector into its components has several practical advantages. One of them is that it makes addition of vectors a good deal easier. This can best be demonstrated by example. Suppose you start at some origin, walk 7 feet northeast, then 7 feet east (Figure 6.24). Where are you? It is a problem in trigonometry to discover your distance and direction from the origin. But if you work with components, it is only a problem of arithmetic. The first 7 feet carried you 5 feet east and 5 feet north. The second 7 feet carried you 7 feet further east and no distance further north. Therefore your final position vector has an *x*-component of 12 feet and a *y*-component of 5 feet. That is all you need to know, since that determines your position exactly. If you still wish to find your distance and direction from the origin, it is an easier problem in trigonometry than before, since now you work only with a right triangle. Its sides are 5 feet and 12 feet. Its hypotenuse is 13 feet, and its angle from the *x*-axis is the angle whose tangent is 5/12, or about 23 degrees.

An excellent example of the usefulness of vector components is afforded by the application of the law of momentum conservation to elementary-particle collisions. Momentum is a vector quantity, and the total momentum (*both* direction *and* magnitude) must be the same after a collision as before. Figure 6.25 illustrates momentum vectors before and after the collision of a high-speed electron with a stationary proton. The *x*-axis is conveniently chosen parallel to the initial direction of the electron. The momentum before the collision is then directed along the *x*-axis; its y-component is zero. The same must be true after the collision for the *total* momentum of the deflected electron and the recoiling proton. The electron after the collision has a positive y-component of momentum and an x-component smaller than before the collision. The recoiling proton has a negative y-component of momentum which must exactly cancel the electron's positive y-component. The x-components of electron and proton after the collision add together to equal exactly the original x-component of the electron alone. Vector analysis of this kind has verified the law of momentum conservation to high accuracy for elementary-particle collisions. Moreover, it serves as a powerful tool to study invisible neutral particles which leave no trails in cloud chambers or bubble

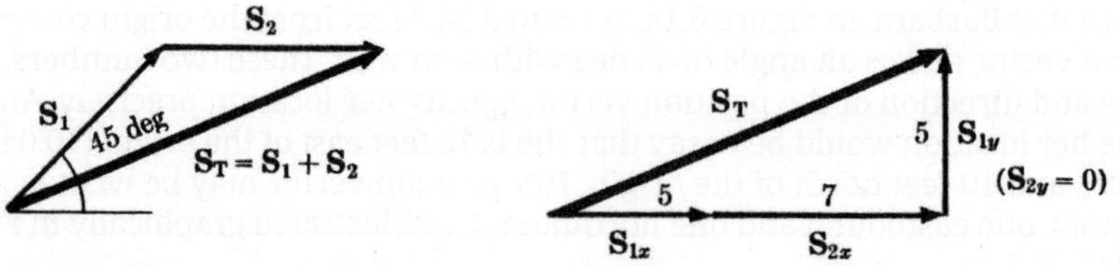

FIGURE 6.24 Vector addition using components. On the left is the arrow diagram for the addition of two vectors without the use of components. On the right the same addition is performed with the help of components. Then only horizontal and vertical directions need be considered.

chambers. If the x and y components of momentum of visible particles in a collision fail to check, it is inferred that another particle, unseen, was also present, with definite known momentum.

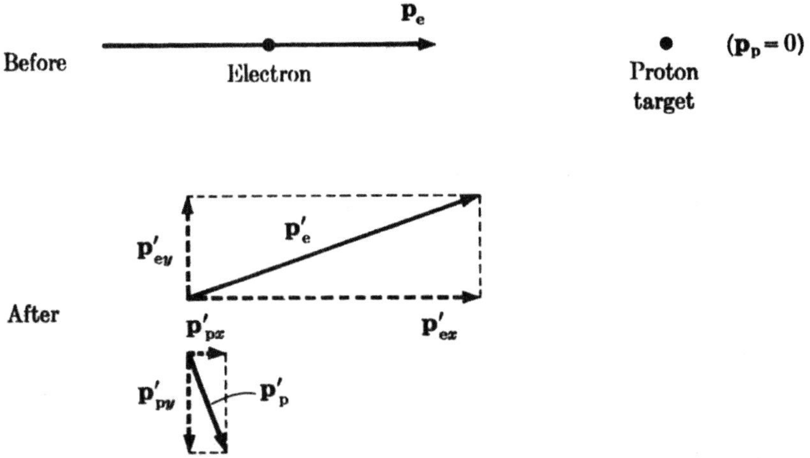

FIGURE 6.25 Momentum conservation in electron-proton collision. The equal and opposite *y* components of momentum after the collision cancel. The *x* components of momentum after the collision add to equal the magnitude of the initial momentum $\mathbf{p}_e$.

6.9 *Vectorial consistency*

Every student learns something about the need for consistency of units by being asked a question such as, "What is the sum of three apples plus two miles?" Not so obviously absurd, but in fact equally inappropriate, is the question, "What is the sum of a scalar and a vector?" Such a sum has no meaning. Two vectors may be added or subtracted. Two scalars may be added or subtracted. But a scalar and a vector can be combined in only one way, by multiplication. Since the physical concepts appearing in the laws of nature are sometimes vectors, sometimes scalars,* there are only certain possible ways in which they can combine in the mathematical statements of these laws. Physical laws must have what is called vectorial consistency. This means simply that if one side of an equation is a vector quantity, the other side must be a vector quantity also. Or both sides must be scalar quantities. For example, the law of momentum conservation may be written

$$\mathbf{p}_1 = \mathbf{p}_2 \, ; \tag{6.8}$$

the momentum before a collision, $\mathbf{p}_1$, must be equal to the momentum after the collision, $\mathbf{p}_2$. Both are vector quantities. A simple example of a scalar equation is the geometric formula for the area of a circle,

$$A = \pi r^2 \, .$$

Both sides are scalar quantities. Newton's law of motion, to be studied in Chapter Eight, has the following mathematical form:

$$\mathbf{F} = m\mathbf{a} \, .$$

* Some physical concepts are represented by mathematical objects more complicated than either scalars or vectors, but these will not be of concern to us in this book.

It states that the force **F** applied to a particle is equal to the mass m of the particle multiplied by its acceleration **a**. This equation has vectorial consistency because, on the right side, a scalar quantity (mass) multiplied by a vector quantity (acceleration) is a vector quantity.

The simplest way to demonstrate the fact that a vector cannot be set equal to a scalar is by example. Suppose, for a particle moving uniformly in a circle, we tried the equation,

$$\mathbf{v} = s\,, \text{ (wrong)}$$

in which the velocity **v** is equated to the speed s. Both might be, for instance, 10 cm/sec. But after some time, the velocity, a vector quantity, has changed. It has a different direction, as important for a vector as a different magnitude. The speed, on the other hand, has not changed. If we had at some time been able to interpret meaningfully the equation $\mathbf{v} = s$, it could not remain true at a later time, for the quantity on the left side is changing, the quantity on the right is not.

The important thing to remember about a vector equation of the form

$$\mathbf{V}_1 = \mathbf{V}_2$$

is that *both* essential aspects of the vectors, their magnitudes *and* their directions, are equal. Therefore a vector equation is a more powerful statement than a scalar equation. It is really several different equalities wrapped in a single package. This fact can be understood most clearly if we think of the vectors resolved into their components. In two dimensions, with an x-axis and a y-axis, the equality of two vectors means that they have equal x-components and separately that they have equal y-components. In three dimensions, a vector equation is really equivalent to three different equations, for each component of one side of the vector equation must be equal tb the corresponding component of the other side of the equation.

The fact that a vector equation can be split into component equations has some important physical consequences. One of these is that in some examples of particle motion, the motion parallel to one axis can be completely "decoupled" from the motion parallel to another axis. Each component of the motion proceeds as if the other component were not present. (This is not always true. It requires that the force acting on the particle have a certain simple form.) The height reached by a thrown baseball, for instance, depends only on the vertical component of its initial velocity, not on the horizontal component. The decoupling of the components that happens to be valid for any trajectory motion near the earth means that it is easy to answer a question such as, "Which goes higher, a baseball thrown vertically upward with an initial speed of 25 ft/sec, or a baseball thrown at an angle 30 deg above the horizontal with an initial speed of 50 ft/sec?" As illustrated in Figure 6.26, both baseballs start out with the same vertical component of velocity. Therefore both reach the same height.

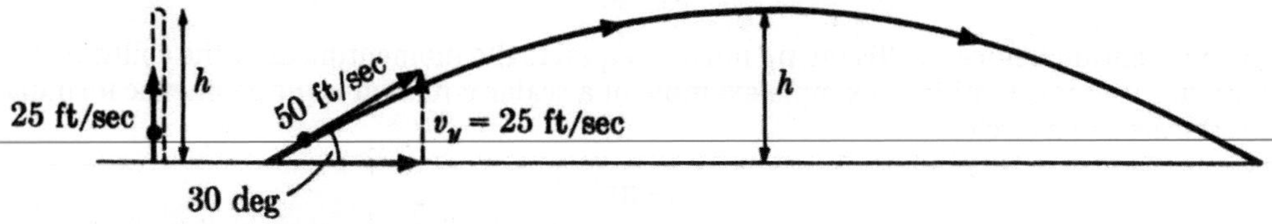

FIGURE 6.26 Significance of velocity components in motion near the earth. Two thrown baseballs with equal initial vertical components of velocity reach the same height, regardless of their horizontal components of velocity.

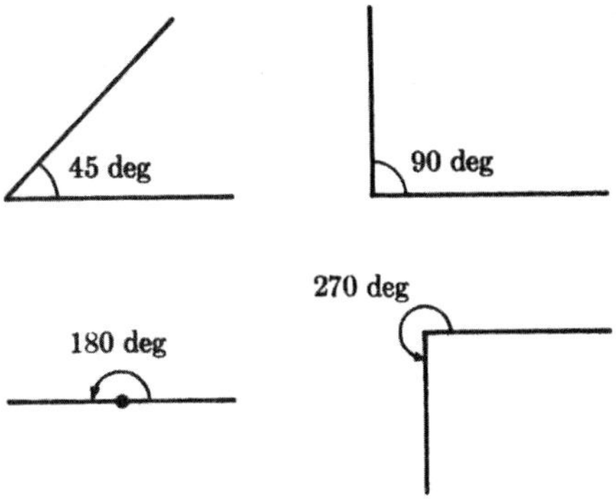

FIGURE 6.27 Several angles measured in degrees.

6.10 Trigonometry

Often useful in working with vectors is the branch of mathematics called trigonometry. Trigonometry (from Greek words meaning measurement of triangles) is the study of the relationships between angles and line segments, especially the sides of triangles. The purpose of this section is to summarize those few aspects of trigonometry that will be needed for the physics presented in this book. The student with no previous training in trigonometry should find this section an adequate guide for study. For the student who has already studied trigonometry, this section may serve as a useful review.

There are two common units of measurement for angles: degrees and radians. The degree (deg) is defined as one 360th part of a complete revolution. A right angle is 90 deg, half a right angle 45 deg, three quarters of a revolution is 270 deg, and so on. Several angles measured in degrees are illustrated in Figure 6.27. In describing rotational motion, it is possible to speak of angles greater than 360 deg. A wheel that has undergone two revolutions, for example, is said to have rotated through 720 deg. One degree is sometimes subdivided into 60 angular minutes, and one angular minute into 60 angular seconds, but these are inconvenient units that we shall avoid.

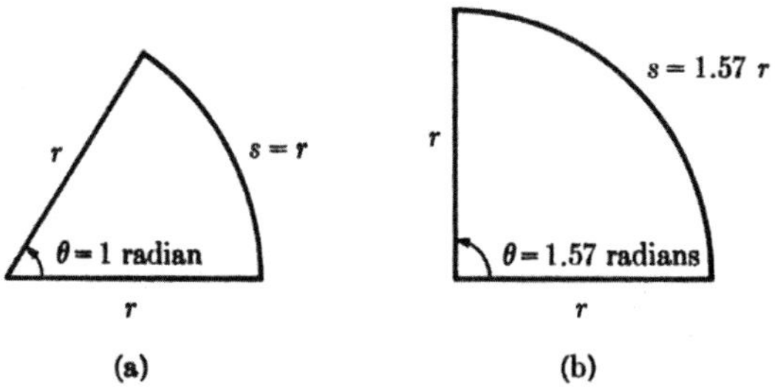

FIGURE 6.28 Definition of the radian. **(a)** If the length of circular arc is equal to the radius, the angular opening is 1 radian, about 57.3 deg. **(b)** For any other angle, the ratio of the length of circular arc to the radius is equal to the angle measured in radians: $\theta = s/r$.

The radian is a larger angular unit than a degree, equal to approximately 57.3 deg. One revolution contains 2π radians, or about 6.283 radians. This seems at first sight a peculiar way to divide up a circle, but actually it is a very sensible way. For most scientific applications, the radian is a more useful angular unit than the degree. The definition of the radian that makes its simplicity more evident is illustrated in Figure 6.28. One radian is that angular opening which intercepts a length of circular arc equal to the radius of the circle. Generally any angle can be measured in radians by taking the length of circular arc it intercepts, and dividing that length by the radius of the circle. This definition does not rest on the size of the circle. Doubling the radius of the circle, for instance, causes the length of arc also to double, so that the ratio of the two remains the same. The radian is called a dimensionless unit. This means that the numerical magnitude of an angle does not depend on the choice of length scale. Whether, in the definition of the radian, the arc length and radius are measured in feet, centimeters, or any other length unit, their ratio remains the same number.

If an angle is changing with time, it is possible to define an "angular velocity." The usual unit for angular velocity is radians/sec, and a frequently used symbol for angular velocity is ω (omega). A wheel moving with an angular velocity of 1 radian/sec will complete one revolution in 6.28 sec. If a line is drawn from the center of a circle to a particle moving uniformly around the circumference of the circle, the line will sweep out angle at some rate ω. If the particle moves with speed v, and the radius of the circle is r, the fundamental relationship among these three quantities is

$$v = \omega r\,, \tag{6.9}$$

provided ω is expressed in radians per unit time. For example, the earth in its orbit around the sun has an angular velocity of 6.28 radians/year. Its distance from the sun is 93 million miles. The product of these two numbers gives the speed of the earth in its orbit: v = 584 million miles per year. In proper scientific notation, this is $v = 5.84 \times 10^8$ miles/yr. (Since there are 31.5 million, or 3.15×10^7 seconds in a year, this speed may also be written: v = 18.5 miles/sec.)

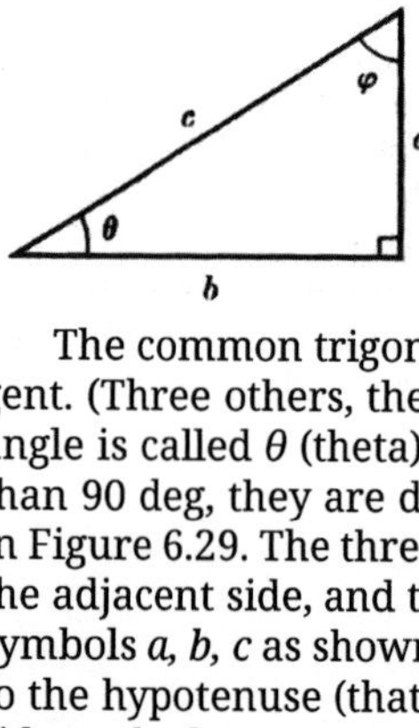

FIGURE 6.29 Right triangle, which defines symbols used in text.

The common trigonometric properties of an angle are the sine, the cosine, and the tangent. (Three others, the secant, the cosecant, and the cotangent, will not be needed.) If the angle is called θ (theta), these quantities are abbreviated $\sin\theta$, $\cos\theta$, $\tan\theta$. For angles less than 90 deg, they are defined geometrically with the help of a right triangle, as illustrated in Figure 6.29. The three sides of the triangle are called the opposite side (from the angle θ), the adjacent side, and the hypotenuse. We represent the lengths of these three sides by the symbols a, b, c as shown in Figure 6.29. The sine of the angle is the ratio of the opposite side to the hypotenuse (that is, the ratio of their lengths). The cosine is the ratio of the adjacent side to the hypotenuse. The tangent is the ratio of the opposite side to the adjacent side. In symbols,

$$\sin\theta = \frac{a}{c}\,, \tag{6.10}$$

$$\cos\theta = \frac{b}{c}\,, \tag{6.11}$$

$$\tan\theta = \frac{a}{b}\,. \tag{6.12}$$

If the angle opposite to θ is called φ (phi), for it the roles of opposite and adjacent sides are interchanged. Therefore

$$\sin\theta = \cos\varphi\,, \tag{6.13}$$

$$\cos\theta = \sin\varphi\,. \tag{6.14}$$

Also, because the sum of the interior angles of a triangle is 180 deg, θ and φ together must add up to 90 deg.

$$\theta + \varphi = 90\ \text{deg} = \frac{\pi}{2}\ \text{radians} \tag{6.15}$$

For angles greater than 90 deg, a slight generalization in the definition of the trigonometric quantities is required. In Figure 6.30, the hypotenuse of a right triangle serves as the radial line in a circle drawn about the origin. If the angle θ is the angle of this radial line measured counterclockwise from the x-axis, the trigonometric quantities are defined by

$$\sin\theta = \frac{y}{r}, \tag{6.16}$$

$$\cos\theta = \frac{x}{r}, \tag{6.17}$$

$$\tan\theta = \frac{y}{x}, \tag{6.18}$$

As is evident from Figure 6.30(a), these definitions coincide with the definitions of Equations 6.10–6.12 if the angle θ is less than 90 deg. For an angle θ between 90 and 180 deg [Figure 6.30(b)], the x-coordinate is negative, so that the cosine and tangent become negative while the sine remains positive.

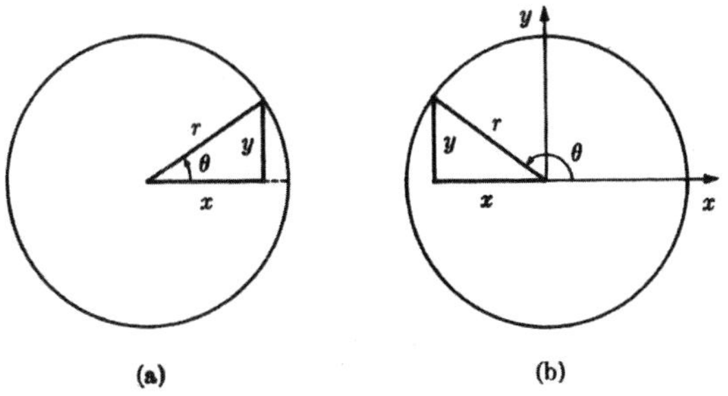

FIGURE 6.30 Diagrams for general definitions of trigonometric quantities.

For dealing with vectors, the cosine is perhaps the most important trigonometric quantity, because of the following simple rule. The component of a vector along any axis is equal to the magnitude of the vector multiplied by the cosine of the angle between the direction of the vector and the direction of the axis. Figure 6.31 illustrates a vector **V** directed 30 deg away from the x-axis, and 60 deg away from the y-axis. If, to be more general for a moment, the angle it makes with the x-axis is called θ, the fundamental definition of the cosine can be written

$$\cos\theta = \frac{V_x}{V};$$

it is the ratio of the x-component of **V** to the magnitude of **V**. Multiplying both sides of this equation by V gives the mathematical expression for the rule stated above:

$$V_x = V\cos\theta\,. \tag{6.19}$$

For the illustrated example

$$V_x = V \cos(30 \text{ deg}) .$$

For the other component of the vector, a similar equation holds,

$$V_y = V \cos(60 \text{ deg}) .$$

As shown by any hand calculator, the cosine of 30 deg is 0.866, and the cosine of 60 deg is 0.500. Referring back to Figure 6.26, we may let **V** be the velocity whose magnitude was 50 ft/sec. Multiplying this magnitude by the respective cosines gives the horizontal and vertical components,

$$V_x = 43.3 \text{ ft/sec} ,$$

$$V_y = 25 \text{ ft/sec} .$$

It is worth noticing that V_y could also have been found from the formula, $V_y = V \sin(30 \text{ deg})$. However, it is probably simpler to remember just a single rule, always using the cosine of the angle between the vector and the axis in question.

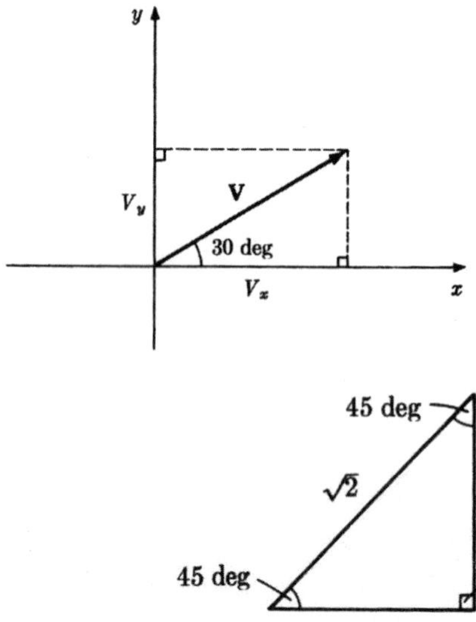

FIGURE 6.31 Diagram for finding the magnitudes of the components of a vector.

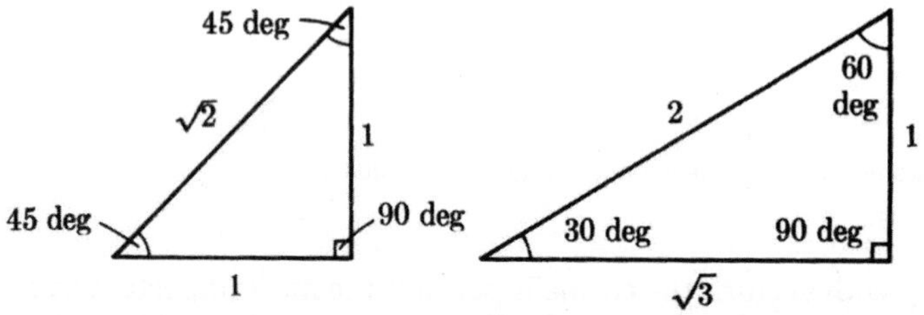

FIGURE 6.32 Two special right triangles, the 45-45-90 triangle and the 30-60-90 triangle.

Two special right triangles, shown in Figure 6.32, have some simple features which make them particularly interesting. The 45-45-90 triangle has two equal sides, and a hypotenuse longer by a factor $\sqrt{2}$ (=1.414) than each side. The trigonometric quantities associated with this triangle are

$$\sin(45 \text{ deg}) = \cos(45 \text{ deg}) = \frac{1}{\sqrt{2}} = 0.707, \tag{6.20}$$

$$\tan(45 \text{ deg}) = 1 . \tag{6.21}$$

The 30-60-90 triangle, with angles in radians $\frac{1}{6}\pi$, $\frac{1}{3}\pi$, $\frac{1}{2}\pi$, has a hypotenuse exactly twice the length of its shorter side. Its trigonometric quantities are

$$\sin(30\text{ deg}) = \cos(60\text{ deg}) = 0.500\,, \tag{6.22}$$

$$\sin(60\text{ deg}) = \cos(30\text{ deg}) = \frac{1}{2}\sqrt{3} = 0.866, \tag{6.23}$$

$$\tan(30\text{ deg}) = \frac{1}{\sqrt{3}} = 0.577, \tag{6.24}$$

$$\tan(60\text{ deg}) = \sqrt{3} = 1.732\,. \tag{6.25}$$

Other angles of special interest are zero and 90 deg;

$$\sin(0\text{ deg}) = \cos(90\text{ deg}) = 0\,, \tag{6.26}$$

$$\cos(0\text{ deg}) = \sin(90\text{ deg}) = 1\,, \tag{6.27}$$

$$\tan(0\text{ deg}) = 0\,, \tag{6.28}$$

$$\tan(90\text{ deg}) = \infty\,. \tag{6.29}$$

The last symbol, ∞, means infinity. We conclude with two formulas connecting sines, cosines, and tangents, which are valid for all angles:

$$\tan\theta = \frac{\sin\theta}{\cos\theta}\,, \tag{6.30}$$

$$(\sin\theta)^2 + (\cos\theta)^2 = 1;\,. \tag{6.31}$$

The aspects of trigonometry covered here amount to a small fraction of the whole subject. Nevertheless, a remarkable power in handling vectors, and in other physical applications as well, can come from a knowledge only of the definitions and simplest properties of angles, and of sines, cosines, and tangents.

EXERCISES

6.1. Sketch diagrams of "geometrical arithmetic" for these operations: **(a)** 3 + 7; **(b)** 3 – 7; **(c)** 3 + 0.5 – 5; **(d)** –4 – 3 – 2 – 1.

6.2. Sketch a diagram of "geometrical arithmetic" for this operation: 2.5 × 4.2.

6.3. Which of the following are numerical quantities, which are vector quantities, and which are neither: **(a)** velocity; **(b)** money; **(c)** momentum; **(d)** mass; **(e)** shape; **(f)** color; **(g)** force; **(h)** age?

6.4. Which of the following physical quantities behave like vectors: **(a)** charge; **(b)** temperature; **(c)** energy? Explain.

6.5. **(1)** Each of two vectors has unit magnitude. Find the magnitude of their sum when **(a)** they are directed parallel, **(b)** they are directed oppositely, and **(c)** one is directed at a right angle to the other. **(2)** What must be their relative direction in order that their sum also have unit magnitude? With each of your four answers, sketch a vector arrow diagram.

6.6. You are driving north on a city street at a steady speed of 10^3 cm/sec. You slow down, round a 90-deg corner to the right, and accelerate back to a steady speed of 10^3 cm/sec.

What is the net change in your velocity vector?

6.7. Three forces whose sum is zero act on a particle. One is directed eastward and has magnitude 3 dynes. One is directed northward and has magnitude 5.2 dynes. What are the direction and magnitude of the third force?

6.8. A railroad train is traveling east at 10^3 cm/sec. A boy on the train throws a baseball straight out the window toward the south at a speed of 3×10^3 cm/sec relative to the train. A man on the ground catches the ball. From what direction does he see it come, and at what speed?

6.9. An airplane with an airspeed of 300 miles/hr is maintaining a northeast heading (that is, the nose of the plane is pointing directly northeast). The air is moving southward at 100 miles/hr. Graphically find the vector giving the airplane's velocity over the ground. What is its ground speed?

6.10. A pilot whose plane flies at 100 miles/hr wants to fly north on a day when a wind of 50 miles/hr is blowing from the west. **(1)** In what direction must he point his plane? **(2)** What is his speed over the ground? **(3)** Draw a vector diagram showing the velocity with respect to the ground as the sum of the velocity with respect to the air plus the velocity of the air.

6.11. Starting from an origin, a man walks 4 meters north, then 12 meters east, then 9 meters south. Draw a reasonably accurate diagram of his trip, and measure the distance from the origin to his final position. What is his final position vector?

6.12. On a whole sheet of paper, sketch a rough map of the state in which you live. At several places on this map indicate by short arrows the position vectors of these places in the state if the city in which you live is chosen as the origin. In drawing the arrows, pay attention to their relative length as well as to their directions.

6.13. Choose as the origin of the schoolroom containing Alice, Barbara, and Charlotte the southeast corner instead of the southwest corner, and redraw Figures 6.14 and 6.15 accordingly, demonstrating that the vectors of relative position, **B** – **A** and **C** – **A**, are unchanged.

6.14. A racing car executes a turn of radius 200 meters at a speed of 90 miles/hr. What is its inward acceleration in cm/sec^2? Is this reasonable? Why?

6.15. An artificial earth satellite moves in a circle of circumference 8×10^9 cm with a period of 1.44×10^4 sec. **(1)** What is the magnitude of its acceleration? How does this compare with the acceleration of gravity at the earth's surface? **(2)** In order to help yourself visualize the characteristics of this satellite, calculate its altitude above the earth's surface in miles and its period in minutes.

6.16. An airplane is flying in a horizontal circle. Its speed is 10^4 cm/sec (about 220 miles/hr). If its acceleration toward the center of the circle is equal to g (980 cm/sec^2), what is the radius of the circle? Express the answer in cm and in miles. *Optional:* What is (or should be) the angle of bank of the airplane?

6.17. Explain in words why doubling the speed of motion around a circle increases the centripetal acceleration fourfold.

6.18. Explain in words why doubling the radius of circular motion at fixed speed halves the centripetal acceleration.

6.19. For the uniform motion of a particle around a circle, what is the direction of the *rate of change* of the acceleration vector?

6.20. Three forces act on an object. Each force has a magnitude of 100 dynes. The directions in which the forces act are: to the east, to the north, and to the northeast. Choose the x-axis in the eastward direction and y-axis in the northward direction. **(1)** Write the x and y components of *each* force. **(2)** Write the x and y components of the *sum* of the three forces. **(3)** What is the magnitude of the sum, and what is its direction?

6.21. Solve Exercise 6.9 using the method of components; specify the east (x) and north (y) components of the velocity over the ground.

6.22. **(1)** What is the angular velocity of an astronaut circling the earth in 90 minutes? **(2)** What is the angular velocity in radians/sec of a crankshaft turning at 5,000 revolutions per minute?

6.23. A large flywheel of radius 200 cm is turning so that a point on its rim is moving with a speed of 6×10^3 cm/sec. What is the angular velocity of the flywheel in radians/sec? How fast is it turning in revolutions per sec?

6.24. What is the angular velocity, ω, in radians/sec of the following: **(a)** the minute hand of a clock; **(b)** the hour hand of a clock; **(c)** a wheel 3 feet in diameter on a car traveling 60 miles/hr; **(d)** the earth in its orbit around the sun?

6.25. What are the following trigonometric quantities: **(a)** sin (360 deg); **(b)** tan (π radian); **(c)** cos (135 deg); **(d)** sin ($\pi/2$ radian) ?

6.26. Draw a right triangle as in Figure 6.29. Using the definitions of sin, cos, and tan prove that

(a) $\tan\theta = \dfrac{\sin\theta}{\cos\theta}$, **(b)** $\sin^2\theta + \cos^2\theta = 1$.

6.27. Prove that

$$\frac{1}{\cos^2\theta} - \tan^2\theta = 1.$$

6.28. **(1)** Which of the following vectors are equal? Which have equal magnitude? Which have the same direction? Which are oppositely directed?

A: $A_x = 2, A_y = 4,$ **B:** $B_x = 3, B_y = 6,$

C: $C_x = -4, C_y = -8,$ **D:** $D_x = -6, D_y = 3,$

(2) Add all four vectors and write the components of the sum. **(3)** Which has greater magnitude: **A** – **C** or **B** – **D**?

6.29. Consider the pair of vectors shown. **(1)** Give the components of **A** and of **B**. **(2)** Find the sum **A** + **B**. **(3)** Find the difference **A** – **B**. **(4)** Sketch vector diagrams for summing the vectors in both orders, **A** + **B** and **B** + **A**.

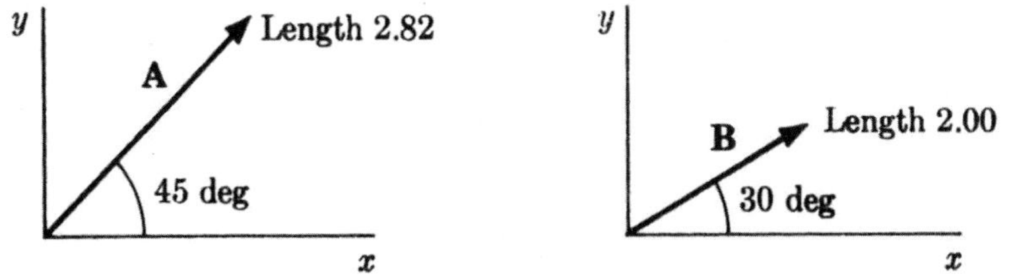

6.30. Consider the pair of vectors **A** and **B** in the preceding exercise. **(1)** Sketch vector sum diagrams for **C** = 2**A** + 3**B** and for **D** = 0.5**A** + 4**B**. (These need not be precise.) **(2)** Use components to find the vectors **C** and **D**. **(3)** Is it possible to multiply the vectors **A** and **B** by scalars α and β in such a way that the sum α**A** + β**B** is equal to zero? If so, what are the values of α and β?

6.31. An eastbound proton strikes a proton at rest. After the collision one proton flies off to the northeast, the other to the southeast, the path of each one at 45 deg to the direction of motion of the original projectile proton. Use the vector character of momentum and the law of momentum conservation to prove that the magnitudes of the two final momenta must be equal. (Use diagrams. They will be helpful.)

6.32. A positive pion strikes a neutron in a nucleus to create a neutral lambda and a positive kaon: $\pi^+ + n \rightarrow \Lambda^0 + K^+$. In a bubble chamber, two tracks are seen, as in the figure. The magnitude of the K^+ momentum is determined to be half the magnitude of the π^+ momentum. Along what direction must the unseen Λ^0 have flown?

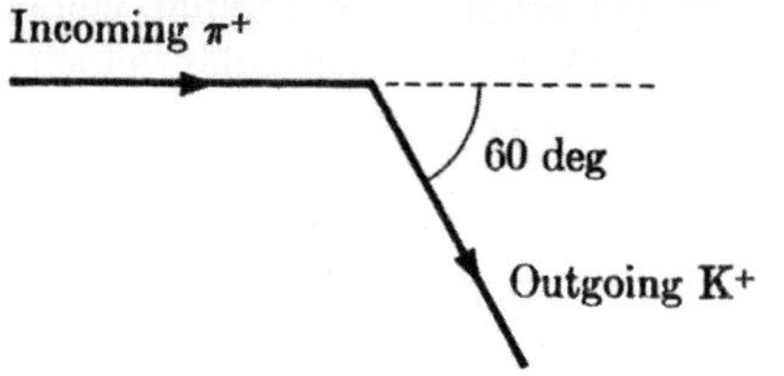

6.33. A long-range cannon is mounted on a railway flatcar (see figure). The shells have a mass of 3×10^5 gm and a muzzle velocity of 10^5 cm/sec. The total mass of flatcar plus cannon is 4×10^7 gm. **(1)** If the cannon is fired horizontally along the track, what is the recoil speed of the flatcar? **(2)** If the barrel is elevated 30 deg above the horizontal, what is the recoil speed of the flatcar after the cannon is fired?

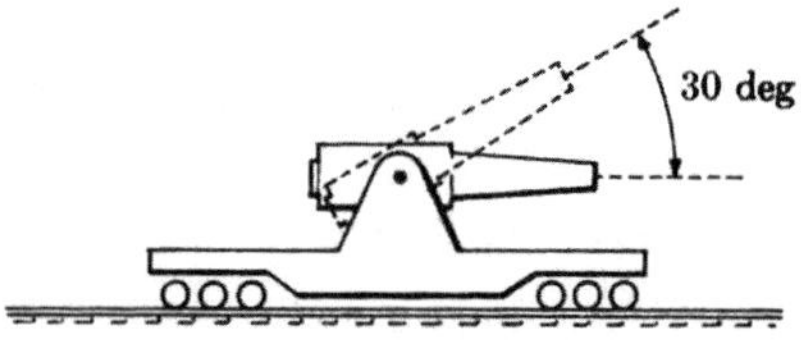

6.34. A cloud-chamber photograph shows the decay of an unstable particle at rest. Two charged particle tracks are seen emerging from the decay point. Measurement shows one particle to have had a momentum of 98 units along the positive x-axis. The other particle had 100 units of momentum in a direction making an angle of 150° with the positive x-axis. A neutral particle would leave no track in the chamber. Do you think one was produced in the decay? Why? If so, what was the magnitude and direction of its momentum? Is it possible that *two* neutral particles may have been produced?

6.35. Explain in what sense a vector equation contains more information than a scalar equation.

6.36. Prove that if three vectors sum to zero, they must lie in the same plane. (HINT: Consider first the sum of two. Then add the third.)

6.37. Consider two vectors, **a** and **b**. Using the laws of vector algebra, write the expression 6(**a** + **b**) – (10**a** – **b**) + 3(**b** – 2**a**) in terms of the vector **b** – **a** alone. Demonstrate graphically the correctness of your result in case **a** is a vector of unit length pointing east, and **b** is a vector of unit length pointing north.

6.38. Check a dictionary for meanings of the word "vector" other than the meaning used in this chapter. Is there any connection between a biological vector and a physical vector?

SEVEN

Units, functions, and graphs

We are concerned in this chapter with several different topics—mainly practical topics—related to the mathematical description of nature: units and dimensions, special functions, and the presentation of data in graphs, tables, and equations. These topics will find application in all of the later chapters.

7.1 *Standard dimensions and units*

That most physical quantities require a unit of measurement is a familiar fact of the everyday world. We reckon distance in inches or feet or miles, time in seconds or minutes or hours, money in cents or dollars, speed in miles per hour or feet per second, and so on. Typical units of measurement for a number of fundamental physical quantities were discussed in Chapter Three. Also discussed there was the distinction between dimensional numbers and dimensionless numbers. Simply stated, a dimensional number is a numerical value (of a physical quantity) that depends upon the choice of units; a dimensionless number is independent of the choice of units. Actually, it would be better to refer to the physical quantity, not the number, as dimensional or dimensionless, but we follow the usual imprecise practice. Dimensionless numbers may result either from counting (the number of pages in this book is a dimensionless number) or from taking the ratio of two quantities expressed in the same unit of measurement. If I am twice as tall as my son, this means that the ratio of my height to his height is 2, a statement that is numerically true (at some time) regardless of the units employed to measure height.

The idea of dimension, and the name, go back to very early thinking about geometry. Length is a concept of one dimension. It is, in geometry, a fundamental idea that cannot be expressed in terms of any others. Therefore length has the dimension of length, and nothing else can be said about it. Area, on the other hand, is a two-dimensional concept. It has the dimension of length multiplied by length. We say that area has the dimension of $(\text{length})^2$. Similarly, volume has the dimension of $(\text{length})^3$. This idea of the dimension of geometrical concepts is obviously related to the units of measurement of the corresponding physical quantities. Distance might be measured in cm, area in cm^2, and volume in cm^3. Or distance in ft, area in ft^2, and volume in ft^3. Whatever the unit of length, there stands behind the unit the more abstract idea of dimension.

In this discussion of geometry, we have made a subtle but important transition from the purely geometric idea of the dimensionality of space to the idea of the dimension of a physically measurable quantity. Having done so, it is an easy next step to generalize the idea of dimension to nongeometric quantities. In kinematics occurs the concept of time as well as the concept of space. What is the dimension of time? In geometric terms it has no dimension. But physically it proves useful to define a dimension of time, a new and independent dimension, not expressible in terms of the dimension of length. Then kinematic quantities such as speed and acceleration can be assigned dimensions that are certain combinations of the new fundamental dimensions, length and time. The dimension of speed is length/time (length divided by time, or length per unit time). The dimension of acceleration is length/$(\text{time})^2$. The rate of expansion of a balloon, to pick a third example, has the dimension of $(\text{length})^3/\text{time}$, or volume per unit time.

The adoption of one fundamental dimension, length, serves as a basis for describing the dimensions of all geometric quantities. The addition of a second, time, provides a basis for assigning a dimension to every kinematic quantity. To go further, and describe the dimensionality of all the physical concepts of mechanics, a third independent dimension is necessary. Interestingly, three fundamental dimensions are sufficient. (These are not to be confused with the three dimensions of space, which are all length dimensions.) Usually mass is chosen as the third independent dimension. Its dimension cannot be expressed as any combination of the dimensions of space and time, but once it is chosen, the dimension of every other mechanical concept can be expressed as some combination of these three—length, mass, and time.

One might suppose that as more and more physical concepts are introduced in new fields of physics, more and more fundamental dimensions would have to be introduced. In fact, it has turned out that three independent dimensions are sufficient to assign a dimension to every known physical quantity. Correspondingly, three independent units of measurement are enough. It is possible, but not necessary, to introduce more than three fundamental dimensions and more than three basic units. In the two systems of measurement commonly employed in physics, four—one more than the most economical number—are chosen. In both of these, systems, called the cgs system and the mks system, length, mass, time, and temperature are chosen as the fundamental dimensions.* Temperature is added only for the historical reason that its conceptual development in physics was independent of mechanical concepts. It still hangs on as an unnecessary independent dimension.

In the cgs, or centimeter-gram-second, system, length is measured in centimeters, mass in grams, time in seconds, and temperature in Centigrade degrees. The units of measurement of every other physical quantity can be expressed as some combination of these four. Speed, to give a simple example, is measured in units of cm/sec. Very often the combination that occurs is sufficiently unwieldy to write and to say that it is convenient to give it a new name. The unit of force, for instance, is gm cm/sec^2, a combination that is given the name *dyne*. Similarly the combination gm cm^2/sec^2 that provides the unit of measurement for energy is given the name *erg*.

$$1 \text{ dyne} = 1 \text{ gm cm/sec}^2,$$

$$1 \text{ erg} = 1 \text{ gm cm}^2/\text{sec}^2 = 1 \text{ dyne cm}.$$

The new names, dyne and erg, must be clearly understood to be abbreviations only, not the names of new units. In later chapters we shall learn how concepts such as electric charge, which seem to have nothing directly to do with mass, length, or time, can also be measured in units that are combinations of centimeters, grams, and seconds.

In the mks, or meter-kilogram-second system, the time unit remains the second and the temperature unit remains the Centigrade degree. The length unit is the meter (100 cm) and the mass unit is the kilogram (1,000 gm). As in the cgs system, new names are introduced for the units of force and energy. The force unit is the *newton*, and the energy unit is the *joule*. They are defined by

$$1 \text{ newton} = 1 \text{ kg m/sec}^2,$$

$$1 \text{ joule} = 1 \text{ kg m}^2/\text{sec}^2 = 1 \text{ newton meter}.$$

One newton is equal to 10^5 dynes; one joule is equal to 10^7 ergs.

Why two sets of basic units? There is no good reason logically, only a historical reason.

* Usually in the mks system, electric current is treated for convenience as a fifth independent dimension. Through its definition in terms of forces, however, it is in fact linked to the mechanical dimensions of length, mass, and time.

Until recently, the cgs system, based on the everyday system of units in continental Europe, was preferred in science. Another set of units, based on the English foot and pound and Fahrenheit degree (and the ubiquitous second), had a lesser following in fundamental science, but a strong following in applied science and engineering. Engineers in the United States still widely use the foot-pound-second system, but this system has largely disappeared from science. It will not be used in this book. (An exception to this rule will occasionally be made for distance, whose unit, for reasons of familiarity, may be chosen to be the inch, the foot, or the mile. Any distance expressed in one of these units should always be re-expressed in centimeters before it is employed in any calculation.) At present the cgs and mks systems both enjoy wide popularity in physics, and are frequently the subject of spirited discussion between their respective adherents. In this book we shall adhere largely to the cgs system, supplemented by certain practical units of electricity. This choice affords some advantages in presenting electromagnetic theory, and it is the set of units most widely used in chemistry and biology.

Besides the "universal" systems of units, there are special systems for special branches of science that will undoubtedly continue to flourish. The nuclear physicist will continue to find the charge on the proton a convenient unit of charge, the quantum physicist will continue to prefer a system of units in which Planck's quantum constant is equal to unity, the astronomer will continue to prefer a unit of length more in keeping with the scale of the universe than is the centimeter or the meter.

7.2 *Dimensional consistency and units consistency*

The simplest equation one can imagine has the form

$$A = B.$$

If this is to be an equation of physics, not merely a mathematical statement, it means that one physical quantity, designated by A, is equal to another physical quantity, designated by B. In particular if A has a certain numerical value, B has the same numerical value. To be physically meaningful in general, the equation must be true regardless of the set of units adopted for measurement. If the dimension of A is not the same as the dimension of B, trouble results. Suppose, for example, that A has the dimension of length and B the dimension of time. In the cgs system, A would be expressed in centimeters and B in seconds. A change to the mks system would result in a change of the numerical value of A because of the change of length scale, but no change in the value of B. Therefore, if the equation were numerically correct in one system of units, it could not be correct in the other. Only if the dimensions of A and B are the same is a change of units irrelevant. If A and B were both lengths, such that the equation might in the cgs system read 200 cm = 200 cm, it would, in the mks system, read 2 m = 2 m. Whenever both sides of an equation are dimensionally the same, both change by the same factor when units are changed, and the equation, if it was true with the first set of units, remains true with the second. It is a common practice, one that we shall follow in this book, to write all basic equations of physics in such a way that their correctness does not depend on a particular choice of units. Equations written in this way are called dimensionally consistent.

Consider the statement about uniform motion in a straight line, distance equals speed multiplied by time. In symbols,

$$d = vt.$$

The dimension of distance is length; the dimension of speed is length/time; the dimension of time is time. The dimensional consistency of this equation may be expressed by another "equation" written in this way:

$$[\text{length}] = \left[\frac{\text{length}}{\text{time}}\right][\text{time}]\,.$$

The square brackets are used to indicate that only the dimensions of the equation are being considered. Dimensional consistency alone, of course, does not prove that an equation is correct. However, dimensional inconsistency can prove that it is incorrect if it is derived entirely from dimensionally consistent equations. Checking each step of a derivation for dimensional consistency is a useful way to look for errors.

The fact that a dimensionally inconsistent equation may nevertheless be correct (with one set of units) can best be seen by example. Suppose the price of steak is one dollar per pound. Then the cost of a steak in dollars is equal to its mass in pounds. If cost is designated by C and mass by M, the equation expressing this fact is

$$C = M.$$

This is a correct but dimensionally inconsistent equation, for the dimension of C is money and the dimension of M is mass. It is true only so long as the unit of money is the dollar and the unit of mass is the pound. The same steak in Germany would be priced about 9 marks per kilogram. There the equation for the cost of a steak would be

$$C = 9M.$$

This again is an equation true only for a particular choice of units. It is clear how to combine these equations into a single dimensionally consistent equation. We must introduce a new symbol—call it P—for the price per pound. The steak cost equation in dimensionally consistent form is

$$C = PM.$$

The dimensional consistency is illustrated by the dimensional equation

$$[\text{money}] = \left[\frac{\text{money}}{\text{mass}}\right][\text{mass}]\,.$$

This example makes clear the great advantage of using only dimensionally consistent equations. First, the equation is correct with any set of units whatever. Second, the equation has greater flexibility and breadth of application. In this example, it allows for the possibility of a change in the price of steak. The price P may, like the cost C and the mass M, be considered a variable instead of a fixed constant. Third, it makes checking the correctness of the equation easier.

Numerical calculation with a dimensionally consistent equation is, if not easier, at least "safer" than with a dimensionally inconsistent equation. This is because the dimensional consistency of the algebraic equation gives way to a units consistency of the numerical equations. It is possible, and almost always desirable, to carry out a kind of "units arithmetic" in parallel with the numerical arithmetic. The idea can be fully illustrated with the simple distance equation considered above,

$$d = vt.$$

If an automobile travels at 30 miles/hr for 3 hr, how far does it travel? Obviously, you respond, 90 miles. Written out, this elementary calculation looks as follows:

$$vt = 30\ \text{miles/hr} \times 3\ \text{hr} = 90\ \text{miles}.$$

Perhaps without thinking about it, when you carry out such a calculation, you are really performing two distinct calculations, one with numbers, one with units. The numerical calculation is 30 × 3 = 90. The units calculation is (miles/hr) × (hr) = (miles). Suppose the original question had been slightly different: If an automobile travels at 30 miles/hr for 40 min, how far does it travel? Obviously the answer is not obtained by multiplying 30 times 40. There are several different ways to get the right answer, and all of them involve the conversion of units to bring about units consistency. First the speed could be converted from 30 miles/hr to 0.5 miles/min. Then the distance is

$$d = (0.5 \text{ miles/min}) \times (40 \text{ min}) = 20 \text{ miles}.$$

Or the time could be converted from 40 min to 0.667 hr. Then the distance is

$$d = (30 \text{ miles/hr}) \times (0.667 \text{ hr}) = 20 \text{ miles}.$$

Another way that may be less familiar is to multiply the originally given numbers 30 and 40 together and to divide their product by a suitable conversion factor. The calculation looks like this:

$$\mathrm{d} = \mathrm{vt} = \frac{30 \text{ miles/hr} \times 40 \text{ min}}{60 \text{ min/hr}} = 20 \text{ miles}.$$

The factor 60 min/hr is a conversion factor introduced to bring about units consistency. The method, a powerful one for avoiding numerical errors, is to write down the given numbers as dictated by the equation, in this case vt = (30 miles/hr) × (40 min). Then one notices that the units arithmetic does not work out properly. The "min" upstairs does not cancel the "hr" downstairs. To get rid of the time units, the conversion factor, 60 min/hr, must be introduced. If it were introduced as another multiplicative factor, time units would still not cancel. The units arithmetic works out correctly to leave the distance unit miles alone only if the conversion factor is introduced in the denominator.

In units arithmetic as in numerical arithmetic, the "upstairs-downstairs" rule applies. A factor downstairs in the denominator is equivalent to a factor upstairs in the numerator. A factor upstairs in the denominator is equivalent to a factor downstairs in the numerator. In symbols,

$$\frac{\frac{a}{b}}{\frac{c}{d}} = \frac{a \times d}{b \times c},$$

true whether a, b, c, and d represent numbers or units. For example,

$$\frac{5 \text{ cm/sec}}{2 \text{ sec/gm}} = 2.5 \text{ gm cm/sec}^2.$$

Sometimes, for reasons of notational convenience, "downstairs units" are written with negative exponents. The speed 5 cm/sec may also be written 5 cm sec^{-1}; the number calculated above might be re-expressed as 2.5 gm cm sec^{-2}. This notation is particularly convenient when the only unit is a downstairs unit. The rate of flow of automobiles through a tunnel, for instance, might be written 120 hr^{-1}, meaning 120 automobiles per hour. Because the number of automobiles is dimensionless, only the time unit appears explicitly.

One more example may be useful to clarify the workings and the value of units arithmetic. How many minutes is required for an astronaut traveling at 24,500 ft/sec to circle the earth, a distance of 25,000 miles? The equation is

$$t = \frac{d}{v}.$$

The calculation, this time requiring two conversion factors, looks as follows:

$$t = \frac{25{,}000 \text{ miles} \times 5{,}280 \text{ ft/mile}}{24{,}500 \text{ ft/sec} \times 60 \text{ sec/min}} = 90 \text{ min.}$$

The number of feet in a mile, 5,280, and the number of seconds in a minute, 60, must both be introduced in order to produce an answer in minutes. Note that in the units arithmetic, feet, miles, and seconds all cancel, leaving only minutes downstairs in the denominator, equivalent to being upstairs in the numerator. Conversion factors may be introduced in this way not only in equations, but for changing the units of a single number. For example, to determine how many ft/sec is 1 mile/hr, one could write

$$\frac{1 \text{ mile/hr} \times 5{,}280 \text{ ft/mile}}{3{,}600 \text{ sec/hr}} = 1.47 \frac{\text{ft}}{\text{sec}}.$$

Some useful conversion factors are listed in Appendix 3.

7.3 *Some dimensional subtleties*

A few hundred years ago algebra was still so strongly tied to geometry that many mathematicians believed that algebraic equations could be meaningful only if each term in the equation had the dimensions of length or area or volume and the equation as a whole had dimensional consistency. A combination such as

$$x + x^2$$

was viewed with suspicion, because if x is a dimensional quantity, x and x^2 have different dimensions and cannot be meaningfully added. This objection no longer has any weight. Either the equation may be regarded as a formal mathematical expression with no physical significance, in which case the dimensions are irrelevant; or, if it does have physical significance, it need not be geometric significance. The symbol x could represent a dimensionless quantity. Then its square is also dimensionless, and the combination $x + x^2$ does have dimensional consistency.

Although this early example of confusion about dimensional and dimensionless quantities has been resolved, there remain a few problems, or at least subtleties, of dimensional analysis that deserve some discussion. There are two important questions that we have not yet properly answered. First, what exactly is the definition of a dimensionless quantity? Second, how many fundamental dimensions are there? The aim of this section is to show that the answers to these two related questions rest upon arbitrary agreement, not upon immutable law.

We stated earlier that a dimensionless quantity either is the result of counting or is the ratio of two quantities with the same dimension. Although correct, this statement does not characterize a dimensionless number with enough precision. The length of a string, for instance, might also be said to be the result of counting—counting the number of centimeters needed to span the string. And the same dimensional quantity, the length of the string, is also a ratio—the ratio of its length to the length of a standard length, 1 cm. The practical definition of a dimensionless quantity is a quantity whose numerical value does not change, no matter what changes are made in the basic units. For example, the number of atoms in a box is dimensionless, for it does not depend on the units of length, mass, or time; whereas the number of length units in a string is dimensional, for it depends on the magnitude of the length unit. Similarly, the ratio of the length of two strings, neither one a standard, is dimensionless, whereas the ratio of the length of one string to a unit length depends on the unit and is therefore dimensional.

These distinctions between dimensional and dimensionless quantities seem straightforward and clear. The subtlety in their application stems from the arbitrariness in the choice of the fundamental dimensions. Usually in physics mass, length, and time are chosen to be the fundamental dimensions. Any quantity that is independent of the units adopted for these three dimensions is, by definition, dimensionless. "Number of atoms," for example, does not depend on the units of length, mass, or time. But the physicist is free to enlarge the number of fundamental dimensions. Mass, length, and time are merely the fewest number that will serve. Additional fundamental dimensions can be chosen, and sometimes it is convenient to add one or two more to the minimal set of three. If this is done, a formerly dimensionless quantity may suddenly become dimensional.

The example of the number of atoms in a box may serve as a basis for illustrating the possible enlargement of the set of fundamental dimensions. To mass, length, and time we could add a new dimension, called an "object" dimension, or a "thing" dimension. Its unit might be an elementary particle or an atom or a molecule or something larger. The answer to the question, "How many things are in the box?" is now a dimensional number, for even though it is the result of counting, its numerical value depends upon the choice of the "thing unit." The contents of the box might be expressed as 50 atoms (of oxygen), or 25 molecules, or 1,200 elementary particles, just as the length of its side might be expressed as 1 ft, or 30.5 cm, or 0.305 m. There are numerous other examples of possible additional fundamental dimensions. Electric charge is a candidate to be chosen as independent of mass, length, and time. Angle is another suitable extra dimension. Temperature is an appendage to mass, length, and time that is usually regarded as independent of them.

These possibilities for extra dimensions are mentioned because it is, in fact, often convenient in calculations to treat some dimensionless quantities as if they were dimensional, assigning new units to them. This procedure is less complicated in practice than it sounds when described. It amounts merely to a way to extend the usefulness and the power of units arithmetic, without actually forsaking the standard cgs units. Suppose that you are told that a box with a volume of 900 cm^3 contains 3×10^{22} oxygen atoms, and you are asked how many elementary particles there are in 1 cm^3 of the box. Number per unit volume is number divided by volume, N/V. If number in this case is treated as dimensional, the correct answer is yielded in a simple calculation with a conversion factor, as in Section 7.2. The calculation may be written

$$\frac{N}{V} = \frac{3 \times 10^{22} \text{ atoms}}{9 \times 10^{2} \text{ cm}^3} \times 24 \frac{\text{particles}}{\text{atom}} = 8 \times 10^{20} \frac{\text{particles}}{\text{cm}^3} .$$

In the units arithmetic, "atoms" cancel to leave particles per cubic centimeter. The conversion factor is obtained from the fact that an atom of oxygen (O^{16}) contains 8 neutrons, 8 protons, and 8 electrons, or 24 particles altogether.*

One dimensionless quantity of physical importance that requires particular care and attention is angle. It is seemingly paradoxical to call it dimensionless, for it is commonly measured in terms of two different units, degrees and radians. Between these there is the conversion factor, 57.3 deg/radian. (The number 57.3 is an approximation to $180/\pi$.) Nevertheless, angle is dimensionless in the same sense as "number of things," which can also be measured in different units. It is dimensionless because its units do not depend upon the basic units adopted for mass, length, and time. But like number of things, angle may frequently be treated as if dimensional. Example: A proton in an accelerator sweeps out an angle of 300 deg in 1 microsecond (μsec†). Through how many radians does it rotate in 5 μsec? Its angular speed is $\omega = 300$ deg/μsec. The time is $t = 5$ μsec. In exact analogy to the equation of uniform straight-line motion, $d = vt$, the equation of uniform rotation is

$$\theta = \omega t ,$$

* A very small fraction of oxygen atoms contain 9 or 10 neutrons in their nuclei instead of 8, but this is an irrelevant complication in this example.

† The prefix micro means one millionth, or 10^{-6}. It is usually abbreviated by the Greek letter μ (mu).

angle equals angular speed multiplied by time. The calculation, including the angular conversion factor, may be written

$$\theta = \frac{300\ \dfrac{\text{deg}}{\mu\text{sec}} \times 5\ \mu\text{sec}}{57.3\ \dfrac{\text{deg}}{\text{radian}}} = 26.2 \text{ radians}\,.$$

It proves useful and simple to treat degrees and radians as units of an independent angular dimension.

Unfortunately, this kind of units arithmetic with angles is not always possible. Some angular equations are true only if the angular unit is the radian. The equation $v = \omega r$ of Section 6.10, relating the speed of a rotating particle to its angular speed and its distance from the center of its circle, is of this type. Therefore it is wise when in doubt always to express angles in radians. One place where the angular unit is irrelevant is in trigonometric quantities. For example,

$$\sin(30 \text{ deg}) = \sin(\tfrac{1}{6}\pi \text{ radian}) = 0.5.$$

It does not matter whether the angle is written 30 deg or ($\pi/6$) radian, because it is the sine of the angle, not the angle itself, that is of numerical interest. All trigonometric quantities are dimensionless, since each is defined as the ratio of the lengths of two sides of a right triangle.

23 Just as the number of fundamental dimensions can be augmented beyond three, it can also, in a certain sense, be diminished below three. The "certain sense" comes about because the decrease rests upon a man-made agreement to regard certain units as fixed. If an absolute monarch decreed that the mks system of units was the only system permitted in his country, the physics in his country would, for practical purpose, be dimensionless. A length would not need to be called 3 meters. The number 3 alone would suffice, since no other length unit could be employed. The number 3 would be the ratio of a certain length to an immutable standard length—1 meter—and could therefore be regarded as a "pure" number, a dimensionless quantity.

The dimensionless physics still sought by physicists will rest not on monarchical decree, but on universal agreement. As discussed at the end of Chapter Three, certain physical quantities have such significance that they are very appealing choices as "natural units," fixed by scientific agreement. So far the smallest number of dimensions found in a "natural" system is one. The speed of light serves as the natural unit of speed; Planck's quantum constant serves as the natural unit of angular momentum. The third natural unit remains to be found.

Exactly how a natural unit reduces the number of fundamental dimensions can best be appreciated through specific example. By providing a natural link between time units and distance units, the speed of light makes it possible to lay aside one of these units and express both times and distances in terms of a single unit. In units where the speed of light has the value of 1, the circumference of the earth is about one-tenth of 1 sec, and the distance from earth to sun is 500 sec. The *time* required for the earth to complete one orbit is 3×10^7 sec (1 year), but the *distance* it travels in that year is 3.2×10^3 sec. The last figure means that light could cover a distance equal to the circumference of the earth's orbit in 3.2×10^3 sec, about one ten-thousandth the time the earth spends in getting around the sun. Once the standard speed unit has been adopted and the meaning of distance measured in seconds is understood, there is no ambiguity. The number of fundamental dimensions has been reduced from three to two. In a similar way, Planck's constant adopted as a unit reduces the number from two to one. Working with fewer than three fundamental dimensions is often convenient for the professional physicist, and is very likely connected with a deeper view of nature. For the beginning student, on the other hand, the use of fewer than three dimensions can sometimes obscure the meaning of new concepts and new laws, and it is not recommended. Nevertheless, it is important to appreciate the meaning, the possibility, and the merit of a dimensionless physics.

7.4 *Functions and graphs*

A key idea in mathematical analysis and in physics is the idea of dependence. One quantity "depends on" another if the variation of one of them is accompanied by a variation of the other. Mathematicians speak of the independent variable and the dependent variable. In physics it is better to think of *interdependent* quantities. The idea that one variable quantity is independent, another dependent, is, like all ideas of cause and effect in science, often useful as a way of describing events, but not strictly necessary. Logically, only the *relationship* of different physical quantities is meaningful.

If one quantity depends on another, it is said to be a *function* of the other. The average stock market price is a function of time. The speed of an accelerating automobile is a function of its position. The brightness of a spot on a television tube is a function of the speed of the electrons. Functional relationships permeate every part of science, for most of science is concerned with connections among different physical concepts. Much of the experimental work in science is devoted to learning in quantitative detail exactly how one quantity depends on another. Reducing experimental relationships to simple mathematical functions is a large part of the theoretical work in science.

The dependence of one quantity on another can be given quantitative expression in three different ways: in tabular presentation, in graphical presentation, and in mathematical equations. Although an equation provides the most concise expression of a functional relationship, tables and graphs are often useful as well. A table of data is appropriate when the precise mathematical relationship is not known, or when it is desired to present numerical values to a high degree of accuracy. Graphs, like tables, can be used to present relationships, such as those discovered experimentally, whose exact mathematical form is not known. Even if the equation *is* known, a graph can do much to bring it alive and make its meaning more evident. We shall frequently employ graphs in this book for just that purpose—to clarify an equation by making it pictorial.

The simplest functional relationship is a direct proportionality. The distance covered by an automobile moving at constant speed, for example, is proportional to the elapsed time. Data for a particular example of such motion could be presented numerically, as in Table 7.1. Each column is headed by the name of the physical quantity listed, and the units in which that quantity is expressed. Because the distance changes as the time changes, we say that the distance is a function of the time. (Alternatively, we could say that the time is a function of the distance, but this is an uncommon way to look at motion. Can you think of a reason why it is more natural to regard time as the independent variable, distance as the dependent variable?) The exact mathematical relationship between time and distance in this example is not immediately obvious when examining the table. This is one of the disadvantages of tabular presentation. Although the numerical values can be precisely specified, they do not at once convey a clear picture of how the variables are related. A graph does this job much better.

TABLE 7.1 Time and Distance Data for Motion at Constant Speed

ELAPSED TIME (min)	DISTANCE (miles)
0	0
2	1.50
4	3.00
6	4.50
7	5.25
9	6.75
10	7.50
15	11.25

In Figure 7.1 the same data are presented graphically. The variable chosen to be regarded as independent—in this case the time—is plotted horizontally; the variable chosen to be regarded as dependent is plotted vertically. Each pair of numbers in the table gives a single point on the graph. It is immediately obvious that the points may all be joined by a single straight line. It is not necessarily true that every point on that line reveals exactly where the automobile was at a particular time. We were supplied with data only for certain selected points. However, the fact that those points lie on a line is very suggestive. It is at least a reasonable supposition that the straight line also represents the location of the automobile at intermediate points. Apparently the motion was carried out with constant speed.

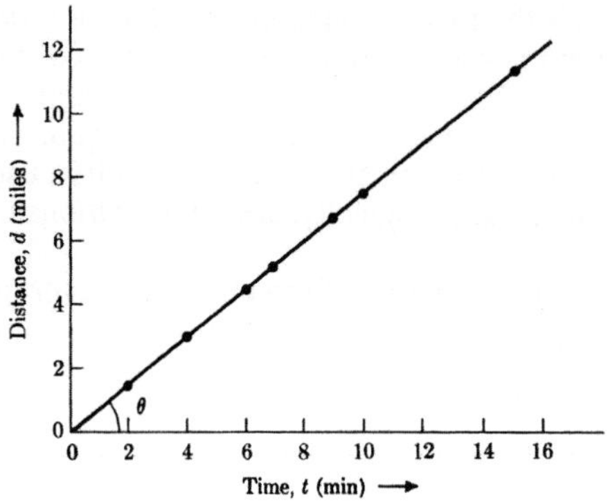

FIGURE 7.1 Graph of distance vs. time for the data of Table 7.1.

The equation that fits these tabular and graphical data is

$$d = 0.75t,$$

where d represents the distance in miles, and t the time in minutes. This equation, lacking dimensional consistency and therefore tied to a particular set of units, is better replaced by

$$d = vt, \tag{7.1}$$

where v stands for a constant whose value in this example is

$$v = 0.75 \text{ miles/min} = 45 \text{ miles/hr}.$$

An equation of this type is called a linear equation, because its graph is a straight line. The distance d is said to be (for constant v) a linear function of the time t. A slightly more general linear equation is

$$d = vt + d_0, \tag{7.2}$$

where d_0 is another constant whose physical meaning in this example is the initial distance from some arbitrarily chosen origin. Thus a direct proportionality is a special case of a linear relationship.

The velocity v, playing the role of a constant of proportionality in Equation 7.1, is also equal to the slope of the straight line in Figure 7.1. The slope of a line in a graph is defined as the tangent of the angle that the line makes with the horizontal axis. This angle is designated by θ in Figure 7.1. Its tangent is the vertical distance to any point on the line divided by the horizontal distance to the same point. For the special example being considered, this ratio is distance divided by time. In symbols,

$$\tan \theta = \frac{d}{t} = v\,.$$

Evidently the trigonometric quantity $\tan \theta$ is the same regardless of what point on the line is chosen to define it.

In general the slope of a straight line on a graph may be interpreted as the rate of change of the quantity plotted vertically with respect to the quantity plotted horizontally. In case the quantity plotted horizontally is time, this rate of change is the same as the everyday meaning of "rate," that is, rate of change with respect to time. More generally, however, rate may be defined with respect to quantities other than time. In Figure 7.1 the slope is rate of change of distance with time, or speed, as already verified above.

As a second example of a functional relationship, consider the motion of a falling ball, involving again a relationship between distance and time. This time we reverse the order, and consider first the mathematical function, then the graph, then a table. Starting from rest, a dropped ball falls a distance d in a time t. The relation between these quantities is

$$d = \tfrac{1}{2}at^2, \tag{7.3}$$

where a is a constant (whose physical meaning is the downward acceleration). Because t^2 appears on the right, the distance d is called a *quadratic function* of the time. This name, like the name "linear function," has a geometric source. However, it is a less evident source, for it has nothing to do with the shape of the graph of the function. If t represents the length of the side of a square, t^2 would be the area of the square—thus the name quadratic, or "four sided."

It would perhaps be more logical to call the quantity $\tfrac{1}{2}at^2$ a parabolic function, for, as illustrated in Figure 7.2, the graph of d vs. t is a parabola.* In each equal interval of time, the distance covered is greater than in the preceding interval, the total distance increasing as the square of the time.

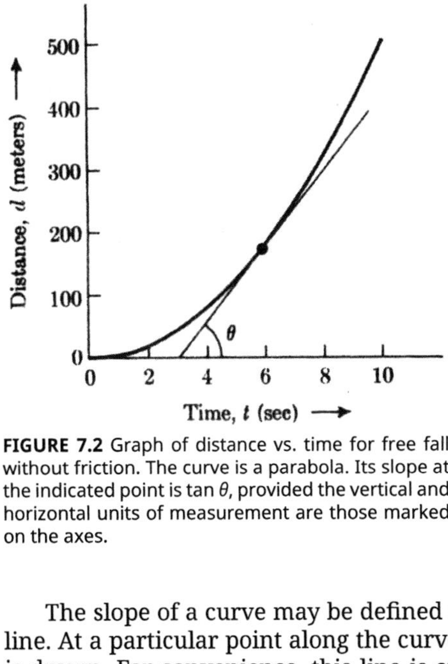

FIGURE 7.2 Graph of distance vs. time for free fall without friction. The curve is a parabola. Its slope at the indicated point is tan θ, provided the vertical and horizontal units of measurement are those marked on the axes.

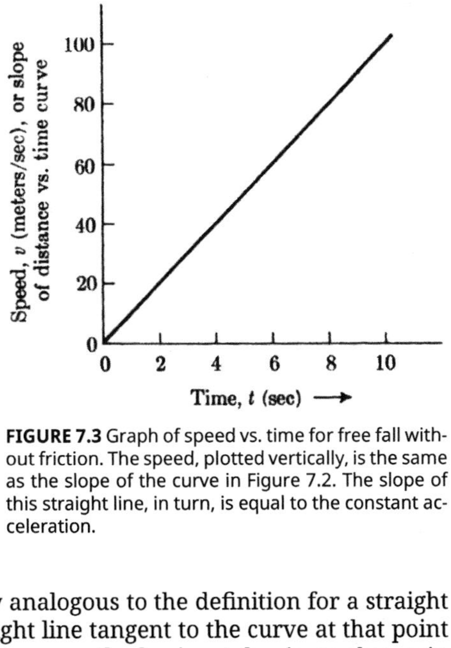

FIGURE 7.3 Graph of speed vs. time for free fall without friction. The speed, plotted vertically, is the same as the slope of the curve in Figure 7.2. The slope of this straight line, in turn, is equal to the constant acceleration.

The slope of a curve may be defined in a way analogous to the definition for a straight line. At a particular point along the curve, a straight line tangent to the curve at that point is drawn. For convenience, this line is extended to cross the horizontal axis, as shown in Figure 7.2, and the angle it makes with the horizontal axis is called θ. The slope of the curve at the chosen point is then defined to be tan θ. Its meaning is the same as for a straight line: rate of change of the quantity plotted vertically with respect to the quantity plotted

* In everyday usage, "parabola" often means any curve with the general shape of the curve in Figure 7.2. Mathematically a parabola is much more precisely defined. A graph of a function is a parabola if one of the variables is a quadratic function of the other.

horizontally. However, the slope is no longer a constant. The rate of change is different at each point. In Figure 7.2, the slope of the parabola at any point is equal to the instantaneous speed of the falling ball at that time. It is obvious from the figure that as time increases, the slope of the parabola increases, and therefore the instantaneous speed increases. Not obvious from the figure, but interesting and important, is the fact that the speed increases at a steady rate. If the slope of the curve in Figure 7.2 is plotted as a function of the time, a straight line results, as shown in Figure 7.3. This line in turn has a slope, which is a constant. This slope is the rate of change of speed with time, or acceleration, the same constant a that appeared in Equation 7.3.

A tabular presentation of time and distance for the falling ball is given in Table 7.2. From the numbers in this table we could verify, by testing each pair of tabular entries, that the time and distance relation follows the mathematical form of Equation 7.3, and that the constant a is equal to 9.8 m/sec^2. A particularly nice way to make this mathematical relationship pictorially evident is to plot d vs. t^2 rather than d vs. t. This procedure, as illustrated in Figure 7.4, "straightens out" the curve of Figure 7.2. It is, of course, much easier to recognize a straight line than a parabola by visual inspection. The slope of the line in Figure 7.4 may be measured. It is 4.9 m/sec^2.

TABLE 7.2 Time and Distance Data for Motion with Constant Acceleration

TIME (SEC)	DISTANCE (meters)
0	0
1	4.9
2	19.6
3	44.1
4	78.4
6	176.4
10	490.0

7.5 *Some special functions*

Besides the straight line and the parabola, two other remarkably ubiquitous curves in physics are the sine curve and the exponential curve. Both can be illustrated with simple physical examples.

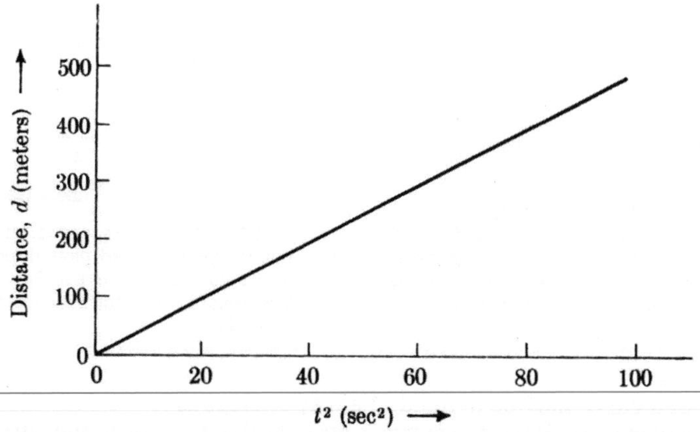

FIGURE 7.4 Graph of distance vs. the square of the time for free fall without friction. This graph makes clear the proportionality of d to t^2, which is not so obvious in the data of Table 7.2 or in the curve of Figure 7.2.

For a particular angle θ, we defined various trigonometric quantities, including sin θ. If the angle θ is regarded not as a constant, but as a variable, the trigonometric quantity sin θ becomes a *function*, simply called the sine function. The angle θ is called the argument of the function (it is also regarded as the independent variable). Moreover, there is no reason why the physical significance of the argument need be an angle. If we write the function

$$\sin x,$$

its argument x could be any dimensionless quantity, not necessarily angle. The expression is nevertheless given definite meaning with the help of an angle. If, for example, x has the numerical value 2, $\sin x$ means the sine of the angle 2 radians. Note that the generalization from angles to other quantities rests on an understanding that the radian shall be the standard unit of angular measurement. The idea of a trigonometric function that has nothing to do with angles and triangles is less mysterious than it might seem at first. It simply means that the mathematical essence has been distilled out of the original physical (or, in this case, geometrical) application. It is not very different than using the quadratic equation $d = \frac{1}{2}at^2$ to describe something other than the motion of a falling ball. In one application, d represents distance and t represents time. In another application, identically the same equation might appear, but with d and t representing completely different physical quantities—the same mathematics, but different physics.

A sine function is plotted in Figure 7.5. That it has the shape it does can be understood by thinking geometrically about the definition of the sine for an arbitrary angle (see Equation 6.16 and Figure 6.30). At zero radian, the sine is zero. As the angle increases to $\frac{1}{2}\pi$ (90 deg), the sine increases to a peak value of 1, after which it starts to decrease. At $\theta = \pi$ (180 deg), the sine has again reached zero. The sines of angles between π and 2π (180 deg and 360 deg) are defined to be negative. A negative peak, with value -1, is reached at $3\pi/2$ (270 deg). Beyond 2π, the pattern repeats, for the sine of, for example 370 deg, is the same as the sine of 10 deg. Therefore, the sine curve is an endlessly oscillating curve. The sine function is called a periodic function, with period 2π.

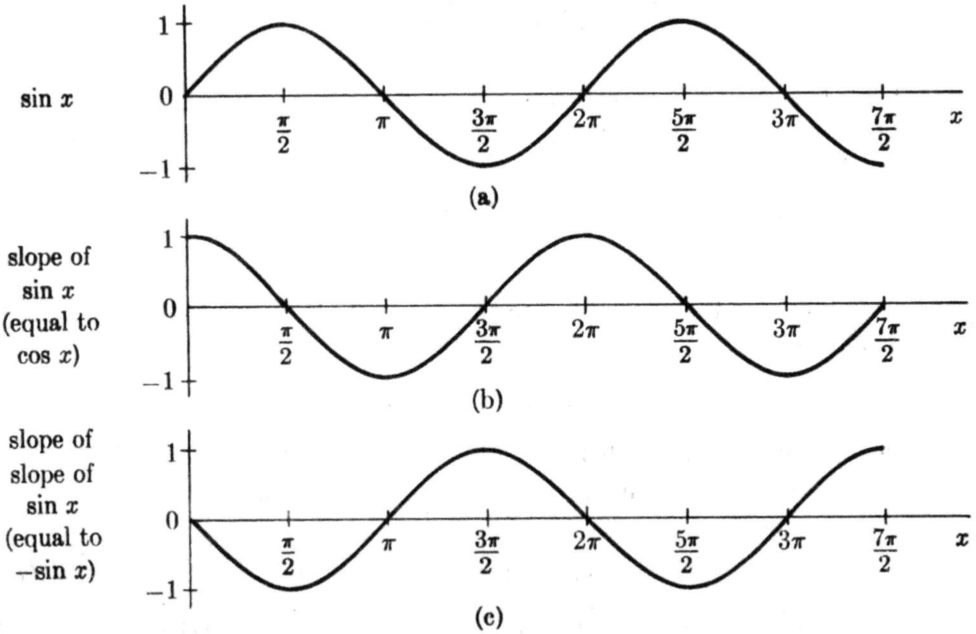

FIGURE 7.5 (a) Graph of $\sin x$ vs. x. The horizontal scale is expressed in radians. **(b)** The slope of the sine function is the same as the cosine function. Its graph is also the same as a sine function displaced horizontally to the left through $\frac{1}{2}\pi$. **(c)** The slope of the slope of the sine function is a negative sine function. Its graph is the same as a sine function displaced horizontally through π. Only the sine function and the closely related cosine function have these simple properties.

The reason for the special significance of the sine curve in physics is that many oscillatory phenomena in nature are described mathematically by the sine function. The curve in Figure 7.6, for example, plots the distance of a pendulum bob from its central position as a function of time. The equation for this curve is

$$d = A \sin \omega t, \tag{7.4}$$

whose two variables are d, the distance, and t, the time. The other two quantities, A and ω are constants. The quantity A, with the dimension of length, is called the amplitude of the oscillation. It is equal to the maximum value achieved by d at the limit of the swing. The constant ω, with the dimension of inverse time (so that the product ωt is dimensionless), is related to period T, the total time for one full oscillation back and forth. A full oscillation of the sine function is completed when the product ωt is equal to 2π. But at that moment, one period has elapsed, so $t = T$. Therefore, $\omega T = 2\pi$, or, solving for ω,

$$\omega = 2\pi/T. \tag{7.5}$$

This equation relates the constant ω to the period T. It means that the pendulum motion equation can also be written

$$d = A \sin (2\pi t/T). \tag{7.6}$$

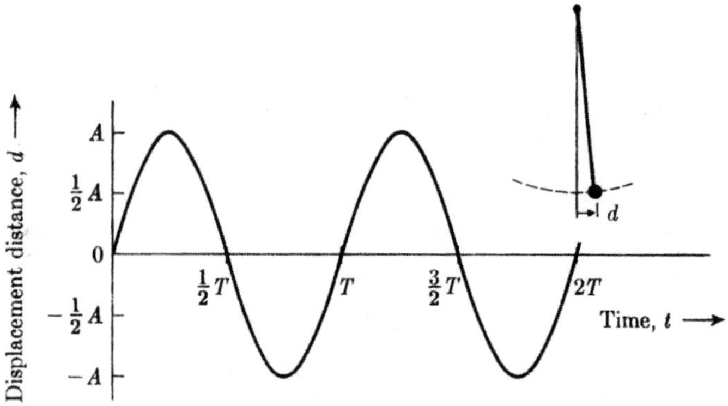

FIGURE 7.6 Graph of displacement distance vs. time for a pendulum swinging through a small angle. This is a sine curve, as expressed by Equation 7.4.

Two special mathematical features of the sine function are worth knowing. The first is a very simple and often useful approximation to the sine function for a small argument. It is

$$\sin x \cong x \text{ if } x \ll 1. \tag{7.7}$$

(The symbol $\cong$ means "approximately equals" and the symbol $\ll$ means "much less than.") An exercise at the end of the chapter asks you to demonstrate this fact. Graphically, this behavior of the sine function means that near $t = 0$, the curve in Figure 7.6 starts out approximately as a straight line following the equation

$$d \cong A\omega t.$$

Comparing this expression with Equation 7.1, you should be able to give the speed of the pendulum bob as it passes through the central point of its swing.

The other interesting feature of the sine curve is its slope. The curve starts out with an upward, or positive, slope. At its peak value, the slope is zero. Then the slope turns downward, meaning that it has a negative value. Wherever the sine curve is rising, its slope is positive; where it is falling, its slope is negative. At the extreme values, +1 and –1, it is flat—that is, its slope is zero. Therefore the slope of the curve is itself an oscillating function, but with its peaks and valleys displaced from those of the sine curve itself. In fact, the slope is also a sine function, displaced from the original sine function by exactly $\frac{1}{2}\pi$. Such a displaced sine curve is the same thing as a cosine curve. Sine curves and cosine curves have identically the same shape. In any graph of distance vs. time, the slope of the curve is equal to the speed of the motion. Therefore the speed of a pendulum bob varies sinusoidally, just as its position does.

An exponential function is one in which the variable appears in an exponent. Whereas t^2 is called an algebraic function, 2^t is called an exponential function. Any other constant raised to a variable power would also be an exponential function. For a particular reason of convenience that will be mentioned later, one particular constant has been adopted as standard to use in exponential functions. It is an irrational number, written *e*, whose approximate numerical value is 2.718. . . . Thus the standard exponential function is written

$$e^x.$$

Since any number raised to the zero power is unity, the value of the exponential function at $x = 0$ is 1. At $x = 1$, it has the value *e*, about 2.72. At $x = 2$, its value is e^2, about 7.39; at $x = 3$, its value is e^3, about 20.1; at $x = 10$, its value is e^{10}, about 22,000. The exponential function is characterized by a very rapid rise as *x* increases. It may also be readily defined for negative values of *x*, because of the rule,

$$e^{-x} = \frac{1}{e^x}. \tag{7.8}$$

Thus at $x = -1$, the exponential function has the value $e^{-1} = 1/2.72 = 0.368$. At $x = -10$, its value is $e^{-10} \cong 4.54 \times 10^{-5}$. Obviously, the exponential function falls very rapidly toward zero as the value of *x* becomes more negative. A graph of the exponential function e^x, simply called an exponential curve, is shown in Figure 7.7. For comparison, the reflected function e^{-x} is also plotted in the graph.

The idea of an exponent that varies smoothly through all values, instead of merely taking on integer values, may not be familiar. However, recall that the square root of *e* may also be written $e^{0.5}$; the fourth root of *e*, $e^{0.25}$; and so on. Then $e^{0.6}$ means the square root of *e* multiplied by the tenth root of *e* ($e^{0.5} \times e^{0.1} = e^{0.6}$). It is an easy generalization to allow an exponent to take on any real numerical value. As illustrated in Figure 7.7, the result is a smoothly varying function, continuously joining the already familiar exponentials e^2, e^3, e^4, and so on.

Since the exponential function ascends ever more rapidly as *x* increases, the numerical value of the slope grows as the numerical value of the function grows. In fact, the exponential function has an important property that it shares with no other function: Its slope is directly proportional to its value. If one point on an exponential curve has twice the magnitude of another point, the slopes at these two points also differ by a factor of two. This is true of any exponential functions, such as 2^x or 3^{5t}. The particular exponential function e^x has an even more special property. Its slope is exactly equal to its value. Thus, in Figure 7.7, at $x = 0$ where the value of e^x is 1, the numerical value of the slope is also 1. It is for this reason that the constant *e* has been adopted as the standard constant for exponential functions.

The exponential function is encountered often in physics, and in chemistry and in biology as well. An example of great importance in the modern world that involves the decreasing exponential is the decay of a sample of radioactive material. The intensity of the radioactivity of a given sample is plotted as a function of time in Figure 7.8. The curve in this figure is a graph of the function

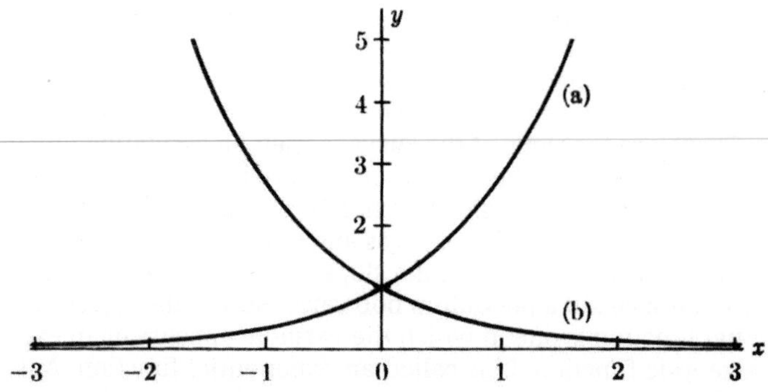

FIGURE 7.7 Graphs of exponential functions. **(a)** Rising exponential, e^x. **(b)** Falling exponential, e^{-x}.

$$I = I_0 e^{-t/\tau}, \tag{7.9}$$

whose variables are the intensity I and the time t. The constant I_0 is the initial intensity, since at $t = 0$, $I = I_0$. The constant τ, with the dimension of time, is called the mean time of the decay. When the time t reaches the value τ, $I_0 e^{-1} \cong 0.368 I_0$. Therefore τ is the time required for the intensity to drop to 36.8% of its initial value. It is actually the mean time or average time of the decay in the following sense. The radioactive sample contains, initially, a large number of radioactive nuclei. Each of these undergoes decay after some time interval that is unpredictable and perhaps different for every nucleus. If the elapsed time until decay were measured for each nucleus separately, and then all of these different times were averaged, their average would be the mean time τ. It should be mentioned that what is meant by intensity is the number of decays taking place per unit time.

Another characteristic time associated with an exponential decay curve is the half life, $t_{½}$. This constant is defined as the time required for the intensity to decrease to one-half its initial value. It is obviously a time somewhat less than the mean time. The exact relationship is

$$t_{½} = 0.694\tau. \tag{7.10}$$

An important mathematical property of the exponential curve is that the time required for the curve to fall from any value to half that value is the same for all parts of the curve. Stating the same property in another way, the mean decay time is the same regardless of the choice of initial time.

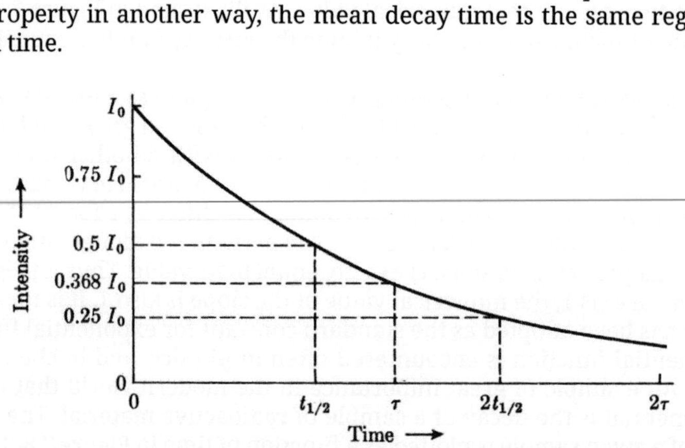

FIGURE 7.8 Exponential decay of intensity for sample of radioactive material. The mean life is τ; the half life is $t_{½}$.

As a final example for this section, we wish to consider a graph whose horizontal coordinate is something other than time, and whose mathematical form is not necessarily known. In Figure 7.9 is plotted the light intensity emitted by a thin phosphor screen (similar to the face of a TV tube) as a function of the speed of the electrons striking the screen. This curve is not based on accurate data for any particular screen; it shows only the general form of such a function. It is typical of what might be called an experimental function, as opposed to a mathematical function. Measurements reveal that the light intensity depends on the electron speed in a certain way, but the physical mechanism of the light emission is sufficiently complicated that the precise mathematical form of the dependence is unknown. Nevertheless, much can be learned about the phenomenon by examining the curve. For small values of the speed, the intensity seems to grow approximately quadratically; the lower part of the curve resembles the parabola of Figure 7.2. As the electron speed grows further, the intensity no longer continues its parabolic rise, but straightens out into an approximately linear dependence. Finally the curve bends in the other direction and seems to approach a horizontal line at very high electron speed. Further increases of electron speed result in very little further increase of light intensity. These conclusions could also have been reached by careful examination of tabular data, but obviously the graphical presentation is very advantageous for giving a clear overall view of the dependence of one variable on another.

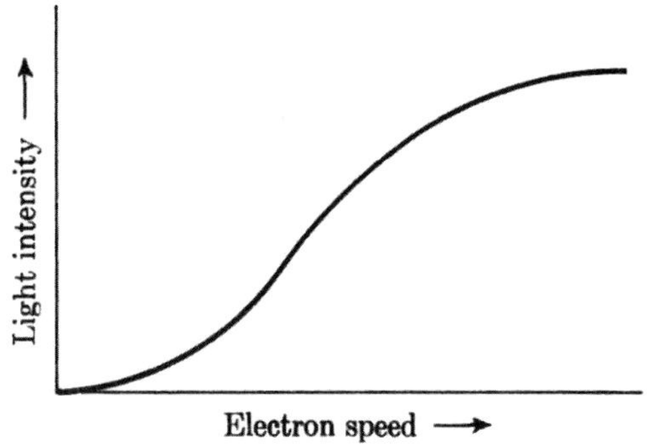

FIGURE 7.9 Approximate graph of light intensity from a thin phosphor screen vs. speed of electrons striking the screen. Graphs can give a quick overall view of a process whose details are complicated.

7.6 *Probability, experimental error, and uncertainty*

Probability is a big subject in mathematics and in physics. Here we wish to discuss only one aspect of probability: that associated with experimental error. Perfect straight lines and perfect parabolas may exist in physical theory, but they do not exist in physical measurement, for perfect precision of measurement is impossible. Where there is measurement, there is error, and where there is error, the concept of probability plays a role. 24

Basically, probability is concerned with ignorance. Probability can be discussed quantitatively because even about ignorance, exact statements can be made. Although we do not know in advance the result of the toss of a coin, we do know that the probability of heads is 0.5. Insurance companies, ignorant of the fate of any policyholder, can nevertheless set premiums intelligently. A gambling casino, ignorant of the outcome of any game, can nevertheless adjust odds to assure a profit. A physicist, ignorant of the lifetime of any particular muon, can nevertheless predict accurately the average lifetime of a collection of muons, and predict the exponential decay of the aggregate. In an experimental measurement, the thing we are ignorant of is the magnitude of the error. By chance a particular measurement might be precisely correct. Or it might have a significant error. In order to make a meaningful comparison between experiment and theory, the physicist must be able to assess the probable magnitude of the error.

Suppose a college catalogue states that the area of a single student's room is 160 ft^2, and a student decides to check this number. He measures the length of the room and its breadth and multiplies these two numbers together to obtain, perhaps, 156 ft^2. Has he demonstrated that the college catalogue is wrong? Not necessarily. It depends on the uncertainty of his measurements. Determination of the uncertainty is usually not easy, and may amount to little more than an educated guess. He might guess, for instance, that his length and breadth measurements are accurate to within about 2%. This would imply that his area determination should be accurate to within 4%. He would then write down for his experimentally determined area,

$$(156 \pm 6)\ \text{ft}^2,$$

indicating his opinion that the true area probably lies somewhere between 150 ft^2 and 162 ft^2. Therefore his measurement would be consistent with "theory," as set forth in the catalogue.

A somewhat better, but still not foolproof, way to determine uncertainty is by repeated measurement. If several measurements of the area of the room agree to within 6 ft^2, the student can be more confident that the correct value lies within 6 ft^2 of his first measurement. However, repeated measurement reveals only random error. Another kind of error, called systematic error, could result from a defective ruler or a consistently faulty measuring technique. These sources of error would act always in the same direction rather than randomly in both directions. Having assessed all sources of error, the physicist tries to assign an uncertainty defined as follows: It should bracket a range of values within which the correct value has a 66% chance to lie, and outside of which the correct value has a 34% chance to lie. Thus, if it had been possible to define the uncertainty with this much certitude for the room area measurement, the expression (156 ± 6) ft^2 would mean that with a probability of 0.66, the true area has a value between 150 ft^2 and 162 ft^2, but with a probability of 0.34, might be less than 150 ft^2 or greater than 162 ft^2.

For only one kind of error, called statistical error, is it actually possible to determine the uncertainty of a measurement accurately. Statistical error results from the measurement of purely random events. It might be determined, for example, that in a particular radioactive sample, exactly 100 nuclei decay in a one-minute interval. In one sense, this measurement is completely free of error, for we may suppose that the number 100 was determined with complete precision. In another sense, however, it does have "error," for it may differ from the average number of decays in any other identical samples. The quantity of real physical importance is the average number of decays per minute in a large number of identical samples. It is this average rather than the actual number of decays for a particular sample that can meaningfully be compared with theory. Thus the idea of error enters in seeking an answer to the question: By how much is the average number of decays for all such samples likely to differ from the actual number measured for this sample? With this understanding of what is meant by the "error" of a precise measurement, a very simple answer to the question can be given. The uncertainty associated with the measurement of n random events is $\sqrt{n}$. If a particular radioactive sample undergoes n decays in a given time interval, the average number for all such samples can be said to be equal to

$$n \pm \sqrt{n}, \tag{7.11}$$

that is, the correct average has a 66% chance to lie within the range $n - \sqrt{n}$ to $n + \sqrt{n}$. If one student in a laboratory measures 100 decays in a minute, he predicts that the average number measured by all of the students working with identical samples will be 100 ± 10. There is a good chance (probability two-thirds) that the true average lies between 90 and 110. This also means that about two-thirds of the students should record values within this range.

The discussion to this point has concerned experimental errors associated with single measurements. Experimental error also plays a very important role in the determination of functional relationships. A curve cannot sensibly be drawn precisely through each experimentally determined point in a graph, for the experimental errors in the measurements are reflected in an uncertainty of location of the points in the graph. Suppose that we wish to determine the functional relationship implied by the following set of time and distance data for a moving particle.

TIME	DISTANCE
1	280
2	450
3	1,000
4	1,300
5	1,450
6	1,840

Suppose further, for the sake of illustration, that the time measurements are so precise that their errors can be ignored and that the distance measurements are rather crude, with the uncertainty of each estimated to be ±70 cm. Figure 7.10 shows the wrong way and the right way to handle these data graphically. In the wrong way of Figure 7.10(a), the measured values are simply joined by straight-line segments. This is wrong first because the experimental errors mean that the plotted points need not represent the actual position of the particle; second because the data provide no evidence about exactly how the motion occurred between the times of measurement. In the right way of Figure 7.10(b), the errors are incorporated into the graph by attaching to each datum point an error line extending 70 cm above and 70 cm below the point. These lines at once provide a picture of where the particle most probably was at each time. Next a smooth line or curve is drawn among the points with no more details of shape and curvature than is justified by the magnitude of the uncertainties. In this example, a straight line adequately fits the data, and it would not be justifiable to seek a "better" fit with a curve. This, of course, does not mean that the straight line necessarily represents the actual motion accurately. We can say only that within the accuracy of the measurements, it is consistent to make the simplest assumption, namely, that the motion occurred with uniform speed such that the distance vs. time graph is a straight line. If the uncertainty of ±70 cm indeed spans a range of values containing the correct distance value

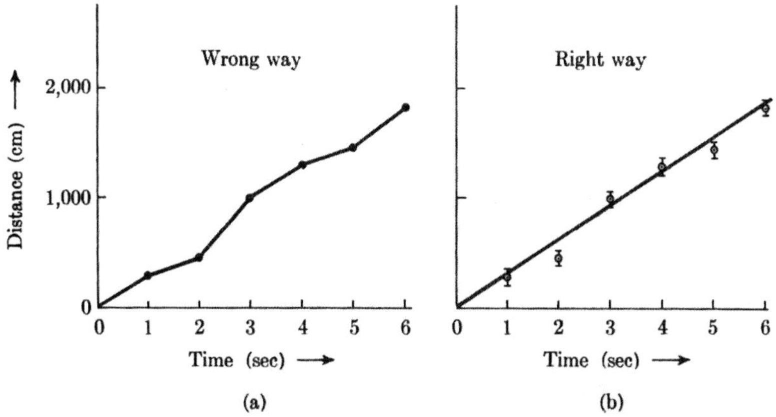

FIGURE 7.10 (a) The wrong way and **(b)** the right way to present experimental data graphically. In this example, the experimental uncertainties are sufficiently large that it is not justified to assume that the graph is anything other than a straight line.

with 66% probability, the line drawn among the points should cut through approximately two-thirds of the error lines and miss one-third of them. The fact that the straight line in Figure 7.10(b) cuts four error lines and misses two tends to support the assumption that the straight line provides an adequate representation of the data. The slope of the straight line then provides a value for the speed of the particle. In Figure 7.10(b), the slope of the line is 300 cm/sec. It is left as a problem to estimate the uncertainty of this speed determination.

We end this chapter with some remarks about practical calculations, and some words of advice. Power and insight in science can come only through the proper balance between qualitative understanding and mathematical description. Mathematics provides, obviously, the power of science. It provides also an important part of the insight into the true nature of science, for much of the beauty of science springs from the remarkable mathematical simplicity of some of its fundamental laws. On the other hand, the mathematics of physical laws is itself only the skeleton of physics, sterile without the flesh and blood of ideas, pictures, and measurements. The special problem for the student, especially the student without a strong background in mathematics, is to separate the difficulties of conceptual understanding from the difficulties of practical calculation. It is wrong to assume that the ideas of physics can be understood and assimilated without calculation. It is equally wrong to assume that a mathematical difficulty in a particular example reflects an underlying difficulty of the basic law being applied. In this book an effort is made to avoid irrelevant or particularly complicated examples and problems. The goal of conceptual understanding is paramount. Equations, algebraic manipulations, and numerical calculations are introduced only where they seem to enhance understanding of the most important laws and ideas.

The advice is this. Think about calculations with the same degree of intensity applied to thinking about ideas expressed verbally. Profit from numbers. Every practicing scientist learns to think pictorially about mathematics, and to think mathematically about pictures (graphs). For the student, learning to bridge the gaps rapidly among tables and graphs and equations is a talent worth developing. A single word or phrase can evoke a host of references to the past, tapping a lifetime of experience in the use of words. For most students, there is nothing like a comparable background of mathematical experience that can enrich the meaning of an equation or a numerical result. That background must accumulate gradually, and at first must be forced into existence by deliberate effort. A number should stimulate comparison, an equation should evoke a picture, a calculation should invite tests of its "reasonableness." To learn that a slow neutron moves at 2,000 m/sec is to learn very little if that speed is not placed mentally in a framework of familiar speeds. How does it compare with the speed of an airplane, or of an astronaut, or of a photon? An equation such as $d = vt$ is a cold fact if not warmed by the image of a moving object, or at least of a straight-line graph. A calculation is a risky undertaking if not done with a wide-awake mind asking whether it makes sense. It would be easy to calculate (mistakenly) that 1 mile is equal to 0.6 km. Does it make sense? No. The foreign traveler might recall that in Europe an automobile covers more kilometers in an hour than it covers miles at home. The sports enthusiast might recall that the 1,000-meter (1 km) run is shorter than a mile. The laboratory student should remember that a meter stick is about one yard long, so that 1,000 meters add up to about 3,000 feet, less than the 5,280 feet in a mile.

Despite the great importance of thinking about equations and their interpretations, there is also a time to turn the mathematical crank. An important part of the power of science stems from the fact that deduction and derivation can be carried out according to a set of mathematical rules without the necessity of careful thinking in physical terms about the meaning of each step. In practice, the student is more likely to err on the side of too little thinking about equations than too much. Nevertheless, worth reflecting upon is Alfred North Whitehead's emphatic statement, quoted in Exercise 1.13, on the merits of turning the crank.

EXERCISES

7.1. **(1)** What is the dimension of momentum? **(2)** What is its cgs unit? (3) Show that the product of force and time has the same dimension as momentum.

7.2. Demonstrate that the dimension of force times speed is the same as the dimension of energy divided by time. Use this equality to express the cgs force unit (the dyne) in terms of the cgs energy unit (the erg) and length or time units.

7.3. What is the speed of an astronaut in miles/hr if he is traveling 24,500 ft/sec? (Introduce conversion factors and carry out the units arithmetic as well as the numerical arithmetic.)

7.4. Re-express each of the following quantities in cgs units: **(a)** 186,000 miles/sec; **(b)** 1 yard; **(c)** 10^{-6} century; **(d)** 1,000 ft^3/hr; **(e)** 32 ft/sec^2.

7.5. Suppose that force, mass, and time are adopted as fundamental dimensions. Express length as a combination of these three.

7.6. In the nineteenth century the English physicist Stokes studied the motion of small spheres moving slowly through liquids. He was able to show theoretically that the "drag force" of the liquid on the sphere is given by the expression

$$F_{\text{D}} = (6\pi)\eta v r,$$

where F_{D} is the drag force, v is the speed of the sphere, and r is the radius of the sphere. The constant η is called the *viscosity* of the liquid. What is the dimension of η? What is its cgs unit of measurement? (Because viscosity is an important characteristic of a fluid, this unit has been given a separate name, the *poise*.)

7.7. The price of gasoline is $0.30 per gallon. Write two equations relating total cost C in dollars to the number of gallons G, first one that is valid only for this particular price, then one that is generally valid for any price per gallon.

7.8. In many universities the credit-hour is a unit of measurement that is approximately equal to 50 minutes of classroom exposure per week for 15 weeks. Re-express the credit-hour in units of classroom-seconds.

7.9. Mention half a dozen different units that you commonly employ (either in everyday life or in other courses of study) which are not cgs units. Find the factors that convert any two of these to the corresponding cgs unit.

7.10. One gallon of water has a mass of 8 pounds. One cm^3 of water has a mass of 1 gm. The cgs to English conversion factor for mass is 454 gm/lb. Express the volume of one gallon in cm^3.

7.11. Sketch on a graph straight lines whose slopes are 0, 0.5, 1, 2, and –1.

7.12. Sketch a curve whose slope varies continuously from zero to infinity.

7.13. On a graph, the surface area of a reservoir is plotted as a function of time. What would be a suitable unit of measurement for expressing the slope of this curve?

7.14. **(1)** Sketch a curve that represents the volume of a cube as a function of the length of one side of the cube. **(2)** What is the physical dimension of the slope of this curve?

7.15. A bug crawls to the right at a speed of 3 cm/sec for 2 sec, then stops for 2 sec, then crawls again to the right at 2 cm/sec for 2 sec, then reverses course and scampers back to its starting point at 5 cm/sec. **(1)** Draw a graph of position vs. time for the bug. **(2)** Draw a graph that presents the slope of the position-vs.-time graph as a function of time.

7.16. What is the physical significance of the slope of the curve in Figure 7.6? Sketch a graph showing this slope as a function of time. (Your graph should be neat and clear, but you need not provide labeled scales. The general shape of the slope-vs.-time curve is what should be displayed.)

7.17. Plot distance vs. time for the data in this table. Describe the motion in words. What is the speed of the motion near t = 8 sec?

t (sec)	s (cm)
0	0
1	250
2	1,000
3	2,250
4	4,000
5	6,000
6	8,000
7	10,000
8	12,000

7.18. Work from the basic definition of an angle expressed in radians (Figure 6.28) and of the sine of an angle (Equation 6.16 and Figure 6.30) in order to demonstrate that a small angle and its sine are nearly equal, as stated in Equation 7.7.

7.19. Sketch the slope of the curve in Figure 7.8 as a function of time. (Pay attention to sign.) What is the nature of the new curve you have drawn? What would be the appearance of *its* slope?

7.20. What is suspicious about the representation of experimental uncertainty in this graph?

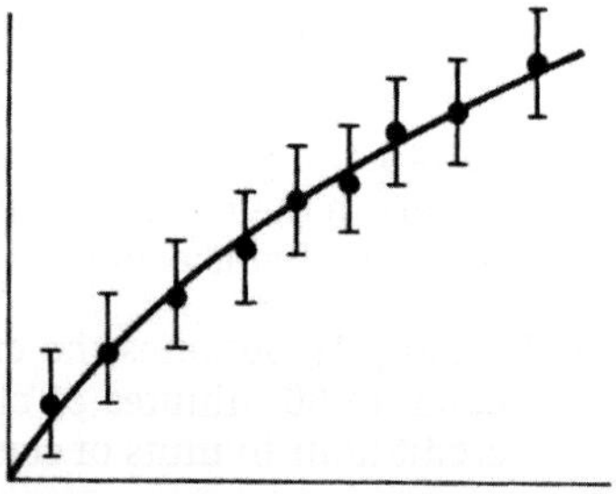

7.21. If $\sqrt{n}$ is the uncertainty associated with the statistical error in the measurement of n random events (Equation 7.11), what is the *fractional* uncertainty? By what fraction must the number of counts n be increased in order to halve the fractional uncertainty?

7.22. Estimate the uncertainty of the slope of the straight line graph in Figure 7.10(b). Use any method you like to do this, but explain how you reach your conclusion.

7.23. In a system of units in which the speed of light is equal to 1, both time and distance can be measured in seconds. Express each of the following distances in seconds: (a) your height; (b) one mile; (c) one light-year.

7.24. Let speed and acceleration be chosen to be the fundamental dimensions of kinematics. (1) Express length and time as combinations of speed and acceleration. (2) Suggest possible units of speed and acceleration that might serve as world standards. (Good standards should be, as nearly as possible, invariably constant, and it should be relatively convenient for scientists to compare measurements with the standards.)

7.25. The table shows the mass of a particular child as a function of his height over the years. **(1)** Graph these data, plotting height horizontally and mass vertically. Join the plotted points by a smooth curve. **(2)** Which part of the curve has maximum slope? What is the approximate magnitude of this slope? **(3)** Extend your curve beyond the range of the data in order to estimate the mass of the unborn child when his height was 35 cm and his later mass should he grow to a height of 175 cm.

HEIGHT (cm)	MASS (kg)
50	3.5
75	11
100	20
125	31
150	45

7.26. The pendulum of a clock swings so that the horizontal displacement of its bob is given by the equation

$$d = A \sin \omega t.$$

The amplitude of the motion is 5 cm, and the period is 1 sec. **(1)** Using graph paper, make a careful plot of d vs. t for t between 0 and 1.5 seconds. **(2)** Use this graph to find approximate value of the speed of the bob at $t = 0$, ⅛ sec, ¼ sec, and ½ sec. **(3)** What is the ratio of speeds at $t = 0$ and $t = 1$ sec?

7.27. The velocity of a racing car during a time trial is given by

$$v = A(1 - e^{-t/\tau}),$$

where $A = 150$ miles/hr and $\tau = 6$ sec. **(1)** Make a plot of v vs. t on graph paper for t from 0 to 10 sec. **(2)** Considering the whole range of t from 0 to $+\infty$ answer the following questions:

(a) What is the minimum speed? When attained?
(b) What is the maximum speed? When attained?
(c) What is the minimum acceleration? When attained?
(d) What is the maximum acceleration? When attained?

(NOTE: Completion of this exercise requires numerical values of the exponential function.)

7.28. An object starting from rest at point A slides down a nearly frictionless track. Its speed is measured at point B one meter from A and at point C two meters from A. The results of four trial slides are shown in the table. **(1)** Estimate the approximate random percentage error of the measurements. **(2)** Find the ratio of speeds v_C/v_B for each trial. Is it consistent to assume that the ratio is a constant within experimental error? (3) Hypothesize an equation of the form $(v_C/v_B) = (x_C/x_B)^n$ relating the speed ratio v_C/v_B to the distance ratio x_C/x_B (which in this instance is 2). If you believe in the simplicity of the laws of nature, the exponent n should be a "simple" number. What value of n fits the experimental results?

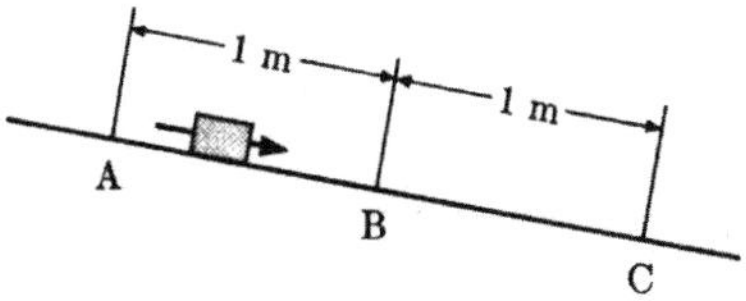

TRIAL NO.	SPEED AT B (cm/sec)	SPEED AT C (cm/sec)
1	24.8	35.0
2	24.2	34.5
3	25.0	34.8
4	24.6	35.2

Part 3
Mechanics

ISAAC NEWTON
(1642 – 1727)

"If we now imagine bodies to be projected in the direction of lines parallel to the horizon from greater heights, . . . those bodies . . . will describe arcs either concentric with the earth, or variously eccentric, and go on revolving through the heavens in those orbits just as the planets do in their orbits."

"Nature is pleased with simplicity, and affects not the pomp of superfluous causes."

EIGHT

The laws of motion

In the foregoing chapters, many of the concepts and ideas of mechanics and some of its laws—the conservation laws—have been presented. It is the task of this part of the book to sharpen the definition of some mechanical concepts, to assemble the concepts and laws into the coherent structure called the theory of mechanics, and to apply this theory and its laws to the description and the explanation of various kinds of motion.

8.1 The conceptual basis of mechanics

Mechanics is usually regarded as the first great theory of physics, whose development in the seventeenth century ushered in the modern scientific era. Indeed this is true. Never before had mathematics and careful observations of natural phenomena been welded together into a single general structure of explanation, prediction, verification, and comprehension. However, no giant forward step in science, not even the "first," can occur without antecedents. Mechanics rested upon two thousand years of mathematics and observational astronomy, upon the Copernican doctrine of a sun-centered universe (1543), upon Stevin's understanding of the nature of force (1605), and—not least—upon the new spirit of intellectual inquiry engendered by the Renaissance and the Reformation. Even aside from its antecedents, the theory of mechanics evolved over a period of many decades. The boundaries in time of this evolution are ill defined, but its major development occurred between the time of Kepler's laws of planetary motion in 1609-1618 and Newton's law of universal gravitation, first published in 1687. Because the history of mechanics is tied so closely to the history of knowledge of the solar system, a discussion of the background of mechanics and of its culminating development in the seventeenth century is postponed to Chapter Twelve, on gravitation. In this chapter and in the next three, we shall be concerned principally with mechanics as a logical structure and as a tool for understanding motion in general.

If the simplicity of a branch of science is measured in terms of the number of its basic concepts, there are branches of science simpler than mechanics. Mechanics is built on the three cornerstone concepts of length, mass, and time. (Three other independent concepts would serve as well—for instance, length, time, and force; or force, energy, and momentum.) Geometry, on the other hand —if regarded as a branch of science, which it has a very good claim to be—can be founded on only a single primitive concept, length. Its derived concepts are area and volume and angle.

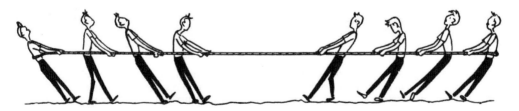

FIGURE 8.1 Example of static equilibrium. The rope experiences no net force.

If to the concept of length the concept of time is added, a basis is formed for the branch of science called kinematics—the description of motion. Its important derived concepts are velocity and acceleration. Sometimes it is said that kinematics *describes* motion, whereas mechanics, or dynamics, *explains* motion. This distinction, although it conveys roughly the right idea, is not very meaningful. As emphasized in Chapter One, it is impossible to distinguish clearly between description and explanation. It is more accurate to say that mechanics provides a deeper, more complete, and more satisfying description of motion than does kinematics. Equally well, one might say that kinematics, limited to the concepts of space and time, *explains* motion much less satisfactorily than does mechanics, which gains its extra power from the introduction of a third basic concept. Much of Galileo's careful work on motion in the first part of the seventeenth century was concerned with the space-time description of motion. He discovered the kinematics of freely falling objects, of balls rolling down inclined planes, of pendulums, and of projectiles. Complementing Galileo's terrestrial kinematics, Kepler, during the same period, was discovering the accurate geometry and kinematics of planetary motion.

Besides kinematics, one other subdivision of mechanics is based on only two primitive concepts. The concepts are force and distance, and the branch of science is called statics. Statics, with origins going back at least to Archimedes in 250 B.C., is concerned with the balancing of forces acting on an object at rest. Figure 8.1, showing two teams exerting equal and opposite forces on a motionless rope, illustrates a simple example of static equilibrium. In modern engineering, the design of a bridge or a dam requires the use of statics, for the engineer wishes to know the magnitude and the direction of the force acting on every part of the structure. The fundamental principle of statics, that the total force acting on a motionless object must be zero,* seems simple enough, but, like the basic facts of kinematics, it was not thoroughly clarified until the seventeenth century. Part of the reason for the lapse of almost 2,000 years between Archimedes' pioneering work and the modern view of statics initiated by Stevin is the fact that force is a vector quantity. The correct law of addition of forces was presented first by Stevin in 1605, but a full appreciation of the vector as a mathematical object came only in the nineteenth century.

The fundamental concepts underlying kinematics and statics together provide an adequate basis for mechanics.† The power and generality of mechanics as compared to kinematics springs in large part from its description of motion in terms of force, and its introduction of mass—the inertial property—as the connecting link between force and motion. Despite the great historical importance of the concept of force and its continuing importance for dealing with macroscopic motion, it is interesting that in modern physics, force has become a very unimportant concept. For describing the interaction among elementary particles, it is entirely unnecessary. Nevertheless, numerous other concepts of mechanics, such as mass, momentum, energy, and angular momentum (besides space and time) remain as important in contemporary physics as they were in the early development of mechanics. We have seen that all of these are key concepts for describing particles.

The nature of scientific theory was discussed in Chapter One, and there mechanics was included in the list of "great theories." In simplest terms, a great theory of physics is a satisfying harmony between mathematical abstraction and experimental observation. It is "satisfying" in two different ways, first quantitatively, second esthetically. The physicist is satisfied with nothing less than precise quantitative accord between prediction and observation. He is also satisfied only with a theory whose basic concepts and equations are simple and

* There is one other principle of statics, that the total torque acting on a motionless object must vanish. Torque is defined and discussed in Section 10.5. Zero total force ensures that a motionless object does not acquire a net bulk motion in any direction. Zero total torque ensures that it will not acquire rotational motion.

† Sometimes the word "dynamics" is used for the theory of motion. In this text we shall use the word "mechanics," now more common among physicists.

few. On both counts, mechanics deserves the title "great." Although we know observational limits in which it ceases to fit the facts, it successfully and accurately describes motion over an enormous range from the atomic to the cosmological; and it does so with an arsenal of concepts and equations simple enough to be grasped completely by the beginning student of physics.

In posing and solving problems in mechanics, we treat them often as mathematical exercises. Such problems are nevertheless an important part of physics. The solution of each one brings into view another aspect of motion that was latent in the laws of mechanics. Moreover, problem solving forms the basis for the experimental foundation of mechanics. It is never a fundamental law that is tested directly, rather some specific implication of that law. For example, the laws of mechanics do not state directly that planets move in elliptical orbits. It was a "textbook problem" solved first by Newton to show that the law of motion and the law of gravitational force together imply elliptical orbits. Another, and simpler, textbook problem is to find the speed acquired by a particle upon which a known force has acted for a known time. Early in this century, experimental observations for high-speed electrons were found to disagree with the theoretical solution to this problem. Thus was discovered the first limitation to the scope of mechanics.

8.2 Kinematics

The primitive concepts of kinematics are two: space and time. Constructed from the space concept is the position vector. The important derived concepts of kinematics are velocity and acceleration. Velocity is the rate of change of position with respect to time; acceleration is the rate of change of velocity with respect to time. Both are vector quantities. Although further derived quantities, such as rate of change of acceleration, could be introduced, they prove to be unnecessary. The complete theory of mechanics requires no kinematic concepts more abstract than acceleration. Acceleration is already far enough removed from the immediacy of sense perceptions to offer some difficulties of visualization. Position and time are primitive concepts not only in the technical sense of their operational definitions, but also in the human sense—they are intuitive ideas directly linked to observation. Velocity we do not sense directly; nevertheless we have learned through common experience to understand it and to think intuitively about it. What we actually observe to give ourselves an idea of velocity are successive positions of an object as time passes, or the position of a speedometer needle, or the flashing lights in an elevator, or the force of wind in an open car. Acceleration, one stage further removed from direct observation, requires practice to think about and appreciate, even though the human body reacts directly to it, as in a turning automobile or a pitching ship. In thinking about the acceleration of an object in the external world, two things are helpful. One is to think graphically, picturing the motion in terms of curves and diagrams. Another is to exploit the fact that acceleration bears the same relation to velocity as velocity bears to position, so that the less familiar relationship can be compared to the more familiar. For example, the fact that an object can be accelerated even though its speed does not change can be understood by noting that an object can have a velocity even though its distance from a fixed point does not change.

In most of this section we shall be concerned with straight-line motion, or one-dimensional motion. For convenience, we suppose that we are dealing with the motion of a particle along the x-axis. Its distance from the origin will be called x, its velocity v, its acceleration a, and the time (also referred to any convenient origin of time) t. Actually, x, v, and a represent the x-components of vector quantities. It might appear that for a vector directed along the x-axis, its x-component and its magnitude are identical. This is not quite true. The magnitude of a vector is always positive, whereas its component may be either positive or negative. This can be seen most easily by examining the equation for a vector component. If $\mathbf{s}$ is the position vector, its x-component is

$$x = s \cos \theta ,$$

where θ is the angle between **s** and the x-axis. If **s** is directed to the right along the x-axis, $\cos \theta = +1$, and $x = s$. If **s** is directed to the left along the x-axis, $\cos \theta = -1$, and $x = -s$. Although the magnitude s is positive, the component x may be positive or negative.

For one-dimensional motion, it is convenient to plot x as a function of t, v as a function of t, and a as a function of t. Figure 8.2 shows such graphs for an example of motion that follows no simple mathematical law. In the graph of x vs. t, the slope of the curve is equal to the velocity. A rising curve means positive velocity (motion to the right); a falling curve means negative velocity (motion to the left). Where the slope of the curve is zero (horizontal), the velocity is zero. This means that examination of the position curve makes it possible to plot beneath it the velocity curve. (To do so exactly, of course, would require precise measurements of slopes. Recall that the slope is defined as a tangent of the angle between the horizontal axis and a line drawn tangent to the curve.) The next step, to the acceleration curve, can be taken in the same way. Where the velocity curve has positive slope, the acceleration is positive, and so on. To summarize: The slope of the position curve is equal to the height of the velocity curve; the slope of the velocity curve is equal to the height of the acceleration curve.

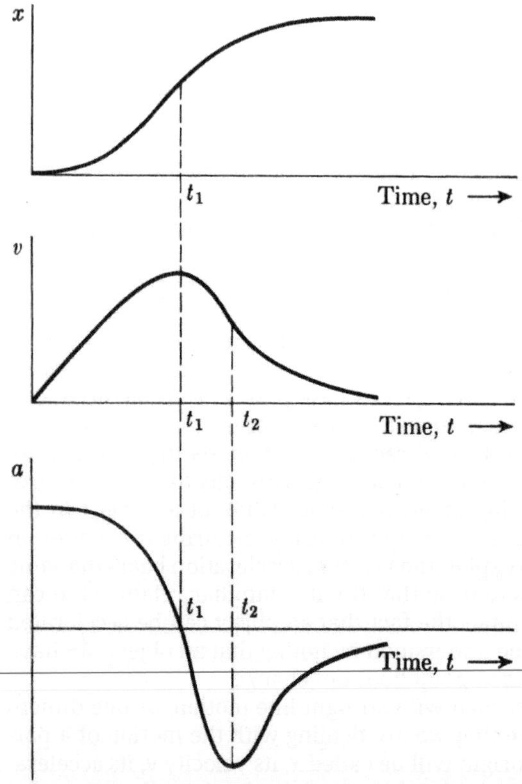

FIGURE 8.2 Graphs of position, velocity, and acceleration for example of motion in one dimension. Each of the quantities plotted vertically is actually the component of a vector quantity. Vertical dashed lines are drawn at t_1, the time of maximum positive velocity and zero acceleration, and at t_2, the time of maximum rate of decrease of velocity (peak negative acceleration).

Having achieved the acceleration curve in two steps, one may compare it directly with the position curve. There is a simple connection. When the acceleration is positive, the position curve has positive curvature. This means it is shaped like part of a right-side-up cup. When the acceleration is negative, the position curve has negative curvature (like an upside-down cup). When the position curve is nearly a straight line, rising or falling at a uniform rate, it has little or no curvature, and the acceleration is near zero. Another way to express the connection is this. Positive acceleration makes the particle "want" to move more rapidly to the right; it makes the slope of the position curve rotate counterclockwise, the direction of approach to a large positive slope. For negative acceleration, the slope of the position curve rotates clockwise, a direction that will, if it proceeds far enough, produce a large negative slope and high speed to the left.

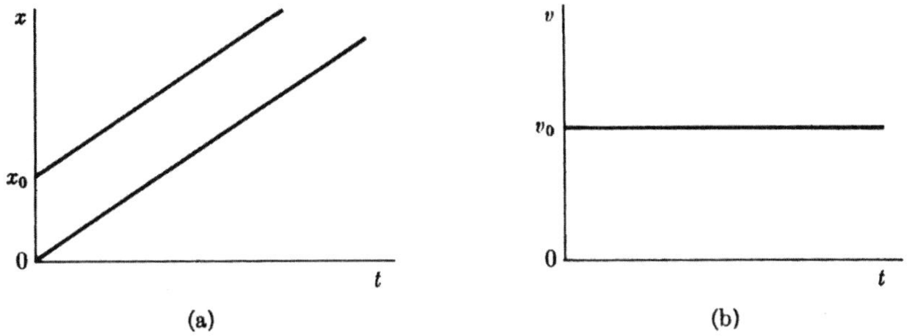

FIGURE 8.3 Constant-velocity motion. **(a)** Graphs of position vs. time for both the special case ($x = 0$ when $t = 0$) and the general case ($x = x_0$ when $t = 0$). **(b)** Graph of velocity vs. time, the same constant for both the special case and the general case.

After this look at one-dimensional motion in general, we return to two special kinds of motion, which were discussed in Section 7.4—motion with constant velocity, and motion with constant acceleration. Here we shall generalize the mathematical expressions for these motions.

For constant-velocity motion, the special case of greater mathematical simplicity occurs when the particle is located at the origin, $x = 0$, at the initial time, $t = 0$. More generally, the particle might be somewhere else, at x_0, when $t = 0$. The equations for the special and general cases are:

Special *General*

$$x = vt\,, \qquad (8.1) \qquad\qquad x = x_0 + vt\,, \qquad (8.2)$$

Both

$$v = v_0\,, \qquad (8.3)$$

$$a = 0\,. \qquad (8.4)$$

As discussed in Section 7.4, and illustrated again in Figure 8.3(a), both $x = vt$ and $x = x_0 + vt$ are examples of linear functions. Both curves* have the same slope, which is the velocity v. The equation $v = v_0$ means the velocity v remains constant, equal to its initial value v_0, so that v and v_0 can be used interchangeably. For this simplest example of motion, the graph of velocity vs. time is a straight horizontal line [Figure 8.3(b)]. Its slope in turn, equal to the acceleration, is zero.

In general, for any kind of motion, the *average* velocity, $\mathbf{v}_{\mathbf{av}}$ during some time interval is defined by

* The graph of any function is usually called a "curve," even if part or all of it may have no curvature. The straight lines in Figure 8.3 are special examples of curves.

$$\mathbf{v}_{\mathbf{av}} = \frac{\mathbf{s}_2 - \mathbf{s}_1}{t_2 - t_1}. \tag{8.5}$$

It is the difference in the position vectors at two times divided by the time interval. For one-dimensional motion, this average velocity may be written

$$v_{\mathbf{av}} = \frac{x_2 - x_1}{t_2 - t_1}. \tag{8.6}$$

As shown in Figure 8.4, the geometrical significance of $v_{\mathbf{av}}$ is that it is the slope of a straight line drawn between two points on a curve of position vs. time. For constant velocity motion, the average velocity between any two points is the same as the velocity at each point. More generally this is not true, and the average velocity is what its name implies, an average of the different velocities that existed between the two points. It is a concept that will be useful in discussing motion with constant acceleration.

For constant acceleration, we distinguish again between a special case and the general case. The special case is the one considered in Section 7.4, characterized by $x = 0$ and $v = 0$ when $t = 0$, motion starting initially from rest at the origin. For the general case, neither the initial position x_0 nor the initial velocity v_0 need be zero. *All* the important equations governing uniformly accelerated motion can be written down in a small space. We shall list them first and then discuss them.

Special		*General*	
$x = \frac{1}{2}at^2$	(8.7)	$x = x_0 + v_0t + \frac{1}{2}at^2$	(8.11)
$v = at$	(8.8)	$v = v_0 + at$	(8.12)
$a = a_0$	(8.9)	$a = a_0$	(8.13)
$v^2 = 2ax$	(8.10)	$v^2 - v_0^2 = 2a(x - x_0)$	(8.14)

The third equation in each column states that the acceleration a remains constant, equal to its initial value, a_0. Stepping up to the second equation in each column, we can easily understand the correctness of these expressions for the velocity by analogy with the previous example. For constant velocity motion, the position coordinate x was a linear function of time. Now it is acceleration that is constant and velocity that is a linear function of time. The second equation in each column states that velocity increases at a uniform rate (or, if the constant a is negative, decreases at a uniform rate).

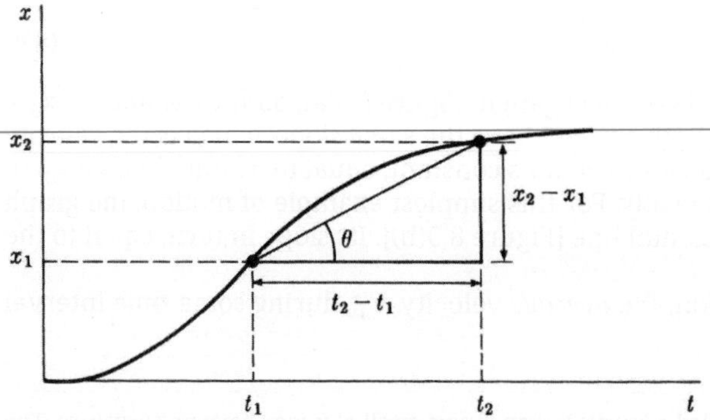

FIGURE 8.4 Graphical determination of average velocity for one-dimensional motion. The average velocity in the time interval t_1 to t_2 is the slope of the straight line drawn between the dots. This in turn is equal to tan θ, which is the ratio $(\mathbf{x}_2 - x_1)/(t_2 - t_1)$. Compare Figures 7.1 and 7.2.

The step up to the first equation in each column is slightly more complicated, and requires the use of the average velocity concept. The average value of a changing quantity is not usually simple to calculate. If, however, the quantity is changing at a constant rate, its average value in any interval of time is simply its mid-point value, halfway from its initial value to its final value. Thus if you walk at constant velocity from point A to point B, your average position during the walk is halfway from A to B. Similarly, for constant acceleration, the average velocity during some interval of time is halfway from the initial velocity to the final velocity. Consider the special case. In a time t, the velocity grows uniformly from zero to a value equal to at. Its average value in this time is $v_{av} = \frac{1}{2}at$. This average value, according to Equation 8.6, is the distance traversed (x in this case) divided by the elapsed time (t in this case). In symbols,

$$v_{\mathbf{av}} = \frac{x}{t};$$

multiplication of both sides by t gives

$$x = v_{\mathbf{av}}t\,.$$

If $v_{\mathbf{av}}$ is replaced by $\frac{1}{2}at$ in this equation, the equation $x = \frac{1}{2}at^2$ results. It is left as an exercise to derive the general formula for x, Equation 8.11, using exactly the same idea of average velocity.

The fourth equations in each column are useful relationships, but they are not really new. They result from eliminating the time variable t between the first two equations. The relations between x and t and between v and t are combined to give a relationship between x and v. It is instructive to compare Equations 8.8 and 8.10. One states that the velocity increases in proportion to time; the other states that the *square* of the velocity increases in proportion to distance, or equivalently, that the velocity increases in proportion to the square root of the distance. The fact that for uniformly accelerated motion velocity has a somewhat simpler relation to time than to distance was first discovered and emphasized by Galileo.* These equations are relevant, for example, to the takeoff roll of a jet, executed approximately with constant acceleration. Suppose, to be specific, that the acceleration is $a = 150$ cm/sec^2, and that the duration of the take-off roll is 40 sec. The equation $v = at$ may be used to find that the take-off speed is 6,000 cm/sec (134 mph). Because there is a direct proportionality between v and t, it may also be concluded at once that half of the take-off speed will have been achieved in half the take-off time, 3,000 cm/sec after 20 sec. From the equation $v^2 = 2ax$, the length of the take-off roll may be calculated. It is 1.2×10^5 cm = 1.2 km. After half of this roll is completed (x = 0.6 km), the speed, according to the same equation is $v = \sqrt{18 \times 10^6 \text{ cm}^2/\text{sec}^2} = 4{,}240$ cm/sec. In half of the take-off distance, the jet has already achieved about 70% of its take-off speed. If a jet pilot finds, when half of the runway is used up, that he has reached 60% of his take-off speed, should he continue the take-off roll or attempt to stop?

Consider now an example requiring the use of the more general equations describing uniformly accelerated motion. A baseball is thrown vertically upward, leaving the hand of the thrower 200 cm above the ground with a speed of 1,500 cm/sec. What is the maximum height above the ground reached by the ball? Before the equations of uniform acceleration can be applied, an origin of the x-coordinate and an origin of time must be chosen, and the values of the three constants, x_0, v_0, and a, must be known. The time origin may conveniently be chosen to be the instant when the ball leaves the thrower's hand; the x-origin may be chosen to be at ground level, with positive x being measured upward from the ground. With

* In *Two New Sciences*, Galileo confesses that when he first began his study of uniformly accelerated motion, he believed that velocity increases in proportion to the distance traversed, an error that he soon corrected.

these choices, the constants are

$$x_0 = 200 \text{ cm},$$

$$v_0 = 1500 \text{ cm/sec},$$

$$a = -980 \text{ cm/sec}^2.$$

The first, x_0, is the initial height above ground, the value of x when $t = 0$. The second, v_0, is the initial velocity, positive because it is directed upward. The acceleration due to gravity, **a**, is a constant vector of magnitude 980 cm/sec^2 directed vertically downward. In this example, its x-component, or upward component, is therefore negative. These numbers may be substituted into Equations 8.11, 8.12, and 8.14 to provide equations describing the subsequent history of the baseball (until it collides with the ground). To answer the particular question posed above, we must utilize the fact that the velocity of the ball is zero at the moment it reaches its maximum height. At that moment the velocity equation, $v = v_0 + at$, becomes

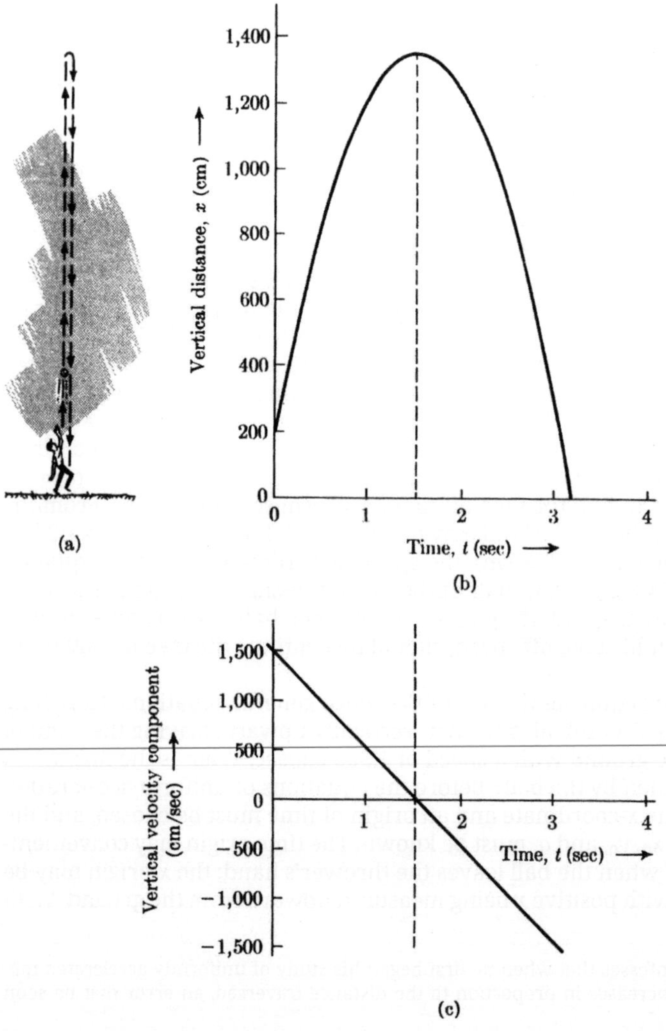

FIGURE 8.5 Example of vertically thrown baseball. **(a)** Pictorial diagram. **(b)** Graph of position vs. time. The curve is a parabola. **(c)** Graph of vertical component of velocity vs. time.

$$0 = 1{,}500 - 980t\,.$$

Its solution, $t = 1.53$ sec, is the time when the ball is at its peak height. The general position equation, $x = x_0 + v_0t + \frac{1}{2}at^2$, is, in this example,

$$x = 200 + 1{,}500t - 490t^2\,.$$

At the time $t = 1.53$ sec, $x = 1{,}350$ cm. This is the maximum height of the ball.

As this example makes, clear, equations are tools that do not provide automatic right answers. It is necessary to choose a frame of reference (in this case the origins of x and t), to find the correct constants, to pick the right equation, to apply the right criterion for the solution (in this case the fact that $v = 0$ at the maximum height), to work with a consistent set of units, and to interpret the solution correctly. Diagrams are frequently helpful in applying laws or equations to specific examples—either pictorial diagrams of the problem in question, or graphs of the variables of greatest interest. Both kinds of diagrams for the example of the thrown baseball are shown in Figure 8.5.

The kinematics of motion in more than one dimension is called vector kinematics. One important example of vector kinematics, the description of uniform motion in a circle, was discussed in Section 6.7. We shall consider other examples of two-dimensional motion in Section 8.10.

8.3 *The concept of force*

Sometimes, with a new concept, the problem for the student is that it represents a wholly 25
new idea, unrelated to anything in his past experience. This is not so with force. The problem of understanding force is rather the opposite. Force is such a familiar idea, used in so many different ways in everyday speech, that difficulty of understanding is more likely to result from the necessity of getting rid of misconceptions about its meaning than from embracing a new idea. In nontechnical usage, it is permissible to speak of the force of an author's words, the force of a man's personality, or the force of troops stationed abroad. We read also that an atomic explosion occurred with enormous force (energy is meant), or that a flywheel is spinning with great force (angular momentum is meant), examples of incorrect technical usage. Instead, force has a single definite—and rather simple—meaning in physics, as it must if it is to be a useful quantitative concept. The physicist takes one of the most common everyday meanings of force, a push or a pull, and sharpens it into a definition precise enough to form a basis of measurement.

Pushes or pulls, as we shall understand forces to mean from here on, have two kinds of effect in the world. They may change the shape or size of an object, or they may influence the motion of an object. A toy balloon subjected to a pair of forces between two hands alters its shape, and if the forces become large enough, it alters its shape drastically by exploding. A wooden plank supported at both ends is pulled downward by the force of gravity and bends into a bow. A soft pillow changes shape according to the forces acting on it. A tree bends from the force of the wind. A spring elongates when pulled and compresses when pushed. This general kind of effect of force we may call deformation. Anyone can, without difficulty, think of dozens of examples of deformation produced by force. Indeed no object is so rigid that it does not undergo at least slight deformation when acted upon by external forces.

The deforming effect of force can be used to define force. Imagine a well-designed standard spring whose elongation when pulled can provide a measure of force (Figure 8.6). One end of the spring is held fixed in position (this requires some force, but it is not the force of interest). The force to be measured is applied to the other end of the spring. We might decide to call the force required to extend the spring by 10^{-3} cm the unit force. The new unit requires a name. We call it one dyne. Two such forces, each of one dyne, acting in the same direction, provide a force of two dynes. If two equal and parallel forces together add to one dyne, each of them is equal to 0.5 dyne. By such combinations any unknown force can in principle be related to the unit force and to the chosen spring extension. If the spring happens to have the happy property that its extension is proportional to the applied force, it is a practical and useful force-measuring device, not just a definer of force in principle. Then, for instance, an extension of 1 cm would measure a force of 1,000 dynes, an extension of 3.2 cm a force of 3,200 dynes, and so on. Such springs, which can in fact be constructed to a good degree of accuracy, are said to obey Hooke's law—deformation is proportional to force.

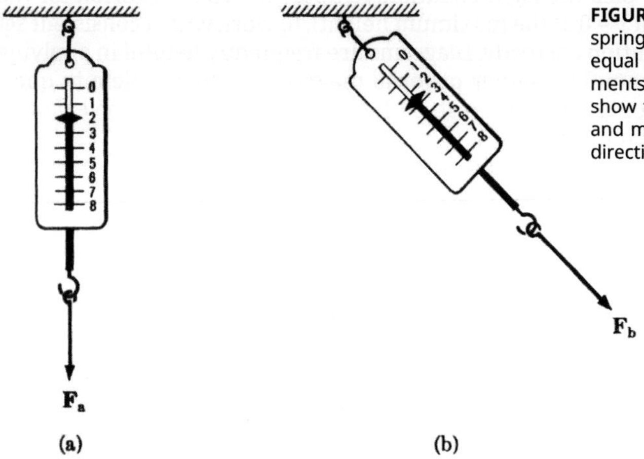

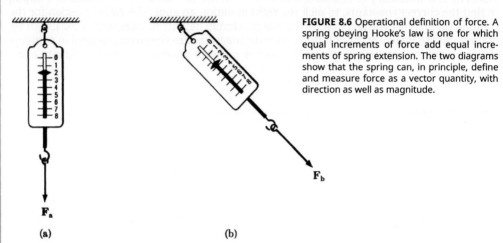

FIGURE 8.6 Operational definition of force. A spring obeying Hooke's law is one for which equal increments of force add equal increments of spring extension. The two diagrams show that the spring can, in principle, define and measure force as a vector quantity, with direction as well as magnitude.

Note that the mere extension of a spring may be considered an "experiment," albeit a very simple one. It is the experiment that operationally defines the concept and the unit of force. The arbitrariness of the unit is reflected in the arbitrary choice of a standard spring and the arbitrary extension of that spring used to specify a force of one dyne. The dyne, incidentally, is, in human terms, quite a small force. A man leaning idly against a wall is pushing the wall with a force of more than one million dynes. The force of gravity on one pound of matter at the earth's surface is 4.45×10^5 dynes.

Although not accurate enough for modern purposes, the spring definition of force is a good tentative definition for use in approaching the laws of motion. It is simple, and it is in familiar terms, placing the quantitative definition of force quite close to our intuitive ideas about the nature of force. Moreover, it has the important merit that it is independent of motion. We wish to learn exactly how force affects motion. This is most easily done if force is first separately defined without reference to motion.* However, *any* action of force can be used to define forces. An alternative approach, in which force is defined through motion, is discussed in Section 8.6.

It might appear that the spring definition of force is limited to contact forces despite the fact that all of nature's fundamental forces are "action at a distance." This is not so. Something must be in contact with the spring to extend it, but the force being measured can reach out across an empty space. An electrically charged object, for example, could be attached to the spring in order to measure an electric force exerted by another charged object some distance away. The ordinary use of a spring balance to weigh a parcel is a measurement of a gravitational force acting from a distance.

* The slight motion associated with the extension of a spring is inconsequential, for, once extended, the spring remains motionless under the action of the force.

Since force is a vector quantity, its definition must include this fact. This means first that force has direction as well as magnitude; both properties can be included in the spring definition (Figure 8.6). But the vector nature of force means more. It means that two or more forces must combine physically in exactly the same way that vectors combine mathematically. As illustrated in Figure 8.7, the standard spring may easily be used to test the laws of combination of several forces to verify their vector character. Two equal-magnitude forces directed at right angles to each other, for instance, can be verified to exert a total force directed halfway between the two combining forces with a magnitude 41% greater than the magnitude of each combining force. Two forces equal in magnitude and opposite in direction can be verified to exert no net force on the spring.

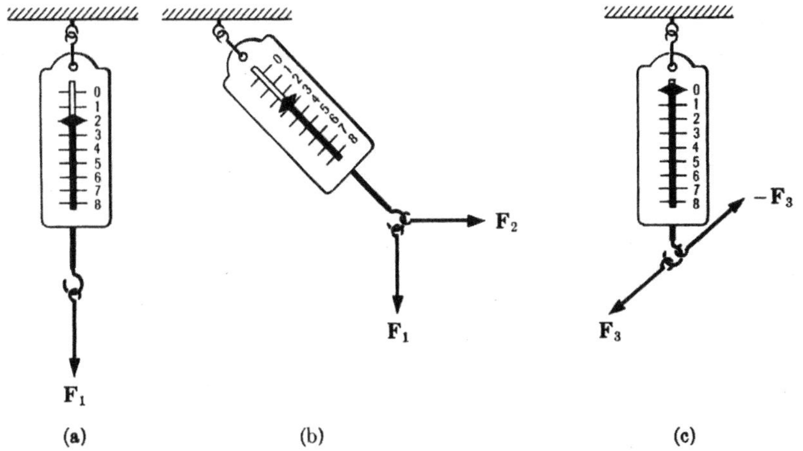

FIGURE 8.7 Direct test of the vector nature of force. **(a)** A force $\mathbf{F}_1$ acting alone has magnitude 2 on this scale. **(b)** Two equal-magnitude forces, $\mathbf{F}_1$ and $\mathbf{F}_2$, acting at right angles, produce a force directed halfway between them with magnitude 2.8. **(c)** Two equal-magnitude forces directed oppositely add to zero.

Having defined force in terms of the deformation of a standard spring, we may study how force affects motion.

8.4 *Force and motion*

The connection between force and motion can be stated very simply: The acceleration experienced by a given object is proportional to the force acting on it. In symbols,

$$\mathbf{F} \sim \mathbf{a}\,. \tag{8.15}$$

Two remarks about the proportionality of Formula 8.15 are worth making at once. First, it is a vector proportionality. Acceleration always has the direction of the net force, regardless of the direction of the velocity. In uniform circular motion, for example, where velocity is directed tangentially around the circle, acceleration and therefore force are directed inward toward the center of the circle. Second, the relationship implies no causal connection. It is usually rather natural (and not harmful) to think of force as the cause, acceleration as the result. Logically, it is as valid to say that acceleration causes force. Sometimes it is in fact easier to think in this way. We can say that the forward acceleration of an automobile "causes" us to feel an extra force from the back of the seat. In any case, as a law of nature, Formula 8.15 states only that force and acceleration always go together in a certain way.

Aristotle, and scholars after him for many centuries, believed that a simple connection existed not between force and acceleration, but between force and velocity. Simply stated, the Aristotelian view of motion was this: Force is required to produce and to sustain motion (that is, velocity). If there is no motion, there is no force (true). If there is no force, there is no motion (false). It is quite easy to see how this incorrect view of motion came to be adopted. The average person untrained in science is likely to hold the same view today. We observe that all our means of locomotion—trains, automobiles, airplanes, or our own legs—seem to require a continual force in order to keep moving. When the motive force ceases, the motion, after a time, ceases too. It appears that the result of zero force, at least eventually, is zero velocity, and that the result of some definite applied force is some definite velocity, not the endlessly growing velocity that would accompany a fixed acceleration. There is one difficulty with this view of motion, and that concerns coasting motion. A hockey puck sliding freely over the ice or an automobile coasting along a flat road both appear to be subjected to no force in their direction of motion, yet they keep moving. Aristotle, recognizing this problem, postulated that the air provides the motive force to sustain coasting motion. Now we know, of course, that the air provides only a frictional force tending to decelerate a coasting object. Ideal coasting, free of all forces, proceeds indefinitely with constant velocity.

The problem of the motion of an object subject to a *constant* force was solved completely by Galileo. There had been no lack of thinking and writing on the subject of motion over the centuries since Greek civilization, but Galileo's approach to the problem contained some new elements—elements very much in the spirit of modern science—that made the difference between philosophical speculation and concrete quantitative results. First, Galileo began simply. Instead of seeking at once a systematic explanation of all motion, he studied with care the kinematics of uniform-velocity motion, then of motion with constant acceleration, then of some more complex motions such as pendulum motion. He neither sought nor found a general theory of motion, but his discoveries were a vital first step toward the theory of mechanics. Second, Galileo was an experimenter, not merely an observer. He deliberately designed experiments to reduce or eliminate unwanted effects and thereby to reveal the simplest properties of motion.* For Galileo, experiments with balls rolling down troughs or swinging pendula required not only insight into the best way to study nature, but courage as well. To most academic scholars of the time, such mundane work was beneath the dignity of intellectual man. Third, Galileo recognized the significance of the quantitative. Because existing clocks were not accurate enough for his purposes, he devised a better method of keeping time.† Finally, Galileo saw the importance of extrapolating away from the realities of friction and other outside influences that disturb an experiment to the ideal domain, approachable but never quite attainable, where simple mathematical laws of nature are precisely correct. In considering the parallel falls of a wooden ball and a metal ball, he realized that the fact that the metal ball reached the earth first was much less significant than the fact that the two balls fell *almost* identically. The not-quite-correct statement that all objects fall with the same acceleration provides far more insight into nature than does the perfectly correct statement that no two objects fall in a precisely identical way.

---

* Some modern scholars question Galileo's skill as an experimenter. It is possible that he pushed his experimental studies only far enough to verify the general correctness of his ideas, and proceeded from there by logical reasoning. It is even possible that in his writings he described some experiments that he had not actually carried through. This is a practice considered indefensible by modern scientists. In the intellectual climate of Galileo's time, with no tradition of experimental science, such a casual approach to the description of experimental results is not surprising. In any event, it is clear that Galileo, much more than most of his predecessors, recognized the importance of appeal to experiment.

† Galileo's "clock" was a large container of water with a small orifice at the bottom which could be opened and closed. The weight of water escaping from the orifice during an interval of time was used as a measure of the duration of that interval.

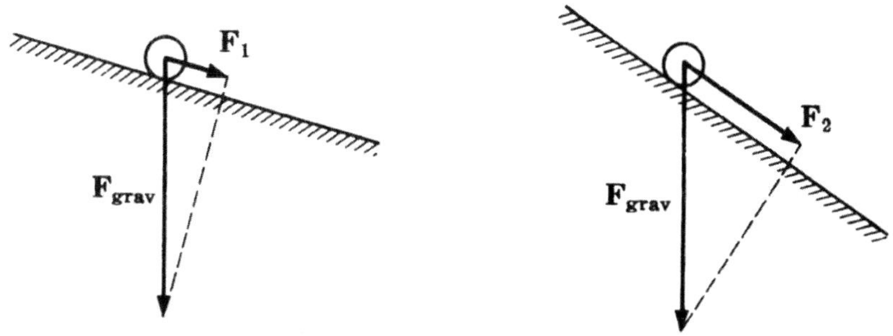

Figure 8.8 Forces in Galileo's experiments. For identical balls on troughs of different slope, the gravitational force Fgrav is the same. The ball's acceleration is proportional to the component of this force directed along the trough. Not shown are force components perpendicular to the trough, which cancel.

If a book is held at shoulder height, a certain force is required to counterbalance the downward force of gravity and keep the book stationary. If the book is held at waist height, the same upward force is required. At ankle height, the same force is still required. At least over man-sized dimensions, we conclude that the weight of the book is independent of its position. To state the result of this and similar experiments with more care: Near the surface of the earth, the force of gravity on a stationary object does not depend on the position of the object. Aware of this law, Galileo guessed, correctly, that the gravitational force on a given object is also independent of its state of motion. He reasoned that a falling object is being drawn downward by the same constant force that would act if it were held stationary at any level. Since his experiments established that an object moving downward under the action of the gravitational force moves with constant acceleration, he concluded that constant force produces constant acceleration.

Because the gain of speed in free fall is so rapid, most of the experiments described by Galileo were carried out with inclined motion rather than with vertical motion. He found that a ball rolls down a smooth inclined trough with constant acceleration, the magnitude of the acceleration increasing as the slope of the trough increases. Moreover, the acceleration along a particular trough is proportional to the component of the gravitational force along the trough (Figure 8.8), verifying the general rule of Formula 8.15 that force is directly proportional to acceleration. Since the component of gravitational force perpendicular to the trough is canceled by the force of the trough itself on the ball, only the component of gravitational force along the trough contributes to the acceleration.

8.5 Motion without force: Newton's first law

An important consequence of Galileo's work is that in the absence of force, it is acceleration, not necessarily velocity, that is zero. Despite the fact that a ball rolling in a horizontal trough will in fact come to rest, Galileo realized that for an ideally smooth trough, the motion would continue with constant velocity forever. Although simple, this result was extremely important, for it ran counter to the prevailing Aristotelian view of motion. It led to what was stated later in more general form by Newton as his first law. In Newton's words: "Every body continues in its state of rest, or of uniform motion in a straight line, unless it is compelled to change that state by forces impressed upon it." This is a statement that the "natural" or "undisturbed" state of a body is unaccelerated motion.

Although the law that $\mathbf{a} = 0$ when $\mathbf{F} = 0$ seems simple enough in the abstract, its application to everyday situations creates some confusion, for it violates our intuitive ideas about motion. We can ask, for example, whether any force acts on a skier going downhill at constant velocity. The natural tendency is to answer yes. He is pulled down by gravity. If there were no slope and no loss of altitude, he could not move without effort. Nevertheless, the fact is that no *net* force acts on the skier. The uphill frictional force precisely cancels the downhill component of gravitational force. If these forces did not add to zero, the skier would not move with constant velocity. If the frictional force were dominant, he would decelerate (that is, he would experience uphill acceleration). If the gravitational force were dominant—as it might become when he starts down a steeper portion of the hill—he would accelerate downward.

The same considerations apply to motion of any kind. While an automobile is accelerating forward, a net forward force is acting. When it slows, its acceleration and the net force it experiences both point rearward. So long as it moves with constant velocity, it is free of external force. In most terrestrial examples of uniform velocity motion, the zero total force arises from the sum of a number of different forces, rather than from the absence of any. Because we sometimes think of one of these forces without thinking of another, it is easy to conclude incorrectly that a net force is acting. An airplane moving with constant velocity, for instance, is subjected to a myriad of forces acting over its whole surface in addition to the downward force of gravity. The net upward component of the forces is called lift; the net rearward component on all parts of the airplane other than the driving surfaces (turbine blades or propellers) is called drag, the net forward component on the driving surfaces is called thrust. As illustrated in Figure 8.9, all of these forces must sum to zero when the airplane moves with constant velocity.

One other example may serve to emphasize the fact that it is acceleration, not velocity, that is related to force. If an object is dropped vertically from a helicopter, the only important force acting on it at first is the gravitational force. It begins to fall with the constant acceleration of gravity, 980 cm/sec^2. As its speed builds up, however, the frictional force of air resistance, directed upward, increases. The net downward force decreases, and the downward acceleration decreases in proportion. So long as there is any downward acceleration, the speed continues to grow. But so long as the speed grows, the frictional force grows. Eventually the frictional force becomes equal in magnitude to the gravitational force. The total force acting on the falling object is then zero. It moves thereafter with constant velocity. The speed of fall required to produce balanced forces and therefore zero acceleration is called the terminal speed. Terminal speed depends on the size and the shape and the mass of an object, but in every case it is the speed for which the total force vanishes. Different objects reach quite different terminal velocities, but for all, the terminal force and the terminal acceleration are zero.

The special significance of Newton's first law, and some subtleties associated with it, are pursued in Sections 8.11 and 8.12.

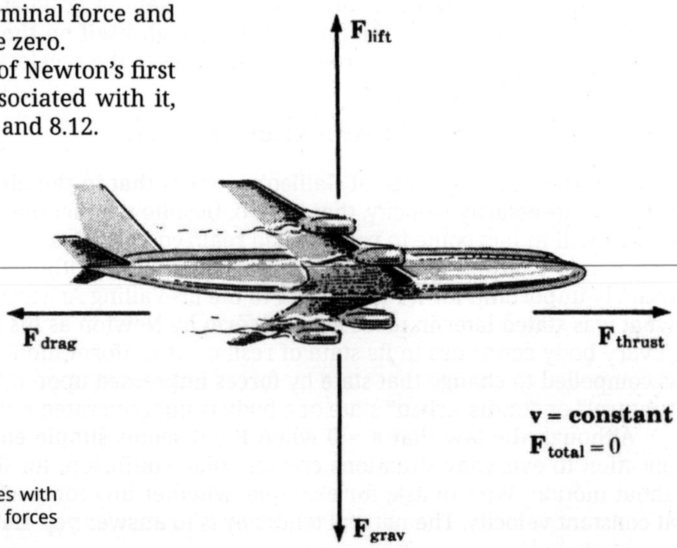

FIGURE 8.9 When an airplane moves with constant velocity, the sum of all the forces acting on it is zero.

8.6 *Inertial mass: Newton's second law*

Properties of uniformly accelerated motion can conveniently be studied in the laboratory 26
with an experiment using apparatus such as that pictured in Figure 8.10. A car rides on a horizontal track with very little friction. On the car can be placed weights of various size. The car and its weights are drawn along the track with a known constant force, and the acceleration is measured. The first result of such an experiment is a verification of the fact that a constant force produces constant acceleration. The second result is that for the car loaded in a particular way, its acceleration is proportional to the force applied. These results are summarized by the proportionality of Formula 8.15.

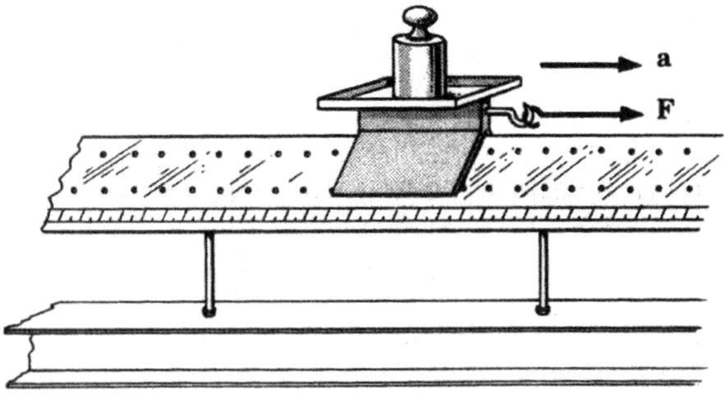

FIGURE 8.10 Experimental arrangement to measure inertial mass. A car and its load, supported on a frictionless air track, responds with constant acceleration **a** to a constant force **F**.

The new point of interest with this experiment concerns the constant of proportionality between force and acceleration. For the car loaded in a particular way, the force may be written as a constant times the acceleration. In symbols,

$$\mathbf{F} = m\mathbf{a}\,. \tag{8.16}$$

This is Newton's second law.* Its implications for motion in general are discussed in later sections. Here we are concerned with its role in fixing the basic concepts of mechanics.

The constant of proportionality in Equation 8.16 is called the *inertial mass,* or often simply the *mass,* of the object under study, in this case the car plus its load. The usual symbol for inertial mass is m. Before discussing the physical meaning of mass, we may note some facts about it at once from its appearance in this equation. We notice that its dimension must be the dimension of force divided by the dimension of acceleration, or force × (time)^2/length. In standard cgs units, the unit of mass, the gram, is equal to one dyne (second)^2/centimeter. With the usual abbreviations, this equality may be written,

$$1 \text{ gm} = 1 \text{ dyne sec}^2/\text{cm}\,. \tag{8.17}$$

Since now mass is usually taken as fundamental, and force as derived, the same connection may be rewritten

$$1 \text{ dyne} = 1 \text{ gm cm/sec}^2\,. \tag{8.18}$$

We notice also from Equation 8.16 that mass appears to be a numerical quantity (or scalar quantity), not a vector quantity. The only thing that can multiply a vector (**a**) to give another vector in the same direction (**F**) is a number. However, as is discussed below, the numerical or scalar character of mass requires experimental verification.

---

* Newton's own version of his second law, although equivalent in content to Equation 8.16, was phrased differently. We use a modernized version suited to our needs in this chapter.

Experiments performed with cars loaded differently reveal different masses. Not surprisingly, the rule is: The greater the load, the greater the mass. From the definition of mass, it follows that a particular force gives more acceleration to a small mass than to a large mass. The greater is the load carried by a car, the more slowly it responds to a given force, and the less velocity it acquires in a given time. This means that mass is a measure of an object's resistance to being set into motion. This "inertia" associated with mass applies equally well to deceleration. A more massive body is more difficult to stop as well as more difficult to start. Most generally, we can say that mass is a measure of an object's resistance to a change in its state of motion. Technically, "state of motion" means nothing more or less than "velocity." Therefore, comparing two objects, we may say that the more massive one has more resistance to a change in its motion because, for a given force, its velocity changes by a lesser amount, that is, it has less acceleration. Anyone who has pushed both a Cadillac and a Volkswagen Beetle by hand knows from experience that greater mass has greater resistance to a change of velocity than does smaller mass. Correspondingly, the Cadillac requires a more powerful engine to achieve comparable acceleration. For the same acceleration, twice as much force is required for twice as much mass.

Among charged particles, the least massive electron is the easiest to accelerate to high speed and the easiest to deflect into a curved path. It responds most readily to the pushes and pulls of electric and magnetic forces. As discussed in Chapter Two, some particles—the photon, the neutrinos, and the graviton— are actually massless. If Newton's second law remained valid in the world of elementary particles, these particles could achieve infinite acceleration and infinite velocity. They would have absolutely no resistance to a change in their state of motion. In fact, because of the existence of a speed limit in nature, they reach only the speed of light rather than infinite speed. However, they can be said to experience infinite acceleration. A photon spends no time reaching the speed of light, but starts immediately with that speed at the moment of its birth.

Returning to the loaded car on the track whose mass is defined as the ratio of the horizontal component of force acting on it to its horizontal component of acceleration, we must inquire about the self-consistency of the definition. Is the mass of two objects taken together equal to the sum of the masses of the separate objects? This is a question that must be answered experimentally. Suppose that it is ascertained by acceleration experiments that the mass of the unloaded car is 0.2 kg, the mass of the car with load A is 0.5 kg, and the mass of the car with load B is 0.7 kg. If masses combine physically in the same way that numbers combine arithmetically, the mass of load B must be 0.5 kg. To test this hypothesis, the car is filled with loads A and B together, and another force and acceleration measurement is made. This experiment reveals that the car carrying loads A and B has a mass of 1.0 kg, a verification of the hypothesis. Many such experiments could be performed to verify the fact that the inertial mass of a group of objects taken together is numerically equal to the sum of the inertial masses of the individual objects. We can therefore say with confidence that mass is a numerical or scalar quantity.

Lurking behind the facts that the mass of a single object is a constant and that the masses of several objects sum numerically is another important law: the law of mass conservation. In the domains of nature where Newton's second law is valid, mass is conserved. In nineteenth-century chemistry, mass conservation played an important role, for, to an extremely high degree of precision, chemical reactions do not alter the mass of a sample of matter. In subatomic domains where Newton's second law fails, mass conservation also fails (to be replaced by mass-energy conservation).

To review: Force has been defined in terms of the static extension of a pulled spring; mass has been defined in terms of the acceleration of an object subjected to a known force. In this formulation, force is a primitive concept, mass a derived concept. Although both concepts are operationally defined, the unit of force is an arbitrary unit referred to an arbitrary standard spring, whereas the unit of mass is defined as a certain combination of the units of force, length, and time. However, because of the experimental connection between force and mass set forth mathematically by Equation 8.16, it is easy to see that mass could have been chosen to be the primitive concept, and force the derived concept. We could, for example, select a standard object and arbitrarily define its mass to be 1 gm. (A thousand-times-larger standard, the kilogram, does in fact define the mass unit for modern scientific work.) The unit force, called one dyne, could then be defined as the push or pull required to give to the standard gram an acceleration of 1 cm/sec^2. The dyne in this case would be regarded merely as a shorthand name for the combination, gm cm/sec^2. Other magnitudes of force than one dyne would also be referred to the standard mass. Three dynes would be that force required to give to 1 gm an acceleration of 3 cm/sec^2, and so on. Finally, the masses of objects other than the standard object could be determined by further acceleration experiments. Most simply, the unknown mass and the standard gram could be subjected to the same force. The ratio of their masses would be defined to be the inverse ratio of their accelerations,

$$\frac{m}{m_s} = \frac{a_s}{a}, \tag{8.19}$$

where a is the acceleration of the unknown mass m, and a_s is the acceleration of the known standard mass m_s, when both experience the same force. For example, if an object subject to a force of 1 dyne experienced an acceleration of 0.4 cm/sec^2, its mass would be 2.5 gm.

There are still other ways to define mass, one of which is discussed in Section 9.8. Which definition is adopted as the world standard has practical importance, for it is related to the degree of precision possible in extremely accurate measurements. So far as the logical framework of mechanics is concerned, however, it is not very important which definition is adopted, or whether mass is regarded as a primitive concept or a derived concept. The important things are only that the theory provides a self-consistent set of equations, that the basic concepts are chosen in such a way that these equations are as simple as possible, and that there is a clearly understood connection between physical measurements and mathematical symbols. When these things are achieved, the theory provides a *simple description* of some part of nature and the *power of prediction*. Nothing else can be asked of it. Modern science neither tries nor succeeds to answer questions such as: What are "really" the most basic concepts? Why does mass behave in the way that it does? The interweaving of definitions and laws in the structure of modern science, although occasionally confusing, is a necessary aspect of a science whose mathematics is tied solidly to nature with the bonds of operational definitions and experimental confirmation.

8.7 Applications of Newton's second law

Newton's second law finds its simplest application in the motion of a particle or an object along a single straight line. An important aspect of any vector equation is that it expresses the separate equalities of the vector components because of the equality of the vector quantities. If three mutually perpendicular directions in space (such as east, north, and up) are called the x-direction, the y-direction, and the z-direction, Newton's second law in vector form, Equation 8.16, may be re-expressed as three different equations for the x, y, and z components,

$$F_x = ma_x, \tag{8.20}$$

$$F_y = ma_y, \tag{8.21}$$

$$F_z = ma_z. \tag{8.22}$$

In these equations, the subscripts designate the component directions. The first of the three, for instance, states that the component of force in the x-direction is equal to the product of the mass and the x-component of the acceleration. Since a vector component is a numerical quantity, the three numerical equations for components replace a single vector equation. In those cases where the motion is known to occur along a single straight line—an automobile on a straight road, a vertically fired rocket, a child on a slide, or a parachutist in still air—the direction of the line, whatever it may be, can be chosen to be the x-direction. Then the acceleration along that direction multiplied by the mass of the object is equal to the component of force along that direction.*

If a constant component of force, F_x acts on a particle moving along the x-axis, the behavior of the particle is easy to ascertain, for its acceleration is a constant:

$$a_x = \frac{F_x}{m} = constant. \tag{8.23}$$

The kinematics of constantly accelerated motion were studied in Section 8.2. Equations 8.11 and 8.12 provide what is called the "solution for the motion," that is, answers to the questions: Where is the particle, and how fast is it moving, at all times? The same force can produce different motions, depending on what are called initial conditions. Suppose, for example, that a constant force of 600 dynes acts in the positive x-direction on an object of mass 2 gm. Its acceleration will be a_x = 300 cm/sec^2. If it starts from rest ($v_x = 0$) at the origin ($x = 0$) at the initial time ($t = 0$), its later position is given by the equation

$$x = ½a_x t^2 ,$$

which, in cgs units, is

$$x = 150t^2 .$$

It moves away to the right with ever increasing speed (Figure 8.11). If, however, it had been located at some other point x_0 at the initial time and had possessed some initial speed v_0 at this time, its subsequent history would be given by the more complicated equation,

$$x = x_0 + v_0 t + ½a_x t^2 .$$

Two examples of such motion are also illustrated in Figure 8.11, one with x_0 = 500 cm and v_0 = 900 cm/sec, one with x_0 = 1500 cm and v_0 = –900 cm/sec. For the initial velocity positive, the object accelerates away to larger values of x, keeping a higher speed than when it started from rest. For the initial velocity negative, the object begins to move to the left (to smaller values of x) but is slowed by the rightward acceleration. After 3 sec it is brought to rest, its direction of motion is reversed, and it accelerates away to larger values of x. The essential points illustrated by this example are these: The one-dimensional motion of an object subject to a constant force proceeds with constant acceleration. Precise details of the motion depend on initial conditions—where the particle was located and how fast it was moving initially. However, for a particular acceleration, all possible motions follow a parabolic graph of position as a function of time; indeed the parabolas are, except for horizontal and vertical shifts on the graph, all the same. Motion begun from rest or in the direction of the acceleration will continue in the same direction. An object moving initially in the direction opposite to the acceleration will be stopped and turned around after some time.

Although one may "know" Newton's second law, it is easy to fall back into the Aristotelian view of motion, thinking that velocity must be in the direction of the force. It is, of

* For such examples of one-dimensional motion, it is often the custom to omit subscripts for convenience and to write $F = ma$, understanding that F and a refer to components in a particular direction. To avoid possible confusion with the full vector equation, we shall retain the component subscripts in this section.

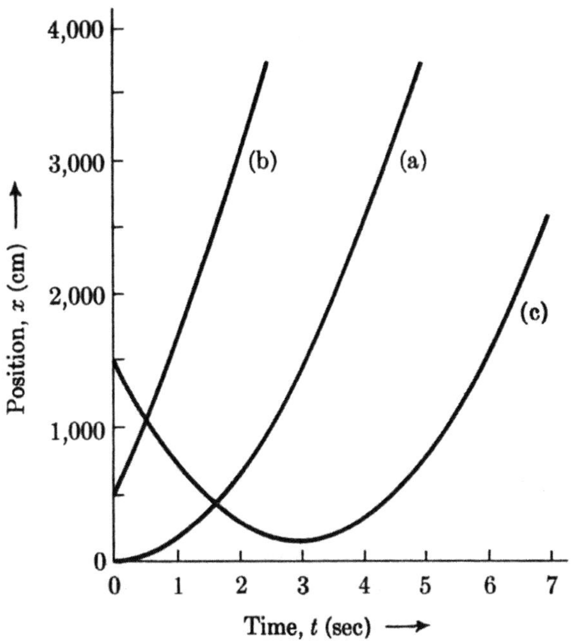

FIGURE 8.11 Three graphs of position vs. time for motion with the same constant acceleration (a_x = 300 cm/sec^2), but with three different initial conditions: **(a)** $x = 150t^2$; **(b)** $x = 500 + 900t + 150t^2$; **(c)** $x = 1500 - 900t + 150t^2$.

course, only acceleration that is in the direction of the force. Velocity may be in a direction opposite to the force (a ball thrown upward, a braking automobile, or the third example above), or indeed in any direction at all relative to the force.

A law of force that occurs rather often for one-dimensional motion in nature is this:

$$F_x = -Kx. \tag{8.24}$$

It says that the force is directed toward the origin (it is negative for positive x, and positive for negative x), and that the magnitude of the force increases in proportion to the distance away from the origin. The quantity K is a constant. At the origin there is no force. This law of force holds true approximately for a pendulum or a swing. A swing at its central equilibrium point (the origin), if at rest, remains at rest. It experiences no horizontal component of force. If displaced from the origin, it is pulled back toward the origin. You know from experience that the horizontal force required to push a swing away from its equilibrium point increases as the distance increases. For not-too-large displacements, the force grows in direct proportion to the distance. Therefore the force law of the swing is the law written down in Equation 8.24.

If m is the mass of a swing (including its occupant, if any), Newton's second law applied to the swing may be written

$$ma_x = -K_x, \tag{8.25}$$

or, equivalently,

$$a_x = -\left(\frac{K}{m}\right)x. \tag{8.26}$$

Since the ratio K/m is a constant for a particular swing, the interesting thing that this equation has to say is that the acceleration is proportional to the negative of the position. A good way to think about what this means is to think pictorially. A graph of acceleration vs. time for a swing will be a curve identical in form to a graph of position vs. time, except that one curve will be an upside-down version of the other. When x is zero, a_x is zero; when x is large and positive, a_x is large and negative; when the x-curve goes up, the a_x-curve goes down; and so on. At first guess, one might think that many different motions have this property that position and acceleration are each other's opposites. In fact, there is only one such kind of motion. It is an oscillating motion whose graph of position vs. time is a sine curve. The sine function was discussed in Section 7.5. The slope of the slope of the sine curve is again a sine curve, but turned upside down (see Figure 7.5).

For the law of force given by Equation 8.24, the motion, most generally, is given by the equation

$$x = A\sin\left[\frac{2\pi(t-t_0)}{T}\right], \tag{8.27}$$

in which A and t_0 are arbitrary constants, and T, called the period of the motion, is given by

$$T = 2\pi\sqrt{\frac{m}{K}}, \tag{8.28}$$

Whenever the time t increases by an increment T, the sine curve completes one full cycle, and the motion completes one full oscillation (Figure 8.12). More important than a detailed understanding of the constants in Equation 8.27 is a recognition of the broad significance of the sine function in describing oscillatory motion. The law of force followed by the swing or by a pendulum is much more common than one might at first suppose. Oscillatory phenomena are very common in nature—a vibrating tuning fork, a child's rocking horse, waves on water, sound transmitted through the air, an entire building swaying in the wind—an almost unlimited number of examples; and for nearly all of them, the simple law of force expressed by Equation 8.24 is a good approximation to the true force acting toward the central equilibrium point of the motion. Therefore a vast variety of swaying, swinging, vibrating, or oscillatory motions are rather accurately described by a sine function.

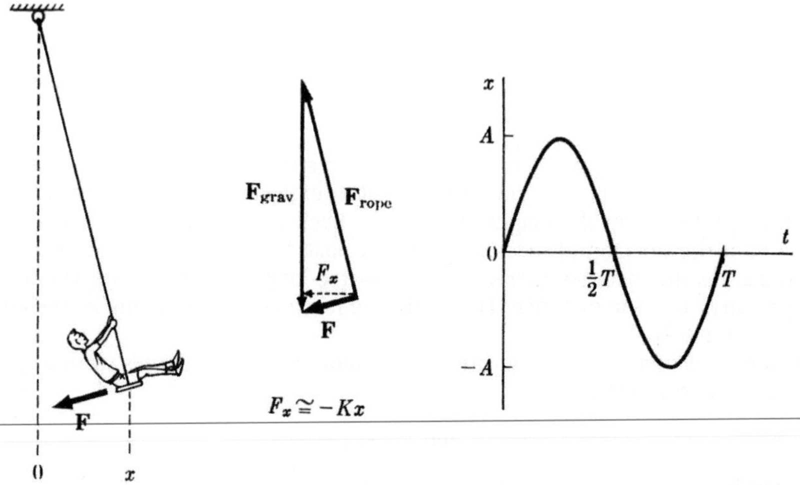

FIGURE 8.12 Simple harmonic oscillation. The restoring force on a swing (and on many other oscillatory systems) is proportional to the negative of the displacement x. Accordingly, the swing executes sinusoidal oscillation. The period T of the oscillation is related to the mass m and the force constant K by Equation 8.28.

Not all sine functions are identical, of course, any more than are all of the parabolas identical that describe uniformly accelerated motion. Besides being displaced one from another as in Figure 8.11, the parabolas may be fatter or thinner, according as the acceleration is less or greater. Similarly, sine functions may be displaced in time; they may have large magnitude or small, according to the amplitude of the oscillation; and they may have the long stately period of a swaying suspension bridge or the frenetic million-times-per-second vibration of a small crystal. Several different sine curves are illustrated in Figure 8.13. They have in common a certain underlying mathematical identity, which can be expressed graphically in this way: Any one sine curve, if stretched or squeezed both horizontally and vertically, can be made identical in shape to any other sine curve. Another way to say this is that by suitable choices of horizontal and vertical scales, the graph of any sine curve can be made to look like any other given sine curve. It is remarkable that a single simple mathematical function describes an enormous range of different types and sizes and speeds of oscillatory motion. The reason, according to Newton's second law, is that forces proportional to displacement and opposite in direction to the displacement are common in all parts of nature.

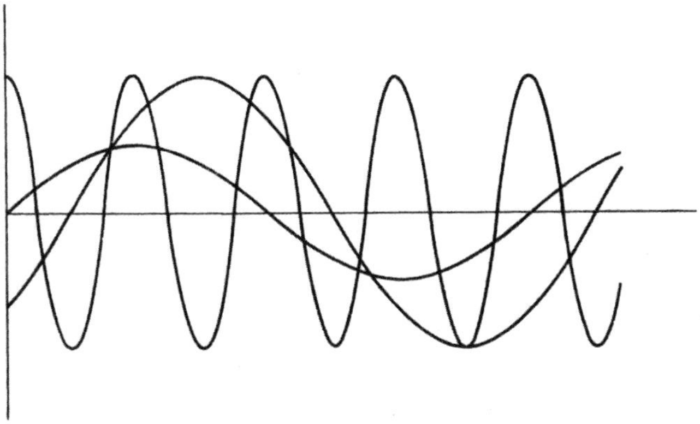

FIGURE 8.13 Sine curves of different amplitudes, different periods, and different initial values.

Newton's second law need not always be used to answer the question: Given the force, what is the motion? It can be equally well used to answer the question: Given the motion, what is the force? Newton himself used it in this way. He made use of the knowledge about lunar and planetary motion supplied by Kepler's laws in order to deduce the law of gravitational force. Here is a more homely example. An automobile with a speed of 10 m/sec rounds a curve whose radius of curvature is 30 m. What is the force experienced by a 75-kg passenger? The study of the kinematics of circular motion in Section 6.7 revealed that the magnitude of the acceleration is equal to the square of the speed divided by the radius of the circle,

$$a = \frac{v^2}{r}.$$

In this example, the acceleration of the automobile is therefore

$$a = \frac{(10^3 \text{ cm/sec})^2}{3 \times 10^3 \text{ cm}} = 333 \text{ cm/sec}^2 .$$

The passenger, following the same path as the automobile, has the same acceleration. The force acting on him must therefore be, according to Newton's second law,

$$F = ma = 75 \times 10^3 \text{ gm} \times 333 \text{ cm/sec}^2 = 2.5 \times 10^7 \text{ dynes}.$$

Force and acceleration are directed along the same line, toward the center of the circular path.

8.8 *The motion of systems: center of mass*

The example of the passenger in the automobile illustrates an aspect of Newton's second law that is of practical importance in solving problems of motion, and of fundamental importance in understanding the motion of systems. By "system" is meant any collection of two or more things, usually interacting with one another, such as the solar system or a collection of molecules making up a solid object. Newton's second law applies separately to any object or particle or system or part of a system. The total force acting on a particular object is equal to the mass of that object multiplied by its acceleration, regardless of how the object may be connected to or interact with other objects. The force on an automobile is equal to its mass times its acceleration. The force on a passenger is equal to his mass times his acceleration. It is equally true that the net force acting on the passenger's little finger is equal to the mass of the little finger times its acceleration.

How to apply Newton's second law to a nonrigid system—the solar system, for instance—is not at once clear. The total force acting on the system and the total mass of the system are clear ideas. But what is "the" acceleration of a system whose individual parts are moving in different directions with different accelerations? This question can best be answered in terms of some simple algebraic manipulations. Suppose the system in question consists of two particles with masses m_1 and m_2, accelerations $\mathbf{a}_1$ and $\mathbf{a}_2$, and subject to forces $\mathbf{F}_1$ and $\mathbf{F}_2$. We call the total mass M ($M = m_1 + m_2$) and the total force $\mathbf{F}$ ($\mathbf{F} = \mathbf{F}_1 + \mathbf{F}_2$). The motion of each particle separately is described by Newton's second law:

$$\mathbf{F}_1 = m_1\mathbf{a}_1\,, \tag{8.29}$$

$$\mathbf{F}_2 = m_2\mathbf{a}_2\,, \tag{8.30}$$

If the sum of the left sides of these two equations is set equal to the sum of the right sides, the result is

$$\mathbf{F} = m_1\mathbf{a}_1 + m_2\mathbf{a}_2\,.$$

On the right side of this equation multiply and divide by the total mass M. Then the sum of the two individual second laws takes the form

$$F = M\left(\frac{m_1\mathbf{a}_1 + m_2\mathbf{a}_2}{M}\right). \tag{8.31}$$

This has the appearance of Newton's second law for the combined system. It states that the total force is equal to the total mass times the quantity within the parentheses, which may be called the "total acceleration." Actually it is an average acceleration, a particular kind of average called a weighted average. We conclude that Newton's second law applies to the nonrigid system as a whole, provided "the" acceleration of the system, $\mathbf{a}$, is defined by the formula,

$$\mathbf{a} = \frac{m_1\mathbf{a}_1 + m_2\mathbf{a}_2}{M}\,. \tag{8.32}$$

For two particles of equal mass, the acceleration of the system is an average in the usual sense, $\mathbf{a} = \frac{1}{2}\mathbf{a}_1 + \frac{1}{2}\mathbf{a}_2$. For particles of unequal mass, the acceleration of the more massive particle is weighted more heavily in the sum. For example, if particle number 1 has a mass of 1.7×10^{-24} gm and particle number 2 has a mass of 3.4×10^{-24} gm, the acceleration of the combined system is $\mathbf{a} = \frac{1}{3}\mathbf{a}_1 + \frac{2}{3}\mathbf{a}_2$. It should be noticed that because this is a vector sum, the system as a whole can have zero acceleration even though its parts are accelerated.

The acceleration of a system of particles is related to an important concept called the center of mass. The center of mass of a system is the average position of all the mass in the system—again a weighted average. Its mathematical definition is this:

$$\mathbf{s}_c = \frac{m_1\mathbf{s}_1 + m_2\mathbf{s}_2 + m_3\mathbf{s}_3 + \ldots}{M}, \tag{8.33}$$

where $\mathbf{s}_1$; $\mathbf{s}_2$, $\mathbf{s}_3$, etc. are the position vectors of the particles that make up the system; m_1, m_2, m_3, etc. are the masses of the particles; M is the total mass; and $\mathbf{s}_c$ is the notation for the position vector of the center of mass. In general, the center of mass need not be at any material point of the system. The center of mass of a pair of equally massive particles is a point in space halfway between them (Figure 8.14). The center of mass of a hydrogen atom (proton plus electron) is a point in space between proton and electron, much closer to the more massive proton than to the electron. Often centers of mass can be accurately guessed without recourse to Equation 8.33. The center of mass of a uniform rod, for example, is at the center of the rod, and of a rectangular slab, at the center of the slab. For nonuniform systems, Equation 8.33 may be required. The center of mass of a 75-kg boy and a 50-kg girl is located at

$$\mathbf{s}_c = \tfrac{3}{5}\mathbf{s}_B + \tfrac{2}{5}\mathbf{s}_G .$$

This is a point 60% of the way from the girl to the boy.

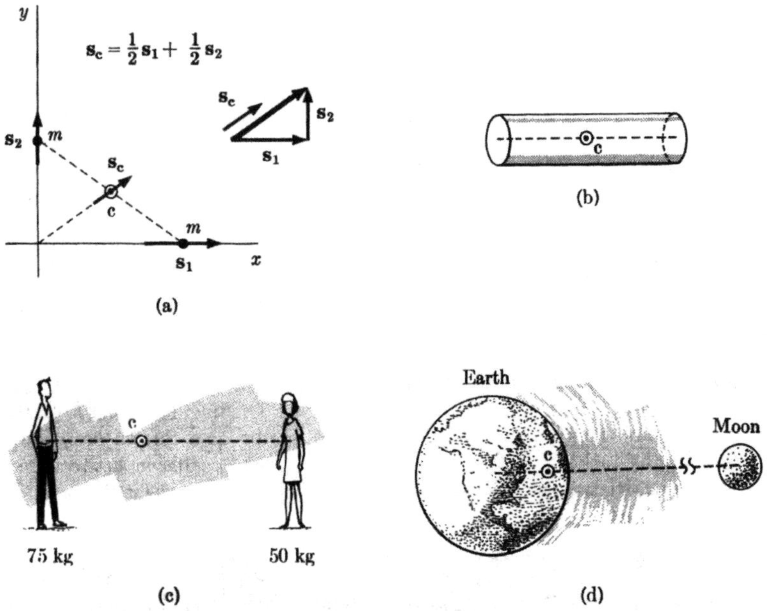

FIGURE 8.14 Centers of mass. **(a)** The center of mass of a pair of protons is a point in space halfway between the particles. To choose a coordinate axis along the line joining the particles would be easier, but it is not necessary. **(b)** The center of mass of a uniform cylinder is located at its geometric center. **(c)** The center of a mass of a 75-kg boy and a 50-kg girl is located 40% of the way from boy to girl. **(d)** The center of mass of the earth-moon system is a point about 2,900 miles from the center of the earth. This is the point that follows a smooth elliptical path about the sun. In all four diagrams, the center of mass is designated by a small circle and the letter c.

As the particles comprising a system move, their center of mass moves. Even though it may be a point in space not attached to any matter, the center of mass is a well-defined point with a well-defined velocity. The foregoing discussion suggests at once what the velocity of the center of mass must be. It is

$$\mathbf{v}_c = \frac{m_1\mathbf{v}_1 + m_2\mathbf{v}_2 + m_3\mathbf{v}_3 + \ldots}{M}, \tag{8.34}$$

the weighted average of the individual velocities. Similarly, the acceleration of the center of mass is the weighted average acceleration,

$$\mathbf{a}_c = \frac{m_1\mathbf{a}_1 + m_2\mathbf{a}_2 + m_3\mathbf{a}_3 + \ldots}{M}, \tag{8.35}$$

This is exactly the same as what was called the acceleration of the system, defined for a two-particle system by Equation 8.32. Therefore, Newton's second law applied to the system as a whole (Equation 8.31) may be rewritten

$$\mathbf{F} = M\mathbf{a}_c. \tag{8.36}$$

We have discovered a point in space, the center of mass, whose acceleration is governed by Newton's second law, using the total force and the total mass of the system. To express this interesting and beautifully simple result another way: The center of mass moves as if all of the mass of the system were concentrated there and all of the forces were applied there (even if, in fact, there is neither mass nor force there). The center of mass is obviously a very special point. It might be called the perfect representative point, the only point whose motion accurately reflects the average motion of the whole system. Figure 8.15 shows a wrench sliding across a smooth horizontal surface. Its center of mass, indicated by a black cross, executes straight line motion at constant velocity, just as a single particle would, while the motion of every other part of the wrench is more complicated. It is the center of mass of our own earth-moon system, not any other point, that follows a smooth elliptical path around the sun. If a tennis racket is tossed into the air with a spinning and twisting motion, its center of mass follows the same uniformly accelerated motion as a thrown ball, while the rest of the racket gyrates. During the violence of a collision between two elementary particles, their center of mass continues its straight-line motion at constant speed.

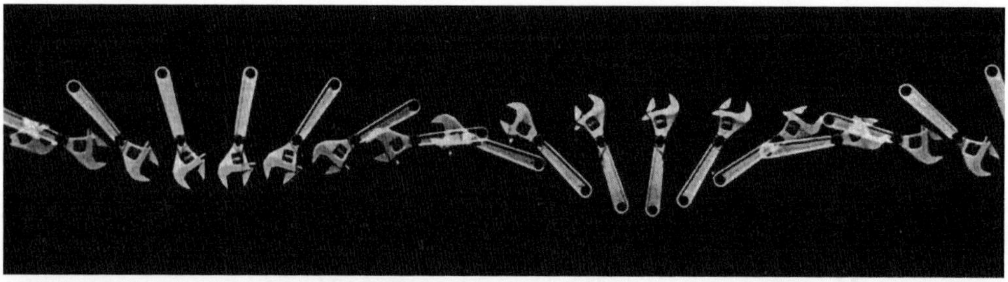

FIGURE 8.15 Motion of wrench without friction. Only the center of mass (black cross) moves with constant velocity. (Photograph from P.S.S.C. *Physics*, D. C. Heath & Company, 1965.)

8.9 *Motion near the earth: gravitational mass*

27 As long as man has been a thinking creature, he must have been aware of two obvious facts about gravity. First, every object near the surface of the earth is pulled downward. Second, any unsupported object falls. Man has not always thought clearly about the difference between these two facts, but if we are to understand gravity properly, it is essential to distinguish the nature of the force from the nature of the motion.

The gravitational force experienced by a body is called its weight. The force acts whether or not the body is in motion, but it is most conveniently measured when the body is at rest. Every scale to record weight is a force-measuring device. The units painted on the dial of a scale are usually not force units, an unfortunate complication that will be clarified below. Nevertheless, the deflection of the needle of the scale is a measure of the gravitational force experienced by the object being weighed. Since the measurement of weight is a static measurement, it provides important information about a fundamental force of nature independent of the laws of motion.

If, armed with a scale and some curiosity, you set about learning all that you could about gravity, you would probably wish to seek answers to the following questions: (1) How does the gravitational force experienced by a given object depend on the location of the object? (2) Does the force on a particular object depend on its state of motion? (3) How does the force vary from one object to another? A little experimentation would reveal that the answer to the first question is, to a good approximation: Not at all. Your weight is the same upstairs as down, the same in one city as in another. Actually precise experiments reveal slight weight difference from one part of the earth to another, and a gradual decrease of weight with increasing altitude. However, if you concluded that every object near the surface of the earth experiences a definite fixed gravitational force regardless of its location, you would be very close to the truth. The dependence of the gravitational force on motion is less easy to measure in general, but experiments reveal that it is independent of motion. You could, for example, weigh yourself on a uniformly moving train or a uniformly moving elevator and verify that your weight is unchanged. In accelerated motion, an ordinary scale would give a different reading. However, this comes about only because the scale, when accelerated, is no longer a suitable device for measuring the gravitational force. Other methods could reveal that even in accelerated motion, the gravitational force is unchanged.

Galileo considered both of these questions and answered them both, the first by measurement, the second by guess. He argued that since an object weighs the same at different heights, it probably continues to feel the same constant force as it falls through these different heights. In this way he connected the observed constant acceleration of a falling object to the hypothesized constant force acting on it.

Much can be learned about the variation of gravitational force from one object to another through static weight measurements, although the full clarification of this subject requires a study of motion as well. Common experience tells us that the weight of an object is somehow related to the amount of matter in it. For a particular kind of material, weight is proportional to volume. One pint of water weighs one pound (lb), one quart weighs two lb, one gallon weighs eight lb.* On the other hand, a gallon of gasoline weighs 6 lb, and, a gallon of lead weighs 109 lb. There is no universal connection between weight and volume. One comes considerably closer to a simple relationship in trying to establish a connection between weight and the number of nucleons (protons plus neutrons) in a piece of material. There is almost a direct proportion, but not quite. The ratio of the weight of an object to the number of its constituent elementary particles varies by as much as eight parts in a thousand from one substance to another. However, weight has been found to be directly proportional to mass. This proportionality, apparently universal for all substances of all sizes, has been tested experimentally for aluminum and gold to an accuracy of three parts in one hundred billion.

Because of the observed proportionality between weight and mass (inertial mass), there is often confusion between these concepts. Recall that inertial mass, defined as the constant of proportionality between force and acceleration in Newton's second law, is a measure of the resistance to a change in motion of an object subjected to a force, *any* force. Weight, on the other hand, is specifically a measure of the intensity of response of an object to a *gravitational* force. Weight is, in fact, the gravitational force experienced by an object, regardless of its motion. There is no obvious reason at all for a connection between these two concepts. Their direct proportionality, discovered by Newton, remained a mysterious coincidence for over two centuries. How Einstein's general theory of relativity made use of this proportionality and removed its mystery is discussed in Chapter Twenty-Two.

---

* The pound is a unit of mass (equal to 454 gm). When we say that an object "weighs" two lb, we mean that its mass has been so determined indirectly by measuring the gravitational force acting on it. See page 175.

The experimental proportionality between gravitational force (weight) and inertial mass can be written in this way:

$$F_{\text{grav}} = mg\,, \tag{8.37}$$

where g is a constant of proportionality. Comparison of this law of force with Newton's second law shows at once that g must have the same dimension as acceleration. Its numerical value in the cgs system is

$$g = 980\ \text{cm/sec}^2.$$

If the gravitational force is the *only* force acting on an object, its acceleration is related to that force by Newton's second law,

$$F_{\text{grav}} = ma\,, \tag{8.38}$$

Since $F_{\text{grav}} = mg$, it follows at once that

$$a = g\,. \tag{8.39}$$

An object near the surface of the earth acted upon only by the earth's gravitational force moves with a constant acceleration whose magnitude is equal to 980 cm/sec^2. On the other hand, if gravity is only one of several forces acting on an object, the acceleration of the object need bear no simple relation to the constant g. You are free, for instance, to accelerate a book in your hand in any direction or to hold it motionless. No matter how it moves in response to the sum of all the forces acting on it, one of those forces, the force of gravity, continues to act vertically downward with the constant magnitude mg.

The equations governing the uniformly accelerated motion of free fall have been discussed before and are set forth in Equations 8.7–8.14. No motion that takes place in the air is truly "free" fall, for air resistance always acts to exert some force in a direction opposite to the motion. For sufficiently low velocity, or for a very dense aerodynamically shaped object such as a bomb, this retarding force may be so small compared with the gravitational force that it can be neglected. Then the motion is successfully described by Equations 8.7–8.10 or 8.11–8.14 as uniformly accelerated motion (with $a = g$). In an opposite limit, when the retarding force of air resistance becomes so great as to be equal in magnitude (but opposite in direction) to the gravitational force, terminal speed is achieved; the acceleration is zero.

The fact that objects near the earth on which the frictional force is negligible all fall with the same constant acceleration is so familiar that its significance is easily overlooked. It ought to be regarded as truly remarkable. Consider a number of different weights. On each the pull of gravity is different. Yet all fall with precisely the same acceleration. How does it come about that a body which experiences twice the gravitational force of another also has exactly twice the inertial mass so that both move in exactly the same way? It is not our purpose here to explain this coincidence, only to re-emphasize that there is no obvious reason for mass and the magnitude of a gravitational force to be proportional. No other fundamental forces in nature are simply related to mass. Proton and positron, for example, experience the same electric force despite a large difference in their mass. Proton and neutron, with nearly the same mass, feel very different electric and magnetic forces.

The puzzle of the proportionality of weight to mass came into prominence in science in the latter part of the nineteenth century. In particular, a question arose whether the ratio of weight to mass at a particular spot on the earth is truly constant or varies slightly from one substance to another. In order to facilitate discussion of this question and to highlight the puzzle, a new concept, gravitational mass, was introduced. The meaning of gravitational mass can best be made clear by considering a hypothetical experiment. Imagine yourself in a laboratory equipped to measure force and velocity and acceleration, with a number of different objects at your disposal, including a standard kilogram. Since the experiment is hypothetical, imagine too that you have available a perfectly flat frictionless horizontal surface upon which to slide the objects. A known horizontal force (*not* a gravitational force) is applied to each object in turn, and the acceleration measured. Through Newton's second law, the force and acceleration provide a measurement of the inertial mass, m_I (the subscript I is introduced to emphasize that it is inertial mass being measured):

$$m_I = \frac{F}{a}. \tag{8.40}$$

Having defined and measured the inertial mass of each object by means of experiments independent of gravity, you turn attention to the nature of the gravitational force acting on each of your test objects. Here a fundamental postulate enters. Its truth will be tested by further experiments. The postulate is this: The magnitude of gravitational force experienced by a particular object depends on only one single property of that object, a scalar quantity whose numerical magnitude is directly proportional to the strength of gravitational interaction of the object. This is obviously a powerful postulate about the simplicity of nature. It says that a single number is enough to reveal how much gravitational force a body will experience, regardless of its shape or color or chemical composition, or whether it contains liquid, solid, or gas. To this numerical quantity the name gravitational mass is given. It is to be regarded as an intrinsic property of a body, completely analogous to electric charge. Whereas charge measures the electric interaction strength of a body, gravitational mass measures the gravitational interaction strength. Indeed gravitational charge might be a better name for the quantity.

According to the postulate, the gravitational force experienced by any one of the test objects should be proportional to its gravitational mass, m_G. The mathematical expression of the postulate is

$$F_{\text{grav}} = g m_G, \tag{8.41}$$

where g is a constant of proportionality.* The new concept, gravitational mass, needs a unit of measurement and a standard for this unit. As a matter of convenience, you may decide to measure gravitational mass in the same unit as inertial mass, the kilogram, and to adopt as its standard the same one-kilogram object. Having done this, you proceed to weigh each of your test objects, that is, to determine by a static measurement the gravitational force on each. The weight of the standard kilogram determines the constant g through the equation

$$g = \frac{F_{grav}}{1\text{ kg}}.$$

The weight of any other object determines its gravitational mass through the equation

$$m_G = \frac{F_{grav}}{g}. \tag{8.42}$$

To review: Using an experiment *with* motion but *without* gravity, you have measured the inertial masses of a number of objects. Using an experiment *without* motion but *with* gravity, you have measured the gravitational masses of the same set of objects. Despite the confusing facts that both quantities are measured in the same unit and referred to the same standard, it is clear that their operational definitions are entirely distinct. Therefore, if you discover the equality of inertial and gravitational mass for every different object, 28

* The quantity g may depend on where the experiment is carried out. It is a constant only at a given place. The more essential fact about g is that it is independent of all properties of the object experiencing the force. The explicit factor m_G depends on the object; g does not.

$$m_I = m_G, \tag{8.43}$$

you have discovered a very significant fact about nature, by no means a mere matter of definition.

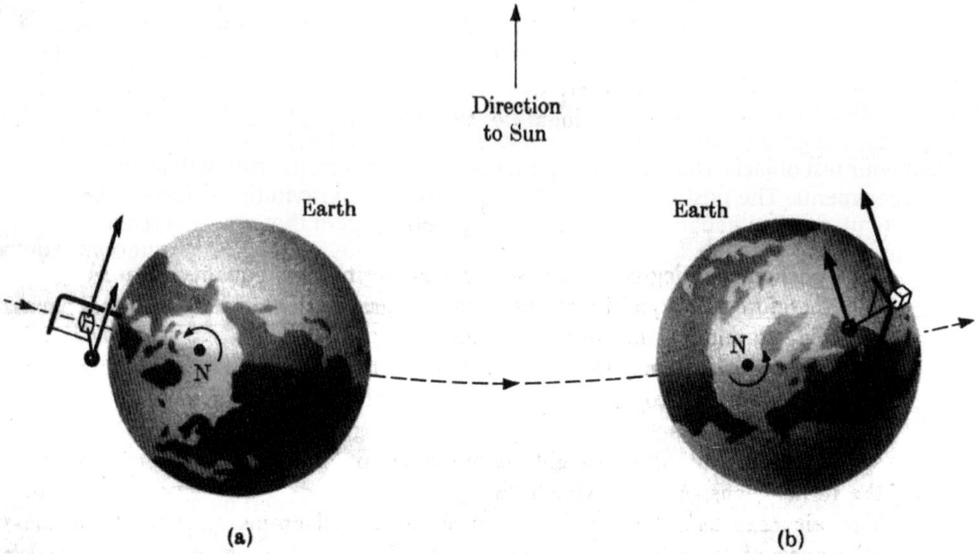

FIGURE 8.16 The Dicke experiment seeks to discover a difference in the gravitational acceleration toward the sun of two objects made of different material, here shown for convenience as a cube and a sphere. Both are suspended from a long fiber. If the two objects are not equally accelerated toward the sun, the fiber will be twisted, one direction at sunset **(a)**, the opposite direction at sunrise **(b)**. The diagrams show a hypothetical greater acceleration of the cube than of the sphere. In fact, no differences were detected.

The success of Newton's law of gravitation in describing planetary motion implies that gravitational and inertial mass are equal. The constant acceleration of all freely falling objects implies the same thing. However, the verification of their equality to an extremely high degree of precision—now to within three parts in 10^{11}—rests on more specialized experiments, carried out first by Roland von Eötvös in the 1880s and refined by Robert Dicke of Princeton University in the early 1960s. Dicke sought to discover whether different materials experience identical gravitational acceleration toward the sun. His sophisticated experiment, although complex in execution, is simple in principle.* Several weights made of different material are suspended from a long fiber in such a way that if one of the weights were more strongly accelerated toward the sun than another, the fiber would twist (Figure 8.16). Because of the rotation of the earth, the sun's pull would twist the fiber (if at all) first in one direction, then in the other direction. Such a daily alternation of the direction of twist would be easier to detect than a twist in only one direction, and could be attributed with reasonable certainty to the action of the sun rather than to unknown perturbing influences. Except for inessential complications produced by the rotation of the earth and the pull of the moon, the suspended weights are, in effect, engaged in "free fall" toward the sun (along with the rest of the apparatus, the experimenter, and the earth itself). The acceleration of each toward the sun is proportional to the ratio of its gravitational mass to its inertial mass,

* A good account of experiments to test the equality of gravitational and inertial mass is to be found in an article by Robert Dicke, entitled "The Eötvös Experiment," in the *Scientific American*, December 1961, p. 84.

$$a \sim \frac{m_G}{m_I}.$$

To a phenomenal degree of precision, Dicke verified the equality of accelerations and therefore the equality of the ratios of gravitational mass to inertial mass for gold, aluminum, and other substances.

Because gravitational force is proportional to inertial mass, an ordinary scale may be calibrated to read either in force units or in mass units. Most scales read in mass units—pounds or grams or kilograms. It is usual to refer to a scale's mass reading as weight, permissible so long as it is understood what the term means. A steak which "weighs" one pound has a mass of 454 grams. The air traveler's "weight" allowance of 20 kilograms is actually a mass allowance of 20 kilograms or 44 pounds. The freedom in choice of a scale's unit comes about, of course, because of the equivalence between inertial and gravitational mass. To consider a fictitious but analogous situation, suppose it were true that every man's weight were proportional to his height. Then scales could be calibrated in inches. In order to determine his height, a man need only step on a scale and read off his height in inches. If asked his weight, he might answer, "I weigh 69 inches." A scale no more measures inertial mass than it measures height. Yet because of the separately established proportionality between gravitational force (or gravitational mass) and inertial mass, it is possible to learn the inertial mass of an object by weighing it.

Throughout most of the rest of this book, we shall accept the equivalence of inertial and gravitational mass as an established fact of nature, and accordingly refer simply to "mass," designating either or both kinds of mass by the same symbol m.

8.10 Motion in two dimensions

Many of the most important examples of motion in nature take place in two dimensions: the collision and interaction of a pair of elementary particles, the deflection of an electron by a nucleus, the orbit of a satellite relative to the earth, the orbit of the earth about the sun. For each of these kinds of motion, the trajectories are traced out in an imaginary plane in space, because there are no forces at work able to deflect the particles or objects out of the plane of their motion. The general principles governing motion in three dimensions are no different from the principles for two-dimensional motion. However, they are more simply illustrated for two-dimensional motion, which is all that we shall consider. Mainly, it is the vector character of Newton's second law that accounts for interesting features of multidimensional motion not found in one-dimensional motion. In this section we shall consider uniform circular motion and the parabolic motion of projectiles near the earth.

Uniform Circular Motion

The kinematics of uniform circular motion was considered in Section 6.7. There we discovered that the acceleration of an object moving at constant speed in a circle is directed toward the center of the circle and has a magnitude given by

$$a = \frac{v^2}{r}. \tag{8.44}$$

Newton's second law, $\mathbf{F} = m\mathbf{a}$, reveals at once what force is required to produce such motion. Since the magnitude of the acceleration is a constant, the magnitude of the force is also a constant, given by

$$F = \frac{mv^2}{r}. \tag{8.45}$$

Moreover, since force and acceleration are vector quantities, both with the same direction, the force must be directed toward the center of the circle. Such a force, directed always toward a fixed center, is called a central force. It is also often known by Newton's original name for it, a centripetal (center-seeking) force. For a stone swung on the end of a rope, the centripetal force is transmitted by the rope. Without the help of a material intermediary, gravitation and electricity act across empty space to produce central forces. The moon is held in an almost circular orbit by a force directed toward the center of the earth. The electron in a hydrogen atom experiences a centripetal force toward the central proton in the atom. Uniform circular motion can occur only when a central force acts, although central forces produce other kinds of motion as well.

Some of the many earth satellites now in orbit are tracing out nearly circular paths at almost constant speed. Since a low-altitude satellite (at an altitude of about 100 miles) experiences nearly the same gravitational force as at the earth's surface, we can easily calculate the time required for it to complete one circle. The inward force is $F = mg$. Because it is in uniform circular motion, this force must also be equal to mv^2/r, so that we can write

$$mg = \frac{mv^2}{r}. \qquad (8.46)$$

The mass of the satellite thus proves to be irrelevant. Its inward acceleration is the same regardless of its mass. This result is the same as that discovered by Galileo for freely falling objects for a very good reason. A satellite *is* a freely falling object, in fact more precisely so than a thrown baseball or a falling stone, since air resistance is negligible for a satellite, or very nearly so. One way to understand how an object can "fall" forever and never strike the earth is this. The downward force of the earth on the satellite causes its velocity vector to rotate toward the center of the earth. By the time the velocity vector has rotated through 90 deg to what would have been a vertical direction where it was, the satellite has moved one-quarter of the way around the earth, so that the new direction of the velocity, far from pointing toward the center of the earth, is directed parallel to the earth's surface at its new location. The velocity vector is continually being changed in direction, always toward the center of the earth, but the orbital speed of the satellite prevents it from striking the earth. A lower-speed projectile would have its velocity vector rotated toward the earth before the projectile could move far enough to avoid striking the earth. Another way to say all this is that the surface of the earth falls away (that is, curves away) from the satellite as fast as the satellite falls toward it.

We seek the period of rotation of a low-altitude satellite. First, its speed may be found from Equation 8.46, rewritten as

$$v^2 = rg.$$

The distance from the center of the earth, r, is about 4,000 miles, or 6.4×10^8 cm, and g near the surface of the earth is 980 cm/sec^2. The product of these numbers yields $v^2 = 6.3 \times 10^{11}$ (cm/sec)^2. Taking the square root, we get the approximate speed of a low-altitude satellite,

$$v = 7.9 \times 10^5 \text{ cm/sec},$$

about 24 times the speed of sound in air, which is 3.3×10^4 cm/sec. In more familiar speed units 7.9×10^5 cm/sec is the same as 17,700 miles/hr. Since the circumference of the earth is 4×10^9 cm (recall that the meter was originally defined as one ten-millionth of the distance from pole to equator), the rotation period is equal to the distance, 4×10^9 cm, divided by the speed, 7.9×10^5 cm/sec. That is,

$$T = \frac{2\pi r}{v} = \frac{4 \times 10^9}{7.9 \times 10^5} \cong 5{,}100 \text{ sec}. \qquad (8.47)$$

This time is about 85 minutes, somewhat less than an hour and a half.

At higher altitudes, satellites require longer to circle the earth. From their rotation period important information can be gained about the diminishing strength of the gravitational force with increasing distance from the earth. A communications satellite, for instance, at a height of 22,236 miles above the earth, completes one circular orbit in a leisurely 24 hours. How does its weight at that altitude compare with its weight on earth? To find out, we must calculate its acceleration, using Equation 8.44. Its distance from the center of the earth is 26,236 miles, or 4.22×10^9 cm. This number multiplied by 2π is the circumference of its orbit, about 2.65×10^{10} cm. Then its speed is this distance divided by its period of 86,400 sec (24 hours):

$$v = \frac{2.65 \times 10^{10}\ \text{cm}}{8.64 \times 10^4\ \text{sec}} = 3.07 \times 10^5\ \text{cm/sec}\,,$$

slower than the low-altitude satellite. The inward acceleration of the satellite in its circular orbit is given by

$$a = \frac{v^2}{r} = \frac{(3.07 \times 10^5)^2}{4.22 \times 10^9} \cong 22.3\ \text{cm/sec}^2.$$

If the satellite's mass were 10^5 gm, the force required to keep it in orbit would be, according to Newton's second law,

$$F = ma = 10^5\ \text{gm} \times 22.3\ \text{cm/sec}^2 = 2.23 \times 10^6\ \text{dynes}.$$

On the other hand, its weight on earth would be

$$F = mg = 10^5\ \text{gm} \times 980\ \text{cm/sec}^2 = 9.8 \times 10^7\ \text{dynes},$$

some 44 times greater than its weight of 2.23 million dynes at its orbital distance. We conclude that the gravitational force exerted by the earth is 44 times weaker at an altitude of 22,200 miles than at the earth's surface.

By way of contrast, it is interesting to compare inner-atomic accelerations with the accelerations experienced by satellites. An electron in its innermost orbit in the hydrogen atom moves at a distance of 5×10^{-9} cm from the proton with a speed of about 3×10^8 cm/sec. Quantum mechanics prevents us from describing its motion as a simple circular orbit. Nevertheless, the formula $a = v^2/r$ provides a good approximate indication of the acceleration that must be experienced by the electron. The numbers just given, when substituted in this formula, give the result,

$$a \cong 2 \times 10^{25}\ \text{cm/sec}^2. \tag{8.48}$$

Obviously, it is in the small-scale world that accelerations are "astronomical." A man cannot withstand lasting accelerations above $10g$, or momentary accelerations above $1{,}000g$. The electron's acceleration is $2 \times 10^{22}g$, unimaginably large compared with any acceleration known in the macroscopic world.

Projectile Motion

Galileo made the first "breakthrough" in the description of projectile motion near the earth when he realized that the horizontal and vertical components of the motion can be treated independently. While the horizontal component proceeds with constant speed, the vertical component experiences a constant downward acceleration. His argument ran like this. Imagine a pebble sliding across a smooth horizontal table. When the pebble reaches the edge of the table, the only thing that changes is that the upward force of the table no longer compensates the downward force of gravity. Under the action of gravity, the pebble begins to experience a constant vertically downward acceleration. However, nothing happens to the horizontal part of its motion. On the table, there was no force in the horizontal direction (if friction was negligible), and off the table there is still no horizontal force. The uniform horizontal motion should continue unchanged after the pebble leaves the table. As Galileo then showed, a parabolic path through space results from the combination of uniform horizontal motion and constantly accelerated vertical motion.

Galileo's argument can now be founded on simple mathematics. According to Newton's second law,

$$\mathbf{F} = m\mathbf{a} .$$

Because this relationship between force and acceleration is a vector relationship, it must be separately true for each component. If we call the horizontal direction the *x*-direction and the vertically upward direction the z-direction (Figure 8.17), the component equations are

$$F_x = ma_x , \tag{8.49}$$

$$F_z = ma_z , \tag{8.50}$$

In the absence of air resistance, a projectile experiences no horizontal force and a vertically downward force equal in magnitude to mg. The same pair of equations may therefore be written

$$0 = ma_x ,$$

$$-mg = ma_z .$$

They may be examined separately. The first states that there is no horizontal acceleration. This means, of course, that the horizontal component of velocity, v_x, is a constant. If we consider in particular the case of the pebble that flies off the edge of the table, and choose the origin of coordinates to be the point where the pebble leaves the table, and the zero of time to be time when the pebble leaves the table, the subsequent increase of the *x*-coordinate is given simply by

$$x = v_x t . \tag{8.51}$$

Since the *z*-component of the motion starts initially from rest (there is neither upward nor downward motion at the instant the pebble leaves the table), the *z*-coordinate is described by the simple special case of Equation 8.7,

$$z = -\tfrac{1}{2}gt^2 , \tag{8.52}$$

negative since the acceleration is in the negative *z*-direction. Thus the horizontal component of the motion is the same as if there were no fall, and the vertical part of the motion is the same as if the fall were straight down.

Equations 8.51 and 8.52 may be combined to yield an equation for the trajectory of the pebble. For this purpose, the time must be eliminated. If, from Equation 8.51, $t = x/v_x$ is substituted into Equation 8.52, the trajectory equation results:

$$z = -\left(\frac{g}{2v_x^2}\right) x^2. \tag{8.53}$$

This is the equation of a parabola (Figure 8.17). The combination in parentheses is a constant; therefore z is proportional to x^2.

The analysis for projectile motion in general is the same in principle—constant horizontal component of velocity combined with constant vertical acceleration—but mathematically slightly more complex. The general x-coordinate equation is

$$x = x_0 + v_{x0}t. \tag{8.54}$$

The general z-coordinate equation is

$$z = z_0 + v_{z0}t - \tfrac{1}{2}gt^2\,. \tag{8.55}$$

These combine again to yield a parabolic trajectory. For a projectile fired upward at an angle from the origin (Figure 8.18) the constants x_0 and z_0 are zero, and some simplification results. In that case, the trajectory equation is

$$z = \left(\frac{v_{z0}}{v_{x0}}\right)x - \left(\frac{g}{2v_{x0}}\right)x^2\,. \tag{8.56}$$

This equation can be used to find, among other things, the range of the projectile. When the projectile again reaches earth, the z-coordinate is again zero; then

$$0 = \left(\frac{v_{z0}}{v_{x0}}\right)x - \left(\frac{g}{2v_{x0}}\right)x^2\,.$$

This quadratic equation has two solutions; the first, $x = 0$, locates the initial point of the trajectory. The second,

$$x = \frac{2v_{x0}v_{z0}}{g}, \tag{8.57}$$

gives the range of the projectile. The horizontal distance covered is twice the product of the initial horizontal and vertical components of velocity divided by the constant acceleration g. Verify that Equation 8.57 has dimensional consistency.

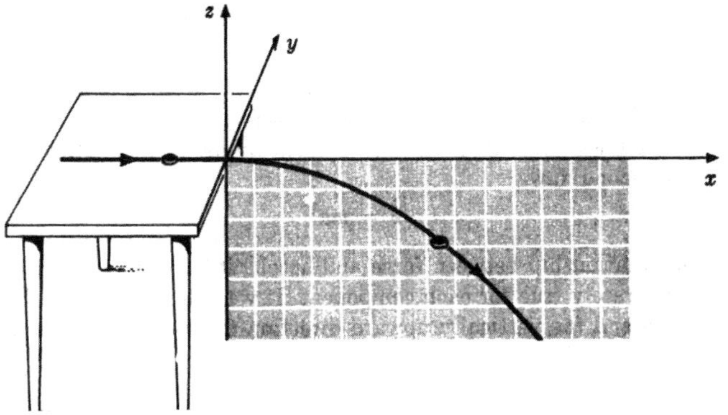

FIGURE 8.17 Coordinate system and path of pebble for motion described by Galileo. The parabolic part of the path is expressed by Equation 8.53.

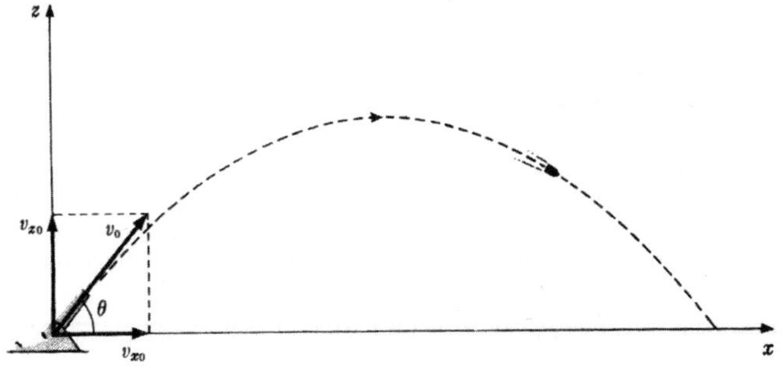

FIGURE 8.18 Parabolic trajectory of a projectile. The resolution of the initial velocity vector into its horizontal and vertical components is superimposed- on the spatial diagram.

8.11 *Assessment of Newton's laws of motion*

Just as modern mathematicians have found no reason to modify Euclid's treatment of Euclidean geometry, modern physicists have found no reason to alter Newton's original formulation of Newtonian mechanics, or classical mechanics, as it is now often called to distinguish it from its twentieth-century offspring, relativistic mechanics and quantum mechanics. Newton based mechanics on three laws, still the cornerstones of classical mechanics. His first law, introduced in Section 8.5, describes the motion of objects free of external forces. His second law, discussed in Section 8.6, relates force to motion. These two laws, to be elaborated in this section, are laws of motion. They make use of the concepts of velocity and acceleration, in addition to the concepts of force and mass. Newton's third law is of a different character. Although it has exceedingly important implications for motion, it is not directly a statement about motion. It is a statement about forces in nature—not about any particular force, but about forces in general. The third law is the subject of Section 9.2.

Newton's three laws form an adequate basis for calculation and prediction of all aspects of motion provided that the forces are known. If they are not known, calculations in mechanics become mere exercises in mathematics. Therefore the full theory of mechanics requires for completeness an understanding of the particular forces that do act in nature as well as an understanding of the general laws governing all motion and all forces. Newton's three laws are of the latter type: general statements about force and motion. The law of gravitational force discovered by Newton,* on the other hand, is a particular statement about one fundamental force. When other fundamental force laws were discovered —the law of electric force in the latter part of the eighteenth century, and the law of magnetic force early in the nineteenth century—these too were found to conform to the three laws of Newtonian mechanics. In addition, Newton's three laws encompassed all of the common and complicated nonfundamental forces, such as the pushes and pulls of physical contact and frictional resistance. In this century we have learned about the strong interactions and the weak interactions, fundamental forces that act in the submicroscopic domain where Newton's laws are not all valid. Nevertheless, it is interesting that after three centuries of scientific progress, the law of gravitational force discovered by Newton and used by him to validate the laws of motion remains in some sense the most fundamental force known to science. It is the only completely universal force, experienced by all matter and all energy.† Also worth noticing is the fact that in the present day Newton's laws, far from having faded into mere historical interest because we have learned of the limits of their validity, probably find more use in day-to-day science than ever they did in the eighteenth or nineteenth centuries. They are still the essential tools of the astronomer, the aerodynamicist, the engineer, and the physicist.

The Significance of Newton's First Law

29 Our special concern in this section is to clarify the significance of Newton's first law, which is the simplest, yet at the same time the subtlest, of the three laws. Examined superficially, Newton's first law appears to be only a special case of the second law, and hardly worth a separate statement. If $\mathbf{F} = m\mathbf{a}$ (Newton's second law), it follows that in the absence of force there is no acceleration (Newton's first law). Why then does the first law deserve the dignity of a separate statement as one of the fundamental laws of mechanics?

* The law of gravitational force was not wholly original with Newton. It had been speculatively suggested earlier by Robert Hooke and possibly others. Nevertheless, Newton does deserve the credit he is usually accorded. His analytical power and insight converted the law of gravitational force from speculation to solid fact, and Kepler's picture of the solar system from mysterious simplicity to mechanical inevitability; see Chapter Twelve.

† The gravitational force is also the only force directly related to the properties of space and time, through Einstein's general theory of relativity; see Chapter Twenty-Two.

Several answers can be given to this question, some of them appreciated by Newton, others the result of later and deeper insights.

(1) Newton's first law *defines* what kind of motion is "natural" or undisturbed motion, and thereby it delimits what kind of motion requires further explanation. According to Aristotle, the forward progress of an arrow through the air required an explanation in terms of some kind of motive power, for to him the "natural" state of motion of the arrow was the state of rest. Newton's first law states instead that the constant horizontal component of the arrow's velocity (in the absence of friction) is what is natural. This component being unaccelerated, it requires no explanation in terms of force. On the other hand, Aristotle and scholars after him accepted circular motion in the heavens as the natural course of events. For Newton, the accelerated motion of the moon and planets *did* require an "explanation," that is, a description in terms of force. The first law makes clear that the planets, if undisturbed by outside influences, would move not in the circles envisioned by the ancients, but in straightline trajectories off into space.

(2) There is a modern view of Newton's first law that makes it truly independent of the second law, a view that was perhaps already implicit in Newton's thinking about the laws of motion. A modern statement of the first law can be phrased this way: The center of mass of an isolated system free of external influences moves with constant velocity. This statement contains two improvements over Newton's original statement of the law, given on page 161. First, it replaces his word "body" by "center of mass." Not all parts of a free body move with constant velocity; only its center of mass does so (Figure 8.15). Second, the new version of the law dispenses altogether with the concept of force. It is this aspect of the new statement that makes it significantly different from the original. As originally phrased by Newton, the first law suggested that one might first have to know what force is in order to know when none is present. Now it seems preferable to refer to "an isolated system free of external influences." True isolation can never be achieved; but it can be approached. The new version of the law states that natural or undisturbed motion is achieved by removing a system to a very great distance from all possible outside influences. The significance of the law does not depend on any understanding of force. The concept of force then enters the theory of mechanics only through Newton's second law.

When Newton's laws of motion are looked at in this way, there is a difference between an automobile moving down a straight road at constant velocity and a meteorite cruising at constant velocity in the depths of empty space. The automobile has no acceleration because the sum of all the forces acting on it happens to be zero. Its constant velocity merely illustrates a special case of Newton's second law. The meteorite, on the other hand, directly illustrates Newton's first law. That it should have constant velocity follows from its isolation, even if we knew nothing about force, or even if force had been inconsistently defined.

(3) In a way that Newton could not have foreseen, the separate formulation of the first law has been justified by history. We know now that Newton's second law ceases to be valid in the submicroscopic world governed by quantum mechanics and in the high-speed world governed by relativity; and that in modern physics, the concept of force no longer plays a central role. The first law, on the other hand, at least when expressed in terms of momentum, remains valid, so far as we know, throughout all of nature. Its universally valid formulation is this: The total momentum of an isolated system free of all external influences is a constant. In the domains where Newtonian mechanics is valid, constant total momentum and constant velocity of the center of mass are equivalent, a relationship which will be explored in the next chapter. In the theory of relativity, however, the law of inertia can be expressed only

in one way; in terms of momentum. Because the law of momentum conservation, the modern equivalent of Newton's first law, is of such great importance, we can say that Newton's first law, far from being a mere special case of the second law, is a pillar of modern physics. In the absence of external influences, a system retains constant momentum.

(4) Finally, Newton's first law is significant in that it defines inertial frames of reference. It is literally true that the acceleration of an object depends on one's point of view. In more technical language, it depends on the frame of reference of the observer. A motionless object seen by an accelerated observer seems to be accelerated. An accelerated object is, relative to an observer accompanying it, at rest. These facts mean that isolated systems are unaccelerated only for certain observers. The frames of reference of these observers are called inertial frames of reference. Only in inertial frames are Newton's laws of motion valid.

Newton's second law may be regarded as a definition of force and at the same time an important law of nature. The first law plays a similar dual role. It provides at one and the same time a definition of an inertial frame of reference and a significant statement about nature. That Newton's first law is not hollow can best be appreciated by stating it in this way: In a frame of reference where one free object is unaccelerated, all other free objects are also unaccelerated. The first half of the statement defines an inertial frame, the second half expresses the law of inertial (undisturbed) motion. Inertial frames of reference are those in which isolated systems move in the simplest possible way (Figure 8.19).

There is more to be said about frames of reference. We devote the last section of this chapter to this tricky but important subject.

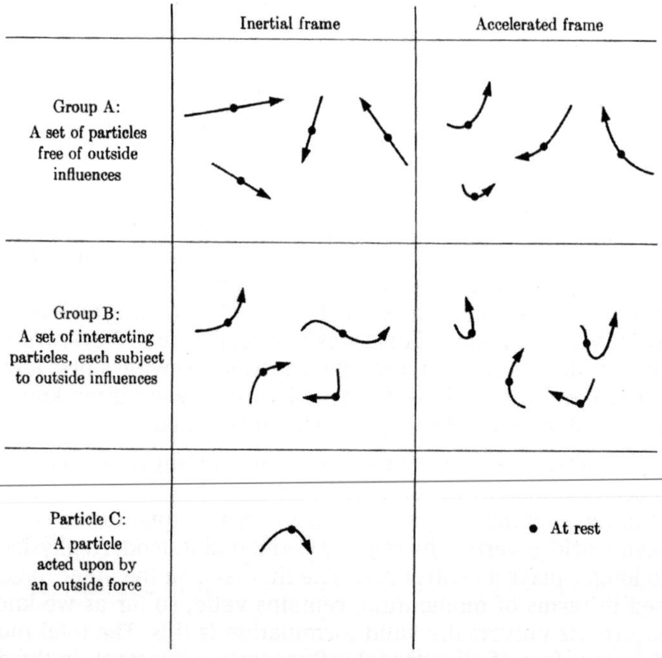

FIGURE 8.19 The motion of particles in two frames of reference. In an inertial frame (left), all objects free of outside influences move with constant velocity. In an accelerated frame (right), the motion of a particular object, such as particle C, might be greatly simplified, but nearly all other motions appear accelerated, including the motion of free particles (Group A).

8.12 *Frames of reference*

To be meaningful, every law of nature must be defined with respect to some frame of reference, or, more often, with respect to some set of frames of reference. By a frame of reference is meant, usually, a set of coordinates with respect to which positions and distances are measured. In classical mechanics, the frame of reference is purely a *spatial* frame of reference; time is assumed to be independent of the frame of reference. The frame may be moving and thereby itself depend on time, but the time coordinate is supposed to be uninfluenced by the motion. To make this point clear, imagine yourself an observer in an airplane en route to San Francisco. In your frame of reference, San Francisco is moving toward you—its position vector is changing with time. To an observer in New York, San Francisco is motionless a fixed distance away. The concept of time is important in distinguishing the two frames of reference. However, the *measurement* of time is presumed to be the same for you and the ground-based observer. Careful comparison would reveal no detectable difference in the two scales of time. In this century we have learned that for two frames of reference with sufficiently high relative speed, there is in fact a perceptible difference in time scales as well as a large difference in position measurements. This circumstance, explained by the special theory of relativity, has led to the idea of a four-dimensional frame of reference, making both space and time measurements depend on the state of motion of the observer. The new insight is of practical importance, however, only at speeds near the speed of light. Even in the frame of reference of an orbital astronaut, very swift by human standards, the measurement of time is so little different than on earth that the difference may be neglected. In the domain of nature where classical mechanics is valid, time may be regarded as absolute, and only spatial measurement is dependent on the frame of reference.

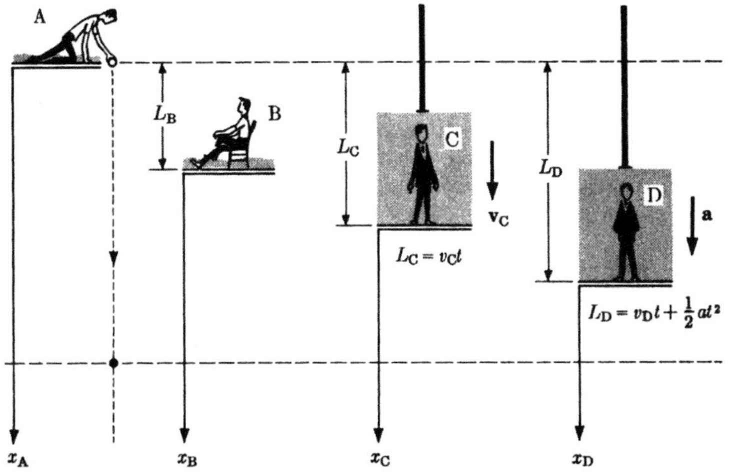

FIGURE 8.20 Frames of reference. The motion of a falling ball is studied by four observers. Observers A, B, and C are in inertial frames of reference. Observer D is in an accelerated frame of reference.

Newton, and Galileo before him, realized that the laws of motion are the same in all inertial frames of reference. This does *not* mean that all aspects of the description of motion are the same in different reference frames. To see the effect of the frame of reference on the description of motion, we can study a particular example. Observer A [Figure 8.20(a)] drops a ball from rest at the origin of his coordinate system. If friction is negligible, the ball falls with constant acceleration g, and he describes its motion by the equation,

$$x_A = \tfrac{1}{2}gt^2 . \tag{8.58}$$

Three other observers examine the same motion. Observer B [Figure 8.20(b)] uses a coordinate system whose origin is a fixed distance L_B below the origin of reference frame A. Observer C, in an elevator descending with constant speed v_C [Figure 8.20(c)], describes the motion relative to his origin whose distance L_C from the starting point of the motion increases uniformly with time:

$$L_C = v_C t . \tag{8.59}$$

Finally, observer D [Figure 8.20(d)], riding a downward accelerated elevator, measures the position of the falling ball relative to the floor level of his elevator. This elevator floor, with downward acceleration a, increases its distance L_D from the origin of frame A according to the equation,

$$L_D = v_D t + \tfrac{1}{2}at^2 . \tag{8.60}$$

(For simplicity, we assume that both elevator floors passed the origin of reference frame A at $t = 0$.) Now we can write expressions for x_B, x_C, and x_D, the position coordinates of the falling ball in the other three frames of reference. Each is equal to the total distance of fall, x_A, minus the distance to the new origin (L_B, L_C, or L_D) :

$$x_B = x_A - L_B ; \tag{8.61}$$

$$x_C = x_A - L_C ; \tag{8.62}$$

$$x_D = x_A - L_D ; \tag{8.63}$$

Substitution from Equations 8.58, 8.59, and 8.60 yields

$$x_B = -L_B + \tfrac{1}{2}gt^2 , \tag{8.64}$$

$$x_C = -v_c t + \tfrac{1}{2}gt^2 , \tag{8.65}$$

$$x_D = -v_D t + \tfrac{1}{2}(g - a)t^2 , \tag{8.66}$$

Although Equations 8.58, 8.64, and 8.65 for x_A, x_B, and x_C are all different, they have this in common: Each describes uniformly accelerated motion with the same acceleration g. Each fits the pattern of Equation 8.11. Frames A, B, and C are inertial frames of reference. In each of them the measured acceleration is the same and the law of force is the same. According to observer D, by contrast, the falling ball has a lesser acceleration, $g - a$ (Equation 8.66), and therefore an apparently lesser force. Because the frame of reference D is not an inertial frame, observer D cannot describe the motion using the same form of Newton's second law as his fellow observers who are in inertial frames.

Another example is illustrated in Figure 8.21. A baseball player throws a ball whose trajectory, in his frame of reference, is a parabola (again ignoring air friction). To another player who runs along under the ball, its trajectory is a straight line, vertically up and down. These two observers assign to the ball different positions and different velocities. However, if both measured the acceleration, they would get the same answer. Since one player is unaccelerated relative to the other, the ball has the same acceleration with respect to both. Newton's second law too is valid for both.

30 The principle that the laws of mechanics are the same in all inertial frames of reference is called the principle of Galilean relativity. It is really a principle of invariance, a statement that the laws governing motion are the *same* in different inertial frames of reference, even though the description of a particular example of motion might be different in the different

reference frames. The significance of Galilean relativity can perhaps best be emphasized by stating the principle in this way: If a particular example of motion is possible (that is, consistent with the laws of motion) in one inertial frame of reference, it is also possible in every other inertial frame of reference. This way of looking at Galilean relativity can be illustrated with the help of Figure 8.21. The motion of the ball as observed by player B is straight-line vertical motion. Although this is *not* the motion observed by player A, it is a possible motion that player A *could* produce in his own frame of reference if he wished to—simply by throwing the ball straight up. Similarly, the parabolic motion observed by player A is a motion that could be observed in player B's moving frame of reference if the ball were thrown differently. The class of all possible motions for observer A is precisely the same as the class of all possible motions for observer B. Therefore they agree about the laws of motion.

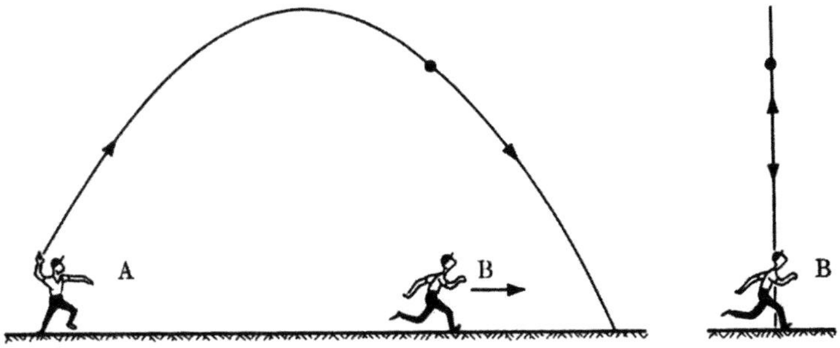

FIGURE 8.21 Description of motion in two inertial frames of reference. Observer A, at rest with respect to the ground, and observer B, running beneath the thrown baseball, assign to the ball different trajectories and different velocities, but they agree about its acceleration and about the law of force governing its motion.

An important consequence of the principle of Galilean relativity is that it is impossible to discover the existence of a truly stationary frame of references, a frame at rest with respect to a hypothetical ether filling all of space. Since all physical phenomena (at least mechanical phenomena) follow the same laws in a frame moving uniformly with respect to this supposed ether as in a frame at rest with respect to the ether, there is no mechanical experiment that can distinguish the two. Galileo himself used the principle to support the Copernican view of the solar system. He argued that the laws of motion would be no different on a moving earth than on a stationary earth, so that one might as well adopt the simpler Copernican view of a moving earth in place of the far more complicated Ptolemaic view of a solar system built around a stationary earth. Before Galileo, almost all scientists and philosophers from Ptolemy onward had argued that the straight vertical fall of a dropped stone demonstrated that the earth must be at rest. According to Galileo's principle of relativity, it proved no such thing.*

It cannot be said that we really know of the existence of any ideal inertial frame of reference, since it is impossible to isolate an object entirely from outside influences and find in what frame it is unaccelerated. However, the ideal can be approached closely. For certain purposes, it is useful to redefine an inertial frame as one in which an object is unaccelerated if it experiences only a gravitational force (instead of no force). Such a frame is provided by a freely falling laboratory, a spaceship coasting in orbit. Although in some ways it is the most fundamental inertial frame, the orbiting laboratory is special in two ways. First, it casts gravity in a favored role. Objects in it respond only to forces *other* than gravity.

* Einstein's principle of relativity is more general. It asserts that inertial frames of reference are equivalent for *all* the laws of nature, not just for mechanics; see Chapter Twenty-One.

Second, it is "local." It acts as an inertial frame only over a region of space small enough so that the gravitational field can be regarded as constant. For motion in the small, the orbiting laboratory is a good inertial frame.

For describing motion on a larger scale, the earth itself is sometimes an adequate approximation to an inertial frame. A sun-fixed frame is better, and a frame in which the galaxy as a whole is unaccelerated is still better. A workable rule is this: If the acceleration of a particular frame of reference with respect to an ideal inertial frame is much smaller than the accelerations being studied, then the frame of reference is a suitable approximation to an inertial frame. The surface of our earth has an average acceleration with respect to a sun-fixed frame of about 2.7 cm/sec^2, less than 1% of the gravitational acceleration g. For any particular application, it must be decided whether this magnitude, 2.7 cm/sec^2, is sufficiently small. If it is, an earth-fixed frame adequately approximates an inertial frame. Elementary particles experience accelerations enormously larger than 2.7 cm/sec^2. For their study, the surface of the earth defines a very close approximation to an inertial frame. For the study of planets, on the other hand, whose orbital accelerations are less than 2.7 cm/sec^2, the earth-fixed frame, far from being an inertial frame, must be regarded as a rapidly accelerated frame. The orbital acceleration of Mars, for instance, is only 0.25 cm/sec^2. For the study of its motion, a sun-fixed frame is a good approximation to an inertial frame, but an earth-fixed frame is not. It is exactly this circumstance, of course, that led to the Copernican revolution which placed the sun, not the earth, as the fixed center of the solar system. Now we know that even the sun is not quite unaccelerated. It orbits about the galaxy with an acceleration of 2×10^{-8} cm/sec^2.

31 One fascinating frame-of-reference problem remains unsolved. If absolute position and absolute velocity are without meaning in empty space, why does absolute acceleration seem to have a meaning? Another way to phrase the question is this: Why are inertial frames of reference a special preferred set? Think of two observers in empty space, one in an inertial frame of reference, one in an accelerated frame. In the inertial frame, an isolated object moves with constant velocity. In the accelerated frame, it does not. Rather, it seems to be responding to forces. In fact, the laws of motion are very different in the two frames of reference, and much simpler in the inertial frame. However, the inertial frame can be said to be accelerated *relative* to the accelerated frame, just as a stationary observer is accelerated relative to a falling object. Each frame is accelerated in exactly the same way relative to the other (equally but opposite). Why, then, is the inertial frame a privileged frame of reference? No satisfactory answer to this question has been found. It seems fairly certain that the distribution of matter spread throughout the universe must in some way be the material framework with respect to which absolute acceleration (but not absolute velocity) has a meaning. From this point of view, an inertial frame is one unaccelerated with respect to the material framework of the universe.

Consider our own rotating earth. In a frame of reference fixed with respect to the earth, the stars in the sky execute circular and therefore accelerated motion. This was the ancient view of the universe. In a frame fixed with respect to the sun, on the other hand, it is the earth's surface that executes circular accelerated motion; the acceleration of the stars is much less than in the earth-fixed frame. Evidently nature prefers a frame of reference in which the stars have little or no average acceleration. Despite the impossibility of defining absolute motion or absolute rest, there does seem to be a material framework in the form of distant matter. It has been speculated that in an otherwise empty universe, a material object would possess no mass—that a body's resistance to acceleration, the property we call its mass, is a resistance to its being moved out of one of nature's preferred inertial frames of reference. Perhaps, continues the speculation, even the existence of measurable intervals of space and time depends upon the existence of matter and energy in the universe. According to this fascinating point of view, the properties and behavior of the elementary particles—nature's tiniest bits of matter—are influenced in an important way by the bulk matter of distant galaxies. No problem in science is more intriguing than the search for links between the submicroscopic and cosmological domains of the physical world.

EXERCISES

8.1. A block of wood sits at rest on a plank inclined at 30 deg to the horizontal. Friction prevents the block from sliding. A gravitational force of 10^5 dynes acts vertically downward on the block. What are the magnitude and direction of the force exerted by the plank on the block?

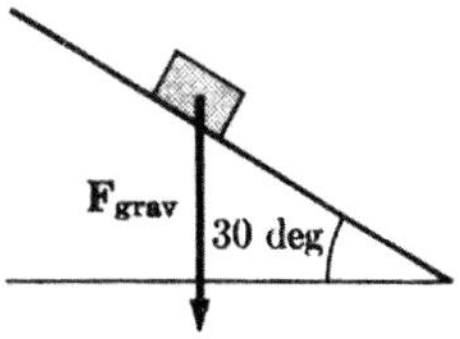

8.2. The motion of a particle moving with constant speed v along the x-axis is described by Equation 8.2. At $t = 1$ sec, $x = 3.5$ cm; at $t = 2$ sec, $x = 5.0$ cm. **(1)** What is its speed v? **(2)** What is the magnitude of the constant x_0? **(3)** Where is the particle at $t = 10$ sec?

8.3. The speed of a runner in the 100-yard dash varies as shown in the figure. **(1)** Sketch a graph showing his acceleration as a function of time. **(2)** Estimate roughly his average speed during the race.

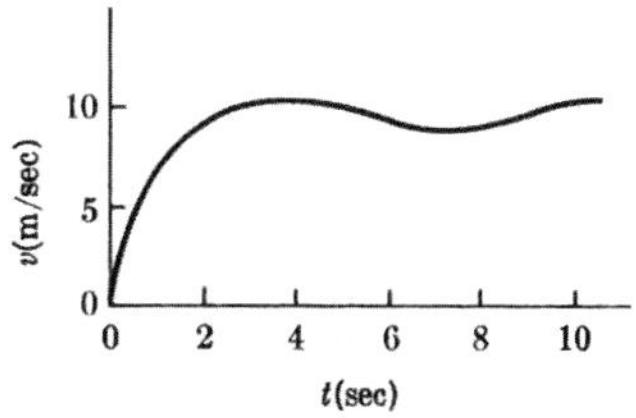

8.4. A particle moving along the x-axis with constant acceleration passes the origin ($x = 0$) at $t = 0$ with velocity $v_0 = 100$ cm/sec. Later, at $t = 4$ sec, it passes the origin again. **(1)** What is its acceleration? **(2)** What is its velocity the second time it passes the origin? **(3)** Does it pass the origin a third time? If so, when? **(4)** What is the greatest value of x reached by the particle? (Recall that at the moment it reaches maximum x, the particle comes momentarily to rest.)

8.5. From Equation 8.12, deduce the average velocity for uniformly accelerated motion in the time interval from zero to t. Combine this with the basic definition of average velocity for one-dimensional motion (Equation 8.6) to derive Equation 8.11 expressing the position as a function of time for constant acceleration.

8.6. To gain more effect, the two men pictured in Figure 6.5 change their direction of pushing to make angles of 30 deg with the long axis of the automobile. Each continues to push with a force of 10^7 dynes. A force of friction, also of magnitude 10^7 dynes, acts to the rear. The mass of the automobile is 1.3×10^6 gm. What is its acceleration?

8.7. The block and plank of Exercise 8.1 are replaced by a small car and frictionless air track also inclined at 30 deg to the horizontal. The vertical gravitational force on the car is 10^5 dynes. **(1)** What is the mass of the car? **(2)** How much time passes as it slides from rest a distance of 200 cm along the track? **(3)** What is its speed at the end of this slide?

8.8. A china cup drops from a height of 5 cm onto a wooden table. **(1)** With what speed does it strike the table? **(2)** If the table yields by 10^{-3} cm as it brings the cup to rest, what is the magnitude of the deceleration of the cup as it is stopped (assuming a uniform deceleration)? **(3)** How long does it take the table to stop the cup? (For calculational convenience, use $g = 1{,}000$ cm/sec^2.)

8.9. A sky-diver drops from a hovering helicopter. After 20 sec, during which time his downward speed has become steady at 6,000 cm/sec, he opens his parachute. In about 2 sec, his speed decreases to a new lower terminal speed of 600 cm/sec. **(1)** Sketch an approximate graph of his vertical component of acceleration vs. time from the moment he leaves the helicopter until he strikes the ground. **(2)** If his mass is 70 kg, what net force acts on him (give magnitude and direction) **(a)** just after he leaves the helicopter; **(b)** just before he pulls the ripcord; and **(c)** just before he reaches the ground?

8.10. A chain of three links, each link of mass 10^4 gm, hangs from a hook. The hook is drawn upward with a constant acceleration of 200 cm/sec^2. What is the net force acting on the center link?

8.11. Two men decide to "weigh" an automobile in the following way: The automobile rests on a level surface with the brakes off. One man pushes it with a steady force of 2×10^7 dynes. After 10 sec the second man measures the speed of the automobile to be 150 cm/sec. If they assume zero frictional force, what will they conclude its mass to be in grams? What would they predict the reading to be in pounds if the same automobile is put on a scale? *Optional:* Since friction cannot be neglected in this experiment, would the actual scale reading be higher or lower than their predicted value? Explain.

8.12. A man whose mass is 175 lb stands on a bathroom scale in an elevator. When the elevator starts upward he notices that the scale registers a constant 200 lb. **(1)** What is his acceleration? **(2)** After 2 sec what is his speed? **(3)** When does he pass the next floor 300 cm above his starting point?

8.13. **(1)** Describe carefully how mass may be defined as a primitive concept, and alternatively as a derived concept. **(2)** Why is there arbitrariness in whether a concept is primitive or derived?

8.14. **(1)** With a string, a shoelace, or a lightweight key chain used to support small weights, carry out a crude experiment to find out how the period of a pendulum of fixed length depends upon the mass of the suspended object. Report briefly on what you did and what results you obtained. **(2)** Interpret your results in the light of the theoretical formula, Equation 8.28. What can you conclude about the quantity K in this formula? Why is this quantity called a constant? *Optional:* With the same "apparatus," find out how the period of a pendulum of fixed mass depends upon the length of the supporting cord.

8.15. Verify by calculation the statement made in the caption of Figure 8.14 that the center of mass of the earth-moon system lies 2,900 miles from the center of the earth.

8.16. In the xy plane a 10-kg mass is placed at the point $x = 5, y = 1$ and a 5-kg mass at the point $x = 5, y = 4$. **(1)** Where is their center of mass? **(2)** A third mass of 30 kg is added at the point $x = 2, y = 2$. Where is the center of mass of the combination of three masses?

8.17. The distance between the centers of the carbon atom and the oxygen atom in a molecule of carbon monoxide (CO) is 1.13×10^{-8} cm (1.13 A). **(1)** Where is the center of mass of the molecule? **(2)** If, at a given instant, the carbon nucleus is moving downward with a speed of 3×10^4 cm/sec and the center of mass is at rest, in what direction and with what speed is the oxygen nucleus moving? (Ignore any contribution of electrons to the molecular mass.)

8.18. Two astronauts leave their space vehicle while in orbit around the earth. They start to throw a medicine ball between them. Describe in general terms how the velocity of each man changes as the game progresses. Describe how the velocity of the center of mass of the two men changes.

8.19. A vernier thrust rocket at one edge of a spacecraft fires accidentally, setting the craft into a twisting and gyrating motion. The rocket pushes the craft, whose mass is 2×10^6 gm, with a force of 4×10^7 dynes for 2 sec, before it is shut off. Describe in words the motion of the center of mass of the craft. Calculate the change of velocity of the craft, assuming that the rocket thrust acts in the same direction for the 2 sec it is applied.

8.20. **(1)** An object subjected to gravitational forces, from whatever sources, experiences a force proportional to its own gravitational mass, m_G. Show that if it experiences no other forces, its acceleration is proportional to the ratio of its gravitational mass to its inertial mass ($a \sim m_G/m_I$). **(2)** Use this fact to explain why a pair of satellites can "fly formation," remaining close together without the use of power.

8.21. Assuming the center of the earth to be at rest in an inertial frame of reference, calculate the centripetal acceleration of a point on the equator, showing it to be approximately 3.4 cm/sec^2. Why is this number larger than the average centripetal acceleration over the whole surface of the earth, stated on page 188 to be about 2.7 cm/sec^2?

8.22. A straight section of model railroad track is joined to a section which is the arc of a circle of radius 40 cm. The model locomotive whose mass is 1,500 gm moves at a constant speed of 120 cm/sec. **(1)** What is the total force on the locomotive on the straight section? Why? **(2)** What is the total force on the locomotive on the curved section? What is the source of the force? (Because the forces on the straight section and the curved section are not equal, actual railroads do not join circular arcs directly to straight tracks. Instead "railroad curves" of gradually changing radius of curvature are used.)

8.23. A man swings a mass of 1 kg in a horizontal circle of radius 200 cm. **(1)** Assuming that he can exert a maximum inward force of 2×10^7 dynes, what is the greatest speed he can impart to the circling mass? **(2)** If he shortens the cord attached to the mass and continues to exert maximum inward force, does the speed of the mass increase or decrease? Does his angular velocity increase or decrease? (Ignore the minor effect of gravity in this exercise and assume that the man pulls horizontally.)

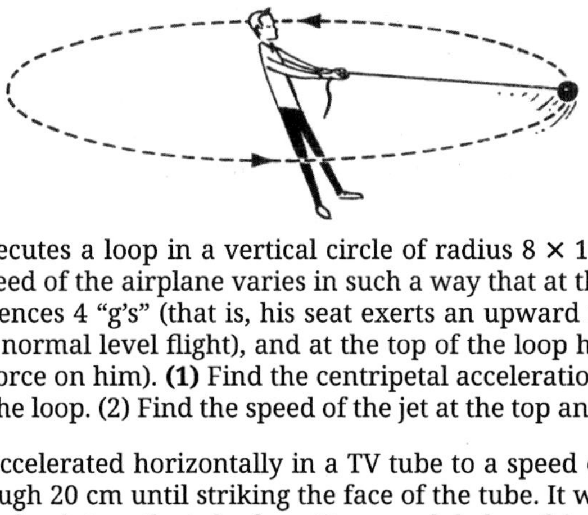

8.24. A jet trainer executes a loop in a vertical circle of radius 8×10^4 cm. Going around the loop, the speed of the airplane varies in such a way that at the bottom of the loop the pilot experiences 4 "g's" (that is, his seat exerts an upward force on him 4 times greater than in normal level flight), and at the top of the loop he is "weightless" (his seat exerts no force on him). **(1)** Find the centripetal acceleration of the jet at the top and bottom of the loop. (2) Find the speed of the jet at the top and bottom of the loop.

8.25. An electron is accelerated horizontally in a TV tube to a speed of 10^9 cm/sec. It then flies freely through 20 cm until striking the face of the tube. It was initially aimed directly at a central point on the tube face. How much below this point does it actually strike because of the downward force of gravity? Is this a significant effect tending to blur television pictures?

8.26. For the projectile motion illustrated in Figure 8.18 and described by Equations 8.54 to 8.57 (with $x_0 = z_0 = 0$), write algebraic expressions for the horizontal and vertical components of velocity, v_x and v_z. Derive an expression for the maximum height of the trajectory reached at its midpoint.

8.27. When the curvature of the earth is taken into account, would you expect the range of an intercontinental ballistic missile to be greater or less than the range given by Equation 8.57? Explain the reason for your answer.

8.28. Prove that a projectile reaches its highest point when it has covered exactly one half of its horizontal range, (HINT: Its horizontal and vertical components of motion are independent.)

8.29. A proton, with a mass of 1.7×10^{-24} gm, moves with a speed of 10^{10} cm/sec. **(1)** It flies down the center of a horizontal evacuated pipe 1 cm in diameter and 10^4 cm long. Does the force of gravity cause it to strike the pipe? **(2)** It enters a bubble chamber where it experiences a magnetic force of 10^{-7} dynes perpendicular to its velocity. What is its radius of curvature? **(3)** It enters a material in which it experiences a deceleration of 10^{14} cm/sec^2. How long does it take to stop?

8.30. At $t = 0$, observer A in Figure 8.20(a) throws a ball downward from the origin of his coordinate system with initial speed $v_0 = 300$ cm/sec. At this same moment, the origin of observer C's coordinate system [Figure 8.20(c)] is passing A's origin with constant downward velocity of 500 cm/sec. **(1)** Write the ball's equation of motion from C's point of view (x_C vs. t). **(2)** At what time does the ball overtake and pass the origin of C's frame of reference? (3) Where is C relative to A at this time? **(4)** How should A throw a second ball in order that his description of its motion be identical to C's description of the motion of the first ball ?

8.31. A metal ball is launched toward a suspended can at the same moment the can is released and starts to fall. Prove that the ball will hit the can in midair (ignoring friction) regardless of the initial speed of the ball, (HINT: Consider as a function of time the distance Δy by which the ball falls below its projected straight line path.)

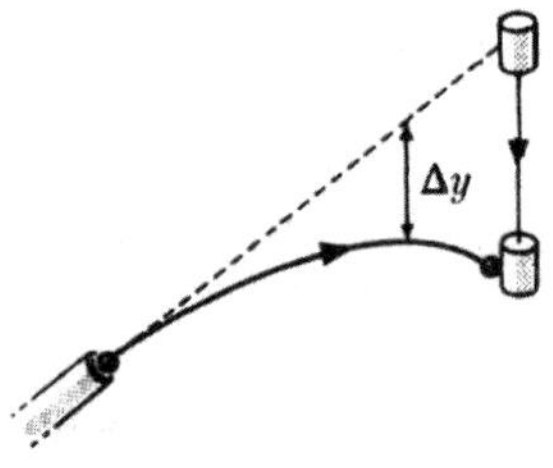

8.32. The motion of a particle on the z-axis is described by the equation

$$x = Ae^{-Bt},$$

where A and B are constants, and t is time. **(1)** What is the physical dimension of B? **(2)** What is the physical meaning of A? **(3)** Plot, as well as you can in freehand sketches, x vs. t, v vs. t, and a vs. t (v is velocity, a is acceleration). **(4)** What is the mathematical form of the law of force (i.e., how does it depend on x, the particle position)?

NINE

Momentum

Chapter Eight was built around four central concepts of mechanics—force, mass, length, and time. 32 The number of important mechanical concepts is not large. Counting such derived concepts as velocity and acceleration, probably less than a dozen suffice for a complete foundation of mechanics. It is not far from the truth to say that when these concepts are understood, the science of mechanics is understood. But understanding a concept involves more than knowing simply a definition or an equation. To "understand" a concept thoroughly, one must know its operational definition (how it is defined and how it is measured), its dimension and unit, its relationship through laws and equations to other important concepts, its role in various parts of the physical world, and its typical magnitudes. In addition, one must acquire something of an intuitive "feeling" for the concept, a kind of immediate recognition and appreciation that can come only after the concept has been looked at from enough different angles and seen at work in enough different examples to make it seem a familiar friend. A word of caution: This worthwhile kind of scientific familiarity founded on quantitative study is very different from the illusory familiarity bred by nontechnical everyday usage of words like force and energy and acceleration.

In a competition for "most fundamental concept" in mechanics, space and time would probably share first prize. Close behind would be the three key concepts: momentum, energy, and angular momentum. Each of these acquired importance in mechanics because of its appearance in a conservation law. Each proved useful in solving problems of motion. Each weathered the twentieth-century revolution of physics and emerged as a key concept in the new theories of relativity and quantum mechanics. Each has found a more solid foundation in physics through its relationship to a principle of symmetry in nature. Although we are concerned in this part of the book only with the classical mechanics of Newton, it is worth knowing which parts of this theory remain valid on the contemporary frontiers of physics (the conservation of momentum, for example), and which parts have been found to have limited scope (the conservation of mass, for example).

In earlier chapters the conservation laws of momentum, energy, and angular momentum have been discussed. In this chapter we seek a deeper insight into the concept of momentum and into the foundation of its conservation law. Angular momentum is the theme of Chapter Ten, and energy is the subject of Chapter Eleven.

9.1 Momentum and Newton's second law

As stated in Chapter Four, the momentum of a particle is defined, in classical mechanics, as the product of the mass and the velocity of the particle:

$$\mathbf{p} = m\mathbf{v}\,. \tag{9.1}$$

The usual notation for momentum is **p**. It is a vector quantity, whose direction is the same as the direction of the velocity. Its physical dimension is mass times length divided by time; its cgs unit is gm cm/sec.

At speeds near the speed of light, the definition of Equation 9.1 has been found to be "wrong"—that is, not useful. In order to have a vector property of motion that enters into a conservation law valid at all speeds, it is necessary to replace Equation 9.1 by what is called the relativistic definition of momentum. This definition is stated in Section 9.8 (Equation 9.60) and explored more deeply in Chapter Twenty-One. It is important too to remember that Equation 9.1 is completely inappropriate for massless particles, which do possess momentum in spite of having no mass. In most of this chapter, however, we shall be concerned with realms of nature where Equation 9.1 adequately defines the momentum of a particle.

In earlier sections we have emphasized several times that momentum is a significant concept worth defining and studying because, under certain circumstances (in isolated systems), it is conserved. There is one other reason to define momentum. It is the property of motion most simply and directly related to force. The rate at which momentum changes is equal to the force applied.

Consider a moving particle which, in some interval of time Δt, changes its velocity from $\mathbf{v}_1$ to $\mathbf{v}_2$, and, correspondingly, its momentum from $\mathbf{p}_1$ to $\mathbf{p}_2$. Using the definition of Equation 9.1 and the fact that the mass of the particle is constant, we can write for the change of momentum,

$$\mathbf{p}_2 - \mathbf{p}_1 = m\mathbf{v}_2 - m\mathbf{v}_1 = m(\mathbf{v}_2 - \mathbf{v}_1) . \quad (9.2)$$

If $\mathbf{p}_2 - \mathbf{p}_1$ is abbreviated $\Delta\mathbf{p}$, and $\mathbf{v}_2 - \mathbf{v}_1$ is abbreviated $\Delta\mathbf{v}$, Equation 9.2 takes on a simpler appearance:

$$\Delta\mathbf{p} = m\Delta\mathbf{v} . \quad (9.3)$$

The change of momentum, $\Delta\mathbf{p}$ (a vector quantity) is equal to the mass of the particle times the change of its velocity. Next we divide both sides of Equation 9.3 by the time interval Δt to obtain

$$\frac{\Delta\mathbf{p}}{\Delta t} = m\frac{\Delta\mathbf{v}}{\Delta t} . \quad (9.4)$$

For a sufficiently short interval, the quotient on the left represents the rate of change of momentum. On the right appears the rate of change of velocity, which is the same as acceleration. Therefore Equation 9.4 can be rewritten,

$$\frac{\Delta\mathbf{p}}{\Delta t} = m\mathbf{a} . \quad (9.5)$$

This equation states that the mass of a particle times its acceleration is equal to the rate of change of its momentum—an equality that follows from the definition of momentum and the assumption that mass is a constant. Since force is equal to mass times acceleration (Equation 8.16: Newton's second law), we can write

$$\mathbf{F} = \frac{\Delta\mathbf{p}}{\Delta t} . \quad (9.6)$$

In words: The rate of change of momentum of a particle is equal to the force acting on it. This is an alternative form of Newton's second law that is simpler than the form used in Chapter Eight in that it makes use of only one property of the particle, its momentum, rather than two, its mass and its acceleration. If a given force $\mathbf{F}$ acts on a particle for a time interval Δt, we can deduce directly from Equation 9.6 that the particle momentum changes by an amount $\Delta\mathbf{p}$ equal to the product of $\mathbf{F}$ and Δt:

$$\Delta\mathbf{p} = \mathbf{F}\Delta t . \quad (9.7)$$

The combination $\mathbf{F}\Delta t$ on the right is called the "impulse." It was in this form that Newton first stated his second law. The change of momentum of a body is equal to the impulse applied to it.

As a simple example illustrating the application of Equation 9.7, consider a particle of mass m moving along the x axis with speed v_0. Then a constant force of magnitude F is applied in the x direction. After a time t, the impulse delivered to the particle in the x-direction is equal to Ft. Therefore the momentum changes by the magnitude of this impulse:

$$p - p_0 = Ft \, , \tag{9.8}$$

where p_0 is the initial x-component of momentum, and p is the x-component of momentum after the force has been acting for a time t. If we replace p_0 by mv_0 and p by mv, Equation 9.8 can be rewritten,

$$m(v - v_0) = Ft \, .$$

Rearrangement of this equation yields

$$v = v_0 + \frac{F}{m} t \, , \tag{9.9}$$

which is an equation of uniformly accelerated motion (see Equation 8.12) with acceleration equal to F/m. We find, as we should, that direct application of Equation 9.7 yields the same result as application of Newton's second law in the form $\mathbf{F} = m\mathbf{a}$.

For a single particle on which no forces act, the conservation of momentum is equivalent to the conservation (or constancy) of velocity. Only for combinations of two or more interacting particles—that is, for systems—does the conservation of momentum take on special interest and significance. Before pursuing conservation in general (Section 9.4), we must study Newton's third law and its implications for the motion of systems (Sections 9.2 and 9.3).

9.2 Newton's third law

Newton phrased his third law of mechanics in this way: "To every action there is always 33
opposed an equal reaction; or the mutual actions of two bodies upon each other are always equal, and directed to contrary parts." In modern terminology, the same statement would be rendered: "For every force in nature there is always an equal and opposite force; or the mutual forces of two bodies upon each other are always equal in magnitude and opposite in direction." The first half of Newton's statement of the third law says that all of nature's forces come in balanced pairs, equal and opposite. The second half of the statement explains that the equal and opposite forces exist between any pair of interacting bodies, and then goes on to emphasize the vector nature of the law. Why Newton chose to use the word "action" instead of the word "force"* is not certain—perhaps to let "action" summarize both the idea of "force" and the idea of "change of momentum," perhaps only for the linguistic reason that "action" has a convenient opposite, "reaction," and "force" does not. In any case, it is adequate to state Newton's third law as a law of equal and opposite forces. In its most compact form, as an equation, the law may be written

$$\mathbf{F}_{AB} = -\mathbf{F}_{BA} \, . \tag{9.10}$$

In words: The force exerted by object B on object A is equal and opposite to the force exerted by object A on object B (Figure 9.1). Note that the third law requires for its application two (or more) interacting bodies. The first and second laws can be phrased in terms of single bodies.

* In his original Latin, Newton refers to *vis* (usually translated "force") in his statements of the first and second laws, and to *actio* (usually translated "action") in his statement of the third law.

The most important aspect of Newton's third law to understand first is that it is a law of force, not directly a law of motion. It is a statement about forces exerted and experienced by particles or objects regardless of how they might be moving. It states, for example, that if a book experiences a downward force of gravity of 10^6 dynes, then the earth is pulled upward by the book with a net force of 10^6 dynes, regardless of whether the book is being lifted or rests on a table or falls to the floor, and regardless of what other forces might be influencing the book and the earth. The motion of the earth and the motion of the book are determined by the *total* force acting on each (Newton's second law). Among all the forces possibly acting on the book, one is the earth's force of gravity. Among all the forces acting on the earth, one is the gravitational pull by the book. These particular two forces, according to Newton's third law, are equal and opposite, a balanced pair.

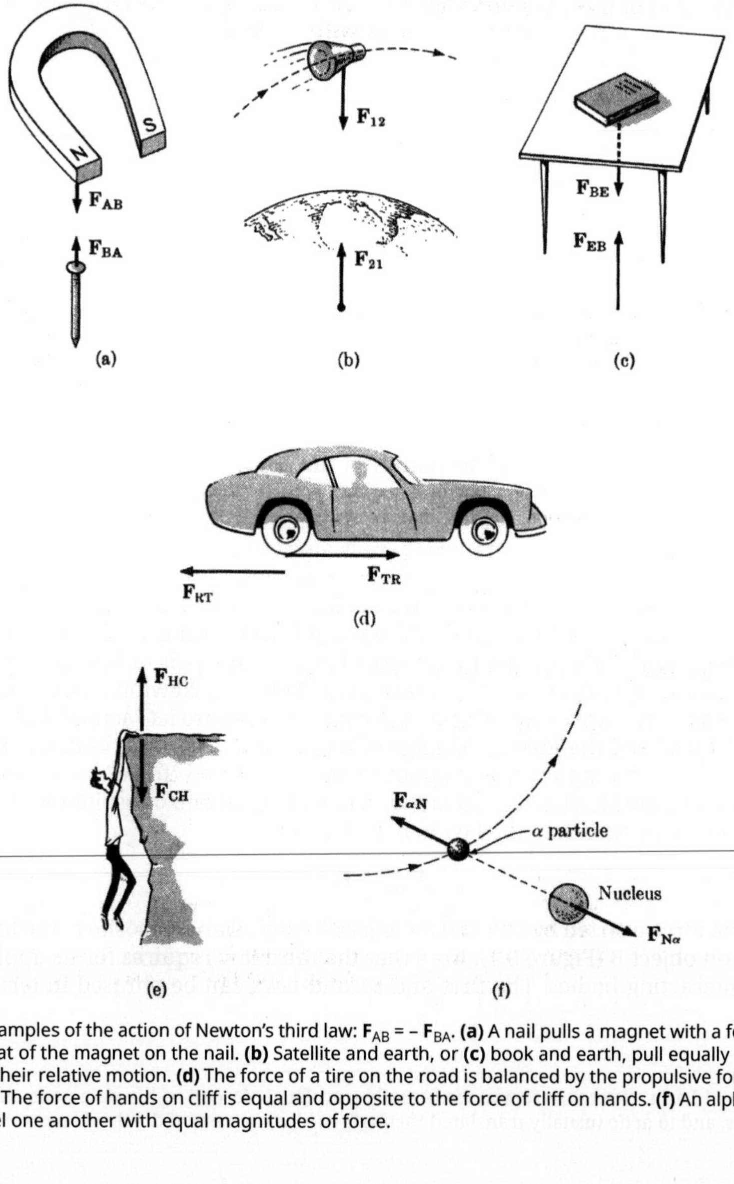

FIGURE 9.1 Examples of the action of Newton's third law: $F_{AB} = -F_{BA}$. **(a)** A nail pulls a magnet with a force equal and opposite to that of the magnet on the nail. **(b)** Satellite and earth, or **(c)** book and earth, pull equally on each other, regardless of their relative motion. **(d)** The force of a tire on the road is balanced by the propulsive force of the road on the tire. **(e)** The force of hands on cliff is equal and opposite to the force of cliff on hands. **(f)** An alpha particle and a nucleus repel one another with equal magnitudes of force.

A second important point to appreciate about Newton's third law is its generality, a 34
generality springing from what it does *not* say. It does not say anything about motion. It does not say anything about forces on object A arising from any source other than object B. It does not say anything about the kind of force acting between A and B. It is a completely general statement about forces in nature, independent of their strength, their source, their kind, or their effect. One cannot, for instance, learn the strength or even the direction of the gravitational force exerted by the earth on a book from Newton's third law. That knowledge comes from a particular law of force, the law of universal gravitation. The third law states only that the force of the earth on the book, whatever its strength, whatever its nature, is balanced by an opposite force of the book on the earth. Moreover, it leaves open the possibility that as yet undiscovered forces may act between bodies. If they do, and if Newton's third law remains valid, the new forces too must come in equal and opposite pairs. When Newton formulated the third law, only the gravitational force was understood quantitatively. Later electric and magnetic forces were found to obey the third law. Ordinary contact forces, basically electric in nature, conform to the same law. In the subatomic domain, the strong and weak interactions have been found to be consistent with the law of momentum conservation, which may be considered to be a modern synthesis of Newton's first and third laws.

Although the idea of equal and opposite forces seems easy enough to grasp, it can some- 35
times be perplexing in practical application, and it has some unexpected implications. Consider the "cart and horse paradox." If a cart pulls backward on a horse with exactly the same magnitude of force that the horse pulls forward on the cart, how is it that cart and horse are able to move at all? In more general terms: If every force in nature is canceled by an opposite force, why is there any motion in the world? Half an answer to these questions is easy to give. Motion does not require force; acceleration does. However, if we rephrase the questions to ask, Why is there any acceleration? the paradox seems to remain. Its resolution rests primarily on one key idea: *The equal and opposite forces of a balanced pair do not act on the same body.* The horse exerts *on the cart* a force $\mathbf{F}_{CH}$ (Figure 9.2). The force $\mathbf{F}_{HC}$ that is equal and opposite to $\mathbf{F}_{CH}$ does not act on the cart, but on the horse. The motion of the cart is determined only by the forces acting on it, not by any forces it may be exerting on other things. If we designate by $\mathbf{F}_C$ the total force acting on the cart (equal to $\mathbf{F}_{CH}$ plus frictional force from the ground plus the force of air resistance), we may apply Newton's second law to the cart alone:

$$\mathbf{F}_C = m_C \mathbf{a}_C . \tag{9.11}$$

Its mass times its acceleration is equal to the total force acting on it. The force of cart on horse which "cancels" the force of horse on cart actually has nothing to do with the motion of the cart. The oppositely directed forces acting on the cart are the forward force supplied by the horse and the backward force of friction. Since these are not action and reaction forces, they need not cancel. The force equal and opposite to that supplied by the horse is a

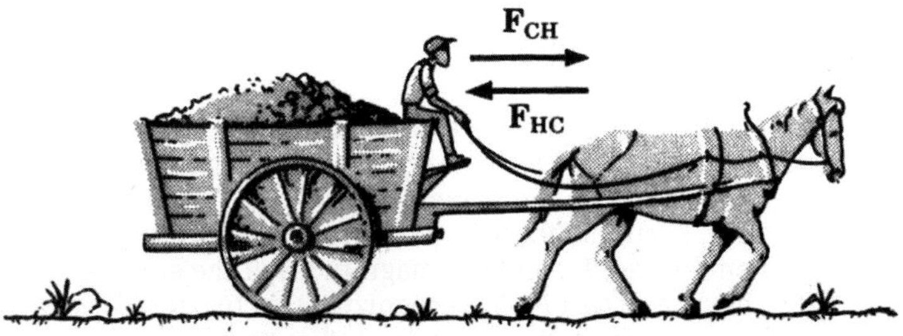

FIGURE 9.2 The "cart and horse paradox." Why does the system accelerate?

force on the horse and affects the horse's motion, not the cart's. The force equal and opposite to the ground's frictional force on the cart is a force exerted on the ground by the cart, which influences the motion of the ground but not of the cart. Newton's third law does not relate the different forces acting *on* the cart. Therefore they need not cancel, and the cart may accelerate.

Similar reasoning applies to the horse. His acceleration is dictated by the vector sum of the backward pull of the cart and the forward push of the ground. Since this pair of forces is not related by Newton's third law, they need not balance. The horse too can accelerate. (A horse and cart example is worked out in the next section.)

Consider now the interesting application of Newton's third law where it does apply; to the interaction between ground and horse. Because the horse pushes backward on the ground, the ground pushes forward on the horse with an equal and opposite force. If the horse accelerates forward, it is because this force supplied by the ground is greater in magnitude than the rearward force supplied by the cart. Friction, far from being the retarding force it is usually assumed to be, is in this case the motive force driving the horse ahead. What drives an automobile ahead? Not its engine, but the road. Rails push a locomotive ahead; the air propels an airplane (even a jet). The engines of automobile, locomotive, or airplane do not push the vehicle forward; they push some material (roadway, rails, or air) backward. The material, constrained by Newton's third law, must react to push the engines, and the vehicles attached to them, forward. The role of the third law in vehicle propulsion is not unlike the "psychological third law" that influences a victim at the edge of a cliff. If he is nudged forward toward the cliff edge, he will react and spring backward. If he is pulled unexpectedly backward, he will react in the opposite direction and may hurl himself over the cliff.

The cart and horse paradox is chosen deliberately to *seem* confusing, the better to clarify the application of Newton's third law. Had we chosen instead a "satellite and earth paradox," the fact that there is no real paradox would have been evident at once. The earth exerts a certain force on a satellite; the satellite exerts an equal and opposite force on the earth. Why does the satellite accelerate? Obviously a net force acts on it. The opposite force acts not on the satellite, but on the earth, and has no effect on the satellite's motion. Similarly the earth accelerates in its orbit around the sun because it is attracted to the sun. The equal and opposite force felt by the sun influences only the sun's motion, not the earth's.

9.3 *The motion of systems*

Two important general principles of great utility in solving practical problems (and in understanding the cart and horse paradox) are these: (1) Newton's second law applies separately to every part of any system, no matter how (or whether) that part is connected to other parts. (2) Newton's third law applies to any two parts of a system, no matter how (or whether) those parts are connected. Actually these two principles are interdependent. Newton's second law applies to any part of a system only because of the action of Newton's third law within that part. For the present, however, let us regard these as independent principles and apply them to two numerical examples in order to clarify further the meaning of Newton's second and third laws.

Example 1. A horse of mass 500 kg pulls a cart of mass 300 kg with an acceleration 200 cm/sec^2. If the frictional force exerted by the ground on the cart is 4×10^7 dynes, what are the forces acting on the horse? As shown in Figure 9.3, we call the magnitude of the forward frictional force on the horse F_1, the magnitude of the backward frictional force on the cart F_2, and the magnitudes of the equal and opposite forces acting between horse and cart F_3. By recognizing that two forces have the same magnitude F_3, we have already taken advantage of Newton's third law and the second of the two principles laid down in the paragraph above. We utilize the first principle by applying Newton's second law in three different ways: to the horse alone, to the cart alone, and to horse and cart together. The equations are:

$$\text{horse:} \qquad F_1 - F_3 = m_H a\,; \tag{9.12}$$

$$\text{cart:} \qquad F_3 - F_2 = m_C a\,; \tag{9.13}$$

$$\text{horse and cart:} \qquad F_1 - F_2 = (m_H + m_C)a\,. \tag{9.14}$$

Each equation states that the net horizontal component of force acting on the particular system named is equal to the product of its mass and its horizontal component of acceleration. Since horse and cart have the same acceleration, the same symbol a appears in all three equations. It is very important to notice that if the horse is chosen as the system of interest, the force F_2 is irrelevant, despite the physical connection that exists between the horse and cart. In thinking of the horse alone, one must mentally erase the cart and the earth from the picture. There stands the horse alone in space, accelerating under the combined action of the forces F_1 and F_3 that he actually experiences. Similarly the cart, despite its physical connection to the horse, can be regarded as a separate system accelerating under the action of forces F_3 and F_2.

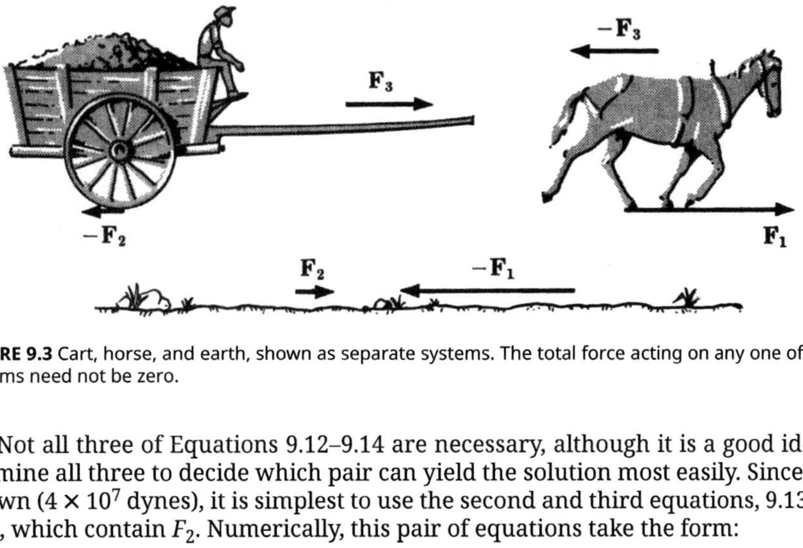

FIGURE 9.3 Cart, horse, and earth, shown as separate systems. The total force acting on any one of these systems need not be zero.

Not all three of Equations 9.12–9.14 are necessary, although it is a good idea to examine all three to decide which pair can yield the solution most easily. Since F_2 is known (4×10^7 dynes), it is simplest to use the second and third equations, 9.13 and 9.14, which contain F_2. Numerically, this pair of equations take the form:

$$\text{cart: } F_3 - 4 \times 10^7 \text{ dynes} = 3 \times 10^5 \text{ gm} \times 200 \text{ cm/sec}^2 = 6 \times 10^7 \text{ dynes;}$$

$$\text{horse and cart: } F_1 - 4 \times 10^7 \text{ dynes} = 8 \times 10^5 \text{ gm} \times 200 \text{ cm/sec}^2 = 16 \times 10^7 \text{ dynes.}$$

Their solutions provide the desired answers, the forces acting on the horse:

$$F_1 = 2 \times 10^8 \text{ dynes,} \tag{9.15}$$

$$F_3 = 10^8 \text{ dynes.} \tag{9.16}$$

These answers may be checked by regarding the horse alone as the system of interest and using Equation 9.12. The left and right sides of this equation are separately found to be equal to 10^8 dynes, verifying the consistency of the answers obtained.

It is left as an exercise to calculate the leftward acceleration of the earth arising from the unbalanced pair of forces F_1 and F_2 acting on the ground (Figure 9.3).

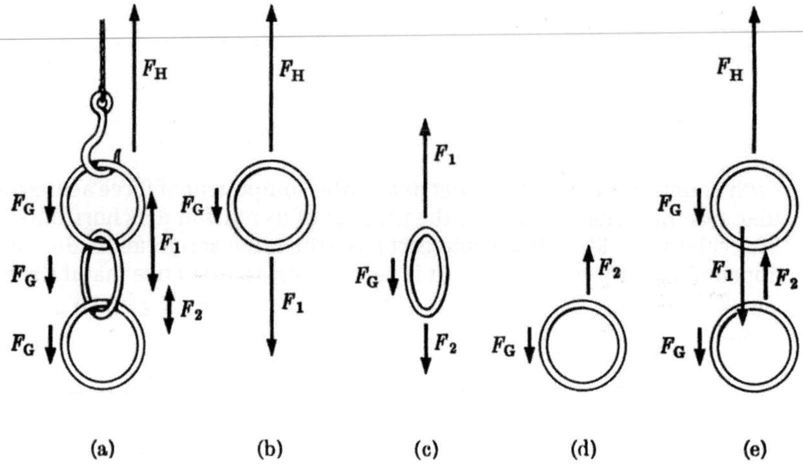

FIGURE 9.4 Acceleration of a three-link chain. **(a)** The whole chain is regarded as the system of interest. **(b)** The upper link alone is considered. **(c)** The middle link alone is considered. **(d)** The lower link alone is considered. **(e)** The upper and lower links together are treated as a single system. Newton's second law applies to every part or combination of parts.

EXAMPLE 2. A three-link chain is being raised by a hook (Figure 9.4). Each link in the chain has a mass of 10 kg, and the upward force supplied by the hook is 4.14×10^7 dynes. What is the acceleration of the chain? What are the forces acting on the middle link? Included in Figure 9.4 are the directions of the various forces acting and the symbols for their magnitudes: F_H for the upward force of the hook, F_G for the downward force of gravity on each link, F_1 and F_2 for the pairs of equal and opposite forces acting between the links. A problem of this kind requires a methodical approach, and it requires a decision—a decision about what to pick as the systems of interest. It can do no harm to pick more systems or parts than are actually needed. We may, for example, apply Newton's second law to the chain as a whole and to each link separately. The equations relating the vertical components of force and acceleration are:

whole chain: $$F_H - 3F_G = 3ma\,; \tag{9.17}$$

upper link: $$F_H - F_G - F_1 = ma\,; \tag{9.18}$$

middle link: $$F_1 - F_G - F_2 = ma\,; \tag{9.19}$$

lower link: $$F_2 - F_G = ma\,. \tag{9.20}$$

The acceleration of the chain is represented by a, and the mass of each link by m. In the first equation the total mass $3m$ appears; in each of the others the mass m of a single link appears. Upward-acting forces appear with plus signs, downward-acting forces with minus signs.

In this example, the knowns are F_H (4.14×10^7 dynes), m (10^4 gm), and F_G ($F_G = mg = 10^4 \times 980 = 0.98 \times 10^7$ dynes). The unknowns are F_1, F_2, and a. From the first of our four equations, the acceleration may be found. That equation—Newton's second law applied to the chain as a whole—is, numerically,

$$4.14 \times 10^7 \text{ dynes} - 3 \times 0.98 \times 10^7 \text{ dynes} = 3 \times 10^4 \text{ gm} \times a.$$

Its solution gives the acceleration,

$$a = 400 \text{ cm/sec}^2. \qquad (9.21)$$

Now the upper-link and lower-link equations may be used to find the forces F_1 and F_2. In numerical form, they are:

$$\text{upper link: } 4.14 \times 10^7 - 0.98 \times 10^7 - F_1 = 10^4 \text{ gm} \times 400 \text{ cm/sec}^2 = 0.4 \times 10^7 \text{ dynes;}$$

$$\text{lower link: } F_2 - 0.98 \times 10^7 = 10^4 \text{ gm} \times 400 \text{ cm/sec}^2 = 0.4 \times 10^7 \text{ dynes.}$$

Their solutions give the two interaction forces,

$$F_1 = 2.76 \times 10^7 \text{ dynes,} \qquad (9.22)$$

$$F_2 = 1.38 \times 10^7 \text{ dynes.} \qquad (9.23)$$

The middle-link equation can now serve as a check of these answers. Its left side is

$$F_1 - F_G - F_2 = (2.76 - 0.98 - 1.38) \times 10^7 = 4 \times 10^6 \text{ dynes.}$$

Its right side is

$$ma - 10^4 \text{ gm} \times 400 \text{ cm/sec}^2 = 4 \times 10^6 \text{ dynes.}$$

To show the power of the principle that Newton's second law may be applied to any part of a system and to obtain a further check on our answers, we may choose as the system of interest the upper link plus the lower link. In Figure 9.4(e), the hook and the middle link have been dissolved away. The remaining system has a mass of 20 kg and, according to our solution, both parts of it have the same upward acceleration, 400 cm/sec^2. What are the forces acting on this system? Acting upward are the forces F_H on the upper link and F_2 on the lower link. Acting downward are the gravitational force $2F_G$ and the force F_1 on the upper link. The net upward force on this two-link system is

$$F_{up} = F_H + F_2 - 2F_G - F_1 = (4.14 + 1.38 - 1.96 - 2.76) \times 10^7 = 0.80 \times 10^7 \text{ dynes.}$$

This is equal, as it should be, to the product of mass and acceleration,

$$2 \times 10^4 \text{ gm} \times 400 \text{ cm/sec}^2 = 0.8 \times 10^7 \text{ dynes.}$$

The Cancellation of Internal Forces

The emphasis in these examples has been on the application of Newton's second law to systems or parts of systems. However, it is Newton's third law in the background that makes such application possible. In Equation 9.17, for example, the forces F_1 and F_2 acting between links did not appear. These were, for the whole chain, internal forces. Because of Newton's third law, these balanced pairs of internal forces canceled. This cancellation is so important in all aspects of the study of motion that it deserves to be stated as a general principle: The sum of all the internal forces in any system or part of a system is always zero. This principle is a direct consequence of Newton's third law. 36

To understand the cancellation of internal forces, we need only understand exactly what is the distinction between an internal force and an external force. An internal force is one whose source and whose action (or whose cause and effect) are both within the system being considered. If cart and horse together are being considered as a single system, the force of the cart on the horse is an internal force. If the whole three-link chain is regarded as a system, the force of the upper link on the middle link is an internal force. For any material object, the forces among its atomic constituents are internal forces. An external force, on the other hand, is one that acts on a system but whose source is outside the system. If the cart alone is treated as a system, both the force supplied by the horse and the force of friction supplied by the ground are external forces. The same force may, of course, be external for one system and internal for another. The sun's force on the earth is an external force if the earth alone is regarded as the system of interest. In the solar system as a whole, the same force is an internal force.

The definitions of internal and external forces together with Newton's third law at once imply that the sum of internal forces must be zero. For every force there is an equal and opposite force. For an internal force, the opposite force also acts within the system and is itself an internal force. In brief, all internal forces come in balanced pairs. Since the sum of each pair is zero, the sum of all pairs is zero. The canceling of internal forces in every system makes for more than convenience in mechanics. One might almost say it makes mechanics possible. We are so accustomed to the fact that rigid objects accelerate only in response to external forces that we may easily overlook the importance of the canceling internal forces. That a billiard ball and an elementary particle follow the same laws of mechanics comes about only because the complexity of structure of the billiard ball, and the myriad of forces within it, have no effect on its motion as a whole. Only because of the action of Newton's third law in causing internal forces to cancel can Newton's second law be applied to the bulk motion of any piece of matter. If internal forces had any effect on the motion of macroscopic objects, Galileo would not have learned about uniform acceleration near the earth, and Newton would not have learned the nature of the gravitational force from the study of planetary motion. In our everyday world, objects upon which no outside forces act would spontaneously accelerate. We have reason to be grateful that Nature's forces obey Newton's third law, which brings to the description of systems of many billions of atoms the same simplicity as the description of elementary particle motion [Figure 9.5 (a)].

Newton's first law, the principle of inertial motion, states that an isolated object moves with constant velocity. This means that the important third law of Newton underlies the first law and the definition of an inertial frame of reference, just as it underlies the second law applied to material objects. Every "object," even a single atom, is really a system of interacting particles. Because interaction forces within the system sum to zero, the system as a whole responds only to external forces. Its internal complexities are irrelevant to its bulk motion. We might suppose that an elementary particle is truly a single entity and not a system. Present theory, however, suggests that even a particle must be regarded as a busy little system, by no means a quiescent speck of matter. In any case, it would be very rash to base any general principles of physics on assumptions about the internal structure of elementary particles, for this domain of nature is still largely a mystery.

An isolated system is one free of external influences. In practice, of course, absolute isolation is an impossibility, but for many systems, the external forces are small enough to neglect, at least for some period of time. Over a few centuries, our solar system may be considered an isolated system. The earth, together with its moon and its artificial satellites, is an approximately isolated system for a duration of a few hours. In this interval of time, the external forces of the sun and other planets have not significantly altered the velocity of the center of mass of the system. The closest approximations to true isolation occur in the cosmological and the submicroscopic worlds, a galaxy on the one hand, an elementary particle on the other. About the universe as a whole, its scope and its laws, we know far too little to speculate whether it may be regarded as an isolated system. Probably the most nearly perfect isolated system is a single neutrino, which can traverse the entire universe with but a small chance to be stopped or deflected.

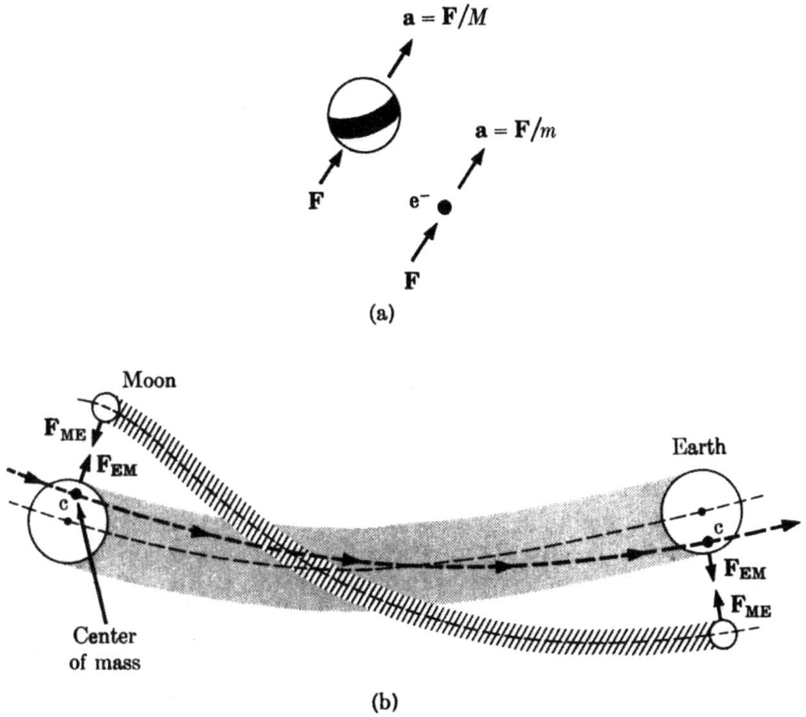

FIGURE 9.5 The cancellation of internal forces. **(a)** A billiard ball responds to an applied force as a single entity, in the same way as a single electron. **(b)** The forces of attraction between moon and earth are internal forces if the earth and moon together are regarded as a single system. These forces have no effect on the motion of the center of mass of the earth-moon system, which follows a smooth elliptical path about the sun. (This drawing is not to scale.)

In Section 8.8 we showed that the center of mass of a system moves in response to the sum total of all the forces acting on the system. Now, with the help of Newton's third law, we can draw a more powerful conclusion about the motion of the center of mass. Divide the forces into internal forces and external forces. Since the sum of the internal forces is zero, the sum of the external forces must be the same as the total sum of all the forces. Therefore we may reword the conclusion of Section 8.8 in this way. The acceleration of the center of mass of a system is proportional to the total external force acting on the system (Figure 9.5). The equation is

$$M\mathbf{a}_{\mathrm{c}} = \mathbf{F}_{\mathrm{ext}}\,; \tag{9.24}$$

the total mass of the system multiplied by the acceleration of its center of mass is equal to the total external force acting on the system. From this improved general statement of Newton's second law follows also a more general version of his first law applicable to any system. If the external force is zero (or negligibly small), the quantity $\mathbf{a}_{\mathrm{c}}$ is zero (or exceedingly small), and the center of mass of the system moves with constant (or nearly constant) velocity. Expressed more succinctly: The center of mass of an isolated system moves with constant velocity.

Now that we have considered the impact of Newton's third law on the motion of isolated systems, it is worthwhile to look once again at the general question suggested by the cart and horse paradox. If every force in nature is canceled by an opposite force, why is there any acceleration in the world? The answer: For an isolated system, indeed there is no acceleration! At least there is no acceleration of the center of mass, the perfect representative point of the system. Individual parts of an isolated system experience acceleration, but that is because those parts are not themselves isolated. Anything in nature, to be accelerated, must be linked to something else outside itself (not necessarily connected by material bonds, of course, but linked through forces). We see acceleration and the interesting complexities of motion only when we fix attention on some part of a total system. If we were to take an Olympian view of nature, noting only the average motion (the center-of-mass motion) of isolated systems, we would see nothing but constant-velocity motion. Looking closely at the solar system, for instance, from a few billion miles away, we would see the individual accelerations of the sun and its planets and moons. If we then backed away into the depths of space until the solar system appeared only as a single dot, those complex motions would fade from view, until all that remained to see would be the almost constant velocity motion of a single isolated "object."* Among the consequences of cancelling internal forces, none is more important than the law of momentum conservation. This is the subject of Section 9.4.

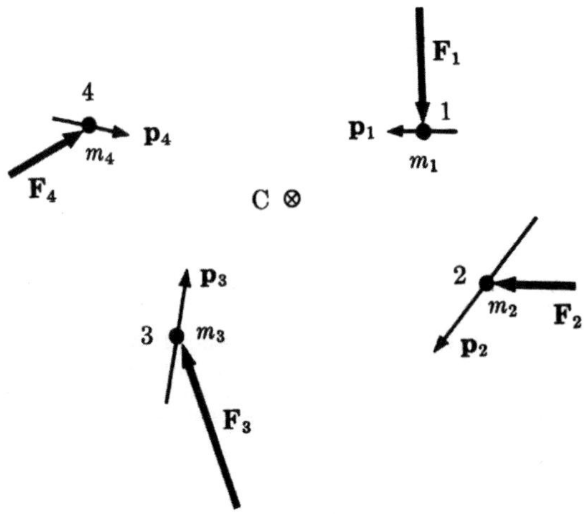

FIGURE 9.6 A system of particles. Each is characterized by its mass and its momentum (as well as other variables), and each is acted upon by a force. The point labeled C is the center of mass of the system.

9.4 *Conservation of momentum*

With the groundwork laid down in Sections 8.8, 9.1, and 9.3, we can very quickly derive the law of momentum conservation and show its connection to center-of-mass motion. Although we may wish to apply the law only to certain simple examples of motion, it is useful to derive it in the most general way in order to appreciate its scope. In so doing, we shall be emulating a typical mathematical technique employed by physicists. Suppose that a system—any system—contains a number of particles. To keep track of them, we suppose that the particles are labeled 1, 2, 3, and so on (Figure 9.6). Newton's second law applied to the first may be written,

$$\mathbf{F}_1 = \frac{\Delta \mathbf{p}_1}{\Delta t}. \qquad (9.25)$$

* The solar system of course is not perfectly isolated, being attracted by the rest of the stars in the galaxy. Its acceleration in the galaxy, however, is extremely small compared with the accelerations of the planets and moons within it. See page 188.

The second particle, subject to force $\mathbf{F}_2$, obeys a similar law,

$$\mathbf{F}_2 = \frac{\Delta \mathbf{p}_2}{\Delta t}. \tag{9.26}$$

According to one of the general principles set forth in Section 9.3, each particle separately is governed by Newton's second law. We can continue to write down equations connecting force and rate of change of momentum for every particle in the system:

$$\mathbf{F}_3 = \frac{\Delta \mathbf{p}_3}{\Delta t}, \tag{9.27}$$

$$\mathbf{F}_4 = \frac{\Delta \mathbf{p}_4}{\Delta t}, \tag{9.28}$$

and so on. Now we add these equations together in a particular way. The sum of all their left sides is equal to the sum of all their right sides:

$$\mathbf{F}_1 + \mathbf{F}_2 + \mathbf{F}_3 + \ldots = \frac{\Delta \mathbf{p}_1}{\Delta t} + \frac{\Delta \mathbf{p}_2}{\Delta t} + \frac{\Delta \mathbf{p}_3}{\Delta t} + \ldots. \tag{9.29}$$

The sum on the left (a vector sum) is the total force acting on the system. Because of the cancellation of internal forces (dictated by Newton's third law), this is the same as the total external force, for which we use the symbol $\mathbf{F}_{\text{ext}}$. On the right is the rate of change of the total momentum of the system, for which we write $\Delta \mathbf{p}/\Delta t$*. Equation 9.29 may therefore be replaced by a beautifully simple equation whose mathematical form is identical to Equation 9.6 relating force and momentum for a single particle:

$$\mathbf{F}_{\text{ext}} = \frac{\Delta \mathbf{p}}{\Delta t}. \tag{9.30}$$

This important result rests equally on Newton's second and third laws. It states that the total momentum of a system (of arbitrary complexity) changes in response to an applied force. In the form of a conservation law, it states that the total momentum of an isolated system does not change. For an isolated system, the external force vanishes ($\mathbf{F}_{\text{ext}} = 0$); therefore the momentum change in any time interval vanishes ($\Delta \mathbf{p} = 0$). A vanishing change of momentum of course means that momentum is constant,

$$\mathbf{P} = \mathbf{P}_1 + \mathbf{P}_2 + \mathbf{P}_3 + \ldots = \textbf{constant}. \tag{9.31}$$

The constant is a vector quantity, fixed in direction as well as in magnitude.

---

* Although this statement seems obviously true, it actually requires a mathematical proof. We cannot supply the rigorous proof here; what follows is a mathematical argument indicative of the lines the actual proof would follow. Suppose that in a particular small interval of time the changes of momentum of the particles are $\Delta \mathbf{p}_1$, $\Delta \mathbf{p}_2$, $\Delta \mathbf{p}_3$, and so on. Their rates of change are $\Delta \mathbf{p}_1/\Delta t$, $\Delta \mathbf{p}_2/\Delta t$, $\Delta \mathbf{p}_3/\Delta t$, ... The sum of these rates of change,

may also be written, more simply,

$$\frac{\Delta \mathbf{p}_1}{\Delta t} + \frac{\Delta \mathbf{p}_2}{\Delta t} + \frac{\Delta \mathbf{p}_3}{\Delta t} + \ldots$$

in which all the momentum changes are summed before dividing by the time interval Δt. However, the sum of all the momentum changes is $\Delta \mathbf{p}$, the change of momentum of the whole system, so that the sum

$$\frac{\Delta \mathbf{p}_1 + \Delta \mathbf{p}_2 + \Delta \mathbf{p}_3 + \ldots}{\Delta t},$$

may be contracted to

$$\frac{\Delta \mathbf{p}}{\Delta t}.$$

Applications of Momentum Conservation

As emphasized in Section 8.12, approximately isolated systems are not as rare as one might at first suppose, especially in the world of elementary particles, where almost every significant interaction event occurs among a small number of particles effectively isolated from the rest of the universe. Because such isolated interaction events are common, the law of momentum conservation has many important applications. One of the simplest is in the decay of one unstable particle into two others, the so-called two-body decay discussed in Section 4.6. If the initial particle is at rest, momentum conservation requires that the two final particles must fly apart with equal and opposite momenta, so that the sum of their momenta is zero, the same as the momentum of the initial particle (Figure 4.2). The observation of two particles emerging from a decay process "back to back" with equal and opposite momenta provides good evidence that no third particle escaped detection. In the decay of a positive pion at rest (Figure 9.7), only a single particle, a positive muon, is observed. Because the muon momentum is always the same in every decay of a stationary pion, the law of momentum conservation suggests that only one other particle is created, whose momentum precisely balances the muon momentum, and whose energy takes up the difference between the pion energy (its mass energy) and the muon energy. This other particle is the muon's neutrino. The decay is written

$$\pi^+ \rightarrow \mu^+ + \nu_\mu .$$

In three-body decays of a particle at rest, the final momenta of the product particles must still sum to zero, but there are infinitely many ways in which the three particles may fan out without violating either the law of energy conservation or the law of momentum conservation. The three-body decay of a positive kaon, denoted by

$$K^+ \rightarrow \pi^+ + \pi^+ + \pi^-,$$

is illustrated in Figure 9.8.

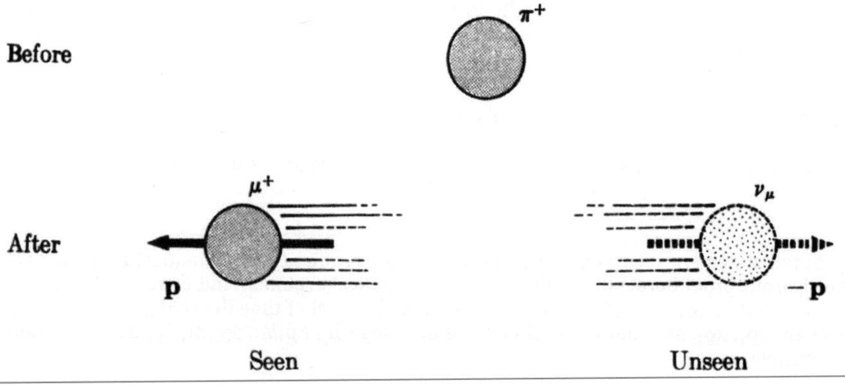

FIGURE 9.7 Momentum conservation in pion decay. The fact that the unseen neutrino must have momentum equal and opposite to that of the muon means that it carries away a fixed fraction of the total energy, leaving to the muon a definite energy, the same in every such decay event.

An event in the macroscopic world not unlike the decay of an elementary particle is the bursting of an artillery shell into two or more fragments. Suppose that a shell fired vertically upward explodes, just as it reaches its peak height, into two fragments. One fragment, with a mass of 5 kg, shoots downward with a speed of 6×10^4 cm/sec. The mass of the other fragment is 10 kg. In what direction and with what speed does it fly off? Since the momentum just before the explosion is zero, so must it be after the explosion, provided air resistance is small enough to enable us to regard the material making up the shell as an isolated system. Here we see clearly the value of the principle of momentum conservation. Regardless of the complexity of the explosive internal forces and regardless of our ignorance of their exact nature (as true for elementary particles as for artillery shells), momentum conservation provides a simple link between "before" and "after." The vertically downward momentum of the lighter fragment in this example is

$$m_1 v_1 = 5 \times 10^3 \text{ gm} \times 6 \times 10^4 \text{ cm/sec} = 3 \times 10^8 \text{ gm cm/sec}.$$

To conserve momentum, the heavier fragment must fly vertically upward with the same magnitude of momentum:

$$m_2 v_2 = 3 \times 10^8 \text{ gm cm/sec}.$$

Since the mass of the second fragment is $m_2 = 10^4$ gm, the speed of the second fragment is

$$v_2 = 3 \times 10^4 \text{ cm/sec}. \tag{9.32}$$

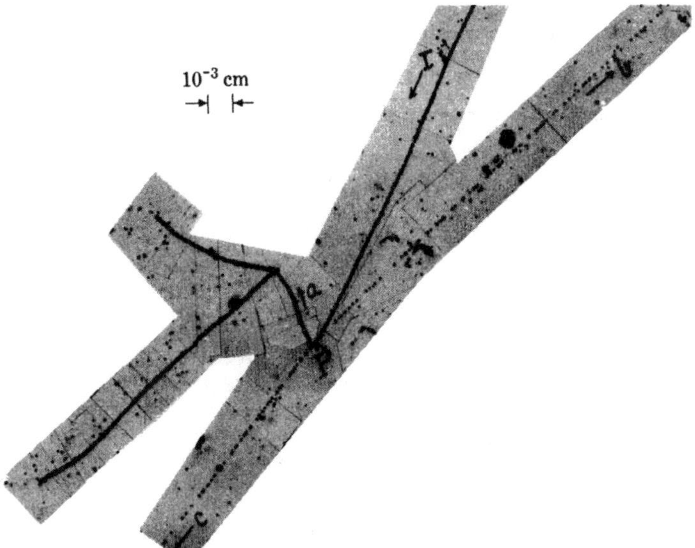

FIGURE 9.8 Three-body decay. A kaon (track labeled τ), after being brought to rest in a photographic emulsion at point P, decays into three pions. A pion of low momentum leaves a heavy track (a); pions of higher momentum leave lighter tracks (b and c). The momenta of the three pions sum to zero. This composite photograph, of historical interest, shows the first such decay event to be observed. [From C. F. Powell, P. H. Fowler, and D. H. Perkins, *The Study of Elementary Particles by the Photographic Method* (New York: Pergamon Press, 1959), p. 289. Reprinted by permission of the authors and publisher.]

An explosion (or decay) into two fragments results in equal and opposite momenta only if the shell (or particle) is initially at rest. Suppose instead that a shell of mass $m_0 = 10$ kg is moving horizontally with a speed $v_0 = 5 \times 10^4$ cm/sec at the moment of its explosion into two equally massive fragments ($m_1 = m_2 = 5$ kg). If one fragment moves at 45 deg above the initial flight path and the other at 45 deg below the initial flight path (Figure 9.9), what is the speed of each? Application of the law of momentum conservation in this case requires the vector addition of the momenta $\mathbf{p}_1$ and $\mathbf{p}_2$ of the fragments, as shown in Figure 9.9(b). Their sum must be equal to the initial momentum $\mathbf{p}_0$, as indicated by the vector equation expressing momentum conservation:

$$\mathbf{p}_1 + \mathbf{p}_2 = \mathbf{p}_0 \,. \tag{9.33}$$

Since the arrow diagram representing the vector addition happens to be a right triangle in this example with equal sides p_1 and p_2 and hypotenuse p_0, the magnitudes satisfy the Pythagorean theorem,

$$p_0{}^2 = p_1{}^2 + p_2{}^2 \,.$$

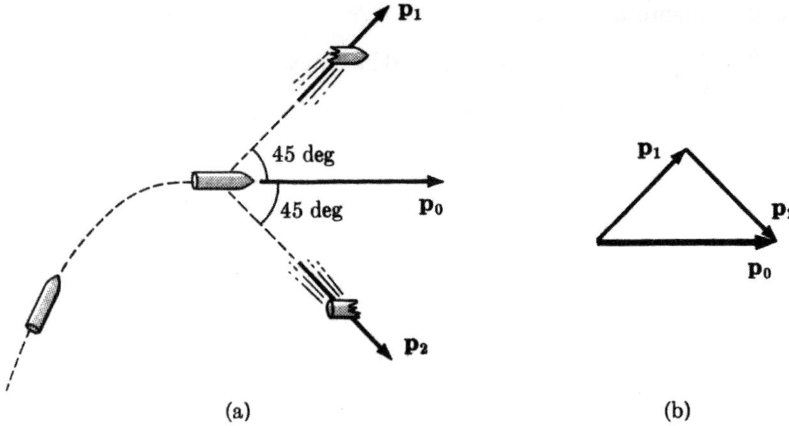

FIGURE 9.9 Momentum conservation in the explosion of a shell. **(a)** The momentum of the shell just before it explodes is $\mathbf{p}_0$; $\mathbf{p}_1$ and $\mathbf{p}_2$ are the momenta of the two fragments. **(b)** Momentum conservation requires that $\mathbf{p}_1 + \mathbf{p}_2 = \mathbf{p}_0$.

Numerically, $p_0 = 10^4$ gm × 5 × 10^4 cm/sec = 5 × 10^8 gm cm/sec. Its square is $p_0{}^2 = 25 \times 10^{16}$ (gm cm/sec)^2, which is equal to the sum, $p_1{}^2 + p_2{}^2$. Since $p_1 = p_2$, the square of either fragment momentum alone is one half of $p_0{}^2$:

$$p_1{}^2 = 12.5 \times 10^{16} \text{ (gm cm/sec)}^2.$$

The square root of 12.5 is approximately 3.54. Therefore the magnitudes of the final momenta are given by

$$p_1 = p_2 = 3.54 \times 10^8 \text{ gm cm/sec.} \tag{9.34}$$

Division by the mass of each, 5 × 10^3 gm, gives the sought-for speeds,

$$v_1 = v_2 = 7.08 \times 10^4 \text{ cm/sec.} \tag{9.35}$$

The same result could be obtained by considering the horizontal component of Equation 9.33. Calling the horizontal direction the *x*-direction, we have

$$p_{1x} + p_{2x} = p_{0x} \,. \tag{9.36}$$

Since $\mathbf{p}_0$ is in the *x*-direction, its *x*-component is the same as its full magnitude, 5 × 10^8 gm cm/sec. However, the component p_{1x} is less than the magnitude p_1 by the factor cos (45 deg) or cos (π/4). This cosine is approximately equal to 0.707. The individual components are therefore given by the expressions,

$$p_{1x} = p_1 \cos(\pi/4) = 0.707p_1,$$

$$p_{2x} = p_2 \cos(\pi/4) = 0.707p_2.$$

Their sum, appearing in Equation 9.36, gives the equation,

$$0.707p_1 + 0.707p_2 = 5 \times 10^8 \text{ gm cm/sec.} \tag{9.37}$$

Consideration of the vertical component of Equation 9.35 (left as an exercise) shows that the magnitudes p_1 and p_2 are equal. Therefore we may equate half of the left side of Equation 9.37 to half of the right side:

$$0.707p_1 = 2.5 \times 10^8 \text{ gm cm/sec.}$$

Division of both sides by 0.707 gives the result,

$$p_1 = 3.54 \times 10^8 \text{ gm cm/sec,} \tag{9.38}$$

the same as Equation 9.34. Since $p_1 = m_1v_1 = (5 \text{ kg}) \times v_1$ it again follows that

$$v_1 = 7.08 \times 10^4 \text{ cm/sec,} \tag{9.39}$$

and that v_2 has the same magnitude as v_1.

There is, of course, nothing fundamentally important about this particular example. It only illustrates a typical application of the law of momentum conservation in the macroscopic world, emphasizing the important facts (1) that details of internal forces are irrelevant to momentum conservation; and (2) that the vector nature of momentum plays an essential role in the law of momentum conservation.

9.5 Rocket propulsion

Ordinary propulsion on earth relies heavily on friction—on the fact that the vehicle, be it 37
boat, locomotive, automobile, or airplane, is *not* an isolated system. For a rocket in space, the situation is entirely different. The rocket has neither earth nor air nor water to push against and must achieve its acceleration another way. Of course, the rocket is not precisely an isolated system either. It experiences the gravitational force of sun, earth, moon, or other astronomical bodies and accelerates accordingly. During its coasting period, which is the major part of any rocket flight, the trajectory is determined by these external gravitational forces. However, there is nothing the rocket pilot—or the ground controller—can do about these. If the rocketeer wishes to accelerate in a way that he can control according to his whim, he must find a substitute for the friction that provides the basis of propulsive control on earth. It is through rocket exhaust and the law of momentum conservation that he finds this control. Imagine now a rocket in empty space that is truly an isolated system. Its center of mass moves with constant velocity. Can anything be done about this? No, nothing can be done. No internal motions or changes or forces whatever can prevent the center of mass of the system from continuing inexorably forward at constant velocity. But one part of the system can be caused to accelerate, that is, to change its momentum, provided another part of the system experiences a compensating opposite change of momentum. The rocketeer must be prepared literally to sacrifice some of his rocket ship for the benefit of the rest of it. *Only* by discarding some mass can he achieve an acceleration of what remains.

In simplest terms, we can say that the rocket ship is divided into two parts, the payload part, and the fuel-exhaust part. Neither of these parts is an isolated system, for it interacts with the other part. Therefore on each part considered alone an external force can act to accelerate that part. What happens is that the payload part pushes the exhaust backward, and the exhaust pushes the payload, with an equal and opposite force, forward. The center of mass of the entire system continues uniformly, unaffected by all this activity. The payload part does, however, achieve the desired acceleration. Usually the payload is a small part of the original total mass of the rocket. For it to acquire a high velocity in one direction, a large mass of exhaust must have been accelerated in the opposite direction.

Suppose that a rocket of mass m is coasting in space with a forward speed v_1. Its pilot wishes to accelerate forward to reach a higher speed v_2. If the exhaust leaves with speed v_{ex} relative to the payload part of the rocket, how much mass must be thrown away in order to achieve the increment of speed, $v_2 - v_1$? The law of momentum conservation provides the answer (Figure 9.10). If the mass of the system before the burst of acceleration is m, its initial momentum is

$$p_1 = mv_1 . \tag{9.40}$$

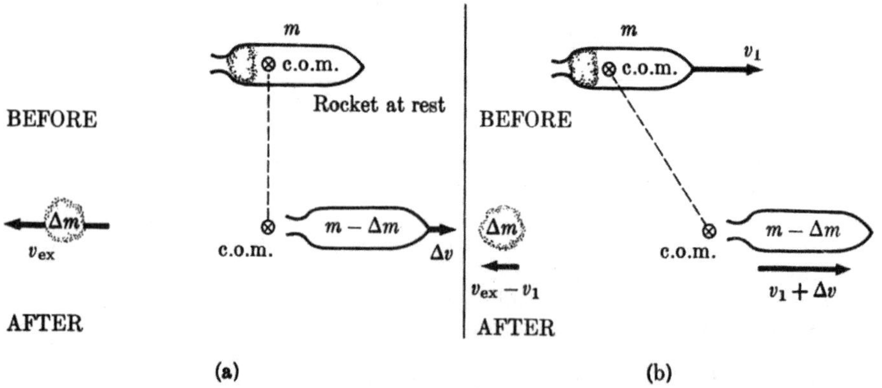

FIGURE 9.10 Rocket propulsion. **(a)** In a frame of reference in which the rocket is initially at rest, the increment of exhaust gas and the remainder of the rocket separate with equal and opposite momenta. Their center of mass remains at rest. **(b)** In a frame of reference in which the rocket is initially moving with speed v_1, the increment of exhaust gas and the remainder of the rocket separate in such a way that the total momentum, equal to mv_1, remains constant. The center of mass of the total system continues to move with constant velocity.

After the burst of acceleration, a lesser mass, $m - \Delta m$, is moving forward with a higher speed v_2, and the mass Δm discarded as exhaust is moving backward with speed $v_{ex} - v_1$. (Recall that the exhaust has a backward speed v_{ex} relative to the rocket, therefore a backward speed of only $v_{ex} - v_1$ relative to the stationary frame of reference.) The net forward momentum of the system finally is given by

$$p_2 = (m - \Delta m)v_2 - \Delta m(v_{ex} - v_1) , \tag{9.41}$$

the first term on the right being the contribution of the payload part, the second term being the contribution of the exhaust part. Since no external forces acted on the system, momentum is conserved (we consider only magnitudes since the motion is one-dimensional) :

$$p_1 = p_2 . \tag{9.42}$$

Therefore the right side of Equation 9.40 may be equated to the right side of Equation 9.41. Algebraic manipulation yields a formula for the amount of mass Δm that must be thrown away:

$$\Delta m = \left(\frac{v_2 - v_1}{v_2 - v_1 + v_{ex}} \right) m . \tag{9.43}$$

Suppose, for example, that the required increment of speed, $v_2 - v_1$ is 10^4 cm/sec (about 225 miles/hr), and the exhaust speed is $v_{ex} = 9 \times 10^4$ cm/sec. For these speeds Equation 9.43 gives

$$\Delta m = 0.10m \,; \tag{9.44}$$

it would be necessary to discard one-tenth of the initial mass as exhaust. It should be remarked that Equation 9.41 is valid only if the change, $v_2 - v_1$, is much smaller than v_{ex}. (Why is this?) This being so, it is permissible to neglect the combination $v_2 - v_1$ relative to v_{ex} in the denominator of Equation 9.43 and to write

$$\Delta m \cong \frac{v_2 - v_1}{v_{ex}} m \,.$$

An instructive way to express this equation is to introduce the notation Δv for the change of speed, $v_2 - v_1$, and to write

$$\frac{\Delta m}{m} = \frac{\Delta v}{v_{ex}} \,. \tag{9.45}$$

This is a fundamental equation of rocket engineering. It states that for each small increment of rocket speed Δv, it is necessary to exhaust a fraction of the remaining mass ($\Delta m/m$) equal to the speed change divided by the exhaust speed relative to the rocket. Since this relation rests on the conservation of momentum, it is valid for an isolated rocket in space, one sufficiently far from all centers of gravitational attraction. It can be applied also to the acceleration of a satellite in an earth orbit provided the momentum change produced by the rocket exhaust is much greater than the momentum change produced by the gravitational force of the earth in the same time interval.

An alternative form of Equation 9.45 is also useful and interesting. This equation may be rewritten

$$v_{ex}\Delta m = m\Delta v \,. \tag{9.46}$$

On the right is the momentum change of the rocket during a small time interval. This momentum change, according to Equation 9.7, is equal to the impulse, $F\Delta t$, delivered to the rocket by the exhaust during this interval. Therefore we may write

$$\textit{Impulse} = v_{ex}\Delta m \,. \tag{9.47}$$

If an increment of mass Δm is ejected with speed v_{ex} relative to the rocket, this exhaust delivers to the rocket an impulse, and therefore a change of momentum, equal to the product $v_{ex}\Delta m$. Since the rocket engine designer wants to throw away as little mass as possible (small Δm), he seeks to achieve the highest possible rocket exhaust speed (large v_{ex}).

9.6 Momentum conservation in systems that are not isolated

Even for systems not isolated from external influences, momentum conservation can play an important role. This can come about in two ways. (1) The external forces may add up to zero. Even though the system is not isolated, it experiences no net external force, and its momentum is therefore a constant. A freight car rolling on a straight horizontal track is such a system, provided friction can be neglected. The downward pull of gravity is balanced by the upward push of the tracks; the total external force is zero. (2) The *component* of the net external force in some direction may vanish. Since momentum is a vector quantity, the law of momentum conservation for an isolated system is really three laws in one. Each component is separately a constant. For a system that is not isolated, the components may be treated separately. Then it may happen that only one or two components of momentum are conserved because only one or two components of the external force vanish. A projectile near the earth, for example, experiences only a vertically downward force (if air resistance is negligible). Therefore its horizontal components of momentum remain constant. The earth in its orbit around the sun experiences no force perpendicular to the plane of its orbit. Therefore the component of its momentum in that direction, initially zero, remains zero, and the earth's orbit does not deviate from its original plane.

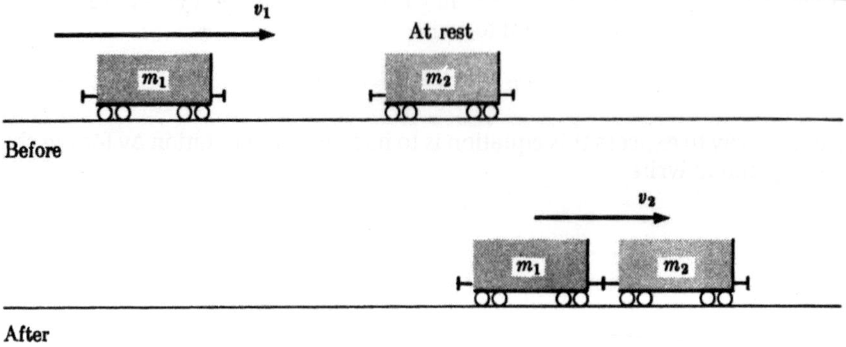

FIGURE 9.11 Momentum conservation in the collision and coupling of freight cars.

Example: A freight car of mass $m_1 = 1.5 \times 10^7$ gm rolling along a horizontal track at a speed $v_1 = 200$ cm/sec collides and couples with another car of mass $m_2 = 1.0 \times 10^7$ gm, after which the pair continue together with a lesser speed v_2 (Figure 9.11). What is the magnitude of v_2? Problems of this kind are best solved algebraically first, because the algebraic solution has a much wider applicability than the particular numerical solution. Before the collision, the second car has no momentum; the total momentum of the two-car system is that of the first car. Its horizontal component, the same as its total magnitude, is

$$p_1 = m_1 v_1 \,. \tag{9.48}$$

After the collision, the entire mass, $m_1 + m_2$, moves with speed v_2. The momentum of the system is

$$p_2 = (m_1 + m_2) v_2 \,. \tag{9.49}$$

Because the forces acting during the collision and coupling are all internal forces for the chosen system of two cars, momentum is conserved: $p_1 = p_2$. The right sides of Equations 9.48 and 9.49 may be equated to give

$$m_1 v_1 = (m_1 + m_2) v_2 \,.$$

The algebraic solution for the final speed v_2 is

$$v_2 = \frac{m_1}{m_1 + m_2} v_1 \,; \tag{9.50}$$

it is equal to the original speed of the first car multiplied by the ratio of that car's mass to the total mass. For the particular numbers of this example,

$$v_2 = \frac{1.5 \times 10^7}{(1.5 + 1.0) \times 10^7} \times 200 \text{ cm/sec} = 120 \text{ cm/sec}. \tag{9.51}$$

The mathematics of this example may be applied without change to other examples. Suppose that a bullet of mass m_1 and speed v_1 strikes a stationary wooden block of mass m_2 and is imbedded in the block. What speed will the block and bullet together acquire? Equation 9.50 provides the answer. This example has practical utility. It provides a convenient way to determine the speed of a bullet. The larger speed v_1 may be discovered by measuring the smaller speed v_2.

9.7 *Momentum conservation and center of mass*

The fact that the center of mass of a system is the perfect representative point of the system has been emphasized before. As expressed by Equation 9.24, it is the point whose acceleration is proportional to the external force acting on the system. It is also the point, as we shall show now, whose velocity is proportional to the total momentum of the system. In both cases, the constant of proportionality is the total mass of the system.

We consider again a system containing an arbitrary number of particles and sum the momenta of the particles to obtain the total momentum **P**:

$$\mathbf{P} = m_1\mathbf{v}_1 + m_2\mathbf{v}_2 + m_3\mathbf{v}_3 + \cdots . \tag{9.52}$$

On the right side are added together (as vectors) the individual momenta. In order to link this sum with the center-of-mass velocity, we must multiply and divide the right side by the total mass M ($M = m_1 + m_2 + m_3 + \cdots$). The resulting equation reads

$$\mathbf{P} = M\left(\frac{m_1\mathbf{v}_1 + m_2\mathbf{v}_2 + m_3\mathbf{v}_3 + \cdots}{M}\right). \tag{9.53}$$

The quantity within the parentheses is the mass-weighted average velocity of all the particles. This is, according to Equation 8.34, the velocity of the center of mass of the system. This means that the total momentum of the system is simply equal to the product of the total mass M and the center-of-mass velocity $\mathbf{v}_c$:

$$\mathbf{P} = M\mathbf{v}_c . \tag{9.54}$$

This and other laws derived earlier show that the center of mass behaves exactly as if all of the mass of the system were concentrated there and all of the external forces acted there. The important equations are summarized here:

Single Particle	*System*
$\mathbf{p} = m\mathbf{v}$	$\mathbf{P} = M\mathbf{v}_c$
$\mathbf{F} = m\mathbf{a}$	$\mathbf{F}_{ext} = M\mathbf{a}_c$
$\mathbf{F} = \dfrac{\Delta\mathbf{p}}{\Delta t}$	$\mathbf{F}_{ext} = \dfrac{\Delta\mathbf{P}}{\Delta t}$

The first line on the left is the definition of particle momentum, mass times velocity; on the right is the result just derived, total momentum equals total mass times center-of-mass velocity. The second and third lines on the left state Newton's second law for a particle in two alternative forms, first as force equals mass times acceleration, second as force equals rate of change of momentum. Equations of mathematically identical form hold true for a system, provided particle mass and particle momentum are replaced by total mass and total momentum, particle velocity and acceleration are replaced by center-of-mass velocity and acceleration, and the force on the particle is replaced by the total external force on the system. The law of momentum conservation follows most directly from the third equation on the right. In the absence of external force, the total momentum is a constant. In the absence of any component of external force, the corresponding component of total momentum is a constant. Finally, constant momentum is reflected in constant center-of-mass velocity.

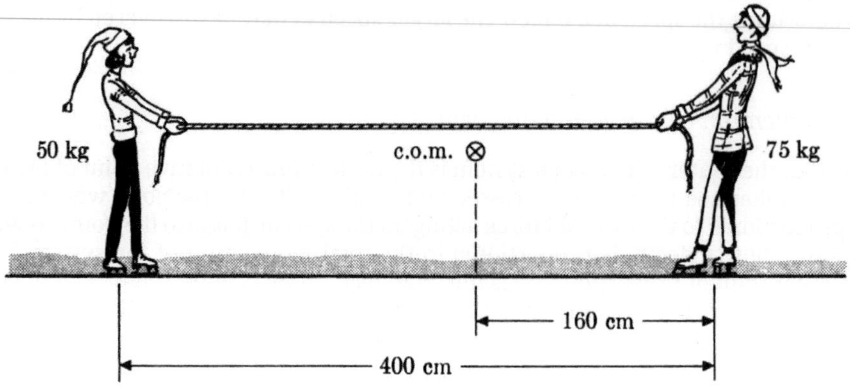

FIGURE 9.12 Conservation of the horizontal component of momentum in a system free of external horizontal force. The center of mass, initially at rest, moves neither to the left nor to the right.

Example: A boy of mass 75 kg and a girl of mass 50 kg face each other standing on ice skates on an ideally smooth horizontal ice surface. They are 400 cm apart and joined by a rope (Figure 9.12). Starting from rest, both the boy and the girl pull toward each other by pulling hand-over-hand on the rope. Where will they meet? It might seem at first sight that their meeting point would depend on which of them pulls harder on the rope. This is not the case, because boy, girl, and rope together form a system whose horizontal component of momentum must be conserved. All forces on the rope are internal forces that cannot influence their total momentum or the motion of their center of mass. The center-of-mass concept provides the easiest solution to the problem. Since the center of mass is initially at rest, it remains at rest. It is a fixed point between the boy and the girl.* No matter how close together they come, that point remains between them and remains stationary. It must therefore be the point where they meet. Then the question to be answered is: Where is the center of mass? The defining equation of the center of mass, Equation 8.33, can be used. Choosing the line joining boy and girl as the x-axis, we have

$$x_c = \frac{m_B x_B + m_G x_G}{m_B + m_G}, \tag{9.55}$$

where subscripts B and G refer to boy and girl. If the origin is chosen midway between the two, so that $x_B = 200$ cm, $x_G = -200$ cm, the center-of-mass coordinate may be calculated:

$$x_c = \left(\frac{75 \text{ kg}}{125 \text{ kg}}\right) \times 200 \text{ cm} - \left(\frac{50 \text{ kg}}{125 \text{ kg}}\right) \times 200 \text{ cm} = 40 \text{ cm}. \tag{9.56}$$

This means that the center of mass is 240 cm from the girl and 160 cm from the boy. That is the point where they meet.

The constant velocity of the center of mass for isolated systems may also be used to good advantage in analyzing particle collisions. Suppose that a neutron moving at 4×10^8 cm/sec

* More accurately stated, the center of mass does not move horizontally. Why can it move vertically up and down?

strikes a proton in such a way that the two particles emerge from the collision with equal speed, one particle 45 deg above the initial flight line of the neutron, the other 45 deg below the initial flight line of the neutron (Figure 9.13). What is the speed of the particles after the collision? For the purpose of this example, we may assume that the neutron and the proton are equally massive (they actually differ in mass by less than one part in a thousand). For the situation before the collision, it is convenient to think of a horizontal line joining the neutron and proton. The right end of this line, at the proton, is stationary. The left end, at the neutron, is moving to the right at the speed of the neutron, 4×10^8 cm/sec. The midpoint of the line, which is the location of the center of mass of this two-particle system, moves to the right at half the neutron speed (as it must to remain halfway between the two particles). According to the law of momentum conservation, the center of mass will continue to move to the right at the same speed after the collision as before, 2×10^8 cm/sec. After the collision, the line joining the particles is vertical. Its midpoint, the center of mass, remains on the same horizontal line it traced out before the collision. Since the speed of this midpoint is 2×10^8 cm/sec, it is a simple matter of geometry to deduce the speed of the recoiling particles. While the center of mass is covering a distance equal to the side of a 45-45-90 triangle, each particle is covering a distance equal to the hypotenuse of the same triangle. Each particle is therefore moving faster than the center of mass, faster by the factor $\sqrt{2}$, or 1.414. We conclude that in order to preserve the constant center-of-mass speed of 2×10^8 cm/sec to the right, neutron and proton must emerge from the collision each with a speed of 2.828×10^8 cm/sec.

9.8 *Reaction definition of mass*

Newton's third law and the law of momentum conservation provide the basis for what is called the reaction definition of mass. In Section 3.5 the example was given of a pair of astronauts floating freely within the cabin of their space ship. Initially at rest, they touch hands and push apart, floating to opposite ends of the cabin. The heavier astronaut recoils more slowly than the lighter astronaut. The ratio of their masses is defined to be the inverse ratio of their recoil speeds. Although not a very practical definition for precision work, this reaction definition of mass has much logical appeal. Here we would like to show that it is a simple consequence of the law of momentum conservation. 38

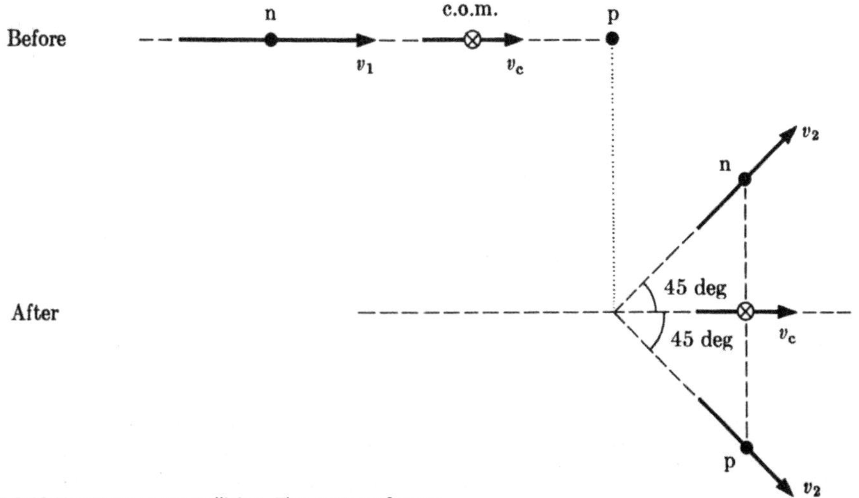

FIGURE 9.13 Neutron-proton collision. The center of moss continues forward with constant velocity.

If the pair of astronauts are originally at rest, their total momentum is zero, and will remain zero so long as they are free of external forces. Since the forces they exert on each other are internal forces, they recoil with equal and opposite momenta to preserve zero total momentum. In a vector equation, their momenta sum to zero:

$$\mathbf{p}_1 + \mathbf{p}_2 = 0 . \tag{9.57}$$

This means that the magnitudes of $\mathbf{p}_1$ and $\mathbf{p}_2$ are equal; the equality can be written

$$m_1 v_1 = m_2 v_2 . \tag{9.58}$$

Collecting masses on one side of the equation and speeds on the other side, we get the quantitative statement of the reaction definition of mass:

$$\frac{m_1}{m_2} = \frac{v_2}{v_1} . \tag{9.59}$$

The attractive feature of this definition is that it is based on a law of constancy—the law of momentum conservation—rather than on a law of change such as Newton's second law. Moreover the definition is economical of concepts. The mass ratio in Equation 9.59 is defined in terms of the kinematic concepts of space and time only, without explicit reference to force. Of course force is not wholly irrelevant. Momentum conservation is intimately related to Newton's third law of equal and opposite forces, and without the mutual internal forces there would have been no recoil and no speeds to measure. However, no forces need be known or measured. As an operational definition, the reaction definition of mass is independent of the concept of force.

Like other definitions of mass, the reaction definition requires an arbitrary standard unit, and tests of self-consistency. Both of these points were discussed in Section 3.5.

In the submicroscopic world of particles, new laws have been discovered. For high-speed motion, force is no longer proportional to acceleration; and momentum, to remain a useful concept entering into a conservation law, must be redefined. The high-speed, or relativistic, definition of momentum for a particle of mass m is this:

$$\mathbf{p} = \frac{m\mathbf{v}}{\sqrt{1 - v^2/c^2}} , \tag{9.60}$$

in which c is the speed of light. The "why" of this definition is discussed in Chapter Twenty-One. Here we wish to show how it alters the mathematical aspect of the reaction definition of mass. For a pair of recoiling particles with equal momenta, Equation 9.58 is replaced by

$$\frac{m_1 v_1}{\sqrt{1 - {v_1}^2/c^2}} = \frac{m_2 v_2}{\sqrt{1 - {v_2}^2/c^2}} . \tag{9.61}$$

As in Equation 9.59, masses may be collected on the left side of an equation, and speeds on the right. The result is

$$\frac{m_1}{m_2} = \frac{v_2}{v_1} \frac{\sqrt{1 - {v_1}^2/c^2}}{\sqrt{1 - {v_2}^2/c^2}} . \tag{9.62}$$

Measurements of the speed of the recoiling particles still provide a determination of the mass ratio, but in a slightly more complicated way than in Newtonian, or classical, mechanics. It is this reaction definition of mass, Equation 9.62, that has been found to have self-consistency in the high-speed world of particles. Note that whenever the speeds v_1 and v_2 are very much slower than the speed of light c, the square-root factors in Equation 9.62 are very nearly equal to 1, and Equation 9.62 "reduces to" the simpler form of Equation 9.59.

We stated at the beginning of this section that the reaction definition of mass has more logical appeal than practical utility. In the macroscopic world, both the difficulty of approximating an isolated system and the difficulty of making precise measurements of speed limit its usefulness. In the particle world, the reaction definition, although not adopted as a standard way of determining mass, can be one of several useful ways to measure mass. Suppose, for example, that a lambda particle at rest decays into a proton and a negative pion, a process indicated by

$$\Lambda \rightarrow p + \pi^- .$$

The proton and pion are observed to fly apart in opposite directions, the proton with a speed $v_1 = 3.19 \times 10^9$ cm/sec, the pion with a speed $v_2 = 1.75 \times 10^{10}$ cm/sec. What is the ratio of proton mass to pion mass? These speeds are quite significant fractions of the speed of light, $c = 3.00 \times 10^{19}$ cm/sec. The proton moves at 10.6% of the speed of light ($v_1/c = 0.106$), the pion at 58.3% of the speed of light ($v_2/c = 0.583$). Therefore the reaction definition of mass requires the relativistic Equation 9.62. Substitution of the given numbers into this equation yields

$$\frac{m_1}{m_2} = \frac{17.5 \times 10^9 \text{ cm/sec}}{3.19 \times 10^9 \text{ cm/sec}} \times \frac{\sqrt{1 - (0.106)^2}}{\sqrt{1 - (0.583)^2}} = 5.49 \times \frac{0.994}{0.813} = 6.71 . \qquad (9.63)$$

The proton is 6.71 times more massive than the pion. The Newtonian formula, Equation 9.59, based on the momentum definition $\mathbf{p} = m\mathbf{v}$, would have yielded the incorrect value 5.49 for the mass ratio.

For the massless particles, not even the relativistic momentum definition, Equation 9.60, holds true. For a photon or a neutrino there is no connection between momentum and speed. Every massless particle moves with the same speed c regardless of its momentum. This means that the momentum of a photon, for instance, is measurable only indirectly via the law of momentum conservation.

When a photon is born, it must be assigned a certain momentum to preserve the law of momentum conservation in the system creating the photon. When the same photon is absorbed, the conservation of momentum in the absorbing system requires that the photon must have contributed some momentum of its own. Since the momentum carried away by the photon at its birth and the momentum it surrenders at its death are found always to be equal, it is verified that a definite momentum can be assigned even to a massless photon; indeed that the photon's momentum is essential for overall momentum conservation.

9.9 *The significance of momentum conservation*

Consider a system of two particles interacting with each other, but isolated from all 39
outside influences. Two fundamental laws of their interaction and motion have been the themes of this chapter. One is Newton's third law of equal and opposite forces, expressed mathematically by

$$\mathbf{F}_{12} = -\mathbf{F}_{21} .$$

The other is the law of momentum conservation, expressed by

$$\mathbf{p}_1 + \mathbf{p}_2 = \textbf{constant} .$$

Is one of these laws more fundamental than the other? There is no unambiguous logical way to answer this question. However, experience can tend to show that one law is more general, more useful, or apparently deeper than another. Originally Newton chose his three laws as the most suitable axioms for the development of mechanics. From the third law (together with the second) the law of momentum conservation was deduced. We have followed the Newtonian approach in this chapter so far: Momentum conservation is a *consequence* of Newton's laws. However, it is important to realize that the law of momentum conservation could supplant Newton's third law as one of the basic foundation stones of mechanics. If it did, the third law would become secondary, a law that could be deduced mathematically from momentum conservation and Newton's second law. The evolution of physics since Newton, and particularly the developments of the twentieth century, suggest that the law of momentum conservation indeed has a good claim to be regarded as a more fundamental statement about nature than does Newton's third law.

Among the reasons for attaching a special significance to the law of momentum conservation is its simplicity. Every conservation law, by picking from the chaos of activity and change in nature a single constant quantity, has a special appeal. Because momentum is a constant for every isolated system or interaction, it is naturally looked upon as a fundamental concept, and its conservation as a particularly important law.

As mathematicians of the eighteenth and nineteenth centuries studied and reformulated mechanics, the concept of force was eclipsed in importance by the concepts of energy and momentum. This was, so to speak, accidental. It simply turned out that the equations with the greatest power for solving mechanical problems were equations in which momentum and energy appeared explicitly, force did not. The demotion of force from its central position in mechanics was completed by the twentieth-century revolutions of relativity and quantum mechanics. In these theories, force, although it may be introduced, need not be. Momentum, on the other hand (suitably redefined, as Equation 9.60), remains a central concept in these modern theories, and still obeys a conservation law.

Force (and with it, Newton's third law) has encountered two main difficulties in modern physics. The first is that in the catastrophic events of annihilation and creation that govern the behavior of elementary particles, we usually do not know what forces act. The conservation of momentum from before to after the event can be verified; the probability of the event can be determined; but exactly what forces are at work during the event (if indeed the event has any duration) cannot be measured. The second difficulty connected with force is more an inconvenience than a fundamental problem. It has to do with the idea of action at a distance. Since forces do not act instantaneously across space, but are propagated with finite speed, Newton's third law must be examined with new care. We must ask: *When* are the pair of forces equal and opposite? It takes some time for particle number 1 to react to the presence of particle number 2 at a distance, and in that time the distance between the particles and the force between them may have changed. The result of this is that Newton's third law is greatly complicated, for it is no longer true that a pair of forces need be precisely equal and opposite at a given instant of time. On the other hand, the law of momentum conservation retains a simple form even when the time lag of propagating forces is taken into account. It is still true that the system has a constant total momentum at each instant of time. The only added feature of the new view is that only part of the total momentum is contributed by the matter in the system. The rest comes from photons or other massless messengers which transmit force, energy, and momentum from one material part to another.

In some ways the most compelling of the reasons for regarding momentum conservation as a profound law of nature is the connection between this conservation law and the uniformity of space. The principle of the uniformity of space might be called the principle of the sameness of nothingness. How a law as general and important as the law of momentum conservation can rest on a principle apparently so innocuous is explained in Chapter Four. As emphasized there, the true beauty of every conservation law may reside in the symmetry principle upon which it is founded. The symmetries of space and time underlie the laws of conservation of energy, momentum, and angular momentum. The search for subtler symmetries of nature underlying some of the other conservation laws is one of the most exciting endeavors of physics.

EXERCISES

9.1. A proton of mass 1.7×10^{-24} gm is moving with a speed of 10^9 cm/sec. **(1)** What is its momentum? **(2)** It is acted upon by a force of 5×10^{-10} dyne. What is its rate of change of momentum?

9.2. A man pushes his stalled automobile along a level stretch of highway. To keep it moving forward with a constant momentum of 1.4×10^8 gm cm/sec, he must exert a steady force of 1.2×10^7 dynes. **(1)** What force of friction acts? **(2)** If he stops pushing and the same force of friction continues to act, how long a respite will he have until the momentum of the automobile is halved?

9.3. A proton moves horizontally to the right with momentum $p_0 = 3 \times 10^{-16}$ gm cm/sec. It then experiences a force of magnitude 1.5×10^{-10} dynes for a time of 1 μsec (10^{-6} sec). Find its momentum after 1 μsec (magnitude and direction) for each of the following three possibilities: **(a)** the force acts horizontally to the right; **(b)** the force acts horizontally to the left; **(c)** the force acts vertically upward.

9.4. A short-order cook slides a plate of spaghetti down a counter. He misjudges the stopping power of the counter; the plate continues off the edge of the counter, arcs through the air, and crashes on the floor. Describe *all* the forces exerted by the plate on its environment in each part of its motion.

9.5. A chain hoist lifts a mass of 10^6 gm at a constant speed of 10 cm/sec. What downward force does the chain hoist exert on the girder that supports it? (Neglect the mass of the chain hoist in comparison with the mass of its load.)

9.6. Referring to Figure 9.3 and Example 1 in Section 9.3, calculate the acceleration imparted to the whole earth by the net leftward force, $F_1 - F_2$, exerted by the horse and cart on the ground. If this acceleration persisted for one hour, how far would the earth move? Compare this distance with the size of an atom, a nucleus, or an elementary particle.

9.7. Show that Equations 9.17 through 9.20 are partially redundant: Derive one of them algebraically from the other three. The physical meaning of the equations can guide you in seeking the right combination.

9.8. A chain of three links, each link of mass 10^4 gm, hangs from a hook. The hook is drawn upward with a constant acceleration of 200 cm/sec^2. Find all the forces acting on the center link.

9.9. What total gravitational force do you exert on the earth **(a)** when you are standing motionless on the ground? **(b)** when you are leaping through the air?

9.10. A head-on collision takes place between a 10,000-kg truck traveling 60 miles/hr and a 1,000-kg passenger car traveling 80 miles/hr in the opposite direction. If the car and truck lock together at the instant of collision, in what direction and at what speed is the wreckage traveling immediately after the collision? What is the change of velocity of the passenger car? What is the change of velocity of the truck? (NOTE: This exercise can be solved without units conversion.)

9.11. Two spheres rolling at the same speed collide head on. After the collision, how would you expect them to be traveling if **(a)** the two spheres are billiard balls; **(b)** the two spheres are balls of putty? Is momentum conserved in *both* collisions or in only one? Explain.

9.12. An artillery shell which is traveling horizontally at a speed of 8×10^4 cm/sec explodes into three fragments. The first fragment has a mass of 15 kg and flies forward in a horizontal direction with a speed of 10^5 cm/sec. The second fragment, with a mass of 10 kg, flies forward in a horizontal direction with a speed of 9×10^4 cm/sec. The third fragment has a mass of 5 kg. What is its velocity?

9.13. An alpha particle with an initial momentum of 7×10^{-15} gm cm/sec approaches a stationary uranium nucleus. In the encounter the alpha particle is turned through 90 deg and emerges with nearly the same magnitude of momentum. What is the recoil velocity (magnitude and direction) of the uranium nucleus, whose mass is 4×10^{-22} gm?

9.14. **(1)** Work out algebraically instead of numerically the exploding shell problem described at the end of Section 9.4 and illustrated in Figure 9.9, and show that $v_1 = v_2 = \sqrt{2}\, v_0$. What is the percent error of the numerical answer given by Equation 9.39? **(2)** Use the vertical component of Equation 9.33 to prove that $p_1 = p_2$.

9.15. A neutral pion with a momentum of 5.10×10^{-15} gm cm/sec decays into two photons, one of which has a momentum of 2.94×10^{-15} gm cm/sec and flies away on a line perpendicular to the original path of the pion. (1) What is the momentum vector of the other photon? Include a diagram with your answer. (2) If the momentum of the pion, whose mass is 2.4×10^{-25} gm, were correctly given by the nonrelativistic formula, $p = mv$ (it is not), what would be its speed before it decayed? *Optional:* Use Equation 9.60 to find the speed of the pion before its decay.

9.16. **(1)** Carry out algebraic manipulation of Equations 9.40 to 9.42 to derive Formula 9.43 for the mass of exhaust gas needed to change the velocity of a rocket. **(2)** Explain why Equation 9.41 is valid only if the change of speed, $v_2 - v_1$, is much smaller than the exhaust speed v_{ex}.

9.17. A group of astronauts are in a space ship whose total mass is 2×10^7 gm. If their rocket engines have an exhaust velocity of 2×10^5 cm/sec, approximately what mass of fuel must they bum in order to change their speed (a) from 8.0×10^5 cm/sec to 8.2×10^5 cm/sec? (b) from 8.0×10^5 cm/sec to 7.8×10^5 cm/sec?

9.18. A certain rifle fires a bullet of mass 25 gm with a muzzle velocity of 7×10^4 cm/sec (about Mach 2). The bullet strikes a wooden block of mass 1,000 gm. The block, with the bullet imbedded in it, acquires a velocity of 1,480 cm/sec. (1) How much loss of momentum did the bullet suffer between leaving the rifle barrel and striking the block? Express your answer both in gm cm/sec and in percentage of initial momentum. (2) What happened to the "lost" momentum?

9.19. A collision slightly less catastrophic than the one described in Exercise 9.10 occurs between a pair of automobiles, each of mass 2×10^6 gm, one traveling north at 15 miles/hr, the other traveling east at 20 miles/hr. What is the velocity of the center of mass of this system (a vector quantity) before, during, and after the collision?

9.20. The masses of the freight cars pictured in Figure 9.11 are $m_1 = 1.5 \times 10^7$ gm and $m_2 = 1.0 \times 10^7$ gm, and the speed of the first car is $v_1 = 200$ cm/sec. **(1)** When the centers of the cars are 25 meters apart, where is their center of mass? **(2)** What is the speed of the center of mass before they collide? **(3)** How does the speed of the center of mass compare with the speed of the cars after they couple?

9.21. Two identical spheres approach one another with equal and opposite velocity (not necessarily directly head on). **(1)** What exactly does the law of momentum conservation say about their motion after the collision? What does it not say? **(2)** Are there any circumstances in which momentum is not conserved in such a collision ?

9.22. An astronaut's rocket on the launching pad is pouring out exhaust gas at the rate of χ gm/sec with exhaust speed v_{ex} cm/sec. **(1)** Consider the momentum of the exhaust gas and use Newton's second law and Newton's third law to show that the thrust (force) on the rocket is the product χv_{ex}. **(2)** If the rocket mass is 10^8 gm at launch and the exhaust speed is 4×10^5 cm/sec, what is the rate of flow of exhaust (in gm/sec) that would be just sufficient to lift the rocket off the launching pad?

9.23. Speculate briefly on possible aspects of the behavior of matter in a world where Newton's third law is not valid.

TEN

Angular momentum

Three basic concepts of mechanics have proved to be hardy perennials, even more deeply rooted in contemporary physics than in the classical physics that gave them birth. Each of the concepts—momentum, angular momentum, and energy—has developed a deeper significance and a wider scope of application than could have been imagined when it was first defined and used in mechanics. Each is the core concept of a fundamental conservation law.

Chapter Nine was devoted to momentum—its conservation and its connection to Newton's third law. In this chapter and in the next we shall be concerned with the key concepts of angular momentum and energy—how each is defined, in what situations each is important, and under what circumstances each is conserved.

10.1 The concept of angular momentum

Angular momentum may be defined mathematically in terms of the already understood concepts of mass, velocity, and distance. The usual symbol for angular momentum is L. The defining equation of its magnitude is

$$L = mv_{\perp}r. \tag{10.1}$$

In words, the magnitude of the angular momentum of a particle with respect to an arbitrarily chosen reference point is equal to the product of its mass, the perpendicular component of its velocity (to be clarified below), and its distance r from the reference point. The velocity component that appears in the definition of angular momentum is the component perpendicular to the line joining the particle and the reference point (Figure 10.1). If the velocity itself is perpendicular to this line, as it would be for circular motion about the reference point [Figure 10.2(a)], the perpendicular component is equal to the full magnitude of the velocity. Then the magnitude of the angular momentum is simply equal to mass times speed times distance from the reference point. If, on the other hand, the velocity is directed exactly toward or away from the reference point (Figure 10.3), the perpendicular component of velocity vanishes ($v_{\perp} = 0$) and the angular momentum is zero. For intermediate directions of the velocity, its perpendicular component is always less than the full speed [Figures 10.2(b) and (c)]. The exact relation, made clear in Figure 10.1, is

$$v_{\perp} = v \cos \theta \,. \tag{10.2}$$

The speed v is multiplied by the cosine of the angle between the velocity vector and the perpendicular to the radial line joining the reference point and the particle. Compare Equation 10.2 and Equation 6.19. In both the angle θ is the angle between the vector direction and the component direction.

There is a simple way to picture this situation. Angular momentum measures a *rotational* property of motion. If a particle flies straight toward (or away from) a point, it has no tendency to rotate about that point. Accordingly, it has no angular momentum with respect to that particular point. If the particle moves transverse to the line joining it to the reference point, it has the maximum tendency to rotate about that point. In that case, the perpendicular component $v_\perp$ takes on its maximum value, the speed v. In between these two extremes, a certain part of the velocity, its perpendicular component, contributes to the rotational tendency, while another part, the parallel component (the component along the line joining particle and reference point) does not. The easiest way to recognize what we here call a rotational tendency is to picture the motion of an imaginary line drawn from the reference point to the particle. If this line rotates as the particle moves, the particle possesses angular momentum. If the line does not rotate, but only stretches or shrinks as the particle moves, the particle possesses no angular momentum. Angular momentum is greater for a greater length of this line, for a faster rate of its rotation, or for greater mass of the particle. The magnitude of the angular momentum depends too on the chosen reference point.

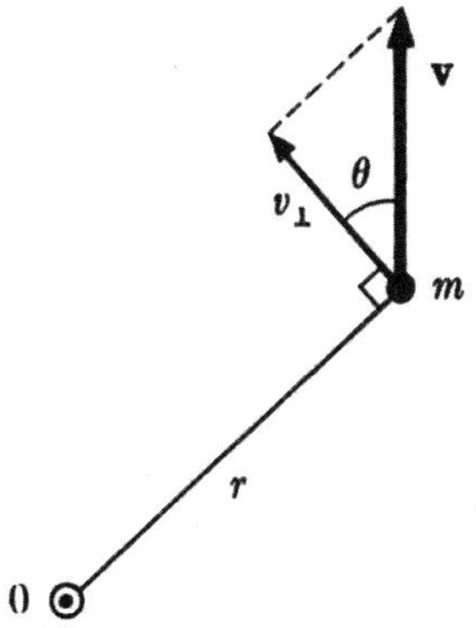

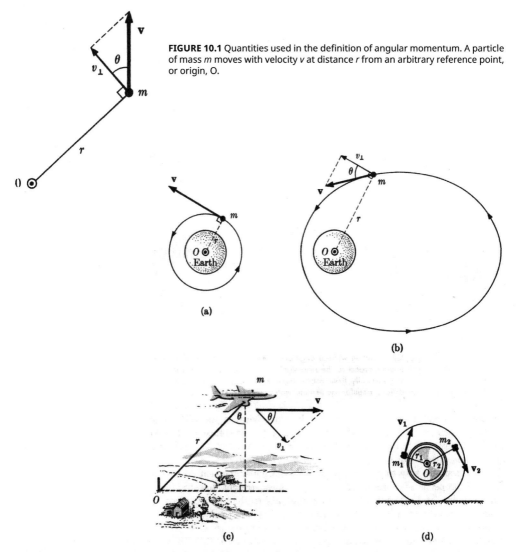

FIGURE 10.1 Quantities used in the definition of angular momentum. A particle of mass *m* moves with velocity *v* at distance *r* from an arbitrary reference point, or origin, O.

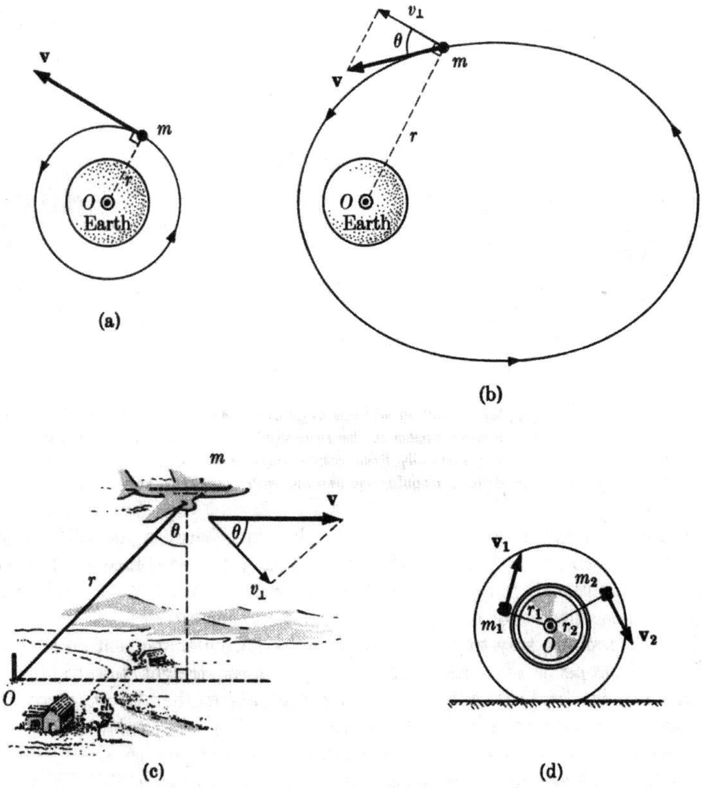

FIGURE 10.2 Examples of motion with angular momentum. **(a)** A satellite in a circular orbit, and **(b)** a satellite in an elliptical orbit. For both, the center of the earth is the most convenient reference point. **(c)** An airplane flying straight possesses angular momentum with respect to a point on the earth. **(d)** For a rotating wheel, each bit of mass contributes some angular momentum to the total. In each example, the reference point is arbitrary. See also Figure 4.4.

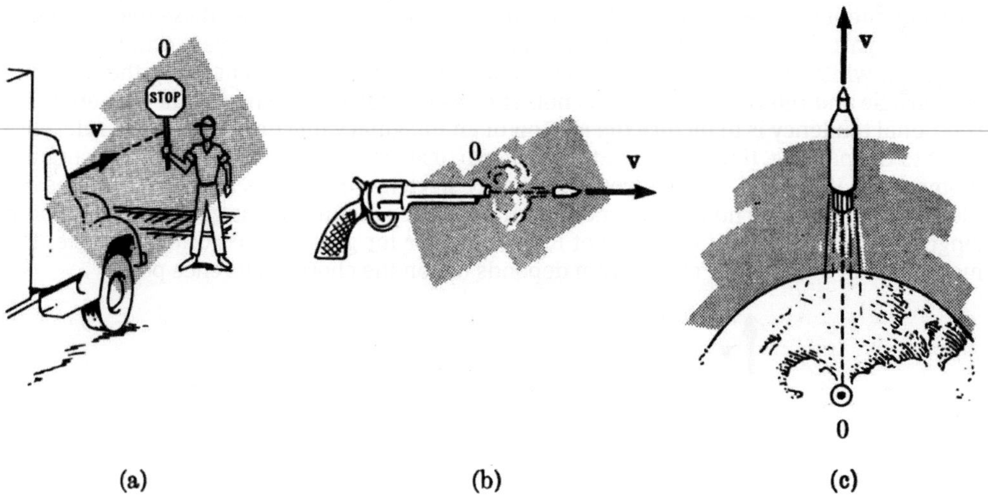

FIGURE 10.3 Examples of motion without angular momentum. The truck possesses no angular momentum with respect to the stop sign, the bullet none with respect to the gun barrel, and the vertically fired rocket none with respect to the center of the earth. Each would have angular momentum with respect to some other origin.

Since momentum is defined as mass times velocity, it is possible to rewrite the definition of Equation 10.1 in even more compact form:

$$L = p_{\perp} r \, . \tag{10.3}$$

The magnitude of the angular momentum of a particle is equal to its distance from a reference point multiplied by the component of its momentum perpendicular to the line joining the particle and the reference point. As the definition makes clear, the physical dimension of angular momentum is the product of the dimensions of mass, velocity, and distance. The cgs unit of angular momentum is gm × cm/sec × cm, or, in more abbreviated notation, gm cm^2/sec. This particular combination of units has received no new name of its own (unlike, for example, 1 gm cm/sec^2, which is 1 dyne). A qualitative discussion of angular momentum is to be found in Section 4.7. We can best think of the "meaning" of angular momentum as the strength of rotational motion, a measure of the effort required to start or stop such motion.

That a particle can possess angular momentum without actually rotating around a point is best made clear through the example of straight-line motion at constant speed. Imagine a particle moving vertically upward with speed v past a point P, coming at its point of nearest approach to within a distance b of the point P (Figure 10.4). At this point of nearest approach, labeled A, the perpendicular component of velocity is the speed v. According to the fundamental definition expressed by Equation 10.1, the angular momentum with respect to point P when the particle is at A is given by

$$L_{\text{A}} = mvb \, . \tag{10.4}$$

Later, at point B, the particle is at a greater distance r from point P. Its perpendicular component of velocity is less than it was at point A, for the particle is not moving transverse to the line joining points B and P. The perpendicular component is $v_{\perp} = v \cos \theta$; the angle θ is illustrated in Figure 10.4. Therefore the angular momentum with respect to point P when the particle is at point B is

$$L_B = mv_\perp r = mvr\cos\theta\,. \tag{10.5}$$

In Figure 10.4 there are two similar right triangles and two equal angles θ. The cosine of θ not only relates the speed v to the perpendicular component $v_\perp$, it also relates the distance r to the distance b according to the formula,

$$b = r\cos\theta\,. \tag{10.6}$$

Therefore, in Equation 10.5 the combination $r\cos\theta$ can be replaced by b to give

$$L_B = mvb\,. \tag{10.7}$$

With respect to the fixed point P, the angular momentum of the particle at point B is the same as at point A. Since point B represents any point at all along the particle trajectory, the angular momentum has constant magnitude, always equal to mvb. In terms of the defining Equation 10.1, this constancy comes about because the increase in the distance r as the particle moves upward from point A is exactly compensated by a decrease in the perpendicular component of velocity $v_\perp$. The product of the two remains unchanged.

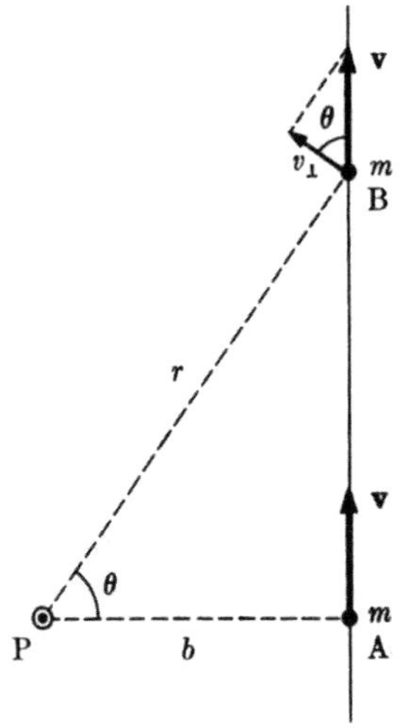

FIGURE 10.4 Straight-line motion with constant velocity. The particle possesses constant angular momentum with respect to the point P.

This example of simple constant-velocity motion reveals two important aspects of angular momentum. First, it shows that the definition of angular momentum, despite its complexity, can be useful, for it leads to the result that in undisturbed inertial motion, the angular momentum is constant. Notice that the magnitude of the angular momentum of the uniformly moving particle is different with respect to different reference points, but that for any particular reference point, it does not change in time. Second, the example shows that angular momentum can accompany nonrotational motion. *Something* is rotating, however, and that something is the imaginary line joining the reference point to the particle. This line sweeps counterclockwise in Figure 10.4 as the particle moves. Without such rotation, there would be no angular momentum.

Among examples of motion with angular momentum, none is mathematically simpler than the uniform rotation of a particle in a circle [Figure 10.2(a)], This simplicity requires that the center of the circle be chosen as the reference point. With that choice, each of the factors in the defining Equation 10.1 is separately constant—the particle mass, the radius of the circle, and the perpendicular component of velocity, which is equal to the total speed v. With respect to the center of the circle, the angular momentum is simply

$$L = mvr = pr \,. \tag{10.8}$$

This equation could be used to calculate with good accuracy the angular momentum of the earth about the sun, an astronaut in a circular orbit about the earth, or a stone swung on the end of a rope. Although the methods of classical mechanics are not all valid in the submicroscopic world governed by quantum mechanics,* Equation 10.8 can also be used to estimate roughly the angular momentum of an electron circling within an atom. An electron half an Angstrom (0.5×10^{-8} cm) from the atomic nucleus moving in a circle with a speed one hundredth the speed of light (3×10^{8} cm/sec) would have, according to Equation 10.8, an angular momentum with respect to the nucleus equal to

$$L = 9 \times 10^{-28}\ \text{gm} \times 3 \times 10^{8}\ \text{cm/sec} \times 0.5 \times 10^{-8}\ \text{cm} = 1.35 \times 10^{-27}\ \text{gm cm}^2/\text{sec}.$$

Actually, in the submicroscopic world (and in the large-scale world as well), all orbital angular momenta come in units of Planck's constant divided by 2π, $h/2\pi = \hbar$ ("h bar"), whose magnitude is $\hbar = 1.05 \times 10^{-27}$ gm cm^2/sec. The electron referred to above could possess an angular momentum equal to $\hbar$, or $2\hbar$, or $3\hbar$, but not in fact equal to the calculated value, which turned out to be about $1.3\hbar$. Planck's constant, with the same dimension as angular momentum, is the fundamental constant of quantum mechanics, governing the magnitudes of other quantum phenomena as well.

10.2 The vector character of angular momentum

40 In order to be dealt with quantitatively and manipulated in equations, every measurable concept in physics must be associated with some mathematical "object," either number or vector or something more abstract. What kind of object successfully describes angular momentum? In the preceding section we studied its magnitude. Does it have a direction as well? It does, but *not* the direction of motion of the particle. The direction that is assigned to angular momentum is an axial direction, the direction of the axis about which the line joining particle and reference point rotates. For circular motion, the axis is evidently a line perpendicular to the plane of the motion. For the straight line motion depicted in Figure 10.4, the axis is a line perpendicular to the page, for it is about this axis that the line connecting the reference point to the particle swings.

Since an axis has two directions, it is necessary to specify a positive direction. The arbitrary convention is based on the right-hand rule. If the curved fingers of the right hand follow the rotational motion or the rotational tendency, the right thumb designates the positive direction along the axis. Thus the angular momentum of the wheel of an automobile moving forward is directed horizontally to the left [Figure 10.5(a)], In reverse, the wheel's angular momentum is directed horizontally to the right. The angular momentum of the earth about its own axis is directed toward the North Star [Figure 10.5(b)], and the angular momenta of the moon with respect to the earth and the earth with respect to the sun are directed approximately in the same way. The angular momentum of the particle moving upward in Figure 10.4 is directed vertically upward from the page. For this example and for the motion of the train shown in Figure 10.5(c), a slightly modified version of the right-hand rule might be helpful. Let the fingers of the right hand point from the reference point to the moving particle in such a way that the palm "pushes" the line joining them in its direction of motion. Then the right thumb indicates the direction of the angular-momentum vector.

* Equation 10.1, for instance, is totally invalid for massless particles, which can possess angular momentum despite the fact that, for them, $m = 0$.

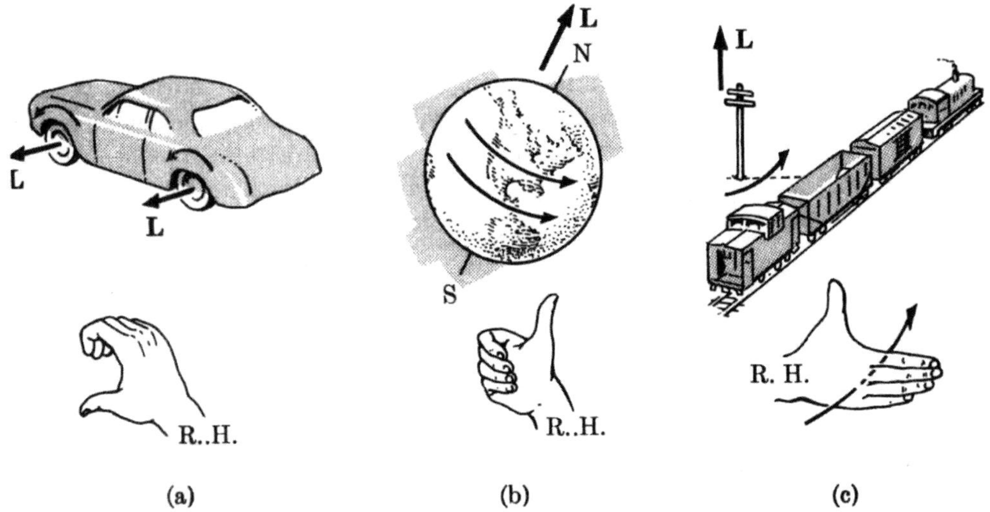

FIGURE 10.5 Direction of the angular-momentum vector. Rotating wheels and rotating earth have angular momentum directed along the axis of rotation. The train has angular momentum directed perpendicular to a line joining it to the reference point.

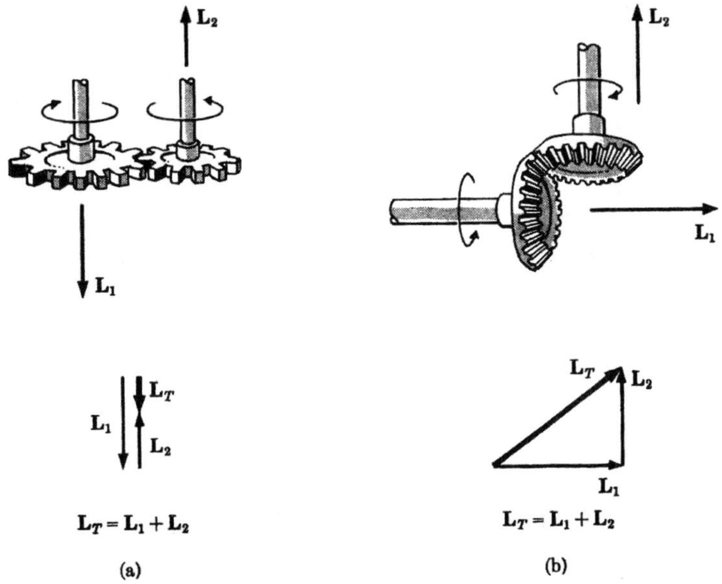

FIGURE 10.6 The vector addition of angular momenta. **(a)** The total angular momentum of this pair of gears is small in magnitude, and directed downward. **(b)** The total angular momentum of this pair of gears is not directed along either axis.

To the question, Why does angular momentum have this directional property?, there is only the solid pragmatic answer that it is fruitful to *define* angular momentum in this way. Then it behaves experimentally like a mathematical vector. If the angular momentum of a system is conserved (as it always is for isolated systems), the angular momenta of its parts combine—or add—as vectors, not as numbers, and the total angular momentum is conserved as a vector, both in magnitude and direction. When angular momentum is not conserved, the law of its change is expressed by a vector equation, the subject of Section 10.5.

These arguments about the vector nature of angular momentum may seem somewhat too abstract to be entirely plausible. They can be supplemented by this plausibility argument. For a particle circling around a point, the only possible fixed direction associated with the motion is along a line perpendicular to the plane of the orbit. Within the plane itself, the velocity and momentum vectors take on all possible directions. Throughout the motion, however, the plane of the orbit remains fixed and therefore the perpendicular direction to the plane remains fixed. It is natural to take this axial direction as the direction characterizing the rotational motion. The direction along the axis chosen as the positive direction is entirely arbitrary. Neither this nor any other argument can show that one direction is preferable to the other. The conventional choice defined by the right-hand rule is merely a historical accident standardized to avoid confusion.

Since angular momentum is a vector quantity, two or more angular momenta add as vectors. The total angular momenta of pairs of coupled gears are illustrated in Figure 10.6. Two equal and opposite angular momenta, just like two equal and opposite forces or momenta, can add to give zero (see Figure 4.5). Examples discussed in Section 10.5 give further insight into the vector character of angular momentum.

10.3 The angular momentum of systems

Most examples of rotational motion encountered in the large-scale world are not orbital motions of particles. They are rotations of objects, systems of many particles or constituent parts. Even in the small-scale world, the spin of a single particle must be viewed as the rotation of a structure. However, there is a deeper reason than the practical one for considering the angular momentum of systems or structures. If the center of mass of a system of particles is at rest, the angular momentum of the system has an important new property. It is the same with respect to all reference points. For a single moving particle, the angular momentum depends on the choice of reference point and might even be variable with respect to one reference point and constant with respect to another. For two or more particles, on the other hand, the angular momentum can be uniquely defined, the same for all reference points, in a frame of reference in which the center of mass is stationary. We shall illustrate this fact, but not prove it in general.

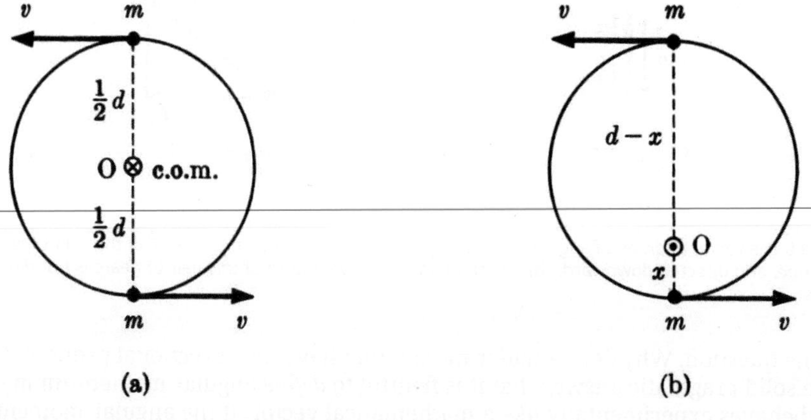

FIGURE 10.7 Equally massive particles rotating about their stationary center of mass have the same total angular momentum with respect to any point.

Consider a pair of equally massive particles rotating about their fixed midpoint (Figure 10.7). Their separation is d; each has mass m, speed v, and is a distance ½d from the stationary center of mass. Accordingly, the magnitude of the angular momentum of each with respect to the midpoint is

$$L_1 = L_2 = \tfrac{1}{2}mvd\,. \tag{10.9}$$

Since the two angular momenta have the same direction (outward from the page in Figure 10.7) as well as the same magnitude, the vector magnitudes add numerically to give the total magnitude,

$$L = L_1 + L_2 = mvd\,. \tag{10.10}$$

This is the total angular momentum of the pair with respect to the point halfway between them. Consider now a reference point located at the position of one of the particles. This particle has no angular momentum with respect to its own location ($r = 0$). The other particle, a distance d away, has angular momentum mvd. The sum of the two is again mvd. Alternatively, pick an intermediate reference point a distance x from one particle and a distance $d - x$ from the other [Figure 10.7(b)], With respect to this reference point, the two angular momenta are

$$L_1 = mvx\,,$$

$$L_2 = mv(d - x)$$

Once again the sum has the same value,

$$L = L_1 + L_2 = mvd\,. \tag{10.11}$$

The angular momentum of the system is the same with respect to all reference points, giving it obviously more significance than it would otherwise have.

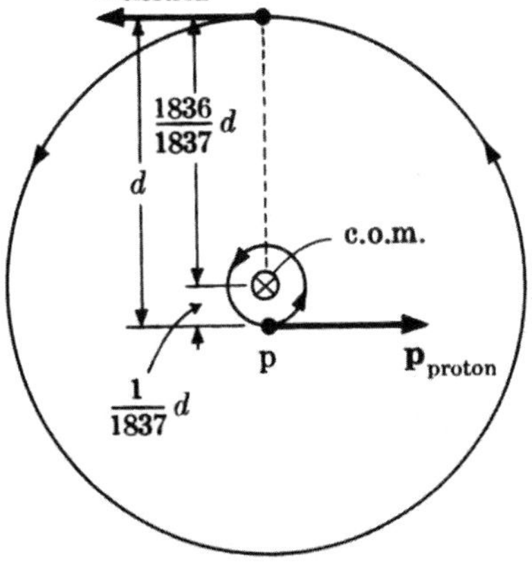

FIGURE 10.8 The hydrogen atom as it might appear if it followed the laws of classical mechanics. The scale is distorted for clarity.

Pretending for the moment that the hydrogen atom is describable in terms of classical particle trajectories, we may examine it as another example of a two-particle system. If the distance between proton and electron is d, the center of mass of the atom is located a distance (1/1837) d from the proton and a distance (1836/1837) d from the electron, very much closer to the massive proton than to the light-weight electron (Figure 10.8). In order that the center of mass of the atom remain at rest, the momenta of the proton and electron must be equal in magnitude and opposite in direction. (Recall from Section 9.7 that the center of mass is stationary only if the total momentum is zero.) Call the magnitude of each momentum p, and suppose at a particular moment that the electron's momentum is directed to the left, the proton's momentum to the right, as in Figure 10.7. Then the individual angular momenta with respect to the center of mass are

$$L_{\text{electron}} = p\left(\frac{1836}{1837}\right)d. \tag{10.12}$$

$$L_{\text{proton}} = p\left(\frac{1}{1837}\right)d. \tag{10.13}$$

Their sum is

$$L_{\text{atom}} = pd. \tag{10.14}$$

It is left as an exercise to show that the atomic angular momentum has the same value with respect to other points.

Any material object—a billiard ball, a top, an automobile wheel—may be looked upon as a system of particles. If the object rotates about an axis, each microscopic speck of matter in the object is a particle tracing out a circular orbit and contributing its microscopic bit to the total angular momentum of the whole object [Figure 10.2(d)], Integral calculus, concerned with summing an infinite number of infinitesimal contributions, is the mathematical tool required to calculate the total angular momentum of an extended object of any shape. However, no mathematical machinery is required in order to understand the principle involved. Each bit of matter behaves like a particle, providing its share of angular momentum according to the definition provided by Equation 10.1, depending on its mass, its speed, and its distance from the reference point. Each contributing angular momentum is directed along the same axis, so that the sum of all contributions is directed along the axis. Moreover, if the axis of rotation passes through the center of mass, so that this point remains stationary, the magnitude of the angular momentum is a definite number, the same with respect to all reference points.

10.4 Spin angular momentum and orbital angular momentum

41 So far in this chapter we have been concerned with the nature of the angular-momentum concept, without reference to the physical laws of its behavior. Still one more aspect of this fundamental attribute of motion needs to be discussed before turning in the next section to the law governing the rate of change of angular momentum. Here we seek to clarify the distinction between spin and orbital angular momentum.

Spin* is the angular momentum of an object or a structure with respect to its own center of mass. If some object—for example, a top or the idling propeller of a parked airplane—is rotating about a stationary axis passing through its center of mass, its spin angular momentum is the same as its total angular momentum. Moreover, as emphasized in the preceding section, this angular momentum has a unique magnitude, the same with respect to all points of reference. It is for this reason that spin may be regarded as an intrinsic property of the rotating structure independent of the observer's point of reference.

More often than not, the center of mass of a spinning object is itself in motion. The earth spins about an axis which moves through space. A billiard ball struck askew spins as it travels across the table. Because the center of mass is moving, the total angular momentum *does* depend on the point of reference. The earth has more angular momentum with respect

* The word "spin" alone is usually used to mean spin angular momentum.

to the sun than with respect to the moon. The billiard ball may have more or less angular momentum with respect to a point on its path than with respect to a point to one side of its path. The complexity of the situation can be appreciated by visualizing the path of a single point as it moves through space—the tip of the Empire State Building attached to the earth, or a speck of dirt on the billiard ball. It follows not a circular path but a more complex path, moving now slower, now faster, with respect to the reference point. Fortunately simplicity can be recaptured by dividing the angular momentum into two parts, a spin part and an orbital part.

The spin of an object continues to be defined as the angular momentum with respect to the center of mass, whether or not the center of mass is in motion. It remains an intrinsic property of the object, independent of the reference point. The spin angular momentum of the earth, for instance, would be calculated as if the earth's axis were stationary, with every part of the earth executing simple circular motion.

Orbital angular momentum is the angular momentum associated with the bulk motion of the whole object. It is the product of three factors: the total mass of the object, the perpendicular component of the velocity of its center of mass, and the distance of its center of mass from the reference point. In symbols,

$$L_{\text{orbital}} = mv_{c\perp}r_c \,. \tag{10.15}$$

The subscript c refers to the center of mass. As illustrated in Figure 10.9, the orbital angular momentum of a single object depends upon the chosen reference point. (For two or more objects in relative motion, the total orbital angular momentum may be, like the spin, independent of the reference point.)

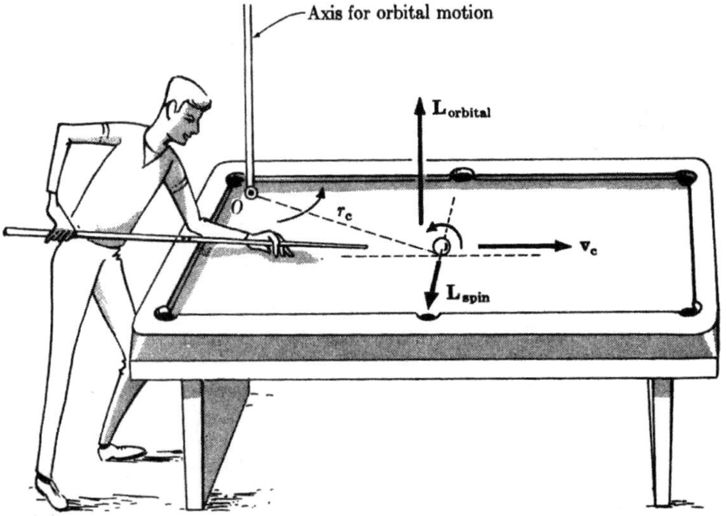

FIGURE 10.9 Orbital angular momentum and spin angular momentum. A billiard ball, hit a low blow, slides and turns across the table, with its spin directed horizontally to the right of its path. With respect to the left rear corner of the table, it possesses orbital angular momentum directed vertically upward.

The division of angular momentum into orbital and spin parts has found its greatest usefulness in the world of particles. Most of the elementary particles possess a spin which is an invariable property of the particle. In addition, any particle may possess orbital angular momentum with respect to any reference point (see Figure 4.6). Because of the quantum rule that orbital angular momentum comes only in units of the constant $\hbar$, an electron might approach a proton with no relative orbital angular momentum, or with one unit $\hbar$, or with two units or three, and so on. For all collisions the electron's spin would have the same value, The proton also has a fixed spin of $\frac{1}{2}\hbar$.

Despite the invariable magnitude of a particle spin, its direction can change. In an interesting class of particle collisions called spin-flip collisions, the direction of the spin angular momentum changes during the collision. Because of the vector nature of angular momentum and the law of conservation of angular momentum that applies in these collisions, a spin flip must be accompanied by a change in orbital angular momentum—in magnitude or direction or both—in order that the total angular momentum remain unchanged. The earth-moon system provides another interesting example of interchange between spin and orbital angular momentum. According to present theory, the moon once possessed more spin and less orbital angular momentum with respect to the earth than at present. Now it spins on its axis only once in each 27-day trip around the earth, and its orbital angular momentum with respect to the earth far exceeds its spin angular momentum.

10.5 *Torque and the law of angular-momentum change*

42 In the preceding four sections, the concept of angular momentum has been defined and elaborated, but not yet placed in the context of a physical law. To summarize the main points emphasized so far: (1) The angular momentum of a particle with respect to a point, a measure of the strength of its rotational tendency about that point, is defined in terms of the mass and velocity of the particle and its distance from the reference point. Even more simply, it may be defined in terms of the particle's momentum and its distance from the reference point. (2) A particle moving at constant velocity in a straight line has constant angular momentum, A particle moving at fixed speed in a circle has constant angular momentum with respect to the center of the circle. (3) Angular momentum is a vector concept. Its direction is the axial direction as defined by the right-hand rule. (4) The angular momentum of a system is the vector sum of the angular momenta of the constituent parts of the system. (5) It is convenient to separate the angular momentum of a material object into a spin part and an orbital part. Spin is its angular momentum with respect to its center of mass. Orbital angular momentum, the additional angular momentum arising from motion of the center of mass, is calculated as if all of the mass of the object were concentrated at the center of mass.

Still to be answered are the important questions: Under what circumstances is angular momentum conserved? What is required to change the angular momentum of a system? What is the law of angular momentum change? If the third question can be answered, the answers to the first two will follow readily. In this section we explore the classical law of angular-momentum change.

Since angular momentum is defined mathematically in terms of previously established concepts, it is not surprising to learn that the law of its change is derivable mathematically from Newton's laws and need not have a new experimental foundation. We shall not carry out this derivation, but rather we shall establish the plausibility of the result through examples.

If a particle is moving in a straight line toward a reference point, it possesses no angular momentum with respect to that point. Acted upon by a force in its direction of motion, it accelerates and changes its momentum, but still acquires no angular momentum since it remains on a collision course with the point. If another particle moves uniformly in a circle, it must be acted upon by a centripetal force in order to maintain its acceleration toward the center of the circle. This force does not change the angular momentum. As these two examples make clear, force alone need not change angular momentum. However, force applied in the right way *can* produce a change in angular momentum.

Consider a wheel attached to a fixed axle. A force applied at the rim of the wheel produces no rotation if it is directed toward the axle, but sets the wheel rotating if it is directed tangentially along the rim. If directed at some intermediate angle, the force is less effective but also produces rotation, thereby changing the wheel's angular momentum. Besides the direction of the force, its distance from the axis is also important. If you grasp the axle of the wheel and twist, it is much more difficult to set the wheel spinning than if you push it along its outer edge. A pump handle or jack handle, hinged to rotate about one of its ends, is much easier to move if held at the opposite end than if held close to its axis of rotation. Other things being equal, a force applied far from an axis changes angular momentum more effectively than a force applied close to an axis.

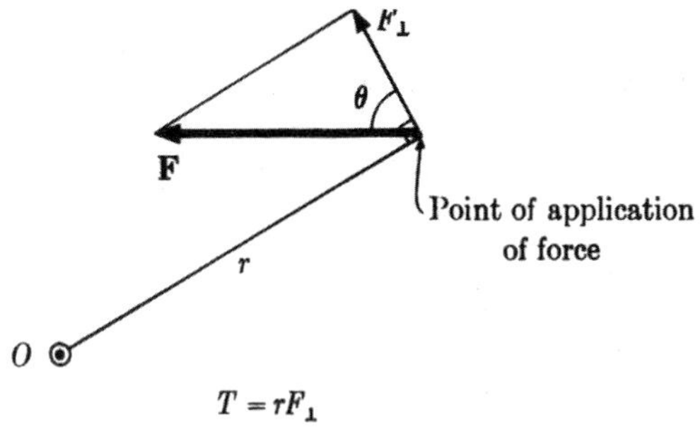

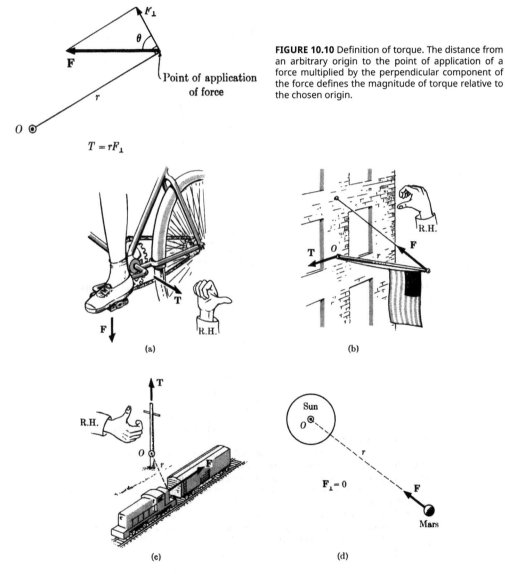

FIGURE 10.10 Definition of torque. The distance from an arbitrary origin to the point of application of a force multiplied by the perpendicular component of the force defines the magnitude of torque relative to the chosen origin.

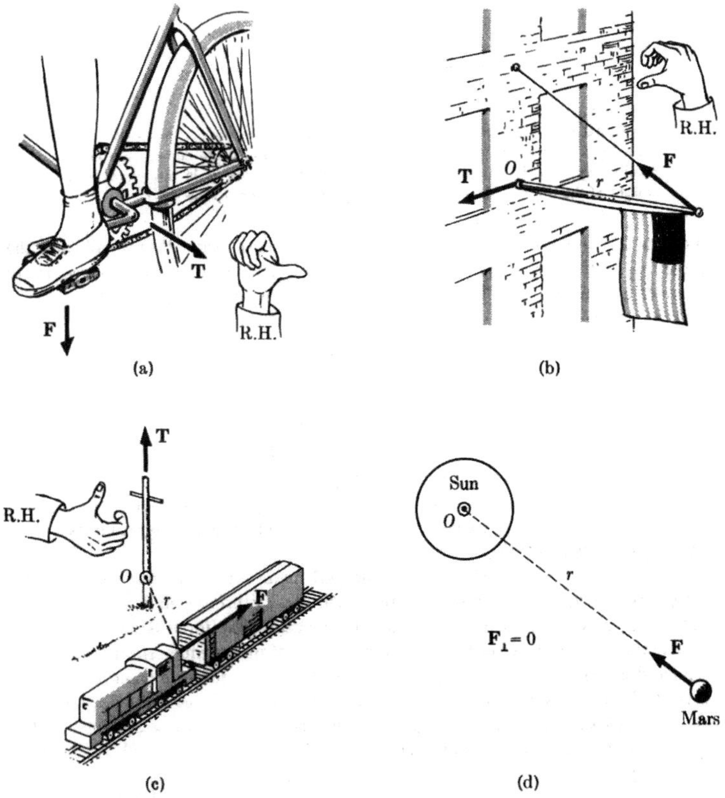

FIGURE 10.11 The vector nature of torque. **(a)** Force on a bicycle pedal produces an axial torque to the left. **(b)** With respect to a point at the base of the flagpole, the force of the guy wire produces a horizontal torque as shown. **(c)** An engine pushing a freight car produces a vertical torque if the origin is chosen to one side of the track. **(d)** With respect to the center of the sun, the gravitational force of the sun on Mars produces no torque.

The particular combination of force magnitude, force direction, and distance that is most simply related to change of angular momentum is called torque. Like angular momentum, torque is measured with respect to a particular point of reference. Its magnitude is defined by this equation:

$$T = F_{\perp} r \,. \tag{10.16}$$

The magnitude of a torque is equal to the perpendicular component of a force multiplied by the distance between the point of application of the force and the reference point (Figure 10.10). In this context, "perpendicular component" means exactly what it meant in the definition of angular momentum: perpendicular to the line joining the reference point to the physical point of interest (in this case the place where the force acts). Also like angular momentum, torque is a vector quantity directly axially. If the curved fingers of the right hand indicate the direction of twist produced by the force, the right thumb indicates the direction of the torque (Figure 10.11). Two or more torques add as vectors to produce a total torque. The cgs unit of torque is the dyne centimeter.*

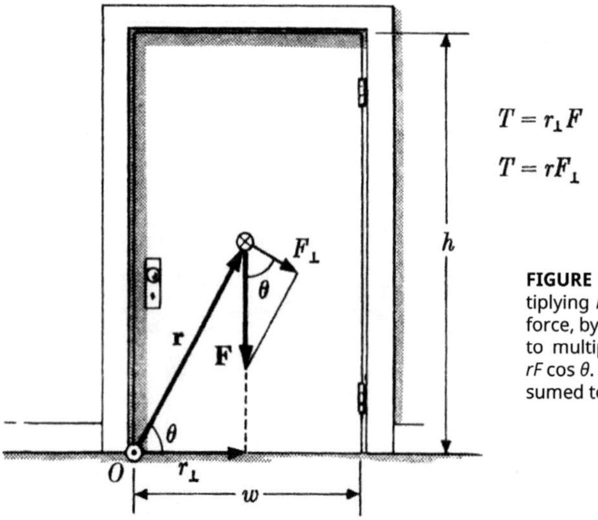

FIGURE 10.12 Alternative definitions of torque. Multiplying $r_{\perp}$, the component of r perpendicular to the force, by F, the total magnitude of force, is equivalent to multiplying r by $F_{\perp}$. Both products are equal to $rF \cos\theta$. In this example, the weight of the door is assumed to act at the center of mass of the door.

A sometimes useful alternative to Equation 10.16 is

$$T = r_{\perp} F \,, \tag{10.17}$$

in which $r_{\perp}$ is the perpendicular distance from the reference point to the line of action of the force. As shown in Figure 10.12, $rF_{\perp} = r_{\perp}F = rF\cos\theta$, so that Equations 10.16 and 10.17 are indeed equivalent. In the example of Figure 10.12, the form 10.17 is somewhat easier to apply, because $r_{\perp}$ can be recognized at once to be ½w, half the width of the door. The weight of the door produces with respect to the lower left corner of the door a torque whose magnitude is ½wmg. (What is its direction?)

Just as force is the idea of a push or pull made into a quantitative concept, torque is the idea of a twist rendered scientifically precise and useful. Torque bears the same relationship to angular momentum that force bears to ordinary momentum. Force is that which changes momentum; torque, similarly, is that which changes angular momentum. In the

* Although torque and energy have the same unit of measurement, they are very different quantities physically. To avoid confusion the word "erg" as a substitute for dyne cm or gm cm^2/sec^2 is used only for energy, not for torque.

absence of force, momentum is conserved; in the absence of torque, angular momentum is conserved. Moreover, there is a close mathematical parallelism that accompanies these statements. The connection between force and the change of momentum is expressed by Newton's second law:

$$\mathbf{F} = \frac{\Delta \mathbf{p}}{\Delta t}.$$

Force is equal to the rate of change of momentum. An exactly similar equation connects torque and the rate of change of angular momentum:

$$\mathbf{T} = \frac{\Delta \mathbf{L}}{\Delta t}. \tag{10.18}$$

This is the physical law governing the change of angular momentum. Although it is of fundamental importance, it has not been dignified by a name such as "Newton's fourth law," because it is simply a consequence of Newton's three laws. Equation 10.18 was in fact unknown to Newton. The concept of angular momentum did not enter mechanics explicitly until the middle of the eighteenth century. Some features of the analogies between force and torque, and momentum and angular momentum, are summarized in Table 10.1.

TABLE 10.1 Some Parallels in Mechanics

Feature	Parallels Between Force and Torque	
An aspect of motion	Momentum, **p**	Angular momentum, **L**
The impetus of change	Force, **F**	Torque, **T**
The law of change	$F = \frac{\Delta \mathbf{p}}{\Delta t}$	$T = \frac{\Delta \mathbf{L}}{\Delta t}$
The requirement for conservation	Absence of external force	Absence of external torque

Note that Equation 10.18 is a vector equation. It states both that the direction of the change of angular momentum is the same as the direction of the torque, and that the magnitude of the change of angular momentum divided by the time interval during which the change occurs is equal to the magnitude of the applied torque. Just as with Newton's second law, it is often convenient to interpret this equation causally and to say that angular momentum changes *because* a torque acts. However, there is no necessity for such an interpretation. The equation expresses only a connection. Torque and change of angular momentum go together. One always accompanies the other. There are even occasions when it seems more sensible to look upon change of angular momentum as a cause, torque as an effect. If a wheel is being slowed "because" of a torque supplied by a brake, the brake drum in turn experiences a torque "because" of the change of angular momentum of the wheel. In either case, the torque is equal to the rate of change of angular momentum.

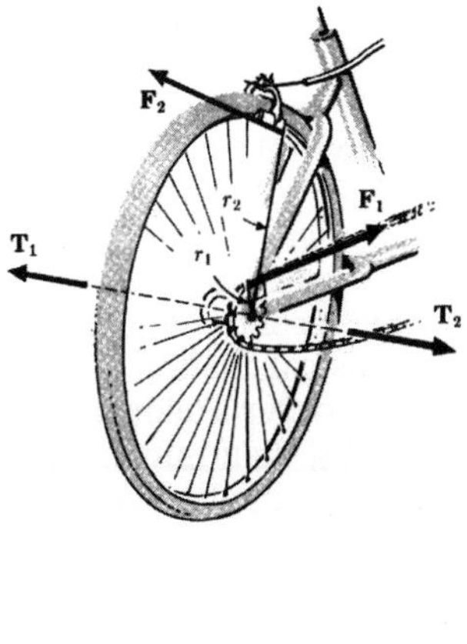

FIGURE 10.13 Forces and torques applied to the rear wheel of a bicycle.

In many physical examples, the conservation of angular momentum associated with the absence of torque is of great importance. That is the subject of Section 10.6. Here we wish to investigate a few examples in which angular momentum *does* change.

The rotation of a wheel about a fixed axis provides a simple example for study. Consider a bicycle suspended so that its wheels can spin freely without touching the ground. We fix attention on the rear wheel as the system of interest (Figure 10.13). A force can be applied at the rear sprocket by the chain and a force can be applied at the rim by the brakes. These two forces are designated $\mathbf{F}_1$ and $\mathbf{F}_2$ in Figure 10.13. The frictional forces at the axle are assumed to be small enough to neglect. With respect to the center of the axle chosen as a reference point, each of the forces $\mathbf{F}_1$ and $\mathbf{F}_2$ gives rise to a torque. Each of these forces is perpendicular to a line drawn from the axis to its point of application. Therefore the perpendicular component of each is equal to the full magnitude of the force. If the distances from the axis to the chain and to the brake are called r_1 and r_2, the respective torques have magnitudes given by

$$T_1 = F_1 r_1 , \tag{10.19}$$

$$T_2 = F_2 r_2 , \tag{10.20}$$

These torques are directed axially, $\mathbf{T}_1$ to the left, $\mathbf{T}_2$ to the right, as indicated in Figure 10.13. This means that any change in the wheel's angular momentum produced by the torque $\mathbf{T}_1$ must be directed axially to the left, any change in its angular momentum produced by the torque $\mathbf{T}_2$ must be directed axially to the right.

Suppose, to be specific, that the axis-to-chain distance is $r_1 = 4$ cm, the axis-to-brake distance is $r_2 = 30$ cm, and the chain can deliver to the sprocket a force $F_1 = 10^7$ dynes. If the wheel starts at rest and the force of the chain is applied for two seconds, what angular momentum will the wheel acquire? First, the magnitude of the torque $\mathbf{T}_1$ must be calculated. From Equation 10.19,

$$T_1 = 4 \text{ cm} \times 10^7 \text{ dynes} = 4 \times 10^7 \text{ dyne cm}.$$

According to the fundamental law governing change of angular momentum (Equation 10.18), this torque is equal to the rate of change of angular momentum, or to the change of angular momentum divided by the change of time. Expressed in terms of magnitudes,

$$\frac{\Delta L}{\Delta t} = T_1 = 4 \times 10^7 \text{ gm cm}^2/\text{sec}^2 .$$

Since the torque acts for an interval $\Delta t = 2$ sec, the angular momentum change ΔL, equal to the product of Δt and T_1, is

$$\Delta L = 8 \times 10^7 \text{ gm cm}^2/\text{sec} . \tag{10.21}$$

Note that the unit of angular momentum is correct, verifying the dimensional consistency of Equation 10.18. Since the angular momentum was initially zero, this *change* of angular momentum is the same as the angular momentum of the wheel after two seconds. It is directed to the left, corresponding correctly to clockwise rotation if the wheel is viewed from the right side.

Imagine now that the wheel spins freely for some time with no change in its angular momentum, since no torques act, after which time the brakes are applied with constant force to decrease the angular momentum of the wheel by half in three seconds. What brake force must be applied to achieve this result? The logical steps of this calculation are the same as those of the preceding paragraph, but executed in the opposite order. Following the generally preferable procedure, we obtain the result first algebraically, then numerically. Let L represent the angular momentum of the wheel before the brake is applied. After an interval of time Δt its angular momentum is to be reduced to $\frac{1}{2}L$, so that the magnitude of the change of angular momentum is also $\frac{1}{2}L$.

$$\Delta L = \tfrac{1}{2}L \,.$$

The fundamental equation connecting torque and rate of change of angular momentum may be written, for this example,

$$T_2 = \frac{\Delta L}{\Delta t} = \frac{1}{2}\frac{L}{\Delta t}\,. \tag{10.22}$$

According to Equation 10.20, this torque is also equal to the product of the force F_2 and the distance r_2. The two expressions each equal to the braking torque T_2 may be equated to give

$$F_2 r_2 = \frac{1}{2}\frac{L}{\Delta t}\,.$$

Since the problem posed was to find the necessary braking force F_2, this equation is solved for F_2:

$$F_2 = \frac{1}{2}\frac{L}{r_2 \Delta t}\,. \tag{10.23}$$

The numerical answer is then readily obtained by substituting $L = 8 \times 10^7$ gm cm^2/sec, r_2 = 30 cm, and Δt = 3 sec. It is

$$F_2 = 4.44 \times 10^5 \text{ gm cm/sec}^2 = 4.44 \times 10^5 \text{ dynes.} \tag{10.24}$$

As illustrated in Figure 10.14, the change in angular momentum produced by the braking torque is directed to the right, so that the final angular momentum, although only half so great in magnitude as the initial angular momentum, is the vector *sum* of the initial angular momentum and the change of angular momentum.

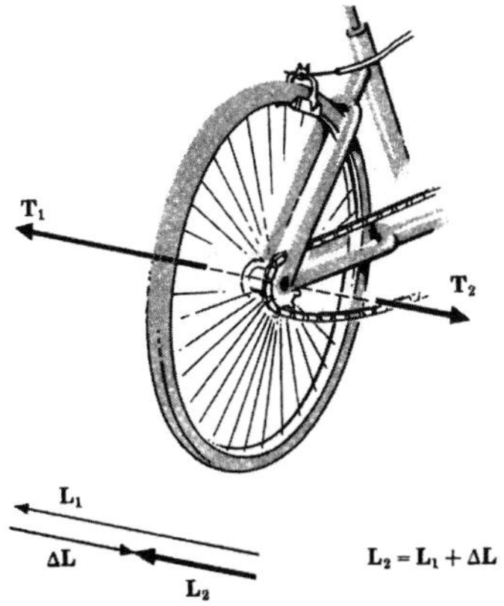

FIGURE 10.14 Changes of angular momentum produced by torques. Set spinning with torque $\mathbf{T}_1$, the wheel has angular momentum $\mathbf{L}_1$. A braking torque $\mathbf{T}_2$ then contributes a change of angular momentum $\Delta\mathbf{L}$. The final angular momentum, $\mathbf{L}_2$, less in magnitude than $\mathbf{L}_1$, is the vector sum of $\mathbf{L}_1$ and $\Delta\mathbf{L}$.

In the example just discussed, the vector nature of angular momentum plays a rather minor role, since all torques and angular momenta were directed left or right along the same fixed axis. When this is not the case—if for instance a torque acts upon a spinning object in a direction other than the direction of its spin—new effects arise that seem at first rather startling, despite the fact that they are direct consequences of Equation 10.18.

Consider a bicycle wheel of the kind often used for lecture demonstrations, mounted on a pair of handles (Figure 10.15). If the wheel is not spinning and one end of one handle is placed on a pointed support, as shown in Figure 10.15, the wheel does what is expected of it. It falls off the support. Its fall can be easily understood in terms of torque and angular momentum vectors. Suppose for definiteness that the handle is initially horizontal in the east-west direction with its west end resting on the support. The downward force of gravity, which can be supposed to act at the center of the wheel, gives rise to a torque with respect to the point of support. The direction of this torque is horizontal to the north. It produces a twist about a north-south axis. Accordingly the wheel acquires rotational motion about this axis. The mass of the wheel starts vertically downward. With respect to the point of support, it gains angular momentum northward. As the wheel falls from the support, it rotates, not about its own axle, but about a north-south axis dictated by the direction of the gravitationally produced torque.

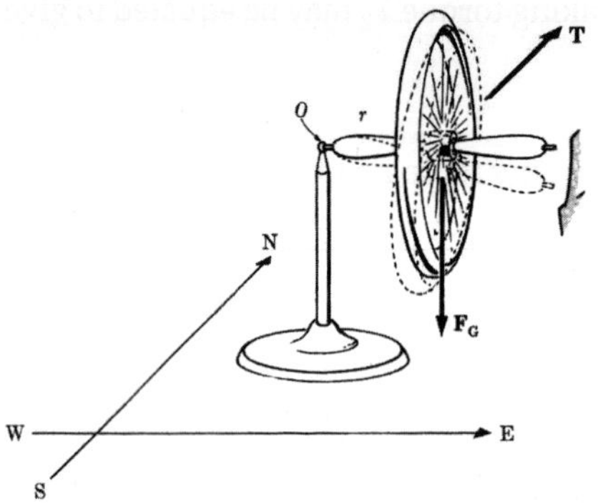

FIGURE 10.15 A nonspinning bicycle wheel falls from its support. In falling, it acquires angular momentum directed northward, parallel to the torque **T** produced by the gravitational force $\mathbf{F}_G$. The chosen origin is the top of the supporting rod.

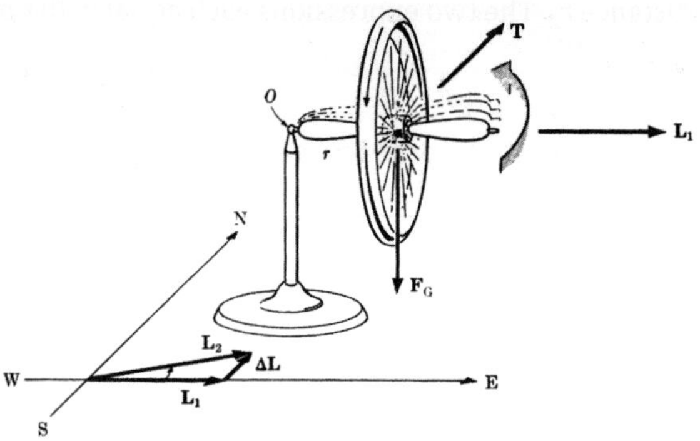

FIGURE 10.16 Precession of a rotating wheel. The angular momentum is directed initially eastward. Torque and the change of angular momentum are directed northward. Therefore $\mathbf{L}_1$ swings around as indicated in the diagram, a motion called precession.

This description of the simple fall of the wheel from its support may seem unnecessarily roundabout. However, it is a very useful preamble to the discussion of an altered situation. Suppose that the wheel is set spinning about its own axle before it is placed as before with the west end of its handle on a point of support. If, viewed from the east, the wheel is rotating counterclockwise, its angular momentum vector is directed eastward. The same gravitational force acts as before, producing a northward torque (Figure 10.16). Now we have an interesting new situation in which torque and angular momentum are not parallel. However, according to the fundamental law expressed by Equation 10.18, the *change* of angular momentum must be parallel to the torque, northward in this case. If the vector $\mathbf{L}_1$ designates the initial angular momentum and the vector $\Delta\mathbf{L}$ designates the change of angular momentum in a brief interval of time, their sum $\mathbf{L}_2$ will designate the angular momentum of the wheel after this time interval:

$$\mathbf{L}_2 = \mathbf{L}_1 + \Delta\mathbf{L}\,. \qquad (10.25)$$

This vector addition is shown pictorially in Figure 10.17. The result of the gravitational torque is not to topple the wheel from the support. Rather it shifts the wheel's axle around to a new direction, in this case slightly north of east. Once the wheel has so shifted, the gravitational torque has also shifted, to a direction slightly west of north. Consequently the new total angular momentum $\mathbf{L}_2$ is further pulled counterclockwise (as seen from above) by a new change $\Delta\mathbf{L}$. The process continues indefinitely, or until friction puts a stop to it, with the angular momentum vector of the wheel swinging through the points of the compass. Physically, this means that the axle of the wheel, as seen from above, rotates counterclockwise. This rotation of the axle, which proceeds at a uniform rate, is called precession. Had the wheel been spinning in the opposite direction, its precession direction would also be opposite. In the design of gyroscopes, whose axles are supposed to remain stationary, precession is something to be avoided. It is the aim of gyroscope designers to eliminate all torques not aligned with the axis of rotation, for a torque in any other direction will cause the gyroscope axis to precess.

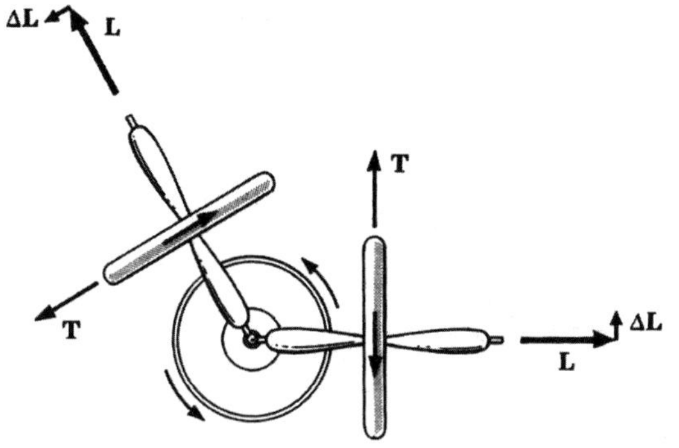

FIGURE 10.17 Top view of precessing bicycle wheel. The magnitude of the angular momentum remains constant as its direction changes.

Actually watching the bicycle wheel precess, supported at the end of one of its handles, is a worthwhile experience. Notwithstanding our ability to understand what is happening, it seems rather surprising that the wheel does not fall from the support. The beauty of this example is the way it shows the power of a mathematical law to penetrate easily a new range of experience. The law connecting torque and the rate of change of angular momentum, based on observations in full accord with common sense, readily accounts for new phenomena whose explanation would otherwise strain common sense.

10.6 The conservation of angular momentum

Under what circumstances does the angular momentum of a system remain constant? Equation 10.18 provides an immediate answer. If the total torque on a system is zero, the total rate of change of its angular momentum is zero. Equation 10.18 can be applied separately to each particular constituent of a complex system. If the vector sum of all the torques acting on and within the system is zero, the rates of change of all the constituent angular momenta also sum to zero. This means simply that the total angular momentum of the system remains constant. Therefore Equation 10.18 applies to the system as a whole as well as to its parts, and can also be written

$$\mathbf{T}_{\text{total}} = \frac{\Delta \mathbf{L}_{\text{total}}}{\Delta t} . \qquad (10.26)$$

Actually a more important and more useful statement than this about total torque and total angular momentum can be made. Consider the division of the total torque into an internal part and an external part, 43

$$\mathbf{T}_{\text{total}} = \mathbf{T}_{\text{int}} + \mathbf{T}_{\text{ext}} . \qquad (10.27)$$

External torques arise from outside the system; internal torques arise from within the system. A sky-diver falling through the air, for example, is subjected to external torques from the air and internal torques from his own muscles. According to excellent experimental evidence—evidence based on the validity of the law of angular-momentum conservation—the internal torques in fact always add up to zero. This is an important and a new statement about nature. It might be called the rotational equivalent of Newton's third law. Because of Newton's third law, which leads directly to the conclusion that internal *forces* always add up to zero, the *momentum* of an isolated system is conserved. Similarly, the vanishing of total internal *torque* leads to the conservation of *angular momentum* for isolated systems. For central forces—those forces acting along the lines joining particles—Newton's third law leads to the vanishing of total internal torque as well as total internal force. In general, however, the requirement that total internal torque vanish is a more powerful restriction on the forces of nature than is Newton's third law.

Since internal torques sum to zero, Equation 10.26 may be replaced by

$$\mathbf{T}_{\text{ext}} = \frac{\Delta \mathbf{L}_{\text{total}}}{\Delta t}. \tag{10.28}$$

The total angular momentum of a system changes only in response to an external torque. If the sky-diver, after falling for a while in a swan dive, begins to rotate, it can only be because of torques delivered by the air. Falling through a vacuum he would be unable, by any contortions whatever, to change his angular momentum.* His own muscles, by changing his shape, can trigger torques in the air which impart angular momentum. No pushing, pulling, or twisting of the muscles, however, can lead to a net internally produced torque.

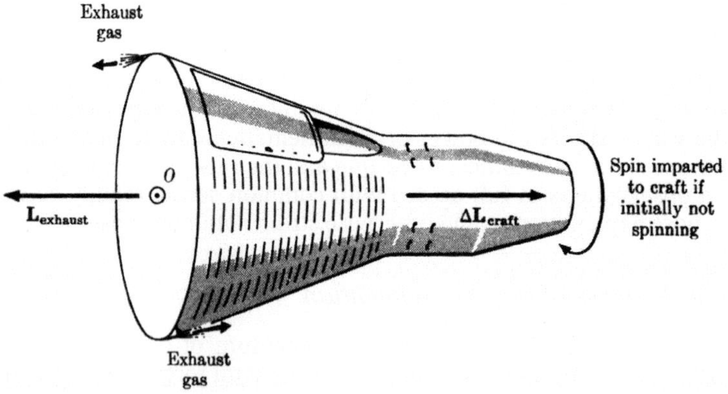

FIGURE 10.18 Application of torque to spacecraft by tangential rocket thrust. The exhaust gas possesses angular momentum with respect to the center of the spacecraft. An equal and opposite change of angular momentum is imparted to the craft.

For the same reason, nothing done internally in a spacecraft can stop or start a uniform rotational motion. This can be accomplished only in the same way that its momentum is changed: by throwing away mass. The exhaust from a tangential nozzle (Figure 10.18) possesses angular momentum with respect to the center of the spacecraft. The craft and what remains in it therefore acquire an equal and opposite change of angular momentum. Spacecraft plus exhaust viewed together as a single system suffer no change of angular momentum since they experience no external torque. If the exhaust is *not* regarded as part of the system, it is part of the external world delivering a torque and a change of angular momentum to what is left behind. What about the exhaust by itself? Is it acted upon by an external torque?

* He could, however, change his orientation without changing his angular momentum. A cat, usually able to land on its feet, is very clever at doing exactly that. For the falling cat, the air contributes nothing important to the turning maneuver, which could, in principle, be performed in a vacuum. The explanation of the cat's feat is left as an exercise.

The easiest way to rid a system of external torque is to rid it of external force. An isolated system experiences no external force and therefore no external torque.* Consequently we have as a fundamental law of nature: The total angular momentum of an isolated system is conserved. The sky-diver without any air and the spacecraft outside the atmosphere illustrated the principle. Examples are easy to find also in the world of atoms and particles. Indeed it is in the submicroscopic world, where true isolation of an individual system from the influence of all other systems is very nearly a reality, not just an idealized abstraction, that the law of angular-momentum conservation can be most precisely tested. Despite the fact that the *quantization* of angular momentum is significant in the submicroscopic world and irrelevant in the macroscopic world, the *conservation* of angular momentum takes exactly the same form and has the same significance in both domains.†

Examples of angular-momentum conservation in particle transformations appear in Chapter Four. Recall in particular the example of lambda decay illustrated in Figure 4.6, in which the orbital and spin angular momenta of the product particles could combine in two different ways to equal the initial angular momentum of the lambda.

The conservation of angular momentum is no less important when there is none of it than when there is some of it. In the decay of a kaon into two pions, denoted, for example, by

$$K^+ \rightarrow \pi^+ + \pi^0 ,$$

all three particles are spinless. With respect to the point occupied initially by the kaon, the total angular momentum is zero. After the decay, angular momentum conservation requires that the two pions fly apart with no relative angular momentum. (Momentum conservation is also at work. It requires that the two pions separate with equal and opposite momenta.)

Another mode of decay of the positive kaon is into muon and neutrino:

$$K^+ \rightarrow \mu^+ + \nu_\mu ,$$

Again the total angular momentum is zero, but this time each final particle has one-half quantum unit of spin. The spinning neutrino has the characteristic property that it is "left-handed." This means that its angular momentum is always directed opposite to its momentum, or equivalently, opposite to its direction of flight (Figure 10.19). To conserve angular momentum in this mode of kaon decay, the muon must also emerge in a left-handed state of motion. Both the momenta and the angular momenta of the two product particles cancel.

10.7 *The law of areas*

For an isolated system upon which no external forces act, both momentum and angular momentum are conserved. However, isolation is not required for angular-momentum conservation. All that is required is the vanishing of external torque. It is quite possible for a system to be acted upon by an outside force, yet be free of torque. For one special yet very important kind of force, a *central* force, the angular momentum of a system is conserved even though the system is not isolated and its momentum is not constant.

Consider Mars, the planet studied most intensively by Kepler, and choose the sun as a reference point. (More precisely, the reference point should be the center of mass of the Mars-sun system, which is not far from the center of the sun.) With respect to this point, Mars possesses orbital angular momentum. However, with respect to the same point, Mars

---

* This statement requires that there be no force of any kind acting, therefore that the system be truly isolated. If the zero force arises from two or more canceling forces, the total torque need not be zero.

† It should be borne in mind, however, that the law of change, Equation 10.28, is a classical law which loses its validity in the quantum world.

experiences no torque [Figure 10.11 (d)]. The external force acting on Mars is the sun's attraction directed along the line joining planet and sun. Since the perpendicular component of this force is zero, the torque is zero. Therefore Mars, as it moves around the sun, retains constant angular momentum. Actually the spin and orbital angular momenta of the planet are separately constant, since it experiences no torque with respect to its own center and none with respect to the sun. We are here concerned with the constancy of its orbital angular momentum.

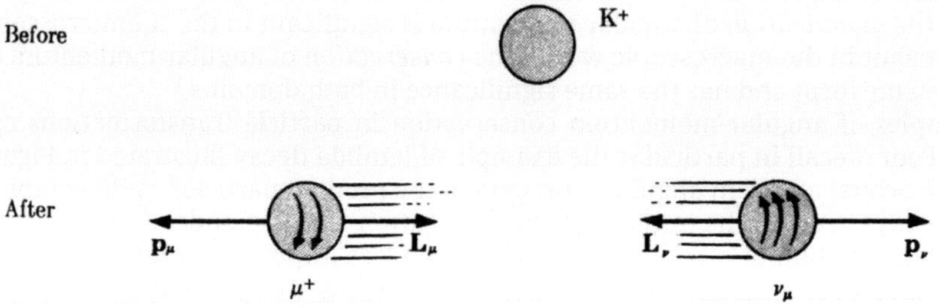

FIGURE 10.19 Two conservation laws in kaon decay. Both momenta and angular momenta of the two product particles add to zero.

A central force is defined as a force acting along the line joining a particle or object to a center of force. By its definition, therefore, a central force has no component perpendicular to this line and can produce no torque with respect to the force center. Any object moving under the action of a central force moves with constant angular momentum. Uniform motion in a circle, which requires a central force, is a special and often cited example of accelerated motion with constant angular momentum. However, an object acted upon by a central force need not move in a circle, or even move around the center of force. A few kinds of motion in central force fields are shown schematically in Figure 10.20. For each, angular momentum with respect to the force center is conserved. Also each such orbit or trajectory has in common an interesting geometric property called the "law of areas."

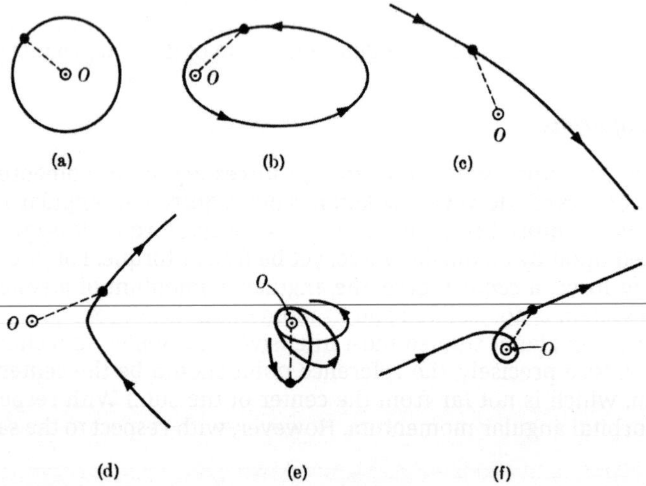

FIGURE 10.20 Possible motions in central force fields. Diagrams **(a)** – **(d)** show circular, elliptical, and hyperbolic paths that are possible when the force varies inversely with the square of the distance ($F \sim 1/r^2$). If the force varies in other ways with distance, an infinite variety of other trajectories are possible. Two such are shown in diagrams **(e)** and **(f)**.

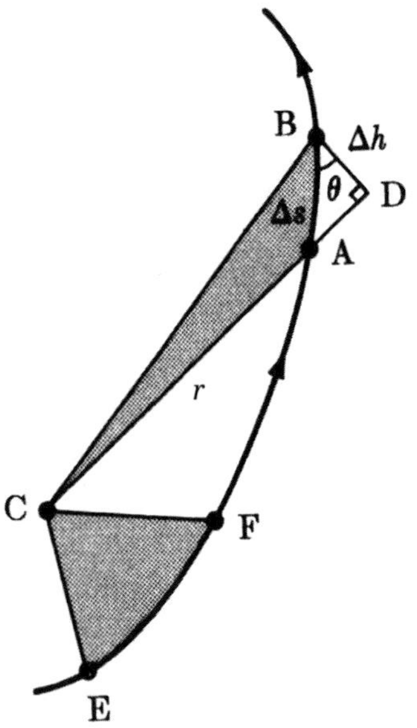

FIGURE 10.21 Geometric construction for the law of areas. Points E, F, A, and B lie along the particle trajectory. Point C is the force center.

Figure 10.21 illustrates the geometric construction that makes clear the meaning of the law of areas. At a particular moment a particle acted upon by a central force is at point A a distance r from the force center C, and is moving with speed v. After the lapse of a time interval Δt it has moved ahead to point B. If the interval Δt is sufficiently brief, the short segment of orbit AB may be approximated as a straight line traversed at constant speed v. These are approximations whose merits will be discussed below. The length of the segment AB may be called Δs. It is equal to the product of speed and time:

$$\Delta s = v\Delta t\,. \tag{10.29}$$

If now a line BD perpendicular to the radial line CA is drawn, it will make some angle θ with the line AB. The length Δs of AB multiplied by the cosine of this angle is equal to the length Δh of BD:

$$\Delta h = \Delta s \cos\theta = v\Delta t \cos\theta\,. \tag{10.30}$$

After the second equal sign, Δs has been replaced by $v\Delta t$. We note now that the combination $v \cos\theta$ is equal to $v_\perp$, the component of velocity perpendicular to the radial line CA. Therefore Δh may be written, more simply,

$$\Delta h = v_\perp \Delta t\,. \tag{10.31}$$

We fix attention now on the area ΔA of triangle ABC, shaded in the figure. This is referred to as the area "swept out" by the particle in the time interval Δt. It is equal to half the product of the base r and the height Δh:

$$\Delta A = \tfrac{1}{2} r\Delta h\,. \tag{10.32}$$

According to Equation 10.31, Δh may be replaced by $v_\perp \Delta t$, so that

$$\Delta A = \tfrac{1}{2} r v_\perp \Delta t\,. \tag{10.33}$$

The reason for these manipulations now begins to appear, for on the right side of Equation 10.33 is the same combination $rv_\perp$ that appears in the definition of angular momentum

(Equation 10.1). This combination is equal to the angular momentum L with respect to the force center divided by the particle mass m. The area ΔA may therefore be written

$$\Delta A = \frac{1}{2}\frac{L}{m}\Delta t\,.$$

As a final step, the factor Δt is taken to the left side of the equation, which becomes

$$\frac{\Delta A}{\Delta t} = \frac{L}{2m}\,. \tag{10.34}$$

The rate of change of area—or the rate at which area is being swept out—is equal to the angular momentum of the particle divided by twice the particle mass. If the force is central, this is a constant for any particular particle, the same at one time as at another. In Figure 10.21, for instance, the areas CAB and CEF are equal if the time intervals for the particle to traverse the orbit segments AB and EF are the same. As the figure makes clear, this means that the particle speed must be greater from E to F than from A to B.

44 The fact that a line drawn from the force center to the particle (or object) sweeps out area at a constant rate is what is called the law of areas. We have derived it as a geometric consequence of the law of angular-momentum conservation. Historically the law of areas was not a consequence but a precursor of angular-momentum conservation. Johannes Kepler, in 1609, before the concept of angular momentum had been invented, before the idea of a central force existed, even before it was understood that planetary motion had anything to do with force, discovered as an empirical fact that the law of areas correctly describes the motion of Mars around the sun. This was one of three important properties of planetary motion discovered by Kepler. Kepler's laws, which tied together a myriad of careful observations of the sun and planets into three neat packages, are discussed further in Chapter Twelve. His law of areas, many decades after its discovery, provided strong support for Newton's contention that the planets were drawn toward the sun by a central force.

The derivation above included two approximations whose validity must now be discussed. First, the curved orbit segment AB was treated as a straight line. Second, a possible change of speed between A and B was overlooked. Both of these approximations are called "first-order approximations." Consider, for example, the constant-speed approximation. The precisely correct expression for the distance AB is

$$\Delta s_{\text{exact}} = v_{\text{av}}\Delta t\,,$$

where v_{av} is the average speed during the time interval Δt. The expression actually used in the derivation was

$$\Delta s_{\text{approx}} = v\Delta t\,,$$

where v is the speed at point A. This expression is in error by an amount $\Delta v\Delta t$, where Δv is the difference between v_{av} and v. The product $\Delta v\Delta t$ is said to be of the second order because it contains two small factors, each of which gets smaller and smaller as the points A and B come closer together. In a tiny segment the speed cannot change much, so that the smaller is Δt, the smaller is Δv. The product $v\Delta t$, on the other hand, is an expression of the first order. For sufficiently small increments Δt and Δv, the second-order expression will be so much smaller than the first-order expression that it may be ignored. In a similar way, it can be shown that the error involved in replacing the curved orbit segment by a straight line is an error of the second order. The use of calculus makes it possible to consider in a meaningful way the limit in which the time interval Δt becomes vanishingly small. In that limit the second-order errors disappear and the law of areas may be proved to be an exact law.

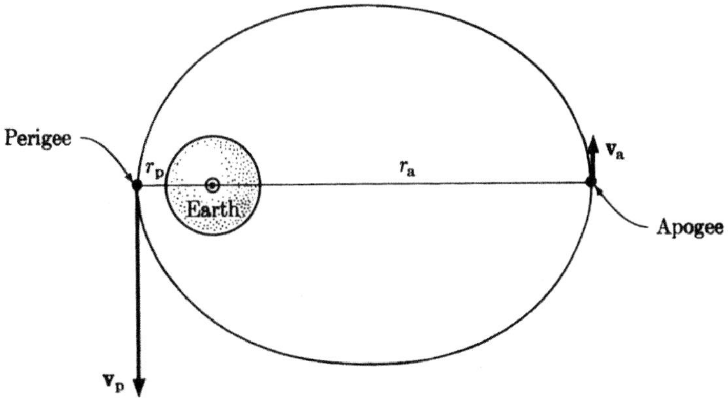

FIGURE 10.22 A satellite at perigee and apogee. The ratio of its speeds at these two points is the inverse ratio of its distances from the center of the earth (Equation 10.37).

At its perigee (nearest point) and at its apogee (farthest point), an earth satellite is moving perpendicular to the line joining it to the center of the earth (Figure 10.22). At these particular points, its perpendicular component of velocity is therefore equal to its total speed.* Accordingly, the angular momentum at perigee and apogee can be written in terms of mass, distance, and speed:

$$L_p = mr_p v_p , \tag{10.35}$$

$$L_a = mr_a v_a , \tag{10.36}$$

The subscripts p and a refer to perigee and apogee. Since the earth's gravitational force acting on the satellite is a central force, angular momentum is conserved, and L_p may be set equal to L_a. This equality leads to a particularly simple relation between speeds and distances at perigee and apogee:

$$\frac{v_a}{v_p} = \frac{r_p}{r_a} . \tag{10.37}$$

The ratio of the speed at apogee to the speed at perigee is equal to the inverse ratio of the distances from the force center at these two points. If apogee is three times as far from the earth's center as perigee, the speed at apogee is one-third the speed at perigee.

The decrease of speed at apogee dictated by the conservation of angular momentum has been used to good advantage in the choice of orbit for some communications satellites. These satellites (Telstar and Relay) have been launched into orbits with apogee about twice as distant (from the earth's center) as perigee. They are used to relay messages principally when they are near apogee and moving slowly enough to allow uninterrupted use for one to two hours. As they disappear over the horizon and approach perigee they speed up, and so waste less time getting around the other side of the earth and back to a useful position than would be the case if they were in nearly circular orbits. Man cannot repeal the law of angular-momentum conservation, but he can use it for his own ends.

* If the satellite possessed an outward radial component of velocity at apogee, it would be at a greater distance from the center a moment later, contradicting the assumption that it was already at its farthest point. If it possessed an inward radial component of velocity at apogee, it would have been at a greater distance a moment earlier. Similar arguments rule out the possibility of an inward or an outward radial velocity component at perigee.

Kepler, in studying the orbits of the six planets known to him, was not favored with the wide variety of orbital shapes now exhibited by artificial earth satellites. As shown in Table 10.2, none of these planets has an orbit markedly different from circular. Mercury, whose orbit is farthest from circular among these six, has an apogee distance 52% greater than its perigee distance (although the longest diameter of its orbit is only 2% greater than the shortest diameter—see Figure 10.23). The orbit of Venus is nearly a perfect circle. Mars, whose orbit is flattened by less than 0.5%, comes about 20% closer to the sun at perigee than at apogee. The data of Tycho de Brahe available to Kepler were of sufficient accuracy to allow him to discover the law of areas for the motion of Mars and to verify its validity for the other known planets.

TABLE 10.2 Some Properties of Planetary Orbits Known to Kepler*

Planet	Average Distance from Sun in Astronomical Units†	Ratio of Apogee Distance to Perigee Distance	Ratio of Longest Diameter of Orbit to Shortest Diameter
Mercury	0.387	1.517	1.022
Venus	0.723	1.014	1.000023
Earth	1.000	1.034	1.00014
Mars	1.524	1.206	1.0044
Jupiter	5.20	1.102	1.0012
Saturn	9.54	1.118	1.0016

* The data in the table are current values of these quantities.

† The astronomical unit (A.U.) is defined to be the average distance of the earth from the sun. It is equal to 1.495×10^{13} cm, about 93 million miles.

The precision tracking of modern satellites provides further tests of the law of areas every day. Of course, for satellites the law of areas is a tool, not a test. At one extreme are the elongated orbits of the Explorer satellites, which brush the fringes of the atmosphere at little more than 100 miles above the earth at perigee and cruise out to distances of more than 100,000 miles—halfway to the moon— at apogee. Explorer 18 rounded its apogee more than 126,000 miles from the center of the earth at a leisurely pace of about 750 miles per hour, to be compared with its speed of over 23,000 miles per hour at perigee. At the other extreme of orbit shapes is the almost perfect circle of a synchronous communications satellite. At the end of 1964, Syncom 3, over 26,000 miles from the center of the earth, was only five miles more distant at apogee than at perigee, and its speed in orbit varied by less than 0.02%.‡

Fortunately for Kepler and the advance of science in the seventeenth century, the law of areas is very nearly an exact law for planetary motion. However, we now know—and indeed Newton knew—that no planet experiences precisely a central force. Mars, for example, although predominantly influenced by the sun, is pulled weakly in other directions by Jupiter, the earth, and every other planet. With respect to the sun's center as reference point, these weaker forces, known as perturbations, are not central forces. Because of the perturbations, Mars experiences some torque, its angular momentum changes slightly in time, and for it the law of areas is not quite an exact law. Because the perturbations have only small effects, the solar system has a pleasing simplicity that it would otherwise lack, a piece of good luck for man in his early efforts to comprehend the world beyond the earth. On the other hand, the perturbations proved eventually to serve science, not impede it.

‡ Since the communications satellites carry small rockets controllable from earth, the orbit of Syncom 3 may now be slightly different. Because it is entirely free of frictional force, it will in any case be a companion of the earth indefinitely.

They made possible sensitive tests of the law of gravitational force; and perhaps most significant, they led to the discovery of two new planets—Neptune and Pluto. How this came about is described in Chapter Twelve.

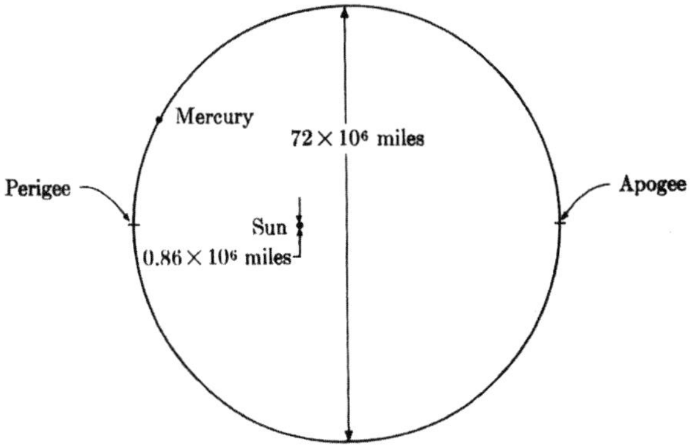

FIGURE 10.23 Orbit of Mercury drawn to scale. The size of the sun is also drawn to scale, but Mercury is greatly enlarged. The speed of Mercury at its perigee is about 52% greater than its speed at apogee.

For satellites circling the earth, the law of areas is also imperfect, an imperfection that can be put to good use to learn more about the earth. If the earth were a perfect sphere with its mass arranged in uniform shells, the force experienced by a satellite would be exactly a central force. But it is not such a perfect sphere. It is flattened at the poles, it is very slightly pear-shaped (a fact first revealed by satellites), and it is rather lumpy in its surface layers. For all of these reasons, the earth can exert small torques on a satellite, changing its angular momentum and causing it to deviate slightly from the law of areas. These orbital deviations have been measured with enough precision for some satellites to reveal the earth's pear shape, and to reveal new details of the distribution of mass within the earth.

10.8 The isotropy of space

The concept of angular momentum has undergone an interesting evolution in physics. 45
Kepler discovered the law of areas as an empirical fact long before angular momentum entered the vocabulary or the tool-kit of physics. Still without the help of the angular-momentum concept, Newton related the law of areas to the action of a central force. Not until the eighteenth century was angular momentum defined and used in mechanics. Not until the nineteenth, with the development of alternative and more powerful mathematical formulations of mechanics, did angular momentum come to be regarded as one of the most fundamental concepts of mechanics. Despite the fact that Newton's *Principia*, published in 1687, provided a complete foundation for mechanics, the theory was far from static over the next two centuries. Refined logically and conceptually, and rendered more powerful with new mathematics, the mechanics of 1887 bore about the same relation to the mechanics of 1687 as gasoline bears to petroleum. Newton struck oil. His successors developed the best that was in it. As the refinement of mechanics proceeded, angular momentum gradually separated itself as a significant concept, the heart of a new conservation law.

Finally in the twentieth century, angular momentum has joined momentum and energy as one of the preeminent mechanical concepts.* There are several reasons for its present

* A reminder: All three of these concepts have required new definitions in the modern theories, definitions that encompass the old.

station. One reason is its conservation; another is its quantization; a third is its relation to a simple symmetry of empty space.

As momentum conservation is related to, and indeed can be founded upon, the *homogeneity* of space (the indistinguishability of one point in space from another), angular-momentum conservation is similarly tied to the *isotropy* of space (the indistinguishability of one direction from another). An isolated object at rest in space is not expected to be self-accelerating in some direction, for that would imply an inhomogeneity of space. Nor is it expected to set itself spontaneously into rotation, for that would imply an anisotropy of space. The absence of spontaneous rotation requires the absence of any net internal torque, which in turn implies that the angular momentum of an isolated system is conserved. The bland sameness of space is at the root of both momentum conservation and angular-momentum conservation.

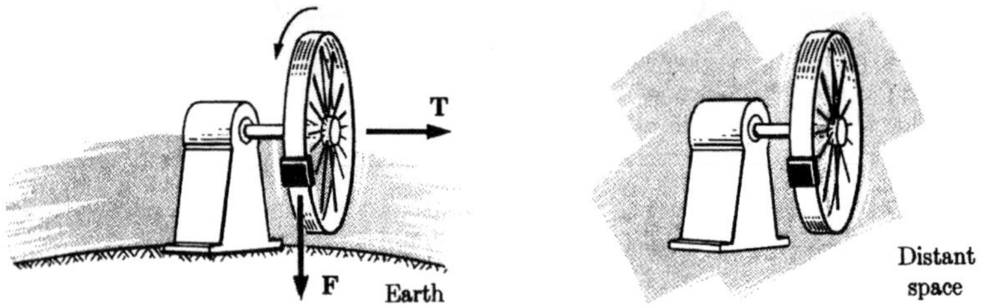

FIGURE 10.24 Angular-momentum conservation and the isotropy of space. **(a)** An unbalanced wheel near the earth rotates "spontaneously." Its angular momentum is not constant. **(b)** The same wheel in the depth of empty space does not start to rotate spontaneously if space is isotropic. This means that it experiences no net torque, which in turn implies that angular momentum is conserved for an isolated system.

The chain of argument here is worth reviewing in somewhat more detail. Consider a wheel that is uniform except for a weight placed at one point of its periphery (Figure 10.24). If suspended near the earth, initially at rest, it will begin to rotate "spontaneously" if the weight starts at any point other than the lowest point. There is nothing surprising about this. An external force, the force of gravity, acts on the wheel, and it produces an external torque. (Another external force, supporting the wheel at its axle, contributes no torque with respect to the axle.) Because of the external torque, the wheel's angular momentum changes. If initially at rest, it begins to rotate. For the wheel, angular momentum is not conserved. Another way to describe this situation is to say that in the neighborhood of the earth there exists a preferred direction, the vertical direction of the earth's gravitational force. The preferred spoke—the one connecting the hub to the weight on the rim—moves in such a way as to align itself with the preferred direction or to oscillate equally about the preferred direction, In empty space, on the other hand, far from the earth or other external influences, there should be no preferred direction and no spontaneous rotation. The same wheel, placed at rest in an ideally remote location, should remain at rest. The "should" in these sentences is based on the fundamental postulate of the isotropy of space. If space possesses the indistinguishability of direction called isotropy, no isolated object should spontaneously begin to rotate. This is the first key step in the argument. The second is to note that no rotation means no torque. Torques, if any, within the isolated object must cancel exactly. Having reached the conclusion that total internal torque must equal zero, we may, for the final step in the argument, allow our isolated object to be rotating initially instead of being stationary. Since it experiences neither external torque nor internal torque, its angular momentum remains constant.

This argument, intended only to be indicative of the existence of a link between the isotropy of space and the conservation of angular momentum, is less rigorous than it may appear. It provides a tight logical link only for rigid objects such as the wheel of Figure

10.24. For looser systems whose parts are in relative motion, the connection between spatial isotropy and angular-momentum conservation is more subtle (but just as real). Then the absence of any preferred direction leads only to the conclusions that *some* rotational property should be conserved. Angular momentum is defined in just such a way that it is the conserved quantity. It is not hard to think of other rotational quantities—angular velocity, for instance—that are not conserved.

Angular-momentum conservation has not been put to the test over domains of space larger than the solar system. It remains a question for the future, and a most intriguing question, whether this conservation law will fail in the galactic and intergalactic domains. If it does, man will have learned that space in the large is not perfectly isotropic, a discovery that would have most important bearing on the structure of the universe as a whole, and on the question of whether the universe is finite or infinite.

EXERCISES

10.1. A proton ($m = 1.67 \times 10^{-24}$ gm) is injected into a circular synchrotron at point C with a speed of 10^9 cm/sec. The radius of the synchrotron is 5×10^3 meters. What is the initial angular momentum of the proton **(a)** with respect to point A? **(b)** with respect to point B? **(c)** with respect to point C? (Give directions as well as magnitudes.)

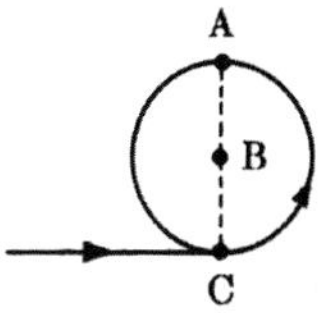

10.2. Work from the definition $L = p_\perp r$ (Equation 10.3) to show that the magnitude of the angular momentum with respect to the origin may also be written $L = pr_\perp$, in which p is the magnitude of the momentum and $r_\perp$ is the component of the position vector of the particle perpendicular to the momentum vector, (HINT: Extend the diagram of Figure 10.1 to show $r_\perp$.)

10.3. A girl is rolling a plastic hoop along the sidewalk at a speed of 150 cm/sec. The hoop has a diameter of 100 cm and a mass of 200 gm. (1) What is the angular velocity in radians per second of the hoop about its center? (2) What is the angular momentum of the hoop with respect to its center?

10.4. A boy tosses a stone into the air at an angle of 45 deg to the horizontal with an initial speed of 1.4×10^3 cm/sec. It lands 2,000 cm away, again with a speed of 1.4×10^3 cm/sec and at an angle of 45 deg to the horizontal. The mass of the stone is 80 gm. **(1)** What is the angular momentum of the stone with respect to the boy at the moment it leaves his hand? **(2)** What is its angular momentum with respect to the boy just before it lands? **(3)** Why does the angular momentum of the stone change?

10.5. Let θ represent the latitude angle of a point on the earth (zero at the equator, +90 deg at the north pole, –90 deg at the south pole), R the radius of the earth, and T the period of rotation of the earth. Find an algebraic expression for the angular momentum with respect to the center of the earth of a man of mass m standing on the surface of the earth at latitude θ. Express the answer in terms of the quantities θ, R, T, and m.

10.6. **(1)** The main rotor of a helicopter, seen from above, rotates counterclockwise. What is the direction of its angular momentum? **(2)** A small tail rotor mounted on a horizontal axis rotates counterclockwise when seen from the right side of the craft. What is the direction of its angular momentum? **(3)** Assuming the main rotor angular momentum to be 10 times the tail rotor angular momentum, sketch a vector arrow diagram for the sum of these angular momenta. Say from what direction this arrow diagram is viewed.

10.7. As in the discussion accompanying Figure 10.8, pretend that the hydrogen atom can be described as an electron and a proton executing circular orbits about their common center of mass. Pick some point other than the center of mass as the reference point and show that with respect to this point, the angular momentum is *pd*, the same as given by Equation 10.14.

10.8. Assume the approximate validity of classical mechanics in describing particle motion and consider a neutron moving along a straight line with a speed of 10^8 cm/sec past a stationary proton located at the origin of a coordinate system. At its point of closest approach, the neutron comes within 3.78×10^{-11} cm of the proton. **(1)** What is the orbital angular momentum of the neutron with respect to the proton? Express your answer in quantum units of $\hbar$. **(2)** Show in a sketch how the spins of neutron and proton must be aligned to maximize the total angular momentum of this system with respect to the chosen origin. With this alignment, what is the magnitude of the total angular momentum?

10.9. A delicate wheel of radius 1 cm requires a minimum torque of 2 dyne cm to set it into motion. At what angle should a force of 4 dynes be applied at the rim of the wheel in order to provide this minimum torque ?

10.10. A heavy flywheel has a radius of 50 cm. In order to start the wheel spinning, a man wraps a rope around its periphery and pulls on the rope for 3 sec with a force of 1.5×10^7 dynes. What is the final angular momentum of the flywheel?

10.11. Give an example of a pair of forces satisfying Newton's third law that produce no net force but do produce a net torque.

10.12. The bicycle wheel pictured in Figure 10.16 has a mass of 2.5 kg. The distance from its support point to its center is r = 20 cm. **(1)** With respect to the support point, what is the torque produced by gravity? **(2)** If the magnitude of the angular momentum of the wheel is $L = 1.2 \times 10^8$ gm cm^2/sec, through what angle does it precess in 0.2 sec? (NOTE: Since this is a relatively small angle, a vector sum diagram like that in Figure 10.16 is appropriate.)

10.13. An alpha particle with initial speed v_0 is deflected by a nucleus as shown. When very far from the nucleus, it moves in a straight-line track which, if extended, would miss the nucleus by a distance b. It actually comes within a distance r_{min} of the nucleus. Its angular momentum with respect to the nucleus is constant. **(1)** What is the direction of the angular momentum vector? (Be careful.) **(2)** What is its magnitude? **(3)** If $r_{min} = 2b$, what is the ratio of the speed v at the point of closest approach to the speed v_0 at great distance?

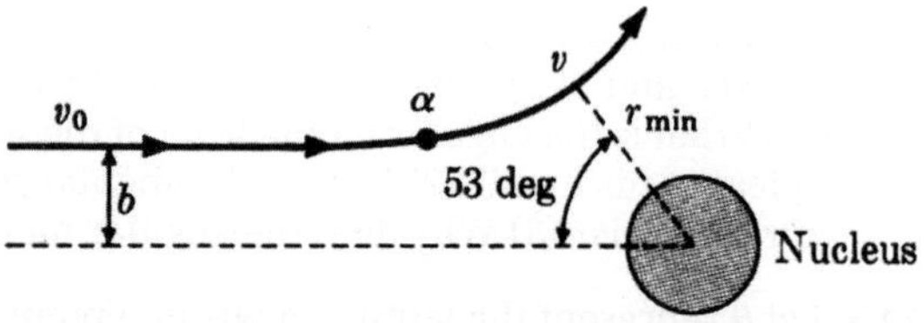

10.14. A particle moving uniformly in a circle under the action of a centripetal force experiences no torque (since the *perpendicular* component of force vanishes) and no change of angular momentum. If you tie a weight to a string, you are able to exert on the weight only an inward force along the string toward your hand. How, then, are you able to set the weight into rotational motion—that is, to impart to it some angular momentum? If puzzled about how you do so, perform the experiment.

10.15. An automobile with its engine idling stands at rest in neutral gear. As the driver "revs" up the engine he notices that his car lurches slightly to one side. Explain. Would it be possible to build an engine that did not have this effect? Explain.

10.16. If a boatload of ore were transported from Alaska to Peru, would the length of the day change? Why or why not?

10.17. Standing at rest on a nearly frictionless turntable an instructor hurls an eraser tangentially, as shown in the figure. The 100-gm eraser leaves his hand with a speed of 1,200 cm/sec at a distance of 80 cm from the axis of the turntable. **(1)** What angular momentum does the system of instructor-plus-turntable acquire? **(2)** Estimate very roughly the instructor's resulting angular velocity. Make reasonable guesses of any required magnitudes.

10.18. A student on a nearly frictionless turntable holds a spinning wheel as shown. Initially the turntable is not rotating. **(1)** The student moves the axle of the wheel further from the axis of the turntable and then brings it back. What happens? **(2)** He turns the wheel over. What happens? **(3)** He turns it back to its original position. What happens? **(4)** He touches the wheel to his jacket and stops it with friction. What happens?

10.19. An earth satellite has a perigee which is 110 miles above the earth's surface and an apogee which is 220 miles above the surface. What is the ratio of the velocity at perigee to that at apogee ?

10.20. A falling cat with zero angular momentum about its own center of mass can turn and land on its feet. It changes its angular position without changing its angular momentum. How can it do so? (HINT: Think of the front and rear halves of the cat as two connected parts of a single system. Only the sum of the angular momenta of the two parts need remain zero.)

10.21. An artificial satellite mounted atop a rocket on the launching pad at Cape Kennedy has a small angular momentum (owing to the fact that it is attached to the rotating earth). After it is injected into orbit it has a considerably larger angular momentum. Carefully reconcile the two statements, "Angular momentum is conserved," and "The angular momentum of the satellite is changed."

10.22. For exercise, a crew of astronauts take along a bicycle. One of the astronauts gets on the bicycle. Initially he and the bicycle float in midcabin, motionless with respect to the spaceship. Then he starts to pedal. Describe the resulting motion. Is angular momentum conserved? What happens when he brakes and the wheels stop turning?

10.23. A boy holds a spinning wheel by handles attached to its axles as shown. He pushes forward with his left hand and pulls back with his right hand. The handles respond by exerting forces on his hands, but not horizontal forces. One handle pushes up, the other down. **(1)** Why is this? **(2)** Does this violate Newton's third law? *Optional:* If the wheel is spinning as shown, which handle seemingly "wants" to go up, which "wants" to go down?

10.24. A pulley turns freely on a fixed horizontal axis. A rope goes over the pulley and hangs down on either side. On one end of the rope is fastened a counterweight of mass m. Clinging to the other side of the rope is a monkey of equal mass m (see figure). The masses of pulley and rope are negligible. **(1)** Let the system of interest consist of pulley, rope, counterweight, and monkey. Name all the external forces acting on this system. Where do they act? **(2)** With respect to the axis of the pulley, what is the total external torque? **(3)** Use the principles developed in this chapter to answer this question: If the monkey starts climbing upward, making good a speed v relative to the pulley, how fast does the rope move, and in which direction?

10.25. In order that a system at rest remain at rest it must experience no total force and no total torque. These are the conditions of static equilibrium. Consider a door of height h, width w, and mass m supported in equilibrium by two hinges as shown. The gravitational force ($F_g = mg$) can be considered to act vertically downward at the center of mass. The force $\mathbf{F}_1$ is known to act horizontally. What is the magnitude of $\mathbf{F}_1$? What is the direction of $\mathbf{F}_2$? (HINT: Since the choice of reference point for the calculation of torque is arbitrary, it is convenient to choose a reference point with respect to which only two of the three forces contribute to the torque.)

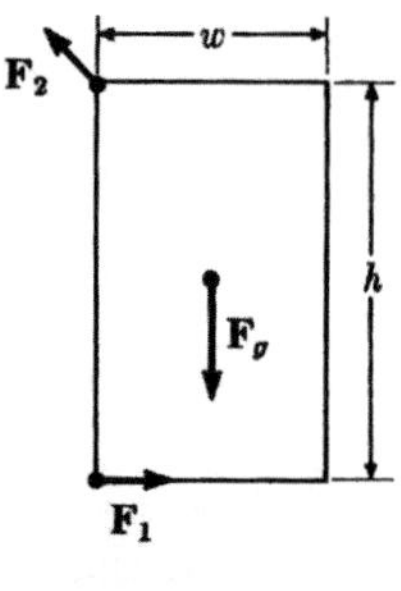

10.26. Consider an idealized solar system containing a fixed center of attractive force and a single spherical planet spinning on its axis and rotating about the force center. **(1)** Explain why the spin angular momentum and orbital angular momentum of the planet are separately constant. **(2)** Imagine now that the planet is deformed into a nonspherical shape (very far from spherical if you like). Explain why its spin angular momentum need not be constant.

ELEVEN

Energy

The number of concepts that can be defined in physics is limitless. The number that are 46
fundamental for the description of nature is remarkably small. Part of the scientist's search for simplicity is his search for economy of concepts. To merit attention, a physical concept must be not merely quantitatively definable—that much is easy—but it must also bring something of special value to the description of nature. Either it must appear in a natural way in the description of many different phenomena, or it must facilitate the application of theory to specific problems, or it must tie together different branches of science, or it must under some circumstances be conserved. In some way it must force itself on the attention of the scientist as something he cannot ignore. By every criterion just named, the concept of energy deserves the attention of the scientist and the student.

11.1 The forms of energy

Ubiquitous, manifold, and constant. These words describe the characteristics of energy that 47
make it seem perhaps the most important single concept in contemporary science, apart from space and time. Energy is a central idea in physics, chemistry, biology—indeed in every area of natural science, as well as in engineering and practical affairs. Chameleon-like, it changes among a variety of forms as it appears in different scientific and technical environments. One need only think of some of the commonly used units of measurement for energy to be reminded of its many roles in many places: the kilowatt hour, the erg, the BTU (British thermal unit), the calorie, the foot pound, the electron volt, the megaton. Finally, and most important, energy is conserved. Only because of its conservation are the widespread and manifold forms of energy of special significance. The transformation of electrical energy to mechanical energy by a motor, or of mass energy to kinetic energy by a particle decay, takes on importance only because of conservation. The loss of one kind of energy is exactly balanced by the gain of another kind of energy. Without conservation, energy would be not one single concept appearing in many phenomena in many guises; it would be many different concepts. From the nuclear energy of the sun's interior to the work performed by a man's muscles runs a long and elaborate path of energy transformations through the sciences of astronomy, physics, chemistry, geology, and biology. Because of its conservation at every step along the way, energy is an unbroken thread tying these disciplines together. From its humble beginning as a secondary concept in mechanics, energy has grown into the most important unifying idea of natural science.

Momentum and angular momentum may be called purely mechanical quantities, or better, properties of motion. Energy, on the other hand, is not exclusively a property of motion. In one of its manifestations, kinetic energy, it is analogous to momentum and angular momentum. However, there is no law of conservation of kinetic energy, only a law of conservation of total energy. Kinetic energy, more elusive than momentum or angular momentum, can vanish to reappear in a different form.

Figure 11.1 shows schematically some of the interconnections that exist among different forms of energy. Mass energy and kinetic energy in particle transformations were considered in Chapter Four. In this chapter the primary emphasis is on the mechanical aspects of energy encircled by a dashed line in Figure 11.1: work, kinetic energy, and potential energy. Energy in these and other forms will recur often in later chapters.

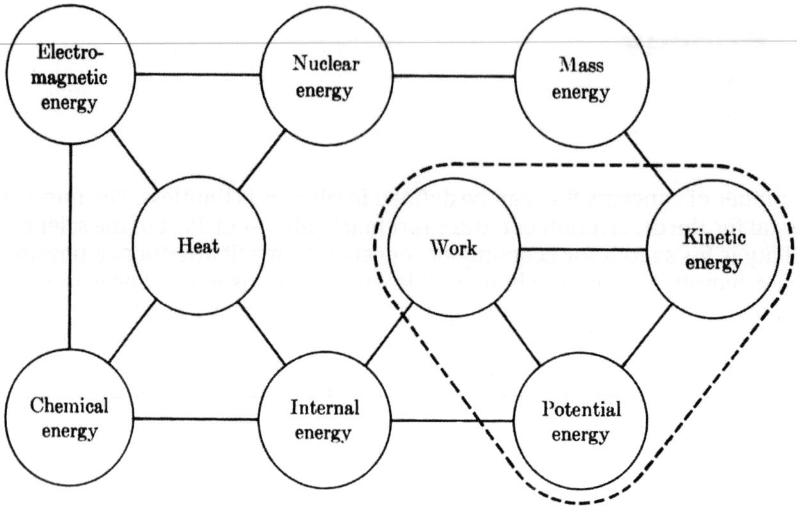

FIGURE 11.1 Some forms of energy, and some of their interconnections. Not all of these forms are clearly distinct: Internal energy is in part molecular kinetic energy; stored chemical energy can be considered to be atomic potential energy; and changes of mass accompany almost all other energy changes. The dashed line encircles the forms of energy relevant to macroscopic mechanics. Heat and work, in the center of the diagram, are modes of energy transfer.

Despite our emphasis on the manifold forms of energy, it should here be stated that the difference in some forms of energy is more difference in appearance than difference in essence. Internal energy is accounted for in terms of microscopic kinetic energy and potential energy. Chemical energy is electric in origin. Potential energy is reflected in changes of mass energy. Heat and work are not strictly forms of energy at all, but measures of energy transfer. At the most fundamental level of elementary particles, only two forms of energy—kinetic energy and mass energy—are required to describe all energy transformations. Nevertheless, there is good reason for regarding the various forms of energy in Figure 11.1 as distinct. In some domains of nature or for some ways of looking at nature, these energy forms manifest themselves in such completely different ways that it only makes sense to treat them as different kinds of energy. When a satellite reenters the atmosphere, its kinetic energy is dissipated and converted via heat energy into internal energy. It is satisfying to know that part of this internal energy is attributable to kinetic energy, the same form of energy characterizing the high-speed motion of the satellite before its reentry. However, the bulk motion of the satellite as a whole bears so little resemblance to the random molecular motion associated with heat and internal energy that the appreciation of their basic similarity is purely an esthetic pleasure, without practical utility. If an astronaut, after riding comfortably in orbit, were burned to a crisp upon reentering the atmosphere, he might well be piqued that some designers had given inadequate attention to the difference between internal energy and kinetic energy.

11.2 Work

48 A good way to approach energy is via the concept of work. Work ties together the already familiar concept of force and the new concept of energy, and it leads directly to the most basic property of energy, its conservation.

Consider the following idealized experiment of great simplicity, which suffices to provide a basis for the definitions of work and energy. On a horizontal and ideally smooth surface, an object is pushed with a constant force $\mathbf{F}_A$ for a time, then allowed to slide freely for a time, then decelerated and brought to rest with a constant force $\mathbf{F}_C$ whose magnitude differs from the magnitude of the accelerating force $\mathbf{F}_A$. For purposes of discussion, we shall divide the motion into three parts, A, B, and C, as illustrated in Figure 11.2. The object begins at rest, moves in a straight line, and ends at rest. Concerning this simple motion we can ask two questions in our search for a new conservation principle: (1) Is there any way to equate the effort required to start the object and the effort required to stop it? (2) Is there any quantity that remains constant throughout the motion? Both questions have an affirmative answer. We consider them in turn.

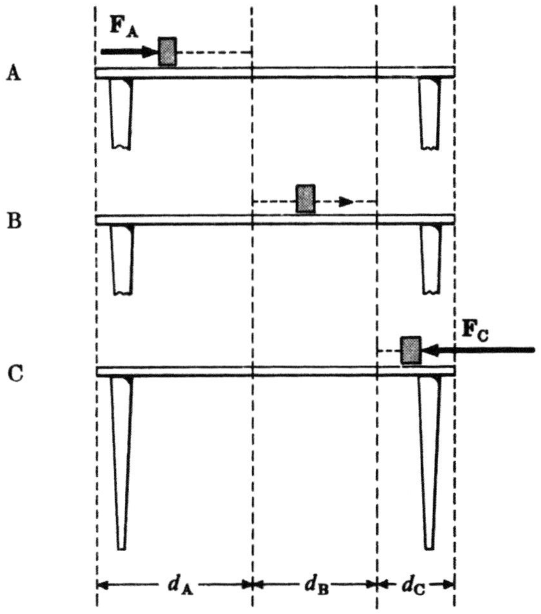

FIGURE 11.2 Idealized example of motion used to introduce the concept of work. An object is accelerated with a constant force $\mathbf{F}_A$ through distance d_A, then allowed to coast with constant velocity a distance d_B, then brought to rest in a distance d_C by a constant force $\mathbf{F}_C$.

Comparing the accelerating phase of the motion, part A, with the decelerating phase, part C, we notice that because the magnitudes of the forces $\mathbf{F}_A$ and $\mathbf{F}_C$ are unequal, the distances traveled and the times elapsed are also unequal. If $\mathbf{F}_C$ exceeds $\mathbf{F}_A$, the object will decelerate in less time and less distance than were required for its acceleration. However, there is one constant linking parts A and C. It is the product of force and distance in each part of the motion:

$$F_A d_A = F_C d_C . \tag{11.1}$$

This important equality can be proved using the properties of uniformly accelerated motion presented in Section 8.2. Most directly it follows from Equation 8.14 relating speed and distance, which, for this example, can be written

$$v_x{}^2 = v_{x0}{}^2 + 2a_x x . \tag{11.2}$$

If an object has x-component of velocity v_{x0} at the initial point $x = 0$ and is accelerated in the x-direction with acceleration a_x, this equation gives its x-component of velocity v_x at a later point x. For our purpose we apply the equation twice, first to connect the beginning and end of part A of the motion, then to connect the beginning and end of part C of the motion. The quantities needed for evaluating the equation, and the form the equation assumes in each part, can best be presented in tabular form, as in Table 11.1. Newton's second law, $\mathbf{F} = m\mathbf{a}$ (or $F_x = ma_x$), in which m is the mass of the object, has been used to relate acceleration to force. The uniform speed v_B of the central part of the motion serves as the final speed in part A and the initial speed in part C. Zero speed also appears twice, as the initial speed in part A and the final speed in part C. Since the two equations show that v_B^2 is equal both to $2F_Ad_A/m$ and to $2F_Cd_C/m$, we have the equality,

$$\frac{2F_Ad_A}{m} = \frac{2F_Cd_C}{m}. \tag{11.3}$$

The common factor $2/m$ on each side may be canceled to yield the equality of Equation 11.1, whose proof was sought. Since the product of force and distance is the same in part C of the motion as in part A, it provides a suitable measure of the starting effort and stopping effort. Since it is independent of the particular force applied and the same for two parts of the motion, it deserves a name. The product of force and distance is called *work*, designated by W:

$$W = Fd. \tag{11.4}$$

Caution: This is a preliminary definition, valid only for a constant force applied in the direction of motion.

TABLE 11.1 Kinematics of Parts A and C of the Motion Illustrated in Figure 11.2

Part A	Part C
$v_{x0} = 0$	$v_{x0} = v_B$
$a_x = F_A/m$	$a_x = -F_C/m$
$x = d_A$	$x = d_C$
$v_x = v_B$	$v_x = 0$
$v_B^2 = 2\frac{F_A}{m}d_A$	$0 = v_B^2 - 2\frac{F_C}{m}d_C$

Up to this point we have been using the word "effort" in a loose way as indicative of the kind of concept toward which we were groping. "Work," by contrast, has a well-defined technical meaning. Before returning in the next section to consider part B of the motion and to define kinetic energy, let us examine some aspects of this new concept, work. First, it would be well to sharpen its definition. If a force acts in some direction other than the direction of motion, only the component of force parallel to the motion contributes to work. The revised definition is

$$W = F_{\parallel}d. \tag{11.5}$$

As indicated in Figure 11.3, the parallel component of force is equal to the total magnitude of the force multiplied by the cosine of the angle between the force direction and the direction of motion. The same definition can therefore be rewritten,

$$W = Fd\cos\theta. \tag{11.6}$$

This revision is significant even in the example of one-dimensional motion just considered, for if the force acts opposite to the motion instead of along the motion, the cosine is −1 instead of +1, and the work is negative instead of positive. We shall return to this point. Finally, if the force is not constant, work can be defined only over an interval so short that the change of force is negligible. The final definition of work that covers all situations is

$$\Delta W = F_{\parallel}\Delta s\,. \tag{11.7}$$

The increment of work, ΔW, is the product of the component of force parallel to the motion and the increment of distance moved, Δs. Despite the fact that work is defined in terms of the magnitudes of vector quantities, it is itself a scalar quantity represented by a single number.

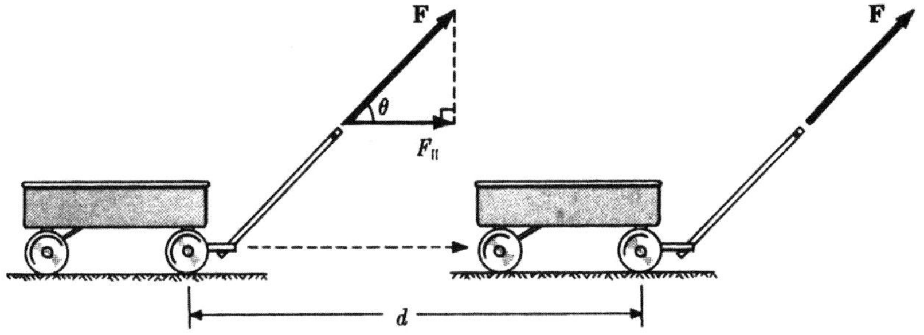

FIGURE 11.3 Example of work performed when force is not in the direction of motion. For a constant force **F** acting on a wagon which moves a distance *d*, the work is $F_{\parallel}d = Fd\cos\theta$.

The physical dimension of work is evidently the product of the dimensions of force and distance. Its cgs unit is the dyne cm, or equivalently the gm cm^2/sec^2. This combination of units, because it is the unit of the all-important energy concept, has been given a name of its own, the erg:

$$1 \text{ erg} = 1 \text{ dyne cm} = 1 \text{ gm cm}^2/\text{sec}^2\,. \tag{11.8}$$

The mks unit of work and energy, the joule, is ten million times larger:

$$1 \text{ joule} = 1 \text{ kg m}^2/\text{sec}^2 = 10^7 \text{ ergs}\,. \tag{11.9}$$

Other energy-conversion factors are found in Appendix 3. A closely related concept worth mentioning here is power. Power is the rate of work, work per unit time (or more generally, energy per unit time),

$$P = \frac{\Delta W}{\Delta t}\,. \tag{11.10}$$

The mks power unit also has a special name, the watt.

$$1 \text{ watt} = 1 \text{ joule/sec} = 10^7 \text{ ergs/sec}\,. \tag{11.11}$$

The commonly encountered kilowatt is 1,000 watts, or 10^{10} ergs/sec.

An important fact to note about work is that it requires motion. Work is defined in terms of a change of position. Regardless of what forces may be acting on a stationary system, there is no associated work. The reason this point requires special emphasis is that the word "work" is so familiar that its many nontechnical meanings cannot lightly be abandoned in favor of a narrow new technical meaning. Actually, of course, its other meanings need not be abandoned; they need only be segregated. The physicist, in borrowing the word "work" to name a special quantitative concept, would be the last to suggest that its use in other contexts should be inhibited. He would still say that his newly repaired radio is working, that thinking is hard work, and that exerting a force through a distance on the playing field is not work, but play. It is important to avoid confusing the technical and nontechnical meanings. A boy supporting a heavy load without moving might legitimately claim to be working hard. However, if he wishes to bring the theory of mechanics to bear on his situation, he must embrace the technical definition and confess that he is doing no work.

If the work associated with a particular motion is positive, work is said to be done *by* the source of the force. If the work is negative, work is said to be done *on* the source of the force. In the example of motion being considered, work is being done *by* the source of the external force in part A and work is being done *on* the source of the external force in part C. Since the direction of the force *exerted* by the moving object is opposite to the direction of the force *experienced* by the object (Newton's third law), the sign of the associated work is also opposite. Work is being done *on* the object as it accelerates and *by* the object as it decelerates. This means that the equality of work in part A and part C can be regarded as the manifestation of a conservation principle. Some work is expended by the source in part A. An exactly equal amount of work is regained by the source of the external force in part C. However, no work is associated with part B. Work itself is not a constant feature of the entire motion. Yet the equality of input work and output work suggests that some aspect of the motion is a constant. In some way the potentiality for doing work is preserved through the intermediate region of part B. During its free motion in part B, the object possesses the ability to do work even though it is doing none.

11.3 Kinetic energy

The view of motion arrived at in the last paragraph comes very close to suggesting a new substance. We say that something is put into the system, is preserved, and is then extracted from the system. This way of looking at the motion is, of course, not necessary, but it proves to be fruitful. The "something" that is added and then taken away is energy, in this case kinetic energy.

Kinetic energy, or energy of motion, for which we use the symbol *K.E.*, was defined by Equation 3.1, which we repeat here:*

$$K.E. = \tfrac{1}{2}mv^2 \,. \tag{11.12}$$

A moving object possesses kinetic energy equal to half its mass multiplied by the square of its velocity. A motionless object possesses zero kinetic energy. Negative kinetic energy is impossible.

The lower left equation in Table 11.1 can be rewritten,

$$F_A d_A = \tfrac{1}{2}mv_B^2 \,. \tag{11.13}$$

In words: The work done *on* the object in part A of its motion is equal to its kinetic energy in part B. In similar manner, the lower right equation in Table 11.1 can be rewritten,

$$\tfrac{1}{2}mv_B^2 = F_C d_C \,. \tag{11.14}$$

This states that the kinetic energy in part B is equal to the work done *by* the object in part C. Both of these equations can be summarized by the single mathematical statement,

$$\Delta K.E. = W \,. \tag{11.15}$$

The change of kinetic energy of an object is equal to the work done on the object (positive or negative). In part A of the motion illustrated in Figure 11.2, positive work is done on the object, and it gains kinetic energy. In part B, no work is done; its kinetic energy does not change. In part C, negative work is done on the object, and it loses kinetic energy.

* Remember that this is a nonrelativistic definition of kinetic energy, valid only at speeds much less than the speed of light. It is not applicable, for instance, to massless particles, which also possess kinetic energy.

The kinetic energy of a system is the sum of the kinetic energies of its constituent parts. 49
Since kinetic energy is a scalar quantity, not a vector quantity, and since it can never be negative, any two kinetic energies always combine to give a larger kinetic energy. A system can have zero kinetic energy only if *all* of its constituent parts are at rest. By contrast, a system can have zero momentum even though its parts are in motion. The kinetic energy of a particle and its momentum, despite certain superficial similarities, are very different concepts which should not be confused. Both are properties of motion depending on the mass and speed of the particle, zero if the speed is zero, and increasing in magnitude as the speed increases. There the similarities stop. One is a vector quantity, the other a scalar quantity. They have different physical dimensions. They enter into different conservation laws. Their laws of change are different. Their dependence on particle speed is different.

Everyone knows that driving an automobile at 60 miles/hr is more than twice as dangerous as driving it at 30 miles/hr. Among other reasons, this can be attributed to the fact that as speed doubles, kinetic energy increases fourfold. How abruptly a car can brake to a stop depends on how much force the tires can exert on the road. Roughly, the maximum stopping force is independent of the speed of the car. If we call this force F_S and the stopping distance d_S, the work done by the car on the road as it stops is their product,

$$W = F_S d_S \,.$$

This work in turn must be equal to the kinetic energy of the car before it begins to brake,

$$F_S d_S = \tfrac{1}{2}mv^2. \tag{11.16}$$

Since the mass of the car is a constant and the force F_S is approximately a constant, the stopping distance is approximately proportional to the square of the speed. The fact that the distance to stop increases as the square of the speed is borne out by measurements and probably is vaguely sensed by most drivers, although not always acted upon. Kinetic energy also plays a role in collision damage. If a car is driven into a wall that stops it in a fixed distance d_S, for example six inches, the stopping force, according to Equation 11.16, is proportional to its kinetic energy. Damage to occupants is determined principally by the magnitude of the deceleration, which is proportional to the force F_S, in turn proportional to the square of the impact speed.

The principle applies no less well to astronauts than to automobile drivers. In a low-altitude orbit, an astronaut is moving at more than 16,000 miles/hr, or about 275 times faster than a driver at 60 miles/hr. The astronaut's kinetic energy is about 75,000 times greater than the driver's, and the work that he must do in coming to rest is therefore 75,000 times greater. Since the driver, if properly strapped in, can withstand about the same force as the astronaut, their stopping distances must be in the same ratio as their energies (see Equation 11.16). If the driver can walk away unscathed from a collision in which his car came to rest in 10 feet, the astronaut, to be equally undamaged, must be stopped in 750,000 feet or about 140 miles. Fortunately, this can be arranged. Even so, the enormous kinetic energy of the returning astronaut means that his transformation of kinetic energy to work must be thought of more as a violent collision than as a pillowed descent.

11.4 Relation of work and energy

One way to define energy in general is that which is capable of doing work. Not all forms of 50
energy are directly or easily converted to work, but all can, through some suitable chain of transformation, manifest themselves as work. If work is done on an object (or a system), it acquires an amount of energy equal to the work done (possibly, but not necessarily, in the form of kinetic energy). This relation of energy change and work can be expressed symbolically by an equation that generalizes Equation 11.15:

$$\Delta E = \Delta W \,. \tag{11.17}$$

If an object does work on something else, ΔW is a negative quantity, and the object loses an amount of energy equal to the work done. The fact that we speak of energy being added to or taken away from an object does not mean that energy is regarded in any sense as an actual material substance. It can at best be called an attribute or a property of the object. Nevertheless, in order to form a mental picture of the difference between energy and work, it is helpful to think of energy as substance, work as action. A system *possesses* energy; it can *do* work. At any instant a system has a certain energy *content*. Part or all of this energy can be transformed into the *activity* of work. In one sense work is just another form of energy because of its transformability without loss to and from energy. In another sense, work is only the active measure of energy, not itself a form of energy. In still another sense (the deepest sense) there is no such thing as work at all. It is only the name for a mode of transfer of energy from one form to another. Work bears the same relation to energy that money bears to material wealth. It can be used to define and to measure energy without itself being a form of energy.

The role of work as a medium of exchange can be elucidated by considering again the sliding object that has occupied our attention in this chapter so far. Initially work is done on it and its kinetic energy accordingly increases. This work must have been supplied by something else. Whatever the source of the work, we may assume that somewhere else a loss of energy occurred to make this work available. This unidentified other loss of energy must equal in magnitude the work done, which in turn equals the kinetic energy gained by our object. Jumping over the intermediate idea of work, we can say simply that our object gains kinetic energy because energy is transferred from outside the system to inside the system (the "system" here is just the sliding object). Looked at in this way, work is merely a transmitter of energy, a medium of exchange. Similarly, in part C of the motion, the kinetic energy loss must be compensated by an equal energy gain in the interacting environment, and work is a measure of the energy transferred. This point of view suggests a powerful as well as practical formulation of the law of energy conservation, not limited to isolated systems. We postulate that any energy change in a system is exactly equal to an opposite energy change outside the system. More simply stated, the loss of energy anywhere is always compensated by an equal gain of energy somewhere else. This law, richly supported by experiment, has required the extension of the energy concept outside of mechanics into every subsequently developed theory and into every branch of science. It is the most important unifying principle in science.

11.5 Potential energy

51 If a pencil is picked up from the floor and set on a table, work is performed (equal to the vertical component of force multiplied by the height of the table top). To stretch a spring requires work. Work is also needed to separate a nail from a magnet. In these examples, the result of the input work is not kinetic energy, but what is called potential energy. As kinetic energy is energy of motion, potential energy is energy of position—more accurately, energy of relative position. The pencil gains potential energy because its distance from the center of the earth increases. The spring gains potential energy because of the slight rearrangement of the relative positions of its atoms. The nail gains potential energy because it moves away from the magnet in the field of magnetic force. There is no question about the reality of potential energy, for in any of the examples just cited, return to the original position releases the potential energy that was added. It can be recaptured as work or as kinetic energy or as some other form of energy.

Potential energy is particularly important (that is, useful) in describing motion influenced by the earth's gravity. The progress of a vertically fired rocket (Figure 11.4) can serve to introduce gravitational potential energy. Suppose that the rocket, because of a malfunctioning engine, is barely able to support its own weight. It staggers 100 meters into the air and there comes to a stop for lack of sufficient thrust. From the moment of launch up to this point work has been done on the rocket in an amount that can easily be calculated. The force produced by the exhaust is nearly equal to the weight of the rocket, *mg*. The distance covered may be called *h* (for height). The product of these two is the work done by the force of the engine:

$$W = mgh\,. \tag{11.18}$$

After this input of work, the rocket is at rest, just as it was initially. Obviously there has been no gain of kinetic energy as there would have been if the same work had been done on the rocket far removed from earth. Yet the work should result in some kind of energy gain by the rocket. Since the only change in the rocket is a change of position, the energy must be an energy of position associated with the altitude gain *h*. This is potential energy.

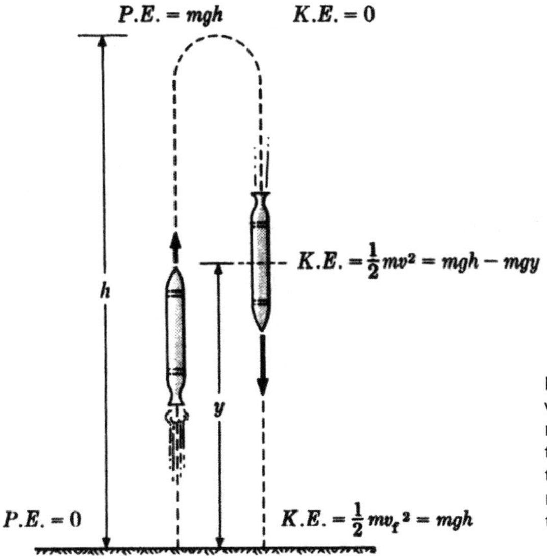

FIGURE 11.4 Gravitational potential energy. Lifted very slowly a distance *h* with a force equal to ***mg***, the rocket gains potential energy equal to *mgh*, which in turn is equal to the work done in lifting it. In free fall through the same distance with the engine off, the rocket gains kinetic energy equal to its loss of potential energy in the fall.

Acting on the general principle that energy change is equal to the work done (Equation 11.17), we can write an equation for the potential energy associated with vertical ascent,

$$P.E. = mgh\,. \tag{11.19}$$

It is worth noticing immediately that this is an energy *change* associated with a *difference* of altitude between the initial and final points. We shall return below to the important fact that potential energy is always defined relatively for two points, never absolutely for one point. If the quantity just defined is to be acceptable as truly a form of energy, it should fit into a law of energy conservation, it should be convertible to other forms of energy, and it should be transformable to work. It is easy to discover that potential energy does satisfy these criteria.

If the rocket engine is extinguished at the moment the rocket comes to a stop at height *h*, the rocket falls back to earth, pulled downward by the force of gravity, of magnitude *mg*. In the descent through the distance *h*, the gravitational force does work *mgh*, and imparts to the rocket a kinetic energy equal to the work done:

$$\tfrac{1}{2}mv_t^2 = mgh\,. \tag{11.20}$$

The symbol v_t denotes the final speed of the falling rocket just before impact with the earth. The rocket's fall may be interpreted as the direct conversion of potential energy to kinetic energy. In dropping through the distance h, the rocket loses potential energy and gains an equal amount of kinetic energy. Moreover, as it falls, its kinetic energy at any instant is equal to the loss of potential energy until that time. Throughout the fall, the sum of kinetic energy and potential energy is constant. This relationship may be written

$$\tfrac{1}{2}mv^2 + mgy = mgh\,. \tag{11.21}$$

The first term on the left is the kinetic energy at any point; the symbol v denotes the variable speed during the fall. The second term is the potential energy at altitude y relative to ground level. On the right is a constant, equal to the initial potential energy. At the start of the fall, this equation takes the form,

$$0 + mgh = mgh\,. \tag{11.22}$$

The rocket has potential energy and no kinetic energy. At the end of the fall (just before impact), the equation takes the form,

$$\tfrac{1}{2}mv_t^2 + 0 = mgh\,, \tag{11.23}$$

the same as Equation 11.20. The rocket then has kinetic energy and no potential energy (relative to ground level). At all points in between, the sum of kinetic energy and potential energy remains constant.

Equation 11.21 is a special case of a more general energy-conservation equation valid for many kinds of motion, which may be written

$$K.E. + P.E. = E \text{ (constant)}. \tag{11.24}$$

The algebraic form $\tfrac{1}{2}mv^2$ gives correctly the kinetic energy of any particle or nonrotating rigid object. No expression of similar generality can be written for potential energy, whose mathematical form depends on the nature of the forces that are present. The form mgy is quite special, appropriate only for motion in the uniform field of gravitational force close to the earth. Even then, the form is not unique, because of the relative rather than absolute nature of potential energy. If, for reasons of our own, we should decide to choose a level a distance d below ground as the reference level for measuring potential energy, the rocket would have a potential energy mgd at ground level and a potential energy $mg(d + y)$ at a distance y above ground. In that case the energy conservation equation would take the form,

$$\tfrac{1}{2}mv^2 + mg(d + y) = mg(d + h)\,. \tag{11.25}$$

The revised energy constant, $mg(d + h)$, compensates for the revised expression for potential energy. Thus total energy E, like potential energy, is without absolute significance. The physical implications of Equations 11.25 and 11.21 are identical.

The solution of physical problems is often facilitated by taking advantage of energy conservation. As a simple example, consider the question: What is the maximum height of a ball thrown vertically upward from ground level with initial speed v_0 = 1,500 cm/sec? If ground level is chosen as the reference plane, the initial potential energy of the ball is zero. Its initial kinetic energy is $\tfrac{1}{2}mv_0^2$. Therefore its initial total energy, the same as its total energy throughout its flight, is $\tfrac{1}{2}mv_0^2$, and the energy conservation equation may be written,

$$\tfrac{1}{2}mv^2 + mgy = \tfrac{1}{2}mv_0^2\,. \tag{11.26}$$

This equation relates the speed v and the height y throughout the ascent and descent of the ball. In order to use the equation to answer the question that was posed, we must use the fact that the maximum height is achieved when the speed v is equal to zero. Is this fact obvious? You may think that it is. But even obvious facts are sometimes worth proving. We mention two proofs, one more physical, one more mathematical. The physical proof goes like this. At first the ball is rising with positive upward component of velocity. It cannot start to fall until the upward component becomes negative. Since the velocity changes smoothly and not in a sudden jump, its upward component can change from positive to negative only by passing through zero. When it is zero, the ball stops rising and starts falling. Therefore, zero speed occurs at maximum height. Alternatively, one may reason mathematically without the aid of a physical picture. Since the sum of kinetic energy and potential energy is a constant, potential energy is maximum when kinetic energy is minimum. Minimum kinetic energy is zero (it can never be negative). Therefore maximum potential energy and maximum height occur when kinetic energy and speed are zero. These arguments validate a method of answering the question about maximum height. Simply substitute $v = 0$ into the energy equation, to get the result,

$$mgy_{max} = \tfrac{1}{2}mv_0^2 . \tag{11.27}$$

The mass m cancels on the two sides, leaving the result for the maximum height,

$$y_{max} = \frac{v_0^2}{2g} . \tag{11.28}$$

Numerically, for $v_0 = 1{,}500$ cm/sec and $g = 980$ cm/sec^2, this gives

$$y_{max} = \frac{2.25 \times 10^6 \text{ cm}^2/\text{sec}^2}{2 \times 0.98 \times 10^3 \text{ cm/sec}^2} = 1{,}150 \text{ cm} . \tag{11.29}$$

Is Equation 11.28 "reasonable"? It is worth looking hard at it as well as substituting numbers into it. The initial speed v_0 occurs "upstairs." More speed means more height. Doubling the initial speed quadruples the initial energy and quadruples the height. That is reasonable. The gravitational acceleration g occurs "downstairs." If it were larger (an imaginary situation), the retarding force would be greater and the height would be less. That is also reasonable. As the numerical example demonstrates, the dimensions are consistent. That is essential. Finally, the absence of mass in the equation is reasonable. More mass means more initial kinetic energy, but it also means more retarding force. The effects cancel and the height does not depend on the mass. Not only is the equation entirely reasonable (it ought to be; it is correct!), it is useful for many situations other than thrown baseballs. What, for instance, would be the maximum vertical height of a shell launched vertically upward with a muzzle velocity of 10^5 cm/sec (about Mach 3), if air friction could be neglected? Substitution into Equation 11.28 gives at once $y_{max} = 5.1 \times 10^6$ cm = 51 km, or about 32 miles.

We have discussed the transformation of work to potential energy (the slowly ascending rocket), the conversion of potential energy to kinetic energy (the falling rocket), and the conservation of kinetic energy plus potential energy (the falling rocket and the thrown baseball). To complete the catalog of properties of potential energy that make it acceptable as a bona fide form of energy, we need only demonstrate its transformation directly back to work. A somewhat fanciful fate of our rocket can serve to illustrate this transformation. Suppose that the rocket, in ascending through the height h, drifts sideways toward a tower of height h. At the moment that the rocket engine stops firing, its tailpipe is resting lightly on top of the tower (Figure 11.5). Then, as the rocket's full weight settles upon the tower, the tower slowly crumples, lowering the rocket gently and majestically back to earth. In this instance, the rocket gains little or no kinetic energy on its way back to earth. Instead its potential energy is transformed to work performed on the tower. Throughout its descent through the distance h, it exerts on the tower a force equal to its own weight mg, so that the work done on the tower is *mgh*, equal to the rocket's loss of potential energy in the descent. It will not be hard for the reader to think of some less far-fetched examples of the transformation of potential energy to work.

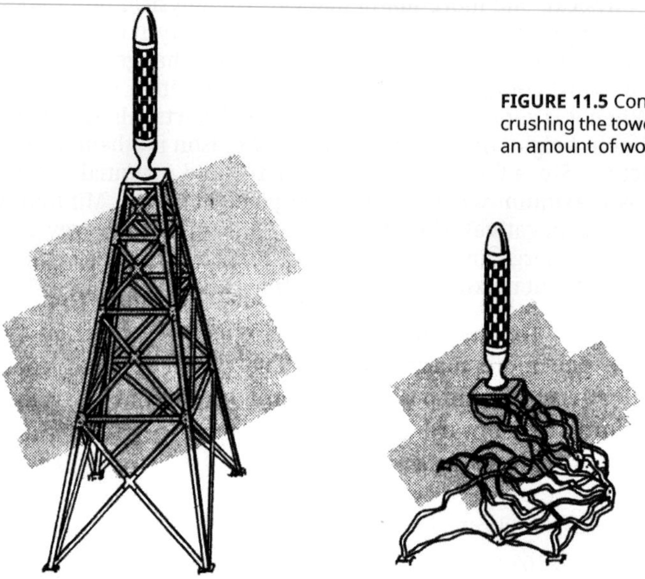

FIGURE 11.5 Conversion of potential energy to work. In crushing the tower as it slowly descends, the rocket does an amount of work equal to *mgh*, its loss of energy.

Work, Force, and Potential Energy

52 In the discussion to this point, one aspect of potential energy has been ignored. Now is the time to acknowledge it. The observant reader may already have noted that potential energy is defined only in the presence of some force, for instance the gravitational force. In fact, potential energy is nothing more nor less than a new name for the work associated with certain forces.

When a stone falls toward the ground, its gain of kinetic energy may be attributed to its loss of potential energy. Alternatively, its gain of kinetic energy may be attributed to the work done on it by the gravitational force. Either description is acceptable, but only *one* must be used. If the gravitational force and its associated work are taken into account explicitly, there is no need for potential energy. If potential energy is introduced, the work done by the gravitational force need not—in fact, must not—be introduced as well.

Among other forces whose effect can be represented by potential energy are the restoring force of a spring, the pull of a magnet on a nail, and the electric force between a proton and an electron. These are forces that produce *recapturable* energy changes. The energy expended in lifting a weight or pulling a spring or separating a nail from a magnet can be regained by lowering the weight or relaxing the spring or allowing the nail to be drawn back to the magnet. By contrast, work performed against a frictional force is not recapturable, at least not by simple mechanical means, for it is transformed into heat and internal energy. A frictional force is called a "dissipative" force. For it, no potential energy can be defined. For nondissipative forces such as the force of gravity, potential energy does have meaning.

When it is possible to use potential energy in describing motion, it is usually useful to do so. Potential energy enters into a conservation law and it often makes easier the solution of practical problems. That is its justification. It needs no other.

11.6 Conservation of work

Certain mechanical devices are transformers of work without the intermediary of energy storage. Work is done on the device and simultaneously work is done by the device. If the heat energy created by frictional forces is sufficiently small, the output work will, to good approximation, be equal to the input work. The study of levers and pulleys, which are devices of this kind, provided the earliest hints of the idea of energy conservation.

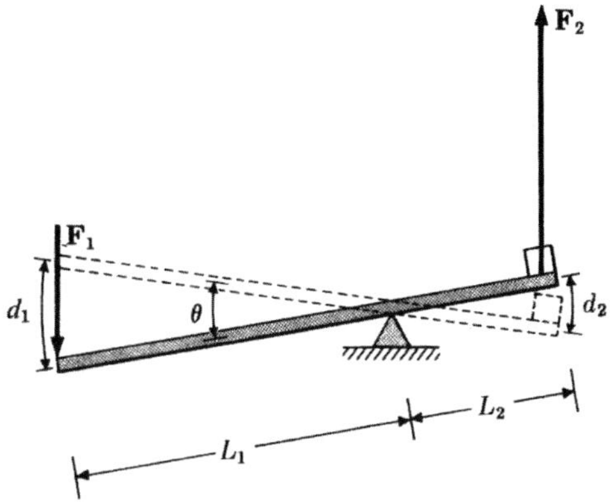

FIGURE 11.6 The simple lever. The input work is equal to the output work.

A simple lever consists of a rigid bar pivoted at some point, not necessarily its center (Figure 11.6). Let us call the distances from the pivot to the two ends of the lever L_1 and L_2. If a force F_1 is applied to move the left end of the lever through a distance d_1 an amount of work F_1d_1 is done on the lever. If the lever is so arranged that its other end applies a force F_2 to some other object as it moves through a distance d_2, the output work is equal to F_2d_2. (We assume that the forces act parallel to the direction of motion.) In the absence of frictional energy dissipation, input work and output work are equal:

$$F_1d_1 = F_2d_2\,. \tag{11.30}$$

The input and output *forces*, on the other hand, are unequal. Instead their ratio stands in the inverse ratio of the distances moved:

$$\frac{F_1}{F_2} = \frac{d_2}{d_1}\,. \tag{11.31}$$

The distances moved, in turn, are proportional to the lengths of the arm of the lever. If the lever turns through angle θ (radians), its left end moves a distance given by

$$d_1 = \theta L_1\,,$$

and its right end moves a distance given by

$$d_2 = \theta L_2$$

(Figure 11.6). Therefore the ratio d_2/d_1 is the same as the ratio L_2/L_1. The normal purpose of a lever is to gain an output force greater than the input force. The ratio of output force to input force is called the *mechanical advantage (M.A.)* of the lever. In symbols

$$M.A. = \frac{F_2}{F_1} = \frac{L_1}{L_2}\,. \tag{11.32}$$

Evidently, because of the conservation of work, the greater force must act through the smaller distance. The lever pictured in Figure 11.6, with the arm L_1 twice the arm L_2, has a mechanical advantage of two. Archimedes, who first discovered the law of the lever represented by Equation 11.32 around 250 B.C., is said to have been so impressed by the unlimited possible magnitude of the mechanical advantage that he declared, "Give me a fulcrum on which to rest, and I will move the earth." It is entirely believable that Archimedes did say this, for the statement reveals two common attributes of the creative scientist: his enthusiasm for his subject, and his ability to extrapolate from the familiar realm in which a law is discovered to a new domain where the law could have new power or receive new tests.

Archimedes' law of the lever can be founded either on the idea of balancing torque or on the idea of equal input and output work. For the lever, these two approaches are equally simple. For understanding most other devices, however, the idea of the conservation of work is decidedly the simpler, more useful, and more general approach. Consider, for example, its application to the operation of a chain hoist. In the chain hoist, a pair of pulleys of slightly different circumference are attached to the same fixed shaft. Below them, as shown in Figure 11.7, is connected a movable pulley which supports the weight to be raised. The input force F_1 is applied to the chain leaving the larger upper pulley. The output force F_2 is supplied by the lower pulley to the load being raised. Of interest is the mechanical advantage, which is the ratio of forces, F_2/F_1. In Figure 11.7 three chain segments are labeled A, B, and C. As segment A is pulled down a distance d_A, segment B moves down a distance d_B, segment C moves up a distance d_C, and the lower pulley moves up a distance $\frac{1}{2}(d_C - d_B)$; the motion of segment B tends to lower it and the motion of segment C tends to raise it. The conservation of work equation may be written,

$$F_1 d_A = F_2(d_C - d_B)/2, \tag{11.33}$$

so that the mechanical advantage is

$$M.A. = \frac{F_2}{F_1} = \frac{2d_A}{d_C - d_B}. \tag{11.34}$$

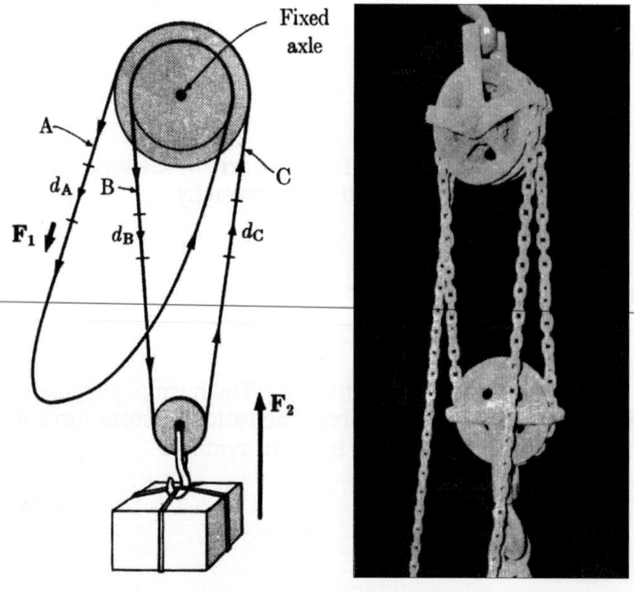

FIGURE 11.7 The chain hoist, a multiplier of force. **(a)** The principal of its operation. While force F_1 acts through distance d_A, the much greater force F_2 lifts the load through a much smaller distance, $\frac{1}{2}(d_C - d_B)$. **(b)** A chain hoist in action. (Photograph courtesy of L. L. Baumunk & Son.)

Consider now the motions that take place during one full revolution of the upper shaft. The segment C moves up and the segment A moves down through the same distance L_1, the circumference of the outer pulley. At the same time the segment B moves down through the distance L_2, the circumference of the inner pulley. Therefore the mechanical advantage may be reexpressed in terms of these two circumferences:

$$M.A. = \frac{2L_1}{L_1 - L_2}. \tag{11.35}$$

If the *difference* of the two circumferences, $L_1 - L_2$, is much less than the outer circumference L_1, the mechanical advantage will be very large. The next time that you are in an automobile garage, watch a chain hoist and notice the large movement of chain segment A required to produce a small movement of the lower pulley and its load.

Among other devices illustrating the principle of conservation of work is the vise common to home work benches. The explanation of its action is left as an exercise. For all such multipliers of force, the principle is the same: the input force multiplied by the distance through which it acts is equal to the output force multiplied by the distance through which it acts. This conservation of work is a very special case of the general law of energy conservation. It requires for its validity that the device store no energy and that it transform no work (or, in practice, very little work) into nonrecapturable energy such as heat energy, so that work input appears immediately as equal work output.

11.7 *Conservation of kinetic energy plus potential energy*

Galileo was aware of a conservation law of height for certain motions in the earth's gravitational field. He knew that on a curved and ideally frictionless slide (Figure 11.8) a sliding block would rise to the same height from which it started, regardless of the shape of the slide. He knew that a pendulum bob, released at a certain height, rises again to that height, even if the string holding the bob is interrupted in its swing by a peg. Although Galileo did not introduce or use the energy concept, we can now understand his rule of constant height as a simple consequence of energy conservation. For the motions he described, kinetic energy plus potential energy is a constant,

$$K.E. + P.E. = E \tag{11.36}$$

(the symbol E denotes the constant total energy). At the points of maximum height, kinetic energy is zero. At these points, therefore, potential energy is equal to the constant total energy. Since potential energy in the earth's gravitational field depends only on height, points of constant potential energy are points of constant height.

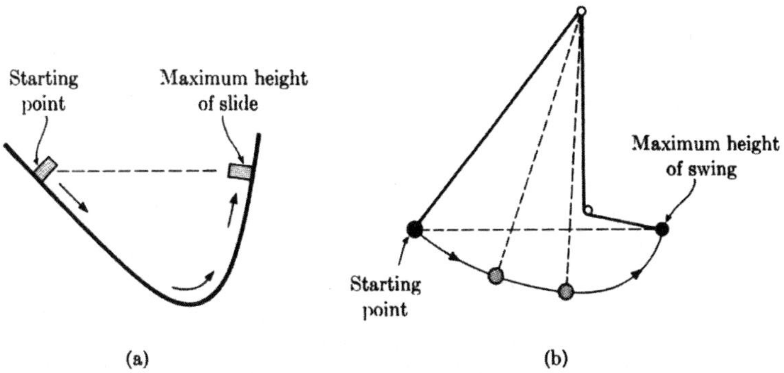

FIGURE 11.8 Examples of motion discussed by Galileo. **(a)** Block on a smooth slide. **(b)** Pendulum. Both rise to the same height as their initial height.

Gravity Near the Earth

Of special value in studying simple motions for which kinetic energy plus potential energy is a constant is the energy diagram. Plotted vertically is the energy, and horizontally the distance along the direction of motion. Consider, for example, the vertical motion of a thrown baseball or a falling stone. First in the diagram is drawn the potential energy as a function of the vertical distance y. If ground level is chosen as the point of zero potential energy, the equation is

$$P.E. = mgy\,, \tag{11.37}$$

and the graph is a straight line passing through the origin (Figure 11.9). Next in the diagram is drawn a horizontal line whose height is equal to the total energy E. It is horizontal because the energy E is the same at all distances y. Finally a vertical line may be drawn connecting any point on the potential energy curve (or in this case, the potential energy line) with the point on the energy line directly above it. The length of this vertical line gives the magnitude of the kinetic energy, since it is the difference between total energy and potential energy.

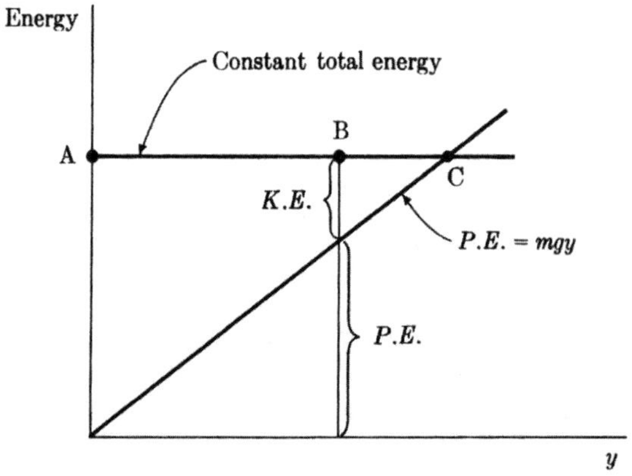

FIGURE 11.9 Energy diagram for vertical motion near the earth.

The energy diagram is considerably more abstract than a physical picture of the actual motion. In this example the fact that the quantity plotted horizontally represents the vertical distance can cause some confusion. Nevertheless the energy diagram has a simplicity of its own and repays careful study. It is in its own way a picture of motion, a picture that reveals more about the motion than does an actual pictorial representation of the motion in space. If a baseball is thrown vertically upward with some total energy E, this energy is initially all in the form of kinetic energy which gradually becomes converted to potential energy as the ball rises. The rise of the ball corresponds to moving to the right in the energy diagram until the potential energy line and the total energy line meet (from A to B to C). This meeting point locates the maximum height possible for the given value of energy. Another motion described by identically the same energy diagram would be that of a ball dropped from this maximum height. It moves to the left in the diagram, from C to B to A. Alternatively a ball with the same total energy (and therefore the same energy diagram) might start at a lower height and be thrown downward with an initial velocity so that it starts out with some kinetic energy and some potential energy. Its progress in the energy diagram is only from B to A.

The different motions just described have different initial conditions, but the same energy. For a given energy, the maximum possible height is defined by the intersection of the energy line and the potential energy line, although this maximum height need not always be reached. For a different energy, the diagram requires but little modification. The horizontal energy line is simply slid up or down to a new level. For greater energy, the maximum height increases in proportion to the energy. To summarize, the advantages of the energy diagram are that it shows at a glance the limit of the motion (in this example the maximum height); it describes simultaneously motions with different initial conditions; it gives pictorially the relative magnitudes of the kinetic energy and the potential energy at any point; and it reveals simply the effect of changing the energy. Its disadvantages are that it is an abstract picture, not a physically visualizable picture, and that it does not show directly how fast the motion occurs from one part of the diagram to another. However, it does give the kinetic energy directly, which is easily related to speed.

The Oscillator

An energy diagram of special interest is that of the oscillator, because oscillatory, or vibrational, motion is such a common feature of the natural world at all its levels. Shown in Figure 11.10 are what might be called the representational oscillator and the abstract oscillator. Part (a) presents a picture of a physical oscillator, a mass suspended between an identical pair of horizontal springs. If displaced from its central position, the mass experiences a force directed toward the central position, a force whose magnitude is approximately proportional to the displacement. If pulled aside and released, the mass oscillates back and forth between equal limits on either side of the central position. Part (b) shows the energy diagram for the oscillator. Its potential energy curve, as we shall show in a moment, is a parabola. For zero energy, the oscillator is quiescent at its midpoint (energy line A). For any greater energy, it moves back and forth between two limits in the symmetrical energy diagram. Doubling the energy (energy line B to energy line C), instead of doubling the amplitude of oscillation, increases it by only 41%.

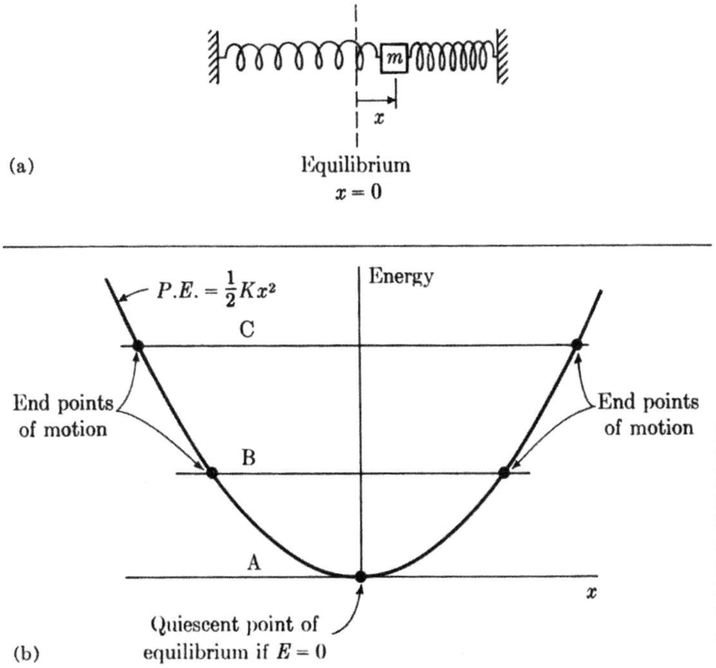

FIGURE 11.10 The oscillator. **(a)** One of many possible physical oscillators, a weight suspended between springs. **(b)** Energy diagram for the oscillator. The potential energy curve is a parabola. Lines A, B, and C designate three different magnitudes of total energy.

To find the oscillator's potential energy function, we must calculate how much work is done against the force of the springs in moving the mass from its central position ($x = 0$) to some other position a distance x from the center. If the springs obey Hooke's law (see page 158), the force function may be written,

$$F_x = -Kx\,. \tag{11.38}$$

This equation (the same as Equation 8.24) states that the force exerted *by* the spring is directed opposite to the displacement and is proportional to the displacement x with proportionality constant K. The constant K is a stiffness constant. A soft spring has a small value of K; a stiffer spring has a larger value of K. If the mass is pushed slowly by hand from its initial position a distance x to the right, at first no force is required; finally a force Kx is required to hold it aside. The *average* force* during the displacement from zero to x is $\frac{1}{2}Kx$ and is directed parallel to the direction of motion. This average force multiplied by the distance moved is the work done,

$$W = \tfrac{1}{2}Kx^2\,. \tag{11.39}$$

This positive work done against the force of the springs is equal to the potential energy:

$$P.E. = \tfrac{1}{2}Kx^2\,. \tag{11.40}$$

The potential energy varies as the square of the displacement x. Its graph is the parabola plotted in Figure 11.10.

For the oscillator, the energy conservation equation reads

$$\tfrac{1}{2}mv^2 + \tfrac{1}{2}Kx^2 = E\,. \tag{11.41}$$

The kinetic energy vanishes at the endpoints of the motion, where the energy line intersects the potential energy curve. At these two points the energy equation takes the form,

$$\tfrac{1}{2}Kx_{\text{limit}}^2 = E\,.$$

Its solution for the limiting displacement is

$$x_{\text{limit}} = \pm\sqrt{\frac{2E}{K}}\,. \tag{11.42}$$

The plus and minus signs are indicated explicitly as a reminder of the two choices possible for the sign of the square root. The plus sign gives the right-hand limit, the minus sign the left-hand limit, of the motion. This result shows that the amplitude varies not in proportion to the energy E, but in proportion to the square root of E. To double the amplitude requires four times the energy. The result also shows how the amplitude depends on the stiffness constant K. As is reasonable, increasing the stiffness constant diminishes the amplitude. The actual time dependence of the position coordinate x is given by Equation 8.27, in which the amplitude A is the same as what we here call x_{limit}.

Gravity in the Large

At high altitude, the force of gravitational attraction toward the earth weakens. At the distance of the moon, the earth's gravity is some 3,600 times weaker than it is at the earth's surface. Because of this decrease of gravitational force with increasing distance from the earth, the gravitational potential energy does not continue to rise indefinitely in proportion to altitude, as might be incorrectly inferred from the energy diagram in Figure 11.9. Instead, as shown in Figure 11.11, the potential energy associated with the earth's gravitational force rises ever more slowly at greater distance from the earth, until finally it approaches a con-

* This is an average over distance, not an average over time.

stant value at infinite distance from the earth. The small circle in Figure 11.11 encloses the part of this energy diagram that corresponds to the "local" energy diagram in Figure 11.9.

As emphasized in Section 11.5, the zero of potential energy is arbitrary. The choices of zero potential energy at the earth's surface in Figures 11.9 and 11.11, and at the midpoint of the oscillator's motion in Figure 11.10, are convenient but not necessary. For gravity in the large, infinite distance is usually adopted instead as the point of zero potential energy, a choice that causes the potential energy at finite distance to be negative. Also, for gravity in the large, it is more convenient to measure distance from the center of the earth than from the surface of the earth. With these changes of scale, the potential energy curve in Figure 11.11 takes the form shown in Figure 11.12. The small circle still indicates the part of the diagram that corresponds to Figure 11.9. Although the mathematical equation describing the potential energy curve in Figure 11.12 is not of primary concern here, it is no secret. The curve is a hyperbola. It is described by Equation 12.36.

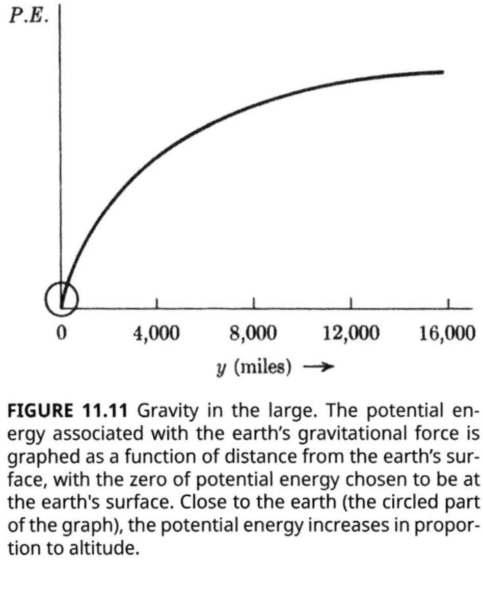

FIGURE 11.11 Gravity in the large. The potential energy associated with the earth's gravitational force is graphed as a function of distance from the earth's surface, with the zero of potential energy chosen to be at the earth's surface. Close to the earth (the circled part of the graph), the potential energy increases in proportion to altitude.

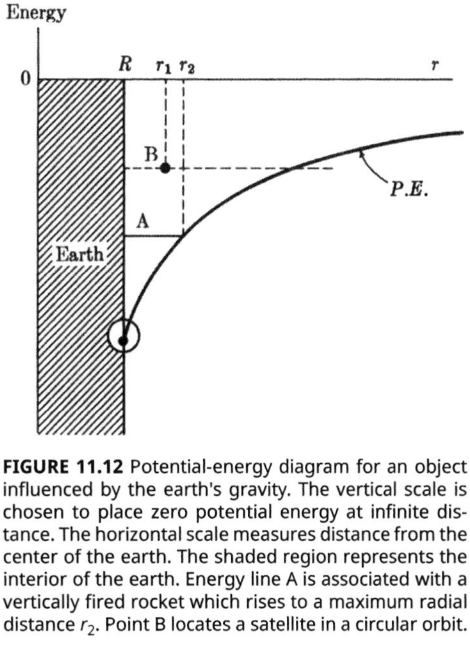

FIGURE 11.12 Potential-energy diagram for an object influenced by the earth's gravity. The vertical scale is chosen to place zero potential energy at infinite distance. The horizontal scale measures distance from the center of the earth. The shaded region represents the interior of the earth. Energy line A is associated with a vertically fired rocket which rises to a maximum radial distance r_2. Point B locates a satellite in a circular orbit.

The horizontal energy line A in Figure 11.12 depicts a vertically fired rocket that ascends to a maximum radial distance r_2, then falls back to earth. At each point in its flight, its potential energy is measured by the distance of the potential energy curve below the horizontal axis (a negative quantity), and its kinetic energy is measured by the distance of the total energy line above the potential energy curve. For a satellite in a circular orbit, neither altitude nor speed change. Therefore both kinetic and potential energies are also constant. In an energy diagram, the circling satellite appears simply as a stationary dot, such as the point B in Figure 11.12. Note that in each of these examples, the total energy, kinetic plus potential, is negative.

The curves in Figures 11.9 through 11.12 can be interpreted pictorially in a way that any skier will find easy to remember. To move from a lower to a higher point (uphill) on a potential energy curve requires that energy be expended. In moving from a higher to a lower point (downhill) on the curve, the given force (gravity or spring tension) does work on the object or system. The steeper the slope, the greater the force. In terms of the potential energy diagram of Figure 11.12, we on the surface of the earth are far down in a potential energy valley. To climb up and out of this valley requires a large expenditure of energy. Only in recent years has man developed the rocket technology to do it.

Submicroscopic Energy Conservation

Finally we mention, without pursuing any of its mathematical details, an example of energy conservation that was of vital importance in leading toward our present picture of atomic structure. Hans Geiger and Ernest Marsden fired energetic alpha particles at metal foils in 1909; they found that a small but unexpected fraction of the alpha particles were deflected through large angles, a few being turned through almost 180 deg. Before long Ernest Rutherford, in attempting to account for the strong force that must be responsible for the large deflections, hit upon an atomic model that proved to be a key to progress. He postulated that all of the positive charge of the atom resides in a tiny nucleus at the center of the atom which creates in its own neighborhood a strong electric force, far stronger than would exist within the atom if the same charge were distributed throughout the entire volume of the atom. The potential energy of the alpha particle moving in the nuclear electric force field has the form shown in Figure 11.13. Like the earth's gravitational potential energy, this electric potential energy varies inversely with distance, and its curve is a hyperbola. It is positive instead of negative because the force is repulsive instead of attractive.

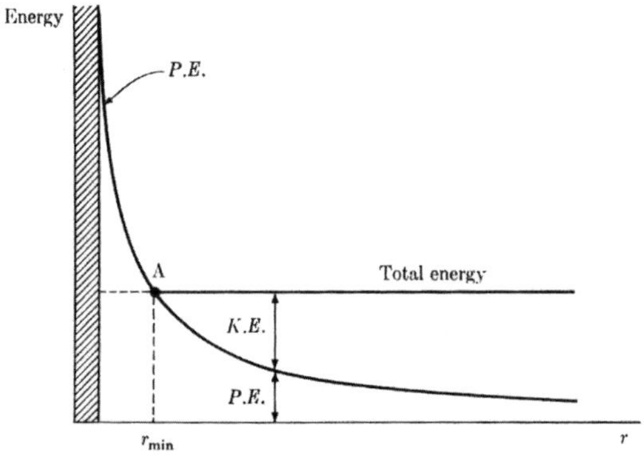

FIGURE 11.13 Potential-energy diagram for an alpha particle in the electric force field of an atomic nucleus. The shaded region indicates the interior of the nucleus.

Subsequent experiments of Geiger and Marsden verified the correctness of Rutherford's atomic model through many measurements of alpha-particle deflection at several different energies. All measurements agreed with Rutherford's calculated deflections. For our discussion here, the important thing about Rutherford's calculation was the fact that he could use the law of energy conservation to determine the distance of closest approach of the alpha particle to the nucleus, even though this distance is extraordinarily small compared to any distance that could be measured directly. Exactly the same condition applies as in the macroscopic examples discussed above. The point in the energy diagram where the energy line intersects the potential energy curve locates a limit of the motion, in this case a small-distance limit. The point A in Figure 11.13 is at the distance r_{min} where an alpha particle headed straight for the nucleus is turned back. For trajectories not headed straight for the nucleus, the distance of closest approach is somewhat greater. Since, in the early experiments, no discrepancies appeared between the calculated and measured deflections, Rutherford could conclude that the nucleus must be smaller than the calculated point of closest approach. Otherwise alpha particles would have penetrated the nucleus and suffered deflections different from those calculated. An alpha particle with an energy of 3 MeV on a collision course for a gold nucleus comes to within 8×10^{-12} cm of the center of the nucleus. This distance, about one thousandth of the atomic radius, is approximately ten times what is now known to be the radius of the gold nucleus.

11.8 *Mass and energy*

There is probably no equation of physics more widely known than that connecting mass 53 and energy, $E = mc^2$, and no scientist better known than the author of this equation, Albert Einstein. Physicists sometimes ask each other why Einstein, among all of the outstanding scientists of the twentieth century, stands preeminent in the public mind as the greatest. The public needs a scientist hero, runs one argument. But this argument, true or false, leaves unanswered the question: Why Einstein? A case can be made out that on the basis of his scientific achievements, Einstein indeed deserves the first position. This is at best a partial answer, for there have been a number of towering figures in twentieth-century physics (as well as in other sciences), and there is no simple yardstick for measuring the relative importance of different contributions. The key to Einstein's unique fame lies more likely in the disconcerting impact of his discoveries on man's view of the world around him. In a single year (1905) Einstein pulverized the solid bedrock of human preconceptions about space, time, and matter, leaving a shifting sand of relativity in its place. The relativity of distance and of time and the mixing together of space and time are subjects of Part 6, on the theory of relativity. There too is discussed the theoretical foundation for the remarkable conclusion that matter may be created and annihilated. Here we are concerned with the impact of the new view of matter and mass on the law of energy conservation.

Einstein's first paper on the subject of mass and energy was entitled, "Does the Inertia of a Body Depend on its Energy?" His discovery—or better, his theoretical prediction—was that mass (inertia) is a reflection of, and a measurement of, the total energy content of a stationary object (or a system with zero total momentum). This is not the same as saying that mass is a form of energy, although under certain circumstances the two statements are equivalent. Consider, for example, the creation of a pion in a proton-proton collision, denoted by

$$\mathrm{p} + \mathrm{p} \rightarrow \mathrm{p} + \mathrm{p} + \pi^0 .$$

This is a commonplace event in a high-energy accelerator. One of the initial protons, the projectile particle, has a certain kinetic energy.* The kinetic energies of the three final particles add up to less than the kinetic energy of the projectile proton, and the difference is equal to the energy equivalent of the pion mass, $m_\pi c^2$. The energy-balance equation may be written,

$$(K.E.)_{\text{initial}} = (K.E.)_{\text{final}} + m_\pi c^2 . \tag{11.43}$$

It does no harm to add the proton mass energies to the left and right sides of this equation and rewrite it in the form,

$$(K.E.)_{\text{initial}} + 2m_p c^2 = (K.E.)_{\text{final}} + 2m_p c^2 + m_\pi c^2 . \tag{11.44}$$

This is a special case of the very general energy-conservation equation that holds true for all elementary particle transformations:

$$[K.E. + Mc^2]_{\text{before}} = [K.E. + Mc^2]_{\text{after}} . \tag{11.45}$$

In this equation, the symbol M stands for the sum of the masses of the participating particles. The total mass M need not be the same before and after, nor need the total kinetic energy be unchanged. But the sum, $K.E. + Mc^2$, is a constant for any particular decay process or reaction.

* Remember that the kinetic energy of a particle is correctly represented by the formula $\frac{1}{2}mv^2$ only if its speed is much less than the speed of light. In this section we will not need to be concerned with the actual relativistic formula for kinetic energy.

A simple way to interpret Equation 11.45 is to accept mass as a form of energy, just one more way in which the slippery concept called energy can manifest itself. In elementary-particle transformations, examined in before-and-after context, no forms of energy other than kinetic energy and mass energy play a role, so that Equation 11.45, much like Equation 11.36, is a simple statement of energy conservation. The energy of motion is kinetic energy. The energy of being is mass energy, mc^2. The sum of these two forms of energy remains constant. Looking at the mass-energy equivalence in this way, we say that the creation of a pion is accompanied by the conversion of energy from one form to another—kinetic energy into mass energy. When the pion later decays into two photons,

$$\pi^0 \rightarrow \gamma + \gamma ,$$

mass energy is converted back into kinetic energy.

54 Einstein's view of mass was deeper and more general. He said not that mass is *one* form of energy, but that localized energy in *any* form possesses mass. Let us examine some implications of this general rule. We restrict attention first to systems with no bulk motion, whose total momentum is and remains zero—a swarm of bees in a stationary box, a single atom at rest. If to such a system an amount of energy ΔE is added—by heating or by any other means—its mass increases by an increment equal to ΔE divided by c^2:

$$\Delta m = \Delta E/c^2 . \tag{11.46}$$

Moreover, the total mass of the system, M_{tot}, measures the total energy content of the system, E_{tot}, through a similar relation,

$$M_{tot} = E_{tot}/c^2 . \tag{11.47}$$

These equations are valid for systems at rest. If the system has bulk motion, it possesses additional kinetic energy that is not reflected in its mass.*

As discussed in Section 3.6, the square of the speed of light that appears in Equations 11.46 and 11.47 is best looked upon as a conversion factor connecting mass units to energy units. In terms of centimeters and seconds, c^2 is equal to 9×10^{20} cm^2/sec^2. Note, however, that 1 cm^2/sec^2 is the same as 1 erg/gm, so that

$$c^2 = 9 \times 10^{20} \text{ ergs/gm} . \tag{11.48}$$

One gram of mass reflects a total energy content of 9×10^{20} ergs. One erg of locally concentrated energy exhibits a mass of only 0.11×10^{-20} gm. On the human scale, the inverse of c^2 is infinitesimally small. For this reason alone, the mass-energy equivalence escaped man's notice until this century.

To appreciate the full significance of Equations 11.46 and 11.47, it is helpful to consider a complex system such as the swarm of bees in a box. How is the mass of the system as a whole related to the masses of its individual constituents? First it is important to decide what is meant by a "constituent." Is it a whole bee, or the wing of a bee, or a molecule within the bee, or a single proton or electron within the molecule? In fact, it can be any of these, or even the whole swarm of bees. The general connection between energy and mass is independent of the way in which the system is subdivided into its parts. It is good that this is so, for a single proton itself is a system whose ultimate constituents we know nothing about. Directions for relating the mass and energy of a system to the masses and energies of its parts can best be given in recipe form. First, decide what you wish to call the parts of the system. Take each of these parts, be it elementary particle or complex structure, and mea-

* Equation 11.47, as we use it, defines the "rest mass" of a system. Some authors use Equation 11.47 also for moving systems to define an "effective mass." This is a less useful concept than rest mass.

sure its mass, either by weighing it (gravitational mass) or by accelerating it with a known force (inertial mass). This procedure yields a set of masses, $m_1, m_2, m_3, \ldots$, and a total mass of parts equal to their sum,

$$M_{\text{parts}} = m_1 + m_2 + m_3 + \cdots . \tag{11.49}$$

The total mass appearing in Equation 11.45 is such a mass of parts. These parts, reassembled to form the total system, may have energy of motion (*K.E.*) and energy of mutual interaction (*P.E.*). If the sun, planets, and moons are chosen as the parts of the solar system, these parts have both kinetic energy and gravitational potential energy. For a system of well-separated elementary particles, the parts (particles) have only kinetic energy. In any case, there is some energy, E_{parts}, associated with the motion and the mutual interaction of the parts of the system. This energy of the parts adds to the intrinsic mass energy of the parts to give the total energy of the system,

$$E_{\text{tot}} = M_{\text{parts}}c^2 + E_{\text{parts}} . \tag{11.50}$$

This is a generally valid equation, whether or not the total momentum of the system is zero. Compare it, for instance, to the left or right side of Equation 11.45. The parts energy, E_{parts}, in Equation 11.50 does *not* include any form of internal energy within any single part, for every contributor to the internal energy of a part has already made itself felt in the mass of that part.

If the system has no bulk motion (zero momentum), its total energy and total rest mass are related by Equation 11.47. In that case, division of each term in Equation 11.50 by c^2 yields an expression for the mass of the system as a whole:

$$M_{\text{tot}} = M_{\text{parts}} + \frac{E_{\text{parts}}}{c^2} . \tag{11.51}$$

Therefore the mass of the system treated as a single unit differs from the sum of the masses of its parts by an amount E_{parts}/c^2 where E_{parts} is what might be called the *external* energy of the parts—the energy associated with the bulk motion of the separate parts and with their interaction, when the system as a whole is not moving.

If you are diet conscious, you might take this discussion to heart and decide that it is better to eat cold apple pie than hot apple pie, since the greater molecular kinetic energy in hot apple pie causes it to have greater mass. But how much greater? Is it really worth the trouble if you happen to prefer the taste of hot apple pie? For a hypothetical analysis of the apple-pie problem, we may choose to subdivide the pie into its individual molecular constituents. Suppose that the sum of the masses of the molecules taken one at a time is 500 gm (about 1.1 lb), that is, $M_{\text{parts}} = 500$ gm. The total molecular energy of a chilly pie is about six million million ergs, that is, $E_{\text{parts}} = 6 \times 10^{12}$ ergs. Dividing this energy by c^2 (9×10^{20} ergs/gm) gives the additional contribution to the mass,

$$\frac{E_{\text{parts}}}{c^2} = \frac{6 \times 10^{12}\ \text{ergs}}{9 \times 10^{20}\ \text{ergs/gm}} = \frac{2}{3} \times 10^{-8}\ \text{gm},$$

hardly big enough to be concerned about. Heating the pie by 100° Fahrenheit (about 44° Centigrade) adds to its energy approximately 9×10^{11} ergs and to its mass approximately 10^{-9} gm. The only dietary solution is to leave the pie in the refrigerator uneaten.

In the apple pie, changes of molecular *kinetic* energy produce slight changes of total mass. *Potential* energy too is a contributor to the mass of the pie, but in a complicated way associated with molecular interaction. To see potential energy at work more clearly as one of the determinants of mass, consider the arrangement illustrated in Figure 11.14. Two carts separated by a spring are pressed together and latched with the spring compressed. The compression of the spring requires work: work which is stored as potential energy ready for transformation back to work, or conversion to kinetic energy when the latch is released. The energy balance is straightforward. If the carts roll without friction or obstruction after the latch is opened, they acquire a kinetic energy equal to the potential energy of the compressed spring, which in turn is equal to the work required to compress the spring. But what of the mass? In prerelativity physics, it was assumed that the mass of the system was a fixed constant. According to Einstein's new view, the system with more energy has more mass. The latched carts and compressed spring have a total mass slightly greater than the sum of the masses of the separated parts. If the work required to compress the spring is, for example, 1 joule or 10^7 ergs (a typical figure for a small laboratory spring), the extra mass amounts to 10^7 ergs divided by c^2, or

$$\Delta M = \frac{10^7 \text{ ergs}}{9 \times 10^{20} \text{ ergs/gm}} = 1.1 \times 10^{-14} \text{ gm} . \tag{11.52}$$

For a macroscopic system such a mass change is far too small to measure. It is easy to see why classical physics contained no hints of the variability of mass.

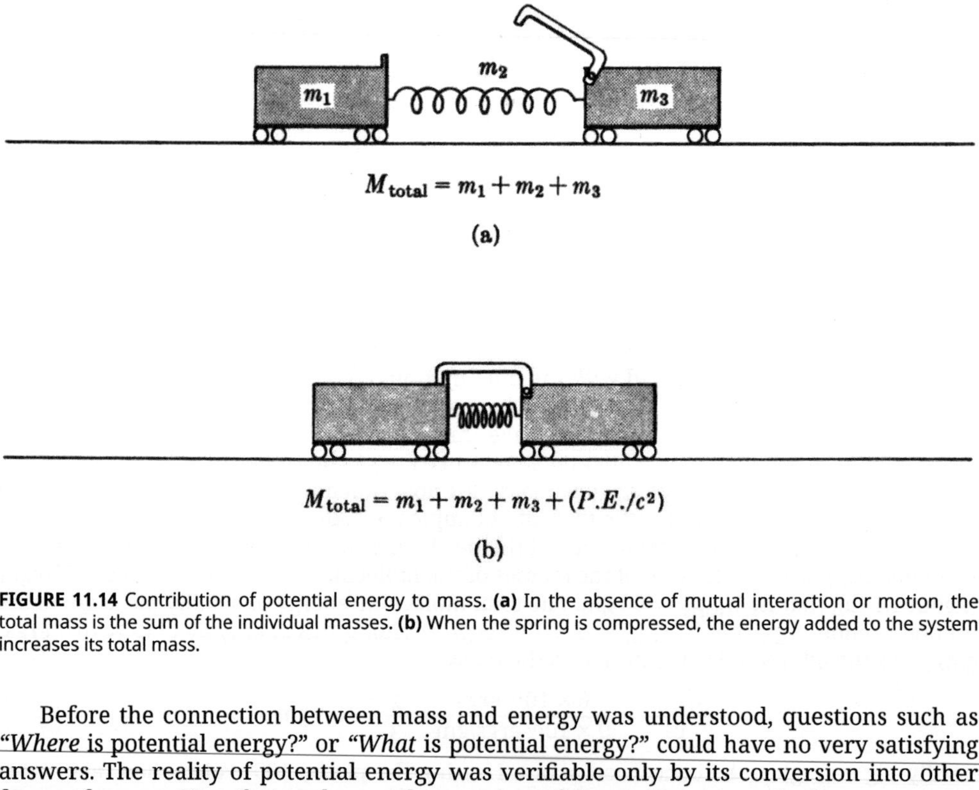

FIGURE 11.14 Contribution of potential energy to mass. **(a)** In the absence of mutual interaction or motion, the total mass is the sum of the individual masses. **(b)** When the spring is compressed, the energy added to the system increases its total mass.

Before the connection between mass and energy was understood, questions such as "*Where* is potential energy?" or "*What* is potential energy?" could have no very satisfying answers. The reality of potential energy was verifiable only by its conversion into other forms of energy. Now that we have a deeper view of the matter, we can look upon mass as the seat of potential energy and the storehouse of potential energy. The amount by which a compressed spring is more massive than the same spring relaxed provides a direct measure (or would provide a direct measure, if the required precision could be achieved) of the potential energy of the compressed spring. Potential energy can be said to have a detectable reality even without being converted to other forms of energy.

The mass of a system need not be greater than the mass of its parts, although this happened to be the situation for the apple pie and for the carts and spring. Attractive forces between the parts of a system can lead to a negative mutual energy for the parts (a negative value of E_{parts} in Equation 11.51), and thereby to a total mass of the system less than the mass of its parts. If two blocks of wood glued together are pried apart, work is done, work that appears as more mass of the separated parts than of the original system. When work is required to separate a system into its parts, the system is said to possess binding energy. Because of its binding energy, an atomic nucleus has less mass than the sum of the masses of its constituent neutrons and protons. Similarly, the earth-moon system taken as a unit has less mass than the sum of the separate masses of earth and moon, because an input of energy would be required to remove the moon from the vicinity of the earth. For a nucleus, the effect of binding energy on mass is measurable. For the earth-moon system, it is not.

The conservation of mass is an intuitively reasonable idea. The mass changes that accompany energy changes in the familiar world around us are so infinitesimally small that we quite naturally suppose the mass (or weight) of every object to be a fixed quantity. Ordinary experience also makes reasonable the idea that the mass of any system or structure is equal to the sum of the masses of its parts. Before relativity, mass conservation was much more than an intuitive idea, however. It was a solid pillar of chemistry, supported by numerous accurate measurements of mass constancy during chemical change, and supporting a world picture of matter built of indestructible and invariable atoms.

Since the form and appearance of matter undergoes marked changes during chemical transformation, it was realized very early in the history of chemistry that mass conservation must be checked experimentally and not taken for granted. Indeed, the verification of mass conservation in chemical change was one of the key experiments that gave birth to modern chemistry in the late eighteenth century. The principle of mass conservation was first clearly stated and used by Antoine Lavoisier in 1789. It was subjected to further experimental tests throughout the nineteenth century. When Einstein first suggested in 1905 that mass could change as energy changes, there was as yet no shred of evidence for his new point of view.

In the most violent chemical changes, little more than 10 eV of energy is released for each participating atom. For comparison purposes, it is convenient also to express the atomic mass in energy units. The energy equivalent of the mass of a hydrogen atom (nearly the same as the mass of a proton) is about one billion eV (1 GeV). Imagine a hypothetical chemical reaction (more vigorous than most) in which 10 eV of energy is released per atom, and the average atomic mass is ten times the proton mass. In this reaction the chemical energy is one billionth of the mass energy. Therefore the mass of the reacting material changes by only one part in a billion as a result of the release of chemical energy.

Since Einstein was aware that mass measurement far more accurate than any ever performed would have been required to detect the mass changes in chemical reactions, the accumulated weight of evidence for mass conservation was no bar to his suggestion that mass was, in fact, not strictly conserved. At the same time he was aware of a possible testing ground for the nonconservation of mass. Only three years earlier, in 1902, Ernest Rutherford and Frederick Soddy had calculated that the energy release per unit mass in the newly discovered radioactive transformations must be at least 100,000 times greater than in chemical reactions. This would imply fractional mass changes of one part in 10^4 instead of one part in 10^9, putting the variability of mass within reach of experimental test. Einstein suggested such a test. Finally, in 1932, John D. Cockcroft and Ernest T. S. Walton first verified that nuclear mass changes are related to nuclear energy changes through Einstein's equation,

$$\Delta E = \Delta M c^2 .$$

We now know that nuclear mass changes considerably larger than one part in 10^4 are commonplace. In nuclear fission, 0.1% of the mass is dissipated as the nuclear energy is released. In fusion reactions, fractional mass changes as large as 0.7% are realized. Not even

Einstein anticipated the 100% mass-to-energy conversion of some particle decays, or the creation of new mass where there was none before. The first explicit theory of mass creation and annihilation was proposed by Enrico Fermi in 1934 to explain the phenomenon of beta decay, in which a newly created electron is ejected from a radioactive nucleus. Similar thinking played a role in Paul Dirac's theory of the positron four years earlier. He proposed that this as yet undiscovered new particle could be created with an electron and could annihilate an electron. Both Dirac and Fermi recognized that symmetry requirements within the theory of quantum mechanics implied that if mass could be created, it could also be annihilated, neither process being possible without the other.

It should be mentioned that there is at present no conceivable way to achieve total conversion of mass to energy on a macroscopic scale. As far as we know now, the fractional change of mass of one part in two hundred that occurs in the thermonuclear burning of a hydrogen bomb is very near the upper limit that is possible. This upper limit, which occurs in stars, is about one part in 120, slightly less than one per cent. Therefore bombs much more destructive per unit mass than those already in existence are not conceivable in the framework of our present understanding of nature.

As mentioned above, every nucleus, because of its binding energy, is less massive than are its individual neutron and proton components. It is through the comparison of nuclear mass measurements and binding-energy measurements that the most accurate tests of the mass-energy equivalence have been achieved. Consider the simplest nucleus with any binding energy, the deuteron, containing one proton and one neutron. The following numbers, based on recent mass measurements, demonstrate the effect of binding energy on mass:

Mass of proton	= 1.00728
Mass of neutron	= 1.00866
Mass of proton + mass of neutron	= 2.01594
Mass of deuteron	= 2.01355
Difference between mass of system and mass of its parts	= 0.00239

The unit of mass employed here is the a.m.u. or atomic mass unit. It is chosen as a matter of convenience* to be exactly equal to one twelfth of the mass of an atom of the most common isotope of carbon, whose nucleus contains six protons and six neutrons. One a.m.u. is equal to 1.6604×10^{-24} gm. As is obvious from the numbers given above, this atomic mass unit is slightly less than the mass of a single neutron or proton.

The deuteron is a system composed of two parts, a neutron and a proton. The mass of the system is less than the mass of the parts by 0.00239 a.m.u., a fractional difference of about one part in a thousand. For the two carts and a spring illustrated in Figure 11.14, the mass of the system is greater than the mass of the parts, because work is required to bring the carts together. For the deuteron, on the other hand, work—or energy—is required to separate the parts. Because of the great strength of the nuclear force, this energy is large enough to have a quite perceptible effect on the mass. The amount by which any nuclear mass is less than the sum of the masses of its constituent neutrons and protons is called the *mass defect*. Because of the mass-energy equivalence, the mass defect is proportional to the nuclear binding energy.

Finally, we can look again at the mass of a single elementary particle. When a particle with mass is created, we usually say that some energy has been transformed into mass energy. Instead, in keeping with the more general view of mass and energy, we could say that a certain amount of energy has been highly concentrated and localized into an entity which we call a particle. This entity has mass *because* it is a concentrated bundle of energy. The exact nature of this energy that clusters together to form the massive particle is unknown.

---

* The convenience is that most atomic masses are very close to integers in this unit.

Perhaps some future theory will reveal more of its details. But whatever its nature, it is the source of the particle's mass.

11.9 *Assessment of energy conservation*

As a law of mechanics alone, energy conservation has some apparent imperfections. If some 55
mass is annihilated, new mechanical energy makes its appearance. If frictional forces are significant, mechanical energy disappears. In short, mechanical energy by itself (kinetic plus potential) is *not* always conserved. Within the framework of the theory of mechanics alone, the law of energy conservation is often nearly correct and often useful; but it is not universally valid. This is in contrast to the conservation laws of the purely mechanical quantities, momentum and angular momentum. Yet the very limitations of energy conservation as a law of mechanics have proved to be the sources of its power and significance as a more general principle of nature. The transformation of mechanical energy via heat to internal energy proved to be basic to the development of the theory of thermodynamics. Concern about the consistent transfer of energy from mechanical form to the form of electromagnetic radiation was an important ingredient of the foundation of relativity theory, and led in turn to Einstein's deep insight into the equivalence of mass and energy.

As nature's most convertible and many-faceted concept, energy plays a special role in science. It prevents man from compartmentalizing the phenomena of nature into rigid, well-separated disciplines. Mechanics cannot be separated from thermodynamics or from electromagnetism, nor can any of these subjects ignore relativity. Energy is the great unifier. As a common concept of every theory describing nature, it joins together the parts of science. As the principal medium of exchange among the phenomena of nature, it literally joins the world together. The sea, the sky, and the land are joined by energy transfer, some of it as subtle as the evaporation of dew, some as violent as a summer thunderstorm. Through nuclear, electromagnetic, and gravitational energy, the galaxies join to form a whole. Through nature's most elaborate and wonderful sequence of energy conversions, the thermonuclear reactions of the sun energize the human body. Civilized man, not content to witness, has devoted much of his talent to converting and transferring energy for his own ends, to which the industrial world's thousands of miles of high-tension lines bear singing testimony.

Because of its conservation, and only because of its conservation, energy is the central unifying concept of science. Without conservation, the disappearance of energy in one place could not be linked with the appearance of energy in a different place or in a different guise. Without conservation, two different manifestations of energy would instead be two distinct concepts. Without conservation, man could not so easily buy and sell and ship energy like a commodity. Without energy conservation, our description of nature would be very different, and for the good reason that nature would be very different.

In this century man's view of energy has deepened in two important ways, first through its connection with mass, second through its connection with the time symmetry of nature. Although it is beyond the scope of this book to discuss in detail this latter connection, it is so fundamental and at the same time so startling that it must at least be mentioned. The conservation of energy has been linked to the uniformity of time, more particularly to the invariance of the laws of nature with respect to change of time. Energy is conserved, according to the modern argument, because the laws of nature are the same yesterday, today, and tomorrow. The enterprising student in search of a hint about the reason for this remarkable connection should look at Section 21.5 and then again at Section 4.8.

EXERCISES

11.1. An automobile with a mass of 10^6 gm is stalled on a horizontal roadway. Starting from rest, a man pushes the car forward with a steady force of 2×10^7 dynes for 5 sec. Assume that the car rolls with negligible friction and calculate the following quantities: **(a)** the speed of the automobile after 5 sec; **(b)** the distance covered; **(c)** the kinetic energy of the automobile after 5 sec; **(d)** the work done by the man.

11.2. A planet executes a circular orbit about a star. **(1)** Explain why the star does no work on the planet. **(2)** Name two quantities that remain constant during the course of this motion, and two that change.

11.3. Verify that for *any* time t during part A of the motion depicted in Figure 11.2, the kinetic energy is equal to the work done up to that time.

11.4. Within a television tube, an electron acquires a kinetic energy of 4×10^{-9} ergs in a distance of 2 cm. **(1)** What is the approximate speed of an electron with this kinetic energy (the nonrelativistic formula suffices)? **(2)** What average force must have acted on the electron to give it this energy?

11.5. A sky diver, after reaching a terminal speed $v = 6{,}000$ cm/sec, exerts on the air a steady force equal to his own weight $F = 7 \times 10^7$ dynes. **(1)** Working from Equation 11.10, show that the power he expends on the air is given by $P = Fv$. **(2)** Calculate this power, expressing the answer in ergs per second and in kilowatts.

11.6. Describe a system composed of two particles which has zero linear momentum, zero angular momentum, but nonzero kinetic energy.

11.7. A shell of mass 10 kg is fired vertically upward with an initial speed of 3×10^4 cm/sec. **(1)** If there were no air friction, so that the sum of kinetic energy plus potential energy remained constant, what would be the maximum height of the shell? **(2)** If, on its way up, the shell dissipates 10^{12} erg in the form of heat, what height does it reach? (In both parts, ignore the variation of gravitational force with height.)

11.8. An automobile whose mass is $m = 1{,}500$ kg coasts with negligible friction on a horizontal road at a speed $v_0 = 20$ miles/hr. **(1)** What is its kinetic energy in ergs? **(2)** What is the greatest vertical height of a hill over which it can coast? **(3)** Solve part **(2)** algebraically as well as numerically and show that the answer is independent of the mass of the automobile.

11.9. A baseball is thrown with initial velocity whose magnitude is v_0 and whose angle to the horizontal is θ. Use energy conservation arguments analogous to those in Equations 11.26 to 11.28 to prove that its maximum height is given by the equation

$$y_{max} = v_0^2 \sin^2 \theta / 2g \, .$$

Discuss the two limits, $\theta = 0$ and $\theta = 90$ deg.

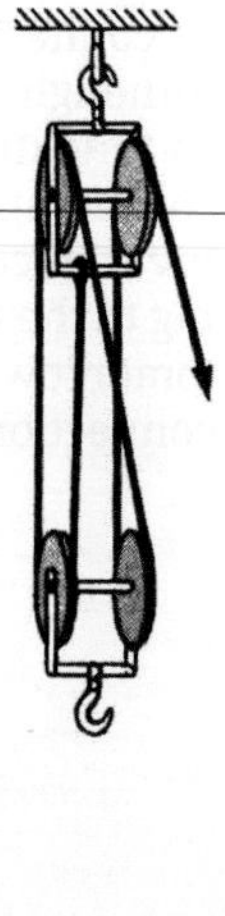

11.10. A common predecessor of the chain hoist was the block and tackle, a combination of multiple pulleys. The combination pictured in the figure, with two double pulleys, has a mechanical advantage of 4. Explain why.

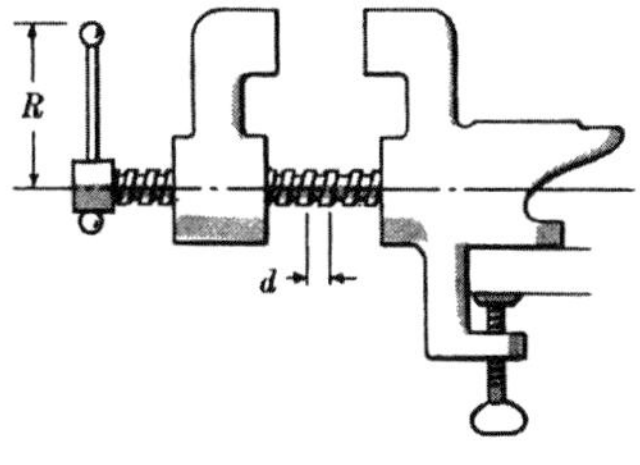

11.11. The input force on a vise is applied a distance R from the axis of the screw. In one turn, the screw thread advances a distance d. The output force is applied at the jaws of the vise. **(1)** Derive a formula for the mechanical advantage (*M.A.*) of the vise, assuming conservation of work. **(2)** If half of the input work is transformed to heat and half appears as output work, what is the mechanical advantage?

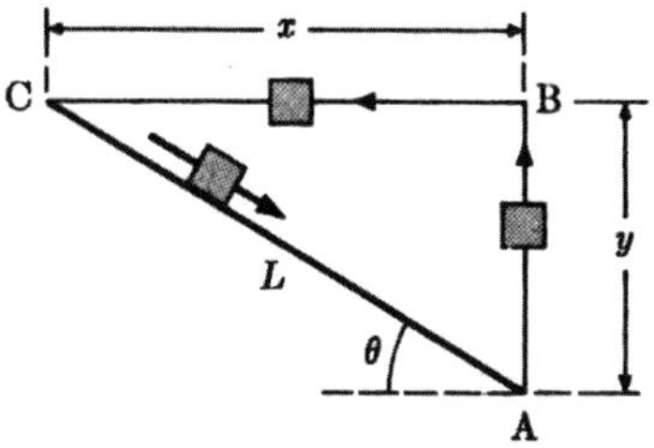

11.12. A block of mass m is lifted vertically a distance y from A to B, which requires an input work equal to mgy. It is then moved horizontally a distance x from B to C, which requires no expenditure of work. Therefore its potential energy at C measured relative to zero potential energy at A is *P.E.* = mgy. Calculate its kinetic energy at A after it slides down a motionless track from C to A in two ways. **(1)** Find the component of gravitational force along its direction of motion, then its acceleration, and its speed at the bottom. **(2)** Apply energy conservation directly. Comment on the relative merits of these two approaches to the problem.

11.13. A rocket as ill-fated as the one pictured in Figure 11.5 rises slowly to a height of 1 km, its thrust being nearly the same as its weight. There it flips over and drives itself vertically downward, still with thrust equal to its weight. **(1)** With what speed does it crash into the ground? **(2)** Give algebraic expressions for its kinetic energy and its potential energy **(a)** just after lift-off; **(b)** at its peak height; and **(c)** just before it strikes the ground.

11.14. An electron of charge $-e$ in a uniform electric field $\mathcal{E}$ experiences a constant force of magnitude $F = e\mathcal{E}$. Suppose that this force is directed to the left and the positive x-axis is directed to the right. ($F_x = -e\mathcal{E}$). **(1)** Obtain an expression for the potential energy of the electron in this field of force. The analogy with gravity near the earth may be helpful. **(2)** Is the location of the zero of potential energy arbitrary in this example?

11.15. An atom in a crystal, if displaced from its equilibrium position ($x = 0$), experiences a restoring force proportional to its displacement, a force given by Equation 11.38. Consider an atom whose force constant is $K = 8 \times 10^5$ dyne/cm vibrating with an energy of 1 eV. **(1)** What is its maximum displacement, x_{limit}? **(2)** If the mass of the atom is 4×10^{-23} gm, what is its maximum speed? **(3)** At what value of x is this maximum speed attained?

11.16. **(1)** In an energy diagram like that of Figure 11.12, draw an energy line representing a vertically fired rocket that escapes the earth and continues on indefinitely. **(2)** The point B in Figure 11.12 represents a satellite in a circular orbit. In your diagram, show how the motion of a satellite in an elliptic orbit must be represented.

11.17. The potential energy curve in Figure 11.13 is a hyperbola; the potential energy varies inversely with the radius, *P.E.* ~ $1/r$. **(1)** If an alpha particle with an initial kinetic energy of 3 MeV can approach to within 8×10^{-12} cm of the center of a gold nucleus, what must be the energy of an alpha particle that can penetrate to the nuclear surface, whose radius is 8×10^{-13} cm? **(2)** At its point of closest approach what is the kinetic energy of this alpha particle and what is its potential energy?

11.18. Calculate and compare the following energies in units of eV: **(a)** Kinetic energy of a hydrogen atom with speed $v = 2 \times 10^5$ cm/sec, typical of normal thermal motion; **(b)** energy of photon of red light emitted by hydrogen, $f = 4.6 \times 10^{14}$ sec^{-1} ($E = hf$); **(c)** energy released per H atom in H bomb if 1 part in 200 of the mass is converted to energy; **(d)** kinetic energy of flea with mass of 0.1 gm lazing along at 0.1 cm/sec.

11.19. Energy released in nuclear explosions is commonly measured in kilotons (10^3 tons) or megatons (10^6 tons). One "ton" is an energy unit equal to 4.2×10^{16} ergs; it is the energy released in the chemical explosion of one ton of TNT. **(1)** Express the kinetic energy of ten thousand cubic kilometers of air (the air over a small county) moving at 500 cm/sec (a light breeze) in kilotons. Take the average air density to be 10^{-3} gm/cm^3. **(2)** How much mass gets converted to energy in the explosion of a 10–megaton H bomb?

11.20. A nuclear bomb containing 10 kg of fissionable material explodes, converting 0.05% of this mass to energy. **(1)** How much energy is released in the explosion in ergs? in kilotons? **(2)** If this energy could be harnessed to lift an ocean liner with a mass of 10^{10} gm, how high could such a ship be lifted against the earth's gravitational field? (Ignore the weakening of the pull of gravity with increasing altitude.)

11.21. The binding energy of a hydrogen atom is 13.6 eV. Is the mass of a hydrogen atom greater or less than the sum of the masses of a proton and an electron? By how much? What fraction is this of the total mass of the atom?

11.22. Two identical safety pins are placed in two identical acid baths. One safety pin is latched, the other unlatched. After the acid has dissolved both pins, is there any difference between the acid baths? What happened to the potential energy of the latched safety pin?

11.23. Consider the following hypothetical system. One billion billion billion photons (10^{27}), each of energy 1 MeV, are flying about within an otherwise empty container whose interior walls are perfectly reflecting. The photons form a system at rest—that is, their total momentum is zero. What is the total energy of this system in ergs? What is its mass in gm? (NOTE: Even though a single photon is massless, two or more photons can comprise a system, which taken as a whole has mass because it has localized energy.)

11.24. Each of the photons in the hypothetical system described in the exercise above is replaced by an electron whose *kinetic* energy is 1 MeV. What is the total energy of this new system in ergs? What is its mass in gm? By how much does the mass of the system differ from the sum of the masses of the electrons? (NOTE: The mass of an electron, 9.1×10^{-28} gm, is equivalent to an energy of 0.511 MeV).

11.25. An earth satellite moves in an elliptic orbit. **(1)** In which part of the orbit does the earth do work on the satellite? **(2)** In which part of the orbit (if any) does the satellite do work on the earth? **(3)** Is the total work done by the earth on the satellite in one revolution positive, negative, or zero? Why?

11.26. When a motionless neutral kaon decays into two pions, the pions leave the scene in exactly opposite directions with equal speed. **(1)** Explain this behavior in terms of energy and momentum conservation. **(2)** If two pions were seen to emerge in a "V" (not oppositely directed) from a place where a kaon was known to be, what possible explanations might be offered (there are several) ?

11.27. Two particles have equal mass m. One is at rest; the other flies toward the first with speed v (see figure). They collide elastically, so that their total kinetic energy is the same after the collision as before, and their total momentum (a vector quantity) is also unchanged. For convenience, define the x direction to be the flight direction of the incident particle. **(1)** If the target particle flies forward in the x direction, what are the velocities of both particles after the collision? **(2)** If the target particle flies off at 45 deg to the x-axis, what are the final velocities of both particles? (HINT: Assign suitable symbols to unknown quantities and write down before-and-after conservation equation for total kinetic energy, and for x and y components of total momentum.) *Optional:* Prove that if the incident particle is deflected, its path after the collision is perpendicular to the path of the recoiling target particle.

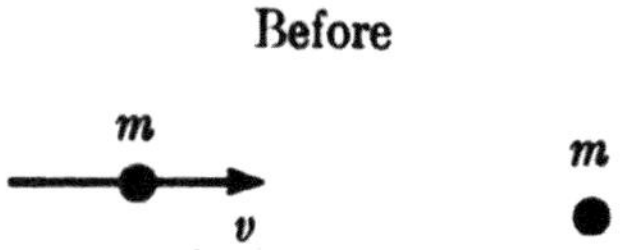

TWELVE

Universal gravitation and the evolution of mechanics

To bring mechanics to bear on nature requires more than Newton's laws. Also required are specific laws of force. One force in particular—gravitation —is singled out for attention in this chapter.

12.1 Laws of force in mechanics

Consider the vast generality of Newton's three laws alone. The first specifies the nature of undisturbed motion for *all* bodies, and defines inertial frames of reference. The second describes how an object reacts to *any* force. The third states that *all* of nature's forces come in balanced pairs. These laws encompass both the experimentally circumscribed domain of physics and the mathematical domain of unfettered abstraction. Newton himself studied the properties of motion associated with some hypothetical laws of force. Students of physics ever since have been working out how objects would move if acted upon by particular forces—forces which may not resemble any of those actually encountered in nature.

Scientists study nature as it might be only to the extent that such study can teach something about nature as it is. To narrow mechanics to the actual, Newton's laws must be supplemented by laws of force governing natural phenomena, in particular the laws of those fundamental forces that act over macroscopic distances*—the gravitational, electric, and magnetic forces.

The fact that Newton's laws do not distinguish real forces from hypothetical forces is not a weakness of mechanics. In fact it is a strength, for it means that mechanics is open, ready and able to accommodate new discovery. The mechanics of Newton accurately describes the deflection of an electron by an electric force, for instance, although neither the existence of the electron nor the law of electric force were known to Newton.

One subtlety must be mentioned. According to a popular point of view about the logical structure of mechanics, Newton's second law should be regarded as a definition of the force concept. This "demotion" of the second law from fundamental law to definition is compensated by a "promotion" of the laws of force. In any event, regardless of the point of view, it is a fact that the description of actual motion requires both Newton's laws and specific laws of force.

The Role of Gravity

56 Among the fundamental forces of nature, gravity is of special interest for several reasons. It is, first of all, the only truly universal force. It acts on every material thing from electron to galaxy, and, as we have learned in this century, it even acts on immaterial things—photons, neutrinos, or energy in any form. Second, gravity has played a uniquely important role in the development and growth of mechanics. Newton, in his definitive formulation of mechanics, drew upon past studies of motion near the earth, influenced by local grav-

* The short-range nuclear forces and the weak interactions act in the subatomic domain where classical mechanics is not valid.

ity, and of planetary motion far from earth, influenced by the sun's gravity. Since Newton's time, motion governed by gravitational force has provided the most stringent tests of mechanics, has served as stimulus for much of the mathematical elaboration of the theory of mechanics, has led to the discovery of distant new planets, and, in our own era of artificial satellites, has revealed new details of the shape and structure of the earth. Through the study of the orbit of Mercury came the first hint of an imperfection in Newtonian mechanics. Mercury's refusal to follow precisely the laws of classical mechanics stands now as one of the experimental supports of the new mechanics of Einstein's general relativity.

Finally, gravity has a special place in mechanics because of its *essential* link with motion. Most forces on earth can be measured statically. In principle, at least, a typical force on earth can be measured with a spring balance or other device that involves no motion. The same is obviously not true of the forces acting on the planets. These can be determined only by studying the motion of the planets. The motion determines the force, but at the same time the force determines the motion. Planetary mechanics must be a self-consistent theory in which the law of motion (Newton's second law) and the law of gravitational force are intertwined.*

12.2 The two threads of mechanics

57 Throughout recorded history, man has responded to the lure of the heavens, examining, recording, and speculating on celestial motion. At least since the time of Aristotle, the study of the motion of sun, stars, and planets has been linked with the study of motion on earth. Aristotle proposed a general theory of all motion that accounted in a single coherent scheme for the motion of every thing, be it wandering planet, falling stone, rolling cart, or rising flame. When Isaac Newton in the seventeenth century proposed an equally general theory of motion, nearly every aspect of Aristotelian physics was overturned.

In the twenty centuries between Aristotle and Newton, some scholars gave more attention to terrestrial motion than to celestial. Others, in greater number and with greater accuracy and sophistication, gave more attention to celestial motion. Despite some disputes over particular Aristotelian points of view about motion, and despite some progressive steps toward modern scientific viewpoints, various of Aristotle's opinions remained alive for over nineteen of those twenty centuries. One in particular that survived was his belief that celestial and terrestrial motions were of basically different natures. This is not to say that celestial motion was regarded as being outside the scope of man's understanding, or that a single system of physics was not believed capable of encompassing both kinds of motion. It means only that celestial motion was assumed to be of a special kind not encountered on earth, and terrestrial motion was of a kind not encountered in the heavens. It was roughly like distinguishing fishes and birds, two very different forms of life, neither living in the realm of the other, but both related by the science of biology. This view of different realms of motion was not challenged by Copernicus when he reintroduced the ancient idea of a sun-centered universe in 1543. The first reputable scientist to suggest that celestial motion was really mundane was probably William Gilbert in 1600. He had discovered that the earth is a giant magnet and suggested that perhaps the earth and other planets are held in their orbits by magnetic forces no different from the force pulling a piece of iron to a hand magnet. Kepler speculated in a similar vein about interplanetary forces. Although Galileo made no prog-

* Actually the form of the law of gravitational force, at least of the earth's gravitational force, can be determined by static measurements, by determining the variation of weight with altitude. Such measurements had not been made in Newton's time. Later the change of weight with altitude and with latitude was used to provide further support for Newton's ideas. In any case, the universality of the gravitational force—the fact that the force producing weight on earth and the force holding the solar system together are the same—could never be inferred from static measurements.

ress toward learning the nature of the planetary forces, his telescopic observations begun in 1609 did more than anything previously had done to break down the conceptual barrier between the celestial and terrestrial domains. His discoveries of a new comet, of four of Jupiter's moons, of spots on the sun, and of the pockmarked surface of the moon served to destroy the prevalent image of an incorruptible cosmos, pure, simple, and perfect.

Finally with Newton the two long and meandering threads of man's concern about motion—the celestial thread and the terrestrial thread—were drawn together in as simple and beautiful and powerful a union as ever occurred in the history of science. Newton formulated mechanics in terms of the three laws that now bear his name, discovered the law of universal gravitation, and showed that these laws accounted for the known features of planetary motion.* This achievement, often referred to as the Newtonian synthesis, capped the intellectual ferment of the seventeenth century, provided a general theory of nature of unprecedented breadth and power, and removed man once and for all from his cherished position at the center of the universe.

12.3 The law of universal gravitation

The historical background of the law of universal gravitation is discussed in later parts of this chapter. Here we wish to state the law and define its terms. Once the target is clearly in view, it will be easier to follow the progress toward it. Consider a pair of particles separated by a distance d, one with mass m_1, and the other with mass m_2 (Figure 12.1). The gravitational force exerted by each on the other is attractive and is central—that is, it acts along the line joining the particles. The magnitude of these equal and opposite forces is proportional to the product of the two masses, and inversely proportional to the square of the distance separating them. It may be written

$$F = G\frac{m_1 m_2}{d^2}; \tag{12.1}$$

the constant of proportionality G is called the gravitational constant. The law of universal gravitation states that precisely this law of gravitational force applies to every pair of particles in the universe, always with the same fixed gravitational constant G. The numerical value of G depends on the units of measurement adopted. In the cgs system, its value is

$$G = 6.673 \times 10^{-8} \text{ dyne cm}^2/\text{gm}^2. \tag{12.2}$$

This small number reflects the extraordinary weakness of the gravitational force. Two equal masses of 1 gm separated by 1 cm attract each other with a force of 6.7×10^{-8} dyne, one million billion (10^{15}) times less than the weight of a man. It is only because of the enormous mass of the earth that its gravitational force seems strong. The action of gravity between any pair of ordinary-sized objects on earth is negligibly small. So far as we know, gravity is also of no consequence in the world of particles. (In the yet-to-be-explored subparticle domain, it is conceivable that the exceedingly small distances may more than compensate for the small masses and small value of G and cause gravity again to become important.)

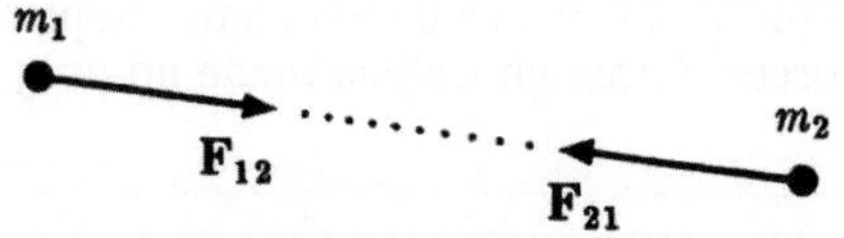

FIGURE 12.1 The gravitational forces acting between a pair of particles are attractive and central (as well as equal and opposite).

* Newton's *Principia* was published in 1687. Some of its key ideas probably go back to 1666, when the 23-year-old Newton left Cambridge to escape a plague and worked at home. He later said of that early year: "In those days I was in the prime of my age for invention and minded mathematics and philosophy more than at any time since."

The most important thing to notice about Equation 12.1 is that it is a law of force, not a law of motion. It specifies the magnitude of the gravitational force independent of how the pair of particles are moving, or even whether they are moving. The masses that appear in the numerator are *gravitational* masses. They are measures of the *strength of interaction.* The gravitational mass of a particle is proportional to the strength of gravitational force it exerts and to the strength of gravitational force it feels. It is a separate experimental fact that gravitational and inertial mass are equal. Actually, it is experimentally meaningful only to say that gravitational and inertial mass are *proportional.* Their *equality* is an arbitrary matter. Expressing the proportionality in more concrete terms, we can say that the resistance of an object to a change in its velocity is always proportional to the strength of its gravitational interaction. Whenever two quantities are always proportional, it is possible to choose their units of measurement in such a way that their numerical values are identical. With the unit of gravitational mass so chosen, the gravitational constant G becomes a measurable quantity. Otherwise the magnitude of G could be selected arbitrarily, a procedure that would constitute a different definition of gravitational mass.

If the factors m_1 and m_2 on the right side of Equation 12.1 are interchanged, the force is unchanged. This mass symmetry is no coincidence. Were it not true, the gravitational-force law would violate Newton's third law. Suppose, for example, we started with the fact that the gravitational force *experienced* by an object is proportional to its own mass. According to Newton's third law, the force exerted by the object is equal in magnitude to the force it feels. Therefore the force it *exerts* is also proportional to its mass. This means that for any pair of objects, the gravitational force must be proportional both to the attracting mass and to the attracted mass. The logic of this argument can be expressed more clearly in a diagram (Figure 12.2).

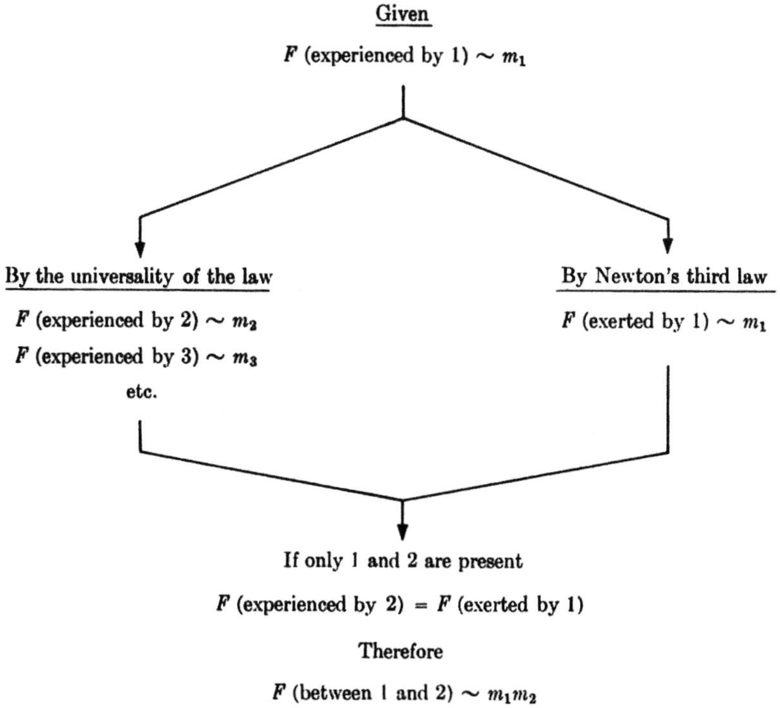

FIGURE 12.2 Logical diagram for the argument that if the gravitational force experienced by an object is proportional to its own mass, and if in addition the force is to be universal and consistent with Newton's third law, then it must be proportional to the product of two masses.

58 Implicit in Equation 12.1, but not obvious, is a superposition principle for gravitational forces. There are two aspects of the superposition principle which can best be explained by example. Consider three particles of different mass, arranged as shown in Figure 12.3. The superposition principle states first that the force exerted on particle 1 by particle 2 ($\mathbf{F}_{12}$) is independent of the position or mass of particle 3, indeed totally independent of the presence of particle 3. This means that Equation 12.1 remains valid for any pair of particles, regardless of their environment. Second, the superposition principle states that the total force acting on particle 1 is the vector sum of the separate contributing forces, in this example $\mathbf{F}_{12} + \mathbf{F}_{13}$. Obviously the superposition principle is a principle of simplicity. It is hard to conceive of any easier way to extend a law of force from two particles to more than two particles.

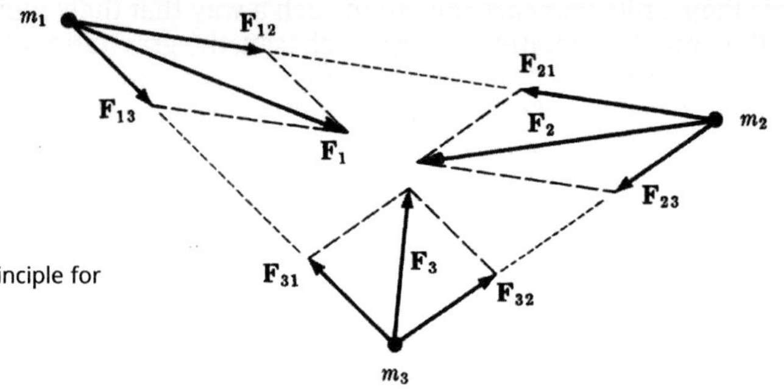

FIGURE 12.3 The superposition principle for gravitational forces.

So far in the discussion of the gravitational force we have limited attention to point particles, idealized objects without spatial extent. Only in this hypothetical limit is the distance d from one object to the other unambiguously defined, and only in this limit is the directional property of the force—along the line joining the particles—unique. Otherwise, we would not know from which point in an extended object to measure distance or direction. In practice nature approximates the idealized situation whenever the size of each of a pair of objects is very much smaller than their separation. Sun and earth, even earth and moon, qualify as very nearly point particles for their mutual interaction, because their separation is vastly greater than their size. On the other hand, man and earth, or satellite and earth, form a system for which the particle approximation is very far from the truth. A man standing on the earth is literally pulled in all directions. Every speck of earth, be it 10 centimeters away or 10,000 kilometers away, is pulling in its own direction, north, south, east, west and down. Above this man the air exerts its own weak gravitational force upward. The myriad of individual gravitational forces, which may as well be considered an infinite number, act in consort. Each alone is an infinitesimal vector quantity; the sum of all of them is the total gravitational force the man calls his weight, directed approximately toward the center of the earth.

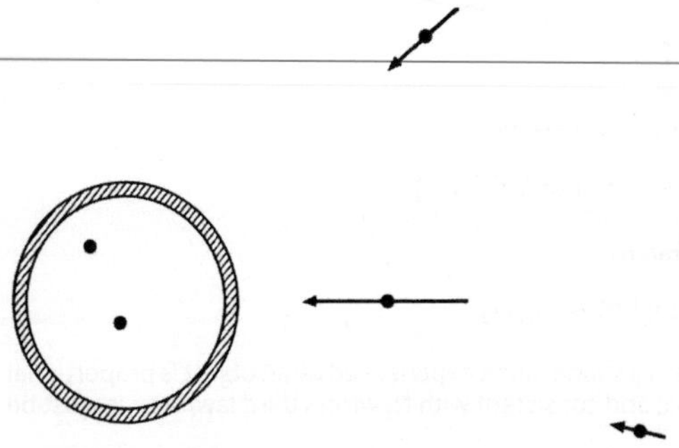

FIGURE 12.4 The gravitational force produced by a spherical shell of mass. Particles outside are attracted as if the mass of the shell were concentrated at its center. Particles inside experience no force.

In the *Principia*, Newton presented the solution to the problem of the gravitating sphere. His exact results are these: (1) A particle outside a uniform spherical shell is drawn toward the shell exactly as if all of the mass of the shell were concentrated at its center. (2) A particle inside the shell is drawn equally in all directions and experiences no force at all (Figure 12.4). Both results are beautifully simple and rather surprising. Particularly interesting is the fact that of all possible laws of force, only the inverse square law obeyed by the gravitational force has such simple properties.

The earth can be likened to an onion of successive spherical shells. A man on the surface is pulled downward by each shell as if all of its mass were at the center of the earth. The force is then calculable by Equation 12.1. The mass m_1 is the mass of the man, the mass m_2 is the mass of the earth, and the distance d is the distance from the man to the center of the earth. The force is the same as if the man were left suspended in space while the earth collapsed to a tiny ball four thousand miles beneath his feet. For calculating the force of gravity on a satellite, the radius of the earth is irrelevant. All that matters is the distance of the satellite from the center of the earth.

Imagine now a hole bored straight through the center of the earth. How would a man's weight vary as he was lowered into the hole? At a depth of 100 m, the outermost spherical shell of the earth, 100 m thick, would cease to exert any net force on him. All the rest of the earth below would still be pulling him downward. At every depth the earth above would be irrelevant, and his weight would be the same as if he were on the surface of a smaller planet. At the center his weight would be zero. In Figure 12.5 is depicted the weight of a man (or of anything else) as a function of altitude and of depth. It is left as an exercise to show that halfway to the center of the earth, his weight is reduced by half.

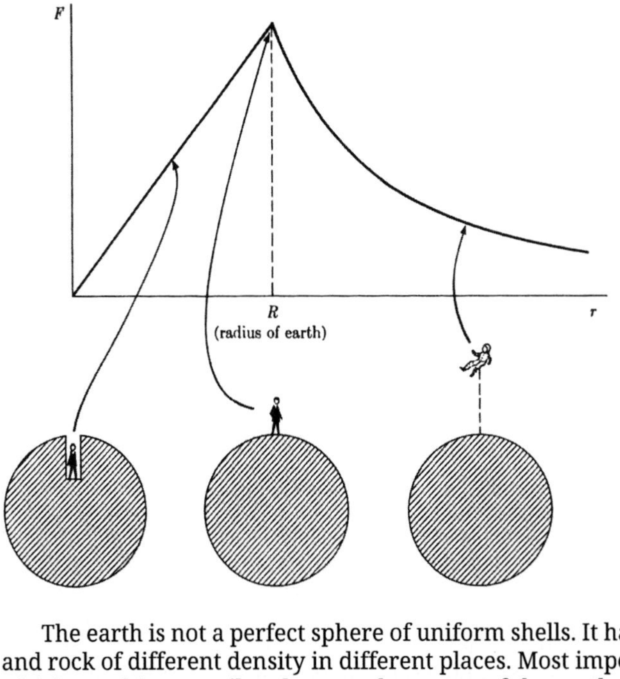

FIGURE 12.5 Weight as a function of distance from the center of the earth. The magnitude of the downward force is designated by *F*, the distance from the center of the earth by *r*.

The earth is not a perfect sphere of uniform shells. It has mountains here, valleys there, and rock of different density in different places. Most important, it is flattened at the poles, which are thirteen miles closer to the center of the earth than is the equator. Newton was aware of these discrepancies between the mathematical calculation and reality, and he actually capitalized on them to provide subtler tests of his theory of gravitation (see Section 12.8). In recent years, satellite orbits have provided new precision in our knowledge of the earth's geography. Because the earth is not a perfect sphere, the force experienced by the satellite is not exactly the same as if the mass of the earth were concentrated at a point. The satellite orbit therefore reflects deviations from sphericity. In 1959 it was discovered that the earth, besides being flattened at the poles, has a very slight pear shape, a minute tendency toward a pointed north-pole head, and the barest hint of jowls about the southern hemisphere (Figure 12.6).

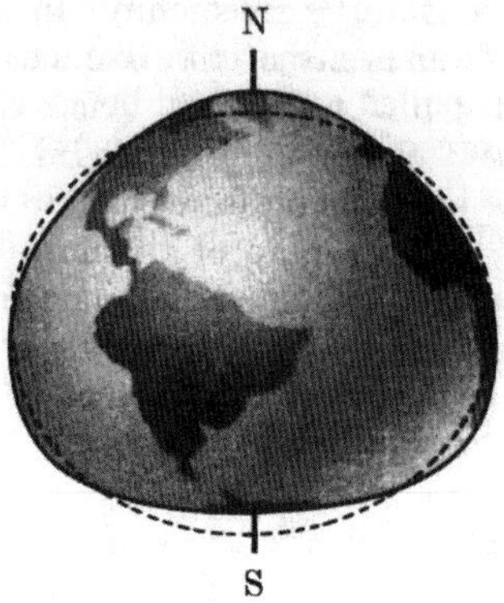

FIGURE 12.6 A highly exaggerated view of the shape of the earth. Besides being flattened at the poles, the earth is very slightly distorted into a pear shape.

12.4 *Humans looked skyward*

59 It is an interesting and at first a seemingly odd fact that the development of mechanics was stimulated more by the study of the heavens than by the study of simple everyday motion on earth. There are several reasons why. First, let it be noted that everyday motion on earth is in fact rarely very simple. The ever present friction accompanying all terrestrial motion complicates matters, and many interesting examples of motion on earth are too short-lived to be easily studied. Celestial motion, by contrast, is durable and frictionless. It was easier for man to accumulate accurate data on the motion of Mars than on the motion of a falling stone. Also the motion of the sun and planets and stars had something to contribute to the practical goals of time-keeping and navigation, whereas the practical fruits of studying terrestrial motions were less obvious. Probably as important as either of these considerations were the psychological and religious factors. The heavens hold for man a natural fascination with which mundane motion cannot compete. A modern man who never thinks twice about the technological wonders that surround him may look skyward with awe, marveling at the structure of the universe. Man has always been concerned about his place in the universe. Even before his conception of the universe expanded to its present enormity, he must have been conscious of himself as a tiny part of the whole. The Greek who believed himself to be at the center of the universe must have been nearly as humbled by thoughts of the giant cosmos surrounding him as is a modern man by thoughts of the inconspicuous corner of the universe assigned to him and his earth.

For the ancients, the study of the heavens had a particular religious significance that it has, by necessity, lost for modern man. They wanted a physical location for the gods and things heavenly. For all ancient civilizations the celestial region was to some extent sanctified. At least until the early seventeenth century the idea persisted that the heavens were made of purer stuff than the earth. Galileo with his telescope struck the most serious blow against the sanctity of the heavens. Newton completed the job of rendering the whole universe earthy.

Even before the time of Aristotle, remarkably accurate data on celestial motion had already been accumulated, and he knew that his system of the world must conform to these data. At the same time there were no quantitative data whatsoever on aspects of terrestrial motion, and Aristotle apparently saw no need to accumulate any. In the centuries after Aristotle, ever more refined measurements of the positions of the sun and moon and planets were made. In step with these, ever more elaborate and complex pictures of the structure of the universe were developed. By the year 1600, the accuracy and extent of astronomical measurements were truly astonishing, and the mechanical structure of man's picture of the universe amazingly complex. At the same time there was as yet no general understanding of uniformly accelerated motion in a straight line, and there were no measurements of motion as simple as that of a swinging pendulum. Galileo was the first to deal carefully with these examples of terrestrial motion.

One other factor may be mentioned that contributed to the remarkable disparity in the degree of advancement of the celestial thread and the terrestrial thread in the study of motion. Before the seventeenth century, science was *observational,* not *experimental.* Men were interested in the planets, so they observed the planets. In modern science, the path to the thing of interest may be much less direct. We approach the subject of mechanics now through spring balances, falling balls, and carts carrying weights not because these things are of much interest by themselves, but because they enable us to formulate concepts and laws which then give us dominion over a vast part of nature, including the planets and other interesting things. Galileo was perhaps the first scientist to see clearly the value of indirection. Before him the path to the planets was direct observation. After him it was indirect, via experiments and theories leading to a scheme of understanding that embraced the planets. Actually, because the knowledge of planetary motion developed sooner and more fully than the knowledge of terrestrial motion, it is equally true to say that the path to an understanding of terrestrial motion was via the planets. When Newton drew the celestial and terrestrial threads together, they reinforced each other. It was the accurate planetary data that provided the best confirmation of the terrestrial laws of motion.

12.5 The system of the world

The modern physicist is seeking knowledge of two kinds. First of all, he wishes to know 60
about the structure of the world—what are its constituent parts, how are they assembled, how do they move and interact? Second, he is looking for theoretical structures that account not only for what is but for what might be. Science before the seventeenth century was concerned almost entirely with knowledge of the first kind. It is true that Aristotle had a grand general scheme of nature, a theoretical framework embracing both the animate and inanimate world. Yet his theory had as its aim only to assemble already known aspects of the world into a coherent picture. What sets apart the knowledge of the second kind gained through modern theories is the power of prediction and power of control that it gives to the scientist. Wholly new kinds of phenomena can be predicted; and the laws of nature can be harnessed for man's own ends.

Aristotle's theory of the solar system was a *specific* theory concerned only with what man can actually see moving in the sky. By contrast, the theory of mechanics together with the law of universal gravitation provides a *general* explanation, embedding the knowledge of our own planetary system in a theoretical explanation of all possible planetary systems. We look upon our planetary system as an accidental special case, one among an infinity of possibilities. There might be systems with more or fewer planets, with different masses, different relative spacings, differently shaped orbits. The same theory that explains the motion of our actual planetary system would explain the motion of any other. This generalization is more than a mere mathematical exercise. Perhaps planetary systems of other suns will actually be discovered. We have created our own planetary system of artificial satellites about the earth, governed by the same laws as are the planets moving around the sun. Because of the new generality, man can *predict* and he can *control.*

Aristotelian physics was not without some powers of prediction. In the Greek explanation of planetary motion, and in the later refined versions of it, the future positions of sun, moon, and planets could be predicted with considerable accuracy. But it was prediction within the closed confines of a single system moving in a certain way. The predictions made possible by modern general theories are of quite a different kind. The behavior of a hypothetical system that has never yet been seen can be predicted. Perhaps man can then arrange to create the system he has predicted, as he has done with earth satellites. Newton, as soon as he had in hand the general theory of mechanics, foresaw the possibility of earth satellites. For technical reasons, of course, not all predictions can be quickly realized; some can never be realized.

Whether the generality of the modern theory is tapped for new inventions or not, it still engenders a characteristic point of view about nature quite different from the ancient view. We look upon our planetary system as something that exists by chance and moves as it does because it happened to get started in a certain way. Conceptually it is in no way unique, because we can imagine other planetary systems and predict their motion, and because the laws governing its motion are the same as the laws governing all motion. According to the Greek view, which persisted for a very long time, the planetary system is a unique system with a unique set of rules to explain its motion. All the emphasis was on finding out by careful observation how the planets do move and then explaining this motion in terms of a mechanism as simple as the observations and human ingenuity and human prejudice permitted. There was no reason why the explanation should explain anything else. It was a closed explanation, with the single goal of explaining a single structure (to be sure, a very elaborate and complex structure). By contrast, the modern explanation is open, enlarged by a general theory to encompass other kinds of motion, real and hypothetical, without limit.

Despite the modern scientist's emphasis on the generality of explanation and on understanding the actual in terms of the hypothetical, he is no less interested than his ancient counterpart in understanding the structure and behavior of the natural world. We think that we understand the solar system more deeply because we see its motion as only one possibility among an infinity of other unrealized possibilities. We are not less interested in the actual solar system on that account.

Aside from satisfying man's curiosity about his celestial neighbors, study of the solar system has contributed in a most important way to the evolution of mechanics. Both for its intrinsic interest and for its role in the development of physics, the solar system deserves our attention.

For probably about 24 centuries man has dared to cope with the problem of understanding the structure of the universe, or, in the somewhat more modest terminology of the seventeenth century, the system of the world. The parts of the world known to the Greeks were the sun, moon, and earth, five planets (Mercury, Venus, Mars, Jupiter, and Saturn), the multitude of stars, and occasional comets. No more constituents of the world were discovered until the late sixteenth century, when a "nova" or new star appeared. For about two thousand years, designers of world systems had an unchanging and seemingly unchangeable list of parts to assemble. The problem of designing a world system was looked upon as a mechanical problem: How can the constituents of the world be fit into a mechanism whose motion reproduces the observed motion? Throughout those two millennia, as the most fashionable or most successful world systems underwent many alterations, four unchanging assumptions dominated almost all thinking on the subject of the system of the world. (1) The earth is at rest at the center of the universe. (2) Celestial motion is based on circular motion. (3) The universe is of finite size with well-defined boundaries. (4) There exists a simple and regular pattern of motion in the heavens which can be described mathematically and understood by man. In short, celestial motion is neither capricious nor unduly complicated.

These assumptions shaped Greek thinking about the system of the world as early as 400 B.C. In 1600, the challengers of some of these assumptions were only beginning to be listened to. By 1700, three of the four assumptions had collapsed.* Only the last, the faith in

* Whether the universe is finite or not is still an open question. Nevertheless, the assumption that it must be finite disappeared from science after Newton.

simplicity, remained, and it remains today, strengthened by three centuries of astonishing progress in science.

What are the raw data of celestial motion? To begin with, everything in the sky seems to move approximately in a circle once each day. Stars near the North Star move in small circles. None of the apparent motion is quite precisely uniform and circular, although the stellar motion is nearly so: The stars move almost as if they were bright spots of paint on the inside of a uniformly rotating spherical bowl. The sun, by contrast, varies its path through the sky and its time of rising and setting each day. Relative to the stars, the sun lags behind by about four extra minutes each day, in addition to its seasonal motion north and south. The moon rises nearly an hour later each day. The near planets, like the sun, move relative to the stars by a few minutes each day; the distant planets change by a lesser amount. If the stars are like paint spots on the inside of a rotating bowl, the planets can be likened to fireflies crawling inside the same bowl. They are carried round and round by the motion of the bowl as they wander slowly among the paint spots. Our word "planet" is derived from the Greek word for wanderer. Usually the slow wandering of the planets among the stars is "upstream"—that is, opposite to the direction of motion of the stars. This means that the planets usually lag a little farther behind the stars each day, as the sun and moon do. But each planet sometimes reverses course relative to the stars and appears earlier each day for a time. This is called retrograde motion. The subtle deviation of the stars themselves from unchanging uniform rotation is an effect known as the precession of the equinoxes. Two points in the sky, one above the north pole, one above the south pole, seem not to move at all as the rest of the sky rotates. This pair of motionless points is not forever fixed, however, but changes slowly over the years. The northern "motionless" point itself moves in a circle among the stars, once around in 26,000 years. In modern terminology, the axis of rotation of the earth precesses so that the ends of the earth's axis point to a gradually shifting pair of points in the sky.

Except for the exact nature of the earth's precession, all the facts of celestial motion just outlined were known to the Greeks. By the fourth century B.C., recorded astronomical data were sufficiently numerous and accurate to inspire the construction of world systems. A useful way to review some of the major systems before Newton is in the context of the basic assumptions about the universe listed on this page. The twenty-centuries-long tenure of these assumptions shaped the history of man's thinking about the universe at large, and thereby also his thinking about the phenomena of his immediate environment.

(1) The Assumption of a Stationary Earth

The motionlessness of the earth is, first of all, "obvious." It does not seem to move. Moreover, for egocentric man, what is more natural than a geocentric universe? Although the Greeks generally preferred a system of the world with a central stationary earth, they did not wholly ignore other possibilities. Heracleides suggested that the earth, although central, was not at rest, but rotated about its axis. Aristarchus, around 280 B.C., proposed a heliocentric (sun-centered) system in which the earth moved about the sun and also rotated about its own axis. However, it was the system with a central stationary earth, whose details were first worked out by Eudoxus (about 370 B.C.) and later refined by Aristotle, that captured the minds of men and overshadowed all other ideas about the universe.

Helped by the perspective of modern knowledge, we might wonder at the psychological pressure that made it seem simpler to assume that thousands of stars moved around the earth every day than that one single earth rotated on its axis. It is like the mother who thinks her marching son is in step while all of his companions are out of step. Undoubtedly, had the Greeks known anything of the distribution of stars in space, they would have concluded that it is the earth that moves, not the stars. Since, in fact, they pictured the stars as being attached to a single sphere with the earth at its center, the choice was between two single rotations. Either the earth rotated, or the stellar sphere rotated. Given this choice, it is not surprising that they picked what the senses directly suggest—a stationary earth and a rotating sphere of stars.

Aristotle did more than cast his vote for a stationary earth. He gave seemingly sound scientific argument to support his view. One argument (first demolished by Galileo) was that on a moving earth a stone thrown straight up should not fall back to the point from which it was thrown, but instead should fall "behind" that point by whatever distance the surface of the earth had moved while the stone was in the air. To put it in human terms, a man throwing a stone vertically upward should not expect to catch it if the earth moves, for while the stone is in the air he is transported by the moving earth to a new point. Arguing against the idea of a rotating earth in particular, Aristotle pointed to the "natural" downward motion of all material near the surface of the earth. If the earth rotated, he said, the natural motion near its surface should be not downward, but around—that is, parallel to the surface. The assumption of a rotating earth is fundamentally inconsistent, he argued, for it requires that a piece of earth on the ground moves naturally *around*, whereas the same piece of earth raised above the ground moves naturally *down*. The reader is afforded an opportunity in the exercises to counter Aristotle's arguments.

These "logical" arguments of Aristotle had a powerful influence on stabilizing the geocentric theory, for they tied his system of the world to his physics, that is, to his ideas on motion and the structure of matter. Of Aristotle's four elements—earth, air, fire, and water—two, he said, have natural motions toward or away from the center of the earth. Earth moves down and fire moves up. This suggested the center of the earth as the center of the universe, the reference point for all motion. The fifth fundamental element, the perfect stuff of the heavens, moved naturally, according to Aristotle, neither toward nor away from the center of the earth, but in circles around this fixed reference point.

Other arguments for a stationary earth were added to Aristotle's. Ptolemy cited the great wind that should rush across the earth if the earth moved. He could see no reason why the tenuous air should be carried along with the earth. Finally there was one truly scientific argument against any orbital motion of the earth, discovered by the Greeks and reemphasized by Ptolemy. This was the absence of any star parallax. If your two eyes are focused on an object near at hand, they are not looking along exactly parallel lines but along lines converging to a point at the object. The nearer the object, the greater the deviation from parallelism. Similarly, if one eye is closed, and the head is moved from side to side, the open eye must alter the direction in which it looks in order to remain focused on a point. Both effects are the same, a change in the angle of looking because of a change in position. This is called parallax. If the object viewed is exceedingly distant compared to the shift in position, the parallax is very small. If a man looking at the moon moves his head one foot to the side, the direction in which he must look to see the moon does not change appreciably. (Try the observation. The moon simply seems to move along with the head.) On the other hand, if two telescopes one hundred miles apart are both pointed at the moon, it is a simple matter to discover that they are not parallel. The Greeks knew that the moon was close enough to exhibit parallax, and from the effect they could discover the distance to the moon, a remarkable achievement of ancient astronomy. Moving from one point to another on the earth was not enough to produce a stellar parallax. But if the earth moved around the sun, they argued, there ought to be a stellar parallax effect over a six-month span of time as the earth shifts its position in space by some two hundred million miles. No such effect could be found. Therefore, argued the Greeks, the earth does not move. We know now that their argument was sound and their observations were good. The trouble came only because the stars, even the nearest ones, are so distant that the semiannual parallax effect is too small to have been detected without modern precision telescopes.* Because the ancients could not accept the idea of stars so enormously distant

* Even today, parallax has been detected only for a few hundred of the closest stars.

compared to the scale of the planetary system, they were forced to embrace the picture of a stationary earth.

There were many reasons why Aristarchus' sun-centered universe withered while Aristotle's earth-centered system took root in the minds of men and flourished. Not least among these was the fact that Aristotle's system was worked out in quantitative detail, while Aristarchus' was not (it could have been, and would have worked as well). Especially after Ptolemy extended and refined the Aristotelian system (A.D. 100), the practical man had nowhere to turn but to the earth-centered system to find exact description and prediction of celestial motion.

Not again until the sixteenth century was a serious effort made to dislodge the earth from its motionless central position. As astronomical observations had improved over the centuries, it was necessary to modify the motion of sun, moon, and planets again and again to fit the facts of observation. With every "improvement" went greater complexity. When taught the latest version of the Aristotelian world system in the thirteenth century, King Alfonso reputedly said, "If the Lord Almighty had consulted me before embarking upon the creation, I should have recommended something simpler." It was the search for something simpler that led Copernicus to try again a sun-centered system of the world. By fixing the sun and allowing the earth to move, Copernicus was able to assign simpler motions to the other planets. He could also account in a natural way for certain observational facts which, in the earth-centered system, were arbitrary. For example, Mercury and Venus never appear far from the sun in the sky, whereas the other planets stray far from the sun. In the Ptolemaic system, there is no particular reason for this difference. In the Copernican system, it could not be otherwise. In 1543, the year of his death, Copernicus' book, *On the Revolutions of the Celestial Spheres,* appeared. Many more decades passed before his ideas were widely accepted.

Galileo accepted the Copernican view of a moving earth, and had some new reasons for doing so. He understood that a moving earth was not inconsistent with known properties of motion on earth. He surveyed the heavens with a telescope and learned that there was no reason to believe in the existence of a more perfect incorruptible substance outside the earth, therefore no reason to believe the earth to be essentially different from any other planet. Interestingly, Galileo did not believe in the elliptical planetary orbits discovered by his contemporary, Kepler. But Kepler's discoveries early in the seventeenth century (we shall return to them below) so convincingly climaxed the simplification begun by Copernicus that the geocentric universe could not hold on any longer. By the middle of the seventeenth century, all thinkers assumed that the earth moved around the sun and turned on its axis. Newton could begin from there.

(2) The Assumption of Circular Motion

Like the motionlessness of the earth, the circular motion of the heavenly bodies is "obvi- 61
ous." The stars seem to move exactly in circles, and the sun, moon, and planets in paths that are very nearly circular. To the Greeks, who had developed geometry to a high degree, it was natural to think about the physical world in geometrical terms. Plato, and Aristotle after him, apparently regarded it as axiomatic that heavenly bodies move in circles, the most perfect (in modern terminology, the most symmetric) of geometric figures. It was a case of rough observations fitting neatly with preconception and so elevating the preconception to the level of absolute truth. This very human tendency to fit observation into a preconceived framework is still common enough in ordinary affairs, and is especially noticeable in political fanaticism. It is no longer so common in science. It must be added, however, that there is a bit of the fanatic in many a successful scientist. Einstein's belief in universal invariance was no less a preconception than Aristotle's belief in perfect circles. A desire to mold nature to man's image of what it should be can be a powerful stimulus to creative science, provided that the scientist remains sensitive to conflicting evidence and even seeks it out, becoming the critic of his own prepared position.

According to Aristotle, every kind of matter had its own kind of natural motion. The natural motion of the perfect celestial matter was the perfect circle, whose center coincided with the center of the earth, which was in turn the center of the universe. Even before Aristotle, a circular motion in the heavens was widely accepted. However, a single fixed circle for the sun, another for the moon, and one for each planet could not fit the facts. Eudoxus, the first person known to develop a world system with quantitative care, assumed that each observed path is compounded of a number of different circular motions. He imagined a set of transparent spheres of different size, nested one within the other, all centered at the earth's center (Figure 12.7). Attached to the innermost sphere was the moon. This sphere rotated at a constant rate about an axis whose ends were attached to the next larger sphere. The motion of the second sphere about its own axis caused the axis of the inner sphere to wobble, so that the moon's motion was not a simple circle. The axis of the second sphere in turn was attached to a rotating third sphere, and so on outward. Only the outermost sphere, to which were attached the stars, executed simple rotational motion about a fixed axis. Most spheres contained nothing visible. Between the moon's sphere and the sun's sphere were two empty spheres; between Mercury and Venus were three empty spheres. Eudoxus found that he could adequately fit the facts with 27 spheres. Even this large a number did not suffice for long, for discrepancies between the model and observations soon appeared. Aristotle accepted the basic picture of the universe used by Eudoxus, but added 28 more spheres to bring the total to 55.

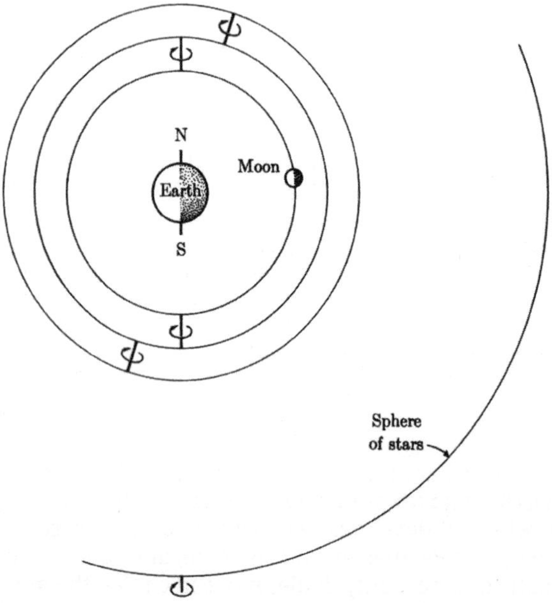

FIGURE 12.7 The world system of Eudoxus. This figure is not to scale, nor does it accurately depict the angles of the various axes. The innermost sphere contains the moon, the outermost sphere the stars. The sun and five planets occupy six of the intervening 25 spheres, the rest being empty.

62 In about A.D. 140 Ptolemy made a departure from the concentric spheres of Aristotle's world system.* His new scheme, which permitted greater flexibility in matching the model to the observations, held on as the accepted system of the world for nearly 1,500 years. Like Aristotle's universe, Ptolemy's had a stationary earth at its center and a uniformly rotating sphere of stars at its outer periphery. In between, things were rather differently arranged in the two pictures of the world. Aristotle's system, conceptually simpler (but not less com-

* Ptolemy's alterations of Aristotle's world system was not entirely new, but Ptolemy was the first to develop the modified system with quantitative care.

plex for calculation) , contained numerous concentric spheres, with the visible objects in the sky attached to only a few of them. Ptolemy, first of all, swept away the invisible spheres between one planet and another, replacing the wheels within wheels by wheels *on* wheels (Figure 12.8). The sphere carrying a planet was centered not at the center of the earth, but on another sphere, whose center in turn might be on still another sphere or might be at a point in space at or near the earth. The major reference sphere for a planet was called the *deferent*. As it rotated, it carried on its back one or more rotating spheres or circles called *epicycles*. The planet moved at a constant rate around a circle whose center in turn moved around another circle. The reference circle whose center did not move (the deferent) was in general *eccentric*, that is, its center did not coincide with the center of the earth. Both the eccentricity and the epicyclic motion, of course, caused the planet (or the sun or moon) to vary its distance from the earth as it moved. Still another innovation of Ptolemy allowed him another source of variation of planetary speed. He imagined a point within the deferent circle, but not at its center, which served as the hub of rotation. A line drawn from this point, called the *equant*, to the moving center of an epicycle, rotated at constant angular speed (Figure 12.8). Despite its complexity, the Ptolemaic system must be admired for its ingeniousness and flexibility. It could and did grow to meet the needs of successively refined observations. Notice that three Aristotelian principles were retained unchanged: (1) The earth is central and stationary. (2) Celestial motion is composed of circular motion. (3) Celestial motion is characterized by uniform rates of rotation. Nothing in the Ptolemaic system was inconsistent with Aristotelian physics or philosophy.

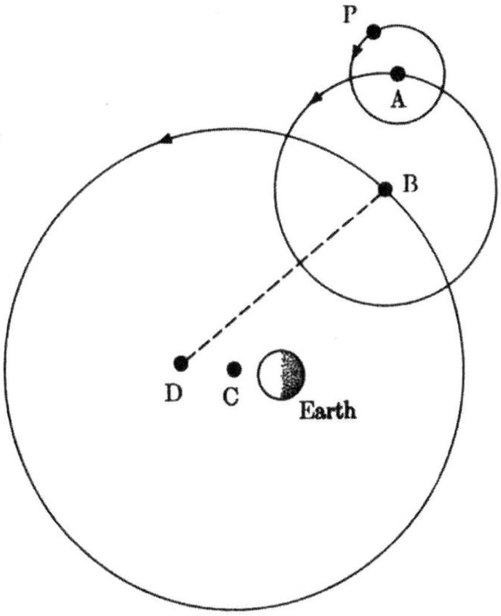

FIGURE 12.8 The mechanism of Ptolemy. A planet P moves at a constant rate in a circle about point A, which in turn might move uniformly in a circle about another point B. The point B might move with constant speed in its circle centered at C, or it might move with constant angular velocity about some other point D. The earth, although not at the center C, remains at the center of the outermost stellar sphere.

Copernicus, despite the revolution he triggered by shifting the center of the universe from the earth to the sun, abandoned neither circles nor uniform rotation. Indeed, one of the motivations of his work was to get rid of the equants of Ptolemy and get the centers of rotation back to the centers of the circles. He did find it necessary to use eccentrics and epicycles; but with fewer and smaller epicycles, Copernicus' planetary system was simpler than the unwieldy and intricate mechanism into which the Ptolemaic universe had evolved. Copernicus' system had certain other appealing features. The correlation between the retrograde motion of the planets and their position relative to the sun received a simple explanation lacking in the Ptolemaic system. And in the Copernican scheme there appeared a new regularity: The farther is a planet from the sun, the longer is its period of rotation about the sun.

Late in the sixteenth century, new standards of accuracy in astronomical observation established by Tycho de Brahe revealed failures in the Copernican scheme which might have led to more patchwork and more complexity, but instead led to the magnificent breakthrough by Kepler to the beautifully simple view of the planetary system that we still have today. Brahe himself, unwilling to abandon either circles or a stationary earth, tried to devise a new system in which Mercury and Venus rotated around the sun and the sun and other planets rotated around the earth. But success came to Brahe's assistant, Johannes Kepler, who had the courage to break loose from the ancient fixation on circles and uniform rotation. By abandoning what had seemed the simplest and most perfect basis of celestial motion, he created a world system far simpler even than the earliest approximation of Eudoxus with its 27 spheres. Gone were the equants and epicycles. Each planet traced a single ellipse in the sky with the sun at one focus, and moved according to a rule that was later recognized as equivalent to the conservation of angular momentum.

Galileo, although he applauded his contemporary Kepler, was unable to give up the circular basis of planetary motion. Both campaigned for the Copernican view of a sun-centered universe, a view that (among other heretical views) brought death by burning to Giordano Bruno in 1600, and trial and imprisonment to Galileo in 1633. The revolution begun by Copernicus moved at a snail's pace for many decades, and with a quickening pace after the work of Brahe, Kepler, and Galileo. By the middle of the seventeenth century, the Keplerian system of the world was widely accepted.

(3) The Assumption of a Finite Universe

The assumption that the stars lay on a sphere beyond which there was nothing (or heaven) was a consistent part of astronomical thinking up to the seventeenth century. In the development of mechanics, this assumption was important only insofar as it tended to support the geocentric system of the world. It did so in two ways. Psychologically, it was easier to think of man as the center of the universe and the whole reason for creation in a world that did not stretch on forever in all directions. Scientifically, the assumed finiteness of the universe, coupled with the absence of any stellar parallax, suggested that the earth did not move, for if it did its changing distance from the stars should be detectable. Actually the finiteness of the universe did not have to be abandoned when the geocentric theory was abandoned. Copernicus simply enlarged the universe to accommodate a moving earth. He supposed that the sphere of the stars, although still finite, was so vast in extent that the motion of the earth around the sun was relatively insignificant. The diameter of the earth's orbit, he argued correctly, must be so small compared to the distance to the stars that the parallax effect is unobservable. So the Copernican universe, although much less cozy than the Ptolemaic universe, was still bounded. Only with Newton did the vista of a possibly endless universe open up.

(4) The Assumption of Simplicity and Regularity

63 Too often we think that science began in the seventeenth century after millennia in which superstition and blind faith reigned. The evolution of man's thinking about the system of the world clearly shows otherwise. The faith in simplicity, the assumption that mathematics describes nature: these essential attributes of modern science were equally parts of ancient science. The concentric spheres of Plato and Aristotle were simple and their motion regular, but they did not hold up in the face of new evidence.

For the development and growth of science, criteria for *rejection* of an idea may be even more important than criteria for *acceptance*. Ptolemy rejected and revised Aristotle's picture of the world in order to get rid of the excessive number of intermediate spheres. The accumulated complexity of the later Ptolemaic picture led Copernicus to reject it in favor of a simpler sun-centered model. Kepler found the Copernican universe still excessively complex (as well as insufficiently accurate) and rejected it even before he had discovered the true laws of planetary motion. The necessary precondition for his great step forward was dissatisfaction with existing explanations. The groundwork for modern science was laid by men who sought simplicity and believed in the mathematical basis of natural phenomena.

12.6 *Kepler's laws*

Newton's laws are statements about motion in general and forces in general. Kepler's laws 64
are of quite a different kind. They are statements about the motion of one single system, the planetary system. Newton's and Kepler's laws differ also in the number of concepts they draw together. In modern terminology, Newton's laws are *dynamic*, connecting mass, force, distance, and time. Kepler's laws are *kinematic*, concerning only distance and time. Kepler's laws are best looked upon as summarized observation. They distill and neatly package a myriad of observations, converting what would otherwise be long tables of numbers with much substance and no form into a few beautifully simple relationships. Newton's laws, instead of *summarizing* a particular set of observations, *generalize* from observation. They connect a few basic concepts for all motion and all systems. Despite the fact that they are both called "laws," the nature and intent of Newton's laws and Kepler's laws are quite different.

Laboring for many years over the astronomical data of Brahe, Kepler was forced in his double search for simplicity and for precision to abandon Copernicus, to abandon circles, and to abandon uniform rotation. His efforts were crowned by the discovery of the three laws of planetary motion that now bear his name. The first two, published in 1609, describe the shape of planetary orbits and the variations of speed of a planet as it executes its orbit. The third, published in 1619, relates one planetary orbit to another. Kepler's laws remain as valid today as when they were discovered.

Kepler's First Law

The path traced out by each planet is an ellipse with the sun at one focus.

An ellipse is an oval-shaped closed curve, roughly describable as a flattened circle. But it is not just any such curve. It is a very special one which can be precisely defined in several ways. First to give some pictorial, but nevertheless exact, definitions: (1) If a plane intersects a circular cone in such a way that the intersection is a closed curve, that curve is an ellipse (Figure 12.9). Because of this definition, the ellipse is called a conic section. Other conic sections, as shown in Figure 12.9, are the parabola, the hyperbola, and the circle. The parabola and hyperbola are open curves. The circle is, in fact, an ellipse, a special ellipse in the same way that a square is a special rectangle. (2) The shadow cast on a flat surface by a circle exposed to parallel rays of light is an ellipse (Figure 12.10). (3) Another definition of the ellipse is easier drawn than said (Figure 12.10). If a pencil point moves in a plane in such a way that the sum of. its distances from two fixed points in the plane is constant, the curve it traces out is an ellipse. These two fixed points within the ellipse are called its foci.

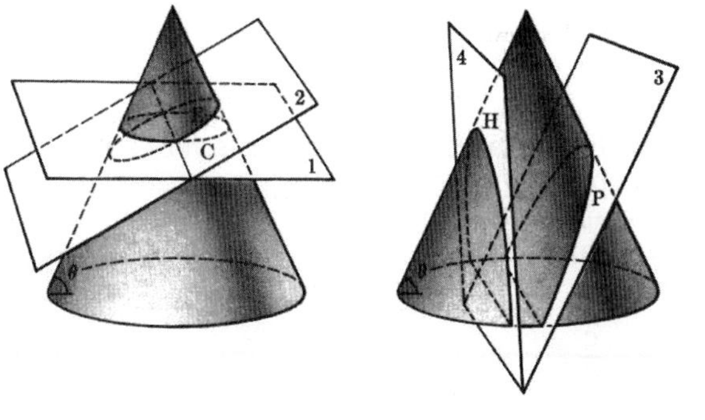

FIGURE 12.9 Conic sections. The edge of the cone makes an angle θ with the horizontal. The intersection of a horizontal plane (1) with the cone defines a circle (C). A plane inclined to the horizontal at less than the angle θ (2) intersects the cone in an ellipse (E). Planes inclined to the horizontal at angles equal to θ (3) and greater than θ (4) define the parabola (P) and the hyperbola (H).

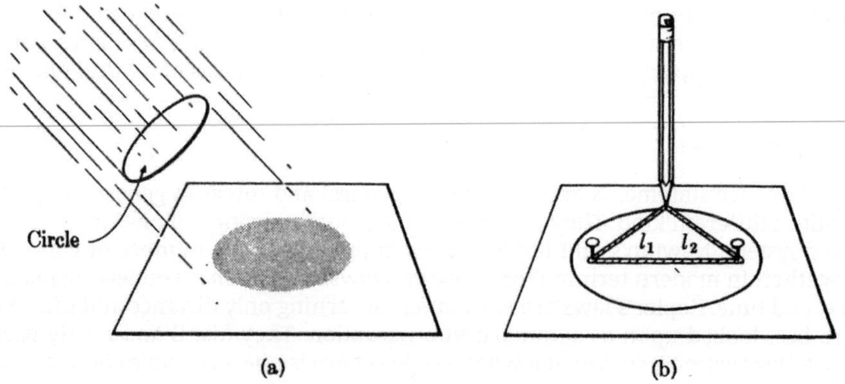

FIGURE 12.10 Other definitions of the ellipse. **(a)** The shadow of a circle intercepting parallel beams of light is an ellipse. **(b)** Points, the sum of whose distances, l_1 and l_2, from fixed points is constant, define an ellipse.

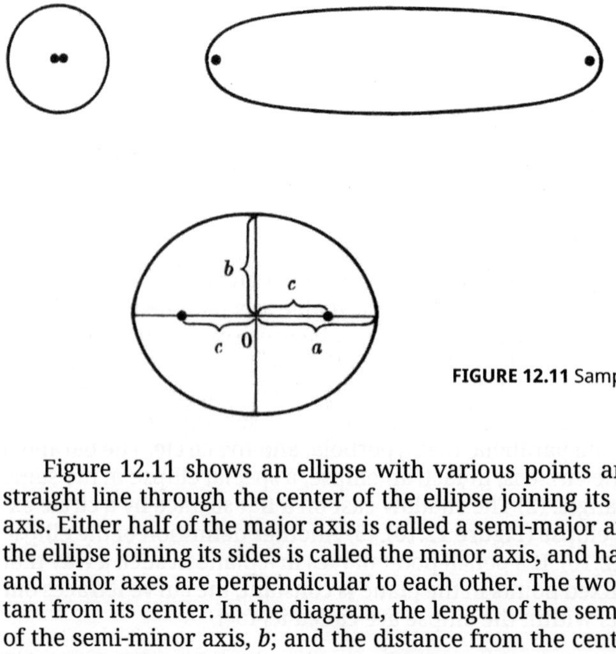

FIGURE 12.11 Sample ellipses. Heavy dots locate the two foci.

Figure 12.11 shows an ellipse with various points and distances labeled. The longest straight line through the center of the ellipse joining its opposite ends is called the major axis. Either half of the major axis is called a semi-major axis. The line through the center of the ellipse joining its sides is called the minor axis, and half of it the semi-minor axis. Major and minor axes are perpendicular to each other. The two foci lie on the major axis equidistant from its center. In the diagram, the length of the semi-major axis is called a; the length of the semi-minor axis, b; and the distance from the center of the ellipse to either focus, c. For every ellipse, there is a simple algebraic connection among a, b, and c:

$$c^2 = a^2 - b^2 . \tag{12.3}$$

For a "fat" ellipse (nearly a circle), a is only slightly larger than b, and c is very small. The two foci are close together. (For a circle, a and b are equal, c is zero, and the two foci coalesce at the center.) For a "thin" ellipse, whose major axis is much longer than its minor axis (a much greater than b), the foci are widely separated near the ends of the ellipse. One other parameter is used to characterize an ellipse. It is the eccentricity e, defined by

$$e = \frac{c}{a} . \tag{12.4}$$

The planetary orbits all have rather small eccentricities. Some comets have large eccentricities (that is, e nearly equal to 1, which is the maximum value) and they sweep very close to the sun at one end of their orbit, very far from it at the other end. What is the eccentricity of a circle?

We add one mathematical definition of the ellipse to the three given on page 299, not because it contributes more to the understanding of Kepler's first law, but because it illustrates the variety of approaches to this interesting curve, the ellipse. (4) If an ellipse is placed with its center at the origin of a coordinate system and its major and minor axes along the x and y axes, respectively, the curve is defined by the equation,

$$\frac{x^2}{a^2} + \frac{y^2}{b^2} = 1 . \tag{12.5}$$

In this equation, a and b are fixed constants, as defined above, characterizing the particular ellipse. The variable coordinates x and y locate any point on the curve. For example, if y is zero, x must be equal to a in order to satisfy the equation. This locates the rightmost point on the curve. At $x = 0$, $y = b$, the topmost point on the curve.

Newton first proved, using his second law and the law of gravitational force, that a planet could, if it happened to be started in the right way, move along an elliptical orbit of any shape or size whatever. In fact he proved even more: that all of the possible trajectories are conic sections. A sufficiently energetic planet or comet could move along a parabolic or hyperbolic path and escape from the solar system. Now that man has created a new planetary system of satellites about the earth, he has created a much wider range of elliptical shapes than were originally recognized in the solar system. He has also sent vehicles into escaping orbits along hyperbolic paths relative to the earth.*

Kepler's first law, it should be noted, is purely a geometrical law. It refers only to the spatial aspect of the planetary orbit, not to time or speed or any other concept. Yet what a great step it was after two thousand years of circles. With the single statement that planets move along ellipses, Kepler swept away epicycles, equants, and deferents, vastly simplifying the picture of the solar system, and at the same time achieved better concordance between observation and calculation than the most elaborate previous scheme had achieved.

Kepler's Second Law

Kepler's second law is the law of areas, discussed in Section 10.7. As a planet moves along its elliptical orbit, its speed varies in such a way that the radial line connecting the sun to the planet sweeps out area at a constant rate. This means that the planet moves more rapidly when close to the sun, more slowly when far from the sun.

Kepler's second law replaces the assumed constant rate of rotation—either about the center of a circle or about an equant—that characterized all of the earlier systems of the world. Kepler's rule for variable speed seems at first more complicated. Yet it does rest on a constant, the constant rate of sweeping out area. As we now know, the really significant underlying constant is angular momentum. The speed of the planet along its orbit varies in just such a way that its angular momentum never changes.

The application of Kepler's second law is especially simple at apogee and perigee, the points of greatest and least distance from the center of force, where the planet's velocity is perpendicular to the radial line. At these two points, the products of speed and radial distance are equal:

$$v_{\mathrm{a}} r_{\mathrm{a}} = v_{\mathrm{p}} r_{\mathrm{p}} . \tag{12.6}$$

The subscripts a and p refer to apogee and perigee. Thus the ratio of speed at perigee to

* No space vehicle has so far escaped from the solar system. Once free of the earth, the interplanetary vehicles settle into elliptical orbits about the sun.

speed at apogee is given by

$$\frac{v_p}{v_a} = \frac{r_a}{r_p} . \tag{12.7}$$

For the earth, r_a = 152 million km and r_p = 147 million km. The earth's greatest speed (at perigee) divided by its least speed (at apogee) is

$$\frac{v_p}{v_a} = \frac{1.52 \times 10^{13}\ \text{cm}}{1.47 \times 10^{13}\ \text{cm}} = 1.034 . \tag{12.8}$$

Kepler's second law deals only in ratios of speeds. It does not reveal the speed at any point along the orbit unless the speed at another point is already known. The measured minimum speed of the earth is

$$v_a = 2.94 \times 10^6\ \text{cm/sec}. \tag{12.9}$$

As previously discussed in Section 10.7, the more general mathematical statement of the law of areas is

$$rv_\perp = constant . \tag{12.10}$$

The subscript ⊥ denotes the component of velocity perpendicular to the radial line. The constant in Equation 12.10 is different for each planet, and there is no special relationship among the constants for different planets. Physically, the constant is equal to the angular momentum of the planet divided by its mass.

Kepler's second law is a *kinematic* statement. It utilizes the fundamental concepts of space and time and is concerned with planetary speed. Like the first law, it is valid for each orbit separately but establishes no connection among different orbits. With earth satellites, Kepler's second law has been verified many times over. Although Kepler did not recognize it as such, his second law of planetary motion was the first great conservation law of mechanics to be discovered.

TABLE 12.1 Planetary Data

Name of Planet	Mean Distance from Sun*		Period		Eccentricity of Orbit	Mass of Planet		Equatorial Radius of Planet		Acceleration of Gravity at Planet Surface‡
	cm	A.U.†	sec	Earth years		gm	Earth masses	cm	Earth radii	
Mercury	5.79×10^{12}	0.387	7.60×10^6	0.241	0.2056	3.25×10^{26}	0.054	2.42×10^8	0.38	360
Venus	1.08×10^{13}	0.723	1.94×10^7	0.615	0.0068	4.87×10^{27}	0.815	6.10×10^8	0.96	870
Earth	1.496×10^{13}	1.000	3.156×10^7	1.000	0.0167	5.98×10^{27}	1.000	6.38×10^8	1.00	980
Mars	2.28×10^{13}	1.524	5.94×10^7	1.881	0.0934	6.46×10^{26}	0.108	3.38×10^8	0.53	375
Jupiter	7.78×10^{13}	5.203	3.74×10^8	11.86	0.0482	1.90×10^{30}	317.8	7.14×10^9	11.19	2450
Saturn	1.43×10^{14}	9.53	9.30×10^8	29.46	0.055	5.69×10^{29}	95.2	6.04×10^9	9.47	1000
Uranus	2.88×10^{14}	19.3	2.65×10^9	84.01	0.051	8.67×10^{28}	14.5	2.38×10^9	3.73	900
Neptune	4.50×10^{14}	30.1	5.20×10^9	164.8	0.008	1.03×10^{29}	17.2	2.22×10^9	3.49	1480
Pluto	5.94×10^{14}	39.7	7.84×10^9	248	0.251	5×10^{27} ?	0.8 ?	3×10^8	0.47	?

* The planet's mean distance from the sun is equal to its semi-major axis.

† Earth's mean distance from the sun is defined to be one astronomical unit (A.U.).

‡ Numbers in this column give average acceleration relative to rotating planet surface.

Kepler's Third Law

If a denotes the length of the semi-major axis of a planetary orbit, and T denotes the time required for the planet to complete one revolution about the sun, then

$$T^2 \sim a^3 \,. \tag{12.11}$$

The square of the period is proportional to the cube of the length of the semi-major axis. Alternatively, Kepler's third law may be written,

$$\frac{T^2}{a^3} = constant\,. \tag{12.12}$$

The ratio, T^2 divided by a^3, is a constant, the same for every planet.

For many years, Kepler had sought to supplement his first two laws with a third that would relate one planetary orbit to another. Since Copernicus, it had been recognized that planets successively more distant from the sun had successively longer periods. Kepler was the first to establish the quantitative relationship between times and distances. Both the form of the law and the numerical value of its constant of proportionality were for Kepler empirical discoveries, unrelated to any deeper rules of planetary motion. It remained for Newton to connect Kepler's third law with the law of universal gravitation. In so doing, Newton showed that the constant in Equation 12.12 is not a universal constant, but is fixed for any particular center of force. It has one value for all of the planets and comets rotating around the sun, another value for all of the moons of Jupiter, and another value for all satellites of the earth, including our moon. For any particular planetary system, Kepler's third law provides a link between different possible orbits.

The principal properties of the sun's nine known planets are summarized in Table 12.1 (see also Table 10.2).

12.7 *Deduction of the law of gravitational force*

In this section we are concerned with Newton's use of Kepler's laws to deduce the law of 65
gravitational force. An important thing to understand first is why Newton was convinced that a mechanical explanation of planetary motion was needed. For Aristotle and for Copernicus, circular motion was the "natural" motion in the heavens. It was the way stars or planets moved when left to themselves (and obviously man had no opportunity to do otherwise than leave them to themselves). It required no further explanation. Kepler and his English contemporary William Gilbert were perhaps the first to think seriously about a force to explain planetary motion. Both imagined that the force might be magnetic. But their speculations could not lead far, for neither understood the principle of inertia, that an undisturbed object continues to move uniformly in a straight line. They imagined that a force should act along the planetary orbit, to impel the planet continually along its path through space. Galileo's attitude toward planetary motion was most interesting, for while he took giant strides forward in understanding mechanical principles and in understanding gravity, he reverted to a view of natural circular motion scarcely more sophisticated than Aristotle's. To a full understanding of inertial motion, now expressed by Newton's first law, Galileo came perilously close. But he did not reach it. He was the first to state that undisturbed motion ("natural" motion) is motion in a straight line with constant velocity. This indeed sounds like Newton's first law. However, Galileo believed this law to be limited in validity to the macroscopic human-sized domain. He pointed out that an object set sliding on an enormous frictionless plane in contact with the earth at one point (Figure 12.12) would not continue indefinitely at constant velocity. As it moved farther from the center of the earth it would be decelerated; what started out as horizontal motion would become uphill motion. On the other hand, on an imaginary perfectly smooth sphere surrounding the earth, an object

set sliding would continue at constant speed in a great circle around the earth. Because he could not free himself mentally from the shackles of the earth's gravitational field, Galileo failed to extend the principle of inertia to the heavens. He succeeded only in advancing new arguments for the ancient view that natural celestial motion is circular motion. Perhaps if Galileo had accepted Kepler's ellipses, he would have reached different views about inertial motion in the cosmos.

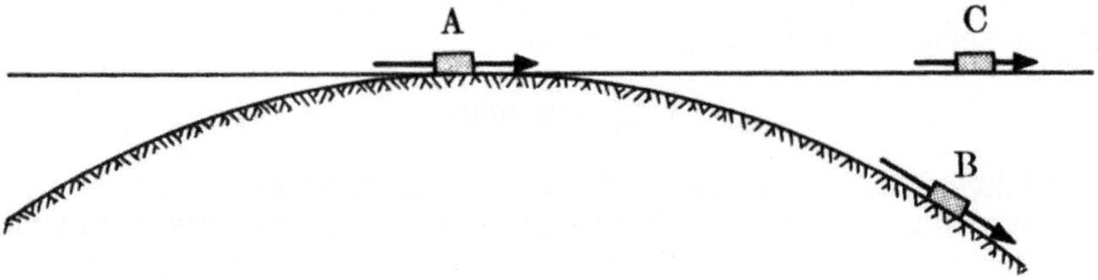

FIGURE 12.12 Galileo's view of frictionless undisturbed motion. An object sliding without friction around the surface of the earth (A to B) maintains constant speed, whereas an object sliding without friction in a straight line (A to C) is decelerated.

Newton's imagination was able to cut loose from the bonds of earthly gravity that had held back Galileo. He unhesitatingly extended Galileo's principle of inertial motion on to infinity and stated it as his first law of mechanics. This was a necessary first step to the discovery of the law of universal gravitation. If the "natural" motion of a planet in the absence of force is straight-line motion, the actual orbital motion requires an explanation. A force must be acting to deflect the planet from its otherwise straight course into its curved path around the sun.

In the five subsections below, we trace a line of reasoning from Kepler's laws to the law of gravitational force. This should be taken as plausible reconstruction of Newton's work, not as authoritative history. There is some evidence about the early development of Newton's ideas in his later correspondence, but Newton himself may have been rewriting the events. Insights in science are much more likely to occur in a patchwork of intuition, deduction, guesswork, and calculation than in any logical orderly chain. Not the least important element is the preconception, the belief that the solution must be found along a certain path.

(1) Kepler's Second Law and the Direction of the Force

Before the time of Newton, the kinematics of uniform circular motion was not understood. As described in Section 6.7, the essential facts are that the acceleration is directed inward toward the center of the circle and has a constant magnitude given by

$$a = \frac{v^2}{r},$$

where v is the speed of the object and r is the radius of the circle. These facts about uniform circular motion were first published by Christian Huyghens in 1673, but they were probably known to Newton in 1666. They mean that if an object moves uniformly in a circle, it must experience a force directed toward the center of the circle. (Here we use the proportionality of force and acceleration.) This conclusion was all-important to the progress in astronomy, for it suggested that the planets move as they do because of a force pulling them toward the sun. Newton put this reasonable guess on a firm footing by proving a theorem not just about uniform circular motion, but about any orbital motion at all that is executed in a plane. In Section 10.7 we proved that an object acted upon by a central force obeys Kepler's second law (the law of areas). Because it experiences no torque with respect to the center of force, it has constant angular momentum with respect to this point, and constant angular momentum in turn implies that the radial line is sweeping out area at a constant rate. To put it in planetary terms: If the sun exerts a central force on a planet, the planet moves in such a way as to satisfy Kepler's second law. This is the theorem that Newton proved, along with its converse: If a line drawn from the sun to an accelerated planet sweeps out area at a constant rate, the force acting on the planet must be directed toward the sun. The essential point is the connection between the central force and Kepler's second law. Using only one of Kepler's laws of planetary motion, Newton could prove the vastly important result that the planets are all acted upon by a central force directed toward the sun.

(2) Kepler's Third Law and the Dependence on Distance and Mass

Kepler's second law (together with Newton's second law) revealed the *direction* of the gravitational force, but no other property of the force. The next problem was: How does the strength of the sun's gravitational pull vary as the distance from the sun varies? There were two ways to get at this problem, both of which Newton employed. The first, our concern in this subsection, was to compare the motion of different planets at different distances from the sun. The second, discussed in the next subsection, was to study the motion of a single planet as its distance from the sun varies during the course of its orbit. For comparing the motion of different planets, Newton had at hand Kepler's third law, stating that the squares of the periods of the planets are proportional to the cubes of the semi-major axes of their elliptical orbits (Formula 12.11). Introducing a constant of proportionality K, we can write this law in the form

$$T_{\mathrm{p}}^2 = Ka_{\mathrm{p}}^3 . \tag{12.13}$$

The subscript p denotes any particular planet. The proportionality constant K carries no such subscript since it is the same for all the planets.

Now we want to make one very important approximation which Newton undoubtedly made in his earliest calculations. We take advantage of the fact that none of the planetary ellipses differs greatly from a circle, and pretend that the planets all move precisely in circles of radius r_{p}. Kepler's third law for these circles may be written

$$T_{\mathrm{p}}^2 = Kr_{\mathrm{p}}^3 . \tag{12.14}$$

This approximation enables us to carry out a simple and significant mathematical derivation. Each planet, moving approximately at constant speed v_{p} in its approximate circle of radius r_{p} has inward acceleration a (not to be confused with the distance a_{p} in Equation 12.13) given by

$$a = \frac{v_{\mathrm{p}}^2}{r_{\mathrm{p}}} . \tag{12.15}$$

In order to take advantage of Kepler's third law, we replace the speed v_{p} in this expression by the orbital circumference divided by the period:

$$v_{\mathrm{p}} = \frac{2\pi r_{\mathrm{p}}}{T_{\mathrm{p}}} . \tag{12.16}$$

(Speed equals distance divided by time.) The formula for the planet's acceleration then becomes

$$a = \frac{1}{r_{\mathrm{p}}}\left(\frac{2\pi r_{\mathrm{p}}}{T_{\mathrm{p}}}\right)^2 = \frac{4\pi^2 r_{\mathrm{p}}}{T_{\mathrm{p}}^2} . \tag{12.17}$$

This is so far a kinematic statement about uniform circular motion in general. Kepler's third law as expressed by Equation 12.14, on the other hand, is an observational fact about the planets. Substitution of Equation 12.14 into the denominator of Equation 12.17 gives a formula for planetary acceleration:

$$a = \frac{4\pi^2}{K}\frac{1}{r_{\mathrm{P}}^2}. \tag{12.18}$$

The combination $4\pi^2/K$ is a constant; the quantities a and r_{p} are variables. Equation 12.18 states that the acceleration of a planet toward the sun is inversely proportional to the square of its distance from the sun. Multiplication of the acceleration of a planet by its mass gives the force acting on it:

$$F_{\mathrm{P}} = \frac{4\pi^2}{K}\frac{m_{\mathrm{P}}}{r_{\mathrm{P}}^2}. \tag{12.19}$$

Compare this with the law of gravitational force stated by Equation 12.1. Here, derived from Kepler's third law, is a major part of the final form of the law of force. Equation 12.19 states that the gravitational force experienced by a planet is proportional to its mass and inversely proportional to the square of its distance from the sun. All that is missing is the proportionality of this force to the mass of the sun.

Two comments about Equation 12.19 are in order. First, recall that the derivation was based on the approximation of circular motion. Therefore we cannot be certain that the result is exactly correct. It turns out, remarkably enough, that the result is unchanged for elliptical orbits. Newton was able to demonstrate the connection between Kepler's third law and the inverse square radial dependence of the acceleration for the true elliptical orbits, not just for the circular approximation.

The second comment is more subtle. It appears that we have proved—at least in circular approximation—that the sun's force on a planet is proportional to its mass and inversely proportional to the square of its distance from the sun. This is not quite true. Suppose, for example, that the mass of a planet were proportional to its distance from the sun. Then the combination $m_{\mathrm{p}}/r_{\mathrm{p}}^2$ would also be proportional to $1/r_{\mathrm{p}}$. For this hypothetical situation, it would be equally true to say that the force varies inversely with the distance and not at all with the mass. Newton probably assumed that there is no particular connection between planetary mass and distance (as indeed there is not), so that the form $m_{\mathrm{p}}/r_{\mathrm{p}}^2$ appearing in Equation 12.19 is not masking some different radial dependence. The clinching evidence for the direct proportionality of gravitational force to mass (and hence for the correctness of Equation 12.19) comes from the idea of the *universality* of the gravitational force. Near the surface of the earth, every object experiences a downward force proportional to its own mass, $F = mg$. Newton assumed this dependence on mass to be a universal feature of gravitation, not limited to gravity on earth. Equation 12.19, concerned with the force of the sun on the planets, confirms and strengthens this assumption. Still another bit of evidence on the same subject appears in the next subsection.

(3) Kepler's First Law: More Evidence on the Radial Dependence

The idea of an inverse square law of gravitational force acting on the planets was in the air around Newton's time. His genius lay not so much in thinking of it as in demonstrating its validity mathematically and weaving it into a coherent theory of universal gravitation. The really crucial test of the inverse square law is provided by Kepler's first law, the statement that planets move in elliptical orbits with the sun at one focus. In any orbit that is not a circle, a planet periodically alters its distance from the sun, sampling stronger and weaker regions of the sun's gravitational field as it moves around. The precise form of its orbit, therefore, depends on the law of force, on exactly how the force weakens as the distance increases. In London in 1684, Edmund Halley, Robert Hooke, and Sir Christopher Wren worked at the problem of connecting the elliptical orbits of the planets to the law of force emanating from the sun. Although they believed in the inverse square law, they failed to connect it to Kepler's first law of elliptical motion. Finally Halley journeyed up to Cambridge to ask Newton about the problem. Here is an account of their meeting written soon afterward by John Conduitt. "Without mentioning either his own speculations, or those of

Hooke and Wren, he [Halley] at once indicated the object of the visit by asking Newton what would be the curve described by the planets on the supposition that gravity diminished as the square of the distance. Newton immediately answered, *an Ellipse*. Struck with joy and amazement, Halley asked him how he knew it? 'Why,' replied he, 'I have calculated it.' " Three years later, in 1687, under the auspices of Halley, Newton's monumental *Principia* was published.

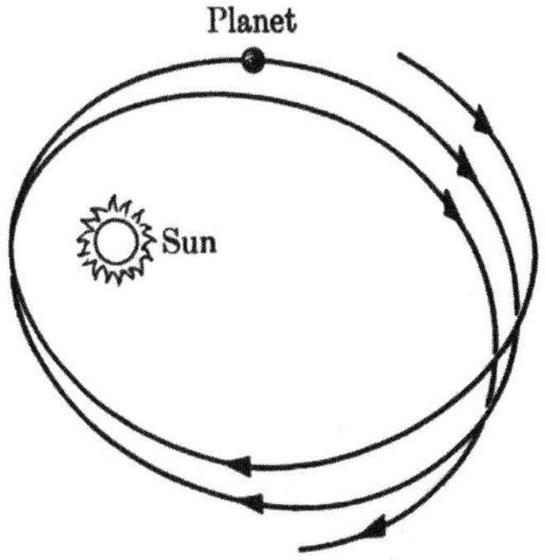

FIGURE 12.13 Hypothetical planetary orbit if the gravitational force depended on distance in some way other than inversely proportional to the square of the distance (see also Figure 10.20). In fact very small deviations of planetary orbits from ellipses do occur, because each planet experiences small forces contributed by the other planets besides the dominant force contributed by the sun.

Before Newton, the inverse square law of gravitational force was an unsupported hypothesis with little more weight than Democritus' belief in atoms. In Newton's hands it became an established law. Not only was he able to carry out the complicated calculation deriving the planetary motion from his law of gravitational force, but he realized correctly that Kepler's first law provided the most sensitive test of the inverse square law. If the force diminished in some slightly different way than inversely with the square of the distance, the orbits would not be ellipses, nor would they close on themselves (Figure 12.13). Instead a planet would follow a somewhat different path on each successive trip around the sun. Any small deviation of the true law of force from the inverse square law would show itself in the geometry of planetary orbits before it would significantly alter Kepler's third law.

About the gravitational force, the planetary ellipses reveal only one thing—but that one thing very accurately—how the force depends on distance. Thus Kepler's first law removes any ambiguity associated with Equation 12.19. Indirectly it strengthens the conclusion based on terrestrial gravity and on Kepler's third law, that the gravitational force is directly proportional to the mass of the body experiencing the force.

(4) Newton's Third Law and the Universality of Gravitation

Insofar as there was thinking about the mechanics of celestial motion before Newton, it was in terms of a force exerted *by* the sun acting *on* the planets. Newton assumed that the planets equally well must pull on the sun and on each other, and indeed that every massive object must exert a gravitational force on every other. This idea of the universality of gravitation rested basically on the principle of action and reaction that we now call Newton's third law. The existence of reaction forces opposing applied forces is obvious in many commonplace examples, but before Newton no one had recognized the general significance of this pairing of forces. Newton obviously grasped its significance, for he adopted as his third fundamental law of mechanics the principle that every force in nature is opposed by an equal and opposite force. Moreover he generalized this principle straight from the force of a man lifting a weight or a horse pulling a cart to the sun pulling the earth. The planets, he argued, if they experience forces to hold them in their orbits, must also exert forces, forces equal and opposite to those they experience. If the sun attracts the earth, the earth equally attracts the sun. But if the earth attracts the sun, it must attract other objects as well—other planets, the moon, or an apple falling from a tree. By this sort of reasoning, Newton was led to think of the *universality* of gravitation—that the local gravity accounting for weight on earth and the cosmic gravity accounting for the orbit of the earth around the sun were all one.

Because of the universality of gravitation, Newton realized that the sun need not be the only center of a planetary system. He noted that the moons of Jupiter obey Kepler's three laws, as do the moons of Saturn. These moons must therefore be held in their orbits by an inverse square central force exactly as are the planets. The only difference is in the total intensity of the force, a difference that shows itself in a different proportionality constant in Kepler's third law (Equation 12.13) for each center of force. The earth, no less than Jupiter and Saturn, could be—and now is—the center of a planetary system, a situation that Newton correctly foresaw, although he surely could not have imagined the hundreds of satellites now circling the earth. Nevertheless, our only natural satellite, the moon, provided for Newton one more crucial test of his law of gravitation. This test is discussed in the next subsection.

Equation 12.19 provides most of the law of gravitational force. To complete it, we need only the idea of universality and Newton's third law. The discussion of Section 12.3 can be reviewed quickly here. If the force *experienced* by a planet is proportional to its own mass, then the equal force it *exerts* is also proportional to its mass (Newton's third law). But if the force a *planet* exerts is proportional to the *planet's* mass, then the force the *moon* exerts is proportional to the *moon's* mass, and the force the *sun* exerts is proportional to the *sun's* mass (universality). Therefore the force experienced by a planet is proportional both to its own mass and to the sun's mass. Calling $F_{\rm p}$ the sun's force on a planet, we can write three different proportionalities:

$$F_{\rm p} \sim m_{\rm S}\,,$$

$$F_{\rm p} \sim m_{\rm p}\,,$$

$$F_{\rm p} \sim 1/r_{\rm p}^{\,2}\,.$$

These express the proportionality of the force on the planet to the sun's mass, to the planet's mass, and to the inverse square of the distance of the planet from the sun. Newton postulated that the gravitational force depends on *nothing else*. The combination of these three proportionalities into a single equation is his law of universal gravitational force:

$$F_{\rm P} = G\,\frac{m_{\rm S} m_{\rm P}}{r_{\rm P}^{2}}\,. \tag{12.20}$$

The constant of proportionality, called *G*, he supposed to be a universal constant, the same for all matter in all places at all times at all distances of separation. There is so far no evidence to the contrary.

Compare Equations 12.20 and 12.1. The subscripts S and p in Equation 12.20 can be replaced by any others to give the gravitational force between any pair of material objects.

Comparison of Equations 12.20 and 12.19 gives a deeper insight into the meaning of the constant *K* in Kepler's third law. What appears in Equation 12.19 as $4\pi^2/K$ must be the same as the combination $Gm_{\rm S}$ in Equation 12.20. Therefore we get a new equality:

$$K = \frac{4\pi^2}{Gm_{\rm S}}\,.$$

This is much more than simply the expression of one constant in terms of another. The constant *K* is an empirical constant from Kepler's third law for the solar system. It is *special* for the sun's planetary system. The constant *G*, on the other hand, is *general*, the same for all systems throughout nature. To give recognition to this difference, we should affix a subscript S to *K* and write

$$K_S = \frac{4\pi^2}{Gm_S}. \tag{12.21}$$

Another value of K relates the periods and distances of the moons of Jupiter. Its magnitude is dictated by the mass of Jupiter and by the same gravitational constant G. We can call the Kepler's-third-law constant for Jupiter K_J and write

$$K_J = \frac{4\pi^2}{Gm_J}. \tag{12.22}$$

Similarly, if Kepler's third law for earth satellites is written

$$T_s^2 = K_E a_s^3, \tag{12.23}$$

the constant K_E may be written

$$K_E = \frac{4\pi^2}{Gm_E}. \tag{12.24}$$

These various constants K can be measured easily. They were quite accurately known to Newton. Not so the constant G. Its magnitude was completely unknown to Newton, and it remains today the least accurately known of the fundamental constants of physics. Notice that in each of the equations above G is multiplied by a large mass—of the sun, of Jupiter, or of the earth. Newton knew the magnitudes of these products, such as Gm_S, but not the magnitudes of either factor.

In order to determine the constant G, it is necessary to measure the force between a pair of objects of manageable size whose individual masses can be separately measured. This experiment, first carried out by Henry Cavendish in the 1790s, is sometimes referred to theatrically as "weighing the earth." Indeed this is what it achieves indirectly, although it could just as well be called weighing the sun. Since the products Gm_S and Gm_E are known, the masses of sun and earth become known as soon as the constant G is separately determined. The difficulty of the Cavendish experiment springs from the extreme weakness of the gravitational force. In order that it be detectable for a pair of human-sized objects, it is necessary to use a highly sensitive balance to measure the force, and it is necessary to get rid of all electric forces, which are intrinsically far stronger. The balance used by Cavendish was a torsion balance (Figure 12.14), a long wire whose twist can respond to a very weak force.

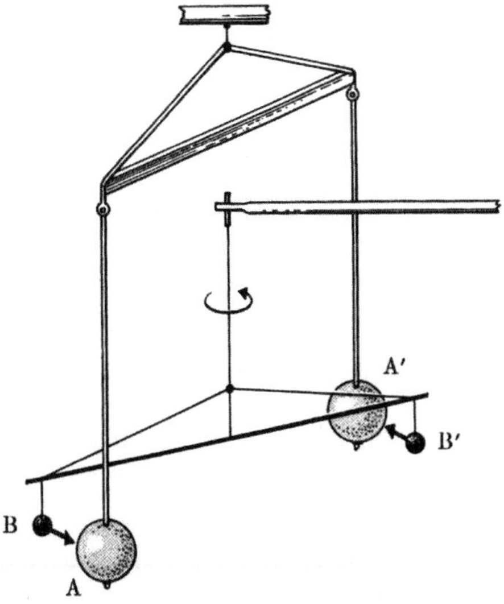

FIGURE 12.14 Schematic diagram of the Cavendish apparatus to measure the gravitational force between objects of ordinary size. Massive lead spheres (A and A') are placed close to smaller spheres (B and B') attached to the ends of a horizontal rod. In response to the torque produced by the gravitational forces, the fiber supporting the rod twists through a measurable angle.

(5) The Gravitational Force Produced by a Sphere

Looking back on his earliest discoveries, Newton later wrote:

> And the same year I began to think of gravity extending to the orb of the Moon, and ... from Kepler's Rule [Kepler's third law] ... I deduced that the forces which keep the Planets in their orbs must [vary] reciprocally as the squares of their distances from the centers about which they revolve: and thereby compared the force requisite to keep the Moon in her orb with the force of gravity at the surface of the earth, and found them to answer pretty nearly.

The moon proved to be the vital link between terrestrial gravity and the cosmic gravitational force. The motion of the moons of Jupiter and Saturn proved that these planets attracted their moons in exactly the same way that the sun attracts the planets. But to our own moon fell the job of proving that this inverse square force which reaches out from every massive body is exactly the same as the familiar force of gravity that we all experience on earth.

To establish this link, Newton had to use the fact that a sphere acts gravitationally exactly as if all of its mass were concentrated at its center.* The problem of the spherical mass provides a beautiful example of a self-consistent circle of reasoning. The equivalence of sphere and point was used in helping to establish the correctness of the inverse square law of force. Yet the sphere is in fact equivalent to a point mass *only* for the inverse square force. Had the gravitational force turned out to have a different form, it would not have been proper to replace the earth in calculations by a point mass at its center. This is only one of the simple properties of the inverse square law that makes it seem so uniquely right. King Alfonso, giving advice to the Creator (see page 295), could not have done better than recommend a gravitational force whose strength varies inversely with the square of the distance.

Every equation of physics is meaningful only in a context of definitions, assumptions, and limitations. This is true of the universal law of gravitation, Equation 12.20, as well as of every other. As a fundamental statement it should be taken to apply to hypothetical point masses. By good fortune it is true also for ideal spheres, in which case the distance r is the distance between the centers of the spheres. It is also true to good approximation whenever the distance between two objects is very much greater than the size of either. But there are circumstances where it is not true. It is not true within a sphere. It is not true in the neighborhood of a nonspherical object. Our own earth differs sufficiently from an ideal sphere that the small deviation of its gravitational field from the inverse square law is not hard to measure, especially with the help of satellites. This, of course, does not mean that Equation 12.20 is wrong—only that it is wrongly applied. It still describes with complete accuracy the force between any speck of matter in a satellite and any other speck of matter anywhere within the earth.

Newton knew that near the surface of the earth all objects fall with the same acceleration. Therefore every object experiences a gravitational force proportional to its own mass. The moon, in its nearly circular orbit, is also "falling" toward the earth, with an easily calculable acceleration equal to the square of its speed divided by the distance of its center from the center of the earth (Equation 12.15). Since the force acting on the moon is presumed also to be proportional to its mass, its inward acceleration is independent of its mass. To find how the gravitational force weakens with increasing distance, it is necessary only to compare the acceleration of the falling moon with the acceleration of a falling apple, both directed toward the center of the earth.

* Recall that the sphere and the point are indistinguishable only outside the sphere; see Section 12.3 and Figure 12.5.

The moon is 384,000 kilometers distant from the center of the earth, and the apple 6,370 kilometers from the center of the earth. If the earth's gravity extends on out into space, diminishing as the inverse square of the distance from its center, it is a simple matter to calculate how much less should be the moon's acceleration than the apple's acceleration. The ratio of these accelerations should be

$$\frac{a_{\text{moon}}}{g} = \left(\frac{6{,}370 \text{ km}}{384{,}000 \text{ km}}\right)^2 = 2.75 \times 10^{-4} . \tag{12.25}$$

Since the moon's speed is known to be 1.02×10^5 cm/sec, its acceleration is easily calculable. It is

$$a_{\text{moon}} = \frac{v^2}{r} = \frac{(1.02 \times 10^5 \text{ cm/sec})^2}{3.84 \times 10^{10} \text{ cm}} = 0.271 \text{ cm/sec}^2 . \tag{12.26}$$

The actual ratio of the accelerations of moon and apple is

$$\frac{a_{\text{moon}}}{g} = \frac{0.271 \text{ cm/sec}^2}{980 \text{ cm/sec}^2} = 2.76 \times 10^{-4} . \tag{12.27}$$

Using the best data available to him, Newton made a similar comparison of the theoretical and actual acceleration ratios, and found them to "answer pretty nearly."

The universal law of gravitational force was the product of Newton's genius and nature's kindness. Nature was kind in arranging for a law of force of such magnificent simplicity. And nature was kind in providing a solar mass so much greater than any planetary mass that the solar force on any planet greatly exceeds the force produced by any neighboring planet. To good approximation every planet responds to the sun and ignores its fellow planets. Only for this reason can planetary motion be well described by Kepler's laws. 66

12.8 *Gravity on earth*

Newton, the obvious hero of this chapter, did not stop when he had dealt with the grand problems of planetary and lunar motion. Perhaps his greatest triumphs came in the coherent clarification of numerous smaller and subtler problems. For example, he used the law of gravitational force to explain the tides, the shape of the earth, and the variation of the acceleration g over the surface of the earth. All of these terrestrial phenomena were already known but had not been explained.

Newton correctly attributed the tides primarily to the pull of the moon, secondarily to the pull of the sun. He could account adequately for the known fact that tides reach their greatest peaks at the new moon and the full moon, when the moon and sun work together (Figure 12.15).

The fact that the earth, moon, sun, and planets are spheres—or very nearly spheres—had been accepted for millennia without ever really being explained. This was hardly a problem of the time, for it is unlikely that more than a few people felt that any explanation was required. Yet Newton realized that there was a good reason for it. The mutual forces among the parts of any object tend eventually to establish the spherical form as the shape of least potential energy. Imagine, for instance, a giant sphere of water suddenly released at one point on the earth (Figure 12.16). We pretend for the moment that the earth is a perfect sphere. Each part of the liquid is attracted to the center of the earth, and consequently flows until it is as close as it can get to the earth's center. It establishes a spherical shell surrounding the earth. The whole earth was probably once hot enough to flow in this way. Even if it were not, after the passage of enough millions of years, solid rock too can "flow" to establish the spherical form. Our older mountain ranges are gradually being worn away and their mass redistributed over the earth. The eventual fate of a nonrotating body would be, in principle, an ideal sphere. In practice, of course, surface irregularities might never be totally erased.

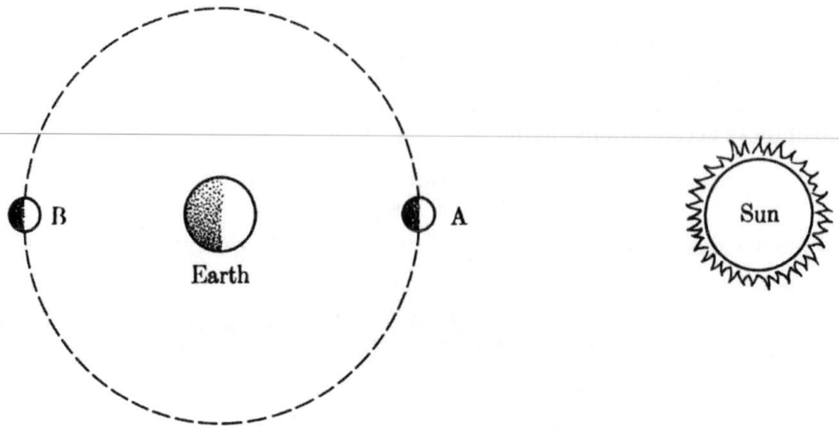

FIGURE 12.15 Earth tides are maximum when earth, moon, and sun are in line at new moon **(A)** and full moon **(B)**.

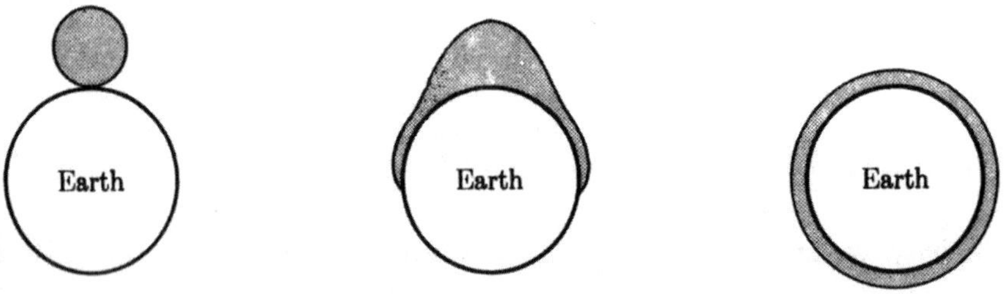

FIGURE 12.16 Imaginary spread of a giant water drop over the surface of the earth. It seeks the almost spherical form of lowest gravitational potential energy.

What if the sphere is rotating? Then a "stationary" bit of matter on the surface of the earth is in fact experiencing acceleration as it moves in a circle once around the axis of the earth each day. Consider a book on a supposedly frictionless table [Fig. 12.17(a)]. The contact force $\mathbf{F}_c$ exerted by the table on the book acts vertically upward. Usually we say that the gravitational force on the book, $\mathbf{F}_G$, has equal magnitude and acts vertically downward, so that the book experiences no net force. Actually, however, these two forces cannot quite cancel. They must add to give a small net force directed toward the axis of the earth ($\mathbf{F}_{net} = \mathbf{F}_c + \mathbf{F}_G$). Since the book is experiencing a net acceleration, it must be experiencing a net force. This means that the gravitational force cannot act vertically downward (except at pole and equator). As indicated in Figure 12.17(b), the gravitational force acts approximately toward the center of the earth, a direction inclined slightly away from the vertical. If the earth remained precisely spherical the vertical direction (perpendicular to the surface) and the direction of the center of the earth would be the same, so that it would be impossible to combine a vertical contact force and a gravitational force to produce the proper net force toward the axis.

Newton was able to prove that the equilibrium shape of a rotating body, if it is capable eventually of flowing, is an ellipsoid, flattened at its poles, bulging at its equator. He confidently predicted such a flattening of the earth, calculated the magnitude of the effect, and produced some striking indirect evidence in support of it. But fifty years went by before direct measurements confirmed the prediction. What is required is a measurement of the distance around the earth along a latitude line at two or more well-separated latitudes.

A flattened earth has a greater equatorial girth than a spherical earth of the same average radius. In the arctic regions the opposite situation prevails. There the flattened earth has smaller circles of latitude than the spherical earth. To establish the circumference of a circle of latitude, it is necessary to measure the speed of rotation of the surface of the earth, which requires accurately timed telescopic observations at two points along an east-west line. In the years 1735-1737 expeditions went out from France to Peru and to Lapland to make the measurements and settle the question of the shape of the earth. Both returned with convincing evidence in support of Newton's prediction. Pierre De Maupertuis, leader of the Lapland expedition, was henceforth known as "the grand flattener."

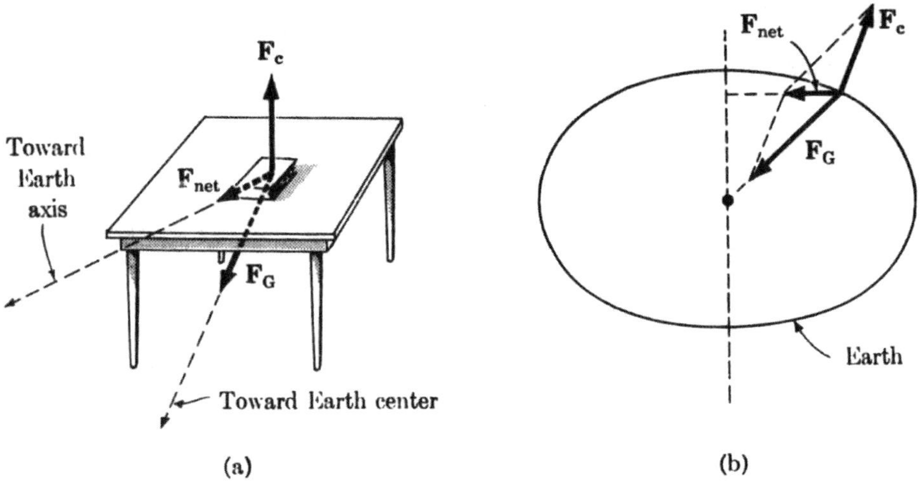

FIGURE 12.17 Explanation of the nonspherical shape of the earth. For a stationary bit of matter at the earth's surface, the gravitational force $\mathbf{F}_G$ and the contact force $\mathbf{F}_c$ must add to give a small net force $\mathbf{F}_{net}$ toward the earth's axis. (The amount of the earth's flattening is greatly exaggerated in the drawing.)

Probably the most dramatic and beautiful aspect of Newton's work on the nonspherical earth was his explanation of the precession of the equinoxes (Section 12.5). Since at least 130 B.C. it had been known that the point in the sky to which the earth's pole points moves gradually among the stars. After the Copernican view of the universe was accepted, it was clear that this effect, called the precession of the equinoxes, must arise from a slow wobbling of the earth's axis, not, as had long been supposed, from a shifting axis of rotation of the stellar sphere. We shall not give Newton's explanation in detail, but only indicate its basis. Because the earth is nonspherical and has a tilted axis, the sun and the moon exert on the earth a torque as well as a force (Figure 12.18). This torque causes the spin angular momentum of the earth to change (Equation 10.18); in particular, the orientation of the earth's axis shifts slowly. Newton realized, long before there was any direct evidence available about the nonspherical shape of the earth, that the precession of the earth's axis is itself compelling evidence for the flattening of the earth at the poles, and that the rate of precession provides good evidence about the amount of the flattening. What a rich variety of images the flattened earth can call to mind: Hipparchus, the discoverer of the precession of the equinoxes, observing the stars on a warm Greek night; Newton, alone in a chilly study in Cambridge, with pages of calculation before him; De Maupertuis, a thousand miles from home, shivering in a frigid arctic wind; a technician in a half-buried bunker in Florida, counting backward to the moment of fiery launch of a satellite destined to reveal new details of the earth's shape.

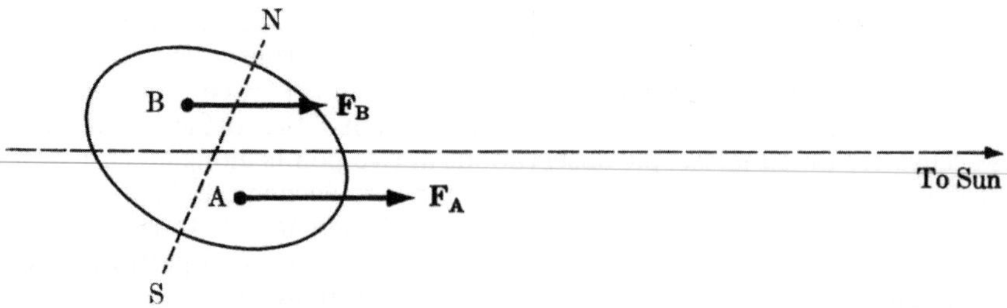

FIGURE 12.18 Explanation of the torque exerted by the sun on the earth. The dashed line represents the ecliptic plane (plane of earth's orbit). The center of mass of the part of the earth below this plane (point A) is slightly closer to the sun than the center of mass of the part of the earth above this plane (point B). Accordingly the sun exerts a force $\mathbf{F}_A$ on the lower half that is slightly greater than the force $\mathbf{F}_B$ which it exerts on the upper half. Because these forces are not precisely equal there is a net torque exerted with respect to the center of the earth (directed out of the page in this diagram). The scale of the figure is of course highly distorted.

The Acceleration of Gravity

For a spherical earth of mass M_E and radius R, the gravitational force exerted on an object of mass m at its surface is

$$F = G\frac{mM_E}{R^2}. \tag{12.28}$$

This force is also equal to the mass of the object multiplied by the acceleration it experiences in free fall:

$$F = mg. \tag{12.29}$$

$$g = \frac{GM_E}{R^2}. \tag{12.30}$$

Comparison of these two equations at once makes clear how the constant acceleration g discovered by Galileo is related to the mass of the earth, the radius of the earth, and the gravitational constant G. The connection is
Notice that here again occurs the product of G and a large mass. Even today this product is known more accurately than either of its factors.

For a stationary and perfectly spherical earth, the acceleration of gravity would be precisely constant over the face of the earth. Not so for a rotating earth. Over the pole an unimpeded object would indeed fall with acceleration g. But over the equator, the earth's surface is itself accelerated toward the center of the earth, so that the falling object has to overtake a downward accelerating surface. Thus the measured value of the acceleration *g relative to the surface* is less at the equator than at the pole. Newton knew the formula $a = v^2/r$ for the inward acceleration in uniform circular motion. Applying it to the surface of the earth at the equator, where the surface speed is $v = 4.65 \times 10^4$ cm/sec (about 1,040 miles/hr), he obtained the acceleration,

$$a_{\text{equator}} = \frac{(4.65 \times 10^4 \text{ cm/sec})^2}{6.37 \times 10^8 \text{ cm}} = 3.4 \text{ cm/sec}^2. \tag{12.31}$$

This is the amount by which the value of g would diminish in going from pole to equator if the earth were spherical. Because it is not spherical, the force of gravity is actually slightly greater at the pole than at the equator, enough greater to account for an additional change in the value of g of 1.8 cm/sec^2. Altogether, the measured value of g is less at the equator than at the pole by an amount Δg equal to the sum of the two contributions:

$$\Delta g = 5.2 \text{ cm/sec}^2 . \tag{12.32}$$

The modern values for g at pole and equator and their difference are:

$$g_{\text{pole}} = 983.217 \text{ cm/sec}^2 \tag{12.33}$$

$$g_{\text{equator}} = 978.039 \text{ cm/sec}^2 \tag{12.34}$$

$$\Delta g = 5.178 \text{ cm/sec}^2 . \tag{12.35}$$

Such an effect (over a smaller span of latitude) had already been noticed in Newton's time, and Newton was able correctly to attribute it to the combined effects of the earth's rotation and the earth's nonspherical shape. The approximate value of g commonly employed, 980 cm/sec^2, is correct to within 0.3% everywhere on earth.

12.9 *Gravitational potential energy: escape speed*

If the zero of potential energy is chosen at infinite distance from the earth, an object of mass m has negative potential energy in the gravitational force field of the earth, given by

$$P.E. = -G\frac{mM_{\text{E}}}{r} , \tag{12.36}$$

in which M_{E} is the mass of the earth and r is the distance from the center of the earth to the object. The potential energy has a mathematical form quite similar to the expression for the magnitude of the gravitational force, Equation 12.1, with the all-important difference that the potential energy varies inversely as the separation distance rather than inversely as the square of the separation distance.* Equation 12.36 was presented graphically in Figure 11.12.

At the surface of the earth, where $r = R$, the potential energy of an object of mass m is (relative to infinite distance)

$$(P.E.)_{\text{surface}} = -G\frac{mM_{\text{E}}}{R} , \tag{12.37}$$

The magnitude of the right side is the "depth of the potential well," the energy required to remove the object completely from the influence of the earth's gravitational force. Comparison of Equations 12.37 and 12.30 shows that this binding energy can also be written

$$(P.E.)_{\text{surface}} = -mgR . \tag{12.38}$$

Consider now a rocket of mass m launched vertically upward which acquires a certain speed v_0 very quickly—while still close to the earth—and thereafter coasts. If friction is negligible, it conserves mechanical energy during its coasting flight:

$$K.E. + P.E. = E \text{ (constant)}. \tag{12.39}$$

If the energy E is negative, the rocket rises to some maximum height and then falls back to earth (energy line A in Figure 12.19). For positive total energy, the rocket escapes (energy line C). The kinetic energy decreases but never reaches zero. The least energy E that permits escape is zero (energy line B). Then at (or near) the surface of the earth,

$$(\text{K.E.})_{\text{surface}} = +mgR \tag{12.40}$$

* Recall that the expressions valid close to the surface of the earth, $F = mg$ and $P.E. = mgh$, also differ only by a factor with the dimension of length.

$$(P.E.)_{surface} = -mgR$$

$$K.E. + P.E. = 0. \quad (12.41)$$

As the rocket rises, kinetic energy and potential energy remain equal and opposite as both decrease in magnitude and eventually approach zero. The minimum initial speed for escape is called the escape speed. It can be calculated from Equation 12.40:

$$\tfrac{1}{2}mv_{esc}^2 = mgR\,,$$

$$v_{esc} = \sqrt{2gR}\,. \quad (12.42)$$

Note that the escape speed is independent of the mass of the rocket. Its numerical magnitude is

$$v_{esc} = 1.12 \times 10^6 \text{ cm/sec}\,. \quad (12.43)$$

Since the speed of sound in normal air is approximately 3.3×10^4, the escape speed is about Mach 34. It is left as an exercise to compare this speed with the speed of a low-altitude satellite.

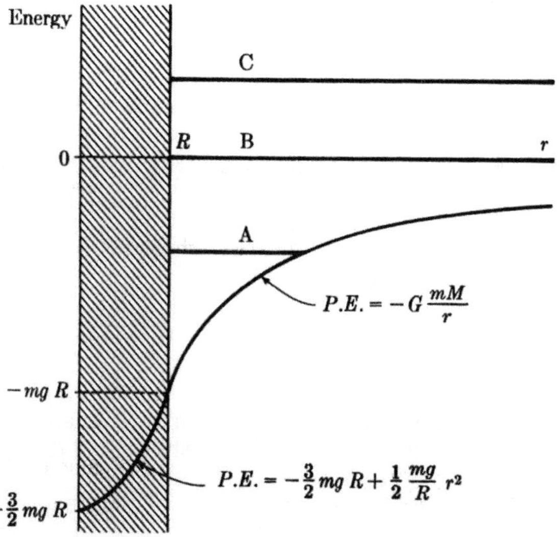

FIGURE 12.19 Energy diagram for rocket or other object interacting with the earth. If the energy is negative **(A)**, the rocket does not escape. If the energy is zero **(B)** or positive **(C)**, the rocket does escape.

Example 1: A vertically fired rocket starts upward with total energy $E = -\tfrac{1}{2}mgR$. What is its maximum height? At the peak of its motion, $K.E. = 0$, and $P.E. = E$, or

$$-G\frac{mM_E}{r_{max}} = -\frac{1}{2}mgR\,.$$

On the right, g can be replaced by GM_E/R^2 (Equation 12.30), so

$$-G\frac{mM_E}{r_{max}} = -\frac{1}{2}G\frac{mM_E}{R}\,.$$

The solution to this equation is

$$r_{max} = 2R\,. \quad (12.44)$$

The peak distance above the earth is

$$h_{max} = r_{max} - R = R \cong 4{,}000 \text{ miles.} \tag{12.45}$$

Example 2: After exhausting its fuel in a very short climb, a vertically fired rocket of mass 10^8 gm has a kinetic energy of 10^{20} ergs. Does it escape? If so, what is its final speed at great distance from the earth? First, determine its total energy. According to Equation 12.38, its potential energy at the start of its trip is

$$\text{P.E.}_{surface} = -mgR = -10^8 \text{ gm} \times 980 \text{ cm/sec}^2 \times 6.37 \times 10^8 \text{ cm} = -6.24 \times 10^{19} \text{ ergs}\,.$$

Its total energy then is

$$\text{E} = 10^{20} \text{ ergs} - 0.624 \times 10^{20} \text{ ergs}$$

$$\text{E} = 3.76 \times 10^{19} \text{ ergs.} \tag{12.46}$$

Since this energy is positive, the rocket does escape (Figure 12.19). At infinite distance, $P.E. = 0$, and $K.E. = E$, or

$$\tfrac{1}{2}mv^2_{final} = E = 3.76 \times 10^{19} \text{ ergs}\,.$$

The solution is

$$v_{final} = \sqrt{\frac{2E}{m}} = 8.67 \times 10^5 \text{ cm/sec}\,. \tag{12.47}$$

The appearance of the gravitational force within the earth is shown in Figure 12.5. Corresponding to this force that is directly proportional to radial distance is a parabolic potential energy curve, as illustrated in Figure 12.19. Also shown in Figure 12.19 is the mathematical form of the gravitational potential-energy function within the earth:

$$P.E. = -\frac{3}{2}mgR + \frac{1}{2}\frac{mg}{R}r^2\,. \tag{12.48}$$

If a hole could be bored through the earth, an object dropped into it would reach a maximum kinetic energy equal to $\tfrac{1}{2}mgR$ at the center.

12.10 Mechanics after 1700

Ending the story of mechanics with the publication of Newton's *Principia* is like ending a romantic novel with the happy couple walking to the altar. Like a real-life marriage, the wedding of celestial and terrestrial mechanics, so happily arranged by Newton, was more a beginning than an ending. It stimulated a century and more of mathematical discovery. It shaped the course of astronomy. It gave birth to hydrodynamics and to the kinetic theory of gases. It profoundly influenced philosophy. In the nineteenth century, physics was dominated by what Einstein and Infeld* have called the mechanical view of nature. In the twentieth century, mechanics is in use as a daily tool of science and technology, and is the fundamental theory underlying such applied areas of research as aerodynamics, seismology, and meteorology.

Perhaps the most dramatic success of mechanics after Newton was the discovery of the planet Neptune in 1846. Newton had recognized that the forces acting directly between one

* A. Einstein and L. Infeld, *The Evolution of Physics* (New York: Simon and Schuster, Inc., 1961).

planet and another, although small, would be significant in causing small deviations from the perfect elliptical orbit that would be followed if only the force of the sun were acting. These extra disturbing forces, which can be ignored for most purposes, are called perturbations. Throughout the eighteenth century, perturbations were studied with increasing care as astronomical observations became more precise and calculational methods more refined. The culminating work on perturbations was Pierre Simon Laplace's *Celestial Mechanics*, published early in the nineteenth century. The perturbing interplanetary forces had been identified and measured, resulting in values of planetary masses which could not be ascertained from the action of the sun's force alone. By this time it was clear that gravitation was indeed a universal force acting among all of the bodies in the solar system. The world system of Laplace was a beautifully self-consistent system of mutually interacting bodies, each influencing and influenced by all the others.

67 Six planets were known to Newton. In 1781, William Herschel in England, after a telescopic survey of the whole sky, had turned up a seventh, Uranus, about twice as distant from the sun as the sixth planet, Saturn. The full machinery of observation and calculation was turned upon Uranus and before long its future orbit was mapped out. With a period of 84 years, Uranus was in no hurry to get on with its job of tracing out the predicted orbit. Nearly fifty years went by before it became clear that Uranus was, in fact, not following precisely the orbit that should have resulted from the action of the sun and other planets. Either Newton's law of gravitation was failing at this great distance from the sun, or still another planet, more remote and still unseen, was perturbing Uranus. As happens so often in science, the more conservative possibility proved to be the correct one. In the 1840s two young men, unknown to each other, John Adams in England and Urbain Leverrier in France, decided to assume that Newton's law of gravitational force was perfectly correct and to calculate where an eighth planet should be located to account for the unexpected perturbation of Uranus. Both completed their calculations at nearly the same time, and both predicted nearly the same position in the sky for the new planet. But Leverrier had the greater luck. While astronomers at Greenwich politely ignored Adams' suggestion to search in a certain area of the sky, the director of the Berlin observatory acted at once upon receipt of a letter from Leverrier and discovered Neptune the very same day.

A somewhat similar but less dramatic chain of events led to the discovery of the ninth and, so far, the last planet, Pluto. Early in this century, an American astronomer, Percival Lowell, completed calculations on the orbits of Neptune and Uranus which suggested that there should exist yet another planet beyond Neptune. It was not so easy to predict the exact location of this new planet, and it escaped detection for 25 years. Finally, in 1930, Pluto was definitely identified at the Lowell Observatory in Arizona. Not until the year 2179 will it have completed one trip around the sun and be back at the point in the sky where it was first seen.

While slight oddities in the behavior of Uranus were very nicely resolved by the discoveries of Neptune and Pluto, another small discrepancy between observation and calculation went unexplained for more than 60 years. Leverrier also found that the motion of the innermost planet, Mercury, failed to conform precisely to prediction. He suggested that a small unseen planet nearer to the sun than Mercury might be held responsible, but this idea bore no fruit. In this case it was the radical and not the conservative assumption which proved correct. Instead of a new planet, a whole new view of space and time was required to account for the extra perturbation of Mercury's orbit. As described in Chapter Twenty-Two, the orbit of Mercury proved to be one of the key supports of Einstein's general theory of relativity in 1915.

Among the important applications of mechanics developed in the eighteenth century, mainly in the hands of mathematicians, were hydrodynamics (fluid mechanics) and rigid-body mechanics. The branch of hydrodynamics called aerodynamics (or gas dynamics) remains an area of active endeavor today. Although these fields in which the theory of mechanics is applied to macroscopic matter in bulk have had important practical value, even more significant for the subsequent development of physics was the application of mechanics to microscopic matter. The mechanical description of large numbers of atoms or molecules in constant random motion with frequent collisions is called the kinetic theory—"theory" because when this description of a gas was first proposed, it was by no means certain that it was the correct description. (Obviously "theory" is here being used in quite a different way than in the "theory of mechanics.") Nearly a century ahead of his time, Daniel Bernoulli, a Swiss mathematician, introduced the kinetic theory of gases in 1738. Not until evidence for the existence of atoms and molecules began to accumulate in the nineteenth century was the subject reopened and pursued. Because the kinetic theory greatly clarified the concepts of heat and temperature and now forms an important part of the theory of thermodynamics, it is a subject belonging to the next part of this book.

After Newton mechanics followed two paths—a practical path and a theoretical path. 68
On the one hand, it was being used to predict and to control, and to describe an ever wider class of motions. On the other hand, it was evolving as a fundamental theory of nature. For Newton, the basic concepts of mechanics were space, time, velocity, acceleration, force, mass, and momentum. To these were added later energy, work, angular momentum, and torque, as well as others which have not been discussed in these pages. As the conceptual basis of mechanics was expanding, so was its mathematical power. Especially in the century between two famous textbooks, that of Leonhard Euler in 1736 and that of William R. Hamilton in 1835, mechanics was being constantly revised and revitalized by many of the leading mathematicians of Europe. From Eulerian angles to Hamiltonian functions, most of what students now learn in courses on advanced mechanics was developed in this span of one hundred years.

From the modern point of view, how do we assess mechanics? Except to a few mathematicians, it is no longer a significant area of research. For over two hundred years, from about 1600 to 1820, mechanics was at the center of research interest in theoretical physics and astronomy. During the nineteenth century it was joined and gradually overshadowed by the great new disciplines of thermodynamics and electromagnetism. In the twentieth century, mechanics is nowhere to be found on the frontiers of physics. This certainly does not mean that mechanics is a dead theory; it means only that the frontiers of physics have moved elsewhere. Mechanics remains very much alive as a theory of magnificent generality and simplicity encompassing a large part of the phenomena in nature. As a tool of research (as opposed to a subject of research) it is in daily use. Engineering and technology would founder without it. One need only think of the problem of landing a man on the moon to see the relevance of mechanics in the modern world.

As a subject of research, the wonder of mechanics is not that it has disappeared from research frontiers, but that it so long remained on the frontier. That Hamilton in 1835 could be reformulating the equations of mechanics in significant new ways shows the remarkable strength and vitality of Newton's ideas. Here indeed was a coherent set of concepts, equations, and links with experience that deserves to be called a great theory. In our own time relativity is in a similar position. So fertile is the theory which Einstein "completed" in 1915 that it remains very much a frontier field of research today, and no one believes its potentiality for new insights to be anywhere near exhausted.

The limitations of mechanics are of two kinds. About some phenomena it has nothing to say; about other phenomena it has the wrong thing to say. Electromagnetic phenomena, for example, are outside the *scope* of mechanics; the motion of high-speed electrons is outside the *range* of mechanics. The theories of nature that have evolved since 1800 have been supplementary to mechanics, extending both its range and its scope. Electromagnetic theory, built around the ideas of fields and waves, provides an entirely new way of looking at nature, a way that still dominates twentieth-century physics. Accompanying the rise of the field point of view was the decline in the mechanical point of view—although electromagnetic theory, in its early history, was still very much tied to the mechanical point of view. In this century the theories of relativity and quantum mechanics extended the *range* of mechanics into domains of high speed and small size where Newtonian mechanics is no longer valid. Thermodynamics, concerned with heat, temperature, and the behavior of bulk matter, is a theory which, like electromagnetic theory, extends the *scope* of mechanics, providing an entirely new way of looking at nature.

Of course the description of nature is not so neatly compartmentalized that each theory takes unto itself a certain class of phenomena and ignores the rest. The links are many, the overlaps are frequent, and the concepts of mechanics permeate every branch of science. All of thermodynamics is based on the mechanics of vast numbers of interacting entities (atoms or molecules). The idea of field rests on the idea of force. Mass, energy, momentum, and angular momentum are key concepts in every branch of physics. Certain phenomena are equally well described from different points of view with different theoretical structures. Others are describable only by using several theories together. The combination of mechanics, thermodynamics, and electromagnetic theory needed to deal with the motion of electrically charged matter in magnetic fields bears the imposing name magnetohydrodynamics.

At the beginning of the twentieth century, both the strengths and weaknesses of mechanics came sharply into focus. As Newton's second law and the conservation of mass were being found wanting in the atomic domain, other aspects of mechanics were emerging as pillars of the new theories. The conservation of energy, already extended as a principle for all of nature in the nineteenth century, continued to hold its own. The conservation of momentum and of angular momentum not only remained valid laws, but moved to a more central position when their connection with invariance principles was demonstrated. Whatever the future may bring in further extending the range and scope of physics, mechanics will remain a useful and a valid theory for comprehending and controlling a large part of nature. It will remain too a monument to the power of the human intellect.

EXERCISES

12.1. A spacecraft whose mass is 1,000 kg is located 10^9 cm from the center of the earth at a particular instant. **(1)** What force of gravity does it experience? (2) What gravitational force does it exert on the earth? **(3)** What is the acceleration of the spacecraft? **(4)** By how much is the earth accelerated because of the pull of the spacecraft?

12.2. **(1)** A man whose mass is 75 kg stands 200 cm from the center of mass of an automobile whose mass is 2,000 kg. Compute the gravitational force of attraction exerted by the car on the man, and then find the ratio of this force to the weight of the man. **(2)** Estimate roughly by any means you can think of the force exerted on the man by a light breeze and compare this with the gravitational attraction of the car.

12.3. According to Formula 12.2, the gravitational constant G is expressed in dyne cm^2/gm^2. Re-express this unit in cm, gm, and sec.

12.4. **(1)** What is your weight in dynes at the surface of the earth (absolute honesty not required)? **(2)** What would be your weight 3,960 miles above the surface of the earth? (3) At what distance from the center of the earth would your weight be 10^6 dynes?

12.5. The force of electrical attraction between a proton and an electron separated by 1 Angstrom (10^{-8} cm) is 2.3×10^{-3} dyne. By what factor is this force greater than the gravitational force between the same two particles? What is the implication of this result for atomic structure?

12.6. A hypothetical satellite skims the surface of the earth—that is, it rotates in a circular orbit whose radius is equal to the radius of the earth. For calculational purposes, assume that the vacuum of space extends down to the surface of the earth and that any interfering mountain ranges have been bulldozed aside. **(1)** What is the acceleration of such a satellite? **(2)** What is its speed? **(3)** What is its period? How much less is this than the actual period of a typical low-altitude satellite, about 90 minutes? (See also Exercise 12.29.)

12.7. A box of sand with a mass of 5×10^4 gm (about 100 lb) is held fixed. A bottle of water with a mass of 100 gm is located 30 cm away on a perfectly smooth horizontal surface. The gravitational force between these two objects causes the bottle to slide toward the sand. If the bottle starts from rest, how far will it have moved after one minute?

12.8. A 50-kg girl and a 75-kg boy are initially at rest 200 cm apart on a horizontal frictionless ice surface. They are attracted to each other by gravitation. Approximately how long does it take for them to get together?

12.9. The center of mass of the earth-moon system moves about the sun in an approximately circular orbit of radius 150 million kilometers. **(1)** Calculate the acceleration of this point toward the sun, showing it to be approximately 0.6 cm/sec^2. **(2)** What force does the sun exert on a 75-kg man? How does this compare with his weight?

12.10. Prove that at the bottom of a hole 2,000 miles deep (halfway to the center of the earth), the weight of a man is half his weight at the earth's surface, (HINT: Work with ratios, and take advantage of the fact that the net force on the man is produced only by the mass at greater depth than he.) *Optional:* Prove more generally that, within the earth, weight is directly proportional to distance from the center of the earth, as is indicated by the straight-line portion of Figure 12.5.

12.11. It is stated in the text that the superposition principle for gravitational forces is "implicit in Equation 12.1." Explain how this is so.

12.12. If a man fell into an 8,000-mile-long hole through the earth, what would be his fate? Describe the nature of his motion in words. (See also Exercise 12.37.)

12.13. Imagine yourself at the center of a hollow spherical shell of matter. **(1)** Why do you experience no force at the center? **(2)** Explain why as you move away from the center and toward the inner wall of the shell you still experience no force. At first thought it might seem that the close matter in the nearer wall should exert a stronger pull than the distant matter in the farther wall. Can you explain why it does not?

12.14. Which two planets can never be seen at midnight? How does the heliocentric theory account for this fact? Illustrate with a diagram. Can the geocentric theory offer an equally plausible explanation? Why or why not?

12.15. Aristotle argued that if the earth rotated, the natural motion of matter in the earth must be in circles, in contradiction to the observed natural motion straight downward of a piece of solid matter that is dropped. Carefully criticize this reasoning. (Do not simply dismiss it as obviously wrong. It deserves careful attention.)

12.16. A Soviet communications satellite, Molniya 1D, when launched in 1966, had a perigee 300 miles above the earth's surface, and an apogee of 24,670 miles above the earth's surface. **(1)** Calculate the three distances a, b, and c characterizing its elliptical orbit (see Figure 12.11). Slide-rule accuracy suffices. **(2)** What is the eccentricity of this ellipse ?

12.17. Intelsat 2A, a commercial communications satellite launched in 1966, was intended for a near-circular orbit (eccentricity near zero). Instead it went into orbit with eccentricity $e = 0.634$. At apogee it is 27,000 miles from the center of the earth. (1) How far above the surface of the earth is it at perigee? **(2)** What is the ratio of its speed at perigee to its speed at apogee ?

12.18. With distances measured in astronomical units (1 A.U. is the average distance of the earth from the sun), and with a suitably chosen coordinate system, the equations of the orbit of Halley's comet can be written: $(x/18.0)^2 + (y/4.6)^2 = 1$. **(1)** Using graph paper, plot the orbit. Label the axes and locate the positions of the foci. **(2)** Compare the dimensions of your plotted ellipse with the dimensions of planetary orbits given in Table 12.1 and state where in the solar system Halley's comet is to be found when it is at perigee and when it is at apogee. (This comet, first observed by Edmund Halley in 1682, is next expected at perigee in 1986.)

12.19. Explain why a "hovering" satellite, one whose orbit causes it to remain always over the same point on the earth, must be above a point on the equator.

12.20. In the labeled ellipse in Figure 12.11, the distances a, b, and c are in the ratios 5 : 4 : 3. Consider an earth satellite moving in an ellipse of this shape and apply Kepler's second law to obtain the following speed ratios: **(a)** the ratio of speed at perigee to speed at apogee; **(b)** the ratio of speed halfway between apogee and perigee (at the end of the semiminor axis) to speed at apogee. Is this intermediate speed closer in magnitude to apogee speed or to perigee speed?

12.21. An unseen planet called Vulcan was once hypothesized to exist in an orbit smaller than the orbit of Mercury. If Vulcan's mean distance from the sun were 0.2 A.U., what would be its period in earth years ?

12.22. Working from the moon's kinematic data given in Appendix 2C, scale downward using Kepler's third law to find **(a)** the distance in miles from the center of the earth of a satellite whose period is 24 hours; and **(b)** the period in minutes of a low altitude satellite whose distance from the center of the earth is 4,100 miles. Assume circular orbits for both satellites.

12.23. Consider planets in circular orbits about a common force center. The squares of their periods are proportional to the cubes of their orbital radii (Kepler's third law). What analogous relationship exists between their speeds in orbit and their orbital radii?

12.24. If a quantity X is changed by a small fraction, the fractional change of X^2 is approximately twice the fractional change of X, and the fractional change of X^3 is approximately three time the fractional change of X. **(1)** Explain the reason for these mathematical facts. **(2)** Assume that a synchronous communication satellite is in a perfect circular orbit. In order that its period of rotation be within one minute of 24 hours, how accurately must its altitude be established? Express answer in miles.

12.25. During part of their Gemini 12 flight, astronauts Lovell and Aldrin were in an orbit whose distance from the center of the earth varied between 4,060 and 4,135 miles. The period of this orbit was 89 minutes. **(1)** What was the ratio of their maximum speed to their minimum speed in orbit? **(2)** Scaling from the Gemini 12 period, find the period of a satellite in an orbit with a semimajor axis of 8,200 miles. **(3)** If the gravitational force varied not as $1/d^2$ but as $1/d^3$, one of the two answers above would be changed, one unchanged. Which would be unchanged, and why?

12.26. Newton derived an expression for the constant of proportionality in Kepler's third law (Equation 12.21). Use Equations 12.21 and 12.24 together with kinematic data (distances and times) on planets and satellites to find the ratio of the mass of the sun to the mass of the earth. Check your answer against the masses given in Appendix 2.

12.27. Suppose we wish to investigate the possibility that the moons of Jupiter are held in their orbits by a force of the following form:

$$F = \frac{GmM_J}{r^n},$$

in which m is the mass of a moon, M_J is the mass of Jupiter, and the exponent n is an unknown. Derive an expression relating the periods and radii of circular orbits. Is there any value of n other than 2 for which the moons would obey Kepler's third law ?

12.28. **(1)** For a point on the earth's surface at 45 deg N latitude, calculate the inward acceleration produced by the rotation of the earth. (Keep in mind that the radius of the circle traced out by this point is not the radius of the earth, but the distance from the point to the earth's axis.) Compare this acceleration with the free-fall acceleration g. Is it a legitimate approximation for most purposes to say that an object at rest on the surface of the earth experiences no net force? **(2)** The gravitational force at this latitude is inclined slightly away from the vertical (Figure 12.17). By approximately what angle?

12.29. **(1)** Derive algebraically an expression for the speed of a low-altitude satellite in a circular orbit, and use Equation 12.30 to cast it into a form containing neither G nor M_E. **(2)** What is the ratio of this speed to the speed of sound in air? (3) Compare this speed with the escape speed for a vertically fired rocket, Equation 12.42.

12.30. **(1)** At what point between earth and moon is an object attracted equally to these two bodies? **(2)** Is its potential energy at this point positive, negative, or zero (if its potential energy at infinity is zero) ? Why?

12.31. Formulas 12.42 and 12.43 for escape speed from the earth assume that the rocket acquires this speed while still close to the earth and then coasts away. **(1)** Assume instead that the rocket begins its vertical coasting flight at some height above the earth where its radial distance from the center of the earth is r. Derive a formula for its escape speed from this position, expressing the answer in terms of r, the gravitational constant G, and the mass of the earth M_E. Reasoning analogous to that in Section 12.9 applies. **(2)** Reexpress Formula 12.42 in terms of R, G, and M_E. Write an expression for the ratio of escape speed from the surface of the earth to escape speed from greater height. **(3)** At what height above the earth's surface is the escape speed half of the escape speed from the surface?

12.32. Let r represent the radius of the earth's orbit (approximated as a circle), M_E the mass of the earth, and M_S the mass of the sun. In terms of these quantities and the gravitational constant G, answer all parts of this exercise algebraically. **(1)** What is the potential energy of the earth in the gravitational field of the sun? **(2)** What is the kinetic energy of the earth? **(3)** How much additional energy would the earth require in order to escape from the solar system? **(4)** Show that the escape speed of the earth from the solar system is

$$V_{esc} = \sqrt{\frac{2GM_S}{r}}.$$

12.33. **(1)** Using the formula given in Exercise 12.32, part (4), compute the escape speed of the earth from the solar system. Note that this solar escape speed does not depend on the mass of the escaping body. A spacecraft, once having "escaped" from the earth, would need this much remaining speed in order to continue on to escape from the sun. **(2)** What is the ratio of the solar escape speed to the average orbital speed of the earth given in Appendix 2C? **(3)** What is the ratio of the solar escape speed to the terrestrial escape speed given by Formula 12.43? What does this imply about the relative difficulty of getting to the moon and getting to Neptune?

12.34. One of the possible ways to define the concept of mass is to make use of the law of gravitational force between all pairs of material objects. Explain how this law might be used to define mass. (Any measurements you suggest should be possible in principle but need not be very practical.) Would this method define gravitational mass or inertial mass?

12.35. **(1)** In orbit an astronaut wants to turn his vehicle around without in any way disturbing the motion of its center of mass. Explain how he can do this. **(2)** Later he wants to change the orbit in order to descend into the atmosphere. Explain how a retro-rocket achieves this aim. At what point in his elliptical orbit should the retrorocket be fired to get the desired effect most easily? (Do not use intuitive arguments. Reason directly from fundamental laws and state the laws that you use.)

12.36. A satellite with a mass of 10^7 gm in a circular orbit 8×10^8 cm from the center of the earth turns on its rocket engine for two minutes, during which time it experiences a force of 10^9 dynes in its direction of motion. **(1)** What is the applied torque relative to the center of the earth? **(2)** What is the change in its angular momentum? **(3)** What were its speed, its angular momentum, and its kinetic energy before the impulse? **(4)** What are each of these same three quantities after the impulse? **(5)** How would you expect its orbit to change? **(6)** In qualitative terms, what would happen had the same impulse been directed **(a)** opposite to the velocity; **(b)** vertically upward away from the earth?

12.37. Estimate by any method the time required for an object to fall the full length of an evacuated tunnel drilled straight through the earth. Explain your mode of reasoning. From the material of Chapters Eleven and Twelve, it is possible to work out an exact answer. However, the object of this exercise is to test your ability to make a reasonable order-of-magnitude estimate.

12.38. **(1)** Why do we speak of Kepler's "laws" instead of Kepler's "theory"? **(2)** Why are Kepler's laws applicable and useful in the description of the motion of artificial earth satellites?

12.39. Suppose that a crank was convinced that the gravitational force of the earth should depend in a simple way not on distance from the center of the earth, but on distance from the surface of the earth. For low altitude satellites he could successfully fit the facts with this equation for acceleration:

$$a = g - \frac{2gh}{R},$$

in which g is the acceleration of gravity at the earth's surface, h is the distance above the surface, and R is the radius of the earth. **(1)** Sketch a graph of a vs. h. **(2)** Compare this graph with the graph of Figure 12.5, and explain in what way the two graphs "agree." **(3)** Suggest an experiment or observation that could easily demonstrate the limitations of this hypothesis for the gravitational acceleration. *Optional:* Prove that

$2g/R$ is indeed the correct coefficient of h to give agreement with Newton's law of gravitational force at low altitude.

12.40. Consult one or more other books to learn more about the retrograde motion of planets. With the help of a diagram, explain how the Copernican theory accounts for retrograde motion.

PART 4
Thermodynamics

JAMES PRESCOTT JOULE
(1818 – 1889)

"Indeed the phenomena of nature, whether mechanical, chemical, or vital, consist almost entirely in a continual conversion of attraction through space [potential energy], living force [kinetic energy], and heat into one another. Thus it is that order is maintained in the universe—nothing is deranged, nothing ever lost, but the entire machinery, complicated as it is, works smoothly and harmoniously."

THIRTEEN

Temperature, heat, and the kinetic theory

Man's imagination has always been challenged by the domains of nature far removed from his immediate human scales of reference. Concerning the limit of the very large, he has pondered the system of the world. Concerning the limit of the very small, he has puzzled over the ultimate structure of matter. Newton's mechanics bridged the gap between the macroscopic (human-sized) and cosmological domains. The theory of thermodynamics links the macroscopic and submicroscopic domains.

13.1 Introduction to thermodynamics

Thermodynamics is the theory of heat and temperature superimposed on the theory of me- 69
chanics. Its great triumph is the explanation of bulk properties of matter, especially thermal properties, in terms of the submicroscopic mechanics of atoms and molecules. This is the aspect of thermodynamics to be emphasized in this chapter and the next—its role as a bridge between two domains of nature.*

Thermodynamics is not directly a theory of the structure of matter, although its development paralleled and supported the development of the atomic theory of matter. The theory of heat and temperature was beginning to be placed on a firm footing late in the eighteenth century, at about the same time that chemical experiments were providing the first solid evidence for the existence of atoms. By the middle of the nineteenth century the evidence of chemistry could leave little doubt about the existence of atoms, and it was at this time that the microscopic and macroscopic threads of physical theory united to create what we now call thermodynamics. By the end of the nineteenth century there was monumental evidence for the existence of atoms, but all of it was indirect. Finally in 1905, Einstein rendered the bridge of thermodynamics visible. His application of thermodynamics to Brownian motion (Section 13.11) made clear that a grain of dust dancing in a liquid is reacting visibly to molecular impact. Before long other and even more direct manifestations of atoms and molecules were being observed.

When an object is set into motion, its structure is normally unaffected. If the same object is electrically charged, it is also unlikely to change its shape or size or appearance. At the macroscopic level, mechanical and electric phenomena may be studied without paying attention to the structure of matter. The same is not true of heat phenomena. If an object is heated, it will, at the very least, change its size slightly. It may also melt or boil or cook or catch fire or explode. How the object reacts will depend very much on its composition. Thermodynamics, being concerned with heat, cannot separate itself from the structure of matter. It must be concerned not only with the fact that matter consists of atoms and molecules, but with how these atoms or molecules are arranged —whether in solids, liquids, or gases—how they move, and how they interact with each other.

For several reasons thermodynamics is mathematically and conceptually more complicated than mechanics. For one reason, it is concerned not with simple systems of a few

* Sometimes the term "thermodynamics" is limited to mean the macroscopic approach to thermal phenomena, reserving "statistical mechanics" to mean the submicroscopic approach. We use the word "thermodynamics" alone to mean the full theory of thermal phenomena, encompassing both approaches.

particles or objects, but with intrinsically complex systems of vast numbers of particles. A drop of water, considered at the molecular level, is a system far more complicated than the solar system. For another reason, thermodynamics looks at nature through a variety of magnifying glasses simultaneously, from the atomic scale to the human scale and beyond. For yet another reason, thermodynamics is concerned with matter in all of its forms—solid, liquid, and gas—and with both its physical and chemical transformations of form.

In this part of the book, we shall attempt neither a panoramic survey of the whole theory of thermodynamics nor a logical presentation of all of its fundamentals. Rather, we shall focus our attention on several of the key ideas of thermodynamics, and in particular on these two:

1. Heat and temperature are both manifestations of molecular energy.
2. Spontaneous change in the large-scale world is governed by laws of probability in the small-scale world.

These two ideas, both concerned with links between the macroscopic and microscopic domains, suffice to support the bridge of thermodynamics. They suffice also to make clear the meaning of the two great principles known as the first and second laws of thermodynamics.

13.2 *The concept of temperature*

Most of the concepts in physics, especially the most fundamental ones, began as somewhat vague ideas used in the description of nature, meaning different things to different people. Gradually their definitions were refined and their measurement made more precise. Force, for instance, springs from the idea of human effort, pulling and pushing. Outside of physics, the word has acquired a myriad of meanings. Within physics, it has become a well-defined quantitative concept which can be accurately measured and mathematically manipulated in the equations of mechanics.

In approaching a new concept such as temperature, it is of the utmost importance to remember that definition and measurement are inseparable. It is perfectly correct to say that temperature is defined by the reading of a thermometer. Then the real essence of the definition is contained in the detailed instructions on how to construct the thermometer. The *usefulness* of the definition, which is another question but a vital one, will depend on whether the concept defined by a particular thermometer can be used to describe natural phenomena in a simple and self-consistent way.

In this section we shall approach temperature from the macroscopic side. With this approach, temperature is a distinct new concept, a primitive concept not directly derivable from the mechanical concepts of mass, length, and time. Later, in Section 13.7, we shall investigate the microscopic aspect of temperature and discover that at the molecular level it does, after all, have a mechanical basis.

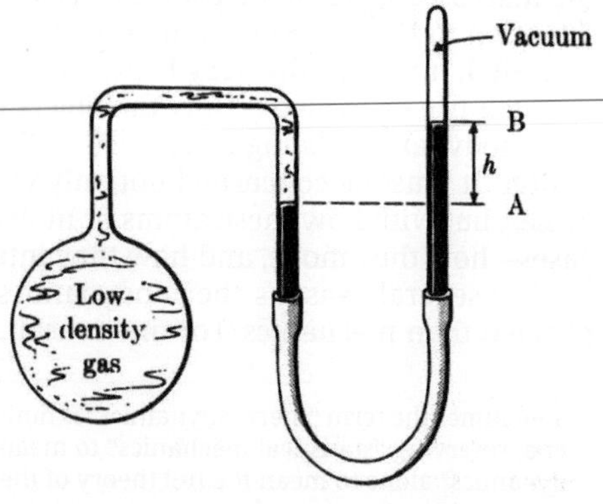

FIGURE 13.1 Constant-volume gas thermometer (idealized). The vertical column on the right can be moved to keep the liquid level A fixed and thereby the gas volume fixed. Then the difference in levels A and B (height h) provides a measure of the gas pressure. Higher pressure means higher temperature (see also Figure 13.2).

A good way to begin thinking about temperature is in human terms. We all feel the difference between a hot object and a cold one, between a warm day and a cool one. But a human reaction to temperature change is too unreliable to form a basis for quantitative measurement. How warm or cold we feel depends on much more than the temperature. It depends on the humidity, the state of our health, the amount of physical exertion, what we have just had to eat or drink, and on whether we have just come from a warm or a cool place.

Looking around us, we have no trouble thinking of reactions of inanimate matter to heat and cold which might serve as a basis for defining and measuring temperature. When heated, a metal bar gets longer; a fluid expands; a gas expands, or its pressure increases, or both. In addition, certain marked changes take place at definite temperatures—freezing, boiling, igniting. In principle, any of these effects could be used to define and measure temperature, and most of them do find a use for temperature measurement. An oven thermometer uses a metal strip. A typical outdoor thermometer uses a column of liquid. The freezing and boiling points of water serve as calibration points. As a standard suitable for the primary definition of temperature, the gas thermometer has been found most suitable. When gas confined in a fixed volume is heated, its pressure rises (Figure 13.1). In an environment of melting ice, the sample of gas has a certain pressure, P_0. Its temperature is defined to be zero degrees Centigrade in this environment. If in some other environment the gas thermometer continues to show the same pressure, P_0, this other environment is assigned the same temperature, 0°C. That is, the pressure P_0 identifies the temperature 0°C. A second calibration point is provided by an environment of boiling water at normal sea-level atmospheric pressure. Here the gas thermometer shows some higher pressure, P_1, which identifies the temperature 100°C. Suppose now that the thermometer is moved into your living room, where after a time it shows a steady pressure just one quarter of the way from P_0 to P_1. Now comes the key point in the temperature definition. The living-room temperature is defined to be one quarter of the way from 0°C to 100°C, or 25°C. If, in a hot-water bath, the gas thermometer pressure stabilizes at the midpoint between P_0 and P_1; the water temperature is defined to be 50°C. If, in an oven, the pressure reaches a value as much greater than P_1 as P_1 is greater than P_0, the oven temperature is defined to be 200°C. And so on. In mathematical language: The temperature is defined to be a linear function of the pressure of a constant-volume gas thermometer. The relationship is made clear by a graph (Figure 13.2). Temperature vs. pressure is a straight line. The pressures P_0 and P_1 exist at the two calibration temperatures. Any other temperature is measured by measuring the gas pressure and reading the corresponding temperature from the graph.

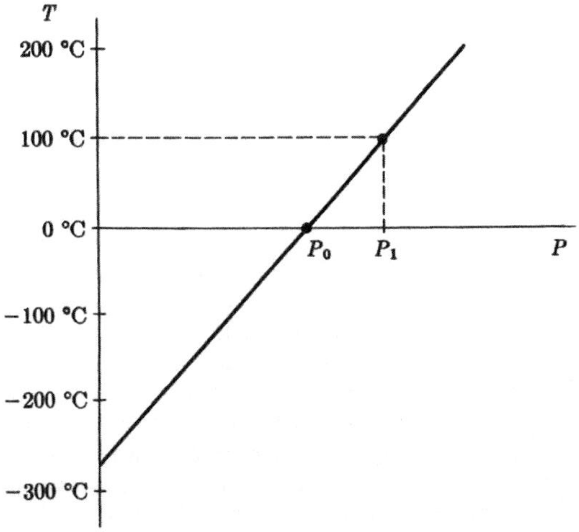

FIGURE 13.2 Temperature defined by pressure of constant-volume gas thermometer. The particular sample of gas has pressure P_0 at the freezing point of water (defined to be 0°C) and pressure P_1 at the boiling point of water at sea level (defined to be 100°C). The straight line drawn between these two points serves to define other temperatures in terms of other pressures.

The linear relationship between the pressure P in the thermometer and the temperature T of its environment is expressed by the equation

$$P = a + bT\,, \tag{13.1}$$

in which a and b designate two constants. For the Centigrade scale, the two calibration points provide the values of the two constants. At 0°C, Equation 13.1 reads

$$P_0 = a\,.$$

At 100°C, the same equation reads

$$P_1 = a + 100b\,.$$

Therefore, the constant a is equal to P_0 and the constant b is equal to $(P_1 - P_0)/100$. For the Centigrade scale, Equation 13.1 takes the special form

$$P = P_0 + \left(\frac{P_1 - P_0}{100}\right)T\,. \tag{13.2}$$

To find the special form of Equation 13.1 for the Fahrenheit scale is left as an exercise. Keep in mind that the mathematical linearity of Equation 13.1 and of the line in Figure 13.1 is a matter of definition. The temperature scale is chosen arbitrarily to bear this simple relationship to the pressure in the gas thermometer.

One reason that the gas thermometer is preferred as a standard of temperature is that its behavior is the same for all gases—at least for all sufficiently dilute gases. Each different liquid and each different solid reacts to temperature change in a somewhat different way. Gases, too, behave differently when compressed to high density. But at low density, gas thermometers containing different gases all define exactly the same temperature scale. This universal feature of the gas thermometer suggests that there is something fundamental about this definition of temperature. In fact, it means more. It means that the linear relation between temperature and pressure (Equation 13.1) is not after all merely a definition. It is a law of nature. This very important point can be appreciated by considering a series of measurements with two different gas thermometers, one, let us say, containing helium, the other hydrogen. Suppose that we choose the helium thermometer to be the primary standard defining temperature. Since only one such standard is needed, the hydrogen thermometer can be regarded not as a thermometer but as a separate experimental device to be studied. Now imagine the helium thermometer and the hydrogen device placed together in an oven which is slowly heated. Every five minutes you, the experimenter, measure the pressure of the calibrated helium thermometer in order to determine the oven temperature. You are curious to know how the hydrogen pressure depends on temperature, so at the same time you measure the pressure in the hydrogen device. To your delight, you discover that the pressure of the hydrogen gas varies linearly with temperature. Since the temperature has been separately defined and measured, this discovery is not tautology. It is an important fact about nature.

Another merit of the gas thermometer is that it leads directly and simply to an important new idea, the idea of absolute zero. If the straight line in Figure 13.2 is extrapolated—that is, extended beyond the range of calibration—it reaches zero pressure at a temperature of −273°C. This suggests that there is a minimum temperature below which it is impossible to go, the temperature at which pressure vanishes. It would be risky to reach this conclusion on the basis of this extrapolation alone, for the predicted absolute zero is a long way from ordinary experience. However, the existence of a real temperature floor is confirmed by many other pieces of evidence, and modern experimenters have been within one thousandth of a degree of it. The idea that temperature is a manifestation of molecular energy (Section 13.7) makes clear that there must be an absolute zero, for when no more energy can be extracted from a system, no further diminution of its temperature is possible. The gas thermometer itself cannot directly verify the existence of absolute zero, even though, by extrapolation, it can locate it. Cooled far enough, the gas liquefies, so that the "gas thermometer" no longer contains gas and no longer functions as a simple thermometer. The liquefaction temperatures of some common gases are shown in Table 13.1. Of all substances, helium condenses to liquid at the lowest temperature, only four degrees above absolute zero. At temperatures close to absolute zero, entirely new methods of temperature measurement must be devised, based on the connection between temperature and energy.

TABLE 13.1 Liquefaction Temperatures of Some Common Gases (at Atmospheric Pressure)

Substance	Liquifaction °K	Temperature °C
Helium	4.3	−268.9
Hydrogen	20.4	−252.8
Neon	27.3	−245.9
Nitrogen	77.4	−195.8
Carbon monoxide	81.7	−191.5
Oxygen	90.2	−183.0
Methane	111.7	−161.5
Chlorine	238.6	−34.6
Ammonia	239.8	−33.4
Steam	373.2	+100.0

Since there is an absolute zero of temperature, it is a sensible procedure to define a new scale of temperature whose zero coincides with absolute zero. This modified scale, called the absolute scale or the Kelvin scale, and designated by the letter K, is now the standard scale for scientific work. It differs from the Centigrade scale simply by the addition of 273.* On the Kelvin scale, water freezes at 273°K and boils at 373°K. Absolute zero is at 0°K. With the Kelvin scale, the linear relationship between pressure and temperature for a gas of constant volume (Equation 13.1) simplifies to become a direct proportionality,

$$P = bT \qquad \text{(at constant } V\text{)}\,. \tag{13.3}$$

Although the existence of an absolute zero was predicted first at the beginning of the eighteenth century, it was not until the middle of the nineteenth century that the absolute scale of temperature came into common use. Actually, the scale introduced at that time by Lord Kelvin did more than recognize the fundamental significance of absolute zero. It also recognized the need for a scale independent of the properties of any particular substance. Kelvin's scale is equivalent to the reading of an "ideal-gas" thermometer. An ideal gas is one which *precisely* satisfies Equation 13.3, as well as a more general ideal-gas law discussed in Section 13.4. A definition in terms of something hypothetical might appear to be meaningless, or at all events useless, but this is not the case. Like infinity, the ideal gas is approachable but not reachable. Like infinity, it is a useful idea. Imagine a series of gas thermometers, some filled with air, one at high density, the next at lower density, the next at still lower density, and so on, and others filled with helium at various densities. No two of these thermometers would yield precisely the same temperature scale (although in practice they would be scarcely distinguishable). However, the low-density air thermometers would tend toward a definite limiting scale. So would the low-density helium thermometers, and both of these limiting scales would be the same. Real gases, when they become exceedingly dilute, tend toward the ideal gas. Thus, the properties of the ideal gas can be measured by extrapolating from the properties of real gases.

* More exactly, the difference is 273.15°.

It should be emphasized that the ideal gas can be so closely approached in practice that the extrapolation from the real to the ideal is hardly necessary. Low-density helium, for all practical purposes, *is* an ideal gas. The reason for its behavior is to be found in the unusual weakness of the interaction between helium atoms. For reasons that are made clear in Section 13.7, a true ideal gas is one whose atoms or molecules fly about freely with no mutual interaction whatever. Helium atoms interact less vigorously with one another than do the atoms of any other gas.

13.3 The mole

Before proceeding with the discussion of temperature, we must mention a few facts about atoms and early chemistry. Gradually, during the nineteenth century, long before the mass of a single atom was known, chemists had deduced the *relative* masses of many atoms. These relative masses, referred to the hydrogen atom as a unit standard, were called atomic weights. Today the same nomenclature is used, although the standard reference atom has changed. The atom of carbon whose nucleus contains six protons and six neutrons (designated ${}_6C^{12}$) is now defined to be exactly 12 atomic mass units (a.m.u.). On this scale, the atomic weight of the lightest isotope of hydrogen (${}_1H^1$) is 1.0078. Usually in chemistry the atomic weight refers to the average mass of the different isotopes of an element in the normal isotopic mixture. The physicist is more likely to be interested in the masses of specific isotopes; to avoid confusion, these are called atomic masses instead of atomic weights. Molecular weights, as the name suggests, are the masses of molecules expressed in atomic mass units.

By the end of the nineteenth century, chemists had a fair idea of the connection between the atomic mass unit and the gram; that is, they knew something about absolute atomic masses as well as relative atomic masses. Accurate values of this conversion factor were not known until early in this century. A more recent precise value for the mass unit expressed in grams is

$$1 \text{ a.m.u.} = 1.6605 \times 10^{-24} \text{ gm} . \tag{13.4}$$

During the nineteenth century, a unit of chemical currency called the mole came into use. One mole of atoms is the number of atoms required to make an amount of material whose mass in grams is numerically equal to the mass of a single atom in atomic mass units. To put it approximately, one mole of hydrogen atoms is the number of hydrogen atoms in 1 gm of hydrogen; one mole of carbon atoms is the number of carbon atoms in 12 gm of carbon. Similarly, one mole of water molecules is the number of water molecules in 18 gm of water. Because two hydrogen atoms unite to form a hydrogen molecule of molecular weight 2, it is also true that one mole of hydrogen *molecules* is contained in 2 gm of hydrogen. Since the mass of a mole of material is chosen to be proportional to the atomic or molecular weight, the *number* of atoms or molecules in a single mole is always the same. The mole, then, is nothing but a convenient way of designating a fixed number of particles. The number of particles in a mole, for which we use the symbol N_0, is called Avogadro's number, and it is a very large number indeed. A little thought will show you that Avogadro's number must be the number of atomic mass units in one gram, which is the inverse of the number in Equation 13.4. It is

$$N_0 = 6.0222 \times 10^{23} \frac{\text{particles}}{\text{mole}} \text{ or } \frac{\text{a.m.u.}}{\text{gm}} . \tag{13.5}$$

The chemist working with macroscopic amounts of matter finds the mole a convenient unit of molecular number in much the same way that the director of a government agency might find one million dollars a convenient unit of dollar number. Actually, there is a difference. The chemist uses the mole not only for reason of contemporary convenience, but for reason of historical precedent. Before the numerical value of Avogadro's number was known even roughly, a mole of matter—that is, 18 gm of water or 12 gm of carbon, or 56 gm of iron—could be defined and used. One of the great triumphs of nineteenth-century chemistry was the deduction of large numbers of atomic and molecular weights by studying the manifold ways in which elements entered into chemical combination with one another. These relative weights provide at once the masses in grams of single moles of material.

One of the earliest important statements about molecular number was the remarkable prediction by Amedeo Avogadro in 1811 that at the same temperature and pressure equal volumes of all gases contain equal numbers of molecules. Several years earlier, Joseph Gay-Lussac had discovered a beautifully simple aspect of chemically reacting gases that is known now as the law of combining volumes. When two gases combine chemically, the volumes that enter into the reaction stand always in simple numerical ratios. If the product is a gas, its volume, too, is simply related to the combining volumes. For example, two liters of hydrogen and 1 liter of oxygen combine to yield two liters of steam. One liter of nitrogen and three liters of hydrogen combine to yield two liters of ammonia. Avogadro guessed correctly that such simple behavior must indicate that all gases have exactly the same number of molecules per unit volume (at a given pressure and temperature). It was nearly half a century later before his prediction was adequately verified and generally accepted. Once the molecular weights of gases had been determined, it was found that the densities of gases were in direct proportion to their molecular weight. This means that the number of molecules per unit volume must be the same. Another way to state Avogadro's hypothesis (still so called despite its ample verification) is this: At fixed temperature and pressure, the volume required to contain one mole of a gas is the same for all gases. Under standard conditions (273°K and 1 atmosphere [atm]), one mole of gas occupies 22.4 liters (22.4×10^3 cm^3).

Note that Avogadro's hypothesis is an important new statement about nature, and a surprising one. Why should molecules of different size and different mass arrange to occupy on the average the same amount of space when confined in a gas? The kinetic theory (Section 13.7) offers an unexpectedly simple answer to this question.

13.4 Two laws of temperature

Temperature enters into the description of nature at many points, and into our own lives nearly every day. Here we would like to discuss two fundamental laws involving temperature.

(1) The Law of Equilibrium

When a roast of beef is put into a freezer, its temperature falls gradually until it is just as cold (that is, until it has the same temperature) as its surroundings in the freezer. When a customer in a hurry pours some cold water into his hot coffee, the water is heated, the coffee is cooled, and the diluted coffee comes to a uniform intermediate temperature. The fact that two things in contact tend toward the same temperature is so familiar that it is easy to overlook its significance. Actually, this equalization tendency is the most basic experimental fact about temperature. Known as the law of equilibrium, or sometimes as the zeroth law of thermodynamics,* it can be stated in this way: Two systems in thermal contact tend toward the same temperature and reach equilibrium at the same temperature. There is both a dynamic and a static aspect to the law. If two systems in contact are *not* at the same temperature, the temperature of one or both will tend to change in the direction of equalization.

* The first and second laws of thermodynamics were named in the nineteenth century before the law of equilibrium was accorded the fundamental status it deserves. In order to give it a position prior to the first law of thermodynamics, it must be called the zeroth law.

The tendency may be countered by outside influences adding or subtracting heat energy from one or the other; for instance, a house furnace prevents the interior of the house from reaching the outdoor temperature. In the absence of such an influence, the direction of spontaneous change is always toward equalization. If two systems in contact *are* at the same temperature, there is no tendency for either to change its temperature spontaneously. They are then said to be in equilibrium.

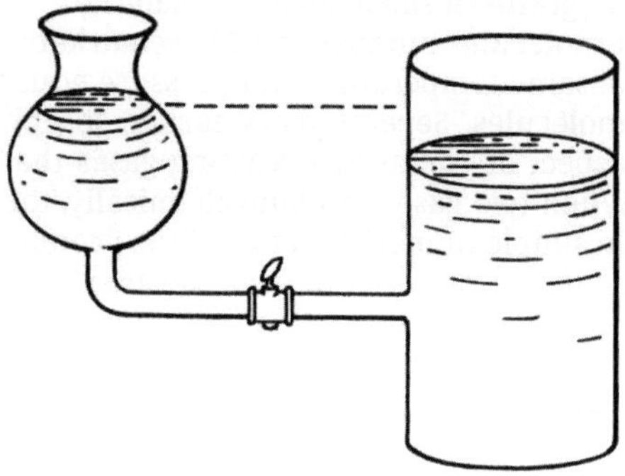

FIGURE 13.3 Mechanical example of equalization tendency. When the valve is opened, water flows until the fluid surfaces reach equal height. Temperature has a similar equalization tendency.

If one thinks of temperature not in familiar everyday terms, but abstractly as a new physical concept, it will be seen that the law of equilibrium is rather special and, therefore, presumably significant. Very few quantities in physics exhibit this spontaneous tendency toward equalization. Masses do not, or forces, or velocities, or accelerations. A mechanical example in which there *is* a tendency toward equalization can be illuminating. Consider two water tanks connected by a pipe, as shown in Figure 13.3. When the valve is opened, water flows until the water level is the same in both tanks. The size and shape of the tanks can be different, the volume of water in each can be different. When equilibrium is established, only one thing is the same: the height of the water surfaces. Although we accept this as familiar and reasonable, we do not regard it as a fact so obvious that it requires no explanation (which was for some time the situation with the law of temperature equilibrium). About the equalization of water level we ask the question: Why? The answer is that the combined system has minimum potential energy when the water levels are equal, and the spontaneous tendency in mechanical systems is toward minimizing potential energy.

For the law of temperature equilibrium, too, it is reasonable to seek a deeper explanation. That explanation has come not from ideas of potential energy, but from quite a different direction. It rests on the microscopic explanation of temperature and on ideas of probability applied to molecular motion. We shall return to it in Section 13.10. It is interesting that a fact of nature once regarded as so obvious as to require no attention found an explanation finally in terms of the most fundamental ideas of thermodynamics. The scientist must approach the familiar with the same care and caution as the unfamiliar.

One important consequence of the law of equilibrium is worth pointing out. Without this law, the measurement of temperature would be much more difficult than it is. What a thermometer reveals directly is its own temperature. It is only because of the law of equilibrium that thermometers are of any use. Once the thermometer comes into equilibrium with its surroundings, we may assume that the temperature of the thermometer is the same as the temperature of its surroundings. Every time that one reads a thermometer, one is making use of the law of equilibrium.

(2) The Ideal-Gas Law

In about 1660, Robert Boyle in England discovered the law of gases that now bears his name. He found that if a quantity of gas is compressed or allowed to expand at constant temperature, it exerts a pressure inversely proportional to the volume it occupies. Boyle's law may be expressed by the equation,

$$PV = constant \quad \text{(at constant } T\text{)} . \tag{13.6}$$

The value of the constant on the right is different for different samples of gas, but for any particular sample, the product of pressure and volume is a constant—provided the temperature is constant. Boyle's law is a statement only about the mechanical properties of a gas—its pressure (force per unit area) and its volume. Temperature plays a subsidiary role: It must be constant, but it does not enter directly into the mathematical statement of the law.

Not until late in the eighteenth century, more than one hundred years later, were the pressure and volume of a sample of gas related quantitatively to the temperature of the gas. Jacques Charles, and after him, Joseph Gay-Lussac, learned that at fixed pressure, the volume of a gas increases in direct proportion to the *absolute* temperature. In symbols,

$$V = constant \times T \quad \text{(at constant } P\text{)} . \tag{13.7}$$

This is now known as Charles's law. As in Boyle's law, the constant is fixed only for a particular sample of gas. A third gas law, basic to the operation of the gas thermometer, is the direct proportionality of pressure to absolute temperature at fixed volume, expressed by Equation 13.3. Actually the gas thermometer was in use long before this law had been accurately tested. The gas thermometer was invented in 1702 by Guillaume Amontons, who was the first to predict the existence of an absolute zero of temperature. He worked only with air. Charles and Gay-Lussac were the first to appreciate the universality of the gas laws, which are as applicable to oxygen or nitrogen or hydrogen as to air.

The three gas laws just discussed can be assembled into a single law, now called the ideal-gas law. The ideal-gas law, in the form it acquired around the middle of the nineteenth century, may be written

$$PV = nRT . \tag{13.8}$$

The three variables are the pressure P (dyne/cm^2), the volume V (cm^3), and the temperature T (°K). The quantity n is the number of moles of gas (dimensionless), and R is a universal constant called simply the gas constant. The ideal-gas law is really a more powerful statement about the behavior of gases than is provided by all three of the special laws just cited. Not until the molecular weights of gases had been measured—after the time of Charles and Gay-Lussac—could the full universality and simplicity of the ideal-gas law be appreciated. The experimental value of the gas constant R is

$$R = 8.3143 \times 10^7 \frac{\text{dyne cm}}{°\text{K}} \text{ or } \frac{\text{erg}}{°\text{K}} . \tag{13.9}$$

(Note that the dimension of pressure times volume appearing on the left of the ideal-gas law is the same as the dimension of energy—see Equation 11.8.) Over a considerable range of pressure, density, and temperature, a large number of different gases do follow the ideal-gas law quite closely.

A good way to look at the ideal-gas law is as a connecting link between the mechanical and thermal aspects of a gas. On the left of Equation 13.8 appear only mechanical and geometric quantities: the volume occupied by the gas, and the force it exerts on each unit area of the containing vessel. On the right appears the temperature. The ideal-gas law unites these aspects. Of all of the statements about nature involving the concept of temperature, this is one of the simplest and most fundamental.

The appearance on the right side of Equation 13.8 of the number of moles n is quite natural, to accompany the appearance of the volume on the left. If the volume of a gas is doubled without changing either its pressure or its temperature, it makes sense that this can be achieved only by doubling the amount of material, that is, by doubling n. It is of some interest to re-express the ideal-gas law in terms of the number of molecules instead of the number of moles. Since there are N_0 molecules in each mole (N_0 is Avogadro's number), there are n times N_0 molecules in n moles. If N denotes the total number of molecules in a sample of gas, the relation is simply

$$N = nN_0 \,. \tag{13.10}$$

This means that N/N_0 may be substituted for n in Equation 13.8 to give

$$PV = N\left(\frac{R}{N_0}\right)T \,. \tag{13.11}$$

Since N_0 is a fixed constant, as is $\boldsymbol{R}$, the ratio R/N_0 is another universal constant. It is called Boltzmann's constant, after Ludwig Boltzmann, a leading contributor to the theory of thermodynamics in the latter part of the nineteenth century. Designated by the letter k, Boltzmann's constant has the value

$$k = \frac{R}{N_0} = 1.3806 \times 10^{-16}\ \text{erg/°K} \,. \tag{13.12}$$

We might call R the macroscopic gas constant, and k the microscopic gas constant. In terms of k, the ideal-gas law has the form

$$PV = NkT \,. \tag{13.13}$$

This is, of course, only a re-expression of the same law, and for many purposes it is inconvenient, because N is such a large number, and k is such a small number. However, it puts the ideal-gas law into closer touch with fundamentals, expressing it in terms of the total number of molecules N in a given sample of gas.

The ideal-gas law has something interesting to say about absolute zero. When the temperature reaches zero on the Kelvin scale, the right side of Equation 13.8 vanishes, so that either the pressure or the volume (of an ideal gas) must also vanish. Negative temperature is unreachable on the absolute scale, for neither negative pressure nor negative volume has any physical meaning. Both the left and right sides of Equation 13.8 must be positive.

Helium gas follows the ideal-gas law quite closely. As a numerical example, consider one liter of helium initially at atmospheric pressure (10^6 dyne/cm^2) and room temperature (300°K). Suppose that it is cooled to 10°K at constant atmospheric pressure. By how much will its volume shrink? Imagine then that it expands back to its original volume at this low temperature. What happens to its pressure? These questions are best answered by using ratios. For constant pressure, volume is proportional to temperature, so that a thirtyfold decrease in temperature must result in a thirtyfold decrease in volume. The gas will shrink to a volume of one thirtieth of a liter. This follows mathematically from the equality,

$$\frac{V_1}{V_2} = \frac{T_1}{T_2} \,, \tag{13.14}$$

which in turn follows from the ideal-gas law at constant pressure. If V_1 = 1 liter, T_1 = 300°K, and T_2 = 10°K, then V_2 = 0.033 liter. Similar reasoning shows that expansion back to a volume of 1 liter at 10°K must result in a thirtyfold decrease of pressure to one thirtieth of an atmosphere. The sequence is

$$P_1 = 1\ \text{atm},\ V_1 = 1\ \text{liter},\ T_1 = 300\text{°K};$$

$$P_2 = 1\ \text{atm},\ V_2 = 0.033\ \text{liter},\ T_2 = 10\text{°K};$$

$$P_3 = 0.033\ \text{atm},\ V_3 = 1\ \text{liter},\ T_3 = 10\text{°K}.$$

Finally we can ask how many moles of helium and how many atoms are contained in the original liter of helium. To give the number of moles, Equation 13.8 may be written

$$n = \frac{PV}{RT}\,.$$

Being careful to use cgs units, and recalling that 1 liter = 10^3 cm^3, we can substitute in this equation to get

$$n = \frac{10^6 \,\dfrac{\text{dyne}}{\text{cm}^2} \times 10^3 \text{ cm}^3}{8.31 \times 10^7 \,\dfrac{\text{dyne cm}}{°\text{K}} \times 300°\text{K}} = 0.040 \text{ moles}\,. \tag{13.15}$$

This number of moles multiplied by Avogadro's number gives the number of molecules:

$$N = nN_0 = 4.0 \times 10^{-2} \times 6.0 \times 10^{23} = 2.4 \times 10^{22} \text{ molecules}\,. \tag{13.16}$$

It is always worth remembering the enormity of molecular number in any macroscopic amount of matter. In this particular example, the number of "molecules" is really the number of atoms, since helium atoms do not join together to form molecules.

13.5 Heat

Heat is now understood as a form of energy. To be more exact, it is a manifestation of energy 70
transfer by molecular collisions. If warm alcohol is added to cold water, the alcohol molecules lose energy and the water molecules gain energy. Energy has been transferred from alcohol to water, and the transfer has been effected by a multitude of molecular collisions. Heat, like work, is more a name for an energy currency than for a "real" energy. It is a mode of energy exchange.

Scientists call the energy content of a substance its *internal energy* and reserve the word *heat* for the gain or loss of internal energy brought about by molecular collisions. In the earliest days of thermodynamics, heat was used to designate both the energy content and the energy exchange. This was quite a natural point of view when heat was itself regarded as a substance, stored in objects and capable of flowing from one object to another. Now, however, it is preferable to separate the two ideas. The main reason for this is that there are a variety of ways in which the internal energy of a system can be caused to change, heat being only one among the ways. Consider a certain quantity of air contained in a glass cylinder with a piston at one end (Figure 13.4). How can we increase its internal energy? One way is to heat the cylinder. The warm inner side of the cylinder transfers some energy to the air as molecules in the air collide repeatedly with molecules in the glass surface. This is heat transfer. Another way is to move the piston inward. The force required to move it multiplied by the distance it moves is work, and this work is transferred to the internal energy of the air. The temperature of the air rises, just as if it had been heated. Colloquially we do say that it has been heated, and that it is hotter. Technically, we should say that its internal energy and its temperature have increased. When *work* is done on the air, it gains energy from the macroscopic bulk motion of the piston. When *heat* is transferred to the air, it gains energy from the microscopic motion of the molecules in the container.

Since heat is a mode of energy exchange, it would be logical for the unit of measurement of heat to be the same as the unit of measurement of energy. Unfortunately, history was not so logical. Heat was defined and used as a quantitative concept long before its connection with energy was known. Nor have subsequent generations of scientists been any more logical. The old practice hangs on, and heat is still usually expressed in a unit different from the standard energy unit. The commonest unit of heat in scientific work is the calorie (cal),* defined as the heat required to raise the temperature of 1 gm of water from 14°C to 15°C. (The heat required to raise the temperature of 1 gm of water by 1 degree at any other temperature between 0°C and 100°C is also very nearly equal to 1 calorie.)

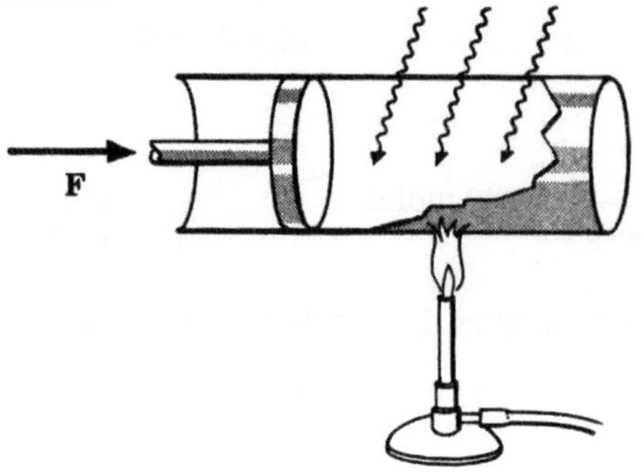

FIGURE 13.4 Air in a cylinder. Its internal energy may be increased by heat (Bunsen flame or radiation) or by work (force moving piston).

Since heat is in fact a manifestation of energy, no historical precedent can prevent its unit from being simply related to the energy unit. The connection is

$$1 \text{ calorie} = 4.184 \text{ joules} = 4.184 \times 10^7 \text{ ergs}. \tag{13.17}$$

In a contemporary setting, this equation is a rather mundane statement about units conversion, much like the equation, 1 foot = 0.305 meter. At the same time, in a historical setting, this equation reports one of the great discoveries of nineteenth-century physics, the equivalence of heat and energy. That heat is a manifestation of energy was suggested first by Julius Mayer and by James Joule in the 1840s. The earliest accurate verification of the equivalence was achieved by Joule in experiments in which he stirred liquids with paddle wheels activated by known external forces. He measured the work done on a liquid by a paddle wheel and the resulting rise in temperature of the liquid. He found that under all circumstances a work input of approximately 4.18×10^7 ergs produced the same temperature rise as a heat input of one calorie. To within the accuracy of the experiment, there was a constant numerical ratio between equivalent inputs of work and heat.

Not only can work and heat produce the same effect (a temperature rise), they are, at least under certain circumstances, interchangeable. The energy put into a liquid by the work of Joule's paddle wheel can be extracted as heat. The liquid is left in the same state as before it was stirred; the net result is a conversion of work to heat. Very often frictional forces act to bring about a similar transformation of work to heat.

Some decades before the work of Mayer and Joule, there was ample evidence pointing to a *connection* between heat and mechanical energy. Based on studies of the production of heat by friction, Count Rumford in 1798 had very forcefully stated his view that heat is a manifestation of microscopic motions. At the same time, steam engines were demonstrating the production of mechanical energy by heat. But before the work of Mayer and Joule, two vital elements in the relationship were missing. These men, first of all, postulated a definite constant ratio for equivalent amounts of heat and work. This alone transformed a qualitative idea into quantitative law. Second, and even more important, they proposed the law of conservation of energy as a single fundamental law, governing all parts of nature. The conservation of mechanical energy and the conservation of heat had been separately recog-

* The food calorie is a unit one thousand times larger than the calorie defined here. Also known as the large calorie or kilocalorie, it is the heat required to raise the temperature of one kilogram of water by one degree.

nized in the eighteenth century as laws valid in certain limiting situations. For the motion of the earth around the sun, mechanical energy is conserved to a high degree of accuracy. For a swinging pendulum that comes gradually to rest, on the other hand, mechanical energy disappears. And in the work produced by a steam engine or by human muscle power, new mechanical energy comes into existence. Heat by itself is conserved if it is transferred without mechanical motion, but it may increase or diminish if its transfer is accompanied by forces and macroscopic motion. New heat is most commonly produced by friction.

The two new elements of Mayer and Joule are of course related. If heat and work were not quantitatively connected by a conversion factor, there could be no basis upon which to build a more general law of energy conservation. Mayer and Joule independently achieved the great vision of energy as a single universal concept capable of changing its appearance drastically without ever changing its total magnitude. Once they had brought about the merger of heat and mechanical energy, the greatest hurdle was surmounted. It was then easy to add electromagnetic energy and chemical energy and eventually nuclear energy. The important thing was their proposal of a general law of energy conservation. If stirring a container of water with a paddle seems a childish pursuit for a grown man, consider its implications. The law of conservation of energy, still a foundation stone of modern physics, has found wider applicability in science and in practical affairs than any other law of nature.

Having learned this much about heat and energy, you might like to put your knowledge to work for some practical purpose. You have been reading this chapter, let us suppose, in the bathtub. Time has slipped by, and the water has become uncomfortably chilly. Why not follow Joule's example and warm it by stirring? You set to work and stir vigorously for a few minutes. Although you feel better for the exercise, nothing seems to have happened to the water temperature. Before you allow suspicion of the law of energy conservation to take shape, it would be wise to do a mental calculation. You estimate that you stirred with a force equivalent to supporting a ten-pound weight. Ten pounds is approximately four kilograms, whose weight is about 4×10^6 dynes. If your hand covered a total distance of 100 meters while stirring, you performed an amount of work equal to the product of 4×10^6 dynes and 10^4 cm, or 4×10^{10} ergs. According to Equation 13.17, this is equivalent to about 10^3 cal of heat. Since you guess that your bathtub contains about 100 kg (10^5 gm) of water, it would require 10^5 cal to raise its temperature by one degree. Your input of mechanical energy has therefore produced an imperceptible temperature rise of only 0.01 °K.

To put this matter in colloquial terms, a little heat is worth a great deal of work. Suppose it were otherwise. What if 42 million calories were equivalent to one erg instead of 42 million ergs to one calorie? It is always interesting to contemplate hypothetical worlds in which physical constants have different values than in our own world. In this *other* world where much heat is equivalent to little work, we can be sure that the general law of energy conservation would have been discovered early. The quantitative link between work and heat would be hard to miss. Life in this world would, at the very least, be hazardous. If a man in a bathtub developed a slight tremor, he might set the water to boiling and scald himself to death. Indeed it is very questionable whether he would exist at all. His ancestral amphibian, trying to climb out of the sea, would probably have been burned to a crisp from friction with the sand.

The actual large magnitude of the mechanical equivalent of heat in our real world has implications for dieters. Suppose you have just finished an ice-cream sundae rated at 500 food calories, and decide to "work it off" by climbing the Empire State Building. If your mass is 60 kg, the 300-meter climb requires a work output of 1.76×10^{12} ergs. (The force is 6×10^4 gm $\times$ 980 cm/sec^2 = 5.88×10^7 dynes.) Dividing this energy by the conversion factor, 4.18×10^7 ergs/cal, gives the equivalent heat—about 42,000 cal. This is a disappointing 42 food calories. You would have to make the climb a dozen times to "work off" one ice-cream sundae. Actually the situation is not quite so bad as this. In the process of performing work, a human being also generates heat; for every erg of mechanical work performed, several ergs of energy are dissipated as heat. Most of the chemical energy released by the oxidation of food makes its appearance as heat, not as work. Nevertheless, this calculation shows why decreasing the intake of calories is much more effective than increasing exercise for losing weight.

13.6 *Internal energy—rival views*

During the late eighteenth and early nineteenth centuries, three views on the nature of what we now call internal energy were current. Recall that at that time the ideas of heat and internal energy were not differentiated. What flows spontaneously from a warmer to a colder object was called heat. What exists within a body to account for its temperature was also called heat.

The first and most widely held view was that heat is a substance. It was given a name, caloric, and was generally assumed to be atomic in nature, like matter. Hot objects were believed to contain more caloric than cold objects, and heat transfer was regarded simply as a flow of caloric from one place to another. According to some chemists, atoms of caloric could unite with atoms of ordinary matter to form new compounds. In some hands, caloric theory became quite elaborate, as indeed it had to be to account for the facts that equal amounts of caloric produced quite different temperature changes in different substances, and sometimes no change of temperature at all, as in melting ice or boiling water. To think of heat as a substance is rather natural, and this point of view does account easily for two main facts: that heat seems to flow from one place to another, and that, at least under certain circumstances, heat is conserved. On the other hand, some powerful arguments were marshalled against the caloric theory. As far as could be determined, an object weighed no more after being heated than before. Friction could apparently produce from a single object an endless supply of heat. And heat could make its way readily through the densest solid matter.

71 One of the earliest and most eloquent critics of the caloric theory was Count Rumford (born Benjamin Thompson in Massachusetts). He performed experiments demonstrating that friction could generate heat in seemingly endless quantity without in any way changing the nature of the material supplying the heat, and that heat had no perceptible effect on the mass of a material object. The following quotations are taken from a paper he published in 1798,* more readable than most scientific papers of any period. His opening paragraphs beautifully summarize an active and pragmatic viewpoint about science that is in some ways characteristically American, although it is a viewpoint that would have been congenial to Galileo as well as to a host of experimental scientists since.

> It frequently happens that in the ordinary affairs and occupations of life opportunities present themselves of contemplating some of the most curious operations of nature; and very interesting philosophical experiments might often be made, almost without trouble or expense, by means of machinery contrived for the mere mechanical purposes of the arts and manufactures.
>
> I have frequently had occasion to make this observation, and am persuaded that a habit of keeping the eyes open to everything that is going on in the ordinary course of the business of life has oftener led, as it were by accident, or in the playful excursions of the imagination, put into action by contemplating the most common appearances, to useful doubts, and sensible schemes for investigation and improvement, than all the more intense meditations of philosophers, in the hours expressly set apart for study. . . .

* "An Inquiry Concerning the Source of the Heat Which Is Excited by Friction," *Philosophical Transactions*, *88* (1798).

> Being engaged, lately, in superintending the boring of cannon, in the workshops of the military arsenal at Munich, I was struck with the very considerable degree of heat which a brass gun acquires, in a short time, in being bored; and with the still more intense heat (much greater than that of boiling water, as I found by experiment) of the metallic chips separated from it by the borer.
>
> The more I meditated on these phenomena the more they appeared to me to be curious and interesting. A thorough investigation of them seemed even to bid fair to give a farther insight into the hidden nature of heat; . . .

Count Rumford goes on to describe *experiments* (as opposed to observations) in which he measures, at least roughly, the work input and heat output during the boring of a brass cylinder, and looks with care for any changes in the brass and its surroundings that might provide a clue about the source of the heat. Finding no material changes whatever associated with the production of heat, he is led to question the existence of caloric. He concludes in this way:

> What is heat? Is there any such thing as an igneous fluid? Is there anything that can with propriety be called caloric?
>
> We have seen that a very considerable quantity of heat may be excited in the friction of two metallic surfaces and given off in a constant stream or flux, in all directions, without interruption or intermission, and without any signs of diminution or exhaustion.
>
> From whence came the heat which was continually given off in this manner, in the foregoing experiments? Was it furnished by the small particles of metal, detached from the larger solid masses, on their being rubbed together? This, as we have already seen, could not possibly have been the case.
>
> Was it furnished by the air? This could not have been the case; for, in three of the experiments, the machinery being kept immersed in water, the access of the air of the atmosphere was completely prevented.
>
> Was it furnished by the water which surrounded the machinery? That this could not have been the case is evident: First, because this water was continually receiving heat from the machinery and could not, at the same time, be giving to, and receiving heat from, the same body; and secondly, because there was no chemical decomposition of any part of this water. Had any such decomposition taken place (which indeed could not reasonably have been expected), one of its component elastic fluids (most probably inflammable air [oxygen]) must, at the same time, have been set at liberty, and in making its escape into the atmosphere would have been detected; but though I frequently examined the water to see if any air bubbles rose up through it, and had even made preparations for catching them, in order to examine them, if any should appear, I could perceive none; nor was there any sign of decomposition of any kind whatever, or other chemical process, going on in the water.
>
> Is it possible that the heat could have been supplied by means of the iron bar to the end of which the blunt steel borer was fixed? Or by the small neck of gun metal by which the hollow cylinder was united to the cannon? These suppositions appear more improbable even than either of those before mentioned; for heat was continually going off, or out of the machinery, by both these passages, during the whole time the experiment lasted.
>
> And, in reasoning on this subject, we must not forget to consider that most remarkable circumstance, that the source of the heat generated by friction, in these experiments, appeared evidently to be inexhaustible.

> It is hardly necessary to add that anything which any insulated body, or system of bodies, can continue to furnish without limitation cannot possibly be a material substance: and it appears to me to be extremely difficult, if not quite impossible, to form any distinct idea of anything, capable of being excited and communicated, in the manner the heat was excited and communicated in these experiments, except it be motion.

The idea that heat is associated with microscopic motions was not new with Rumford. Indeed it was, until the 1780s, the most widely held idea about heat. With origins in antiquity, it was a view supported by Newton and other seventeenth-century scientists. Nevertheless, at the time of Rumford's work, the new idea of heat as a substance was ascendant, because it seemed to mesh nicely with the new discoveries in chemistry, then at the center of scientific interest. Rumford was in any event the first to perform careful experiments tending to support the idea that heat is a manifestation of microscopic motion, and he did so at a time when he had a popular caloric target to shoot at.

Although Count Rumford's marksmanship was good, his target survived the attack. For the next fifty years, the caloric theory hung on, supported by some scientists, while two other theories, both attributing internal energy to molecular motion, also gained currency. The rival doctrines existed in parallel for the good reason that there was insufficient evidence upon which to base a definite choice. Those inclined to believe in caloric could claim that every material contained so much caloric that an apparently inexhaustible supply could be set free by friction without seeming to change the material. In addition, they had to argue that caloric must be far less massive than ordinary matter, perhaps even massless. Caloric, like electricity, became known as an imponderable (i.e., massless) fluid. There were of course some scientists who withheld support from any definite view of heat and internal energy, recognizing that the subject was a mystery still awaiting clarification.

The other two popular views of heat can best be discussed together. Both attributed heat to molecular motion. They differed in the kind of molecular motion that was postulated. According to one view, which has survived to become the basis of our modern understanding, molecules of a gas move about randomly, striking each other and the walls of the container, but otherwise not inhibited in their travels. The quantitative elaboration of this picture of gas behavior is called the kinetic theory. According to the simplest version of the kinetic theory, heat added to a gas goes into the kinetic energy of molecular motion. Greater molecular kinetic energy in turn implies greater temperature.

The second popular view of heat as molecular motion attributed rotational but not translational motion to gas molecules. The minds of some scientists must have recoiled at the randomness and chaos implied by the picture of gas molecules in constant motion and frequent collision. According to their more orderly view, gas molecules remained in fixed relative positions but could rotate more or less rapidly as heat was added or taken away. This picture of gases was in fact the more popular in the early nineteenth century.

It would be simple and neat to say that in the course of time the conception of molecules moving from place to place was proved right, and that of molecules rotating at fixed points was proved wrong. Unfortunately this oversimplifies. True, in liquids and gases, molecules are not constrained to stay at fixed points, but in solids atoms are indeed tied to fixed locations (except that each vibrates about its home base). In a gas, on the other hand, a molecule may rotate as well as move from place to place, and heat added to a gas increases the rotational as well as the translational energy of molecular motion. Nevertheless it is correct to say that the picture of random molecular motion was the most fruitful idea of the early nineteenth century for clarifying and for building a bridge between the microscopic and macroscopic domains.

13.7 The kinetic theory of gases

Even after Joule and Mayer had demolished the idea of caloric by showing heat to be a manifestation of energy, the decision about the molecular behavior of matter remained to be made. The new general law of energy conservation by itself shed no light on one important problem of the structure of matter. As of 1850, the rival pictures of a gas as randomly moving molecules and as molecules rotating in fixed positions both appeared to have a good chance of surviving. Within the next decade, the successes and the simplicity of the kinetic theory had strongly tipped the balance in favor of the picture of randomly moving molecules.

According to the kinetic theory, molecules in a gas fly about in all directions, frequently colliding with each other and with their containing walls (Figure 13.5). Conceptually simple though it may be, this picture is not something suggested directly by any of our experience with matter in bulk. However, it does account at once for the tendency of a gas to expand without limit. In a brilliant piece of work a century before the scientific community was ready to appreciate its significance, Daniel Bernoulli, a Swiss mathematician, worked out the consequences of the kinetic theory in 1738. His work apparently had no impact whatever on the course of science, being noticed only after James Joule in England, and two German physicists, August Krönig and Rudolph Clausius, had developed the same theory independently in the 1840s and 1850s. In the intervening century, Charles and Gay-Lussac had clarified the macroscopic laws of gas behavior, the gas thermometer had been perfected, chemical evidence for the existence of atoms and molecules had become overwhelmingly convincing, the steam engine was utilizing heat energy for mechanical work, thermodynamics was well launched as a quantitative theory, caloric had been abandoned, and the general law of energy conservation had been formulated. By the time these strides in the science of temperature and heat had caught up with the forward leap of Bernoulli, he was long dead and this part of his work forgotten. Yet, when the time was ripe, the same ideas were regenerated by several scientists in different places at nearly the same time in a more typical pattern of scientific development. Only rarely can a long leap into the unknown come down on solid ground, for the infinitude of possible error is so vast compared with the infinitesimal islands of truth.

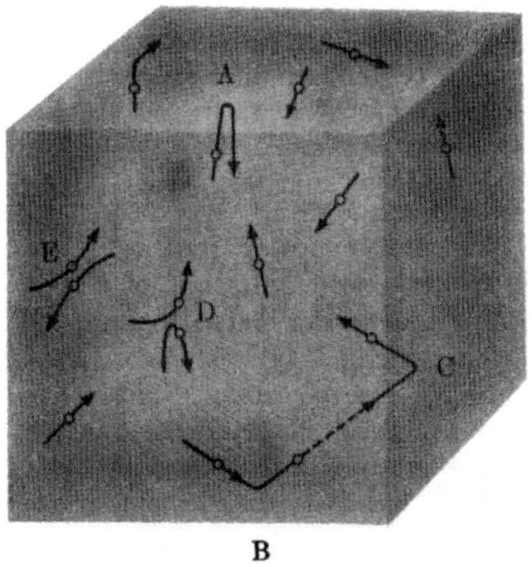

FIGURE 13.5 Random motion of molecules in a gas. Collisions of molecules with the container walls are indicated at points A, B, and C, and collisions of molecules with each other at points D and E.

It might seem surprising that any mathematical derivation based on the kinetic theory could be simple. At the submicroscopic level, a gas appears to be an incalculable chaos of flying molecules, no two moving in exactly the same way, and none moving far between vigorous collisions with its neighbors. Yet from this chaos flows a beautifully simple derivation of the ideal-gas law which illuminates the content of the kinetic theory and, even more important, which establishes the connection between temperature and molecular energy.

What this derivation sets out to do is to relate the pressure exerted by a gas on the walls of its container with the mechanical properties of the molecular motion—that is, with the mass and speed of the molecules. This is a relation between the macroscopic and the submicroscopic, therefore a focus of our concern with thermodynamics. When a gas molecule strikes a wall and is turned back it delivers a tiny hammer blow of force to the wall. The multitude of these incessant hammer blows produces an average outward force on the container, the manifestation of the gas pressure. If a molecule strikes perpendicular to a wall (such as at A in Figure 13.5), it delivers a relatively large hammer blow. If it strikes only a glancing blow (such as at B in Figure 13.5), its hammer blow is weaker, but in that case it is in a position to deliver a stronger blow (as at C) to another wall. On the average, each molecule contributes the same share to the total outward push of the gas on the walls, regardless of how it is moving. This means that we can successfully calculate the outward force by pretending, quite unrealistically, that all of the molecules do move perpendicular to the container walls, as shown in Figure 13.6.

The originators of the kinetic theory learned three simplifying facts about random molecular motion which we shall take advantage of but not prove.

(1) As just discussed, a molecule exerts the same average force on its containing vessel whether it strikes glancing or perpendicular blows on any particular wall. Therefore we can artificially assume simple paths for all molecules.

(2) The outward pressure of a gas in a given volume does not depend on the shape of its container. We take advantage of this fact by choosing, for calculational purposes, a cubical container.

(3) The gas pressure does not depend on the frequency of collisions among molecules. Like the two other simplifying facts above, this one is (after a little thought), if not obvious, at least reasonable. If a molecule leaves the wall numbered 1 in Figure 13.6 and travels unimpeded to wall 2, it delivers its hammer blow to wall 2 when it gets there. On the other hand, if it collides with another molecule on the way, it delivers its hammer blow not to wall 2 but to the intermediate molecule. This molecule in turn is propelled toward wall 2. It may also experience collisions on the way. Eventually some molecule will strike wall 2, conveying to it the pulse of force that the first molecule set on the way. Whether this pulse reaches the opposite wall in one step or in many steps does not matter. This means, again as a calculational simplification, that collisions among molecules can be ignored in our derivation.

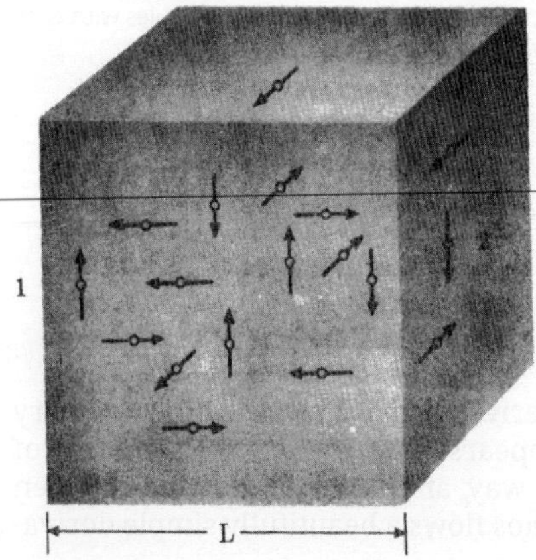

FIGURE 13.6 Idealized model for kinetic theory calculation. All molecules are assumed to move perpendicular to the walls of a cubical box without mutual interaction.

Consider then the idealized gas in a cubical box whose molecules travel in orderly armadas back and forth perpendicular to the walls without mutual collision. To equalize the pressure on all walls, we must assume that one third of the molecules move in each of the three perpendicular directions. To reduce the mathematics to simplest form, one more assumption must be made: that all of the molecules move at the same speed v.

Let us fix our attention now on a single molecule moving back and forth between a pair of walls, and define the *x*-axis to lie parallel to its path. Just before striking the left wall (Figure 13.7), its momentum in the *x*-direction is

$$p_x \text{ (before)} = -mv \, .$$

Just after rebounding, now moving to the right, its x component of momentum is

$$p_x \text{ (after)} = mv \, .$$

The *change* of momentum of the molecule, Δp_x, is p_x (after) – p_x (before), or

$$\Delta p_x = 2mv \, . \qquad (13.18)$$

To produce this change of momentum, the wall has exerted on the molecule a short sharp force, and, according to Newton's third law, the molecule must have exerted on the wall an equal and opposite force. The precise magnitude of this hammer blow of force is not itself of interest. Instead the important quantity is the average force exerted on the wall over a period of time by the myriad of successive hammer blows. For our idealized gas without collisions, the particular molecule on which we fixed attention will return to strike the left wall after an interval of time Δt equal to the round-trip distance $2L$ divided by the speed v:

$$\Delta t = \frac{2L}{v} \, . \qquad (13.19)$$

In each interval of time Δt this molecule will experience a momentum change Δp_x at the left wall. Now we turn to Newton's second law, which states that force is equal to the rate of change of momentum,

$$\mathbf{F} = \frac{\Delta \mathbf{p}}{\Delta t} \, . \qquad (13.20)$$

(Note that we are using the fundamental laws of mechanics directly and *assuming* their validity at the submicroscopic level.) In order that the molecule experience a momentum change Δp_x at the left wall once in each time interval Δt, it must be subjected to an average x component of force given by

$$\mathbf{F}_x = \frac{\Delta p_x}{\Delta t} = \frac{2mv}{\left(\frac{2L}{v}\right)} = \frac{mv^2}{L} \, . \qquad (13.21)$$

This is the average force of the wall on the molecule, directed to the right. The molecule therefore exerts the same average magnitude of force on the wall, but to the left, that is, outward from the container. If the cubical container contains N molecules, one third of them moving left and right parallel to the x axis, the total average outward force on the left wall is

$$F_{\text{total}} = \frac{1}{3} N \frac{mv^2}{L} \, . \qquad (13.22)$$

The pressure of the gas on this wall is simply the force on the wall divided by its area L^2:

$$P = \frac{F_{\text{total}}}{L^2} = \frac{1}{3} N \frac{mv^2}{L^3} \, . \qquad (13.23)$$

Since the quantity L^3 appearing downstairs on the right is the volume V of the cube, this result may be rewritten in the form

$$PV = \tfrac{1}{3}Nmv^2 . \tag{13.24}$$

This important equation, a key result of the kinetic theory, is worth examining closely. It provides a link between the large-scale and small-scale worlds. On the left appear macroscopic quantities, pressure and volume; on the right appear molecular quantities, the mass and speed of a single molecule and the number of molecules in the container. Across the equal sign are joined two worlds of vastly different scale. In terms of a simple equation derived from the fundamentals of mechanics, the bulk properties of a gas—its pressure and volume—are related to the motion of its invisible multitude of careening constituents.

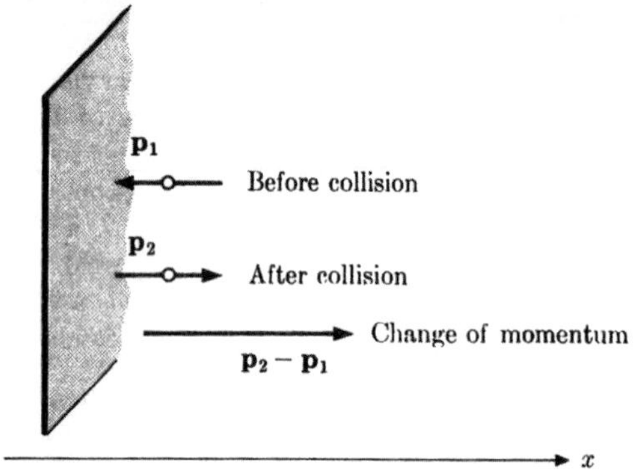

FIGURE 13.7 Perpendicular collision of molecule with container wall.

Bernoulli, and his followers a century later, showed this equation to hold true under much more general conditions: for any shape and size of container, for molecules moving randomly in all directions, and with or without collisions among the molecules. If the molecules do not all have the same speed, only a small correction is required: The quantity v^2 is replaced by $(v^2)_{\text{aver}}$, the average value of the square of the molecular speed—or, more succinctly, the mean square speed. We can then write as the fundamental equation of the kinetic theory,

$$PV = \tfrac{1}{3}Nm(v^2)_{\text{aver}} . \tag{13.25}$$

This equation describes an ideal gas, but still falls short of perfection in describing real gases. Two aspects of molecules have been overlooked: their finite size and the forces they reach out to exert on each other at a distance. In exercises at the end of the chapter you will be afforded the opportunity to discuss how and why molecular size and molecular force alter the ideal-gas law somewhat. For the rest of this discussion, we assume the validity of Equation 13.25.

By the time the kinetic theory of gases was reincarnated around 1850, the ideal-gas law was well known, and it was immediately obvious that the theoretical result expressed by Equation 13.25 has something important to say about the nature of temperature. To make this clear, we write down side by side Equations 13.8 and 13.25:

$$PV = nRT ,$$

$$PV = \tfrac{1}{3}Nm(v^2)_{\text{aver}} .$$

The first is an experimental fact about the thermal behavior of gases. The second is a theoretical result derived from the assumptions of the kinetic theory. If this theory is correct, the right sides of these two equations must be equal. Note that the right side of Equation 13.8, the ideal-gas law, contains three macroscopic quantities: the number of moles n, the gas constant R, and the temperature T. The right side of Equation 13.25, by contrast, is expressed in terms of molecular quantities. The equality of these expressions will provide a vital new link between the macroscopic and submicroscopic worlds and a fundamental definition of temperature.

Since the total number of molecules N in a sample of gas is equal to the number of moles n multiplied by the number of molecules N_0 in a single mole, the equality of interest can be written

$$nRT = \tfrac{1}{3}nN_0m(v^2)_{\text{aver}}\,. \tag{13.26}$$

Division of both sides of this equation by nR yields the new definition of temperature:

$$T = \frac{1}{3}\frac{N_0}{R}\,m(v^2)_{\text{aver}}\,. \tag{13.27}$$

Two small changes of notation enable us to cast this result into even simpler form. First, we recall that R/N_0 is defined as Boltzmann's constant, usually written $\boldsymbol{k}$. Second, we note that mv^2 is twice the kinetic energy of a molecule. Therefore, the molecular definition of temperature may be written more compactly as

$$T = \frac{2}{3k}\,(K.E.)_{\text{aver}}\,. \tag{13.28}$$

This equation makes the important statement that the temperature of a gas is proportional to the average kinetic energy of its molecular constituents. Suddenly temperature is neither a new nor a mysterious concept. It is simply expressed in terms of a familiar concept of mechanics, kinetic energy. The quantity $(2/3k)$ in Equation 13.28 is a constant of proportionality, not itself of fundamental significance. Its role is to tie together the unit of temperature and the unit of energy. The numerical value of Boltzmann's constant k depends on both of these units. In the cgs-Kelvin system, its value is given by Equation 13.12. Just as J, the mechanical equivalent of heat (4.184×10^7 ergs/cal), is a conversion factor joining the units of heat and energy, the constant k is (except for the factor 2/3) a conversion factor joining the units of temperature and energy. Boltzmann's constant could equally well be called the mechanical equivalent of temperature. Turning Equation 13.28 around, we can express molecular kinetic energy in terms of temperature:

$$(K.E.)_{\text{aver}} = \tfrac{1}{2}m(v^2)_{\text{aver}} = \tfrac{3}{2}kT\,. \tag{13.29}$$

Multiplication of the absolute temperature T by the constant $\tfrac{3}{2}k$ gives the average kinetic energy of the molecular constituents of a gas.

If the kinetic theory and its definition of temperature are correct, there are numerous implications for the behavior of matter. We mention here only a few of the myriad of tests that have by now established beyond question the correctness of the kinetic theory. The ideal-gas law is implied, as is Avogadro's hypothesis. Notice in Equation 13.25 that if pressure, volume, and mean molecular energy are the same in two different samples of gas, the number of molecules must also be the same in both. Therefore, if equal molecular energy means equal temperature, Avogadro's hypothesis is true.

More direct tests of the connection between temperature and molecular energy are provided by diffusion and by sound propagation. Diffusion is the intermingling of two different kinds of gas. Sound propagation (in a gas) is the transmission of a pulse of pressure through the gas. Both of these effects occur only because the molecules are in motion, and both proceed at a rate dependent on the molecular speed. From Equation 13.29 it follows that the mean square molecular speed is proportional to temperature and inversely proportional to molecular mass:

$$(v^2)_{\text{aver}} \sim \frac{T}{m}. \tag{13.30}$$

One way to test the correctness of the kinetic definition of temperature is to find out whether molecular speed indeed follows this proportionality. Numerous tests verify that it does. Helium gas, for instance, diffuses through air four times as rapidly as does sulfur dioxide. This suggests that at a given temperature a helium atom moves at four times the speed of a sulfur dioxide molecule. Since a molecule of sulfur dioxide, with a molecular weight of 64, is sixteen times as massive as a helium atom, the proportionality of the square of the speed to the inverse of the mass is verified (at fixed temperature). The important thing is that the kinetic energy of the two molecules is the same. Therefore, the more massive one must move more slowly. A perfume manufacturer who wishes his product to act more swiftly at a distance should seek out a scent with low molecular weight.

As temperature changes, molecular kinetic energy changes, an effect that can be inferred, for example, from studies of the speed of sound in air. Sound propagates through air at a speed proportional to the square root of the absolute temperature. This suggests, according to Equation 13.30, that the speed of sound is proportional to the speed of the gas molecules. Indeed the kinetic theory predicts that this is the case. It is obvious that a sound wave should not make its way through a gas at a speed greater than the molecular speed, for the sound must be propagated by a series of molecular collisions. Somewhat surprisingly, however, the pressure pulse travels at not a great deal less than the molecular speed. In nitrogen (approximately the same as air), the relationship is

$$\text{speed of sound} \cong 0.7 \times \text{speed of molecule}. \tag{13.31}$$

A general rule to remember is that molecules are always somewhat supersonic.

It is always useful to have a few specific physical magnitudes in mind to provide a framework for quantitative thinking—the speed of light, for instance, the size of an atom, the mass of a proton, the location of absolute zero. One of these useful quantities is the speed of sound in air, because it is also indicative of typical molecular speeds. At normal temperature, sound travels at 3.3×10^4 cm/sec in air (about one millionth of the speed of light). This is also about 1,100 ft/sec or 750 miles/hr. A nitrogen molecule in air moves at about 5×10^4 cm/sec, or Mach 1.5. From this it is easy to scale to other temperatures and other molecules. How fast, for example, does a hydrogen molecule move at the surface of the sun where the temperature is 6,000°K? A nitrogen molecule raised from 300° to 6,000° would increase its kinetic energy by a factor of 20 and its speed by a factor of $\sqrt{20}$. Hydrogen, being 14 times less massive than nitrogen, must have at the same energy a speed greater than nitrogen by a factor of $\sqrt{14}$. Altogether, a hydrogen molecule at 6,000° moves faster than a nitrogen molecule at 300° by a factor of $\sqrt{280}$, or 16.7. Its speed is $16.7 \times 5 \times 10^4$ cm/sec $= 8.4 \times 10^5$ cm/sec.

13.8 The molecular basis of heat and internal energy

The most important result of the kinetic theory is the simple connection it establishes between temperature and molecular energy. But if temperature is proportional to energy, is temperature perhaps after all not distinct from the concepts of heat and internal energy? It *is* distinct, as some further discussion of molecular energy can make clear. First, let us eliminate heat from the discussion. Heat is energy transferred by molecular collisions. Temperature and internal energy, on the other hand, are both quantities referring to energy stored in a material. These are the quantities to be distinguished.

For ease of discussion, we continue to fix our attention on nearly ideal gases. For these, temperature measures the mean *kinetic* energy of the molecules. But kinetic energy is only one form of molecular energy. The internal energy of a gas consists of all forms of molecular energy summed together. Internal energy is exactly what its name suggests, the total energy stored within a gas in its molecules. Only part of this energy, that associated with translational motion of the molecules, contributes to the temperature.

When energy is added to air, part of the energy goes into increased kinetic energy of the oxygen and nitrogen molecules. Part increases the rotational energy of these molecules. At high temperatures, part of the energy goes into vibrational motion and other internal motions of the molecules. All of these parts together comprise the internal energy of the air. The temperature of the air, however, is determined only by the translational kinetic energy of the molecules. Recall that the derivation which led up to the kinetic definition of temperature was concerned with the pressure exerted by a gas in a box. Energy in forms other than kinetic energy does not increase the capacity of a molecule to exert pressure. Therefore it does not contribute to the temperature.

The fraction of internal energy in the form of kinetic energy varies from one material to another and from one temperature to another. It depends on details of molecular structure that need not concern us here, for they are not directly related to the fundamental laws of nature. For any particular material at any particular temperature, the relationship between temperature and internal energy is summarized by an experimentally measured quantity called the specific heat. The specific heat of a substance is the amount of heat that must be added to one gram of the substance to raise its temperature by one degree Kelvin. As a general rule, the specific heat of a simple gas at low temperature tends to be small. A *small* specific heat means that little heat is required to produce a given temperature change, which in turn means that a large fraction of the added heat is going into molecular kinetic energy. For a complicated material at high temperature, on the other hand, only a small fraction of the added energy goes into molecular kinetic energy. This implies a *large* specific heat, for more total energy must be supplied to produce the same temperature change.

The specific heat of water at 14°C, although certainly not a fundamental quantity itself, has, as a matter of convenience, been adopted as the definition of the calorie. Using the symbol s for specific heat, we have, as a matter of definition,

$$s(\text{water, } 14°\text{C}) = 1\,\frac{\text{cal}}{\text{gm °K}}\,. \tag{13.32}$$

Compared to most materials at ordinary temperature, water has a high specific heat. Note in Table 13.2 that the specific heats of other common materials are all less than 1 cal/gm °K.

It is left as an exercise to derive the formula

$$\Delta H = sM\Delta T\,, \tag{13.33}$$

which states that the amount of heat ΔH required to raise the temperature of a mass M through a temperature interval ΔT is equal to the product of the specific heat of the material, its mass, and the temperature interval.

TABLE 13.2 Specific Heats of Some Common Materials at Constant Volume and Room Temperature

Substance	Specific Heat (cal/gm °K)
Air	0.17
Carbon dioxide	0.15
Water	1.00
Aluminum	0.22
Iron	0.11
Copper	0.093

Suppose that a gas had a particularly simple structure, such that *all* of its internal energy was in the form of molecular kinetic energy. What would be its specific heat? From the fundamental connection between temperature and kinetic energy it follows that a temperature increment ΔT and a molecular kinetic energy increment $(K.E.)$ are related by

$$\Delta(K.E.) = \tfrac{3}{2}k\Delta T\,. \tag{13.34}$$

Therefore the energy needed *per molecule* to raise the gas temperature by one degree is $\tfrac{3}{2}k$. This quantity divided by the mass of a molecule gives the energy requirement per unit mass for a temperature rise of one degree. Except for the conversion from energy units to heat units, this is the specific heat. Including the conversion factor J (4.184×10^7 ergs/cal), we can write for specific heat of our idealized simple gas

$$s = \frac{1}{J}\frac{3}{2}\frac{k}{m}\,. \tag{13.35}$$

Actually this idealization is not far from the truth for the monatomic gases such as helium, neon, and argon. For helium, whose atom has a mass $m = 6.64 \times 10^{-24}$ gm, this formula predicts

$$s = \frac{1.5 \times 1.38 \times 10^{-16}\,\dfrac{\text{erg}}{°\text{K}}}{4.18 \times 10^{7}\,\dfrac{\text{erg}}{\text{cal}} \times 6.64 \times 10^{-24}\,\text{gm}} = 0.746\,\frac{\text{cal}}{\text{gm °K}}\,.$$

This compares favorably with the experimental value, $s = 0.753$ cal/gm °K.

13.9 *The first law of thermodynamics*

From classical mechanics emerged a fundamental conservation law, the conservation of energy, valid in the absence of heat flow and temperature change. The extension of this law of energy conservation to include heat and internal energy is called the first law of thermodynamics—a powerful principle of nature and an important bridge between the large-scale and small-scale worlds. When bulk motion occurs without significant thermal effects, as in the coasting flight of a space ship, mechanical energy—or macroscopic energy—is conserved. When thermal effects occur without bulk motion, as in the heating of a saucepan by an electric grille, molecular energy—or submicroscopic energy—is conserved. Of even greater interest are examples of energy transformations back and forth between the macroscopic and submicroscopic worlds. For understanding such examples, it is useful to distinguish between ordered energy and disordered energy. Consider a single molecule in your little finger. Suppose at a particular moment it is moving at a speed of 2.5×10^4 cm/sec, or about 550 miles/hr. How hot is your little finger? This is not a question than can be answered, even if you know the mass of the molecule in question. It depends on what the other molecules are doing. Perhaps you are on a jet plane and all of the molecules in your little finger are comfortably flying along together in the same direction at the same speed. This would be an example of *ordered* energy. The ordered kinetic energy of a large group of molecules means simply that the group as a whole are moving together. The ordered motion of molecules does *not* contribute either to temperature or to internal energy. It is counted as macroscopic mechanical motion.

If the same molecule in your little finger were moving in the same direction with the same speed but were not joined in concert by all of its fellow molecules, your sensations might be quite different. Perhaps you have just inadvertently stuck your finger into a flame and the heat transfer has set the molecules into an increased frenzy of random motion. The molecular speeds and the total kinetic energy are the same, but now all of that kinetic energy is *disordered* energy, contributing to internal energy and to increased temperature. Obviously the physical manifestations of ordered and disordered energy are very different, as different as pleasure and pain. Energy transformations back and forth between the macroscopic and submicroscopic worlds are neither more nor less than changes of the form of the energy between ordered and disordered.

Some such energy transformations can be classified with the help of the diagram in Figure 13.8. On the right are two forms of energy: internal energy and bulk kinetic energy. On the left are two modes of energy exchange: heat and work. The arrows indicate schematically four kinds of energy transformation. In every transformation, of course, total energy is conserved.

(1) Heat Flow Without Bulk Motion

When an experimenter, copying a primitive cooking method, drops a hot stone into a container of water, the internal energy of the stone decreases and the internal energy of the water increases. Heat is the mechanism of energy exchange. Via the transformation indicated by arrow 1, the water is warmed as the stone is cooled.

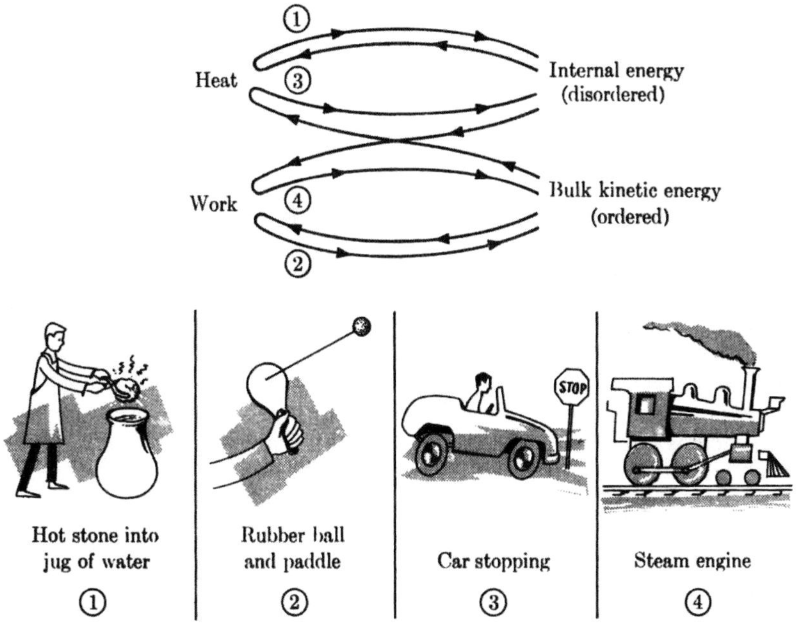

FIGURE 13.8 Some modes of energy transfer. Examples discussed in the text are illustrated by arrows in the diagram and by the corresponding numbered drawings.

(2) Bulk Motion Without Heat Flow

A rubber ball attached to a paddle by an elastic string loses energy (bulk kinetic energy) as it flies away from the paddle. In doing so, it does work on the elastic string. The string in turn does work on the ball as it draws it back to the paddle. As indicated by arrow 2, kinetic energy is lost and then gained, with work as the intermediary.

(3) Frictional Dissipation

A driver steps on the brake pedal. His automobile loses kinetic energy. The brake drums and surrounding parts are heated and gain internal energy. Probably no other form of energy transformation is more familiar in everyday experience than loss of mechanical energy accompanied by gain of internal energy. Since in most applications concerned with the smooth running of machinery, the aim is to preserve mechanical energy, this transformation is usually considered a "loss" of energy. Friction dissipates energy in the same way that a profligate dissipates a fortune, by converting it from ordered to disordered form.

Such energy transformation is also known as energy degradation. In Figure 13.8, the arrow labeled 3 oversimplifies, for work as well as heat transfer may be involved in the energy transformation process. (At what point is work done as an automobile is braked?)

(4) The Ordering of Disordered Energy

When steam in a cylinder expands to push a piston outward, work is done to move the piston and it gains kinetic energy. This ordered bulk energy is extracted from the internal energy of the steam; the steam must therefore be cooled as it expands. Arrow 4 indicates the transformation of the disordered to ordered energy via work. That the steam expanding in the cylinder *must* lose internal energy is a necessary consequence of energy conservation. It is nevertheless interesting to inquire into the exact mechanism of this energy loss. When a molecule of the water vapor rebounds from the retreating cylinder, the molecule loses speed and kinetic energy, tending thereby to lower the temperature of the steam [Figure 13.9(a)], Conversely, a molecule striking an advancing piston recoils with increased speed, like a tennis ball struck by a moving racket. So a gas is cooled by expansion and heated by compression.

A rocket provides a particularly interesting example of the conversion of internal energy to bulk kinetic energy, for in the propulsive action of a rocket, a gas pushes on itself and changes the form of its own energy. An idealized rocket motor is shown in Figure 13.9(b). On the right is a chamber containing hot gas. On the left is the nozzle through which the escaping gas streams. In the chamber the gas has large internal energy and (relative to the rocket) no bulk kinetic energy. As the gas streams out to the left, it acquires ordered energy and must therefore lose some of its disordered energy. Note that in this important transformation process, the gas molecules need suffer no change of speed at all. The change occurs in the relationship of different molecular velocities to each other. Within the hot gas chamber, molecules are moving randomly in all directions and the average velocity (the *vector* average) is zero. In the exhaust, the average molecular velocity is directed to the left, since the gas as a whole is moving to the left. There remains superimposed on this bulk motion some random motion, but less in magnitude than it was before the gas started to expand. (Suppose that the molecules in the exhaust gas moved exactly parallel to one another with equal velocity. What would be the temperature of the exhaust gas?)

13.10 The equipartition theorem and the zeroth law of thermodynamics

73 Like the first law of thermodynamics, the zeroth law finds an explanation in the application of Newtonian mechanics to molecular motion. Although the law of thermal equilibrium—the tendency of systems in contact to come to the same temperature—was accepted as self-evident for a long time before it was dignified with the title of the zeroth law of thermodynamics, its molecular basis is subtler than the molecular basis of energy conservation. It is by no means self-evident that the result of a myriad of collisions among molecules of different mass should be to equalize their average kinetic energy, and thereby to equalize their temperature, yet this is what happens. The understanding of this equalization is rooted in a fundamental statement about energy sharing called the equipartition theorem. The equipartition theorem in turn is a theoretical result derived from the ideas of mechanics and the ideas of probability and statistics applied to molecular collisions.

Imagine a hollow container evacuated except for two lonely molecules of equal mass (a practical impossibility, of course, but a useful idealization for purposes of discussion). Suppose that one of these molecules is moving rapidly, the other slowly. After a time they will collide. In the collision process, energy will be conserved and momentum will be conserved. The precise angles of deflection of the two molecules in the collision cannot be predicted, nor can the energy exchanged between them be predicted. However, a statement about *probability* can be made. The faster molecule is more likely to lose than to gain energy in the collision; the slower molecule is more likely to gain energy than to lose it. This probability implies a tendency toward equalization. In a particular collision, the less energetic molecule might actually lose energy and enhance the discrepancy between the two energies. But after many collisions, the greater probability will win out. Over a long span of time, one molecule will be moving now faster, now slower than the other, in such a way that each one has, on the average, the same energy as the other. This is one example of the equipartition theorem at work. Unequal molecular energies tend to equalize, and once equalized, they remain equal, *on the average.*

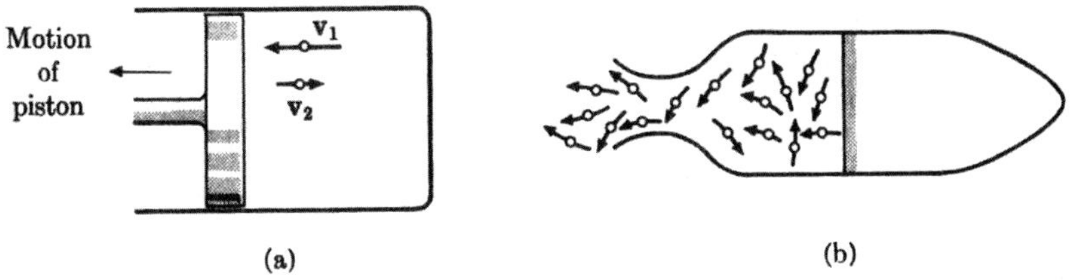

FIGURE 13.9 The molecular mechanism of energy transfer from disordered to ordered form. **(a)** A molecule rebounding from a retreating piston loses energy in the collision. **(b)** The molecules of a hot gas streaming through the nozzle of a rocket convert part of their random motion into ordered bulk motion. The exhaust is at lower temperature.

For two molecules of identical mass, the symmetry of the situation means that it is unthinkable that anything other than equal average energy (and equal average speed) could result from a series of random collisions. There is no possible reason for one of the equal pair to pull ahead of the other. Therefore the prediction of the equipartition theorem that the total available energy is shared equally by the two molecules is hardly surprising. In its full generality, however, the equipartition theorem goes far beyond this example. The theorem describes energy sharing for any number of molecules, for any mixture of molecules of different kind and different mass, and including all kinds of internal energy, not only kinetic energy.

To understand fully the equipartition theorem, it is necessary to grasp one new idea, the concept of degree of freedom. This is not an easy concept to define straightaway, so we shall first circle around it by discussing how to count the number of degrees of freedom. Just as it is easier to count the number of people in a room than it is to say what a person is, it is easier to find out how many degrees of freedom a system possesses than it is to say exactly what one degree of freedom is.

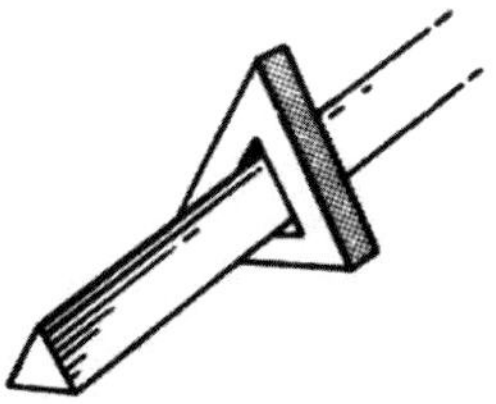

FIGURE 13.10 A bead on a wire, a system with one degree of freedom.

The number of degrees of freedom of a system is the number of coordinates required to specify completely the spatial arrangement of the system. Consider a single bead moving on a fixed straight wire. To simplify matters further, suppose that the bead is prevented from rotating around the wire, as in Figure 13.10. This is a system with just one degree of freedom, for only one single coordinate is required to locate the position of the bead. This coordinate could be the distance of the bead from the left end of the wire or from any other point on the wire. There are various possibilities, any one of which suffices. Evidently the fact that a bead on a wire has only one degree of freedom is associated with the fact that the bead can move in only one dimension. This is the second aspect of "one-ness" for a bead on a wire. It has only one mode of motion. For eight of these beads strung on the same straight wire (Figure 13.11), the number of degrees of freedom of the collection is eight, one for each bead. Eight coordinates are required to locate the beads (that is, to specify the spatial arrangement of the system).

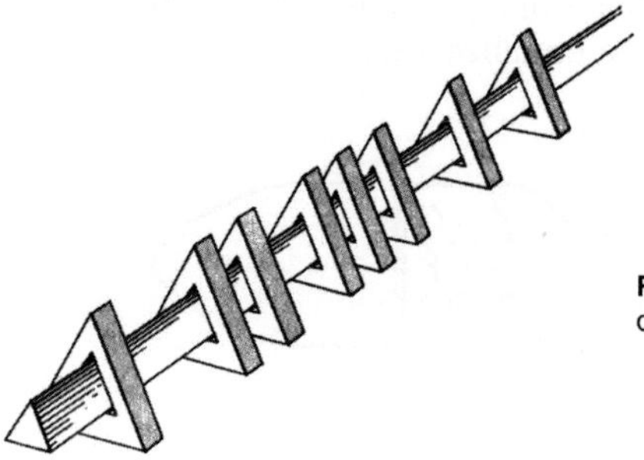

FIGURE 13.11 Eight beads on a wire, a system with eight degrees of freedom.

Consider now another single bead on a straight wire, this one free to rotate. To keep track of its rotation, a stripe is painted on the bead (Figure 13.12). How many degrees of freedom has this system? To find out, we must add up the number of coordinates required to locate precisely the position and the orientation of the bead. Two suffice to do the job. First is a distance coordinate, specifying the location of the bead on the wire. Second is an angular coordinate, specifying how far the stripe on the bead has rotated from some reference position. Once these two facts are known, the spatial arrangement of the system is determined.

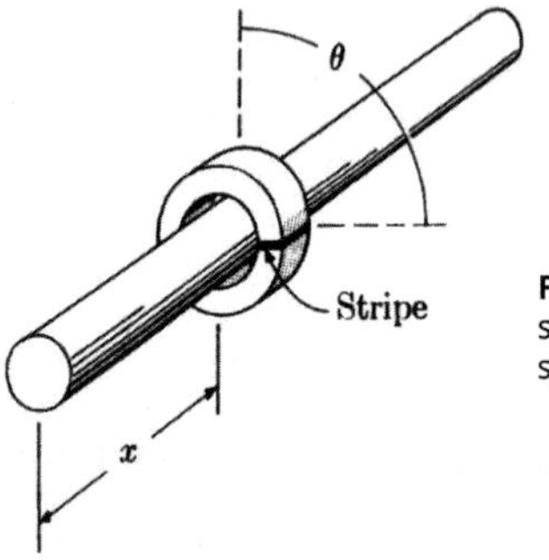

FIGURE 13.12 A bead free to rotate and to slide on a wire. This system has two degrees of freedom. Two coordinates, x and θ, can specify its position exactly.

Counting the independent coordinates of a system in order to learn the number of degrees of freedom focuses attention on the static aspect of a system, the arrangement of its parts at a particular instant. Exactly correlated with the static aspect is a dynamic aspect. The number of degrees of freedom is also equal to the number of independent modes of motion of the system. This is no coincidence. It is merely a different way of looking at the same thing, for an independent mode of motion is defined as a particular motion (possible in principle but perhaps not realized in practice) in which one coordinate changes and all the rest remain fixed. The bead on the wire can move along the wire without rotating, or it can rotate without moving along the wire. These are two independent modes of motion. There are no more, for any other motion must be compounded of these two. In general, the actual motion of a system is a complex intermixing of the independent modes of motion.

Based on this discussion, we can now define a degree of freedom. It is used in two closely related senses. First, it means the *possibility* of an independent mode of motion. The stationary bead on the wire pictured in Figure 13.12 has two possibilities for motion. It can slide or it can rotate. These freedoms exist, whether or not they are exercised. On the other hand, the bead is not free to move away from the wire. A man halfway up a flagpole has more degrees of freedom. He can move up and down; he can move around; or, if he is foolhardy, he can leap away from the pole.

Second, degree of freedom is sometimes used to mean the actual independent mode of motion, not merely the possibility of it. Thus we might say of the bead that half of its energy is in its translational degree of freedom and half in its rotational degree of freedom. This means that its sliding motion accounts for half of its energy and its simultaneous turning motion accounts for the other half.

With the concept of degree of freedom in hand, we can now return to the equipartition theorem. It is literally an equal-division or equal-sharing theorem. It states that after a system comes to equilibrium, its disordered energy is, on the average, equally divided among all of the degrees of freedom of the system. The parcel of energy assigned to each degree of freedom is

$$\text{(Energy/degree of freedom)} = \tfrac{1}{2}kT\,. \tag{13.36}$$

When the energy is not equally divided, the direction of spontaneous change is toward equal division.

Note the limitation in the statement of this theorem to disordered energy. For a space vehicle hurtling toward the moon, the *ordered* energy is by no means equally divided among the degrees of freedom. A molecule of nitrogen (N_2, molecular weight 28) and a molecule of carbon dioxide (CO_2, molecular weight 44) in the air within the vehicle are both being carried along at the same velocity toward the moon. The CO_2 molecule, being more massive, has more ordered energy. Its degree of freedom associated with motion toward the moon is more richly supplied with energy than is the same degree of freedom of the N_2 molecule, and there is no tendency for these to equalize. But superimposed on the headlong flight of these two molecules through space is the random motion of each within the space vehicle. Associated with this random motion is disordered kinetic energy. It is this disordered energy that determines the temperature and that is equally shared by the translational degrees of freedom of the molecules. Therefore in its random motion relative to the space vehicle, the speed of the heavier CO_2 molecule is less than the speed of the lighter N_2 molecule.

Among the implications of the equipartition theorem, none is simpler or more important than the zeroth law of thermodynamics. The spontaneous tendency of matter toward uniform temperature is merely a consequence of the spontaneous tendency of molecules in interaction toward equal division of the available energy among their degrees of freedom. Not all molecules have the same number of degrees of freedom. However all have the same number of *translational* degrees of freedom, namely, three. Three coordinates are required to specify the position of a molecule, and an additional number, which will be described in Section 13.12 are required to specify the orientation and arrangement of the parts of the molecule. Associated with each coordinate is a mode of motion and a degree of freedom. In whatever way unlike molecules might differ internally, all have in common three degrees of freedom for their motion from place to place. Therefore, according to the equipartition theorem, they tend toward equal kinetic energy. Since temperature is determined only by molecular kinetic energy and not by any other forms of molecular energy, the trend toward equal kinetic energy is the same as a trend toward uniform temperature.

Our beads on a wire can serve to render this idea of energy sharing simple and concrete. Imagine a stretched wire on which are threaded six beads, three of them free to rotate, three of them able only to slide (Figure 13.13). These bear no resemblance to molecules, but they form a "system," one to which the equipartition theorem can be applied if we pretend that in collisions of the beads with each other and with the end plates no energy is degraded or dissipated. Initially five of the beads are at rest and the sixth (one of those free to turn) is sent sliding and turning down the wire with an energy of 90 ergs. At every collision energy is exchanged. According to the equipartition theorem, the result of numerous interactions (collisions) among the parts of the system is to divide the energy equally—on the average—among the degrees of freedom. Each gets 10 ergs. After enough time has elapsed for the originally concentrated energy to be spread among the beads—that is, after equilibrium is established—we can imagine a series of energy measurements. No two would give exactly the same distribution of energy. The first bead at one moment might have a total energy of 16 ergs, at a later time 21 ergs, and so on. Its average energy after many measurements would be 20 (or very nearly 20) ergs because of its two degrees of freedom. Similarly beads 2 and 3 would each have average energies of 20 ergs, while the last three beads, each restricted to a single mode of motion, would equally share the last 30 ergs. Although some of the beads have twice as much energy as others, all have the same average *translational* energy. Ten ergs is associated with the sliding motion of each of the six beads. The rotational energy of three of the beads accounts for the remaining energy. If a bead "temperature" were defined proportional to translational kinetic energy, the system would come to uniform temperature despite its asymmetric distribution of total energy. In just the same way, a pan and the water it contains tend toward the same temperature despite the fact that the internal energy of a kilogram of water is much greater than the internal energy of a kilogram of pan.

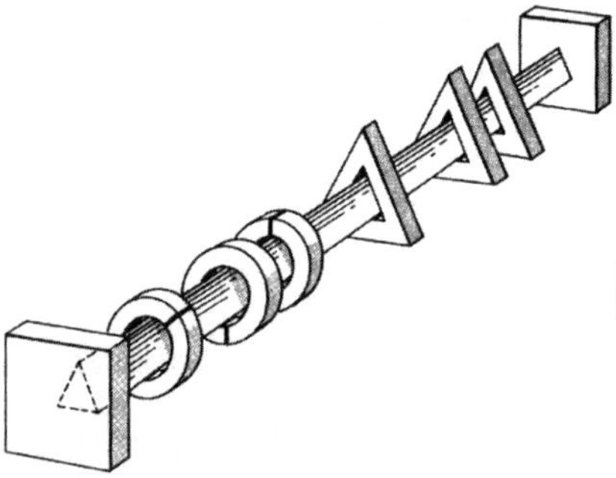

FIGURE 13.13 A system with nine degrees of freedom. Three of the beads are free to rotate; three are not.

In this discussion we have used the word "equilibrium." In macroscopic mechanics, equilibrium refers to a static situation, one in which forces and torques vanish and there is no motion. In thermal equilibrium, on the other hand, a system is far from static. Its molecular constituents are in constant violent agitation. It is only the average properties that remain constant.

13.11 Brownian motion

The equipartition theorem played two very important roles in physics at the turn of this century. Max Planck and others had attempted to apply the theorem to the sharing of energy by matter in the walls of a container and radiation trapped within the container. For the good and simple reason that radiation adds to the system an infinite number of degrees of freedom, they ran into trouble. Planck concluded that the classical equipartition theorem could not be applied to radiant energy. Yet he was unwilling to abandon the search for an understanding of this energy sharing. Instead, in 1900, he abandoned classical physics. Eventually, in the quantum theory that he introduced, was found a new form of the equipartition theorem valid for radiation as well as matter.

Five years later, in a less revolutionary but still dramatic application of the equipartition theorem, Albert Einstein showed (in a theoretical paper) how simple observations with a microscope could reveal the masses of molecules. His key idea was simple but new. If the equipartition theorem describes the sharing of energy between two molecules of quite different size and mass, why should it not also describe energy sharing between a molecule and a larger bit of matter, of any size whatever?

In 1827, an English botanist, Robert Brown, had noticed and had duly reported that grains of pollen suspended in a liquid appear under a microscope to be in a constant state of agitation, moving erratically from place to place. This phenomenon, which came to be known as the Brownian motion, went unexplained for 50 years. Brown at first attributed the motion to a mobility possessed by the pollen grains by virtue of their animate nature. Later he found the motion to be shared by all microscopic bits of matter, animate or inanimate. Until Einstein showed how to extract vital information about molecules from the Brownian motion, the phenomenon was not studied quantitatively.

Einstein made a simple assumption and worked out the mathematical consequence of his assumption. He assumed that the microscopically visible speck of matter is a participant in the incessant dance of collision, energy exchange, and energy sharing on exactly the same footing as an invisible molecule. This means, first of all, that in thermal equilibrium, the speck acquires an average translational kinetic energy equal to that of each molecule. If the energy of the speck could be measured, the energy of the molecule could be inferred. It should be remarked that in 1905 molecular masses were not well known. Therefore molecular energies were not well known. The kinetic theory was well established, but one vital constant, Avogadro's number, was known only in rough approximation. This constant, equal to the number of particles in a mole, links the macroscopic gas constant R and the submicroscopic gas constant (Boltzmann's constant) k:

$$R = N_0 k \quad \text{(equivalent to Equation 13.12).} \tag{13.37}$$

Boltzmann's constant in turn links temperature and molecular energy:

$$\tfrac{1}{2}m(v^2)_{\text{aver}} = \tfrac{3}{2}kT \qquad \text{(same as Equation 13.29).} \tag{13.38}$$

Finally, according to Einstein's assumption, a molecule and a speck of dust share translational kinetic energy equally:

$$\tfrac{1}{2}M(V^2)_{\text{aver}} = \tfrac{1}{2}m(v^2)_{\text{aver}}\,. \tag{13.39}$$

In this equation, M and V denote the mass and speed of the speck, m and v denote the mass and speed of the molecule.

Even with the help of microscopic observation, it is not possible to measure the average kinetic energy of a speck of dust in a liquid or a gas. Its changes of speed and direction are too frequent. Nevertheless, let us pretend for a moment that such a measurement is possible, for then we can understand the nature of Einstein's reasoning in simplified form. A measurement of the kinetic energy of the speck would at once reveal the kinetic energy of the molecule, according to Equation 13.39. This energy in turn would yield the value of Boltzmann's constant k from Equation 13.38. Then, since the macroscopic gas constant is well known from measurements of gas properties, Equation 13.37 would give a value for Avogadro's number. With this constant in hand, the mass of any molecule can be determined from the molecular weight of a substance. Through these steps a single molecule could be "weighed" by measuring the average energy of a visible bit of matter.

The procedure suggested by Einstein to do the job of weighing molecules is somewhat more indirect than the procedure just described, but its basic idea is the same—to use the speck of dust as a visible link to the submicroscopic world. Since the speck is rapidly bombarded from all sides by molecules, its migration through the liquid depends on slight irregularities, or statistical fluctuations. During a particular interval of time, more molecules happen to strike it from one side than another. If molecules were much smaller even than they actually are, and the total number of them striking the speck correspondingly greater, the fluctuations would be reduced. In one second, the impacts of the molecules with the speck would be nearly equal in all directions, and the speck would show little tendency to move. If, on the other hand, molecules were much more massive, and therefore less numerous, the speck would be jarred into more erratic motion. The idea that fluctuations become smaller as numbers become larger is familiar from populations. The ratio of males to females is very nearly the same in England as in the United States. But the ratio of males to females fluctuates greatly from one family to another. The fluctuations in Brownian motion afforded Einstein the key to discover molecular number and therefore molecular mass.

By making use of fluctuations, Einstein's analysis of Brownian motion goes beyond the equipartition theorem and provides a more searching test of the kinetic theory of matter. However, the essential point is not whether one or another aspect of the kinetic theory is tested, but that the key ideas of random motion, collisions, and energy sharing are extended from molecules to any and all bits of matter. Many a forward stride in science has resulted when a scientist took seriously some already existing law or equation or theorem and pushed it into a new domain. The work of Planck and the work of Einstein were both of this kind. The results were different. When Planck tried to enlarge the equipartition theorem to encompass radiation, he failed at first and had to introduce an entirely new idea, the quantum idea. When Einstein tried to extend the equipartition theorem—and fluctuations therefrom—to encompass microscopically visible bits of matter, he succeeded brilliantly, accounting for the Brownian motion, providing a new and rather accurate way to determine Avogadro's number, and putting man into closer touch with individual molecules than he had been before.*

13.12 Specific heat of simple gases

74 For 13 beads strung on the same straight wire, each one capable of sliding and turning, the number of degrees of freedom of the collection is 26, two for each bead. Consider now 13 molecules, and suppose that these occupy an otherwise empty cubical box. How many degrees of freedom has the system? To answer the question, we must add up the number of coordinates required to specify the spatial arrangement of the system precisely. Since the molecules are all alike, it is sufficient to find out how many coordinates are required for each one and multiply that number by 13.

To locate a particular molecule in the box, we must first of all specify three coordinates, which might be the x, y, and z coordinates of its center of mass with respect to a corner of the box (Figure 13.14). Corresponding to these three coordinates, the molecule has three translational modes of motion or three translational degrees of freedom. Having located the spot in the box occupied by the molecule, we need some more coordinates to specify its orientation with respect to the sides of the box, since orientation is part of the total spatial arrangement. In Figure 13.15, the hydrogen molecule is depicted as a dumbbell-shaped object. Three angular coordinates serve to specify its orientation. The first of these (called θ) is the angle that the molecular axis makes with the z-axis. The second (called φ) is the angle that the projected molecular axis† makes with the x-axis. The third (called ψ) is the angle of rotation of the molecule about its own axis, analogous to the rotation of a bead on a wire. It would appear that there are three more degrees of freedom associated with molecular orientation, to make a total of six. Now we come face to face with the first of two peculiarities added to the degree of freedom concept by the theory of quantum mechanics. This is

* As luck would have it, entirely different avenues of science were leading at the same time to still closer touch with single atoms. By 1904 the bursts of nuclear energy released by radioactive atoms were being individually detected. The Geiger counter, invented in 1907, is still in use today along with a variety of other tools that are sensitive to single particles.

† Imagine a light shining vertically downward on the molecule. Its shadow in the x-y plane is called its projection onto the x-y plane. The angle φ measures the angle between the x axis and the axis of this imaginary shadow.

that rotation about the molecular symmetry axis, indicated by the angle ψ does not in fact correspond either to a mode of motion or to a degree of freedom. The molecule actually has only two orientational degrees of freedom.

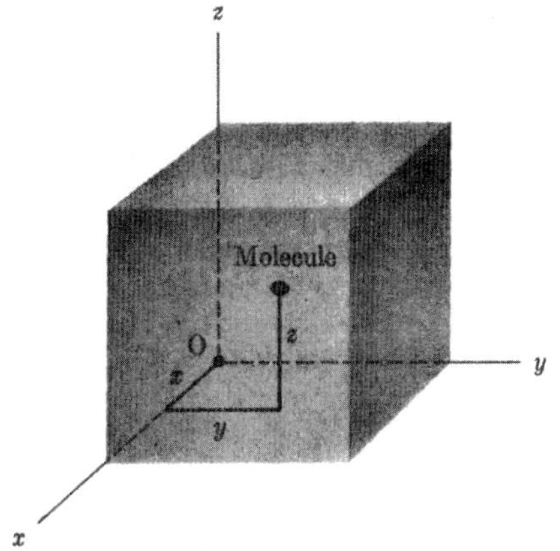

FIGURE 13.14 Three coordinates are required to specify the position of the center of mass of a molecule in space. These correspond to the three translational degrees of freedom of the molecule.

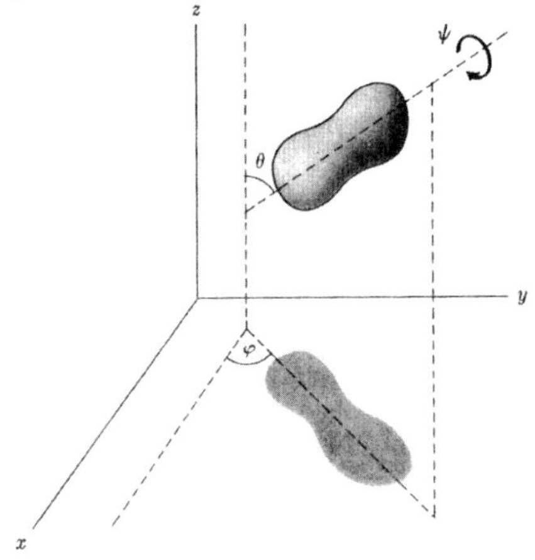

FIGURE 13.15 Three coordinates are required to specify the angular orientation in space of an object. A diatomic molecule, depicted here as a fattened dumbbell, in fact has only two rotational degrees of freedom, associated with angles θ and φ, because it is unable to rotate about its own axis (angle ψ).

A real dumbbell, held in the hand, can certainly be rotated about its symmetry axis, and this is undoubtedly an independent mode of motion of the dumbbell. To keep track of this rotation, a thin stripe can be painted along the dumbbell. The hydrogen molecule differs from the dumbbell in one all-important respect. There is no such thing as painting a stripe on it or in any other way keeping track of this kind of rotation. After rotation through any angle ψ, the hydrogen molecule is identically and indistinguishably the same as before the rotation. The irrelevance of the angle ψ is a very practical consequence (practical since it changes the number of degrees of freedom) of an almost philosophical question: If there is no experimental way to observe a rotation through an angle ψ, how can this rotation have any meaning? The answer provided by quantum mechanics is that no energy can go into this mode of motion. If a mode of motion has no energy and is unable to acquire any, then for practical purposes it does not exist.

The second peculiarity brought to the degree-of-freedom concept by quantum mechanics is the "frozen" degree of freedom. We shall return to that idea in the next section.

We conclude that our 13 hydrogen molecules have 65 degrees of freedom, 5 each. For this and other simple gases, specific heat is simply related to the number of degrees of freedom per molecule. For the collection of hydrogen molecules, three-fifths of any energy added is apportioned to the translational degrees of freedom and therefore contributes to an increase of temperature. The remaining two-fifths of the added energy is apportioned to the rotational degrees of freedom, where it does not influence the temperature. Evidently the larger the number of degrees of freedom, the larger the specific heat, for more degrees of freedom require more energy input to achieve the same temperature rise.

The specific heat of molecular hydrogen can be worked out mathematically with ease. Since each degree of freedom is accorded an average energy equal to $\frac{1}{2}kT$ in thermal equilibrium, a single hydrogen molecule has an average energy equal to $\frac{5}{2}kT$, and a collection of N hydrogen molecules has a total energy given by

$$E = \tfrac{5}{2}NkT\,. \tag{13.40}$$

An increase of the energy E by an amount ΔE results in a temperature increase ΔT governed by a similar equation,

$$\Delta E = \tfrac{5}{2}Nk\Delta T\,. \tag{13.41}$$

Division by the conversion factor J changes energy units to heat units, so that the increment of heat ΔH required to produce the temperature increase ΔT is given by

$$\Delta H = \frac{5}{2}\frac{N}{J}k\Delta T\,. \tag{13.42}$$

This relationship, a proportionality between heat increment and temperature change, is similar to Equation 13.33, which defines specific heat. Comparison of Equation 13.33 and Equation 13.42 yields a new equality,

$$sM = \frac{5}{2}\frac{N}{J}k\,. \tag{13.43}$$

Some simplification results if the mass of the sample of gas is written as the product of the mass to of a single molecule and the number N of molecules:

$$M = Nm\,. \tag{13.44}$$

After this substitution, Equation 13.43 yields

$$s = \frac{5}{2}\frac{1}{J}\frac{k}{m}\,. \tag{13.45}$$

The quantity s is the specific heat of molecular hydrogen at constant volume. If the volume of the gas were allowed to increase as heat is added, additional energy would be expended in the work of bulk motion, and the specific heat would be larger. To make clear the restriction to constant volume, a subscript V is sometimes added to the symbol for specific heat, making it, in our notation s_V.

In Section 13.8, an idealized gas with only three degrees of freedom per molecule was discussed, and its specific heat (at constant volume) was given to be (Equation 13.35)

$$s_V = \frac{3}{2}\frac{1}{J}\frac{k}{m}\,.$$

This formula in fact describes accurately the specific heat of the rare gases (helium, neon, argon, xenon, and radon). Each of these gases is monatomic—that is, its "molecule" is a single atom. Moreover, the rare-gas atom is spherically symmetric, so that after any rotation, the atom is indistinguishable from the same atom before the rotation. Just as rotation of a hydrogen molecule about its symmetry axis is an irrelevant mode of motion that acquires no energy, the rotation of a rare-gas atom about *any* axis is irrelevant and acquires no energy. Therefore, a rare-gas atom has only three degrees of freedom; to be more precise, it has three degrees of freedom that acquire energy and participate in the sharing dictated by the equipartition theorem. Notice that Equation 13.45 for hydrogen (and other diatomic gases) and Equation 13.35 for helium (and other monatomic gases) differ only in the numerical factor, $5/2$ in the first and $3/2$ in the second.

13.13 Frozen degrees of freedom

If two marbles collide, they rebound with less total kinetic energy than they had before the 75
collision. Each is warmed slightly by the collision, since the lost energy has spread as internal energy within the marbles. Some of the original macroscopic ordered energy has been converted to microscopic disordered energy. This is a simple example of the equipartition theorem at work. Energy concentrated initially in one mode of motion tends to distribute itself evenly over all of the available modes of motion.

The situation is usually very different for atomic collisions. A helium atom, like a marble, is a system composed of smaller constituents. Yet when two helium atoms collide at low energy, there is no energy dissipation. The total kinetic energy of the atoms after the collision is exactly the same as before, and the internal structure of the atoms is completely unaffected by the collision. This behavior is a consequence of *quantization* in the atom and is something that could not have been foreseen or explained in the framework of classical physics. If such collisions without dissipation could be arranged in the macroscopic world, it would make possible perpetual-motion machines. Indeed, on the atomic scale, perpetual motion is a reality.

The internal energy of a single atom is quantized; that is, it can take on only a certain definite set of values. For each atom there is a lowest energy state, called the ground state. At higher energies lie the "excited states" of the atom. Between the ground state and the first excited state is an energy gap that represents the smallest amount of energy the atom can absorb. This minimum energy required to excite the atom is called an energy quantum. If offered less than the required energy quantum, the atom cannot accept it. Instead it remains unchanged in its ground state.

Picture, for instance, a collision between two helium atoms, one initially at rest, the other having an initial energy of 0.2 eV. The collision results in energy sharing. One atom gains energy, the other loses energy. Because the minimum energy quantum to excite a helium atom is 20 eV, neither atom can be disturbed internally by the collision. The kinetic energy of a whole atom, on the other hand, is not restricted by such a quantum condition. The energies of motion of the atoms after the collision might be 0.12 eV and 0.08 eV, or 0.195 eV and 0.005 eV, or any other pair of values adding up to 0.2 eV. Energy sharing takes place only among the translational degrees of freedom of the whole atoms. The internal degrees of freedom associated with the motion of the electrons within the atom (and the protons and neutrons within the nucleus), because of their all-or-nothing requirement for energy gain, get none of the energy. These degrees of freedom we call "frozen." Since they do not ordinarily participate in energy sharing, they can be ignored. Indeed they must be ignored. To assign more than three degrees of freedom to a helium atom would lead to incorrect predictions (for example, of the specific heat).

This idea of frozen degrees of freedom was important at the birth of quantum theory. When physicists had tried to apply the equipartition theorem to the sharing of energy among a collection of molecules and electromagnetic waves trapped in a hollow box, They came up against the apparent infinite number of degrees of freedom of the electromagnetic radiation field. To explain why the waves did not take all the energy, leaving none for the matter, Planck postulated that a minimum quantum of energy was required to excite an electromagnetic wave. In order to fit the already known facts on how the energy distributed itself in equilibrium, Planck had to assume that the minimum energy quantum increased in proportion to the frequency of the radiation, according to the simple formula,

$$E = hf, \tag{13.46}$$

in which E is the minimum energy quantum, f is the frequency of the radiation, and h is a constant of proportionality now known as Planck's constant. By this revolutionary quantum postulate, Planck succeeded in "freezing out" the high-frequency radiation. Whenever less than the photon quantum energy is available, no energy at all can go into radiation of a particular frequency. The infinitely many degrees of freedom associated with infinitely high frequencies became frozen degrees of freedom. The equipartition theorem was saved, the facts of electromagnetic energy sharing were explained, and the quantum theory was born.

The full description of Planck's quantum theory is more complicated than is indicated by this discussion. Degrees of freedom, like butter, freeze only gradually. A particular mode of motion (or a particular electromagnetic wave) is never totally excluded from energy sharing, although its exclusion may be very nearly complete. Consider a sample of helium in which the *average* kinetic energy of each atom is 0.2 eV. In almost all collisions, the available energy will be less than the 20 eV required to produce internal excitation of an atom. Occasionally, however, a single atom will by chance acquire a kinetic energy greater than 20 eV. When it collides with a neighboring atom, one of them may emerge from the collision in an excited state of internal motion. The internal degrees of freedom are almost, but not completely, frozen out of the energy sharing. When the average kinetic energy of the atoms approaches the quantum energy of internal excitation, such collisions producing internal excitation occur more often, and the internal degrees of freedom become gradually unfrozen.

At normal temperatures near 300°K, the average molecular kinetic energy of thermal motion is about 0.04 eV. Typical quantum energies of electronic excitation in atoms and molecules amount to several eV, about one hundredfold greater than average thermal energies. (Helium is not typical. Its quantum excitation energy of 20 eV is greater than that of any other atom.) This means that the internal electronic degrees of freedom are quite effectively frozen out. When heat is added to a substance at normal temperature, a negligible part of it goes into atomic excitation. At much higher temperatures the situation changes. When collision energies become large enough to excite atoms internally, more degrees of freedom come into play, and much of the energy of the system is in the form of atomic excitation, a smaller fraction residing in the translational kinetic energy that defines temperature. As a consequence, the specific heat of most substances increases as temperature increases; the greater the number of "unfrozen" degrees of freedom, the more energy is required to produce a given temperature rise. Conversely, near absolute zero, where most degrees of freedom are frozen, specific heats are very small.

The vitally important factors governing the properties of a substance are its characteristic quantum energies relative to the characteristic thermal energy, $\frac{3}{2}kT$. The behavior of a piece of matter at any particular temperature depends more than anything else on which of its degrees of freedom are open for energy sharing, and which are frozen. As an originally solid piece of matter is heated, it experiences successive marked changes—melting, boiling, dissociation, ionization— as successively higher quantum energies topple under the impact of increasing thermal energy. Ice melts at 273°K, when its thermal energy of molecular vibration becomes comparable to the energies binding one molecule to another. At the boiling point of water, 373°K (where the kinetic energy $\frac{3}{2}kT$ is about one twentieth of an eV), the translational degrees of freedom become completely unfrozen. Melting and boiling are examples of "phase changes" occurring at well-defined temperatures. Transformations of form occurring at still higher temperature are gradual, requiring thousands of degrees or more to be completed. Water molecules dissociate (break apart into their constituent atoms) in the range 1,000 to 10,000°K. Ionization (the separation of electrons from their parent atoms) begins in this temperature range and is completed only above one million degrees. At such temperatures, which occur within stars, the gas of protons, electrons, and oxygen nuclei obviously bears little resemblance to ice or water or steam. This ionized gas is called a plasma. Because its particles are electrically charged, it has properties radically different from those of ordinary neutral gases. In the stellar plasma at a temperature of millions of degrees, the energy of thermal motion, $\frac{3}{2}kT$, is large compared to most atomic quantum energies, but is still small compared to the quantum excitation energy of the oxygen nucleus. Despite the intense heat, the internal degrees of freedom of the protons and neutrons in the nucleus remain frozen. Collisions leave the nucleus undisturbed in its ground state. Only if the temperature were to rise to many billions of degrees would the nuclear degrees of freedom thaw, allowing the individual nuclear particles to share energy. It is likely that such enormous temperatures are actually reached in the late stages of the evolution of a star.

TABLE 13.3 Some Temperature-Energy Comparisons

Temperature (°K)	$\frac{3}{2}kT$	Remarks
0	0	
273	0.035 eV	Water freezes
1,000	0.13 eV	Most substances gaseous
6,000	0.78 eV	Sun's surface temperature
10^6 (one million)	130 eV	Light elements largely ionized
10^7 (ten million)	1.3 keV	Sun's central temperature
10^9 (one billion)	130 keV	
10^{10} (ten billion)	1.3 MeV	Thermal excitation and disintegration of nuclei

As a help for thinking panoramically over the temperature range from absolute zero to billions of degrees, Table 13.3 shows some temperature-energy comparisons. It is worth noticing that the energies in this table are all small compared to those commonly achieved in particle accelerators. In the explosion of an atomic bomb a temperature of 40 million degrees might be reached, giving to each particle an energy of about 5 keV. Yet a very modest accelerator produces particles a thousand times more energetic, and some accelerators concentrate six million times more energy on one particle than does a nuclear explosion. The difference of course is in numbers of particles. Vastly many more particles are heated by a nuclear explosion in a microsecond than are accelerated by all the synchrotrons on earth in a year.

EXERCISES

13.1. Find the special form of Equation 13.1 relating pressure and temperature in a constant-volume gas thermometer for the Fahrenheit scale. At the freezing point of water when the gas pressure is P_0, the temperature is defined to be 32°F. At the boiling point of water (at sea level) where the thermometer gas pressure is P_1, the temperature is defined to be 212°F.

13.2. **(1)** Derive a conversion equation for translating a temperature measured in degrees F to the same temperature expressed in degrees C. **(2)** With the help of this equation, fill in the second column of this short table:

°F	°C
0	
32	
68	
80	
98.6	

(3) What is the Fahrenheit temperature of absolute zero?

13.3. The Rankine temperature scale used in some engineering work is defined by these two conditions: **(1)** At absolute zero the Rankine temperature is zero; 0°R = 0°K. **(2)** The difference between two temperatures is the same on the Rankine scale and the Fahrenheit scale, $\Delta T(°R) = \Delta T(°F)$. From these facts, derive the conversion equation linking the Kelvin and Rankine scales. Explain your mode of reasoning. (The Rankine scale belongs to the British system of units, no longer common in scientific work.)

13.4. When you read a household thermometer, you actually measure a length—the height of a column of liquid, or the distance of a pointer from the edge of a scale. **(1)** What physical quantity is measured to determine temperature with a gas thermometer? **(2)** Explain how it is that temperature can depend for its measurement on mechanical quantities yet be a new concept independent of mechanical quantities. Give an example within mechanics of two independent concepts, one of which utilizes the other in its definition.

13.5. Explain carefully why the number of a.m.u./gm is equal to the number of particles per mole, or Avogadro's number. This identity rests on the definition of the mole.

13.6. **(1)** Calculate the number of water molecules in one cup of water, whose mass is 227 gm. How many moles of water molecules is this? **(2)** If the water is dissociated into its atomic constituents, two hydrogen atoms and one oxygen atom from each molecule, how many moles of atomic hydrogen and how many moles of atomic oxygen result? **(3)** Suppose that every atom in this collection could be disintegrated into its elementary particle constituents: electrons, protons, and neutrons. Each hydrogen atom would contribute one electron and one proton. Each oxygen atom would contribute 8 electrons, 8 protons, and 8 neutrons. How many moles of neutrons would the cup of water yield? (This total fragmentation of matter could actually occur, but only at a temperature well in excess of the central temperature of the sun.)

13.7. One mole of an ideal gas occupies 22.4 liters (2.24×10^4 cm^3) at standard conditions. **(1)** Use the density of air given in Appendix 2B to calculate the mass of air in 22.4 liters. **(2)** Find the average molecular weight of air, which is 78% nitrogen, of molecular weight 28; 21% oxygen, of molecular weight 32; and 1% argon, of molecular weight 40. (This is a *weighted* average. Multiply the molecular weight of each component by the fraction of that component, then add.) **(3)** Compare the answers to parts (1) and (2). Does this comparison provide evidence for or against the idea that air behaves like an ideal gas?

13.8. Call the molecular weight of a substance *M.W.*, and the mass of a single molecule of the substance m. Derive a formula for m in terms of *M.W.* and Avogadro's number N_0. This simple formula is worth saving.

13.9. Let *M.W.* represent the molecular weight of a substance, ρ its mass density in gm/cm^3, and u its number density in molecules/cm^3. Derive a formula for u in terms of ρ, *M.W.*, and Avogadro's number N_0. This formula is worth saving.

13.10. **(1)** For an ideal gas at standard conditions (0°C and 1 atmosphere), calculate the number of molecules per cm^3. From this number calculate approximately the average separation between two molecules. **(2)** Calculate the same two quantities for water, whose density is 1 gm/cm^3 and whose molecular weight is 18. **(3)** Calculate the number of atoms per cm^3 and the approximate average separation of atoms in aluminum, whose density is 2.7 gm/cm^3 and whose atomic weight is 27. **(4)** Discuss briefly the significance of the relative spacing of atoms or molecules in these three substances: solid, liquid, and gas.

13.11. The law of equilibrium has nothing to say about the time required to establish equal temperature in two systems which are in thermal contact. Give one example of each of the following: **(a)** a system that comes into equilibrium with another system (or with its surroundings) in 1 sec or less; **(b)** a system that requires a few hours to establish equilibrium with its surroundings; and **(c)** a system that does not reach equilibrium with its surroundings over a time span of millions of years. Do you expect the last-named system to reach equilibrium eventually?

13.12. Explain how the ideal-gas law in the form of Equation 13.8 requires the validity of Avogadro's hypothesis.

13.13. From the ideal-gas law, derive the result that the molar volume at standard conditions (273°K and one atmosphere) is 2.24×10^4 cm^3. (The pressure of one standard atmosphere is 1.013×10^6 dyne/cm^2.)

13.14. Most airplanes operate noticeably better on a cool day than on a warm day. If the air pressure is the same on a winter day at 0°C and on a summer day at 30°C, by what fraction does the air density differ on these two days? Which day has denser air?

13.15. Three sealed containers of gas, all of different volume, lost their identifying tags. When found, the tags bore these labels: **(a)** 2 moles of helium at 300°K and 5×10^6 dyne/cm^2; **(b)** 6×10^{23} neon atoms at 0°C and 2 atmospheres pressure; and **(c)** 7 gm of N_2 at room temperature and 20 atmospheres pressure. Which label belongs to the largest container, which to the one of intermediate size, and which to the smallest container? Would the smallest container fit in your pocket?

13.16. What is the expansion factor when water turns to steam at 373°K and one atmosphere pressure? Use a water density of 1 gm/cm^3, air pressure of 10^6 dyne/cm^2, and assume that the steam behaves like an ideal gas.

13.17. A tire gauge reads pressure above atmospheric. Before a trip, an automobile tire containing 6 moles of air at 300°K shows a gauge pressure of 1.5 atmospheres; this means that the pressure within the tire is 2.5 atmospheres. Later, after running for a time, the tire shows a gauge pressure of 2.0 atmospheres. **(1)** If the tire volume has not changed, what is the new temperature of the air within it? **(2)** If, at this higher temperature, the tire is deflated back to its original pressure, how many air molecules escape from the tire?

13.18. A furnace raises the air temperature in a house from 10°C to 20°C. It does so without changing the volume of the house or the pressure of the air in the house. How can this be accomplished without violating the ideal-gas law? What, besides the temperature, must also change, and by what fraction?

13.19. If the energy required to vaporize one gram of boiling water, 539 calories, were used instead to lift the gram of water in the earth's gravitational field, to what altitude could it be raised? Ignore the variation of gravitational force with height, but comment on whether it is a good approximation to do so.

13.20. Use Equation 13.29 to verify that the average kinetic energy per molecule of a gas at room temperature is about 0.04 eV.

13.21. Nitrogen molecules in the air enclosed within an orbiting space vehicle are traveling with the vehicle at a speed of 17,000 miles/hr (7.6×10^5 cm/sec). What would be the temperature of a sample of nitrogen gas at rest if its molecules were moving randomly at this speed?

13.22. Estimate numerically the speed of sound in molecular hydrogen (H_2) at 6,000°K (the surface temperature of the sun), and compare it with the speed of sound in air at normal temperature.

13.23. Explain in terms of the kinetic theory why the speed of sound in air at high altitude is nearly the same as the speed of sound at sea level, despite a big difference in air density.

13.24. Explain in terms of molecular speeds and molecular collisions why the speed of sound in water is considerably greater (actually more than four times greater) than the speed of sound in air at the same temperature.

13.25. **(1)** Derive a formula for the average speed of a gas molecule in terms of its molecular weight *M.W.*, the absolute temperature *T*, and the gas constant *R*. Equation 13.29 is a good starting point. **(2)** Find the average speed at 0°C of the principal molecular constituents of air: nitrogen, *M.W.* = 28; oxygen, *M.W.* = 32; and argon, *M.W.* = 40.

13.26. A molecule of water vapor ($m = 3.0 \times 10^{-23}$ gm) in a rocket exhaust nozzle has a kinetic energy of 0.2 eV. **(1)** Find its speed and compare this with the speed of an astronaut in a 90-minute orbit. **(2)** If the average random molecular kinetic energy in a sample of material were 0.2 eV, what would be its temperature?

13.27. Explain the following phenomena from the microscopic (i.e., kinetic theory) viewpoint. **(1)** A gas released in free space would expand indefinitely; a solid or liquid would not. **(2)** A swimmer emerging from a heated pool into warm air feels chilly. **(3)** The pressure in an automobile tire increases after the automobile has been driven awhile. (Do more than cite the ideal-gas law; explain in molecular terms.)

13.28. Derive Equation 13.33, which expresses the heat required to raise the temperature of a sample of material through a given interval. Your "derivation" may be a careful verbal argument. Under what circumstances would you expect this formula to be invalid?

13.29. The specific heat of copper is 0.092 cal/gm °K. **(1)** How much heat is required to raise the temperature of a 10-gm piece of copper from 0°C to 100°C? **(2)** How does this energy compare with the energy required to accelerate the same piece of copper from rest to a speed of Mach 1 (3.4×10^4 cm/sec) ?

13.30. Into a 200-gm saucepan initially at 20°C is poured 800 gm of boiling water at 100°C. When equilibrium is established, water and pan are at 93°C. If half of the internal energy lost by the water went into heating the pan (the other half being dissipated to the surroundings), what is the specific heat of the metal in the pan?

13.31. When a space vehicle is decelerated upon reentering the atmosphere, it is heated. Discuss this phenomenon of reentry heating in molecular terms, basing your discussion on the idea of ordered and disordered energy, the kinetic definition of temperature, and the first law of thermodynamics.

13.32. Suppose that a rocket like the one pictured in Figure 13.9(b) is propelled by an ideal monatomic gas, one whose internal energy is entirely in the form of translational kinetic energy of the molecules (see Equations 13.34 and 13.35). Initially the gas, with all of its energy in disordered form, is at 2,000°K. As it leaves the exhaust nozzle it is at 1,000°K. Its total energy has not changed. **(1)** If the atomic weight of the gas is 4, what was the average speed of its molecules initially? **(2)** What is the exhaust velocity, that is, the velocity of the ordered bulk motion of the gas leaving the nozzle?

13.33. An automobile with a mass of 1.5×10^6 gm traveling at a speed of 1,500 cm/sec (about 33 miles/hr) is braked to a halt. As it stops, 80% of its energy is dissipated as heat in the brake linings and surrounding parts. **(1)** What is the magnitude of this heat in calories? **(2)** If this heat is distributed uniformly through 10 kg of iron whose specific heat is 0.1 cal/gm °K, by how much is the temperature of the iron raised? **(3)** What is the eventual fate of the energy that was initially in the form of bulk kinetic energy of the automobile?

13.34. If the wire in Figure 13.10 is bent into a closed circle, around which the bead can slide, how many degrees of freedom has the system? Explain the basis of your answer.

13.35. Show carefully, by mathematical reasoning, that Equation 13.41 follows from Equation 13.40.

13.36. Make the assumption (incorrect) that for ordinary water all the internal energy is in the form of translational kinetic energy of the molecules. What then would be the specific heat of water? Contrast your calculated value with the actual value for water, and discuss the meaning of the difference.

13.37. At sufficiently high temperature a diatomic molecule adds to its translational and rotational motion another mode of motion, vibration along the line joining its two atoms. **(1)** How many relevant degrees of freedom has a molecule when all these modes of motion are available to it? **(2)** By how much does the number of degrees of freedom change when the molecule is dissociated? **(3)** Suggest a reason why the vibration of the diatomic molecule is unimportant at low temperature.

13.38. Sometimes specific heat is expressed in cal/mole °K instead of cal/gm °K. Call this molar specific heat s'. **(1)** Show that Equation 13.33 can be rewritten $\Delta H = s'n\Delta T$, where n = number of moles. **(2)** Show that for an ideal monatomic gas,

$$s' = \frac{3}{2}\frac{R}{J},$$

where R is the gas constant. Work from Equation 13.35.

13.39. The fundamental equation of the kinetic theory, Equation 13.25, is valid if the dimension of a molecule is much less than the average distance between molecules, and if each molecule is, most of the time, free of forces exerted by other molecules. Explain why the finite size of molecules should alter somewhat the idealized kinetic theory. Do you think this effect should cause the actual gas pressure to be greater or less than the ideal pressure given by Equation 13.23? (HINT: Think about what would happen in the limit that the molecules are always close together.)

13.40. The fundamental equation of the kinetic theory, Equation 13.25, is valid if the dimension of a molecule is much less than the average distance between two molecules, and if each molecule is, most of the time, free of forces exerted by other molecules. Explain why the forces acting between molecules should alter somewhat the idealized kinetic theory. Do you think this effect should cause the actual pressure to be greater or less than the ideal pressure given by Equation 13.23? (HINT: Think about a solid, in which no atom is ever free of the forces of neighboring atoms.)

13.41. **(1)** Why is the "temperature of a single particle" a meaningless phrase? **(2)** Name a concept other than temperature which also has meaning only for an assembly of many particles. Explain why.

13.42. At a fixed temperature T, a container of volume V contains n_1 moles of gas 1, n_2 moles of gas 2, and n_3 moles of gas 3. If the n_1 moles of gas 1 were present alone in the container, the pressure would be p_1. Similarly, with gases 1 and 3 removed, the remaining gas 2 would exert pressure p_2; and gas 3 alone at this temperature and in this volume would exert pressure p_3. These are called partial pressures. Use the approach of the kinetic theory to prove that the actual pressure with all three gases present is the sum of the partial pressures: $P = p_1 + p_2 + p_3$.

13.43. Consider a particular molecule moving back and forth between two walls of a cubical container, as in Figure 13.6. According to Equation 13.21, this molecule exerts on one wall an average force given by mv^2/L. **(1)** Pick reasonable values of m, v, and L and calculate a typical magnitude of this average force. **(2)** By any method, estimate the actual magnitude of the force exerted by this molecule on one wall during the brief instant of its collision. (3) What is the approximate ratio of the actual force in one collision to the average force over many collisions? Compare the estimated actual force with some known force such as the weight of an object in order to make clear how large it is. (NOTE: Only order of magnitude calculations are called for here.)

13.44. This exercise is intended especially for chemistry students. An outside reference is required for some of the needed data. **(1)** Derive this conversion equation: 1 eV/molecule = 23 kilocalories/mole. **(2)** What is the heat of vaporization of water in eV/molecule? **(3)** Compare this with the mean kinetic energy of a water molecule at 373 K, also expressed in eV. **(4)** What is the binding energy of a water molecule in eV? **(5)** At about what temperature would you expect water vapor to "boil," that is, to dissociate into hydrogen and oxygen?

FOURTEEN

Entropy and the second law of thermodynamics

As profound as any principle in physics is the second law of thermodynamics. Based on uncertainty and probability in the submicroscopic world, it accounts for definite rules of change in the macroscopic world. We shall approach this law, and a new concept, entropy, that goes with it, by considering some aspects of probability. Through the idea of probability comes the deepest understanding of spontaneous change in nature.

14.1 Probability in nature

When a spelunker starts down an unexplored cavern, he does not know how far he will get 76
or what he will find. When a gambler throws a pair of dice, he does not know what number will turn up. When a prospector holds his Geiger counter over a vein of uranium ore, he does not know how many radioactive particles he will count in a minute, even if he counted exactly the number in a preceding minute. These are three quite different kinds of uncertainty, and all of them are familiar to the scientist.

The spelunker cannot predict because of total ignorance of what lies ahead. He is in a situation that, so far as he knows, has never occurred before. He is like a scientist exploring an entirely new avenue of research. He can make educated guesses about what might happen, but he can neither say what will happen, nor even assess the probability of any particular outcome of the exploration. His is a situation of uncertain knowledge *and* uncertain probability. The gambler is in a better position. He has uncertain knowledge but certain probability. He knows all the possible outcomes of his throw and knows exactly the chance that any particular outcome will actually occur. His ignorance of any single result is tempered by a definite knowledge of average results.

The probability of atomic multitudes, which is the same as the probability of the gambler, is at the heart of this chapter. It forms the basis for the explanation of some of the most important aspects of the behavior of matter in bulk. This kind of probability we can call a probability of ignorance—not the nearly total ignorance of the spelunker in a new cave or the researcher on a new frontier, but the ignorance of certain details called initial conditions. If the gambler knew with enough precision every mechanical detail of the throw of the dice and the frictional properties of the surface onto which they are thrown (the initial conditions) he could (in principle) calculate exactly the outcome of the throw. Similarly, the physicist with enough precise information about the whereabouts and velocities of a collection of atoms at one time could (with an even bigger "in principle"*) calculate their exact arrangement at a later time. Because these details are lacking, probability necessarily enters the picture.

The prospector's uncertainty is of still a different kind. He is coming up against what is, so far as we now know, a fundamental probability of nature, a probability not connected with ignorance of specific details, but rather connected with the operation of the laws of

* Because classical mechanics does not suffice to calculate exactly the outcome of an atomic collision, this hypothetical forecast of future atomic positions and velocities could be extended but a moment forward in time.

nature at the most elementary level. In atomic and nuclear events, such as radioactivity, probability plays a role, even when every possible initial condition is known. This fundamental probability in nature, an essential part of the theory of quantum mechanics, is pursued in Chapter Twenty-Three. In thermodynamics—the study of the average behavior of large numbers of molecules and of the links between the submicroscopic and macroscopic worlds—the fundamental probability in nature is of only secondary importance. It influences the details of individual atomic and molecular collisions, but these details are unknown in any case. Of primary importance is the probability of ignorance stemming from our necessarily scant knowledge of precise details of molecular motion.

The triumphs of thermodynamics are its definite laws of behavior for systems about which we have incomplete knowledge. However, it should be no surprise that laws of probability applied to large enough numbers can become laws of near certainty. The owners of casinos in Nevada are consistent winners.

14.2 *Probability in random events*

77 We turn our attention now to a system that at first sight has little to do with molecules, temperature, or heat. It is a tray of coins (Figure 14.1). For the purposes of some specific calculations, let us suppose that the tray contains just five coins. For this system we wish to conduct a hypothetical experiment and make some theoretical predictions. The experiment consists of giving the tray a sharp up-and-down motion so that all the coins flip into the air and land again in the tray, then counting the number of heads and tails displayed, and repeating this procedure many times. The theoretical problem is to predict how often a particular arrangement of heads and tails will appear.

FIGURE 14.1 A tray of coins, a system governed by laws of probability.

The experiment you can easily carry out yourself. Be sure that the tray is shaken vigorously enough each time so that at least some of the coins flip over. Here let us be concerned with the theory. To begin, we enumerate all possible ways in which the coins can land. This is done pictorially in Table 14.1. There are 32 possible results of a tray shaking.* If all we do is count heads and tails without identifying the coins, the number of possible results is 6 instead of 32 (Table 14.1). Ten of the ways the coins can land yield three heads and two tails. There are also ten different ways to get three tails and two heads. Both four heads and one tail and four tails and one head can be achieved in five ways. Only one arrangement of coins yields five heads, and only one yields five tails. These numbers do not yet constitute a prediction of the results of the experiment. We need a postulate about the actual physical process, and a reasonable one is a postulate of randomness: that every coin is equally likely to land heads up or tails up and that every possible arrangement of the five coins is equally likely. This means that after very many trials, every entry in Table 14.1 should have resulted about $\frac{1}{32}$ of the time. Note, however, that equal probability for each arrangement of coins is

* Since each coin can land in two ways, the total number of ways in which five coins can land is $2 \times 2 \times 2 \times 2 \times 2 = 2^5 = 32$. Three coins could land in 8 different ways (2^3), four coins in 16 ways (2^4), and so on. In how many ways could 10 coins land?

not the same as equal probability for each possible number of heads or tails. After 3,200 trials, for example, we would expect to have seen five heads about 100 times, but three heads and two tails should have showed up ten times more frequently, about 1,000 times. The exact number of appearances of five heads or of three heads and two tails or of any other combination cannot be predicted with certainty. What *can* be stated precisely (provided the postulate of randomness is correct) are probabilities of each such combination. Shown in Table 14.2 are the basic probabilities for all the possible numbers of heads and tails that can appear in a single trial. It is interesting to present these numbers graphically also, as is done in Figure 14.2. The probability of a certain number of heads plotted vs. the numbers of heads gives a bell-shaped curve, high in the middle, low in the wings.

TABLE 14.1 Possible Arrangements of Five Coins

Coin 1	Coin 2	Coin 3	Coin 4	Coin 5	
H	H	H	H	H	1 way to get 5 heads
H	H	H	H	T	
H	H	H	T	H	
H	H	T	H	H	5 ways to get 4 heads and 1 tail
H	T	H	H	H	
T	H	H	H	H	
H	H	H	T	T	
H	H	T	H	T	
H	T	H	H	T	
T	H	H	H	T	
H	H	T	T	H	10 ways to get 3 heads and 2 tails
H	T	H	T	H	
T	H	H	T	H	
H	T	T	H	H	
T	H	T	H	H	
T	T	H	H	H	
H	H	T	T	T	
H	T	H	T	T	
H	T	T	H	T	
H	T	T	T	H	
T	H	H	T	T	10 ways to get 2 heads and 3 tails
T	H	T	H	T	
T	H	T	T	H	
T	T	H	H	T	
T	T	H	T	H	
T	T	T	H	H	
H	T	T	T	T	
T	H	T	T	T	
T	T	H	T	T	5 ways to get 1 head and 4 tails
T	T	T	H	T	
T	T	T	T	H	
T	T	T	T	T	1 way to get 5 tails

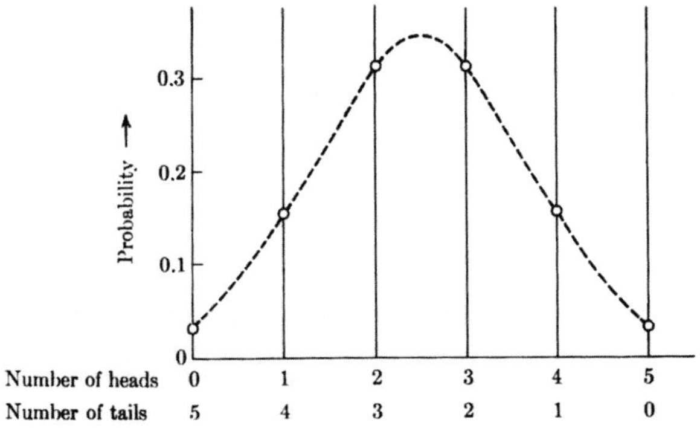

FIGURE 14.2 Probabilities for various results of tray-shaking experiment with five coins.

TABLE 14.2 Probabilities for Different Numbers of Heads and Tails When Five Coins Are Flipped

No. Heads	No. Tails	Probability
5	0	1/32 = 0.031
4	1	5/32 = 0.156
3	2	10/32 = 0.313
2	3	10/32 = 0.313
1	4	5/32 = 0.156
0	5	1/32 = 0.031
		Total probability = 1.000

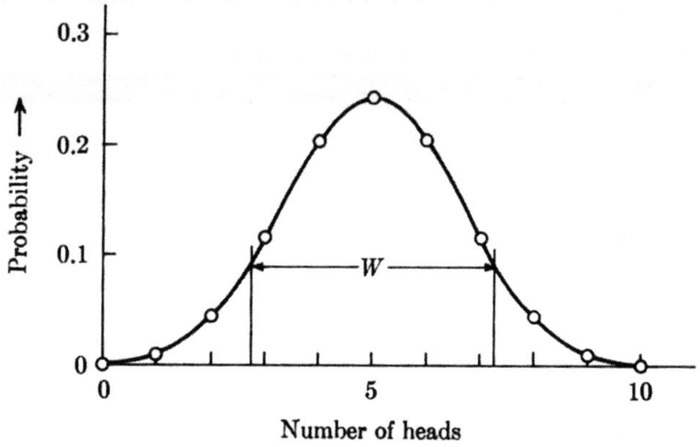

FIGURE 14.3 Probabilities for various results of tray-shaking experiment with ten coins.

TABLE 14.3 Probabilities for Different Numbers of Heads and Tails When Ten Coins Are Flipped

No. Heads	No. Tails	Probability
10	0	1/1024 = 0.0010
9	1	10/1024 = 0.0098
8	2	45/1024 = 0.0439
7	3	120/1024 = 0.1172
6	4	210/1024 = 0.2051
5	5	252/1024 = 0.2460
4	6	210/1024 = 0.2051
3	7	120/1024 = 0.1172
2	8	45/1024 = 0.0439
1	9	10/1024 = 0.0098
0	10	1/1024 = 0.0010
		Total probability = 1.0000

The same kind of calculation, based on the postulate of randomness can be carried out for any number of coins. For ten coins, the basic probabilities are given in Table 14.3 and in Figure 14.3.* Two changes are evident. First, the probability of all heads or all tails is greatly reduced. Second, the bell-shaped probability curve has become relatively narrower. The

* The reader familiar with binomial coefficients may be interested to know that the number of arrangements of n coins to yield m heads is the binomial coefficient

$$\binom{n}{m} = \frac{n!}{m!(n-m)!}.$$

Thus the probabilities in Table 14.3 are proportional to

$$\binom{10}{0}, \quad \binom{10}{1}, \quad \binom{10}{2},$$

and so on.

greater the number of coins, the less likely is it that the result of a single trial will be very different from an equal number of heads and tails. To make this point clear, the probability curve for a tray of 1,000 coins is shown in Figure 14.4. The chance of shaking all heads with this many coins would be entirely negligible even after a lifetime of trying. As Figure 14.4 shows, there is not even much chance of getting a distribution as unequal as 450 heads and 550 tails.

The tendency of the probabilities to cluster near the midpoint of the graph, where the number of heads and the number of tails are nearly equal, can be characterized by a "width" of the curve. The width of the curve is defined to be the distance between a pair of points (see Figures 14.3 and 14.4) outside of which the probabilities are relatively small and inside of which the probabilities are relatively large. Exactly where these points are chosen is arbitrary. One convenient choice is the pair of points where the probability has fallen to about one third of its central value—more exactly to $1/e = 1/2.72$ of its central value. The reason for defining a width is this: It spans a region of highly probable results. After the tray is shaken, the number of heads and the number of tails are most likely to correspond to a point on the central part of the curve within its width. The distribution of heads and tails is unlikely to be so unequal as to correspond to a point on the curve outside of this central region. When the number of coins is reasonably large (more than 100), there is a particularly simple formula for the width of the probability curve. If C is the number of heads (or tails) at the center of the curve, the width W of the curve is given by

$$W = 2\sqrt{C}. \tag{14.1}$$

The half-width, that is, the distance from the midpoint to the $1/e$ point of the curve, is equal to $\sqrt{C}$. This simple square root law is the reason for the particular factor $1/e$ used to define the width. With this choice the probability for the result of a tray-shaking to lie within the width of the curve is 84%.

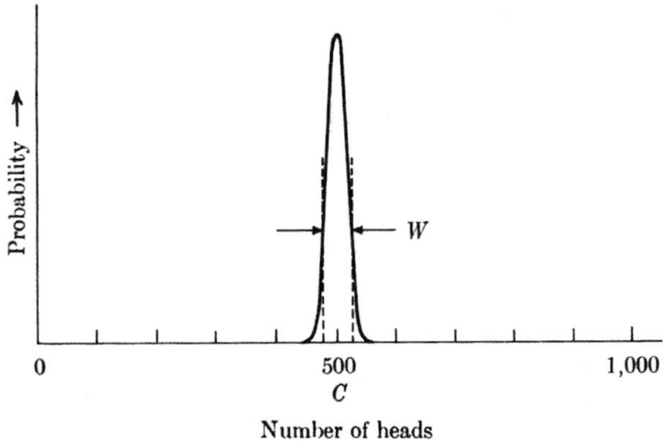

FIGURE 14.4 Probabilities for various results of tray-shaking experiment with 1,000 coins.

In Figure 14.4 the value of C, the midpoint number of heads, is 500. The square root of C is roughly 22. Thus the width of the curve is about 44, extending from $500 - 22 = 478$ to $500 + 22 = 522$. The total chance for a result to lie within this span is 84%; to lie outside it, 16%.

An important consequence of the square-root law is to sharpen the probability curve as the number of coins increases. The ratio of the width to the total number of coins N ($N = 2C$) is

$$\frac{W}{N} = \frac{2\sqrt{C}}{2C} = \frac{1}{\sqrt{C}}. \tag{14.2}$$

This ratio decreases as C (or N) increases. For 200 coins, the width-to-number ratio is about 1/10. For 2,000 coins, it is about 1/32. For 2,000,000 coins, it is 1/1,000. If the number of coins could be increased to be equal to the number of molecules in a drop of water, about 10^{22}, the width-to-number ratio of the probability curve would be $1/10^{11}$. Then the result of vigorous shaking of the coins would produce a number of heads and a number of tails unlikely to differ from equality by more than one part in one hundred billion. The probability curve would have collapsed to a narrow spike (Figure 14.5).

Two more points of interest about these head-and-tail probabilities will bring us closer to the connection between trays of coins and collections of molecules. First is the relation between probability and disorder. Ten coins arranged as all heads can be considered as perfectly orderly, as can an array of all tails. Five heads and five tails, on the other hand, arranged for example as HHTHTTTHTH or as TTHTHTHHHT, form a disorderly array. Evidently a high state of order is associated with low probability, a state of disorder is associated with high probability. This might be called the housewife's rule: Order is improbable, disorder is probable. The reason this is so is exactly the same for the household as for the tray of coins. There are many more different ways to achieve disorder than to achieve order.

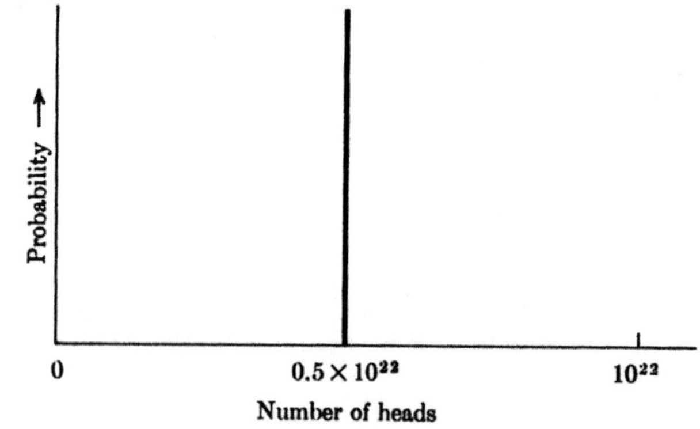

FIGURE 14.5 For 10^{22} coins, the probability curve is a spike much narrower even than the line on this graph.

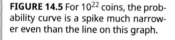

The second point of special interest concerns the way probabilities change in time. If a tray of 1,000 coins is carefully arranged to show all heads, and is then shaken repeatedly, its arrangement will almost certainly shift in the direction of nearly equal numbers of heads and tails. The direction of spontaneous change will be from an arrangement of low probability to an arrangement of high probability, from order to disorder. The same will be true whenever the initial arrangement is an improbable one, for instance 700 tails and 300 heads. If instead we start with 498 heads and 502 tails, no amount of shaking will tend to move the distribution to a highly uneven arrangement. This can be considered an equilibrium situation. Repeated trials will then produce results not very different from the starting point. Clearly there is a general rule here—a rule of probability, to be sure, not an absolute rule: Under the action of random influences, a system tends to change from less probable arrangements to more probable arrangements, from order to disorder. The generalization of this rule from trays of coins to collections of molecules, and indeed to complex systems of any kind, is the *second law of thermodynamics*—a law, as we shall see, with remarkably broad and important consequences.

14.3 Probability of position

78 Most of the large-scale properties of substances are, when examined closely enough, probabilistic in nature. Heat and temperature are purely macroscopic concepts that lose their meaning when applied to individual atoms and molecules, for any particular molecule might have more or less energy than the average, or might contribute more or less than the average to a process of energy exchange. Temperature is proportional to an *average*

kinetic energy; heat is equal to a *total* energy transferred by molecular collision. Because of our incomplete knowledge about the behavior of any single molecule, and the consequent necessity of describing molecular motion in probabilistic terms, neither of these thermal concepts is useful except when applied to numbers so large that the laws of probability become laws of near certainty. The same can be said of other concepts such as pressure and internal energy.

A single molecule is characterized by position, velocity, momentum, and energy. Of these, position is the simplest concept and therefore the one for which it is easiest to describe the role of probability. Consider, for instance, an enclosure—perhaps the room you are in—divided by a screen into two equal parts. What is the relative number of molecules of air on the two sides of the screen? Not a hard question, you will say. It is obvious that the two halves should contain equal, or very nearly equal, numbers of molecules. But here is a harder question. Why do the molecules divide equally? Why do they not congregate, at least some of the time, in one corner of the room? The answer to this question is exactly the same as the answer to the question: Why does a tray of coins after being shaken display approximately equal numbers of heads and tails? The equal distribution is simply the most probable distribution. Any very unequal distribution is very improbable.

The mathematics of molecules on two sides of a room proves to be identical to the mathematics of coins on a tray. By the assumption of randomness, every single molecule has an equal chance to be on either side of the room, just as every coin has an equal chance to land as heads or as tails. There are many different ways to distribute the molecules in equal numbers on the two sides, but only one way to concentrate them all on one side. If a room contained only five molecules, it would not be surprising to find them sometimes all on a single side. The probability that they be all on the left is 1/32 (see Table 14.1), and there is an equal probability that they be all on the right. The chance of a 3-2 distribution is 20/32, or nearly two thirds. Even for so small a number as five, a nearly equal division is much more likely than a very uneven division. For 10^{28} molecules, the number in a large room, the distribution is unlikely to deviate from equality by more than one part in 10^{14}. The probability for all of the 10^{28} molecules to congregate spontaneously in one half of the room is less than

$$10^{-(10^{+27})} .$$

This number is too small even to think about. Suddenly finding ourselves gasping for breath in one part of a room while someone in another part of the room is oversupplied with oxygen is a problem we need not be worried about.

The second law of thermodynamics is primarily a law of change. It states that the direction of spontaneous change within an isolated system is from an arrangement of lower probability to an arrangement of higher probability. Only if the arrangement is already one of maximal probability will no spontaneous change occur. Air molecules distributed uniformly in a room are (with respect to their position) in such a state of maximal probability. This is an equilibrium situation, one that has no tendency for spontaneous change. Nevertheless it is quite easy through external actions to depart from this equilibrium to a less probable arrangement. Air can be pumped from one side of the room to the other. In a hypothetical vacuum-tight room with an impenetrable barrier dividing it in half, almost all of the air can be pumped into one half. When the barrier is punctured, the air rushes to equalize its distribution in space. This behavior can be described as the result of higher pressure pushing air into a region of lower pressure. But it can equally well be described as a simple consequence of the second law of thermodynamics. Once the barrier is punctured or removed, the air is free to change to an arrangement of higher probability, and it does so promptly.

It is worth noting that frequent molecular collisions play the same role for the air as tray-shaking plays for the coins. A stationary tray displaying all heads would stay that way, even though the arrangement is improbable. If molecules were quiescent, they would remain on one side of a room once placed there. Only because of continual molecular agitation do the spontaneous changes predicted by the second law of thermodynamics actually occur.

14.4 *Entropy and the second law of thermodynamics*

79a There are a variety of ways in which the second law of thermodynamics can be stated, and we have encountered two of them so far: (1) For an isolated system, the direction of spontaneous change is from an arrangement of lesser probability to an arrangement of greater probability; and (2) for an isolated system, the direction of spontaneous change is from order to disorder. Like the conservation laws, the second law of thermodynamics applies only to a system free of external influences. For a system that is not isolated, there is no principle restricting its direction of spontaneous change.

A third statement of the second law of thermodynamics makes use of a new concept called *entropy*. Entropy is a measure of the extent of disorder in a system or of the probability of the arrangement of the parts of a system. For greater probability, which means greater disorder, the entropy is higher. An arrangement of less probability (greater order) has less entropy. This means that the second law can be stated: (3) The entropy of an isolated system increases or remains the same.

Specifically, entropy, for which the usual symbol is S, is defined as Boltzmann's constant multiplied by the natural logarithm of the probability of any particular state of the system:

$$S = k \log P \,. \tag{14.3}$$

The appearance of Boltzmann's constant k as a constant of proportionality is a convenience in the mathematical theory of thermodynamics, but is, from a fundamental point of view, entirely arbitrary. The important aspect of the definition is the proportionality of the entropy to the logarithm of the probability P. Note that since the logarithm of a number increases when the number increases, greater probability means greater entropy, as stated in the preceding paragraph.

Exactly how to calculate a probability for the state of a system (a procedure that depends on the energies as well as the positions of its molecules) is a complicated matter that need not concern us here. Even without this knowledge, we can approach an understanding of the reason for the definition expressed by Equation 14.3. At first, entropy might seem to be a superfluous and useless concept, since it provides the same information about a system as is provided by the probability P, and S grows or shrinks as P grows or shrinks. Technically these two concepts *are* redundant, so that either one of them might be considered superfluous. Nevertheless both are very useful. (For comparison, consider the radius and the volume of a sphere; both are useful concepts despite the fact that they provide redundant information about the sphere.) The valuable aspect of the entropy concept is that it is additive. For two or more systems brought together to form a single system, the entropy of the total is equal to the sum of the entropies of the parts. Probabilities, by contrast, are multiplicative. If the probability for one molecule to be in the left half of a container is ½, the probability for two to be there is ¼, and the probability for three to congregate on one side is ⅛. If two containers, each containing three molecules, are encompassed in a single system, the probability that the first three molecules are all on the left side of the first container *and* that the second three are also on the left side of the second container is $\frac{1}{8} \times \frac{1}{8} = \frac{1}{64}$. On the other hand, the entropy of the combination is the sum of the entropies of the two parts. These properties of addition and multiplication are reflected in the definition expressed by Equation 14.3. The logarithm of a product is the sum of the logarithm of the factors:

$$S_{\text{total}} = k \log P_1 P_2 = k \log P_1 + k \log P_2 = S_1 + S_2 \,. \tag{14.4}$$

The additive property of entropy is more than a mathematical convenience. It means that the statement of the second law can be generalized to include a composite system. To restate it: (3) The total entropy of a set of interconnected systems increases or stays the same. If the entropy of one system decreases, the entropy of systems connected to it must increase by at least a compensating amount, so that the sum of the individual entropies does not decrease.

Even though the second law of thermodynamics may be re-expressed in terms of entropy or of order and disorder, probability remains the key underlying idea. The exact nature of this probability must be understood if the second law is to be understood. Implicit in our discussion up to this point but still requiring emphasis is the a priori nature of the probability that governs physical change. The statement that physical systems change from less probable to more probable arrangements might seem anything but profound if the probability is regarded as an after-the-fact probability. If we decided that a uniform distribution of molecules in a box must be more probable than a nonuniform distribution because gas in a box is always observed to spread itself out evenly, the second law would be mere tautology, saying that systems tend to do what they are observed to do. In fact, the probability of the second law of thermodynamics is not based on experience or experiment. It is a before-the-fact (a priori) probability, based on counting the number of different ways in which a particular arrangement could be achieved. To every conceivable arrangement of a system can be assigned an a priori probability, whether or not the system or that arrangement of it has ever been observed. In practice there is no reason why the state of a system with the highest a priori probability need be the most frequently observed. Consider the case of the dedicated housewife. Almost every time an observant friend comes to call, he finds her house to be in perfect condition, nothing out of place, no dust in sight. He must conclude that for this house at least, the most probable state is very orderly state, since that is what he most often observes. This is an after-the-fact probability. As the housewife and the student of physics know, the orderly state has a low a priori probability. Left to itself, the house will tend toward a disorderly state of higher a priori probability. A state of particularly high a priori probability for a house is one not often observed, a pile of rubble. Thus an arrangement of high probability (from here on we shall omit the modifier, a priori) need be neither frequently observed nor quickly achieved, but it is, according to the second law of thermodynamics, the inevitable destination of an isolated system.

In comparison with other fundamental laws of nature, the second law of thermodynamics has two special features. First, it is not given expression by any mathematical equation. It specifies a direction of change, but not a magnitude of change. The nearest we can come to an equation is the mathematical statement,

$$\Delta S \geqslant 0\,. \tag{14.5}$$

In words: The change of entropy (for an isolated system or collection of systems) is either positive or zero. Or, more simply, entropy does not spontaneously decrease.

Every fundamental law of nature is characterized by remarkable generality, yet the second law of thermodynamics is unique among them (its second special feature) in that it finds direct application in a rich variety of settings, physical, biological, and human. In mentioning trays of coins, molecules of gas, and disorder in the house, we have touched only three of a myriad of applications. Entropy and the second law have contributed to discussion of the behavior of organisms, the flow of events in societies and economies, communication and information, and the history of the universe. In much of the physics and chemistry of macroscopic systems, the second law has found a use. Only at the submicroscopic level of single particles and single events is it of little importance. It is a startling and beautiful thought that an idea as simple as the natural trend from order to disorder should have such breadth of impact and power of application.

In most of the remainder of this chapter we shall be concerned with the application of the second law of thermodynamics to relatively simple physical situations. In Section 14.9 we return to some of its more general implications.

14.5 *Probability of velocity: heat flow and equipartition*

Since the velocities as well as the positions of individual molecules are generally unknown, velocity too is subject to considerations of probability. This kind of probability, like the probability of position, follows the rule of spontaneous change from lower to higher probability. It should not be surprising to learn that for a collection of identical molecules the most probable arrangement is one with equal average speeds (and randomly oriented velocities). This means that available energy tends to distribute itself uniformly over a set of identical molecules, just as available space tends to be occupied uniformly by the same molecules. In fact, the equipartition theorem and the zeroth law of thermodynamics can both be regarded as *consequences* of the second law of thermodynamics. Energy divides itself equally among the available degrees of freedom, and temperatures tend toward equality, because the resulting homogenized state of the molecules is the state of maximum disorder and maximum probability. The concentration of all of the energy in a system on a few molecules is a highly ordered and improbable situation analogous to the concentration of all of the molecules in a small portion of the available space.

The normal course of heat flow can also be understood in terms of the second law. Heat flow from a hotter to a cooler body is a process of energy transfer tending to equalize temperature and thereby to increase entropy. The proof that equipartition is the most probable distribution of energy is complicated and beyond the scope of this book. Here we seek only to make it plausible through analogy with the probability of spatial distributions.

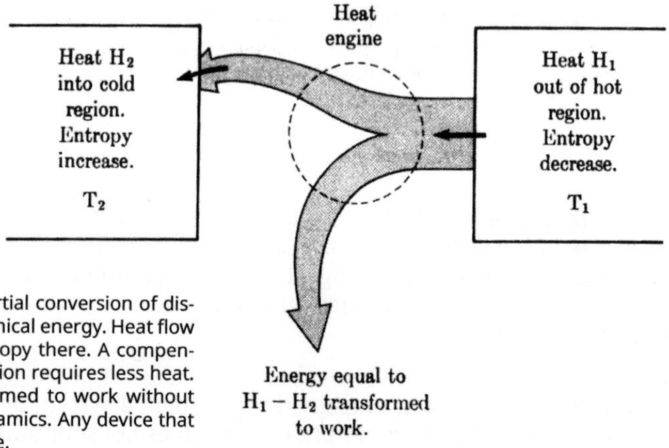

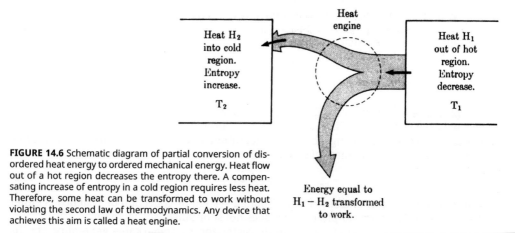

FIGURE 14.6 Schematic diagram of partial conversion of disordered heat energy to ordered mechanical energy. Heat flow out of a hot region decreases the entropy there. A compensating increase of entropy in a cold region requires less heat. Therefore, some heat can be transformed to work without violating the second law of thermodynamics. Any device that achieves this aim is called a heat engine.

79b Heat flow is so central to most applications of thermodynamics that the second law is sometimes stated in this restricted form: (4) Heat never flows spontaneously from a cooler to a hotter body. Notice that this is a statement about macroscopic behavior, whereas the more general and fundamental statements of the second law, which make use of the ideas of probability and order and disorder, refer to the submicroscopic structure of matter. Historically, the first version of the second law, advanced by Sadi Carnot in 1824, came before the submicroscopic basis of heat and temperature was established, in fact before the first law of thermodynamics was formulated. Despite a wrong view of heat and an incomplete view of energy, Carnot was able to advance the important principle that no heat engine (such as a

steam engine) could operate with perfect efficiency. In modern terminology, Carnot's version 79c
of the second law is this: (5) In a closed system, heat flow out of one part of the system cannot be transformed wholly into mechanical energy (work), but must be accompanied by heat flow into a cooler part of the system. In brief, heat cannot be transformed completely to work.

The consistency of Carnot's form of the second law with the general principle of entropy 80
increase can best be appreciated by thinking in terms of order and disorder. The complete conversion of heat to work would represent a transformation of disordered energy, a replacement of random molecular motion by orderly bulk motion. This violates the second law of thermodynamics. As indicated schematically in Figure 14.6, a *partial* conversion of heat to work is possible because a small heat flow into a cool region may increase the entropy there by more than the decrease of entropy produced by a larger heat flow out of a hot region. At absolute zero, the hypothetically motionless molecules have maximum order. Greater temperature produces greater disorder. Therefore heat flow into a region increases its entropy, heat flow out of a region decreases its entropy. Fortunately for the feasibility of heat engines, it takes less heat at low temperature than at high temperature to produce a given entropy change. To make an analogy, a pebble is enough to bring disorder to the smooth surface of a calm lake. To produce an equivalent increase in the disorder of an already rough sea requires a boulder. In Section 14.7, the quantitative link between heat flow and entropy is discussed.

The reverse transformation, of total conversion of work to heat, is not only possible but is commonplace. Every time a moving object is brought to rest by friction, all of its ordered energy of bulk motion is converted to disordered energy of molecular motion. This is an entropy-increasing process allowed by the second law of thermodynamics. In general, the second law favors energy dissipation, the transformation of energy from available to unavailable form. Whenever we make a gain against the second law by increasing the order or the available energy in one part of a total system, we can be sure we have lost even more in another part of the system. Thanks to the constant input of energy from the sun, the earth remains a lively place and we have nothing to fear from the homogenizing effect of the second law.

14.6 *Perpetual motion*

We have given so far five different versions of the second law, and will add only one more. 79d
Of those given, the first three, expressed in terms of probability, of order-disorder, and of entropy, are the most fundamental. Worth noting in several of the formulations is the recurring emphasis on the negative. Entropy does *not* decrease. Heat does *not* flow spontaneously from a cooler to a hotter region. Heat can *not* be wholly transformed to work. Our sixth version is also expressed in the negative. (6) Perpetual-motion machines cannot be constructed. This statement may sound more like a staff memorandum in the Patent Office than a fundamental law of nature. It may be both. In any event, it is certainly the latter, for from it can be derived the spontaneous increase of probability, of disorder, or of entropy. It is specialized only in that it assumes some friction, however small, to be present to provide some energy dissipation. If we overlook the nearly frictionless motion of the planets in the solar system and the frictionless motion of single molecules in a gas, everything in between is encompassed.

A perpetual-motion machine can be defined as a closed system in which bulk motion persists indefinitely, or as a continuously operating device whose output work provides its own input energy. Some proposed perpetual-motion machines violate the law of energy conservation (the first law of thermodynamics). These are called perpetual-motion machines of the first kind. Although they can be elaborate and subtle, they are less interesting than perpetual-motion machines of the second kind, hypothetical devices that conserve energy but violate the principle of entropy increase (the second law of thermodynamics).

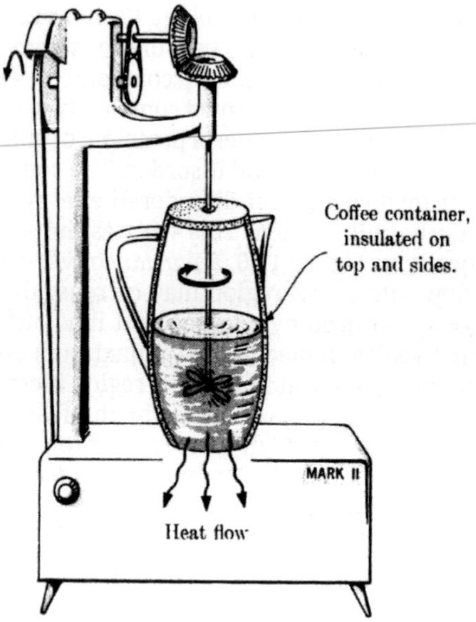

FIGURE 14.7 A perpetual-motion machine of the second kind. The device labeled MARK II receives heat energy from the coffee and converts this to mechanical energy which turns a paddle wheel, agitating the coffee, returning to the coffee the energy it lost by heat flow. It is not patentable.

As operating devices, perpetual-motion machines are the province of crackpot science and science fiction. As *inoperable* devices they have been of some significance in the development of science. Carnot was probably led to the second law of thermodynamics by his conviction that perpetual motion should be impossible. Arguments based on the impossibility of perpetual motion can be used to support Newton's third law of mechanics and Lenz's law of electromagnetic reaction, which will be discussed in Chapter Sixteen. Any contemporary scientist with a speculative idea can subject it to at least one quick test: Is it consistent with the impossibility of perpetual motion?

Suppose that an inventor has just invented a handy portable coffee warmer (Figure 14.7). It takes the heat which flows from the coffee container and, by a method known only to him, converts this heat to work expended in stirring the coffee. If the energy going back into the coffee is equal to that which leaks off as heat, the original temperature of the coffee will be maintained. Is it patentable? No, for it is a perpetual-motion machine of the second kind. Although it conserves energy, it performs the impossible task of maintaining a constant entropy in the face of dissipative forces that tend to increase entropy. Specifically it violates Carnot's version of the second law (No. 5, page 381), for in one part of its cycle it converts heat wholly to work. Of course it also violates directly our sixth version of the second law.

One of the chief strengths of the second law is its power to constrain the behavior of complex systems without reference to any details. Like a corporate director, the second law rules the overall behavior of systems or interlocked sets of systems in terms of their total input and output and general function. Given a proposed scheme for the operation of the automatic coffee warmer, it might be quite a complicated matter to explain in terms of its detailed design why it cannot work. Yet the second law reveals at once that no amount of ingenuity can make it work.

14.7 *Entropy on two levels*

81 The mathematical roots of thermodynamics go back to the work of Pierre Laplace and other French scientists concerned with the caloric theory of heat in the years around 1800, and even further to the brilliant but forgotten invention of the kinetic theory of gases by Daniel

Bernoulli in 1738. Not until after 1850 did these and other strands come together to create the theory of thermodynamics in something like its modern form. No other great theory of physics has traveled such a rocky road to success over so many decades of discovery, argumentation, buried insights, false turns, and rediscovery, its paths diverging and finally rejoining in the grand synthesis of statistical mechanics which welded together the macroscopic and submicroscopic domains in the latter part of the nineteenth century.

In the long and complex history of thermodynamics, the generalization of the principle of energy conservation to include heat stands as probably the most significant single landmark. Joule's careful experiments on the mechanical equivalent of heat in the 1840s not only established the first law of thermodynamics, but cleared the way for a full understanding of the second law, provided a basis for an absolute temperature scale, and laid the groundwork for the submicroscopic mechanics of the kinetic theory. Progress in the half century before Joule's work had been impeded by a pair of closely related difficulties: an incorrect view of the nature of heat, and an incomplete understanding of the way in which heat engines provide work. To be sure, there had been important insights in this period, such as Carnot's statement of the second law of thermodynamics in 1824. But such progress as there was did not fit together into a single structure, nor did it provide a base on which to build. Not until 1850, when the great significance of the general principle of energy conservation was appreciated by at least a few scientists, was Carnot's work incorporated into a developing theoretical structure. The way was cleared for a decade of rapid progress. In the 1850s, the first and second laws of thermodynamics were first stated as general unifying principles, the kinetic theory was rediscovered and refined, the concepts of heat and temperature were given submicroscopic as well as macroscopic definitions, and the full significance of the ideal-gas law was understood. The great names of the period were James Joule, William Thomson (Lord Kelvin), and James Clerk Maxwell in England, Rudolph Clausius and August Krönig in Germany.

One way to give structure to the historical development of a major theory is to follow the evolution of its key concepts. This is particularly instructive for the study of thermodynamics, because its basic concepts—heat, temperature, and entropy—exist on two levels, the macroscopic and the submicroscopic. The refinement of these concepts led both to a theoretical structure for understanding a great part of nature and to a bridge between two worlds, the large and the small. Of special interest here is the entropy concept.

Like heat and temperature, entropy was given first a macroscopic definition, later a molecular definition. Being a much subtler concept than either heat or temperature (in that it does not directly impinge on our senses), entropy was defined only after its need in the developing theory of thermodynamics became obvious. Heat and temperature were familiar ideas refined and revised for the needs of quantitative understanding. Entropy was a wholly new idea, formally introduced and arbitrarily named when it proved to be useful in expressing the second law of thermodynamics in quantitative form. As a useful but unnamed quantity, entropy entered the writings of both Kelvin and Clausius in the early 1850s. Finally in 1865, it was formally recognized and christened "entropy" by Clausius, after a Greek word for transformation. Entropy, as he saw it, measured the potentiality of a system for transformation.

The proportionality of entropy to the logarithm of an intrinsic probability for the arrangement of a system, as expressed by Equation 14.3, was stated first by Ludwig Boltzmann in 1877. This pinnacle of achievement in what had come to be called statistical mechanics fashioned the last great thermodynamics link between the large-scale and small-scale worlds. Although we now regard Boltzmann's definition based on the molecular viewpoint as the more fundamental, we must not overlook the earlier macroscopic definition of entropy given by Clausius (which in most applications is easier to use). Interestingly, Clausius expressed entropy simply and directly in terms of the two already familiar basic concepts, heat and temperature. He stated that a change of entropy of any part of a system is equal to the increment of heat added to that part of the system divided by its temperature at the moment the heat is added, provided the change is from one equilibrium state to another:

$$\Delta S = \frac{\Delta H}{T}. \tag{14.6}$$

Here S denotes entropy, H denotes heat, and T denotes the absolute temperature. For heat gain, ΔH is positive and entropy increases. For heat loss, ΔH is negative and entropy decreases. How much entropy change is produced by adding or subtracting heat depends on the temperature. Since the temperature T appears in the denominator in Equation 14.6, a lower temperature enables a given increment of heat to produce a greater entropy change.

There are several reasons why Clausius defined not the entropy itself, but the change of entropy. For one reason, the absolute value of entropy is irrelevant, much as the absolute value of potential energy is irrelevant. Only the change of either of these quantities from one state to another matters. Another more important reason is that there is no such thing as "total heat." Since heat is energy transfer (by molecular collisions), it is a dynamic quantity measured only in processes of change. An increment of heat ΔH can be gained or lost by part of a system, but it is meaningless to refer to the total heat H stored in that part. (This was the great insight about heat afforded by the discovery of the general principle of energy conservation in the 1840s). What is stored is internal energy, a quantity that can be increased by mechanical work as well as by heat flow. Finally, it should be remarked that Clausius' definition refers not merely to change, but to *small* change. When an otherwise inactive system gains heat, its temperature rises. Since the symbol T in Equation 14.6 refers to the temperature at which heat is added, the equation applies strictly only to increments so small that the temperature does not change appreciably as the heat is added. If a large amount of heat is added, Equation 14.6 must be applied over and over to the successive small increments, each at slightly higher temperature.

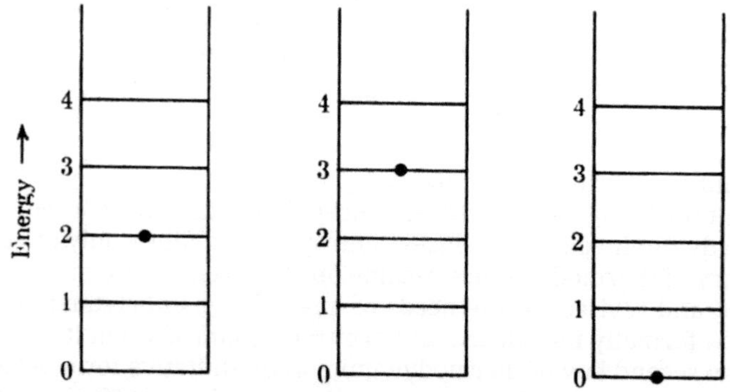

FIGURE 14.8 Idealized energy diagram for a system of three molecules, each with equally spaced energy states. Each ladder depicts the possible energies of a particular molecule, and the heavy dot specifies the actual energy of that molecule.

To explain how the macroscopic definition of entropy given by Clausius (Equation 14.6) and the submicroscopic definition of entropy given by Boltzmann (Equation 14.3) fit together is a task beyond the scope of this book. Nevertheless we can, through an idealized example, make it reasonable that these two definitions, so different in appearance, are closely related. To give the Clausius definition a probability interpretation we need to discuss two facts: (1) Addition of heat to a system increases its disorder and therefore its entropy; (2) The disordering influence of heat is greater at low temperature than at high temperature. The first of these facts is related to the appearance of the factor ΔH on the right of Equation

14.6; the second is related to the inverse proportionality of entropy change to temperature.

Not to prove these facts but to make them seem reasonable, we shall consider an idealized simple system consisting of just three identical molecules, each one capable of existing in any one of a number of equally spaced energy states. The overall state of this system can be represented by the triple-ladder diagram of Figure 14.8, in which each rung corresponds to a molecular energy state. Dots on the three lowest rungs would indicate that the system possesses no internal energy. The pictured dots on the second, third, and bottom rungs indicate that the system has a total of five units of internal energy, two units possessed by the first molecule, three by the second, and none by the third. The intrinsic probability associated with any given total energy is proportional to the number of different ways in which that energy can be divided. This is now a probability of *energy* distribution, not a probability of spatial distribution. However, the reasoning is much the same as in Section 14.3. There the intrinsic (a priori) probability for a distribution of molecules in space was taken to be proportional to the number of different ways in which that distribution could be obtained. Or, to give another example, the probability of throwing 7 with a pair of dice is greater than the probability of throwing 2, because there are more different ways to get a total of 7 than to get a total of 2.

TABLE 14.4 Internal Energy Distribution for Idealized. System of Three Molecules

Total Energy	Distribution of Energy			Number of Ways to Distribute Energy
0	000			1
1	100	010	001	3
2	200	020	002	
	110	101	011	6
3	300	030	003	
	210	201	012	
	120	102	021	
	111			10
4	400	040	004	
	310	301	031	
	130	103	013	
	220	202	022	
	211	121	112	15
5	500	050	005	
	410	401	041	
	140	104	014	
	320	302	032	
	230	203	023	
	311	131	113	
	122	212	221	21

Table 14.4 enumerates all the ways in which up to five units of energy can be divided among our three idealized molecules. The triplets of numbers in the second column indicate the occupied rungs of the three energy ladders. It is an interesting and instructive problem to deduce a formula for the numbers in the last column, (HINT: The number of ways to distribute 6 units of energy is 28.) However, since this is a highly idealized picture of very few molecules, precise numerical details are less important than are the qualitative features of the overall pattern. The first evident feature is that the greater the energy, the more different ways there are to divide the energy. Thus a higher probability is associated with greater internal energy. This does not mean that the system, if isolated and left alone, will spontaneously tend toward a higher probability state, for that would violate the law of energy conservation. Nevertheless, we associate with the higher energy state a greater probability and a greater disorder. When energy is added from outside via heat flow, the

entropy increase is made possible. This makes reasonable the appearance of the heat increment factor, ΔH, in Equation 14.6.

Looking further at Table 14.4, we ask whether the addition of heat produces a greater disordering effect at low temperature than at high temperature. For simplicity we can assume that temperature is proportional to total internal energy, as it is for a simple gas, so that the question can be rephrased: Does adding a unit of heat at low energy increase the entropy of the system more than adding the same unit of heat at higher energy? Answering this question requires a little care, because of the logarithm that connects probability to entropy. The relative probability accelerates upward in Table 14.4. In going from 1 to 2 units of energy, the number of ways to distribute the energy increases by three, from 2 to 3 units it increases by four, from 3 to 4 units it increases by five, and so on. However, the entropy, proportional to the logarithm of the probability, increases more slowly at higher energy. The relevant measure for the increase of a logarithm is the *factor* of growth.* From 0 to 1 unit of energy, the probability trebles, from 1 to 2 units it doubles, from 2 to 3 units it grows by 67%, and so on, by ever decreasing factors of increase. Therefore the entropy grows most rapidly at low internal energy (low temperature). This makes reasonable the appearance of the temperature factor "downstairs" on the right of Equation 14.6.

82 This example focuses attention on a question that may have occurred to you already. Why is it that energy addition by heat flow increases entropy, but energy addition by work does not? The definition, Equation 14.6, makes reference to only one kind of energy, heat energy. The difference lies basically in the recoverability of the energy. When work is done on a system without any accompanying heat flow, as when gas is compressed in a cylinder (Figure 14.9), the energy can be fully recovered, with the system and its surroundings returning precisely to the state they were in before the work was done. No entropy change is involved. On the other hand, when energy in the form of heat flows from a hotter to a cooler place, there is no mechanism that can cause the heat to flow spontaneously back from the cooler to the hotter place. It is not recoverable. Entropy has increased. In a realistic as opposed to an ideal cycle of compression and expansion, there will in fact be some entropy increase because there will be some flow of heat from the compressed gas to the walls of the container.

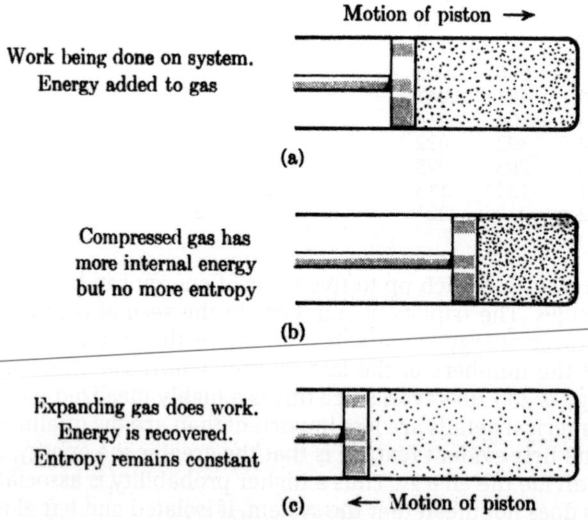
Motion of piston →
Work being done on system.
Energy added to gas
(a)
Compressed gas has
more internal energy
but no more entropy
(b)
Expanding gas does work.
Energy is recovered.
Entropy remains constant
(c)
← Motion of piston

FIGURE 14.9 Idealized cycle of compression and expansion of gas, accompanied by no change of entropy. If any heat flow occurs in the cycle, entropy does increase.

* Logarithms are defined in such a way that the logarithms of 10, 100, 1,000, and 10,000 or of 5,10, 20, 40, and 80 differ by equal steps. It is this feature which makes the multiplication of a pair of numbers equivalent to the addition of their logarithms.

Another useful way to look at the difference between heat and work is in molecular terms, merging the ideas of position probability and velocity or energy probability. If a confined gas [Figure 14.9(b)] is allowed to expand until its volume doubles [Figure 14.9(c)] what we learned about position probability tells us that, so far as its spatial arrangement is concerned, it has experienced an entropy increase, having spread out into an intrinsically more probable arrangement. In doing so, however, it has done work on its surroundings and has lost internal energy. This means that, with respect to its velocity and energy, it has approached a state of greater order and lesser entropy. Its increase of spatial disorder has in fact been precisely canceled by its decrease of energy disorder, and it experiences no net change of entropy. Had we instead wanted to keep its temperature constant during the expansion, it would have been necessary to add heat (equal in magnitude to the work done). Then after the expansion, the unchanged internal energy would provide no contribution to entropy change, so that a net entropy increase would be associated with the expansion—arising from the probability of position. This would match exactly the entropy increase $\Delta H/T$ predicted by the Clausius formula, for this change required a positive addition of heat.

Although the macroscopic entropy definition of Clausius and the submicroscopic entropy definition of Boltzmann are, in many physical situations, equivalent, Boltzmann's definition remains the more profound and the more general. It makes possible a single grand principle, the spontaneous trend of systems from arrangements of lower to higher probability, that describes not only gases and solids and chemical reactions and heat engines, but also dust and disarray, erosion and decay, the deterioration of fact in the spread of rumor, the fate of mismanaged corporations, and perhaps the fate of the universe.

14.8 Application of the second law

That heat flows spontaneously only from a warmer to a cooler place is a fact which can itself be regarded as a special form of the second law of thermodynamics. Alternatively the direction of heat flow can be related to the general principle of entropy increase with the help of the macroscopic definition of entropy. If body 1 at temperature T_1 loses an increment of heat ΔH, its entropy change—a decrease—is

$$\Delta S_1 = -\frac{\Delta H}{T_1}. \tag{14.7}$$

If this heat is wholly transferred to body 2 at temperature T_2, its entropy gain is

$$\Delta S_2 = -\frac{\Delta H}{T_2}. \tag{14.8}$$

The total entropy change of the system (bodies 1 and 2) is the sum,

$$\Delta S = \Delta S_1 + \Delta S_2 = \Delta H\left(\frac{1}{T_2} - \frac{1}{T_1}\right). \tag{14.9}$$

This entropy change must, according to the second law, be positive if the heat transfer occurs spontaneously. It is obvious algebraically from Equation 14.9 that this requirement implies that the temperature T_1 is greater than the temperature T_2. In short, heat flows from the warmer to the cooler body. In the process, the cooler body gains energy equal to that lost by the warmer body but gains entropy greater than that lost by the warmer body. When equality of temperature is reached, heat flow in *either* direction would decrease the total entropy. Therefore it does not occur.

A heat engine is, in simplest terms, a device that transforms heat to mechanical work. Such a transformation is, *by itself*, impossible. It is an entropy-decreasing process that violates the second law of thermodynamics. We need hardly conclude that heat engines are impossible, for we see them all around us. Gasoline engines, diesel engines, steam engines, jet engines, and rocket engines are all devices that transform heat to work. They do so by incorporating in the same system a mechanism of entropy increase that more than offsets the entropy decrease associated with the production of work. The simple example of heat flow with which this section began shows that one part of a system can easily lose entropy if another part gains more. In almost all transformations of any complexity, and in particular in those manipulated by man for some practical purpose, entropy gain and entropy loss occur side by side, with the total gain inevitably exceeding the total loss.

The normal mechanism of entropy gain in a heat engine is heat flow. Carnot's great insight that provided the earliest version of the second law was the realization that a heat engine must be transferring heat from a hotter to a cooler place at the same time that it is transforming heat to work. How this is accomplished varies from one heat engine to another, and the process can be quite complicated and indirect. Nevertheless, without reference to details, it is possible to discover in a very simple way what fraction of the total energy supplied by fuel can be transformed into usable work. This fraction is called the efficiency of the engine. Refer to Figure 14.6, which shows schematically a process of partial transformation of heat to work. From the hotter region, at temperature T_1, flows an increment of heat H_1. Into the cooler region, at temperature T_2, flows heat H_2. The output work is W. The first and second laws of thermodynamics applied to this idealized heat engine can be given simple mathematical expression.

1. Energy conservation: $$H_1 = H_2 + W\,. \tag{14.10}$$

2. Entropy increase: $$\Delta S = \frac{H_2}{T_2} - \frac{H_1}{T_1} > 0\,. \tag{14.11}$$

If this heat engine were "perfect"—free of friction and other dissipative effects— the entropy would remain constant instead of increasing. Then the right side of Equation 14.11 could be set equal to zero, and the ratio of output to input heat would be

$$\frac{H_2}{H_1} = \frac{T_2}{T_1}\,. \tag{14.12}$$

From Equation 14.10 follows another equation containing the ratio H_2/H_1,

$$\frac{W}{H_1} = 1 - \frac{H_2}{H_1}\,. \tag{14.13}$$

Substitution of Equation 14.12 into Equation 14.13 gives for the ratio of output work to initial heat supply,

$$\frac{W_{\text{max}}}{H_1} = 1 - \frac{T_2}{T_1}\,. \tag{14.14}$$

Here we have written W_{max} instead of W, since this equation gives the maximum possible efficiency of the idealized heat engine. If the temperatures T_1 and T_2 are nearly the same, the efficiency is very low. If T_2 is near absolute zero, the theoretical efficiency can be close to 1—that is, almost perfect.

The modern marvels of technology that populate our present world—automobiles, television, airplanes, radar, pocket radios—all rest ultimately on basic principles of physics. Nevertheless they are usually not instructive as illustrations of fundamental laws, for the chain of connection from their practical function to the underlying principles is complex and sophisticated. The refrigerator is such a device. Despite its complexity of detail, however, it is worth considering in general terms. Because it transfers heat from a cooler to a warmer place, the refrigerator appears at first to violate the second law of thermodynamics. The fact that it must not do so allows us to draw an important conclusion about the minimum expenditure of energy required to run it. The analysis is quite similar to that for a heat engine. Suppose that the mechanism of the refrigerator is required to transfer an amount of heat H_1 out of the refrigerator each second. If the interior of the refrigerator is at temperature T_1, this heat loss contributes an entropy decrease equal to $-H_1/T_1$. This heat is transferred to the surrounding room at temperature T_2 (higher than T_1), where it contributes an entropy increase equal to H_1/T_2. The sum of these two entropy changes is negative. Some other contribution to entropy change must be occurring in order that the total change may be positive, in consonance with the second law. This extra contribution comes from the degradation of the input energy that powers the refrigerator. The energy supplied by electricity or by the combustion of gas eventually reaches the surrounding room as heat. If the external energy (usually electrical) supplied in one second is called W, and the total heat added to the room in the same time is called H_2, energy conservation requires that H_2 be the sum of H_1 and W:

$$H_2 = H_1 + W. \tag{14.15}$$

The energy flow is shown schematically in Figure 14.10. At the same time the total entropy change is given by

$$\Delta S = \frac{H_2}{T_2} - \frac{H_1}{T_1}. \tag{14.16}$$

Since ΔS must be zero or greater, the ratio H_2/H_1 [= (heat added to room)/(heat extracted from refrigerator)] must be at least equal to T_2/T_1. If the energy conservation equation is written in the form

$$W = \mathrm{H}_1\left[\frac{H_2}{H_1} - 1\right],$$

we can conclude that

$$W \geqslant H_1\left[\frac{T_2}{T_1} - 1\right]. \tag{14.17}$$

The right side of this inequality gives the minimum amount of external energy input required in order to transfer an amount of heat H_1 "uphill" from temperature T_1 to temperature T_2. As might be expected, the input energy requirement increases as the temperature difference increases. If the temperature T_1 is near absolute zero, as it is in a helium liquefier, the external energy expended is much greater than the heat transferred.

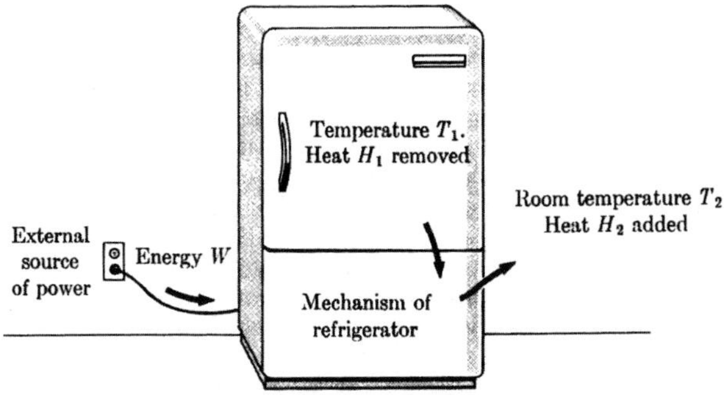

FIGURE 14.10 Energy and heat flow in the operation of a refrigerator.

The real beauty of the result expressed by Equation 14.17 is its generality for all refrigerators regardless of their construction and mode of operation. The input energy W could be supplied by an electric motor, a gas flame, or a hand crank. It is characteristic of the second law of thermodynamics, just as it is characteristic of the fundamental conservation laws, that it has something important to say about the overall behavior of a system without reference to details, perhaps without knowledge of details. In the small-scale world, our inability to observe precise features of individual events is one reason for the special importance of conservation laws. In the large-scale world, the elaborate complexity of many systems is one reason for the special importance of the second law of thermodynamics. Like the conservation laws, it provides an overall constraint on the system as a whole.

83 In many applications of the second law, the concept of available energy is the easiest key to understanding. In general, the trend of nature toward greater disorder is a trend toward less available energy. A jet plane before takeoff has a certain store of available energy in its fuel. While it is accelerating down the runway, a part of the energy expended is going into bulk kinetic energy (ordered energy), a part is going into heat that is eventually dissipated into unavailable energy. At constant cruising speed, all of the energy of the burning fuel goes to heat the air. Thermodynamically speaking, the net result of a flight is the total loss of the available energy originally present in the fuel. A rocket in free space operates with greater efficiency. Being free of air friction, it continues to accelerate as long as the fuel is burning. When its engine stops, a certain fraction (normally a small fraction) of the original available energy in the fuel remains available in the kinetic energy of the vehicle. This energy may be "stored" indefinitely in the orbital motion of the space vehicle. If it re-enters the atmosphere, however, this energy too is transformed into the disordered and unavailable form of internal energy of the air. To get ready for the next launching, more rocket fuel must be manufactured. The energy expended in the chemical factory that does this job is inevitably more than the energy stored in the fuel that is produced.

In general the effect of civilization is to encourage the action of the second law of thermodynamics. Technology greatly accelerates the rate of increase of entropy in man's immediate environment. Fortunately the available energy arriving each day from the sun exceeds by a very large factor the energy degraded by man's activity in a day. Fortunately too, nature, with no help from man, stores in usable form some of the sun's energy—for periods of months or years in the cycle of evaporation, precipitation, and drainage; for decades or centuries in lumber; for millennia in coal and oil. In time, as we deplete the long-term stored supply of available energy, we shall have to rely more heavily on the short-term stores and probably also devise new storage methods to supplement those of nature.

14.9 *The arrow of time*

84 Familiarity breeds acceptance. So natural and normal seem the usual events of our everyday life that it is difficult to step apart and look at them with a scientific eye.

Men with the skill and courage to do so led the scientific revolution of the seventeenth century. Since then, the frontiers of physics have moved far from the world of direct sense perception, and even the study of our immediate environment more often than not makes use of sophisticated tools and controlled experiment. Nevertheless, the ability to take a fresh look at the familiar and to contrast it with what would be the familiar in a different universe with different laws of nature remains a skill worth cultivating. For the student, and often for the scientist as well, useful insights come from looking at the familiar as if it were unfamiliar.

Consider the second law of thermodynamics. We need not go to the laboratory or to a machine or even to the kitchen to witness its impact on events. It is unlikely that you get through any five minutes of your waking life without seeing the second law at work. The way to appreciate this fact is by thinking backward. Imagine a motion picture of any scene of ordinary life run backward. You might watch a student untyping a paper, each keystroke erasing another letter as the keys become cleaner and the ribbon fresher. Or bits of hair clippings on a barber-shop floor rising to join the hair on a customer's head as the barber unclips. Or a pair of mangled automobiles undergoing instantaneous repair as they back apart. Or a dead rabbit rising to scamper backward into the woods as a crushed bullet re-forms and flies backward into a rifle while some gunpowder is miraculously manufactured out of hot gas. Or something as simple as a cup of coffee on a table gradually becoming warmer as it draws heat from its cooler surroundings. All of these backward-in-time views and a myriad more that you can quickly think of are ludicrous and impossible for one reason only—they violate the second law of thermodynamics. In the actual sequence of events, entropy is increasing. In the time reversed view, entropy is decreasing. We recognize at once the obvious impossibility of the process in which entropy decreases, even though we may never have thought about entropy increase in the everyday world. In a certain sense everyone "knows" the second law of thermodynamics. It distinguishes the possible from the impossible in ordinary affairs.

In some of the examples cited above, the action of the second law is obvious, as in the increasing disorder produced by an automobile collision, or the increasing entropy associated with heat flow from a cup of coffee. In others, it is less obvious. But whether we can clearly identify the increasing entropy or not, we can be very confident that whenever a sequence of events occurs in our world in one order and not in the other, it is because entropy increase is associated with the possible order, entropy decrease with the impossible order. The reason for this confidence is quite simple. We know of no law other than the second law of thermodynamics that assigns to processes of change in the large-scale world a preferred direction in time. In the submicroscopic world too, time-reversal invariance is a principle governing all or nearly all fundamental processes.* Here we have an apparent paradox. In order to understand the paradox and its resolution, we must first understand exactly what is meant by time-reversal invariance.

The principle of time-reversal invariance can be simply stated in terms of hypothetical moving pictures. If the filmed version of any physical process, or sequence of events, is shown backward, the viewer sees a picture of something that could have happened. In slightly more technical language, any sequence of events, if executed in the opposite order, is a physically possible sequence of events. This leads to the rather startling conclusion that it is, in fact, impossible to tell by watching a moving picture of events in nature whether the film is running backward or forward. How can this principle be reconciled with the gross violations of common sense contained in the backward view of a barber cutting hair, a hunter firing a gun, a child breaking a plate, or the President signing his name? Does it mean that time-reversal invariance is not a valid law in the macroscopic world? No. As far as we know, time-reversal invariance governs every interaction that underlies processes of change in the large-scale world. The key to resolving the paradox is to recognize that possibility does not mean probability. Although the spontaneous reassembly of the fragments of an exploded bomb into a whole, unexploded bomb is wildly, ridiculously improbable, it is not, from the most fundamental point of view, impossible.

At every important point where the macroscopic and submicroscopic descriptions of matter touch, the concept of probability is crucial. The second law of thermodynamics is basically a probabilistic law whose approach to absolute validity increases as the complexity of the system it describes increases. For a system of half a dozen molecules, entropy decrease is not only possible, it is quite likely, at least some of the time. All six molecules might cluster in one corner of their container, or the three less energetic molecules might lose energy via collisions to the three more energetic molecules ("uphill" heat flow). For a system of 10^{20} molecules, on the other hand, entropy decrease becomes so improbable that it deserves

* For the first time in 1964, some doubt was cast on the universal validity of time-reversal invariance, which had previously been supposed to be an absolute law of nature. In 1968 the doubt remains unresolved. Even if found to be imperfect, the principle will remain valid to a high degree of approximation, since it has already been tested in many situations. In particular, since all interactions that have any effect on the large-scale world do obey the principle of time-reversal invariance, the discussion in this section will be unaffected.

to be called impossible. We could wait a billion times the known lifetime of the universe and still never expect to see the time-reversal view of something as simple as a piece of paper being torn in half. Nevertheless, it is important to realize that the time-reversed process is possible in principle.

Even in the world of particles, a sequence of events may occur with much higher probability in one direction than in the opposite direction. In the world of human experience, the imbalance of probabilities is so enormous that it no longer makes sense to speak of the more probable direction and the less probable direction. Instead we speak of the possible and the impossible. The action of molecular probabilities gives to the flow of events in the large-scale world a unique direction. The (almost complete) violation of time-reversal invariance by the second law of thermodynamics attaches an arrow to time, a one-way sign for the unfolding of events. Through this idea, thermodynamics impinges on philosophy.

In the latter part of the nineteenth century, long before time-reversal invariance was appreciated as a fundamental law of submicroscopic nature, physicists realized that the second law had something quite general to say about our passage through time. There are two aspects of the idea of the arrow of time: first, that the universe, like a wound-up clock, is running down, its supply of available energy ever dwindling; second, that the spontaneous tendency of nature toward greater entropy is what gives man a conception of the unique one-way direction of time.

The second law of thermodynamics had not long been formulated in a general way before men reflected on its implications for the universe at large. In 1865, Clausius wrote, without fanfare, as grand a pair of statements about the world as any produced by science: "We can express the fundamental laws of the universe which correspond to the two fundamental laws of the mechanical theory of heat in the following simple form.

"1. The energy of the universe is constant.

"2. The entropy of the universe tends toward a maximum."

These are the first and second laws of thermodynamics extended to encompass all of nature. Are the extensions justifiable? If so, what are their implications? We know in fact no more than Clausius about the constancy of energy and the steady increase of entropy in the universe at large. We do know that energy conservation has withstood every test since he wrote, and that entropy increase is founded on the very solid principle of change from arrangements of lesser to those of greater probability. Nevertheless, all that we have learned of nature in the century since Clausius leaped boldly to the edge of existence should make us cautious about so great a step. In 1865, the single theory of Newtonian mechanics seemed to be valid in every extremity of nature, from the molecular to the planetary. A century later we know instead that it fails in every extremity—in the domain of small sizes, where quantum mechanics rules; in the domain of high speed, where special relativity changes the rules; and in the domain of the very large, where general relativity warps space and time.

The logical terminus of the universe, assuming it to be a system obeying the same laws as the macroscopic systems accessible to experiment, is known as the "heat death," a universal soup of uniform density and uniform temperature, devoid of available energy, incapable of further change, a perfect and featureless final disorder. If this is where the universe is headed, we have had no hints of it as yet. Over a time span of ten billion years or more, the universe has been a vigorously active place, with new stars still being born as old ones are dying. It is quite possible that the long-range fate of the universe will be settled within science and need not remain forever a topic of pure speculation. At present, however, we have no evidence at all to confirm or contradict the applicability of thermodynamics to the universe as a whole. Even if we choose to postulate its applicability, we need not be led inevitably to the idea of the ultimate heat death. The existence of a law of time-reversal invariance in the world of the small and the essential probabilistic nature of the second law leave open the possibility that one grand improbable reversal of probability could occur in which disorder is restored to order. Finally, we can link this line of thought to the second aspect of the arrow of time, the uniqueness of the direction of man's course through time, with this challenging thought. If it is the second law that gives to man his sense of time's direction, the very construction of the human machine forces us to see the universe running down. In a world that we might look in upon from the outside to see building order out of disorder, the less probable from the more probable, we would see creatures who remembered their future and not their past. For them the trend of events would seem to be toward disorder and greater probability and it is we who would seem to be turned around.

In the three centuries since Newton, time has evolved from the obvious to the mysterious. In the *Principia*, Newton wrote, "Absolute, true, and mathematical time, of itself, and from its own nature flows equably without regard to anything external, and by another name is called duration." This view of time as something flowing constantly and inexorably forward, carrying man with it, persisted largely intact until the revolution of relativity at the beginning of this century. The nineteenth century brought only hints of a deeper insight, when it was appreciated that the second law of thermodynamics differentiated between forward and backward in time, as the laws of mechanics had failed to do. If time were run backward, the reversed planetary orbits would be reasonable and possible, obeying the same laws as the actual forward-in-time orbits. But the reversal of any entropy-changing transformation would be neither reasonable nor possible. The second law of thermodynamics points the way for Newton's equable flow.

Relativity had the most profound effect on our conception of time. The merger of space and time made unreasonable a temporal arrow when there was no spatial arrow. More recently, time-reversal invariance has confirmed the equal status of both directions in time. Relativity also brought time to a stop. It is more consistent with the viewpoint of modern physics to think of man and matter moving through time (as they move through space) than to think of time itself as flowing.

All of the new insights about time make clear that we must think about it in very human terms—its definition, its measurement, its apparently unique direction stem not from "absolute, true and mathematical time" but from psychological time. These insights also reinforce the idea that the second law of thermodynamics must ultimately account for our sense of time.

It is a stimulating idea that the only reason man is aware of the past and not the future 85
is that he is a complicated and highly organized structure. Unfortunately, simpler creatures are no better off. They equalize future and past by remembering neither. An electron, being precisely identical with every other electron, is totally unmarked by its past or by its future. Man is intelligent enough to be scarred by his past. But the same complexity that gives him a memory at all is what keeps his future a mystery.

EXERCISES

14.1. Section 14.1 describes three kinds of uncertainty, associated respectively with a spelunker, a gambler, and a uranium prospector. Which of these kinds of uncertainty characterizes each of the following situations? **(1)** A pion of known energy enters a bubble chamber. The number of bubbles formed along its first centimeter of track is measured. The number of bubbles along its second centimeter of track can then be predicted approximately, but not exactly. **(2)** Another pion is created in the chamber. How long it will live before decaying is uncertain. **(3)** Still another pion, of energy higher than any previously studied, strikes a nucleus. The result of the collision is uncertain. Which, if any, of these examples of uncertainty is governed by thermodynamic probability (the probability of atomic multitudes)?

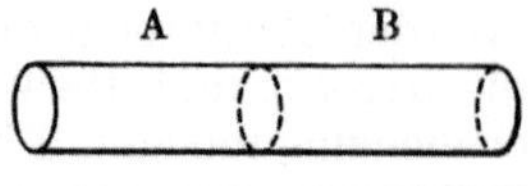

14.2. Suppose that a small cylinder (see figure) could be so nearly perfectly evacuated that only 100 molecules remained within it. **(1)** Using Figures 14.3 and 14.4 and Equation 14.1 as guides, sketch a curve of relative probability for any number of these molecules to be found in region A, which is half of the container. (2) If you placed a bet at even money that a measurement would reveal exactly 50 molecules in region A, would this be, from your point of view, a good bet or a poor bet? **(3)** If you bet, also at even money, that a series of measurements would show less than 60 molecules in region A more often than not, would you be making a good bet or a poor bet?

14.3. **(1)** What is the probability that a single throw of a pair of dice will yield "snake eyes" (a pair of ones) ? Explain how you arrived at your answer. **(2)** Is this an a priori probability? Why or why not? **(3)** What is the most probable total number resulting from the throw of a pair of dice? Why?

14.4. What if the number of air molecules in a room were much smaller than usual, perhaps equaling only a few thousand? Describe any interesting effects that might occur.

14.5. A house is in complete disarray. A housewife spends the day straightening it up—creating order out of disorder, converting a more probable arrangement to a less probable arrangement. Explain how she can do so without violating the second law of thermodynamics.

14.6. Within a hypothetical container is a perfect vacuum. Then 10 molecules are admitted into the left half of the container. At some later time 5 molecules are found in the left half and 5 in the right half. (1) What has been the change of entropy from the first arrangement to the second arrangement of these 10 molecules? (For this calculation you will need logarithms.) **(2)** Is this an entropy increase or entropy decrease? Could the change have occurred spontaneously?

14.7. **(1)** Which arrangement of 10 molecules between the two halves of a container is the arrangement of maximum entropy? **(2)** Suppose that this maximum-entropy arrangement has been reached and that the container and its contents remain free of outside influences. Would you expect any later spontaneous decrease of entropy? Why or why not?

14.8. If the probabilities of two events are 0.15 and 0.35, their *relative* probabilities are 15 and 35, or 3 and 7, or any other pair of numbers with the same ratio. Except as a matter of convenience for dealing with phenomena near absolute zero, only relative probabilities are important in thermodynamics. Explain how the irrelevance of the absolute value of entropy stated in connection with Equation 14.6 is equivalent to a statement that only relative probability plays a role in the definition expressed by Formula 14.3. (HINT: Consider the mathematical properties of the logarithm.)

14.9. Each of the following events is possible but unlikely. **(1)** A glass of water in a warm room freezes, and the room gets warmer. **(2)** A judge believes testimony of a convicted criminal when it conflicts with the testimony of an officer. **(3)** After being shaken, a tray of coins that previously showed 50 heads and 50 tails shows 65 heads and 35 tails. **(4)** Heat flows spontaneously in interstellar space from a cool region containing 1,000 atoms to a warm region containing 1,000 atoms. **(5)** A pion lives 10 minutes. Which of these events violate the second law of thermodynamics? Which of them might reasonably be expected actually to happen? Explain briefly the reasons for each of your answers.

14.10. Name any system not isolated from outside influences whose entropy "spontaneously" decreases. For the system you have named, where and how does a more than compensating increase of entropy take place?

14.11. Discuss the incubation of a hen's egg in general terms from the point of view of the second law of thermodynamics. What part of the system is tending toward a more disorderly arrangement of higher a priori probability?

14.12. A proton approaches the earth from outer space. It experiences the earth's magnetic force and gravitational force. It enters the atmosphere and is slowed down. Discuss the energy exchanges involved in this sequence of events. How is the initial energy of the proton finally distributed? Does the proton come finally to rest? Does entropy increase?

14.13. One pint of water (454 gm) at 20°C is mixed with another pint of water at 40°C. This system is isolated and comes to equilibrium at 30°C. **(1)** Explain how the trend toward uniform temperature in the mixture is a consequence of the second law of thermodynamics. **(2)** Show that the approximate magnitude of the entropy increase in this change is 0.5 cal/°K.

14.14. The quart of water at 30°C which resulted from the mixing described in the preceding exercise is divided into two pints. One of these pints is placed in contact with a large block of ice at 0°C until its temperature falls to 20°C. The other pint is placed in contact with a large container of hot water at 60°C until its temperature rises to 40°C. **(1)** What is the approximate total magnitude of entropy change in these processes? (HINT: Four entropy increments must be considered.) **(2)** Compare this entropy change with the entropy change of 0.5 cal/°K resulting from the process described in the preceding exercise. Would there be a way to get the two pints of water back to their original temperatures of 20°C and 40°C with less total increase of entropy?

14.15. One gram of steam experiences the following changes. For each process, calculate the entropy change of the steam, paying attention to sign. **(1)** The steam is heated from 375°K to 385°K with the input of 4.8 cal of heat as it expands under constant pressure. **(2)** It is maintained at a temperature of 385 °K by the addition of further heat as it continues to expand, doing 10^8 ergs of work. **(3)** By means of 10^6 ergs of external work without heat flow, the steam is compressed to a smaller volume. (Its temperature is increased only slightly in this process, by much less than 1°.) **(4)** Extraction of 4.8 cal of heat from the steam at constant pressure lowers its temperature from 385°K to 375°K. Sum the entropy changes of the steam and explain why the sum is not zero. How does the steam at the end differ from the steam at the beginning ?

14.16. A heat engine, as represented schematically in Figure 14.6, takes in heat at 800°K and gives out heat at 300°K. **(1)** What is its maximum theoretical efficiency (the greatest ratio of output work to input heat that does not violate the second law of thermodynamics)? **(2)** If its actual efficiency is half of this theoretical limit, how many ergs of work does it produce for each calorie of input heat?

14.17. Consider an idealized heat engine (Figure 14.6 and Equation 14.14) exhausting its excess heat to a reservoir whose temperature is very close to absolute zero. Explain in your own words why this makes possible a theoretical efficiency of almost 1, that is, an almost total conversion of heat to work.

14.18. A helium liquefier with an inside temperature of 4°K operates in an environment at 300°K. For each erg of energy extracted from the helium at 4°, how many ergs of energy (at least) must be added to the environment as heat?

14.19. To freeze a tray of water, a household refrigerator extracts 4×10^4 cal of heat from the freezing compartment at 260°K. The temperature of the room is 300°K. **(1)** What is the minimum energy input in ergs required from an outside power source to do this? **(2)** If this refrigerator is able to extract 10^4 cal of heat from its freezing compartment in 1 minute and operate at near maximum efficiency, what is its electric power requirement in watts?

14.20. **(1)** Equation 14.14 relates the maximum output work of a heat engine to the input heat H_1 and the temperature ratio T_2/T_1. Prove that the maximum output work may also be expressed in terms of the output heat H_1 by means of this equation:

$$W_{\text{max}} = H_2\left(\frac{T_1}{T_2} - 1\right)$$

(2) If the source of heat is at temperature $T_1 = 500$°K, for what range of exhaust temperatures T_2 will the "wasted" energy H_1 be no greater than the useful work W_{max}?

14.21. As indicated schematically in the figure, two heat engines are run in series. Engine 1 takes heat H_1 from a source at temperature T_1 and exhausts heat H_2 at a temperature T_2. The input of engine 2 is the heat H_2 at temperature T_2 and it exhausts heat H_3 at temperature T_3. If engine 1 does work W_1 and engine 2 does work W_2, calculate the maximum possible efficiency of these two engines working together—that is, the maximum value of $(W_1 + W_2)/H_1$. How does it compare with the maximum efficiency of a single engine operating between temperatures T_1 and T_3?

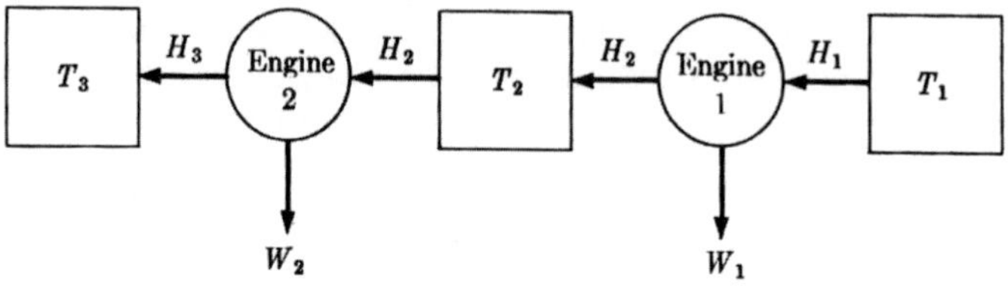

14.22. **(1)** Prove that the engine-refrigerator combination indicated schematically in the figure is possible if both the engine and the refrigerator operate at maximum efficiency (zero entropy change). The energy W required to operate the refrigerator is supplied by the engine. The heat H_2 exhausted by the engine into the region at temperature T_2 is the same as the heat removed from that region by the refrigerator. (*NOTES:* 1. The equation given in Exercise 14.20 may be useful in addition to those given in the text. 2. Take note of the notational differences between this figure and Figure 14.10, differences which will change the appearance of Equations 14.15 to 14.17.) **(2)** Prove that $H_3 = H_1$. **(3)** State in general terms how reality must differ from the idealization of this example.

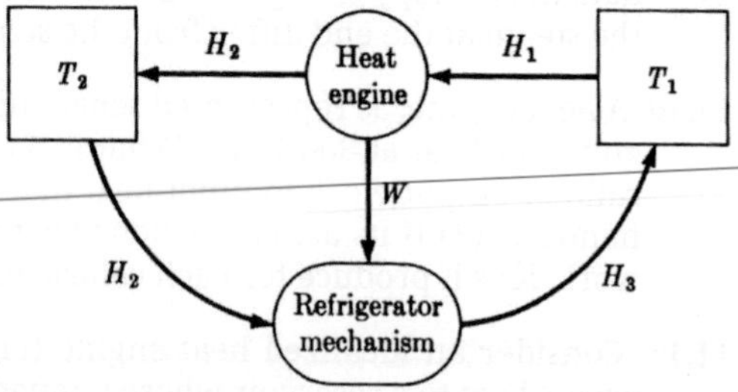

14.23. Why can the second law of thermodynamics be regarded on the one hand as one of the most general and profound laws of nature and on the other hand as not a fundamental law at all?

PART 5
Electromagnetism

JAMES CLERK MAXWELL
(1831 – 1879)

> *"It seems we have strong reason to conclude that light itself (including radiant heat, and other radiations if any) is an electromagnetic disturbance in the form of waves propagated through the electromagnetic field according to electromagnetic laws."*

FIFTEEN

Electrostatics and magnetostatics

More than a century elapsed between Newton's discovery of the law of gravitational force and Coulomb's discovery of the law of electrical force. This seems, in retrospect, a very long time, once the spirit and methods of modern science had come alive. However, it is not difficult to find some reasons for the lag of electricity behind mechanics. Mechanics was concerned with the grand sweep of the cosmos and linked earthly laws with universal laws. Electricity, thought at first to be an attribute of only certain substances, seemed to have less to do with the structure and behavior of matter. As a practical matter, electrical phenomena, in spite of ready accessibility, were not easy to deal with quantitatively. An electrified object gradually lost its charge; charges in metals moved about in a manner the experimenter could not control; electric currents persisted only for a moment. Planetary motion was not subject to such vagaries. Whatever the reasons, the fact is that while mathematicians and scientists were perfecting the techniques and the tests of mechanics, electricity remained, literally, a sideshow attraction. Around the middle of the eighteenth century, the sparks and shocks of electricity were more often used by entertainers seeking personal profit than by scientists seeking knowledge.

This is not to say that electricity had been completely ignored by men of science. By about 1750, a date roughly marking the beginning of electricity and magnetism as serious branches of science, a certain amount of rudimentary knowledge about electricity had been accumulated. That a mineral called amber could be electrified—or, in modern terminology, charged—was known to the Greeks. Since for many centuries electrification was believed to be a property exclusively of this one substance, the Greek word for amber, *elektron*, has furnished the root of our "electricity." Not until the time of Brahe and Kepler did knowledge of electricity take a significant step forward from the Greeks of 2,000 years earlier. A treatise by William Gilbert, *De Magnete*, published in England in 1600, revealed a number of new facts about electricity (and about magnetism), setting the stage for true scientific studies of these fields. In particular, Gilbert discovered that electrification is by no means limited to amber, but is a general phenomenon. The middle of the eighteenth century marks, then, not so much the time when studies in electricity were initiated, as the time when the tempo of effort and exchange of ideas reached a point that led to rapid progress.

15.1 Early knowledge of electrical force

We can summarize most of what was known of electricity in the mid-eighteenth century in terms of the following simple experiments, which may easily be performed by the reader. The only equipment required is a few small rubber balloons, a few feet of string, and some wool or fur (Figure 15.1). A balloon rubbed with wool (or fur or any of numerous other materials) becomes "electrified." What exactly does this mean? If we pretend that we know nothing of electric charge, and proceed only on the basis of our experimental findings, we can say only that this process called electrification has altered the properties of the balloon in some way. Before electrification, it had no properties of attraction for anything else. After electrification, it will readily stick to a wall or a ceiling, and it will attract to itself bits of paper or dust. Next we might discover that electrification can give rise to repulsion as well as attraction. Two electrified balloons suspended from threads and held not too far apart will repel one another, as evidenced by the fact that the strings take up a slanting position instead of hanging vertically. If the balloons are held a great distance apart, there seems to be little or no force; if they are brought closer together, the force gradually increases in intensity. Moreover, Newton's third law can be verified for this new kind of force. Regardless of the degree of electrification of the balloons or their distance apart, the force of the first on the second is equal and opposite to the force of the second on the first. In a similar way the attractive electrical force, say between a balloon and the piece of wool used to electrify it, can be studied and found to have the same properties.

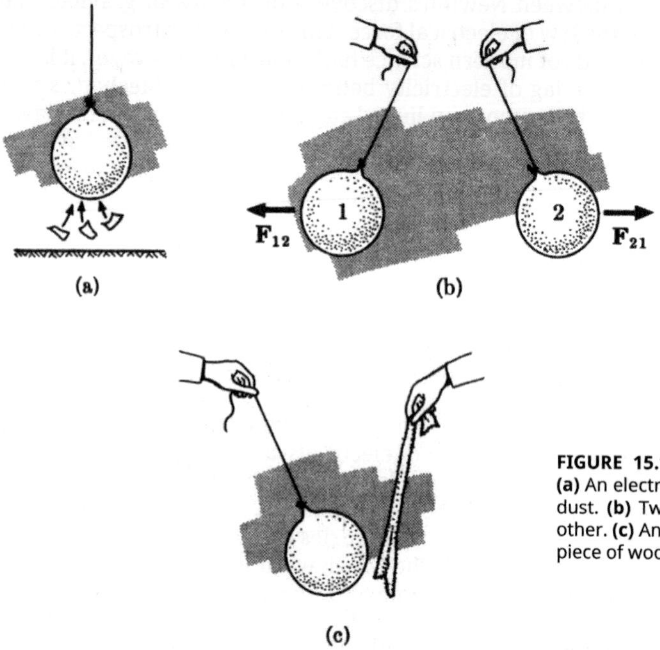

FIGURE 15.1 Simplest facts of electrical force. **(a)** An electrified balloon attracts bits of paper or dust. **(b)** Two electrified balloons repel one another. **(c)** An electrified balloon is attracted to the piece of wool that was used to electrify it.

The fact that an electrified balloon is attracted to almost any other object led at first to the erroneous conclusion that electrical forces are always attractive. Not until 1733 was it discovered (by Du Fay) that electrification can lead both to attraction and to repulsion.

86 It should be noticed that in order to approach an understanding of a new phenomenon, electricity, we make use of the already well-established mechanical concept of force. We learn about electricity first by tying it together with something already known. (The concept of distance, also fundamental to mechanics, is employed as well in considering how the force becomes stronger with decreasing separation of the balloons.) Even an entirely new field of study, electricity, rests upon previous scientific knowledge. Attempts to understand electricity by studying "purely" electrical phenomena—sparks jumping between electrified objects, for example, or shocks administered to human subjects—would not fare nearly so well. Eventually electricity developed into a complete and self-consistent theory with an independent status. But even in its final form, the theory preserves its ties with mechanics. Force, energy, and momentum remain essential concepts of electromagnetism. No theory stands entirely alone, simply because no part of nature is wholly independent of any other part.

Several other important facts about electrical force were known before 1750, which, like those already mentioned, may readily be verified with balloons and wool. The direction of the electrical force is of interest as well as its magnitude. We find that it acts along the line joining the electrified objects. It is, like the gravitational force, a central force. (This fact could, of course, be more accurately verified using objects much smaller than balloons.) Also the electrical force, again like the gravitational force, can act through a vacuum. It requires no material medium for its transmission. One very vital fact about electrical force, yet a fact readily overlooked, is that the electrical force is exceedingly powerful. Since a balloon sticking to a wall can readily be pulled off by a child, it is not so obvious that the force is strong. Yet if it is compared with the gravitational force, it becomes obvious that the electrical force is enormous. The gravitational force between a pair of balloons is entirely negligible compared to the electrical force. Finally we can readily discover that the electrical force an object is capable of exerting or of feeling seemingly has nothing to do with the mechanical or the geometrical properties of the object. The electrification of a balloon depends on how much it is rubbed, and what it is rubbed with, but it does not depend on the size or shape or weight or color or orientation of the balloon. It is apparently an entirely new and different property of the balloon.

We may summarize our qualitative knowledge of electrical force based on a few simple experiments, knowledge roughly the same as that possessed by scientists in the middle of the eighteenth century. The electrical force is (1) either attractive or repulsive, (2) central, (3) weaker at greater distance, (4) capable of acting through a vacuum, and (5) exceedingly powerful compared to gravity. It (6) obeys Newton's third law, and (7) its strength is independent of mechanical or geometrical properties of an electrified object.

With this much knowledge about electrical force assimilated, scientists began to ask two major questions: (1) What exactly is the nature of the electrification phenomenon? That is, can the property of a body causing it to exert and to feel electrical force be quantitatively defined and measured? (2) How does the strength of electrical force vary as the distance between electrified objects is varied? Both of these questions were answered with scientific precision, and answered together, in the latter half of the eighteenth century. Before that happened, the concept of electrical charge evolved gradually as a qualitative idea.

15.2 The concept of charge: charge conservation

Early studies of electricity involved not only electrical force, but also—indeed to a greater extent—the phenomenon of electrification itself. Gilbert proposed in 1600 that electrical effects arise from an electrical fluid. With some substantial modifications, the doctrine of the electrical fluid held sway until late in the nineteenth century. Even our present concept of charge does not differ very drastically from old ideas about the electrical fluid. 87

Basically the idea of the electrical fluid is very simple. Matter was supposed to contain, besides its material constituents, a rather ethereal fluid. According to Gilbert, friction could release a part of the fluid into the space surrounding an object, and the gradual flow of the fluid back to its parent object accounted for electrical attraction. According to this view, a neutral balloon would contain its normal quota of electrical fluid. After being rubbed with wool, it suffered a deficiency of the fluid, the lost fluid being distributed in the space around the balloon. A bit of paper drawn to the balloon was being borne on the tide of returning fluid. The eventual neutralization of the balloon occurred when all the fluid had returned to the balloon. That was an appealing theory, perhaps, for the then known facts, but a theory that could not survive new experiments of the early eighteenth century. Gilbert was unaware of electrical repulsion.

In brief, the subsequent development of the idea of electrical fluid was this. When Gray discovered the conduction of electricity through metals in 1729, Gilbert's idea of a separate electrical fluid attached to each material object gave way to the idea of a single electrical fluid which could flow from one body to another. Then the discovery of electrical repulsion led to the introduction of a two-fluid theory to rival the one-fluid theory. The two fluids corresponded to what we now call positive and negative charge. Benjamin Franklin held to the one-fluid theory, in 1747 making the inspired suggestion that the total amount of electrical fluid remained forever constant, any loss of fluid by one body being exactly compensated by an equal gain of fluid by another body. In his view, neutrality represented a "normal" amount of fluid, what we now call positive charge represented an excess of fluid, and what we now call negative charge, a deficiency of fluid. Although Franklin correctly guessed the law of charge conservation, he and his contemporaries had not yet abandoned Gilbert's original idea that the electrical fluid exists in the space surrounding an electrified object as well as in the object itself. Experiments of Franz Aepinus reported in 1759 showed that there was, in fact, no evidence for any electrical fluid occupying the space outside an electrified body. At this point, the fluid, having retreated to the interior of material objects (except in certain phenomena such as sparks), scarcely differed from our modern idea of charge. Gradually toward the end of the eighteenth century the two-fluid theory reemerged as the dominant theory of electricity, in part because it proved to be more convenient in calculations of electrical force, in part because it was championed by influential scientists such as Charles Coulomb. The two fluids came to be called charge, and for a century the source of all electrical (and magnetic) phenomena was successfully described as a pair of fluids which could flow through some materials, be held fast on other materials, which acted as the source of electrical force, and which, when present in equal quantities, canceled each other's effects.

The idea that electric charge was not a continuous fluid, but that instead definite fixed amounts of charge were associated with each atom, was suggested by Faraday as early as 1840. That charge is indeed concentrated in quantized lumps on individual particles was finally verified beyond doubt by Thomson's discovery of the electron in 1897. Obviously charge in tiny discrete bundles appears to be a very different picture from charge as a fluid spread throughout a substance. Most important, it renders irrelevant the idea of a *separate* fluid, distinct from the material basis of the substance. Charge becomes a property of matter, not an additional substance added to matter. Yet, viewed more carefully, our modern idea of charge is perhaps not so significantly different from Gilbert's original idea of an electrical fluid. We do not really have any deeper understanding of the nature of charge. We can say no more than Gilbert could have said, only that charge is that certain something possessed by material objects that makes it possible for them to exert electrical force and to respond to electrical force. Although we speak of charge as an intrinsic property of a particle, it is still a property distinct from the purely mechanical properties, for example mass and size. It is true that we cannot change the charge of a particle without destroying the particle altogether, yet our mental picture of charge is still a picture of something extra, a "substance" carried by charged particles, a substance that neutral particles lack. The concept of charge has undergone a steady evolution, and several revolutions, since the electrical fluid of Gilbert. Yet it remains a mystery at the deepest level. The fact that we "understand" electric and magnetic phenomena—that we can support a vast range of experimental results on a simple and powerful theoretical framework and can successfully predict new phenomena—should not delude us into thinking that we really understand electric charge.

Benjamin Franklin was first introduced to electricity in 1746. By the following year the results of a few simple experiments involving the transfer of electrification from one object to another had inspired him to suggest the law of charge conservation, a law that has survived and that stands as one of the few absolute conservation laws of nature. However, Franklin's brilliant guess must be recognized as a guess. He made the guess before anyone knew how to put it to the test, for in 1747 charge had not yet actually been defined and measured as a meaningful quantitative concept. Qualitative understanding of the idea of charge was developing, and Franklin had an accurate "feeling" for the nature of charge. Yet an operational definition of charge that could convert a qualitative idea into a quantitative concept had to await the discovery of the exact law of force between electrified objects several decades later.

15.3 *The definition of charge: Coulomb's law*

In his monumental work begun in 1666, Newton discovered three vital facts about the gravitational force. First, the strength of the force varies inversely as the square of the distance between objects. Second, the force can act through empty space. Third, the gravitational force exerted by an object (and experienced by an object) is proportional to its mass. A century went by before it was discovered that the electrical force shares with the gravitational force the first two of these fundamental properties. It must be remembered that in the manifestations most evident to man, these two kinds of force seemed then to have nothing whatever in common. Gravity acted over interplanetary distance; electricity was known only to act over short distances within the laboratory. The source of gravity, mass, was an invariable property of an object that no amount of heating, distorting, or rubbing could alter; the source of electrical force, charge, on the other hand, could easily be altered on a given object in a variety of ways.

That electricity, like gravity, stimulates "action at a distance" was not appreciated until 1759. One simple experiment performed in that year by Aepinus that strongly suggests this conclusion is the following (Figure 15.2). Two metal plates, one negatively charged, one positively charged, are placed close together with a thin layer of air between, like a sandwich with no meat. It is found that the rate at which each plate gradually loses its charge is little affected, if at all, by the presence of the other plate. If the charge (or the electrical fluid) of one plate extended into the space outside the plate, the presence of the second plate should have provided a place for the fluid of the first plate to flow. Since the plates were oppositely charged and attracted each other, the electrical fluid should have flowed from one to the other and quickly neutralized the electrification. It did not. Aepinus concluded, correctly, that the electrical fluid (or charge) was contained only within the plates, not in the space between them. This view, supported as well by other experiments, required that the electrical force arose without contact of electric charge, and therefore represented action at a distance. Actually Robert Boyle had observed much earlier the action of electric force through a vacuum, but he and his contemporaries continued to believe in an electrical fluid that flowed out from an electrified body.

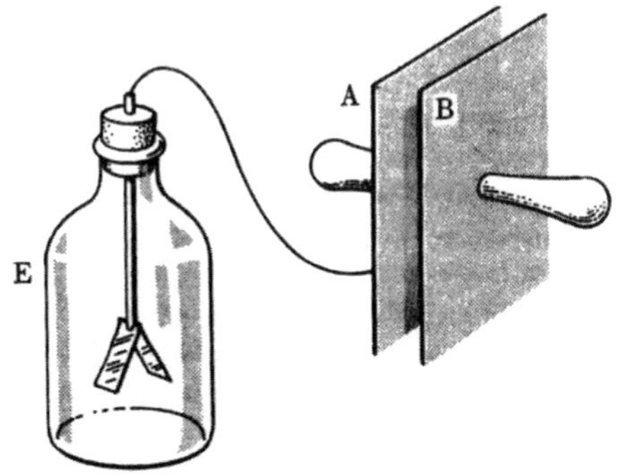

FIGURE 15.2 Schematic representation of an early experiment (Aepinus, 1759) which suggested that charge (or electrical fluid) exists only within matter, not in the space surrounding an object. The rate at which plate A loses its electric charge is unaffected by the presence of plate B. The mutual repulsion of the metal foil leaves in the electroscope E provides a measure of the degree of electrification of plate A.

The inverse square law of electrical force was discovered in 1785 by Coulomb. Expressed in symbols, it is

$$F \sim \frac{1}{d^2}, \tag{15.1}$$

where F is the magnitude of the force exerted by either of a pair of electrified objects on the other (by Newton's third law, they must be equal), and d is the distance between the two objects. The symbol $\sim$ means "proportional to." This relationship tells us that if the distance between two charged bodies is doubled, the force each exerts on the other is made four times smaller; if the distance is trebled, the force is nine times smaller; if the distance is halved, the force is four times larger; and so on. But the relationship does not yet tell us exactly how big the force is at any particular separation. It only provides the *ratio* of forces at different separations. The discovery of this law was a turning point in the history of electromagnetism. It provided for the first time a way to define and measure charge exactly. Equally important, by demonstrating that electrical and gravitational forces vary in the same way with distance, it suggested that the electrical force is perhaps also a fundamental force of nature, and worthy of as much attention as gravitation had received.

To say that Coulomb "discovered" the inverse square law of electrical force is an oversimplification of the facts. His achievement was to measure the electrical force to a hitherto unobtainable precision and so establish beyond doubt the correctness of the inverse square law. This variation of electrical force with distance had, in fact, been suggested on a number of previous occasions. A particularly interesting earlier suggestion was that of Joseph Priestley in 1767. Franklin, whose interest in electricity was still very much alive twenty years after his first contributions to the subject, noticed that a charged metal pail apparently directed all of its electrical effects outward and none inward. A small charged object outside the pail experienced a force; when held inside (but not touching) the pail, it experienced little or no force. Franklin was at a loss to explain this effect, and he wrote about it to his friend Priestley in England. Priestley was aware of a mathematical result from gravitation theory that seemed to bear on the problem. If a sphere (the earth, for example) had a hollow spherical center, there would be no gravitational forces anywhere within the hollow. Moreover, he knew that this was true of gravity only because of the inverse square law of gravitational force. The analogy seemed compelling to Priestley, and he wrote: "May we not infer from this experiment that the attraction of electricity is subject to the same laws with that of gravitation, and is therefore according to the squares of the distances; since it is easily demonstrated that, were the earth in the form of a shell, a body in the inside of it would not be attracted to one side more than another?"

There are two footnotes to this story of Priestley and the inverse square law. The first concerns his error, the second his "vindication." Priestley's argument was incorrect, for a rather subtle reason. It is true that inside a hollow earth there would be no gravitational force. It is also true, as Franklin observed, that there is little or no electrical force within a charged pail. But it is *not* true that a pail-shaped earth would be devoid of gravitational force in its interior. The analogy Priestley drew between electricity and gravity is imperfect: The mass responsible for gravitation is fixed rigidly in position; the charge responsible for electrical force can move around within the interior of a piece of metal. If, for instance, the pail has an excess of electrons, these will move about under the influence of their own mutual repulsion, distributing themselves finally in a non-uniform way over the outside surface of the pail. Now for the "vindication." More modern knowledge has shown that the mobile electrons will indeed distribute themselves in such a way as to produce little or no electrical influence within the pail. Moreover, they do so because of the inverse square law of electrical force. For reasons somewhat more technical than Priestley considered, the hollow interior of a piece of metal will be free of electrical force regardless of the shape of the metal or of the hollow. For gravitation, on the other hand, the only force-free region is a spherical hollow centered within a spherical mass. Today the most exact proof of the inverse square law of electrical force rests not on a direct measurement of force such as Coulomb's, but on a verification of the *absence* of force within a hollow piece of metal. Although Priestley's reasoning was slightly shaky, his insight was flawless.

We come now to a crucial point, the precise definition of charge. It is here that electricity departs from mechanics and from gravitation. The concepts employed so far—at least those employed rigorously—force and distance, are mechanical concepts. The laws of action at a distance, and of inverse square dependence on distance, are identical to laws of gravitation.

But charge is a new concept, a purely electrical idea. We have discussed the idea already, and we have seen that Franklin had a sufficiently clear idea of its meaning to suggest the law of charge conservation. But how do we define charge exactly? This is a question we wish to examine with some care.

To understand the definition of charge, let us imagine some simple experimentation, 88
this time with a number of small plastic pellets, each one electrified, each suspended by a thread (Figure 15.3). How the pellets became charged need not concern us—perhaps each was touched by one of the balloons or pieces of wool left over from the previous studies. In any event, they are surely charged, for they exert electrical forces on one another. Consider any particular pair. Let us call them pellet A and pellet B. If they are placed 10 cm apart, they will exert on each other a certain measurable force. Then if the distance between them is changed, the force will change according to the inverse square law. Another pair, C and D, will also follow the inverse square law, but at a given distance of separation, such as 10 cm, will not experience the same force as did pellets A and B. Now the variation with distance is, for our present purposes, no longer of interest. We wish to study the actual strength of the force and learn how it varies from one pair of pellets to another. (For these idealized experiments, we shall ignore the fact that in practice the charge on a pellet gradually leaks off into the air.) Upon what does the electrical force depend? Not on the shape or size or mass of the pellets, we could easily learn by experiment. A series of experiments and trial-and-error guesses could lead us to the right answer, but instead of describing false starts, let us begin by making what appears to be the *simplest possible* assumption about the source of the force. Suppose that the force depends on one single new property of the pellet, a property to be called *charge* that can be described by a single number. Charge is in this way like mass. The mass of an object is a certain number of grams or pounds, one single quantity. This is to be contrasted with a concept like shape. To specify shape requires many quantities—length, breadth, height, and angles. Charge and mass are scalar quantities, mathematically the simplest possible concepts. Second, suppose that the strength of the force is proportional to the charge. Doubling the charge doubles the force, and so on. Again, it is difficult to imagine a simpler assumption. These two assumptions prove to be all that are needed, and they work. They lead to a self-consistent definition of charge and to a successful description of electrical force, and the charge so defined proves to satisfy Franklin's law of charge conservation. It is no wonder that scientists emphasize the simplicity of the fundamental laws of nature.

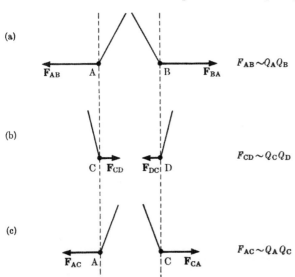

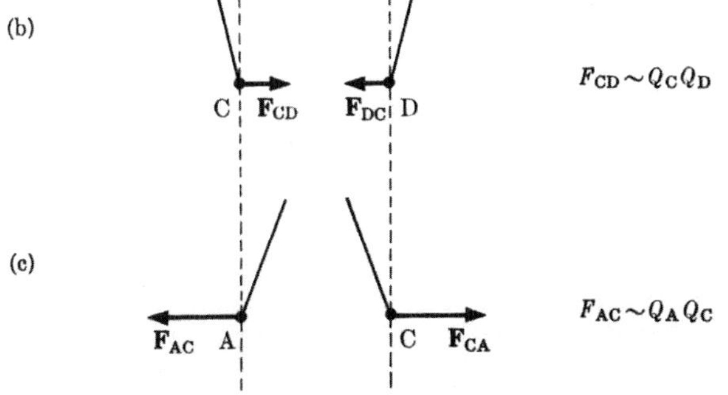

FIGURE 15.3 Idealized experiments with electrified pellets. The magnitude of the force is assumed to be proportional to the product of charges. The self-consistency of this assumption is tested by many experiments with different pairing of given charges.

Armed with these assumptions about charge, let us return to the pellets. Consider again pellets A and B placed close together, well separated from all other pellets, and apply the general idea that the force experienced by any pellet is proportional to its own charge. Then the force on A is proportional to the charge on A, and the force on B is proportional to the charge on B. But since these two forces are equal to each other in magnitude (Newton's third law), it follows that the force on A must also be proportional to the charge on B, and the force on B proportional to the charge on A. Introducing the symbol Q to designate charge, we may express this result symbolically as follows:

$$F_{AB} \sim Q_A Q_B \,. \tag{15.2}$$

In words, the force exerted on A by B (and the equal force exerted on B by A) is proportional both to the charge on A and to the charge on B, that is, to the product of the two charges Q_A and Q_B. Note that this result stems only from making the assumptions of the preceding paragraph consistent with Newton's third law.

The mathematical expressions (15.1) and (15.2) may now be joined. The dependence of electrical force both on charge and on distance is given by

$$F_{AB} \sim \frac{Q_A Q_B}{d^2} \,, \tag{15.3}$$

where d is the distance between the two charges Q_A and Q_B. So far Expression 15.3 is only a proportionality, not an equality. But by the very simple device of choosing the unit of charge to suit our convenience, we can convert it into an equality, and write

$$F_{AB} = \frac{Q_A Q_B}{d^2} \,. \tag{15.4}$$

This is the fundamental law of electrical force, usually called Coulomb's law. (For completeness, it must be supplemented by the statement that the electrical force is a central force, acting along the line joining the pair of objects.) In order to understand the last step, the sudden introduction of the equal sign, let us consider an analogy. Suppose that potatoes cost 9 cents per pound. Two pounds cost 18 cents, three pounds cost 27 cents, and so on. The cost, C, is proportional to the mass of potatoes, M. This may be written mathematically,

$$C \sim M \,.$$

This expression does not give the cost, it says only that the cost is proportional to the mass purchased. Usually a proportion is converted to an equation by introducing a constant of proportionality. In this case the constant of proportionality is 0.09 dollars/pound, and the equation reads

$$C = 0.09M \,.$$

However, there is another possibility—not very practical in this case, to be sure, but possible. We might decide that potatoes, being such a fundamental staple of diet, deserve a new unit of mass of their own. We could *define* 11.1 pounds of potatoes to be one "Bag." One "Bag" of potatoes would be the mass of potatoes that cost one dollar. Three "Bags" cost \$3.00; five and a half "Bags" cost \$5.50. In terms of the new unit of mass, the equation would read

$$C = M \,.$$

Another way to describe this result is to say that the unit of mass has been so chosen that the constant of proportionality is exactly equal to 1. This is what has been done with charge. In fact, since charge is a new concept, we have full freedom to choose the most convenient unit, and the unit is chosen to make Coulomb's law, Equation 15.4, have the simplest form.

This unit of charge is called the e.s.u. (electrostatic unit). The same unit is sometimes known as the statcoulomb.*

One problem remains with Coulomb's law and the definition of charge. That is the problem of measuring the charge on a single object. Suppose that our two pellets A and B, when 10 cm apart, are repelled, each with a force of 1 dyne. This means, according to Equation 15.4, that the product of their two charges (in e.s.u.) is equal to 100. This could come about in an infinite number of ways. Pellet A might have a charge of 25 e.s.u., pellet B, 4 e.s.u.; or A could have –2 e.s.u. and B, –50 e.s.u.; or both could have 10 e.s.u. With these two pellets alone, we have no way of knowing how the charge is divided. To remove the ambiguity, we bring pellet A near a third pellet C with B removed, then bring B near C with A removed. By letting A and B interact separately with the same pellet C, we can discover the relative charges on A and B. Suppose, for instance, that at a given distance from C, A experienced four times the force experienced by B in the same position. Then the charge on A must be four times that on B. Since the product of the two charges is 100, we could with a little algebra (or trial and error) conclude that the magnitude of charge on A must be 20 e.s.u., the magnitude of charge on B, 5 e.s.u.

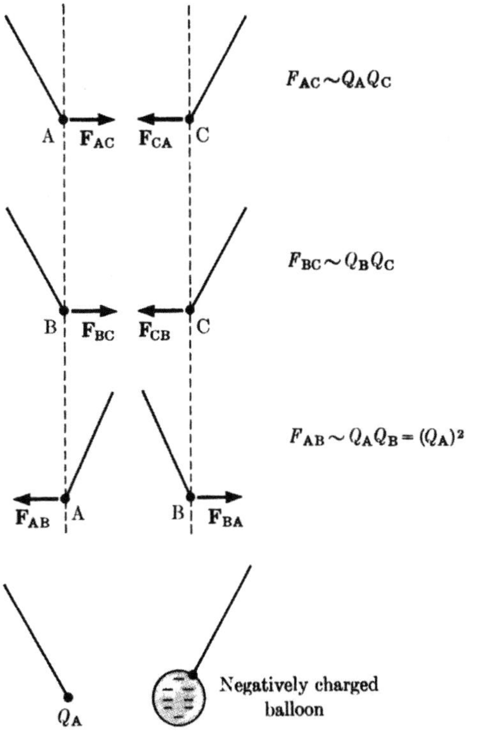

FIGURE 15.4 Idealized experiments to establish the unit charge. Since charges Q_A and Q_B are equally influenced by charge Q_C, the two charges Q_A and Q_B are equal: $Q_A = Q_B$. If the mutual repulsive force between these equal charges is 1 dyne when they are 1 cm apart, each is a unit charge: $Q_A = Q_B = 1$ e.s.u. or –1 e.s.u. Since the charge on an electrified rubber balloon is defined to be negative, attraction between Q_A (or Q_B) and a charged balloon shows the sign of Q_A and Q_B to be positive.

The signs of the charges must be separately determined by reference to an arbitrarily chosen standard of sign. When a rubber balloon is rubbed with a piece of wool, the charge acquired by the balloon is *defined* to be negative. The sign of any other charge can be determined in principle by finding out whether it is attracted or repelled by a charged rubber balloon. Defining something as important as the sign of electric charge in terms of a toy balloon might

* There is another unit of charge in common use, the coulomb. When this unit is employed, the constant of proportionality in Equation 15.4 is not equal to 1.

sound rather childish. In fact it is perfectly acceptable. By an unbroken chain of experiment, the charge on the balloon can be shown to have the same sign as the charge of an electron.

Finally, we may give a rigorous operational definition of the charge unit, one e.s.u. (Figure 15.4). Any two charges equally influenced by a third have equal charge. If two equal charges are placed 1 cm apart and experience a force of 1 dyne, the magnitude of charge on each is 1 e.s.u. With our large collection of charged pellets, for example, we could first, with the help of a rubber balloon, determine the sign of the charge on every pellet. Then we could select all those pairs with equal positive charge. Next, measuring the force between every equal pair, we could determine the charge on each member of the pair. One particular pair might turn out to have one e.s.u. of charge on each pellet. If so, one of this pair could be selected as the standard unit of charge, and the charge on every other pellet, equal or not, could be measured by finding how much force it experienced when near the standard pellet. In practice, of course, much more elaborate procedures than this are required. Our pellets do not retain their charge long enough for accurate work. But the principles involved in more exact measurements are the same.

In the elementary-particle world, the identity of like particles can be used to advantage. If we measure the force between two protons, we do not suffer from ambiguity about how much charge is on one proton, how much on the other. We know that the two charges are identical. The force between two protons cannot be measured directly because it is far too small, but an indirect method can yield the charge on the proton. If one proton, a "projectile," is fired at another proton, the "target," the force of electrical repulsion between them causes the first proton to be deflected and the second proton to be accelerated. Study of the angles at which the two protons emerge from the point of collision* can reveal the strength of the force that acted between them and thereby the magnitude of the charge on each proton. The proton charge is 4.80×10^{-10} e.s.u., the electron charge, -4.80×10^{-10} e.s.u.. Most other elementary particles also carry this same fundamental magnitude of charge unless they are neutral. About two billion proton charges are required to make up one e.s.u. The mks unit of charge, the coulomb, is much larger than the e.s.u.:

$$1 \text{ coulomb} = 3 \times 10^9 \text{ e.s.u.}$$

In this unit, the fundamental quantum of charge is $e = 1.60 \times 10^{-19}$ coulomb.

Since charge may be either positive or negative, the numerical magnitude of force between charges calculated with Coulomb's law, Equation 15.4, may turn out to be either positive or negative. If the calculated force is positive, this means repulsion; if it is negative, it means attraction. Actually, Equation 15.4 is an expression for the component of a vector quantity—the component of force on charge A directed *away* from charge B. A negative component indicates an opposite vector direction, a force on A directed toward B.

15.4 Microscopic and macroscopic manifestations of charge

Electricity began as a curiosity associated with the mineral amber. Then it came to be known as a property of many materials. Next charge and electrical force were recognized as manifestations of a new fundamental theory comparable in simplicity and generality to the theory of gravitation. Charge, it was later learned, resided in fundamental units on electrons, protons, and other elementary particles of nature. Now we know electrical force to be the ruling interaction over a wide domain of nature from the size of atoms to the size of man. In this section we shall examine some of the ways in which electric charge dictates the structure and properties of matter throughout this domain.

* In the particle world, "collision" means an event in which a high-speed particle is deflected or otherwise influenced by another particle. It does not mean that the particles "touched" one another. Touching, or contact, is in any case not a well-defined concept for particles.

The simplest atom, an atom of hydrogen, consists of one proton and one electron. These two constituents of the atom are held together by their force of mutual electrical attraction (Figure 15.5). Because of the laws of quantum mechanics, these two particles do not draw together into the same spot. Instead the light-weight electron takes for its domain of motion a spherical region of space, centered at the heavier proton. The average separation of the two particles is about 0.5 Angstrom (0.5×10^{-8} cm). Under certain conditions, the electron may temporarily have some extra energy and extend its motion to greater distance, but the diameter of the atom can never be less than about 1 Angstrom. If sufficient extra energy is given to the electron, it may escape the attractive force of the proton altogether and move away to a great distance. In this case the atom is said to be "ionized"; the proton left behind is called a positive ion. For this proton the symbol H^+ is sometimes used, designating a positively charged hydrogen atom. Under conditions of extreme temperature, above several thousand degrees Kelvin, such as in a star, hydrogen gas may be largely ionized, consisting mostly of free protons and electrons. Such a gas is called a plasma.

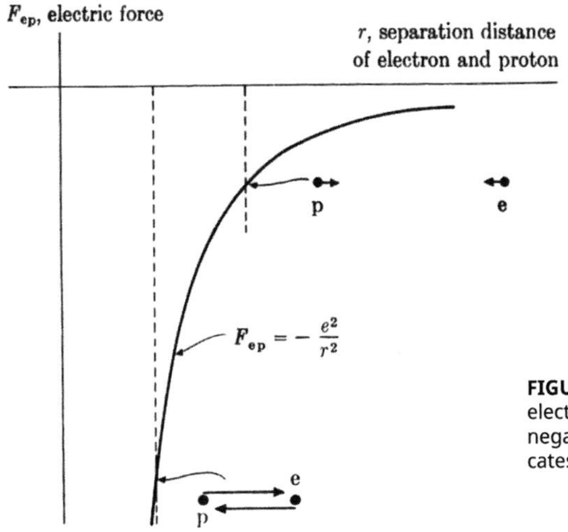

FIGURE 15.5 Graph of electrical force between proton and electron as a function of their separation. By convention, a negative sign attached to the magnitude of the force indicates attraction (see Equation 15.4). Compare Figure 17.5.

In a plasma electrical forces reach out to relatively great distance, and each ion "feels" the presence of many other ions. At more moderate temperatures, however, each electron will be attached to its own proton, forming a neutral hydrogen atom, designated simply H. Because electron and proton have equal and opposite charge, the atom as a whole is neutral and its electrical effects do not extend very far beyond its own boundaries. The electrical effects have, so to speak, been confined within the atom. Two hydrogen atoms a few diameters apart are oblivious of one another. Two protons or two electrons at the same distance, on the other hand, would feel a strong electrical force. A gas of ordinary atoms is called a neutral gas, because each atom as a whole is neutral, although electrical forces are of dominant importance within each atom. One might be led to guess that in a neutral gas, electrical force has been confined entirely within the atom and is of no consequence outside it. This is not true. Rather the neutrality of the atom has sharply limited the range of the force. If two hydrogen atoms get quite close together, for example to a separation of their centers of one Angstrom, they are, roughly speaking, in contact, or touching. All this means is that they are so close together that the proton and electron of one atom can separately feel the electrical forces of the proton and electron of the other atom. The net force felt by the atom is not a Coulomb force; that is, it does not vary inversely with the square of the distance of separation. It is considerably more complicated, its exact form depending on the laws of quantum mechanics as well as on the presence of electrical forces. At great separation there is no force at all. As the atoms approach one another, they experience a weak attractive force, but when they get so close together that they are "touching," this changes to a strong repulsive force that prevents the two atoms from overlapping significantly. The weak attraction is not unimportant. It is what causes pairs of hydrogen atoms to join together to form hydrogen molecules (H_2). The still weaker but analogous force attracting hydrogen molecules to one another is in turn responsible for the condensation of hydrogen gas to a liquid at low temperature, and the freezing of hydrogen into a solid at still lower temperature.

Within every atom, electrical force is the ruling interaction. Consider carbon, whose nucleus contains 6 protons. An electron will be attracted 6 times more strongly to a carbon nucleus than to a proton and can orbit the nucleus at a distance one-sixth its distance from the proton in a hydrogen atom. This structure has a total charge of +5 (in atomic units) and can attract to itself five more electrons. The neutral carbon atom contains 6 electrons in motion about the central nucleus. Because the last electron to be added "sees" a net charge of +1 (six protons and five electrons), its orbit is about the same size as in hydrogen. Thus the carbon atom is no larger than the hydrogen atom. In fact, all atoms are about the same size. An atom of uranium, with 92 protons and 146 neutrons packed into its central nucleus, and 92 electrons in orbit, occupies no more space than the single proton and single electron of the hydrogen atom. Likewise the forces between atoms are all generally of the same character, weakly attractive when the atoms are somewhat separated, strongly repulsive when the atoms begin to overlap (Figure 15.6). Yet the differences are sufficient to introduce a rich variety into the structure of collections of atoms. Between carbon atoms or tungsten atoms, the attractive forces are strong enough to hold these substances in solid form up to temperatures of several thousand degrees Kelvin. For helium, on the other hand, the thermal agitation a few degrees above absolute zero is sufficient to break the interatomic bonds and permit the liquid to boil. When we turn to consider the forces not only between atoms of the same kind, but between atoms of different kinds, the richness of possibilities grows even greater. This is the foundation of chemistry, the study of the combination of atoms to form molecules. It belongs properly to the theory of quantum mechanics; it is a large subject by itself which we shall not pursue further here.

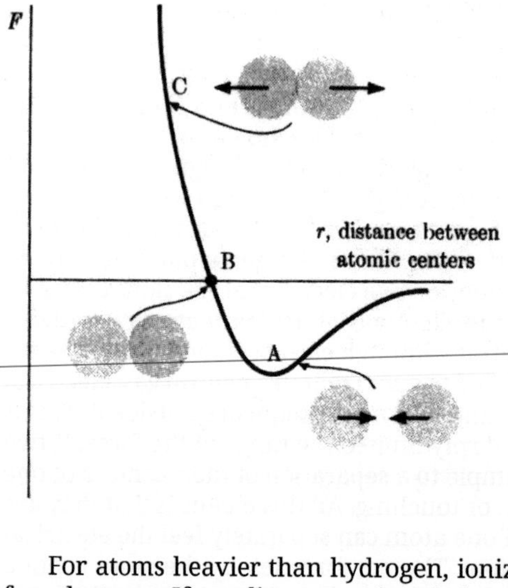

FIGURE 15.6 Approximate form of interatomic force. It is weakly attractive when the atoms are close together but not overlapping (A), and strongly repulsive when the atoms overlap significantly (C). Point B indicates the equilibrium distance where the force vanishes.

For atoms heavier than hydrogen, ionization usually involves the loss of only one or a few electrons. If a sodium atom (symbol Na) loses one of its eleven electrons, the remaining structure, a sodium ion, designated Na^+, consists of the nucleus with its eleven protons surrounded by ten electrons. To remove successively larger numbers of electrons requires more and more energy, for the electrons in inner orbits are held more tightly than those in outer orbits. Nevertheless, total ionization can occur; it is common in stars and rare on earth.

Cosmic rays arriving from outer space consist primarily of totally ionized atoms, most of them protons (H^+), some of them heavier bare nuclei. A nucleus of iron, for example, containing 26 protons, might strike the atmosphere at high speed. As it slows down, it gradually accumulates electrons until finally it acquires its full complement of 26 and becomes a normal neutral atom of iron. The slowing down process itself is accomplished mostly by further ionization. The fast charged particle, in flying through matter, interacts via electrical force with atomic electrons, accelerating some of them sufficiently to eject them from their parent atoms. In this process the electron gains energy at the expense of the fast particle, which accordingly is gradually slowed down. This kind of process strips away electrons, leaving positive ions. Less violent processes may produce negative ions. A hydrogen atom, for instance, can acquire a second electron, very weakly bound in a large orbit, becoming a negative hydrogen ion (H^-). Many other atoms can do the same. One of the simplest kinds of chemical binding is called ionic binding and involves the transfer of an electron from one atom to another. If a cesium atom and a fluorine atom encounter each other, the fluorine atom may "steal" an electron from the cesium, creating the ions Cs^+ and F^-. In symbols,

$$\mathrm{Cs} + \mathrm{F} \rightarrow \mathrm{Cs}^+ + \mathrm{F}^- .$$

The oppositely charged ions then attract each other in large part by the ordinary Coulomb force of electrical attraction and stick together to create a molecule of cesium fluoride.

Within the atom is a nucleus, some ten thousand times smaller than the atom itself, with a structure of its own. Here new forces, more powerful than electrical forces, dominate. In a nucleus of carbon 12 (C^{12}), 6 neutrons and 6 protons are packed tightly together, bound by the nuclear forces acting between each pair of them. This force pays no attention to charge. Protons attract protons, neutrons attract neutrons, protons attract neutrons. The force of electrical repulsion between the proton pairs is also acting, but it is outweighed by the attractive nuclear force. In this subatomic domain, electrical forces are present, and are even important (they prevent the formation of long-lived nuclei heavier than the uranium nucleus); but they are not dominant. Only for atoms and larger structures do electrical forces rule. Electrons are uninfluenced by nuclear forces. Because of the orbital electrons, one nucleus is prevented from getting near enough to another nucleus to interact with it.* Consequently, except at high energies when electrons are stripped away, the nuclear forces are bottled up at the center of each atom and have no influence on how atoms interact with each other.

Air and other normal gases on earth are neutral. Accordingly, they are not good conductors of electric charge. Under certain conditions, however, the gas can "break down," permitting a sudden surge of charge to flow through it. This is a spark. In the Geiger counter and in the spark chamber (Figure 2.11), the ionization produced in the gas by a high-energy particle is the trigger releasing the spark. For the instant that a spark exists, a portion of the gas becomes highly ionized. After the discharge, the ions recapture electrons and the gas settles back to neutrality. In the upper part of our atmosphere, known as the ionosphere, molecules of air are being ionized by ultraviolet light from the sun, and, to a lesser extent, by cosmic rays from space. Although the air in the ionosphere is still mostly neutral, enough charged particles are present to alter its properties significantly. In particular, it is capable of reflecting some radio waves, making long distance radio communication around the earth possible.

* Thermonuclear reactions are exceptions to the rule. They result from the interaction of one nucleus with another via nuclear forces. However, they can be ignited only at very high temperature, as in an atomic explosion or in the sun.

In the liquid state of matter, atoms or molecules are held close together, but they are not fixed in position. A molecule of water may shoulder its way about from one part of the liquid to another. Because many liquids contain a substantial number of ions, they are rather good conductors of electricity. In water, for example, some of the molecules, H_2O, are separated into a proton, H^+, and the combination OH^-, one oxygen atom, one hydrogen atom, and one extra electron bound together. Because water is a reasonably good conductor of charge, it is poor practice for an electrician to work in wet bare feet.

A remarkable fact about solids is the vast range of different electrical conductivities encountered in ordinary material. If a rubber balloon is rubbed on one side with a piece of wool, and that same side is touched to a wall, the balloon will stick. If the side not rubbed is touched to the wall, the balloon will not stick. This simple experiment shows that the extra electrons acquired by the balloon remain immobile on the side of the balloon that was rubbed. The other side acquires no electrification. A material such as the rubber of the balloon through which charge does not flow readily is called an insulator. In other materials, particularly metals, charge flows with ease from one point to another. Such a material is called a conductor. There is no absolute distinction between a conductor and an insulator, for charge can to a greater or lesser extent move through any material. However, differences in the ease with which charge moves through different materials are enormous. A quantity called conductivity measures how readily charge can move from place to place. The conductivity of a metal can be more than 10^{19} times greater than the conductivity of an insulator such as glass. Consider the fact that in a high-tension line charge flows much more easily through hundreds of miles of metal wire than through the few inches of insulating material separating the wire from its supporting tower. In an ordinary lamp cord in the home, charge moves from the outlet through perhaps six feet of wire, through a light bulb, and back through the other wire in preference to flowing across from one wire to the other through a small fraction of an inch of rubber insulation.

How do such remarkably great differences in the electrical conducting powers of different solids come about? The answer to this question has been provided only recently, for it depends on rather subtle differences in the atoms and the interatomic forces in the solid. Simply stated, an insulator is a material all of whose electrons are attached to particular atoms, as in a neutral gas. A conductor is a material in which each atom has contributed one or a few electrons to a general sea of electrons attached to no particular atoms, free to move long distances through the material. It is thus analogous to a plasma of ions and electrons. It may take only a small difference in the binding between atoms to decide whether any electrons are freed to move about inside the solid. But that small difference in atomic forces can be reflected in a difference in conductivities measured in millions or billions. In the early studies of electricity it was learned that metals are useful for transferring charge from one place to another. This knowledge acquired great practical significance only after man learned how to create continuous streams of electric current instead of the short burst of current flow generated by static charges. ("Current" means simply a flow of charge.) In the mid-twentieth century, we can attribute the serious depletion of the world's reserves of copper ore mainly to the fact that copper is a material possessing a particularly high conductivity (second only to silver among the elements).

In the modern world the controlled flow of charge through wires to provide light and heat and power far outweighs in importance the effects of accumulations of static charge. However, static charge has acquired some new uses as well. The spark chamber and the Geiger counter make use of the electrical force of static charge. A particle accelerator called the Van de Graaf accelerator builds up a large static charge on a metal sphere whose electrical force then accelerates protons or other particles to high energy. A device called a capacitor (Figure 15.7) stores electric charge on metal plates as a convenient way of saving electrical energy for later use. And in everyday life, like it or not, we continue to be subjected to the effects of static charge. The most dramatic example is the lightning bolt of a thunderstorm, whose basically electrical nature was first demonstrated by Franklin. Anyone who lives in a dry climate knows also the sometimes annoying manifestation of charge, the sharp shock experienced upon touching a metal object such as a door knob after generating a charge, perhaps by friction of shoes on a rug. It is interesting to speculate that if Athens had a drier climate, the Greeks might have learned that electricity is not a phenomenon restricted to amber.

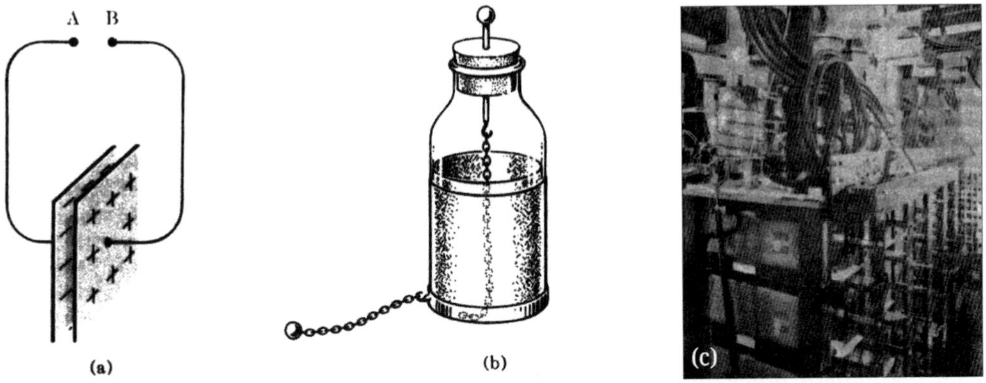

15.7 CAPACITORS. (a) Basic idea of capacitor. Opposite charge on parallel metal plates represents stored energy. Electrical connection between A and B allows charge to flow, tapping the stored energy. **(b)** Early version of capacitor, the "Leyden Jar," invented in 1745. Glass separates layers of metal foil. **(c)** A modern "capacitor bank" at the Lawrence Radiation Laboratory, Livermore, California. The discharge of such a bank can release more than 10^{11} ergs of stored energy in a few microseconds, [(c) Photograph courtesy of Lawrence Radiation Laboratory, University of California at Livermore, under the auspices of the U.S. Atomic Energy Commission.]

The enormous strength of the electric force can best be illustrated by example. A balloon briskly rubbed with wool might acquire a charge (negative) of several thousand e.s.u. To make the calculation easy, let us suppose the charge is –4,800 e.s.u. Dividing this amount of charge by the charge on a single electron, -4.8×10^{-10} e.s.u., we conclude that the balloon has acquired an excess of 10^{13} electrons. This certainly seems to be a large number of electrons, but when compared with the number of electrons already present in the balloon, it becomes insignificant. The balloon has a total of some 10^{24} electrons, 10^{11} times more than were added. The electron population of the balloon increased by only one part in 100 billion. The birth of a single individual changes the world's population by a fraction substantially more than this tiny change in the number of electrons. Yet it is just this tiny imbalance of positive and negative charge in the balloon that is sufficient to create quite marked electrical forces. In the large-scale world, electrical forces are always manifestations of a very slight disturbance in the normally perfect balance of positive and negative charges. In the atomic world, bare fundamental units of charge interact as the raw power of electrical force comes into its own, holding atoms and molecules together, preventing the dissolution of solid matter, and providing the source of all chemical energy.

We end this section with a few remarks on ionizing radiation and living matter. One solid generalization seems to be agreed upon by biologists: Ionization of the complex molecules of living matter is usually harmful to the organism. Nevertheless mankind and all other life have successfully evolved in the face of a steady bombardment of particles whose effect is to tear electrons loose from their parent molecules. Anything capable of removing electrons from atoms or molecules, leaving positive ions behind, is called ionizing radiation. In the language of modern physics, we recognize such "radiation" as particles of one kind or another, either charged material particles or photons. Photons of visible light are not energetic enough to produce significant ionization. Ultraviolet photons, X rays, and gamma rays can do so, as can any charged particle with more than a few eV of energy. The ionizing radiation to which man was subjected before the nuclear age consisted mostly of radioactive emanations, especially beta and gamma rays; cosmic radiation, principally energetic muons; and medical and dental X rays. To this, nuclear explosions have added additional radioactive materials (usually called fallout) beyond those present naturally in the earth and the air. The trail of ions left behind a high-energy particle in a bubble chamber or spark chamber can be utilized for particle detection. The same trail of ions in a living cell is potentially, although not necessarily, dangerous. An ion may simply recapture an electron after a time and return to normal. Alternatively the loss of an electron may sometimes induce a chemical change in a molecule, that is, alter its structure. Such a change in most molecules within a cell represents no hazard, but if the ionization happens to alter the structure of a genetic molecule involved in reproduction, a hazard exists. As a rule, a man ought to avoid unnecessary exposure to charged particles or photons capable of producing ionization. On the other hand, the idea of the calculated risk, an essential concomitant of all of man's activities, is as valid for the unseen risk of a high-energy muon as it is for the seen risk of an oncoming automobile on the highway. A chest X ray, a skiing holiday in the more intense cosmic radiation of high altitude, or, for a physicist, the extra exposure associated with working near an accelerator are risks deemed worth taking.

15.5 Magnetic poles and magnetic force

89 That magnetism and electricity have something to do with each other was first definitely established by Oersted in 1820, although such a connection had been suspected at least since the publication of Gilbert's book in 1600. Franklin believed that the phenomena of magnetism and electricity were related, and in 1751 thought that he had succeeded in proving it by producing magnetism with an electric current. However, a later repetition of his experiment failed to produce the same result; his work on this particular problem apparently had no impact on the subsequent histories of electricity and magnetism. It is approximately true to say that until the time of Oersted's work, electricity and magnetism developed in parallel as distinct nonoverlapping branches of science. Following Oersted's work, which welded the ideas and the words of electricity and magnetism into the single science of electromagnetism, came half a century of progress and consolidation. Between 1820 and 1870, all of the detailed connections among the phenomena of electricity and magnetism were worked out, light and other radiations were merged with electromagnetism, and the entire theory was formulated in terms of a few equations of remarkable simplicity and beauty.

The qualitative facts of magnetism can be stated briefly. Certain objects called magnets, usually (but not necessarily) containing iron, can attract and repel other magnets, and can attract pieces of iron that are not magnets. Unmagnetized iron, placed near a magnet or touched by a magnet, can itself become magnetized. That magnetism is distinct from electricity is demonstrated simply by the fact that a charge and a magnet placed near one another exert no mutual forces on each other. Magnets influence magnets, charges influence charges, but magnets and charges at rest coexist without interaction. Each magnet acts as if it had two centers of force. In a long thin magnet, these two centers of force may be near the ends of the magnet well separated from each other (Figure 15.8). These are called the "poles" of the magnet, and poles are found to attract or repel other poles according to an inverse square law, in exact analogy to Coulomb's law of electrical force (or Newton's law of gravitational force). A "pole strength" may be defined (analogous to charge) such that two equal poles, each having a pole strength of 1 m.s.u. (magnetostatic unit), exert a force of 1 dyne on each other when separated by 1 cm. The force law for poles may be written

$$F_{\mathrm{M}} = \frac{P_1 P_2}{d^2}, \qquad (15.5)$$

where P_1 and P_2 are the pole strengths of two poles and d is their separation in centimeters. The subscript M on the force F serves to remind that the force (measured in dynes) is magnetic, not electric. The peculiarity of the poles of a magnet is that they can never be isolated. The total pole strength of any object, magnetized or not, is always found to be exactly zero. By convention, the pole of a magnet attracted to the north end of the earth is called a north pole and its pole strength is called positive. The opposite pole, attracted to the south end of the earth, is called a south pole; its pole strength is called negative. In a simple magnet with two poles, the pole strengths are equal and opposite, so that the total pole strength of the whole magnet vanishes. If such a magnet is broken in two, each half acquires equal and opposite poles of its own. Like poles repel and unlike poles attract. Since the earth itself is a magnet, its south pole is actually located near its northern end, its north pole near its southern end. This unfortunately confused notation can be alleviated somewhat by using the words "north-seeking pole" instead of "north pole" to describe the end of a magnet drawn toward northern latitudes. The facts summarized in this paragraph comprise most of what was known of magnetism in 1800—in truth, not a great deal.

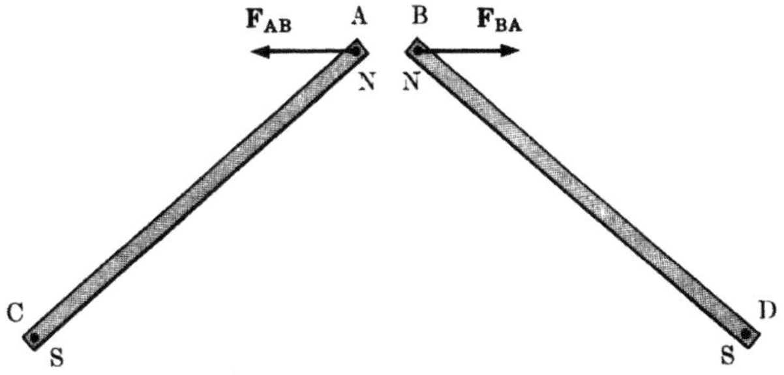

FIGURE 15.8 Long thin magnets act as if they have two well-separated centers of force, called north and south poles. In the arrangement shown here, two north poles repel one another, following a law exactly analogous to Coulomb's law (Equation 15.5). Weaker forces on the more distant south poles (points C and D) can be ignored.

Magnetism, like electricity, was known to the Greeks. As they attributed electricity to a single substance, amber (*elektron*), they also attributed magnetism to a single substance, magnetite (*e lithos magnetis*). Magnetite is a form of iron ore frequently found to be magnetized in its natural state in the earth. A magnet, unlike a charged object, does not lose its power to exert forces after a short time. A magnet may remain magnetized indefinitely. This fortunate property makes it well suited to serve as a compass. When the earth's magnetism was first tapped for navigational aid is unknown. It seems certain that the tendency of one end of a pivoted magnet to swing to the north was discovered independently in Europe and in China, and that this property was put to use for navigational purposes in Europe at least as early as the twelfth century, and in China in the thirteenth century. It remained for Gilbert in 1600 to point out that this property of magnets could be understood simply as arising from the fact that the earth itself is a giant magnet. It is interesting that in the twentieth century, the captain of an ocean liner or of a jet airliner still relies heavily on the magnetic compass as a primary source of information on his direction of motion.

The idea of north and south poles of magnets was originated by Gilbert. Although Newton failed in his effort to determine exactly the nature of the force between magnetic poles, success came to John Michell in 1750 some years before the corresponding law of electrical force had been established. Besides determining that the magnetic force between poles varies inversely as the square of the distance of separation, Michell learned that the total pole strength of a magnet adds up to zero. Both facts were later verified to higher precision, Coulomb for example verifying that magnetic and electrical forces vary in the same way with distance. The fact that the law of magnetic force was discovered before the law of electrical force can very likely be attributed to the fact that magnetization of an object is stable and long-lasting, in contrast to the electrification of an object, which gradually dissipates.

Efforts to understand magnetism in terms of magnetic fluids followed a course in parallel with similar thinking about electric fluids. Because the final theory of magnetism differs so markedly from such theories (we now recognize magnetism as a subsidiary effect of electricity), we shall not discuss these provisional theories. So far no isolated magnetic poles (sometimes called "monopoles") have been discovered, although efforts to find monopoles in the form of transitory elementary particles are still continuing. Up to the present, Michell's discovery that the total pole strength of every object is zero remains exactly true. This implies a law of conservation of pole strength, but it is a law neither so interesting nor so useful as the law of charge conservation. One object may gain or lose charge, with a compensating loss or gain elsewhere to preserve the constancy of the total charge. For pole strength, on the other hand, no object ever gains or loses pole strength, so that the constant value of zero total pole strength is not of great significance. Constancy during change is of fundamental importance; constancy without change is much less interesting. Of course the negative fact that there are no isolated poles is fascinating, and still remains puzzling.

As a practical matter, magnetism and electricity behave in quite different ways. The facts that both follow inverse square laws of force and that charge and pole strengths, defined in similar ways, obey conservation laws, should not be allowed to obscure their differences. Magnetism is a property only of iron and a few other substances; electrification can be imparted to any substance. Charge can flow from one place to another; magnetic pole strength cannot. Magnetism is stable; electrification is not. The earth is magnetized, but is not charged, at least not as a whole. (Large local charge separations do occur; they are the source of lightning.)

Scientists of the early nineteenth century were faced with the facts of electricity and magnetism that have been outlined so far in this chapter. It was widely believed that the sciences of electricity and magnetism were related, but no one knew how. On the one hand, the laws of electrical and magnetic force had the same form, and both forces were enormously large compared to gravity. On the other hand, there were many differences in the ways electric and magnetic phenomena manifested themselves, and there existed no direct force between magnets and charged objects. How these two subjects were merged into the single science of electromagnetism is the subject of the next chapter. Before discussing that important merger, we examine two new concepts, the electric field and the magnetic field. The idea of "field" was actually introduced into scientific thinking after the discovery of the connecting links between electricity and magnetism, but the field idea has become so central in modern thinking about electromagnetism that it is convenient to introduce it at this point.

15.6 *Electric fields and magnetic fields*

Occasionally a new concept makes its way into physics with a deceptively simple definition but remarkably fruitful consequences. Electric field and magnetic field are concepts of this type.

Electric Field

The electric field at a point is defined operationally as the electrical force on a charge placed at that point divided by the magnitude of the charge placed there (provided that all other charges in the vicinity remain undisturbed). In symbols,

$$\mathcal{E} = \frac{\mathbf{F}}{Q'}, \tag{15.6}$$

$\mathcal{E}$ being the usual symbol for electric field; $\mathbf{F}$ designates force and Q' designates charge. (The reason for adding a prime to the symbol Q will be explained below.) For example, the assertion, "there is an electric field at the center of this room," could be tested by placing a charge at the center of the room and observing whether or not it experienced a force. If it did, we would say that by definition an electric field exists at that point. The magnitude and direction of the field could then be measured by measuring the force and dividing by the magnitude of the test charge (presumed known already). Note that the electric field is, according to its definition, a vector quantity. It is chosen to point in the direction that a positive charge would be pushed. The unit of electric field is evidently the unit of force divided by the unit of charge. In the system of units we adopt in this book, the unit of electric field is dynes/e.s.u. According to its definition, electric field is simply force per unit charge.

The definition of electric field poses no conceptual problem. Wherever a charged particle experiences a force, we are quite free to divide the force by the charge and give this quotient a new name, "electric field." To be sure, the newly defined quantity is a vector, but this is not a serious complication. If a charge of +3 e.s.u. located at a certain point experiences a force of 24 dynes to the north, we say that an electric field of 8 dynes/e.s.u. pointing north exists at that point in space. If an electron with a charge of -4.8×10^{-10} e.s.u. is attracted toward a proton with a force of, for example, 0.096 dynes, the electric field at the location of the electron is the force (-9.6×10^{-2} dynes, called negative because it is attractive) divided by the charge, or $+2 \times 10^{8}$ dynes/e.s.u. Its direction is away from the proton, the direction in which a positive charge would be pushed. (It is equally permissible to calculate with positive magnitudes only, and decide afterward in what direction the electric field must point.)

To understand the definition of electric field is not difficult. The real problem is to understand why this new concept is useful. What was the motivation for introducing the idea of electric field? What have been the benefits of thinking about electrical (and magnetic) phenomena in terms of fields? The first essential fact to appreciate about a field is that it can be said to exist at a point in space whether a charge is located at that point or not, whether any force is being experienced at that point or not. The field can be thought of as a latent force, or potential force, describing what force would act at any point if a charge were placed there. Moreover, the field, one single quantity, can provide information about the force that would act on *any* charged particle placed at the given point. According to the definition of the field, the force on a charge Q' placed where the electric field is equal to $\mathcal{E}$ would be given by

$$\mathbf{F} = Q'\mathcal{E}. \tag{15.7}$$

At this point we wish to comment on the use of the symbol Q' for charge. In general, the charge of a particle measures both how much electrical force the particle can exert and how much electrical force the particle can feel. Of course one and the same charge plays both roles. But as a matter of convenience, it can help to avoid confusion if one symbol is used for charge when its active role is emphasized, another symbol when its passive role is emphasized. Throughout the remainder of this book, we shall use Q to designate a charge that is exerting a force (the active role) and Q' to designate a charge that is experiencing a force (the passive role). In the definition of electric field, it is important to notice that we utilize only the passive role of charge. This means conveniently that an electric field can be defined and measured without knowledge of its source. To summarize: The first advantage of the electric-field concept is that it can be thought of as a property of a particular point in space, whether or not the source of the field is known, and whether or not any charge is actually located at that point to experience a force. The field is a latent force.

If the electric field can be defined at any particular point in space, it can of course be defined at all points in space. It can be regarded as a continuous distributed property over any region of space or over all of space. This idea is actually more familiar than it might seem. Inside a room, for example, we may imagine that there is a certain temperature at every point. There cannot be a thermometer at every point, but we know that a thermometer moved to any location in the room will reveal the temperature at that location. Even without moving the thermometer around or filling up the room with thermometers, we have no trouble visualizing the idea of a "temperature field," that is a distribution of temperature throughout all points of the space in the room. The electric field is a similar idea, only made slightly more complicated by the fact that it is a vector quantity rather than, like temperature, a scalar quantity.

Related to the idea of the distributed electric field is the great pictorial convenience resulting from use of the field concept. Consider a single isolated positive charge Q at a fixed point. What is the electric field in the space surrounding this charge? To answer this question, we imagine a second test charge Q' moved about from place to place in the neighborhood of the fixed charge Q. If Q' (presumed to be positive) is placed above the charge Q, it will experience an upward force. This means that an electric field pointing upward away from Q exists at that point. Below Q, the force on Q' is downward. Again the electric field points away from Q. At any other point where Q' is placed, it will experience a repulsive force directed away from the fixed charge Q. If at each point in space, we imagine a little arrow representing the electric field at that point, all of the arrows will be pointing outward like the spines of a porcupine away from the charge Q [Figure 15.9(a)]. Now the test charge Q' has served its purpose to define the electric field. We take it away, leaving the single charge Q surrounded by its electric field. If the separate arrows are joined together by straight lines radiating outward from Q, the resulting porcupine quills, extending on to infinity, provide a three-dimensional "picture" of the electric field created by a single charged particle. As in Figure 15.9(b), we may attach arrowheads to the radiating lines to indicate the direction of the field, the direction of force that would be exerted on a positively charged particle.

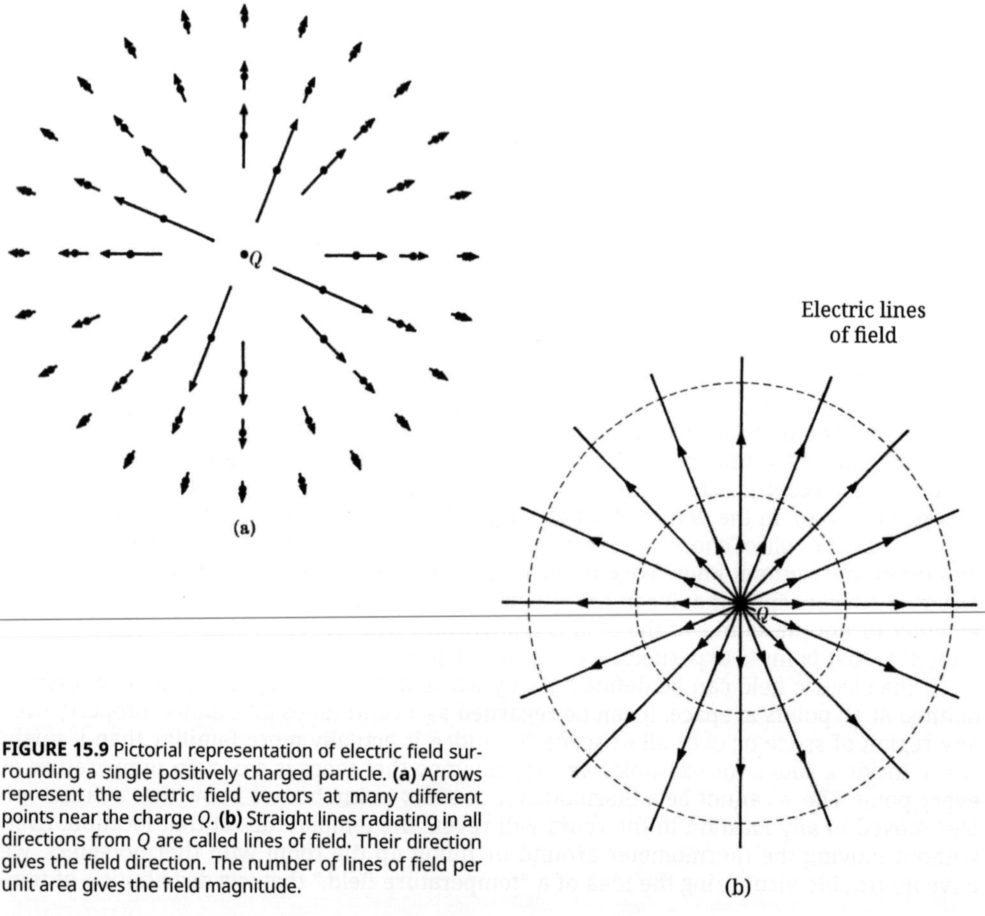

FIGURE 15.9 Pictorial representation of electric field surrounding a single positively charged particle. **(a)** Arrows represent the electric field vectors at many different points near the charge Q. **(b)** Straight lines radiating in all directions from Q are called lines of field. Their direction gives the field direction. The number of lines of field per unit area gives the field magnitude.

The lines constructed in this way are called "lines of field" or "lines of force." They conveniently pictorialize the electric field. Actually they do even more than this. If interpreted in the right way, they provide a picture of the strength of the field as well as its direction. In the example just cited, with straight lines of field radiating from a point, imagine two concentric spheres, one with twice the radius of the other, each being pierced by the lines of field radiating from their common center [Figure 15.9(b)]. The larger sphere, with twice the radius of the smaller, has four times the surface area of the smaller. Since the number of lines piercing each sphere is the same, the number of lines per unit area is four times less for the larger than for the smaller sphere. If each cm^2 of area of the smaller sphere were pierced by four lines of field, each cm^2 of the larger sphere would be pierced by only one line. A test charge placed on the outer sphere would experience only one-quarter as much force as the same test charge placed on the inner sphere because of the inverse square dependence on distance of Coulomb's law. Correspondingly, the electric field is only one-quarter as intense at the outer sphere as at the inner. Thus the number of lines of field per unit area diminishes at the same rate as the strength of the field. Where the lines of field are numerous and closely packed, the electric field is large; where the lines of field are sparse, the field is small. At the location of the point charge Q, the radiating lines of field become infinitely dense and the field becomes infinitely strong.

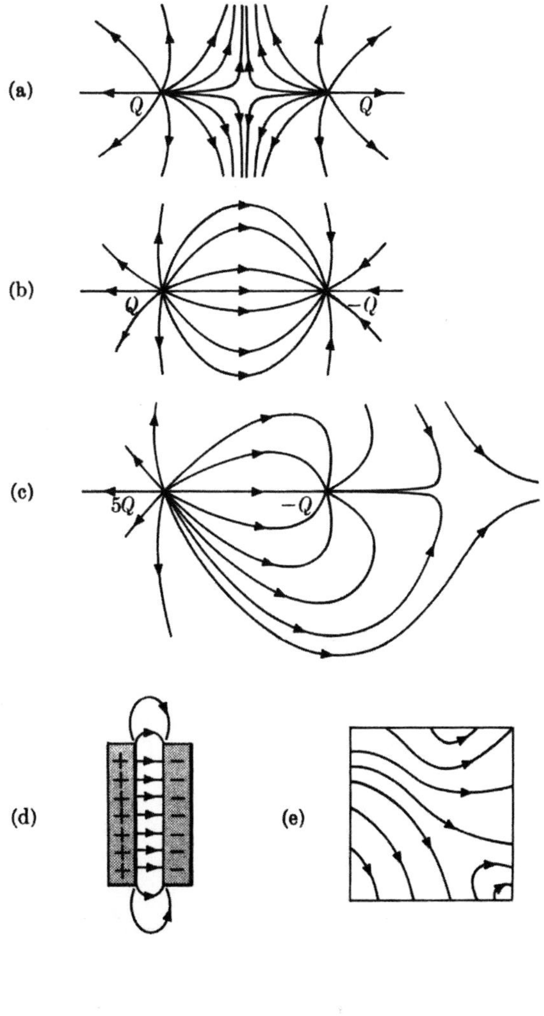

FIGURE 15.10 Lines of field associated with various distributions of electric charge. **(a)** Two equal positive charges. **(b)** A pair of equal and opposite charges. This combination is called a dipole. **(c)** Charges +5*Q* and −*Q*. **(d)** Two parallel metal plates, as in a capacitor, one charged positively, the other with equal negative charge. **(e)** A "black box," a field distribution within an enclosure that can be defined and measured even if the sources of the field are unknown.

As illustrated by several examples in Figure 15.10, the lines of field are of great help in visualizing distributions of electric field. It is left as an exercise to show that the fields created by two like charges and by two unlike charges are as shown in this figure. The last example in Figure 15.10 shows a possible distribution of electric field inside a "black box." It illustrates the fact that the field idea and the pictorial representation of fields can be very useful even when the sources of the field are unknown. In this example, there may be charge distributed over the walls of the box and other charge outside it. Regardless of how many charges may be contributing to the field at a point within the box, the field at that point is still one single quantity, determining the net force that would act on a test charge placed at that point. The field "automatically" sums and summarizes the many contributions of outside charges contributing to the force.

So far we have considered two advantages attending the introduction of the new concept of electric field. First, the field describes a latent force that may be regarded as existing in empty space. Second, the field, distributed throughout space, lends itself to graphic presentation through the idea of lines of field. A third and no less significant advantage has to do mostly with human psychology. Despite the clear and inescapable fact that forces can act across empty space, the idea of "action at a distance" has consistently been repugnant to scientists. Through the introduction of one intermediary or another, man has sought to account for the transmission of force from one point to another, without embracing the idea of action at a distance. Newton regarded as obviously absurd the idea that the sun's gravitational force could act on the earth without an intermediate agent to transmit the force. This agent was given the name ether; it persisted in scientific thought until early in this century. With similar motivation, Gilbert had introduced the idea of the electric fluid flowing out into the space surrounding an electrified body to transmit the electrical force to neighboring objects. After the electric fluid was forced by the weight of experimental evidence in the mid-eighteenth century to retreat back into the electrified body, the doctrine of electrical action at a distance gained a partial, and a temporary, victory. The victory was partial because, although no electric fluid occupied the space between charged objects, the ether was still there and could be assigned the job of transmitting electrical force as it was supposed to transmit gravitational force. Alternatively, in the view of some, two ethers could coexist in space, one to transmit gravitational force, one to transmit electrical force. The victory was temporary because the concept of electric field, introduced in the nineteenth century, again replaced the idea of distant action by the idea of local action. For this reason, as well as the two reasons already cited, the electric field gained ready acceptance among scientists as a useful concept for describing electrical phenomena.

According to the view of the electric field that finally evolved, the force exerted by one charge on another can be regarded as a two-stage process in which the field plays an intermediate role. Suppose there is a charge Q at point A and a charge Q' at a point B, a distance d away (Figure 15.11). Then according to Coulomb's law, the force experienced by Q' due to the presence of Q is

$$F_{\mathrm{E}} = \frac{QQ'}{d^2}. \tag{15.8}$$

This formula may be regarded as a statement of distant action. With the help of the field concept, it may be broken into two pieces, one a statement of the creation of a field, one the statement of the action of the field. Expressed mathematically, these statements are (omitting their vector character)

$$1.\quad \mathcal{E} = \frac{Q}{d^2}, \tag{15.9}$$

$$2.\quad F_{\mathrm{E}} = Q'\mathcal{E}. \tag{15.10}$$

The first says that at a distance d away from itself, the charge Q creates an electric field $\mathcal{E}$. The second says that this electric field causes the charge Q' to experience a force F_E. Evidently if the first formula for $\mathcal{E}$ is substituted into the second formula, Coulomb's law results again. What the field idea has effected is not any change in Coulomb's law, nor in the calculated force; it has only altered the way man looks at the phenomenon of electrical force.

The second of our pair of formulas above is obviously concerned with local action. An electric field $\mathcal{E}$ at the point B acts on a charge Q' at the same point B to produce a force F_E, again at the point B. But what about the first formula? It might appear that the field idea has not got rid of action at a distance. It has only changed "force at a distance" to "field at a distance," for the charge Q creates an electric field at a distant point. However, the change from force to field is conceptually an important change. If the space between points A and B is empty, there are no forces in this space. Without the field concept, we imagine the force as reaching out to appear at point B without existing anywhere between A and B. But the *field* does exist at all points in space. The field at B is not isolated. It is connected to its "parent" charge Q by a continuous distribution of field occupying all the space between A and B. Thus the "reaching out" of the field from A to B is not regarded as a leap across empty space.

FIGURE 15.11 The two-stage view of electric force. **(a)** A charge at point A creates a field at point B. **(b)** A charge Q' placed at B experiences a force because of the existence of the field at that point.

Is the field "real"? Logically speaking, it is as real as any other quantitative concept 90
in physics. It is operationally defined, it can be assigned a definite, unambiguous, and verifiable value and a direction at any point in space. This is an unsatisfying answer to an important question. It says no more than that scientists have defined the electric field in a logically self-consistent way. An endless series of useless concepts could be just as accurately defined. The significant question is not whether the field is real, but whether it is useful. In the end, the reality of nature—that is, the image of the physical world to which scientific discovery leads us—is based on those concepts that have proved useful in science. We quite naturally attribute a deeper reality to the few most useful concepts of science than to numerous peripheral concepts, even though the latter may be equally "real" from a logical point of view. Even though the definition of electric field as force per unit charge may appear at first not to add much to one's understanding of electrical phenomena, in fact, from our present perspective we can say without doubt that the electric field has been an exceedingly useful and fruitful concept in science. Including value judgments as well as mere logic, then, we must conclude that the electric field represents reality.

Magnetic Field

So far only the electric field has been defined, but most of the discussion of this section applies with equal force to the magnetic field. Conceptually the two fields are quite similar. Nevertheless, it is important to realize that they are distinct; both are necessary for describing electromagnetic phenomena. The magnetic field may be thought of as a potentiality for exerting a force on a magnetic pole. It is defined (in exact analogy to the electric field) as the force per unit pole strength:

$$\mathcal{B} = \frac{\mathbf{F}}{P'} . \tag{15.11}$$

Conventionally, the symbol $\mathcal{B}$ represents magnetic field (a vector quantity); **F** stands for force, and P' is the pole strength experiencing the force. Again in analogy with the electrical case, the component of magnetic field directed away from a stationary pole of strength P is given by

$$\mathscr{B} = \frac{P}{d^2}, \tag{15.12}$$

where d is the distance from the location of the pole to the place where the field is measured. Combining this equation with the statement of magnetic force, $F = \mathscr{B}P'$, one obtains Coulomb's law of force between magnetic poles,

$$F = \frac{PP'}{d^2}. \tag{15.13}$$

The magnetic field plays the intermediate role as the transmitter of magnetic force. Its units are dynes/m.s.u. (force/pole strength). This particular unit has acquired another more commonly used name, the gauss. A magnetic field of one gauss exerts a force of one dyne on a pole strength of one m.s.u. (A similar name has not been adopted for the unit of electric field, which remains dynes/e.s.u.). Despite the differences in nomenclature and despite their real physical differences, magnetic and electric fields do have the same physical dimension in the cgs system. To see this, compare Equation 15.13 with Equation 15.8, and Equation 15.12 with Equation 15.9.

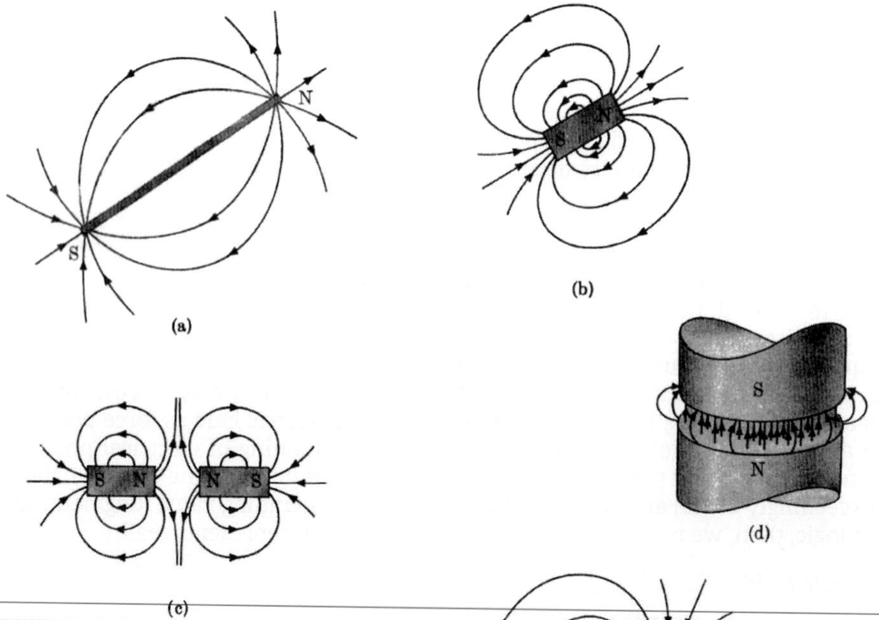

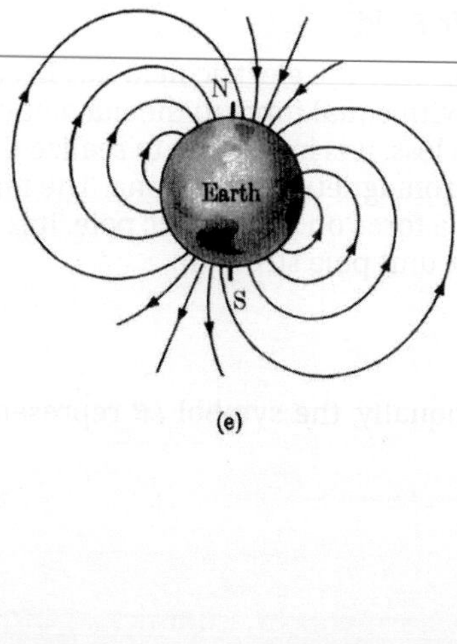

FIGURE 15.12 Examples of magnetic field distributions. **(a)** Long skinny magnet. The field near each end approximates the field of an isolated pole. **(b)** Short fat magnet. The field differs significantly from the field of two simple poles. **(c)** Two magnets repelling one another. **(d)** A nearly uniform field in the space between a pair of large flat parallel pole faces. **(e)** The magnetic field of the earth. Note carefully the direction of the field lines. (All of these diagrams are approximate.)

The direction of the magnetic field is taken to be the direction in which a north pole would be pushed. Recall that a north pole is one drawn to the northern end of the earth. Therefore the earth's lines of magnetic field point in a generally northerly direction. Several examples of lines of magnetic field, including those of the earth, are illustrated in Figure 15.12. In view of the fact that isolated poles do not exist in nature (or, if they do exist, have not yet been detected), it might seem inappropriate to base the definition of magnetic field on magnetic poles. An alternative definition in terms of dipoles (paired pole strengths, equal and opposite) is possible but slightly more complicated. We shall not pursue it. In any event, isolated poles can be approximated, as mentioned earlier, by long thin magnets (Figure 15.8).

How exactly do the electric field and the magnetic field differ? Operationally, it is easy to distinguish them. A stationary charge experiences a force in an electric field but is uninfluenced by a magnetic field. A stationary pole, conversely, feels no force in an electric field but is pushed by a magnetic field. We may say that an electric field is that which can exert a force on a charge, a magnetic field is that which can exert a force on a magnet. So long as no motion is involved, electric and magnetic phenomena are totally independent; electric and magnetic fields coexist in space without mutual effect. With motion or change, they do mingle. Even then, however, electric and magnetic fields retain their identity as two distinct concepts. Actually, the theory of relativity has effected a merger of the ideas of electric field and magnetic field into the single concept of electromagnetic field. But the two ideas have not lost their individuality and distinctness through the merger, any more than man and wife cease to be individuals when married. To draw a simpler physical analogy, we may say that the ideas of "east" and "north" and "up" may be merged into the single idea of "displacement in space." The three directions are obviously closely related ideas, but they are still distinct and distinguishable. In the same way, although modern developments have drawn electric and magnetic fields together into a larger whole, they remain distinct concepts. Their individuality will be further emphasized in the next chapter.

15.7 *The law of superposition*

We end this chapter with a discussion of a law of fields that is sometimes taken for granted, 91
but which is so important that explicit notice needs to be taken of it. This is the law of superposition. Its meaning can best be illustrated by example. First, in the everyday world, suppose that Arthur is pushing with a certain force F_A on a stalled car. He gives up and is replaced by Benjamin, who pushes with a force F_B. Now Arthur and Benjamin join forces and push together on the car. What will be the total force exerted (if they cooperate and push in the same direction)? "Obviously," you might say, "it will be the sum, $F_A + F_B$." Indeed, this is true, but it is something that cannot be taken for granted in science. It needs to be verified by experiment. We regard it as obviously true only because normal experience conditions us to believe it. To a limited extent, at least, most of us have experimented enough to verify roughly the law of superposition for forces encountered in everyday life. Now consider electrical forces at a more fundamental level. Suppose that a charge Q_A is located at point A, and a charge Q_C' is located at point C, this pair of charges being far removed from all others [Figure 15.13(a)]. The charge Q_C' will experience a force $\mathbf{F}_{CA}$ (a vector quantity) because of the presence of the neighboring charge Q_A. Now remove charge Q_A to a great distance, and bring up another charge Q_B to a point B near C. The force on Q_C' will now be contributed by Q_B; call the new force $\mathbf{F}_{CB}$. Finally, leaving the charges Q_B and Q_C' at their locations B and C, return charge Q_A to its original location A. What now will be the force exerted on charge Q_C'? Experiment shows that the combined force is the vector sum of the forces $\mathbf{F}_{CA}$ and $\mathbf{F}_{CB}$ contributed by the two other charges separately. The result holds true for any number of contributing forces. The fact that individual contributions may simply be summed (as vectors) to give the total force is what is called the law of superposition.

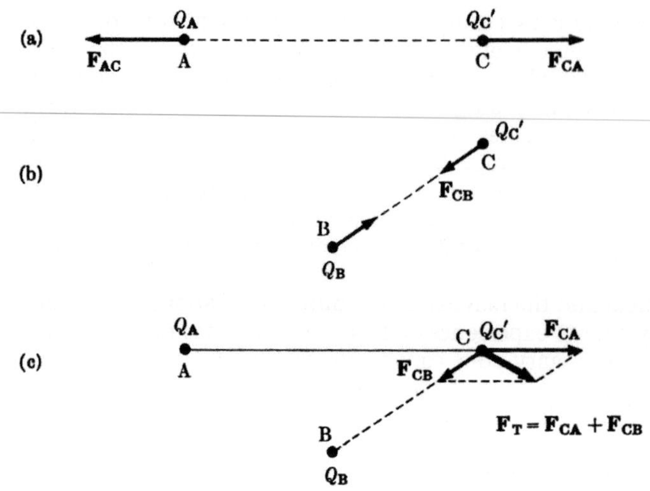

FIGURE 15.13 Illustration of the law of superposition of electric forces. In the presence of charges Q_A and Q_B together, charge Q_C' feels a force equal to the vector sum of the forces that would act if the charges Q_A and Q_B were separately present one at a time.

It may seem self-evident that if a charge is being pushed by two other charges, the force it feels is the sum of the forces it would feel if each of the other charges were present one at a time. Actually this superposition of forces is an extremely important fact about electric forces that has no right at all to be called self-evident. Elsewhere in the physical world, the idea of superposition is not always valid. If a pint of water is added to a pint of alcohol, the total volume of liquid turns out to be less than two pints. If a neutron and a proton join to form a deuteron, the mass of deuteron is less than the sum of the masses of neutron and proton. If three nuclear particles are close together, the force on one of them is not equal to the sum of the forces that would be exerted by the other two separately. Perhaps the best proof that the superposition law of electrical forces is not self-evident is the fact that it is not quite precisely true. In recent years we have learned that electrical forces are not exactly additive. Even in the atomic world, the deviation from additivity—that is, the extent to which the law of superposition is "violated"—is exceedingly small. In the large-scale world, the effect is entirely negligible, and the law of superposition is precisely correct to within the limits of human skill in measurement. The almost exact validity of this law for electric (and magnetic) forces further affirms our faith in the simplicity of nature. It is difficult to imagine any simpler way to combine two forces than by adding them together.

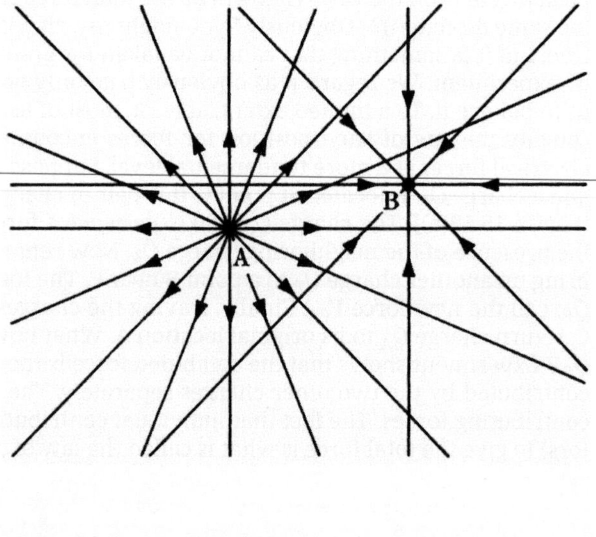

FIGURE 15.14 Superposition in terms of fields. The law of superposition states that fields created by different particles can be assumed to coexist in space without mutual interaction.

If electric (and magnetic) forces obey the law of superposition, then electric (and magnetic) fields do so too. Indeed the significance of the law can better be appreciated in terms of its meaning for fields. Imagine two charges Q_A and Q_B at the points A and B, this time without a third charge present (Figure 15.14). If either Q_A or Q_B were present alone, it would establish in its neighborhood a "Coulomb field," directed away from itself (for positive charge) and diminishing in strength as the inverse square of the distance. According to the law of superposition, the field in space with both charges present is the vector sum of the two different Coulomb fields centered at the points A and B. This means that we may think of the two fields as coexisting in space, each entirely uninfluenced by the presence of the other. The physical meaning of the law of superposition is that the field created by a given charge does not depend at all on how many other fields may also be present in the same space. In view of our modern view of the field as a "substance," the possibility of superposing fields in the same space without mutual interaction is a remarkable fact which clearly illustrates the simplicity that the law of superposition has contributed to our view of nature.

EXERCISES

15.1. An electrified rubber balloon has an excess of negative charge. It is attracted to most objects (walls, tables, blackboards, people, etc.). Does this mean that most objects are positively charged? If so, why this imbalance that favors positive charge? If not, why is the balloon more often attracted than repelled? *Optional:* Test a charged balloon by bringing it near numerous objects. Do you verify the assertion made at the beginning of this exercise? Do you find any objects that repel the balloon?

15.2. A rubber balloon is initially neutral. After being rubbed with wool it is electrically charged. **(1)** Explain why this change of charge does not violate the law of charge conservation. **(2)** Describe an experimental procedure (it need not be very practical) which could test the conservation of charge in this process.

15.3. Suppose that both of the strings supporting balloons 1 and 2 in Figure 15.1(b) make an angle of 30 deg with the vertical. If the hands supporting the strings are moved apart until the distance between the centers of the balloons is doubled, what will be the new inclination of each string to the vertical?

15.4. The charge of a proton is 4.80×10^{-10} e.s.u. The charge of a lead nucleus is 82 times as great. **(1)** When 10^{-8} cm from a lead nucleus, what electrical force does a proton experience? **(2)** If the proton, initially at rest, experienced approximately this force for 10^{-16} sec, what momentum would it acquire? What velocity?

15.5. A uranium nucleus splits into two equal fragments, each with 46 quantum units of charge. **(1)** When these fragments are separated by 10^{-12} cm as they start to fly apart, what electric force acts on one of them? **(2)** What electric force does this fragment feel an instant later after each fragment has moved 10^{-12} cm and they are separated by 3×10^{-12} cm? **(3)** Multiply the approximate average force experienced by one of the fragments by the distance it has moved to obtain the work done on it (roughly). Express this energy in eV.

15.6. Two charges are near one another and isolated from the rest of the universe. If the force experienced by the first is proportional to its own charge, $F_1 \sim Q_1$, prove, using Newton's third law and the assumption that the law of force is valid for any charge that this force must also be proportional to the second charge, $F_1 \sim Q_2$, and therefore that it is proportional to the product of both charges, $F_1 \sim Q_1Q_2$.

15.7. Three pellets with charges of +10 e.s.u., +3 e.s.u., and −10 e.s.u. are arranged as shown in the figure. Find the magnitude and direction of the force on each.

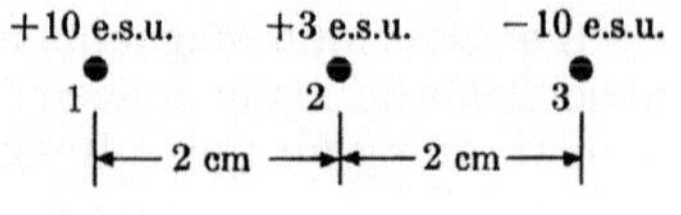

15.8. A pair of protons are located 2 Angstroms apart, as shown in the figure. **(1)** What is the vector force on an electron located at point C? **(2)** If the electron is displaced from point C to point D, what is the vector force acting on it? **(3)** What is the acceleration of the electron when at point D ?

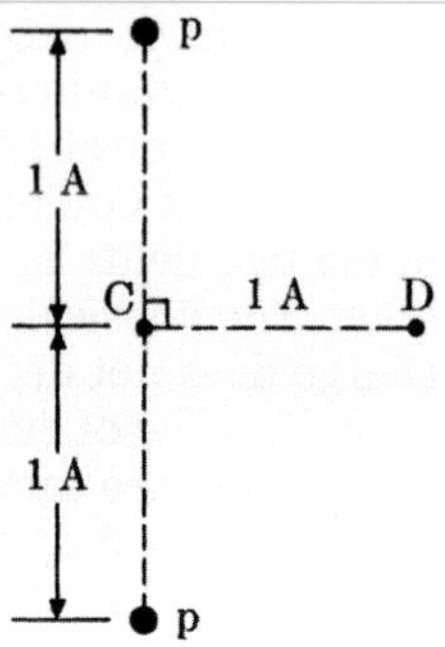

15.9. An arrangement of two equal and opposite charges such as Q_1 and Q_2 in the diagram is called a dipole. A test charge of +1 e.s.u. is moved in the vicinity of this dipole. **(1)** What force does it experience when at point A? **(2)** What force does it experience when at point B? **(3)** Is there any point where it experiences no force? If so, what is the approximate location of this point?

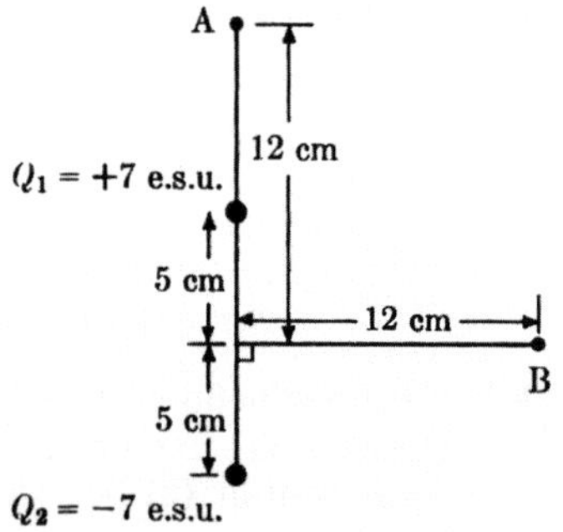

15.10. From Coulomb's law, Equation 15.4, work out an expression for the cgs unit of charge (the e.s.u. or statcoulomb) in terms of the gm, cm, and sec.

15.11. A charge Q_1 is located at the origin of a coordinate system. Another charge Q_2, located at (x, y) a distance r away, experiences an electric force F_2 produced by charge Q_1. **(1)** Show that the x and y components of the force F_2 are given by

$$F_{2x} = \frac{Q_1 Q_2 x}{r^3} \qquad \text{and} \qquad F_{2y} = \frac{Q_1 Q_2 y}{r^3}.$$

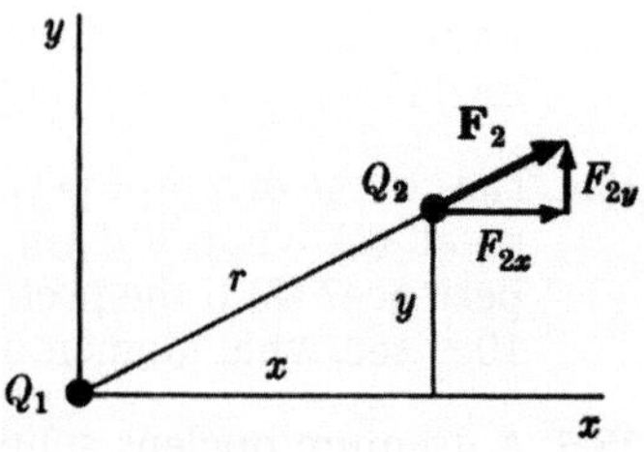

(2) Write down a single expression for the vector $\mathbf{F}_2$ whose components are F_2x and F_2y. (HINT: Consider what other physical vector is parallel to F_2.)

15.12. An eccentric scientist, unwilling to follow the crowd, defines charge (for which he uses the symbol ξ) in such a way that the law of electric force takes the form,

$$F = K \frac{\xi_1^2 \xi_2^2}{d^2},$$

where K is a constant of proportionality. **(1)** Explain why this law of force is perfectly compatible with experiment. **(2)** Is there still a conservation law analogous to the law of charge conservation? If so, what is the conserved quantity? (On purely logical grounds this law of force and Coulomb's law are equally acceptable. Yet it is easy to see why scientists prefer Coulomb's law.)

15.13. A proton in an evacuated container is pulled downward by the earth's gravity. **(1)** What magnitude of electric field would be required to counter this gravitational force? **(2)** How distant would an electron need to be to provide the required electric field?

15.14. In attempts to produce controlled thermonuclear reactions, experimenters have restricted themselves to working with isotopes of hydrogen. Suggest one reason why hydrogen is preferred.

15.15. Two identical bar magnets are arranged as shown. Each acts approximately as a pair of magnetic poles, $P_1 = +100$ m.s.u., $P_2 = -100$ m.s.u., separated by 6 cm. The distance between the magnets is 8 cm. **(1)** Do the magnets repel or attract one another? **(2)** Find the *total* force acting on one of the magnets. **(3)** Suppose that the magnet on the right is inverted so that the positions of its poles P_1 and P_2 are interchanged. What then is the force acting on it? (As this exercise demonstrates by example, the force between a pair of magnets depends on their relative orientation as well as their distance apart.)

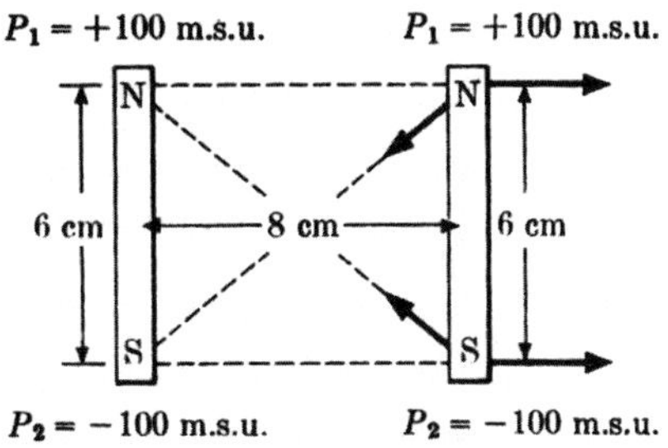

15.16. At a certain point in space the force on an electron is 2×10^{-4} dyne, directed vertically upward. **(1)** What is the electric field vector at this point? **(2)** The electron is removed. What then is the electric field vector at the unoccupied point in space? **(3)** The electron is replaced by a bit of matter whose charge is $Q = +3 \times 10^{-3}$ e.s.u. What is the force on this object?

15.17. Two charges Q_1 and Q_2 known to have equal magnitude are separated by 10 cm. Midway between them the electric field is found to be directed toward the charge Q_2 and to have magnitude 1.6 dynes/e.s.u. Find Q_1 and Q_2.

15.18. Give one example of a vector field quantity other than the electric field and the magnetic field. Explain why the quantity you have picked has the properties of a vector field.

15.19. An electron is a distance of 5×10^{-9} cm away from a proton and consequently is attracted to the proton with a force of 0.92×10^{-2} dyne. **(1)** Give the magnitude and direction of the electric field that the electron "feels." **(2)** The electron is removed from point B and replaced by an alpha particle. How does the electric field at B change? How does the force change?

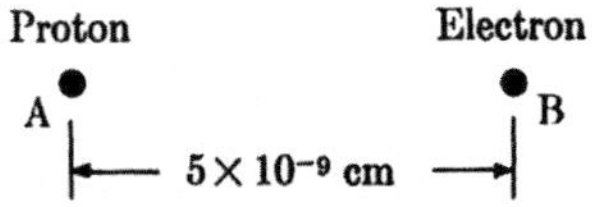

15.20. Consider a spherical surface of radius r surrounding a positive charge Q as shown in the figure. **(1)** What is the product of electric field intensity at this surface and the surface area? This product is called the *electric flux*, for which the usual symbol is Φ_E. **(2)** How does the electric flux vary as the radius r varies? **(3)** Can you think of any other hypothetical form of Coulomb's law for which the electric flux would vary in as simple a manner? (NOTE: More generally, electric flux is defined as the product of a surface area and the component of electric field perpendicular to that surface.)

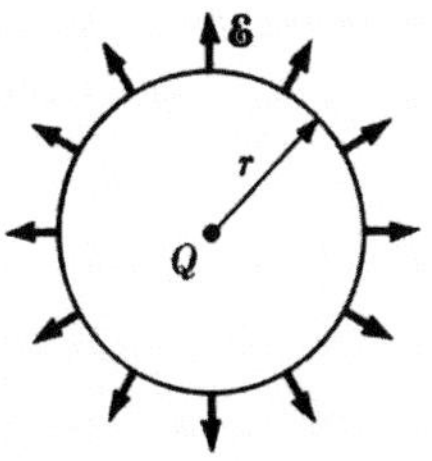

15.21. Three charges, Q_1, Q_2, and Q_3, are arranged at three corners of a square as shown. The first two charges are known to be $Q_1 = 10$ e.s.u. and $Q_2 = 28$ e.s.u.; the electric field at the fourth corner of the square is known to be directed horizontally to the right as shown. **(1)** What is the charge Q_3? **(2)** What is the magnitude of the total electric field $\mathscr{E}$ at the fourth comer?

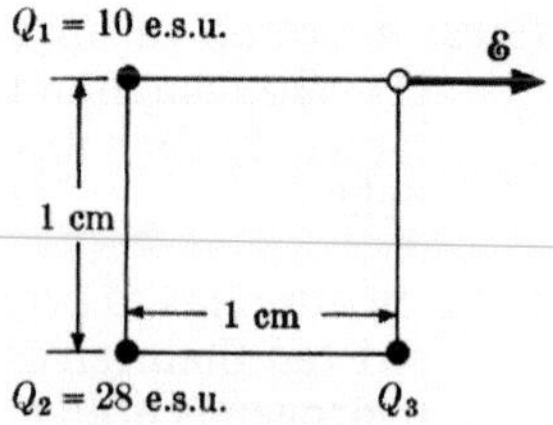

15.22. Work out the field line diagram for two equal positive charges or for an equally and oppositely charged pair of particles [Figures 15.10(a) and 15.10(b)]. Method: At each of a number of points sketch the magnitudes and directions of the two contributing fields and complete a vector sum diagram ($\mathscr{E} = \mathscr{E}_1 + \mathscr{E}_2$). Do this as accurately as you can by eye without precise measurement. The resulting pattern of arrows will suggest how to draw the smooth field lines.

15.23. A compass needle, pivoted at its center, is oriented perpendicular to the direction of the earth's magnetic field (see figure). It experiences a torque of 300 dyne cm. If the compass needle can be approximated as a pair of poles, $P_1 = 250$ m.s.u. and $P_1 = -250$ m.s.u., separated by 3 cm, what is the magnitude of the earth's magnetic field $\mathscr{B}$ at this location ?

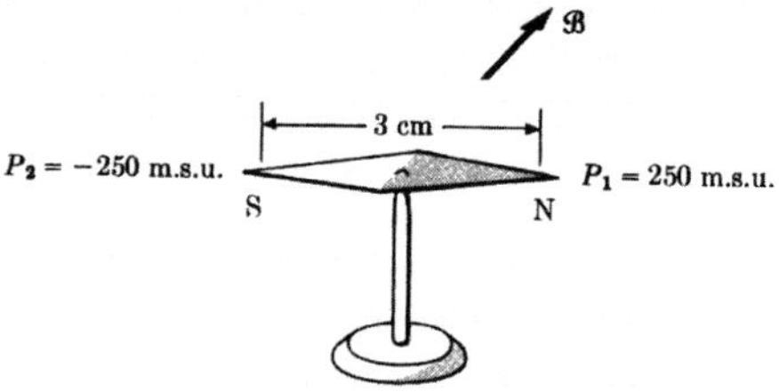

15.24. Give arguments based on fundamental equations presented in this chapter to show that **(a)** electric charge and magnetic pole strength have the same physical dimension; and **(b)** electric field and magnetic field have the same physical dimension.

15.25. Give a quantity, or combination of quantities, that has the same physical dimension as energy, and the same cgs unit as energy (gm cm^2/sec^2), yet is not an energy. (This helps to explain why electric and magnetic fields are distinct quantities, despite having the same cgs unit.)

15.26. Suppose the unit of mass had been chosen such that two equal masses, each of 1 g.u. (gravitational unit), separated by 1 cm, exerted on each other a force of 1 dyne. **(1)** Derive the conversion factor relating gm to g.u. **(2)** What is the mass of an 80-kg man in g.u.? **(3)** In a system of units employing dyne, cm, and g.u., what would be the appearance of Newton's law of gravitational force?

15.27. The electrical potential energy of two electrons, each of charge e, a distance r apart, is

$$P.E. = \frac{e^2}{r}.$$

At what separation is this potential energy equal to the mass energy of the electron, mc^2? This distance is called the "classical electron radius." How big is it? That is, compare it with something else.

SIXTEEN

Electromagnetism

Why is "electromagnetism" such a ponderously long word? There is a simple scientific-historical reason for it. In the first half of the nineteenth century, the theories of electricity and of magnetism drew together into a single theory embracing both. As the theories merged into one, their names merged into one. Electromagnetism is not alone in showing its divided past in its modern name. Ideas of heat and of mechanics (or dynamics) drew together in the theory of thermodynamics. Mechanics and the quantum theory were merged in quantum mechanics. More recently, a successful merger of the theories of quantum mechanics and electromagnetism has been christened "quantum electrodynamics." It is unfortunate that great strides in simplifying and unifying our view of nature must be accompanied by increasing complexity of scientific nomenclature. But perhaps there is merit in allowing history to tinge the names of the fundamental theories of physics. In this chapter we shall explore the ideas and experiments that brought electricity and magnetism together.

16.1 Key discoveries in electromagnetism

A definite connection between electricity and magnetism was first established by Hans 92
Christian Oersted, a professor in Copenhagen, in 1820. At the end of a lecture on the subjects of electricity and magnetism, he happened to place a compass needle—a small magnet—near a wire carrying an electric current. The compass needle swung to a new direction, indicating the existence of a magnetic force, a force that could be attributed only to the nearby electric current. Apparently the flow of electric charge in the wire created magnetic effects (in later terminology, it created a magnetic field), although stationary charge had long been known to have no magnetic effects. Whether this experiment was carried out by chance or by design is uncertain. In any case, Oersted was prepared to capitalize on the discovery. As the subjects of his lecture suggest, he already believed in the existence of a connection between electricity and magnetism. Upon the discovery of the connection, he quickly established the pattern of what we now call the lines of magnetic field. They trace circular paths about a current-carrying wire (Figure 16.1), very different indeed from the radial lines of magnetic force emanating from a magnetic pole at rest. Other scientists, especially in France, were drawn to investigate Oersted's discovery. Through the efforts of. Jean Baptiste Biot, Félix Savart, and André-Marie Ampere, many aspects of the link between electric currents and magnets had been worked out in mathematical detail within a few months of the arrival in Paris of the news of Oersted's discovery.

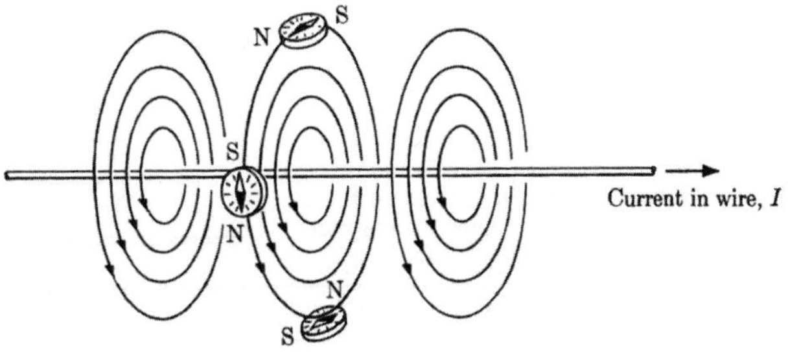

FIGURE 16.1 Oersted's discovery. A current-carrying wire exerts a force on a magnet. The modern interpretation: Moving charge creates a magnetic field. The arrow attached to the wire indicates the conventional direction of current flow, opposite to the actual direction of motion of the mobile negative electrons.

What exactly is the content of Oersted's discovery? We can best express it from a modern point of view. (1) A moving charge, or equivalently, an electric current, creates a magnetic field. Therefore a magnet placed near a current will experience a force. Because of Newton's third law, verified to be valid for this new effect, the magnet in turn must cause the moving charge or current to experience a force, equal and opposite to that felt by the magnet. Thus a corollary to the creation of a magnetic field by a moving charge is this: (2) A magnetic field exerts a force on a moving charge (Figure 16.2). Finally, because a moving charge both creates a magnetic field and feels a force in a magnetic field, no magnets are necessary at all to demonstrate the new effect: (3) One current will exert a force on another current (Figure 16.3). Since a wire carrying current contains moving charge but normally has no net electrification, any force exerted by one current-carrying wire on another is necessarily distinct from the electrostatic force of Coulomb. These were the important new discoveries of 1820 that launched the combined science of electromagnetism. They are pursued in later sections of this chapter.

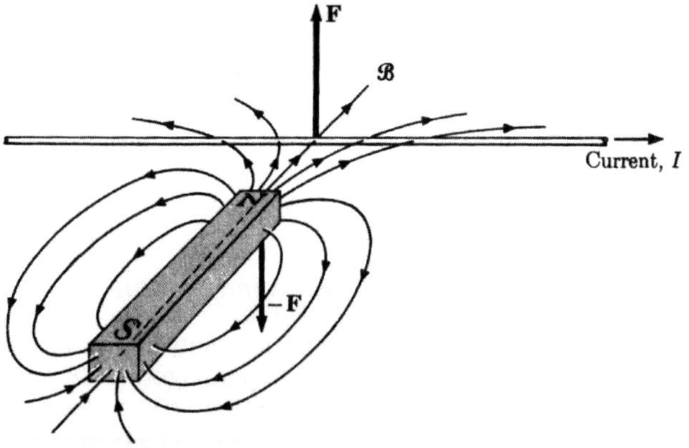

FIGURE 16.2 A corollary to Oersted's discovery. A magnet exerts a force on a current-carrying wire (equal and opposite to that exerted by the wire on the magnet). The modern interpretation: Charge moving in a magnetic field experiences a force. Only the magnetic field created by the magnet is shown.

Eighty-five years before Oersted's discovery, the following description of some evidence for a link between electricity and magnetism was published in England in the *Philosophical Transactions:** "A tradesman of Wakefield, having put up a great number of knives and forks in a large box, and having placed the box in the corner of a large room, there happen'd in July, 1731, a sudden storm of thunder, lightning, etc., by which the corner of the room was damaged, the Box split, and a good many knives and forks melted, the sheaths being untouched. The owner emptying the box upon a Counter where some Nails lay, the Persons who took up the knives, that lay upon the Nails, observed that the knives took up the Nails." The electrical discharge had created magnetism. Despite this and earlier indirect pieces of evidence for a connection between electricity and magnetism, the scientific study of this link had to await the development of an important new tool, the chemical cell.

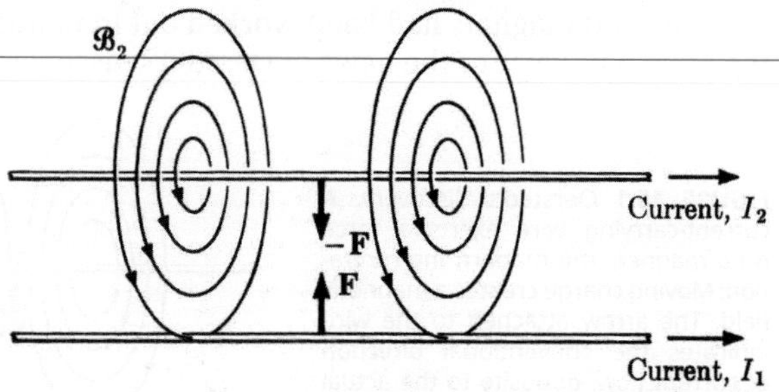

FIGURE 16.3 Ampere's discovery based on Oersted's work. One current-carrying wire exerts a force on another. The modern interpretation: Each current creates a magnetic field; each current experiences a force in the field created by the other. Only the magnetic field created by current I_2 is shown.

* *Philosophical Transactions, 39,* **74** (1735).

Electromagnetic effects require moving charge. Before the end of the eighteenth century, scientists had no means at their disposal to produce a steady stream of moving charge, that is, a steady current. Such flow of electric charge as did exist was transitory and difficult or impossible to reproduce—a spark or a bolt of lightning or the discharge of an electrified body through a wire. The magnetic effects arising from charge in motion had little chance to be investigated before a way was found to keep charge flowing for a long period of time. The discovery of the chemical cell, or battery,* which made this possible, evolved in a most unlikely way from a chance observation in an anatomy laboratory in Bologna in 1780. Luigi Galvani, professor of anatomy, noticed that when the nerve of a frog's leg lying on a metal plate was touched with a metal scalpel, the leg muscle convulsed. The peculiar hand of chance was present at this observation to steer Galvani's thinking in the right direction for the wrong reason. Since a machine for generating electric charge by friction happened to be present on the same table as the frog's leg, he at once supposed that the phenomenon was electrical in nature. Indeed it was, but his subsequent observations led him finally to conclude that the electrostatic machine or any other external electrical effects were entirely irrelevant to the observed convulsion of the frog's leg. All that was required was that the two ends of the nerve be connected through two dissimilar metals. With the aid of the metals, the frog's leg seemed to possess within itself the power to cause a muscle contraction.

That Galvani's chance observation led as quickly as it did to the invention of the battery is a quite remarkable fact attributable to the clear thinking and careful experimenting of Galvani and his countryman, Alessandro Volta. Two beguiling false clues associated with Galvani's original observation might well have led investigators astray for a longer time. The first was the electrostatic machine and its suggestion that electricity was supplied to the frog's leg from outside. Galvani himself proved this to be a false lead. The second clue was the frog's leg itself. Most scientists assumed that the muscle contraction arose from a property associated with living or once living matter, and had nothing to do with the inanimate world. Volta took the opposite view, seeking to explain the phenomenon without recourse to a new and mysterious "animal electricity." In a series of experiments carried out in the 1790s, he showed that the frog's leg had merely acted as a convenient detector of electric current. The source of the electric current was the pair of dissimilar metals, which caused a charge to flow through the frog's leg. The leg responded to the current by contracting.† Actually the leg was an important link in the circuit, but Volta found that it could as well be replaced by a piece of moist cardboard. His first successful battery, capable of generating substantial current for an extended period of time, was completed in 1800. It consisted of a series of cells, each made of one layer of copper, one layer of zinc, and one layer of moist cardboard. The atomic explanation for the action of the cell, which converts chemical energy directly to electrical energy, came a century later. About the workings of the cell, Volta wrote that it "disturbs the electric fluid, or gives it a certain impulse. Do not ask in what manner: it is enough that it is a principle, and a general principle." Often it is wisdom on the part of a scientist to recognize that the time is not ripe for a deeper explanation of some phenomenon, despite his curiosity to know more. The important thing for the history of electromagnetism is that the battery (or Voltaic pile, as it was then called) made possible the generation of steady electric currents. In the words of Volta: "This endless circulation or perpetual motion of the electric fluid may seem paradoxical, and may prove inexplicable; but it is nonetheless real, and we can, so to speak, touch and handle it."

In the series of experiments that proved the unity of electricity and magnetism, two stand out as crucial. The first was Oersted's discovery in 1820 that a moving charge pro-

---

* Strictly speaking, a battery is a series of chemical cells. A typical automobile battery consists of three or six cells. What is often called a flashlight battery is in fact a single cell.

† Galvani's discovery had important implications for physiology, since it linked nerve action to electricity, but we shall not pursue them here.

duces a magnetic field, with its corollary that a magnetic field exerts a force on a moving charge.* The second was Michael Faraday's discovery in 1831 that a moving magnet generates a current in a nearby wire.† One of Faraday's experiments was the following (Figure 16.4). A magnet is moved near a coil of wire, which is closed through a current-sensitive meter. When the magnet moves, a current is observed to flow in the wire. The strength of the current is proportional to the speed of the magnet. It is also greater if the magnet is more powerful, and greater if the magnet is moved closer to the wire. When the magnet stops moving, the current stops flowing. This effect is usually called induction. The moving magnet "induces" a current.

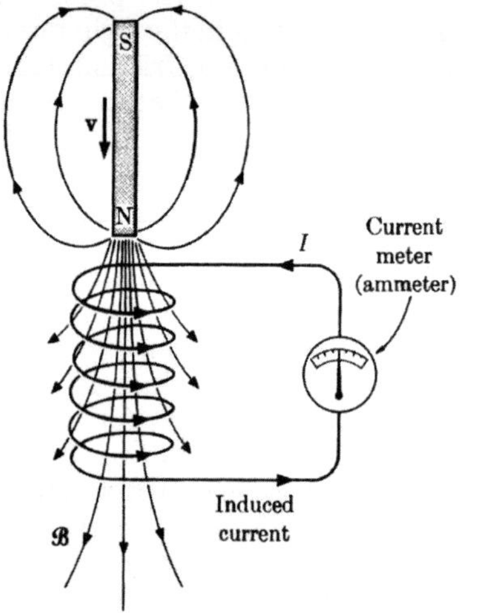

FIGURE 16.4 Faraday's discovery. A moving magnet generates a current in a nearby wire. The modern interpretation: The moving magnetic field creates an electric field which accelerates charge within the wire to cause the current. Arrows attached to the wire show the direction of conventional current flow, opposite to the actual direction of motion of electrons through the wire.

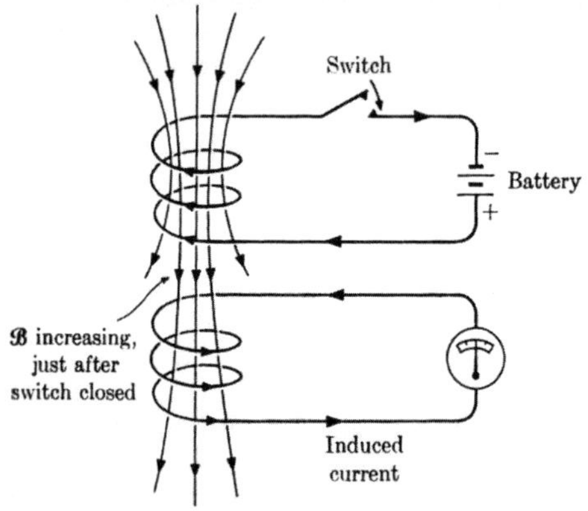

FIGURE 16.5 A corollary to Faraday's discovery. A changing current in one wire creates a current in another wire. The growing magnetic field created by the increasing current in the upper loop just after the switch is closed influences the lower loop in the same way that the loop would be influenced by a moving magnet (compare Figure 16.4).

To describe the new phenomenon, Faraday invented the concept of magnetic field and moved part way toward our modern interpretation of magnetic induction. He imagined lines of magnetic field penetrating the wire and there being able to induce current whenever the lines of field moved. Now we add to this picture of local action the electric field as an intermediary. We say that a moving (or changing) magnetic field creates an electric field. Within the wire, this electric field—which exists only so long as the magnetic field is changing—accelerates charge and thereby induces current.

The magnetic-field concept proved at once to be fruitful, for it suggested new possibilities to Faraday. There were other ways to cause lines of magnetic field to move than moving a magnet. One could move a current-carrying wire near the wire that had no current. Or

* We must emphasize that this statement of the essence of Oersted's discovery makes use of more recently acquired insights. Oersted did not use the concept of magnetic field; and, although he knew electric current to be a flow of charge (or of electric fluid), he did not recognize its equivalence to the motion of a charged object.

† This discovery was made independently at about the same time by Joseph Henry in the United States.

one could start or stop a current in a neighboring wire without moving it at all (Figure 16.5). Indeed both of these procedures resulted in an induced current, as Faraday shortly verified.

Volta's battery had supplied current from chemical energy. Faraday's discovery of induction provided a way to produce current from mechanical energy. All modern generators are based on this principle, deriving current from the relative motion of conducting wires and magnetic fields. Faraday clearly saw the important implications for man of his discovery, but he chose to continue to pursue fundamental research rather than to devote his talents to the development of practical generators.

In the decades following Faraday's discovery, insight into electromagnetism deepened as new implications of the electromagnetic laws were worked out mathematically, and as new applications were discovered. Long-distance telegraphy, initiated in 1844, set off the communications revolution that has been accelerating to the present day. Motors and generators, directly harnessing the bonds between electricity and magnetism, evolved into practical devices of commerce in the period 1850-1880. In this period of consolidation and application, the electric and magnetic fields moved gradually to a more central position in thinking about the fundamentals of electromagnetism. Both fields showed up as important concepts in the work of James Clerk Maxwell in the 1860s. By 1885, Oliver Heaviside had cast the equations governing the fields into a form that revealed the beautiful symmetry that exists in nature between electricity and magnetism.

Three of Maxwell's achievements deserve special mention. First is his unification of all of electromagnetism. He drew together the loose ends of past work into a single theoretical structure. From his equations could be derived all that was already known of the subject, plus much that was still to be learned. Second, he discovered (theoretically, not experimentally) a new connecting link between electric and magnetic fields. The nature of this link is illustrated in Figure 16.6. Current flowing toward one metal plate and away from another produces an increasing electric field between the plates. Surrounding this region of growing field appears a magnetic field, much like the field created by a current—even though the region between the plates contains neither current nor charge. Maxwell thought of the region between the plates as containing something he called a displacement current. Now, however, we simply describe the phenomenon in this way: A changing electric field creates a magnetic field. This makes Maxwell's law an exact parallel of the law discovered by Faraday and Henry, which, in modern terms, is: A changing magnetic field creates an electric field.

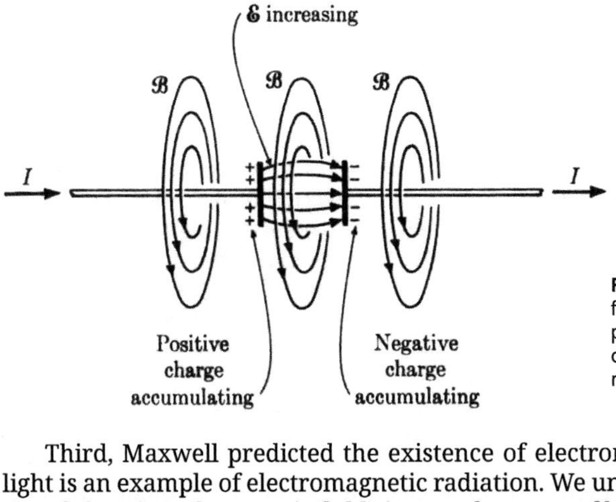

FIGURE 16.6 Maxwell's discovery. An electric field that is changing, such as that between a pair of plates in the process of being charged, creates a magnetic field analogous to the magnetic field created by a current.

Third, Maxwell predicted the existence of electromagnetic waves, and suggested that light is an example of electromagnetic radiation. We understand such radiation now to consist of electric and magnetic fields in a perfect state of balance, each field continually changing and continually interacting with the other field (Figure 16.7). The wave requires the oscillatory motion of charge for its creation, but once created, it propagates without further stimulus from charge or matter. The continually changing electric field in the wave creates a magnetic field; the continually changing magnetic field in turn creates an electric field. Like two freezing trappers in the wilds who huddle together to stay alive, the electric and magnetic fields hang together and keep each other alive through the laws linking electricity and magnetism. The wave terminates only when it meets another charge which can be set into motion and absorb the energy in the wave. Ordinary light is "atom-made" electromagnetic radiation. The first "man-made" radiation was achieved in the 1880s. Now our atmosphere and the space near the earth is filled with the man-made electromagnetic waves of radio, television, and radar. Thanks to the law of superposition cited at the end of the last chapter, all of these waves coexist in space without mutual interference.

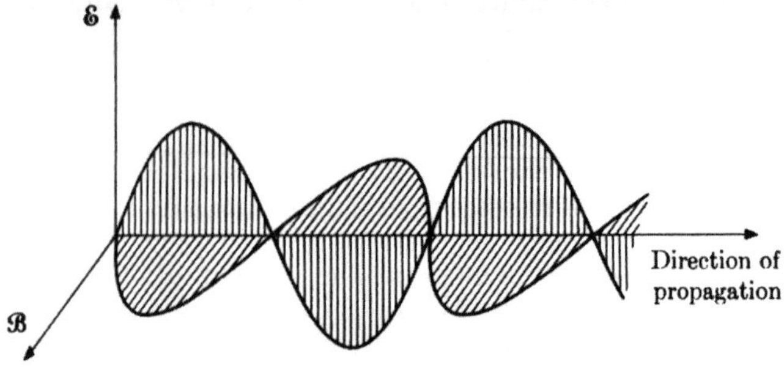

FIGURE 16.7 Maxwell's prediction. Electromagnetic waves, far removed from charges or currents, should be able to propagate through space. We understand these waves now to consist of oscillating electric and magnetic fields of equal intensity (in the cgs system of units).

In the remainder of this chapter we shall describe the few fundamental laws of electromagnetism in more quantitative terms, and illustrate them with examples.

16.2 The magnetic force on a moving charge

A modern device that can conveniently be used to illustrate the essentials of Oersted's discovery is the cathode-ray tube. In this tube a narrow beam of electrons flies through the evacuated space within the tube, then hits the fluorescent end of the tube to produce a spot of light. If the beam is deflected—that is, if the moving charges feel a force—the spot of light moves to a different point on the face of the tube. If the spot is moved about the face of the tube faster than the eye can follow it, the whole tube face seems to be illuminated. This is what happens in a television tube.

In the cathode-ray tube (Figure 16.8), electrons "boil off" from the surface of a hot metal filament not unlike the filament of a light bulb. These free electrons are drawn to a positively charged plate a little way down the tube. Most of the electrons strike the plate. A few, after being accelerated by the attractive force of the positive plate, fly through a small hole in the center of the plate. It is these few that continue in a narrow beam to strike the face and create a fluorescent spot. The hot filament is called a "cathode," the positively charged plate an "anode," names that go back to the time before the electron was discovered. When the electron beam in a cathode-ray tube was first studied in the 1890s, little was known about the beam except that it emanated from the cathode. Hence the name "cathode ray." In 1897, J. J. Thomson reported careful investigations of these cathode rays. He studied their deflection in electric and magnetic fields and their attenuation in passing through rarefied gases. Through these studies he could identify the cathode rays as beams of a new negatively charged subatomic particle, to which he gave the name corpuscle. Now we call Thomson's corpuscle the electron.

If a positively charged object is held above the electron beam in a cathode ray tube, the spot on the tube face will move upward, indicating an attractive upward force on the electrons. This illustrates Coulomb's law and is not surprising. Careful experiments would reveal that the force on the moving electrons in an electric field is equal to the product of the electron charge and the electric field strength,

$$\mathbf{F}_E = Q'\mathscr{E}, \tag{16.1}$$

the same as if the electrons were at rest. What we are concerned with here is how the moving electrons react to a *magnetic* field. If the charged object above the electron beam is replaced by a long thin magnet with its north pole nearer the electron beam (Figure 16.9), the spot on the tube face, surprisingly, will move neither toward nor away from the magnet. It will swing to its own right (to the left side of the tube face from the vantage point of a "viewer"). The force is at right angles to a line joining the magnet and an electron. If the magnet is swung around to the viewer's left, the spot will move downward. For any position of the magnet, the electrons are observed to experience a force perpendicular to their direction of flight (their velocity vector) and perpendicular to the direction of the magnetic field. This is decidedly more complicated than the action of an electric field on a charge. The electric force does not depend on the electron velocity at all, and it acts along the lines of electric field.

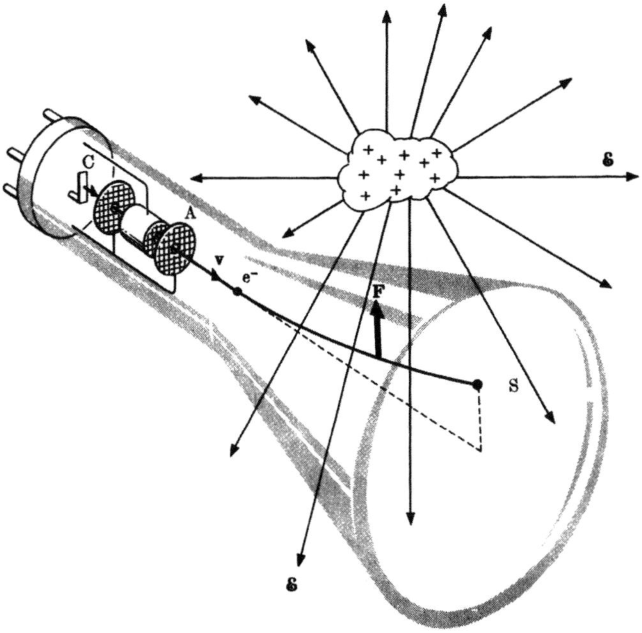

FIGURE 16.8 The cathode-ray tube (simplified, with inessential detail omitted). Electrons boil from the hot cathode (C) and are attracted to the positively charged anode (A). Those passing through the central hole in the anode fly on to cause a fluorescent spot (S) on the tube face. Illustrated is an upward deflection produced by a positively charged object above the beam.

If the magnet is turned around, exchanging the positions of its north and south poles, the direction of the electron deflection is reversed. If the magnet is moved closer, the deflection is greater. If the electrons are caused to move more rapidly, the force is found to be greater. A series of experiments with electrons of various speeds moving through known magnetic fields would lead to the following law:

$$F_M \sim v_\perp Q'\mathscr{B}. \tag{16.2}$$

The magnetic force on a moving charge is proportional to the magnetic field strength $\mathscr{B}$, to the charge Q', and to the component of velocity of the charge perpendicular to the magnetic field, $v_\perp$. If a charge is moving transversely across the lines of magnetic field, $v_\perp$ is equal to v, the entire velocity is perpendicular to the magnetic field. If a charge is moving along a line of magnetic field, $v_{\perp}$ is equal to zero. Then the charge experiences no force. If the charge cuts across lines of magnetic force at an intermediate angle, $v_{\perp}$ has a value less than the full velocity v; only part of the velocity is effective in producing magnetic force.* As shown in Figure 16.10, the direction of the force is perpendicular both to the velocity vector and to the magnetic field vector. In view of the three-dimensional complexity of this situation, it is worth committing to memory a simple rule for determining the direction of the force.

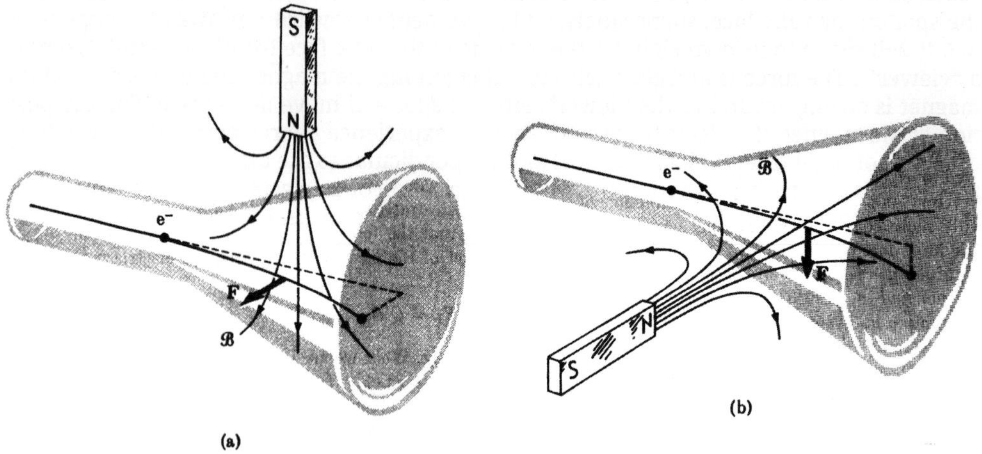

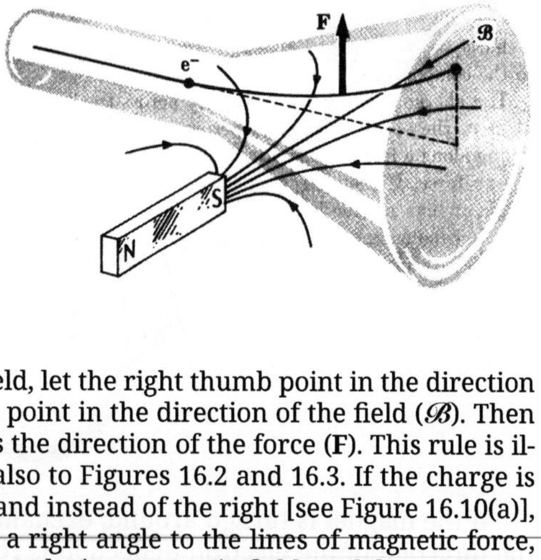

FIGURE 16.9 Deflection of moving electrons by a magnet. **(a)** With magnetic field lines running downward through the tube, electrons are deflected toward the right as shown. **(b)** With a horizontal field as shown in the tube, the beam is deflected downward. **(c)** Reversing the magnet polarity and changing its distance from the tube reverses the direction of the force and changes its magnitude.

Right-and-Left Rule 1

If a positive charge moves in a magnetic field, let the right thumb point in the direction of motion (**v**) and the fingers of the right hand point in the direction of the field ($\mathscr{B}$). Then the direction in which the right palm pushes is the direction of the force (**F**). This rule is illustrated in Figure 16.10(b). It can be applied also to Figures 16.2 and 16.3. If the charge is negative, use the same rule, but with the left hand instead of the right [see Figure 16.10(a)],

In the simplest case of a charge moving at a right angle to the lines of magnetic force, each of the three important vector quantities—velocity, magnetic field, and force—is perpendicular to the other two. Contrast this with the simpler cases where Coulomb's law applies, to the magnetic force on a magnetic pole, or the electric force on an electric charge. Then the force is parallel to the field. Perpendicularity enters in the link between electricity and magnetism.

Any proportionality can be transformed into an equation by the introduction of a constant of proportionality. For the law of magnetic force on a moving charge this constant

* If the velocity vector makes an angle θ with the magnetic field, the exact relation between $v_{\perp}$ and v is: $v_{\perp} = v \sin\theta$.

cannot be unity as it is in Coulomb's law of force, because each of the quantities appearing in Formula 16.2 has already been separately defined. We are no longer free to adopt a new unit for charge or for magnetic field (or for velocity). We are dealing with a new law connecting already defined quantities. The measured force determines the constant of proportionality. This constant is now written $1/c$, so that the new law of force becomes

$$F_{\mathrm{M}} = \frac{v_{\perp}}{c} Q'\mathcal{B}. \tag{16.3}$$

This equation gives the magnitude of the magnetic force on a moving charge; its direction is given by right-and-left rule 1. This is a modern statement of a fundamental law discovered by Ampere not long after Oersted's observation of the magnetic effect of moving charge. Ampere, of course, worked with currents within conducting wires, not charge flying through space. Beams of charged particles in a vacuum were not available until near the end of the nineteenth century. Also Ampere was not able to measure the value of the constant c with precision. Note that this new constant has the dimension of velocity. Since charge times field has the dimension of force (compare Equation 16.1), the extra factor $v_{\perp}/c$ in Equation 16.3 must be dimensionless. The constant c in the electromagnetic-force equation was first measured accurately in 1856 by Wilhelm Weber and Rudolph Kohlrausch in Leipzig. Its magnitude is

$$c = 3 \times 10^{10} \text{ cm/sec}. \tag{16.4}$$

The fact that this constant has exactly the same value as the speed of light created a puzzle in physics finally resolved by Maxwell.

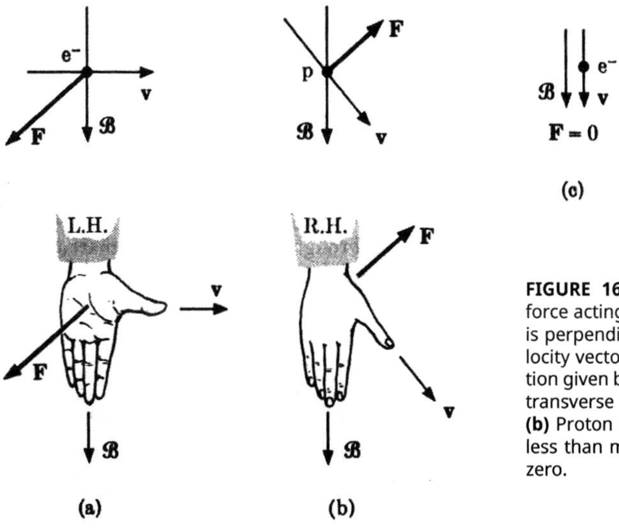

FIGURE 16.10 Directional properties of magnetic force acting on a moving charged particle. The force is perpendicular to the plane determined by the velocity vector and the magnetic field vector, its direction given by right-and-left rule 1. **(a)** Electron moves transverse to magnetic field; force is maximum. **(b)** Proton moves at a lesser angle to field; force is less than maximum. **(c)** Motion along field; force is zero.

Inspection and comparison of Equations 16.1 and 16.3 is worthwhile. Notice that in each a force is proportional to a charge times a field. The magnetic force on a charge is proportional to the extra factor $(v_{\perp}/c)$ not contained in the equation for the electric force on a charge. A charge moving at a speed much less than the speed of light should normally experience magnetic forces much less than electric forces. A charge moving near the speed of light would react equally (except in direction) to equal electric and magnetic fields. Roughly speaking, the link between electricity and magnetism is strong for speeds near the speed of light and weak for speeds much less than the speed of light. The link vanishes if the speed is zero.

In a number of ways in nature, but more especially in man-made devices, the magnetic force on moving charges is at work. It is a guiding principle in every electric motor and in every television set. In the motor, a current-carrying wire in a magnetic field experiences a force. Since the wire is rigidly attached in a coil to a rotor, the force causes the rotor to turn. In a television tube, precisely controlled rapidly changing magnetic fields deflect the electron beam according to a schedule, drawing its end-point in successive horizontal sweeps across the tube face,* making (in the U.S.A.) 525 sweeps per picture and 30 pictures per second.

The magnetic deflection of electrons was used by Thomson to identify the electrons as light negatively charged particles. In the modern bubble chamber Thomson's technique is still employed to advantage for particle identification (Figure 16.11). If a charged particle is moving at right angles to lines of magnetic field, it experiences a force perpendicular to its own velocity. In this respect it is like a stone twirled at the end of a string or like the moon in its orbit round the earth. The effect of a force perpendicular to velocity is to change the direction of motion but not the speed. If the magnetic field is uniform, the charged particle, having swung to a new direction at the same speed, will continue to experience the same magnitude of force, continually altering the direction of its motion in the same way. As a result, it will execute circular motion, just as the stone and the moon do, with constant speed and an acceleration directed toward the center of the circle. It is left as an exercise to prove that a charged particle moving perpendicular to the lines of a uniform magnetic field moves along a circular path with radius of curvature given by

$$r = \frac{pc}{Q'\mathscr{B}}, \tag{16.5}$$

where p is the momentum of the particle and c, Q', and $\mathscr{B}$ are the quantities appearing in Equation 16.3. For elementary particles observable in a bubble chamber, the magnitude of Q' is always the same, 4.80×10^{-10} e.s.u. The known field strength $\mathscr{B}$ is typically about 10^4 gauss, and the constant c is, of course, also known. Therefore, a measurement of the radius of curvature of a bubble-chamber track reveals the momentum of the particle that left the track. Sometimes the combination pc/Q' of a moving charged particle is known as the "magnetic rigidity" of the particle, for it is a measure of the difficulty of bending the trajectory. Particles of high rigidity are barely deflected as they pass through the bubble chamber; particles of low rigidity may be deflected into a full circle within the bubble chamber (Figure 16.11).

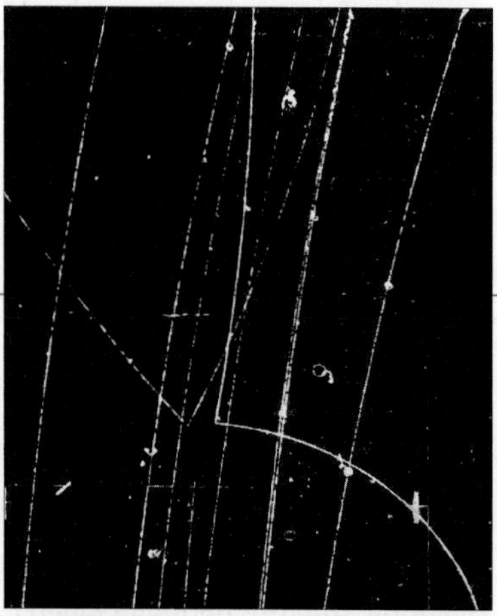

FIGURE 16.11 Magnetic deflection of charged particles in a bubble chamber. The magnetic field in this chamber was directed inward, away from the camera. The incoming particles are flying upward, and two new pairs of charged particles are created in the chamber. Can you determine the sign of the charge on each particle? Assuming that the magnitude of the charge is the same on all the particles, why are some tracks curved more than others? (Photograph courtesy of Lawrence Radiation Laboratory, University of California, Berkeley.)

* The television beam intensity, which makes possible a picture instead of a uniformly illuminated screen, is controlled electrically, not magnetically.

If a charged particle is moving not directly across lines of magnetic field, but has some component along the lines, its motion will be a helix rather than a circle. Its trajectory will look like a bedspring, a combination of circular motion across the lines of field and uniform drifting motion along the lines of field. If such a drifting particle encounters a more intense field, its motion will tighten into a smaller helix and the drift will be slowed down. Enough intensification of the field will cause the drifting motion to reverse its course. The particle is said to have encountered a "magnetic mirror." It spirals toward the region of stronger field, then away from it. A so-called "magnetic bottle" has a magnetic mirror at both ends.

The earth's magnetic field provides a natural "magnetic bottle" in the space outside the earth's atmosphere. The Van Allen radiation belt consists of protons and electrons trapped in this "bottle" (Figure 16.12). A typical proton in the Van Allen radiation belt has an energy of about 1 MeV, or a velocity of 1.4×10^9 cm/sec. Using the proton's mass, 1.7×10^{-24} gm, its charge, 4.8×10^{-10} e.s.u., and the earth's magnetic field in the Van Allen belt, about 0.1 gauss, we can calculate with the help of Equation 16.5 the radius of curvature of the proton's motion. It turns out to be about 1.5×10^6 cm. This is 15 km, or about 9 miles. This seems a great deal, but it is small compared to the size of the Van Allen belt. In addition to this circular motion, the protons drift along the lines of field, being reflected near the north and south poles of the earth (but still above the atmosphere), and accordingly remain trapped for long periods. When disturbances in the earth's field allow the Van Allen protons to dip into the atmosphere, they contribute to auroral displays.

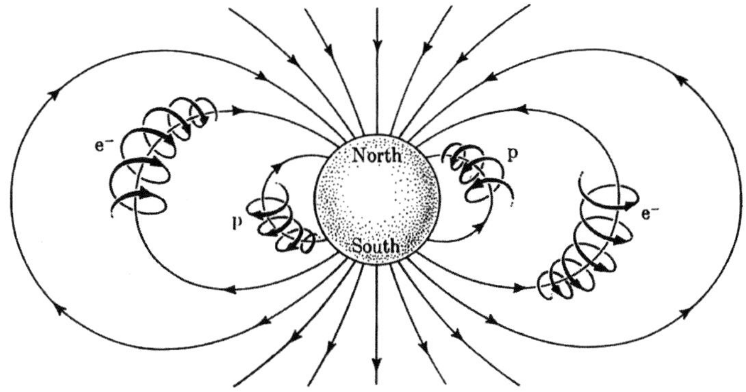

FIGURE 16.12 Trapping of charged particles in the Van Allen radiation belt. Protons spiral one way, electrons the other, and both are reflected by the magnetic mirrors near the poles. Helical paths are not to scale.

If a 1-MeV proton encounters a laboratory magnetic field of 10^4 gauss, 10^5 times stronger than the earth's field in space, it will execute a circle with a radius 10^5 times smaller than the radius of motion of the Van Allen protons. This means 15 cm instead of 15 km. On the other hand, if the proton is much more energetic than 1 MeV, its trajectory has greater rigidity, and not even the laboratory field of 10^4 gauss can curve it in a small circle. In early cyclotrons, protons of a few MeV were held in circles of a few tens of centimeters. In a large modern accelerator like the synchrotron in Brookhaven, New York, 33-GeV protons are magnetically guided around a "race track" 840 feet in diameter.

Another interesting consequence of the magnetic force on moving charged particles occurs in interstellar space, where weak magnetic fields are probably responsible for the acceleration of charged particles of the cosmic radiation to enormous energies. A stationary field deflects a particle without changing its speed. But if the field itself, attached to a tenuous plasma in space, is in motion, the particle may gain energy as well as change direction. It is like a ball gaining energy when it is struck by a moving racket, in contrast to a ball rebounding from a stationary wall without any gain in speed. According to the present theory of cosmic rays, the drifting magnetic fields in our galaxy gradually accelerate protons and some heavier nuclei to energies as great as 10^{16} eV or more. The largest accelerator on earth falls short of peak cosmic-ray energies by a factor of more than one million.

16.3 *The magnetic field created by a moving charge*

Based on Oersted's original discovery, we know that lines of magnetic field describe circles about a straight current-carrying wire (Figure 16.1). This behavior of the magnetic field created by moving charge, obviously very different from the radiating lines of electric field from a charge, can be discovered in a different way, by considering the cathode-ray tube referred to in the previous section. If a north pole held above the beam of a cathode ray causes the beam to be deflected to the viewer's left (Figure 16.13), it is required by Newton's third law that the magnet itself experience an equal and opposite force to the viewer's right. To be more accurate, we should say that *if* Newton's third law is valid for this new force, the equal reaction force to the right on the magnet is expected. Indeed, the measurement of this force verifies the applicability of Newton's third law. Since a stationary north pole experiences a force to the right, this means a magnetic field exists at that location pointing to the right. The moving beam of electrons has created a magnetic field. Above the beam, the field is to the viewer's right; to the right of the beam, the field is directed downward; and so on. From the vantage point of the viewer in front of the tube face, the magnetic field describes clockwise circles about the electron beam (Figure 16.13).

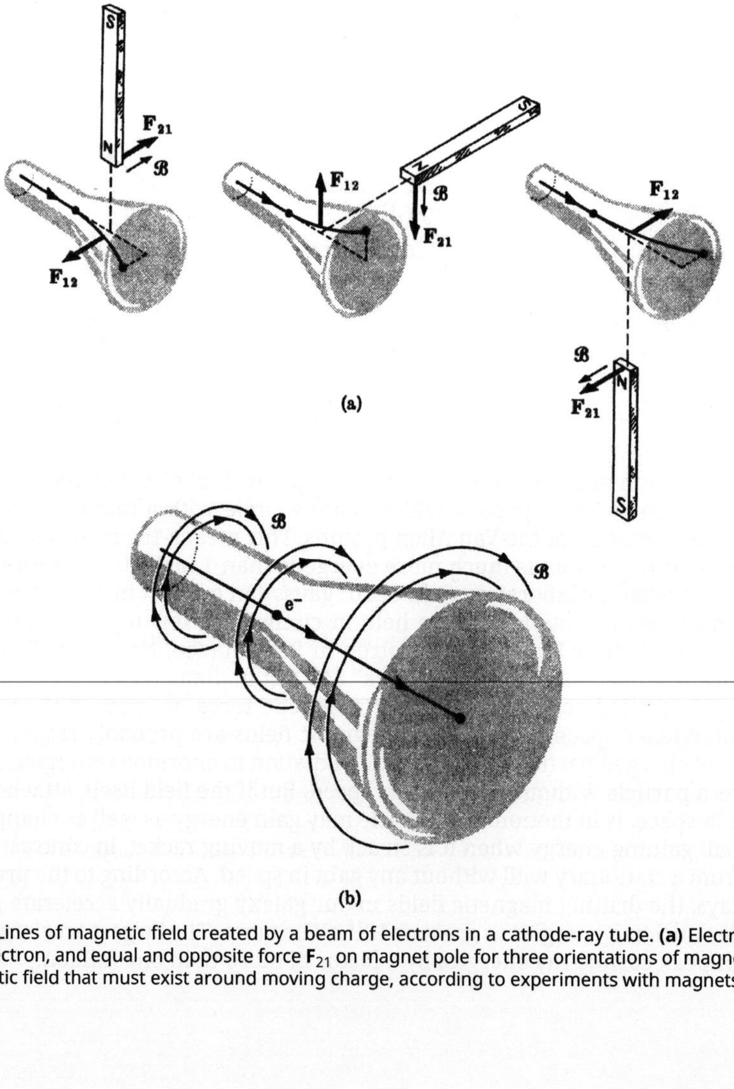

FIGURE 16.13 Lines of magnetic field created by a beam of electrons in a cathode-ray tube. **(a)** Electron deflection, force $\mathbf{F}_{12}$ on electron, and equal and opposite force $\mathbf{F}_{21}$ on magnet pole for three orientations of magnet. **(b)** Circular lines of magnetic field that must exist around moving charge, according to experiments with magnets.

Next we must inquire about the quantitative law. How big is the magnetic field created by a moving charge? The equation of the previous section together with Newton's third law can provide the answer. The force exerted on the moving charge by the magnet was

$$F = \frac{v_\perp}{c} Q'\mathcal{B},$$

as given by Equation 16.3. If the field $\mathcal{B}$ at the location of the moving charge Q' arises from a hypothetical pole P a distance d above the beam, its strength is given by

$$\mathcal{B} = \frac{P}{d^2}. \tag{16.6}$$

Therefore the force on the moving charge arising from the presence of the pole is

$$F = \frac{v_\perp}{c} Q' \frac{P}{d^2}. \tag{16.7}$$

This force, according to Newton's third law, must be the same in magnitude as that experienced by the pole because of the presence of the moving charge. However, the magnetic force on a pole is related to the magnetic field simply by

$$F = P\mathcal{B}. \tag{16.8}$$

Comparing these last two equations, 16.7 and 16.8, we see that the strength of the magnetic field created by a moving charge is given by

$$\mathcal{B} = \frac{v_\perp}{c} \frac{Q}{d^2}. \tag{16.9}$$

Here we have replaced Q' by Q, following our practice to designate a charge creating a field by Q, and a charge reacting to a field by Q'.

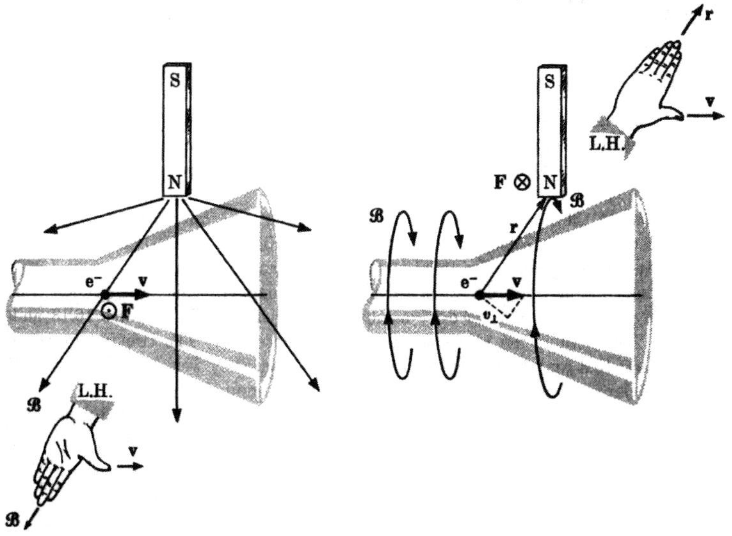

FIGURE 16.14 Two points of view. **(a)** The moving electron in its passive role experiences a force (outward from the page) as it moves through the field produced by a north pole above the cathode ray tube. Equation 16.3 is relevant. **(b)** The same electron in its active role creates a magnetic field which, above the tube, is directed into the page. Equation 16.9 is relevant. The first diagram illustrates the application of right-and-left rule 1, the second an application of right-and-left rule 2. In this instance the negative electron calls for the use of the left hand.

Application of Newton's third law has enabled us to shift viewpoint from the passive role of moving charge experiencing a force (Equation 16.3) to the active role of moving charge creating a magnetic field (Equation 16.9). These two viewpoints are illustrated for this example in Figure 16.14. As shown in each of the diagrams (a) and (b) in this figure, the quantity $v_\perp$ is the component of the velocity v perpendicular to a line joining the charge and the pole. To cover other situations, we must generalize somewhat. In Equation 16.3, $v_\perp$ is the component of $\mathbf{v}$ perpendicular to the magnetic field at the location of the charge. This applies to Figure 16.14(a). In Equation 16.9, $v_\perp$ is the component of $\mathbf{v}$ perpendicular to the radial line from the charge to the point where the field is evaluated. This applies to Figure 16.14 (b).

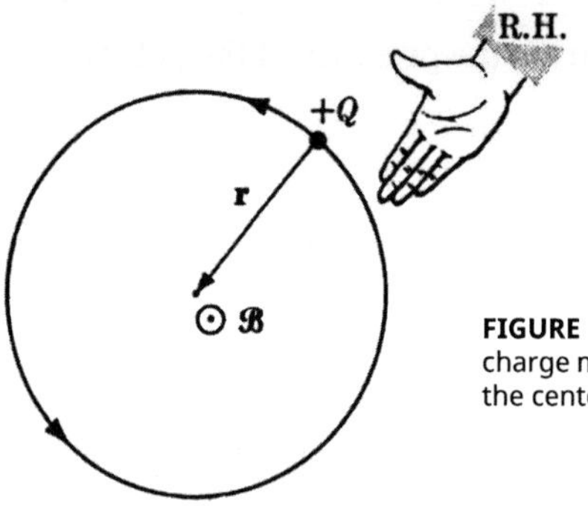

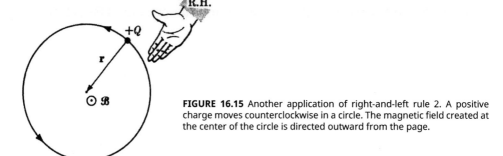

FIGURE 16.15 Another application of right-and-left rule 2. A positive charge moves counterclockwise in a circle. The magnetic field created at the center of the circle is directed outward from the page.

Both the magnetic force experienced by a moving charge and the magnetic field created by a moving charge have directional properties governed by a rule of perpendicularity in three dimensions. To cover the latter (Equation 16.9), a second right-and-left rule is required.

Right-and-Left Rule 2

Let the right thumb point in the direction of motion (**v**) of a positive charge and the fingers of the right hand point in the radial direction (r) from the charge to the point of observation. Then the direction in which the right palm pushes is the direction of the magnetic field ($\mathscr{B}$) created at the point of observation. If the charge is negative, use the same rule, but with the left hand instead of the right. This rule is illustrated in Figure 16.14(b) and in Figure 16.15. It can be applied also in Figures 16.1, 16.3, and 16.5.

To derive Equation 16.9 we assumed that a magnetic pole was present, influencing and being influenced by a moving charged particle (Figures 16.13 and 16.14). This assumption was convenient but unnecessary. The two fundamental equations, 16.3 and 16.9, may be regarded as separately verified statements of physical laws governing the electromagnetism of moving charges.

It is instructive to compare Equation 16.9 with the electric-field equation,

$$\mathscr{E} = \frac{Q}{d^2}, \tag{16.10}$$

given before as Equation 15.9. A charge, whether moving or not, creates an electric field that decreases in strength inversely as the square of the distance from the charge.* If the charge is moving, it creates in addition a magnetic field, weaker than its electric field by the factor $v_\perp/c$, the ratio of a component of its velocity to the speed of light. Apart from this difference in strength, the two fields retain their characteristic individuality of direction. The electric

* Equation 16.10 is valid only for "slowly moving" charge (slow compared to the speed of light). We shall not pursue the modifications to Equation 16.10 required at very high speed.

field lines radiate outward from the charge. The magnetic field lines execute circles about the path of the charge.

16.4 Review of links between charge and electric field and between charge and magnetic field

We have encountered so far four fundamental laws linking charge and fields. Each can be given a qualitative and a quantitative statement. 93

(1) $Q \rightarrow \mathcal{E}$

In words: Charge creates electric field.
In symbols:

$$\mathcal{E} = \frac{Q}{d^2}\,. \tag{16.11}$$

Direction: Electric field is directed radially away from positive charge and toward negative charge.

(2) $\mathcal{E} \rightarrow Q'$

In words: An electric field exerts a force on a charge.
In symbols:

$$F = Q'\mathcal{E}\,. \tag{16.12}$$

Direction: The electric force on a positive charge is parallel to the acting electric field. On a negative charge, the force is opposite to the direction of the field.

(3) $Q \rightarrow \mathcal{B}$

In words: Moving charge creates magnetic field.
In symbols:

$$\mathcal{B} = \frac{v_\perp}{c}\frac{Q}{d^2}\,. \tag{16.13}$$

Direction: Right-and-left rule 2 (page 442).

(4) $\mathcal{B} \rightarrow Q'$

In words: A magnetic field exerts a force on a moving charge.
In symbols:

$$F = \frac{v_\perp}{c}Q'\mathcal{B}\,. \tag{16.14}$$

Direction: Right-and-left rule 1 (page 436).

In Equation 16.13, $v_\perp$ designates the component of velocity perpendicular to the direction associated with the distance d that appears on the right side of the equation. In Equation 16.14, $v_\perp$ designates the component of velocity perpendicular to the direction of the magnetic field $\mathcal{B}$ that appears on the right side of the equation. The close similarity of these two statements makes it easier to remember them.

A close look at the two right-and-left rules shows that they too can be unified. On the right sides of Equations 16.13 and 16.14 appear two quantities with associated directions: velocity and distance from charge to observation point in the first, velocity and magnetic field in the second. In the application of both right-and-left rules, the thumb points in the direction of the velocity, the fingers point along the direction of the other quantity on the right side of the equation that has a directional property, and the palm pushes in the direction of the vector quantity on the left side of the equation.

It is not far from the truth to say that for the effects connecting moving charge and magnetism, everything is perpendicular to everything else. Associated with Equation 16.13: Only the part of the velocity perpendicular to the distance vector is relevant, and the magnetic field created is perpendicular both to the distance vector and to the velocity vector. Associated with Equation 16.14: Only the part of the velocity perpendicular to the magnetic field direction is relevant, and the force is perpendicular both to the magnetic field vector and to the velocity vector. For the connections between charge and electric field, parallelism rather than perpendicularity is the rule.

16.5 Electric currents

In the modern world, we are familiar with many examples of charge moving through space: the electrons in a television tube or in an electronic vacuum tube; the protons in an accelerator, in the Van Allen belt, or in the cosmic rays of outer space; the elementary particles created in a high-energy laboratory. However, the study of the properties of charge in motion got its start not with the study of freely moving charged particles, but with the study of the flow of electric current in wires.* The reader need hardly be reminded that current in wires remains today a vitally important aspect of electromagnetism. It is probably fair to say that through the utilization of current more than in any other single way, physical science has had its impact on the everyday life of man. In the more technically advanced countries, electricity is available in nearly every home for heat, for light, for refrigeration, and for turning the motors of vacuum cleaners or mixers or power saws or a dozen other gadgets. Equally important, power supplied by electric current does most of the heavy work in the industries turning out manufactured goods. Those societies least advanced technically are the ones with the least utilization of electric power. In those areas of the world little touched by the implications of scientific discovery, the dominant impact of physical science in the future will likely come through the introduction of electric current and electric power.

According to the fundamental laws of electromagnetism summarized in the preceding section, the magnetic field created by a moving charge is always weaker than its electric field. This should be especially true in wires, where the electrons move at speeds far less than the speed of light. However, in a wire the total positive charge is almost exactly equal to the total negative charge. Outside the wire the electric effects cancel because of the wire's neutrality. But because the electrons in the wire are moving along the wire in one direction and the positively charged ions are not moving, the magnetic effects do not cancel. Outside the wire the countless tiny magnetic contributions add together to produce a possibly very strong magnetic field. In this important way the current in a wire differs from the flow of charged particles in a beam through space. The electrons in a cathode-ray tube produce electric as well as magnetic effects. The electrons in a wire produce purely magnetic effects.

Surrounding a straight current-carrying wire is a magnetic field whose lines describe circles with the wire as an axis (Figure 16.16). As indicated in the figure, an alternative to moving the palm according to right-and-left rule 2 is curving the fingers to show the direction of the magnetic field. (This is *not* a new rule, but a slight variant of an old rule. Use it only if you find it comfortable and convenient to do so.) Reference to the actual motion of the negatively charged electrons along the wire calls for use of the left hand. From here on, we shall follow instead the conventional practice of indicating current flow in the direction in which positive charge would move (if it were free to move). Then the right hand is used.

* Not until late in the nineteenth century was the equivalence between electric current and moving charged particles fully appreciated.

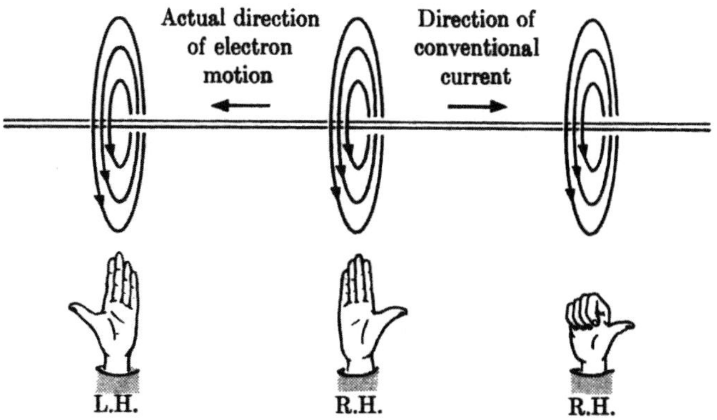

FIGURE 16.16 Magnetic field surrounding a straight current-carrying wire, illustrating the application of right-and-left rule 2. Alternatively, as shown, the fingers of the right hand may be curved to show the direction of the magnetic field.

Now what happens if this current-carrying wire is bent into a circle? Imagine the current flowing in a clockwise direction as seen from above, as in Figure 16.17. The magnetic field lines loop around the wire in such a way that they all dive into the center of the loop from above and point upward everywhere around the outer periphery of the loop. The simple current loop provides the most elementary example of an electromagnet. Within the loop, all the field lines are pointing in the same direction, and this concentration of magnetic field could be used to attract magnetized or magnetizable objects through the center of the loop.

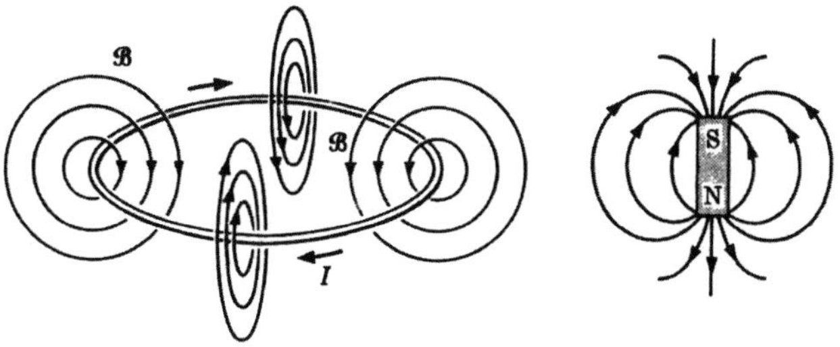

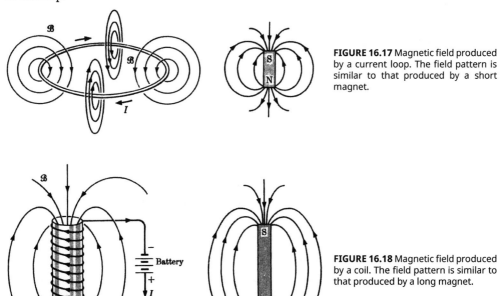

FIGURE 16.17 Magnetic field produced by a current loop. The field pattern is similar to that produced by a short magnet.

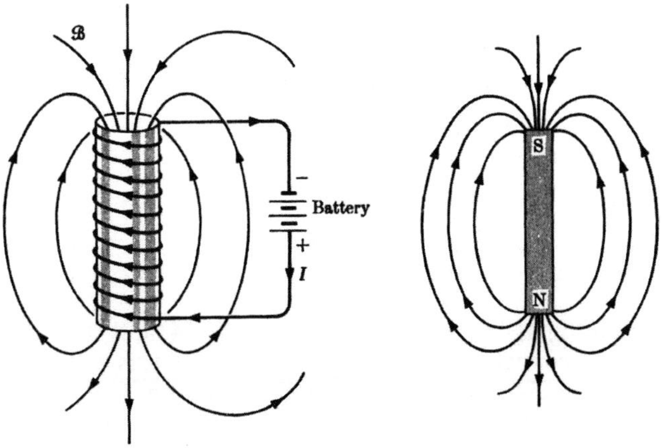

FIGURE 16.18 Magnetic field produced by a coil. The field pattern is similar to that produced by a long magnet.

Next, think of a helical coil of current-carrying wire (Figure 16.18). The current, instead of executing the same loop repeatedly, spirals upward along a cylinder. If again, as seen from above, the current flow is clockwise, each loop will contribute a downward-pointing magnetic field within the coil and an upward-pointing field outside the coil. Because of the concentrating effect of the loops, the field will be most intense within the coil. At the lower end of the cylindrical coil, field lines will emerge; at its upper end, field lines will point downward into the coil. The lower end of the coil behaves much like the north pole of a magnet; its upper end behaves much like a south pole. The coil of wire, usually helical, is the basic design of all electromagnets. It is also the most efficient way to generate intense magnetic fields.

Returning to the single circular current loop, we can learn how to express quantitatively the magnetic field generated by a current. The technical definition of current is charge per unit time. It is the charge flowing past a point in some time divided by that time interval (the usual symbol for current is I):

$$I = \frac{Q}{\Delta t}. \tag{16.15}$$

For the circular loop of radius r, the time required for each electron to complete one trip around the loop is the circumference divided by the speed v:

$$\Delta t = \frac{2\pi r}{v}. \tag{16.16}$$

In this length of time, each electron participating in the flow passes a given point in the wire once. Therefore the current is the total charge in motion divided by this time:

$$I = \frac{Q}{\Delta t} = \frac{Qv}{2\pi r}. \tag{16.17}$$

This result may be written alternatively as

$$I\Delta l = Qv, \tag{16.18}$$

where Δl, replacing $2\pi r$, stands for the length of the wire. This last equation is the key to transforming our previous fundamental Equations 16.13 and 16.14 from the description of charged particles moving in space to the description of currents flowing in wires. Charge multiplied by velocity is equivalent to current multiplied by the length of a segment of wire. Since the same rules of perpendicularity apply to currents as to moving charges, the more general correspondence is

$$I\Delta l_{\perp} \leftrightarrow Qv_{\perp}. \tag{16.19}$$

The double arrow means "is equivalent to" or "may be replaced by." For currents, Equation 16.13 is replaced by

$$\Delta\mathcal{B} = \frac{I\Delta l_{\perp}}{cd^2}, \tag{16.20}$$

and Equation 16.14 is replaced by

$$\Delta F = \frac{1}{c} I\Delta l_{\perp}\mathcal{B}. \tag{16.21}$$

The subscripts $\perp$ have the same meaning as in Equations 16.13 and 16.14, and the same right-and-left rules apply. (So long as the conventional direction of current flow is used, only the right hand is needed.) The delta symbols on the left sides of Equations 16.20 and 16.21 indicate that these equations provide only the contribution to the field or to the force from a particular segment of wire, not the total field or total force.

From Equation 16.13 it follows directly that the magnetic field at the center of the loop carrying a total charge Q at velocity v is

$$\mathscr{B} = \frac{v}{c}\frac{Q}{r^2}. \tag{16.22}$$

The velocity **v** at each part of the loop is perpendicular to the radius vector from the center of the loop to the periphery. The magnetic field $\mathscr{B}$, directed along the axis of the loop, is perpendicular to all of the radii and to the velocity vectors at all points about the loop. The transition to the current description is provided by Equation 16.20. In this special example, for which all parts of the wire are equidistant from the point of observation, the length of segment, Δl, may be replaced by the total length of wire l to give the total field at the center of the loop,

$$\mathscr{B} = \frac{Il}{cr^2} = \frac{2\pi I}{cr}, \tag{16.23}$$

the latter equality resulting when l is set equal to $2\pi r$.

The magnetic field produced by a long straight wire is also of interest, although it is less easy to derive. We state the result without proof:

$$\mathscr{B} = \frac{2I}{cr}, \tag{16.24}$$

where r is the perpendicular distance from the wire to the point of observation. This equation expresses what is known as the Biot-Savart law, discovered shortly after Oersted first established the link between current and magnetism. Its special feature of interest is the inverse dependence on distance, in contrast to the inverse square dependence of the electrostatic field.

In the system of units employed so far, current is expressed in e.s.u./sec, the unit of charge divided by the unit of time. This is a satisfactory scientific unit, and is the unit that should be employed in the fundamental equations of this chapter. In practical applications, a much larger unit of current, the ampere (amp), is more convenient. The ampere is equal to one coulomb per second, and the coulomb is equal to 3×10^9 e.s.u. Accordingly the ampere is 3×10^9 e.s.u./sec. Through a 100-watt light bulb flows a current of about one ampere, or three billion e.s.u./sec. The charge of a single electron is 4.8×10^{-10} e.s.u., less than one billionth of one e.s.u. In terms of the larger unit, the coulomb, the electron's charge is only 1.6×10^{-19} coulomb. This means that about 6×10^{18} electrons flow each second through the filament of the 100-watt bulb.

One current exerts a force on another current. This consequence of the laws of electromagnetism was discovered and studied in mathematical detail by Ampere not long after Oersted's original discovery of the magnetic force of a current. If two current-carrying wires are placed near to each other, the first will generate a magnetic field at the location of the second. Since the second current is charge moving through a magnetic field, it will experience a force. Similarly, the magnetic field created by the second current will cause the first to experience a force. Both right-and-left rules must be used in order to learn that two parallel wires with currents in the same direction will attract each other (Figure 16.3). With currents in opposite directions, they will repel. Recall that current-carrying wires, because they are neutral, produce purely magnetic effects. The mutual force experienced by currents therefore represents an example of magnetism without magnets.

The chief application of the current-current force is in the electric motor. One current-carrying coil is wound on an immovable element, usually the outer part of the motor. Another coil is wound on a rotor attached to an axis. The magnetic force between the two coils is so directed that the rotor is caused to turn, and the rotational motion of the axis may be harnessed to do work.

16.6 *Magnetism in nature*

94 In the world as we know it, electricity without electric charge is impossible. But, as we have just learned, magnetism without magnets is quite possible. A moving charge, for example, generates a magnetic field. In a magnetic field, a moving charge experiences a force. Currents exert forces on each other. These are examples of magnetic phenomena created not by magnets, but by moving charge. Perhaps *all* magnetic phenomena, including the magnetism of magnets, is only a manifestation of charge in motion. This idea arose quite naturally from the work of Ampere on the magnetic effects of currents, and was immediately appealing.* In particular, it could explain at once why no one had succeeded in separating and isolating the north and south poles of magnets. If poles do not, in fact, exist but are only fictions used to describe the approximate properties of magnetic fields generated by currents within magnets, they can of course not be isolated.

Consider the magnetic field generated by a long thin cylindrical coil, called a solenoid (Figure 16.19). Lines of magnetic field draw together in a region of intense field at one end of the solenoid, they extend through the interior of the solenoid, then they emerge and fan out at the other end. A true north pole would be a spot from which magnetic lines of field emerge, a source of field. It would be analogous to a positive charge which literally creates lines of electric field emerging from itself. Looked at not too closely, the right end of the solenoid in Figure 16.19 would appear to be a north pole. Magnetic field lines emerge from it and spread out. In fact, however, they do not all emerge from a single point, nor are they created at that end of the solenoid. It is only an approximate north pole. Similarly the left end of the solenoid behaves somewhat like a south pole, a point where lines of field coalesce at a point and terminate. In this sense it is analogous to a negatively charged particle, a terminus of lines of electric field, or a "sink" of electric field. Again, however, this end of the solenoid is only approximately a pole.

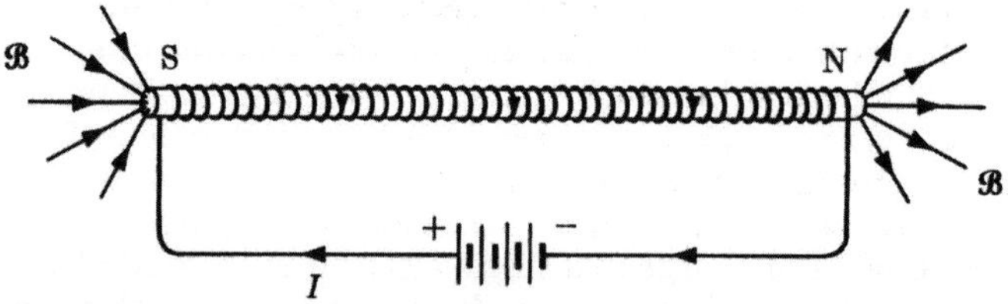

FIGURE 16.19 Magnetic field produced by a long thin solenoid. Near each end of the coil, the magnetic field approximates the radial inverse square field that would be associated with pure magnetic monopoles.

The difference between the electric and magnetic fields found in nature can be expressed in this way. Lines of electric field can begin and end. Lines of magnetic field never begin or end. An electric charge is a primitive source or sink of electric field. From a positive charge emerge lines of electric field and into a negative charge disappear lines of electric field, regardless of the state of motion of the charges. As we now know, charge is carried by electrons, protons, and other elementary particles. No such primitive sources and sinks of magnetic field have ever been discovered, as elementary particles or in any other form. Instead, it seems that all of nature's magnetism arises from the motion of charge. Although this fact simplifies the theory of electromagnetism, accounting for all phenomena in terms of electric charge alone, it does introduce into electromagnetism a puzzling asymmetry that has so far escaped a basis of theoretical understanding.

* Not all of Ampere's contemporaries accepted his idea that microscopic currents within magnets accounted for their magnetism. Indeed the full verification of Ampere's theory had to await the twentieth-century developments of atomic theory.

If a rod had a true north pole at one end, and a true south pole at the other end, it would create a magnetic field scarcely different from that created by the solenoid. Yet at the microscopic level there would be important differences. At the ends of the rod, field lines would begin and end. At the ends of the solenoid, field lines only seem to begin and end. In fact, the lines of field thread their way through the core of the solenoid unbroken. Likewise, in an ordinary magnet, field lines disappearing into one end pass continuously through the interior of the magnet and appear at the other end. Ampere had no way of knowing the exact nature of the internal currents within a magnet that produce its magnetic field. Since no effort to measure a current within a magnet succeeded, he could only assume that the magnetism was produced by tiny microscopic current loops within the material, not by any bulk flow of charge over a measurable distance.

Nearly a century had to elapse before the correctness of Ampere's hypothesis of microscopic current loops could be verified. According to modern atomic theory, electrons within every atom whirl in orbits of radius 10^{-8} cm or less. In addition, every electron spins about its own axis, a further and even smaller-scale manifestation of moving charge. Because of these rotational motions of charge within every atom, no material is free of some magnetic properties. However, for most materials, the individual current loops are oriented in all directions at random, and the magnetic fields they create cancel each other out, leaving no large-scale magnetic effect. In a few materials, called ferromagnets, the current loops become aligned and remain so. Then their fields reinforce each other and a large-scale magnetic field results.* Iron, the best-known ferromagnetic, is commonly employed for magnets, although alloys of iron and other metals are superior for producing strong permanent magnets. Probably the most remarkable fact about ferromagnetism is that it arises from the spin motion of electrons. Not even the vision of Ampere could have foreseen this. Each spinning electron is truly an elementary magnet. For those few materials whose energy is lowered when the axes of electron spin array themselves in parallel, there occurs a natural tendency for self-reinforcement of the atomic magnetism, resulting in the large-scale magnetism discovered more than 2,000 years ago by the Greeks.

In spite of the beautiful explanation of magnetism afforded by modern atomic theory, many physicists remain troubled by the asymmetry of electromagnetism. If elementary poles as well as elementary charges existed, electric and magnetic phenomena would be in perfect balance. Both electric and magnetic fields would have their own primitive sources. As moving charge produces magnetic field and reacts to magnetic field, so moving poles would produce electric field and react to electric field. For every electric phenomenon there would be a precisely equivalent magnetic phenomenon. However, it seems that nature has chosen to be economical rather than symmetric. Charge alone is sufficient to produce both electric and magnetic phenomena. Only charge has been found.

Nevertheless, interest in the possible existence of magnetic poles has not died. Indeed, the recent discoveries of new families of elementary particles has rekindled interest in the possibility of an as yet undiscovered elementary particle carrying pole strength instead of charge. Careful searches for such a particle have been carried out in the debris from high-energy nuclear collisions at the largest accelerators, and in the still higher-energy cosmic radiation arriving from outer space. The results: no poles.

As techniques are refined, the search will go on, for the physicists' faith in the symmetry of nature is hard to shake. There can be no doubt that if magnetic poles exist at all, they are exceedingly rare, at least in our part of the universe. Antiparticles are also rare (fortunately), yet their existence confirms another symmetry of nature. If poles do not exist, the physicist must seek to answer the question: Why not? Our hope would be that the apparent asymmetry of electromagnetism is a reflection of some deeper still undiscovered symmetry.

* A ferromagnetic material need not always be a magnet. Within one domain of microscopic size, all of the atomic magnets align to create a large field. But unless the domains also line up to reinforce one another, a large piece of the material will not act as a single coherent magnet.

16.7 The interaction of electric and magnetic fields

95 More often than not in scientific history, the full latent power of a new discovery required a considerable time to be appreciated. Indeed the more fundamental the discovery, the greater is the realm of nature it encompasses, and therefore the less likely is it that its full ramifications are quickly grasped, even though the fact that it is important may be sensed at once. Over a period of two centuries, for example, Newton's laws of motion, besides spawning countless developments in mathematics, came to encompass an ever larger part of nature, in particular being extended into the molecular domain to account for heat and temperature. In the development of electromagnetism too, a few key discoveries have had to percolate through the minds of many men, being correlated and examined in new ways, before their full content could be assimilated and electromagnetic theory brought to final form. The evolution of electromagnetism as a quantitative theory—not counting its twentieth-century confrontation with the quantum theory, a confrontation still existing in a state of uneasy and uncertain harmony— can be said to have begun about 1785 with Coulomb's law of electrical force and ended about 1905 with Einstein's theory of relativity. Its most active period of development, when the links between electricity and magnetism were discovered and welded into a concise mathematical form, extends from Oersted's discovery of the magnetic force produced by an electric current in 1820 to Heaviside's formulation of the electromagnetic laws in 1885.

Central figures in this active period in the history of electromagnetism were Michael Faraday and James Clerk Maxwell. As already noted, Faraday was responsible for the concepts of fields and lines of force, which have come to be the most important ideas in the mathematics and in the imagery of electromagnetism. It was also Faraday who arrived at the view that the atomic basis of matter is essentially electrical in nature, and that a definite amount of electricity is associated with one mole of a substance.* To Maxwell we owe the prediction of electromagnetic radiation, and the first general mathematical formulation of electromagnetic theory. In this section we are concerned with the key discoveries of Faraday and Maxwell that are illustrated by Figures 16.4 and 16.6. These laws of induction, expressed in modern form, state that a changing magnetic field creates an electric field, and that a changing electric field creates a magnetic field. The great significance of these laws lies in the fact that they express connections between fields alone, with no reference to charge (or to poles). Accordingly, they govern the interaction of fields even in space far removed from matter.

Although the laws of induction stand properly as separate significant statements about electric and magnetic fields, both have interesting close connections with the earlier discoveries of Oersted and Ampere. One remarkable fact is that, from a strictly logical point of view, the experiment of Faraday (and of Henry) illustrated by Figure 16.4 need not have been performed at all. With the benefit of our present comprehensive knowledge of electromagnetism, we recognize that the result of the experiment was to have been expected. This was of course not known at the time. Faraday's result was rightly regarded by his contemporaries as a significant new discovery. Without it the path to the unification of electromagnetism would have been longer and harder.

Consider now a slight modification of the experiment depicted in Figure 16.4. A bar magnet, oriented vertically, is held stationary, and a coil of wire, connected to its current-meter, is moved vertically toward the magnet (Figure 16.20). This represents the motion of charge in a magnetic field. Within the wire are electrons. They are not flowing in either direction along the wire, but are being caused to move upward by the bulk motion of the wire. They will therefore experience a force. To find the direction of this force, we must

* For a substance whose atoms gain or lose a single electron in chemical combination, the molar amount of electricity is the charge carried by 6×10^{23} electrons. This amount of charge, now known as the faraday, is 2.9×10^{14} e.s.u., or 96,500 coulombs.

apply right-and-left rule 1, and it may be helpful to think of the directions in geographical terms. Suppose that the lower end of the magnet is its north pole, from which radiate lines of magnetic field. Consider a particular electron in the part of the coil east of the north pole. Its velocity vector **v** is upward in a magnetic field $\mathscr{B}$ directed toward the east. If the left thumb points in the direction of **v** and the left fingers in the direction of $\mathscr{B}$ (see the figure), the left palm pushes in a southerly direction, the direction of the force acting on the electron. The same rule applied north of the pole gives an easterly force; south of the pole it gives a westerly force; and so on. Each electron experiences a force that is along the direction of the wire, and all are set into motion in such a way that a clockwise flow of electrons (seen from above) results. If the direction of motion of the coil were reversed, or the direction of the magnetic field were reversed, the electron-flow direction would also be reversed. (Remember that the direction of conventional current flow indicated in Figures 16.4 and 16.20 is opposite to the direction of electron flow.) To review: If a wire is at rest in a magnetic field and electrons are flowing along the wire, the force on the electrons is transverse to the direction of the wire, and the wire as a whole tends to be set into motion (Ampere's result); if electrons are at rest within a wire and the wire is moved through a magnetic field, the force on the electrons is transverse to the motion, accelerating the electrons along the wire to create a current (Faraday's result).

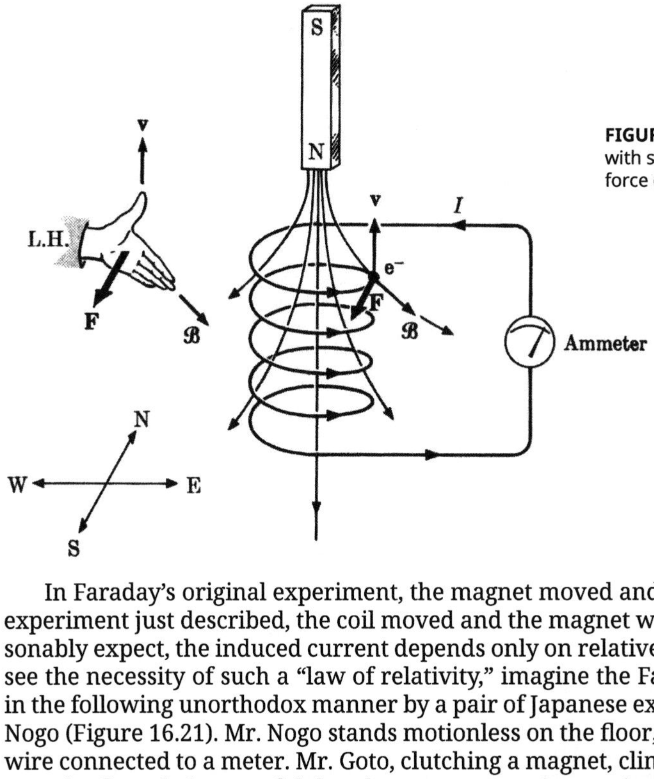

FIGURE. 16.20 Faraday induction experiment with stationary magnet and moving coil. The force on a particular electron is indicated.

In Faraday's original experiment, the magnet moved and the coil was stationary. In the experiment just described, the coil moved and the magnet was stationary. As we might reasonably expect, the induced current depends only on relative motion of magnet and coil. To see the necessity of such a "law of relativity," imagine the Faraday experiment carried out in the following unorthodox manner by a pair of Japanese experimenters, Mr. Goto and Mr. Nogo (Figure 16.21). Mr. Nogo stands motionless on the floor, at rest with respect to a coil of wire connected to a meter. Mr. Goto, clutching a magnet, climbs onto a table and then leaps onto the floor, being careful that the magnet remains motionless with respect to himself. As Goto's magnet passes near the coil, a current is induced. From Nogo's point of view, a moving magnet generated a current in a stationary wire. From Goto's point of view, a stationary magnet generated a current in a moving wire. Since both see the same current recorded by the same meter, they cannot fail to agree that a current was generated. Hence both of their ways of stating the law of induction must be correct.

FIGURE 16.21 Induction experiment in which one observer sees a stationary coil and a moving magnet, the other observer sees a stationary magnet and a moving coil.

There can be power in a point of view. Because he looked upon the induction effect in the way that he did, replacing the idea of "moving magnet" by the idea of "changing magnetic field," Faraday was led to try another important experiment, the one depicted in Figure 16.5. Without a magnet, and without visible motion, a changing current in one wire generates a current in another wire. An even simpler arrangement to demonstrate the same effect is illustrated in Figure 16.22. One straight segment of wire is connected through a switch to a chemical cell; a parallel segment of wire is connected to a current-meter. When

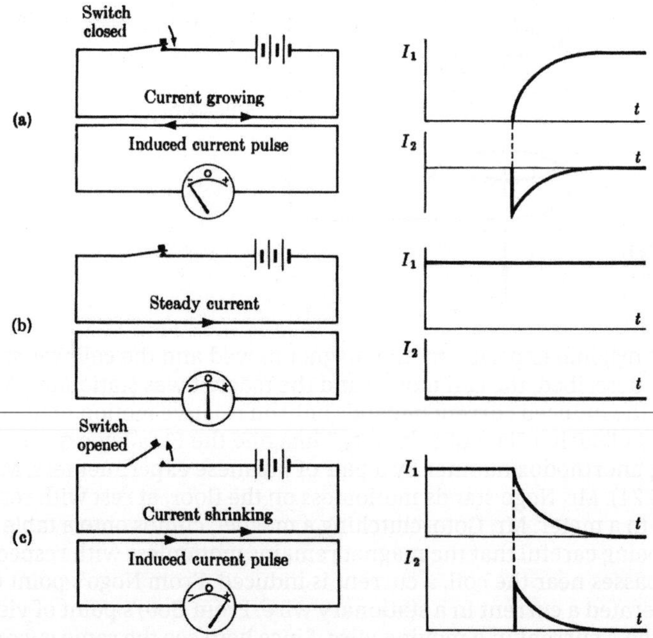

FIGURE 16.22 Principle of the transformer. Whenever current in one wire changes, current is induced in a neighboring wire. The graphs accompanying the diagrams show how the currents I_1 and I_2 vary with time. Coils enhance the induction effect, but do not change the principle.

the switch is closed, current flows in the first wire. During the brief interval of time in which the magnetic field about the first wire is growing to its full strength, a pulse of current is recorded in the second wire. Then, with current in the first wire steady, and no further change in the magnetic field, no current flows in the second wire. When the switch is opened and current in the first wire ceases, there is again a momentary pulse of induced current in the second wire, this time in the opposite direction. Whenever the magnetic field at the position of the second wire changes, for whatever reason, an electric field is created and current is caused to flow. This phenomenon of induction without motion underlies the action of the modern transformer, a device discussed in Section 17.5.

We turn attention now to the second fundamental law linking fields (Figure 16.6): A changing electric field creates a magnetic field. Like Faraday's law of induction, it is a law of fields only, and it is a local law, having to do with the fields at a particular point in space. Like Faraday's law also, it is implicit in the earliest laws of the interaction between electricity and magnetism, but it was not fully appreciated for many years—not until thinking about electromagnetism had shifted away from the material and mechanical concepts of charge, current, and force to the immaterial and ethereal concepts of fields in empty space. That a changing magnetic field creates an electric field is often known as the law of electromagnetic induction. That a changing electric field in turn creates a magnetic field is sometimes called the law of magnetoelectric induction. Regardless of what long names might be attached to these laws, the beautiful symmetry they introduce into electromagnetism is evident.

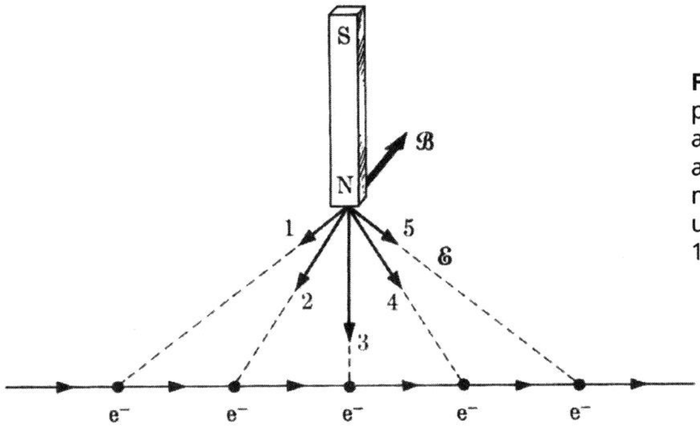

FIGURE 16.23 Reinterpretation of magnetic force produced by moving charge in terms of fields alone. As an electron moves by, the electric field at the location of the magnetic pole changes. The magnetic field produced at that point can be attributed to the changing electric field (compare Figure 16.14).

Recall the experiment illustrated by Figures 16.13 and 16.14. When a beam of electrons fly through a cathode-ray tube, a nearby magnet experiences a force. In Section 16.3 we attributed this to a law expressed qualitatively by the statement: A moving charge creates a magnetic field; and quantitatively by Equation 16.13. It may be looked upon in a different way. Focusing attention on the part of space occupied by the magnet, we may inquire what the magnet "sees" at its own location. As an electron sweeps by in the distant beam, the intensity of its electric field at the magnet changes, rising to a maximum when the electron comes abreast of the magnet, then diminishing (besides changing its direction). Since the magnet is stationary and feels a force, a magnetic field must also be present (Figure 16.23). We may assign a cause-and-effect relationship, and say: The changing electric field creates the magnetic field. A more logically accurate statement is that a magnetic field always accompanies a changing electric field. However, any mode of thought that helps us to appreciate and understand nature can be justified, be it analogy, or pictorial schemes, or assignment of cause and effect, provided only that we bear in mind that these modes of thought may not correspond exactly to any part of the mathematical description of nature. A purely mathematical description of nature without any illogical trimmings is more likely to sterilize progress than are the logical inaccuracies of the fabric of thought about nature that extends beyond pure mathematics.

16.8 *Summary of electromagnetic laws*

96 The theory of electromagnetism rests upon some "old" concepts, drawn from mechanics—length, time, force, energy. To these have been added three new "pure" electromagnetic concepts—charge, electric field, and magnetic field. Actually a fourth, pole strength, was introduced and then laid aside, for it is unnecessary so long as isolated poles are not found in nature. Tying these new concepts together, and linking them with mechanical quantities are the set of laws that have been introduced in this chapter and in the previous chapter. A convenient way to catalogue the laws of electromagnetism is in terms of the Q-$\mathcal{E}$-$\mathcal{B}$ triangle in Figure 16.24. Each of the three electromagnetic quantities—charge Q, electric field $\mathcal{E}$, and magnetic field $\mathcal{B}$—is capable of influencing the other two. Associated with each of the six arrows in Figure 16.24 is one fundamental law of electromagnetism. The arrow from Q to $\mathcal{B}$ designates the influence of charge on magnetic field, the arrow from $\mathcal{E}$ to Q designates the influence of electric field on charge, and so on. Four of the laws were summarized in Section 16.4, in both qualitative and quantitative terms. Here we add the last two.

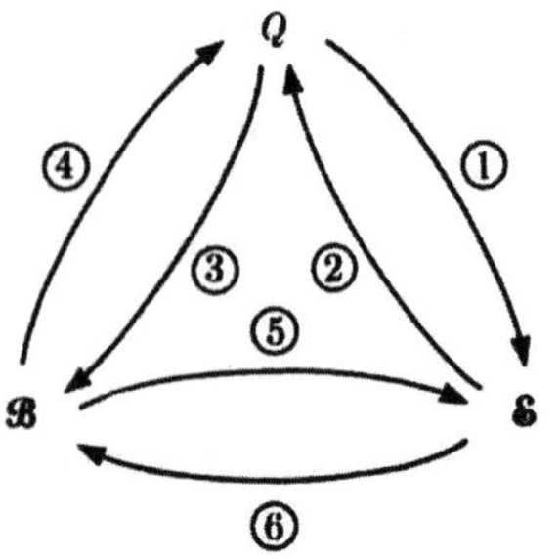

FIGURE 16.24 The Q-$\mathcal{E}$-$\mathcal{B}$ triangle, a schematic reminder of the fundamental laws connecting the concepts of charge, electric field, and magnetic field.

(5) $\mathcal{B} \rightarrow \mathcal{E}$

In words: A changing magnetic field creates an electric field.

(6) $\mathcal{E} \rightarrow \mathcal{B}$

In words: A changing electric field creates a magnetic field.

We shall not give mathematical expressions for this last pair of laws, but remark that here as in all other ways in which electricity and magnetism interact, perpendicularity is the rule. The direction of the induced field is perpendicular to the direction of change of the inducing field.

A pair of laws outside of our Q-$\mathcal{E}$-$\mathcal{B}$ triangle that require explicit statement are concerned with the intrinsic properties of charge and pole strength.

(7) Charge Conservation

The total amount of charge in an isolated system remains constant. An alternative statement: In an isolated event, the total charge after the event is the same as the total charge before the event. In applications of this law one must remember that equal and opposite charges are equivalent to zero charge.

(8) Nonexistence of Poles

Magnetic poles do not exist. One might question the value of calling such a negative statement a fundamental law. There are, after all, countless other quantities one might define that also do not exist. For example, the statement, "A race of men eighteen feet tall does not exist," would hardly be regarded as an important anthropological truth. Nevertheless, the fact that all attempts so far to find magnetic poles have been fruitless is important, for magnetic field does exist. The fact that the known sources of magnetic field are moving charges only and not poles determines in part the nature of the magnetic field. Most important, the absence of poles in nature means that lines of magnetic field never terminate.

The first complete summary of the laws of electromagnetism was given in 1862 by Maxwell. Later, in the hands of Oliver Heaviside and Heinrich Hertz, Maxwell's theory was brought into the elegant and concise mathematical form that we still use today. The Heaviside-Hertz equations,* just four in number, contain electric and magnetic field and charge and, except for the variables of space and time, nothing else. They are local equations, free of the idea of action at a distance, linking what happens at a given point in space only to what is happening in the immediate neighborhood of that point. Despite the great importance of these equations, we can unfortunately not state them here in meaningful form. Because they are differential equations rather than algebraic equations, their meaning and their basic simplicity are evident only to someone with adequate mathematical training. An algebraic equation, the kind we encounter in this book, is a relationship among the magnitudes of a set of quantities. A differential equation involves not only magnitudes of quantities, but their rates of change as well. For example, an algebraic quantity associated with a skier might be his altitude above the valley floor. Differential quantities associated with his descent would include his vertical component of velocity (rate of change of altitude with time) and the slope of the hill (rate of change of altitude with distance). Maxwell's equations involve both the temporal and spatial rates of change of electric and magnetic field. As we have already emphasized, motion and change are key aspects of electromagnetism, and these ideas naturally find their way into the fundamental equations of the theory.

16.9 Energy in electromagnetism

The field concept brought to electromagnetic theory many valuable dividends beyond ease of visualization. It made possible a local theory without action at a distance; it brought a deeper understanding of the laws of induction linking electricity and magnetism; it provided a basis for a concise mathematical theory of electromagnetism; and it introduced a way of thinking about nature that proved to be fruitful in the modern theories of relativity and quantum mechanics. Finally, the fields provided a seat of electromagnetic energy. To the question, "Where is the electromagnetic energy?" we can now give the answer, "In the electric and magnetic fields." The field is literally a physical repository for energy.

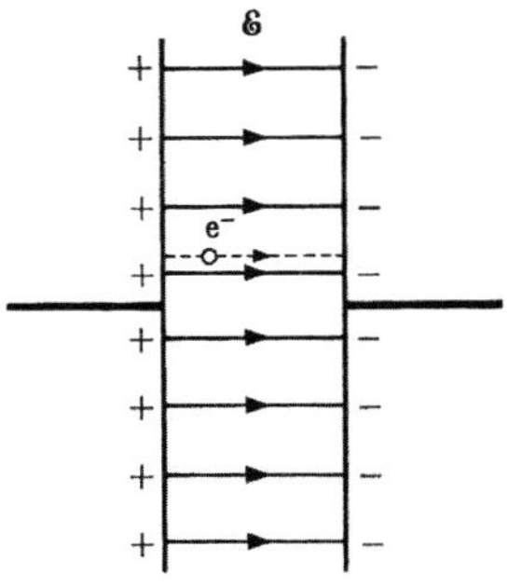

FIGURE 16.25 Schematic representation of a capacitor, a repository of stored charge and stored energy. Moving an electron from the positive to the negative plate increases the stored charge and takes work, work that appears as extra energy stored in the electric field between the plates.

A capacitor (Figure 16.25; see also Figure 15.7) illustrates in a simple way the idea of field energy. One version of the capacitor consists simply of a pair of parallel metal plates, on which charge may be stored, usually equal and opposite charges on the two plates. Then there exists an electric field between the two plates, the surrounding region containing little or no field. Lines of field emerge from the positively charged plate and terminate on

* Physicists usually call these "Maxwell's equations," more out of honor to the great achievements of Maxwell than out of historical accuracy. In what follows, we shall adhere to this now well-established tradition.

the negatively charged plate. To consider the energy associated with the capacitor, we think about the process of charging its plates. Once some positive charge exists on one plate, work is required to add more positive charge to the same plate, for the charge already there exerts a repulsive force on the charge being added. Likewise, work—the expenditure of energy—is required to increase the negative charge on the negative plate. The energy required to charge a capacitor may be visualized in another way, by considering the process of removing an electron from the positive plate and transferring it across the intervening space to the negative plate. Throughout its trip across, it is being subjected to a combined force from both plates tending to push it back to the positive plate. Therefore work must be done on it to overcome this force and transfer the electron from the positive to the negative plate, thereby slightly increasing the net charge on both plates.

The energy required to charge a capacitor is not lost. It stands ready to be used. For example, the two plates might be connected together through a lamp or a motor. The flow of current associated with the discharge of the capacitor could provide energy in the form of heat or light or mechanical work, an amount of energy exactly equal to the energy expended in charging the capacitor. In its charged state of readiness, we may say that the capacitor possesses potential energy. Logically this is adequate and it preserves the law of energy conservation, but it is unsatisfying. The field provides an answer to the question: Where *is* the energy?

That the capacitor is a repository of stored energy was suggested first by Hermann von Helmholtz in 1847.* Now we know that the quantity $\mathscr{E}^2/8\pi$ (where $\mathscr{E}$ is the electric field between the plates) multiplied by the volume of the space between the plates is equal to the stored energy in the capacitor. It is as if an amount of energy equal to $\mathscr{E}^2/8\pi$ is contained in each cubic centimeter of space where the field & is present. This view of stored energy in the field is in fact successful for any arrangement of charge and electric field, not just for the capacitor. The total energy calculated by assigning an energy per unit volume of $\mathscr{E}^2/8\pi$ is always the same as the potential energy calculated from the work required to move the charges and create the fields.

Not surprisingly, a similar expression, $\mathscr{B}^2/8\pi$, gives the energy per unit volume stored in the magnetic field. The general connection between fields and energy is

$$\text{Stored energy per unit volume} = \frac{\mathscr{E}^2}{8\pi} + \frac{\mathscr{B}^2}{8\pi}. \qquad (16.25)$$

If $\mathscr{B}$ is expressed in gauss and $\mathscr{E}$ in dynes/e.s.u. (or, equivalently, statvolts/cm), the right side gives the energy density in ergs/cm^3. In a propagating electromagnetic wave, with electric and magnetic fields in perfect balance, the two terms on the right are equal in magnitude. The energy of the wave is half electric, half magnetic.

At the most fundamental level, we recognize now only three kinds of energy: kinetic energy, field energy, and mass energy. Electric and magnetic and gravitational potential energy have been accounted for as field energy. Heat energy is kinetic energy. Light energy is electromagnetic field energy. Chemical energy is a complicated mixture of the kinetic energy of atomic electrons and the field energy associated with the atomic electrical forces. There is some reason to speculate that the future may see a reduction to only one kind of energy—field energy. If mass proves to be only an intense localized bundle of field energy, as it may, then kinetic energy too (mass in motion) will be a form of field energy. Already we know that changes of field energy are reflected in changes of mass, so that in balancing the energy books in elementary-particle processes, it is usually sufficient to consider only mass and kinetic energy.

* In the same paper Helmholtz suggested that the conservation of energy was a universal and absolute law of nature. This paper, too radical for his older colleagues, was rejected by a leading German scientific journal and had to be printed privately by Helmholtz.

Lenz's Law

An interesting general statement about electromagnetic phenomena is Lenz's law, not itself really an independent law, but a useful way to formulate the implications of energy conservation in electromagnetism. It is this: If a change 1 induces an effect 2, the effect 2 will create a reaction 3 opposing the effect 1. Lenz's law is sometimes loosely expressed by the phrase, "Nature fights back." Picture, for example, the north pole of a magnet moving downward past a horizontal wire (Figure 16.26). The motion of the magnet causes a current to be induced in the wire. The induced current in turn produces a magnetic field ($\mathscr{B}_2$), and this field exerts a force on the magnet, opposing its motion. You may verify, by using both right-and-left rules in turn and by referring to Figure 16.26, that indeed the reaction field opposes the motion of the magnet. In doing this, recall that, relative to the magnet, the electrons in the wire are moved upward. The implication of Lenz's law in this example is that an external force must be applied to the magnet to keep it moving. Work must be done. This is necessary for energy conservation, since the induced current can itself expend energy or do work. If no work were required to induce the current, there would be energy output with no energy input. In a modern generator, current is produced by the relative motion of wires and magnetic fields. The motion is maintained by the continual input of mechanical energy, for example, from a steam engine. When you switch on a light in your home, you draw more current from the generator. This extra current increases the magnetic field created by the induced current in the generator, increases the force opposing the motion of the generator, and thereby calls on the steam engine to supply just enough extra mechanical energy to equal the electrical energy expended in your light.

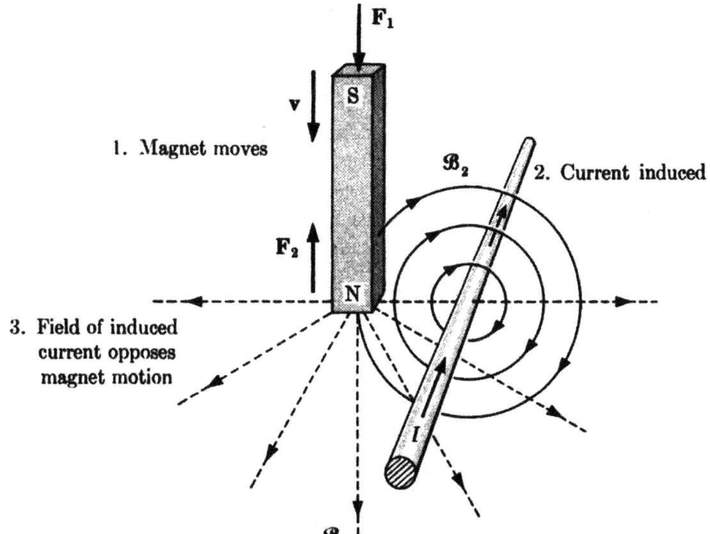

FIGURE 16.26 Lenz's law at work. A change 1 (moving magnet) produces an effect 2 (induced current), which produces a reaction 3 (force $\mathbf{F}_2$ on magnet) opposing the original change. In the figure, $\mathscr{B}_1$ designates the field of the magnet and $\mathscr{B}_2$ the field of the induced current. A force $\mathbf{F}_1$ is required to keep the magnet moving. (To keep the diagram simple, field lines are shown in only one plane.)

Other examples of Lenz's law will be left to the exercises. Note that Lenz's law must be carefully distinguished from Newton's third law of equal and opposite forces. If a magnet is at rest with respect to a current-carrying wire, the wire will experience a force, and the magnet will experience an equal and opposite force. This is not an application of Lenz's law, for no exchange of energy is involved. If the magnet were moved, energy would become relevant and Lenz's law could be invoked to help determine whether the current flow would be increased or decreased by the motion.

From three new concepts and half a dozen fundamental laws has flowed a technology so widespread and so intimately bound up with a myriad of aspects of modern life that we can now scarcely visualize civilized society without electromagnetism, although it existed only a century ago. In the next chapter we shall examine selected applications of electromagnetism, especially those that illustrate the basic laws in a simple way.

EXERCISES

16.1. Is the magnetic field between a pair of parallel wires carrying equal current stronger or weaker than the field at the same place if only one wire were carrying current? Answer for both cases: **(a)** currents in the same direction; and **(b)** currents in the opposite direction. State briefly the reason for each answer.

16.2. Although not indicated in the diagram, a force also acts on the south pole of the magnet in Figure 16.2. **(1)** In what direction does this force act? Very roughly, how does this force compare in magnitude with the force acting on the north pole? **(2)** Does the magnet experience a net torque? If so, what is the direction of the torque vector?

16.3. After the switch in Figure 16.5 has been closed for a time, no current flows in the lower circuit. Then the upper circuit, with the switch still closed, is pulled away from the lower circuit. Does this motion cause a current to be induced in the lower circuit? If so, in what direction? Give reasons for your answers.

16.4. If you hold in your hand a rod with a net charge of 1,000 e.s.u. and swing it briskly at a speed of 900 cm/sec across the lines of magnetic field of a powerful magnet whose field is 3×10^4 gauss, what force acts on the rod? Would this force be perceptible? Compare it with some more familiar force.

16.5. An electron in a TV tube moves with a speed $v = 2 \times 10^6$ cm/sec through a magnetic field $\mathscr{B} = 800$ gauss. Find the magnitude of the force on the electron for each of three angles between the vectors **v** and $\mathscr{B}$: **(a)** 90 deg, **(b)** 45 deg, **(c)** 0 deg. For the largest of these three forces, what is the acceleration of the electron?

16.6. A beam of protons (charge e) moves horizontally to the east with speed v through a region containing a magnetic field $\mathscr{B}$ directed horizontally to the north. In the same region exists an electric field $\mathscr{E}$ chosen so that the total force, magnetic plus electric, on each proton is zero. **(1)** Derive an algebraic expression for the required magnitude of electric field. **(2)** What must be the direction of this electric field? **(3)** If the speed of the protons is $v = 10^6$ cm/sec and the intensity of the magnetic field is $\mathscr{B} = 1{,}000$ gauss, what is the electric field $\mathscr{E}$ in dynes/e.s.u.?

16.7. A charged particle moving in a plane perpendicular to the lines of a constant magnetic field follows a circular path. Prove that the radius of the circle is given by Equation 16.5. You will need to combine what you know of magnetic force and what you know of circular motion from mechanics.

16.8. A proton moving from west to east near the equator in the earth's magnetic field of 0.3 gauss experiences an upward magnetic force. With what speed must it move in order that this force be equal and opposite to the force of gravity acting downward on the proton?

16.9. Suppose a free isolated magnetic pole exists. **(1)** Sketch the magnetic field created by a pole. **(2)** Discuss briefly the motion of an electron **(a)** as it approaches a pole head on; and **(b)** as it passes by a pole at some distance.

16.10. Consider a low energy electron whose speed is 2×10^8 cm/sec moving in a circle above the earth's atmosphere in a magnetic field of 0.1 gauss. **(1)** What force acts on the electron? **(2)** What is the radius of its circular path? **(3)** An electron in a hydrogen atom moves with about the same speed. Its motion can be approximately described as circular with a radius of 5×10^{-9} cm. What is the factor of difference between the force on the electron in the hydrogen atom and the force on the electron of equal speed in the Van Allen belt?

16.11. Residents of northern Canada are bombarded by more intense cosmic radiation than are residents of Florida. Why? (HINT: Consider the effect of the earth's magnetic field on charged particles approaching the earth from different directions.)

16.12. Inside a certain laboratory room there is said to be a magnetic field but no electric field. What experiments might be performed to check the correctness of the assertion?

16.13. An electron moving at high speed is about to strike the central point on the circular face of a cathode-ray tube. Give the directions and the relative magnitudes of the magnetic fields at the points A, B, and C. (The three points are equidistant from the center of the tube face, C being in line with the path of the electron.)

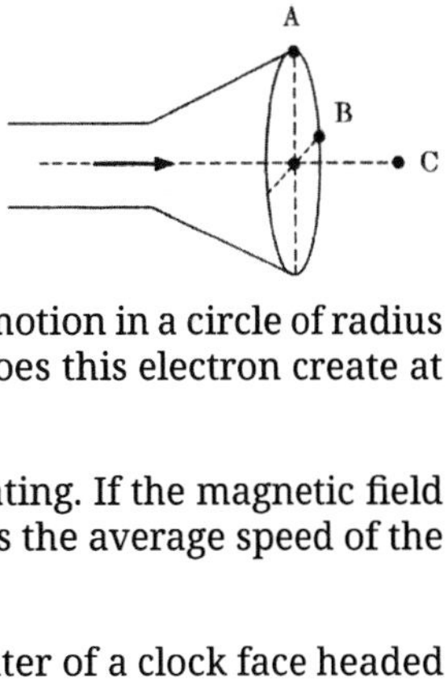

16.14. If the motion of an electron in an atom is approximated as motion in a circle of radius 10^{-8} cm at a speed of 2×10^8 cm/sec, what magnetic field does this electron create at the center of the atom?

16.15. In a loop of wire of radius 10 cm, 10^{22} electrons are circulating. If the magnetic field at the center of the loop is measured to be 100 gauss, what is the average speed of the electrons around the loop?

16.16. A pellet carrying a net charge of +50 e.s.u. flies past the center of a clock face headed toward the 3-o'clock position at a speed of 3,000 cm/sec. The radius of the clock face is 5 cm. At this instant, find the magnitude and direction of the magnetic field created by the moving pellet at **(a)** the 9-o'clock position; **(b)** the 12-o'clock position; and **(c)** the 1-o'clock position.

16.17. In a magnetic field of 200 gauss, an electron executes a circular path whose radius is 0.3 cm. **(1)** What is the momentum of the electron? **(2)** What is its speed? Is the formula $p = mv$ approximately valid? **(3)** What is the magnitude of the magnetic field created by the electron at the center of its circular path? **(4)** Compare this field in magnitude and direction with the original field of 200 gauss which caused the deflection.

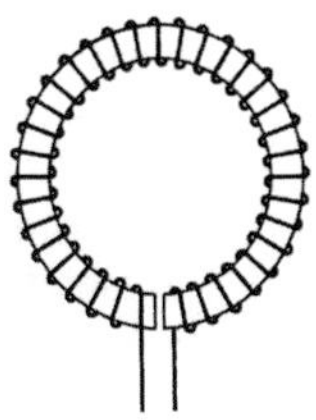

16.18. If the long thin solenoid in Figure 16.19 were bent until its two ends joined and it formed a torus (see figure), what would be the pattern of magnetic field lines ?

16.19. An electron executes a circular orbit of radius 5×10^{-9} cm at a speed of 3×10^{8} cm/sec. What is the electron "current" **(a)** in e.s.u./sec, and **(b)** in amperes?

16.20. A long straight wire carries current I_1. Formula 16.24, known as the Biot-Savart law, gives the magnetic field created by such a current. At a distance r from this wire is a parallel segment of wire of length l carrying current I_2. Derive an algebraic expression for the magnitude of the force on the wire segment of length l. What is the direction of this force?

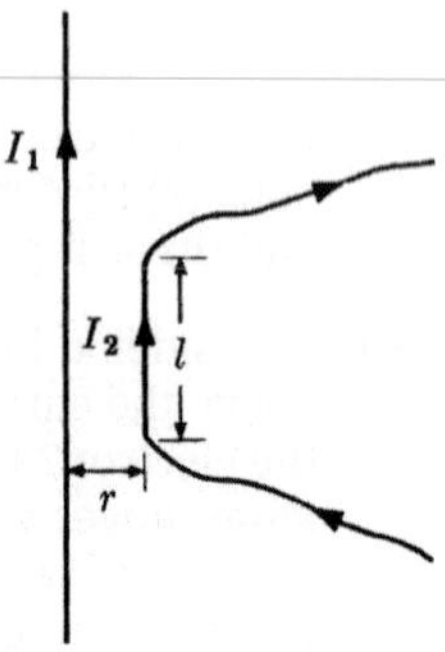

16.21. A proton in a cyclotron has a kinetic energy of 10 MeV. **(1)** What is its speed? **(2)** What is the radius of its circular path in a magnetic field of 10^4 gauss? **(3)** If 10^{12} protons/sec emerge from the cyclotron, what is the current of the beam in amperes? **(4)** What is the power of the beam in watts? Do you think it would be comfortable to stand in the way of the beam?

16.22. A circular loop of wire of radius 6 cm carries a current of 1 amp. **(1)** What is its current in e.s.u./sec? **(2)** What is the magnetic field at the center of the loop? **(3)** Where should a long straight wire carrying 20 amps be placed, and how should it be oriented, in order that its magnetic field cancel the field at the center of the loop? Include a sketch with your answer.

16.23. A circular loop 10 cm in diameter carries a current of 50 amps. Two protons start from the center of the loop, both with a speed of 10^4 cm/sec. **(1)** The velocity of the first proton is directed initially along the axis of the loop. **(2)** The velocity of the second proton is directed initially toward the periphery of the loop. Describe the motion of both in words.

16.24. A wire segment 200 cm long with a mass of 100 gm is located at the equator, where the earth's magnetic field is horizontal and points north, with a magnitude of 0.3 gauss. How should the wire be placed, and what current should flow through it (in which direction) in order that the magnetic force on it can support its own weight?

16.25. A long straight wire carries a current I. At a distance r from the wire are placed **(a)** a stationary charged particle of charge Q'; and **(b)** a stationary magnetic pole of pole strength P'. Calculate the force on each.

16.26. A rectangular loop of n turns carrying current I in a magnetic field $\mathscr{B}$ is suspended so that it can rotate about the vertical axis indicated by the dotted line in the figure. **(1)** Find expressions for the force on each of the four sides of the loop. From these, derive a formula for the torque on the loop. **(2)** Calculate the torque if $I = 10^{10}$ e.s.u./sec, $n = 100$ turns, $\mathscr{B} = 10^3$ gauss, $I = 10$ cm, and $r = 5$ cm. **(3)** In which direction does the loop tend to turn? This exercise illustrates the principle of the electric motor.

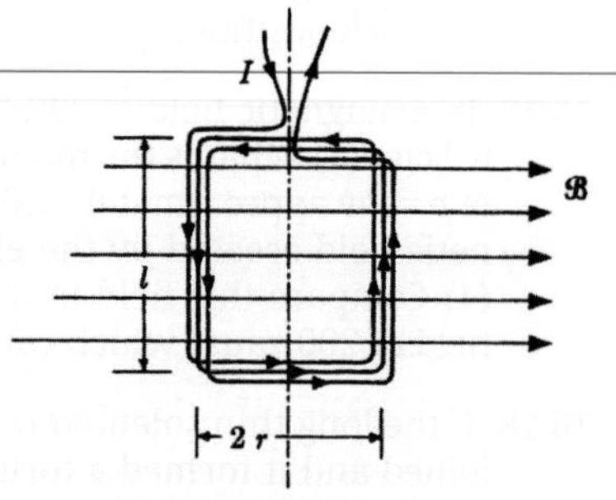

16.27. A uniform magnetic field $\mathcal{B}$ points into the page. A wire of length l slides vertically upward with speed v along stationary wires joined at the top as indicated in the diagram. An electron in the moving segment of wire experiences a force. **(1)** In what direction is this force? **(2)** Derive a formula for the magnitude of the force on the electron. **(3)** Explain why current circulates around the loop as long as the wire segment is moving. This exercise illustrates the principle of the electric generator.

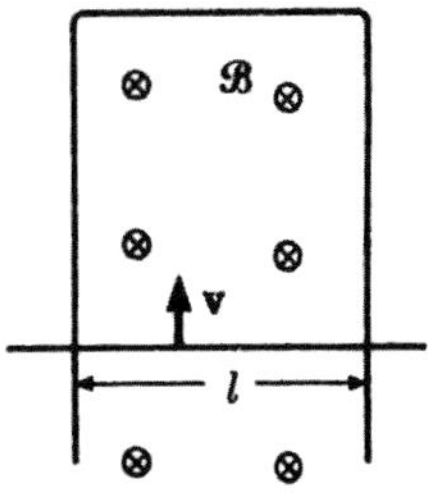

16.28. An airplane flies north with its wings perpendicular to the downward-sloping lines of the earth's magnetic field. **(1)** What is the direction of the magnetic force on an electron in the metal wing? **(2)** What is the direction of the induced current (conventional current)? **(3)** Why does this induced current flow only for a limited time and then stop?

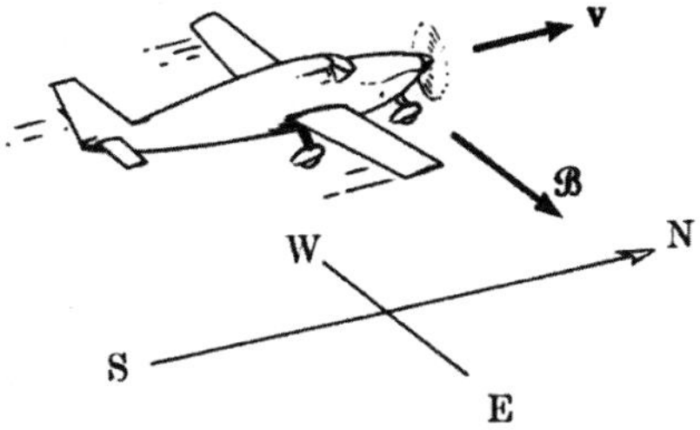

16.29. The diagram of Exercise 16.27 shows a wire segment moving through a magnetic field. Discuss the resulting phenomenon of current generation from the point of view of an observer moving upward at speed v, for whom the wire segment is stationary. In this frame of reference is there an electric field in the wire segment? If so, in what direction?

16.30. The electromagnetic radiation within a laser possesses an energy density of 10^5 ergs/cm^3, divided equally between electric and magnetic fields. **(1)** Find the average $\mathcal{E}$-field and the average $\mathcal{B}$-field in cgs units. **(2)** At what distance from a proton is the energy density in the electric field the same as in this laser?

16.31. In the space between a pair of capacitor plates (Figure 16.25) is stored 10 ergs of energy in an electric field which is nearly constant over a volume of 25 cm^3. **(1)** What is the magnitude of the electric field in dynes/e.s.u. (or statvolts/cm)? **(2)** What is the acceleration of an electron released in this region?

16.32. Describe the course of events if Lenz's law acted oppositely in Figure 16.26, to produce a downward force $\mathbf{F}_2$ on the downward-moving magnet. What fundamental law would be violated?

16.33. In Mr. Goto's leap (Figure 16.21), will electromagnetic forces act to make his net downward acceleration greater or less than the gravitational acceleration g? Explain in terms of Lenz's law.

16.34. The conservation of energy requires that work must be done on the wire segment in the diagram of Exercise 16.27 to move it through the magnetic field. Explain this same fact **(a)** in terms of appropriate right-and-left rules; and **(b)** in terms of Lenz's law.

16.35. Two identical insulated circular loops of wire are laid on a table, one on top of the other. A switch is closed that starts a current flowing clockwise in the upper loop. Use Lenz's law to determine in which direction a current will be induced to flow in the lower loop.

16.36. When current starts to flow in the loop pictured in Exercise 16.26, the loop starts to rotate. Use Lenz's law to find out in what direction it starts to rotate. Explain why Lenz's law can be applied.

16.37. The north pole of the magnet depicted in Figure 16.2 experiences a downward force. If it moves downward in response to this force, it induces in the wire a current I' in addition to the current I already flowing. **(1)** In what direction is this induced current I'? Why? **(2)** Would you expect I' to be greater in magnitude than I, equal to I, or less than I'? Why?

16.38. A switch is closed, causing current to flow upward as shown, through a thin flexible wire parallel to a strong bar magnet. **(1)** What shape does the wire assume as a result of the magnetic forces acting on it? **(2)** At the first instant the current flows, before the wire has had time to deform, what is the direction of the total force acting on the magnet? **(3)** At this same instant, what is the direction of the total torque acting on the magnet?

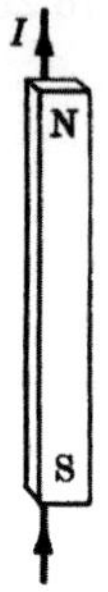

16.39. A charged particle moving in an arbitrary magnetic field never changes its speed. Why? (HINT: Consider its change of *velocity* over a very short interval of time.)

16.40. A pair of long straight parallel wires separated by a distance d carry currents I_1 and I_2. **(1)** Derive an expression for the force per unit length acting on either one of the wires because of the presence of the other wire. **(2)** Do both wires experience the same magnitude of force per unit length?

16.41. A circular current loop of radius r carries current I. Derive a formula for the magnetic field $\mathscr{B}$ on the axis of the loop at point P a distance d from the center of the loop, (HINT: Take advantage of the fact that the net field is directed along the axis. Consider the component of field in this direction produced by each small segment of the current loop.) Verify that if $d = 0$, your result duplicates Formula 16.23.

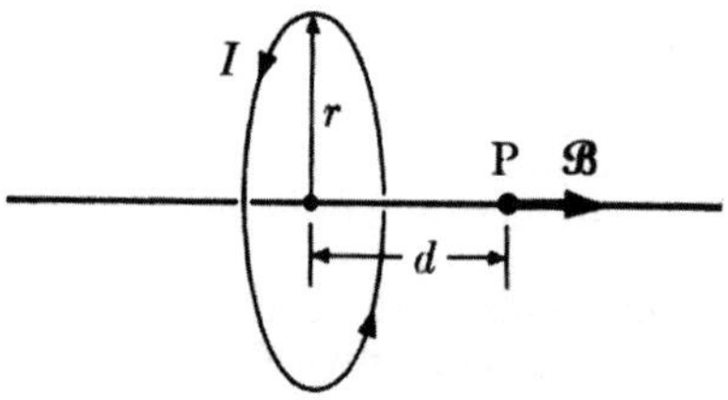

16.42. In an arrangement known as Helmholtz coils, two coaxial coils, each of radius a, are separated by the distance a. Each coil has n turns of wire carrying the same current I. What is the magnetic field strength at the midpoint of the line joining the centers of the two coils ?

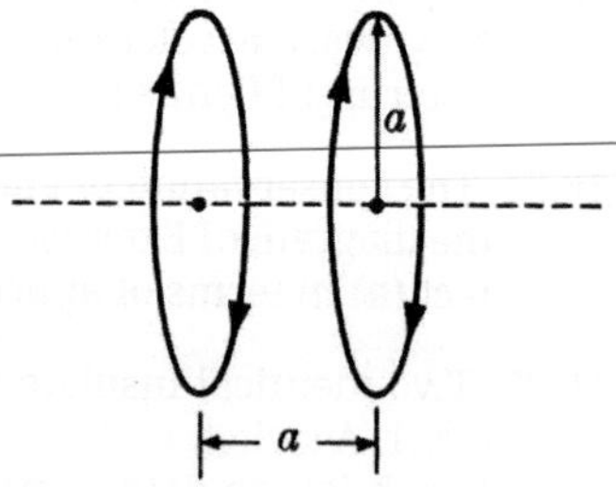

SEVENTEEN

Applications of electromagnetism

The workhorse of modern society is its smallest member, the electron. By good fortune, a certain class of materials called metals contain "free" electrons—about 10^{23} of them in each cubic centimeter. Pushed and pulled through wires, they transmit power and intelligent communication from place to place. In antennas, they create electromagnetic radiation. In the coils of motors, they interact with magnetic fields to do work. Released from the surface of hot metals, electrons can be guided through the space within vacuum tubes to perform tasks of amplification and control and computation. In semiconductors—materials whose electrons are movable but not free—they provide further units of control and of memory.

In modern research not only electrons but other charged particles as well produce fields and react to fields in ways that serve to reveal more about nature. In the following sections, we shall examine several applications of the laws of electromagnetism, some of practical importance, some of research interest.

17.1 Potential

Imagine an electron placed at the negative plate of a capacitor and then allowed to fly across the evacuated space between the plates to strike the positive plate. How much kinetic energy will it have when it reaches the positive plate? To answer the question, we must suppose that we know the magnitude of the electric field $\mathcal{E}$ existing between the plates (Figure 17.1). If the magnitude of the electron's charge is designated by e, the force that the electron experiences will be given by the product of charge and field,

$$F = e\mathcal{E}\,. \tag{17.1}$$

(This is an equation for the magnitude of **F**. The vector equation is $\mathbf{F} = -e\mathcal{E}$.) This electric force continues to act on the electron through a distance d, the separation of the plates. Since force multiplied by distance is work, the work done by the field on the electron will be

$$W = Fd = e\mathcal{E}d\,. \tag{17.2}$$

Now this work makes its appearance in the form of kinetic energy. The energy gained by the electron will be

$$K.E. = e\mathcal{E}d\,. \tag{17.3}$$

Conversely, if the electron at the positive plate were pushed "uphill" against the retarding electric force until it again reached the negative plate, the work input would be the same amount, $e\mathcal{E}d$. We conclude that, for the electron, there is a definite energy associated with the difference between the two plates.

If the electron in this example were replaced by another particle or object with a different charge, for example Q, the energy associated with its trip between the plates would be $Q\mathcal{E}d$ instead of $e\mathcal{E}d$. For every particle there is a characteristic energy proportional to its own charge. If the energy is divided by the charge, there results a quantity independent of charge. For the electron,

$$\frac{\text{Energy}}{\text{Charge}} = \frac{e\mathcal{E}d}{e} = \mathcal{E}d\,. \tag{17.4}$$

For the other object,

$$\frac{\text{Energy}}{\text{Charge}} = \frac{Q\mathcal{E}d}{Q} = \mathcal{E}d\,. \tag{17.5}$$

Energy divided by charge is more universal than energy or charge alone. It is a quantity associated with the capacitor itself, or to be more exact, with the space between the plates of the capacitor, not with any particular charge being considered. To this quantity, energy per unit charge, we give the name *potential.* Like work and energy and charge, it is a scalar quantity.

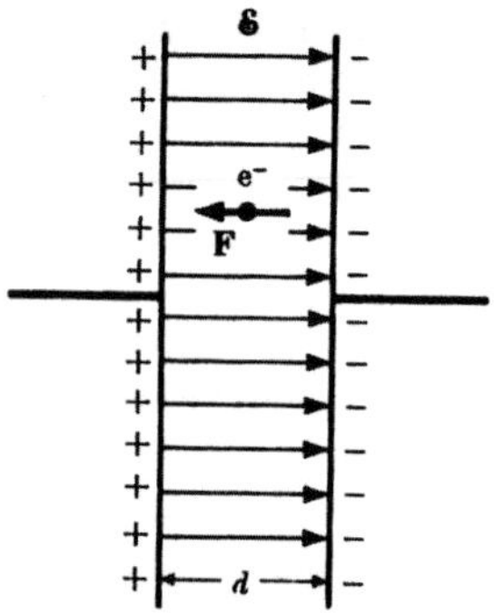

FIGURE 17.1 An electron being accelerated from one plate of a capacitor to the other. The energy it gains divided by its charge is equal to the potential difference between the plates.

Since potential is connected with the all-important idea of energy, it is not surprising that it is one of the central concepts of electromagnetism. We postponed its introduction to this chapter in part because it is not a necessary concept for understanding the meaning of the fundamental laws of electromagnetism, in part because it is of special value in considering a practical application of electromagnetism, electric circuits. Under the name "voltage," potential is actually a familiar idea in the everyday world. We know that auto batteries are 12 volts and that our 110–volt appliances will not function on Europe's 220 volts.

An analogy between potential and electric field is worth noting. Electric field was defined as force per unit charge (a vector quantity). Potential is defined as energy per unit charge (a scalar quantity). The simple act of dividing force by charge led to a powerful new concept because the field could be associated with a point in space whether or not any charged object was there to experience a force. Dividing energy by charge performs very much the same function. We may assign a potential difference to two points in space whether or not a particle actually moves between the two points to release or expend energy. It is important to notice that potential is defined for a pair of points, whereas a field can be defined at one point. Only potential *difference* has meaning.

Since potential is energy per unit charge, the unit of potential is the unit of energy divided by the unit of charge. In the cgs system, the name given to the unit of potential is the statvolt:

$$1 \text{ statvolt} = \frac{1 \text{ erg}}{1 \text{ e.s.u.}}\,. \tag{17.6}$$

This means that an energy of one erg is associated with the motion of a charge of one e.s.u. between two points whose potential difference is one statvolt. A more common unit of potential is the volt (V), defined as an energy of one joule (10^7 ergs) divided by a charge of one coulomb:

$$1 \text{ volt} = \frac{1 \text{ joule}}{1 \text{ coulomb}} . \tag{17.7}$$

The statvolt is a unit 300 times larger than the volt:

$$1 \text{ statvolt} = 300 \text{ V} . \tag{17.8}$$

Thus the common potential of 110 V is slightly more than one third of a statvolt.

In the example of the capacitor, the potential was measured between the two plates of the capacitor. Other examples are illustrated in Figure 17.2.

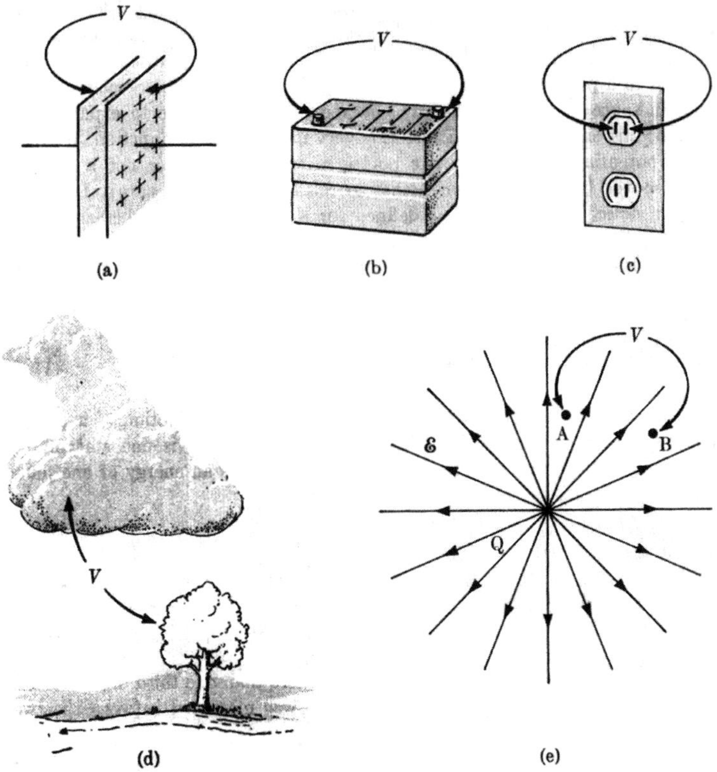

FIGURE 17.2 Examples of potential difference. **(a)** Between plates of charged capacitor. **(b)** Between terminals of storage battery. **(c)** Between slots of household outlet. **(d)** Between storm cloud and tree. **(e)** Between two points in space near a charged particle.

In a household electrical circuit, the potential alternates at 60 hertz. In a "110–volt" circuit, the potential actually swings back and forth between peak values of +156 volts and –156 volts (Figure 17.3). The root-mean-square potential* is 110 V, less than the peak potential by a factor of $\sqrt{2}$, or 1.414. An a.c. circuit is usually designated by its root-mean-square potential, because it is this potential that figures most simply in equations relating potential to current, resistance, and power. Unless otherwise stated, the root-mean-square value of alternating quantities may be used correctly in all equations in this chapter.

* The root-mean-square potential is calculated as follows: Square the potential. Find the average value of the squared potential. Take the square root of this average. For any sinusoidally varying quantity, the root-mean-square is 0.707 times the peak value.

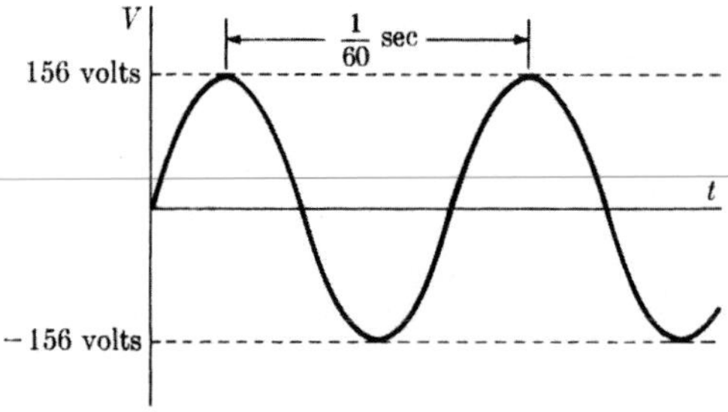

FIGURE 17.3 Potential in "110 volt" household a.c. circuit. The root-mean-square potential is less than the peak potential by a factor of $\sqrt{2}$, or 1.414.

The usual symbol for potential is V. Evidently from the definition of potential

$$E = VQ\,. \tag{17.9}$$

Energy is equal to the product of potential and charge. We leave purposely vague the nature of the energy. It may manifest itself in many ways—in the heat of an electric iron, the light and heat of an electric bulb, the mechanical work of a motor, the kinetic energy of an electron flying between two points in space. Or it may be external work performed against electrical force as in the example of an electron moved from the positive to the negative plate of a capacitor.

A simple modification of the equation above connecting energy and potential leads to a relation between power and current. Since power is defined as energy per unit time, and current is defined as charge per unit time, division of both sides of Equation 17.9 by time leads to the relation,

$$P = VI\,. \tag{17.10}$$

Power is equal to the product of potential and current. In the cgs system current is measured in e.s.u./sec and power in ergs/sec. In the mks system, the unit of current is the ampere (one coulomb/sec), and the unit of power is the watt (W) (one joule/sec). In terms of units, volts times amperes equals watts. Thus a current of 2 amp flowing in a circuit of 110 V expends a power of 220 W (220 joules of energy each second). A toaster of 550 W is so designed that when it is plugged into a 110-V circuit, 5 amp of current will flow through it, and the 550 W of power will be dissipated as heat.

The reason potential or voltage is a particularly important concept in the practical utilization of electricity is that very often the generators of electric current develop a particular fixed potential regardless of how much current, if any, flows. This applies both to chemical cells and to mechanical generators. Between the terminals of an automobile battery there might exist a potential difference of 12 V. This potential difference remains very nearly constant whether or not charge moves from one terminal to the other. Anything from zero to perhaps 25 amp can flow between the terminals, depending on what is connected between them, without much change in the potential. Likewise the generators supplying household circuits supply nearly the same alternating potential at all times. Current and power vary, depending on what is switched on or plugged in by the users.

The algebraic sign of potential has a meaning which, like the sign of electric charge and the direction of electric field, is determined by arbitrary convention. If a point A is at a higher potential than a point B, a positive charge moved from A to B ("downhill") will be capable of doing work or expending energy. A positive charge would need outside energy to be moved from B to A. The opposite conditions would prevail for negative charge. The central core of a flashlight cell, for example, is called its positive terminal, and its outer case is called its negative terminal. This means that the central core has the higher potential. The "natural" direction of flow of positive charge is from higher to lower potential, and of negative charge from lower to higher potential. In the simple circuit illustrated in Figure 17.4, the direction of conventional current flow is from positive terminal to negative. The actual direction of electron flow is from negative to positive.

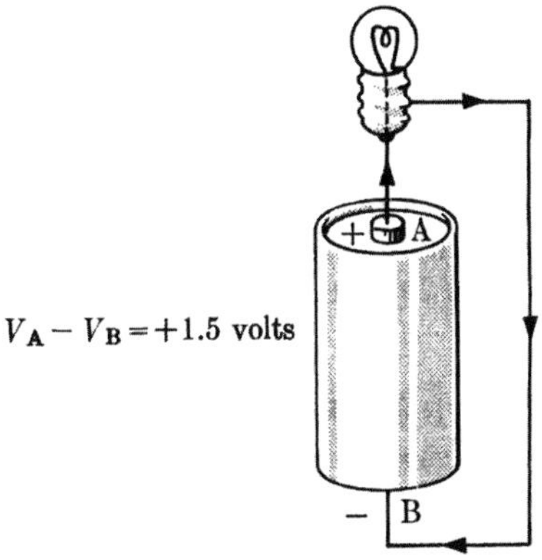

FIGURE 17.4 Significance of the sign of potential. Positive charge flows spontaneously from higher to lower potential. This is the same as the direction of conventional current flow.

A particularly simple and fundamental application of the potential concept is to the region of space surrounding a fixed point charge. Suppose for definiteness that a positively charged nucleus is located at a fixed point. (For this discussion, we may ignore the finite size of the nucleus.) Consider then the energy associated with the motion of another positive charge, for example a proton, in the vicinity of the nucleus. Since forces act, there is of course an energy associated with moving the proton from one point to another. If it is moved from a point 3 Angstrom units (A) from the nucleus to a point 2 A from the nucleus, a certain amount of work is done against the oppositely directed repulsive force. The potential difference between the two points is the work done divided by the proton charge, and the point nearer the nucleus is assigned the higher potential value. To move the proton an equal distance again from 2 A away to 1 A away requires more work, for the repulsive force is now stronger. Accordingly the potential difference is greater. On the other hand, the potential difference between two points 200 and 201 A away is far less, for much less energy is associated with that shift of position of the proton. Alternatively the potential may be investigated by considering the kinetic energy acquired by the proton if it is released near the nucleus and allowed to fly away. The gain in its kinetic energy between any two points divided by its charge is equal to the decrease of potential between those two points. Its final kinetic energy as it coasts away to an infinite distance measures the potential between its starting point and an infinitely distant point. It is convenient and usual to assign a zero value to the potential at infinity. Then any point closer to the nucleus has a higher potential. If the charge on the nucleus is Q, the potential defined in this way turns out to be

$$V = \frac{Q}{d}, \tag{17.11}$$

where d is the distance from the nucleus to any particular point in space. The similarity to gravitational potential energy (Equation 12.36) is evident. The analogy is even closer if we convert from potential to potential energy. This requires that the potential, Equation 17.11, be multiplied by the proton charge (Equation 17.9). Call the charge of the proton Q'. Then the potential energy associated with the proton-nucleus interaction is

$$P.E. = \frac{QQ'}{d}. \tag{17.12}$$

In the example being discussed, the potential and the potential energy are both positive quantities. Since charge may take either sign, so may the potential and the potential energy.

For oppositely charged particles, attracting one another, the right side of Equation 17.12 is negative, like the right side of Equation 12.36 for gravitational potential energy. Figure 17.5 shows a graph of potential vs. distance for a positively charged nucleus, and graphs of potential energy vs. distance for this nucleus interacting first with a proton, second with an electron.

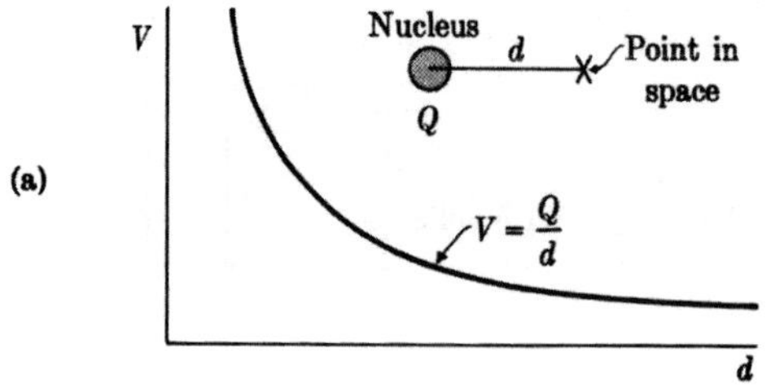

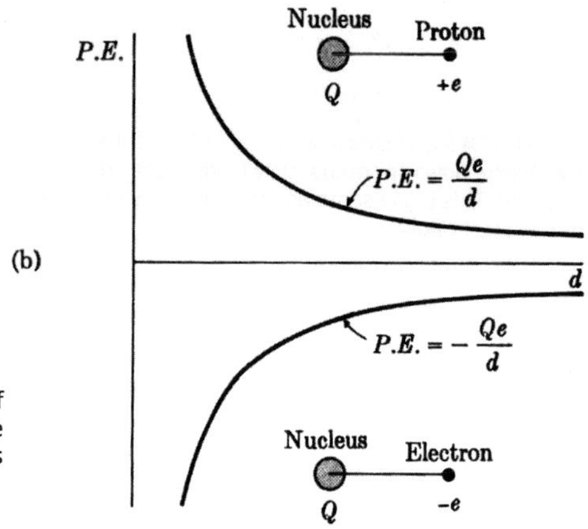

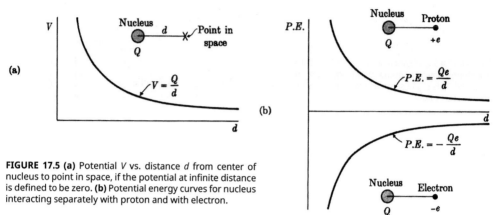

FIGURE 17.5 (a) Potential V vs. distance d from center of nucleus to point in space, if the potential at infinite distance is defined to be zero. **(b)** Potential energy curves for nucleus interacting separately with proton and with electron.

Equation 17.11 seems to specify the potential at a point, but in fact its real meaning is the *difference* in potential between that point and another infinitely remote point. Note that if d is set equal to infinity in this formula, $V = 0$ results. It is instructive to compare this formula with that giving the magnitude of the electric field surrounding the nucleus,

$$\mathcal{E} = \frac{Q}{d^2}. \tag{17.13}$$

The potential decreases as the inverse of the distance, the electric field as the inverse square of the distance. (Keep in mind too that one is a scalar quantity, the other a vector quantity.) For more complicated arrangements of charge, there need be no such simple relationship between potential and field.

The connection between energy and potential has led to the introduction of a new energy unit, the electron volt, now the most commonly used energy unit in the small-scale world. Despite the word "volt" appearing in its name, the electron volt, usually abbreviated eV, is a unit of energy, not of potential. It is the energy gained or lost by an electron (or any other particle with the same magnitude of charge) in moving between two points with a potential difference of one volt. The magnitude of the electron volt expressed in ergs or joules may easily be calculated from the fundamental relation between energy and potential,

$$E = QV.$$

Substituting on the right the charge of an elementary particle, $Q = 1.6 \times 10^{-19}$ coulomb, and the potential difference 1 V, we find the number of joules of energy in 1 eV:

$$1 \text{ eV} = 1.6 \times 10^{-19} \text{ joules}. \tag{17.14}$$

Evidently the eV is an exceedingly small amount of energy compared with the macroscopic unit, the joule. In the cgs system, the elementary unit of charge is 4.8×10^{-10} e.s.u., and a one-volt potential difference is 1/300 statvolt. The product of these two numbers is the same energy expressed in ergs:

$$1 \text{ eV} = 1.6 \times 10^{-12} \text{ ergs}. \tag{17.15}$$

This result could of course also have been obtained from the knowledge that one joule is equal to 10^7 ergs.

One thousand electron volts is usually abbreviated keV, one million electron volts, MeV, and one billion electron volts, GeV. Some typical energies of the submicroscopic world (in round numbers) are the following:

Translational kinetic energy per molecule at normal temperature	0.04	eV
Translational kinetic energy per molecule at 7,500°K	1	eV
Typical energies of chemical reaction	1-10	eV per atom
Typical energies of nuclear reaction	1-10	MeV per nucleus
Mass energy mc^2 of electron	500	keV
Mass energy mc^2 of proton.	1	GeV
Energy of electron in TV tube	1	keV
Energy of proton in large accelerator	30	GeV

17.2 Ohm's law: simple circuits

Prior to Volta's discovery of the chemical battery in 1800, very little had been learned about the flow of current through material substances. So long as charge could be caused to flow only in momentary pulses as electrified objects were discharged, the difficulties of measurement were too great to permit any accurate quantitative results. It was known only that different materials had vastly different conducting properties. Metals allowed charge to flow readily from point to point, and accordingly were called conductors. Through some other substances, called insulators, the flow of charge was so greatly inhibited that it was not possible to detect any current at all. Substances of intermediate conducting power, such as salt water, were also known.

Volta's battery made possible for the first time* the creation of steady currents of long duration, a technical advance of very great significance. Probably the most important consequence of the generation of steady currents, as emphasized in the last chapter, was that it opened the way to the discovery of the link between electricity and magnetism. Secondarily, it made possible a study of the nature of current flow itself, a necessary first step toward the practical utilization of electricity. During the first few decades of the nineteenth century came the few key discoveries that provided the foundation for the electrical technology that burgeoned half a century later.

From 1800 onward to the present time, the flow of current in circuits has been a subject of substantial research interest. In electrical terminology, a "circuit" is any continuous path or array of paths along which current may flow. A circuit usually contains a battery or other source of potential to create the current. In addition it may contain anything from a single wire to a complicated collection of wires, tubes, transistors, capacitors, and other "circuit elements." The path from one terminal of a flashlight cell through the lamp and back to the other terminal of the cell (Figure 17.4) is a simple circuit. A string of Christmas-tree lights plugged into a wall socket forms part of a circuit, or it may be considered a circuit by itself. The maze of parts within a television receiver are formed into a circuit.

We wish to consider here only simple circuits consisting of a battery whose terminals are joined by wires or other pieces of solid matter. Such circuits were the first to be studied. A remarkably simple law of their behavior was discovered by Georg Ohm in 1826. He found that the current flowing between any two points in a circuit (or in a single wire) is proportional to the potential difference between those two points. In symbols,

$$I \sim V, \tag{17.16}$$

* The single chemical cell, or Galvanic cell, although available before 1800, became a practical research tool only when combined by Volta into the multicell battery.

where I designates the current, and V the potential difference (Figure 17.6). If the potential is doubled, the current is doubled, and so on. A simple way to look at the law is to picture the potential as the "motive power" (the British word for potential, "tension," is suggestive), the current as the resulting effect. Doubling the potential across a circuit element causes a doubling of the amount of charge that flows through it in one second. This assignment of cause and effect of course is only an aid to visualization, and has no deep meaning.

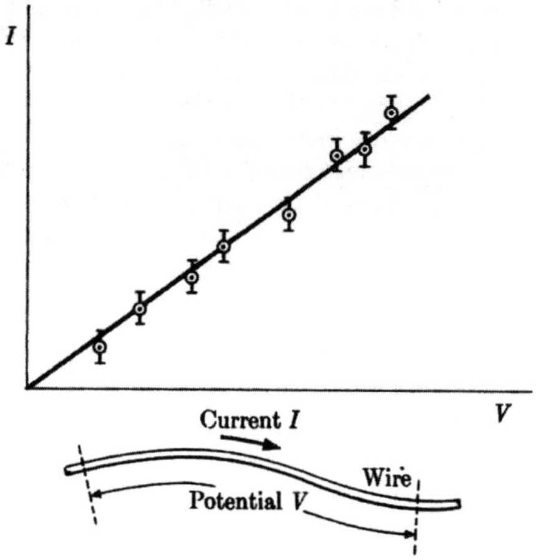

FIGURE 17.6 Ohm's law. If the current I flowing through a wire or other circuit element is directly proportional to the potential difference V across the circuit element, the material obeys Ohm's law. Its resistance is the inverse of the slope of the graph of I vs. V.

To transform the proportionality of Formula 17.16 into an equation, a constant of proportionality must be introduced. This constant is written $1/R$, and R is called the resistance of the circuit or of the piece of material being considered. Ohm's law then takes the form,

$$I = \frac{1}{R}V. \tag{17.17}$$

Alternatively, the same equation may be re-expressed in the form,

$$V = IR. \tag{17.18}$$

This is the form of Ohm's law most frequently encountered. In words: The potential difference between two points in a wire or in a circuit is equal to the current flowing between the two points multiplied by the resistance of whatever lies between the two points. The new quantity that has been introduced, resistance, has the dimensions of potential divided by current. In the mks system of units, the only set we shall consider in dealing with Ohm's law, the unit of resistance is volts/amperes. This particular combination has been given a new name, the ohm.

$$1\text{ ohm} = \frac{1\text{ volt}}{1\text{ ampere}}. \tag{17.19}$$

In a single equation, three giants of the early nineteenth century are immortalized with small letters.

It must be emphasized that the new quantity called resistance is a constant only in a certain sense. For any particular circuit element (at least any one that obeys Ohm's law), the ratio of potential to current is a constant, independent of the potential. But each different wire or circuit element may have a different resistance. The resistance of a particular wire is a constant for that wire, but need have nothing to do with the resistance of anything else. The significance of resistance and the reason for its name can be appreciated by examining Ohm's law. For a given potential difference, the product of current times resistance is constant. The greater the resistance, the less the current. Resistance is exactly that, a measure of the extent to which the given circuit element impedes the flow of current. Alternatively, one may say that if one wire has more resistance than another, more potential is required to force a given current through the wire of greater resistance. A short length of ordinary wire has a resistance of less than one ohm. A typical light bulb has a resistance of about 100 ohms, an iron or electric toaster a resistance of 15 or 20 ohms. Inside radio and television receivers are used many circuit elements called resistors whose resistance varies from a few ohms to millions of ohms.

Perhaps the most important fact to realize about Ohm's law is that it is not a fundamental law at all. For that reason it makes its first appearance in this chapter on applications of electromagnetism. Coulomb's law of force between charged particles, to choose an example that *is* regarded as fundamental, is precisely correct, so far as we know, for all charges at all separations. Ohm's law, on the other hand, is not precisely true for any circuit element, and for some it is very far from the truth. Nevertheless it is quite a good approximation to the truth for almost all solid matter (therefore in particular for almost all wires), provided the matter is not subjected to extremes of temperature. At very low temperatures, some metals become superconductors; their resistance vanishes and current can flow with zero potential difference. Many materials also show significant change of resistance when they become very hot. For certain special circuit elements such as vacuum tubes, the ratio of current to potential difference is far from constant, and Ohm's law is not applicable at all. However, it is an approximate law of considerable utility for a wide range of substances and a wide range of temperatures. A circuit element that obeys Ohm's law is called a linear resistor, because of the appearance of the graph in Figure 17.6. A nonlinear resistor is one that does not obey Ohm's law. In the remainder of this section we shall deal only with linear resistors.

$$V_1 = I_1R_1,$$

$$V_2 = I_2R_2.$$

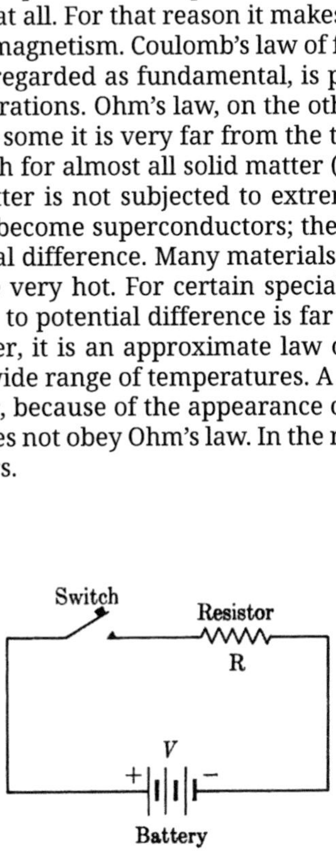

FIGURE 17.7 Simple circuit.

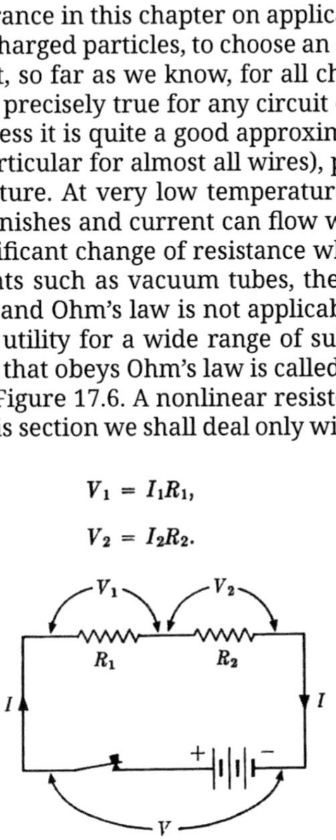

FIGURE 17.8 Series circuit.

If the terminals of a battery of potential V are connected through a resistor of resistance R, Ohm's law tells us that the current flow will be given by the formula

$$I = \frac{V}{R}.$$

This simple circuit is diagrammed in Figure 17.7; this figure also illustrates the usual symbols that are employed to indicate a battery, a resistor, and a switch in circuit diagrams.

The simplest series circuit consists of two resistors, the first with resistance R_1, the second with resistance R_2, joined together in series and then connected to a battery (Figure 17.8). We can apply Ohm's law to each of the resistors separately:

$$V_1 = I_1R_1, \tag{17.20}$$

$$V_2 = I_2 R_2 . \tag{17.21}$$

The potential difference across each resistor is equal to the product of the current through it and its resistance. Now a little fundamental thinking about the nature of current and potential reveals two useful facts about the series circuit. First, the currents through the two resistors are equal,

$$I_1 = I_2 . \tag{17.22}$$

Second, the potential difference across the pair of resistors is equal to the sum of the potential differences across each one,

$$V = V_1 + V_2 . \tag{17.23}$$

(The symbol V denotes the applied potential of the battery.) Consider a point between the two resistors. The charge arriving there each second is equal to the current I_1 through resistor 1. The charge leaving the same point each second is equal to the current I_2 through resistor 2. Since we postulate a steady current flow with no accumulation or depletion of charge at this point, the two currents must be equal. This is quite analogous to saying that for a river in a state of steady flow whose level is neither rising nor falling, the amount of water passing under one bridge in a second must be the same as the amount of water passing under another bridge in a second. From the most fundamental point of view, Equation 17.22 is a manifestation of the law of charge conservation.

The second fact to be proved concerns the sum of the potentials. Here we must think of the meaning of potential. It is energy per unit charge. An electron in passing through the first resistor expends an energy eV_1, where e is the magnitude of the electron's charge. The same electron in passing through the second resistor expends an energy eV_2. The total energy it has expended is the sum of these two energies:

$$\text{Total energy} = eV_1 + eV_2 = e(V_1 + V_2). \tag{17.24}$$

Since energy divided by charge is potential, the total potential difference is $V_1 + V_2$, as stated above in Equation 17.23. From the most fundamental point of view, this summability of potential is a manifestation of the law of energy conservation. The energy supplied to the electron by the battery (eV) is equal to the total energy dissipated by the electron in the resistors.

Now we may assemble these facts about the simple series circuit. Since $I_1 = I_2$ (equal currents), we may drop the subscripts and call the current simply I. Then the two potential differences are given by

$$V_1 = IR_1 ,$$

$$V_2 = IR_2 .$$

The sum of these is the total potential:

$$V = V_1 + V_2 = IR_1 + IR_2 . \tag{17.25}$$

This equation for the total potential difference may also be written,

$$V = I(R_1 + R_2) . \tag{17.26}$$

Now we have Ohm's law for the circuit as a whole. The total resistance of the series circuit is given by

$$R = R_1 + R_2 . \tag{17.27}$$

This simple result states that two resistors in series behave like a single resistor with a resistance equal to the sum of the two resistances. One might almost have guessed this result, but its derivation has required the use of some algebra and a knowledge of the exact meaning of current and potential. For three resistors in series, the total resistance is

$$R = R_1 + R_2 + R_3\,, \tag{17.28}$$

and similar addition applies to any number of resistors in series. An example of a series circuit is the "old-fashioned" string of Christmas-tree lights, in which the failure of a single bulb causes all the lights to go out.

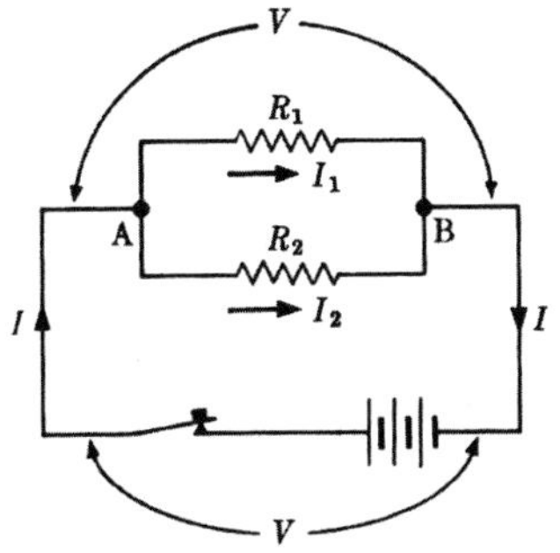

FIGURE 17.9 Parallel circuit.

The rule for combining resistors in parallel is a bit more complicated, but it can be derived with exactly the same kind of reasoning. Figure 17.9 illustrates the simplest example of a parallel circuit, in which two resistors are joined in parallel, their ends being connected to a battery. Again Ohm's law may be applied to each circuit element separately:

$$V_1 = I_1 R_1\,, \tag{17.29}$$

$$V_2 = I_2 R_2\,. \tag{17.30}$$

This time, however, it is the potential differences and not the currents that are equal. The two ends of the first resistor are at points A and B. The two ends of the second resistor are at the same points A and B. Since there is some definite potential difference between these points A and B (recall that potential always refers to a difference between two points), it is the same potential across both resistors. In this example, this is the same as the potential of the battery. In symbols,

$$V = V_1 = V_2\,. \tag{17.31}$$

The currents through the two resistors need not be equal. The first is given by

$$I_1 = \frac{V}{R_1}\,, \tag{17.32}$$

the second by

$$I_2 = \frac{V}{R_2}\,, \tag{17.33}$$

The total flow of current from the battery is divided, some going through the first resistor, some through the second, so that the sum of the two currents is equal to the total current (conservation of charge),

$$I = I_1 + I_2\,. \tag{17.34}$$

Equations 17.32 and 17.33 may be substituted into this equation to give

$$I = \frac{V}{R_1} + \frac{V}{R_2},$$

which in turn may be rewritten as

$$I = V\left(\frac{1}{R_1} + \frac{1}{R_2}\right), \tag{17.35}$$

For the parallel circuit as a whole we expect Ohm's law to hold true:

$$I = V\frac{1}{R}. \tag{17.36}$$

Comparison of the last two equations shows that the total resistance of the parallel circuit (most conveniently written as an inverse) is given by

$$\frac{1}{R} = \frac{1}{R_1} + \frac{1}{R_2}. \tag{17.37}$$

Notice that this relationship implies that R is less than either R_1 or R_2. This is not surprising. The presence of the second resistor provides an alternate path for current flow; the current is less impeded by the pair of parallel resistors than it would be by either alone. For example, two resistors, each of 2 ohms, in parallel have a total resistance of 1 ohm. For four resistors in parallel, the inverse of the total resistance is given by

$$\frac{1}{R} = \frac{1}{R_1} + \frac{1}{R_2} + \frac{1}{R_3} + \frac{1}{R_4}, \tag{17.38}$$

and a similar rule holds for any number of parallel resistors. Ordinary household circuits are always parallel circuits. All of the lamps, motors, and heaters plugged into sockets in a house are in parallel with each other. The potential across each lamp or appliance is then always the same, and the current flowing through it does not depend on what else is connected. As more things are turned on or connected, the total resistance of the entire household circuit decreases and the total current flow increases, the potential remaining unchanged.

An alternative expression of Ohm's law is somewhat more fundamental than Equation 17.16 or Equation 17.18. In a given kind of material, the current flow at one place is proportional to the electric field at that place. In symbols,

$$I \sim \mathcal{E}. \tag{17.39}$$

This formulation of the law refers only to a single location, not to a pair of points. It will be left as an exercise to show that for a wire the proportionality of current to electric field is equivalent to the proportionality of current to potential difference.

We conclude this section with some statements about the connection of resistance to energy expenditure. According to Equation 17.10, power (energy per unit time) is given by the product of potential difference and current:

$$P = VI.$$

This is a precise equation, for it depends simply on the definition of potential.* Since, according to Ohm's law, $V = IR$, power may also be expressed as

---

* For a.c. circuits, the validity of Equation 17.10 requires that potential and current must alternate in phase (that is, both must reach peak values at the same instant of time), and that the root-mean-square values of both V and I be used. However, it does not require the validity of Ohm's law or any other definite connection between V and I.

$$P = I^2R\,. \tag{17.40}$$

Alternatively, writing $I = V/R$, power may be expressed as

$$P = \frac{V^2}{R}\,. \tag{17.41}$$

In these equations, if the units of V are volts, the units of I are amperes, the units of R are ohms, and the units of P are watts (joules/sec). In a simple resistor, this power expenditure appears as heat. In lamps, it appears partly as light.

We must reemphasize that the concept of resistance is not a fundamental concept. Each equation in this section that includes resistance is an approximate equation, very nearly true for certain materials in certain temperature ranges, but not a statement of a fundamental law of nature. Nevertheless, the concept of resistance and Ohm's law have proved to be exceedingly fruitful for circuit analysis and practical applications of electric currents.

17.3 *Electric acceleration of charged particles*

The law of electric force,

$$F = Q'\mathcal{E}\,, \tag{17.42}$$

finds applications in solids, liquids, gases, and in empty space. Inside a metal wire, for instance, exists an electric field which exerts forces on the electrons in the wire. (It also exerts forces on the positively charged nuclei, but these are unable to move.) The electrons are set into motion along the wire. Their frequent encounters with crystal imperfections and with individual ions in the wire subject them to a retarding force in a direction opposite to the force of the electric field. This is the origin of resistance. When these two forces come into balance the electrons are, on the average, subject to no force. They are unaccelerated and move along the wire with a constant velocity, on the average. This average velocity along the wire is usually called the drift velocity, for it is quite small compared with the actual speed of the electrons. Each electron darts this way and that within the wire at a speed of almost 10^8 cm/sec. In addition, if subjected to an electric field (or equivalently, to a potential difference between the ends of the wire), it drifts along the wire with a drift velocity of perhaps 1 cm/sec or less. Only the net drifting motion contributes to the current flow, to the transmission of power, and to the creation of magnetic fields.

In a conducting liquid or gas, a similar situation prevails, except that positive and negative ions as well as electrons can drift through the material and contribute to the current flow. The phenomenon called breakdown can occur in gases, resulting in the surges of current we call sparks. In a normal gas, there are very few free electrons and very few ions. If the electric field becomes sufficiently intense, a chain reaction can occur which releases temporarily a vast number of free electrons. Between two successive collisions, an electron can be accelerated by the electric field to such a high speed that it has sufficient energy to ionize the next atom it hits and detach an electron from it. Each of these electrons in turn is accelerated and can make another ionizing collision, releasing more electrons. Soon there is a flood of electrons pouring through the gas. The flood subsides when the electric field strength diminishes below the critical value. Then ions recapture electrons and the gas returns to its original almost neutral state. A spark can be maintained indefinitely if the intense electric field is maintained, but normally the surge of current tends to neutralize the source of the electric field, so that the spark is quenched after a short time. So far nature has managed to produce, in lightning bolts, more intense and more awesome sparks by far than any produced by man.

The law of electric force finds its simplest application in devices that accelerate electrons through empty space. Within a cathode-ray tube (or television tube), electrons boiling off the hot cathode are subjected first to an electric field that accelerates them a short distance down the neck of the tube (see Figure 16.8). They then fly through a hole in the anode into a region nearly free of electric field and complete their journey to the tube face at nearly constant speed. During their acceleration from cathode to anode, a force, $e\mathcal{E}$, acts over a distance d, to contribute to the electrons a kinetic energy equal to the product of force times distance:

$$K.E. = e\mathcal{E}d\,. \tag{17.43}$$

Since the potential difference V between cathode and anode is equal to the product of electric field times distance, the energy may also be written,

$$K.E. = eV\,. \tag{17.44}$$

For a typical cathode-ray tube, $V = 1{,}000$ V, and $K.E. = 1{,}000$ eV.

The simplest vacuum tube, the diode, contains only a cathode and an anode separated by empty space. The next most complicated vacuum tube, the triode, contains three elements, a cathode, an anode, and between them a screen called a grid. Figure 17.10 shows the actual typical construction of diodes and triodes together with the schematic representations of diode and triode used in circuit diagrams.

In a diode the hot cathode boils off electrons into the space around it. If the anode is positively charged, these electrons are accelerated to the anode and current flows through the tube. If the anode is negatively charged, the electric field forces the electrons back toward the cathode and no current flows. In Great Britain a vacuum tube is more descriptively called a valve, because it can be used to turn on or shut off the flow of current. One common use of the diode is to "rectify" alternating current, that is, to convert alternating current to direct current. The anode is alternately charged positively and negatively with respect to the cathode. Half the time (when the anode is negative) no current flows. Half the time (when the anode is positive) current does flow through the tube. Although the resulting direct current is not steady, it flows in only one direction.

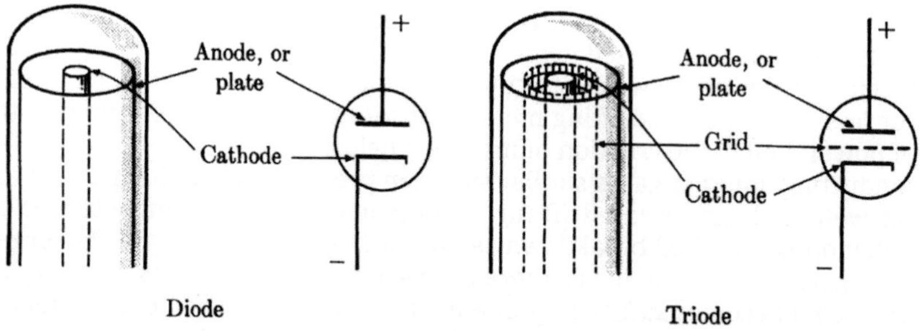

FIGURE 17.10 The diode and the triode, simple vacuum tubes used to control current flow. Not shown in either the pictorial or schematic representations of the tubes are the filament wires used to heat the cathode.

The triode finds its chief applications in control and amplification. Typically the grid of the triode is maintained at a small negative potential with respect to the cathode, while the anode potential is large and positive with respect to the cathode. This means that electrons between cathode and grid feel a weak force repelling them back toward the cathode. Any electron beyond the grid, on the other hand, feels a strong force attracting it to the anode. Most of the electrons boiling from the cathode are prevented by the repulsive electric force from reaching the grid. A few emerge from the cathode with sufficient energy to overcome the repulsive force and make their way past the grid, where they are drawn on to the anode.

Now if the grid potential is made slightly more negative, far fewer electrons will be able to penetrate past the grid, and the current flow through the tube will be greatly reduced. If the grid potential is made slightly less negative, more electrons will surmount the repulsive force between cathode and grid, and the current through the tube will increase. This situation can be pictorialized by a potential diagram like that in Figure 17.11. The grid represents a small hill between cathode and anode. A small change in the height of this hill can result in a large change in the flow of electrons reaching the anode. The grid might be compared to the accelerator pedal in an automobile. A small force applied to the pedal results in a much larger force delivered to the wheels of the automobile. Similarly, a small charge delivered to the grid of a triode results in a far larger change in the charge flowing to the anode.

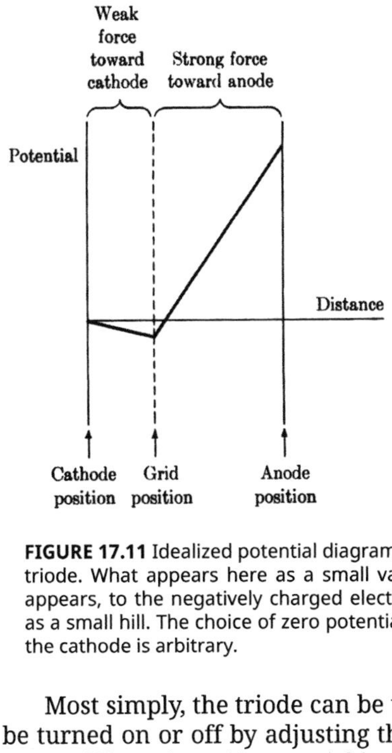

FIGURE 17.11 Idealized potential diagram for triode. What appears here as a small valley appears, to the negatively charged electron, as a small hill. The choice of zero potential at the cathode is arbitrary.

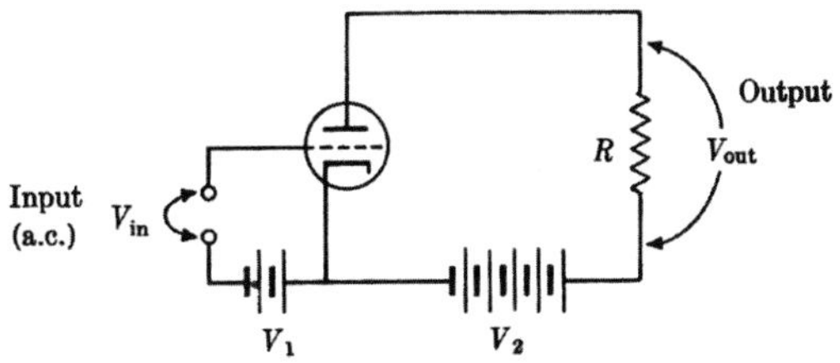

FIGURE 17.12 Simple amplifier. Relative to cathode, the grid of a triode is held at a slightly negative potential V_1, and the anode is held at a high positive potential by a battery of potential V_2. Additional small alternations of potential V_{in} at the grid control the current flow and produce larger but proportional changes in the potential V_{out} across the resistor. (The circuit to heat the cathode is not shown.)

Most simply, the triode can be used as a switch. A flow of current through the tube can be turned on or off by adjusting the potential on the grid. More often the triode serves as an amplifier. The grid potential is varied up and down about some prechosen neutral value that permits a moderate flow of current through the tube. As the grid potential increases, the tube current increases; as the potential decreases, the tube current decreases. To each alternation of the grid potential corresponds an alternation of the current through the tube. This current in turn may be allowed to flow through a resistor attached in series to the tube (Figure 17.12). Because of Ohm's law, the potential across this resistor, given by

$$V = IR\,,$$

will be proportional to the current through the tube (the same as the current through the resistor). Hence every change of grid potential will be reflected in an exactly corresponding, but usually much larger, change of potential across the series resistor. The triode has acted to amplify potential. One or more triodes form the heart of many phonograph amplifiers. In such an amplifier, the potential fed to the grid is rapidly oscillating in synchronism with the vibrations of sound that were recorded. If the amplified potential is precisely proportional to the input potential, the amplifier is said to be linear. A high-fidelity amplifier is one that is very nearly linear.

More and more in modern circuits, vacuum tubes are being supplanted by transistors. However, the path from fundamental laws to practical applications is neither so short nor so simple for transistors as it is for vacuum tubes. Accordingly, despite their growing practical importance, we shall not describe the operation of transistors.

As early as 1897, J. J. Thomson constructed a cathode-ray tube that provided electric acceleration of electrons perpendicular to the axis of the tube as well as parallel to the axis. His method is still in use in oscilloscopes, devices that use cathode-ray tubes to display graphically a potential that varies rapidly with time. The bare essentials of a tube arranged to give transverse acceleration are shown in Figure 17.13. An electron reaches the deflection plates with horizontal component of velocity v_x, and then spends between the plates a time t given by

$$t = \frac{l}{v_x}, \tag{17.45}$$

where I is the length of the deflection plates. During this time, the electron experiences a vertical force (downward in the illustrated example) given by the product of its charge and the electric field strength:

$$F_y = e\mathscr{E}. \tag{17.46}$$

Since the electric field is equal to the potential between the plates V_d divided by their separation d, this force can also be written

$$F_y = \mathrm{e}\frac{V_d}{d}. \tag{17.47}$$

Newton's second law relates this force to the vertical component of acceleration:

$$F_y = ma_y. \tag{17.48}$$

The downward acceleration can therefore be expressed in terms of properties of the electron (m and e) and properties of the deflection plates (d and V_d):

$$a_y = \frac{F_y}{m} = \frac{e}{m} \times \frac{v_d}{d}. \tag{17.49}$$

Since this acceleration is constant during the time t, the electron acquires a vertical component of velocity given by (recall Equation 8.8):

$$v_y = a_y t. \tag{17.50}$$

Substitution from Equations 17.45 and 17.49 into 17.50 yields

$$v_y = \frac{e}{m}\frac{v_d}{d} \times \frac{l}{v_x}. \tag{17.51}$$

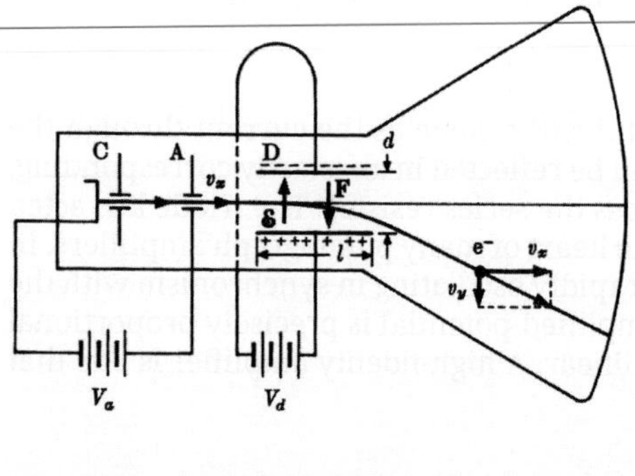

FIGURE 17.13 Cathode-ray tube with internal plates to provide electric deflection (idealized and not to scale). Electrons are accelerated from cathode C to anode A by an accelerating potential V_a, and acquire a horizontal component of velocity v_x. Passing between deflection plates D, the electrons are accelerated vertically by a deflection potential V_d, and acquire a vertical component of velocity V_y. A second pair of plates to provide deflection left and right on the screen is not shown.

The number of mathematical steps required to get this result should not obscure the basic simplicity of the process. Between the deflection plates the electron moves in a uniform field of force. Its horizontal component of velocity remains constant; its vertical component increases in proportion to time. Its motion is equivalent to that of a baseball thrown horizontally. Electron and baseball follow parabolic paths. The physical magnitudes associated with these two examples of motion are of course vastly different. The electron enters the deflection region with a speed of perhaps 2×10^9 cm/sec. For definiteness, let us suppose that the length of the plates is $l = 3$ cm, their separation is $d = 1$ cm, and their potential difference is $V_a = 100$ volts = 0.33 statvolt. The required electron properties are $e = 4.80 \times 10^{-10}$ e.s.u. and $m = 9.1 \times 10^{-28}$ gm. According to Equation 17.49, the electron experiences an acceleration

$$a_y = \frac{4.8 \times 10^{-10}\text{ e.s.u.}}{9.1 \times 10^{-28}\text{ gm}} \times \frac{0.33\text{ statvolt}}{1\text{ cm}} = 1.8 \times 10^{17}\text{ cm/sec}^2\,, \tag{17.52}$$

or about $1.8 \times 10^{14} g$. This acceleration persists for a time given by Equation 17.45:

$$t = \frac{3\text{ cm}}{2 \times 10^9\text{ cm/sec}} = 1.5 \times 10^{-9}\text{ sec}\,. \tag{17.53}$$

In this time, the vertical component of velocity increases to a value given by Equation 17.50 or 17.51:

$$v_y = 1.8 \times 10^{17}\frac{\text{cm}}{\text{sec}^2} \times 1.5 \times 10^{-9}\text{ sec} = 2.7 \times 10^8\frac{\text{cm}}{\text{sec}}\,. \tag{17.54}$$

One must marvel that laws of force and laws of motion discovered in the sluggish macroscopic world remain valid in the frenetic domain of electron motion (provided, as we now know, that the electron speed is not too close to the speed of light).

Equation 17.51 shows that the electron gains a vertical component of velocity proportional to the deflection potential V_d. This means in turn that the displacement of the beam spot on the tube face is proportional to V_d. In effect this cathode-ray tube is a pictorial voltmeter. If another pair of plates swings the beam from left to right while an unknown potential is applied to the vertical deflection plates, the beam traces out on the tube face a graph of the unknown potential (Figure 17.14). This is the principle of the oscilloscope.

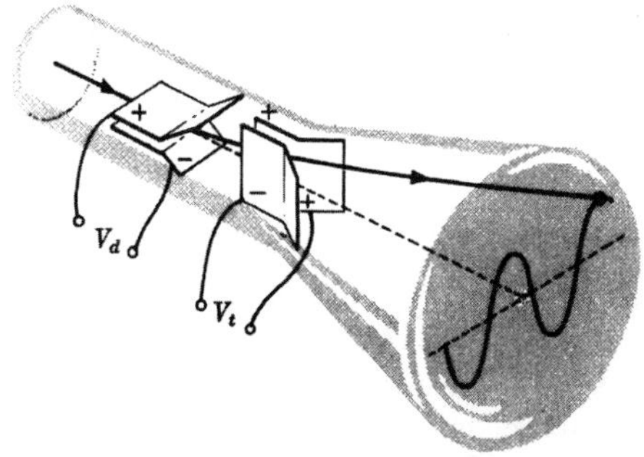

FIGURE 17.14 The principle of the oscilloscope. While a "time-base" potential V_t is swinging the electron beam from left to right at a steady rate, the potential to be studied, V_d, is swinging the beam up and down. As a result, the moving spot of the beam on the tube face traces out a graph of V_d vs. time.

17.4 *Magnetic deflection of charged particles*

Properties of the force experienced by a charged particle moving in a magnetic field were studied in the preceding chapter. The formula for the force is

$$F = \frac{v_\perp}{c} Q'\mathscr{B}\,, \tag{17.55}$$

where Q' is the charge, $\mathcal{B}$ is the magnetic field, $v_\perp$ is the component of the particle velocity perpendicular to the magnetic field, and c is the speed of light. The direction of the force is determined by right-and-left rule 1. A particular consequence of this force law is that a charged particle moving in a uniform magnetic field follows a helical path, circular motion about the lines of field together with uniform drifting motion along the lines of field. The drifting velocity can, as a special case, be zero. Then the particle executes simple circular motion with a radius of curvature given by Equation 16.5:

$$r = \frac{pc}{Q'\mathcal{B}}.$$

In this section we shall examine how magnetic deflection of charged particles can be used to determine the ratio of their charge to their mass, and how magnetic deflection plays a role in cyclotrons and synchrotrons.

In a cathode-ray tube, electrons are first accelerated by an electric field through a known potential difference. Each electron thereby acquires an energy equal to the product of its charge e and the potential difference V (Equation 17.44):

$$\tfrac{1}{2}mv^2 = eV. \tag{17.56}$$

J. J. Thomson, when he studied cathode rays in 1896, was aware of this equation, but he knew neither the mass of the electron, m, nor its charge, e. In order to learn something about these quantities, he subjected the cathode-ray beam to a known magnetic field $\mathcal{B}$ and measured the radius of curvature of the beam, which was deflected into a circular arc. He knew also that this radius of curvature was given by the equation

$$r = \frac{mvc}{e\mathcal{B}}. \tag{17.57}$$

(Here Q' has been replaced by e, and p by mv.) Now by some simple algebra, the unknown velocity v can be eliminated from Equations 17.56 and 17.57. (This will be left as an exercise.) Once v is eliminated, there results the following equation for the ratio e/m:

$$\frac{e}{m} = \frac{2Vc^2}{\mathcal{B}^2 r^2}. \tag{17.58}$$

This equation, first used by Thomson to determine the ratio of charge to mass of the electron, clearly separates the knowns from the unknowns. On the left appear the two unknown properties of the particle. On the right appear the knowns: the potential difference used to accelerate the electrons, the speed of light, the magnetic field used to deflect the electrons, and the measured radius of curvature of the deflected beam. If the units on the right are statvolts for V, cm/sec for c, gauss for $\mathcal{B}$, and cm for r, the charge to mass ratio on the left is expressed in e.s.u./gm.

Notice that the Thomson experiment reveals only the ratio of two unknowns, not either one alone. This ratio Thomson measured to lie between 3×10^{17} and 9×10^{17} e.s.u./gm for electrons,* some thousands of times larger than the ratios of charge to mass then known for some ions. Thomson correctly guessed that his measured ratio of e/m was large not because the electron had an unusually large charge, but because it had an unusually small mass. (This conclusion was actually based on more than guesswork. It was supported by evidence on the manner in which high-speed electrons penetrate through gases.) Subsequently Thomson's technique was greatly improved in accuracy and also extended to the measurement of the charge to mass ratio of other particles—protons and ions. In 1913 Robert A. Millikan was able to measure the charge of the electron alone rather accurately (within

* We now know the correct value to be 5.2728×10^{17} e.s.u./gm.

1%). His measurement combined with the e/m determination then led to an accurate value for the mass of the electron. Also, at about the same time, Bohr's theory of atomic structure made it clear that all ions possess a charge equal in magnitude to that of the electron or an integral multiple of it. Therefore the masses of all atoms could be determined from e/m measurements. The modern technique of precision mass measurements, called mass spectroscopy, stems directly from Thomson's original cathode-ray experiment. It led, in 1913, to the first observation of isotopes, atoms of like nuclear charge and like chemical properties, but different mass. All of these developments, it must be remembered, rest upon direct application of the laws of electric and magnetic force.

Most modern particle accelerators make use of magnetic deflection to keep the particles trapped within a fairly small domain of space while they are being accelerated.* In the 33-GeV Brookhaven accelerator, for instance (Figure 2.9), each proton travels about 150,000 miles during its acceleration cycle, but because of magnetic deflection it is held within a circle 840 feet in diameter while making this trip. The first accelerator designed to hold particles in circular orbits while they were gaining energy was the cyclotron, invented in 1931 by Ernest O. Lawrence.

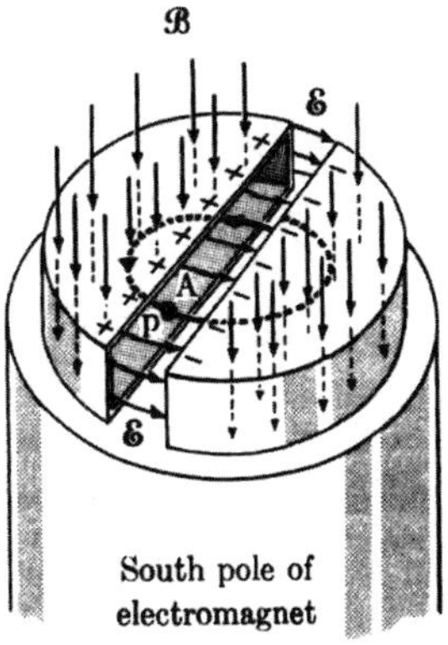

FIGURE 17.15 Early version of cyclotron. The nearly uniform magnetic field points vertically downward. (The upper pole face of the electromagnet is not shown.) At the instant shown, the proton at point A is being electrically accelerated to the right, so that its half-circle in the right dee will have a slightly larger radius than its previous half-circle in the left dee.

In its original form, the cyclotron contained two hollow metal containers placed between the pole faces of a large and powerful electromagnet (Figure 17.15). The metal containers can be thought of as the slightly separated halves of a large pillbox sliced through the middle. Because of their D shape, the containers are called the "dees" of the cyclotron. Each dee is electrically charged, one positively, one negatively. In the gap between the dees there is therefore an electric field, although there is very little electric field within each dee. The lines of magnetic field are vertical and nearly uniform, passing through the dees as well as through the gap. As the acceleration cycle proceeds, the magnetic field remains steady and unchanging, while the electric field is periodically reversed, pointing first from left to right, then from right to left, with several million reversals per second. Projectile particles, for example protons, are released near the center of the pair of dees and spiral outward in circles of gradually increasing radius until they are near the outer periphery of the dees. There they may strike a target or be deflected through a thin window to form an energetic external beam. The dees themselves are enclosed within an evacuated chamber.

The operation of the cyclotron can best be understood by considering some intermediate time during the acceleration cycle. The magnetic field causes the proton to execute a circle with radius given by Equation 17.57. Twice in each revolution the proton crosses the gap between the dees. As it crosses from left to right, the left dee is positively charged, the

* Some accelerators, called linear accelerators (see Figure 2.3) do not use magnetic deflection.

right dee negatively charged, so that the electric field in the gap points from left to right. The proton feels a momentary force in the direction of its motion. It gains energy and it gains speed. During the time it executes its half circle in the right dee (with a slightly larger radius than in the preceding half circle), the charges on the dees are reversed. Therefore at its next passage across the gap, this time from right to left, the proton again feels an electric accelerating force, and gains another increment of energy, of speed, of momentum, and of radius of curvature.

One key to the success of the cyclotron is the stability of the particle orbits. The protons or other particles being accelerated must not drift off course and strike the inner walls of the dees. This stability is achieved by letting the magnetic field be very slightly weaker near the outer edge than at the center. It is left as an exercise to show how the weakening of the field may be related to orbit stability.

A second key to the success of the cyclotron is the interesting fact that the time required for a proton (for example) to execute each half circle within a dee is a constant, the same for low energy and small radius as for high energy and large radius. The time for a half circle is half the circumference, πr, divided by the speed of the particle.

$$t = \frac{\pi r}{v}. \tag{17.59}$$

If Equation 17.57 is divided on the left and right sides by v, it gives

$$\frac{r}{v} = \frac{mc}{e\mathscr{B}}.$$

This value of r/v may be substituted into Equation 17.59 to give

$$t = \frac{\pi mc}{e\mathscr{B}}. \tag{17.60}$$

The time to execute a half circle depends on the mass of the particle, the speed of light, and the charge of the particle, which are constants, and on the magnetic field strength, which is very nearly a constant. Therefore the time to complete each half circle is nearly a constant. This was important in early cyclotrons. Because the rate of alternation of charge on the dees is so high, it is easiest to keep the pulses of electric acceleration synchronized with the passage of the particles across the gaps if these recur at equally spaced time intervals.

In order to obtain typical magnitudes of time and of size for a cyclotron, let us suppose that the projectile particles are protons, whose mass and charge are given by

$$m = 1.7 \times 10^{-24}\ \text{gm},$$

$$e = 4.8 \times 10^{-10}\ \text{e.s.u.};$$

and that the magnetic field strength is

$$\mathscr{B} = 10{,}000\ \text{gauss}.$$

In addition, for substitution into Equation 17.60 we need only the value of π and $c = 3 \times 10^{10}$ cm/sec. Evaluating Equation 17.60 numerically, we obtain a value for the time required for the proton to execute one half circle of its orbit:

$$t = \frac{\pi mc}{e\mathscr{B}} = \frac{(3.14)(1.7 \times 10^{-24})(3 \times 10^{10})}{(4.8 \times 10^{-10})(10^4)} = 3.3 \times 10^{-8}\ \text{sec}, \tag{17.61}$$

less than one ten-millionth of a second. In order to find the required size of the cyclotron, we may use Equation 17.57. Suppose that the final energy of the proton is 30 MeV (a low energy by modern accelerator standards, but larger than the energy of the earliest cyclotrons).

Since 1 eV = 1.6×10^{-12} erg, this final energy in ergs is $(30 \times 10^6) \times (1.6 \times 10^{-12}) = 4.8 \times 10^{-5}$ erg. This may be equated to the kinetic energy $\frac{1}{2}mv^2$ in order to find the final speed of the proton. It is

$$v = \sqrt{\frac{2E}{m}} = \sqrt{\frac{2 \times 4.8 \times 10^{-5}}{1.7 \times 10^{-24}}} = \sqrt{56.5 \times 10^{18}} = 7.5 \times 10^9 \text{ cm/sec}, \tag{17.62}$$

about one-fourth of the speed of light.* The largest radius of curvature reached by the protons in this 30-MeV cyclotron can now be calculated. It is

$$r = \frac{mvc}{e\mathscr{B}} = \frac{(1.7 \times 10^{-24})(7.5 \times 10^9)(3 \times 10^{10})}{(4.8 \times 10^{-10})(10^4)} = 80 \text{ cm}. \tag{17.63}$$

The diameter of the cyclotron is 160 cm, or about five feet. This seems a moderate amount, but even this "low-energy" cyclotron requires quite a large number of tons of iron for its magnet.

One of the simple features of the cyclotron, the equal time of orbits as given by Equation 17.60, is lost at particle speeds near the speed of light.† Moreover, at energies beyond a few hundred MeV, the size and weight of the cyclotron magnets becomes prohibitively large. The modern successor to the cyclotron is called the synchrotron; it may be pictured roughly as a "hollow cyclotron." Instead of starting at the center and spiraling outward, the particles start at the periphery and stay there, orbiting round a doughnut-shaped container at a fixed radius. At one or more points around the circumference, they receive synchronized pulses of electric force, roughly as in the cyclotron (although the dees have vanished). Since the orbital radius is constant, the magnetic field deflecting the particles around their racetrack can no longer be constant in time. This is best understood by reference to the equation,

$$r = \frac{pc}{Q'\mathscr{B}}. \tag{17.64}$$

Since the radius on the left side is constant, and the quantities c and Q' on the right are also constant, the *ratio* $p/\mathscr{B}$ must remain constant. As the particle gains momentum, the magnetic field must be increased in exact proportion to the increasing momentum in order to maintain a constant radius of curvature. At the end of the acceleration cycle the magnetic field has reached its maximum value (usually about 12,000 gauss), and the momentum and energy of the particles are at their maximum. In the proton accelerator at the Brookhaven National Laboratory, the magnetic field is increased from a small initial value to 13,000 gauss in slightly less than one second, during which time the protons complete about 300,000 circuits around a track 840 feet in diameter to reach a final energy of 33 GeV. The practical advantages of the synchrotron that make possible its extension to energies higher than any reached by cyclotrons are first that it requires a magnetic field only around its peripheral track, not in its hollow center; and second that its peak magnetic field does not need to be maintained continuously. It should be obvious from the description given here that the synchrotron as a useful research tool rests heavily not only on the fundamental laws of electromagnetism, but on much of the electrical technology that has flowed from those laws in the last hundred years.

---

* At a particle speed this close to the speed of light, the definition of Newtonian mechanics, *K.E.* = $\frac{1}{2}mv^2$, is no longer quite correct, but it is accurate enough for this example.

† Expressed in the form $r = pc/Q'\mathscr{B}$, the equation for the radius of curvature remains valid in the theory of relativity at particle speeds near the speed of light, but the relation of momentum to velocity, $p = mv$, is no longer true.

17.5 *Interaction of fields*

In the last chapter we expressed the laws of interaction of field in qualitative form: (1) A changing electric field creates a magnetic field; and (2) a changing magnetic field creates an electric field. The most subtle laws of electromagnetism and the last to be discovered, these laws are in some sense also the most fundamental, for they most clearly suggest the reality of the field concept, they are perfectly symmetric between electric and magnetic fields, and they underlie the "pure" electromagnetic phenomenon of radiation in free space. Yet they also find a wealth of practical application—quite directly, for instance, in an everyday device found in many homes and on many telephone poles, the transformer.

The transformer is a device for transferring electrical energy from one circuit to another via fields with no physical contact between the circuits. In its simplest form it might consist of two neighboring wires, one connected to a source of alternating current, the other connected to a motor or any other device that requires current for its operation (Figure 17.16). The circuit containing the source of alternating current is called the primary circuit. The other circuit, containing no source of current, is the secondary circuit. The current flow in the primary circuit creates a magnetic field which embraces the secondary circuit. Each moving electron in the secondary wire experiences a force from this magnetic field, but because the electrons move randomly, there is no net force, and therefore no net secondary current is created by the magnetic field alone. However, the fact that the primary current is alternating brings a law of changing field into play. The continually changing primary current produces a continually changing magnetic field. At the location of the secondary wire, the changing magnetic field creates an electric field. This electric field accelerates free electrons in the secondary wire all in the same direction at any instant, producing a secondary "induced" current. As the magnetic field oscillates, the electric field oscillates, and so does the secondary current. Therefore the secondary induced current is also an alternating current. A direct current in the primary circuit would induce no secondary current. Alternating current makes available to the electrical engineer laws of field that would not be at work with direct current. Because transformers are very convenient in many applications, alternating current has become standard in the United States and in most other technically advanced countries.

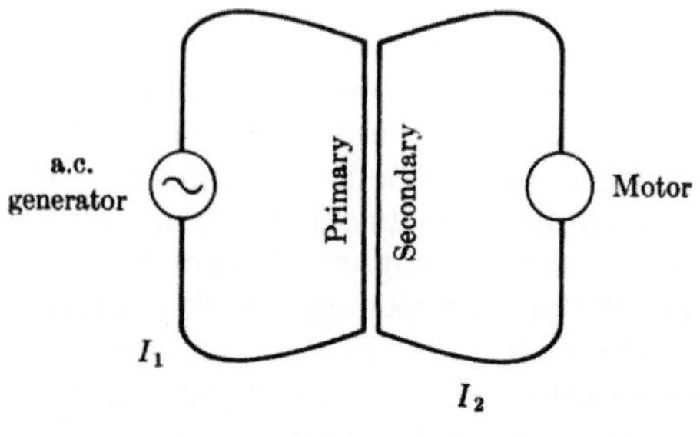

FIGURE 17.16 Principle of the transformer. The alternating current in the primary wire induces an alternating potential in the secondary wire.

In practice the primary and secondary wires of a transformer are not simply placed parallel to one another in a straight line, but are wound in coils around a common piece of iron. (When very high frequency alternating current is used, the iron is omitted.) The primary coil serves to intensify the magnetic field, and the secondary coil serves to expose a long length of secondary wire to the changing field. The compact design also results in remarkably high efficiency. In a well-designed transformer, more than 99% of the energy supplied to the primary circuit is transferred to the secondary circuit.

In addition to these merits, the coil design of the transformer makes it possible conveniently to have primary and secondary wires of different length. This is achieved simply by having more turns round the coil for one or the other wire (Figure 17.17). For a given primary circuit, there is a certain electric field generated at each point in the secondary coil. Since force times distance is equal to energy, force per unit charge (electric field) times distance is equal to energy per unit charge (potential). In symbols,

$$\mathcal{E}d = V. \tag{17.65}$$

The total potential difference across the secondary coil at any instant is equal to the electric field created in the secondary coil at that instant multiplied by the length of the secondary wire in the coil. The potential in the secondary circuit can therefore easily be adjusted to have any desired value by suitably choosing the length of the wire in the secondary coil. If primary and secondary coils have the same number of turns (which usually means about the same length of wires), the induced potential is the same as the primary potential. If the secondary coil is shorter than the primary coil, the secondary potential is less than the primary potential. Such a transformer is called a step-down transformer. Conversely, a step-up transformer is one in which the secondary coil is longer than the primary coil, and the secondary potential is larger than the primary potential. Although potentials can be stepped up or down, energies cannot. For an ideal transformer of 100% efficiency, the power (energy per unit time) delivered to the secondary circuit is the same as that supplied to the primary circuit. Since power is equal to potential times current, this means (for the ideal transformer) that

$$I_1V_1 = I_2V_2, \tag{17.66}$$

where the subscripts 1 and 2 denote primary and secondary circuits. Therefore if V_2 is larger than V_1 (a step-up transformer), I_2 must be correspondingly less than I_1. This equal-power formula also explains one of the main reasons why transformers are useful. For transmitting power over long distances, small currents are desirable in order to minimize the loss of energy arising from the resistance of the transmitting wires (Equation 17.40). Small current calls for large potential. A common potential used in high-tension lines is 110,000 V. For household use, on the other hand, a much smaller potential is desirable in order to minimize the danger to human life and the danger of fire. The standard household potential in the United States is 110 V. Typically the output of an electric generator is stepped up for long-distance transmission, and then stepped down in several stages to the various industrial and household needs of a community.

(a)

(b)

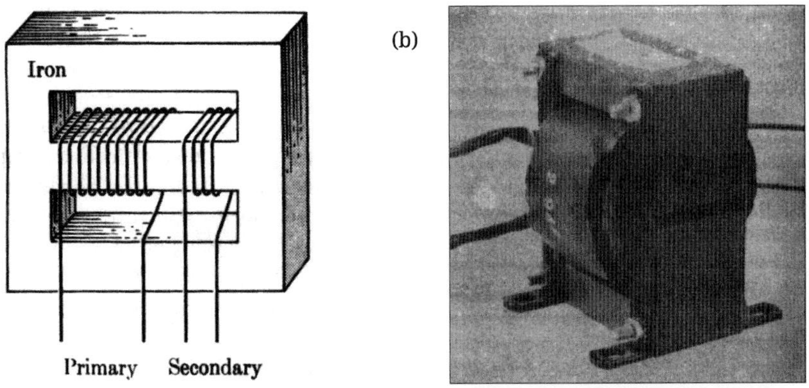

FIGURE 17.17 Transformer. **(a)** Schematic diagram of step-down transformer. Lines of magnetic field thread around through the iron. **(b)** Photograph of small transformer. Its primary and secondary voltages differ by a factor of about 18. (Photograph courtesy of John H. Atkinson, Jr.)

97 Another device that quite directly illustrates the law that a changing magnetic field creates an electric field is the betatron, a special kind of particle accelerator. Although by far less common than the transformer, it is an interesting example of the direct utilization of a law of interacting fields for a practical purpose. In the cyclotron and synchrotron, a magnetic field is employed for particle deflection, and a separate electric field is used for increasing the speed of the particles. In the betatron, a single magnetic field is harnessed for both purposes. Electrons (beta particles) move within an evacuated doughnut-shaped container called a toroid (Figure 17.18). A magnetic field passes through the toroid as well as through its hollow center. The electrons, deflected by the magnetic field, make their way around the toroid in a circular path. Acceleration of the electrons to higher speed is accomplished simply by increasing the strength of the magnetic field. As the magnetic field changes from weaker to stronger, it creates an electric field within the toroid (as well as elsewhere). This electric field pushes the electrons to higher speed. The remarkable fact is that by a suitable choice of the magnetic field in the center relative to that in the toroid, the increasing deflecting power of the magnetic field can be made to keep perfect pace with the increasing energy of the electrons, so that the electrons continue to move in a circle of constant radius. The induced electric field adds momentum to the electron at just such a rate that the ratio of momentum to magnetic field ($p/\mathscr{B}$) is a constant, whence the radius of curvature is a constant (Equation 17.64).

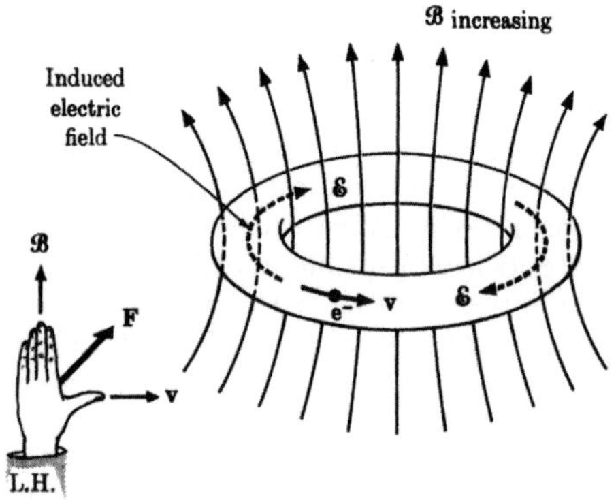

FIGURE 17.18 Principle of the betatron. The magnetic field serves a dual purpose. It deflects the electrons in a circular path. By changing, it creates an electric field that accelerates the electrons to higher speed.

The betatron affords also an interesting application of Lenz's law. How do we know that the electric field created by the changing magnetic field tends to speed up the electrons, not to slow them down? To be definite, imagine the magnetic field directed upward. Application of right-and-left rule 1 tells us that an electron moving horizontally in this field will be deflected to its own left (Figure 17.18). Seen from above, the electrons move in a counterclockwise direction. Next, right-and-left rule 2 may be applied to reveal that the magnetic field created by the moving electrons is directed downward at the center of the toroid. (This field is, of course, weaker than the original upward-pointing field.) Finally, Lenz's law may be applied to find out what happens if the upward-pointing magnetic field is increased. The electrons must react in such a way as to oppose the increase. This opposition takes the form of an increase in the speed of the electrons, for this increases their own downward-pointing magnetic field, counter to the increase of the applied field.

17.6 Inductance and capacitance: the oscillator

One of the simplest applications of a law of interacting fields occurs whenever a current is switched on or off. Imagine a straight wire in which a steady direct current is flowing (Figure 17.19). Around the wire are lines of magnetic field created by the current. What happens when the source of the current is suddenly disconnected, for example by throwing a switch in the circuit to the "off" position? One might suppose that the current simply stops with equal suddenness. It does not. Although the current must, of course, stop eventually, the laws of electromagnetism prevent this from happening suddenly. As the current begins to diminish in strength, the intensity of the surrounding magnetic field also diminishes. This changing magnetic field creates in the wire an electric field. By Lenz's law, we know that this electric field must act in a direction to keep the current flowing, for this will be the direction opposing the change. On the other hand, we know that the current cannot continue unabated, for then there would be no changing magnetic field to create an electric field to keep the current flowing. The current decreases, but gradually, not suddenly. Similar arguments show that when a circuit is switched on, the current must rise gradually to its final value. Actually in a practical case, the time required for the current to fade away or to build up is too short to notice without sensitive measuring apparatus. When a light is switched on, for instance, we are unaware of the time lag between throwing the switch and seeing the light, since the lag amounts to a very small fraction of a second.

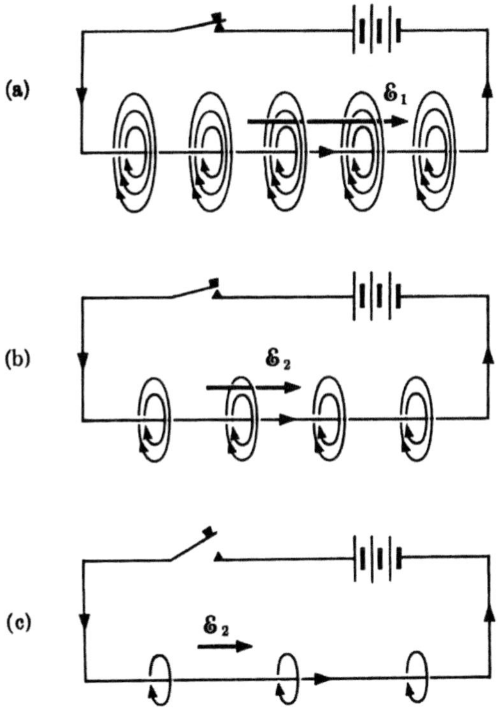

FIGURE 17.19 Self-induction. As a switch is opened, the diminishing magnetic field associated with the diminishing current induces an electric field in the wire that prevents the current from stopping suddenly. In diagram **(a)**, $\mathcal{E}_1$ designates the field in the wire produced by the battery. In diagrams **(b)** and **(c)**, at later times, $\mathcal{E}_2$ designates the self-induced electric field. The time scale is much less than one second.

This effect tending to oppose any change in the magnitude of a steady current is called *self-induction*. Any current brought about by an electric field that in turn is created by a changing magnetic field is called an induced current. If the changing magnetic field is created by one circuit and the induced current is in another circuit, as in a transformer, the phenomenon is called *mutual induction* (Figure 17.16). Exactly the same laws operate in a single circuit to cause the persistence of current known as self-induction.

It is possible to define for a wire or for a circuit a quantity called inductance. Just as resistance is the measure of opposition to flow of current, inductance is the measure of opposition to a *change* in the flow of current. If current is changing at a certain rate in a circuit, an electric field is induced in the circuit to oppose the change. This means in turn that some potential difference is induced across the circuit. Inductance is defined as the ratio of the induced potential to the rate of change of current. In symbols,

$$L = \frac{V}{\left(\frac{\Delta I}{\Delta t}\right)}. \qquad (17.67)$$

where L is used to designate the new concept, inductance. Since a more rapid rate of change of current I produces a proportionately larger induced potential, the ratio defining L is a constant for a given circuit. Notice the close similarity of this equation to the equation defining the resistance of a circuit,

$$R = \frac{V}{I}. \qquad (17.68)$$

Here the denominator contains the current rather than the rate of change of current. Here also the ratio is constant for a given circuit. In the mks system, with potential in volts, current in amperes, and time in seconds, the unit of inductance is given the name henry, after Joseph Henry, the American contemporary of Faraday who independently discovered the phenomenon of current induction by a changing magnetic field.

98 Just as a circuit element designed to add resistance to a circuit is called a resistor, a circuit element designed to add inductance is called an inductor. An ordinary straight wire has some inductance, but far more inductance can be achieved by winding a wire into a coil (Figure 17.20). Then the changing magnetic field created by one part of the wire has a chance to influence many other parts of the wire. In some circuits it is desired to minimize the so-called transient effects that arise from the time lag in current changes. Then the circuit must be designed with as little inductance as possible. Equally or more often, however, the inertial effect of an inductor serves some beneficial purpose in a circuit. Probably the most important role of inductance is in a circuit designed to produce electric oscillations. Before explaining the operation of the oscillator, we must examine its other vital circuit element, the capacitor.

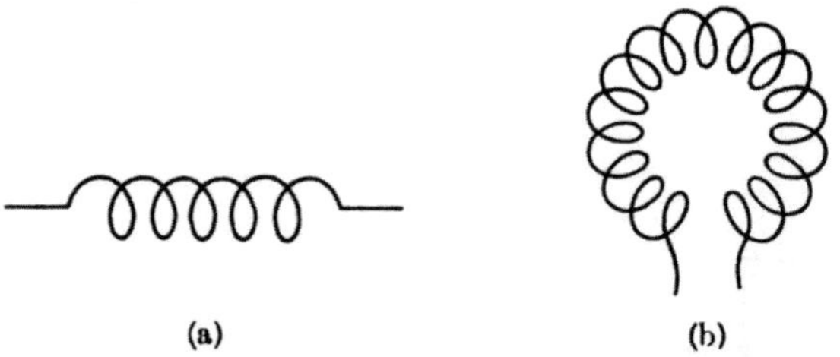

FIGURE 17.20 The inductor. **(a)** Schematic representation of an inductor used in circuit diagrams. **(b)** Typical physical arrangement of actual inductor. The coil bent into a toroid allows the maximum induction effect of one part of the wire on another with the minimum magnetic disturbance of other nearby circuit elements.

The physical structure of a capacitor as a pair of parallel plates separated by air or other insulating material was mentioned in Section 16.9 (see also Figures 17.21 and 15.7). One function of a capacitor is as a storage depot for charge. In some applications, a charged capacitor is connected to an external circuit when a short burst of intense current is desired. It can be charged slowly and discharged rapidly. Pulsed electromagnets, for example, receive their bursts of current from charged capacitors, enabling them to create momentary magnetic fields of high intensity.

When a capacitor is charged, there exists a potential difference between its plates (the energy associated with moving a unit charge from one plate to the other). If the charge on the plates is increased, the potential is increased in proportion, so that the ratio of these two quantities is a constant. This ratio is defined as the capacitance of the capacitor:

$$C = \frac{Q}{V}, \qquad (17.69)$$

C being the standard symbol for capacitance. In this equation, Q stands for the magnitude of charge on either plate, since the two plates have equal and opposite charge. In the mks system, the unit of capacitance is given a new name, the farad (after Michael Faraday). One farad is equal to one coulomb divided by one volt.

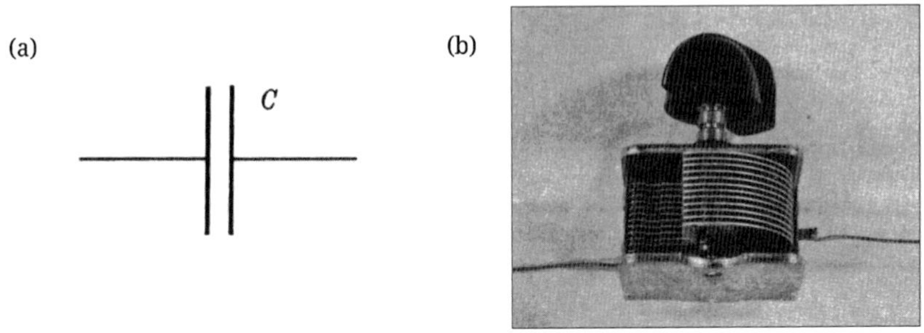

FIGURE 17.21 The capacitor as a circuit element. **(a)** Schematic representation of capacitor used in circuit diagram. **(b)** Actual capacitor used in radio-frequency oscillator. Other capacitors vary enormously in size and in physical construction. (Photograph courtesy of James Curtis.)

In one sense a capacitor as a circuit element acts oppositely to an inductor. Whereas an inductor tends to perpetuate a flow of current, a capacitor tends to stop a flow of current. Since the plates of a capacitor do not touch, the capacitor seems like an open switch that ought to stop current dead. In fact it stops it only gradually. As indicated in Figure 16.6, one may think of current flowing "through" a capacitor when electrons flow onto one plate and away from the other. This process causes a buildup of charge on the plates that gradually produces a potential opposing further flow of current. For example, if a capacitor is connected to a chemical cell, current flows until the charge on the plates has produced a capacitor potential equal and opposite to the cell potential. Then current flow ceases.

The simplest oscillator circuit consists of an inductor and a capacitor connected in se- 99
ries and joined into a closed loop (Figure 17.22). Imagine at a particular instant that there is a clockwise flow of current in the circuit and that the capacitor is uncharged. From this starting point we can analyze what will happen next. The inductor will tend to keep the current flowing, and charge will start to accumulate on the capacitor plates. Gradually the opposition to current flow brought about by the capacitor will reduce the current to zero. At some later instant, there will be no current, but the upper plate of the capacitor will be negatively charged, the lower plate positively charged. At this instant the charged capacitor will begin to discharge, with current now flowing counterclockwise around the loop. Because of the presence of the inductor, it will take some time for the discharging current to build up. At the instant that the capacitor is fully discharged, a current will be flowing, and the inductor will not allow it to stop. The counterclockwise current will persist until the lower plate of the capacitor is sufficiently negatively charged, and the upper plate correspondingly positively charged, to stop the current flow again. The process continues, with regular alternations of current flow like the swings of a pendulum.

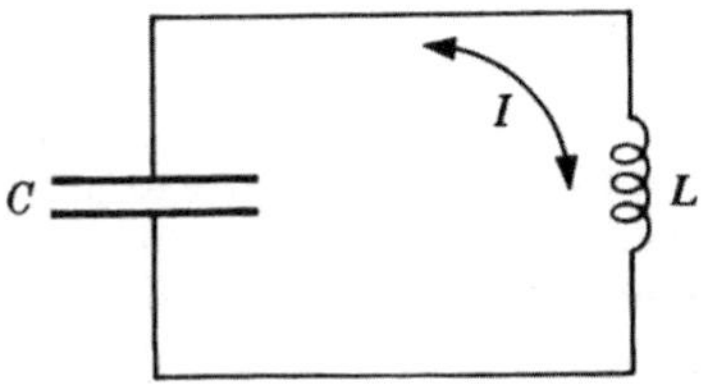

FIGURE 17.22 Simplest oscillator circuit, a loop containing a capacitor and an inductor.

The analogy of the electric oscillator to a swinging pendulum is more than superficial. The equations describing the two kinds of oscillation are mathematically identical. The inductance of the circuit plays the same role as the mass of .the pendulum (the inertial property). The inverse of the capacity acts like the "force constant" opposing the displacement of the pendulum away from equilibrium. Finally, the charge on one plate of the capacitor (positive or negative) follows the same law as the displacement of the pendulum (left or right). In fact, the analogy goes a bit further. Just as a frictionless pendulum would swing forever, an oscillator circuit with no resistance would continue to oscillate forever. In practice, of course, every pendulum has some friction and every circuit has some resistance. A clock pendulum is kept swinging by a light tap once each cycle. Similarly, an oscillating circuit needs to be "tapped" gently to keep it going. This needed extra stimulation might come, for example, from an adjacent circuit via mutual induction.

As a final point of similarity between pendulum and oscillator circuits, both are used to keep time. This does not mean that all oscillators drive clocks (although some do). Rather the oscillator provides a very rapid regularly repeating signal of known frequency that can be used for a number of applications. Chief among them is the generation of radio waves of definite frequency. Every radio transmitter contains an oscillator circuit, and so does every radio receiver. In the receiver it is the weak signal from the antenna that "taps" at the oscillator circuit within the radio to set it into electrical oscillation. To choose an interesting but less common example, electrical oscillators are used to create the sounds of "electronic music." Some piano tuners have also found it expedient to replace tuning forks with electrical oscillators.

Just as each pendulum has its own characteristic frequency, so has each oscillator. The time required for one complete oscillation of the circuit is given by a simple formula:

$$t = 2\pi\sqrt{LC}, \tag{17.70}$$

where L and C are the inductance and capacitance of the circuit, and t is the time for one oscillation, or the period. Note the close resemblance of this equation to Equation 8.28 for the period of a mechanical oscillation. If L were 1 henry, and C were 1 farad, t would be 6.28 sec. Actually 1 farad is by normal standards an enormous capacitance, and it is a rare circuit that would oscillate so slowly. However, practical oscillators of the type described here span an exceedingly great range of frequencies, from a few cycles per second to a few hundred million cycles per second. The frequency, or number of oscillations per second, is of course the inverse of the period:

$$f = \frac{1}{2\pi\sqrt{LC}}. \tag{17.71}$$

One of the important technical problems in the creation of practical radar was the design of an oscillator with a frequency of about 10^{10} cycles/sec, or a period of 10^{-10} sec. Since even a straight wire has inductance, and even a severed wire without plates has some capacity, it proved impractical to lower the values of L and C in a standard oscillator circuit sufficiently to achieve a period of only 10^{-10} sec. This problem found its solution in the invention of a new kind of oscillator called a magnetron, in which electrons oscillate in an evacuated cavity rather than back and forth in a wire.

We are so accustomed to the ubiquitous electrification of our modern world that we may find it difficult to picture a world not full of the applications of electromagnetism. The illustrations in this chapter form the merest handful from the thousands of applications that have been developed in the past hundred years. Yet these examples suffice to show that electromagnetism is far more than a theory of some special phenomena in nature. Each of the most fundamental laws of electromagnetism—the same laws that govern atoms and the cosmos—has found direct application in numerous practical devices of the everyday world and in tools for further research.

EXERCISES

17.1. Express the dimension of potential in terms of mass, length, and time.

17.2. A parallel-plate capacitor with vacuum between the plates is charged to a potential of 100 volts. If an electron leaves the negative plate with zero initial speed, how fast is it traveling when it strikes the positive plate?

17.3. The potential between the plates of a capacitor separated by 0.2 cm is 300 volts. What is the magnitude of the electric field in this region?

17.4. In a certain television tube electrons "fall" through a potential of 20,000 volts on their way to the screen. **(1)** What energy do the electrons acquire in eV? **(2)** With what speed do they strike the screen? **(3)** If the acceleration is accomplished in a distance of 4 cm, what average force acts on the electrons as they are being accelerated?

17.5. A giant spark is created by bringing near together two metal spheres with a potential difference of 10^6 volts. The spark carries a current of 10^{-3} amp and persists for 0.5 sec. **(1)** What is the power of the spark? **(2)** What total energy is associated with the spark? **(3)** How many electrons pass from one sphere to the other?

17.6. A capacitor has plates of area 20 cm^2 separated by an air gap of 0.3 cm. Stored in the electric field between the plates is an energy of 2 ergs. What is the potential between the plates?

17.7. **(1)** What is the difference in potential between a point 1 Angstrom (10^{-8} cm) from a proton and a point 2 A from the same proton? **(2)** What is the difference in potential between the point 2 A from the proton and a point infinitely remote? **(3)** In moving from a great distance to within 1 A of a proton, does an electron gain or lose potential energy? How much? Express the answer in eV.

17.8. An alpha particle ($Q_1 = 2e$) located 2×10^{-12} cm from the center of a uranium nucleus ($Q_2 = 92e$) has zero kinetic energy. **(1)** Describe in words its subsequent motion, if any. **(2)** What is its final kinetic energy in eV?

17.9. What is the physical dimension **(a)** of current; **(b)** of resistance; **(c)** of power?

17.10. An electric space heater, when plugged into a 110–volt outlet, dissipates 1,460 watts. What current flows through the heater? (NOTE: The answer is the same for a d.c. circuit and for root-mean-square values of an a.c. circuit.)

17.11. A light bulb is rated for 500 watts at 110 volts. If it is connected to a 110–volt source of potential, **(1)** how much power does it dissipate? **(2)** How much current flows through it? **(3)** What is its resistance? (Assume the resistance to be constant.)

17.12. One light bulb whose resistance is 100 ohms and another whose resistance is 200 ohms are connected in parallel across a 120-volt d.c. line. How many electrons pass through each bulb in 1 sec?

17.13. The derivation of Equation 17.28 expressing the resistance of a combination of resistors in series rests in part on the law of charge conservation. Explain how it does so.

17.14. **(1)** If three appliances having resistances of 10, 20, and 30 ohms respectively are connected in series, what is the total resistance of the combination? **(2)** If they are connected in parallel, what is their total resistance? **(3)** Draw circuit diagrams showing each of these combinations connected to a battery and a switch.

17.15. A 12-volt storage battery can deliver a total charge of 100 ampere-hours. **(1)** Express this charge in coulombs. **(2)** What is the total stored energy in the battery? **(3)** For how long could this battery light a pair of headlamps, each of 60 watts?

17.16. Estimate approximately **(a)** the resistance of an electric iron designed to operate at 110 volts; **(b)** the total current being supplied to an apartment with half a dozen lights burning and a hot plate operating; and **(c)** the power available in a laboratory designed to accommodate 20 students. Explain how you arrive at each answer.

17.17. Prove that this combination of four equal resistors has the same total resistance R as one of the resistors alone.

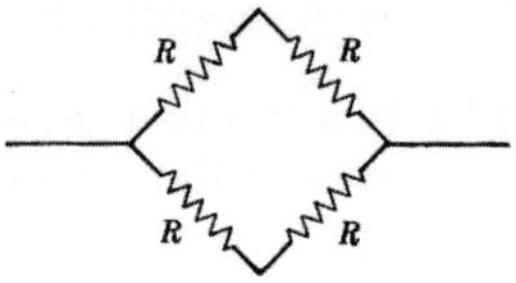

17.18. The battery potential in this circuit is 12 volts. The resistances are R_1 = 10 ohms, R_2 = 5 ohms, and R_3 = 20 ohms. **(1)** Find **(a)** the current in each resistor, and **(b)** the power dissipated by each resistor. **(2)** How long must this circuit operate before 1 kilowatt-hour of energy has been supplied by the battery? In what form is this energy manifested?

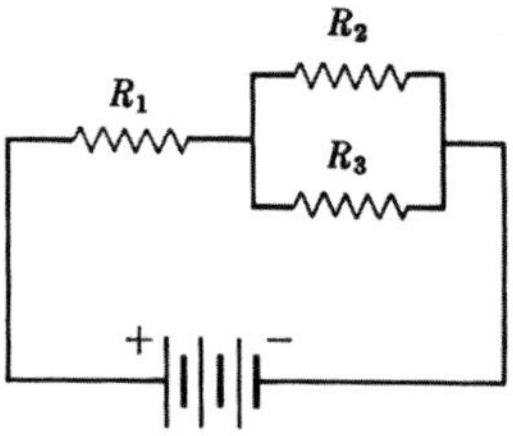

17.19. The diagram represents schematically a long-distance power-transmission line joining a generating station to a city. If the potential at the generating station is V_1 = 50,000 volts, the current is I = 2,000 amp, and the resistance of each wire of the transmission line is 0.3 ohm, **(1)** what power is dissipated in the transmission line? **(2)** What fraction is this of the total power supplied by the generating station? **(3)** What is the potential V_2 at the city? **(4)** If the potential V_1 is doubled and the current I halved, so that the total power supplied by the generating station is unchanged, how does the power loss in the transmission line change?

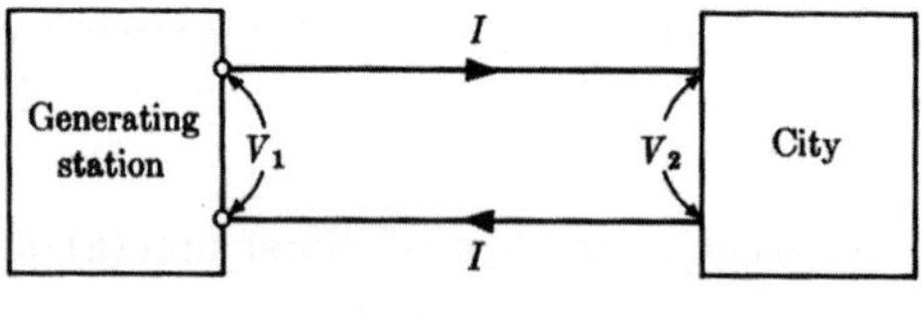

17.20. In a wire of length l there exists an electric field $\mathcal{E}$ directed along the wire. A current I flows in the wire, and the potential difference between its ends is V. Show that if the current is proportional to the field (Formula 17.39), the current is also proportional to the potential (Formula 17.16).

17.21. Ohm's law is sometimes written $I/A = \sigma\mathcal{E}$, where I is current, A is cross-sectional area of a wire, $\mathcal{E}$ is electric field in the wire, and σ is a constant called the conductivity of the wire. For a wire of length l and cross-sectional area A, with potential difference V between its ends, prove that conductivity and resistance are related by the equation,

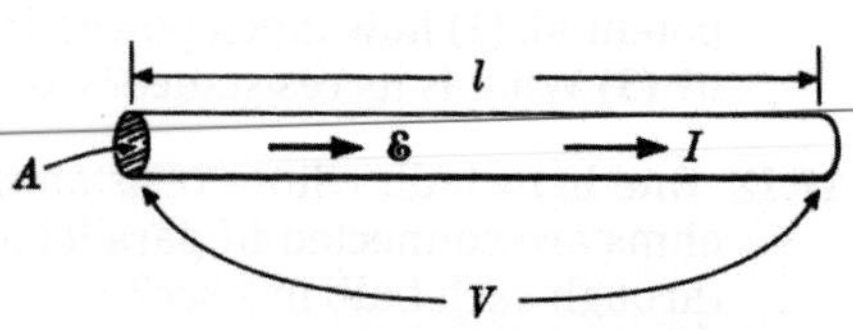

$$R = \frac{l}{\sigma A}.$$

17.22. In a diode, anode and cathode have a potential difference of 500 volts and are separated by 0.6 cm. A current of 3×10^{-3} amp flows through the tube, which is evacuated. **(1)** With what speed do electrons strike the anode? **(2)** The diode does not obey Ohm's law. Does the formula $P = VI$ correctly give its power? **(3)** If the electron energy is transformed to heat at the anode, calculate the heat produced in cal/sec.

17.23. Through the triode in Figure 17.12 flows a current of 1 milliamp (10^{-3} amp). The resistance R is equal to 50,000 ohms, and the potential V_2 is 450 volts. **(1)** What is the potential V_{out} across the resistor? **(2)** What is the effective resistance of the vacuum tube (the ratio of its potential to its current)? **(3)** An increase of the input potential V_{in} by 1 volt causes the current flowing through the triode to double. What is the increase of V_{out}?

17.24. In the cathode-ray tube diagrammed in Figure 17.13, an electron, in moving from cathode to anode, gains kinetic energy equal to the product of its charge and the accelerating potential, $K.E. = eV_a$. **(1)** Use this fact together with the result expressed by Equation 17.51 to prove that the ratio of vertical to horizontal components of velocity is given by

$$\frac{v_y}{v_x} = \frac{l}{2d}\frac{V_d}{V_a}.$$

(2) Substitute numbers you consider to be reasonable on the right side of this equation. What angle of deflection do you calculate?

17.25. Electrons in the cathode-ray tube of Figure 17.13 are accelerated through a potential $V_a = 1{,}000$ volts, then pass between deflection plates of length $l = 2$ cm and separation $d = 0.4$ cm. If the distance from deflection plates to screen is 20 cm, what deflecting potential V_d is required to move the electron beam spot 2.5 cm away from the center of the tube face?

17.26. **(1)** From Equations 17.56 and 17.57, derive Equation 17.58. **(2)** If an electron beam is accelerated by a potential V = 2,000 volts (6.67 statvolts) and deflected by a magnetic field $\mathcal{B} = 1{,}000$ gauss, what is its radius of curvature?

17.27. In Section 17.4, a 30-MeV proton cyclotron is "designed." Suppose physicists decide to use this same cyclotron to accelerate deuterons. The charge of a deuteron is the same as the charge of a proton; its mass is twice the proton mass. **(1)** What is the period of rotation of the deuterons in the cyclotron? **(2)** What is the maximum kinetic energy reached by the deuterons? *Optional:* For any nucleus of charge $Q = Ze$ and mass $M = Am_p$ used as a projectile in this cyclotron, develop scaling laws for the dependence of rotation period and maximum energy on the quantities Z and A.

17.28. A toy train transformer is plugged into a 110–volt household line. **(1)** If the house circuit furnishes a total power of 140 watts, of which 20 watts is dissipated as heat in the transformer, what current flows in the 6-volt secondary circuit? **(2)** Does the primary circuit or the secondary circuit require heavier wires? Why?

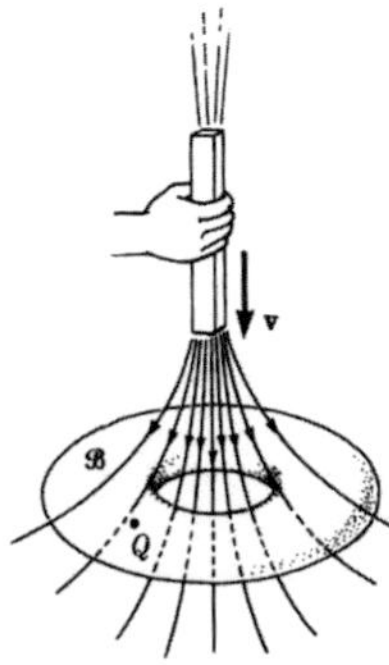

17.29. The figure shows a hypothetical hand-powered device like a betatron used to accelerate positively charged particles around a ring. With the help of Lenz's law, **(1)** deduce the direction of acceleration of the charged particles; and **(2)** explain why work is required to move the magnet downward.

17.30. A capacitor has charge $Q = 10^{-8}$ coulomb on each plate (one positively charged, the other negatively charged). **(1)** If the potential between the plates is 50 volts, what is the capacitance of the capacitor in farads? If the charge on each plate is doubled, **(2)** how does the capacitance change? **(3)** How does the potential change?

17.31. Sketch a graph of potential V across a capacitor vs. charge Q on one plate of the capacitor. What is the significance of the slope of the line in your graph?

17.32. A capacitor is discharging through a resistor as shown. **(1)** Find an expression for the ratio of the current I through the resistor to the charge Q on the capacitor. **(2)** As the remaining charge Q decreases, what happens to the current I? **(3)** Do you expect the current flow to "overshoot" so that the capacitor starts to charge again with opposite sign?

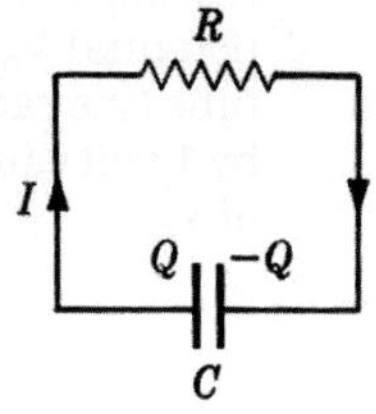

17.33. In a simple circuit containing an inductor and a resistor but no source of potential, a current is flowing at a particular instant. Will this current stop abruptly, diminish gradually, continue unchanged, or increase in magnitude? Justify your answer in terms of fundamental laws of electromagnetism.

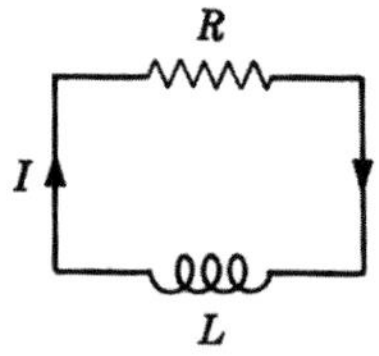

17.34. An inductance $L = 10^{-3}$ henry and a capacitance $C = 2.6 \times 10^{-11}$ farad form a simple oscillator circuit like that in Figure 17.22. **(1)** At what frequency does this circuit oscillate? **(2)** If the peak potential across the capacitor is 100 volts, what is the approximate average magnitude of current in the circuit?

17.35. In a cyclotron the magnetic field (into the page in this diagram) weakens slightly in going from the center to the edge of the pole face. The circle through A and B is a "proper" trajectory for a proton partway through its acceleration phase. If the field were exactly constant, a displaced circle through A' and B' would also be possible. Show that the gradual weakening of the field from center to edge tends to "stabilize" the orbits. Consider first a proton that is accidentally at A' instead of A (but with the same velocity), then a proton that is accidentally at B' instead of B. Stability means that these displaced protons should tend to return to the "proper" trajectory.

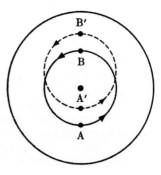

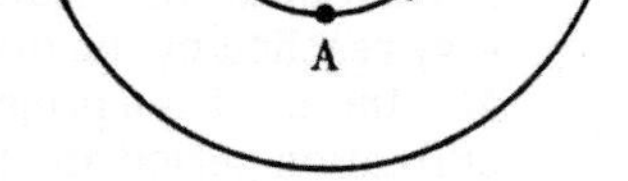

EIGHTEEN

Electromagnetic radiation and wave phenomena

Interest in the nature of light must go back to prehistory. It is undoubtedly as old as any problem in science, and probably one of the most rewarding scientific problems ever tackled. Interwoven with the whole history of modern physics, from the seventeenth century to the present, is the study of light, which has been closely tied to the development of the theories of electromagnetism, relativity, and quantum mechanics, as well as to the practical science of optics, to numerous discoveries in mathematics and, in the modern period, to such things as radar and lasers. From Römer's discovery that the speed of light is finite, based on observations of the times of appearance of the moons of Jupiter (1675), to modern measurements of quasi-stellar red shifts and high-energy photon reactions, the history of the study of light has closely paralleled the overall history of physics. In those 300 years the frontiers of physics have diverged from the solar system upward to the universe at large and downward to the world of elementary particles.

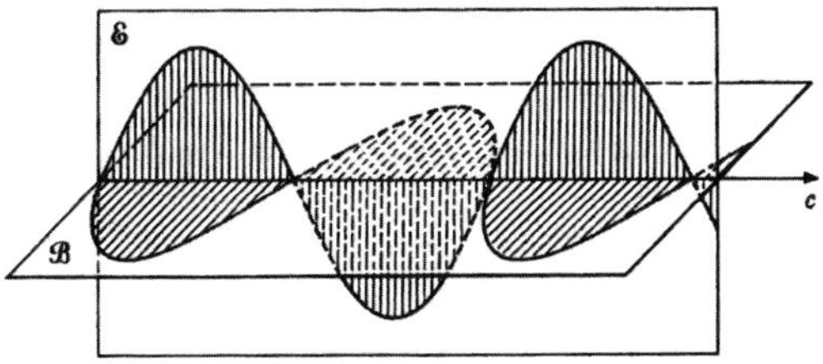

FIGURE 18.1 Electromagnetic wave. In this example, the wave is sinusoidal. Other waveforms are possible. Whatever the shape of the wave, it consists of transverse electric and magnetic fields changing in space and in time, and propagating at the fixed speed *c*.

According to the brilliant prediction of James Clerk Maxwell in 1865, light is a propagating and oscillating combination of electric and magnetic fields—an electromagnetic wave (Figure 18.1). Although this insight was the greatest milestone in the history of light, it was by no means the end of the road. At the beginning of this century the photon was discovered. Then came the special theory of relativity, assigning to the speed of light a new central role in physics. Going beyond the special theory, Einstein predicted connecting links between light and gravitation, which have been verified in the deflection of star light passing near the sun and in the gain of energy of photons descending in the earth's gravitational field. The emission and absorption of light by atoms was a topic central to the development of quantum mechanics. In recent years precise details of the electromagnetic properties of electrons and muons have been explained in terms of transitory bits of light called virtual photons. There remain unsolved puzzles concerned with these virtual photons, and it will not be surprising if a deeper knowledge of the nature of light accompanies still unforeseen advances in physics to come.

18.1 *The nature of light*

The fact that light moves at a definite, albeit enormous, speed, and not infinitely rapidly, has been known since the work of Römer, and many successive measurements have determined this speed with great precision. By 1700 several important facts about light were 100 known. White light was known to consist of a mixture of colors. Light was known to be deflected upon passing from one medium to another (refraction), the amount of deflection depending on the color. The speed of light in empty space was known to be the same for all colors (otherwise a moon of Jupiter would appear one color upon first appearance from behind Jupiter, later other colors, or white). Light was known to travel in straight lines in a uniform medium and to experience a slight deflection in passing an obstacle. It was known to carry energy. And a peculiar phenomenon called double refraction was known: upon entering some crystals, light undergoes not one deflection but two at once, splitting into two separate beams.

In spite of this knowledge, and more accumulated during the eighteenth century, the nature of light remained a mystery for another hundred years. The fundamental question to be decided was: Does light consist of particles or is it a wave phenomenon? When the wave theory triumphed early in the nineteenth century, the defeated adherents of the particle theory could scarcely have imagined that after another hundred years light would turn out to consist of particles after all, leading to the modern resolution of the argument: Light is both wave and particle.

Given the facts that light travels at a fixed speed in straight lines through empty space carrying energy from one place to another, a quite natural first guess is that light must consist of a stream of particles. This is the view commonly believed to have been championed by Newton, although, in fact, Newton recognized that the evidence was inadequate to decide one way or the other about the nature of light. The idea of light particles raised two problems in particular. (1) Why does an object giving off light apparently lose no weight, and an object absorbing light gain no weight? No one had yet imagined massless particles. (2) Why is the speed of light invariable? One might suppose that the particles of different colors of light should travel at different speeds, or that a more intense source of light should give off faster particles of light.

The wave theory, whose early champion was Newton's contemporary Christian Huyghens, accounted nicely for these two difficulties. A wave can transmit energy without conducting mass from one place to another, and it is a common feature of waves to have a fixed speed, independent of their strength and of their wavelength. The speed of sound, for instance, does not depend on the intensity or the pitch. Both wave and particle theories can give an account of refraction, but with an important difference. The wave theory requires that light move more slowly in the denser medium, the particle theory that it move more quickly in the denser medium. When, in the mid-nineteenth century, experiment finally decided in favor of the slower speed, the wave theory was already rather firmly established.

One difficulty with the wave idea seemed to be that it required an all-pervasive substance filling space that could transmit the light vibrations. The ether invented for this purpose had to be indeed a very ethereal substance, for, unlike water and air, it had to be totally transparent and frictionless, offering no impediment to the passage of material objects through it. (Otherwise, the earth would be slowed down and would spiral into the sun.) In spite of these unlikely properties, the ether was acceptable to most scientists. The idea of action at a distance with no transmitting agent seemed more repugnant to most men than did the idea of a mysterious ether.

Quite independent of arguments and speculation about the ether, a series of experiments in the first two decades of the nineteenth century so strongly supported the wave idea that there could no longer be any doubt that light is a wave motion. It is true that the diehard exponents of the light particles offered some tortured and implausible explanations of these phenomena, but the same phenomena were so simply and beautifully explained with the light waves that there could be little doubt that the wave nature of light offered the "right" explanation. It is important to note at this point that these experiments, which led to the triumph of the wave theory, are as valid today as they were 150 years ago. In spite of our deeper knowledge which leads us now to say that light does, after all, consist of particles—photons—the old experiments cannot be rejected. They may be repeated today with the same results and with the same interpretation in terms of waves. We must accept the new evidence that light is particle-like, but we must retain the old evidence that light is wavelike. Fortunately, the theory of quantum mechanics has come along to explain how photons can exhibit both properties, and how all other particles can do so too.

The conclusive evidence for the wave nature of light came through the phenomena of diffraction and interference. A wave passing an obstacle does not leave a precise sharp shadow, but is deflected a little into the dark region, giving the shadow a slightly fuzzy edge. This is diffraction. Two waves arriving at the same point may strengthen each other if they are crest to crest and trough to trough, or they may cancel each other out if the crest of one coincides with the trough of the other. This is interference. A little thought will show that each of these phenomena would be rather difficult to explain in terms of light particles. The wave theory, of course, did more than give a qualitative explanation of the existence of these phenomena. It provided a quantitative theory of diffraction and interference that accorded perfectly with the experimental facts. The exact way in which the light intensity varies smoothly through the region of the fuzzy shadow edge, for example, is predicted, as is the pattern of interference resulting from two slit sources of light. These and other characteristic wave phenomena are discussed in later sections.

The same phenomena that proved the existence of light waves provided a tool for measuring wavelength, and it was soon learned that different colors are distinguished by different wavelengths. Visible light spans about one octave* of wavelength, from shortwave violet, 3.5×10^{-5} cm, to long-wave red, 7×10^{-5} cm. Although small, these wavelengths are still several thousand times larger than the size of an atom, which is about 10^{-8} cm.

Besides wavelength, a wave can be characterized by its frequency, or number of vibrations per second. Light waves vibrate extremely rapidly, more than 10^{14} cps. The highest pitched sound audible to the human ear is about 10^4 cps. Radio waves vibrate at about 10^6 cps in the ordinary band on up to near 10^9 cps in what is called the UHF (ultra high frequency) band. The vibrations of light are about one million times more rapid than the UHF vibrations, and the wavelength, correspondingly, a million times shorter. Frequency and wavelength are related by a very simple equation:

$$\lambda f = v\,, \tag{18.1}$$

where λ is the wavelength (for example, in centimeters), f is the frequency (for example, in vibrations per second†), and v is the speed of the wave (for example, in centimeters per second). The formula can be used for any wave at all. Thus, for the standard musical A, $f = 440$ cps. The speed of sound in air is 3.3×10^4 cm/sec, so the wavelength in air of the tone A is its speed v divided by its frequency f, which turns out to be about 75 cm. We can use the same equation to calculate the frequency of light of known wavelength, for example green light with $\lambda = 5 \times 10^{-5}$ cm. Its speed of 3×10^{10} cm/sec divided by this wavelength gives its frequency of 6×10^{14} sec^{-1}.

Equation 18.1 is easy to derive. Imagine yourself in a boat at anchor watching an evenly spaced series of wave crests roll by. If the wave speed is v and the wavelength (crest-to-crest distance) is λ, how much time elapses between successive crests? This time (let us call it T) is the time required for the wave to move ahead a distance λ, and is equal to the distance divided by the speed:

$$T = \frac{\lambda}{v}\,. \tag{18.2}$$

* An octave represents a factor-of-two change in wavelength or frequency. If, for example, middle C is tuned to 262 cycles per second (cps), the C one octave higher is 524 cps, and the next C (high C) is 1,048 cps.

† Note that "cycles" or "vibrations" are dimensionless, so that the unit of frequency can be written simply: sec^{-1}.

This time T, the period of the wave, is the inverse of the frequency: $T = 1/f$. If your anchored boat is lifted on a wave crest every 4 sec, the frequency of the wave is 0.25 wave (or cycle) per second. In Equation 18.2, replacement of T by $1/f$ gives

$$\frac{1}{f} = \frac{\lambda}{v},$$

which is equivalent to Equation 18.1. The all-important equation governing wave propagation, $\lambda f = v$, is in fact nothing more than a restatement of the basic equation of constant-speed motion (Equation 8.1): distance equals speed multiplied by time.

By 1820, light was known to consist of waves, the phenomena of interference and diffraction had been explained mathematically, the wavelength of light had been measured, at least approximately, and the phenomenon of double refraction had been explained in terms of the transverse vibration of light waves. Thomas Young in England was the first to propose a wave explanation for interference, in 1801. He went on to design specific experiments to reveal interference fringes (look ahead to Figure 18.34) and to determine the wavelength of light. Subjected to ridicule by the then dominant adherents of the particle theory of light, Young abandoned his researches after a few years, returning to them only after 1815 when Augustin Fresnel rediscovered optical interference effects and received a more sympathetic hearing for his wave ideas in France than Young had received in England. In the years 1815-1820, Fresnel and Dominique Arago in France and Young in England carried out a series of experiments and mathematical analyses that won over most scientists to the wave theory. By 1821 Joseph von Fraunhofer in Bavaria had invented the diffraction grating (Section 18.11) and measured some wavelengths of light to high precision.

The riddle of double refraction was first correctly solved in 1817 when Young suggested that the vibrating motion in light waves might be transverse to the direction of propagation of the wave. A water wave is a simple example of a transverse vibration. The motion of the water is up and down, transverse to the motion of the wave, which is horizontal. A swimmer is lifted up and down as a wave passes by. But sound waves or shock waves that pass through a medium rather than over its surface are longitudinal—the material vibrates in the direction in which the wave is proceeding. A man struck by a shock wave from an explosion is first pushed away from the explosion, then sucked back toward it; thus he is caused to vibrate along the direction of the wave motion. It was quite natural to think of light passing through the ether as analogous to sound passing through air, vibrating along its direction of motion.

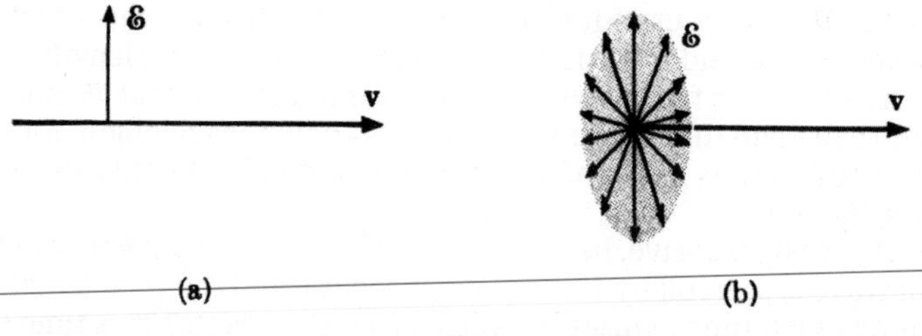

FIGURE 18.2 (a) Polarized wave, represented schematically by single electric field vector perpendicular to the direction of propagation. **(b)** Unpolarized wave, actually a mixture of many different polarized waves, represented by 6 vectors in all possible transverse directions. Every component of the unpolarized wave can be resolved into just two independent polarization directions.

From the time of Christian Huyghens' pioneer work on waves in the seventeenth century, it had been assumed that light, if it is a wave, is a longitudinal wave. Nevertheless, Young's contrary suggestion was not long in finding acceptance. Two important puzzles were ripe for solution, and the hypothesis of transversatility of light waves nicely accounted for both.

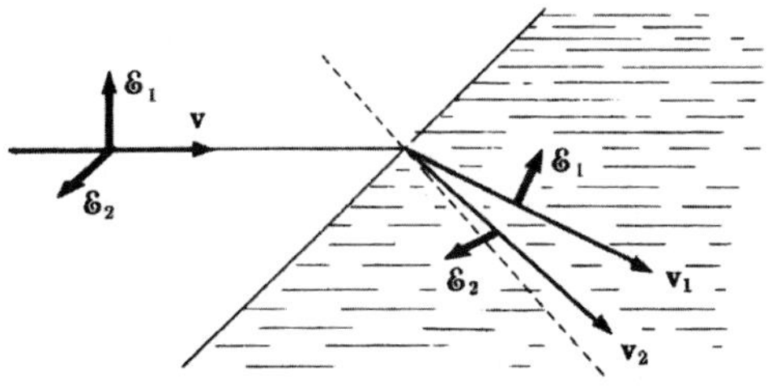

FIGURE 18.3 Double refraction. Upon entering some crystals, an unpolarized beam is split into two differently polarized beams. This behavior was not given a satisfactory explanation with either the particle theory of light or the longitudinal wave theory.

The first of these puzzles was the phenomenon of double refraction, which had gone unexplained since its discovery by Erasmus Bartholinus in 1669. It was necessary only to assume that the speed of light in the crystal depends on the orientation of the direction of vibration. Light vibrating in one direction will then be refracted differently from light vibrating in another direction. For a longitudinal wave there is only one possible direction for the vibration; hence there is no basis for explaining double refraction. It would seem that a transverse wave should have infinitely many different possible directions of vibration (Figure 18.2). Indeed it does. However, it has only two *independent* directions, perpendicular to each other (and to the direction of propagation). Any intermediate direction of vibration can be resolved into the two independent directions, just as any direction on earth can be resolved into a north-south component and an east-west component. For this reason, crystals are at most doubly refracting, not triply or multiply refracting (Figure 18.3).

Apart from double refraction, light was known to have a dual character in general. In 1808 Etienne Malus discovered that light could be polarized by reflection from glass and many other substances. Eight years later Arago and Fresnel learned to their surprise that light beams with different polarization could not be made to cancel each other through interference. This was the second important puzzle that was quickly clarified by Young's suggestion of transverse vibration. No matter how arranged, two beams whose electric field vectors are not parallel cannot cancel one another in destructive interference.

To summarize: Between 1800 and 1820, understanding of the nature of light took giant strides forward. The wave nature of light was established beyond reasonable doubt, and wavelengths and frequencies were measured. Polarization phenomena were successfully explained by the transversality of light waves. One other discovery of importance not so far mentioned, was also made in this period. The existence of waves shorter and longer than light (ultraviolet and infrared) was established, giving the first hint of the vast electromagnetic spectrum that we know today.

Even after these advances, there remained deep mysteries of light that required another century to plumb. Why does light travel at the speed it does, and why is its speed independent of its wavelength? What exactly is the nature of the vibrating medium? What is the mechanism of emission and absorption of light? Why is light polarized by reflection and refraction? What is the connection between light and thermal radiation? What is the connection between light and electricity and magnetism? In 1820 not all of these questions had been asked, much less answered. Eventually all of them proved important, and all found answers.

18.2 The electromagnetic theory of light

James Clerk Maxwell was born in 1831, the same year in which Faraday in England and Henry in America discovered that a changing magnetic field creates an electric field (magnetic induction). Thirty years later, building on the work of Faraday, Maxwell introduced the symmetrical principle that a changing electric field creates a magnetic field. With these two basic laws of changing fields in hand, Maxwell went on to predict the existence of electromagnetic waves—combinations of electric and magnetic fields that, through continually oscillating change, reinforce each other and propagate through space, freed of charge and current and matter. From the laws of electric and magnetic interaction, Maxwell could predict that these waves should be transverse, as shown in Figure 18.1, and should travel at a speed equal to the constant c in the law of magnetic force (Equation 16.3). Since this constant was by then known to be equal to the speed of light, it was a short step for Maxwell to go on and suggest that light was such an electromagnetic wave. "We can scarcely avoid the inference," he wrote, "that light consists in the transverse undulations of the same medium which is the cause of electric and magnetic phenomena." So were united the science of optics and the science of electromagnetism.

101 There are two keys to understanding electromagnetic waves physically—that is, to acquiring some intuitive grasp of their reason for being. One is the idea of *perpendicularity*, the other is the idea of *mutual reinforcement*. As we learned in Chapter Sixteen, perpendicularity characterizes all of the links between electricity and magnetism. Whereas the electric force on an electric charge is parallel to the electric field, the magnetic force on a moving charge is perpendicular to the magnetic field. The magnetic field generated by moving charge is perpendicular both to the direction of motion of the charge and to the line connecting the charge and the point of observation. This principle of perpendicularity extends to the laws of disembodied fields. A changing magnetic field creates an electric field perpendicular to the direction of *change* of the magnetic field (not necessarily perpendicular to the magnetic field itself). In a betatron, for instance (Figure 17.18), the induced electric field is perpendicular to the changing magnetic field. (In this device the direction of the magnetic field and the direction of change of the field are the same.) Similarly, a changing electric field creates a magnetic field perpendicular to the direction of change of the electric field. Through these principles of perpendicularity in electromagnetic interaction the transverse character of electromagnetic waves becomes comprehensible. Note that in Figure 18.1 the electric and magnetic fields are perpendicular to each other. Since each field is undergoing continual change in its own direction, each field is also perpendicular to the direction of change of the other.

This kind of reasoning explains why the two fields are perpendicular to each other in an electromagnetic wave. It does not yet explain why both are perpendicular to the direction of motion of the wave. This fact is understood best in terms of the deep symmetry between electricity and magnetism. In the absence of electric charge, electric and magnetic fields are in perfect balance in Maxwell's equations. Whatever effect the electric field can have on the magnetic field, the magnetic field can have precisely the same effect (sometimes with the opposite sign) on the electric field. The mathematical symmetry of the fundamental equations implies a physical symmetry in nature that is clearly illustrated in the balanced appearance of an electromagnetic wave.

The symmetry of electromagnetism in empty space brings us to the second key required to understand electromagnetic waves—the idea of mutual reinforcement. The propagation of the wave can be looked at as a continual regeneration and reinforcement. A changing electric field creates a magnetic field. This changing magnetic field recreates an electric field. This electric field, still changing, produces another magnetic field. And so on, back and forth. Neither field could propagate alone. Together they can propagate in equal strength provided they never stop changing. Endless propagation requires endless change. And only at the speed of light, it turns out from Maxwell's equations, do the propagating fields maintain themselves in a state of balance.

102 Why the speed of light and the constant *c* in the equation for the magnetic field created by a moving charge (Equation 16.9) are exactly the same number is a fact that cannot be accurately explained without complicated mathematics. However, the following example will serve to make it reasonable and will also reveal something about the way in which electromagnetic waves are created. Suppose that you have, concentrated on the end of a

stick, a certain amount of charge Q. If you hold the stick stationary [Figure 18.4(a)], the static charge is surrounded by an electric field, but no magnetic field. This is very nonsymmetrical between electricity and magnetism, because electric charge is present, but a magnetic pole is not. Now move the stick up and down at an average speed v [Figure 18.4(b)]. A magnetic field is created, but it is weaker than the electric field. The electric field magnitude is

$$\mathscr{E} = \frac{Q}{d^2}.$$

The magnetic field magnitude is

$$\mathscr{B} = \frac{v}{c}\frac{Q}{d^2},$$

less by the ratio v/c. If you could shake the stick at a speed close to the speed of light, you could create magnetic and electric fields of comparable intensity. (The changing magnetic field you have created acts back to induce an electric field opposing your motion. This manifestation of Lenz's law is a complication we can overlook in this discussion.) Note that the inverse of the speed of light acts as a "coupling constant" linking magnetism and electricity. If the speed of light were a much larger constant, a particular moving charge would create a smaller magnetic field. For a hypothetical smaller value of the constant c, the magnetic effects of moving charge would be greater.

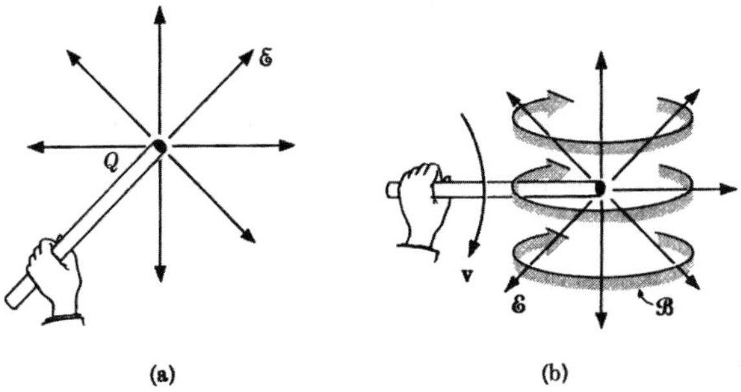

FIGURE 18.4 (a) Surrounding a stationary charge is an electric field, but no magnetic field. **(b)** The same charge in motion creates a magnetic field as well. If the stick is shaken back and forth at frequency *f*, a small fraction of the field intensity escapes to propagate away as an electromagnetic wave of this frequency.

For disembodied fields in space, as well as for those tied to charges, the inverse of the speed of light plays the same role as a coupling constant. If an oscillating electric field of average magnitude $\mathscr{E}$ has wavelength λ and frequency f, it creates a magnetic field with average magnitude $\mathscr{B}$ given by

$$\mathscr{B} = \frac{\lambda f}{c}\mathscr{E}. \tag{18.3}$$

This, in simplest form, is the law of magnetic induction in free space. In this equation we wish to regard c as the constant of electromagnetic linkage, pretending for the moment that we do not know it is also the speed of light. If the combination $\lambda f/c$ is less than unity, the induced magnetic field is less than the electric field. This magnetic field will induce in turn a still smaller electric field, and the whole disturbance will die out. A value of the combination $\lambda f/c$ greater than unity is impossible, for that would imply a crescendo of increasing reinforcement that would violate energy conservation. Only if $\lambda f/c$ is equal to unity can the fields propagate as a wave with equal magnitudes of electric and magnetic fields. For a propagating wave, the product of wavelength and frequency is equal to the speed of the wave, so we conclude that electromagnetic waves travel at the speed c. In symbols,

$$\frac{\lambda f}{c} = \frac{v_{\text{wave}}}{c} = 1. \tag{18.4}$$

This is an oversimplified and inadequate version of Maxwell's derivation, but it emphasizes two main points of his discovery: (1) Propagating waves require electric and magnetic fields in equal strength.* (2) To maintain equal strength, the fields must change at such a rate that their speed of propagation is equal to the constant *c*. Therefore *c* is the speed of light. Right at the heart of electromagnetic radiation is the balanced coupling of electricity and magnetism, and the constant that measures the strength of this coupling is the speed of light.

18.3 *The electromagnetic spectrum*

Although Maxwell's theory provided a speed restriction—that all electromagnetic waves must travel in empty space at the same speed *c*—it provided no restriction on frequency or wavelength. In fact, far from restricting frequency, it opened up the possibility of a limitless range of frequencies and wavelengths. At the time of Maxwell's work, the known band of frequencies extended from infrared to ultraviolet, including visible light and extending somewhat beyond the visible both to higher and lower frequencies. The vast panorama of electromagnetic radiation, of which visible light forms only a narrow slice (Figure 18.5), remained to be discovered.

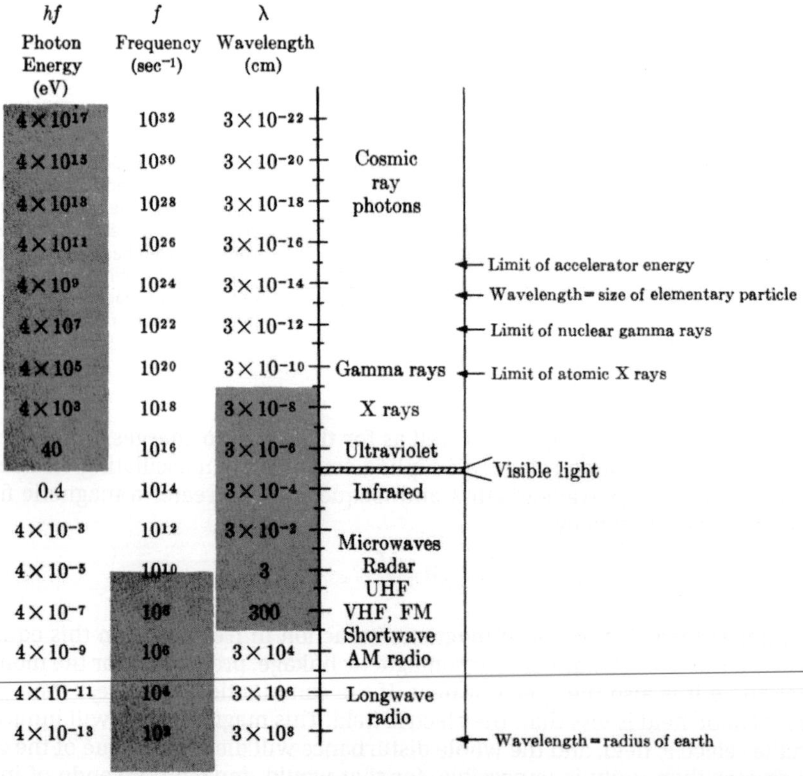

FIGURE 18.5 The electromagnetic spectrum. As indicated by the shading, waves are usually characterized by frequency at the low-frequency end of the spectrum, by photon energy at the high-frequency end, and by wavelength in an intermediate region.

* This equality of strength holds true in the cgs system of units, in which $\mathcal{E}$ and $\mathcal{B}$ have the same physical dimension.

When a narrow beam of white light passes through a prism, the fact that different wavelengths are differently refracted causes the beam to diverge, so that the beam emerging from the prism produces a spread-out image showing the distinct colors from red to violet (Figure 18.6). To this band of colors Newton gave the name *spectrum* (the Latin word for appearance or visible manifestation). By now the term spectrum has evolved to mean any orderly arrangement according to wavelength or frequency,* be it visual, photographic, graphical, or tabular.

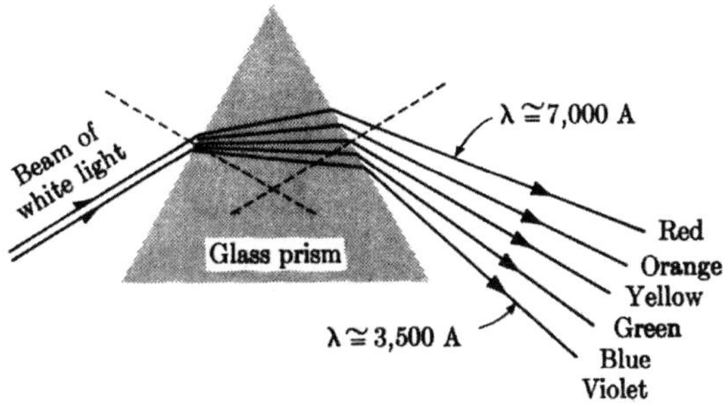

FIGURE 18.6 The spectrum of visible light. Since a prism refracts different frequencies differently, it serves to separate a beam of white light into its component frequencies. This separation is called dispersion.

The fact that the speed predicted for electromagnetic radiation was equal to the known speed of light was by itself strong evidence in favor of Maxwell's suggestion that light is electromagnetic radiation. To this piece of evidence were added the facts that the new theory explained the transverse character of light waves, that it predicted an equal speed for all colors (all wavelengths), and that it predicted that light should be slowed in material media. Nevertheless, a *direct* proof of the electromagnetic nature of light was lacking, for no one knew how to create light by electromagnetic means, or how to detect electric and magnetic fields in a light beam. The first direct demonstration that waves of electric and magnetic fields can propagate through space came not through the study of light but through the generation of radio waves.

In 1887, Heinrich Hertz in Germany succeeded in creating and detecting electromagnetic waves.† His transmitter consisted of the oscillatory discharge of a capacitor across a spark gap. His detector (receiving antenna) consisted simply of a wire bent into a loop with its two ends not quite touching. A visible spark across the gap of the detector signaled the presence of an oscillatory current in the detecting loop, which in turn revealed the presence of an oscillatory electric field in the vicinity. The sensitivity of the detector was enhanced by designing it to have a natural oscillation frequency equal to the frequency of the transmitting circuit. Exactly this principle is used in modern radios. Turning the dial of a radio changes the natural frequency of the detecting circuit and thereby changes the frequency of the electromagnetic radiation to which the radio is sensitive.

Hertz was able to reflect his radio waves, to diffract them, to demonstrate that they were influenced by intervening matter, to show that they traveled at finite speed, to prove them to be transverse, and to measure their wavelength by means of interference effects. The frequencies utilized by Hertz were around 100 megacycles, in what is now called the VHF band

* Since photon energy is proportional to frequency, modern physicists refer also to energy spectra. Led further by the mass-energy equivalence, we speak of the mass spectrum of elementary particles.

† David Hughes in England was probably the first person to create electromagnetic radiation by means of an oscillatory circuit and detect it at a distance. His work of 1879-1880 went unreported and unknown for many years. To Hertz belongs the priority of publication. The credit due Hertz, however, is not so much for the accident of first publication as for his achievement over an extended series of experiments in conclusively demonstrating the wave properties of the new radiation and discovering its points of similarity to light.

(now used for television and aircraft communication, among other things). The wavelength of a 100-megacycle wave ($f = 10^8\ \text{sec}^{-1}$) is

$$\lambda = \frac{c}{f} = \frac{3 \times 10^{10}\ \text{cm/sec}}{10^8\ \text{sec}^{-1}} = 300\ \text{cm}\,. \tag{18.5}$$

Hertz's oscillator frequencies were a happy choice, for they provided macroscopic wavelengths of convenient size. This was not altogether a coincidence. A macroscopic antenna usually emits a macroscopic wavelength. A single atom, by contrast, is a submicroscopic "antenna" whose principal radiation is much shorter in wavelength than radio waves.

Beginning around 1800, knowledge of the radiant spectrum expanded gradually from the visible into the infrared and ultraviolet, until by 1900 the region accessible to optical experiments had grown from one octave to eight. After 1887, the longer wave radio spectrum was rapidly filled in. By the early years of this century wavelengths from less than 1 cm to many kilometers had been successfully created and detected. Practical utilization of radio for communication began in 1900 at the lower frequencies, with the VHF and UHF regions being exploited only in recent decades.* Radar operating with wavelengths from 1 to 10 cm was a development of the 1940s. A commonly encountered wavelength in student laboratories now is 3 cm, quite convenient for studies of electromagnetic wave properties. Waves in this part of the spectrum are called (somewhat confusingly) microwaves, because their wavelengths are very much less than the wavelengths of normal radio waves.

Like the earliest transcontinental railroad being pushed simultaneously from east and west toward a meeting point, the electromagnetic spectrum was attacked by electronic means at low frequencies and optical means at higher frequencies until a meeting point was reached in the 1930s around 0.01 to 0.1 cm, the frontier between microwaves and infrared waves. At the long wavelength end of the spectrum, commercial broadcasting in the United States extends down to 500 kilocycles ($5 \times 10^5\ \text{sec}^{-1}$), with aircraft and marine stations down to 100 kilocycles. Government use of the spectrum continues down to less than 1 kilocycle, where a single wavelength extends more than 300 kilometers.

At the other end of the spectrum, X-rays were discovered by Wilhelm Roentgen in 1895, and nuclear gamma rays by Paul Villard in 1900. Still higher energy gamma rays showed up in the cosmic radiation in the 1920s. Man-made gamma rays at accelerators extend now to frequencies as great as $5 \times 10^{24}\ \text{sec}^{-1}$ or wavelengths as short as 6×10^{-15} cm, yet cosmic rays easily hold the frequency record at about $10^{32}\ \text{sec}^{-1}$.

The discovery of X-rays is a particularly interesting chapter of physics, for it was an accident and it triggered an even more famous accident, the discovery of radioactivity (Chapter Twenty-Five). Roentgen found by chance that radiation emerging from the place where an energetic electron beam (then still called a cathode ray) struck the glass wall of a cathode-ray tube could produce fluorescence in certain salts. Because the radiation was new and its properties unknown, he called it X-radiation. It was soon guessed that X-radiation might be electromagnetic radiation. The proof of its wave nature came in 1912, when Max von Laue successfully produced X-ray interference patterns and measured X-ray wavelengths. Von Laue's technique was simple and ingenious. He let the regular arrays of atoms in a crystal serve as centers of diffraction to produce measurable interference patterns. Microwaves can be significantly diffracted by strips of metal or wire, visible light waves by finely etched lines on a glass plate. X-rays, with wavelengths one thousandth or less of the wavelength of light, requires a correspondingly finer mesh of lines or points to produce much diffraction. Fortunately, nature, in assembling crystals in row upon row of perfect order, provides exactly the array needed, with spacing between atoms of about one Angstrom (10^{-8} cm). Von

* The long-distance communication record of 200 million miles established between earth and a Mariner space vehicle in 1966 utilized frequencies above 2000 megacycles in the UHF band.

Laue's method is still in use, but now X-rays are the tools and crystals the objects of study instead of the other way around.

An electromagnetic wave can be characterized by its frequency, or by its wavelength, or by the energy of one of its photons. All three of these quantities are in common use (and all appear in Figure 18.5), each suited to a particular domain of the spectrum. At the low-frequency end of the spectrum, where the electric circuits used to create and detect radio waves are characterized by certain natural oscillation frequencies, the frequency of the wave is the convenient way to catalogue it. In the optical region, where diffraction and interference effects separate the spectrum into its components, wavelength is the usual measure. Gamma rays, detected individually as single photons, are most often specified by their energy. Because of the vast range of the electromagnetic spectrum there is a proliferation of units used to describe it. Frequencies are measured in cycles,* kilocycles (10^3 sec^{-1}), megacycles (10^6 sec^{-1}), and gigacycles (10^9 sec^{-1}). You may encounter wavelengths expressed in meters, centimeters, microns (μ) (10^{-4} cm), or Angstroms (A) (10^{-8} cm). The electron volt (eV) and its multiples, the keV (10^3 eV), the MeV (10^6 eV), and the GeV (10^9 eV), are the common units of measurement of photon energy.

Being infinite in extent, the electromagnetic spectrum can never be fully explored. However, with known wavelengths extending from one billionth the size of an elementary particle to thousands of kilometers, the exploration has been rather complete, with no uncharted gaps remaining. The ratio of the highest to the lowest frequency included in Figure 18.5 is 10^{30}. In musical language, this is about 100 octaves. By comparison, a piano covers about 5 octaves, and the human ear about 10 octaves (of sound waves), an AM radio receiver two octaves, and an FM radio receiver less than one octave (of electromagnetic waves). Visible light accounts for only one of these 100 octaves of electromagnetic radiation.

18.4 Emission and absorption

An electromagnetic wave can travel a billion light-years through empty space, or it can jump an Angstrom from one atom to another in solid matter. Its wavelength can span the earth or it can fit comfortably inside an elementary particle. Whatever its history, whatever its intensity, whatever its wavelength, there is one sure constant feature of its origin and its termination: accelerated charge. An electromagnetic wave can be created only by the acceleration of charge; it can be absorbed only when it causes charge to be accelerated.†

Energy change must accompany the acceleration. An emitting charged particle loses energy, an absorbing charged particle gains energy. These energy changes in matter are of course precisely balanced by energy changes in the electromagnetic field. It is interesting to examine the entire electromagnetic spectrum in terms of the single idea of accelerated charge. For each part of the spectrum, what charged particle creates a particular wave? In what system does the charge move and over how great a distance? What causes the acceleration of the emitting particle? What charged particle is most likely to absorb the wave? To answer these questions in a survey of the entire spectrum, it is useful to discuss separately three kinds of accelerated motion: (1) oscillation of charged particles in nature; (2) man-made oscillation of charge; (3) accelerated motion that is not oscillatory (Figure 18.7).

Natural Oscillation of Charge

Long before Hertz—several billion years before—certain parts of the electromagnetic spectrum must have been already well filled with radiation. Every atom in nature has as the very basis of its structure accelerated charge in the form of circulating electrons. Therefore

---

* One cycle per second is also known as one hertz.

† If magnetic monopoles should exist, they too could create and absorb electromagnetic waves.

every atom is potentially a radiator. There was a brief period of consternation in physics in the early part of this century, in the interval between the discovery that atoms contain electrons and Niels Bohr's theory of the hydrogen atom. At that time the question was asked: Why do all atoms not radiate continuously? Bohr's answer was that they are prevented from doing so by the conservation of energy.

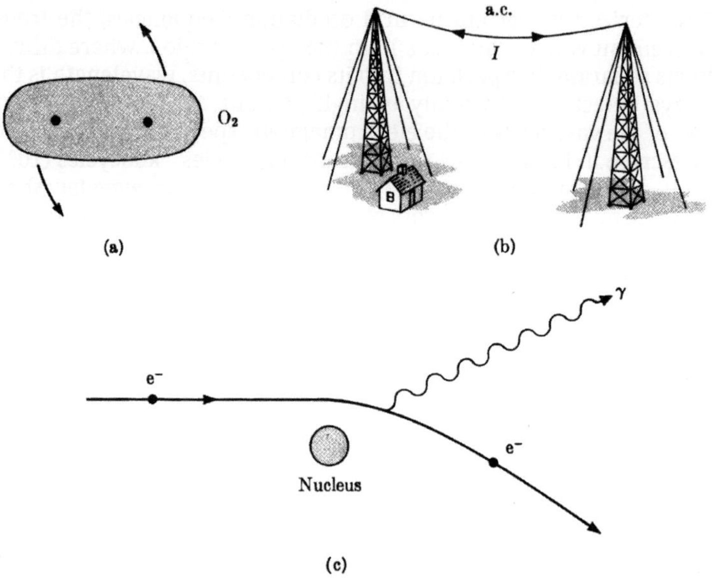

FIGURE 18.7 Sources of electromagnetic radiation. **(a)** Natural oscillation of charge, as in a rotating molecule, **(b)** Man-made oscillation of charge, as in a radio transmitter. **(c)** Acceleration of charge without oscillation, as in the deflection of a high-speed electron by a nucleus.

He postulated that every atom has a lowest energy state, or "ground state" (now a well-established fact). In its ground state an atom can lose no further energy and can therefore not emit radiation, despite the continual acceleration of its oscillating electrons. In a higher energy state, or "excited state," however, the electrons can give rise to electromagnetic radiation. According to classical theory, the frequency of the emitted radiation should be equal to the frequency of the oscillating electron. Quantum mechanics has changed this view. The radiated frequency is a "transition frequency" f related to the emitted energy E by the Planck-Einstein equation,

$$E = hf. \tag{18.6}$$

Under certain circumstances the oscillating electron can undergo a transition to a nearby energy state so that the energy E is small. Then the emitted frequency f, being proportional to E, is also small, possibly much less than the actual oscillation frequency of the electron. As a general rule, however, the radiated frequency from an atom or any other system is expected to be of the same order of magnitude as the oscillation frequency of the charged particles in the system.

Consider an electron circling a proton at a distance of 1 A ($r = 10^{-8}$ cm). The electron experiences a Coulomb centripetal force given by the product of the charges divided by the square of the distance:

$$F = \frac{e^2}{r^2}.$$

According to Newtonian mechanics, which will suffice for this approximate calculation, its acceleration is v^2/r. The mass of the electron multiplied by this acceleration is equal to the force:

$$\frac{mv^2}{r} = \frac{e^2}{r^2}. \tag{18.7}$$

This equation can be solved for the speed v to give

$$v = \sqrt{\frac{e^2}{mr}}. \tag{18.8}$$

The oscillation frequency of the electron is equal to its speed divided by the circumference of its circular orbit, or

$$f_{osc} = \frac{v}{2\pi r}. \tag{18.9}$$

Substitution from Equation 18.8 for v gives a final result for the oscillation frequency of the electron:

$$f_{osc} = \frac{1}{2\pi}\sqrt{\frac{e^2}{mr^3}}. \tag{18.10}$$

Evaluation of the right side of Equation 18.10 in cgs units gives

$$f_{osc} = \frac{1}{2\pi}\sqrt{\frac{(4.80 \times 10^{-10}\ \text{e.s.u.})^2}{9.1 \times 10^{-28}\ \text{gm} \times (10^{-8}\ \text{cm})^3}} = 2.5 \times 10^{15}\ \text{sec}^{-1}. \tag{18.11}$$

An electromagnetic wave of this frequency lies in the ultraviolet, and indeed most atoms emit readily in the ultraviolet. The outermost, or valence, electrons of atoms, typically about 1 Angstrom from the nucleus, are principally responsible for the emission of visible and ultraviolet light. Electrons closer to the nucleus, experiencing greater acceleration, are responsible for the higher frequency X-rays. Molecules have a more complex range of possible emission frequencies, for they have, besides electronic motion similar to atoms, bulk motion of rotation and vibration. Since the atoms composing a molecule are much more massive than electrons, they move at a more stately pace and emit correspondingly lower frequency radiation than do electrons. Molecular rotation and vibration are important sources of infrared radiation.

Within a nucleus, a proton, although massive, experiences such a strong nuclear force that its oscillation frequency is actually much greater than the oscillation frequency of atomic electrons, in fact about one million times greater. A typical nuclear gamma-ray photon, the result of the oscillation of nuclear protons,* has an energy of about 1 MeV, whereas a photon of visible light has an energy of about 1 eV.

To summarize the emission of radiation by naturally occurring oscillation of charge: Molecular vibration and rotation produce infrared radiation; oscillation of the outermost electrons in atoms and molecules produces primarily visible light and ultraviolet radiation; more tightly bound electrons near the nucleus of an atom produce X-radiation; within the nucleus, the accelerated motion of protons and neutrons produce still higher frequency gamma rays. Some quantum transitions are also known in atoms and molecules with energy changes so small that the emitted radiation is in the radio and microwave region. However, these radiations are weak. Electrons "prefer" to emit radiation with a frequency not greatly different from their own frequency of oscillation.

A good emitter of radiation is a good absorber of the same radiation. Molecules readily absorb infrared radiation, atoms absorb visible light and ultraviolet light, nuclei preferen-

* Neutrons in the nucleus also contribute to gamma-ray emission because of their magnetism.

tially absorb gamma rays. X-rays emitted by a particular atom are readily absorbed by the same kind of atom. However, the very high frequency X-rays emitted by heavy atoms are not easily absorbed by lighter atoms, for in the lighter atom, not even the innermost electron has a sufficiently high frequency of oscillation to match the X-ray frequency. This fact underlies X-ray photography and explains why high-frequency or "hard" X-rays easily penetrate organic matter composed mostly of hydrogen, carbon, nitrogen, and oxygen (atomic numbers 1, 6, 7 and 8), but penetrate calcium-rich bone with difficulty (the atomic number of calcium is 20), and heavy metals scarcely at all.

Note in the catalogue of natural emitters that the larger the system, the larger the wavelength (or the lower the frequency). Molecules are larger than atoms. Valence electrons occupy more space than inner electrons, which in turn occupy more space than nuclei. A longer wavelength (or lower frequency) can be achieved either by decreasing the speed of the oscillating charge, or by increasing the distance it traverses in each cycle of oscillation. An approximate formula relating the emitted wavelength to the speed of the emitting particle and to the size of its orbit is

$$\lambda \approx \frac{c}{v} d . \tag{18.12}$$

Here λ is the wavelength, c is the speed of light, v is the average speed of the oscillating particle, and d is the distance covered by the particle in each cycle. It is left as an exercise to derive this equation. Notice that, since the ratio c/v must be greater than 1, the emitted wavelength must be as large or larger than the emitting system.

No systems of particles smaller than nuclei are known. Therefore no radiation with wavelength less than that of nuclear gamma rays (down to about 10^{-12} cm) is known to result from the natural oscillation of charge. A single particle may, however, be considered a "system" whose inner details still remain a mystery. Since a neutral pion decays explosively into two photons, the only assumption consistent with all else that we know about the emission of photons is that within the structure of a single neutral pion resides oscillating charge. (This is not inconsistent with the neutrality of the particle. Recall that an atom is also neutral.) Each photon emerging from the decay of a neutral pion has an energy of about 70 MeV and a wavelength of about 2×10^{-12} cm.

One other naturally occurring oscillation of charge remains to be mentioned, one that occurs on a much grander scale than the molecular, atomic, or nuclear. Free electrons and protons in space experience magnetic fields and accordingly are deflected by magnetic force. In a constant magnetic field, a particle with a charge Q follows a circular or helical path whose radius r is related to the particle momentum p and the magnetic field strength $\mathscr{B}$ by Equation 16.5,

$$r = \frac{pc}{Q'\mathscr{B}} .$$

Such a particle is experiencing an acceleration,

$$a = \frac{v^2}{r} = \frac{v^2 Q'\mathscr{B}}{pc} , \tag{18.13}$$

and accordingly it emits electromagnetic radiation. This radiation from the cosmos, detected only in recent years with the advent of radioastronomy, is named after a modern invention, the synchrotron. For the designers of circular accelerators, synchrotron radiation is an unfortunate nuisance, for it drains energy from the circulating particles. For the astronomer, on the other hand, cosmic synchrotron radiation is an important source of information about conditions in other galaxies and in other parts of our own galaxy. Most of the low-frequency radio waves striking the earth from outer space are probably synchrotron radiation. Their study has made it possible to map the magnetic field in the spiral arms of our galaxy, and to identify a new class of distant objects that radiate intensely in the radio frequency part of the spectrum, the quasi-stellar objects, or quasars.

Man-Made Oscillation of Charge

Between the extremes of atoms and galaxies, man has created oscillators in order to produce controlled radiation, primarily for communication. A radiating antenna of a radio broadcasting station is simply a wire in which a current of electrons is caused to oscillate at a particular frequency. Information is superimposed on the "carrier wave" by varying its intensity (amplitude modulation, AM) or by varying its frequency slightly above and below the standard frequency (frequency modulation, FM). Television uses a combination of AM and FM.

A receiving antenna can be thought of simply as a collection of charged particles—the mobile conduction electrons in the metal antenna—ready and able to be accelerated by an arriving electromagnetic wave. Because the receiving antenna is connected to a circuit with a characteristic natural frequency of oscillation, a large current will be generated in the antenna only by a wave whose frequency is equal to the characteristic frequency of the receiver circuit (Figure 18.8). The electric field in the incoming wave taps the electrons in the wire back and forth, perhaps two million times per second. If the receiver circuit oscillates naturally at this same rate, it is said to be in *resonance* with the incoming wave. This electrical resonance is closely analogous to the more familiar mechanical resonance. If a child on a swing is pushed repeatedly at a frequency different from the natural frequency of the swing, very little motion results. However, if the pushes are in resonance with the natural frequency of the swing, each push adds a little more to the amplitude until the swing gains a large energy. Without the phenomenon of resonance, radio and TV receivers would be useless, for they would receive only a very weak jumble of the myriad of different frequencies impinging on their antennas.

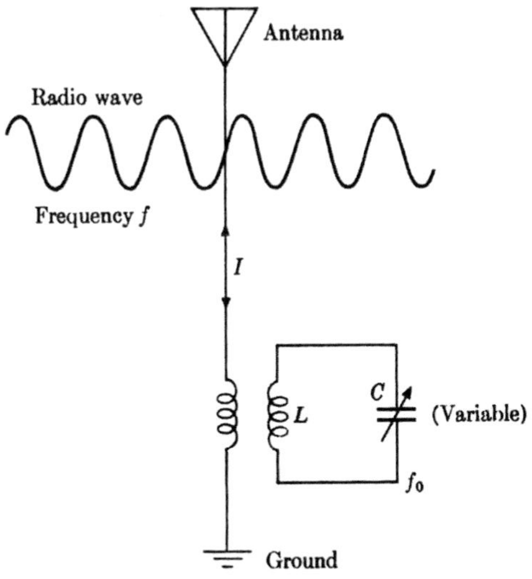

FIGURE 18.8 Part of the circuit in a radio receiver. In the antenna wire, connected to ground through an inductor, electrons are set into oscillation at the frequency f of the arriving radiation. Mutual induction couples the antenna to an oscillator circuit whose natural frequency f_0 is adjustable by means of a variable capacitor. Only when $f = f_0$ will a large current be built up in the oscillator circuit. This is the phenomenon of resonance.

In a typical AM broadcasting antenna, the current reverses direction two million times per second (one million full cycles per second, or 1 megacycle). This seems rapid, but viewed in proper perspective, it is slow. In one millionth of a second, radio waves can travel 3×10^4 cm, about one-fifth of a mile. In much less than one millionth of a second, signals can travel from one part to any other part of a normal-sized electric circuit. Man-made electric oscillations of 1 megacycle or even 100 megacycles are not difficult to create with electrons in wires. At higher frequencies difficulties do arise, associated with the physical size of electric circuits. For infrared radiation, the "oscillator circuit" is a single molecule. As man pushed radio frequencies higher and higher toward the infrared, he had to devise ever smaller circuits, in which electric signals could get from one part of the circuit to another in less than the time of a single cycle of oscillation. In the microwave region, where wavelengths are measured in centimeters or millimeters, the oscillator is contained entirely within a single evacuated tube, a klystron or a magnetron. In these tubes, electrons oscillate over distances of a few millimeters in space, in contrast with their oscillation over many meters of wire in radio transmitters. Developed over the past few decades, these tubes have successfully bridged the gap between radio and infrared.

Acceleration Without Oscillation

At extremely high frequencies beyond the range of nuclear gamma rays, where neither man nor nature has constructed oscillators, single or nonrepeating accelerations are the only source of radiation. One form of such radiation is called bremsstrahlung (one of many German words that have entered the international vocabulary of science: it means "deceleration radiation"). When a high-energy charged particle is deflected, it can emit one or more photons and so lose energy. Aside from the limitation imposed by the conservation of energy, there is no restriction on the frequency of the photon. The particle could give up almost all of its energy to a single high-frequency photon, or it could lose but a tiny fraction of its energy to a low-frequency photon. However, the modern quantum theory of electromagnetism successfully predicts the relative probability for the emission of different frequencies. Synchrotron radiation can be regarded as one form of bremsstrahlung. It differs from the usual bremsstrahlung in that the acceleration is produced by a magnetic field rather than by a collision. Like ordinary bremsstrahlung, it covers a broad frequency range.

Irregular and nonrepeating acceleration is also an important source of lower frequency radiation. In the sun, electrons and protons move randomly in frequent collision and emit a broad spectrum of radiation. A metal heated to incandescence emits light in a similar way, from random accelerations of electrons in the material. This type of radiation, called thermal radiation, contains a wide range of frequencies, but is particularly strong in photons whose energy is not very different from the average energy of the randomly accelerated charged particles. Mathematically, this statement reads

$$hf \approx \tfrac{3}{2}kT\,, \tag{18.14}$$

in which the left side is the energy of a "typical" photon and the right side is the average kinetic energy of the particles. For ideal thermal radiation, called "black-body radiation," resulting from perfectly established thermal equilibrium between photons and material particles, the average photon energy is somewhat less than twice the average particle energy:

$$hf_{\text{av}} = 2.70kT\,. \tag{18.15}$$

The distribution of intensity in black-body radiation is shown in Figure 18.9. Solar radiation and radiation from the tungsten filament of a light bulb approximate this distribution. The yellow light from a sodium vapor lamp, by contrast, is predominantly of a single frequency (monochromatic). It results not from random accelerations of electrons, but from a particular quantum transition of the valence electron in the sodium atom.

Every emission process has its corresponding absorption process. Since a single deflection of a charged particle can cause a photon to be emitted, photon absorption can give to a charged particle a single pulse of acceleration. Such an absorption process is called a free-free transition, since the absorbing particle is in a free state both before and after the absorption.* This is a common process in the sun, where photons are constantly being absorbed as well as emitted.

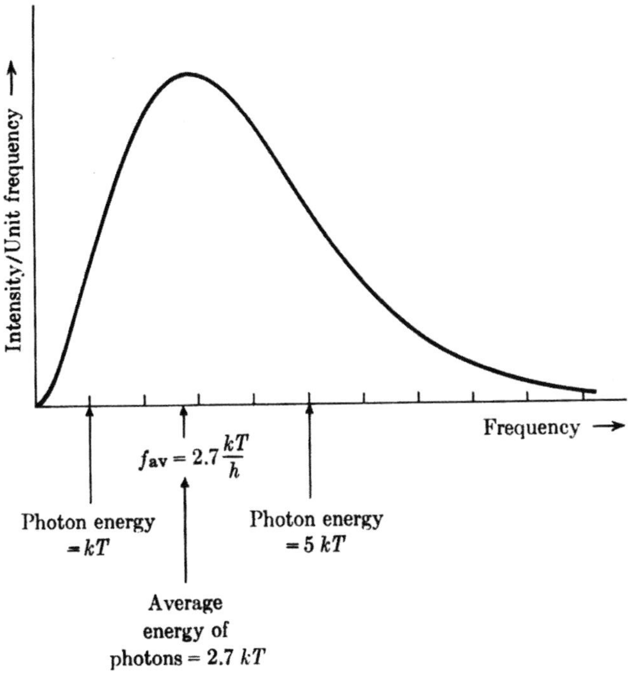

FIGURE 18.9 Black-body radiation. This distribution of intensity vs. frequency results from the random acceleration of atoms or molecules at a particular temperature *T*.

One single important idea has been presented in this section: The emission and absorption of electromagnetic radiation is associated with the acceleration of charge. The acceleration can be the leisurely oscillation back and forth of a current in an antenna, or the frenetic oscillation of an electron in an atom, the gentle deflection of a proton in a galactic magnetic field, or the violent deflection of a pion in a nuclear collision. From one situation to another the frequency may be vastly different, the physical system unrelated, the physical manifestations of the radiation unrecognizably altered; but the underlying idea of an electromagnetic wave coupled to an accelerated charge remains the unchanged unifying thread tying all parts of nature together.

18.5 Properties of waves

Of all the known kinds of waves, electromagnetic waves are the most universal, the fastest moving, and the most strikingly varied in their manifestations across the spectrum. Many of their properties, however, are common to waves in general and can be discussed in a wider context. In sections to follow, we shall examine half a dozen properties of electromagnetic

* A bound-bound transition is one in which an electron absorbs a photon and jumps from one bound state to another in an atom or molecule. In a bound-free transition, an atomic or molecular electron absorbs enough energy to be ejected from a bound state to a free state.

waves that are shared with some or all other forms of wave motion:

1. Polarization.
2. Superposition.
3. Reflection.
4. Refraction.
5. Diffraction.
6. Interference.

The fundamental differentiation among waves is the nature of whatever it is that vibrates. A wave is not by itself a material thing, but a name for a particular oscillatory state of motion of something else. This is an obvious statement, but it is well to keep in mind how sound waves and light waves, with so many properties in common, do differ in the most fundamental way. One is the mechanical vibration of matter, the other the coupled vibration of electric and magnetic fields free of matter. Because matter is granular, divided into atoms and molecules, a sound wave or other mechanical vibration requires a domain of many atoms in order to exhibit an average wave behavior. Electromagnetic waves, on the other hand, can exist, so far as we know, in arbitrarily small volumes and with arbitrarily short wavelengths.

The speed of a wave is governed by the nature of the vibrating medium. The speed of light is a constant associated with the coupling between electric and magnetic fields. The speed of sound is determined by the interaction between atoms or molecules. In a gas, the speed of sound depends on the temperature and the molecular weight, the quantities that affect molecular speed (see Equation 13.30). In air at normal temperature, sound waves travel at 3.3×10^4 cm/sec, about one millionth the speed of light. In a solid, the closely packed atoms conduct sound more rapidly. Some speeds of sound in common substances are shown in Table 18.1.

TABLE 18.1 The Speed of Sound in Common Substances

Material	Speed of Sound at Room Temperature (cm/sec)	Ratio to Speed in Air
Air	3.43×10^4	1.0
Hydrogen	1.32×10^5	3.9
Chlorine	2.13×10^4	0.6
Water	1.49×10^5	4.4
Kerosene	1.34×10^5	3.9
Iron	6.0×10^5	17.5
Aluminum	6.4×10^5	18.7
Glass (Pyrex)	5.6×10^5	16.3
Rubber	$(1.5 \text{ to } 1.8) \times 10^5$	4.4 to 5.3

Maxwell and his followers in the late nineteenth century thought of the ether filling space as analogous to a solid. They thought of each point in the ether being coupled to its neighboring points, the strength of this coupling determining the speed of light. This mechanical conception of the ether gave way before the theory of relativity, yielding to the more abstract picture of interacting fields.

18.6 Polarization

If a wave is transverse, it is said to be polarizable, meaning that it can be caused to vibrate in a particular direction perpendicular to its direction of propagation. A water wave traveling horizontally is vertically polarized, for it is vibrating up and down. In a light wave, the direction of polarization is *defined* to be the direction in which the *electric* field is vibrating. (This is an arbitrary matter. The magnetic field could equally well have been chosen to define the direction of polarization.) Vertically and horizontally polarized light waves are illustrated in Figure 18.10.

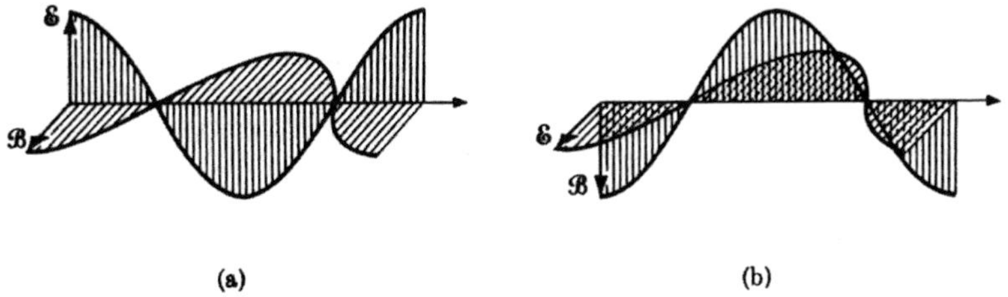

FIGURE 18.10 **(a)** Vertically polarized light wave. **(b)** Horizontally polarized light wave.

A wave can be unpolarized for two reasons. It might be longitudinal and therefore, by its very nature, incapable of being polarized. This is true of sound waves. Or it might be a mixture of many different directions of polarization so that its average polarization is zero [Figure 18.2(b)], Light coming from the sun or from an electric bulb is unpolarized for this reason. Its myriad of photons stream out with randomly oriented polarizations. Any device that selects from this array a particular direction of polarization is called a polarizer. A doubly refracting crystal is a polarizer, for it refracts different polarizations differently and permits two polarized beams to be separated (Figure 18.3). The material used in Polaroid sunglasses is an efficient polarizer working on a different principle, selective absorption. Polaroid is nearly transparent to light whose electric field vibrates parallel to an axis of molecular alignment in the material, and is nearly opaque to light whose electric field vibrates perpendicular to this axis (Figure 18.11).

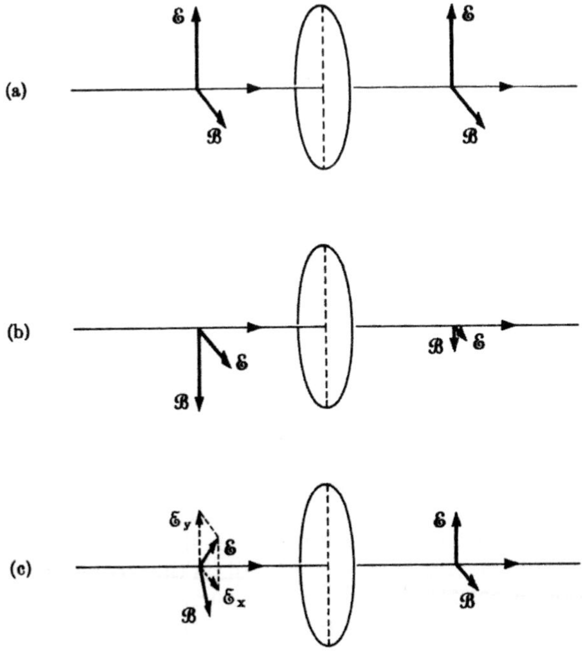

FIGURE 18.11 Action of Polaroid. **(a)** Light polarized in one direction is transmitted with little absorption. **(b)** Light polarized perpendicular to the favored direction is almost completely absorbed. **(c)** Light polarized at an intermediate angle is partially absorbed and emerges with the favored direction of polarization. (Dotted line indicates Polaroid axis.)

The use of Polaroid in sunglasses is related to another mechanism of polarization, ordinary reflection. In about 1810, Etienne Malus discovered that light reflected from glass and water is partially polarized, and at a certain critical angle for each substance, completely polarized. The reflected beam is strongest in light polarized parallel to the surface of the reflector; the refracted beam is strongest in light polarized oppositely (Figure 18.12). The axis of the Polaroid in sunglasses is therefore oriented in order to absorb horizontally polarized light and thereby eliminate some of the glare of reflected light. A sunbather lying on her side (Figure 18.13) foregoes this advantage, for her rotated sunglasses preferentially transmit the reflected light instead of preferentially absorbing it.

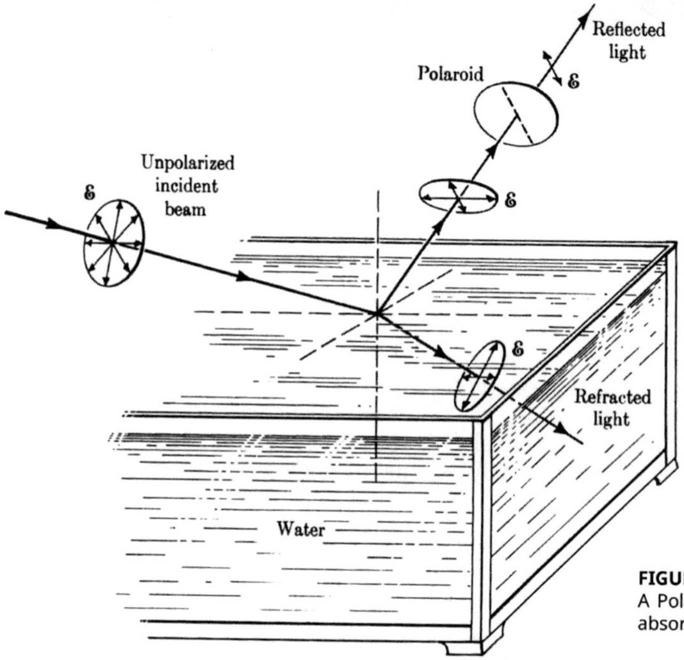

FIGURE 18.12 Polarization by reflection. A Polaroid oriented as shown strongly absorbs the reflected light.

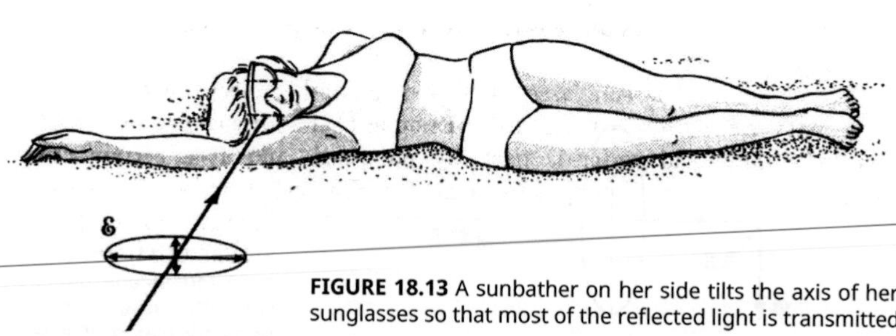

FIGURE 18.13 A sunbather on her side tilts the axis of her Polaroid sunglasses so that most of the reflected light is transmitted.

A microwave polarizer, acting on macroscopic wavelengths, is particularly easy to understand (if approached warily). It consists of a grid of parallel wires (Figure 18.14), the length of each wire being greater than the wavelength of the radiation, the spacing of the wires being less than the wavelength. If unpolarized microwaves fall on this grid with its wires running vertically, which direction of polarization will pass through? Here is where caution is required. One might think uncritically of a skinny prisoner slipping between two bars of his cell and answer: Vertically polarized waves pass through. This is wrong. The vibrating electric field in the vertically polarized wave sets the electrons in the wires into up-and-down oscillation. Energy is transferred from the field to the electrons, so that the wave is absorbed. For the horizontally polarized wave, on the other hand, the vertical wire is a very poor absorber, for its electrons are prevented from swinging back and forth horizontally through more than the small diameter of the wire. Therefore it is the horizontally polarized wave that sneaks through the vertical grid almost undiminished, while the vertically polarized wave is absorbed.

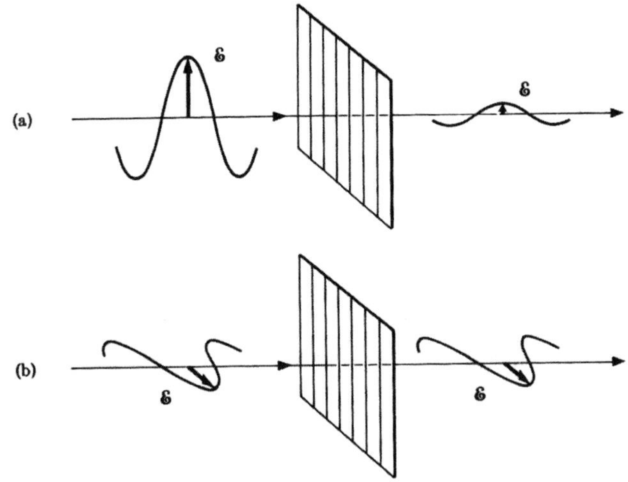

FIGURE 18.14 Microwave polarizer. **(a)** With the $\mathcal{E}$ vector parallel to the wires of the grid, electrons are set into oscillation along the wires and absorb the wave energy. **(b)** With the $\mathcal{E}$ vector perpendicular to the wires, the electrons cannot move back and forth far enough in the thin wires to absorb much energy.

This microwave polarizer illustrates how marked can be the effect of changing the wavelength of electromagnetic radiation. For microwaves, the grid is wholly transparent to one polarization, largely opaque to another. For light, it has no sensitivity whatever to polarization. For microwaves, it acts as a single unit. For light, it is a series of alternate transparent and opaque stripes acting separately. Similar illustrations could be drawn from many parts of nature. Always the important quantity is the scale of the object relative to the wavelength of the radiation.

A vitally important aspect of polarization that we have hinted at but not yet properly discussed is its vector character. This means that a wave polarized in any direction can be regarded as the sum of two or more waves polarized in other directions—that is, it can be resolved into components. Also two or more differently polarized waves propagating in the same direction precisely in phase act in consort as a single wave with a single polarization. The idea of component polarization appeared in Figure 18.11. A second example is illustrated in Figure 18.15. An electromagnetic wave propagates straight upward from a radar antenna, polarized in the northeast-southwest direction. This wave can also be treated as two equally intense waves, one polarized north-south and one polarized east-west. If a wire grid polarizer is superposed with its wires running east-west, what gets through? The common-sense answer (and the correct answer) is a north-south polarized wave half as intense as the original wave. Although both sensible and correct, some subtlety actually lies behind this answer. By the cosine law for vector components, if the peak electric field to the northeast is $\mathcal{E}$, the peak fields to north and east are

$$\mathcal{E}_{\text{north}} = \mathcal{E} \cos (45 \text{ deg}) = 0.707\mathcal{E}, \tag{18.16a}$$

$$\mathcal{E}_{\text{east}} = \mathcal{E} \cos (45 \text{ deg}) = 0.707\mathcal{E}. \tag{18.16b}$$

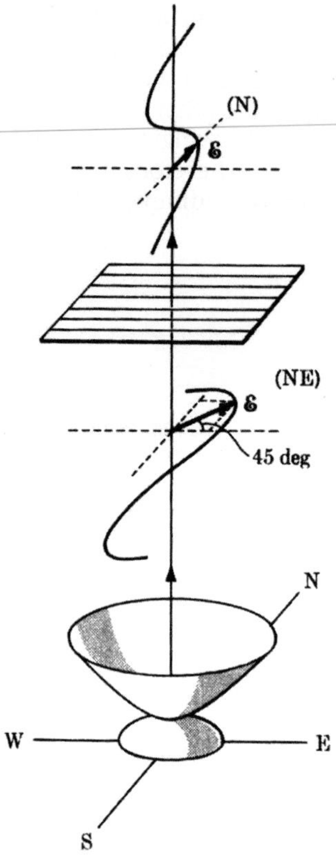

FIGURE 18.15 Effect of microwave polarizer oriented at 45 deg to polarization direction of radar wave.

The field strength slipping through the polarizer is not half of the original field strength, but more than 70% of it. However, the intensity of the wave—the energy in the wave—is proportional to the square of the electric field strength (see Equation 16.25). Therefore the energy in the transmitted wave is indeed half the original energy.

18.7 Superposition

103 The vectorial combination of differently polarized waves illustrates one aspect of the fundamental idea of superposition. As discussed in Section 15.7, the superposition of electric or magnetic fields means that the field at a point arising from several sources is the vector sum of the separate fields that would be present if each source were acting alone. Physically, it means that fields coexist in space without influencing each other. If a second field is added where a field exists already, the first is unchanged by the second, and the second is uninfluenced by the first. The total field is the sum of the two. Consider, by contrast, a law of nonsuperposition, the graduated income tax law. If a man earns $10,000 and then $10,000 more, the tax on the second $10,000 is influenced by the presence of the first $10,000. The total tax is not the sum of the taxes that would be paid on the separate amounts.

Since electric and magnetic fields can be superposed, it follows that electromagnetic waves can be superposed. Think of light from two stars intersecting somewhere in the depths of space. Each beam continues inflexibly on its way, undeviated and unchanged by having interpenetrated the other. Of more vital importance on earth is the superposition of radio waves, without which radio communication would be impossible. Think of the electromagnetic disturbances in the space around you. If they were visible you would see a gaudy and ever changing splash of every conceivable color. If they were audible you would hear a cacophony of sound. Yet every component of this turmoil, with its own frequency, polarization, and direction of propagation, retains its unique identity, undisturbed by sharing the same space with thousands of other signals.

In the seventeenth century, long before the full import of superposition was appreciated for electromagnetic waves, Christian Huyghens made use of the idea of superposition in describing the propagation of a single wave. He asserted that each point on an advancing wave front could be regarded as the source of a new wavelet, and that the subsequent position of the wave front was determined by the superposition of all of the tiny wavelets. This application of the idea of superposition, summing individual contributions to an advancing

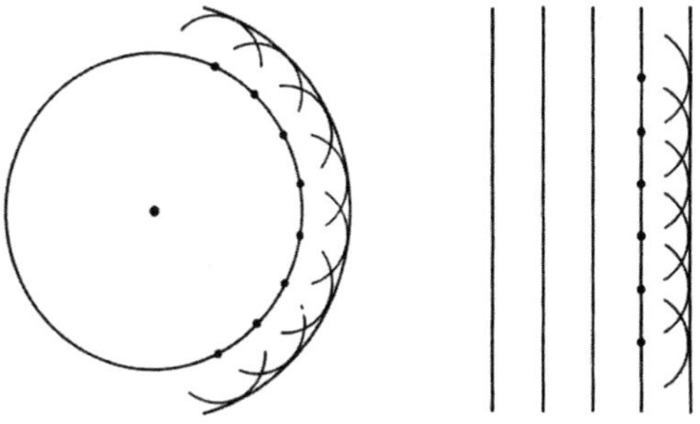

FIGURE 18.16 Huyghens' principle. A broad advancing wave front can be regarded as the superposition of many tiny wavelets emanating from points along the wave front at an earlier instant.

wave to get the total wave, is known now as Huyghens' principle. In a modernized form that takes proper account of the phase of the wavelets, it provides a very useful approach for understanding many wave phenomena, including interference and diffraction.

The simplest application of Huyghens' principle is to the unimpeded advance of a broad wave front. Figure 18.16 shows how the superposition of wavelets accounts for the advance of a spherical wave and a plane wave. The physical justification of Huyghens' principle can be understood best by considering water waves. If a single point on an otherwise calm water surface is caused to move up and down, a circular wave propagates outward from that point. Now consider straight waves rolling across a water surface and focus attention on any single point on the surface. This point oscillates up and down exactly as at the center of a circular wave. If we pretend blindness to the rest of the wave, we would say that the motion of this particular point should produce a circular wavelet. But so should its neighboring point and the point next to that. The sum of all these circular wavelets, said Huyghens, should be the straight-line wave propagating forward. In order to apply Huyghens' principle in the simplest way, it is necessary to consider the wavelets propagating from points along a wave front. These are points of equal phase that rise and fall together. Only for forward propagation of the wave front do the wavelets reinforce one another. For other directions, destructive interference among the wavelets eliminates any net propagation. Even for a single wave traveling straight ahead, the idea of interference is useful.

For water waves and other waves in material media, the principle of superposition is usually valid to good approximation, but it is never precisely valid. Figure 18.17 shows superposed water waves interpenetrating with no noticeable mutual effect. The condition required for superposition is that the wave disturbance be relatively weak—gentle waves on water, for example, but not breaking waves at the beach; or mild sound waves, but not shock waves. If two shock waves cross one another, the resulting wave at their point of intersection is not the sum of the two separate waves, and they do not continue on their way uninfluenced by the encounter.

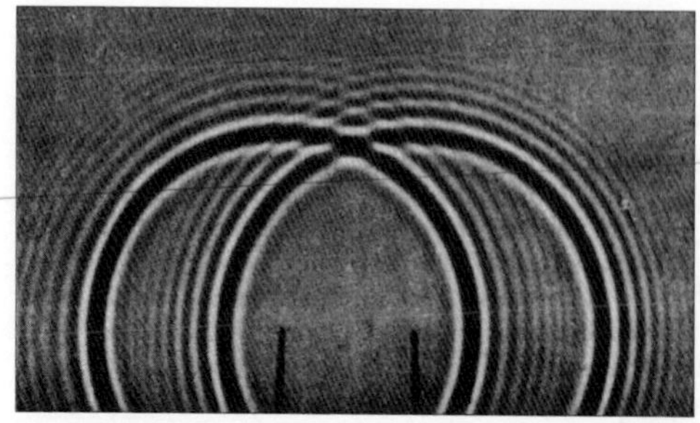

FIGURE 18.17 Superposed water waves. (Photograph courtesy of Film Studio, Education Development Center.)

The development of the quantum theory of radiation in the 1930s revealed that not even for electromagnetic waves is the principle of superposition precisely valid. If two photons meet in space there is a chance—an exceedingly small chance—that they will interact and be deflected by the encounter. This violates the principle of superposition, for the resulting radiation flow is not the same as if each photon had passed separately by. This process, known as the scattering of light by light, has been measured in the laboratory, but it is too improbable to have any perceptible effect on radio communications or on any of the other practical examples of superposed electromagnetic waves (or photons). It is interesting, however, that if two extremely intense beams of light (more intense than any known) intersected, they could interact strongly, as shock waves do, and seriously violate the principle of superposition. Another way to say this is that photons, if crowded together densely enough, would begin to act like bits of matter.

18.8 Reflection

A tennis ball thrown against a wall is "reflected." If it is not spinning, it obeys approximately the law of reflection: The angle of incidence is equal to the angle of reflection. Light striking a perfectly flat mirror obeys this law exactly, but the mechanism of its reflection is much more complicated than the mechanics of the tennis ball's reflection. Light waves or photons do not simply bounce from the reflecting surface. They are absorbed and reemitted. They are diminished in intensity and are partially polarized. These details need not concern us here, however. We wish only to understand the law of reflection in terms of Huyghens' principle of superposed wavelets. The diagram in Figure 18.18 is useful for this purpose. We consider a particular wave front, initially at AA', later at BB', and an equal time later at C. As each point on this wave front reaches the reflecting surface, it gives rise to a new wavelet propagating in all directions. We are interested particularly in the wavelets propagating back outward from the reflecting surface.

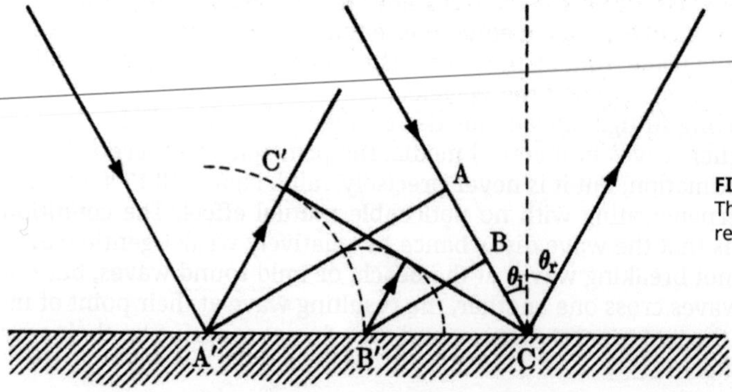

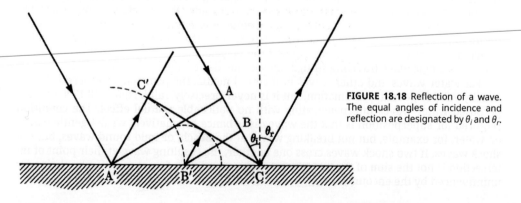

FIGURE 18.18 Reflection of a wave. The equal angles of incidence and reflection are designated by θ_i and θ_r.

In the time it has taken the incoming wave front to move from A to C, the wavelet beginning at A' has expanded to a radius A'C' equal to the distance AC. The wavelet beginning at B', starting later, has propagated back only half as far.

The dashed curves outline these two wavelets at the instant the incoming wave front reaches the point C. At this instant, the wavelet emanating from C has not yet moved. It is indicated by a heavy dot at C. The reflected wave front is then the line CC', touching the two circles and the dot at C. The geometry of the diagram makes clear that the angle of reflection is equal to the angle of incidence. This comes about basically because of the equal speed of the incident wave and the reflected wave. Because the radius of the circle A'C' is equal to the distance AC, the angles of incidence and reflection are equal.

After this demonstration and others like it were given by Huyghens, there was no theoretical bar to the acceptance of light waves. He showed that waves could account as well as particles for the known properties of light. Crucial tests of the wave theory came more than a century later.

18.9 Refraction

Refraction, the deflection of a wave passing from one medium to another, is caused by the difference in the speed of the wave in the two media. Why light is slowed in passing from air to glass is by no means a simple matter. The speed of light, c, is no less a universal constant in glass than in air or in empty space. However, in making its way through glass, light is continually being absorbed and reemitted. Thinking in terms of photons, we can picture a jerky process in which the light is alternately moving at the constant speed c and not moving at all. This start-and-stop progression through the glass results in an average speed less than the constant c. Light moves most rapidly in a vacuum. In any material medium, its average speed is less. In air at ordinary density the average speed is about 0.3% less than c. In glass the average speed may be 30% less than c. Fortunately, the erratic progress of photons through matter, when averaged over the myriad of photons involved, results in a smooth wave behavior, much as in empty space, but with a slower wave speed in the medium.

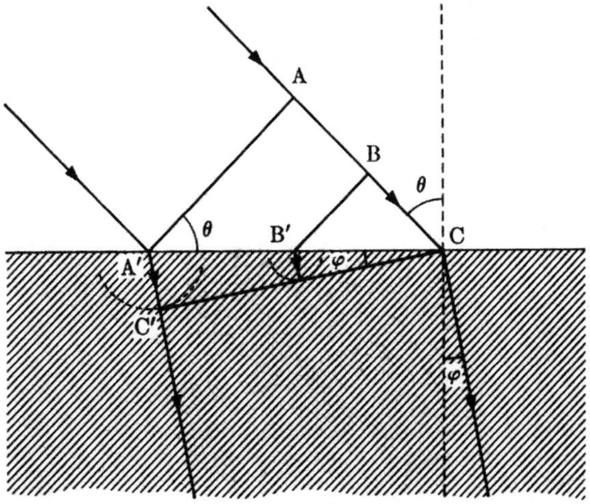

FIGURE 18.19 Refraction of a wave. The angle of incidence is θ, the angle of refraction is φ. The speed is slower in the lower medium.

Huyghens accounted for refraction much as he accounted for reflection, by considering wavelets emanating from the interface between two media. His geometrical construction is still a good way to gain an understanding of the wave aspect of refraction. Consider again a plane wave front approaching a flat interface between two media (Figure 18.19). Let us suppose for convenience that the wave travels more slowly in the lower medium, and that the ratio of speed in the upper medium to the speed in the lower medium is a number n (greater than 1). At the moment the wave front reaches the line AA', a wavelet begins to spread out into the lower medium from A', but at a speed less than the speed of the incident wave. While the incident wave front advances from A to C, the wavelet initiated at A' has propagated a lesser distance A'C'. The ratio of these two distances is equal to the ratio of the speeds:

$$\frac{AC}{A'C'} = n\,. \tag{18.17}$$

The new wave front in the lower medium extending from C to C' has been deflected. (It is left as an exercise to show that the midpoint wavelet beginning at B' has advanced just far enough to touch the line CC'.) If θ is the angle of incidence, it is also, as shown in Figure 18.19, the angle AA'C. The angle of refraction φ is also the angle C'CA'. Using the definition of the sine of an angle for these two right triangles AA'C and C'CA', we can write

$$\sin\theta = \frac{AC}{CA'}\,, \tag{18.18}$$

$$\sin\varphi = \frac{A'C'}{CA'}\,. \tag{18.19}$$

Taking the ratio of these two sines causes the distance CA' to cancel from the two expressions and gives

$$\frac{\sin\theta}{\sin\varphi} = \frac{AC}{A'C'}\,. \tag{18.20}$$

The ratio on the right is, according to Equation 18.17, the speed ratio *n*, so that we may write more simply

$$\frac{\sin\theta}{\sin\varphi} = n\,. \tag{18.21}$$

This law of sines for refraction was discovered experimentally by Willebrord Snell early in the seventeenth century, and is now referred to as Snell's law. Snell discovered that the ratio of the sines is a constant for any particular pair of media —such as air and glass—but he was probably unaware of the physical significance of the constant *n* as a speed ratio.

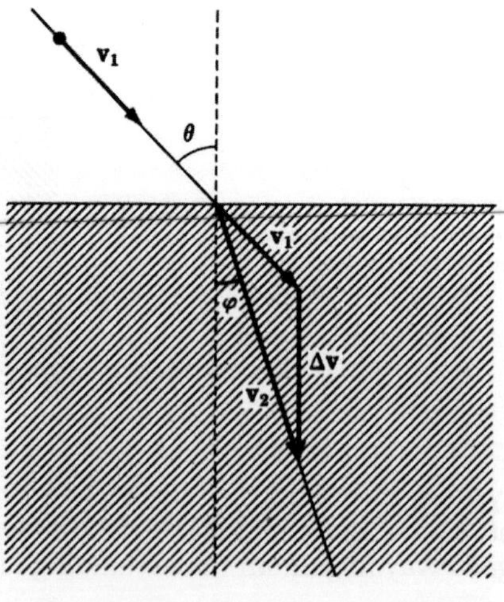

FIGURE 18.20 Refraction according to the particle theory of light. The particle moves faster in the lower medium.

According to the particle view of light adopted by Descartes and explored by Newton, refraction is accounted for by an *increase* of speed of light in passing from air to a denser medium. Figure 18.20 illustrates this alternative conception of refraction. A particle of light entering glass from air is supposed to be attracted by the glass surface, thereby being given an inward acceleration and deflected to a path more nearly perpendicular to the surface. Thus before the end of the seventeenth century one crucial test of the wave vs. the particle theory of light was clear. Does light travel more rapidly or more slowly in dense matter than in air? The answer to the question remained beyond the means of experimental physicists for more than 150 years. Finally in 1850 Jean Leon Foucault succeeded in measuring the speed of light in water and found it to be less than in air.

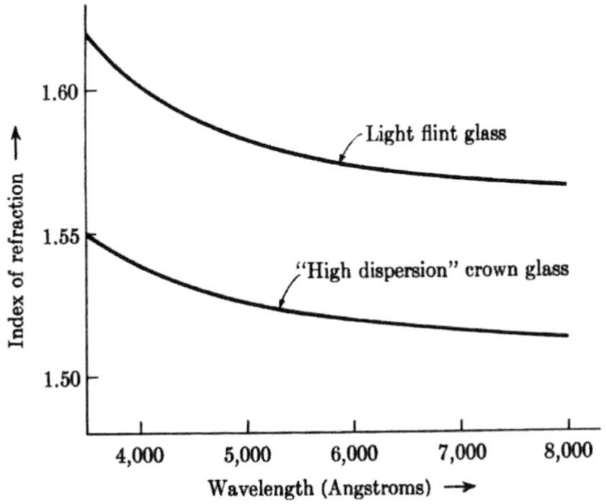

FIGURE 18.21 Variation of index of refraction of two kinds of glass with wavelength of light.

The ratio of the speed of light in vacuum to its average speed in a particular substance is called the index of refraction of the substance. (The quantity n in Equation 18.21 is the ratio of two indices of refraction.) The index of refraction of air at normal density is 1.00029. The index of refraction of water is about 1.33. Indices of refraction for different kinds of glass are in the range 1.4 to 1.8. Besides the variation from one substance to another, the average speed of light in matter varies slightly with the wavelength of the light. In Figure 18.21 is shown a graph of index of refraction vs. wavelength for two kinds of glass. The greater the index of refraction, the greater the refraction. Therefore, upon entering glass at an angle, blue light (shorter wavelength) is deflected more than red light (longer wavelength). This dependence of refraction on wavelength explains why a prism spreads out a beam of white light into a rainbow of colors* (see Figure 18.6).

The separation of white light into colors, or, more generally, of any mixture of wavelengths into separate wavelengths, is referred to as *dispersion*. The dispersion of light by a prism was for a long time the chief tool for analyzing the visible and near-visible parts of the electromagnetic spectrum. Although still used in spectroscopes and spectrographs, the prism has now been largely replaced by the diffraction grating as the means of dispersing light in these instruments.

Any equation is likely to yield interesting insights if examined in one or more limiting situations. This is true of Equation 18.21 expressing Snell's law. You might consider the

* The colors of a rainbow in the sky are similarly accounted for by the dependence on wavelength of the index of refraction of water droplets.

implications of Snell's law if n is very close to unity (the two media almost the same) or if n is much greater than unity. Here we wish to consider another limit, the limit of maximum angle of incidence. This consideration leads to a wholly new idea, total internal reflection. To be specific, consider light passing from air to glass with a speed ratio $n = 1.5$. For zero angle of incidence, it enters the glass undeflected, perpendicular to the surface. As the angle of incidence increases, so does the angle of refraction, but the latter less rapidly. When the angle of incidence has reached its maximum, so that the light beam grazes the surface, $\theta = 90$ deg $= \pi/2$ radians, $\sin\theta = 1$, and Equation 18.21 reads

$$\frac{1}{\sin\varphi_{max}} = n\,. \tag{18.22}$$

The angle of refraction has reached a maximum value φ_{max}, whose sine is given by

$$\sin\varphi_{max} = \frac{1}{n}\,. \tag{18.23}$$

If $n = 1.5$, $\sin\varphi_{max} = 0.667$, and $\varphi_{max} = 41.8$ deg. This particular situation is illustrated in Figure 18.22. All light entering the glass, from whatever direction, is confined within an allowed range of angles within the glass. The other angles, comprising the "forbidden" region, receive no light.

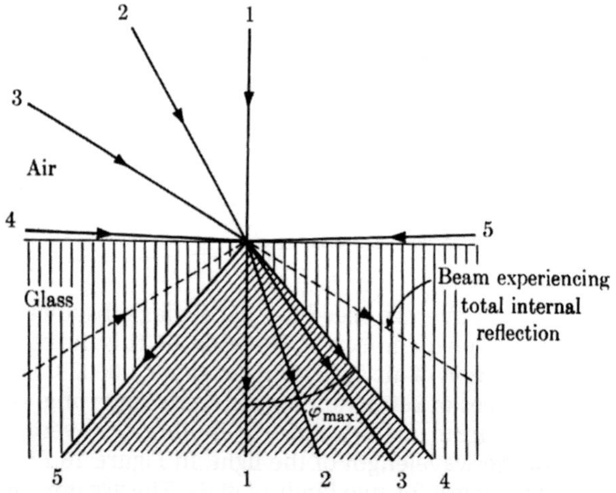

FIGURE 18.22 Refraction at air-glass surface with speed ratio $n = 1.5$. Light entering the glass from the air is confined in angle within the diagonally hatched cone. Angles greater than φ_{max} are "forbidden." A beam of light (dashed line) moving at an angle in the forbidden region cannot escape the glass. It experiences total internal reflection.

In order to extract more from this analysis, we must now consider the reverse process, light passing from glass to air. It is left as an exercise to show that Equation 18.21 remains valid, with φ, the angle in the glass, being now the angle of incidence, and θ, the angle in the air, being the angle of refraction (see Figure 18.19). What happens as the angle φ increases from 0 to $\pi/2$? According to Snell's law, the angle of refraction of the light beam in the air is given by

$$\sin\theta = n\sin\varphi. \tag{18.24}$$

At the critical angle φ_{max}, $n\sin\varphi_{max} = 1$ and $\theta = \pi/2$. The refracted light just escapes into the air, grazing the glass surface. What if we insist on going further, to larger values of φ, causing the light within the glass to strike the surface a more nearly glancing blow? The incident light is then in the forbidden zone, and mathematical trouble results. According to Equation 18.24, $\sin\theta$ becomes greater than one. But there is no angle whose sine exceeds one. Which is wrong, the mathematical assumption that Equation 18.24 remains valid, or the physical assumption that light can exist in the "forbidden" region? Actually neither assumption is wrong. Mathematically, Equation 18.24 tells us correctly that there is *no* angle of refraction θ; no light escapes from glass to air. Physically, it is perfectly possible to illuminate the surface internally within the "forbidden" region (it is forbidden only for light refracted from air to glass). Since this light must go somewhere but cannot cross the surface into the air, it reflects from the surface and remains within the glass. This phenomenon is called total internal reflection. It can occur whenever light is incident on a surface where the average speed of light outside the surface is greater than within the surface.

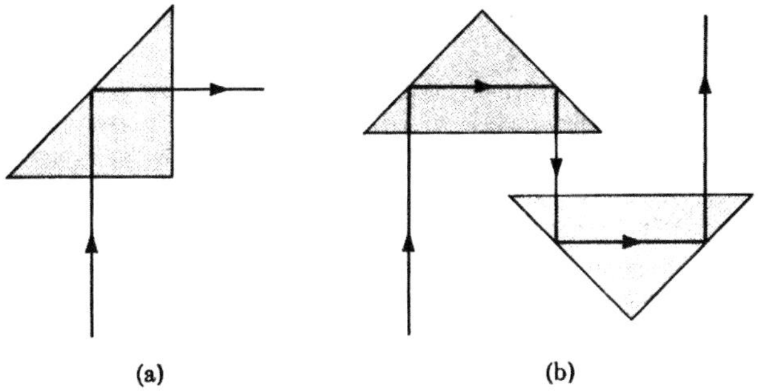

FIGURE 18.23 Practical utilization of total internal reflection in prisms. **(a)** Deflection of light through 90 deg. **(b)** Quadruple reflection to lengthen light path in binoculars.

Total internal reflection can be harnessed for several useful purposes. One is to change the direction of light more efficiently than a mirror can do it. A mirror absorbs some light and is spoiled by dirt. An internally reflecting surface is almost perfectly efficient and its action is less marred by dirt outside the surface. Prisms reflecting light through 90 deg and 180 deg are shown in Figure 18.23. In binoculars, pairs of prisms arranged as in Figure 18.23(b) are used in order to lengthen the light path between the lenses without inconveniently lengthening the barrels of the binoculars. Another interesting use of total internal reflection is in light pipes (Figure 18.24). Lucite light pipes are used by experimental physicists to conduct light from a scintillating crystal to a light-sensitive photomultiplier tube, and by some dentists to conduct light to a localized region in the patient's mouth. A light pipe is literally what its name implies. It "pipes" light from one point to another by a series of total internal reflections. A bundle of many fine fibers, each acting as a separate light pipe, can be used to transmit images along paths of almost any shape (Figure 18.25).

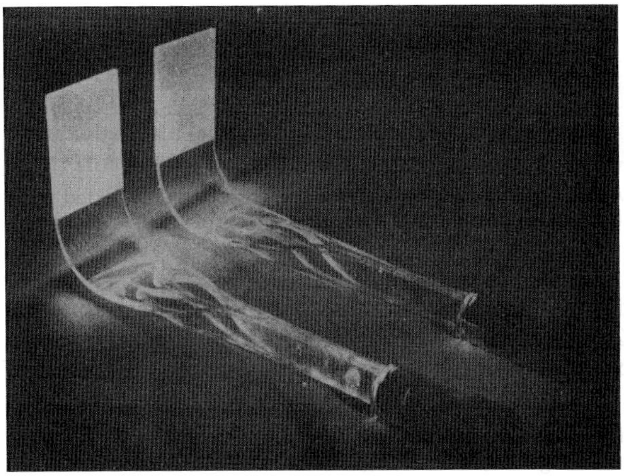

FIGURE 18.24 Light pipes. Light originating in a scintillator plate (painted white) makes its way via total internal reflections through the curved and twisted lucite to a light-sensitive photomultiplier tube. (Photograph courtesy of Lawrence Radiation Laboratory, University of California, Berkeley.)

that touches almost everyone. Yet it is an elaborate area of application based on two simple physical ideas—that waves can be refracted and reflected. The equations appearing in this section, despite their practical importance, must not be confused either in exactness or in fundamental significance with equations such as those expressing Coulomb's law of electric force or Newton's laws of mechanics.

Although geometrical optics rests ultimately on the wave nature of light, it overlooks two key properties of waves, diffraction and interference. These properties, which established the wave nature of light in the early part of the nineteenth century, are the subjects of the next two sections.

18.11 *Diffraction*

Diffraction can be defined as any deviation of wave motion away from straight-line propagation. It manifests itself particularly when waves pass through apertures or around obstacles. Because of diffraction, shadows are not sharp. Instead there is a gradual transition of intensity from the illuminated to the shadowed region. The breadth of this transition region depends on the wavelength (increasing as the wavelength increases) and may also depend on the size of the obstacle or aperture. As shown in Figure 18.31, the shadow transition zone may also contain fringes of alternately greater and lesser intensity.

The diffraction of light, discovered around 1660 by Francesco Grimaldi, was known to Newton and to Huyghens. Somewhat surprisingly, this discovery had little impact on the controversy over the nature of light. Not until after Young's discovery of interference fringes in 1803 was diffraction studied with enough care to provide definitive evidence in favor of the wave nature of light. Until then it was assumed that the deflection of light particles in passing near matter could also account for the smudging of shadows.

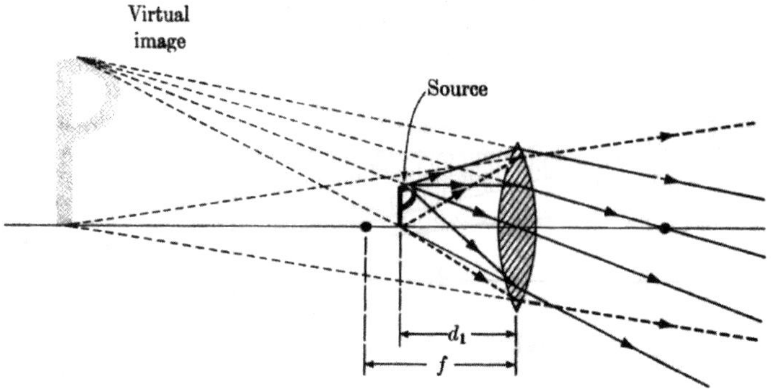

FIGURE 18.30 Formation of a virtual image by a thin lens. When the distance d_1 is less than the focal distance f, the rays seem to come from an enlarged image on the same side of the lens as the object. A hand magnifier forms a virtual image of this kind. Note that the image is not inverted.

According to Huyghens' principle, every point on an advancing wave front acts like a source of a new wavelet spreading out in all directions. This is really a principle of diffraction, in fact a principle of maximum diffraction. In seeking a wave explanation of diffraction, we are faced not so much with the problem of explaining why diffraction *does* occur as with the problem of explaining why it does *not* occur always and obviously for all propagating waves. Why does sunlight not bend around the earth and keep the far side bright in the middle of the night? Why does a searchlight beam seem to travel in a perfectly straight line into the sky? Just such questions were asked by Newton. He was dubious about the wave theory because he doubted its ability to explain the simple fact of straight-line propagation of light.

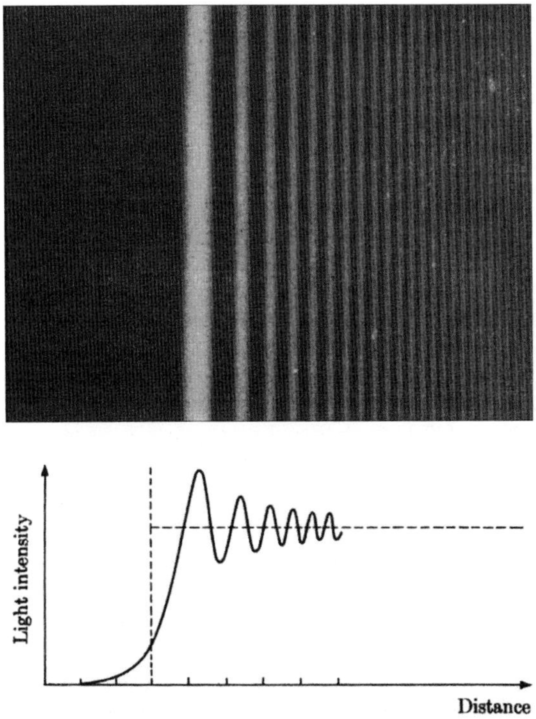

FIGURE 18.31 Diffraction produced by the sharp edge of a screen. A well-defined but not abrupt change from darkness to light at the shadow boundary is followed by a series of fringes of gradually decreasing intensity extending into the illuminated region. [Photograph from M. Cagnet, M. Françon, and J. C. Thrierr, *Atlas of Optical Phenomena* (Berlin: Springer-Verlag, 1962). Courtesy of Professor Françon.]

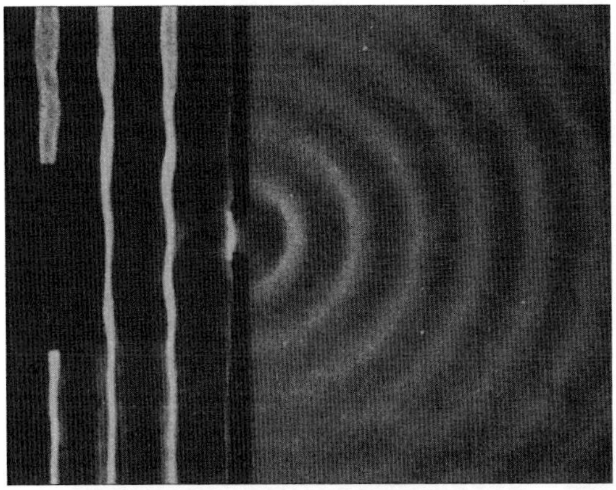

FIGURE 18.32 The reality of the Huyghens wavelet. When all but a narrow segment of an advancing wave front is stopped by a screen, a circular wave spreads in all directions from this small part of the wave front. (Photograph courtesy of Film Studio, Education Development Center.)

The resolution of the apparent conflict between Huyghens' principle and straight-line propagation is to be found in the ideas of superposition and interference. It was mentioned in Section 18.7 that the superposed wavelets emanating from different parts of a wave front can interfere destructively in some directions, thereby eliminating propagation in those directions. For the ordinary propagation of a broad straight wave front, this interference keeps the wave energy moving forward with almost no deviation from straight-line flow. There is a way, however, to see that the extreme diffraction predicted by Huyghens' principle is fact and not fiction. If a screen with a small hole or slit in it blocks all but a narrow portion of a wave front, the Huyghens wavelet will appear on the far side of the screen spreading in all directions, unperturbed by interference with other wavelets. Figure 18.32

shows an example of such a diverging wave coming from a narrow part of a straight wave front on water. The effects of widening the aperture are shown in Figure 18.33. The requirement for extreme diffraction—the "pure" Huyghens spreading wavelet—is that the width of the hole or slit be small compared to the wavelength. (Notice that this is a *relative* requirement. A 1-mm slit is narrow for radar waves but wide for light.) When the width of the wave front becomes comparable to a single wavelength, interference effects between different parts of the wave front set in and begin to narrow the angle of diffraction. Nearly straight-line propagation is restored when the width of the aperture becomes much greater than the wavelength, although residual diffraction effects at the shadow boundary always remain.

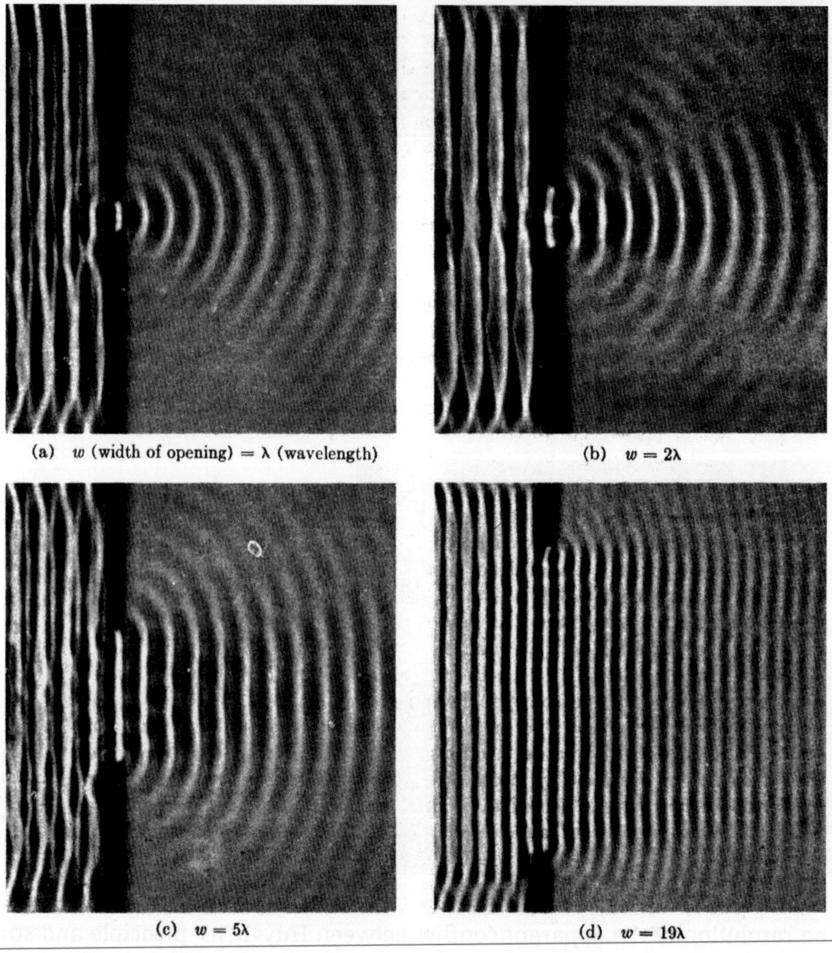

FIGURE 18.33 The effect of slit width on diffraction of water waves. As the slit gets wider, wavelets from different parts of the opening interfere to produce a wave more nearly like the incident plane wave. (Photograph courtesy of Film Studio, Education Development Center.)

Probably the most important practical consequence of diffraction is the limitation it imposes on microscopy. Because of the deflection of light passing an obstacle, the shadow of a small object becomes less and less well defined as its size shrinks toward the wavelength of the light illuminating it. If it is smaller than a single wavelength, it looks at best like a structureless dot—if it is visible at all. No amount of magnification or perfection of microscope design can defeat this fundamental diffraction limit. It can be defeated—or better, circumvented—only by abandoning light and substituting an illumination of shorter wavelength. X-rays have not proved suitable for microscopy because of the difficulty of focusing them and producing magnification. (They have, however, proved to be very useful for studies of crystal structure.) The most valuable extension of the light microscope has been the electron microscope, developed in the 1930s. Electrons, like photons, have a wavelength and are also subject to their own diffraction limit. However, it is easy to produce electrons of wavelength very much less than the wavelength of light, and by means of electric and magnetic forces these electrons may be focused and caused to produce magnified images. The connection between the wavelength of an electron and its momentum is dealt with in Section 23.5.

18.12 Interference

Young's classic two-slit experiment remains the simplest and most direct demonstration of wave interference. He illuminated one side of a sheet containing a pair of narrow parallel slits and examined the pattern of intensity projected onto a screen on the other side (Figure 18.34). The alternating dark and light fringes that appeared gave the first wholly convincing evidence for the wave nature of light. Such interference was already known for sound waves and water waves. Since then, interference has been demonstrated for waves in most other parts of the electromagnetic spectrum as well as for electron waves and neutron waves. Figure 18.35 compares interference patterns produced by X rays and by electrons.

(a)

(b)

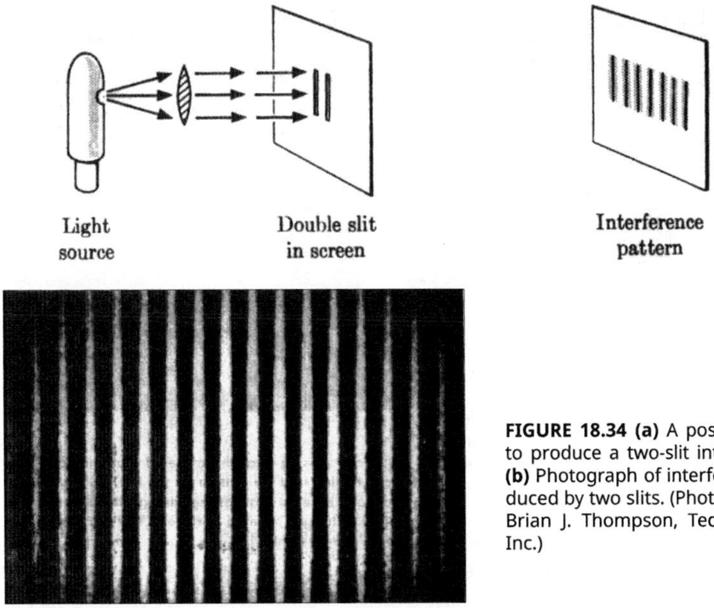

FIGURE 18.34 (a) A possible arrangement to produce a two-slit interference pattern. **(b)** Photograph of interference fringes produced by two slits. (Photograph courtesy of Brian J. Thompson, Technical Operations, Inc.)

In an idealized version, the two-slit experiment can be analyzed quantitatively in a simple way. This analysis is worth attention not just because it is easy, but because it makes clearer the physical reason for interference fringes and provides a basis for the most important practical use of interference, the measurement of wavelength. The idealization consists in supposing that the width of each slit is much less than a single wavelength. (For microwaves, this condition is easy to satisfy in practice; for light, it is very difficult.) Because of the narrowness of the slits, waves diverge from both of them in all directions. Each slit acts as a source of a Huyghens wavelet, as in Figure 18.32. Where these wavelets are in phase, they are said to interfere constructively, giving a point of illumination. Where they are completely out of phase (the crest of one coinciding with the trough of the other), they interfere destructively, giving a point of darkness. Consider, in particular, the intensity pattern on a screen placed parallel to the sheet containing the slits and perpendicular to the original direction of the incident light (Figure 18.36). Call the distance from the midpoint between the slits to the observation screen L, and the distance between the two slits d. Directly opposite the midpoint between the slits is a point on the screen equidistant from the two slits. This point, labeled O in Figure 18.36, is a point of constructive interference, for the two wavelets arriving there have traveled the same distance and are in phase. The point O is both the center of the interference pattern and the center of a bright fringe. Notice that it is also the center of what would be a shadow if waves were not diffracted. Regardless of the slit separation d, the center of the "shadow" is brightly illuminated. Consider next a point A on the screen a distance y from the central point O. The wavelet from the upper slit S_1 travels a distance x_1 to reach A. The wavelet from the lower slit S_2 travels a distance x_2 to reach A. If the distances x_2 and x_1 differ by exactly half a wavelength (or by 1.5, 2.5, 3.5 .. . wavelengths) the point A will be a point of maximum *destructive* interference, the center of a dark fringe. If the distances x_2 and x_1 differ by exactly one wavelength (or by 2, 3, 4 . . . wavelengths) the point A will be a point of maximum *constructive* interference, the center of a bright fringe.

(a)

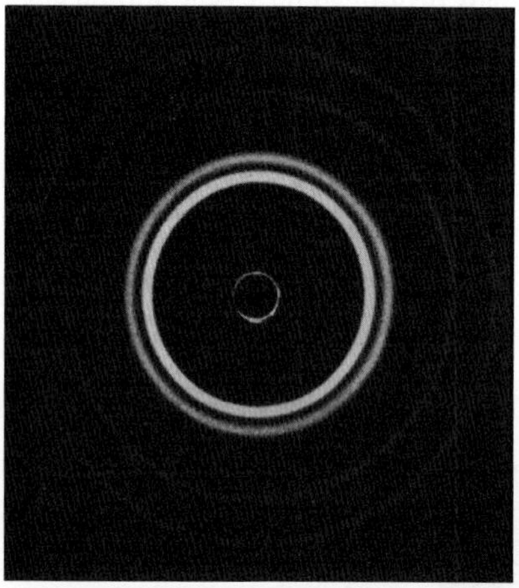

(b)

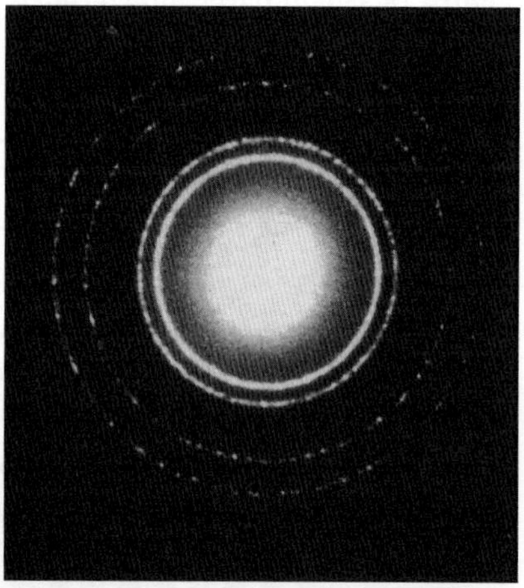

FIGURE 18.35 (a) X-rays passing through powdered aluminum crystals were diffracted and produced the interference pattern on the left. **(b)** A very similar interference pattern produced by electrons passing through the same material clearly demonstrates the wave nature of material particles. (Photographs courtesy of Film Studio, Education Development Center.)

We need now the mathematical relationship between the path difference $x_2 - x_1$ and the distance y on the screen in order to identify the position of bright and dark fringes. This can be obtained by considering the two right triangles S_1BA and S_2CA shaded in Figure 18.36. For the first of these, the Pythagorean theorem states:

$$x_1^2 = L^2 + (y - \tfrac{1}{2}d)^2 . \qquad (18.39)$$

For the second, the Pythagorean theorem states:

$$x_2^2 = L^2 + (y + \tfrac{1}{2}d)^2 . \qquad (18.40)$$

Taking the difference between Equations 18.40 and 18.39, we get

$$x_2^2 - x_1^2 = (y + \tfrac{1}{2}d)^2 - (y - \tfrac{1}{2}d)^2 .$$

The left side factors into the product of $(x_2 - x_1)$ and $(x_2 + x_1)$. The right side simplifies after expansion to the single term $2yd$. Then

$$(x_2 - x_1)(x_2 + x_1) = 2yd . \tag{18.41}$$

Although this is an exact equation connecting the path difference $x_2 - x_1$ and the distance y on the screen, the appearance in it of the variable quantity $x_1 + x_2$ makes it inconvenient for practical use. However, a very simple and useful approximation can be derived from it if the size of the interference pattern being studied is much smaller than the distance from slits to screen—that is, if y is much less than L. Then neither x_2 nor x_1 will differ significantly from L, and the sum $x_2 + x_1$ can be replaced by $2L$. Using this approximation, we can write Equation 18.41 in the form

$$x_2 - x_1 = \frac{yd}{L} . \tag{18.42}$$

Bright fringes will be pinpointed by the condition of constructive interference, $x_2 - x_1 = n\lambda$, where λ is the wavelength, and n is an integer. For dark fringes, the condition for destructive interference is $x_2 - x_1 = (n - \tfrac{1}{2})\lambda$, where again n is an integer. These conditions substituted into Equation 18.42 yield the following equations for the fringe positions:

$$y \text{ (bright fringes)} = n\lambda L/d . \tag{18.43}$$

$$y \text{ (dark fringes)} = (n - \tfrac{1}{2})\lambda L/d . \tag{18.44}$$

In the first of this pair of equations, $n = 0$ locates the central bright fringe at O ($y = 0$), $n = 1$ locates the first bright fringe off center, and so on. In the second equation, $n = 1$ locates the dark fringe adjacent to the central bright fringe, $n = 2$ locates the next dark fringe, and so on.

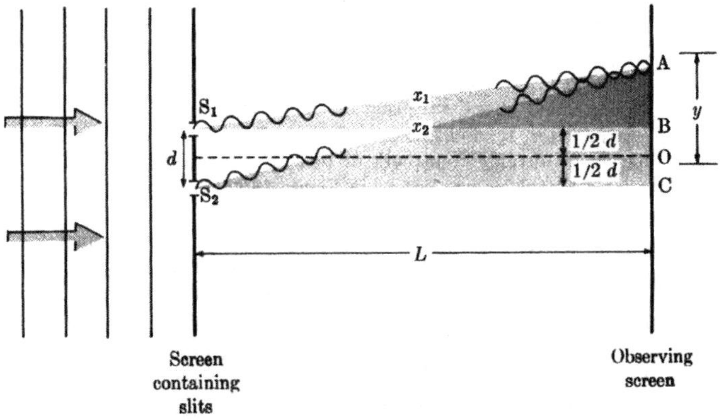

FIGURE 18.36 Diagram for analyzing two-slit interference patterns. A bright fringe is centered at point O. In this example, a dark fringe is located at point A because x_2 is greater than x_1 by half a wavelength.

A simple numerical calculation will bring out the content of Equations 18.43 and 18.44. Consider green light, whose wavelength is about 5,000 A ($\lambda = 5 \times 10^{-5}$ cm) falling on a pair of slits separated by one millimeter ($d = 10^{-1}$ cm). What will be the interference pattern on a screen two meters distant ($L = 200$ cm)? At the center of the pattern will be a bright fringe. According to Equation 18.43, the first bright fringe on either side of the center will be located at a distance $\lambda L/d$ from the center:

$$y\text{ (first bright fringe)} = \lambda\frac{L}{d} = 5\times10^{-5}\text{ cm}\times\frac{200\text{ cm}}{10^{-1}\text{ cm}} = 10^{-1}\text{ cm} = 1\text{ mm}\,. \qquad (18.45)$$

The next bright fringe will be 2 mm from the center, the next 3 mm, and so on, with equal spacing. Between each pair of bright fringes is a dark fringe, the one closest to the center being located by the equation

$$y\text{ (first dark fringe)} = \frac{1}{2}\lambda\frac{L}{d} = 0.5\text{ mm}\,. \qquad (18.46)$$

Here is an interesting number. In this example there is darkness exactly opposite each slit (half a millimeter above and below the center), where straight-line propagation would predict light. There could be no clearer proof of interference. Where either slit alone would produce light, the two acting together produce darkness.

The simple pattern of equally spaced fringes is valid so long as Equations 18.43 and 18.44 are valid. Their validity requires that the distance y on the screen be small compared to the distance L. This condition is well satisfied in this example. A pattern of 20 bright fringes occupies only 2 cm, which is 1% of the distance L.

Fringe separations of 1 mm are readily visible to the unaided eye or may be photographed. One interesting way to think about the two-slit interference pattern is as a wavelength magnifier. In Equation 18.45 the wavelength λ is multiplied by a "magnification factor" L/d which, in this example, is equal to 2,000. The fringe separation is 2,000 times larger than the wavelength; thereby the microscopic wavelength is, so to speak, rendered visible. Either by increasing the distance L or by decreasing the slit separation d we can increase the magnification factor and further spread out the interference pattern.

Interference does more than demonstrate the existence of waves. It provides a valuable tool for measuring wavelength. Young followed up his earliest experiments on interference with measurements of the wavelengths of light, and interference effects still provide the basis of all accurate wavelength measurements. In the simple two-slit interference pattern, the fringe separation is directly proportional to wavelength. Equation 18.43 implies that the separation of adjacent bright fringes is $\Delta y = \lambda L/d$. Turning this around, we can express the wavelength in terms of the fringe spacing:

$$\lambda = \frac{d}{L}\Delta y\,. \qquad (18.47)$$

Now the observed spacing Δy is "demagnified" by the ratio of known distances, d/L, to give the unknown wavelength.

For a given arrangement of slits and screen, the distances d and L remain fixed, and different wavelengths reveal themselves through differently spaced interference patterns. If two or more wavelengths illuminate the slits simultaneously, their separate interference patterns will overlap on the screen. The two slits and screen constitute a crude version of a spectroscope—a device that makes it possible to determine the components of the spectrum present in some unknown radiation. The key to the action of any spectroscope is dispersion, the fragmenting of radiation into its separate constituent wavelengths. As discussed in Section 18.9, dispersion can be achieved by the variable index of refraction of a glass prism. It can also be achieved by the phenomenon of interference. White light passing through a pair of slits produces a different fringe spacing for each wavelength and therefore a pattern of many rainbows. (The central fringe remains white. Why?)

If the slits giving rise to an interference pattern are not narrower than a single wavelength, the analysis is more complicated, but the basic principle is unchanged. Then there is interference not only between light passing through one slit and light passing through the other slit, but interference among the wavelets emanating from different parts of the same slit. This "self-interference" from a single slit is, unimportant for directions nearly straight ahead toward the screen, but it is important for directions away from the initial direction of propagation. As a result, the interference pattern differs from that of the idealized arrangement discussed above only in the "wings" of the pattern. Near the center of the pattern, the fringes produced by wide slits are nearly the same as those produced by narrow slits. In particular, the locations of the centers of the fringes, as given by Equations 18.43 and 18.44, are completely unchanged. The connection between wavelength and fringe spacing is the same.

As the heart of a spectroscope, a double slit has a serious disadvantage: The fringes it produces are broad (Figure 18.34). As a result, it is impossible to locate the centers of fringes precisely enough to make very accurate wavelength measurements, and it is impossible to discriminate the patterns produced by two wavelengths that differ very little from one another. In addition to suffering from these shortcomings on account of fringe breadth, the double slit allows but a small fraction of the incident light to reach the observer. These problems are all beautifully cured by the diffraction grating, which in its simplest form is a picket fence of alternating opaque and transparent strips, an array of many parallel slits. The first gratings for interference studies and accurate wavelength measurements were constructed by Joseph von Fraunhofer around 1820. These consisted of fine wires strung on a frame like the warp of a rug before weaving begins. His best wire gratings had as many as 19 wires per millimeter. Fraunhofer went on to construct finer gratings by scratching parallel lines on glass, and in this way achieved up to 300 lines in a millimeter. This method of making gratings grew in technical perfection during the remainder of the nineteenth century, until in the 1880s Henry A. Rowland succeeded in making near-perfect glass gratings with as many as 1,000 lines per millimeter. (The spacing between lines in such a grating, 10^{-4} cm or 10,000 A, is about equal to one and a half wavelengths of red light.) Some of Rowland's gratings, and replicas of them, are still in use.

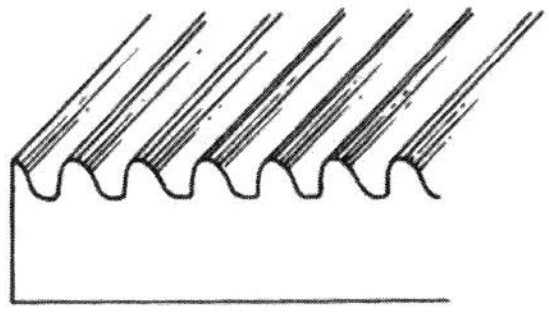

(a)

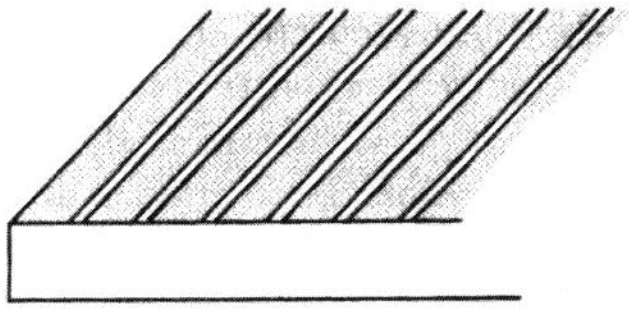

(b)

FIGURE 18.37 Diffraction grating. **(a)** Highly magnified view of grating consisting of equally spaced array of identical grooves in glass. **(b)** Idealized version of grating suitable for analysis: thin screen with narrow parallel slits separated by wider opaque stripes.

A glass grating need not have any opaque strips or any specified shape for its grooves. All that is necessary is that its successive grooves be equally spaced and nearly identical—that the grating, viewed end on, show a regularly repeating structure [Figure 18.37(a)]. The analysis of the grating, however, is much simplified—without losing the essence of its action—if it is idealized as a set of very-narrow slits separated by wider opaque stripes [Figure 18.37(b)], Then each slit may be regarded as the source of an elementary Huyghens wavelet. In a typical grating spectroscope (Figure 18.38), the light diffracted through an angle θ is gathered by a lens and brought to a focus. Therefore the interference between wavelets from different slits may be dealt with by considering the path difference between successive parallel tracks. A subtle but important point requires mention here. If the wavelets reaching points A, B, C, D, and E in Figure 18.38 are in phase, they will still be in phase at the focal point F. Despite the shorter distance from C to F than from A to F, the *time* from C to F is the same as the time from A to F because the wave going from C to F traverses more glass and is delayed just long enough to permit it to be overtaken by the wave going from

A to F at the instant when both reach the focal point. This equality of time for all the tracks corresponds also to an equality of *number* of wavelengths. Waves vibrating in phase at A, B, C, D, and E also vibrate in phase when they converge at F. Now we ask: For what angle of diffraction θ will the wavelets arriving at A, B, C, D, and E indeed be in phase in order to produce a bright image at F? This condition of reinforcement requires that the path difference Δx between successive tracks be an integral number of wavelengths. In particular, for what is called first-order diffraction, the lag of each wavelet behind its neighboring wavelet is exactly one wavelength:

$$\Delta x = \lambda . \tag{18.48}$$

According to the geometry of each little right triangle in Figure 18.38 whose hypotenuse is the slit separation d, the path difference is

$$\Delta x = d \sin \theta . \tag{18.49}$$

Therefore wavelength and diffraction angle for the first bright image are related by

$$\lambda = d \sin \theta \qquad \text{(first-order diffraction).} \tag{18.50}$$

This is the fundamental equation of the grating spectroscope. Since d and θ can be measured accurately, the wavelength λ can be determined accurately.

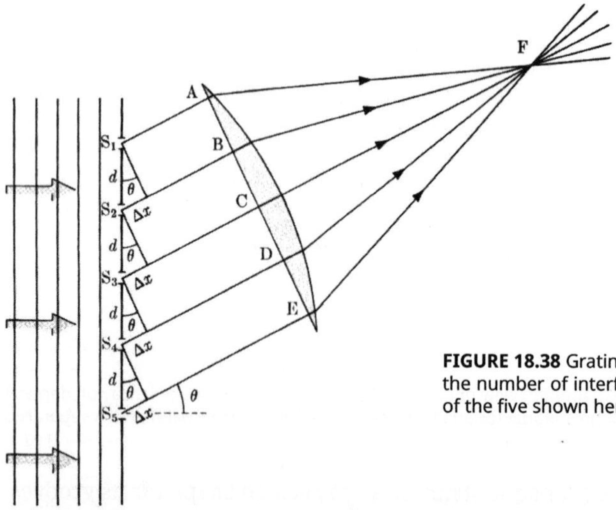

FIGURE 18.38 Grating spectroscope. In an actual spectroscope, the number of interfering wavelets is in the thousands instead of the five shown here.

The advantage of many slits (tens of thousands in practice) instead of only two becomes evident if we consider what happens when the angle θ is slightly changed away from the angle given by Equation 18.50, which locates the center of a bright fringe. For a slight change of θ, the path difference Δx between two successive tracks, such as S_1A and S_2B, changes slightly so that the wavelets reaching A and B are no longer perfectly in phase. If these were the only two slits acting, the intensity at F would diminish only slightly, for the two wavelets would still be far from a condition of destructive interference. With a very large number of slits acting, however, the intensity at F drops sharply. Although the wavelet from S_2 differs in phase only slightly from the wavelet from S_1, the wavelet from S_3 differs slightly more in phase, that from S_4 slightly more, and so on. Eventually there will be a slit whose wavelet is exactly out of phase with the wavelet from S_1. Summing the contributions of thousands of slits, there will be almost complete destructive interference and darkness at

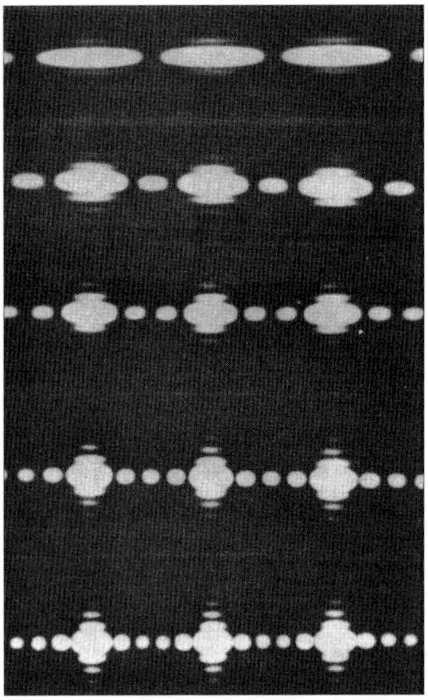

FIGURE 18.39 Interference patterns produced by gratings with (from top to bottom) 2, 3, 4, 5, and 6 slits. These photographs were made using a point source rather than the usual line source. With further increase in the number of slits, the principal maxima become narrower and the subsidiary maxima become dimmer. (Photograph courtesy of Brian J. Thompson, Technical Operations, Inc.)

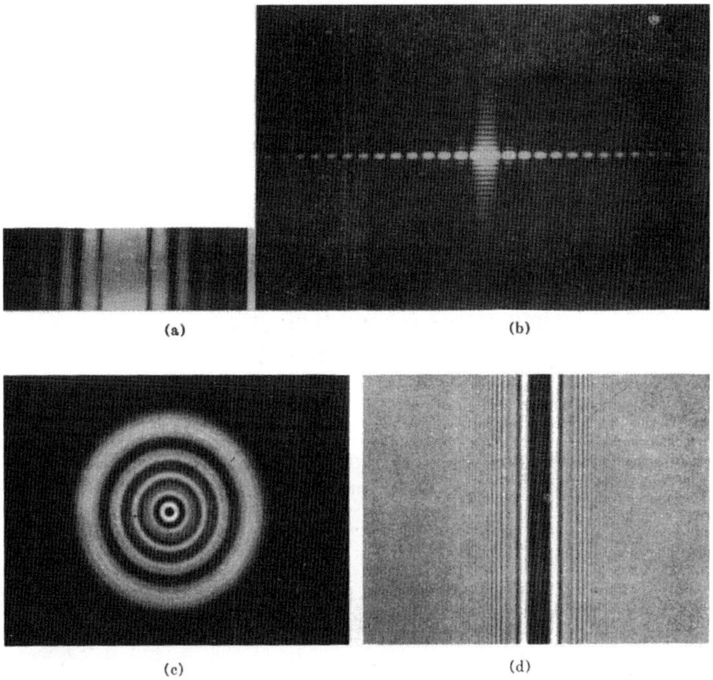

FIGURE 18.40 Patterns of interference resulting from diffraction produced by single apertures and a single obstacle. **(a)** Single slit. **(b)** Rectangular hole. **(c)** Circular hole. **(d)** Wire obstacle. [Photographs (a) and (d) courtesy of Brian J. Thompson, Technical Operations, Inc. Photographs (b) and (c) from M. Cagnet, M. Françon, and J. C. Thrierr, *Atlas of Optical Phenomena* (Berlin: Springer-Verlag, 1962). Courtesy of Professor Françon.]

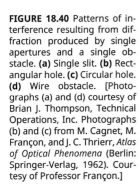

F despite the fact that the angle θ has shifted only slightly from a fringe center. As a result, the broad fringes characteristic of the two-slit interference pattern collapse in the grating spectroscope into narrow lines separated by expanses of darkness. Figure 18.39 shows the beginning of this effect as the number of slits is increased from two to six. Because of the appearance of the narrowed interference fringes in a grating spectroscope, light of a particular wavelength is often referred to as a spectral line.

Oddly enough, the pattern of interference resulting from diffraction by a single slit or a single obstacle is more complex than the pattern from a large number of slits. Since these patterns reveal no new physical principles, it is unnecessary to discuss them here. Nevertheless, many of them are agreeable to look at. With this in mind, we reproduce several in Figure 18.40.

18.13 Line spectra

104 Much of what the ancients knew about nature, and a great part of what we now call the classical physics of the seventeenth, eighteenth, and nineteenth centuries, rests ultimately on the quantum mechanics of the submicroscopic world. The quantum nature of atoms dictates the density of ordinary matter; the behavior of solids, liquids, and gases when heated; the transparency, opacity, and color of materials; the existence of physical change such as freezing and boiling; and all aspects of chemical change. Because of the quantum structure of atomic nuclei, nuclear burning lights the sun and stars and nourishes life on earth. The facts that metals are conductors, that chemical cells generate electricity, that a crystal can be doubly refracting, that light travels more slowly in matter than in empty space—all find their explanation now in quantum theory.

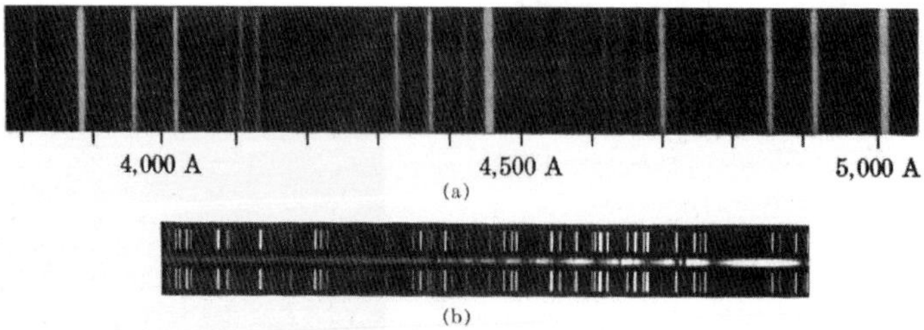

FIGURE 18.41 Line spectra. **(a)** Light emitted by helium in the wavelength range 3,800 A (violet) to 5,000 A (green). **(b)** Violet spectral lines of iron in the wavelength range 3,600 A to 3,800 A. (Photographs courtesy of Mount Wilson and Palomar Observatories.)

One of the important precursors of quantum theory was the discovery of line spectra. Under certain circumstances, the light emitted by matter is not continuously distributed over a region of the spectrum but instead consists of a definite set of discrete wavelengths (Figure 18.41). Such a mixture of wavelengths is called a line spectrum because of its appearance when examined with a spectroscope. Even the sun, which emits a continuous spectrum, reveals certain dark spectral lines arising from selective absorption in its outermost envelope. A portion of the sun's spectrum is shown in Figure 18.42.

FIGURE 18.42 Portion of the sun's spectrum in the region of green and yellow light. The strongest pair of dark lines, near 5,900 A, are contributed by the absorption of light by sodium atoms. (Courtesy of Mount Wilson and Palomar Observatories.)

As a general rule, when light is emitted by electrons within individual atoms, a line spectrum results, because only certain quantum energy changes are permitted and therefore only photons of certain definite energies and frequencies are emitted (Equation 18.6).

This connection between energy and frequency is the key to the quantum interpretation of line spectra that was missing before 1900. Earlier it was assumed that atoms had certain preferred modes of vibration at particular resonant frequencies, just as a violin string, the air in an organ pipe, or the current in an oscillator circuit has certain characteristic vibration frequencies. Line spectra were taken as evidence for the existence of atoms with a fairly complicated internal structure containing charged particles; but before the work of Einstein and Bohr, spectra were not recognized as direct evidence of the quantum structure of atoms.

If electrons or other charged particles emitting radiation are not restricted to a set of discrete energy changes, they produce a continuous spectrum instead of a line spectrum. The bremsstrahlung emitted by high-energy charged particles is in the form of a continuous spectrum. Electrons in the sun, freed from bondage to any particular atom, emit a continuous spectrum. Thermal radiation in general, as from an incandescent bulb, is continuous, usually with superposed lines. Not surprisingly, continuous spectra were the first to be discovered, since the earliest optical work was done with sunlight or the light of a candle flame.

The first known observation of a spectral line was made by Thomas Melvill in 1752, who found that salts of sodium dropped in a flame produce a yellow light that is of "one determined degree of refrangibility [is refracted through a particular angle, is not dispersed]; and the transition from it to the fainter color adjoining is not gradual, but immediate." This momentous discovery went largely unnoticed and apparently had no impact on the subsequent history of science. Only after Fraunhofer's study of dark lines in the solar spectrum (1814) and his accurate determinations of wavelengths with diffraction gratings (1821-1823) did spectroscopy establish itself as an important branch of experimental physics. The solar lines, seen first by William Hyde Wollaston in 1802, were measured by Fraunhofer and catalogued as the A line, the B line, the C line, and so on. Even after 150 years, we still speak of the Fraunhofer lines. The D line is Melvill's yellow line of sodium, which Fraunhofer found to be a pair of close lines, or a doublet (see Figure 18.42). The A line and the B line are red lines of oxygen; the C line is a red-orange line of hydrogen.

Line spectra afford an interesting example in the history of science of extensive and accurate data waiting a long time for a theory. Ninety-nine years elapsed between Fraunhofer's first spectral analysis of solar lines and Niels Bohr's theory of the hydrogen spectrum. In between, thousands of spectral lines were measured to high precision. What kept interest from flagging after many decades of accumulation of data was the practical utility of spectroscopy. As early as the 1820s it was suggested that spectral lines might provide evidence on the composition of materials. This fact was definitively established by Gustav Kirchhoff and Robert Bunsen in the 1850s and 1860s. They made of spectroscopy a primary tool of chemical analysis. Since every atom emits a characteristic line spectrum, and the spectra of no two atoms are the same, the line spectrum emitted by any substance provides unambiguous evidence about the atomic constituents of the substance. Through spectral analysis we know much about the distribution of elements in the sun and throughout the universe. One element, helium, was actually "seen" in sunlight before it was discovered on earth. In laboratory measurements, spectroscopy can reveal tiny traces of elements that would be difficult or impossible to recognize by other means. A number of rare elements have been discovered first through their spectra.

In the study of line spectra we find complementary ideas that have been crucial in the development of quantum mechanics and in the evolution of our contemporary picture of nature: the idea of wave and particle, of continuity and discreteness. Nature presents these two faces to us, and in line spectra they were there for all to see. Continuity shows itself in the spread of a wave through space and in its diffraction. At the same time the wave shows discreteness, or particle-like properties, in its definite wavelength and frequency, and, as we now know, in its composition of photons of definite energy. Line spectra show us also the beautiful unity of nature. They have tied together the electromagnetic theory of radiation and the quantum theory of matter, and they have tied together physically the vastness of the universe into a single community of laws. We can look at a spectral line from a distant galaxy and know that there, a billion years ago, existed a hydrogen atom indistinguishable from a hydrogen atom on earth today, restricted in its behavior by the same laws of nature that govern man and his environment now.

EXERCISES

18.1. An audible sound wave in air has a wavelength of 10 cm. **(1)** What is its frequency? **(2)** What is the frequency of an electromagnetic wave with the same wavelength? In what part of the spectrum is this wave ?

18.2. Calculate approximately **(1)** the wavelength of 118.3 megacycle radiation (in the VHF band) from an airport control tower; **(2)** the frequency of radiation in the 30-meter shortwave radio band; and **(3)** the speed of a wave whose wavelength is 3 cm and whose frequency is 10 gigacycles (1 gigacycle = 10^9 sec^{-1}). What might this last wave be?

18.3. A radio transmitting antenna is linked to an oscillator circuit containing an inductance of 2 millihenry and a capacitance of 10^{-10} farad. Find the frequency and the wavelength of the emitted radiation.

18.4. A certain harpsichord spans 5 octaves. **(1)** What is the ratio of its highest to its lowest frequency? **(2)** Estimate its highest and its lowest frequency if a note at the center of the keyboard has a frequency of 250 sec^{-1}.

18.5. An organ pipe 10 feet long emits sound whose wavelength in air is 40 feet. If this note is sounded for 1 sec, how many full vibrations of the wave are generated?

18.6. A sound wave in air strikes a metal surface. It is partially reflected and partially transmitted into the metal. **(1)** Explain in terms of the physical mechanism of sound propagation why it is the frequency and not the wavelength that is the same in air and metal. **(2)** Is the wavelength in the metal greater or less than the wavelength in air? **(3)** Is the wave in air longitudinal or transverse? **(4)** Is the wave in metal longitudinal or transverse ?

18.7. A radar antenna transmits electromagnetic radiation of 3 cm wavelength for a time duration of 0.5 microsecond, after which it can receive reflected radiation. **(1)** What is its frequency? **(2)** How long is the emitted wave train, and how many wave cycles does it contain? **(3)** If the transmitter remains off for 1 millisec, how far away are the nearest and farthest objects whose reflections can be received? *Optional:* If the radiated power is 10^9 erg/sec (100 watts), how many photons are emitted in 0.5 microsec?

18.8. From the moving charged rod in Figure 18.4(b) a weak electromagnetic wave propagates to the right. **(1)** Is this wave polarized? If so, in what direction? **(2)** Make a rough estimate of the frequency of this wave and explain the basis of your estimate.

18.9. **(1)** Approximately how many photons of an AM radio wave are required to equal in energy one photon of visible light? **(2)** Convert to photons/sec a power of 1 watt of **(a)** radio waves of frequency 10^8 sec^{-1}, and **(b)** gamma rays of frequency 10^{20} sec^{-1}.

18.10. A proton in a nucleus oscillates in a domain of about 10^{-12} cm at about 10% of the speed of light. What is its approximate frequency of oscillation? How does this compare with the frequencies of nuclear gamma rays?

18.11. Explain in your own words why different regions of the electromagnetic spectrum are characteristically radiated by different kinds of physical systems, and why the highest frequency radiation is emitted by the smallest systems.

18.12. A charged particle oscillates through a distance d at average speed v. It radiates electromagnetic radiation whose frequency is equal to its own frequency of oscillation. Show that the wavelength of the radiation is given approximately by Formula 18.12.

18.13. According to Formula 18.12, the wavelength of electromagnetic radiation is at least as great as the dimension of the radiating system. Is this condition consistent with your own casual observations of various kinds of radiating antennas (broadcast stations, radar, taxicabs, police cars, ham radios, aircraft, microwave relays, etc.)? Justify your answer for a few of these and attempt to explain any apparent discrepancy.

18.14. **(1)** A straight wire acting as a transmitting antenna radiates polarized radiation. In what direction is it polarized? Justify your answer. **(2)** Would you expect a receiving antenna to absorb one polarization preferentially?

18.15. Air flow over mountains in California produces a pattern of rising and falling air known as the Sierra wave. **(1)** Is this wave longitudinal or transverse? **(2)** Is it standing or propagating?

18.16. **(1)** Explain why a microwave polarizer (Figure 18.14) does not polarize light. **(2)** Describe a hypothetical physical system that acts on microwaves in the same way that the microwave polarizer acts on light.

18.17. Prove carefully that the angles of incidence and reflection, θ_i and θ_r, in Figure 18.18 are equal.

18.18. **(1)** If a reflected wave had less speed than the incident wave, would the angle of reflection be greater or less than the angle of incidence? To answer this question, consider a modified version of Figure 18.18. **(2)** When a ball bounces obliquely from a wall with some loss of energy, is its angle of reflection greater or less than its angle of incidence? Answer with the help of experiment if necessary. **(3)** Briefly contrast the different bases underlying the reflection of waves and the reflection of particles.

18.19. Prove that the Huyghens wavelet emanating from point B' in Figure 18.19 just touches the refracted wave front line CC'.

18.20. If n_1 is the index of refraction of medium 1 through which an incident wave approaches a surface and n_2 is the index of refraction of medium 2 into which the wave is refracted, show that Snell's law may be written

$$n_1 \sin \theta = n_2 \sin \varphi \, .$$

Refer to Figure 18.19 for the definitions of θ and φ.

18.21. Use Huyghens' principle and a geometrical construction analogous to that in Figure 18.19 to show that Equation 18.21 remains valid when φ is the angle of incidence, θ the angle of refraction, and the speed of light *increases* by the factor n in crossing the surface.

18.22. A beam of light falls on the surface of a sheet of glass having an index of refraction of 1.33. **(1)** If the angle of incidence of the light is 30 deg, what is its angle of refraction within the glass? **(2)** What is the average speed of light inside the glass?

18.23. A pencil of light approaches the surface of a piece of glass from within the glass. The space outside the surface is evacuated. For each of several indices of refraction from 1.0 to 1.8, calculate the maximum angle of incidence of the light beam in the glass, φ_{max}, for which the light can escape through the surface. Present the results of these calculations in graphical form.

18.24. A ray of light is incident on the interface between two sheets of glass with different indices of refraction, as shown in the figure. **(1)** Is the light deflected toward or away from the line perpendicular to the surface? **(2)** By how many degrees is it deviated from its straight line course?

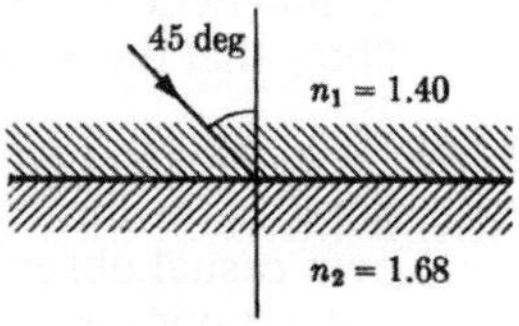

18.25. **(1)** What is the average speed of light in a glass prism that is just barely able to produce total reflection of the kind shown in this diagram? (Take the index of refraction of air to be exactly 1.00.) **(2)** Would the prism still be totally reflecting if submerged in water?

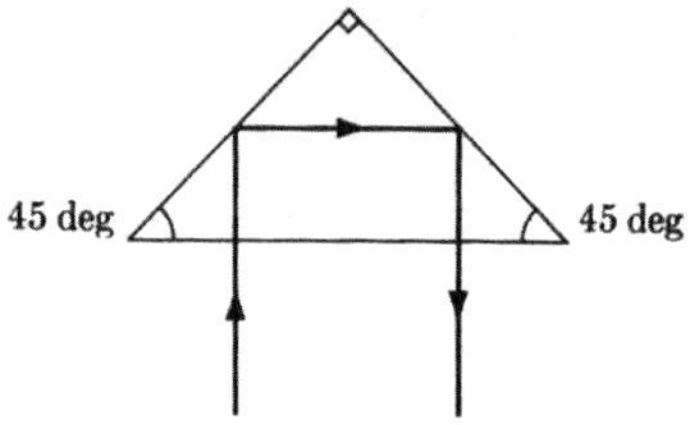

18.26. If you believed in the particle theory of light, how would you explain total internal reflection?

18.27. Parallel rays of light are incident, as shown, on a concave mirror whose surface is a small part of a large sphere. Show that if the angles of reflection are small, the mirror brings the light approximately to a focus at a focal distance f equal to half the radius of curvature of the mirror.

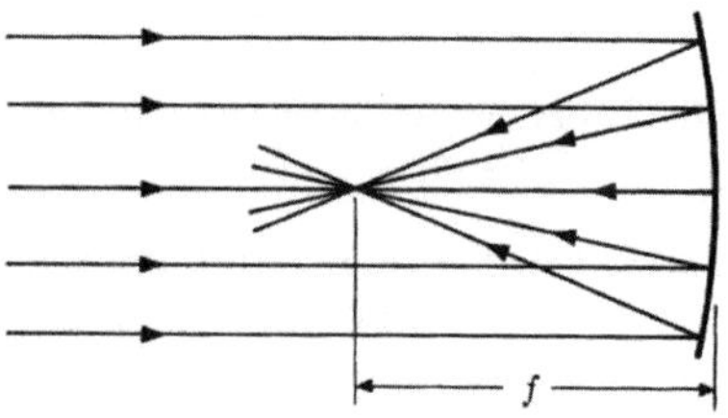

18.28. For a lens of fixed focal distance f = 10 cm, prepare a graph of magnification M vs. object distance d_1 (Equation 18.38). Extend the graph into the region of negative M and suggest a possible interpretation of its meaning in this region.

18.29. Figure 18.30 shows an example of virtual image formation by a thin lens. If the distance from virtual image to lens is called d_2, the equation given in Exercise 18.41 relates d_1, d_2, and f. Work from this equation to derive a formula for the magnification of the virtual image:

$$M = \frac{1}{1 - \frac{d_1}{f}}.$$

Note the similarity of this result to Formula 18.38 for the magnification of a real image.

18.30. Borrow the reading glasses of a far-sighted friend and measure approximately the focal distance of the lenses. With the help of a diagram, describe how you carried out the measurement. (If suitable eyeglasses are not available, use any converging lens or hand magnifier.)

18.31. At the center of the dark shadow of a circular disk is a bright spot. Explain this fact in terms of diffraction and interference.

18.32. The length of a grain in a photographic emulsion is 0.5 micron (1 micron = 10^{-4} cm). How does this dimension compare with the wavelength of visible light? Should it be possible to determine in detail with a microscope the shape of the grain?

18.33. Orange light of wavelength 6,000 A is incident on a screen containing a pair of narrow slits separated by 0.2 cm. **(1)** Calculate and sketch the interference pattern produced on a screen 400 cm beyond the slits. Would this pattern be visible to the unaided eye? **(2)** In a companion sketch beside the first one, indicate the pattern that would be observed if light traveled in straight lines without diffraction.

18.34. Why can the difference $x_2 - x_1$ in Equation 18.41 not be approximated in the same way as the sum $x_2 + x_1$? To answer this, think about the distance from New York to Istanbul (x_1) and from Washington to Istanbul (x_2). They are nearly equal and their sum is well approximated by twice the distance from Wilmington to Istanbul. Yet the distance from New York to Washington ($x_2 - x_1$) is not zero. Convert this kind of reasoning to quantitative form by discussing percentage errors introduced by approximations in sums and differences.

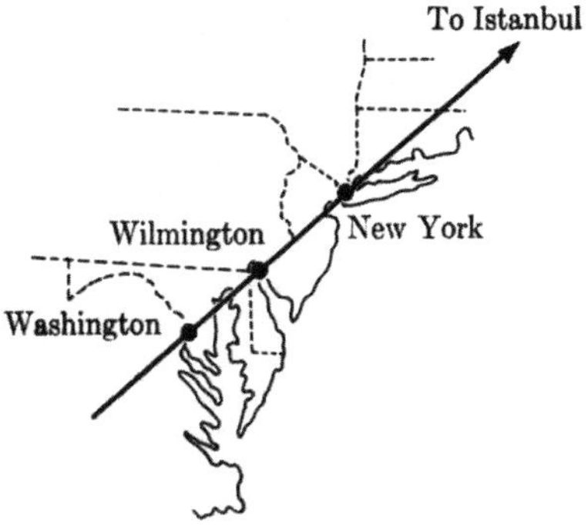

18.35. You have been given a source of light labeled "Guaranteed to emit radiation of wavelength 5×10^{-5} cm." Design an interference experiment to test the claim, specifying the numerical magnitudes of all significant dimensions, and calculating the expected distance to be measured between two points of constructive interference.

18.36. Red light from three different sources is mixed and illuminates a spectroscope whose grating has 2,500 lines per centimeter. Find the angle of first-order diffraction for each of these three spectral lines. Would you expect these three lines to be clearly distinct and easily identifiable?

Source	Wavelength
Hydrogen	6,563 A
Neon	6,402 A
Argon	6,965 A

18.37. Light of known wavelength, $\lambda = 6{,}328$ A, from a helium-neon laser is used to calibrate a grating spectroscope. If the first-order diffraction line is found to make an angle of 18 deg with the incident beam, **(1)** what is the grating spacing d? **(2)** At what angle will the second-order diffraction line be observed?

18.38. **(1)** Prove that, if a pair of spectral lines have a wavelength difference $\Delta\lambda$, their frequency difference is given approximately by $\Delta f = -c\Delta\lambda/\lambda_{av}^2$, where λ_{av} is the average wavelength of the pair. **(2)** Use this formula to find the frequency difference and then the photon energy difference of the yellow D lines of sodium, whose wavelengths are $\lambda_1 = 5{,}890$ A and $\lambda_2 = 5{,}896$ A. (This energy difference is equal to the energy required to invert the spin of one electron in the sodium atom.)

18.39. A spectral doublet radiated by potassium has wavelengths $\lambda_1 = 7{,}665$ A and $\lambda_2 = 7{,}699$ A. (1) In what part of the spectrum does this doublet lie? **(2)** Are the photons radiated by potassium to make these spectral lines of greater or less energy than the photons of the sodium D lines, whose wavelengths are stated in the preceding exercise? **(3)** Is the difference in the two potassium photon energies greater or less than the difference in the two sodium photon energies? By about what factor?

18.40. Apply Snell's law and the thin lens approximation to the light beam deflection pictured in Figure 18.28, and prove that Equation 18.29, $\alpha \cong b/f$, remains valid. (This is more difficult than the discussion of Figure 18.26 in the text, because refraction at both surfaces of the lens must be considered.)

18.41. A light ray emanating from a point on the axis of a lens inside the focal point is deflected in such a way that it appears to come from a more distant point on the same side of the lens. Prove that the distances d_1 and d_2 (defined in the diagram) and the focal distance f are related by

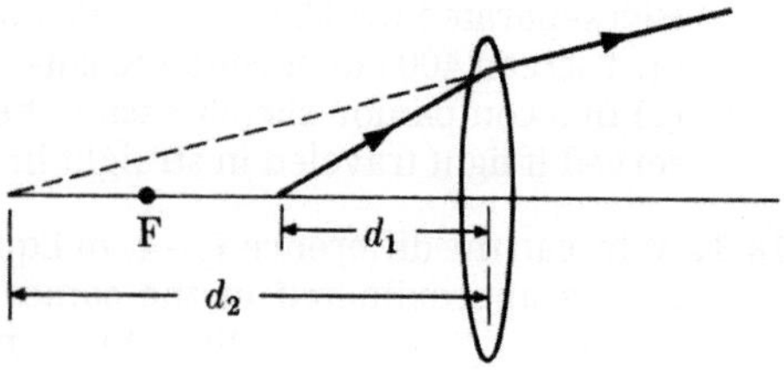

$$\frac{1}{f} = \frac{1}{d_1} - \frac{1}{d_2}.$$

This result rests on the thin lens approximation and on the fact that the angle of deflection of the light ray, α, is approximately equal to b/f (Equation 18.29).

PART 6
Relativity

ALBERT EINSTEIN
(1879 – 1955)

"We can attach no absolute significance to the concept of synchronism; two events which are synchronous when viewed from one system will not be synchronous when viewed from a system moving relative to this system."

"Science is not just a collection of laws, a catalogue of unrelated facts. It is a creation of the human mind, with its freely invented ideas and concepts.... The reality created by modern physics is indeed, far removed from the reality of the early days."

NINETEEN

Galilean relativity and the speed of light

In the year 1887, two notable experiments concerned with light took place, one a success, one apparently a failure. In Karlsruhe, Heinrich Hertz demonstrated that the oscillations of electric charge in an ordinary spark give rise to radio waves which travel a great distance through air. In so doing, he provided a key support for electromagnetic theory and bolstered Maxwell's contention that light is electromagnetic radiation. At the same time, in Cleveland, Ohio, Albert A. Michelson and Edward W. Morley set out to measure what they believed to be another prediction of electromagnetic theory—that the speed of light with respect to the earth should depend upon the motion of the earth through the all-pervasive substratum known as the ether—and they failed. Michelson by himself, as a young visitor in Berlin six years earlier, had tried the same experiment with a similar negative result. Now with the help of Morley, an older professor with a worldwide reputation for accurate measurements, he repeated the experiment with a new and much more precise apparatus. Again no effect. Michelson and Morley were disappointed, indeed almost incredulous. They published their negative findings and went on to other things, little realizing that they had chipped at the foundations of physics.

19.1 *Physics at the end of the nineteenth century*

The closing years of the nineteenth century were, for the most part, years of consolidation 105
in physics. Thermodynamics, statistical mechanics, and electromagnetic theory were being molded into seemingly final form. Together with mechanics, these subjects comprise what we now call classical physics. At the time, the structure of physics seemed to stand, in the words of Sir James Jeans, "foursquare, complete, and unshakable." Eminent scientists saw the future of science as a working out of further consequences of these theories and the application of them to new fields. Fortunately, a younger generation were unwilling to accept this verdict against the excitement of future discovery in science. Within a few years after the turn of the century, relativity and quantum theory had arrived to revolutionize science, philosophy, and, eventually, the life of man.

To the inquiring eye of the curious scientist, the structure of physical science in the 1890s was not quite perfect. There were a few loose bricks, a few things that did not quite fit into place. In 1856, Urbain Leverrier in France had calculated with great care the influence of the other planets on the motion of Mercury. He found that the elliptical orbit traced out by Mercury should itself rotate slowly about the sun, the point of Mercury's closest approach to the sun (its perihelion) advancing at 527 seconds of arc per century, or once around in about 2,500 centuries. (There are 60 seconds of arc in one minute, 60 minutes in one degree, and 360 degrees in a full circle.) But Mercury's orbit was known to shift around by 565 seconds per century. The tiny but significant discrepancy of 38 seconds per century lurked unexplained in the background of science for sixty years. Improved measurements and calculations had succeeded only in making the discrepancy grow slightly, to 42 seconds, when in 1916 Einstein predicted from the general theory of relativity an extra rotation of exactly 42 seconds.*

* More recently the discrepancy between observation and the prediction of classical physics has shrunk back to 39 seconds per century (see Section 22.5), providing a new challenge for the ingenuity of the physicist.

Another puzzle of physical science at the end of the nineteenth century was the amazing shyness of the ether, which refused to show itself to its most persistent pursuers. To Michelson and Morley, as well as to several of their contemporaries with alternative stratagems of their own, the ether refused to reveal its presence. It is not the first time in history that what was difficult to conquer took on a special fascination, and some of the most inspired scientific work of the period 1890 to 1905 was dedicated to explaining how the ether could exist yet defy detection. Hendrik A. Lorentz in Holland and Henri Poincaré in France developed some of the ideas of relativity in this period from the desire to explain the ether's anonymity. But it took the genius of Einstein to reject the ether altogether and to appreciate the revolutionary nature of the vacuum left by its departure.

Still another puzzle whose solution led to revolutionary new ideas at the turn of the century had to do with light waves and energy. A drama of physics was played out in what seems to be a most uninspiring arena, just an empty container, or hohlraum. Within the hohlraum the classical theories of mechanics, thermodynamics, and electromagnetism should have met and united in a final triumph of the classical description of nature. The mechanical vibrations of molecules in the wall and the electromagnetic vibrations of waves in the cavity should have shared energy according to the equipartition theorem of thermodynamics. But they refused to cooperate. The high-frequency radiation received less than its fair share of the energy. In the first years of this century, Max Planck saved the equipartition theorem, but he had to introduce the energy quantum and topple a good deal of the structure of classical physics to do it.

Finally, in the late 1890s, the atom was beginning to act up. Few people doubted the existence of atoms. The role of atoms in chemistry was appreciated, and a good deal about the external properties of atoms was known. The existence of characteristic line spectra suggested also a complex inner structure for the atom, but so far there was no evidence that this structure would not yield to analysis with the tools of classical physics. When Henri Bequerel discovered in Paris in 1896 that certain atoms (which we now call radioactive) emit penetrating radiation, the inside of the atom forced itself upon our attention in a more insistent way, and we have not been able to ignore it since. At about the same time, in Cambridge, J. J. Thomson learned that the negative electricity which can be torn from atoms consists of small swift particles beside which atoms appeared suddenly large and clumsy. So was discovered the first and most important elementary particle, the electron, whose existence revealed the presence of a new and unknown world within the atom. In the meantime the electron has gone on to independent glories of its own, now doing most of the brain work and some of the manual labor in the world's millions of machines.

From our present vantage point in history, it is easy to select these examples of weak spots and gaps in classical physics at the end of the century. But it should be borne in mind that to most observers these appeared as rather small and inconsequential flaws in a magnificent and well-tested structure. To Einstein, Planck, Lorentz, Poincaré, Rutherford, and the relatively small group of scientists throughout the world who fixed their attention on the flaws without being blinded by the beauty of the rest of the structure do we owe the creation of the new structure of twentieth-century physical science.

19.2 Relativity and invariance

The quantum theory and the theory of relativity, both born within the first five years of the twentieth century, together immensely deepened our understanding of nature. The same could of course be said for any important theory of nature. But there was something unique about the new theories, symbolic of a changing relationship between science and human experience. Each of them contributed to a complete revolution in the scientists' and philosophers' ways of looking at the world. For the first time in science, concepts were introduced that violated common sense and defied visualization. Sixty years later, man is no better able to picture a four-dimensional space-time or the wave-particle nature of a photon than he was when these ideas were introduced. The reason is not hard to find. We humans are classical creatures, and pretwentieth-century science suffices quite well to describe us in the large. The quantum theory is relevant to us only at the microscopic biochemical level below the range of our perception, and relativity is likewise unimportant in normal life because we move about, even in jet travel, at speeds far less than the speed of light. In short, we are very large (compared with an atom) and very slow (compared with light). Twentieth-century science has been mainly a science of the small and the fast.

Do these twentieth-century theories, with their odd concepts, then have any relevance to modern man, other than to professional scientists? All of us, scientists or not, do have some curiosity about the nature of the elementary material from which we are constructed and how it is put together. Aside from that, the new theories have had a direct impact. It is through our increased knowledge of the very small and the very fast that have arisen, to name a few things, electronics, pocket radios, synthetic fibers, and atomic bombs.

The theory of relativity is concerned with two seemingly opposite ideas, relativity and 106
invariance. Relativity calls to mind phrases such as, “Well, that depends.” In the theory of relativity it means relativity of observation—I see a given phenomenon one way, you see it another. It refers to *disagreement*.

Invariance appears to mean just the opposite: definiteness, absence of relativity. In the theory of relativity, it refers to areas of *agreement*, those aspects of a phenomenon (and even more important, those laws of phenomena) that are the same for different observers. Instead of relativity and invariance, we could just as well have used two more familiar words, subjectivity and objectivity, except perhaps that these words are too familiar. The subjectivity of relativity is a definite kind of physical subjectivity, not referring merely to differences in human perception. And the objectivity of relativity is not the philosophers' “objective reality.” It is rather objectivity by definition, an agreement among observers to accept as real the common aspects of their measurements. (Invariance actually means a bit more than this; we shall return to it later.)

The theory of relativity has, surprisingly, added both more relativity (or subjectivity) and more invariance (or objectivity) to science. Einstein showed that a number of quantities previously thought to be invariant are in fact relative, most notable of these being time. But at the same time he showed how to extract from the increased relativity of observation new invariant quantities. More important, he raised to the level of a fundamental postulate of science the principle that, despite the relativity of the raw observations of phenomena, the *laws* governing these phenomena must be invariant. From the increased subjectivity of observation that relativity brought to the world came a new and deeper view of the objectivity of physical laws.

19.3 Agreements and disagreements in Newtonian relativity

The ideas of relativity and invariance exist in ordinary Newtonian mechanics, and can be 107
illustrated with simple mechanical experiments. Consider a child in a uniformly moving train who releases a ball and allows it to fall straight down. From his point of view it starts from rest, and falls vertically downward with uniform acceleration. Someone watching from outside the train would have a different view (Figure 19.1). The “actual” path of the ball in space, the outside observer would say, is a parabola. The ball had a forward motion when released, and fell like a projectile through a parabolic arc. He would concede, however, that the child was right in assigning to the ball a uniform downward acceleration, and indeed both would agree on the magnitude of the acceleration. We may summarize the areas of agreement and disagreement between the two observers in a short table. (Bear in mind that this table refers to Newtonian mechanics, and will have to be altered soon when we come to the new mechanics of Einstein.)

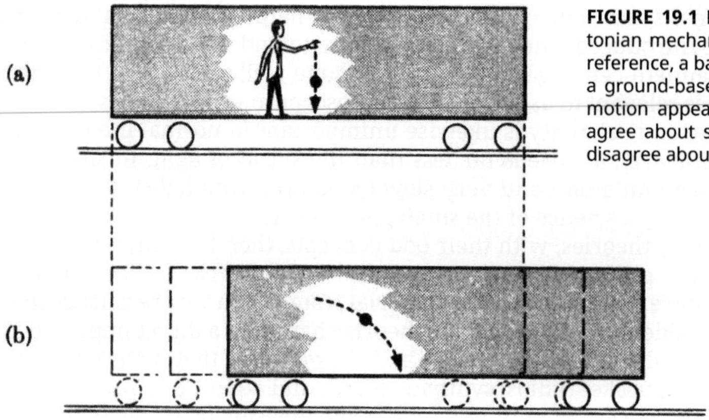

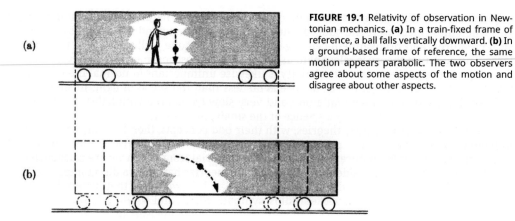

FIGURE 19.1 Relativity of observation in Newtonian mechanics. **(a)** In a train-fixed frame of reference, a ball falls vertically downward. **(b)** In a ground-based frame of reference, the same motion appears parabolic. The two observers agree about some aspects of the motion and disagree about other aspects.

Agreement	Disagreement
Acceleration	Position
Mass	Velocity
Force	Coordinates
Time	
Laws of motion	

There is obvious disagreement about position, since there is disagreement about the shape of the trajectory. There is also obvious disagreement about velocity, since the outside observer considers the ball to have some horizontal component of velocity, whereas the inside observer considers the ball to have only vertical velocity. There is also possible disagreement about coordinates, something which is not necessarily connected with the relative motion of the observers. Any two observers are always free to choose different coordinate systems to which to refer their measurements.

Despite the disagreements, there remains a large area of agreement between the observers. Both would measure the same acceleration. This may not be immediately obvious. It comes about because there is no relative acceleration between the two observers. Since each has zero acceleration with respect to the other, both agree about the acceleration of the moving objects. Mass and time are *assumed* in mechanics to be definite invariant quantities, and the theory of mechanics would be considerably upset if it were not so (as indeed it *is* upset by the theory of relativity). Force would be measured to be the same by both observers, since experimentally train travelers do not change their weight.* Most important of all, if the train traveler (who had better now be a scientist and not a child) supplements the experiment of dropping a ball with many other experiments, he would arrive at Newton's laws, exactly those that the outside observer already knows to be valid at rest outside the train.

The fact that the observers moving relative to each other agree about the laws of motion is known as Galilean invariance. It means that the earth is neither a better nor a worse laboratory than the train, and that the traveling scientist has as much right as the earthbound scientist to claim that he is at rest and the other is moving. The invariance eliminates any

* In one possible logical formulation of mechanics, force is defined as mass times acceleration. In that case, the invariance of mass and acceleration automatically implies the invariance of force. In a different possible logical formulation, where force is separately defined, it must be experimentally verified to be invariant.

single preferred frame of reference for mechanics, and thereby eliminates absolute motion. The essence of this invariance principle, as its name implies, was already appreciated by Galileo. He used it to support the Copernican view that the earth moves around the sun. As passengers on a moving earth, he argued, we cannot detect the motion of our vehicle earth through space.

The implications of the fact that the same laws of motion hold in the train and outside it can be appreciated by considering possible types of motion in the two frames of reference. The train defines the traveler's coordinate system or his frame of reference; the earth defines the frame of reference of the fixed observer. Although the traveling observer and the fixed observer disagree about the trajectory of the dropped ball—one says it fell straight down, the other that it followed a parabolic arc—each concedes that what the other claims to see is a *possible* motion according to Newton's laws. The outside observer could easily duplicate the straight vertical motion that the train traveler saw, and the traveler could without difficulty cause the ball to follow a parabolic arc with respect to the railroad car, if he chose to throw it in a different way. Invariance of the laws of motion means just that both observers agree about the possible paths of moving particles, but not necessarily about the motion of a particular object which they are both observing.

Galilean invariance is in fact well known, at least qualitatively, to everyone. We all know that in a uniformly moving elevator, or in an airplane in smooth air, there is no sensation of motion. We feel neither heavier nor lighter; there are no unexpected physiological sensations. If we drop an object, it falls to the floor. Everything seems "normal" and that normalcy means only that the same laws of mechanics apply to us and our surroundings in the elevator or in the airplane as at rest on the earth.

19.4 The Galilean transformation

The curious scientist is not satisfied to know that the traveling observer and the observer at rest measure different positions and velocities for the same moving ball. He wishes to know just how they differ, i.e., to express the difference quantitatively. This is easy to do. In general, a set of mathematical expressions relating one observer's measurements to another observer's measurements is called a transformation. In mechanics, the transformation of space and time measurements from one observer to another is called a *Galilean transformation*. For definiteness, suppose that our traveling observer measures horizontal distances x' forward from the rear of the train and vertical distances y' upward from the floor of the train (Figure 19.2). The observer on the ground measures horizontal distances x from some fixed pole A on the ground, and vertical distances y upward from a platform at the floor level of the train. Then the measurements y and y' will agree and the time measurements t and t' must be assumed to agree if both observers carry accurate clocks (an assumption that will soon have to be changed). Only x and x' will differ. If the rear of the train passed the pole A at $t = 0$, it will later have moved forward a distance $d = vt$, where v is the speed of the train. It is clear from Figure 19.2 that x and x' differ by this distance d. The Galilean transformation may therefore be written as a set of three simple equations:

$$x = x' + vt' , \tag{19.1}$$

$$y = y' , \tag{19.2}$$

$$t = t' . \tag{19.3}$$

On the right side of each of these three equations is a measurement (or measurements) of the observer on the train. On the left side of each equation is a measurement of the observer on the ground. Thus we have a *transformation* from x, y, t to x', y', t'. It is of course also equally well a transformation in the other direction.

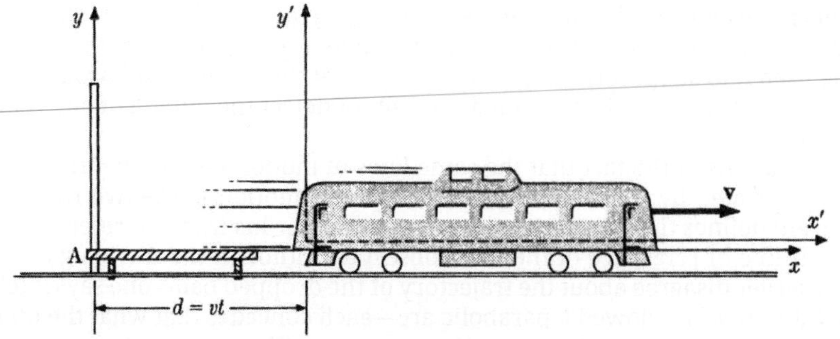

FIGURE 19.2 Definition of coordinates in two frames of reference.

Consider again the two fundamental ideas in the theory of relativity: relativity and invariance. The Galilean transformation laws give quantitative expression to the *relativity* of mechanics, for they relate the different measurements of different observers. The *invariance* of mechanics is contained in the principle of Galilean invariance, the statement that the laws of mechanics are the same in the two frames moving uniformly with respect to each other.

We have so far not been very careful about stating in which frames of reference the laws of mechanics remain invariant. In fact there is no way to tell exactly how to find even *one* frame of reference in which Newton's laws are valid. But if one can be found, then it is easy to find many more, for every frame moving uniformly with respect to it will also be a frame in which Newton's laws are valid. Such frames are inertial frames (Section 8.11). Any frame moving uniformly with respect to an inertial frame is itself an inertial frame. The task of finding the first inertial frame is not difficult (in principle), for it is a frame in which Newton's first law is true. It is necessary only to observe a particle upon which no forces act, and then choose a frame of reference in which that particle is at rest or moves uniformly. For many practical purposes, the surface of the earth is a good approximation to an inertial frame of reference. A rapidly decelerating automobile, on the other hand, is obviously not an inertial frame, for a passenger on whom no horizontal forces are acting is "thrown" forward, i.e., he is accelerated with respect to the car.

19.5 Electromagnetism and the Lorentz transformation

The fact that the laws of mechanics are the same in all inertial frames illustrates what Poincaré and Einstein called the Principle of Relativity.* We have seen that the principle is true for mechanics if the space and time measurements of different observers are related by the Galilean transformation. We might be tempted to try out the same ideas on electromagnetism, testing the equations of this theory for invariance. Disappointingly, such an effort meets at first with failure. If the space and time measurements of one observer are transformed to the space and time measurements of another observer by means of Equations 19.1 to 19.3, the equations of electromagnetism are altered in form. Indeed, equations true in one frame of reference are transformed into equations not true in the other frame of reference. The detailed reasons for this apparent failure of the Principle of Relativity are subtle, but we can get a hint about the difficulties by referring back to the table on page 550. Recall that the two observers in relative motion disagreed about the velocity of the ball whose motion they were watching. This disagreement did not prevent them from agreeing

* Poincaré stated this as a principle of classical physics that seemed to be in trouble. Einstein stated it as a principle of nature to which classical physics, if necessary, would have to yield.

about the laws of motion because the law $\mathbf{F} = m\mathbf{a}$. does not contain the velocity explicitly. It contains instead the acceleration, about which the two observers agreed. In the theory of electromagnetism, however, the fundamental equations do contain velocity. The magnetic field generated by a moving charge depends on its velocity; the magnetic force on a moving charge depends on its velocity; and the basic equations contain the speed of light. All of these velocities would be changed by a Galilean transformation. This is one reason why the laws of electromagnetism do not demonstrate Galilean invariance. They seem to call for a preferred frame of reference.

This difference between the theory of mechanics and the theory of electromagnetism 108
can be viewed in several alternative ways. (1) The Principle of Relativity happens coincidentally to be satisfied by mechanics but it is not a general principle of nature and is not important. (2) The theory of electromagnetism is incorrect and must be changed to conform to the Principle of Relativity. (3) The Principle of Relativity is correct, but the Galilean transformation (and therefore Newtonian mechanics) must be discarded and a new transformation found which will permit the laws of electromagnetism to be invariant. Einstein adopted the third bold view, which led to the theory of relativity, with its enormously successful consequences, including a new mechanics, a new view of the world, and incidentally even a deeper new insight into electromagnetism. Although the equations of electromagnetism weathered the revolution of relativity unchanged, the interpretation of these equations was somewhat altered.

Einstein's transformation law relating the space and time measurements of different observers (all in inertial frames!) is called the Lorentz transformation. Lorentz had shown in 1904 that this transformation law would preserve the invariance of the theory of electromagnetism and account for the shyness of the ether. Over the course of the next five years, as scientists reluctantly but finally discarded the ether altogether, the more general validity of the Lorentz transformation and some of its startling consequences came to be realized. These form the subject of the next chapter.

The Lorentz transformation plays a vital role in the theory of relativity because it is the essential quantitative link between the measurements of different observers. Without it, science would be reduced to a chaos of subjectivity; one observer measures one thing, another measures something different, and no two observers could get together and agree about what is "really" happening. The Lorentz transformation is a world court of arbitration, reconciling to everyone's satisfaction the divergent parochial viewpoints of scientists who insist on referring their measurements to frames of reference that are in states of relative motion. Because it accounts quantitatively for the relativity of observation, it makes possible the invariant "objective" agreement among different observers. (Actually, human observers do not move fast enough to reveal the special features of the Lorentz transformation. However, the scientist can cast himself mathematically into a high-speed frame of reference and can describe physical processes in that frame without climbing aboard it himself.)

Along with other scientists of the time, Einstein was interested in problems posed by the treatment of electromagnetic phenomena in moving frames of reference. Even more than most of his contemporaries, he regarded these as theoretical problems, problems of the internal self-consistency of electromagnetic theory and the consistency of electromagnetism with general principles such as energy conservation. In his search for harmony, Einstein asked the question: Since the Galilean transformation changes the laws of electromagnetism, is there some other transformation that leaves these laws unchanged? The answer is yes; it is the Lorentz transformation. This route to relativity, although it has a certain appeal for the trained scientist, is not the easiest route for the beginning student of physics. We shall start instead from an experimental base. One experiment in particular, the Michelson-Morley experiment, mentioned at the beginning of this chapter, is useful because it makes clear the physical basis of relativity and leads very directly to the notion of the relativity of time, the key to understanding the new ideas of the theory of relativity. Even though Einstein took no notice of the Michelson-Morley experiment in his first work on relativity,* it was later recognized as an experimental cornerstone of the new theory. In order to understand why the Michelson-Morley experiment was carried out, and why its results are significant, it is necessary to appreciate the late nineteenth-century view of the ether and of the nature of light.

19.6 *The ether and electromagnetic radiation*

109 It is to Aristotle that we owe the statement, "Nature abhors a vacuum" (a statement that beautifully and succinctly illustrates what was wrong with Aristotelian physics, but is no less memorable on that account). A much better case could be made for the proposition, "Human beings abhor a vacuum," if we understand by a vacuum truly empty space—nothingness. So abhorrent was the vacuum to Newton that he wrote, "that one body may act upon another at a distance through a vacuum, without the mediation of anything else, ... is to me so great an absurdity, that I believe no man who has in philosophical matters a competent faculty for thinking, can ever fall into it." Newton believed in the existence of an all-pervading substance filling up space. This substance, which came to be called the ether, was supposed to transmit gravitational forces and light, which would otherwise have to make their way through nothingness. In the two centuries between Newton and Einstein, the belief in the reality of the ether became more, not less, firmly fixed in the minds of scientists. In a panegyric to the ether in 1873, Maxwell said: "The vast interplanetary and interstellar regions will no longer be regarded as waste places in the universe, which the Creator has not seen fit to fill with the symbols of the manifold order of His kingdom. We shall find them to be already full of this wonderful medium, ... It extends unbroken from star to star; and when a molecule of hydrogen vibrates in the dog-star, the medium receives the impulses of these vibrations; and after carrying them in its immense bosom for three years, delivers them in due course, regular order, and full tale into the spectroscope of Mr. Huggins, at Tulse Hill."

Newton rejected the empty vacuum on philosophical grounds. He was well aware that his law of gravitation was sufficiently justified by its empirical success; his failure to provide a detailed picture of the mechanism of transfer of gravitational force from one place to another in no way detracted from that success. At the time of Newton, the ether was just an "idea," and no crucial experiment to prove or disprove the idea could be imagined. To its proponents in the late nineteenth century, the ether was more than an idea; it was a substance with definite properties. One of these properties was its characteristic rate of propagation of "disturbances" (i.e., electromagnetic waves). Just as every material substance—air, water, iron, and so on—has a characteristic speed of propagating sound waves, so the ether was supposed to have a characteristic speed of propagating light waves. Maxwell's successful prediction of the speed of light was undoubtedly a dominant factor behind his belief in the ether.

So long as the ether escaped definite experimental detection, it had its detractors, as well as its proponents. Some scientists held fast to the belief in "action at a distance," for they felt that the all-pervasive frictionless ether had properties too strange to be believed. How could one conceive of a medium that offered resistance to the flow of something as tenuous as light, yet permitted material objects to move through it unimpeded and unaffected? Clearly it was time to try to pin down the ether or give it up, as scientists had given up one by one in the past the mysterious fluids of phlogiston—a substance supposedly contained in combustible materials—caloric, and electricity.

* He was probably unaware of it. It was said of Einstein at the time that he read little, and thought much.

19.7 The Michelson-Morley experiment

That a definitive experimental test for the existence of the ether should be possible was first realized by Albert Michelson, who in 1879 as a 26-year-old American naval ensign had already established a name for himself by measuring the speed of light with great precision. His first effort to discover the motion of the earth through the ether came two years later in Potsdam, Germany. First as an ensign on leave, then as an ex-ensign, Michelson had been spending his time at the leading scientific centers of Europe, where he had come to see that the demonstration of the existence of the ether was one of the most important and challenging problems in physics. In Berlin he devised a method of coming to grips with the ether, but traffic vibration forced him to move the sensitive apparatus to the quieter suburb of Potsdam. Here he found surprisingly that he was unable to detect any relative motion of earth and ether. But the accuracy of the measurement was not high, and no significant stir in the world of science resulted.

When Michelson and Morley repeated the experiment six years later in Cleveland, Ohio, with much improved apparatus, the significance of the negative finding could no longer be overlooked. No apparent motion of the earth through the ether could be detected. Many repetitions of the experiment in succeeding decades have given agreement with the 1887 result—there is no detectable relative motion of earth and ether.*

In order to understand the Michelson-Morley experiment, we turn temporarily from light waves to sound waves. If passengers on an airplane in flight had a way of measuring the speed of sound waves encountered in the air, they would observe quite different values for the speed of sound relative to the airplane according to the direction of motion of the sound. For example, on a plane traveling at half the speed of sound, a sound wave going the opposite direction would seem to move past the airplane at a speed 50% greater than the normal speed of sound. A sound wave overtaking the plane, on the other hand, would pass the plane at a relative speed of only half the normal speed of sound. According to Maxwell's view, we are all passengers on the earth which is flying through the ether. If we encounter light waves we should find them to be moving at different speeds relative to the earth, according to whether the light is encountering us head-on in our passage through the ether, or whether it is overtaking us in our motion. The reason is the same in both examples. Sound travels at a definite speed with respect to the air, and will accordingly have different speeds with respect to different observers moving through the air. Light, according to Maxwell, was supposed to move at a definite speed with respect to the ether and should therefore have a different speed with respect to the earth, which is in motion through the ether.

Imagine now a rather unusual airspeed indicator mounted on the airplane. It consists 110
of a sound generator—a drum would do—and a sound receiver (such as a microphone) mounted on top of the plane amidships, and small plates to reflect sound, one attached to the tail and one to the wingtip, equidistant from the sound generator-receiver combination (Figure 19.3). A noise is created, some time elapses, and echoes return from the tail and the wingtip. If the plane is in motion, the wingtip echo will arrive slightly sooner than the tail echo, and the time difference of the echoes could be displayed in the cockpit to let the captain know how fast the plane is moving. Suppose, for instance, that the speed of the plane is 1.5×10^4 cm/sec (about 340 mph), half the speed of sound, and that the reflecting plates are 3×10^3 cm from the source of sound. The sound wave traveling to the tail would have a speed with respect to the airplane of 50% more than the speed of sound, that is, 4.5×10^4 cm/sec. Its journey to the tail would therefore require 0.067 sec. On its return trip, it would be mak-

* A more recent version of the experiment was carried out at Columbia University in 1958 by J. P. Cedarholm, G. F. Bland, B. L. Havens, and C. H. Townes, and reported in the *Physical Review Letters, 1*, 342 (1958). The expected negative result was found, but to higher precision than heretofore. Since any departure from the null result, no matter how small, would be exceedingly important, the experiment will undoubtedly continue to be repeated as more accurate techniques are developed.

ing good a speed with respect to the plane of only 1.5×10^4 cm/sec, and would require three times as long to get back, 0.2 sec. Altogether 0.267 sec would elapse before the echo from the tail returned. We can express this algebraically. The outbound time from the source A to the tail reflector M_1 (Figure 19.3), a distance L away, is

$$t_{\text{out}} = \frac{L}{v_s + v_a}, \tag{19.4}$$

where v_a is the speed of the airplane and v_s is the speed of sound. The sum $v_s + v_a$ in the denominator is the speed of the backward-propagating sound wave relative to the airplane. The time for the echo to return is

$$t_{\text{return}} = \frac{L}{v_s - v_a}, \tag{19.5}$$

The round-trip time for the tail echo is the sum of these two expressions,

$$t_{\text{out}} + t_{\text{return}} = t\,(\text{tail echo}) = \frac{2L}{v_s} \times \frac{1}{1 - (v_a/v_s)^2}. \tag{19.6}$$

Analysis of the trip to the wingtip and back is somewhat complicated by the forward motion of the plane. Derivation of an equation for the round-trip time is left as an exercise. It is

$$t\,(\text{wingtip echo}) = \frac{2L}{v_s} \times \frac{1}{\sqrt{1 - (v_a/v_s)^2}}. \tag{19.7}$$

The important point is that this latter time is shorter, 0.231 sec in our example. Both echoes are delayed by the motion, but the tail echo is delayed more than the wingtip echo, and the difference is easily interpreted in terms of the speed of the airplane.

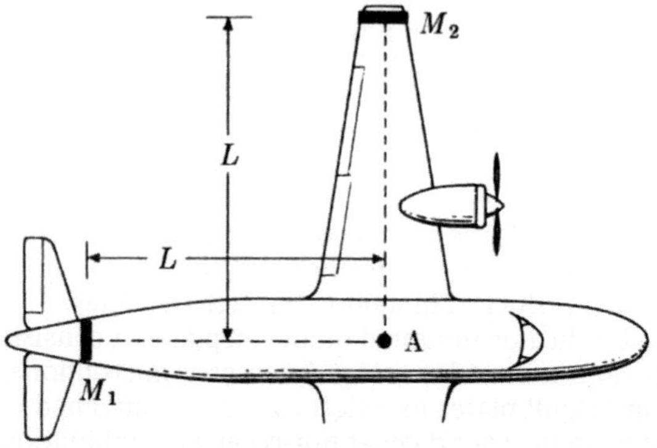

FIGURE 19.3 Unusual airspeed indicator. Sound waves start from point A and return to point A after reflecting from the equidistant plates M_1 and M_2. The time difference between the returning echoes provides a measure of the airspeed.

The Michelson-Morley experiment is an etherspeed indicator designed in exact analogy to our unorthodox airspeed indicator. Light is sent simultaneously in two perpendicular directions and then reflected back. The source of light and the apparatus are fixed with respect to the earth, and are therefore being carried through the ether by the earth. The time difference between the two reflected light signals is too small to measure directly, but it is inferred from the interference of the two waves. The design of the experiment is indicated in Figure 19.4. A light wave arriving from the left is split by a lightly silvered glass plate into two waves, one running upward to mirror M_1 and the other continuing on to mirror M_2. This wave-splitting action ensures that the two waves get underway at exactly the same instant. Some of the light returning from mirror M_1 passes through the plate and some returning from M_2 is reflected from the plate to the eye of the observer (or better, to a light-sensitive device more accurate than the eye). If the mirrors M_1 and M_2 are exactly the same distance from the glass plate (this is not necessary, but it simplifies matters slightly), and if the apparatus is at rest with respect to the ether, the two waves arriving at the observer will be exactly in phase and will reinforce each other to give a bright light. If instead the apparatus is moving through the ether, for example in the upward direction in the diagram, the light from M_1 will be slightly more delayed in its round trip than the light from M_2, and the two waves will no longer be exactly in phase. There will be partial interference of the two waves, and the light seen by the observer will be less bright.

Of course the earth cannot be stopped and started at will to look for a change of light intensity. But what is simple and just as effective is to change the orientation of the apparatus. Michelson rotated the whole apparatus through 90 deg so that the path to M_1, if initially "upwind" and "downwind," became "cross-wind." A change in intensity should have been observed as the rotation proceeded. Michelson and Morley also repeated the experiment at various times of the year to catch the earth's motion in various directions through space as the earth swung around the sun.

The magnitude of the effect that Michelson hoped to observe is extremely small, because of the leisurely pace of the earth round the sun, and it is small wonder that the Berlin traffic, even in 1881, was sufficient to disturb his experiments. The orbital speed of the earth is about 3×10^6 cm/sec (67,000 mph), which is one ten-thousandth of the speed of light. We may suppose, as Michelson did, that the earth also moves through the ether at this speed. A typical distance from the lightly silvered plate to the reflecting mirrors (Figure 19.4) was 120 cm. If these numbers are used in Equations 19.6 and 19.7, one finds that the upwind-downwind beam should have been retarded with respect to the crosswind beam by about 4×10^{-17} sec. In this exceedingly short time light travels 120 A (or 1.2×10^{-6} cm). A single wavelength of visible light is about fifty times greater, or 6,000 A. Therefore as the apparatus is rotated, the shift of the two waves with respect to each other is much less than the 3,000 A that would be required to convert bright constructive interference into dark destructive interference. Nevertheless, a slight change of intensity should occur, and Michelson felt confident that he could observe this tiny effect of the ether without difficulty.

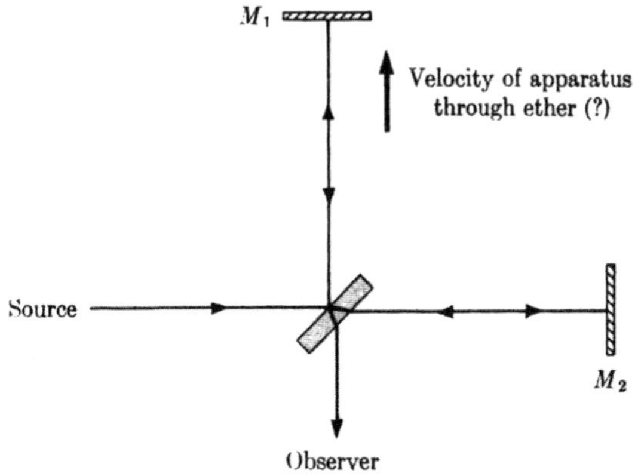

FIGURE 19.4 Schematic arrangement of Michelson-Morley experiment.

No matter what the time of day or time of year, no matter what source of light was employed, a rotation of the apparatus never produced any observable change of light intensity. The Michelson-Morley etherspeed indicator always gave a reading of zero, indicating no motion of the earth through the ether. Since the earth is rotating about its own axis and traveling about the sun, it seems inconceivable that the ether should execute such gyrations as to remain always at rest with respect to the earth. It is easier even to give up the ether than to imagine this!

19.8 Efforts to keep the ether

111 Yet it was far from easy to give up the ether. For nearly two decades after the Michelson-Morley experiment, many efforts were made to retain the ether yet explain the experimental result. Michelson himself gave no thought to discarding the ether. He believed that in some inexplicable way nature had tricked him, making his experimental arrangement unsuitable for detecting the ether. The simplest such trickery to imagine is that the earth drags some of the ether along with itself. According to this ether-drag hypothesis, the Michelson-Morley experiment shows no effect because just near the surface of the earth, the ether is motionless with respect to the earth. This kind of drag is well known for airplanes, which drag along a very thin shield of motionless air known as a boundary layer. An airspeed probe which penetrated only into the boundary layer would record no motion. It must stick out through the boundary layer into the air-stream beyond in order to detect the motion of the airplane through the air. Today satellites would make it possible in an analogous way to put the Michelson-Morley etherspeed probe out into space beyond the ether boundary layer. But in the meantime the ether-drag hypothesis has had to be abandoned: first, because it should cause some bending of starlight arriving at the earth, which has never been seen; second, because the successful theory of relativity has led us to discard the ether entirely, with or without drag. Nevertheless scientists are accustomed to expect the unexpected, and the Michelson-Morley experiment will no doubt one day be repeated far from the earth's surface.*

Another suggestion which suffered the same fate as the ether-drag hypothesis, and for the same reasons, was the proposal that light travels at a fixed speed not with respect to the ether, but with respect to the source of the light. Before the passage of many years, this hypothesis was laid to rest by astronomical evidence. Some stars occur in pairs which rotate about each other. As one of the pair moves toward the earth, the other moves away. Yet careful observations showed that the light from the approaching star must be traveling no faster toward earth than light from the receding star.

A more radical suggestion made by George F. FitzGerald in 1892 also came to nought as an explanation of the Michelson-Morley results, but like so many wrong suggestions in science, it gave rise to some fruitful consequences before it died. Suppose, said FitzGerald, that the experimental apparatus shrinks in the direction of motion through the ether by just enough to compensate exactly for the slower average speed of light upwind and downwind than crosswind. Since the required shrinkage would be, on earth, only about one part in one hundred million, this subtle effect would not contradict any previous measurements. The FitzGerald contraction hypothesis was formulated mathematically in the same year by Lorentz, and his mathematical expressions found their way eventually into the theory of relativity, but with a markedly different interpretation. At every point where mathematics is used to describe nature, the mathematical skeleton is meaningless without interpretation and definition of the symbols used. The formulas of Lorentz underwent a marked evolution of interpretation between the time of FitzGerald's hypothesis in 1892 and the time of Einstein's theory of relativity in 1905.

Other unsuccessful efforts to detect the ether followed the work of Michelson and Morley. These experiments contributed, around the beginning of this century, to a heightened interest in theoretical problems of electromagnetism, especially problems concerned with moving frames of reference. Einstein, although unaware of the Michelson-Morley experiment in 1905, provided, through his special theory of relativity, a simple reason for its "failure." The expected result of the Michelson-Morley experiment was based on the supposition that light travels at a fixed speed with respect to the ether, and therefore at a greater or lesser speed with respect to some observer, depending on how that observer is moving through the ether. Einstein abandoned this point of view and abandoned the ether. He pos-

* In 1926-1928 Auguste Piccard repeated the Michelson-Morley experiment in a balloon about 8,000 feet above the earth, with the expected negative result.

tulated instead that the speed of light is a universal constant, the same with respect to all observers. This hypothesis, which appears simple at first glance, and appears impossible after a little thought, is the source of the revolution of relativity. In the next chapter we shall explore some of its consequences. Two chapters later, in Chapter Twenty-Two, we shall find that the ether is not so easy to kill after all. In a new and sophisticated form it is with us still.

EXERCISES

19.1. Section 19.1 mentions several unexplained observations that seemed to mar the structure of physics at the close of the nineteenth century. **(1)** Very briefly describe three of these in your own words. **(2)** Which of those you have named were in conflict with the then known principles of physics, and which merely lacked an explanation?

19.2. Does the magnetic field created by a moving charge depend upon the state of motion of the observer? Illustrate your answer with a simple example.

19.3. If the table of agreement and disagreement in Section 19.3 is extended, into which column should each of the following concepts go: **(a)** momentum; **(b)** kinetic energy; **(c)** temperature; and **(d)** charge? Explain the reason for each choice.

19.4. A physicist and his equipment are shut inside a railroad car with no windows. Describe a simple experiment that would enable him to determine a *change* of velocity of the car even though he is unable by any means to measure its velocity when it moves with constant velocity.

19.5. **(1)** According to Newtonian mechanics, does the principle of Galilean invariance remain valid for observers whose relative motion is accelerated? **(2)** For such a pair of observers, how must the table of agreement and disagreement in Section 19.3 be changed?

19.6. An experimenter on a train tosses a ball toward the rear of the train. It leaves his hand moving horizontally with initial velocity components $v_{x0}' = -v$ (where v is the speed of the train over the ground) and $v_{y0}' = 0$. Its initial position is $x_0' = 0, y_0' = h$. **(1)** Write equations for the subsequent position coordinates x' and y' at time t' in the coordinate system of the train-based observer. **(2)** With the help of the Galilean transformation, write equations for the coordinates x and y as a function of t, as measured by a ground-based observer. **(3)** Is the elapsed time from leaving the hand of the man on the train until striking the floor of the car the same or different for the two observers? **(4)** Is the distance covered by the ball in its fall the same or different for the two observers?

19.7. A railroad car moves over straight tracks at a constant speed $v = 1{,}000$ cm/sec. A block on the floor of the car is pushed in such a way that it experiences a constant forward acceleration of 50 cm/sec^2 relative to the car. **(1)** What is the acceleration of the block according to an observer on the ground? **(2)** If the block starts from rest in the car, write expressions for x' vs. t' (train-based coordinates) and x vs. t (ground-based coordinates), assuming the validity of the Galilean transformation. **(3)** Compare $x_2 - x_1$ with $x_2' - x_1'$ and $t_2 - t_1$ with $t_2' - t_1'$ for any chosen pair of instants t_1 and t_2.

19.8. Table 18.1 reveals in an approximate way a correlation between the speed at which sound propagates through a medium and the hardness or solidity of the medium. Based on this correlation, what would you expect to be the nature of a medium that propagates a wave at the speed of light? Is this extrapolation from material media consistent with your conception of the ether?

19.9. Explain in your own words why sound requires a material medium for its propagation, but light requires neither matter nor even an ether.

19.10. **(1)** Make sketches, approximately to scale, of the successive positions of the airplane in Figure 19.3 when **(a)** a sound pulse leaves source A, **(b)** the pulse reaches mirror M_1, and **(c)** the pulse returns to point A, if the speeds and dimensions are those stated in the example preceding Equation 19.4. **(2)** By what distance does the airplane move forward between sketches (a) and (b)? By what distance does it move forward between sketches (b) and (c)? **(3)** For what speed of the airplane would the sound pulse never return to point A?

19.11. With the help of a diagram like this one, derive Equation 19.7 for the round trip time, t (wingtip echo). The speed of the airplane relative to the air is v_a; the speed of sound relative to the air is v_s. Note that for the wingtip echo, in contrast to the tail echo, the time from source to mirror is the same as the time from mirror to source.

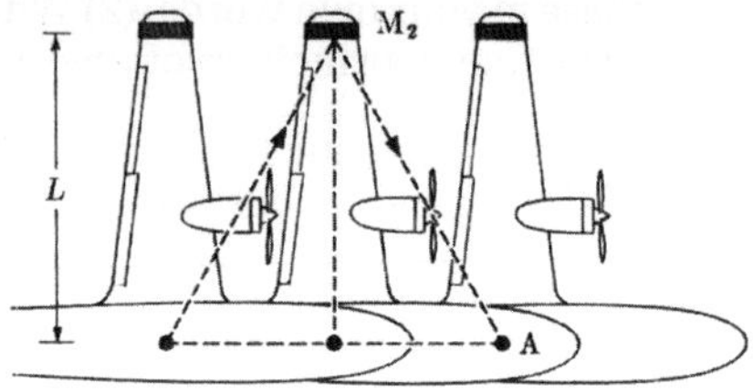

19.12. ETHER ELUDES MICHELSON AND MORLEY, SCIENTISTS PERPLEXED. Write the lead paragraph of a news story to follow this headline.

19.13. Suppose that the Michelson-Morley experiment had yielded a positive result, showing absolute motion of the earth and revealing a preferred frame of reference. **(1)** How could the direction and speed of the sun in the preferred frame of reference be determined? **(2)** In a hypothetical world, with such a preferred frame of reference, where Newtonian mechanics remained valid, it would be possible to bring a photon to rest. Explain why.

19.14. The Michelson interferometer (Figure 19.4), apart from its use in the Michelson-Morley experiment, became a valuable tool for precision measurements of wavelength. As one of its mirrors is slowly moved to lengthen one arm of the interferometer, what should happen to the observed pattern of light? Suggest a precise method of wavelength determination that makes use of such mirror motion.

19.15. In the Ritz emission theory of light it was assumed that the velocity of light is constant relative to the emitter. Thus if an object emitted light while approaching the observer, that light would be traveling at a speed $c + v$ relative to the observer. If the light was emitted when the source was receding from the observer, the light was supposed to move toward the observer with speed $c - v$. Suppose that a star of small mass is going in a circle around its massive companion. On the basis of the Ritz theory show that an observer on earth might see the star in a later position before he saw it in an earlier position. Illustrate your argument with a diagram. The absence of such anomalies in observations of binary stars is one of the pieces of evidence against the Ritz emission theory.

TWENTY

The Lorentz transformation

Because it is simple in conception, and because its result disagrees with the classically expected result, the Michelson-Morley experiment is a useful springboard into relativity theory. The result of the experiment can be conservatively stated in this way: The speed of light seems to be constant with respect to the earth at all times. Since the earth moves now one way relative to the rest of the universe, now another, an earth-fixed observer at two different times can be regarded as two different observers in states of relative motion, and it is not too big a jump to the statement: The speed of light seems to be the same for observers in different states of motion. Some of the early attempts to explain the results of the Michelson-Morley experiment concentrated on the "seems to" in this statement. The explanation offered by the theory of relativity rests instead on a powerful new hypothesis: The speed of light *is* the same for all observers in inertial frames of reference. (That the earth itself is not a perfect inertial frame of reference is true, but it is unimportant in the interpretation of the Michelson-Morley experiment and in most other terrestrial experiments. The acceleration of the earth may be ignored if the *change* of velocity of the earth is small during the time a given experiment lasts.)

20.1 Constancy of the speed of light

The statement

$$c = constant \tag{20.1}$$

appears innocent enough, but it leads to some remarkable, and seemingly paradoxical, consequences. First of all, it requires that the ether be rejected altogether. Recall that the *only* way the ether was supposed to make its presence known was through electromagnetic effects. It was entirely frictionless and devoid of any interaction with matter or with heat. If now its presence can also not be detected by means of an experiment with electromagnetic radiation, such as the Michelson-Morley experiment, the ether becomes entirely unobservable and therefore scientifically meaningless. It remains at best a philosophical crutch for those who still abhor a vacuum. It must not be imagined, of course, that Einstein said dogmatically, "There is no ether." Rather he pointed out that science could get along very well without the nineteenth-century conception of the ether, which proved to be neither necessary nor useful. It led, in fact, to some predictions that were in disagreement with experiment, and to no successful predictions that could not have been made as well without the ether.

Vanishing along with the ether were the notions of a preferred frame of reference and of absolute velocity. If empty space is truly empty, a rocket ship can at best measure its speed with respect to other objects, not with respect to "space," whose very emptiness implies the impossibility of any measurement with respect to it. If rocket ship A drifts past rocket ship B in empty space, spaceman A and spaceman B will each observe the relative motion, but each will be quite unable to decide which is moving and which is at rest, if either. The two frames of reference are not equal, but are equally "good," or, more exactly stated, they are physically equivalent. Actually the Michelson-Morley experiment only suggests this possibility of the equivalence of different (inertial) frames of reference; it does not imply such equivalence unambiguously. The experiment tells only that absolute motion cannot be detected with light waves, but leaves open the possibility that some other physical phenomena might be sensitive to absolute motion. However, according to Einstein's second fundamental postulate, inertial frames of reference are equivalent for *all* physical phenomena. This postulate is called the Principle of Relativity; a few of its consequences will be explored in the next chapter.

20.2 The relativity of time

112 The direct and startling consequence of the assumption that light travels at an invariable speed with respect to all observers (in inertial frames) is that time must be relative—observers in states of relative motion must have different ideas about the measurement of time and the time interval between events.

To see just what is paradoxical about the statement c = *constant*, imagine observer A standing beside a road and observer B traveling along the road in a supercharged rocket car (Figure 20.1). Just as B comes abreast of A, a light wave passes them both by. Later on A meets B and remarks, "Did you see that light wave come by us this afternoon? I happened to notice that it was traveling at exactly 300 meters per microsecond." "Nonsense," replies B. "You must be in error. It passed me at a relative speed of 300, and I was doing 100 myself." "No, no," retorts A, "I clocked you both. It's true that you were doing 100, but the light wave passed you at a relative speed of only 200." Our common-sense point of view is that either A or B (or both) must be in error. But according to Einstein's postulate, both are indeed correct. Yet something has to give. If we are prepared to admit that the light wave was moving at the same speed c relative to both A and B, we must admit the possibility that there is some intrinsic difference in the way they are defining speed. Since a measurement of speed involves measurements of both distance and time, perhaps they disagree about length measurements or about time measurements. In fact they must disagree about both, as we will see presently.

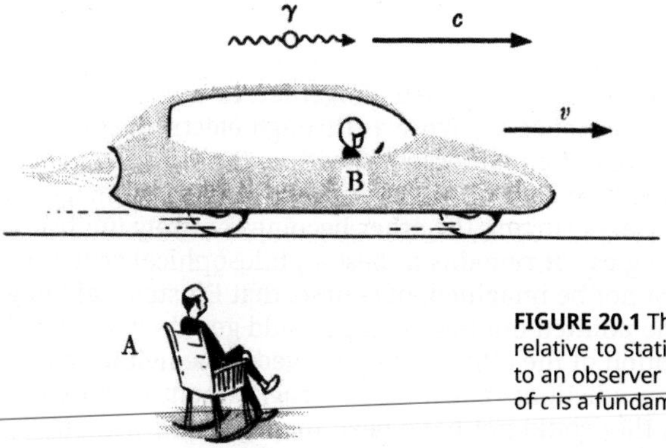

FIGURE 20.1 The "paradox" of relativity. A photon has a speed c relative to stationary observer A and the same speed c relative to an observer B in high-speed motion. The universal constancy of c is a fundamental postulate of relativity.

It must be remarked that the hypothetical conversation reported above is farfetched, not merely because B is reported to have traveled at a quite phenomenal speed, but because in a world where such speeds are commonplace, the constancy of the speed of light and the relativity of time would be so well known in everyday experience that no such controversy would arise. But suppose that the rocket car is a recent invention and that men sometimes fall back into their old ways of thinking. "Sorry," says A. "I can't stay to argue. It's nearly midnight and I am quite tired." "What!" replies B. "Why it's only 10:00 by my watch, and I feel quite fresh after cruising about all day."

The conclusion that we are forced to is that the constant speed of light is paradoxical unless observers moving with respect to each other disagree about distance and/or time measurements. It is easy to construct innumerable thought experiments which demonstrate this fact. As a slight variant on the observations of A and B reported above, suppose that on some other occasion B switches on his headlights just as he passes A. "Well," he reports later to A, "I verified that light travels at 300 meters per microsecond with respect to my rocket car, for exactly 10^{-8} seconds after switching on my lights, the light beam had reached a point just 3 meters ahead of me." "Quite impossible," replies A wearily. "I took some measurements on that beam myself. After 10^{-8} seconds, it had progressed exactly 3 meters from *me* and was only 2 meters ahead of you." "But wait," says A as an afterthought. "Wasn't your watch running slowly the other day? Perhaps what you took to be 10^{-8} seconds was really longer, giving the light beam some extra time to get three meters ahead of you."

The discrepancies of observations of A and B reported so far could be explained either by differences in their length scales, or, as A suggests, by difference in their time scales. Another simple thought experiment will show that their time scales (at least) must be different (Figure 20.2). Suppose that B switches on the interior light in his rocket car as he speeds past A. If the light is located just at the center of the car, B will conclude that its illumination reaches the front and the back of his car simultaneously, for he consistently finds that light travels at a fixed speed with respect to his car. But A as adamantly claims to see light always traveling at the same fixed speed with respect to *him*. From A's point of view the light will spread at equal speed in each direction and reach the back of the car, which is approaching the light, before it reaches the front of the car, receding from the light. Two events (the arrival of light at the front and at the back of the car) are judged by B to be simultaneous and by A to occur at different times. Clearly there must be something different about time itself for the two observers.

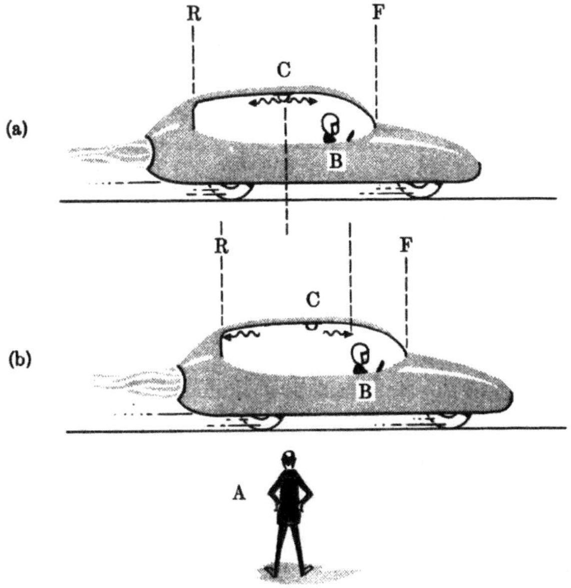

FIGURE 20.2 The relativity of time. **(a)** At some instant the central light C in the rocket car is switched on. **(b)** Sometime later, according to the roadside observer A, the light has reached the rear point R but has not yet reached the front point F. But according to observer B in the car, the light reaches the equally distant points R and F simultaneously.

20.3 *Time dilation*

We will permit A and B, with their remarkable powers of observation, one more experiment, this time with a view to learning by exactly how much their time scales differ. They agree that it would be useful to measure the round-trip time for a light wave starting on the floor of the car, going vertically upward with respect to the car, reflecting from a mirror on the ceiling, and returning to its starting point on the floor of the car. Using the equation for constant-speed motion, distance equals speed times time, B can write down

$$2L = ct_B\,, \tag{20.2}$$

if L is the distance from floor to ceiling and t_B is the round-trip time that B measures. According to A, watching from beside the road, the light beam has executed a sawtooth pattern, and traveled not a distance $2L$, but a greater distance $2H$, where H stands for the hypotenuse of the right triangle in Figure 20.3. A's equation connecting distance, speed, and time, is

$$2H = ct_A\,, \tag{20.3}$$

Since H is greater than L, but both observers measure the same speed c, it is already clear that t_A must be greater than t_B, and that A will consider B's time scale to be running too slowly. To get the quantitative relation, A makes use of the Pythagorean theorem for the right triangle in Figure 20.3,

$$H^2 = L^2 + (\tfrac{1}{2}vt_A)^2\,, \tag{20.4}$$

where v is the speed of the car with respect to A. For H in this expression he substitutes $\tfrac{1}{2}ct_A$ from his distance equation, Equation 20.3, and for L he substitutes $\tfrac{1}{2}vt_B$ from B's distance equation, Equation 20.2. The Pythagorean relation then becomes

$$\tfrac{1}{4}c^2t_A{}^2 = \tfrac{1}{4}c^2t_B{}^2 + \tfrac{1}{4}v^2t_A{}^2\,. \tag{20.5}$$

Some simple algebra, which is left as an exercise, converts this equation to the form

$$t_A = \frac{t_B}{\sqrt{1-(v^2/c^2)}}\,. \tag{20.6}$$

This is one of the most famous equations of relativity, and has by now received ample experimental verification (see Section 20.6). It represents what has come to be called the time-dilation phenomenon. According to A, B's time is "dilated," that is, B's clocks run slowly, and the factor giving the ratio of the rates of the clocks is $\sqrt{1-(v^2/c^2)}$. For normal human speeds, this factor is so close to one that we are quite unable to detect the time-dilation effect in ordinary life.

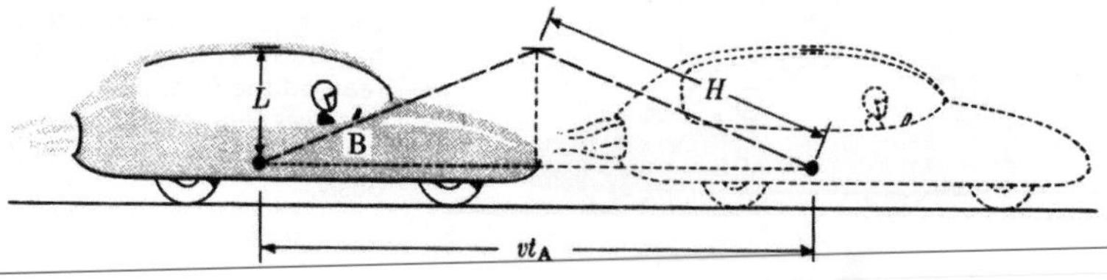

FIGURE 20.3 Experiment to determine relative time measurements. A light beam going from floor to ceiling and back travels a distance $2L$ in a time t_B, according to driver B; and a distance $2H$ in time t_A, according to the roadside observer A. In half the round-trip time, observer A sees the car move forward a distance $\tfrac{1}{2}vt_A$.

Since time scales (and distance scales too) are different for different observers, we must become suddenly quite cautious, indeed suspicious, in approaching the processes of measurement. For example, dare we draw the "obvious" conclusion that since B's clocks seem to A to be slow, A's clocks must seem to B to be fast? No, B will be equally convinced that it is A's clocks that are slow. Without this reciprocity of observation, indeed, we would be able to identify absolute motion and return to a preferred frame of reference. This reciprocal relationship means that B can respond to any slurs that A might cast about his qualifications as an observer, "It's mutual, I'm sure." In the experiment just described, the apparent absence of reciprocity comes about because the point to which the light returns is the same in B's frame of reference, but different in A's frame. Reciprocity would reenter if this experiment were compared with one in which it is A on the ground who shines a light vertically up and down, and B who observes the sawtooth pattern. (In view of the reciprocity of measurement, how could it have happened, as reported in an earlier conversation between A and B, that B returned from a day of fast cruising to find his watch lagging behind A's? This fascinating question, which has stirred controversy for decades, is discussed in the next chapter.)

In our description of the time-dilation experiment above, we tacitly assumed that A and B agreed about the vertical distance L from floor to mirror. It is quite natural to be suspicious about this, but fortunately the principle of reciprocity tells us that about this measurement, at least, A and B will agree. If A judges B's car to have shrunk vertically, then B must judge A to have shrunk vertically. That these two points of view are inconsistent can be demonstrated by A if he holds out a wet paintbrush at a height equal to the height of the car when at rest. If he believes that the moving car has diminished its height, his brush will be above the car and leave no trace. But according to B, the brush should be below the roofline of the car and will leave a stripe. Since the car cannot emerge from the encounter both with a stripe and without one, it must be true that A and B do agree about vertical dimensions, and the brush would just graze the top of the car.

Another implication of the reciprocity principle is that the two observers agree about their relative speed. If they disagreed, this would introduce into their observations an asymmetry inconsistent with the assumption that all inertial frames are equivalent. In terms of vector quantities, if the velocity of B as measured by A is $\mathbf{v}$, the velocity of A as measured by B is $-\mathbf{v}$.

There is one other essential area of agreement in the theory of relativity that needs to be stated explicitly, in the spirit of extreme caution about accepting common-sense notions. This is that all observers *do* agree about events at the same place at the same time. If at a particular place at a particular time a pion is annihilated and a muon and a neutrino are created in its place, all observers will agree that the annihilation and creation events did occur at the same place at the same time. Although different observers might have different ideas about where this pair of events occurred and at what time, there will be no disagreement about the facts that the events were simultaneous and at the same place. This much of common sense, at least, is retained in the theory of relativity. It means that two or more things that happen at the same place at the same time may be called one single event. All disagreements in the theory of relativity are concerned with measurements of differences between different events.

20.4 *The Lorentz transformation equations*

The time-dilation experiment of A and B reveals a part of the total relationship between the space and time measurements of A and B. Numerous other experiments could be carried out to reveal other facets of the relationship, all resting on just a single key assumption, that the speed of light is always the same with respect to each observer. A and B might for example shine light waves horizontally along that direction of their relative motion; they might synchronize and compare different clocks in their separate frames of references. Combining the results of such measurements, they would arrive at the general relationship between their space and time measurements, which is the Lorentz transformation. Actually this transformation rests not exclusively on the constancy of the speed of light. One other assumption is necessary, which may be taken to be either the assumption of reciprocity, or the assumption of agreement about events at the same place at the same time. The Lorentz transformation equations may be written in a rather general way to cover all possible situations, but it is much easier to refer them to a particular experimental arrangement. We will imagine a train moving with constant speed v along a straight track (Figure 20.4). An observer within the train measures distances $\Delta x'$ along the train, distances $\Delta y'$ vertically, distances $\Delta z'$ transversely, and time intervals $\Delta t'$. The ground-based observer measures distances Δx, Δy and Δz with respect to the ground also along the track, up and down, and across the track, and he measures time intervals Δt. The relationship among the measurements are given by the transformation equations, which we here state without proof:*

$$\Delta t = \frac{\Delta t' + (v\Delta x'/c^2)}{\sqrt{1-(v^2/c^2)}}, \tag{20.7}$$

$$\Delta x = \frac{\Delta x' + v\Delta t'}{\sqrt{1-(v^2/c^2)}}, \tag{20.8}$$

$$\Delta y = \Delta y', \tag{20.9}$$

$$\Delta z = \Delta z'. \tag{20.10}$$

On the right side of each equation is a measurement or measurements of the moving observer, and on the left of each equation is a measurement of the stationary observer. (We are of course using the words "moving" and "stationary" only as convenient labels for our two observers, not from any conviction about which one is "really" moving.) The Lorentz transformation ties together the subjectivity of measurement and provides a basis for extracting "objective reality" from subjective measurement. Without the transformation there would be as many different views of nature as there are different states of relative motion of observers.

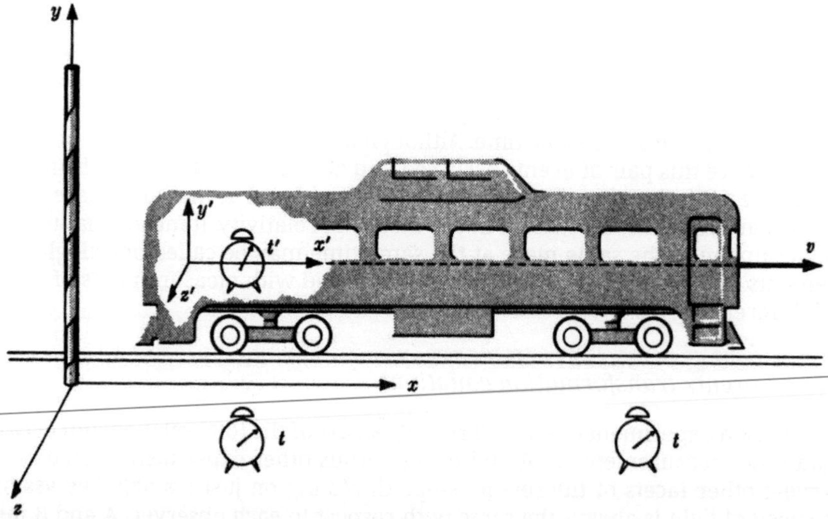

FIGURE 20.4 Coordinate systems of observers in relative motion. The Lorentz transformation equations relate the space and time interval measurements of the ground-based observer to the space and time interval measurements of the train- based observer.

* Their "proof" rests of course ultimately on their success in describing what happens in nature. Their derivation rests on thought experiments similar to the one illustrated in Figure 20.3.

Inspection of the Lorentz transformation equations reveals at once two interesting facts. (1) Space and time can be *mixed*. One man's space is some combination of another man's space *and* time. This mixing has led to the concept of "spacetime" as a single four-dimensional entity. (2) No speed can exceed the speed of light. This unexpected prediction of a speed limit in nature follows from the mathematical form of the transformation equations. If v were to exceed c, the quantity $\sqrt{1-(v^2/c^2)}$ would become an imaginary number and we would not know how to interpret the equations physically. If we accept the fact that the Lorentz transformation relates all possible physical measurements, we must accept the conclusion that no two objects can have a relative speed greater than c. There is now ample evidence that this speed limit exists. Electrons have been accelerated (in the Stanford Linear Accelerator in California) to a speed of $0.9999999997c$, but cannot be pushed beyond c.

20.5 *Consequences of the Lorentz transformation*

In order to appreciate other consequences of the Lorentz transformation, it is well to specialize to particularly simple conceptual experiments.

Time Dilation

Suppose that we choose $\Delta x' = 0$ in the transformation equations. From Equation 20.7 we get at once the time-dilation equation discovered earlier (Equation 20.6):

$$\Delta t = \frac{\Delta t'}{\sqrt{1-(v^2/c^2)}} \,. \tag{20.11}$$

This relates the two observers' measurements of a time interval between events *at the same point in the train*, such as the time duration of a passenger's nap (Figure 20.5).

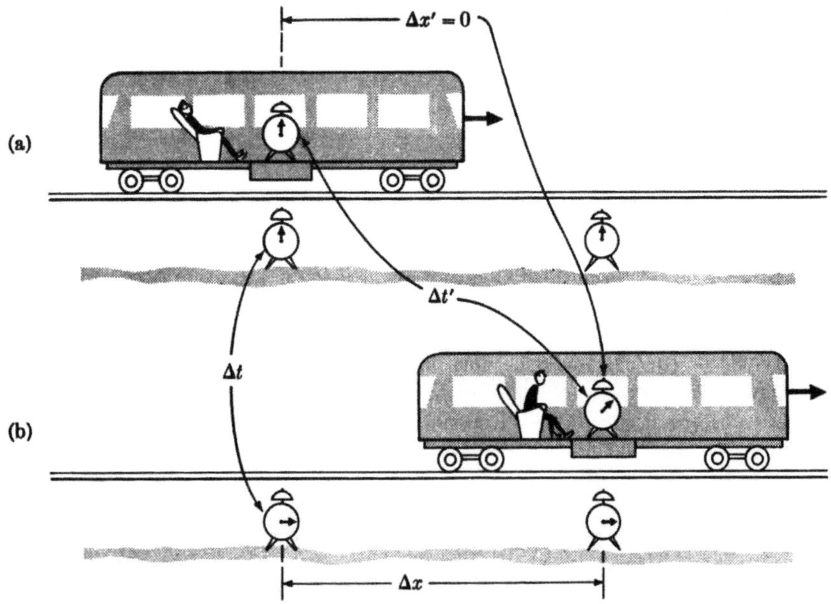

FIGURE 20.5 Time dilation. A passenger's catnap lasts a time $\Delta t'$ in his frame of reference, and a longer time Δt in the ground-based frame of reference. Between the events of going to sleep and waking up, there is no spatial separation in the train ($\Delta x' = 0$), but a spatial separation Δx on the ground. (The difference between Δt and $\Delta t'$ is of course too small to observe with an actual train.)

When the ratio v/c is much less than 1, a very useful approximate equation is this:

$$\frac{1}{\sqrt{1-(v^2/c^2)}} \cong 1 + \frac{1}{2}\left(\frac{v}{c}\right)^2 . \qquad (20.12)$$

Its deduction is left as an exercise. Suppose that a supersonic transport pilot moves with respect to the earth at twice the speed of sound, or $v = 6 \times 10^4$ cm/sec. The ratio of this speed to the speed of light is $v/c = 2 \times 10^{-6}$; so, for the relative motion of supersonic transport and earth, the right side of Equation 20.12 is equal to $1 + 2 \times 10^{-12}$, scarcely different from 1. In order for his clock to fall one second behind earth-based clocks, the pilot would have to log 0.5×10^{12} sec of supersonic flight time, or about 16,000 years aloft. In a transcontinental flight of two hours, his clock would lag by 14 nanoseconds (1 nanosecond = 10^{-9} sec).

The Lorentz Contraction

To bring out another implication of the Lorentz transformation, we can set $\Delta t' = 0$, and inquire about the spatial separation of events that are simultaneous in the train. For such pairs of events, Equation 20.8 takes the form

$$\Delta x = \frac{\Delta x'}{\sqrt{1-(v^2/c^2)}} . \qquad (20.13)$$

This defines what is known as the Lorentz contraction. Why contraction and not expansion? We must imagine two observers in the train who simultaneously (in their own frame of reference) look out of windows and see telephone poles just opposite themselves (Figure 20.6). Their measured distance, $\Delta x'$, between the poles is *less* than Δx, the ground-based measurement of the separation of the poles. The relative motion has caused the observers in the train to think that the outside world has become compressed (and, of course, the outside observer, noting the positions of the front and rear of the train at the same time in *his* frame of reference, concludes that it is the train that has contracted in its direction of motion). Notice that the factor of difference in distance scales, $\sqrt{1-(v^2/c^2)}$, is the same as the factor of difference in time scales. When the ratio v/c is small, the approximation given by Equation 20.12 is again useful.

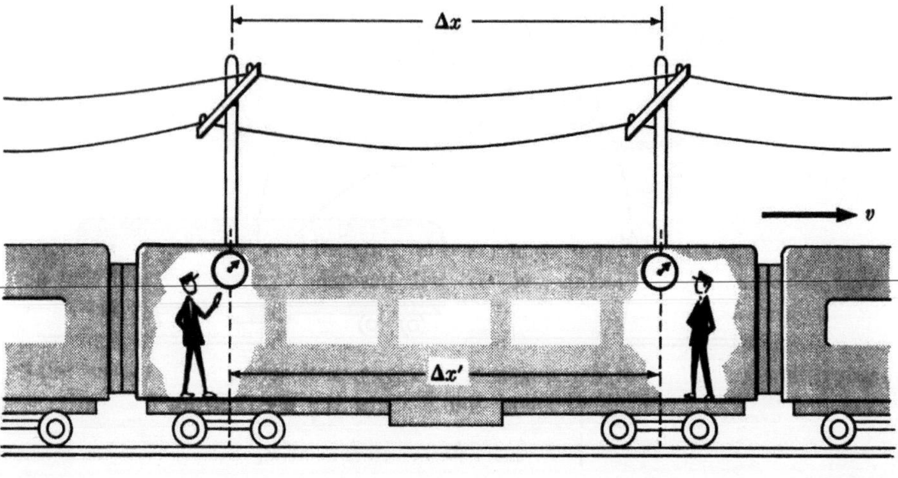

FIGURE 20.6 The Lorentz contraction. A pair of observers on the train find themselves opposite a pair of poles on the ground at the same time according to train-based clocks ($\Delta t' = 0$). They measure a separation of the poles, $\Delta x'$, less than the separation x measured by a ground-based observer.

Although usually called the Lorentz contraction, the *apparent* contraction described by Equation 20.13 must be carefully distinguished from the *material* contraction hypothesized by FitzGerald and discussed by Lorentz in the early 1890s. According to FitzGerald, the contraction occurs for objects moving through the ether, quite unconnected with the way an observer might be moving. And no principle of reciprocity applies to the FitzGerald contraction. Most important, the Lorentz contraction of relativity (which might better be called the Einstein contraction) arises from the properties of space and time, while the FitzGerald contraction was supposed to be associated with the properties of matter. FitzGerald pictured a physical compression of material objects; Einstein imagined a compression and stretching of space and time itself.

By how much does the supersonic transport mentioned above appear to be shortened by its motion? The fractional effect is again $\frac{1}{2}(v^2/c^2)$, or 2×10^{-12}. If the length of the airplane is 5,000 cm, the apparent contraction at Mach 2 is $(2 \times 10^{-12})(5 \times 10^3 \text{ cm}) = 10^{-8}$ cm, less than the diameter of a single atom.

An example that illustrates the "paradoxical" character of the Lorentz contraction is shown in Figure 20.7. A train which, at rest, is 300 meters long, approaches a tunnel of length 250 m at 80% of the speed of light. A band of hijackers have sealed the exit of the tunnel (B), and they intend to block the entrance (A) as soon as the train is inside the tunnel, in order to trap it there. At the rather swift pace of the train, the Lorentz contraction factor, $\sqrt{1 - (v^2/c^2)}$, is equal to 0.60. This means that the hijackers, being ground-based observers, measure the length of the moving train to be 0.60×300 m = 180 m. They believe it will fit comfortably within the 250-meter-long tunnel. To the passengers on the train, the length of the train is still 300 m, but the tunnel is shrunk to 0.60×250 m = 150 m, only half the length of the train. From their point of view, the train is far too long to fit within the tunnel.

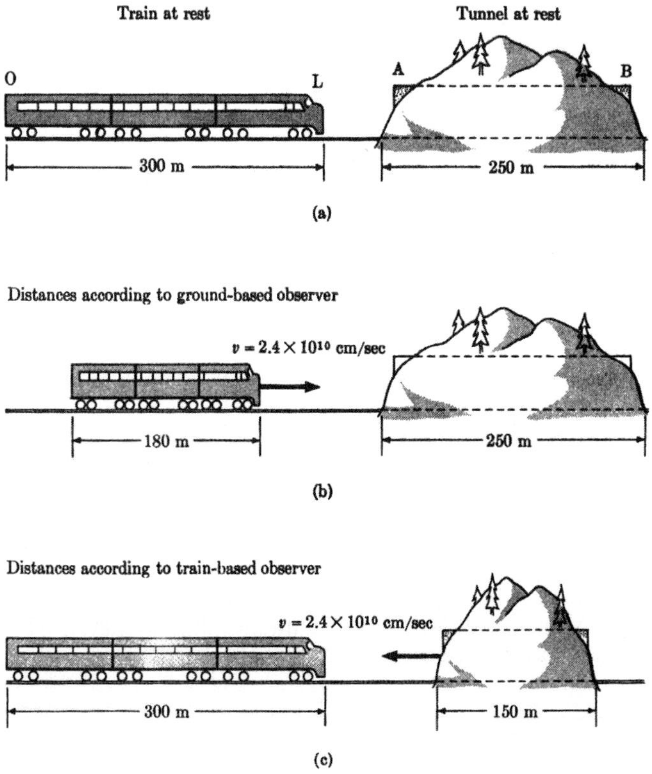

FIGURE 20.7 Will the train fit in the tunnel? **(a)** With both at rest, the train is 300 m long, the tunnel 250 m long. **(b)** As the train approaches at $v = 0.80c$, it appears to the ground-based observer to be 180 m long. **(c)** At the same time the tunnel appears to the train-based observer to be 150 m long.

Will the hijackers succeed in their nefarious plan? Instead of answering this question directly, let us consider what happens if the exit B is not blocked and the train continues at constant velocity. One thing all observers can agree about is that at *some* time the locomotive (L) will pass the exit point B and that at some time the observation car (O) will pass the entrance point A. What observers may differ about is when these events occur. That is the key to the apparent paradox. The relativity of distance measurement is closely related to the relativity of time measurement. Indeed, almost all of the peculiarities of relativity can be traced to the fact that time, once considered absolute, has become relative. In this example, the hijackers find that the locomotive reaches the exit point B *after* the observation car has passed the entrance point A, but the passengers find, according to their clocks, that the locomotive reaches the exit point well *before* the observation car reaches the entrance. Since the two sets of observers disagree about the time sequence of events, it is not surprising that they disagree about the length of moving objects. (To make this clear, think about what length you would measure your moving car to have if you marked a spot in the pavement under its rear bumper at one moment, then five seconds later marked a spot under its front bumper, and measured the distance on the pavement between these two spots.)

Now what happens if the exit B is blocked and the train is forced to stop? Will it first be trapped and then burst through the blockade or be crushed in trying to do so? Or will the observation car never reach the entrance point A? We leave these questions for the reader to think about. They are deliberately tricky; it is possible to understand much of relativity without being able to answer them. However, they are good exercise for those who like to wrestle with challenging questions.

The Relativity of Simultaneity

The problem of the hijackers and the train shows the important connection between length measurements and time scales when motion is involved. This connection is revealed also by another simplified form of one of the Lorentz transformation equations. In Equation 20.7, set $\Delta t = 0$. This gives

$$\Delta t' = -v\Delta x'/c^2\,, \tag{20.14}$$

which defines the relativity of simultaneity. Events judged by the ground observer to be simultaneous ($\Delta t = 0$) will be judged by the observer on the train to occur with a time separation $\Delta t'$ that depends on their separation in space.

The Addition of Velocities

Another relation of great interest can be derived from the Lorentz transformation, but less directly. It is the law of addition of velocities, which puts a firm foundation under the idea of nature's speed limit. We imagine the train traveling now with speed v_1 relative to the ground, and inside the train a passenger running forward with a speed v_2 relative to the train (Figure 20.8). What is the speed of the runner with respect to the ground? By now we should not be surprised to learn that the answer is *not* the common-sense answer $v_1 + v_2$. The runner's speed is instead less than $v_1 + v_2$ and is given by

$$v = \frac{v_1 + v_2}{1 + (v_1 v_2/c^2)}\,. \tag{20.15}$$

Note that if the runner is replaced by a light wave ($v_2 = c$), then automatically $v = c$. *Any* speed added to c gives c again. This velocity-addition law also implies that no matter what speeds v_1 and v_2 we choose (if each is less than c), their "sum" will also be less than c.

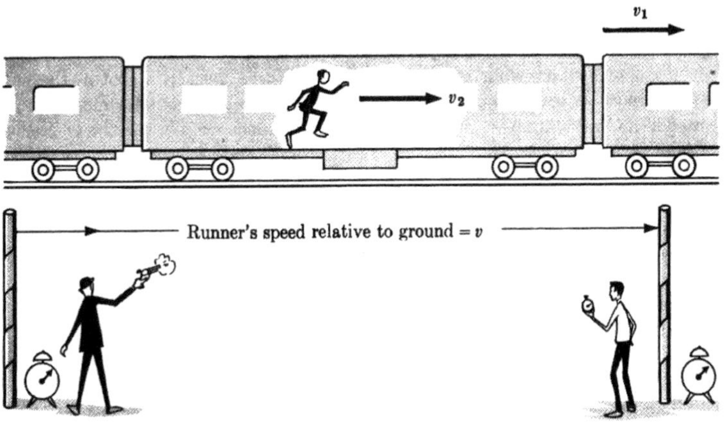

FIGURE 20.8 The addition of velocities. A passenger runs at speed v_2 relative to the train, which is moving at speed v_1 relative to the ground. The ground-based observer measures the runner's speed to be not $v_1 + v_2$, but something less (Equation 20.15). Reciprocally, the runner finds the speed of the ground relative to himself to be given by the same equation.

These four results—time dilation, Lorentz contraction, the relativity of simultaneity, and the addition of velocities—constitute all of the essential kinematic consequences of special relativity, those consequences having to do with space and time. All violate common sense, and all for the same reason. Man never directly perceives motion with speed greater than a tiny fraction of the speed of light. No evidence comes directly to our senses about the remarkable properties exhibited by space and time when observers move relative to one another at very great speed.

20.6 Tests of the Lorentz transformation

Human beings are clearly too slow to be significantly affected by the new effects of relativity. This will remain true for a long time to come. Speeds of space travel in the foreseeable future will remain quite small compared with the speed of light. But fortunately the elementary particles are less phlegmatic than we, and they achieve speeds near the speed of light with great ease, making possible detailed quantitative tests of the theory of relativity. Time dilation and the velocity-addition law have each been thoroughly tested in innumerable experiments, although no direct way has ever been found to measure the Lorentz contraction or the relativity of simultaneity.

113 Time dilation is revealed most directly by the characteristic lifetime of unstable particles. Pions at rest, for example, have a well-defined and easily measured mean lifetime of about 10^{-8} sec. A pion emerging from an accelerator with an energy of 1 GeV is traveling at a speed 99% of the speed of light and has a mean life in the earth-fixed frame about seven times longer than the lifetime of its fellow pion at rest with respect to the earth. Time-dilation effects this large and even much larger are commonplace occurrences in modern experimental work with accelerators and with cosmic rays. Indeed the dilation of time has a marked effect on the intensity of cosmic radiation at the earth's surface. Protons striking nuclei near the top of the atmosphere create pions. Even with the benefit of time dilation, most of these pions decay into muons (and neutrinos) while still high above the earth's surface. At rest the mean lifetime of muons is about 2 μsec (2×10^{-6} sec). Moving near the speed of light, a muon covers only 6×10^4 cm or about a third of a mile in 2 μsec of earth time. If high-speed muons were characterized by the same lifetime as muons at rest, they too would largely disappear in the atmosphere before reaching the earth. Instead, their lifetime is so

stretched by their motion (according to earth-based clocks) that many reach the earth before decaying.* A 10–GeV muon, for instance, experiences a time-dilation factor of 95 and flies, on the average, about 35 miles before it decays.

Accelerators reveal in a direct way the law of addition of velocities. In the Stanford Linear Accelerator, electrons flying down a two-mile-long evacuated tube are repeatedly "pushed." At each push, some speed is added to what speed the electron had before. But as its speed gets closer and closer to the speed of light, the added speed gets less and less, as the electron inches its way toward, but never reaches, the speed of light. At 10 GeV, halfway to its final energy of 20 GeV, the electron is traveling at 0.999999999*c*, or just 39 cm/sec less than the speed of light (39 cm/sec is about 0.9 miles/hr). In a frame of reference moving at this speed, the addition of another 10 GeV of energy adds nearly 3×10^{10} cm/sec to the electron speed. In the earth-fixed frame, it adds instead only 29 cm/sec, bringing the electron to within 10 cm/sec (0.2 miles/hr) of the speed of light. In a frame of reference moving with the electron at its final speed, the length of the accelerator, two miles, is contracted to three inches. Under conditions of this kind, every common-sense notion about space, time, and speed becomes ludicrously inappropriate.

20.7 *Intervals, space-time, and world lines*

The Lorentz transformation provides a rational account of the relativity of observation. It contains also hidden within itself an unexpected statement of invariance. If each of the transformation Equations 20.7 to 20.10 is squared, and the squares are combined in the following particular way,

$$\Delta x^2 + \Delta y^2 + \Delta z^2 - c^2\Delta t^2 ,$$

the equations reveal that this combination is exactly equal to

$$\Delta x'^2 + \Delta y'^2 + \Delta z'^2 - c^2\Delta t'^2 .$$

Apparently this combination of length and time measurements is something about which the observers agree, in spite of their disagreement about length and time separately. It is natural to attach significance to an invariant quantity, and this combination is named the square of the "interval," for which we use the symbol I:

$$I^2 = \Delta x^2 + \Delta y^2 + \Delta z^2 - c^2\Delta t^2 , \tag{20.16}$$

$$I^2 = \Delta x'^2 + \Delta y'^2 + \Delta z'^2 - c^2\Delta t'^2 . \tag{20.17}$$

Between two simultaneous events in one frame of reference, the interval is just the spatial separation of those events. Then $\Delta t = 0$, and

$$I^2 = \Delta x^2 + \Delta y^2 + \Delta z^2 = (\textit{distance})^2 , \tag{20.18}$$

in this frame of reference. (In a frame moving with respect to this one, $\Delta t'$ is not zero—the events are not simultaneous—and $\Delta x'$ is not equal to Δx, but the combination on the right side of Equation 20.17 remains equal to the combination on the right side of Equation 20.16.) Two events occurring in the same place ($\Delta x = \Delta y = \Delta z = 0$) but at different times in one frame of reference are separated by an interval that is proportional to the elapsed time between the events:

$$I^2 = -c^2\Delta t^2 . \tag{20.19}$$

* The flux of cosmic-ray muons at the earth's surface is about 0.05 per cm^2 per sec.

The interval is a generalization and a combination of the ideas of spatial and temporal separation. The reason why the square of a physically meaningful quantity can be negative is something we shall not go into. Suffice it to say that it causes no trouble. If I^2 is positive, the interval is said to be "space-like." If I^2 is negative, the interval is called "time-like."

The Lorentz transformation has already shown that space and time have to be thought of as mixed together. Now the invariance of the interval further strengthens that view of a kind of "interchangeability" of space and time. Space and time are still not quite equivalent, for in the definition of the interval the space terms are preceded by plus signs and the time term by a minus sign, yet it is possible and useful to think of a four-dimensional space-time labeled by the four coordinates x, y, z, and t, just as ordinary space is three-dimensional and can be labeled by three coordinates x, y, and z. The interval can then be pictured as a distance in four-dimensional space-time, an invariant quantity about which all observers agree, in spite of their disagreements about distances in space and time separately.

Although four-dimensional space-time defies visualization, it is still quite instructive to picture a slice of space-time and examine simple events in that slice. For example, consider the "Minkowski diagram" in Figure 20.9, with one spatial dimension plotted horizontally and the time dimension plotted vertically. The complete history of any object restricted to move only in the x-direction can be represented by a line in this diagram. A man at rest is represented by the straight vertical line A. A man who walks to the right, stops for awhile, and walks to the left leaves the trace B. A light wave traveling to the right has a history represented by C. Such lines are called world lines, and any point on a world line is an event. The most interesting events are those in which two or more world lines intersect. The dashed line D could not qualify as the world line of anything physical, for it corresponds to a speed greater than the speed of light.

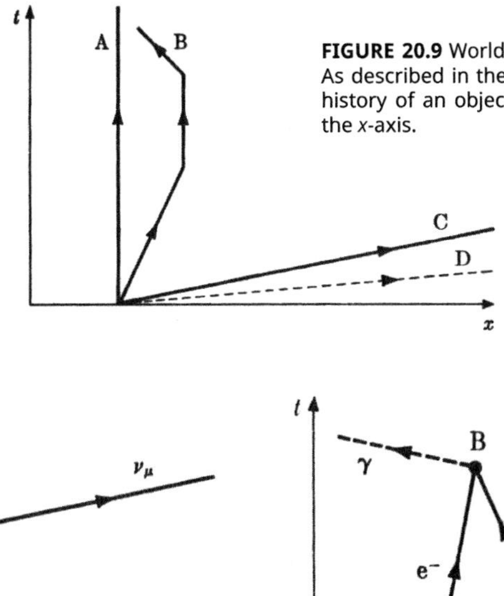

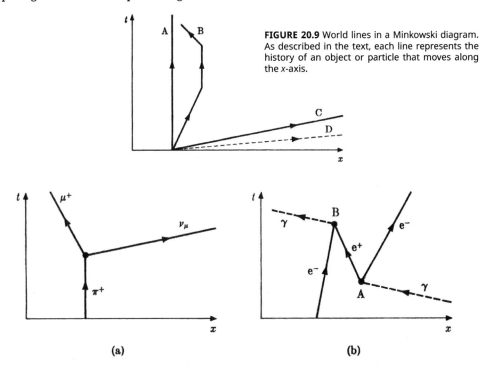

FIGURE 20.9 World lines in a Minkowski diagram. As described in the text, each line represents the history of an object or particle that moves along the x-axis.

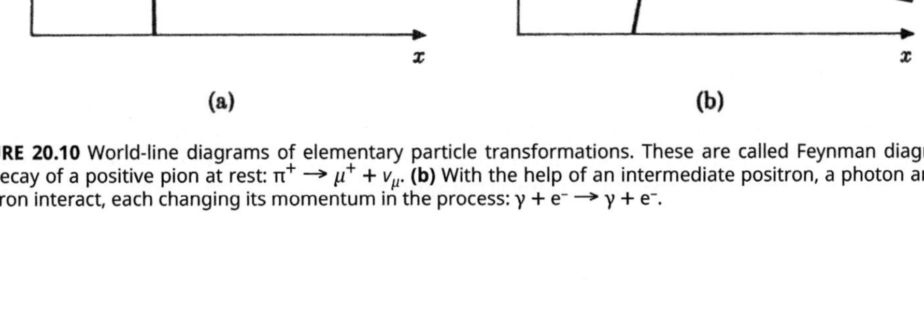

FIGURE 20.10 World-line diagrams of elementary particle transformations. These are called Feynman diagrams. **(a)** Decay of a positive pion at rest: $\pi^+ \rightarrow \mu^+ + \nu_\mu$. **(b)** With the help of an intermediate positron, a photon and an electron interact, each changing its momentum in the process: $\gamma + e^- \rightarrow \gamma + e^-$.

Elementary-particle transformations can be conveniently represented by world lines. The decay of a pion at rest is illustrated in Figure 20.10(a). The world line of the motionless pion is a straight vertical line until the moment of its annihilation. Then the massless neutrino that is created moves off in one direction at the speed of light, while the newly created muon moves away in the other direction at a lesser speed. Fascinating world lines also result from the annihilation and creation of positron-electron pairs. In the diagram of Figure 20.10(b) a positron-electron pair is created at A as two new world lines come into existence. At a later time at point B the positron encounters another electron, both annihilate, and two world lines terminate. But dare we take a more radical view of these events? Does the electron at B perhaps not annihilate at all, but merely turn around and move backward through time until it reaches point A, where once again it turns around and resumes its course forward through time? An answer to this question has been given in 1949 by Richard Feynman. He showed that either view of the events is equally acceptable, and that an entirely consistent mathematical theory of positrons can be constructed in which they are regarded as electrons moving backward in time. World-line portrayals of particle transformations, such as those in Figure 20.10, are called Feynman diagrams. They receive more attention in Chapter Twenty-Seven.

20.8 Significance of the speed of light

114 The speed of light is the culprit in special relativity. Its constancy with respect to different observers led to the Lorentz transformation, to the relativity of time, and to all the odd consequences of the Lorentz transformation. It then showed itself as nature's speed limit. One is tempted to probe deeper and ask why light happens to travel at just the speed 3×10^{10} cm/sec and not some other speed. Before answering the question, we consider another: Why is the mechanical equivalent of heat 4.18×10^7 ergs/cal? The answer is that the unit of energy (the erg) and the unit of heat (the calorie) were separately defined before it was known that there was a relation between heat and energy. Had it been known soon enough that heat is itself just a form of energy, then heat would have been measured in ergs and there would have been no need even for the idea of a mechanical equivalent of heat. The number 4.18×10^7 is just a way of correcting past error and joining together two ideas that were believed to be different but are not.

The idea of space-time leads us to a similar view of the speed of light. Space and time are joined together into a single entity, space-time. Properly, space and time should have the same unit of measurement. Because the centimeter and the second were independently defined before any connection between space and time was known, the speed of light appears as the correction factor relating the two different units, which should have been the same. On this view, the numerical value of c has no particular meaning. Yet the analogy with the mechanical equivalent of heat should not be pressed too far. It is true that with "properly" chosen units, the numerical value of c would be 1, yet we cannot dispense with the *idea* of the speed of light. Whatever its numerical value, it is important in setting a speed limit in nature. Even to a cosmic creature not bound by time who could take in the full panorama of space-time at a glance, to whom human history is just a map laid before him, the idea of the speed of light would be important, for it would provide a boundary line between possible and impossible world lines in the space-time map.

Granted that there is a limiting speed in nature, why is it light that sets this limit, and not something else? The answer is to be found in the modern particle concept of light. It is the masslessness of photons that enables them—indeed requires them—to move at nature's greatest speed. This means that light, although special, is not unique. Neutrinos and the hypothetical gravitons, also massless particles, move at the same speed as light. When the theory of relativity was formulated, neutrinos were unknown and unsuspected. The theory developed primarily from studies of electromagnetism. Contributing to the point of view about nature that helped stimulate its development were experiments with light such as the Michelson-Morley experiment. For these reasons, light was regarded as the special radiation of special relativity. In bare outline, the chain of reasoning about the speed of light that approximates the actual historical development is this:

(1) Because of the desire to extend the Galilean type of invariance from the laws of mechanics to the laws of electromagnetism, it is postulated that the speed of light is the same for all observers (in inertial frames of reference). This postulate is backed by experiments with light that failed to reveal the ether.

(2) The postulate of constant light velocity (together with the idea of reciprocity) leads to the Lorentz transformation equations, which supplant the Galilean transformation equations.

(3) The Lorentz transformation equations become meaningless for speeds greater than the speed of light, suggesting that light sets the speed limit for nature.

This reasoning should be capped by an experimental finding:

(4) Nothing has ever been observed to move faster than light.

An alternative chain of reasoning, which from the modern point of view has a little more appeal than the historical sequence outlined above, runs like this:

(1) Suppose that there exists in nature a maximum speed v_{max}, which is the same for all observers.*

(2) The existence of the constant speed v_{max} leads to the Lorentz transformation equations and to other consequences of the theory of relativity.

(3) Massless particles, if any, will achieve the speed v_{max}. Massive particles will move more slowly.

This logic does more than merely replace the symbol c by the symbol v_{max}, for it leaves open the possibility that nothing actually achieves the limiting speed.

Suppose that there were in nature no massless particles to move at the greatest speed. Would there be a theory of relativity? If so, could it have been discovered? The answer to both questions is emphatically yes. Its discovery might have been delayed a few years, but surely for not very long, for the behavior of high-energy electrons would have provided ample evidence for the inadequacy of previous theory. It is a bit of luck that light achieves nature's greatest speed. However, the careful study of any high-speed particle motion would have led physicists inescapably to the same theory of relativity.

* If the speed v_{max} were not the same for all observers, there would be one observer with the largest v_{max}; his would be a "maximum maximum speed." According to the other fundamental postulate of the theory of relativity (Chapter Twenty-One), all inertial frames of reference are equivalent and there are no such privileged observers.

EXERCISES

20.1. Make up a thought experiment other than those mentioned in Section 20.2 that shows the "paradoxical" nature of the postulate c = *constant*. Does your experiment imply a relativity of time? or of space? or of both? Does it violate common sense?

20.2. Working from Equation 20.4, fill in the algebraic details to derive the time dilation formula, Equation 20.6.

20.3. Anthony states: "At 12:01 precisely two photons simultaneously struck the same point in my frame of reference. One microsecond later, an electron was emitted from this same location." Bartholomew, in high-speed motion relative to Anthony, observed the same events. With which parts of Anthony's description does he agree, and with which parts does he disagree?

20.4. In terms of differences of coordinates, two of the Galilean transformation equations, Equations 19.3 and 19.1, can be written in the form

$$\Delta t = \Delta t' ,$$

$$\Delta x = \Delta x' + v\Delta t' .$$

Compare these with the corresponding Lorentz transformation equations, Equations 20.7 and 20.8. What condition must be satisfied in order that the Lorentz transformation equations, are accurately approximated by the Galilean transformation equations?

20.5. **(1)** Find the following square roots approximately: $\sqrt{1.1}, \sqrt{1.01}, \sqrt{0.9}, \sqrt{0.96}$. From your results arrive at approximate formulas for $\sqrt{1+x}$ and $\sqrt{1-x}$, if x is a small number. **(2)** Find the following inverses approximately: 1/1.01, 1/0.98, 1/0.92, 1/0.94. From your results arrive at an approximate formula for $1/(1-x)$, if x is a small number. **(3)** Combining the results of (1) and (2), obtain the following approximate formula:

$$\frac{1}{\sqrt{1-\left(\frac{v}{c}\right)^2}} \cong 1 + \frac{1}{2}\left(\frac{v}{c}\right)^2 ,$$

if (v/c) is a small number. *Optional:* Use the binomial theorem to obtain this result. What third term can be added to the right to improve the accuracy of the approximation?

20.6. Muons at rest have a mean lifetime of 2.2 microsec. Muons emerging from a cyclotron have a speed of 2.4×10^{10} cm/sec (80% of the speed of light). What should the mean lifetime of these moving muons be observed to be?

20.7. A passenger on an interplanetary express bus catnaps for five minutes by his watch. If the bus is maintaining a speed of $0.99c$ through the solar system, how long do observers fixed on the planets judge the nap to be?

20.8. Rocket taxis of the future move about the solar system at half the speed of light. Drivers receive $10 per hour as measured by clocks in the taxis. The taxi drivers' union demands that the pay be based on earth time instead of taxi time. If their demand is met, by what fraction will their wages rise?

20.9. An astronaut is circling the earth at a speed of 7×10^3 cm/sec. He remains in orbit for exactly two weeks according to earth-based clocks. By how much less than two weeks does his flight last according to a clock in the space capsule?

20.10. Some pions at rest in the laboratory were observed to have an average lifetime of 2.55×10^{-8} sec. Another group of high-speed pions were found to have an average lifetime of 3.19×10^{-8} sec according to laboratory clocks. **(1)** How fast were the second group of pions moving? **(2)** If the moving pions had been carrying clocks, how long would they have recorded their own lifetime to be? **(3)** What average distance in the laboratory was covered by the moving pions?

20.11. Meter stick A moves at half the speed of light past meter stick B at rest in the laboratory. At the instant of time in the laboratory when the front of the moving meter stick is even with the front of the stationary meter stick, where, according to the laboratory observer, is the rear of the moving meter stick?

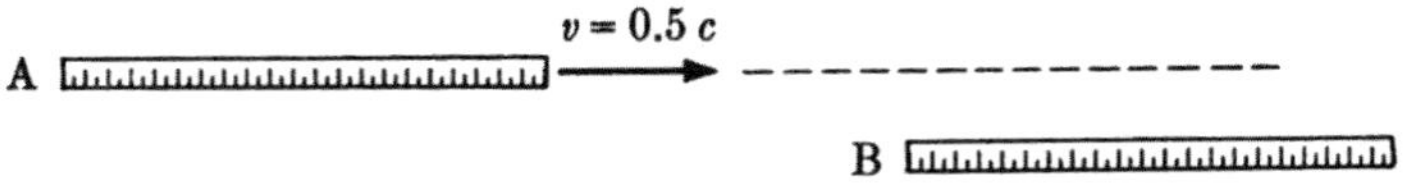

20.12. A fictitious, relativistic railroad car, when at rest, has a length of 14 meters. **(1)** How long does it appear to be to ground-based observers as it streaks by at 99 per cent of the speed of light? **(2)** Observers at rest on the ground determine its length by noting the position of its front at a particular time and the position of its rear at the same time. According to observers in the car, how much time elapsed between these two measurements that the ground-based observers call simultaneous?

20.13. A spaceman in a satellite traveling at a speed of 6×10^5 cm/sec sees events in Los Angeles and New York (4×10^8 cm apart) which he judges to be simultaneous. How much time elapses between these events according to earthbound observers?

20.14. Suppose that the hijackers described in Section 20.5 determine the length of the moving train by placing marks on the ground opposite the front and rear of the train at the same time and measuring the distance between the marks. By this means they find its length to be 180 meters. What criticism would observers on the train have of this procedure? Give a quantitative answer.

20.15. **(1)** With the help of Equation 20.7, explain how it is that one observer can judge an event X to occur before an event Y, while another observer thinks event Y occurred before event X. **(2)** Under certain circumstances, Δt and $\Delta t'$ have the same sign, regardless of the sign or magnitude of v (within physically allowable limits). This means that all observers agree on the time sequence of the pair of events. What condition must be satisfied to bring about this much agreement between observers?

20.16. Explain carefully in words the exact meaning of Equation 20.14 expressing the relativity of simultaneity. To be definite, consider a pair of events in space-time, one at x_1, t_1 (and x_1', t_1'), the other at x_2, t_2 (and x_2', t_2').

20.17. **(1)** Show from Equation 20.15 that the "sum" of c and any other speed v is c. **(2)** Find the relativistic velocity sums for these pairs: **(a)** $v_1 = v_2 = 0.5c$; **(b)** $v_1 = v_2 = 0.9c$; **(c)** $v_1 = 0.99c$, $v_2 = 0.10c$.

20.18. Rocket ship 1, cruising from Earth to Jupiter at half the speed of light relative to the solar system, passes rocket ship 2, at rest with respect to the solar system. Experimenters in ship 1 observe two cosmic ray particles, one moving toward Jupiter, one moving toward Earth, both at the same speed, 0.75c, in the frame of reference of ship 1. What velocities do observers in ship 2 assign to each of these particles? (HINT: The symbols v_1 and v_2 in Equation 20.15 refer to components of velocity in a particular direction. Signs are important.)

20.19. (1) Prove that for speeds near the speed of light, the square root factor that appears in many of the equations of relativity may be approximated as follows:

$$\sqrt{1 - v^2/c^2} \cong 1.41\sqrt{\frac{c-v}{c}}.$$

(2) Use this result to determine the Lorentz contraction factor in a frame of reference moving with a 20-GeV electron, for which $c - v = 10$ cm/sec.

20.20. Explain in your own words why we are unaware of relativistic effects in our everyday lives. Illustrate the point with a numerical example. Can you draw any conclusion about the probable relation between human perceptions and the fundamental theories of nature?

20.21. Use Equations 20.7 to 20.10 to derive the following equation of invariance for the space-time interval:

$$\Delta x^2 + \Delta y^2 + \Delta z^2 - c^2\Delta t^2 = \Delta x'^2 + \Delta y'^2 + \Delta z'^2 - c^2\Delta t'^2 .$$

(HINT: Deal first with $\Delta x^2 - c^2\Delta t^2$, then add $\Delta y^2 + \Delta z^2$.)

20.22. Use the interval concept to answer the following question: If observer A sees two events as simultaneous and 4 cm apart, and observer B sees the same pair of events separated by 5 cm, what is the time separation of the events according to B? Note that a knowledge of the relative velocity of A and B is not required.

20.23. The square of a four-dimensional interval may be negative (see, for instance, Equation 20.19). The square of a speed cannot be negative and still be physically meaningful. Explain this difference by discussing the difference in the way a speed is measured and the way an interval is measured.

20.24. Pick any two simple occurrences, either in the elementary particle world or in your own life, and sketch the world-line diagrams for these occurrences.

20.25. Event E_1: A photon leaves the rear of a train, headed forward. Event E_2: The photon arrives at the head of the train. The length of the train, according to the train-based observer, is $L = 3 \times 10^3$ cm. The speed of the (rather relativistic) train is 2.4×10^{10} cm/sec. Find $\Delta x'$, $\Delta t'$, Δx, and Δt, the space and time differences between the two events for train-based observer and ground-based observer.

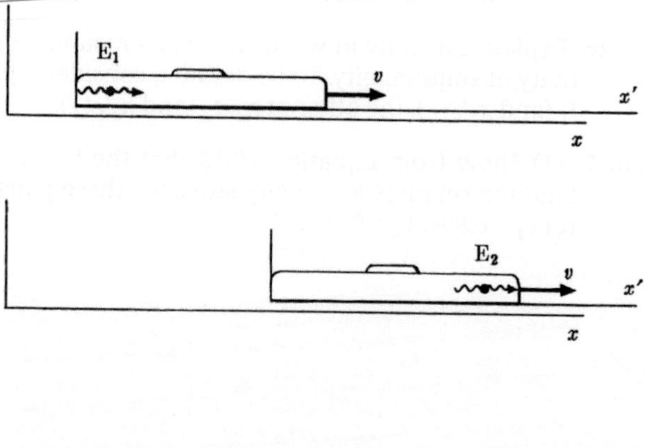

20.26. What is the four-dimensional interval between the pair of events E_1 and E_2 described in Exercise 20.25: **(a)** according to the train-based observer; and **(b)** according to the ground-based observer?

20.27. A roadside observer A sends a light beam vertically upward a distance L, where it is reflected back to its starting point. He measures the round-trip time to be t_A. A moving observer B measures its round-trip time to be t_B. Draw a diagram analogous to Figure 20.3 to help analyze this example, and derive an equation analogous to Equation 20.6 relating the two time measurements.

20.28. Prove that if v_1 and v_2 both lie between 0 and c, their relativistic "sum" as given by Equation 20.15 is less than c. (HINT: Proceed as follows. Write Equation 20.15 in the form

$$\frac{c}{v} = \frac{1 + (v_1/c)(v_2/c)}{(v_1/c) + (v_2/c)}$$

and seek to show that the right side of this equation is greater than 1 if v_1/c and v_2/c are both positive and less than 1.) *Optional:* Extend the proof to include negative values of v_1 and/or v_1.

20.29. A light in the center of a rocket car is switched on. According to an observer in the car, photons reach the front and the rear of the car simultaneously, each after covering a distance of 300 cm in 10^{-8} sec. Answer all of the following questions from the point of view of an observer on the ground, who measures the speed of the rocket car to be 70.7% of the speed of light. **(1)** What is the length of the rocket car (between the points reached by the photons)? **(2)** What is the speed of the forward-going photon? **(3)** Which arrives at an end of the car sooner, the forward-going photon or the rearward-going photon? How much sooner? **(4)** What is the four-dimensional interval between the two events of photons reaching the front and the rear of the car?

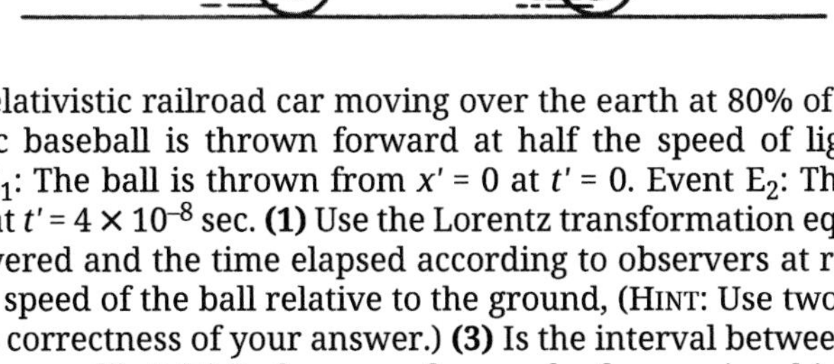

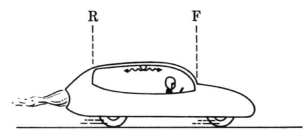

20.30. Within a relativistic railroad car moving over the earth at 80% of the speed of light, a relativistic baseball is thrown forward at half the speed of light relative to the car. Event E_1: The ball is thrown from $x' = 0$ at $t' = 0$. Event E_2: The ball is caught at $x' = 600$ cm at $t' = 4 \times 10^{-8}$ sec. **(1)** Use the Lorentz transformation equations to find the distance covered and the time elapsed according to observers at rest on the ground. **(2)** Find the speed of the ball relative to the ground, (HINT: Use two methods in order to check the correctness of your answer.) **(3)** Is the interval between events E_1 and E_2 time-like or space-like? **(4)** Is there any frame of reference in which the ball appears to be caught before it is thrown? Justify your answer mathematically.

20.31. Regard Equations 20.7 and 20.8 as two equations for two unknowns, $\Delta x'$ and $\Delta t'$. **(1)** Solve for these unknowns. **(2)** Explain how the resulting formulas for $\Delta x'$ and $\Delta t'$ support the principle of reciprocity.

20.32. Measurements in the runner's frame of reference in Figure 20.8 (x'', t'') are related to measurements in the train-based frame of reference (x', t') by the Lorentz transformation equations

$$\Delta t'' = \frac{\Delta t' - v_2 \Delta x'/c^2}{\sqrt{1 - v_2^2/c^2}},$$

$$\Delta x'' = \frac{\Delta x' - v_2 \Delta t'}{\sqrt{1 - v_2^2/c^2}}.$$

A similar pair of equations (with v_2 replaced by v_1) relates train-based measurements to measurements in an earth-fixed frame of reference (x, t). **(1)** Write down the Lorentz transformation equations which express $\Delta x'$ and $\Delta t'$ in terms of Δx and Δt. **(2)** Substitute from one pair of equations into the other to express $\Delta x''$ and $\Delta t''$ in terms of Δx and Δt. Show that the resulting equations are equivalent to a single Lorentz transformation to a frame of reference moving at a speed given by Equation 20.15. Although the algebra is complicated, the result is important, for it shows that two successive Lorentz transformations can be replaced by a single Lorentz transformation, and it yields the relativistic law of addition of velocities.

TWENTY-ONE

The special theory of relativity

By 1900 it had become clear that the ether was not easily detectable. Scientists were beginning to feel like Aesop's fox who could not reach the grapes. Henri Poincaré, perhaps the first to see clearly the potential significance of the failures to discover the ether, was led to say, "Our ether, does it really exist? I do not believe that more precise observations could ever reveal anything more than *relative* displacements." Einstein independently rejected the ether. In his first paper on relativity in 1905 (developing what we now call the special theory of relativity), he wrote: "The introduction of a 'luminiferous ether' will prove to be superfluous inasmuch as the view to be developed here will not require an 'absolutely stationary space' provided with special properties."

21.1 Einstein's two postulates

As the two basic postulates of the special theory, Einstein adopted the Principle of Relativity and the constancy of the speed of light. From these flow a host of implications, of which the irrelevance of the ether is only one. Quite aside from its remarkable and general consequences, the special theory of relativity has a unique allure because of the simple elegance of these two postulates upon which the whole theory rests:

(1) The laws of nature are the same in all inertial frames of reference.

(2) The speed of light is the same in all inertial frames of reference.

From the postulates Einstein derived the Lorentz transformation equations. Within the next few years Einstein and Planck discovered the relativistic mechanics which had to replace Newtonian mechanics, Minkowski achieved the mathematical synthesis of space and time and discovered the usefulness of the world-line diagrams, and Poincaré studied in depth the mathematical properties of the Lorentz transformation. By 1909 the special theory of relativity stood as a complete fundamental theory of physics, developed with explosive speed in but a few years. However, its seeds were planted at least as early as the 1880s, and its fruit is not all harvested yet.

We have seen in the preceding chapter that Einstein's second postulate—the constancy of the speed of light—is responsible for the Lorentz transformation, for the relativity of time, for nature's speed limit, and indeed for all of the strange new effects associated with space and time, as well as for a complete reorientation of our view of the nature of space and time. The innocent-looking statement, $c = constant$, an abbreviated way of expressing this postulate, is the keystone of the idea of *relativity*. Because of it, much more relativity of observation is introduced into science than was suspected before. But at the same time it leads to the Lorentz transformation, which accounts for the disagreements of different observers, and provides a rational basis for eventual agreement about a background objectivity.

Einstein's first postulate, the Principle of Relativity, may be regarded as the keystone of 115
the idea of *invariance* (and indeed might better have been named the Principle of Invariance). The theory of relativity has brought to science both more relativity *and* more invariance. According to the old idea that the Galilean transformation represented the right way

to relate the observations of two observers in relative motion, mechanics was invariant (its laws were the same in all inertial frames), but electromagnetism was not invariant. The Principle of Relativity asserts that *all* laws of nature, those known and those yet to be learned, are the same in all inertial frames of reference. In this sense relativity is a theory of theories, imposing requirements even on theories not yet formulated. The straitjacket of the Principle of Relativity has been a valuable tool over the last half century in the search for new laws of the elementary-particle world. It has so far met only with success, and will, like the principles of energy conservation and momentum conservation, remain a pillar of physical science until experimental evidence forces its rejection or modification.

The postulate of the constancy of the speed of light is itself an invariance law. Indeed if one chose to call the statement, "Light travels at the constant speed *c*" a law of nature, then the second postulate could be dispensed with and the entire theory of relativity would follow from the Principle of Relativity alone. This is entirely a matter of taste. It is probably better to let postulate 2 stand as a reminder of the particular significance of the speed of light in relativity. We have encountered one other invariant quantity so far, the four-dimensional space-time interval, which replaces the old ideas of separately invariant distances and times. We must consider now what impact the Principle of Relativity has had upon the classical theories of mechanics and electromagnetism.

21.2 The impact of relativity on classical theories

Since the laws of mechanics are invariant if the measurements of different observers are related by the Galilean transformation, it is not surprising to learn that the use of the Lorentz transformation to relate different observations causes the theory of mechanics to lose its invariance. But with the use of the Lorentz transformation, the theory of electromagnetism, as expressed in Maxwell's equations, "miraculously" remains invariant in all inertial frames. Since one theory is lost and one gained for the Principle of Relativity, the net advance produced by relativity might appear questionable. Fortunately, Einstein did not take such a negative view of matters. He and Planck showed that Newtonian mechanics could be modified in such a way that it too became invariant if the Lorentz transformations were used to relate one frame of reference to another. Some experiments with fast electrons that had been performed before 1905 tended to confirm the correctness of this modification of mechanics, and innumerable experiments since that time have verified to very high accuracy the correctness of relativistic mechanics and the incorrectness of Newtonian mechanics. In the meantime Maxwell's equations, unmodified, have been repeatedly tested and found to describe electromagnetic phenomena without error. Clearly the Principle of Relativity has scored a sweeping success.

That electromagnetic theory survived relativity unscathed should, of course, be regarded as neither accidental nor miraculous. In studying consequences of the Lorentz transformation, we learned that the new and interesting effects of relativity are significant only at high speeds. Newtonian mechanics had weathered over two centuries of tests and applications and appeared to be entirely correct only because, until Thomson's experiments with electrons, it had never been applied to the description of very high-speed motion. The equations of electromagnetism, on the other hand, had already described successfully the very *highest* speed motion, the propagation of light itself. To be correct at all, it *had* to be relativistic from the start, even without a theory of relativity to give significance to its properties of invariance.

Even though relativity did not change the equations of electromagnetism, it did produce new insight into the meaning of the concepts of electromagnetism. In particular, it merged the electric and magnetic fields into a single entity called the *electromagnetic field tensor*. Until now we have encountered two kinds of mathematical "objects," numbers and vectors. A number has one "component," its numerical value; a vector in space has three components, and cannot be characterized by just a single magnitude. A tensor is a mathematical object one stage more complicated than a vector, having more than three components. It is difficult to explain the meaning of a tensor in words, or to render it visualizable. Nevertheless the main idea here can be grasped—that two different concepts, electric field and magnetic field, have been merged by relativity into one single concept, the electromagnetic field. It is roughly analogous to drawing together the three ideas of forces in the x, y, and z direction (each a *number*) into the single idea of a force *vector*. Numerous other mergers of electromagnetic concepts have been effected by relativity. For example, electric charge and electric current have been united as a single four-dimensional vector. Even without mathematical change, electromagnetism has profited by relativity.

21.3 Relativistic energy

Relativistic mechanics is a subject as vast as Newtonian mechanics, and somewhat more complicated in applications. Instead of attempting a superficial survey of the whole theory, we shall concentrate on what relativity has to say about the ideas of mass, momentum, and energy. In this section we encounter relativity's most famous formula, $E = mc^2$.

From the contemporary point of view, it is easiest to say that the equivalence of mass and energy is an experimental fact. Numerous experiments have demonstrated that mass is convertible to energy, and energy to mass, and that the conversion factor between the two is the square of the speed of light. However, it is of interest to know how Einstein, long before the existence of any such experimental evidence, was led from ideas of relativity to the idea that mass should be a form of energy.

Just as the constancy of the speed of light leads to some remarkable and unexpected 116
conclusions about the nature of space and time, the Principle of Relativity leads to some equally remarkable and unexpected conclusions about the nature of energy (and of mass and momentum). The law of energy conservation, said Einstein, as one of the fundamental laws of nature, ought to be true in all inertial frames of reference. Yet this assumed invariance of the law of energy conservation—an application of the first postulate of the theory of relativity— has as a consequence that mass is energy. In order to describe how this comes about, we can do no better than discuss Einstein's original "thought experiment," slightly modernized.

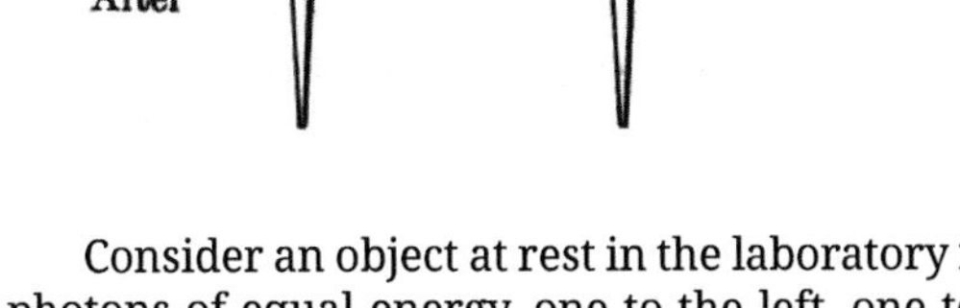

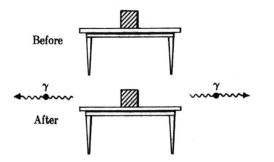

FIGURE 21.1 Laboratory frame of reference. An object emits oppositely directed photons of equal energy.

Consider an object at rest in the laboratory frame of reference (Figure 21.1). It emits two photons of equal energy, one to the left, one to the right. (Our use of photons is the modernization. Einstein originally spoke of light waves of equal energy. Photons do not change the essence of the argument.) Since equally energetic photons have equal magnitudes of momentum, the total momentum carried away by the photons is zero (a vector sum). The motionless object, having lost no momentum, remains motionless. It is as if a rifle could fire bullets simultaneously in opposite directions and thereby suffer no recoil. Our object does lose energy, however. Its energy loss is the sum of the energies of the two photons, or twice the energy of each. The fact that some energy has been carried away from the object does not by itself reveal that the object must have diminished its mass, for the energy might have been supplied by chemical change, nuclear transformation, or the random kinetic energy of molecular vibrations. There is no shortage of possible internal sources of energy to account for the emission of the photons. That a mass change accompanies the energy change can be discovered only by reexamining the same process from a moving frame of reference.

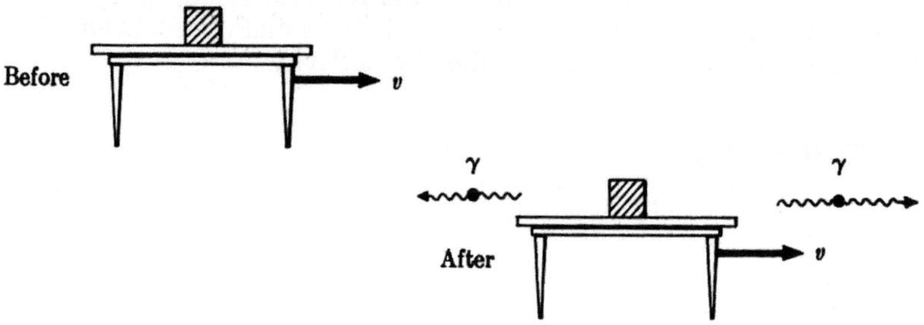

FIGURE 21.2 Moving frame of reference. The object does not change speed as it emits photons of different energy.

Imagine yourself now moving through the laboratory at constant speed v, westbound, let us say, to be definite. From your point of view the object (and, of course, the whole laboratory, but that is irrelevant) is moving eastward at a fixed speed v before it emits the pair of photons, and at the *same* speed v after it emits the photons (Figure 21.2). Since it experienced no recoil in the laboratory frame of reference, it experiences no change of velocity in the moving frame of reference. This is a key point. However, in the moving frame of reference the photons do not possess the same momenta and energies as in the laboratory frame. This is not surprising. If a baseball pitcher threw the ball from a platform moving toward the batter, the ball would arrive over the plate with more energy and more momentum than if he threw the ball from a stationary platform or from a receding platform. Similarly, the eastbound photon, thrown "forward" from the moving observer's point of view, has more energy than the same photon in the laboratory frame of reference, more by a factor

$$\frac{1 + (v/c)}{\sqrt{1 - (v^2/c^2)}}. \tag{21.1}$$

The "backward"-thrown photon has less energy than in the laboratory frame, less by a factor

$$\frac{1 - (v/c)}{\sqrt{1 - (v^2/c^2)}}. \tag{21.2}$$

These factors were discovered by Einstein in his study of Maxwell's equations of electromagnetism.* We state them here without proof. Their importance lies in the fact that their sum is *not* unity. Suppose that the energy of one photon in the laboratory frame is called $\frac{1}{2}\Delta E$. Then, in the moving frame, the energies of the two photons are

$$\frac{1}{2}\Delta E\frac{1 + (v/c)}{\sqrt{1 - (v^2/c^2)}} \quad \text{and} \quad \frac{1}{2}\Delta E\frac{1 - (v/c)}{\sqrt{1 - (v^2/c^2)}}.$$

* These factors of energy change hold true in electromagnetic theory for *any* amounts of electromagnetic radiation. They are not tied to the photon concept. However, Einstein noted that the factor of energy change between two frames of reference is the same as the factor of frequency change. This discovery may have had an impact on his important paper published in the same year, in which he proposed that all electromagnetic energy is emitted and absorbed in quantum bundles, photons, whose energy is proportional to their frequency.

The sum of these two energies is

$$\frac{\Delta E}{\sqrt{1-(v^2/c^2)}} \quad \text{(total energy loss in moving frame)}\,. \tag{21.3}$$

By contrast, the total photon energy in the laboratory frame is

$$\Delta E \quad \text{(total energy loss in laboratory frame)}. \tag{21.4}$$

The same pair of photons have more energy in the moving frame than in the laboratory frame. Accordingly the object has lost more energy from the moving observer's point of view than from the laboratory observer's point of view. Apparently the law of energy conservation does not possess the invariance required by the Principle of Relativity. Can it be saved? Yes, but only at the expense of what was one of the solidest laws of nineteenth-century physical science, the law of mass conservation.

Einstein argued that the place to find the extra energy needed to make up the apparent discrepancy between the two observers is in the bulk kinetic energy of the object that emits the photons. In the laboratory frame the object remains motionless, with zero kinetic energy both before and after the photons are emitted. In the moving frame, it possesses kinetic energy which, according to old ideas, remains constant because its speed remains constant. But, said Einstein, suppose that a decrease of mass accompanies the emission of the photons. Then, in the moving frame of reference, there is a loss of kinetic energy *even though there is no change of speed.* This change of kinetic energy can provide the extra energy content of the photons in the moving frame of reference and preserve the invariance of the law of energy conservation. It was Einstein's genius to recognize the more fundamental nature of energy conservation than of mass conservation, and to sacrifice the one in order to save the other.

Once the change of mass has been pinpointed conceptually as the key to preserving energy conservation, it is not a difficult matter to derive mathematically the relationship between the energy ΔE of the emitted photons and the mass change Δm of the emitting object. It is left as an exercise to show that this relationship is

$$\Delta m = \Delta E/c^2\,. \tag{21.5}$$

In words: If the object suffers a loss of mass equal to the total radiated energy divided by the square of the speed of light, the law of energy conservation will be preserved in all inertial frames of reference.

In his first paper on mass and energy, Einstein suggested that the energy changes observed in radioactive transformations might be great enough to produce perceptible mass changes, a suggestion finally verified in 1932. Not even Einstein was bold enough to suggest in 1905 that the total conversion of mass to energy might be an experimental possibility. Yet it is now recognized to be commonplace in the world of elementary particles. It is useful to reexamine Einstein's thought experiment with this possibility in mind.

Suppose that the "object" in the thought experiment is a neutral pion, which after an average lifetime of about 10^{-16} sec emits a pair of photons and disappears, transforming its entire mass into radiant energy. The decay process is indicated symbolically by

$$\pi^0 \rightarrow \gamma + \gamma\,.$$

In the laboratory frame of reference, where the pion is supposed to be at rest, the two photons have equal and opposite momenta, and each carries away an energy equal to $\frac{1}{2}mc^2$ (m is the pion mass). The energy dissipated in the laboratory frame is mc^2. In a moving frame of reference, on the other hand, the energy dissipated, according to Equation 21.3, is greater. It is

$$\frac{mc^2}{\sqrt{1-(v^2/c^2)}}\,.$$

Since, in both frames of reference, the pion disappears, giving up all of its energy, the energy carried away by the photons must be equal to the total energy content of the pion before the decay. Thus the total energy content of a particle may be expressed by

$$E = mc^2 \qquad \text{(stationary particle)}, \tag{21.6}$$

$$E = \frac{mc^2}{\sqrt{1-(v^2/c^2)}} \qquad \text{(moving particle)}\,. \tag{21.7}$$

A particle at rest has an intrinsic energy, an "energy of being," equal to mc^2. In motion, its energy is greater; it is mc^2 divided by the ubiquitous factor $\sqrt{1-(v^2/c^2)}$. In these equations, the symbol m represents what is sometimes called the rest mass. It is a fixed constant for a particular object, regardless of the speed of the object, and must not be confused with the variable mass or "relativistic mass" used by some authors.*

The difference between the total energy of a particle in motion (Equation 21.7) and its total energy at rest (Equation 21.6) may be interpreted as its kinetic energy. Thus the relativistic equation for kinetic energy, which, like the energy-mass equivalence, can be deduced from Einstein's original thought experiment, is

$$K.E. = \frac{mc^2}{\sqrt{1-(v^2/c^2)}} - mc^2 \qquad \text{(relativistic)}\,. \tag{21.8}$$

At first glance this appears very different indeed from the equation of Newtonian mechanics,

$$K.E. = \tfrac{1}{2}mv^2 \qquad \text{(nonrelativistic)}. \tag{21.9}$$

In order to perceive their similarity, we must examine the mathematical behavior of the relativistic equation for low speed, that is, when the ratio v/c is a small fraction of one. This requires use of the approximation given by Equation 20.12:

$$\frac{1}{\sqrt{1-(v^2/c^2)}} \cong 1 + \frac{1}{2}\frac{v^2}{c^2}\,.$$

Substitution of this approximation into Equation 21.8 yields

$$K.E. = mc^2\left[1 + \frac{1}{2}\frac{v^2}{c^2}\right] - mc^2\,, \tag{21.10}$$

a version of the relativistic formula for kinetic energy valid for slow particles (slow compared with the speed of light). The right side of Equation 21.10 reduces to $\tfrac{1}{2}mv^2$, exactly the Newtonian expression for kinetic energy. This is not a piece of luck; it is a necessity. More than two centuries of precise tests of classical mechanics, especially in describing the motions of the planets, had shown the correctness of the equation, $K.E. = \tfrac{1}{2}mv^2$, at low speed. Relativistic mechanics, to be acceptable at all, must duplicate the equations and the predictions of classical mechanics in the domains of their validity. At the same time, it must be remembered that Equation 21.8, like Equation 21.9, is a *definition*. The revised definition is required to save the law of energy conservation.

It is in the realm of high-speed motion that the relativistic kinetic energy diverges completely from the Newtonian formula. This divergence is illustrated in Figure 21.3. At the "speed-of-light barrier" the energy of a material particle approaches infinity. Here all similarity between Newtonian mechanics and the facts of nature disappears. New equations, new concepts, and new ways of thinking are required.

* The relativistic mass m_r is defined by $m_r = m/\sqrt{1-v^2/c^2}$. In terms of this concept, Equation 21.7 takes the form $E = m_r c^2$. We shall use only the rest mass m, which, because of its invariance, is a more significant quantity than the variable mass m_r.

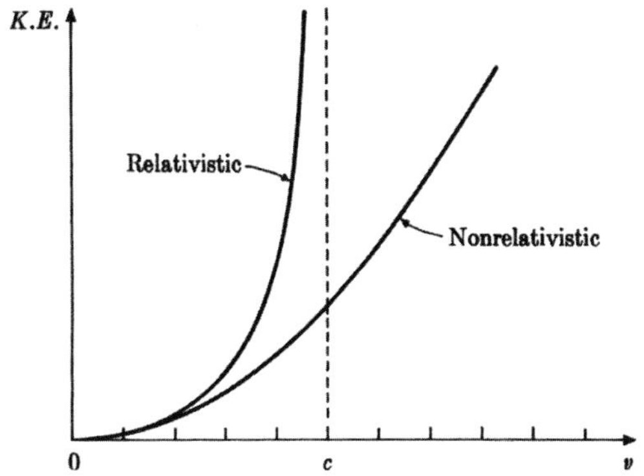

FIGURE 21.3 Graphs of kinetic energy vs. speed according to the relativistic definition, Equation 21.8, and the nonrelativistic definition, Equation 21.9. The nonrelativistic curve is a parabola. The relativistic curve approximates a parabola at low speed, but curves more rapidly upward to approach infinity at the speed-of-light barrier.

No concept of physics has weathered more winds of change than the concept of energy. And from every storm it has emerged a stronger and a broader concept, with its branches now extending into every field of physical and biological science. First potential energy was added to work and to kinetic energy. Then came electrical energy, chemical energy, heat energy, electromagnetic energy, and nuclear energy. To the list of the manifold forms of energy the theory of relativity added mass, the most fundamental form since kinetic energy, for at the level of particles every energy is manifested either as energy of motion or as energy of being.

The speed of light, which showed up as the link between the unit of distance and the unit of time, appears again (or rather its square appears) as the link between mass and energy. In the equation $E = mc^2$ the factor c^2 is not fundamentally important (this is, of course, not to say that it can be omitted!). It may be thought of as a correction factor to convert the units of mass to the units of energy, and is of no more importance than that. The modern interpretation of the equation would be simply: mass *is* energy. Indeed modern scientists are beginning to develop a picture, still very hazy, of a material particle as just a highly concentrated and localized bundle of energy, which has mass only *because* it has energy.

21.4 *Relativistic momentum*

The requirement of self-consistency, a powerful tool of the mathematician, has its uses in 117
physics too, and especially in relativity theory, a theory whose origins are to be found in some clear thinking about the consistency of measurements by different observers. Another aspect of the self-consistency of the theory of relativity is the interesting way in which the nonconservation of mass can be linked either to the conservation of energy or to the conservation of momentum. Consider again the arrangement of Einstein's thought experiment—an object at rest in the laboratory emits equal-energy photons to the left and to the right. In a moving frame of reference, the photons have neither equal energy nor equal momentum; nevertheless, the object that emits them continues at constant speed. Since the photon momenta are unbalanced and do not add to zero, the object must suffer a change of momentum when the photons are emitted. Now if speed is constant, the only way to change momentum is to change mass. In this way, the loss of mass accompanying the photon emission can be based on momentum conservation instead of energy conservation, with the same result as before: $\Delta M = \Delta E/c^2$. Here is a valuable hint, that perhaps relativity will have more to say about a connection between energy and momentum.

Another example may show even more simply the connection between momentum conservation and the mass-energy equivalence. Imagine two identical objects, each of mass m, a distance d apart (Figure 21.4). Object A emits a photon of momentum p to the right, and recoils with equal momentum p to the left. If the mass m is large and the recoil velocity v

small, the Newtonian formula for momentum, $p = mv$, may be used with confidence (even though we have every reason to expect this formula to be incorrect at high speed). After a time $t = d/c$, the photon will have reached object B, and object A will have recoiled a distance $x = vt = pt/m$, so that at this moment objects A and B have increased their separation to the distance $d + x$. According to a general consequence of momentum conservation in classical mechanics, the center of mass of an isolated system, if once at rest, remains at rest. Yet here apparently, the initially motionless center of mass has shifted leftward through a distance ½x. To keep the center of mass riveted in place—at first halfway between the two objects, but later farther from object A than from object B—it is necessary to shift some mass. It is left as an exercise to show that the mass Δm lost by object A and gained by object B is equal to the photon energy divided by c^2: $\Delta m = E_\gamma/c^2$. It seems that the photon has "carried" mass from A to B. This is true in a way, but the photon is nevertheless massless. It is better to say that some of the mass of object A has been transformed into the electromagnetic energy of the photon. Later this energy, upon being absorbed by B, is transformed back to mass.

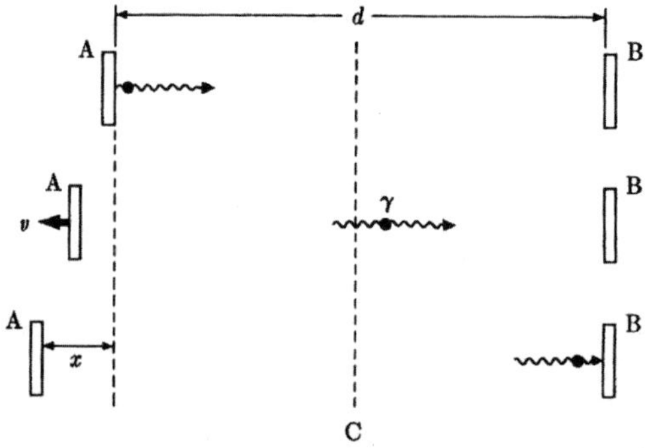

FIGURE 21.4 Thought experiment relating momentum conservation to the mass-energy equivalence. If the center of mass of the system (indicated by the dotted line C) is to remain at rest, object A must lose some mass and object B gain some mass.

Since the relativistic definition of kinetic energy implies that energy becomes infinite as the speed of a material object approaches the speed of light, it will not be surprising to learn that relativity offers a new definition of momentum with this same property. Instead of deriving the relativistic formula for momentum (which can be done in a variety of ways), we shall state the definition and then demonstrate its consistency with two other equations of relativity, the mass-energy relationship and the law of addition of velocities. Remember that momentum, like energy, is defined in such a way that it satisfies a conservation law, even at high speed, even for the interaction of matter and radiation.

Momentum is a vector quantity with the same direction as velocity and with magnitude given by

$$p = \frac{mv}{\sqrt{1 - (v^2/c^2)}} . \tag{21.11}$$

The numerator is the same combination of mass and speed that defines momentum nonrelativistically. The denominator is the now familiar factor of relativity whose value is imperceptibly different from one at low speed but becomes very small and produces an important effect at high speed. Since the denominator is always less than one, the relativistic momentum is always greater than the product mv: only slightly greater at low speed, but very much greater as the speed nears the speed of light. This behavior of the relativistic momentum is illustrated in Figure 21.5. There appears the same "speed-of-light barrier" preventing motion at speeds greater than the speed of light. No matter how much momentum a particle acquires, its speed can never quite reach the limiting speed c (unless its mass is zero, a special case that will be discussed at the end of this section).

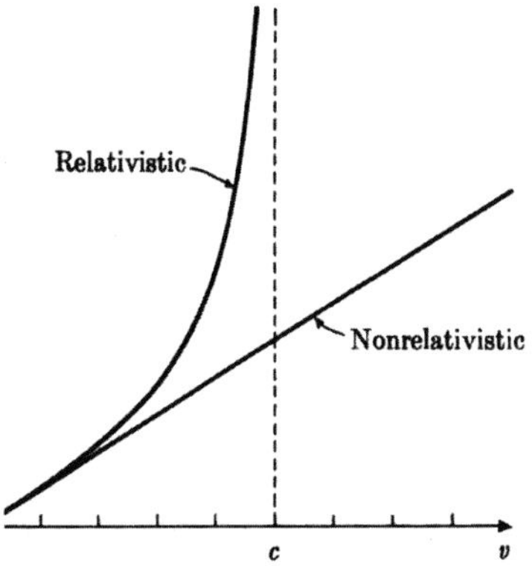

FIGURE 21.5 Graphs of the magnitude of momentum of a particle according to the relativistic definition, Equation 21.11, and according to the nonrelativistic definition, $p = mv$. At low speed, the relativistic graph closely approximates the straight-line classical graph.

The relativistic momentum increase was what Thomson noted when he discovered that high-speed electrons were more difficult to deflect than he thought they should be. This "extra momentum" is sometimes described by saying that a particle when moving is more massive than when stationary. But it is better not to think of the particle's mass as changing, for the factor m appearing in the definition of momentum is a fixed number, regardless of how fast the particle might be moving. The apparent mass increase comes about because a relativistic particle with fixed mass when set in motion "behaves like" a Newtonian particle with increased mass. The mass increase is an illusion arising from thinking in old-fashioned Newtonian terms about a relativistic problem. It is quite unlike, for example, time dilation, which is a real, verifiable effect of motion.

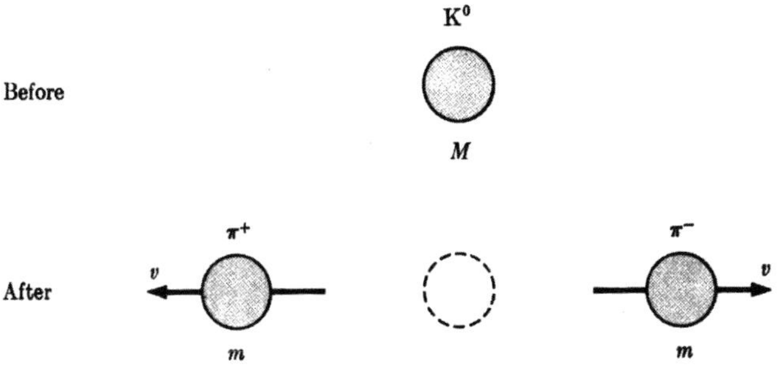

FIGURE 21.6 Decay of stationary kaon in the laboratory frame. Pions fly apart with equal and opposite momenta, and each with half the total energy of Mc^2.

To turn now to an application of the relativistic momentum formula and a test of its self-consistency, consider the decay of a kaon at rest in the laboratory into a pair of pions, indicated by

$$K^0 \rightarrow \pi^+ + \pi^- .$$

Call the mass of the kaon M, the mass of each pion m, and the speed of each pion v. To preserve the initial zero momentum, the pions must fly apart at equal speed in opposite directions (Figure 21.6). The initial energy is Mc^2. The final energy of each pion is

$$E_\pi = \frac{mc^2}{\sqrt{1-(v^2/c^2)}} . \tag{21.12}$$

Energy conservation requires the equality of energies before and after the decay:

$$Mc^2 = \frac{2mc^2}{\sqrt{1-(v^2/c^2)}} . \tag{21.13}$$

Now suppose that you choose to examine the same decay process in a frame of reference moving to the left at speed v. In this frame (Figure 21.7), the kaon before its decay moves to the right with speed v and possesses momentum

$$p_K = \frac{Mv}{\sqrt{1-(v^2/c^2)}} . \qquad (21.14)$$

After the decay, the pion that moved leftward in the laboratory frame is at rest in the moving frame, so that all of the initial momentum of the kaon in this frame must be acquired by the right-going pion. Its speed is v_1 (*not* twice v) and its momentum is

$$p_\pi = \frac{mv_1}{\sqrt{1-(v^2/c^2)}} . \qquad (21.15)$$

Momentum conservation (and herein lies the test of our new definition of momentum) requires the equality of these two momenta:

$$\frac{Mv}{\sqrt{1-(v^2/c^2)}} = \frac{mv_1}{\sqrt{1-(v^2/c^2)}} . \qquad (21.16)$$

The correctness of this equation (by no means,obvious as it stands) can be verified in several ways. Suppose, for instance, that we solve it for v_1. On the left may be substituted, from the energy-conservation equation, Equation 21.13,

$$M = \frac{2m}{\sqrt{1-(v^2/c^2)}} ,$$

so that the momentum-conservation equation (Equation 21.16) reads

$$\frac{2mv}{1-(v^2/c^2)} = \frac{mv_1}{\sqrt{1-(v^2/c^2)}} . \qquad (21.17)$$

The factors m cancel on the two sides, and the algebraic solution for v_1 (left as an exercise) is

$$v_1 = \frac{2v}{1+(v^2/c^2)} . \qquad (21.18)$$

This is exactly what would have been predicted by the velocity addition formula of Chapter Twenty. The speed v_1 is the relativistic "sum" of v (the speed of the right-going pion with respect to the laboratory) plus v again (the speed of the laboratory with respect to the observer). The relativistic velocity-addition law gives

$$v_1 = \frac{v+v}{1+(vv/c^2)} = \frac{2v}{1+(v^2/c^2)} . \qquad (21.19)$$

Through this example we have demonstrated the mutual consistency of (a) the relativistic definition of energy, (b) the relativistic definition of momentum, and (c) the relativistic equation for the addition of velocity. From any two of these, the third may be derived. Indeed

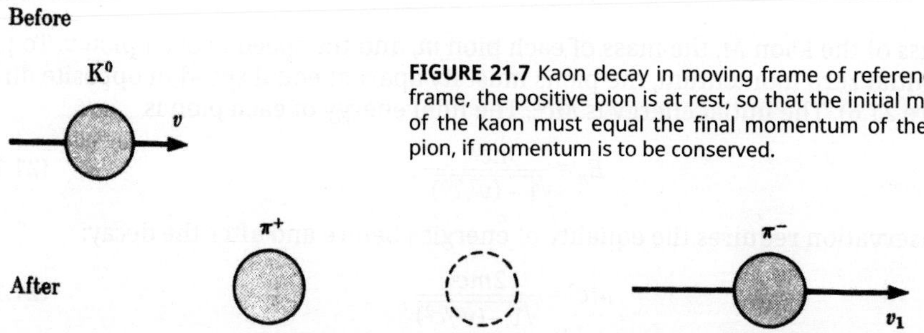

FIGURE 21.7 Kaon decay in moving frame of reference. In this frame, the positive pion is at rest, so that the initial momentum of the kaon must equal the final momentum of the negative pion, if momentum is to be conserved.

there is no other possible definition of momentum that is consistent with the already given definition of energy and law of velocity addition. The requirement of invariance of the conservation laws imposes a powerful restriction on the mathematical form of the quantities conserved.

Massless particles occupy a special place in relativity theory. As pointed out at the end of Chapter Twenty, there is no reason known why they *must* exist, yet several are known in nature. There is excellent experimental evidence that the photon and the electron's neutrino are truly massless. For a massless particle, the already given definitions of momentum and energy lose their meaning. Consider, for instance, the expression for the momentum of a massive particle,

$$p = \frac{mv}{\sqrt{1 - (v^2/c^2)}} .$$

If the mass m is zero, the numerator is zero. However, since in that case the particle moves at the speed of light, $v = c$, the denominator is also zero. The equation is said to be "singular." It is simply meaningless, for the form $0/0$ is undefined. Nevertheless, some insight into the reason why massless particles travel at the fixed speed c can be gained by examining the behavior of the relativistic momentum as the mass m becomes exceedingly small. Graphs of momentum vs. speed are shown in Figure 21.8 for various values of mass. A particle with very little mass need acquire only a small momentum in order to move near the speed of light. Thereafter further increases in its momentum increase its speed only slightly as it inches closer to the limiting speed. For a massless particle, the momentum "curve" becomes the backward-L indicated by the heavy lines in Figure 21.8. Any momentum at all is sufficient to cause the particle to move at the maximum speed c. No further increase of momentum can make it move any faster. For this reason, there exists no equation linking the momentum of a massless particle to its speed. There exists only a relationship (a beautifully simple one) between its momentum p and its energy E:

$$E = pc . \tag{21.20}$$

Energy is directly proportional to momentum. This energy E may be regarded as pure kinetic energy, since the rest energy mc^2 vanishes. A massive particle moving near the speed of light also behaves "like" a massless particle, that is, its energy and momentum satisfy approximately the same relationship:

$$E \cong pc . \tag{21.21}$$

For such a particle, its total energy so greatly exceeds its rest energy mc^2 that the fact that it has mass becomes unimportant.

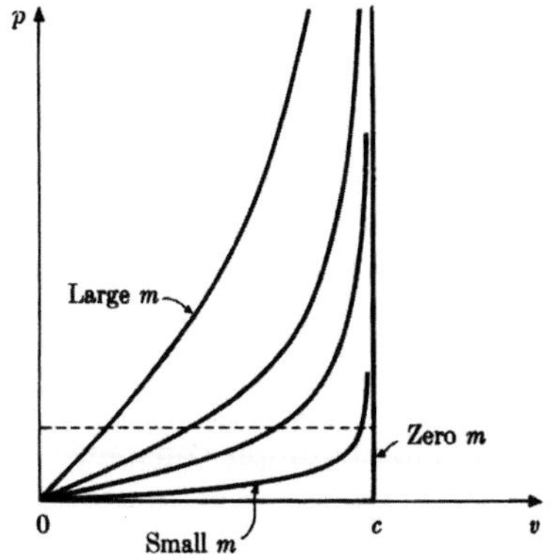

FIGURE 21.8 Variation of momentum graphs with particle mass. For a given momentum, as indicated by the horizontal dashed line, less mass means more speed. Massless particles, regardless of their momentum, move at the maximum speed c.

21.5 Four-vectors

118 In Chapter Twenty, the invariance of the interval and the mixing together of space and time in the Lorentz transformation led us to introduce the idea of a four-dimensional space-time, filled with world lines and events. Each event is a point in space-time located by three space coordinates and one time coordinate, and any pair of events is separated by a space-time "distance," the interval. With this picture of space-time, we may quite naturally think of the four coordinates locating an event as being the four components of a single vector which, to avoid confusion with old-fashioned three-dimensional vectors, we shall call a four-vector. This space-time four-vector is a generalization of the idea of a position vector in ordinary space. Instead of locating a point in space, it locates an event in space-time. Its components are x, y, z, and ct, and the square of its "length" is $x^2 + y^2 + z^2 - (ct)^2$. (The minus sign preceding the contribution of the fourth component is important and distinguishes the fourth component from the first three.) The length of the four-vector is a constant for all observers, for it is just the interval under a new name.

In Newtonian mechanics, several different vector quantities appeared: position, velocity, acceleration, momentum, force. We have been able to extend the idea of the position vector to the four-dimensional world by appending ct to it as a fourth component. This extension has several technical advantages in the theory of relativity, the most important of which, for our present purpose, is that the "length" of the four-vector is an invariant quantity, whereas the ordinary three-dimensional length is not invariant, but depends on the state of motion of the observer. This process of generalizing vectors from three to four dimensions can be carried out not only for the position vector but for all of the vectors of mechanics, and some unexpected and interesting partnerships turn up. Next in interest to the joining together of space and time is the union of energy and momentum.

Momentum, because it is a three-dimensional vector, may be expressed in component form:

$$p_x = \frac{mv_x}{\sqrt{1-(v^2/c^2)}}\,, \tag{21.22}$$

$$p_y = \frac{mv_y}{\sqrt{1-(v^2/c^2)}}\,, \tag{21.23}$$

$$p_z = \frac{mv_z}{\sqrt{1-(v^2/c^2)}}\,, \tag{21.24}$$

Since space has acquired a fourth dimension—time—momentum needs a fourth component, or "time component." Remarkably, energy (or more exactly, energy divided by the speed of light) proves to be the time component of momentum:

$$p_t = \frac{E}{c} = \frac{mc}{\sqrt{1-(v^2/c^2)}}\,, \tag{21.25}$$

This merger of two of the key concepts of mechanics into a single four-dimensional entity was achieved by H. Minkowski in 1908. Shortly other mergers were recognized, such as electric charge with electric current, and electric field with magnetic field, so that soon all of classical physics became imbedded in the new space-time.

The four-dimensional momentum-energy vector has a length, whose square is defined in much the same way as the square of the space-time interval. It is

$$p_x^2 + p_y^2 + p_z^2 - p_t^2\,. \tag{21.26}$$

This combination, as a little algebra reveals, is exactly equal to the simple constant,

$$-m^2c^2\,,$$

the same for all observers. Although both the energy and momentum of a particle depend on the state of motion of the observer, here is a certain invariant combination about which all observers agree.

Although space-time vectors connecting pairs of events may be either spacelike or time-like, the momentum-energy vector of a material particle is always time-like, as indicated by the minus sign on the right of the equation

$$p_x^2 + p_y^2 + p_z^2 - p_t^2 = -m^2c^2 . \tag{21.27}$$

For a massless particle, always standing apart as a special case, the momentum-energy vector has zero length (although momentum and energy are of course not separately zero). Such a vector is called a *null vector*. A briefer and often useful way to write Equation 21.27, valid for particles with or without mass, is

$$E^2 = p^2c^2 + m^2c^4 . \tag{21.28}$$

This expresses the total energy of a particle in terms of its momentum and its mass.

In general, what does the fourth component of any vector mean? It scarcely seems an adequate answer to say it is a component pointing in the time direction. We can appreciate the meaning of north, south, east, west, up, and down. But which direction is the time direction? Unfortunately man does not seem able to visualize four dimensions, yet that is just what the theory of relativity is asking him to do. The best one can do is try and extrapolate one's pattern of thought from two dimensions to three and on to four, to form analogies, and to make use of what has already been learned about space and time. By way of analogy, we might think of a two-dimensional worm living out his life in a plane. He would "know" that space has two dimensions. If a more learned worm tried to explain to him the meaning of the third dimension, he might become irritated and exclaim, "But what can you *mean* by a third dimension? It is nothing you can point to; it is not anything we can experience." We three-dimensional creatures must likewise be content to accept the fourth dimension as something at best vaguely visualizable, yet vitally important in deepening our view of the world.

To summarize, relativity has revealed two important new aspects of energy. First, there exists a new kind of energy, the energy of mass. Second, energy is linked to momentum in the same way that time is linked to space. Just as space and time become mixed in relating the measurements of two different observers, so do momentum and energy become inextricably mixed, and the laws of conservation of momentum and of energy combine into a single more general conservation law. All this because the ether was rejected. How beautiful and how unexpected are the consequences of Einstein's two simple postulates!

21.6 The twin paradox

So contrary are the ideas of relativity to ordinary ways of thought that numerous puzzles 119
and "paradoxes" have been proposed over the years, some of the best of which provide challenging tests of one's ability to think in relativistic terms. Probably the most famous of these, one which has excited a great deal of interest and even stirred up considerable controversy, is called the "twin paradox." Anton (twin A) stays at home on earth, while Bertram (twin B) cruises about the universe at nearly the speed of light. Since Bertram's rapid motion has slowed down his clocks and his life processes along with them, he (1) returns home to find himself much younger than Anton. But from Bertram's point of view, it is Anton who was traveling and Anton who was living more slowly. Could we not with equal justification say (2) that Anton should be younger than Bertram when they rejoin? Or, in view of the reciprocity of observation, (3) should they instead still be equal-aged twins when they reunite? The first of these three possibilities is the correct one. Bertram *will* be younger than Anton.

The problem is to reconcile this result with the principle of reciprocity and the absence of preferred frames of reference. The solution can, of course, be given mathematically, but we shall instead discuss a closely analogous problem that illuminates the twin paradox and makes it seem not paradoxical.

Suppose that Anton is standing motionless on earth as Bertram speeds past in a train. Both note as they pass that their watches point to the same time, but, making rapid observations, each concludes that the other's watch is running slowly. Each therefore predicts that later on the other's watch will be behind his own. Can they both be right? Oddly enough, they can. It depends on how they go about checking up on their watches later on.

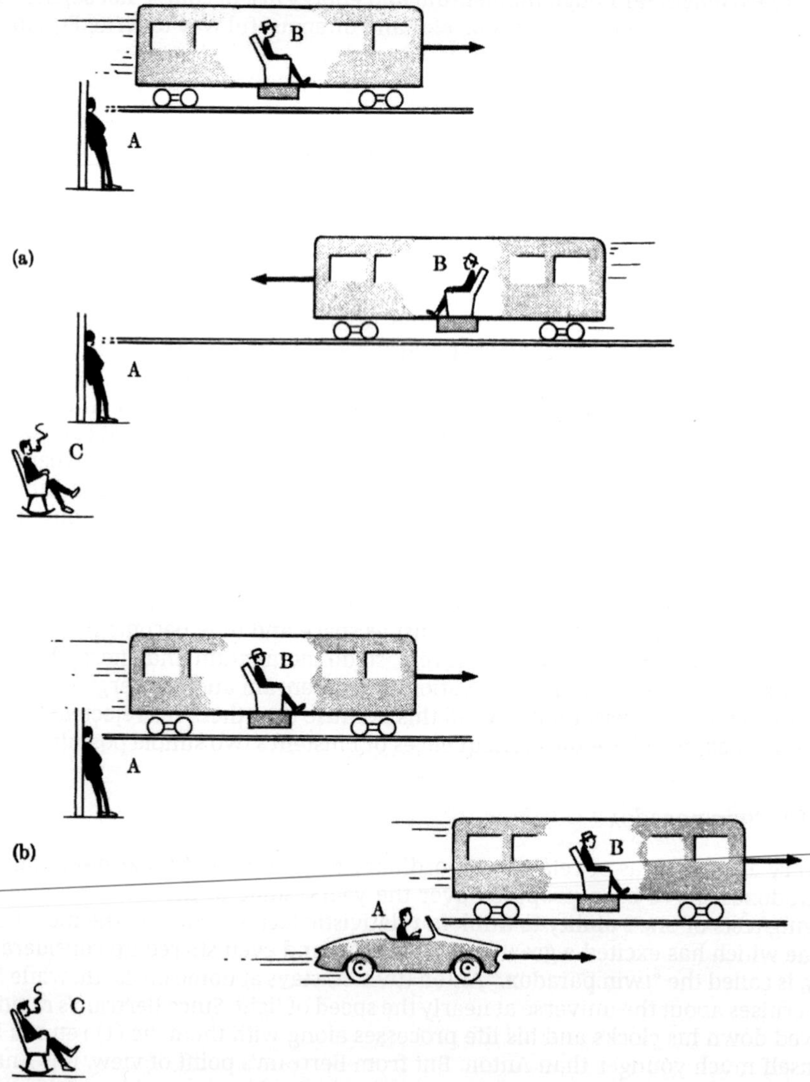

FIGURE 21.9 The twin paradox. **(a)** Anton waits for Bertram to return. Bertram is younger when they meet. **(b)** Anton overtakes Bertram, who waits patiently in his inertial frame. Anton is younger when they meet. Charles understands this, for he sees Bertram moving more rapidly than Anton in case (a) and Anton moving more rapidly than Bertram in case (b).

Let us suppose that the twins pass each other in this manner on two occasions (Figure 21.9). On the first occasion, Bertram leaves the train at the next station, catches the next train back, and rejoins Anton to find his watch behind Anton's. "That's odd," says Bertram, "I could have sworn it was your watch that was slow. Oh, well," he adds philosophically, "at least I haven't aged as much as you." On the second occasion, Anton rents a fast motorcar, overtakes the train and boards it at the next station. Now it is Anton's turn to be surprised, but to take what comfort he can from his lower rate of aging. Each observer's prediction has been borne out, provided that he remains patiently in his own inertial frame of reference and lets the other observer come to him. "All quite simple," says an impartial observer, Charles, who takes the human point of view that the earth is really *the* frame of reference. "Bertram's watch was of course running slowly because he was moving, as was proved when he came back to Anton. But if Anton is so foolish as to travel even faster than Bertram in his effort to overtake him, small wonder that Anton's watch slowed down even more than Bertram's."

The general rule to cover these experiments as well as the first statement of the twin paradox is that any observer who leaves an inertial frame, moves with respect to it, and then rejoins it, will find that his time is behind the time of observers who remained in the frame. There is actually nothing paradoxical about the twin paradox. Indeed, the explanation given above has even been verified—with twin nuclei, not twin people! Nevertheless it raises some fundamental questions to which no satisfactory answers have yet been provided. Why are there any inertial frames at all? Given that there are inertial frames, what determines *which* frames of reference are inertial? We can at best define an inertial frame operationally. It is one in which Newton's first law holds true—objects on which no force acts are unaccelerated. And, of course, any frame moving uniformly with respect to an inertial frame is inertial. The most appealing hypothesis about the "true nature" of inertial frames is due to Ernst Mach; it is usually called Mach's principle. Mach suggested that the distribution of matter in the universe determines preferred frames of reference, and that any frame unaccelerated with respect to the "fixed stars," that is, with respect to the average distribution of matter in the universe, is an inertial frame.* This would mean that matter throughout the universe would determine the properties of space-time at the earth—a challenging idea, but one which has so far defied a mathematical formulation, and which has not been fruitful in leading to the prediction of new experiments, although it probably played a role in shaping the general theory of relativity. (Actually, the more modern view would replace Mach's "matter" by matter plus energy, but since most of the energy in the universe seems to be in the form of matter, the difference is not significant.)

21.7 *Experimental tests of special relativity*

By 1910 relativity was still far from a well-tested theory. Although it accounted at once for experiments that had failed to uncover the ether, most of its experimental tests came much later. Einstein himself in justifying the theory placed great weight on the internal logic and simplicity of the theory, which to him lent to it a kind of necessity at least as great as that provided by the supporting evidence of experiments.

The first important success of relativity beyond its ability to explain the shyness of the ether was in accounting for the momentum of fast electrons. The apparent mass increase of high-speed electrons first noticed by Thomson was measured quantitatively by W. Kaufmann in 1901 and succeeding years. Still more careful measurements of A. H. Bucherer in 1908 confirmed completely Planck's relativistic definition of momentum. By now,

* Mach actually suggested even more, that inertial mass exists at all only because of the presence of the rest of the matter in the universe. According to this principle, a body alone in an otherwise empty universe would possess no mass.

of course, it has been verified up to incredibly great energies and to speeds within one mile per hour of the speed of light. If the relativistic definition of momentum were inconsistent with other definitions of relativity theory, or did not lead to a law of momentum conservation, the latest particle accelerators would fail to work, resulting, at the very least, in red-faced embarrassment for the designers of these giant machines.

Not until 1941, a few years after the discovery of the muon, did a direct test of time dilation become possible. Now it is commonplace in spark chambers and bubble chambers to see unstable particles whose lifetimes are increased tenfold or a hundredfold by the relativistic dilation of time. What in our everyday world is a tiny effect below the limit of measurability becomes for energetic particles an enormous correction. Somewhat earlier, in 1937, a group in Essen, Germany, confirmed indirectly both the time dilation and Lorentz contraction effects by studying with extreme precision the dependence of the natural frequency of vibration of a rod on the orientation of the rod. Aside from this indirect test, the Lorentz contraction effect has not been verified, but it is such an integral part of relativity that it is inconceivable that the effect does not exist.

Passengers flying between El Paso and Albuquerque can still today see clearly in the New Mexico desert a black circle of glazed sand attesting to the correctness of Einstein's mass-energy equivalence. Actually this equivalence was well tested in the laboratory a decade before the explosion of the first atomic bomb. During the 1920s, the masses of many nuclei were accurately measured, and it was known that these masses were close to, but not exactly equal to, integral multiples of the mass of the proton. The first direct test of the mass-energy equivalence came in 1932 when John D. Cockcroft and Ernest T. S. Walton in England bombarded nuclei of lithium with protons that had been speeded up in a newly designed accelerator. Some of the time a nuclear reaction occurred in which the proton and lithium nucleus were transformed into two alpha particles,

$$p + \mathrm{Li}^7 \rightarrow \alpha + \alpha\,.$$

(The nucleus of Li^7 contains three protons and four neutrons; each alpha particle contains two protons and two neutrons.) Cockcroft and Walton found that the emerging alpha particles had more kinetic energy than the incoming proton. This excess energy could be exactly accounted for by Einstein's mass-energy relation, for the two alpha particles had less total mass than did the proton plus lithium nucleus. The relation $E = mc^2$ was exactly what was needed to save the principle of energy conservation in the reaction. Since then the equivalence of mass and energy has been tested even more dramatically by the creation and total annihilation of particles such as muons, pions, and positrons.

What must stand as the most satisfying success of the theory of relativity resulted in 1928 from the union of relativity and quantum mechanics. This synthesis was achieved by a young Englishman, Paul Dirac, who at the age of 26 was already a world-renowned mathematical physicist. In attempting to reconcile quantum mechanics with the invariance requirement of relativity, Dirac was led to an equation (still called the Dirac equation) from which he derived two startling and wholly unexpected conclusions. The first was that the electrons described by the equation must be spinning. The fact that all electrons do behave like tiny tops spinning continuously about an axis had been discovered a few years earlier by the Dutch physicists Samuel Goudsmit and George Uhlenbeck. Now Dirac revealed a hidden and unsuspected power of relativity. The theory, like black magic, required that electrons *must* spin.

Dirac's second startling conclusion was that there should exist two kinds of electrons. No one knew what to make of this prediction, since only one type of electron was known. But the Dirac equation could not be given up, for it had succeeded in accounting perfectly for subtle details of atomic structure as well as for the electron spin. The matter hung indecisively in the air for four years until Carl Anderson, a 26-year-old American physicist, discovered in a photograph of a cloud chamber exposed to cosmic radiation unmistakable evidence for a particle in every way like an electron except that it was positively charged. Anderson's positron (as it is now called) was Dirac's second electron. It verified beyond doubt Dirac's successful application of the theory of relativity to quantum mechanics.

21.8 *The significance of special relativity*

Any successful theory in science contributes to man's understanding in two ways. First it 120
provides a satisfying explanation of a set of empirical facts of nature, and shows how these facts can all be understood in terms of a few simple ideas. Its success in quantitative prediction is the acid test of any theory. But a good theory does more. It alters more or less profoundly man's view of the world around him. It may do so by demonstrating the utility and therefore the importance of some new concept. The idea of field, for example, which was provided by the theory of electromagnetism, remains today a central concept of modern physics, having grown enormously in importance during the last hundred years. From relativity has come the idea of invariance, a pillar of nearly every branch of contemporary physical science. Other examples from the classical theories will occur readily to the reader.

Besides introducing new concepts, a theory may revolutionize our view of nature by drawing together old concepts or old facts in an entirely new and unsuspected way. The revolution of relativity has come about in large part from syntheses of ideas previously thought to be unrelated. Space and time have been drawn together into a single concept of space-time; the electric and magnetic fields have been amalgamated into the electromagnetic field tensor; momentum and energy have been united; mass has been shown to be just another form of energy. When relativity was joined to quantum mechanics, more unsuspected connections appeared. The spin of the electron and the existence of antiparticles were miraculous consequences of the relativistic invariance requirement.

A particular significance of special relativity has been its role in physics as a kind of supertheory. *All* laws of nature, said Einstein, must be the same in all inertial frames of reference. This invariance requirement has been a straitjacket on the free invention of scientists, and a surprisingly powerful one. In probing into the unknown world of the interactions among elementary particles, physicists have learned that the rejection of all laws of interaction that do not conform to the Principle of Relativity leaves only few possibilities. Besides the relativistic invariance (in inertial frames), other kinds of invariance principles have been discovered. It remains a challenging open question whether all the invariance principles taken together will so rigidly circumscribe the conceivable laws of nature that the actual laws will be uniquely determined by invariance requirements alone. If so, the idea of invariance introduced by special relativity will have been promoted to the central position in all of physics.

More than any other theory since mechanics, relativity has had a powerful impact upon man's view of his place in the universe. This has come about in part because relativity has removed finally the last remaining vestiges of the pre-Copernican view of the privileged status in nature of man and his earth. When the ether vanished, there vanished with it any hope that the earth, or even our solar system or our galaxy, could be a preferred frame of reference in the universe. The empty space of Einstein which replaced the ether of Newton and Maxwell cast man adrift from any anchor in the world. And with the vanishing of a spatial anchor went the vanishing of a temporal anchor, as absolute time gave way to relative time and to the mixing of space and time.

The other vital aspect of relativity's impact on man's view of his place in the universe can be summed up by the phrase: Relativity violates common sense. Common sense, or intuition, or self-evident truth, derives from man's perceptions of the world about him. The lesson of relativity has been that man's perceptions are an inadequate guide to the description of the scheme of nature. Pretwentieth-century physical science, no matter how revolutionary, no matter how unexpected, remained always visualizable and not out of harmony with common sense. The earth swinging around the sun is not hard to picture, even if our more naive impression is that the sun moves around the earth. A spherical earth is not hard to accept, even if we cannot readily perceive its curvature. Even electric and magnetic fields are visualizable with a little effort, and nothing about them need cause incredulity. But in relativity we encounter both the incredible and the nonvisualizable. The relativity of time, the addition of velocity, the speed-of-light barrier remain incredible because nothing in our experience has ever raised the strange effects of relativity above the threshold of observability. Four-vectors and space-time remain nonvisualizable because we are, like it or not, three-dimensional creatures.

In 1920 it was argued that the next generation of school children would learn about relativity at an early age and would come to accept its concepts and its peculiarities as quite everyday matters not out of harmony with common sense. A generation has passed with no noticeable progress in this direction. This optimistic prediction seems very unlikely of fulfillment at any time, for no amount of early exposure to the theory of relativity can alter the fact that all examples of important application of the theory are outside the domain of human perception. The difficulties of visualization and the noncommon-sense aspects of relativity are as real to the trained scientist as to the layman. The situation is similar for the other great theory of this century, quantum mechanics. Together they suggest that we may be on the threshold of a complete divergence between the fundamental concepts needed to describe the basic laws of nature and the everyday concepts used to describe the world around us. If this is so, the progress of fundamental science may be slowed as man gropes for the concepts needed to probe deeper. On the other hand, the power of mathematics together with the evidence of new experimental findings may lead man on to the new concepts, no matter how strange. Or—and this seems quite unlikely—physics may be passing through a difficult phase to be replaced again by one featuring concrete and familiar concepts, as still deeper and more general theories evolve.

The special theory of relativity has raised a host of questions which for the present must be called philosophical, although they may one day become scientific. Why, for example, does man perceive space and time in such different ways if they are really only parts of the same entity, space-time? Why, if we can move this way or that in space, do we find ourselves moving inflexibly in only one direction through time? Are there other parts of the universe or other eras in the history of the universe when the direction of motion through time was different? If we are three-dimensional creatures imbedded in a four-dimensional world, are there still more dimensions still further removed from our perception? Such questions, apparently idle today, give some indication of the kinds of questions that might be answerable within the framework of science tomorrow.

EXERCISES

21.1. The theory of relativity has brought to our description of nature both more relativity and more invariance. Name **(a)** two quantities considered invariant in classical physics that turned out to be relative (that is, different for different observers); and **(b)** two new invariant quantities not recognized in classical physics.

21.2. **(1)** State in words the meaning of a Galilean transformation and of a Lorentz transformation. **(2)** What does it mean to say that a law of nature is invariant under a transformation? **(3)** Under which transformation is Newton's second law invariant?

21.3. How much mass gets converted to energy in the explosion of a 10–megaton H-bomb? (See Appendix 3 for a conversion factor.)

21.4. An electron emerges from a certain betatron with a total energy of 1 MeV. **(1)** What is its kinetic energy? **(2)** What is its speed?

21.5. **(1)** By how much does the speed of a particle differ from the speed of light if its energy is 1,000 times its rest energy? **(2)** What is the energy in eV of a muon moving at this speed (for a muon, $mc^2 = 105$ MeV)? **(3)** What is the average lifetime in the laboratory of muons moving at this speed (refer to Table 2.1 for the average lifetime of muons at rest)?

21.6. A proton emerging from the Berkeley Bevatron has a total energy of 7 GeV. For calculational ease in this exercise, approximate its rest energy as 1 GeV (a 6% error). Find **(a)** its kinetic energy; **(b)** its momentum; and **(c)** its speed. What is its "effective mass," defined as the ratio of its momentum to its speed? How does this compare with its rest mass?

21.7. The wavelength of light from a certain laser is 6,328 Angstroms. What is the momentum of each of its photons?

21.8. An earth satellite in a low orbit travels at a speed of about 5 miles/sec. What fractional error in a calculation of its momentum results from use of the classical formula $p = mv$? Express your answer first algebraically, then numerically.

21.9. From the definitions of energy and momentum, Equations 21.7 and 21.11, obtain an expression for the ratio E/p. Compare this with the ratio E/p for a massless particle as implied by Equation 21.20. Comment on the connection.

21.10. Two electrons, each of total energy 300 MeV, collide with equal and opposite velocity. What is the maximum possible mass of newly created particles in this collision that is consistent with the conservation laws of energy, momentum, and electron family number? Explain how each of these three conservation laws is relevant.

21.11. Equations 21.7 and 21.11 define relativistic energy and momentum. **(1)** Combine these two equations to eliminate the speed v and obtain a formula for E in terms of m, p, and c. (HINT: Square both equations and regard v^2 as the variable to be eliminated.) **(2)** Under what circumstances is $E = pc$ a good approximation to your result?

21.12. **(1)** Assume that the object in Figure 21.2 is moving slowly so that the nonrelativistic definition of kinetic energy suffices. Its kinetic energy before emitting the photons is $\frac{1}{2}m_1v^2$ and afterwards is $\frac{1}{2}m_2v^2$. Make use of Equations 21.3 and 21.4 to show that its loss of mass, $\Delta m = m_1 - m_2$, is given *approximately* by Equation 21.5, $\Delta m = \Delta E/c^2$. **(2)** Describe the kinetic energy of the object in Figure 21.3 by the relativistic definition, Equation 21.8. Again using Equations 21.3 and 21.4, show that its loss of mass is given *exactly* by Equation 21.5.

21.13. Answer in physical terms: *Why* can no material particle be brought from rest to a speed which exceeds the speed of light?

21.14. After emitting a photon, object A in Figure 21.4 has mass $m - \Delta m$. After absorbing the photon, object B has mass $m + \Delta m$. **(1)** Show that if the center of mass of the pair of objects remains fixed, $\Delta m = xm/(x + d)$, where x and d are distances defined in Figure 21.4. **(2)** Explain why, for sufficiently massive objects, it is a good approximation to write $\Delta m \cong xm/d$. **(3)** The momentum of the slowly recoiling object A is $p = mv$. The momentum of the photon is related to its energy E_γ by $p = E_\gamma/c$. With the help of the law of momentum conservation and the results of parts (1) and (2), show that $\Delta m \cong E_\gamma/c^2$.

21.15. Fill in the algebraic details omitted in the text to verify that Equation 21.18 follows from Equation 21.17.

21.16. A neutral pion moving eastward decays into two photons; one flies to the east, one to the west. The eastbound photon has twice the energy of the westbound photon. Prove that the pion moved at one third the speed of light before its decay, (HINT: Both energy conservation and momentum conservation are relevant.)

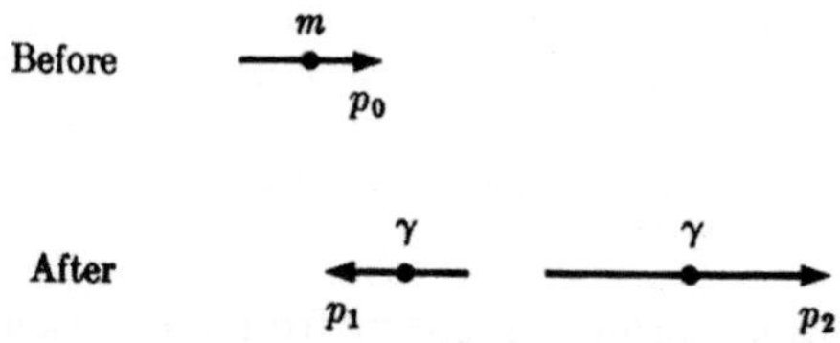

21.17. **(1)** As shown in Figure 21.6, a neutral kaon at rest decays into two charged pions. Refer to Table 2.1 for the masses of these particles, and calculate the speed v of the created pions. **(2)** If a neutral kaon moves through the laboratory at such a speed that one of the two pions resulting from its decay remains at rest in the laboratory (see Figure 21.7), with what speed v_1 does the other pion fly away?

21.18. Just as a gun recoils when fired, a flashlight recoils when it is shining. Consider a flashlight whose mass is 1 kg, initially at rest in an inertial frame of reference, switched on by remote control. What recoil speed does it acquire after 5 sec if, during that time, the power in the narrowly directed light beam is 0.1 watt?

21.19. A 20-GeV electron is "superrelativistic" because its total energy is very much greater than its rest energy, and it moves very nearly at the speed of light. **(1)** What is the momentum of such an electron in gm cm/sec? **(2)** What is the speed of a 2-gm ping-pong ball with the same momentum as the electron? **(3)** What is the kinetic energy of this ping-pong ball? **(4)** What is the total energy of the ping-pong ball? Express the last two answers in GeV in order to compare them with the energy of the electron.

21.20. **(1)** Prove that the momentum-energy four-vector of a massless particle is a null vector. **(2)** What other null vector is associated with a propagating photon or neutrino?

21.21. From Equations 21.22 to 21.25, derive Equation 21.27, which expresses the invariance of length of the momentum-energy four-vector.

21.22. For the kaon decay illustrated in Figures 21.6 and 21.7, evaluate the combination $E_T{}^2 - p_T{}^2c^2$ in the laboratory frame, where E_T and p_T designate total energy and total momentum. Evaluate the same combination in the moving frame and show that the combination is invariant, although neither of its terms is separately invariant.

21.23. Anton (twin A) travels about the galaxy at 80% of the speed of light while Bertram (twin B) remains on earth. When Anton rejoins Bertram after 20 years of earthtime, which twin is older, and by how much?

21.24. An astronaut is in a circular orbit around the earth at a speed of 7.7×10^5 cm/sec. As he passes over a ground station he synchronizes his clock with an identical one on the ground. After traveling around the earth in 90 minutes he compares his clock with the one on the ground. **(1)** Is his clock ahead or behind the earth-based clock? By how much? **(2)** If the clocks are reliable to within 0.1 sec per year, would the difference be experimentally detectable?

21.25. Write a description of the two twin experiments illustrated in Figure 21.9 (1. B changes trains and returns to A who waited at the station; 2. A overtakes the train in a fast motorcar and joins B) from the point of view of a train conductor T, who considers his train to be the true preferred frame of reference. On both occasions the conductor remains on the train moving to the right at constant velocity.

21.26. A proton ($mc^2 \cong 1$ GeV) has a total energy of 7 GeV in the laboratory. **(1)** Show that the product of its laboratory momentum and the speed of light (pc) is also nearly equal to 7 GeV. **(2)** This proton approaches another proton at rest. In a certain frame of reference—called the center-of-momentum frame—these two protons have equal and opposite momentum. Show that this frame of reference is moving with respect to the laboratory at a speed of approximately 87% of the speed of light, (HINT: How is the speed of the center-of-momentum frame related to the speed v_1 of the protons in this frame?)

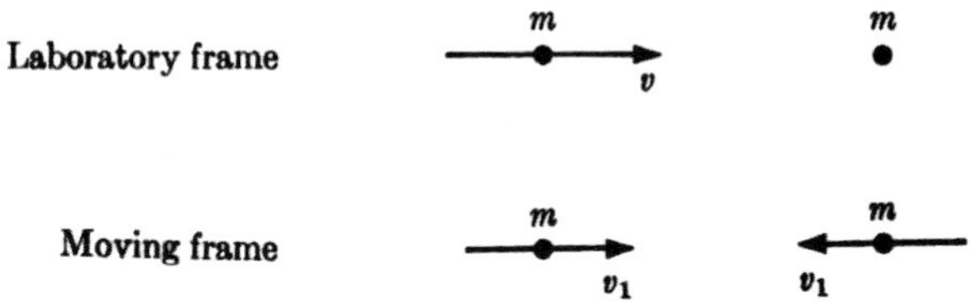

21.27. The high-energy proton of Exercise 21.26 strikes a stationary proton, and several new particles are created. The whole collection of particles moves forward together at the same speed, $0.87c$. **(1)** What is the total mass of all the particles moving together after the collision? **(2)** Compare the energy equivalent of the newly created mass with the kinetic energy of the original proton. Why are these two energies not equal?

21.28. Momentum and energy are relative quantities which obey Lorentz transformation equations almost identical to those obeyed by space and time coordinates. If a particle moves along the x-axis with momentum p_x and energy E in a "stationary" frame of reference, its momentum p_x' and total energy E' in a frame moving in the positive x direction with speed v are given by

$$p_x' = \frac{p_x - vE/c^2}{\sqrt{1 - v^2/c^2}},$$

$$E' = \frac{E - vp_x}{\sqrt{1 - v^2/c^2}}.$$

(Note that the speed v in these equations is the speed of one frame of reference relative to the other, not the speed of the particle in either frame.) **(1)** Verify the dimensional consistency of both equations. **(2)** Combine the equations to show that they yield the invariance equation, $p_x^2c^2 - E^2 = p_x'^2c^2 - E'^2$.

21.29. A particle of mass m is at rest in the laboratory. In a frame of reference moving to the right at speed v (positive x direction), it appears to move to the left at speed v. **(1)** What are its total energy E and its momentum p_x in the laboratory frame? **(2)** Apply the Lorentz transformation equations given in Exercise 21.28 to find its energy E' and momentum p_x' in the moving frame.

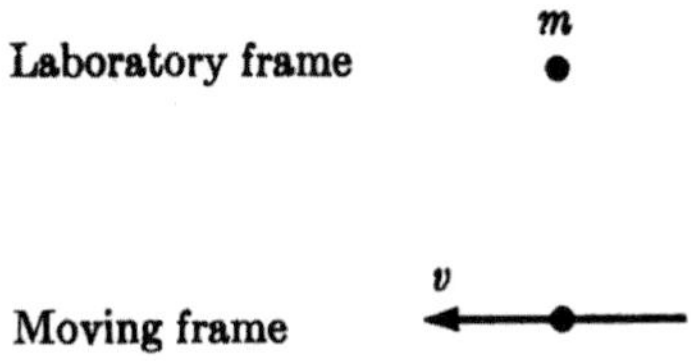

21.30. In the laboratory frame of reference, a photon moves in the positive x direction with momentum p and energy $E = pc$. Its energy and its frequency are related by $E = hf$. In a frame of reference moving along the x-axis with speed v, it has momentum p', energy E', and frequency f'. With the help of one of the Lorentz transformation equations given in Exercise 21.28, prove that the frequencies in the two frames of reference are related by

$$f' = f\frac{1 \pm v/c}{\sqrt{1 - v^2/c^2}}.$$

When would the plus sign be appropriate, when the minus sign? Such a shift of frequency from one frame of reference to another is called a Doppler shift.

21.31. Certain photons emitted by hydrogen atoms on earth have an energy of 10.2 eV. Presumably the same kind of photons with the same energy are emitted by hydrogen atoms on quasistellar objects (the Principle of Relativity). Yet the energy of such photons from the cosmos, measured on earth, may be as little as 3.4 eV. **(1)** According to the relativistic Doppler shift formula given in Exercise 21.30, what is the relative velocity of two frames of reference in which the energy of a given photon differs by a factor of three? **(2)** Is the quasistellar object moving toward or away from the earth? **(3)** A photon of energy 10.2 eV has a wavelength of 1,216 A. What is the wavelength of the "red-shifted" photon of energy 3.4 eV? In what part of the spectrum do these photons lie?

21.32. Write a brief essay answering the question: What do the theories of electromagnetism and relativity have to do with each other? Consider in particular the historical impact of electromagnetism on relativity and the theoretical impact of relativity on electromagnetism.

TWENTY-TWO

General relativity

The special theory of relativity, in spite of its all-embracing scope as a theory of theories, is indeed "special" in the sense that it describes transformations of observations only among inertial frames of reference, and imposes an invariance requirement on physical laws only in inertial frames. This circumstance, of course, need not be regarded as a *defect* of the theory, any more than Newton's laws of motion were regarded as defective for the same reason—since they too apply directly only to motion in inertial frames. Nevertheless, Einstein did regard it as a defect, and his conviction that the laws of nature should be expressed in a form invariant in *all* frames of reference, accelerated or not, was the primary motivating force which led to the general theory of relativity, a structure of magnificent beauty and simplicity from the mathematician's point of view, yet more difficult to understand, interpret, and apply than any other theory in the history of science.

22.1 Inertial forces and gravitational forces

In order to generalize relativity, it is necessary to think about observations in non-inertial, 121
that is, accelerated, frames of reference. In fact we are all experienced observers in noninertial frames, and we know that things are not quite normal when our frame of reference is accelerated. A high-speed elevator when starting or stopping produces an odd but unmistakable feeling in the passengers. An automobile rounding a corner "throws" the passengers to one side. The same car slowing down throws the passengers forward. The pushing of passengers this way and that inside an accelerated car is accomplished by "inertial forces." If we are trained in Newtonian mechanics, we know that inertial forces are not real forces at all, but only apparent forces arising from the car's motion. Our being thrown forward is only a manifestation of our inertial tendency to keep moving as the car stops.

FIGURE 22.1 Inertial force. In a decelerating car, all passengers are thrown forward with equal acceleration. In the car-based frame of reference, each experiences an inertial force proportional to his own inertial mass.

Yet what if we were inside a car with blackened windows and could not see the action of the driver or hear the engine? How could we distinguish real forces from inertial forces? The first thing to note about inertial forces is that they are proportional to mass. If an adult and a child are in an automobile that is suddenly slowed (Figure 22.1), they will be thrown forward side by side with the same acceleration relative to the car, a quite obvious fact to an outside observer who sees only the car being decelerated and all its occupants, large or small, tending to continue forward with constant velocity. But to the passengers within, the force is real enough, and they would decide that inertial forces are proportional to mass. For if the child received, let us say, the *same* inertial force as the adult, the child would be catapulted much more violently forward, owing to its smaller inertial mass. In fact the apparent inertial forces are just equal to the *inertial masses* of the passengers multiplied by the negative of the acceleration of the vehicle:

$$\mathbf{F}_{\text{inertial}} = -m_{\text{I}}\mathbf{a}_{\text{car}}\,. \tag{22.1}$$

The proportionality of the apparent inertial force to mass calls to mind the real gravitational force, which has the same property, except that for gravity it is the gravitational mass and not the inertial mass to which the force is proportional. However, gravitational mass and inertial mass had been known for a long time to be equal. The experiment of Roland von Eötvös in 1890 had established the equality to within one part in 10^8, making even more remarkable what in pre-relativity science had to be regarded as a coincidence. (Of course no one believed that the equality of the two kinds of mass really was a coincidence, but no explanation for the equality had been advanced.) The Eötvös experiment played for general relativity a role similar to that of the Michelson-Morley experiment for special relativity. In each case the *absence* of a measurable effect or difference was the key result, and in each case the null result is directly related to a central postulate of the theory. The Michelson-Morley experiment was accounted for by the second postulate of special relativity, the constant speed of light. The Eötvös experiment underlies the Principle of Equivalence.

22.2 *The Principle of Equivalence*

122 The idea that the inertial property of matter is related in some way to gravity predates even special relativity. The first attempt to draw from this possible connection a definite quantitative prediction was made by Max Planck in 1907, after special relativity had linked energy and inertial mass. Perhaps, said Planck, the same link exists between energy and gravitational mass. If so, pure energy, even without mass, should experience gravitational force and exert gravitational force. In the same year Einstein formulated similar ideas in a bolder way. Inertial effects and gravitational effects, said Einstein, are not merely closely related, they are identical and indistinguishable. The observers in the closed car should be unable by any means whatever to ascertain whether the forces they feel are true gravitational forces or apparent inertial forces. This hypothesis of Einstein, called the Principle of Equivalence, formed a cornerstone of the theory of general relativity which evolved over the next eight years. An immediate consequence of this postulate is the equality of inertial and gravitational masses, for otherwise the difference in these two kinds of mass would afford a means of distinguishing gravitational and inertial forces.

A few examples may help to clarify the meaning of the Principle of Equivalence. Consider first an elevator falling freely down a very long elevator shaft without friction (Figure 22.2). To observers within the elevator, there are no external forces at all. A ball released from the hand will not fall to the floor of the elevator, for it will fall toward the earth just as the elevator itself and the passengers are falling, and to the passengers it will appear to float in air. Relative to the elevator, the ball satisfies Newton's first law, and the passengers, pursuing this and other experiments as they fall, will conclude that they are in an inertial frame of reference. Not at all, say the ground-based observers. The elevator is an accelerated system. It just happens that upward-acting apparent inertial forces cancel out the downward-acting real gravitational forces, and the passengers are deluded into believing themselves to be in an inertial frame with no forces acting. The "weightless" condition of astronauts in orbit about the earth arises exactly from this cancellation of inertial and gravitational forces.

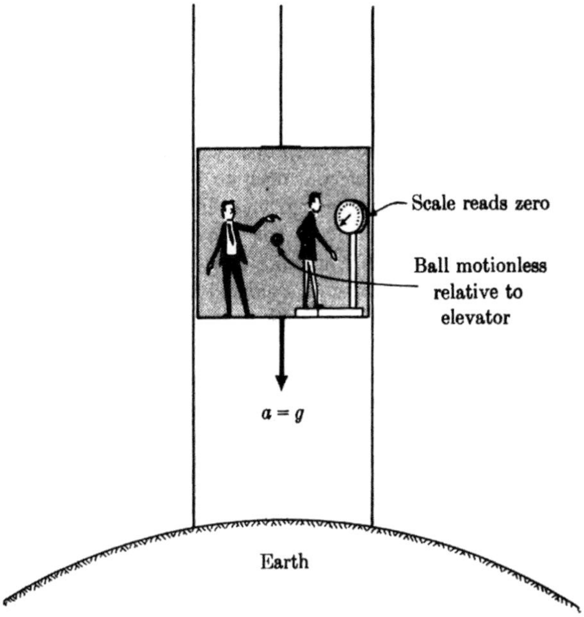

FIGURE 22.2 A freely falling elevator is an inertial frame to its occupants. The cancellation of upward inertial forces and downward gravitational forces in the accelerated frame of reference makes it indistinguishable from a true inertial frame floating free in gravity-free space.

Or consider a space ship, far from any sources of gravitational force, being uniformly accelerated by its rocket engine [Figure 22.3(a)]. This time a ball dropped from the hand will fall with uniform acceleration to the floor of the cabin. From the point of view of an outside observer, the ball remains motionless in space, but the cabin floor is accelerated upward to meet it. He recognizes that there are no real forces acting, only apparent forces arising from the acceleration of the spaceship. To observers within the spaceship, however, these forces seem quite real. The crew might decide that they are motionless, parked on some planet [Figure 22.3(b)]. According to Einstein's Principle of Equivalence, there is no experiment which the occupants of the falling elevator or the accelerating rocket could carry out that would unambiguously decide in favor of their point of view or the viewpoint of the outside observers. That gravity and acceleration through space produce identical physical effects is the startling assertion of the Principle of Equivalence.

FIGURE 22.3 An accelerated spaceship in gravity-free space **(a)** is, to its occupants, indistinguishable from a spaceship sitting motionless in the gravitational field of a planet **(b)**. (The planet, of course, must be chosen so that the acceleration of free fall at its surface, g, matches the acceleration, a, of the first spaceship.)

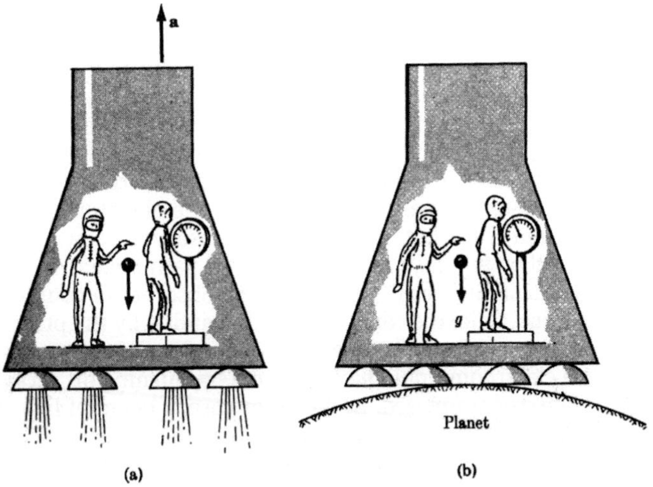

To return to the passengers being thrown forward in the slowing automobile, Einstein's principle means that an enormous mass placed in front of a stationary car would pull the passengers forward in a way indistinguishable from the effect of suddenly slowing a moving car (Figure 22.4). But we know that inertial forces are unreal. They are properties of the motion of our frame of reference. We only think we are experiencing a force because of the acceleration of our frame of reference. In short, the inertial force is not a force at all, but an effect associated with the behavior in space and time of our frame of reference. If we accept the Principle of Equivalence, we must accept the idea that gravitational forces also are "unreal," and are in some way merely properties of space and time. This is Einstein's point of view, to which we shall return after discussing several predictions of the Principle of Equivalence that have been verified experimentally.

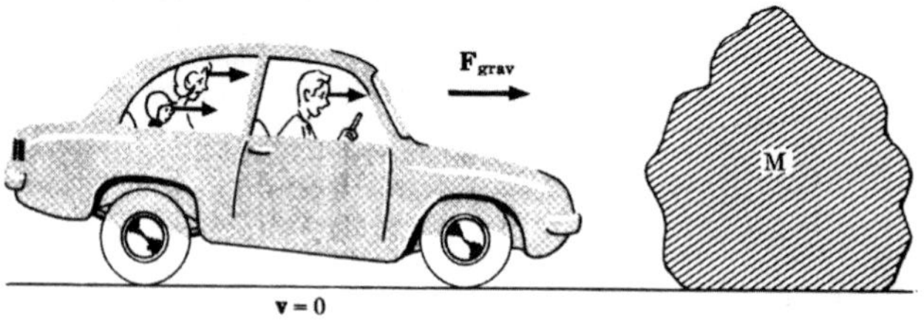

FIGURE 22.4 If a hypothetical enormous mass is placed in front of a stationary automobile, its passengers can be accelerated forward in a way that, to them, is indistinguishable from the effect of slowing a moving automobile (Figure 22.1).

22.3 The bending of light by gravity

123 Returning to the example of the elevator that was falling freely down its shaft (Figure 22.2), let us suppose that the outside observer, in his zeal to enlighten the elevator passengers and convince them that they are not in an inertial frame of reference, hits upon a clever stratagem. He arranges with the passengers to make a pinhole in one side of the elevator and to erect within the elevator several equally spaced translucent screens which will enable the passengers to see what route a narrow beam of light follows through the elevator. He then shines a narrow beam in horizontally through the pinhole. It makes its way across the elevator at the speed of light, leaving a visible spot on each screen as it passes. Since the elevator is falling with ever increasing speed, the distance it falls as the light travels from the second to the third screen is greater than the distance it falls as the light moved from the first to the second screen, and so on for the successive intervals. According to the outside observer, the light, executing straight line motion through space, should leave a track within the elevator which seems to be an upward curving parabola (Figure 22.5). Since light could hardly travel in a curved path in an inertial frame, this should convince the passengers that they are in an accelerated frame. But according to the Principle of Equivalence, there is no physical way for the elevator passengers to determine whether their frame is truly inertial or one in which gravitational and inertial forces are both present but cancel out. The only light path consistent with the principle is a straight line within the elevator (Figure 22.6). But a straight line to the passengers is a downward-curving arc to the outside observer. The result of the experiment will not be that the elevator passengers become convinced of the error of their ways, but that the outside observer learns to his surprise that his beam of light is attracted by gravity, following a curved path instead of a straight line.

This surprising result for the outside observer of course does not violate the Principle of Equivalence. Quite the contrary, it is demanded by the principle. The equivalence of gravitational and inertial effects means that any effect of the earth's gravity observed, for example, within a classroom, could be duplicated by "turning off" gravity and instead accelerating the whole classroom upward. In the latter case, a beam of light moving horizontally should appear to be curving downward. Therefore in the former case (gravity turned on again) the light beam should also "appear" to be curving downward, that is, it should be attracted by gravity.

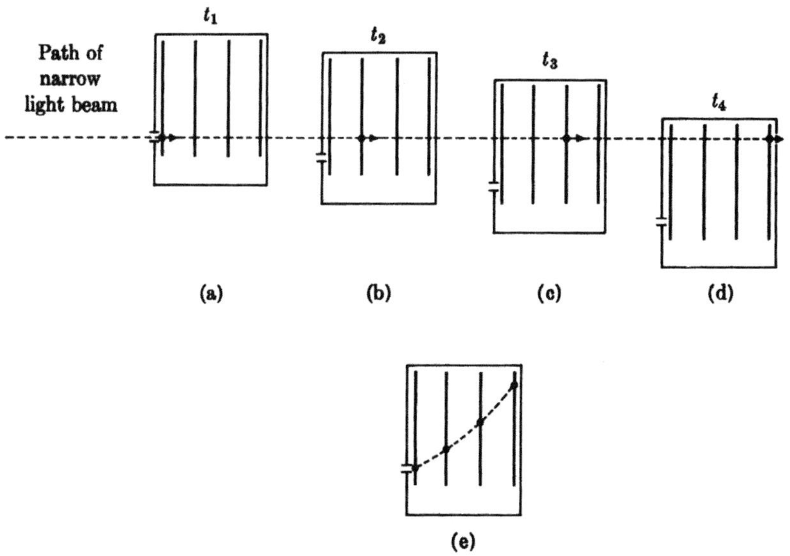

FIGURE 22.5 Path across translucent screens in falling elevator predicted by outside observer who believes that light travels in a straight line in his frame of reference. **(a)** through **(d)**: Successive positions of elevator at equally spaced time intervals as light beam crosses screens. **(e)** Apparent curved path seen by passengers if the outside observer's contention is correct.

The gravitational deflection of light was implied by Planck's hypothesis in 1907 that all energy should experience gravitational force (and exert gravitational force). Yet the phenomenon was apparently not appreciated until 1911, when Einstein pointed out that it was to be expected on the basis of the Principle of Equivalence. Moreover, he suggested a particular method of detecting the effect. Starlight passing close to the sun should experience some deflection, and the apparent position of the star should accordingly be slightly shifted relative to its position at some other time of the year when its light received at the earth does not pass near the sun (Figure 22.7). Unfortunately the sunlight itself is so intense that it was impossible at the time to observe stars whose position in the sky is near the sun. Not until 1919 did a total eclipse of the sun make possible the brief but decisive measurements of apparent positions of stars near the sun that verified Einstein's prediction that light (energy without mass) is deflected by gravity. Neither these measurements nor those that followed at later eclipses were accurate enough to test with precision the exact magnitude of the effect. Now at last, after more than half a century, new techniques are available to measure accurately the apparent positions of stars near the sun. These may soon yield greatly improved data on the bending of light by gravity.

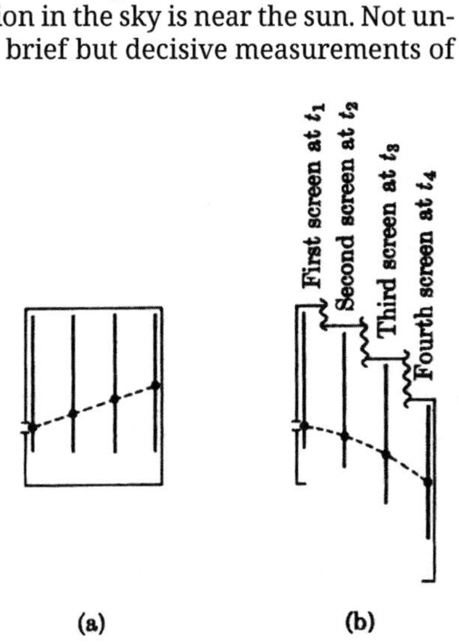

FIGURE 22.6 (a) Actual straight path of light seen by elevator passengers is consistent with other evidence that their frame is equivalent to an inertial frame. **(b)** Superimposed view of elevator position at equally spaced time intervals shows the downward deflection of the light beam seen by the outside observer.

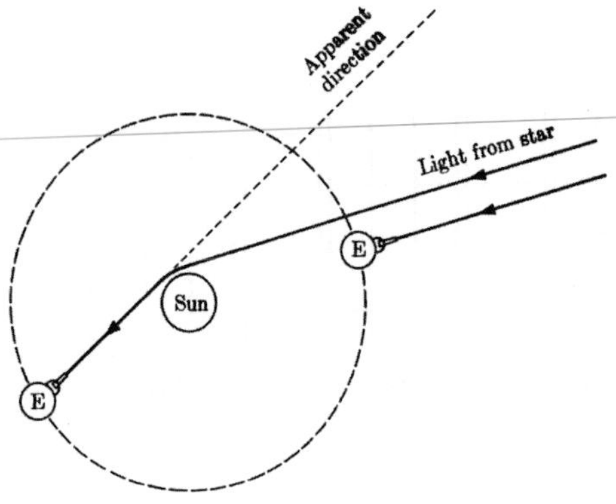

FIGURE 22.7 Deflection of starlight by the sun. The diagram enormously exaggerates the effect, which in fact amounts to less than 2 seconds of angle (5.5×10^{-4} deg).

22.4 *Gravity and time: the gravitational red shift*

Two other famous predictions of general relativity can be approximately understood in terms of the equivalence principle as applied to length and time measurements in accelerated frames of reference. In order to understand these we will consider a rotating frame of reference (which is, of course, an accelerated frame) instead of a frame uniformly accelerated in a straight line. To stay with familiar objects, let us replace the elevator by a merry-go-round. Consider two clocks on the merry-go-round, one at its center and one at its edge. From special relativity we know that the clock at the center, since it is not moving with respect to the ground, should run at the same rate as ground-based clocks. But the clock at the edge is in motion with respect to the earth and should therefore be observed to be running more slowly than ground-based clocks. This tells us at once that the two clocks on the merry-go-round, although attached to the same frame of reference, do not run synchronously, but the outer clock loses time with respect to the inner clock (Figure 22.8). This difference would be noticeable to observers on the merry-go-round as well as to those on the ground. A ground-based observer attributes the slower rate of the peripheral clock to its motion. To an observer on the merry-go-round, however, neither clock is in motion. He must attribute the difference to something else. In his frame of reference there does exist one marked difference in the environments of the two clocks. At the edge of the merry-go-round is a strong "centrifugal force" (an inertial force arising from the accelerated motion), but no such force at the center. He therefore decides that in a region of stronger force, clocks run more slowly. Applying the Principle of Equivalence, we must conclude that clocks in strong gravitational fields of force run more slowly than clocks in weak fields of force. Einstein predicted such an effect at the same time he formulated the Principle of Equivalence in 1907, and suggested a way to observe it. All atoms emit light of certain well-defined characteristic frequencies associated with the rate of vibration of the electrons within the atom. Each atom is therefore itself a clock, and the slowing down of clocks—that is, the slowing of all physical processes—would be manifested by a slowing down of atomic vibrations. A particular kind of atom on the sun, where the gravitational force is strong, should therefore emit light of a lower frequency (slower vibration) than that emitted by the same kind of atom on the earth. This effect was indeed observed in sunlight, but various disturbing influences made it hard to measure, and the gravitational red shift,* as this effect is called, was in fact never reliably verified until 1960, when an entirely new technique made possible an accurate measurement on earth.

* Since red light is at the low-frequency end of the visible spectrum, a lowering of frequency of visible light shifts the color toward the red. The term "red shift" has come to be used for any lowering of frequency even if the radiation is not visible light.

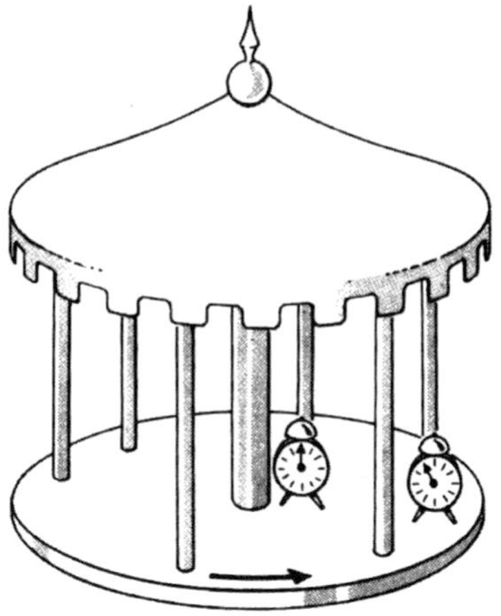

FIGURE 22.8 Clocks in an accelerated frame of reference. The moving clock at the periphery runs slower than the clock at the center. In the moving frame of reference the peripheral clock experiences an inertial force. Since gravity can duplicate the effects of acceleration of the frame of reference, a clock must run slower in a strong gravitational field than in a weak field.

Usually when an atom shoots out a photon of light, it recoils like a gun so that some of the energy that was scheduled for the photon is drained off by the recoiling atom. But less energy for the photon means less frequency. The act of emitting the photon has lowered the photon frequency, producing a "recoil red shift." In the comparatively weak gravitational field on the surface of the earth, the recoil red shift is so much greater than the gravitational red shift that it was never possible to measure the tiny effect of gravity superposed on the big effect of recoil. However, in 1957, Rudolf Mössbauer in Heidelberg, Germany, discovered a way to eliminate the recoil red shift entirely. Under certain conditions, atoms held within solid crystals at low temperature may be so tightly glued in place that they are unable to recoil when they emit a photon. Atoms in this condition are arranged shoulder to shoulder so that the outer electrons usually responsible for emitting the atom's characteristic photons are no longer able to do so without serious interference from their neighbors. This would negate the benefits of having the atoms held rigidly in place were it not for the fact that the isolated and well-protected nucleus deep within the atom is also capable of emitting photons (gamma rays). The nucleus is fortunately held rigidly in place without being disturbed, so that it may emit its characteristic photons at their full frequency, free of any recoil red shift. 124

Robert Pound of Harvard, and simultaneously a group in England, realized that this "Mössbauer effect" should make possible at last an accurate test of Einstein's 1907 prediction. If a nucleus identical to the emitting nucleus is used as a receiver of the photon, the receiving nucleus will strongly absorb the photon only if the time scales of the two nuclei, emitter and absorber, are the same and if no effect has altered the photon frequency. An exceedingly minute change of frequency (or of time scales) will be at once noticeable because the receiving nucleus will be less well able to absorb the photon. Pound and an assistant, G. A. Rebka, Jr., were the first to achieve reliable results, and announced their findings in the summer of 1960. To within an accuracy of 3%, Einstein's prediction was exactly verified. Pound and Rebka worked in a shaft within their laboratory, separating the receiving nucleus and the sending nucleus vertically by 70 feet. According to the Principle of Equivalence, the time scale for the lower nucleus (in the slightly stronger gravitational field) should be dilated with respect to the time scale for the upper nucleus by about two parts in 10^{15}. This almost inconceivably small effect corresponds to the lower clock losing 1 second with respect to the upper clock in fifteen million years. At no time before in the history of experimental science has such a small effect been measured. If a Texas millionaire lost a penny, the fractional change of his fortune would be ten million times greater than the fractional change of time scales measured by Pound and Rebka. 125

It is interesting that this red shift can be looked at from an entirely different point of view, making use of Einstein's equation relating the energy, E, of a photon to its frequency, f: $E = hf$, where h is Planck's quantum constant. Since a photon has energy, it will be influenced by gravity. A photon leaving the sun will be "retarded" by the sun's gravity and will lose energy as it flies away, just as a ball thrown straight up loses kinetic energy because of the pull of the earth. The photon's loss of energy will be equivalent to a loss of frequency, and it will arrive at the earth with less frequency than a brother photon emitted on the earth. This different viewpoint serves to illustrate the inner consistency of general relativity, and the fascinating way in which the concepts of space-time, gravity, and energy are linked together.

22.5 *Gravity and space: the motion of Mercury*

Returning to the merry-go-round, we may notice an effect on distance measurements quite analogous to the effect on time measurements. A ruler laid along the circumference of the merry-go-round will appear contracted to the observer on the ground, while a ruler near the center, moving much more slowly, will be nearly unaffected. Again making use of the Principle of Equivalence, we must conclude that distance measurements will be dependent upon the strength of a gravitational field, even if no relative motion is involved. Recall that the rotating merry-go-round is equivalent to a stationary merry-go-round with a strong gravitational field near its edge and a weak gravitational field near its center. This dependence of distance scales on the strength of gravitational force has an effect on the motion of planets about the sun. Exactly what the effect is is by no means easy to see, and Einstein was not able to predict the corrections to planetary motion based on the equivalence principle until he had arrived at a complete mathematical formulation of gravity and of the properties of space-time in 1915. He then found that only the innermost planet, Mercury, which experiences the strongest gravitational force, should be measurably influenced. Mercury moves in an elliptical orbit which carries it periodically into regions of weaker and stronger gravitational force, so that, according to the consequences of the equivalence principle that have just been outlined, it experiences periodically varying properties of space and time. The effect on its motion is exceedingly small, but thanks to the great precision of astronomical observation, quite easily measurable. It is predicted that the entire ellipse representing Mercury's orbit will itself slowly rotate about the sun, so that successive trips about the sun will trace out slightly different paths in space (Figure 22.9). Actually the rotation of Mercury's ellipse is caused only in part by the effect of general relativity. Venus, Earth, and other planets also perturb Mercury's motion and contribute to the slow rotation of its orbit. The general-relativity contribution amounts to about one one-hundredth of a degree per century, or one extra revolution in three million years. This effect had been known, and unexplained, for many decades. Einstein's prediction was therefore immediately verified by past observation.

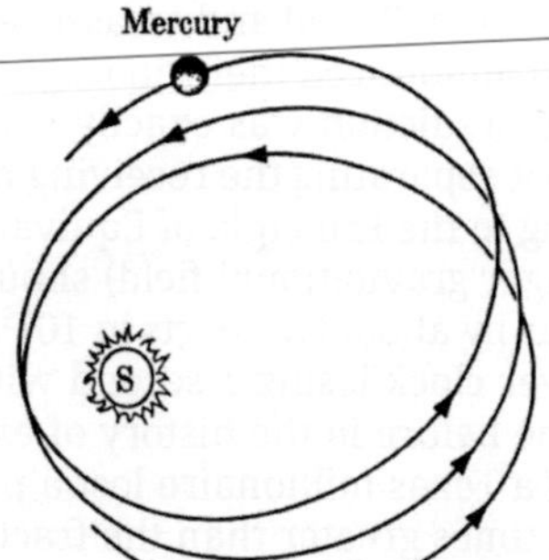

FIGURE 22.9 Precession of the orbit of Mercury. One of the important contributors to the slow rotation of Mercury's orbit is the effect of the sun's gravitational field on the properties of space-time. In this diagram, both the eccentricity of the orbit and its rate of turning are greatly exaggerated.

Until 1967 the rate of precession of Mercury's orbit was believed to be the most accurate and most crucial test of general relativity. Then a discovery by Robert Dicke and H. Mark Goldenberg in Princeton that the sun is not precisely round introduced a slight blemish in the picture. Because the sun is not quite a perfect sphere, its gravitational field does not follow exactly an inverse square law. This in turn contributes to the precession of Mercury's orbit. Including this new effect along with all other known effects and the prediction of general relativity, there remains a tiny discrepancy between theory and observation—about one thousandth of a degree per century. Perhaps the "known" effects are not well enough known, and improved calculations will wipe out the discrepancy. Perhaps general relativity will require modification. Its important contributions to our understanding of gravity and space-time will surely not be abandoned, however, nor will the now solid Principle of Equivalence.

Another interesting point appears in considering length measurements on the merry-go-round—geometry itself must be altered in an accelerated frame of reference (or in a fixed frame with a gravitational field). A distance measurement along a radius of the merry-go-round should be unaffected by the motion, because the motion is perpendicular to the radius. But distance measurements around the circumference *are* affected. The *ratio* of circumference to diameter will no longer be the fixed constant π (3.14159 . . .) but some other number (Figure 22.10). This surprising conclusion is another in the list of violations of common sense that relativity has introduced. It just means that gravity causes space to be non-Euclidean; the ordinary laws of Euclidean geometry taught in high school are no longer valid. For example, circles need not all have the same ratio of circumference to diameter, and right triangles need not satisfy Pythagoras' theorem. It must be borne in mind, of course, that the deviations from "common sense" predicted by relativity in our everyday world are entirely too small to be observed. (If that were not so, common sense might be quite different!)

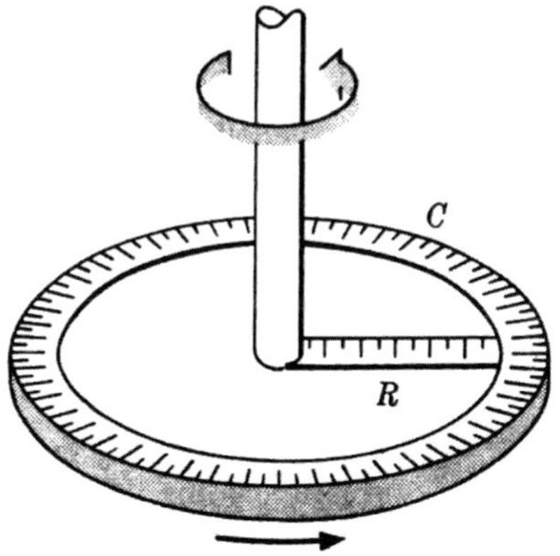

FIGURE 22.10 Rulers in an accelerated frame of reference. The equation of Euclidean geometry, $C = 2\pi R$, is not precisely valid in a rotating frame of reference, or in a region of space containing a nonuniform gravitational field.

22.6 *The new geometry and the new mechanics*

Einstein's final formulation of general relativity in 1915 placed geometry in a central role. The mathematics of the theory is too formidable to present, even sketchily, and we shall have to be content with pointing out a few interesting features of the theory, and indicating some of its implications, both for how we think about the world around us, and for possible future theories of nature. But first let us review in logical order (not so very different from the actual historical order) the crucial steps that lead from special relativity to general relativity. First is the desire to extend the Principle of Relativity to embrace invariance of physical laws in all frames of reference, not just inertial frames. This leads to the consideration of accelerated frames, in which apparent forces, or inertial forces, exist. Because of the equality of gravitational mass and inertial mass, the inertial forces are indistinguishable from real gravitational forces. The Principle of Equivalence makes of this indistinguishability a fundamental postulate. Applying it to the propagation of light, it is seen that gravity must deflect light, a phenomenon independently suggested by the equivalence of mass and energy. Coupling the equivalence principle with the time dilation and Lorentz contraction phenomena of special relativity leads to the conclusion that the properties of space-time depend on the strength of the gravitational force, and leads to the prediction of an effect on the motion of Mercury, and to the prediction of the gravitational red shift (which is independently suggested by the photon concept of light). Finally, since gravity influences the properties of space and time, it is possible and preferable to regard gravity itself as nothing more than a property of space and time. This reduces gravity to a manifestation of space-time geometry and gives a rational basis to the equivalence principle, since it has always been clear that the "unreal" inertial forces are just properties of space and time.

General relativity demands of the imagination more than the imagination is able to give. Special relativity had already demanded that we give up the notion of absolute time and that we try to picture a four-dimensional world of spacetime. To these hurdles general relativity adds a non-Euclidean geometry, or "curved" space-time, and asks us to visualize objects not being acted upon by gravitational forces, but rather responding in their motion to the "curvature" or "warping" of space in their own neighborhood. In classical physics, space was the stage upon which physical phenomena acted out their parts as absolute time rolled independently on. In special relativity, space and time had to be merged, but still were the stage, or perhaps better the canvas, upon which phenomena traced out their world-line histories. In general relativity, space-time is more than the canvas upon which history is painted. It becomes itself an active participant in history, its hills and valleys and warping *being* the phenomena—or at least the phenomena of gravitation. Although the ether was banished by the revolution of special relativity, general relativity has reintroduced a new and subtler "ether"—space-time itself.* Among the challenging questions raised by relativity are: Are space and time manifestations of matter and energy? Would space and time have any meaning in an empty universe?

In order to attempt to visualize curved space-time, we must retreat from four dimensions to two. The surface of the earth is a good example of a "curved space." In any small region, it seems quite flat, and ordinary geometry works quite well. But if we consider figures of large size, it is obviously non-Euclidean. An equilateral triangle with sides about 6,000 miles long (Figure 22.11) has three angles of 90° instead of three of 60°. The Equator is a circle with a circumference exactly twice its diameter instead of more than three times its diameter. These are extreme cases, and a surveyor with sufficiently precise measurements could determine the non-Euclidean character of the surface with much smaller figures.

A two-dimensional worm on the earth's surface would be unable to visualize the curvature, since it requires a third dimension to describe the earth's surface. Nevertheless he could survey the surface with great care and deduce that it was not a plane because the laws of Euclidean geometry were not precisely correct. He might also journey forever away from home in a straight line, only to find himself home again after 25,000 miles. Man in space is like the worm on the earth's surface, able to determine by measurement whether his space is "flat" or "curved" but quite unable to visualize a curved space. According to relativity, our space is not curved uniformly like the earth's surface, but has more curvature near matter and energy and less curvature far from matter and energy. Yet it probably has an overall

* In 1920 Einstein said: "According to the general theory of relativity, space is endowed with physical qualities; in this sense, therefore, there exists an ether." Compare this statement with the quotation on page 682 from Einstein's 1905 paper.

average curvature. It is possible (but not yet shown to be so) that space is closed in upon itself, just as the earth's surface is closed, so that the space traveler flying away forever in a straight line would reappear at the earth some billions of years later.

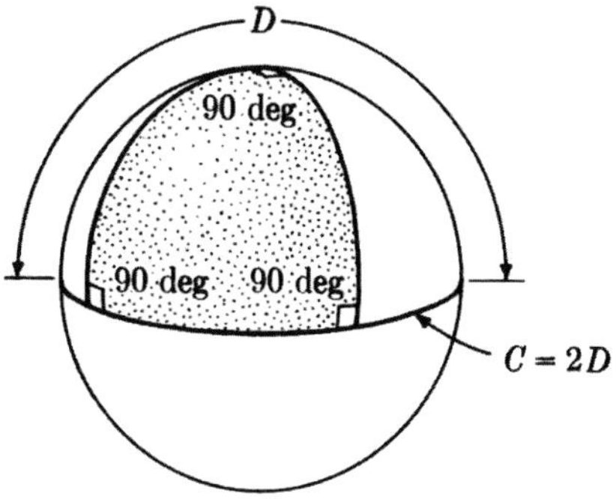

FIGURE 22.11 The surface of the earth, a curved two-dimensional space. Its geometry differs from the Euclidean geometry of a flat space.

This picture of the space traveler executing a closed circuit without ever turning raises the question: What is a straight line? This is a key question, for the law of mechanics in general relativity is (taking into account only gravitational forces): All objects (including photons and neutrinos) move in straight lines. Returning to the surface of the earth, we might define a "straight line" as the shortest distance between two points. On the earth this is usually called a great-circle route. The technical name for it is a geodesic. Einstein's definition of a straight line in space-time is the same. It is the shortest four-dimensional distance between two events. The law of mechanics is therefore a law of cosmic laziness! All particles take the shortest route. In a hilly space, of course, the shortest route might not look like a straight line. The worm on the earth would not get from 34th Street and Broadway to 33rd Street and Fifth Avenue by climbing up one side of the Empire State Building and down the other. He would go around, obeying his own instinct to minimize his path. Looking down from high above, unaware of the building, we would see that the worm did not follow a straight path and might incorrectly conclude that some force had acted upon him, deflecting his path. Three-dimensional man, unable to perceive the hills and valleys of space-time, attributed curved paths to forces until Einstein showed how they might much more simply be accounted for as routes of shortest distance through space-time.

22.7 *The new view of nature*

The view of the world to which general relativity has led us was foreseen in 1870 by Wil- 126
liam Clifford, a British mathematician, who displayed the following remarkably prophetic insight: "I hold in fact (1) That small portions of space *are* in fact of a nature analogous to little hills on a, surface which is on the average flat; namely, that the ordinary laws of geometry are not valid in them. (2) That this property of being curved or distorted is continually being passed on from one portion of space to another after the manner of a wave. (3) That this variation of the curvature of space is what really happens in that phenomenon which we call the *motion of matter*, whether ponderable or etherial. (4) That in the physical world nothing else takes place but this variation, subject (possibly) to the law of continuity." It remained for Einstein, forty-five years later, to bring this vision to fruition as a successful mathematical theory. Not even Clifford could foresee that this success would require the merging of time with space into a single entity. The view of the world represented by Clifford's vision and by Einstein's interpretation of general relativity has recently been given by John Wheeler the name "geometrodynamics" to symbolize the merging together of geometry and the dynamics of motion—the fusion of actor, stage, and action. Whether *all* physi-

cal phenomena will ultimately be described as merely manifestations of the properties of space and time remains today still an open question. So far only the force of gravity has been simply and convincingly merged with space-time.

Since the original motivation for the general theory of relativity was the desire to formulate *all* laws of nature in a manner invariant for observers in all frames of reference, accelerated or not, how is it that *one* law, the law of gravitation, assumed in general relativity a central role? This came about because of the dual nature of mass, on the one hand as inertia, creating apparent forces in accelerated frames of reference, and on the other hand as the source of gravitational force. No other properties of elementary particles seem to have such a dual role. Electric charge, for instance, is a seat of electric force, but it has no "space-time" property like inertial mass which would lead to a geometrical interpretation of electricity as simple as the geometrical interpretation of gravity. This circumstance troubled Einstein and many other scientists who have sought a "unified field theory" in which electromagnetic and gravitational phenomena were tied equally simply and directly to the hills and valleys of space-time. These efforts have not been successful so far. This is not to say that electromagnetism is excluded from general relativity. Space-time curvature responds to *any* energy, including electromagnetic, and a photon will create its little disturbance in space-time as surely as will a particle with mass. General relativity is indeed general, but gravity stands in the theory in a privileged position.

22.8 The universe in the large

127 General relativity is undoubtedly the least tested theory of nature that is widely accepted. Its three famous predictions—the correction to the motion of Mercury, the deflection of light by the sun, and the gravitational red shift—have been verified, to varying degrees of accuracy, but none to very high precision. It has been subjected to no other crucial tests. Yet the beauty and economy of the theory coupled with these few tests have been sufficient to convince most scientists of the correctness of general relativity. Man is rather like the worm on the surface of the earth who has discovered tiny but significant discrepancies between Euclidean geometry and observation, and on this basis asserts that the earth is a sphere. A pragmatic worm could declare: "Since a sphere is nothing I can visualize anyway, and since these discrepancies you report are entirely too small to be important, I think your theory is irrelevant. The discrepancies could probably be explained in some simpler way anyhow." To which the scientist-worm might reply: "The discrepancies, small or not, are nevertheless present. Since the sphere explains them, and since the sphere is such a magnificently beautiful and simple figure, I choose to believe in it."

The sphere of the worms is the universe of man. So minute is the average curvature of man's space that he is normally unaware of it. In our everyday world, and on downward into the submicroscopic world, the manifestations of general relativity are too small to notice (although extrapolation to distances shorter than any measured so far is dangerous, for there new surprises, including a possible significant role of space-time curvature, may await us). But over the enormous distances of intergalactic space, the cumulative effects of space curvature become large—just as to the worm traveling an appreciable fraction of the distance around the earth, the earth's curvature would become significant. For this reason, the most interesting speculations on the consequences of general relativity have been in the cosmological domain. Various "models" of the universe have been proposed. According to one model, the universe is continually expanding away from a highly condensed state which existed some ten or twenty billion years ago. According to another model, the universe is only apparently expanding, and is in a "steady state" in which new matter is continually coming into existence. Each of these models is consistent with general relativity and with astronomical observation. It has not been possible so far to decide whether either of these models or any other model describes the actual universe. Many fascinating questions remain unanswered: Is the universe closed or open? Is it finite or infinite? Is the amount of matter (and energy) in the universe truly a constant? Is there a connection between the world of the very large and the world of the very small? Man has dared to grapple with these questions—as questions of science—over the past few decades, and answers to them may appear within the next half century. It should be remembered that when the "true" picture of the universe emerges, it will be a truth of simplicity, convenience, consistency, and successful predictions, just as is the truth of every other theory of physical science.

EXERCISES

22.1. "Centrifugal force" is an example of an apparent force or inertial force. If a child of mass m stands a distance r from the axis of a merry-go-round turning with angular velocity ω, what is the centrifugal force seeming to act on the child in this accelerated frame of reference? Give its direction and a formula for its magnitude.

22.2. An airplane can perform a maneuver that causes its passengers to experience "weightlessness." **(1)** What is the general nature of the maneuver? Can it last very long? **(2)** How does this method of inducing weightlessness illustrate the Principle of Equivalence?

22.3. **(1)** An upward force is applied to the elevator in Figure 22.2 just sufficient to reduce its acceleration to zero; it continues downward at constant velocity. **(2)** The upward force is then increased for a time to reduce its velocity to zero. **(3)** Having stopped, the elevator hangs motionless in its shaft. How would passengers in the elevator who believe themselves to remain at rest at all times interpret these three situations?

22.4. Imagine a hypothetical world in which the ratio of gravitational mass to inertial mass is not the same for all bodies. Describe an experiment (possible in principle, but not necessarily practical) which would show that the Principle of Equivalence is not valid in this world.

22.5. In each of the two rocket ships depicted in Figure 22.3, the passengers erect a series of translucent screens and an outsider shines in a narrow beam of light. Apply the Principle of Equivalence and discuss the light path in each of the rocket-ship frames of reference.

22.6. A passenger in the rear of a railroad car which is accelerating forward sends photons of a particular frequency in the forward direction. A passenger at the front of the same car measures the frequency of the photons reaching him. If he can measure with unlimited precision, does he measure a frequency which is less than, equal to, or greater than the frequency measured by the passenger at the rear. Why? Invoke the Principle of Equivalence to justify your answer.

22.7. Machines 1 and 2 in the diagram on the following page are hypothetical mass-energy converters of perfect efficiency. Machine 1 absorbs a photon of energy E_1 and from it manufactures a particle of mass $m = E_1/c^2$. The particle, starting from rest, drops a distance h with acceleration g in the earth's gravitational field into machine 2, where its total energy is converted into the energy of a photon. This photon makes its way back to machine 1, where it supplies the energy to make another particle, and so on. Prove that the conservation of energy requires that the photon lose energy on its upward trip, and that its fractional change of energy be given by

$$\frac{E_2 - E_1}{E_1} = \frac{gh}{c^2}.$$

Assume that h is small compared to the radius of the earth and that a nonrelativistic description of the fall of the particle suffices.

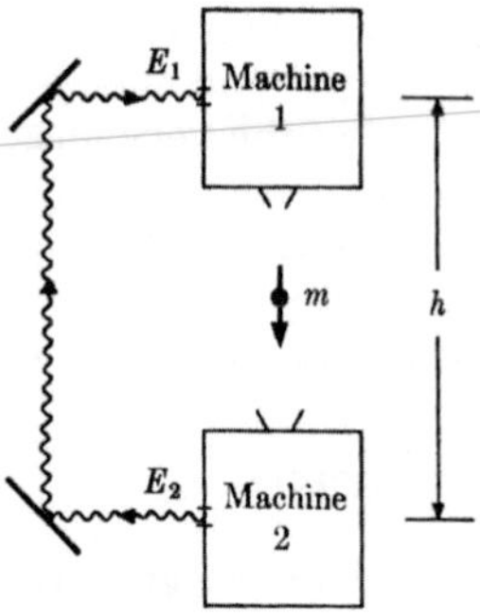

22.8. The implication of the thought experiment described in Exercise 22.7 is that a photon of original energy E loses an increment of energy $\Delta E = Egh/c^2$ in ascending a distance h near the earth. **(1)** Express its change of momentum in terms of E, g, h, and c. **(2)** If force is defined as rate of change of momentum ($F = \Delta p/\Delta t$), what force acts on a photon near the earth? **(3)** Calculate and compare the earth's gravitational force on a visible photon of energy 2 eV and on an electron.

22.9. The equation given in Exercise 22.7 shows the dependence of photon energy on altitude in the gravitational field near the earth. **(1)** Write an analogous equation for photon frequency instead of photon energy. **(2)** In the experiment of Pound and Rebka (Section 22.4), the effect of a 70-foot change of altitude on photon frequency was measured. By what fraction did the frequency change?

22.10. **(1)** The quantity gh/c^2 which gives the fractional change of photon energy with an altitude change h (see Exercise 22.7) also gives the fractional change of time scale with the same altitude change. Why are photon energy and time related in this way? **(2)** A clock suspended in a balloon 10 km above the earth runs faster than a clock below it on earth. How much time does the balloon-borne clock gain in a year?

22.11. An astronaut moves slowly away from the earth toward much more massive Jupiter. Before his departure his transmitter and a receiver on earth are tuned perfectly to the same frequency. Because of the effect of gravity on time scales, he must alter the frequency of his transmitter from time to time in order that its radiation reach earth with the proper frequency for the receiver. In which part of his trip from the surface of Earth to the surface of Jupiter must he transmit at a slightly increased frequency? In which part at a slightly decreased frequency? In which part at the original frequency? Consider the effects of gravity only, not the effects of motion. (NOTE: In practice the effects considered here would not be large enough to be measurable.)

22.12. Explain in a brief paragraph why gravitational force turns out to be the "special" force of general relativity.

22.13. A photon passing near the earth is imperceptibly deflected. An astronaut near the earth is very perceptibly deflected. If both are following the same law, tracing out the "shortest distance in space-time," why are their paths so different? (NOTE: Only a plausible answer to this question is expected. There are two routes to an answer: [1] an argument by analogy, considering low-speed and high-speed motion over a two-dimensional surface that is not flat; [2] a semimathematical argument, based on considering the relative importance of the spatial and temporal parts of the four-dimensional interval for low-speed motion and for high-speed motion.)

22.14. The potential energy associated with the gravitational interaction of a pair of electrons is $-Gm^2/d$, where m is the mass of an electron, and d is the separation of the two electrons. **(1)** At what separation d does this energy have a magnitude of 1 eV? **(2)** At what separation does it have an energy equivalent to the mass of the two electrons, about 1.0 MeV? At this distance, general relativity predicts a drastic alteration in the properties of space and time. **(3)** Compare these distances with characteristic distances now known in the submicroscopic world, and comment.

22.15. Explain why **(a)** the special theory of relativity is vitally important in understanding elementary particle phenomena, and **(b)** the general theory of relativity has appeared so far to be irrelevant in the submicroscopic world.

PART 7
Quantum Mechanics

NIELS BOHR
(1885 – 1962)

"Every scientist is constantly confronted with the problem of objective description of experience, by which we mean unambiguous communication."

WERNER HEISENBERG
(1901 – 1976)

"The two processes, that of science and that of art, are not very different. Both science and art form in the course of the centuries a human language by which we can speak about the more remote parts of reality."

TWENTY-THREE

The key ideas of quantum mechanics

In December, 1900, Max Planck introduced an idea destined to shake the foundations of physics, the idea that material energy can be transformed into radiant energy only in units of a certain size, "quantum units." A dozen years later, Niels Bohr generalized Planck's idea into a quantum principle of nature and used the principle with astonishing success to account for the structure of the hydrogen atom (Chapter Twenty-Four). Bohr's principle is this: Some of nature's variables can take on only discrete values, and accordingly can change only in finite jumps. Building on this principle of discreteness or quantization in nature, Werner Heisenberg, Erwin Schrödinger, Max Born, Wolfgang Pauli, Paul Dirac, and Bohr created in the years 1925-1928 the edifice that we now call the theory of quantum mechanics. The giant stride of physics in those few years has not been equaled since.

23.1 The revolution of quantum mechanics

Like relativity, quantum mechanics is a true revolution in intellectual history. These two theories gave new life to a science that some thought was dying, and their impact on philosophy and literature can be compared only to the impact of the Newtonian revolution more than two centuries earlier. Accompanying each of these twentieth-century theories have been a number of new ideas that are completely contrary to those of classical physical and to everyday experience. Man has been forced to look at nature in a new way, and to recognize that the "common sense" distilled from his sense experiences may have little to do with the deeper sense of nature's design.

Quantum mechanics is, in simplest terms, the theory of the submicroscopic world. It is well to remember that every fundamental theory in physics is associated with a particular domain of nature or with a particular set of phenomena. Classical mechanics describes the motion of material objects, provided, as we now know, that the objects are neither too small nor too swift. Its greatest early triumph was the successful explanation of planetary motion. Electromagnetic theory achieved its greatest success in accounting simultaneously for electricity, magnetism, and light. Quantum mechanics scored first in providing an explanation of atomic structure. Describing the motion of particles, the forces between particles, the union of particles to form atoms and molecules, and the processes of creation and annihilation of particles, quantum mechanics has extended man's powers of prediction and quantitative description downward from the molecular to the subatomic and subnuclear domains. Because the submicroscopic world is vitally important within stars, quantum mechanics has, interestingly, also pushed back the frontiers of astronomy. Together, quantum mechanics and relativity have revealed the boundaries of validity of classical mechanics. So far, no definite limit of validity is known for either quantum mechanics or relativity, but these theories are, in the broad sweep of scientific history, still young.

For reasons both practical and fundamental, we shall focus attention in this part of the book more on ideas and less on equations and logical organization than in what has gone before. More advanced mathematics is required for quantum mechanics than for relativity or for classical physics, mathematics not accessible to most beginning students. We should not on this account abandon the effort to grapple with the ideas of quantum mechanics. We will in fact find it possible to introduce quantitative description at certain key points, the essence of the ideas of Planck and Bohr, for example, being expressible with simple algebra. The practical reason aside, the focus on general ideas and principles is in keeping with the spirit of contemporary physics. The professional physicist himself is given more to emphasizing general principles in the modern period. Some of these take the form of conservation laws (Chapter Four). In Part 6 we encountered a principle of invariance and a principle of equivalence. An important part of quantum mechanics is summarized by the uncertainty principle (Section 23.6).

128 In the remainder of this chapter we shall discuss and illustrate five key ideas of quantum mechanics:

(1) Granularity. Much of nature in the small is granular—its physical variables as well as its material pieces.

(2) Probability. The fundamental laws of quantum mechanics are laws of probability, not laws of certainty.

(3) Annihilation and Creation. Any particle may be annihilated or created; stable matter is a happy chance.

(4) Waves and Particles. Units of matter and units of radiant energy both share wave properties and particle properties.

(5) The Uncertainty Principle. Apart from limitations on human ingenuity, nature imposes fundamental limitations on the precision with which some physical quantities can be measured.

In the chapters that follow, these ideas will be applied in more detail to atoms, nuclei, and elementary particles.

23.2 Granularity

129 No single fact about nature in the small is more striking than its granularity. Not only are the atomic building blocks of matter granular, so are numerous physical variables: energy, angular momentum, mass, and electric charge, among others. In the macroscopic world of our senses, all physical quantities *seem* to be continuous. A piece of iron pipe can apparently be cut to any length whatever, not just six feet or eight feet or ten feet. A child's top as it is slowing down seems to decrease its rate of rotation smoothly as it gradually comes to rest. Yet we know that, viewed sufficiently closely, the pipe is composed of separate atoms, and the spin of the top is changing jerkily from one discrete value to another. An invisible speck of iron might be six atoms long or eight, but not six and a half. The spin of the top, as it slows down, makes "quantum jumps" from one allowed value to another, like a base stealer who vanishes from first and appears on second without having covered the distance between. Only because the increment between adjacent quantum values is so extremely small on the human scale does smooth continuous variation seem to be the rule in the large-scale world.

In connection with his original suggestion that energy transfer from matter to radiation is quantized, Planck introduced a numerical constant, now known as Planck's constant, for which the usual symbol is h. Its value is $h = 6.6255 \times 10^{-27}$ gm cm^2/sec (or erg sec). The dimension of h is the same as the dimension of angular momentum (mass × distance × speed), which in turn is the same as the dimension of energy × time. Planck's constant has turned out to be the fundamental constant of quantum mechanics, in much the same way that the speed of light is the fundamental constant of relativity. It is obviously, by macroscopic standards, an extremely small quantity. In the domain where it rules, it is neither small nor unimportant. Planck's constant determines the entire scale of the submicroscopic world—the energy of photons, the spin of particles, and the size of atoms.

Albert Einstein related the constant h to the energy carried by electromagnetic radiation. The particle of light, the photon, is the subject of Section 24.1. Niels Bohr and John Nicholson were among the first to see a wider significance of Planck's constant as the basis of a general quantum principle in nature. Both recognized angular momentum as a granular variable. In 1915 William Wilson and Arnold Sommerfeld independently gave the general rule for angular-momentum quantization,

$$L = n\frac{h}{2\pi}, \tag{23.1}$$

in which L represents angular momentum and n is any integer. The combination $h/2\pi$ has proved to be so ubiquitous in quantum theory that it deserves its own symbol, $\hbar$ (pronounced "h-bar"). Its numerical value is

$$\hbar = 1.05457 \times 10^{-27} \text{ erg sec.} \tag{23.2}$$

Thus $\hbar$ is the quantum unit of angular momentum in nature.

The discovery by George Uhlenbeck and Samuel Goudsmit in 1925 that the electron has a spin angular momentum equal to $\frac{1}{2}\hbar$ upset the Sommerfeld-Wilson rule somewhat, but it proved to be possible to incorporate this discovery into quantum mechanics without difficulty. We understand now that particle spin may be either an integral or a half-integral multiple of $\hbar$, although orbital angular momentum is restricted, as Nicholson and Bohr first guessed, to integral multiples of $\hbar$.

Not all of nature's granularity is understood. The reasons for discontinuity of energy and of spin are provided by the quantum theory. But why mass and charge come only in lumps of certain sizes remains a mystery and a challenge. Perhaps a new quantum constant, analogous to Planck's constant, is waiting to be discovered.

23.3 Probability

One of the most important insights contributed by the theory of quantum mechanics is this: 130
The fundamental laws of nature are laws of probability, not laws of certainty. If a hydrogen atom is put into an "excited state" (a state of greater than normal energy), and the atom is left to itself, it will, after a time, get rid of its excess energy by spontaneously emitting a photon of light. The length of time the atom remains in its excited state of motion before emitting the photon is entirely uncertain and cannot be calculated. But the *probability* that the photon will be emitted within any particular time interval can be calculated exactly. Quantum mechanics is an unambiguous and quantitative theory in the sense that probabilities can be calculated precisely. It is indefinite in that one can calculate only the chance that something will happen, never what will, in fact, happen for a particular atom or system. To check the calculated probability of photon emission against experiment, a single atom will never do. One must study a great many atoms and infer the probability for a single one by observing the average behavior of the whole collection. Similarly, one could never prove with a single toss of a coin that the probability of "heads" is exactly one half. Verification would require a great many tosses.

The decay of an unstable particle affords a simple and direct test of the working of probability at the fundamental level. If a great many pions are created under identical conditions at the target point in a particle accelerator, those which move away at a certain speed in a certain direction may be photographed in a bubble chamber. It will be observed that some decay into muons and neutrinos after moving a very short distance; some, after moving a greater distance; a few, after moving very much farther. There will be some average distance and correspondingly some average time for the occurrence of the decay process. If this experiment is repeated time after time, each time with a very large number of pions, the *average* lifetime of the pions will be exactly the same for every group. This average is a perfectly definite quantity, a measure of the decay probability of the pion, and it can be measured to arbitrarily high precision if a large enough group of pions is used. Nevertheless, the length of time that any single pion will live is indeterminate. It might die long before most of its fellow pions, or it might outlive them all.

The idea that the fundamental processes of nature are governed by laws of probability should have hit the world of science like a bombshell. Oddly enough, it did not. It seems to have infiltrated science gradually over the first quarter of this century. Only after the quantum theory had become fully developed in about 1926 did physicists and philosophers sit up and take note of the fact that a revolution had occurred in our interpretation of natural laws.*

As early as 1899, Ernest Rutherford and others studying the newly discovered phenomenon of natural radioactivity noticed that the decay of radioactive atoms seemed to follow a law of probability. Yet Rutherford and his fellow physicists did not shout from the housetops that the fundamental laws of nature must be laws of probability. Why was this? The answer is very simple. They did not recognize that they were dealing with *fundamental* laws. Probability in science was, after all, nothing new. What was new, but not yet recognized, was that for the first time probability was appearing in simple elementary phenomena in nature.

Classical physics was built solidly on the idea that nature follows exact deterministic or causal laws. If enough is known about a particle or a light wave or any system at one time, its future behavior is, according to pretwentieth-century physics, exactly calculable. There is no question about where the earth or moon will be at some future date. A bridge can be built or an electromagnet designed with solid assurance that it will not collapse or fail to work because of some unpredictable fluctuation in the laws of mechanics or electromagnetism. Indeed, the exact determinism of physical laws had a powerful effect on philosophy in the nineteenth century; one popular view was that the universe can be regarded as a giant mechanism—the "world-machine"—inevitably and relentlessly unrolling history according to a predetermined plan.

Nevertheless we are all well aware of probability in everyday life and do not need to go to the elementary-particle world to find it. As life-insurance actuaries and gamblers know, life and death and roulette wheels are governed by laws of probability. So is the behavior of bulk matter as it follows the second law of thermodynamics. The probability of the macroscopic world is a probability of ignorance (lack of detailed information); the probability of the submicroscopic world is a fundamental probability of nature. One contributor to macroscopic probability is ignorance of initial conditions. According to quantum mechanics it is impossible, in principle, as well as in fact, to calculate the exact cause of an atomic event, no matter how precisely the initial conditions are known. One can know every possible thing there is to know about a pion, and still not be able to predict when it will decay.

We can now understand why Rutherford must not have been unduly surprised to discover a law of probability in radioactive decay. He assumed that he was dealing with a probability of ignorance. The interior of the atom was, as far as he knew, a complicated structure, and the apparently random nature of the decay process could be attributed to unknown differences in the internal state of different atoms. Yet even before quantum mechanics was developed as a fully acceptable theory (in 1925) there were hints that the probability of the atomic world might be of a more fundamental kind. Rutherford himself discovered (with Frederick Soddy, in 1902) that radioactivity represented a sudden catastrophic change in an atom, and was not the result of a gradual process of change. This, in itself, made the radioactive transmutation appear to be a rather fundamental event. Einstein's photon theory

* The *fundamental* role of probability in nature seems to have been first clearly emphasized by Niels Bohr, Hendrik Kramers, and John Slater in 1924. Their effort to build a new quantum theory failed, but success came the following year to Werner Heisenberg, and in 1926 Max Born gave to the new theory the probability interpretation that remains as a keystone of quantum mechanics to the present day.

of light in 1905 and Bohr's theory of the hydrogen atom in 1913 also contained hints of a new fundamental role of probability, but we shall not go into the details of these discoveries here.

Probability in the submicroscopic world shows itself in several ways. First, and most directly, it manifests itself through a randomness of microscopic events. Anyone with a luminous-dial wrist watch and a Geiger counter can perform a simple experiment to demonstrate this. The Geiger counter should be of the common type arranged to give an audible click when a high-speed particle triggers the device. The watch is held at such a distance from the counter that the individual clicks can be heard. It will be obvious to the listener that the clicks are not coming in a regular sequence like the ticks of the watch, but occur in an apparently random fashion. Indeed, a mathematical analysis would show that they are exactly random. The time at which a given click occurs is completely unrelated to the time elapsed since the previous click or to the time at which any other click occurs.

In doing an experiment like this, one feels in unusually close touch with the submicroscopic world. The single audible click means that somewhere among the countless billions of atoms on the watch face, one nucleus has suddenly spontaneously ejected a particle at high speed and transmuted itself into a different nucleus. Very literally, a nuclear explosion has occurred and, in the private world of the nucleus, the time at which the explosion occurred has been governed exclusively by a law of probability. An identical neighboring nucleus may have long since exploded, or it may be destined to live yet a long time.

Probability manifests itself in another way that is not so obvious to the eye or ear, but that is equally convincing to someone with a little mathematical training. This is through the exponential law of decay. Rutherford, in fact, discovered the role of probability in radioactivity in this way, for in 1899 he did not yet have any way to observe single transmutation events (it was several years later that his research student, Hans Geiger, invented the Geiger counter). Rutherford noticed that when the total intensity of the radioactivity was graphed as a function of time, a curve like that in Figure 23.1 resulted. The most marked characteristic of an exponential curve is that it falls vertically from any value whatever to half that value in a fixed horizontal distance. This meant in Rutherford's experiment that a definite fixed time was needed for the radioactivity to diminish in intensity by half, regardless of the initial intensity. This fixed time is called the half life of the material, designated $t_{1/2}$. The curve in Figure 23.1 is described mathematically by the function

$$I = I_0 e^{-t/\tau}, \tag{23.3}$$

where τ is the mean time, or average lifetime of the radioactive nuclei. When $t = \tau$, the intensity is $I = I_0 e^{-1} = 0.368 I_0$. When $t = t_{1/2} = 0.694\tau$, $I = I_0 e^{-0.694} = 0.5 I_0$. In Figure 23.1, the initial intensity I_0 is set equal to one.

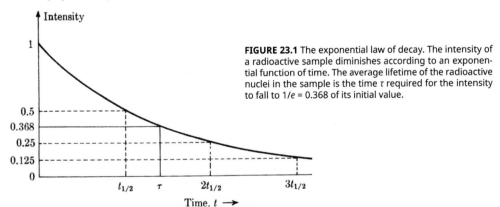

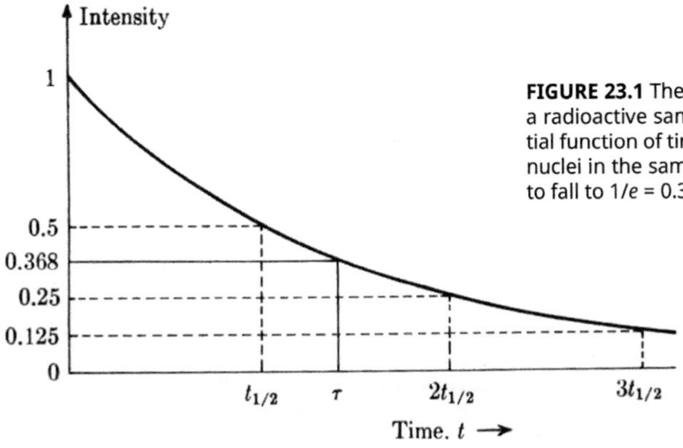

FIGURE 23.1 The exponential law of decay. The intensity of a radioactive sample diminishes according to an exponential function of time. The average lifetime of the radioactive nuclei in the sample is the time τ required for the intensity to fall to $1/e = 0.368$ of its initial value.

What Rutherford knew and what we here state without proof is that the exponential curve results from the action of a particular law of probability on the individual radioactive-decay events. For each single nucleus, the probability of decay per unit time is constant, and the half life represents a halfway point in probability. The chance that the nucleus will decay in less time is one half; the chance that it will decay after a longer time is one half. When this probabilistic law acts separately on a large collection of identical nuclei, the total rate of radioactive decay falls smoothly downward along an exponential curve. The same is true in the particle world. Each of the mean lives* appearing in Table 2.1 was measured by studying the exponential curve of decay for particles of that kind. (The times themselves are usually not measured directly, but they can be inferred by measuring the speeds of the particles and the distances traveled.)

The span of known half lives from the shortest to the longest is unimaginably great. At one extreme are the super-short-lived particles, or resonances, with half lives of 10^{-20} sec or less. The more respectable particles live from 10^{-10} sec up to several minutes—except, of course, for the stable particles, which as far as we know live forever. Radioactive nuclei are known with half lives ranging from about 10^{-3} sec up to more than 10^{15} years. Regardless of the half life, the decay of every kind of unstable particle or nucleus proceeds inflexibly along an exponential curve.

The radioactivity created by nuclear explosions consists of a mixture of many different radioactive species, so that the fallout subsequent to bomb tests does not follow a single simple exponential curve. Some of the radioactive nuclei decay so quickly after the explosion—within seconds or minutes—that they contaminate only a local region and are not a public-health problem. Other species have such a long life—millions of years—that their rate of decay remains always very small. In between are the nuclei with half lives of a few years up to a few thousand years; these constitute the potential hazard of fallout. The often-discussed Co^{60} (cobalt 60) has a half life of 5.3 years, and Sr^{90} (strontium 90) a half life of 25 years. The isotope of carbon, C^{14}, which has been so useful for dating archaeological finds because of its half life of 5,700 years, is also formed in the atmosphere by bomb tests. This will greatly complicate, perhaps invalidate altogether, the C^{14} dating method for archaeologists of future millennia.

We have discussed so far only the probability of time which determines the characteristic decay pattern of unstable particles and nuclei. Probability also manifests itself in other ways in the submicroscopic world. There is the probability of "branching ratio." A kaon may decay in various ways: into two pions or into a muon and a neutrino, among other possibilities. Which branch any particular kaon will choose to follow is completely indeterminate, but the probability for each branch is readily measurable (given enough kaons). There is also a probability of position and probability of angular deflection in scattering. A particularly fascinating aspect of probability at the fundamental level is the phenomenon of "tunneling." If a particle is held on one side of a wall which, according to classical physics, is totally impenetrable, there is a chance that it will emerge on the other side. In certain transistors, electrons tunnel through potential-energy barriers. The alpha decay of nuclei can also be explained as a tunneling phenomenon (Section 25.6). As with so many other aspects of quantum mechanics, tunneling is of significance only in the submicroscopic world. A man leaning idly against the wall of his hotel room need have no fear that he will suddenly find himself in the next room. Nor should a student in a dull lecture put any hope in the tunneling phenomenon as a way out.

* In general a mean life, or average life, is not the same as a half life. In 1950, for example, the average life expectancy (mean life) of an American male was 66.3 years, but his half life was about four years greater. He would need to reach an age of 70.7 years in order to outlive half of his contemporaries. In the particle world, on the other hand, the half life is considerably less than the mean life. A neutron, with a mean life of 17 minutes, has outlived half of its contemporary neutrons after 12 minutes. The ratio of half life to mean life is 0.694 whenever a law of constant probability per unit time operates to produce an exponential change.

Not every aspect of nature is uncertain and probabilistic. Many of the properties of stable systems—for example the spin of an electron or its mass—are precisely defined. Even where a law of probability is at work, the probability of an event may be so close to zero or to one that its nonoccurrence or occurrence can be regarded in practice as a certainty. The chance that the tunneling phenomenon will be experienced by a man can be said to be effectively zero. The chance that a proton will decay in a billion years is essentially zero (this has been measured). The chance that a pion will live for two hours is as good as zero. Because quantum-mechanical probabilities in the macroscopic world are always so close to zero or one, the deterministic laws of classical physics are completely adequate and accurate for describing large-scale phenomena.

Is the probability of the submicroscopic world really a fundamental probability of nature, or is it perhaps, after all, a probability of ignorance, arising out of a complicated deeper, as yet undiscovered, substructure of matter? The simplest answer that can be given is: "No one knows." Most scientists regard it as not a very interesting question. Since nothing of a deeper substructure is known, it is not fruitful at this moment in history to discuss it. So far as we know now, the probability is indeed fundamental, but one need cling to this idea no more firmly than to any other in science.

Nevertheless, some of the greatest scientists of this century did find the question interesting and have discussed it. Those arguing for the fundamental nature of the laws of probability have a bit better time of it, for they have all of the successes of quantum mechanics on their side. Those favoring the view that the probability of quantum mechanics is really a probability of ignorance can adduce at best philosophic, and not scientific, arguments. Einstein, for example, liked to remark that he did not believe in God playing dice, and in 1953 he wrote,* "In my opinion it is deeply unsatisfying to base physics on such a theoretical outlook, since relinquishing the possibility of an objective description . . . cannot but cause one's picture of the physical world to dissolve into a fog."

The arguments in favor of the truly fundamental role of probability in nature are more subtle, being based on the theory of quantum mechanics. We shall give just one here. The manufacturers of baseballs try hard to make all of their balls identical. It is, of course, an impossible task. No two can ever be precisely alike down to the last microscopic detail, because each ball is a complicated structure with many constituents—more than 10^{25} constituents if we count atoms. On the other hand, there is quite good evidence that any two electrons are, in fact, truly identical, and that relatively few parameters are required to specify an electron completely. In short, the electron seems to be a decidedly much more elementary structure than a baseball. This is not a trivial conclusion. If there were infinitely many layers of nature to uncover, the electron could just as well be about as complex as the baseball. Since the electron follows laws of probability, one is led to suspect that these laws are themselves of an elementary and fundamental kind, not merely a reflection of the fact that a complicated unknown structure resides within the electron.

Although arguments of this kind sound scientific, they are no more rigorous than Einstein's statement of belief. We can only wait and see.

23.4 Annihilation and creation

Nineteenth-century chemistry was based solidly on two laws of conservation: the conserva- 131
tion of mass and the conservation of energy. The theory of relativity showed that mass is convertible into energy and energy into mass. It did not say that matter *must* be created or annihilated, only that it could be. But nature, like a dog on a leash, has a way of doing everything not absolutely forbidden to it. The discovery of the positron in 1932 provided the first clear evidence of the creation and annihilation of matter, and Fermi's theory of beta decay

* A. Einstein, in *Scientific Papers Presented to Max Born* (New York: Hafner, 1953), p. 40. Original in German.

shortly afterward showed that the electrons emitted in beta radioactivity must be created on the spot. Quantum mechanics had, in the meantime, provided a theoretical framework for dealing with mass creation, in particular, showing that the creation of a photon of light energy does not differ in any essential way from the creation of a material particle. According to quantum mechanics, emission and absorption of light is precisely equivalent to creation and annihilation of particles. By the mid-thirties, creation and annihilation of material particles was a well-established fact. Now we recognize that any and all particles may be created or annihilated. All of the unstable particles undergo spontaneous annihilation; the stable particles can be annihilated by coming into contact with their antiparticles. When enough energy is at hand, any particle, stable or unstable, may be created. Figure 23.2 shows the creation and annihilation of several particles in a bubble chamber.

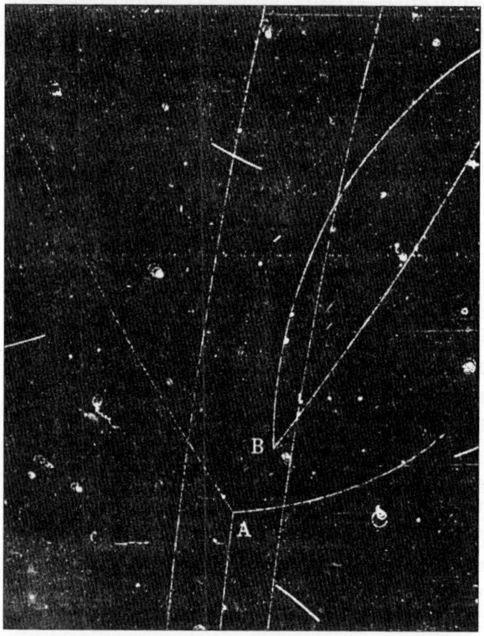

FIGURE 23.2 Creation and annihilation of material particles. Three kaons enter the chamber from below. At point A, one kaon strikes a proton. In the collision, both the kaon and the proton are annihilated, and three new particles are created: $K^- + p \rightarrow \Lambda^\circ + \pi^- + \pi^+$. The pions, being charged, leave tracks as they fly away, the negative pion to the left, the positive pion to the right. Although the neutral lambda leaves no track, it reveals its presence at point B, where it decays into a proton and a pion: $\Lambda^\circ \rightarrow p + \pi^-$. (Photograph courtesy of Lawrence Radiation Laboratory, University of California, Berkeley.)

The decay of an unstable particle is the simplest example illustrating both annihilation and creation of mass. Some typical decay modes are listed in Table 2.1. In the beta decay of the neutron, written

$$n \rightarrow p + e^- + \overline{\nu_e},$$

a neutron is annihilated, and a proton, an electron, and an antineutrino are created. This is the only way in which a neutron may decay, except for the rare case in which it also emits a photon. A kaon, on the other hand, has a variety of decay modes. Positively charged kaons can vanish in more than half a dozen different ways, producing electrons, muons, neutrinos, charged and neutral pions, and gamma rays. Each mode of decay is constrained by conservation laws and controlled by probability.

Spontaneous decays and particle-antiparticle annihilation are "downhill" events. The mass of the product particles is always less than the mass of the initial particle or particles, and the difference is converted into energy of motion imparted to the product particles. "Uphill" events, in which new mass is created, can be stimulated by the use of high-speed projectile particles, either furnished free in the cosmic radiation, or furnished at great expense by man-made accelerators.

When relativity swept away the law of mass conservation, it swept away the idea of a solid and reliable material basis of the universe. The modern view, based on the mass-energy equivalence predicted by relativity and the routine annihilation and creation of matter predicted by quantum mechanics, is much more tenuous. Most of the material particles do not live long enough to be of any use for building the world. Even those that do can be annihilated if struck by other energetic particles or if brought into contact with their antiparticles. The modern view might be described as follows: Because of certain conservation laws, a very few of nature's particles happen by chance to be stable. Even these are not indestructible, but because where we live the flux of projectile particles is very low, and because our corner of the universe happens to contain a great deal of matter and very little antimatter, the stable particles have time enough to build a durable material world.

23.5 Waves and particles

In the world of the very small, waves and particles appear to be not merely closely related, 132
but actually one and the same thing—or, more accurately stated, different aspects of one and the same thing. This remarkable fact was first implied by Einstein's theory of the photon in 1905, and came to be fully appreciated after the work of de Broglie, Schrödinger, and others on the quantum theory twenty years later. Now we recognize the wave nature of matter as the factor that gives atoms their size, "explains" the uncertainty principle (Section 23.6), elucidates the role of probability in nature, and thwarts man's efforts to study the interior of elementary particles.

The concept of a particle, however small the particle may be, is easy to grasp. We can picture a golf ball and imagine it shrunk down to elementary-particle size, about 10^{-13} cm. We see in our mind's eye a tiny spherical lump of matter. It has mass; it is located at some definite point; it can move from one place to another at some measurable speed. Energy is required to set it in motion, and it gives up energy when slowed down or stopped. Picturing a massless particle that always shoots about at the speed of light is a bit harder, and we postpone discussion of the photon (Section 24.1) and the neutrino (Section 26.4). Here we shall keep to more conventional material particles to illustrate the wave-particle duality. At first thought, waves seem to be different from particles in every way. A particle has mass, can be located at a definite point, and can be imagined to have a definite size; a wave is massless, is necessarily spread out, and has an ill-defined size. Moreover, the quantities used to characterize a wave—its amplitude, wavelength, and frequency—are quantities that seem to have no meaning for particles.

In spite of these obvious differences, quantum mechanics has succeeded in merging the ideas of waves and particles. To make the merger reasonable, let us consider those properties waves and particles do have in common, even in our macroscopic world. First of all, both obviously can travel from one place to another, and at a definite speed. But here, too, there is a difference. The speed of a wave usually depends very little on its wavelength or its amplitude. Fortunately for the listener at the back of the second balcony at a symphony concert, sound travels at a nearly fixed speed, regardless of its loudness (amplitude) or its pitch (frequency). Particles, on the other hand, can easily be caused to travel at different speeds, dependent on their energy. Most important as a point of similarity, waves and particles can do the same job. Each can receive energy, carry it elsewhere, and transmit it to something else. If two boys hold the ends of a long rope (Figure 23.3), one can supply energy by shaking his end. A wave will run along the rope and transmit an impulse to the hand of the other boy. Alternatively, the same amount of energy could have been transferred by means of a "particle" such as a baseball thrown from one boy to the other.

The merging of the ideas of wave and particle was made possible only by some changes in our view of waves and of particles. Both concepts have had to yield a little in order to grow more alike. Usually we think of a wave as a vibration of something. Water waves need water, sound waves need air, the rope wave needs a rope. Before relativity made the idea untenable, it had naturally been assumed that light waves too need a material capable of vibrating: the ether. The modern idea is rather that light is the wavelike propagation of electric and magnetic fields through empty space. Quantum mechanics deals with the emission, propagation, and absorption of fields, fields associated with material particles as well as the electromagnetic field associated with photons. Because of the shift of emphasis from ether to field, we gain a new perspective on waves. A wave becomes a more material thing, an entity by itself. It is still spread out, still characterized by wavelength and frequency and amplitude, but it is something by itself, not just the name given to a vibration of an underlying medium. It is as if the boys could transmit a rope wave without a rope. This is obviously a big step in the direction of making a wave more particle-like, an unattached bundle of energy.

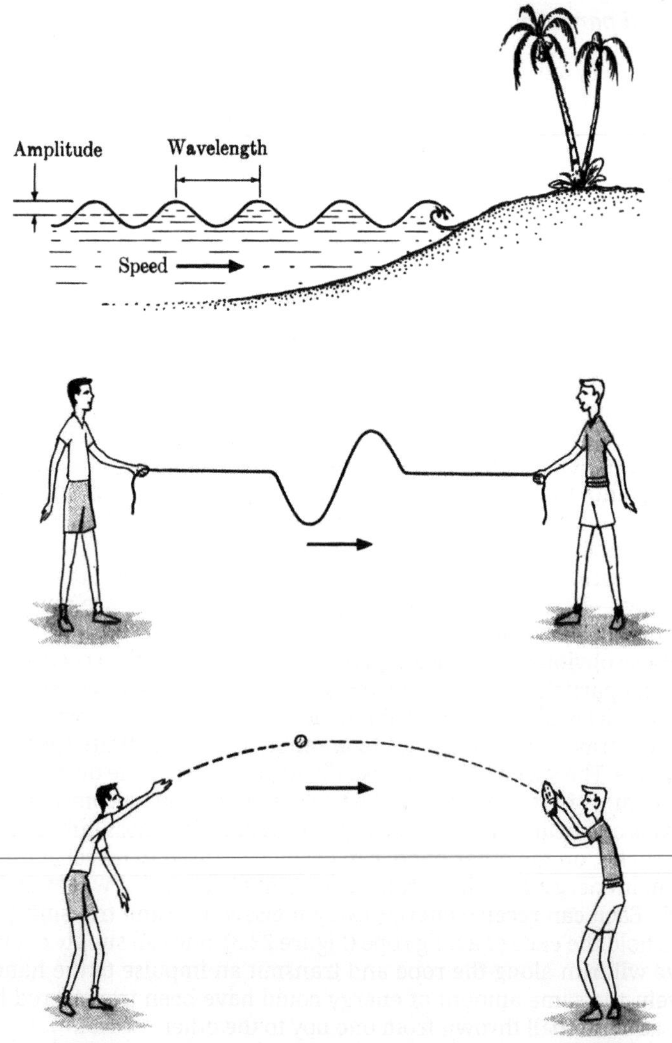

FIGURE 23.3 Waves and "particles" in the macroscopic world. Water waves rolling toward a beach are characterized by wavelength, amplitude, speed, and frequency. A rope wave has the same characteristics but is more localized. It can transmit energy from one boy to the other just as a thrown ball (or "particle") can.

Quantum mechanics, then, brought with it some necessary changes in our view of particles which rendered them less distinct and therefore more wavelike. The most essential feature is the nonlocalizability of particles. According to the uncertainty principle, which lies at the core of quantum mechanics, the location of a particle can never be precisely specified. The particle therefore loses its distinctness, becoming spread out and a bit fuzzy, like a wave. The bigger the particle, the less important is this fuzziness, so that in the world of our senses all "particles" appear to be perfectly localized and have well-defined, sharp boundaries. In the world of the very small the fuzziness becomes all-important. The fact that the hydrogen atom is 100,000 times larger than the proton at its center comes about entirely because of the nonlocalizability of the very lightweight electron, which refuses to sit quietly alongside the proton, requiring instead all of this extra space for its domain of motion.

We can say that particles are nonlocalizable because they are wavelike or that they are wavelike because they are nonlocalizable. It does not really matter which we say. In practice it is somewhat simpler to take the wave nature of the particle as basic, and derive from that the other new features of quantum mechanics. The single key equation that specifies the wave nature of a particle was first postulated by Louis de Broglie in 1924; shortly afterward it was incorporated into the full theory of quantum mechanics. That the photon was in some sense both wave and particle had been known since 1905. De Broglie was the first to suggest that *every* particle should have a wave nature.

The de Broglie relation may be written

$$\lambda = \frac{h}{p}. \tag{23.4}$$

This equation looks simple enough but has consequences as significant as Einstein's famous $E = mc^2$. Here λ is wavelength, p is momentum, and h is Planck's constant.

Momentum is a particle-like property. Wavelength is obviously a wavelike property. The de Broglie equation links these two properties, and the link tying them together is Planck's constant. It is the small size of this quantum constant h that makes the wave properties of particles irrelevant except in the world of the very small. Just as with Einstein's equation, $E = mc^2$, where the heart of the equation is the proportionality of the energy E to the mass m, the factor c^2 being a constant of proportionality, so with de Broglie's equation, the heart of the equation is the proportionality of wavelength λ to $1/p$, the *inverse* of the momentum p, Planck's constant h being the constant of proportionality. Because p is in the denominator, larger p implies smaller λ. For the enormous momenta of macroscopic objects, the associated wavelength is so small that the wave property is completely unobservable. A man walking at 3 miles/hr has a wavelength of less than 10^{-33} cm. If he tried to move more slowly in order to have a larger wavelength it would not help much. Progressing at one centimeter per century, he would have a wavelength of less than 10^{-21} cm, still one hundred million times smaller than the size of an elementary particle. On the other hand, a single electron moving at about 3×10^8 cm/sec in the hydrogen atom has a wavelength of 2×10^{-8} cm, just about the diameter of the hydrogen atom. Not until man extended his powers of observation far beyond the normal range of human perception did he have a chance to discover quantum mechanics.

The most convincing direct evidence of waves comes through the phenomena of diffraction and interference. Through these effects the wave nature of light was established beyond question early in the nineteenth century. More than a century later, the same effects proved that material particles too exhibit wave behavior. Following de Broglie's suggestion that all particles should show wave properties, George P. Thomson, Clinton Davisson, and Lester Germer discovered diffraction and interference in electron beams (1925). More recently, neutrons have proved to be the best particles for demonstrating the phenomena of diffraction and interference. The particular merit of the neutron is that it carries no electric charge. The most marked wave effects occur for the greatest wavelength, which in turn—according to the de Broglie equation—requires the least momentum. Electrons of low momentum are easily disturbed by any small electric forces which they encounter; thus they cannot penetrate solid matter. But neutrons can be slowed down to a walk, 10^5 cm/sec (about 2,200 miles/hr) or less, without being readily subject to disturbing influences. These slow neutrons, with a relatively long wavelength, pass easily through thin layers of solid material. As a benchmark for the wavelength-momentum relationship, we note that a neutron moving at 9,000 miles/hr or 4×10^5 cm/sec—about half the speed of an orbital astronaut—has a wavelength of one Angstrom (10^{-8} cm).

A surprising number of facts about the submicroscopic world can be understood in terms of the wave nature of particles. Perhaps the most essential fact about a wave, as far as the world of particles is concerned, is its nonlocalizability. A wave cannot be said to be exactly at this point or exactly at that point. At best it is known to be in this region or that region. It can be *approximately* localized, but the crucial distance below which it makes no sense to speak of the position of a wave is its own wavelength. Crudely speaking, a wave has to go through at least one cycle of oscillation in order to be a wave at all, and therefore it must occupy a space at least as big as its own wavelength. Picture a long rope that has been given a single shake at one end. A wave in the shape of a single hump will run along the rope; the position of the hump marks the position of the wave. But the wave occupies a region, not a point, and the size of the region is the length of the hump, which is roughly the wavelength of the disturbance.

The relevance of this nonlocalizability in the submicroscopic world is simply that the position of a particle can never be known, even in principle, to an accuracy much greater than the wavelength of the particle. The wave nature of matter introduces an essential fuzziness into nature; the particle wavelength defines a region of uncertainty, within which the whereabouts of the particle is unknown and unknowable. One might think this should permit us to dispense with particles altogether and say that there are only waves. This cannot be done, for the particle property is still evident in processes of change, that is, in events of annihilation and creation. The birth and death of particles is "particle-like," occurring suddenly at one point in space and time; the life of particles between creation and annihilation is "wavelike," characterized by a wavelength and diffused over a region of space.

133 Let us apply these ideas to the size of the hydrogen atom. A hydrogen atom consists of one proton, a heavy particle which we may think of for the moment as being fixed at a certain point, and one electron, a light particle moving about the proton. Between them acts an attractive electric force. According to classical physics, the electron should emit light waves, gradually lose energy, and spiral down into the proton, so that the size of the atom would finally be about the size of the proton, 10^{-13} cm.* It is the wave nature of the electron that prevents this collapse. If the electron spiraled into the proton, it would be confined to a smaller and smaller region of space, which means that its associated wavelength would have to become smaller and smaller. According to the de Broglie equation, smaller wavelength means larger momentum, which in turn means more energy of motion (kinetic energy). This is the crux of the matter. The wave nature of the electron means that it can be confined to a small region only if it has a high kinetic energy. Because of the electrical attraction, the electron "wants" to be near the proton. But in order to have the smallest possible energy, it "wants" to have a very large wavelength and be spread over a large region of space. These two opposing influences—the proton's force tending to pull it in, its wave nature tending to push it out—reach a point of balance for the electron wave spread over a certain distance; this distance happens to be about 10^{-8} cm and determines the size of the atom (Figure 23.4).

* The thoughtful reader might translate this situation to the solar system, and wonder why the earth, which is described by classical laws, does not spiral into the sun. The answer: It does! The difference, an all-important one, is a matter of time. The electron should spiral into the proton in about 10^{-8} sec, the earth into the sun in about 10^{24} years (the age of the earth is less than 10^{10} years).

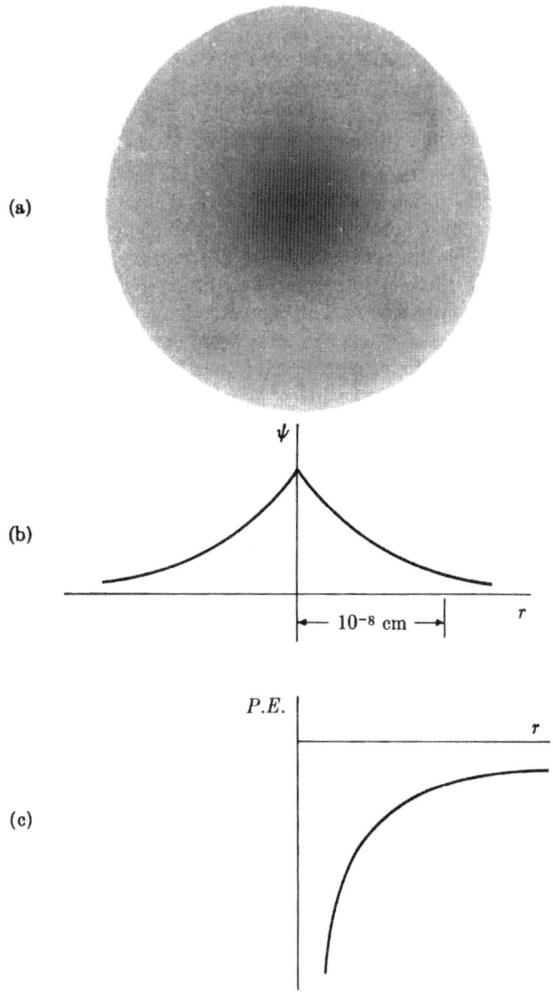

FIGURE 23.4 Electron wave in hydrogen atom. **(a)** Physical picture of wave occupying a spherical region around the proton. **(b)** Graph of wave amplitude ψ. **(c)** Potential energy of electron as a function of distance from proton. If the electron wave were to shrink to smaller dimension, the potential energy would decrease, but the kinetic energy associated with the shorter wavelength would increase even more.

The argument can be expressed mathematically in the following way. The electrical potential energy of the electron-proton system, negative because the force is attractive, is given by

$$P.E. = -\frac{e^2}{r}, \tag{23.5}$$

the product of the charges divided by their separation. The electron kinetic energy is

$$K.E. = \frac{1}{2}mv^2 = \frac{p^2}{2m}. \tag{23.6}$$

According to the de Broglie equation, the electron momentum p can be replaced by h/λ, giving for the kinetic energy

$$K.E. = \frac{h^2}{2m\lambda^2}. \tag{23.7}$$

Now we assume that the wavelength of the electron is about equal to the diameter of the atom (the idea of nonlocalizability), or $\lambda = 2r$. This means that the electron kinetic energy can be expressed in terms of the atomic radius by

$$K.E. = \frac{h^2}{8mr^2}. \tag{23.8}$$

This equation states that as the atomic size decreases, the kinetic energy necessarily increases (because of the wave nature of the electron). At a very large radius, this kinetic energy is negligible and the potential energy dominates. Here the electron is drawn inward. At a very small radius, the kinetic energy overwhelms the potential energy and the electron tends to fly outward. The electron takes for its domain of motion a distance such that its kinetic and potential energies are comparable in magnitude:

$$\frac{h^2}{8mr^2} \cong \frac{e^2}{r}. \tag{23.9}$$

Solution of this approximate equation for the atomic radius gives

$$r \cong \frac{h^2}{8me^2}. \tag{23.10}$$

Numerically, the right side of this formula is equal to 2.6×10^{-8} cm. It is important to be aware that this derivation, although mathematical, is only qualitative. It is an order of magnitude calculation. In truth the electron's average kinetic and potential energies are not exactly equal, nor is the electron wavelength exactly twice the atomic radius. Nevertheless, the derivation is significant, for it shows that the size of an atom is determined by a certain combination of fundamental constants, h^2/me^2. (Because of the approximate nature of the derivation, the particular numerical factor 8 appearing in Formula 23.10 is without special significance.) Until Planck's constant h entered physics, there was no possible way to explain the size of atoms.

In atoms heavier than hydrogen, the distance of the outermost electrons from the nucleus is determined by similar considerations. All atoms in fact have diameters of about 10^{-8} cm. The wave nature of the electron, incidentally, was used as an argument to banish electrons from the nucleus. An electron confined to the small size of the nucleus would have too much kinetic energy and could not be held there. An electron that shoots out of the nucleus in beta decay, therefore, must have been formed at the moment of decay, and not been supplied from a reservoir of electrons already there.

The wave nature of particles is intimately connected with the fundamental role of probability in nature. The simplest aspect of particle probability is probability of position; the hydrogen atom can serve again to illustrate the idea. In the atom, the electron cannot be thought of as existing at any particular point, but must be visualized in terms of a spread-out wave. In a particular state of motion, the electron is described by a wave amplitude or wave function, for which the usual symbol is ψ. Such a wave function is graphed in Figure 23.4(b). Where ψ is large, the electron is likely to be found; where ψ is small, the electron is unlikely to be found. Specifically, the probability that the electron will be found in a small volume V, if an experiment designed to reveal the particle aspect of the electron is carried out, is

$$P = \psi^2 V. \tag{23.11}$$

The square of the wave function is interpreted physically as probability per unit volume.

So long as the electron is left to itself in a particular state of motion in the atom, we need be concerned neither with its particle aspect nor with the idea of probability. We may simply visualize it as a wave occupying a region of space. Yet it is possible to do an experiment to reveal the electron as a particle at a specific location (Figure 23.5). If a high-speed positron is fired at the atom it may strike the electron; if it does so, both will vanish and a pair of photons will emerge. These photons could be studied to reveal the place within the atom where the electron was at the moment of its annihilation. (This does not happen to be a practical experiment, but it is all right in principle.) Through the concept of probability the apparent paradox of the electron wave and the electron particle is resolved. If the same experiment with the positron is repeated with a number of hydrogen atoms, each one *exactly* the same as each other one in every property we know how to specify, the results of the different experiments will not be the same. Sometimes the electron will be found in one part of the atom, sometimes in another; sometimes close to the nucleus, sometimes far away; but it will almost always be found within about 10^{-8} cm of the proton, within the region where the electron wave is large. The relative probability to reveal the electron at any place is proportional to the square of the wave function at that place for the state of motion of the electron that existed before the annihilation event.

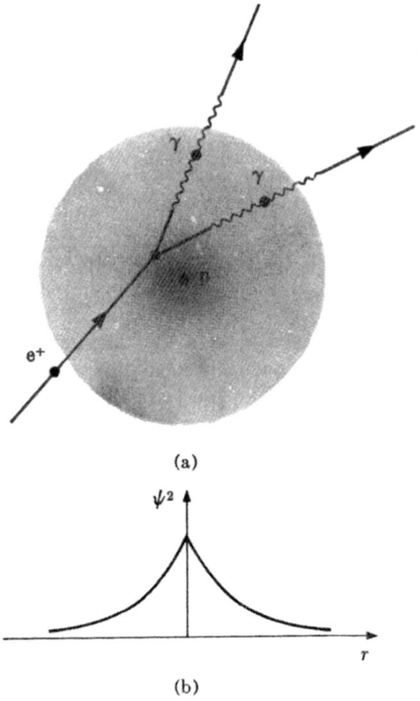

FIGURE 23.5 Probability in the hydrogen atom. **(a)** A high-energy positron can annihilate with the electron in the atom to produce a pair of gamma rays which can, in principle, locate the part of the atom where the annihilation event occurred. **(b)** The square of the wave function gives the probability per unit volume that the electron will manifest itself as a particle in such a process.

Surprisingly at first thought, the wave nature of matter is also intimately connected with the granularity of the submicroscopic world. The very fact that an electron is not localized accounts for some of its quantized properties. The connection between waves and discreteness is pursued in Section 24.7. Suffice it to note here that such a connection is in fact familiar in the macroscopic world. Piano strings have discrete fundamental frequencies because of their characteristic wave vibrations.

23.6 *The uncertainty principle*

One of the most important insights into nature revealed by quantum mechanics is the 134
Heisenberg uncertainty principle. This is a general principle that takes many forms, but it will be sufficient to consider one of its forms, which can be written

$$\Delta x \Delta p = \hbar \,. \qquad (23.12)$$

On the right side is $\hbar$, the ubiquitous Planck constant (here divided by 2π) which turns up in every equation of quantum mechanics. Momentum is represented by p, and position (distance) by x. The Δ symbols are used here to mean "uncertainty of" (not "change of"): Δx is the uncertainty of position, Δp is the uncertainty of momentum. The product of these two uncertainties is equal to the constant $\hbar$. Since $\hbar$ is, on the human scale, a very small quantity, Δx and Δp can be so close to zero in the macroscopic world that there is for all practical purposes no uncertainty whatever in the position and momentum of large objects. If we wish to specify a man's position to within the size of a single atom, his speed could, in principle, be determined to an accuracy of about 10^{-24} cm/sec. Needless to say, the inaccuracies of measurement take over long before the inherent fundamental limitation of accuracy implied by the uncertainty relation can play any role. But in the world of particles, this is not so. Masses and distances are so small that the uncertainty principle is of vital importance. An electron, in order to be localized within a distance 10^{-8} cm (which is equivalent to assigning it this great an uncertainty of position), has an inherent uncertainty of speed of about 10^{8} cm/sec.

The uncertainty principle, when discussed by itself outside the framework of quantum mechanics, is often assigned a profundity and depth that are probably unjustified. It has obvious philosophic implications and it is especially popular with those who wish to attack science, for it shows that even the "exact" scientist is prohibited by nature from measuring things as exactly as he might like. One might also argue that nature is shielding its innermost secrets by allowing man to proceed only so far and no farther in his downward quest. In truth, the uncertainty principle is fundamental and presents in capsule form an important part of the physical content of quantum mechanics. Nevertheless, it may be viewed as just one more aspect of the wave nature of matter, in which case it seems considerably less mysterious.

135 In order to understand the uncertainty principle, one must come to grips with a somewhat difficult but extremely important idea about the localizability of waves: A wave can be localized only if different wavelengths are superposed. Consider first a "pure" wave of a single wavelength. Figure 23.6 shows such a wave, spread out in space, a perfect sine wave repeating itself indefinitely with a precisely defined constant wavelength. If this is an electron wave, it represents a particle moving along at constant speed, with a well-defined momentum given by the de Broglie equation: $p = h/\lambda$. Where is the electron? It is everywhere, or it is equally likely to be anywhere (along the infinite dimension of the wave).* Its uncertainty of position is infinite; its uncertainty of momentum (and of wavelength) is zero. This is an extreme situation that is at least consistent with the uncertainty principle. If the product of two uncertainties is constant, one of the two can become vanishingly small only if the other becomes infinitely great.

FIGURE 23.6 A pure wave of a single frequency. This sine wave has a single wavelength and— if it is a particle wave— a definite momentum free of uncertainty. Since it oscillates indefinitely in space, its uncertainty of position is infinite.

Consider now the result of superposing several wavelengths. The superposition of two waves differing by 10% in wavelength is shown in Figure 23.7. Five cycles from the point of maximum reinforcement, the waves interfere destructively. Five cycles further on, they

* At an instant of time, the square of the wave function pictured in Figure 23.6 has hills and valleys, so that the particle has greater probability to be in some places than other places. If it is a propagating wave, the regions of high and low probability change with time, resulting in an equal average probability for the particle to be found anywhere along the wave.

again reinforce. This alternation of constructive and destructive interference produces partial localization of the wave, a bunching together of the wave into regions each about 10 wavelengths in extent. There is already an important hint about the uncertainty principle in this result. Partial localization has required a mixing of wavelengths, which in turn implies an uncertainty in momentum.

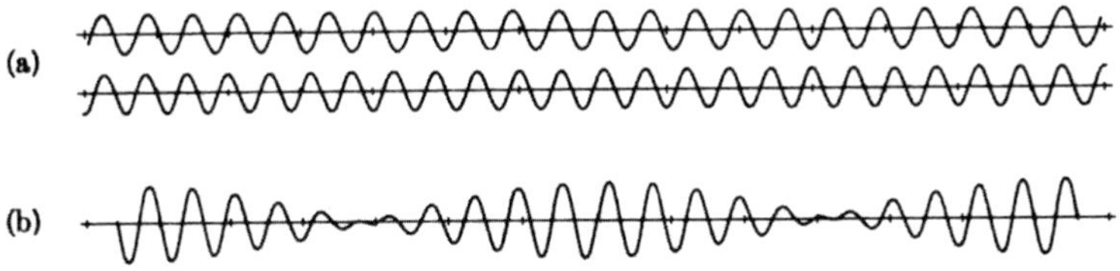

FIGURE 23.7 Superposition of two waves differing by 10% in wavelength. **(a)** The two pure sine waves. **(b)** The wave formed by superposing these two.

Greater localization is achieved by superposing a large number of different wavelengths. Figure 23.8 shows the result of combining 6 different pure waves, whose wavelengths differ successively by 2%. The total spread of wavelengths remains 10%, or 5% on either side of the average wavelength. We can therefore assign an uncertainty of momentum of about 5% of the dominant momentum:

$$\Delta p \cong 0.05p\,. \tag{23.13}$$

Here p represents an average momentum, related to an average wavelength λ by $p = h/\lambda$. In Figure 23.8, the sixfold superposition produces a better localization of the wave. (If all possible wavelengths within the given span were superposed, the wave could be completely localized. Outside of a given region of space its amplitude would fall to zero, never to rise again.) The uncertainty of position is roughly the extent of the wave on either side of its center—about five wavelengths from the figure:

$$\Delta x \cong 5\lambda\,. \tag{23.14}$$

The product of position and momentum uncertainties is, in this example,

$$\Delta x \Delta p \cong 0.25\lambda p\,. \tag{23.15}$$

Since the product λp on the right is equal to Planck's constant h, this equation can be written

$$\Delta x \Delta p = 0.25h\,. \tag{23.16}$$

Because we have taken rough-and-ready estimates of the uncertainties, the numerical factor 0.25 in this equation has no special significance. The technical definition of uncertainty used in quantum mechanics makes the right side of the uncertainty equation turn out to be exactly $h/2\pi$, or $\hbar$, as stated in Equation 23.12.

FIGURE 23.8 Superposition of six waves spanning 10% in wavelength. The two waves shown in Figure 23.7(a) plus four others intermediate between these two are superposed to form the wave shown here. (The vertical scale is altered.)

Since a 10% spread in momentum "crushed" the wave packet from infinite extent (Figure 23.6) to a little more than 10 wavelengths (Figure 23.8), it is a reasonable guess that momenta varying by nearly 100% would be required to narrow the packet down to a single cycle. This is true. Maximum localization, $\Delta x \cong \lambda/2$, requires that wavelengths (and momenta) differing by as much as a factor of two be superposed. Then $\Delta p \cong p/2$. In these equations, p and λ refer to average values of momentum and wavelength, since neither of these variables is well defined for a localized wave. Again the product $\Delta x \Delta p$ is of the same order as the product λp, which in turn is equal to Planck's constant. The electron wave function in Figure 23.4 shows such a highly localized wave, completing only one cycle of oscillation. Note that the shape of this wave function differs markedly from the shape of a pure sine wave, a reflection of the fact that it is a superposition of many different wavelengths.

This detour into wave superposition has demonstrated the essential point that the Heisenberg uncertainty principle is a consequence of the wave nature of particles. It is no more profound, and in this form says no more and no less than the de Broglie equation giving the wavelength of material particles. Uncertainty of measurement arises essentially from the nonlocalizability of waves.

One practical consequence of the uncertainty principle is that it makes life hard for the elementary-particle physicist, and expensive for the governments of the world. Our knowledge of the world of the very small comes mainly from the results of "scattering" experiments, in which projectile particles are fired at target nuclei, where they are deflected, or scattered, and reemerge to be observed with Geiger counters, bubble chambers, or other detectors. More complicated things than deflection may occur in these collisions; for instance, new particles may be created and also emerge. Scattering has come to be used to describe the general process in which two or more particles come together and, after a brief interaction, various particles fly apart. Man must infer what happened in the brief moment of interaction and over the tiny distances when the particles were close together by studying in detail the particles that emerge from the collisions—what particles they are, how fast they travel, and which way they go. One important limitation on the accuracy with which he can do this is imposed by the wave nature of the particles.

Suppose that we wish to study ships in a harbor by analyzing waves that pass them. Waves rolling past a large ship at anchor would be strongly affected (Figure 23.9). The ship would leave a "shadow" of calm water and the waves rounding the ends of the ship would be diffracted in a characteristic way. We could learn the shape and size of the ship fairly accurately by studying waves that had passed it in various directions. If, on the other hand, the same waves roll by a piece of piling sticking out of the water (Figure 23.10), they would scarcely be affected and would at most show that some small thing was there, without revealing its size or shape. But we should have no difficulty analyzing the piling with light waves, that is, by looking at it. The essential point is this. Waves provide a good method of analysis only when the wavelength is comparable to or smaller than the dimension of the object being studied. If the wavelength is much greater than the size of the object, no details can be revealed. Therefore, if one wishes to study an object with waves, the wavelength chosen should not be greater than the dimension of the object.

FIGURE 23.9 Diffraction of waves by a ship could reveal details of the size and shape of the ship if the wavelength of the waves is smaller than the dimension of the ship.

FIGURE 23.10 A piece of piling can be accurately analyzed with light, whose wavelength is small compared with the dimension of the piling, but not with water waves of long wavelength.

The "objects," or better, the small regions of space, which the physicist wishes to study are now about 10^{-13} cm or less—he would be delighted if he could study much smaller regions. He therefore wishes to use, as projectiles for his scattering experiment, particles whose wavelength is as small as possible. The difficulty is that, according to the de Broglie equation, small wavelength means large momentum. To probe smaller and smaller distances, the physicist must use more and more energetic particles with more and more momentum. This requires the construction of large particle accelerators; in recent years, these have become enormous and costly.

At an energy of 600 MeV, an electron has a wavelength of 2×10^{-13} cm, roughly the diameter of a proton (Figure 23.11). Recall, for comparison, that the electron in the hydrogen atom has a wavelength of 2×10^{-8} cm, one hundred thousand times larger. In the 30-GeV accelerators at Brookhaven, New York, and Geneva, Switzerland, protons are accelerated until their wavelengths are shrunk to 4×10^{-15} cm. Still larger accelerators, and smaller wavelengths, are planned for the future, but man is up against the hard fact that he must solve the puzzles of the elementary particles, if he is to solve them at all, by using wavelengths of not much less than 10^{-15} cm.* The 200-GeV machine planned for Weston, Illinois, whose protons will have a wavelength of 0.6×10^{-15} cm, will cost hundreds of millions of dollars to construct. In this sense, nature, through the uncertainty principle, is impeding man's effort to discover its deepest secrets.

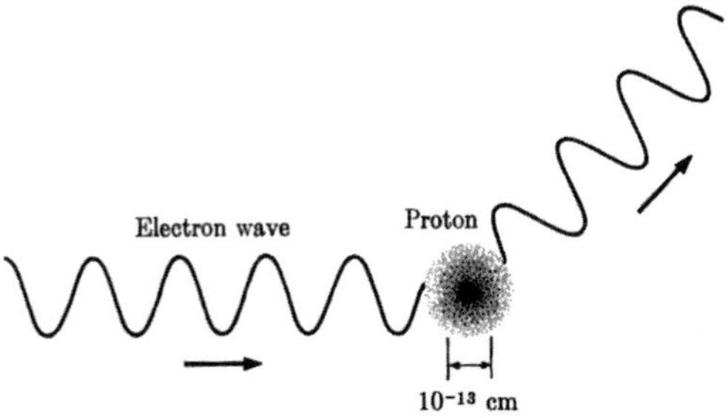

FIGURE 23.11 Details of the inner structure of the proton have been revealed by using electrons whose wavelength is comparable to the dimension of the proton. (The wavelength shown, 2×10^{-13} cm, corresponds to an electron energy of 600 MeV.)

The key ideas of quantum mechanics are strange ideas: discreteness instead of continuity, probability instead of certainty, matter that is neither solid, localizable, nor permanent. These are very likely only a foretaste of stranger ideas to come.

Man's direct perceptions are limited, and as methods of observation and accompanying theories extend beyond the range of this direct perception, it should not be surprising if the ideas and ways of looking at the world which result conflict with the evidence of the senses. Just as a pilot learning to fly on instruments must learn to believe the instruments and forget his sensations, the scientist (and eventually the nonscientist, too) must learn to think in

* Some cosmic-ray particles strike the earth with energies far greater than 30 GeV, and wavelengths much less than 10^{-15} cm. Their number is so small, however, that they have not so far proved valuable for detailed studies of the subnuclear domain.

new ways and give up the preconceived notions based on past experience. The evidence of history so far suggests that man's ability to accommodate to strange new ideas and, even more important, to generate such ideas, will enable him to proceed still a long way in his quest for fundamental understanding.

EXERCISES

23.1. **(1)** The speed of light is the fundamental constant of relativity. In what sense is it "large"? A comparative answer is required. **(2)** Planck's constant h is the fundamental constant of quantum mechanics. In what sense is it "small"?

23.2. **(1)** A 1-kg rock is swung in a circle of radius 100 cm with a speed of 210 cm/sec. How many quantum units of angular momentum does it possess relative to the center of its circle? **(2)** A proton with a speed of 6.3×10^8 cm/sec misses a nucleus by a distance of 3.0×10^{11} cm as it flies by. What is the angular momentum of the proton with respect to the nucleus in units of $\hbar$?

23.3. Name one physical variable that is, so far as we now know, infinitely divisible, free of granularity. Speculate very briefly on the possible alteration of our view of nature that might result, should this quantity later prove to be granular.

23.4. Two particles believed to be muons are created at the same point in a laboratory, and both move at half the speed of light. One covers a distance of 3.30×10^4 cm before it decays, the other a distance of 3.81×10^4 cm before it decays. Do these distance measurements provide any evidence that one or both of the particles may not be a muon? Why or why not?

23.5. A prospector hiking in the desert with a Geiger counter hears a regular succession of clicks from his counter, one every half second for 10 sec. **(1)** Is it possible that he is passing over a vein of radioactive material? Why or why not? **(2)** Should he next **(a)** hold the Geiger counter over the same place for a longer time? **(b)** keep walking? **(c)** start digging? **(d)** check the Geiger counter for malfunction?

23.6. The number of radioactive nuclei remaining in a sample, if graphed as a function of time, yields an exponential curve similar to the one shown in Figure 23.1. **(1)** Sketch such a curve of number vs. time with labeled axes. **(2)** What is the physical significance of the slope of this curve? **(3)** Recall that the slope of an exponential curve is proportional to the height of the curve. How is the proportionality of slope to height of your curve related to the action of a law of probability in the decay process? **(4)** Name one other process in nature that is represented by a law of exponential change (either increasing or decreasing).

23.7. In an Arizona cave archaeologists find a cache of prehistoric sandals. From a knowledge of the relative abundance of C^{14} in the world, they can deduce the approximate number of C^{14} nuclei that must have been contained originally in a sample of this material. Their measurements of radioactivity show that only one quarter of these radioactive nuclei remain in the sample. How old are the sandals?

23.8. **(1)** Name two ways in which a negative kaon can be annihilated. **(2)** Name two ways in which an electron can be created.

23.9. **(1)** Neutrons are cooled by passing them through liquid helium at 3°K. If they emerge with kinetic energy equal to $\frac{3}{2}kT$, what is their wavelength? **(2)** Ordinary thermal neutrons have 100 times as much kinetic energy ($T = 300$°K). What is their wavelength?

23.10. Calculate the wavelength of an automobile of mass 1.5×10^6 gm **(a)** traveling at 60 miles/hr; and **(b)** traveling at 1 Angstrom per century. What is the wavelength of the automobile if its velocity is zero? Compare this answer with the answer to part (b) and discuss.

23.11. A high-energy neutron interacts with a nucleus only if it strikes the nucleus. The "cross section" of the nucleus is then equal to the projected area of the nucleus (see figure). For low-energy neutrons, however, the effective cross section of the nucleus may be much greater than its projected area. A low-energy neutron can interact with a nucleus even if its classically calculated trajectory misses the nucleus by a substantial margin. The lower the energy of the neutron, the greater the effective nuclear cross section can be. Explain these facts.

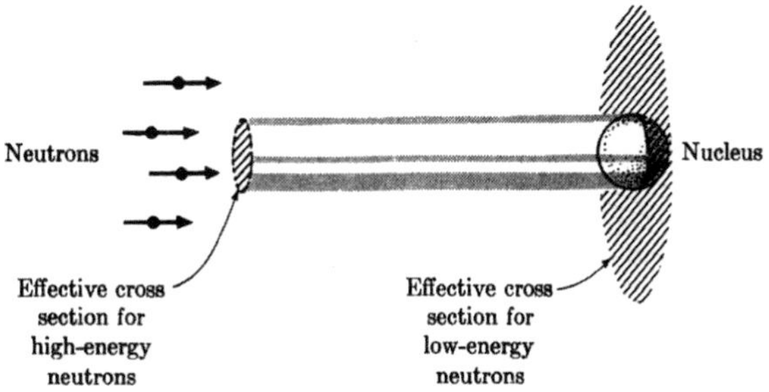

23.12. A neutron, an electron, and a photon all have the same wavelength, $\lambda = 10^{-8}$ cm (1 Angstrom). **(1)** Find the speed of the neutron in cm/sec. Is this "slow" or "fast"? **(2)** Find the energy of the neutron in eV. **(3)** Find the energy of the electron in eV. **(4)** Find the energy of the photon in eV. In what part of the electromagnetic spectrum does this photon lie? (For neutron and electron, nonrelativistic formulas suffice.)

23.13. **(1)** Electrons in the Stanford Linear Accelerator reach an energy of 20 GeV. What is their wavelength at this energy? How does this compare with the size **(a)** of an atom? **(b)** of a nucleus? **(2)** Protons with an energy of 20 GeV have a wavelength that is very nearly the same as the wavelength of 20 GeV electrons. Why? (HINT: Relativity is important for both particles.)

23.14. Show that the combination of fundamental constants appearing in Equation 23.10, h^2/me^2, has the physical dimension of length.

23.15. **(1)** The combination h/mc is known as the Compton wavelength of a particle. What is the Compton wavelength of **(a)** an electron? **(b)** a proton? **(2)** At what speed is the wavelength of a particle equal to its Compton wavelength?

23.16. The wave nature of matter is irrelevant in the macroscopic world, yet the whole structure of the macroscopic world depends upon the wave nature of matter. Interpret this statement and discuss it briefly.

23.17. Discuss very briefly why we are unaware in everyday life of **(a)** the wave nature of particles; **(b)** the particle nature of electromagnetic radiation; **(c)** the creation and annihilation of particles.

23.18. **(1)** An electron has an approximate speed of 10^8 cm/sec. What is its approximate wavelength? **(2)** Its uncertainty of speed is 3×10^7 cm/sec. What is its uncertainty of position? **(3)** What kind of electromagnetic radiation has the same wavelength?

23.19. The momentum of a 1-gm pellet is determined to within an uncertainty $\Delta p = 10^{-6}$ gm cm/sec. To what accuracy could its position in principle be determined? Make an approximate guess of the maximum accuracy to which its position could be determined in practice.

23.20. An electron in its lowest state in the hydrogen atom has an uncertainty of position of about 10^{-8} cm (see Figure 23.4). What is its uncertainty **(a)** of momentum? **(b)** of velocity? How does this compare with the speed of light? Explain how the uncertainty principle "prevents" the collapse of the hydrogen atom.

23.21. An electron is located in space within an uncertainty $\Delta x = 10^{-4}$ cm. Its speed is approximately 10^8 cm/sec. It is desired to measure its momentum to an accuracy of one part in a thousand. Is this possible? Justify your answer with an appropriate calculation.

23.22. A proton is known to be located within a cubical box 1 meter on a side. **(1)** Explain why the proton cannot be precisely at rest within the box. **(2)** Estimate the minimum speed of the proton that is consistent with its localization within the box.

23.23. About what fractional range of wavelengths must be superposed to produce each of the wave patterns shown in the figure?

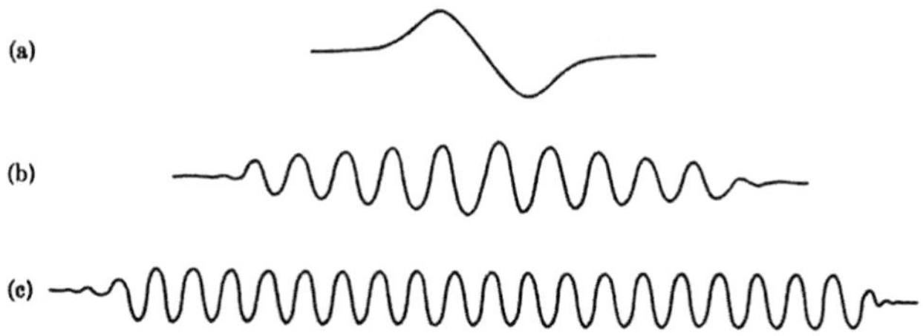

23.24. **(1)** Sketch the approximate appearance of a wave formed by superposing many waves which span a 20% range of wavelength: $\Delta\lambda \cong 0.2\lambda_{av}$. **(2)** If what you have sketched is the wave function of a material particle and $\lambda_{av} = 10^{-9}$ cm, what are Δx and Δp, its uncertainties of position and momentum? (Only approximate answers are required.)

23.25. The superposition of different wavelengths can localize a wave. Conversely, a wave that is localized must contain components of different wavelengths. Consider the sound wave emanating from one of the long pipes of an organ, whose frequency is approximately 30 sec^{-1}, if the note is sounded for only one third of a second, long enough for 10 cycles of oscillation. **(1)** What is the length of the wave train ($v = 3.4 \times 10^4$ cm/sec)? **(2)** What is the average wavelength? **(3)** What is the range of wavelengths that must be mixed to form this wave? Optional: Discuss the musical implications. Why do composers never call for short-duration notes of very low frequency?

23.26. A combination of constants that proves to be useful and important in quantum mechanics is $e^2/\hbar c$. This combination is called the fine structure constant; its usual symbol is α. **(1)** Calculate the numerical magnitude of α and verify that it is a dimensionless number. **(2)** If an international commission increased by a factor of 10 the standard unit of measurement of mass, by what factor would α change?

23.27. Explain briefly in your own words why high-energy projectile particles are needed in order to probe the interior of nuclei and particles.

23.28. Just as relativity introduced into our description of nature both more relativity and more invariance, quantum mechanics introduced both more discreteness and more fuzziness. **(1)** Name, and illustrate with a suitable example, a quantity once believed to be continuously variable which proved to be discrete, or granular. **(2)** Do the same for a quantity once believed to be sharply definable which proved to be "fuzzy." **(3)** How does the constant h enter into each of your examples?

TWENTY-FOUR

Atoms

The photon was "discovered" by a theoretical physicist, Albert Einstein, in 1905.* Five years earlier another theoretical physicist, Max Planck, had introduced the quantum idea on which Einstein capitalized. Because these two men took the vital first quantum steps down the road that led eventually to a full theory of atomic structure, we place the photon first in a chapter on atoms.

24.1 The photon

The photon is, from our modern point of view, one of the elementary particles—a rather special particle, to be sure, but not a unique sort of entity. It has no mass, no charge, and one quantum unit of spin (equal to $\hbar$). It interacts with all charged particles (as well as with some neutral ones); it is the carrier of electric and magnetic forces; and it *is* light—or, more generally, electromagnetic radiation. In the universe at large, it is the currency of the light and power company and of the FCC rolled into one, for photons bring to earth almost all of our energy and all of our information about the rest of the world. On earth, photons serve most of man's communications needs. They also provide—and this is what concerns us here—a link between man and atoms. Every atom emits and absorbs (creates and annihilates) photons of particular energies and frequencies. Studies of these photons have revealed most of what we know about the structure of atoms.

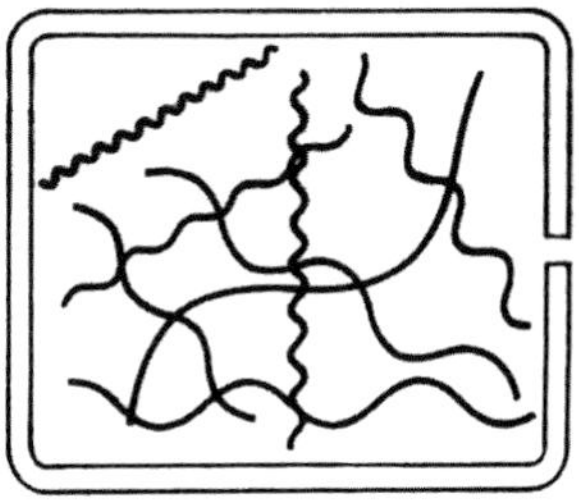

FIGURE 24.1 The "hohlraum," or hollow cavity, studied by Planck. Indicated schematically are different electromagnetic waves within the cavity. These waves are in thermal equilibrium with the matter in the walls. Radiation emerging from a small hole in the cavity can be studied to reveal the distribution of energy among waves of different frequency.

Planck was led to the quantum idea by studying the way in which energy is distributed 136
inside a closed box within which electromagnetic waves are bouncing back and forth (Figure 24.1). The waves themselves carry energy that is continuously being exchanged with the energy of atoms in the walls, and the problem Planck set out to solve was how to explain the way in which all the available energy was distributed—some of it in the walls, some of it in the radiation, and that in the radiation divided among waves of different frequencies. This is a complex system of many atoms and many wavelengths. Planck found that he could account for the behavior of this system and successfully explain how the available energy was shared among the various parts of the system only if he postulated that energy could

* It was for this discovery, rather than for the theory of relativity, that Einstein received the Nobel Prize in 1921.

be transferred from matter to radiation in bundles of a certain size, given by the equation

$$E = hf, \tag{24.1}$$

where E is the energy in one bundle of radiation—or one photon, as we would say now—and f is the frequency of the radiation. Actually Planck thought only in terms of quantized energy exchange, not in terms of quantized radiation. He continued to believe in a continuous pool of radiant energy.* By adjusting the numerical value of his new constant h to get the best agreement between the calculated and measured distribution of energy among different wavelengths, Planck was able to determine h to an accuracy of about 2%.

Einstein took the second radical step. He assumed that the quantum bundle of energy, after being transferred from matter to radiation, retained its quantum identity, behaving as an identifiable "particle" of light, a photon, such that it could be reabsorbed by matter only as a whole unit or not at all. Einstein realized that this granular concept of electromagnetic radiation could account beautifully for a different and very much simpler phenomenon than the one studied by Planck. Planck had considered a system composed of a vast number of atoms and of photons; Einstein considered an experiment involving the elementary event of the absorption of a single photon.

A few years earlier, it had been noted that when ultraviolet light shone upon the surface of some metals, electrons were emitted from the surface (Figure 24.2). This phenomenon, christened the photoelectric effect, could be understood qualitatively in terms of Maxwell's wave theory of light, but that theory failed entirely to account for the quantitative details of the process. According to the wave theory, the electromagnetic radiation striking the surface set electrons near the surface into motion, and some of these were caused to move so rapidly that they could escape and fly off from the surface.

The wave theory had two main predictions to make, and both were in disagreement

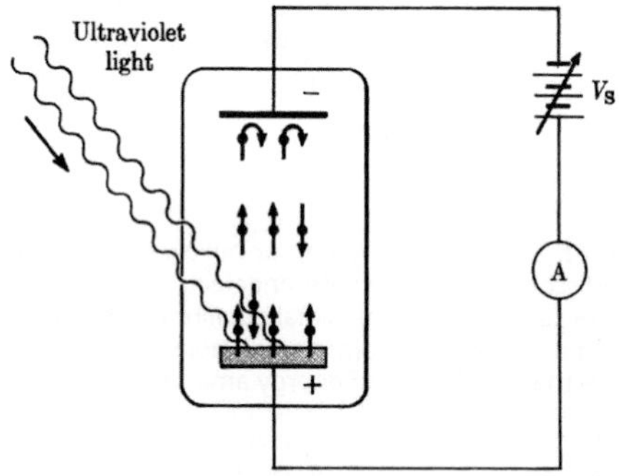

FIGURE 24.2 The photoelectric effect. Ultraviolet light incident upon a metal surface releases electrons. The stopping potential V_S measures the energy of the electrons.

with the facts. First, more intense radiation should have pushed harder on the electrons, and caused them to fly off with more energy. Instead, the energy of the departing electrons did not vary as the light was made more intense. The only change was that a larger number of electrons escaped. Second, according to the wave idea, the energy should not depend particularly upon the frequency of the light, as long as sufficient light intensity shines on the surface. In fact, higher-frequency radiation caused the electrons to fly off more energetically, even if the intensity was lowered. Below some particular frequency, no electrons at all escaped. Ordinary visible light was incapable of ejecting any electrons, however intensely it shone on the surface.

Einstein noted that the photon theory of light could explain the observed facts of the photoelectric effect very simply and elegantly. According to the wave idea, an electron gradually absorbs energy from the wave, and can absorb any amount, large or small. Suppose

* In fact, Planck never accepted the photon idea of Einstein.

instead, said Einstein, that energy can be absorbed from the incoming light only in bundles of a certain definite size. An electron either absorbs exactly one whole photon, or it absorbs none. Increasing the intensity of the light increases the number of photons, but it does not change the energy of each one. The higher intensity results in more electrons absorbing photons, but it does not increase the amount of energy absorbed by any one. If the energy of each photon follows Planck's equation, $E = hf$, the energy absorbed by a given electron depends on the frequency of the impinging light but not on its intensity. The chance that one electron absorbs more than one photon is negligible, because the number of photons is much lower than the number of electrons. Having absorbed a photon, an electron either flies from the surface of the metal or dissipates its energy within the metal in a time so short that it has no significant chance to absorb a second photon.

The two basic facts about the photoelectric effect—that the number of electrons depends on the intensity of the light, and that the energy of each electron depends on the frequency of the light—were simply explained by the photon hypothesis, while being entirely unexplainable by the wave theory. In the photoelectric effect, one was witnessing the single events of photon absorption by electrons, much simpler than the complex system considered by Planck. For most physicists, there was no recourse but to accept the photon as part of reality, in spite of all of the accumulated evidence for the wave nature of light.

The verification of Einstein's theory of the photoelectric effect came through a careful study of the maximum kinetic energy of the ejected electrons as a function of the frequency of the incident radiation. According to the photon idea (and energy conservation), an electron that absorbs a photon of frequency f acquires an energy equal to the entire energy of the photon, hf. The electron may lose some of this energy before escaping from the metal and therefore fly off with a kinetic energy of less than hf. Or it may lose all of its energy within the metal and not escape at all. Even if the photon absorption occurs at the metal surface, the electron will have to expend some of its energy in order to escape, for there is a force of attraction holding the electrons within the metal. The energy required to overcome this attractive force is called the work function of the metal, for which the usual symbol is W. For a particular metal with a smooth clean surface, W is a constant. Because of the work function, the kinetic energy of an ejected electron will be at most the photon energy minus W:

$$(K.E.)_{\max} = hf - W\,. \tag{24.2}$$

If the maximum kinetic energy of escaping electrons is measured at each of several frequencies, and the results plotted as a function of frequency, a straight-line graph should result, as shown in Figure 24.3. This graph shows a "threshold," that is, a minimum frequency below which no electrons escape. The threshold frequency, determined by the condition $\boldsymbol{hf = W}$, is shown for several metals in Table 24.1, along with values of W for these metals. For comparison with these values, note that the high-frequency end of the visible spectrum (violet light) has a frequency of about 8×10^{14} sec^{-1}, and a photon energy of about 3.3 eV.

Table 24.1 Photoelectric Threshold Frequency and Work Function for Several Metals.

Metal	Threshold Frequency (sec-1)	Work Function (eV)
Cesium*	4.6×10^{14}	1.9
Beryllium	9.4×10^{14}	3.9
Titanium	9.9×10^{14}	4.1
Mercury	1.09×10^{15}	4.5
Gold	1.16×10^{15}	4.8
Palladium	1.21×10^{15}	5.0

* The threshold frequency for Cs lies in the visible part of the spectrum. For the other metals, ultraviolet light is required to eject electrons.

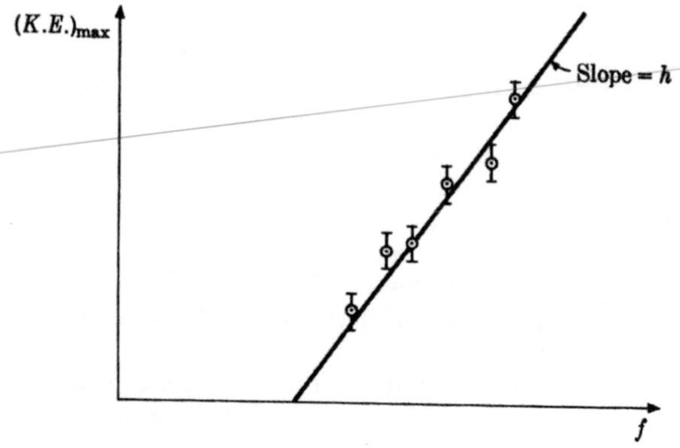

FIGURE 24.3 Maximum kinetic energy of ejected electrons vs. the frequency of the incident radiation.

The most significant feature of the graph in Figure 24.3 is the slope of the line, which is exactly Planck's constant h. Experimental measurements leading to such a graph therefore provide a way to measure h quite independently of Planck's method of determining this constant. In practice the photoelectric effect provides a value for the ratio of two fundamental constants, h and the electron charge e, from which h can be found if e is known. The actual property of the electron that is easy to measure is not energy but energy per unit charge. If a "stopping potential" V_S is applied, as shown in Figure 24.2, just sufficient to turn back the most energetic electrons, it satisfies the energy equation,

$$eV_S = (K.E.)_{max} \,. \tag{24.3}$$

Therefore $(K.E.)_{max}$ can be replaced by eV_S in Equation 24.2 to give

$$eV_S = hf - W \,,$$

or

$$V_S = \frac{h}{e} f - \frac{W}{e} \,. \tag{24.4}$$

A graph of the directly measured quantity V_S vs. frequency f is a straight line with slope h/e.

137 It is well to remember that in 1905 the equation $E = hf$ was a *law* of nature, required to account for several experimental facts, but not yet resting on any underlying general theory. Representing as it does a fundamental statement about electromagnetic energy, this simple equation deserves careful attention. One obviously important fact about it is its statement that photon energy is proportional to frequency. The photon of a high-frequency nuclear gamma ray carries much more energy than the photon of a low-frequency radio wave, in fact about 10^{14} times more. A single gamma ray has so much energy that it is very easily detected by itself. A single photon leaving the antenna of a radio broadcasting station has so little energy that it is undetectable. Only the combined effect of a vast army of radio photons can be detected, and when many photons act in consort, their individual-particle aspect is overwhelmed by the overall wave aspect. The radio engineer never has to be concerned with individual photons and can think always in terms of waves. The nuclear physicist, on the other hand, thinks of gamma rays primarily as particles, and is less concerned with their wave aspect. In the intermediate region of visible light, single photons can be detected with sensitive photocells, although in most optical experiments, large numbers of photons are involved.

Another point about the photon energy equation worth special notice is the role of Planck's constant, appearing as a constant of proportionality between a particle-like variable and a wavelike variable, much as in the de Broglie equation. If h were much smaller, the amount of energy in one photon would be smaller and the quantum aspect, or particle aspect, of light would be less noticeable. If we imagine h disappearing altogether, becoming zero, then there would be no photons and light would be purely waves again. On the other hand, if h were much larger, the quantum aspect of light would be more noticeable. Single photons would carry so much energy that they might be separately observed as flashes of light by the eye of man. (This is pure fiction, for if h really *were* larger, atoms would be larger, and man would be larger too. This discussion is only for the purpose of clarifying the meaning of h, whose size, of course, cannot be manipulated.) What Planck's constant does is to determine the scale of the quantum world. We live in a classical world because the energies of our everyday experience—the energy of raising a hand, the energy of reading a page—are very large compared to the energy of a single photon of light.

The similarity between the Planck-Einstein equation $E = hf$ and the de Broglie equation 138
$p = h/\lambda$ is no accident. For light they are in fact equivalent equations. Even before the advent of the quantum theory, it was known that electromagnetic radiation carries both energy and momentum, and that these quantities are simply related in a propagating wave by the equation,

$$E = pc\,. \tag{24.5}$$

(We now know this to be the energy-momentum relationship for any massless particle.) The photon equation may therefore be rewritten in terms of photon momentum:

$$pc = hf\,. \tag{24.6}$$

Since $c = f\lambda$ for a wave moving at speed c, this equation in turn can be written $p = h/\lambda$, identical to the de Broglie equation. This equivalence is true only for photons (or other massless particles). De Broglie's contribution was the suggestion that the momentum-wavelength relationship should be true for *all* particles.

24.2 Atomic spectra

Every atom has its own signature. Through the set of characteristic frequencies that comprise its line spectrum (see Section 18.13), it identifies itself uniquely. This fact was known and used many decades before it received a quantum interpretation. Since musical instruments and many other mechanical systems were known to possess characteristic discrete vibration frequencies, it did not seem surprising that atoms too should be capable of vibratory motion, or that every different kind of atom should have its own particular set of frequencies. Before the photon forced a new look at line spectra, these atomic signatures revealed only that atoms contained electric charge and that this charge could oscillate. Physicists assumed, in accordance with classical electromagnetic theory, that the radiated frequencies coincided with the oscillation frequencies of electric charge within the atom.

It is an important fact of physics and an agreeable attribute of nature that great insights flow often from simple systems or simple experiments. Think of Galileo rolling balls down troughs, Faraday plunging a bar magnet into a coil, or Joule stirring a container of water. In atomic physics, the hydrogen atom, lightest and simplest of all atoms, has been the inspiration and the testing ground for most of the important forward steps, from Bohr's first atomic theory in 1913 to the quantum electrodynamics of Feynman, Schwinger, and Tomonaga* in 1949. Even before the first hint of a quantum theory of atoms, the hydrogen atom made

* In 1965 Richard P. Feynman, Julian Schwinger, and Sin-itiro Tomonaga were honored with the Nobel Prize for their success in refining the quantum theory of matter and radiation to account for certain minute but significant differences between the observed line spectrum of hydrogen and the line spectrum predicted by earlier quantum mechanics.

its special simplicity evident through its line-spectrum signature. In 1884 Johann Balmer discovered that the frequencies of the known lines of hydrogen could be expressed mathematically by a very simple empirical law. Since it is not frequency but wavelength that is measured directly, we divide frequency by the speed of light and write Balmer's law in the form

$$\frac{f}{c} = \frac{1}{\lambda} = \mathcal{R}\left(\frac{1}{2^2} - \frac{1}{m^2}\right), \tag{24.7}$$

in which $\mathcal{R}$ is a constant and m is an integer greater than two. The constant $\mathcal{R}$, called the Rydberg constant in honor of Johannes Rydberg, who discovered other approximate spectral formulas a few years later, is an exceedingly accurately known constant, equal to 109,677.58 cm^{-1}. Even at the time of Balmer's work, this constant could be determined with considerable precision, for spectral wavelengths were among the best measured quantities in physics. Even to fit the four visible lines of the hydrogen spectrum accurately with so simple a law was an interesting achievement. Balmer then found that his formula accounted perfectly for five more lines in the ultraviolet. Counting another five lines observed in starlight and attributed to hydrogen, Balmer's formula simply summarized fourteen accurately measured wavelengths, all that were known of the hydrogen spectrum at the time. There could be no doubt of its significance. Like Kepler's laws of planetary motion, Balmer's law neatly packaged a collection of data, although not providing any explanation for the data.

The set of spectral lines expressed by Balmer's law is known as the Balmer series. Since spectral lines known in 1885 corresponded to every integral value of m from 3 through 16, it was natural to assume that additional lines corresponding to higher m values also existed, awaiting discovery. More were soon found. By the time of Bohr's work in 1913, 33 lines of the Balmer series had been observed (to $m = 35$), all accurately described by Balmer's formula. The higher members of the Balmer series cluster ever closer together, as shown in Figure 24.4, with a definite series limit being determined by letting m become infinite in Balmer's formula. It is instructive to calculate the wavelengths of the first few members of the Balmer series:*

$$\frac{1}{\lambda_1} = \mathcal{R}\left(\frac{1}{4} - \frac{1}{9}\right) = (6{,}565\ \text{A})^{-1}, \text{ red}, \tag{24.8}$$

$$\frac{1}{\lambda_2} = \mathcal{R}\left(\frac{1}{4} - \frac{1}{16}\right) = (4{,}863\ \text{A})^{-1}, \text{ green}, \tag{24.9}$$

$$\frac{1}{\lambda_3} = \mathcal{R}\left(\frac{1}{4} - \frac{1}{25}\right) = (4{,}342\ \text{A})^{-1}, \text{ blue-violet}, \tag{24.10}$$

$$\frac{1}{\lambda_4} = \mathcal{R}\left(\frac{1}{4} - \frac{1}{36}\right) = (4{,}103\ \text{A})^{-1}, \text{ violet}, \tag{24.11}$$

The series limit is located in the ultraviolet at about twice the frequency (half the wavelength) of the first line of the series:

$$\frac{1}{\lambda_\infty} = \frac{1}{4}\mathcal{R} = (3{,}647\ \text{A})^{-1}, \text{ ultraviolet}, \tag{24.12}$$

Balmer was sufficiently impressed by the success of his law to speculate that other hydrogen series might exist described by similar laws in which the inverse squares of other integers replaced the ¼ in his formula. His guess for a more general spectral law can be written

* These calculated wavelengths are vacuum wavelengths. In practice, wavelengths are usually measured in air, not in vacuum. Because of the index of refraction of air, the Balmer wavelengths in air are about 0.03% less than in vacuum.

$$\frac{1}{\lambda} = \mathscr{R}\left(\frac{1}{n^2} - \frac{1}{m^2}\right), \tag{24.13}$$

in which both n and m are integers (m greater than n). His speculation (it can hardly be called a prediction) was first confirmed by Friedrich Paschen, who in 1908 discovered the first two members of a new hydrogen series in the infrared. This series, now called the Paschen series, is characterized by $n = 3$, with $m = 4, 5, 6 \ldots$ The Brackett series ($n = 4$), and the Pfund series ($n = 5$), which, like the Paschen series, lie entirely in the infrared, were first identified in the 1920s.

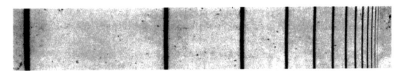

FIGURE 24.4 The Balmer series of spectral lines emitted by hydrogen. This photograph spans about one octave of the electromagnetic spectrum, from 6,600 A on the left (red) to 3,600 A on the right (near ultraviolet). [From Gerhard Herzberg, *Atomic Spectra and Atomic Structure* (New York: Dover Publications, 1944).]

The highest frequency (shortest wavelength) series of hydrogen lines, characterized by $n = 1$ (the Lyman series), lies entirely in the ultraviolet, with inverse wavelengths running from $\frac{3}{4}\mathscr{R}$ to $\mathscr{R}$. Although Theodore Lyman observed the first member of this series in 1906, he did not attribute it to hydrogen until Bohr predicted such a line seven years later.

The hydrogen spectrum, although the simplest known, is not the only one to exhibit regular series. Johannes Rydberg, in the 1890s, succeeded in finding approximate empirical laws to describe various spectral series recognized in the radiations of other elements. Figure 24.5 shows two series of sodium lines. Rydberg's laws were neither so simple nor so accurate as Balmer's law, but one feature of them proved to be of eventual theoretical significance. All of Rydberg's expressions for wave number (inverse wavelength) had one thing in common near the series limits: They contained a term approximately equal to the fixed constant divided by the square of an integer. Bohr was later able to attribute this common feature to the fact that for all elements, the higher members of spectral series arise from the motion of a single electron far from the atomic nucleus.

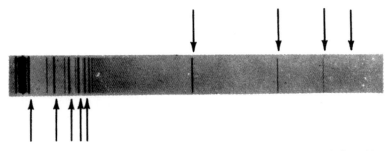

FIGURE 24.5 Series of lines in the spectrum of sodium. One series in the visible region is indicated by arrows below the photograph, another series in the ultraviolet by arrows above the photograph. [From Gerhard Herzberg, *Atomic Spectra and Atomic Structure* (New York: Dover Publications, 1944).]

Another empirical law of spectra discovered in the early part of this century proved to be significant in the development of atomic theory. First noticed by Rydberg and later advanced as a general principle by Walther Ritz (1908) was the fact that the wave number of a spectral line of a particular element is often equal to the sum or difference of two other wave numbers in the spectrum of the same element. This "combination principle" is evident in Balmer's law for hydrogen. The wave numbers of the first two lines of the Balmer series, for instance, are

$$\frac{1}{\lambda_1} = \mathcal{R}\left(\frac{1}{4} - \frac{1}{9}\right),$$

$$\frac{1}{\lambda_2} = \mathcal{R}\left(\frac{1}{4} - \frac{1}{16}\right).$$

Their difference,

$$\frac{1}{\lambda_2} - \frac{1}{\lambda_1} = \mathcal{R}\left(\frac{1}{9} - \frac{1}{16}\right). \tag{24.14}$$

is equal to the wave number of the first line of the Paschen series. The Ritz combination principle proved to be valid for many lines in many elements, and helped greatly to organize complex spectra into simple patterns. Note that the principle applies not to wavelengths but to inverse wavelengths (wave numbers), or, equivalently, to frequencies. Because of the proportionality of spectral frequency to atomic-energy change, we now recognize the Ritz combination principle as nothing more than a statement of the law of energy conservation. At the time of its discovery, however, the principle was, like Balmer's law, a useful but mysterious fact of nature.

By the time that Ernest Rutherford literally opened up the atom in 1911, atomic spectroscopy was in an advanced state of development. The light that came from atoms had been accurately measured, catalogued, and found to be governed by several empirical laws. It had been used for practical ends, to identify elements and to unravel radioactive decay chains. Yet the mechanism within the atom responsible for this light was entirely mysterious. A wealth of data awaited explanation.

24.3 The electron

That electrically charged particles reside within atoms had been surmised since the time of Faraday. Not until 1897 was definitive evidence produced for the existence of such a subatomic particle, the electron. In a series of ingenious experiments J. J. Thomson succeeded in proving that cathode rays are beams of negatively charged particles very probably much smaller in size and considerably less massive than atoms. Even today we often refer to beams of electrons as cathode rays, a name dating from a time when little for sure was known about them except that they emanated from a negatively charged terminal, or cathode.

Figure 24.6 shows the main evolutionary steps in the family history of the cathode-ray tube. In the earliest version,* a high voltage applied to terminals in a rarefied gas caused a current to flow and the gas to glow. Later, improved vacuum pumps made it possible to lower the gas pressure to a point such that the gas no longer glowed although current still flowed. Interposition of a metal obstacle between cathode and anode (positive terminal), and alteration of the placement of the anode, made clear that something was emanating from the cathode. This "something," which came to be called cathode rays, could be deflected by a magnetic field, but otherwise seemed to move in straight lines, and caused glass to fluoresce. In one of Thomson's cathode-ray tubes [pictured in Figure 24.6(c)], a narrow beam of cathode rays was defined by means of a pair of anodes, each containing a narrow slit. The beam then passed between a pair of parallel plates which could be electrified, and struck the end of the tube where a scale permitted deflections to be measured. The parallel plates,

* The glow of rarefied gases stimulated by the passage of electric current through them was observed as early as 1748. Not until the 1850s were gas-discharge tubes used as subjects of serious scientific study. Modern fluorescent bulbs and neon lights are direct descendants of these early gas-discharge tubes.

when charged, could deflect the beam electrically; electromagnets placed near the tube could deflect it magnetically. The modern cathode-ray tube does not differ in principle from Thomson's tube, although it has been refined in many details. The cathode is now coated with special materials and heated in order to make it emit electrons copiously. The anode, or pair of anodes, is replaced by a series of plates and cylinders designed to produce a sharply focused beam of controllable intensity. The glass of the screen is coated with phosphor for more intense fluorescence. In oscilloscope tubes two pairs of parallel plates permit electric deflection, both vertical and horizontal. In a TV tube, the position of the electron beam on the screen is controlled magnetically by means of coils placed next to the neck of the tube.

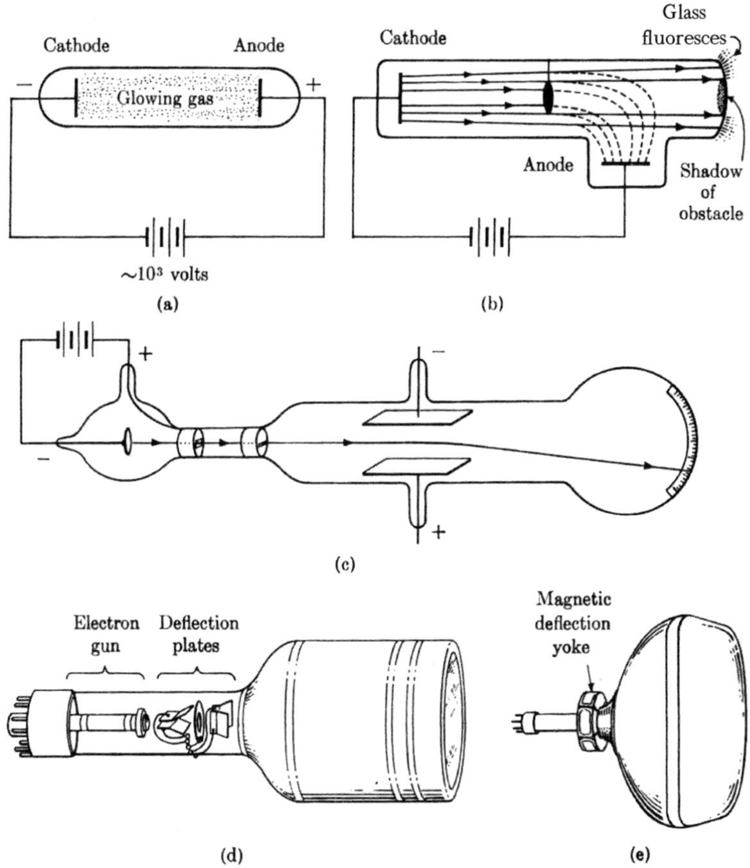

FIGURE 24.6 Evolution of the cathode-ray tube. **(a)** Gas discharge tube, within which rarefied gas glows when high voltage is applied to its terminals. **(b)** With higher vacuum, residual gas no longer glows, but glass fluoresces. (Solid lines indicate electron paths; dashed lines indicate lines of electric field.) **(c)** Cathode ray tube of J. J. Thomson, containing anode slits to define narrow beam and parallel plates to deflect beam electrically. **(d)** A modern cathode-ray tube of the type used in oscilloscopes has a more elaborate series of electrodes to produce a sharply focused beam, two pairs of plates for deflection both horizontally and vertically, and a phosphor screen. **(e)** A television tube uses magnetic instead of electric deflection.

By means of electric deflection, Thomson showed the cathode rays to be negatively 139
charged. The small size of the negative particles he inferred from the ease of their passage through a gas; the distance required for the beam to dissipate its energy was much greater than the distance in which a beam of ionized atoms or molecule of comparable energy

would be stopped. Most significant among Thomson's measurements was his approximate determination of the ratio of the charge to the mass of the cathode-ray particles. Through this measurement he came into closest touch with the subatomic world, for the measurement revealed a fundamental property of a single electron. As outlined in Section 17.4, the charge-to-mass ratio is determined by accelerating electrons through a known potential difference, then deflecting them in a known magnetic field. A modern value of the charge-to-mass ratio of the electron is

$$\frac{e}{m} = 5.2727 \times 10^{17} \text{ e.s.u./gm} . \tag{24.15}$$

With appropriate scientific caution, Thomson attributed the large magnitude of his measured *e/m* ratio to "the smallness of *m* or the largeness of *e*, or to a combination of these two." Then he went on to argue that the electrons (or corpuscles, as he called them) were very likely much less massive than atoms. He wrote in summary:

> The explanation which seems to me to account in the most simple and straightforward manner for the facts is founded on a view of the constitution of the chemical elements which has been favorably entertained by many chemists: this view is that the atoms of the different chemical elements are different aggregations of atoms [particles] of the same kind. . . . Thus on this view we have in the cathode rays matter in a new state, a state in which the subdivision of matter is carried very much further than in the ordinary gaseous state: a state in which all matter—that is, matter derived from different sources such as hydrogen, oxygen, & c.—is of one and the same kind; this matter being the substance from which all the chemical elements are built up.

In the decade following Thomson's work, much additional evidence supported the soundness of his conclusions. Electrons were found to be ejected from metals by ultraviolet light (the photoelectric effect). Beta particles shooting out of radioactive substances were identified as electrons. The momentum of very high-speed electrons was found to depart from the classical value, *mv*, in the way predicted by relativity. Ions were interpreted as atoms with an excess or a deficiency of electrons. The existence of electrons within atoms provided a simple reason why atoms can emit and absorb electromagnetic radiation. Very shortly few doubts remained that electrons were indeed subatomic particles found in all atoms. Seventy years later, the electron remains the most primordial bit of matter we know, whose internal structure, if any, is a complete mystery.

The charge of the electron was first separately determined with some degree of precision by Robert A. Millikan in 1913. Beautifully simple in conception, Millikan's oil-drop experiment (Figure 24.7) goes right to the heart of the problem by providing a way to measure the force acting on a single electron or a small number of electrons. Tiny droplets of oil are suspended in air between plates that can be charged to provide a known electric field in the region under study. Each droplet, although microscopic in size, contains billions of atoms, yet the net imbalance of charge on a droplet may be equal to the charge of a single electron, or a few electrons, or zero. Since the droplet may have either a slight excess or slight deficiency of electrons, its charge, if any, may be either negative or positive.

In the absence of an electric field, a droplet is acted upon by two forces, a gravitational force acting downward and a frictional force of air drag acting upward. These two forces quickly come into balance as the droplet reaches terminal speed. The fall of the droplet under these conditions can be watched through a microscope and timed. The heavier the droplet, the more rapidly it falls. If the exact law of air friction is known, the mass of the droplet can be calculated from its terminal speed. Once its mass is known, each of the two forces acting on it can also be calculated.

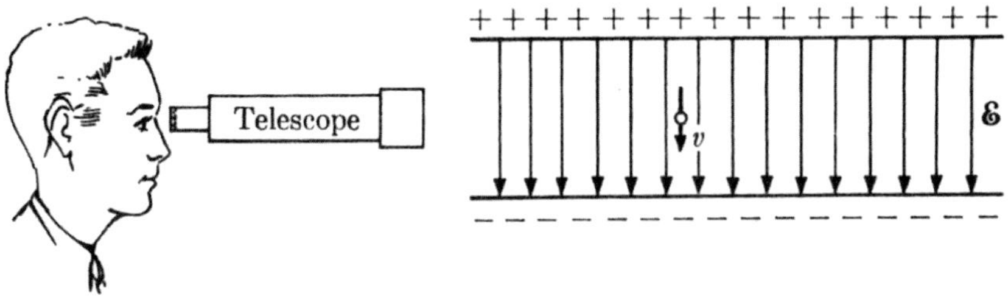

FIGURE 24.7 Schematic diagram of Millikan oil drop experiment. The terminal speed of a tiny charged droplet is measured as it moves under the combined influence of gravity, frictional air drag, and a known electric field.

Now the electric field is turned on. If the droplet has any net charge, a third force is added to the other two. This force is simply equal to the product of the charge on the droplet and the electric field ($F_E = Q'\mathcal{E}$). The droplet responds by accelerating to some new terminal speed. It might reach a higher downward speed, or a lower downward speed, or it might reverse direction and move upward. In any case, it settles at a new velocity where the sum of all three forces is zero. From the new velocity (and with a knowledge of the mass of the droplet), the electric force can be calculated. Since the electric field is known, the electric force at once determines the electric charge.

In a long series of measurements, Millikan produced direct and unambiguous evidence for the quantization of charge. The net charge on a droplet proved always to be an integral multiple (positive or negative) of a smallest unit of charge. This unit he assumed to be the magnitude of the charge on a single electron. A modern value for this fundamental constant is

$$e = 4.803 \times 10^{-10} \text{ e.s.u.} = 1.6021 \times 10^{-19} \text{ coulomb.} \tag{24.16}$$

Given the charge of the electron and its charge-to-mass ratio, the mass of the electron can of course be inferred at once. A modern value for the electron mass is

$$m = 9.109 \times 10^{-28} \text{ gm.} \tag{24.17}$$

In energy units, this is

$$mc^2 = 0.5108 \text{ MeV.} \tag{24.18}$$

An interesting sidelight on the Millikan oil-drop experiment is the essential role played by aerodynamics in the determination of a fundamental property of the electron. At the time he began his work, Millikan did not know with sufficient precision the exact law of frictional air drag on small spheres. Much of the work required to obtain an accurate value of e went into refining and testing the law of drag force, for this was the yardstick against which the electric force had to be measured.

24.4 Key postulates of Bohr's atomic theory

In hundreds of places in the modern world—in books and newspapers, on billboards and television and army uniforms—we see atoms represented symbolically by a small central nucleus surrounded by the intersecting oval orbits of several electrons (Figure 24.8). This is a tribute to the impact of Niels Bohr on the modern world. At the same time it is an unfortunate distortion of the contemporary scientist's conception of the atom. Bohr's planetary model has evolved into what might better be called a nebular model, with electrons spread as waves of probability over the whole volume of the atom. This development notwith- 140 standing, Bohr's original atom model deserves a place in a survey of the fundamentals of physics, for, apart from its historical importance as a giant step toward quantum mechanics, it contained several basic ideas that have withstood the test of time and which remain as essential aspects of modern atomic theory. In this section and the next two we shall develop the Bohr model of the hydrogen atom. Later sections of the chapter are concerned with some of the subsequent developments of quantum mechanics that created a revised picture of hydrogen and made possible a successful theory of heavier atoms as well.

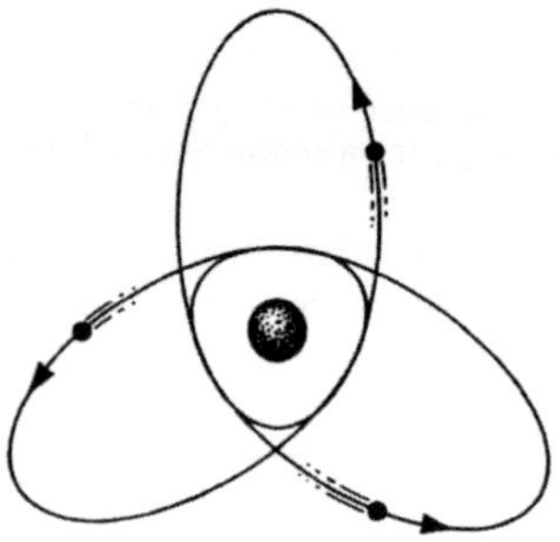

FIGURE 24.8 An all-too-common contemporary view of the atom. In fact it is not possible to assign classical trajectories to electrons in atoms.

With a fresh Ph.D. in his pocket, Niels Bohr journeyed from Denmark to England in 1911 and came under the influence of two of the giants of atomic science, J. J. Thomson of Cambridge, and Ernest Rutherford in Manchester. Back in Copenhagen the following year, Bohr began to work out the details of a quantum mechanical model of the atom. His 1913 paper, describing what we now call the Bohr atom, represented the first successful application of quantum ideas to the structure of matter.

Bohr drew upon three main threads of earlier work which, to that time, had not been successfully woven together. First was the basic quantum idea of Planck and Einstein that radiation is emitted and absorbed only in discrete energy packets, with the energy of a single packet and its frequency being related by the equation $E = hf$. Second was the highly developed empirical science of spectroscopy, particularly the simple regularities of the hydrogen spectrum and the general combination principle of Ritz (Section 24.2). Third was the new physical picture of the atom advanced by Rutherford—a heavy positively charged nucleus of small dimensions surrounded by electrons. Bohr recognized that in some way Planck's quantum constant had to play a role in the mechanics of the atom, for only then could the motion of the electrons and the processes of emission and absorption of radiation be drawn together in a unified theory. With what one colleague has called daring conservatism, Bohr abandoned only as much of classical mechanics as necessary, holding firm to those classical ideas that still seemed to work within the atom.

Four of Bohr's ideas deserve special mention, for they have survived to be part of modern quantum mechanics.

(1) The Idea of the Stationary State

141 Bohr postulated that electrons in atoms can exist in various states of motion, each state of motion being distinct and characterized by a fixed energy. Such a state is called a stationary state. (Note, of course, that the electrons themselves are not stationary in a stationary state.) A corollary of this idea of stationary states is that energy is a quantized variable in atoms, limited to certain discrete values.

(2) The Idea of the Quantum Jump

142 It was already known that a quantum principle governed radiation. This led Bohr to abandon all hope of a classical description of the processes of emission and absorption of light. According to classical radiation theory, an electron in a Rutherford atom should radi-

ate continuously; the radiated frequency should change steadily to create spectral bands, not spectral lines; and the atom should collapse finally to the size of the nucleus. None of these things happened. Bohr postulated that an atom undergoes sudden transitions (quantum jumps) from one stationary state to another. He replaced the idea of smooth steady radiation by an idea of occasional tiny explosions, in each of which the electrons suddenly alter their state of motion and a photon is emitted. In a letter to Bohr commenting on a prepublication copy of Bohr's paper, Rutherford wrote, "There appears to me to be one grave difficulty in your hypothesis, which I have no doubt you fully realise, namely, how does an electron decide what frequency it is going to vibrate at when it passes from one stationary state to the other? It seems to me that you would have to assume that the electron knows beforehand where it is going to stop." Rutherford's question, although a very good one (it had to wait many years for an answer), betrayed one mode of classical thinking that Bohr was willing to abandon. Classically, in order to emit radiation of frequency f, an electron must vibrate at this same frequency f. According to the idea of stationary states connected by quantum jumps, there can be no such simple connection between radiated frequency and electron frequency. An electron may vibrate at one frequency in its initial state, a different frequency in its final state. The frequency of the radiated photon, although related to these frequencies, is equal to neither of them. By the Planck-Einstein equation, the photon frequency is determined only by the energy of the photon. Fortunately, Bohr let neither the unpredictability nor the nonvisualizability of the quantum jump deter him from advancing the idea. Quantum mechanics subsequently made possible calculations of the *probability* of transitions (the problem that worried Rutherford). These transitions, or quantum jumps, remain as nonvisualizable as they were to Bohr, and any particular one nearly as unpredictable.

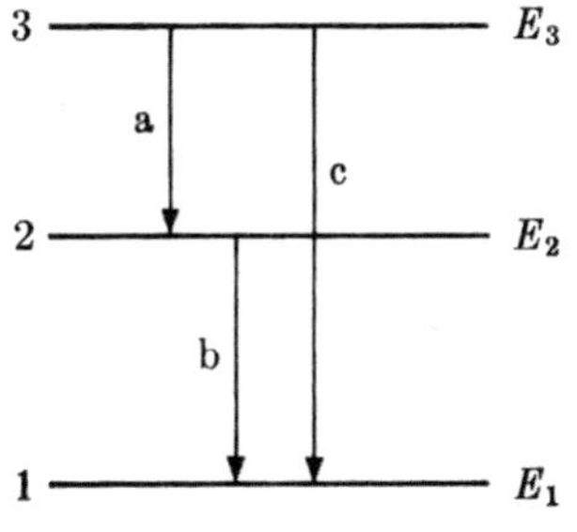

FIGURE 24.9 The Ritz combination principle and submicroscopic energy conservation. The horizontal lines represent energies of three stationary states. The arrows indicate quantum jumps accompanying photon emission. The combination principle for frequencies, $f_a + f_b = f_c$, rests on energy conservation, $\Delta E_a + \Delta E_b = \Delta E_c$.

(3) Submicroscopic Energy Conservation

We have mentioned already that Bohr held onto as much of classical physics as seemed 143
to work in the atomic domain. In particular he assumed that the law of energy conservation remained rigorously valid for electrons and photons. This assumption made possible a simple explanation of the Ritz combination principle. Consider, for instance, three stationary states of an atom with energies E_1, E_2, and E_3 (Figure 24.9). If the atom undergoes a quantum jump from the third to the second state, energy conservation requires that the emitted photon carry away exactly the energy difference between these two states:

$$hf_a = E_3 - E_2 . \tag{24.19}$$

For a quantum jump from the second to the first state, energy conservation imposes a similar condition,

$$hf_b = E_2 - E_1 . \tag{24.20}$$

Similarly, conservation of energy for the leapfrog transition from the third to the first state can be expressed,

$$hf_{\mathrm{c}} = E_3 - E_1 \,. \tag{24.21}$$

It is obvious from these equations that the frequencies of the light for these three possible transitions are simply related by the equation,

$$f_{\mathrm{c}} = f_{\mathrm{a}} + f_{\mathrm{b}} \,. \tag{24.22}$$

This is a mathematical statement consistent with the Ritz combination principle. The sum of two frequencies in the spectrum of the atom is equal to a third frequency in the spectrum. The same is of course true of wave numbers:

$$\frac{1}{\lambda_{\mathrm{c}}} = \frac{1}{\lambda_{\mathrm{a}}} + \frac{1}{\lambda_{\mathrm{b}}} \,. \tag{24.23}$$

The additive rule does *not* apply to wavelengths, and energy conservation explains why. Bohr was not the first to assume that energy is conserved in atomic systems, but he was the first to make good use of the assumption. The significance of the Ritz combination principle as a simple manifestation of energy conservation could be appreciated only after Bohr had broken loose from the idea that radiated frequency is equal to electron vibration frequency.

(4) The Correspondence Principle

144 The three ideas outlined above—stationary states, quantum jumps, and submicroscopic energy conservation—are evidently intertwined. Together they provide a simple description (not, to be sure, a classically visualizable description) of the processes of emission and absorption of radiation, and they relate atomic spectra to atomic mechanics. Bohr's fourth key idea provided an essential bridge between the classical and quantum worlds. His bridge is the correspondence principle, the idea that quantum mechanics must have a "classical limit." This idea of the classical limit was already a part of relativity theory, where it took a simple form. When particles move slowly compared with the speed of light, the relativistic description of their motion reduces to the classical description. To put it simply, the new theory (be it relativity or quantum mechanics) must agree with the old where the old is known to be correct.

In his theory of the hydrogen atom, Bohr used the correspondence principle in this way. He postulated that classical mechanics should approximately correctly describe an atomic transition when that transition takes place between two stationary states that differ very little in energy and in other properties. Through the correspondence principle he could get hold of the mysterious quantum jump. The new and unknown was tied to the old and familiar.

24.5 The Bohr atom

We turn now to the specifics of Bohr's model of hydrogen. He knew of the Balmer series and the Paschen series, described by Equation 24.13, or, in terms of frequency, by

$$f = c\mathscr{R}\left(\frac{1}{n^2} - \frac{1}{m^2}\right). \tag{24.24}$$

Multiplication of both sides of this equation by Planck's constant h gives the radiated energy per photon, hf, which Bohr assumed to be equal to the energy difference, ΔE, between two stationary states:

$$\Delta E = hc\mathscr{R}\left(\frac{1}{n^2} - \frac{1}{m^2}\right). \tag{24.25}$$

A single set of energy values to account for these energy differences is 145

$$E_n = -\frac{hc\mathcal{R}}{n^2}. \qquad (24.26)$$

The integer n identifies a particular stationary state. The first (and lowest) is characterized by $n = 1$, the second by $n = 2$, and so on. The negative sign in Equation 24.26 reflects the fact that the electron is bound to the positive nucleus, like the moon to the earth, or the earth to the sun. It is often convenient to use the binding energy W, a positive quantity,

$$W_n = +\frac{hc\mathcal{R}}{n^2}. \qquad (24.27)$$

As shown in Figure 24.10, the Balmer series is produced by quantum jumps ending at the second stationary state, and the Paschen series by quantum jumps ending at the third stationary state. Without any evidence for the lowest energy state ($n = 1$), Bohr confidently predicted its existence. Before long, the Lyman series, produced by quantum jumps ending at the first stationary state, was identified, vindicating the prediction.

So far we have used the ideas of stationary states, quantum jumps, and energy conservation, yet Equation 24.27 for the binding energies of the stationary states is still an *empirical* formula, chosen to fit the facts of the spectrum (in the framework of these ideas). The Rydberg constant is a quantity determined by experiment, so far unrelated to other constants in nature. The real triumph of Bohr in his 1913 paper was his “explanation” of the Rydberg constant. He was able to write $\mathcal{R}$ as a combination of other fundamental constants—the electron mass, the electron charge, Planck’s constant, and the speed of light. This great achievement required the use of his fourth essential idea, the correspondence principle.

Classically, an electron far from the nucleus radiates continuously at a frequency equal to its own frequency of revolution about the nucleus. It spirals inward, radiating at ever higher frequencies. Quantum-mechanically, the electron moves in one stationary state, then jumps to a lower state, then to a still lower state, and so on, cascading toward the nucleus while emitting a series of photons which contribute to discrete spectral lines. According to the correspondence principle, these two seemingly very different descriptions of the atom should merge into one when the stationary states are close together in energy, and the successively emitted photons are close together in frequency. Then the granularity of the quantum description gives way to the continuity of the classical description. It is evident from Figure 24.10 that this classical limit could be approached in the hydrogen atom only where the energy levels cluster together, at high values of n.

Classically, an electron in a planetary orbit about a fixed proton rotates with a frequency f_e given by

$$f_e = \frac{2W}{2\pi e^2}\sqrt{\frac{2W}{m}}, \qquad (24.28)$$

where W is the binding energy, e is the magnitude of the electron charge (and also the magnitude of the proton charge), and m is the mass of the electron. The derivation of this expression for circular orbits is left as an exercise. It is also true for elliptical orbits. In order for a quantum description of a cascading electron to correspond to a classical description of a spiraling electron, the cascading electron must pass successively through every stationary state. In a transition from state n to state $n - 1$, the electron emits a photon whose frequency is given by

$$f_r = c\mathcal{R}\left(\frac{1}{(n-1)^2} - \frac{1}{n^2}\right). \qquad (24.29)$$

Here the subscript r designates radiation. This equation can be rewritten

$$f_r = \frac{c\mathcal{R}(2n-1)}{n^2(n-1)^2} \,. \tag{24.30}$$

For very large n, the factor $2n - 1$ in the numerator can be replaced by $2n$, and the factor $(n - 1)^2$ in the denominator replaced by n^2, to yield

$$f_r \cong \frac{2c\mathcal{R}}{n^3} \qquad (n \text{ very large}) \,. \tag{24.31}$$

According to the correspondence principle (Figure 24.11), the radiated frequency f_r at very large n should equal the electron frequency f_e. If Equation 24.27 is substituted into Equation 24.28, this correspondence condition can be written

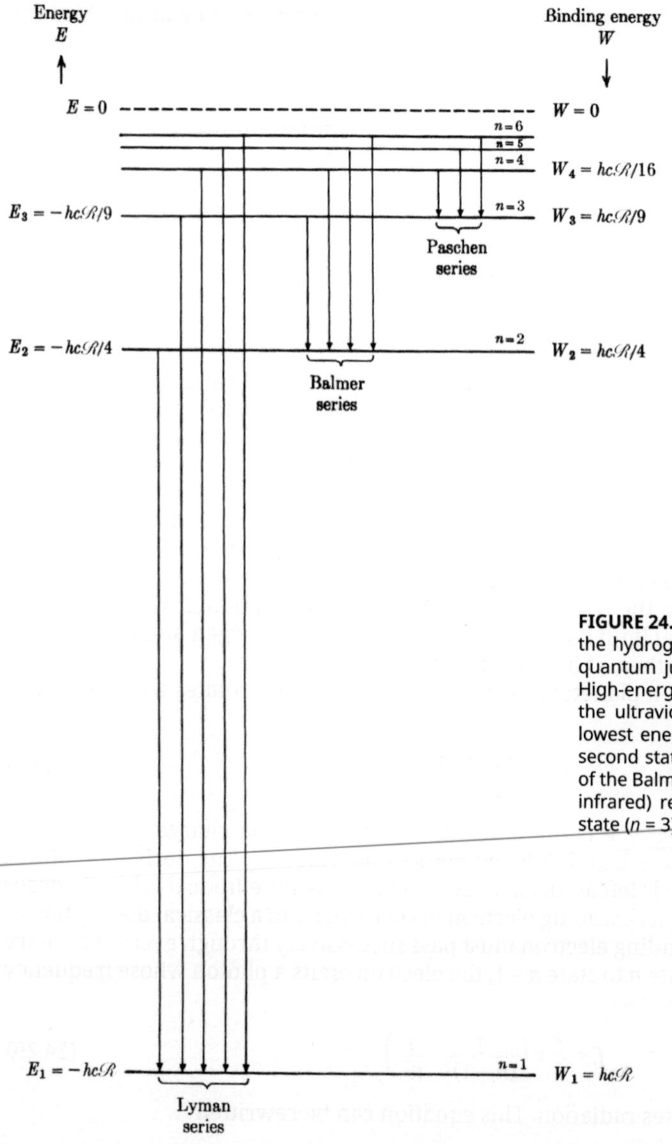

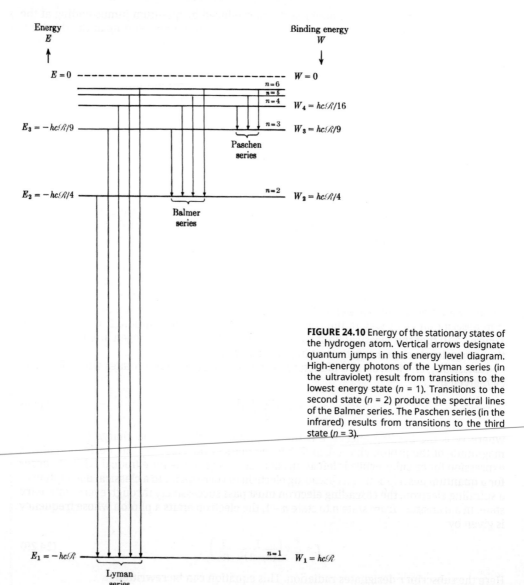

FIGURE 24.10 Energy of the stationary states of the hydrogen atom. Vertical arrows designate quantum jumps in this energy level diagram. High-energy photons of the Lyman series (in the ultraviolet) result from transitions to the lowest energy state ($n = 1$). Transitions to the second state ($n = 2$) produce the spectral lines of the Balmer series. The Paschen series (in the infrared) results from transitions to the third state ($n = 3$).

$$\frac{2c\mathcal{R}}{n^3} = \frac{2hc\mathcal{R}}{2\pi e^2 n^2}\sqrt{\frac{2hc\mathcal{R}}{mn^2}}. \tag{24.32}$$

The most important thing to note first about this equation is that the factor n^3 on the left cancels against an equal power of n on the right. This confirms that the correspondence principle holds generally for all large values of n, not just for a particular transition. It is in fact possible to prove by a somewhat more general argument that the requirement of the correspondence principle can be met *only* if the binding energy varies in proportion to $1/n^2$ at large n, as given in Equation 24.27. Next we notice that Equation 24.32 can be solved for $\mathcal{R}$ in terms of other fundamental constants. The result is

$$\mathcal{R} = \frac{2\pi^2 e^4 m}{h^3 c}. \tag{24.33}$$

This splendid unification of electron constants (e and m), quantum constant (h), and spectral radiation constants (c and $\mathcal{R}$) remains valid today. The Rydberg constant, now known to 2 parts in 10,000,000, helps, through this equation, to determine an accurate value for Planck's constant. With the accuracy of constants available to him, Bohr was able to verify the correctness of his equation for $\mathcal{R}$ to within 6%. A modern value of the Rydberg constant is given on page 648. If we substitute Equation 24.33 for $\mathcal{R}$ into Equation 24.27, we get an alternative expression for the binding energies of the stationary states of the hydrogen atom:

$$W = \frac{me^4}{2\hbar^2}\frac{1}{n^2}, \tag{24.34}$$

expressed in terms of Planck's constant (recall $\hbar = h/2\pi$) and two basic properties of the electron, m and e.

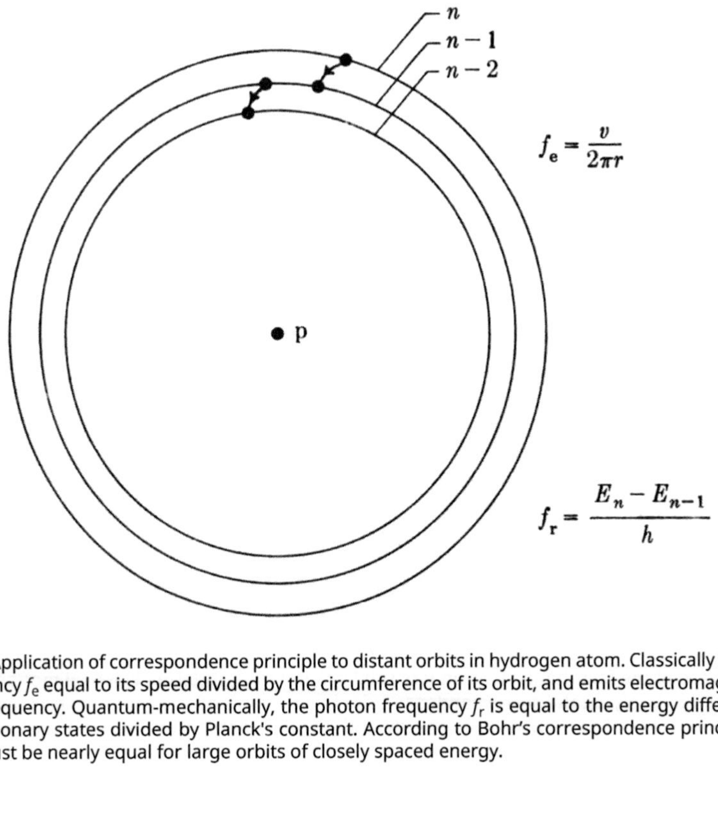

FIGURE 24.11 Application of correspondence principle to distant orbits in hydrogen atom. Classically the electron rotates at frequency f_e equal to its speed divided by the circumference of its orbit, and emits electromagnetic radiation of the same frequency. Quantum-mechanically, the photon frequency f_r is equal to the energy difference between successive stationary states divided by Planck's constant. According to Bohr's correspondence principle, these two frequencies must be nearly equal for large orbits of closely spaced energy.

24.6 Bohr orbits

With caution and with some equivocation, Bohr took a further step and assumed that a classical description of the electron motion might be valid in each stationary state separately.* On this view, the electron moves in a Keplerian ellipse when not engaged in emitting or absorbing radiation. The wave nature of particles has taught us that this picture of sharply defined orbits is wrong. Nevertheless, it is instructive to examine Bohr's allowed orbits, for they yield the correct order of magnitude for the size of the atom and the speed of the electron. They also suggest a principle of angular-momentum quantization in nature that has proved to be central in quantum mechanics.

Since Bohr orbits are approximate at best, it is appropriate to simplify them to the greatest extent possible by considering only circular orbits. The potential energy of an electron moving about a proton at a distance r is

$$P.E. = \frac{e^2}{r}. \tag{24.35}$$

Its kinetic energy in this orbit is

$$K.E. = \frac{e^2}{2r}. \tag{24.36}$$

The sum of these two energies is the total energy, $E = -e^2/2r$, and the negative of E (again a positive quantity) is the binding energy W:

$$W = \frac{e^2}{2r}. \tag{24.37}$$

This equation may be simply rearranged to express r in terms of W,

$$r = \frac{e^2}{2W}. \tag{24.38}$$

On the right we substitute the quantum equation for the binding energies of stationary states in hydrogen, Equation 24.34, to get

$$r = \frac{\hbar^2 n^2}{me^2}. \tag{24.39}$$

This equation can be written

$$r = an^2, \tag{24.40}$$

in which a, now called the Bohr radius, is a combination of constants with the dimension of length:

$$a = \frac{\hbar^2}{me^2}. \tag{24.41}$$

The magnitude of the Bohr radius is

$$a = 5.29 \times 10^{-9}\ \text{cm}, \tag{24.42}$$

about half an Angstrom. This magnitude agreed with what was known to Bohr about the size of the hydrogen atom in its normal state, or ground state. According to Equation 24.40,

* Bohr was aware of the fact that he was on less solid ground in postulating classical motion in all stationary states than in applying the correspondence principle to states of large n. He expressed the hope that at least the constants of motion such as energy and angular momentum might be correctly calculated classically.

the radius of the orbit of the lowest state, if it is circular, is a, the radius of the second orbit is $4a$, of the third orbit $9a$, and so on. This rapid growth in the predicted size of the atom with increasing excitation energy nicely explained why very high members of the Balmer series were seen only in starlight, not in laboratory-produced light. In the atmosphere of stars, the density of matter is so low that enormously inflated hydrogen atoms can exist without being seriously disturbed by neighboring atoms.

The electron speed in a circular Bohr orbit is also interesting. According to Equations 24.36 and 24.37, the electron kinetic energy is equal to its binding energy, or

$$\tfrac{1}{2}mv^2 = W\,. \tag{24.43}$$

From this equation follows at once an expression for the electron speed

$$v = \sqrt{\frac{2W}{m}}\,. \tag{24.44}$$

Substitution from Equation 24.34 for W on the right gives

$$v = \frac{e^2}{\hbar}\frac{1}{n}\,. \tag{24.45}$$

The speed is equal to $e^2/\hbar$ in the lowest state ($n = 1$), and successively smaller values in higher states. It is illuminating to cast this equation into dimensionless form by dividing both sides by the speed of light:

$$\frac{v}{c} = \frac{e^2}{\hbar c}\frac{1}{n}\,. \tag{24.46}$$

The dimensionless constant on the right, $e^2/\hbar c$, known as the fine-structure constant,* has fascinated physicists for decades, in large part because it *is* dimensionless. We recognize it now as a "coupling constant," measuring the strength of interaction between an electrically charged particle and the electromagnetic field. It can also be looked upon as the magnitude of e^2 measured in "natural" units, that is, in units of $\hbar c$. Numerically, the fine-structure constant is

$$\frac{e^2}{\hbar c} = 0.007297 = \frac{1}{137}\,. \tag{24.47}$$

Most physicists believe that the fine-structure constant requires an explanation, that there ought to be a deeper reason why three fundamental constants of nature combine to give one particular pure number. So far not a glimmer of light has been cast on this tantalizing possibility.

In Equation 24.46, the fine-structure constant plays the role of "nonrelativistic constant." Because $e^2/\hbar c$ is small compared with one, the ratio v/c is small compared with one in every orbit of hydrogen, and the effects of relativity are small. Relativistic corrections are nevertheless easily measurable in spectra, and account for small corrections to Bohr's energy formula, Equation 24.26. In heavy atoms, with highly charged nuclei, the innermost electrons experience a much stronger force, move at higher speed, and are more strongly influenced by relativity.

The integer n that has appeared throughout this section and the last is called a *quantum number*. In general, when a physical variable (such as the binding energy of the electron in

---

* In the spectra of many atoms, what appears with coarse observation to be a single spectral line proves, with finer observation, to be a group of two or more closely spaced lines. The spacing of these fine-structure lines relative to the coarse spacing in the spectrum is proportional to the square of $e^2/\hbar c$, for which reason this combination is called the fine-structure constant. We now know that the significance of the fine-structure constant goes far beyond atomic spectra.

the hydrogen atom) can take on only discrete values, it is convenient to label the possible values by an integer. The quantity n, called the principal quantum number in hydrogen, labels the energy values. For the state of lowest energy, $n = 1$; for the next lowest state, $n = 2$; and so on. At the same time, n does more than label, for it appears explicitly in many equations. Like most quantum numbers, it is both a label and a variable.

Bohr orbits, with properties expressed by Equations 24.40 and 24.46, only roughly approximate the state of the atom for small values of the principal quantum number n. For large n, however, the classical limit is approached, the correspondence principle can be applied, and orbits more nearly portray the true situation. Figure 24.12 compares the modern wave description and the Bohr orbit description for $n = 1$ and $n = 10$. Evidently, for larger n, the electron wave is beginning to collapse into a well-defined ring. At $n = 100$, the classical and quantum correspondence is very close. One reason that Bohr's energy formula, Equation 24.26, remains valid today is that in deriving it he relied on a classical description of the electron motion only at very large n. It should incidentally be remarked that the equations for radius and speed, Equations 24.40 and 24.45, although accurate at large n, tell only part of the story. For simplicity we considered circular orbits. Elliptical orbits are also possible.

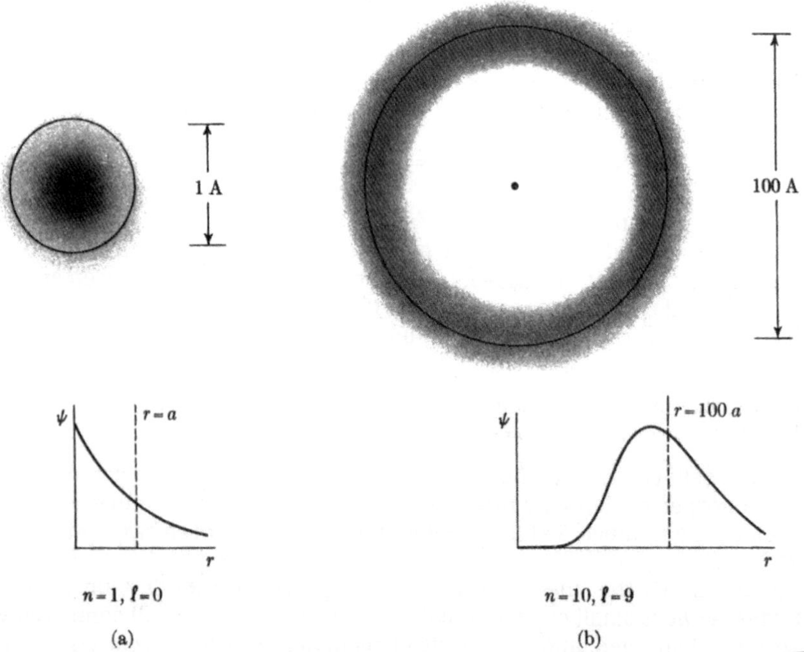

FIGURE 24.12 Comparison of Bohr orbit picture and modern wave picture for electron motion in the hydrogen atom. **(a)** Lowest stationary state, $n = 1$. **(b)** Most nearly circular orbit at $n = 10$. Below each sketch of the atom is a graph of the wave function. The distance scales are not the same in the two graphs, since the atom is 100 times larger at $n = 10$ than at $n = 1$.

146 We have left until last in this discussion of the Bohr atom a quantity that was peripheral in Bohr's original theory of hydrogen but that has become increasingly central in quantum mechanics—angular momentum. Bohr noticed that if (and only if) the electron in the hydrogen atom was assumed to move in a circle, a remarkably simple rule of angular-momentum quantization emerged. To get this condition, we need only combine Equations 24.45 and 24.40:

$$L = mvr = m\left(\frac{e^2}{\hbar}\frac{1}{n}\right)\left(\frac{\hbar^2}{me^2}n^2\right), \tag{24.48}$$

or

$$L = n\hbar\,. \tag{24.49}$$

The angular momentum (in circular orbits) is an integral multiple of $\hbar$. Here is a rule for hydrogen that is both erroneously based (the orbits are neither sharply defined nor necessarily circular) and wrong (actually, the lowest state of hydrogen has zero orbital angular momentum). Yet, the luck and genius of Bohr combined to produce a rule that, reexpressed only slightly, is a valid law of quantum mechanics. Orbital angular momentum is always equal to an integral multiple of $\hbar$. (For hydrogen, the integer need not be the same as the principal quantum number.)

This simple law of angular-momentum quantization was "in the air" at the time. It had been suggested first by J. W. Nicholson in 1912, and again by Paul Ehrenfest as well as by Bohr in 1913. It is only a slight caricature of the modern approach to physics to say of this law, "It is so simple, it must be right." In introducing the idea, Bohr wrote: "If we therefore assume that the orbit of the electron in the stationary states is circular, the result of the calculation can be expressed by the simple condition: that the angular momentum of the electron round the nucleus in a stationary state of the system is equal to an entire multiple of a universal value, independent of the charge on the nucleus." With a no more solid base than this, Bohr went on to use angular-momentum quantization as a basic principle for explaining the structure of heavier elements. In fact, he made little headway with the theory of multiple-electron atoms, but he had touched on the quantization law that was to prove basic for explaining the entire periodic structure of the elements.

24.7 *De Broglie waves*

The energy of a photon and its momentum are very simply related:

$$E = pc\,.$$

This means that the equation for the energy of a photon can be reexpressed as an equation for the photon's momentum:

$$pc = hf\,. \tag{24.50}$$

Since the frequency, wavelength, and speed of an electromagnetic wave are connected by $c = f\lambda$, this equation in turn can be cast into the form

$$p = \frac{h}{\lambda}\,. \tag{24.51}$$

We now call this the de Broglie equation because Louis de Broglie first suggested that it should govern all particles, not merely photons. As emphasized in Chapter Twenty-Three, this proposed generalization had revolutionary consequences, for it implied a wave nature for all matter. De Broglie's idea came at a time when physics was ripe for new ideas. Suddenly, a quarter of a century after Planck had introduced a new quantum constant into the description of nature, things began to fall into place. Here are some of the major developments in a short span of time:

1924 Wave nature of matter predicted (de Broglie)

1925 Electron spin (Uhlenbeck and Goudsmit)

1925 Exclusion principle (Pauli)

1925 Mathematical theory of quantum mechanics (Heisenberg)
1925 Wave nature of matter verified (Thomson; Davisson and Germer)
1926 Wave theory of quantum mechanics (Schrödinger)
1926 Probability interpretation (Born)
1927 Uncertainty principle (Heisenberg)
1928 Relativistic quantum theory of the electron (Dirac)

Before the wave nature of the electron was proposed, the Bohr quantization rules in hydrogen had seemed exactly that, *rules*, necessary and successful, but not derived from any deeper theory. It is of course true that any theory, no matter how deep, can eventually be described as a set of rules. Yet it is pleasing when we have a physical picture which makes the rules seem reasonable or even necessary. That is what the de Broglie waves provided for the Bohr orbits. Consider, most simply, circular orbits. An electron circling with speed v at a distance r from the nucleus has, according to the de Broglie equation, an associated wavelength λ given by

$$\lambda = \frac{h}{p} = \frac{h}{mv}. \tag{24.52}$$

Suppose that we imagine this wave following around the electron orbit and closing on itself (Figure 24.13). Like a snake biting its own tail, the wave may interfere destructively with itself. If the head and the tail of the wave are only slightly out of phase after one revolution, they will be further out of phase after two, and so on. After a very large number of revolutions, the wave will have totally destroyed itself. The sum of all the superposed waves will be zero. If, on the other hand, the wave after one revolution is precisely in phase with itself, interfering constructively, it would remain in phase after two or any number of revolutions', and remain alive as a "standing wave." The condition for reinforcement is that the circumference contain an integral number of wavelengths:

$$2\pi r = n\lambda, \tag{24.53}$$

where n is an integer. On the right, λ may be replaced by h/p or h/mv to give

$$2\pi r = n\frac{h}{mv}. \tag{24.54}$$

This equation may be rewritten

$$2\pi mvr = nh.$$

On the left mvr may be replaced by L, angular momentum. After dividing both sides by 2π we get

$$L = n\hbar, \tag{24.55}$$

exactly the rule found by Bohr for circular orbits. Now the rule is "explained" as the result of a wave closing on itself and interfering constructively only for certain discrete radii.

This de Broglie wave picture of the electron orbits in hydrogen can be described as lying about halfway between the classical orbit theory and the eventual quantum-mechanical theory. It is incomplete in two major respects. First, it still retains the sharply defined Bohr orbit as a guiding line for the wave. Second, it ignores the three-dimensional aspect of the wave, paying attention only to the wavelength along the supposed line of travel. It adds the wave without abandoning the particle. Eventually, in the Schrödinger wave theory of quantum mechanics, the particle and its orbit had to be thought of as spreading out to *be* the wave in a stationary state.

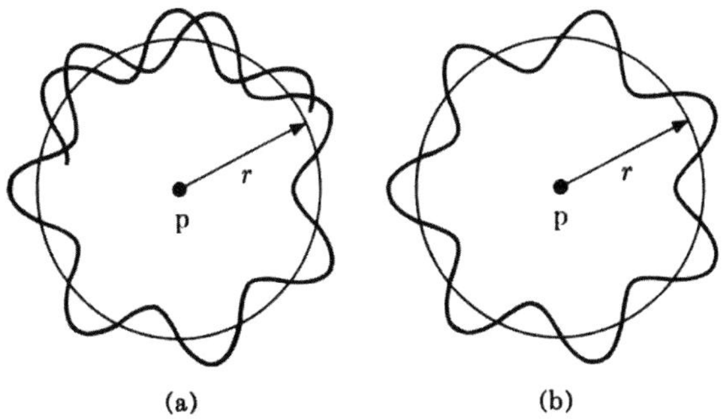

FIGURE 24.13 Simplified de Broglie picture of self-reinforcement of waves as an explanation of discrete stationary states of electron motion. **(a)** The wave interferes with itself destructively. **(b)** The wave satisfies the condition of constructive interference, $n\lambda = 2\pi r$.

Despite these shortcomings of the simple de Broglie description, it contains some correct ingredients that reveal how we can understand energy quantization and angular-momentum quantization in terms of quantum waves. Only those wave distributions are permitted, according to the modern model, that are self-reinforcing, neither canceling to give zero, nor building up into waves of infinite amplitude. In mathematical terminology, the waves must be well-behaved and must satisfy certain boundary conditions.

In the hydrogen atom, an electron of fixed energy has more momentum near the nucleus, less momentum far from the nucleus. Correspondingly the electron wave has shorter wavelength near the nucleus, longer wavelength far from the nucleus. In fact, its wave "length" may change by a large fraction in less than one oscillation of the wave. As a result, the appearance of the wave in the radial direction is quite distorted relative to a simple sine wave. Some examples of electron waves in hydrogen are discussed below. It is instructive first to examine a much simpler system, one in which a particle moves with constant speed, and therefore constant wavelength.

Consider a particle in a box, bouncing back and forth between parallel impenetrable and perfectly elastic walls (Figure 24.14). Classically the description of the motion is very simple. The particle has some fixed speed v (any speed is possible), a constant magnitude of momentum, mv, and a constant kinetic energy, $\frac{1}{2}mv^2$. If the walls are separated by a distance d, the particle executes periodic motion, completing one round trip in time $T = 2d/v$. Quantum-mechanically, the description of this same motion provides insight into the implications of the wave nature of matter. We use the same principle of self-reinforcement that was applied to circular orbits in hydrogen. In hydrogen, the principle was imperfectly applied because the spread of the wave in the radial direction was ignored. For the particle in the box, the principle is in fact precisely correct, because the actual motion proceeds in only one dimension. According to this principle of self-reinforcement, only those states of motion are permitted for which the de Broglie wave, after closing upon itself, is precisely in phase with itself to produce constructive interference. All other waves are self-destructive. For self-reinforcement, an integral number of wavelengths must fit into the round trip distance $2d$:

$$n\lambda = 2d\,. \tag{24.56}$$

As before, n denotes an integer—a quantum number. If λ is replaced by h/p, this equation reveals the allowed magnitudes of momentum,

$$p = \frac{h}{2d}\,n\,. \tag{24.57}$$

Here then is a granularity introduced by the wave nature of the particle. The momentum is restricted to integral multiples of a unit $h/2d$. Kinetic energy is likewise quantized. We can write

$$K.E. = \frac{p^2}{2m} = \frac{h^2}{8md^2}\, n^2\,. \tag{24.58}$$

The energy diagram in Figure 24.15 shows the set of allowed energies—the stationary states—for the particle in the box. If the particle were electrically charged, it could emit a photon and drop from one of these states to another, just as the electron makes quantum jumps in hydrogen.

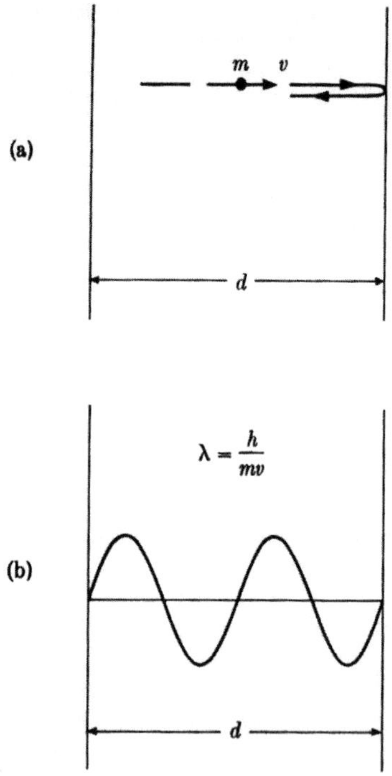

FIGURE 24.14 Particle in a box, moving in one dimension between perfectly reflecting walls. **(a)** Classical illustration of motion. Any momentum is possible. **(b)** Quantum wave illustration. The momentum is restricted to values that produce self reinforcement of the wave.

The energy unit in Equation 24.58 is $h^2/8md^2$. How big is this? For an object of mass 1 gm moving between walls 1 cm apart, it is 5.5×10^{-54} erg, or 3.4×10^{-42} eV, a phenomenally small energy, illustrating again the irrelevance of quantum granularity in the macroscopic domain. For an electron ($m = 9.1 \times 10^{-28}$ gm) held between hypothetical walls 1 A apart ($d = 10^{-8}$ cm), on the other hand, the same energy unit is 6×10^{-11} erg, or 38 eV, which is the energy of an ultraviolet photon, a "large" energy on the atomic scale.

147 One special feature of Figure 24.15 deserves mention, because it is a common feature of all quantum systems that are restricted in space. This is the absence of a quiescent state. At the very least, the particle in the box has an energy equal to $h^2/8md^2$. This is called its zero-point energy. Like a dollar bill glued to the bartender's wall, the zero-point energy is useless currency. There is no question about its reality, but it cannot be drawn upon. Nature has established a minimum balance of energy in every system. In an isolated atom the electrons move tirelessly and forever after the last photon has been radiated away. In a collection of atoms, a zero-point energy of vibrational motion remains even at absolute zero. This is but one of the many new features of the behavior of matter predicted by quantum mechanics.

De Broglie waves provide a visualizable model to explain the granularity of energy, momentum, and angular momentum. The concept of probability in turn gives physical meaning to the waves. Consider, for instance, the second stationary state of motion of a particle in a box. The wave function ψ for this state is

$$\psi = \sin\left(2\pi \frac{x}{d}\right), \tag{24.59}$$

where x is the distance measured from the left wall toward the right wall (Figure 24.16). Since the square of a wave function is proportional to probability of position (Equation 23.11), the probability distribution for the second state of motion in the box is

$$P \sim \sin^2\left(2\pi \frac{x}{d}\right) \tag{24.60}$$

(The exact constant of proportionality need not concern us. It is determined by the requirement that the probability of finding the particle *somewhere* is 100%.) As shown in Figure 24.16(b), this probability—which, in this example, is actually probability per unit distance—has the interesting attribute that it is zero at the midpoint as well as at both walls. The wave nature of the particle produces a bunching effect, making the particle more likely to be at some places, less likely to be at others. Classically, in a random look, the particle should be found with equal probability at any point between the walls.

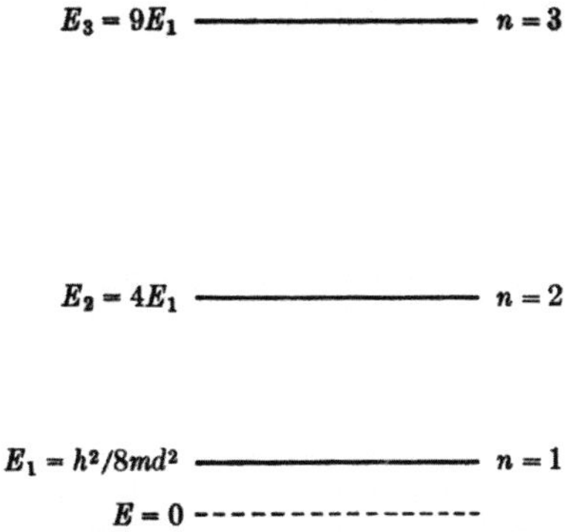

FIGURE 24.15 Energy-level diagram for particle in box. Because of the condition of self-reinforcement applied to the de Broglie wave, the energies are discrete.

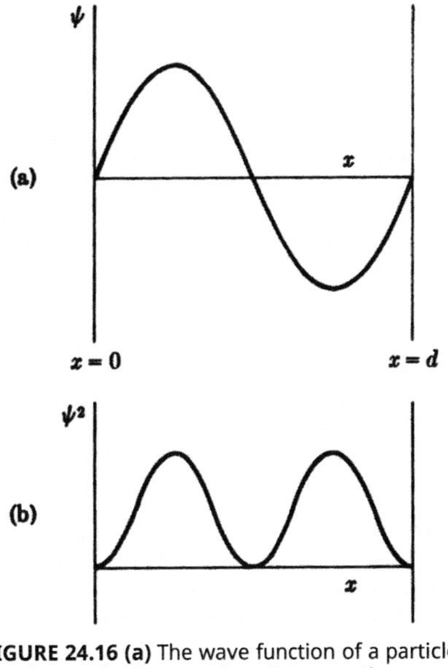

FIGURE 24.16 (a) The wave function of a particle in a box for n = 2. Two wavelengths cover the round trip distance $2d$. **(b)** The square of the wave function, proportional to the probability per unit distance to find the particle at any distance x from the left wall.

By suitable stretching of the mind and imagination, many quantum phenomena, such as probability waves, can be adequately visualized. Nevertheless our usual vocabulary is not quite a match for the nuances of quantum mechanics. In an undisturbed stationary state, the wave of probability *is* the particle. If this state is violently disturbed in such a way as to pinpoint the particle by its interaction with another particle or other outside influence, the wave suddenly shifts its role to that of probability signpost. It is easy to ask about quantum mechanics unanswerable questions, such as: If the particle is never to be found halfway between the walls, how can it get from one half to the other half? The only answer that can be given to this question is that the "particle" exists simultaneously in both halves of the available space. Yet when an interaction causes it to materialize, it shows itself in one half or the other.

24.8 *The hydrogen atom*

149a In the Bohr theory of the hydrogen atom, each stationary state is characterized by a single quantum number, n, the principal quantum number.

(1) Principal Quantum Number

$$n = 1, 2, 3, \ldots, \text{all positive integers.}$$

This quantum number and Bohr's energy equation, Equation 24.34, survived in the full theory of quantum mechanics, although the physical picture of the atom underwent radical revision. Besides the fact that wave distributions superseded particle orbits, two more quantum numbers were added to the description of stationary states. First was an angular-momentum number L, defined by

$$L = \ell\hbar. \tag{24.61}$$

Bohr's tentative suggestion that orbital angular momentum might be quantized according to this simple rule was found to be correct, not only for hydrogen, but for all physical systems. In hydrogen, the ground state has zero angular momentum, rather than one unit as Bohr's circular orbit model suggested. For states of higher energy, an interesting multiplicity appears. For the energy defined by $n = 2$ (the final energy for Balmer transitions), there are actually two distinct stationary states, one with zero angular momentum, one with one unit of angular momentum. These two states are said to be "degenerate," meaning that they have equal energy. Therefore they do not show themselves separately in the spectrum. For $n = 3$, there are three degenerate states of motion, with angular momentum quantum numbers $\ell = 0, 1,$ and 2. These states *correspond* to classical elliptical orbits of equal energy but different eccentricity (Figure 24.17). The general rule for this second quantum number is:

(2) Orbital Angular-Momentum Quantum Number

$$\ell = 0, 1, \ldots, n-1 \qquad (n \text{ different values}).$$

In order to display both quantum numbers n and ℓ, physicists draw "term diagrams" in which the energy of atomic states is graphed vertically, and the angular momentum is graphed horizontally. Figure 24.18 shows the term diagram for hydrogen. The arrows in the diagram indicate typical quantum jumps accompanying photon emission. Partly because photons have one unit of angular momentum (the full story is more complicated), most atomic transitions take place between states differing by one unit of orbital angular momentum.

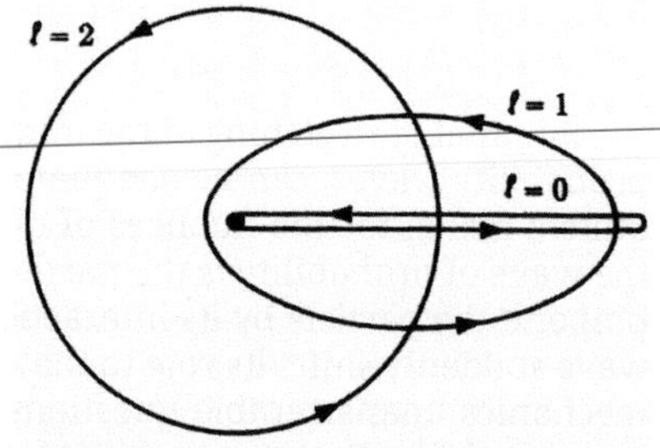

FIGURE 24.17 Three classical orbits that correspond to the three degenerate states of motion in the hydrogen atom with $n = 3$ and $\ell = 0, 1,$ and 2. Classical orbits of equal energy have equal semi-major axes. Actually, of course, the wave functions for these three states of motion are spread out and only roughly correspond to the classical orbits.

Quantum mechanics showed angular momentum—a vector quantity—to be granular not only in magnitude but also in direction. A third quantum number characterizing states of motion in hydrogen (and states of motion generally) is an orientation quantum number, m.

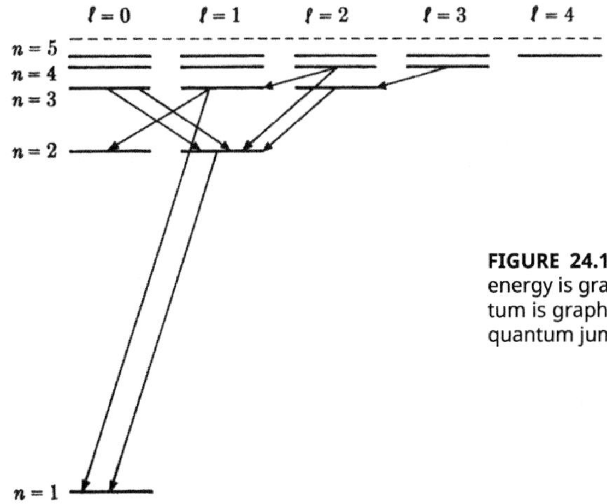

FIGURE 24.18 Term diagram of the hydrogen atom. Binding energy is graphed vertically downward; orbital angular momentum is graphed horizontally to the right, Arrows indicate typical quantum jumps.

(3) Orientation Quantum Number

$$m = -\ell, -\ell + 1, \ldots, \ell \qquad (2\ell + 1 \text{ different values}).$$

The component of angular momentum along a fixed direction can, like the total angular momentum, take on only discrete values, with the separation between allowed values being $\hbar$. If this fixed direction is called the z-axis, the quantization of orientation can be expressed by the equation,

$$L_z = m\hbar\,, \tag{24.62}$$

where L_z is the component of the vector **L** along z. It may seem remarkable, and indeed it is remarkable, that this rule holds true for any direction in space. In practice, the quantization of orientation takes on special importance when a particular direction in space is physically defined, as by a magnetic field pointing in a certain direction. In the absence of such a special direction, differently oriented states are degenerate. This is not surprising. In empty space, with all possible directions equivalent, it should make no difference to the atom in which direction its angular momentum is pointing. This irrelevance of orientation in the absence of a special direction is true of all atomic systems.

In a term diagram such as the one in Figure 24.18, the differently oriented states are not separately displayed. Each rung in the diagram must be imagined to represent $2\ell + 1$ distinct states of motion differing only in orientation. For the ground state, with no angular momentum, there is indeed only one orbital state of motion. (The effect of the spin of the electron remains to be considered.) For a state with one unit of angular momentum, three orientations are possible, with $m = -1$, 0, and +1 (Figure 24.19). Altogether, at the energy corresponding to $n = 2$, the electron can choose among four different states of motion—one with $\ell = 0$, three with $\ell = 1$. At $n = 3$, nine separate states of motion are available to the electron. In terms of quantum numbers, the electron in the hydrogen atom is as restricted as a freight car in a switching yard. At the same time, because of its wave nature, the electron ranges free as a cloud over the whole volume of the atom in every one of its restricted states.

Figure 24.20 shows cross sections in the radial direction of electron waves for several stationary states in the hydrogen atom. In addition, the waves show oscillation in the angular direction. Pictorial representations of the three-dimensional waves are given in Figure 24.21 by means of differential shading. The intensity of shading corresponds approximately to the probability per unit volume, or the square of the wave function.

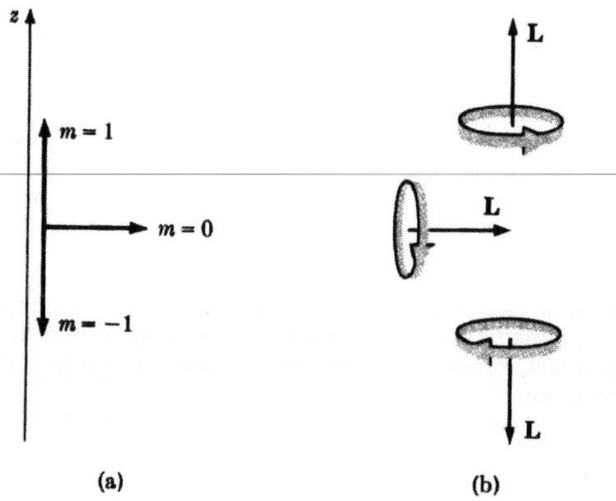

FIGURE 24.19 Quantization of orientation. **(a)** Three possible orientations of orbital angular momentum vector if ℓ = 1. **(b)** Corresponding classical motion. If a magnetic field is directed along the *z* axis, each of the three states of motion has a different energy. The splitting of spectral lines caused by the dependence of energy on orientation in a magnetic field is called the Zeeman effect.

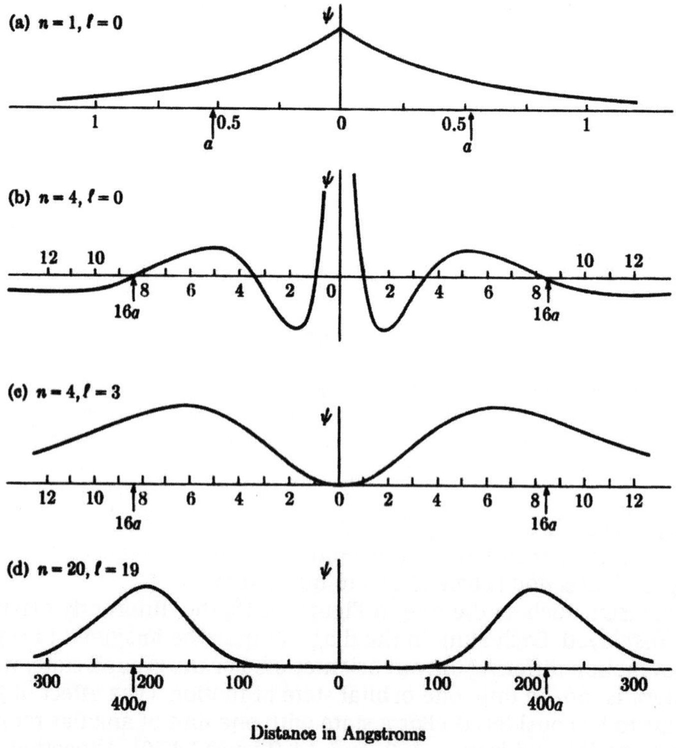

FIGURE 24.20 Cross sections of electron waves in the radial direction for four states in the hydrogen atom. **(a)** The ground state, $n = 1$, $\ell = 0$. **(b)** The state $n = 4$, $\ell = 0$, corresponding to a classical straight line trajectory through the nucleus. **(c)** The state $n = 4$, $\ell = 3$, corresponding to a circular orbit. **(d)** The state $n = 20$, $\ell = 19$, corresponding to a very large circular orbit. Radial scales are not the same in the different graphs.

147 The four states chosen for pictorial representation illustrate significant features of quantum waves. (a) The most featureless wave is that of the ground state ($n = 1$, $\ell = 0$). Lacking any angular momentum, this state also lacks a preferred direction in space, and is characterized by a spherically symmetric wave. In the radial direction the wave behaves in the simplest possible manner, rising to a single maximum at the center of the atom, then

falling again. (b) For the higher energy state with $n = 4$ and $\ell = 0$, the absence of angular momentum is again reflected in the spherical symmetry of the wave. The classical analogue of this state of motion is a straight-line trajectory back and forth through the nucleus (think of a long narrow ellipse squeezed to become a straight line). In the quantum wave, this to-and-fro motion shows itself in the radial oscillation of the wave. Having a greater distance to cover at $n = 4$ than at $n = 1$, the electron wave undergoes a number of cycles of oscillation from one side of the atom to the other.

(c) At the same energy level ($n = 4$), the wave function of a state of higher angular momentum ($\ell = 3$) also shows oscillatory behavior, but directed around the nucleus in orbital fashion rather than toward and away from the nucleus. This wave, representing one of Bohr's circular orbits, is concentrated in a doughnut around the nucleus.

(d) As quantum numbers increase in magnitude, and the granular effects of quantum mechanics become less significant, the wave concept begins to resemble closely the classical orbit concept. Figures 24.20(d) and 24.21(d) illuminate the correspondence principle. For $n = 20$ and $\ell = 19$, the quantum wave has narrowed into a shell with a large hollow center. The wave oscillates as it follows around a circular Bohr orbit, closing upon itself with constructive interference in much the manner visualized by de Broglie (Figure 24.13). The gradual transition from quantum to classical behavior is one of the beautiful aspects of the submicroscopic world.

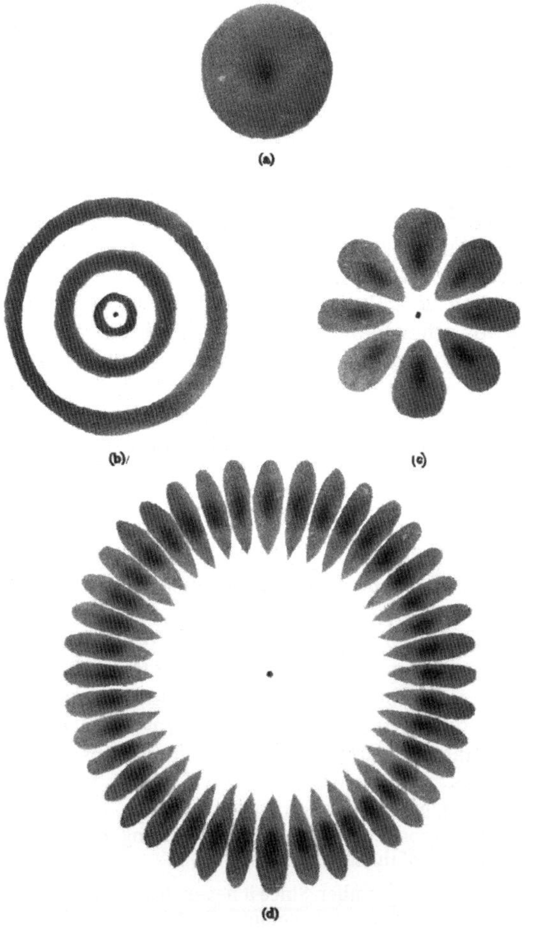

FIGURE 24.21 Pictorial representation of probability density for electron waves of the same four states as in Figure 24.20. Radial scales are not the same in the four diagrams.

24.9 Electron spin

Two big discoveries of the year 1925 provided the keys to unlock the mysteries of atoms containing more than one electron. One was the discovery of electron spin by George Uhlenbeck and Samuel Goudsmit. The other was the discovery by Wolfgang Pauli of a most remarkable principle, the exclusion principle. Together with what was already known of hydrogen, these two ideas made possible a grand structural scheme of all atoms. Much of atomic structure had already been deduced from the experimental evidence of spectra in the decade 1915-1925. Suddenly in 1925-1926, all the facts and knowledge fell into place as the quantum-mechanical theory of spinning electrons provided a simple foundation of understanding for all atoms, and therefore all of the material world constructed of atoms. It was a victory of the human intellect comparable to Newton's illumination of the solar system. Referring to this giant step, Paul Dirac wrote matter-of-factly in 1929, "The underlying physical laws necessary for the mathematical theory of a large part of physics and the whole of chemistry are thus completely known." Our concern in Sections 24.9 through 24.12 is atomic structure and the periodic system of the elements.

In 1921 Arthur Compton suggested that the electron might be a spinning particle. This was, at the time, a hypothesis unsupported by any solid evidence. The evidence was provided several years later by Uhlenbeck and Goudsmit. Their "discovery" of the spin of the electron consisted of a theoretical analysis showing that several aspects of spectra could be explained if the electron was assumed to be spinning with angular momentum equal to $\frac{1}{2}\hbar$ and magnetic moment equal to $e\hbar/2mc$. Atoms in a magnetic field emit a spectral pattern differing slightly from the pattern emitted without a field. The difference, known as the Zeeman effect, can be attributed to the intrinsic magnetism of the atoms. Before the analysis of Uhlenbeck and Goudsmit, it had been assumed that this atomic magnetism is created exclusively by the orbital motion of electrons, although many details of the known Zeeman patterns went unexplained. These men showed that an orderly explanation was possible if it were assumed that the electron generates a magnetic field both by orbital rotation and by spin rotation. Moreover, the known Zeeman patterns required for their explanation specific magnitudes of electron spin ($\frac{1}{2}\hbar$) and magnetic moment* ($e\hbar/2mc$).

The spin of the electron also manifests itself directly in the doublets observed in the spectra of sodium (Figure 24.22) and numerous other elements. The most famous spectral doublet, the D lines of sodium, is attributed to a quantum jump—or, more exactly, a pair of quantum jumps—to the ground state from excited states in which the spin of the electron has two different orientations. In one of these states, the spin is directed parallel to the orbital angular momentum; in the other, it is directed antiparallel to the orbital angular momentum. Regardless of the details, the doublet character of the lines can be readily understood in terms of the magnitude of the electron spin. Since the quantization of orientation separates successive components of spin by one whole quantum unit, $\hbar$, a spinning system (or particle) whose angular momentum is $\frac{1}{2}\hbar$ can be oriented in only two possible directions. With respect to a chosen direction in space, its component must be $\frac{1}{2}\hbar$, or $-\frac{1}{2}\hbar$. No other possibilities are separated by $\hbar$.

The discovery of the spin of the electron meant that the states of motion of the electron in the hydrogen atom had to be characterized by four quantum numbers, not three. Besides the principal quantum number, the orbital-angular-momentum quantum number, and the orbital-orientation quantum number, there is another orientation quantum number, that of spin.† To distinguish these two orientation quantum numbers, we introduce subscripts ℓ

* If a magnet is aligned perpendicular to a magnetic field, it experiences a torque tending to twist it into alignment with the field. Its magnetic moment, which is a convenient measure of the strength of its magnetism, is defined as the ratio of this torque to the strength of the magnetic field.

† The magnitude of the spin is, strictly speaking, also a quantum number. Since it never changes, there is no reason to keep track of it. It is the same for all states of all electrons in all atoms.

and s and write for the components of **L** and **S**:

$$L_z = m_\ell \hbar , \tag{24.63}$$

$$S_z = m_s \hbar . \tag{24.64}$$

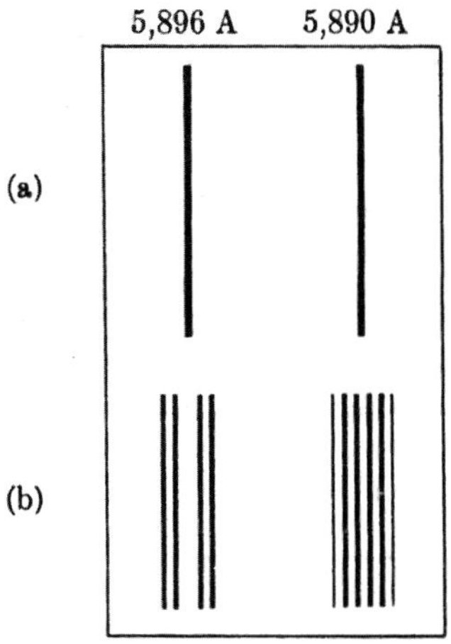

FIGURE 24.22 (a) The D lines of sodium, a pair of closely spaced spectral lines (a doublet). The transitions responsible for these lines are alike in all respects except that in one initial state, the electron spin is aligned parallel to its orbital angular momentum; in the other initial state, the electron spin is aligned antiparallel to its orbital angular momentum. **(b)** For sodium atoms in a magnetic field, states of motion with different orientations have slightly different energies. Then the D lines are transformed into this more complex Zeeman pattern. [Drawing after photograph in E. Back and A. Landé, *Zeeman Effect and Multiplet Structure* (Berlin: Springer-Verlag, 1925).]

For the third quantum number, the summarizing statement on page 669 must be modified slightly: 149b

(3) Orbital Orientation Quantum Number

$$m_\ell = -\ell, -\ell + 1, \ldots, \ell \qquad (2\ell + 1 \text{ different values}).$$

For the fourth quantum number, we have the summarizing statement:

(4) Spin Orientation Quantum Number

$$m_s = -\tfrac{1}{2}, +\tfrac{1}{2} \qquad (2 \text{ different values}).$$

Altogether, the state of motion of a single electron in the hydrogen atom is characterized by the set of quantum numbers,

$$n, \ell, m_\ell, m_s.$$

Since m_s can take on two distinct values, the fact that the electron is spinning doubles the number of states of motion. Table 24.2 lists the possible combinations of quantum numbers for the lowest two energy levels in hydrogen. The $n = 2$ floor of the Balmer transitions actually consists of eight different states of motion. The energies of these eight states are not all identical. They divide into three slightly different energies, contributing to what is called the fine structure of the spectrum. In a magnetic field the energy is further subdivided into eight separate closely spaced levels, since no two of the states have identical magnetic properties.

An alternative set of quantum numbers for the electron corresponds more closely to the actual physical situation in the hydrogen atom. Instead of imagining the two angular momenta, orbital and spin, to be separately oriented in space, we can first combine these two into a total angular momentum (designated by **J**), this total then being oriented in space, with orientation quantum number m_j. The alternative set of quantum numbers is

$$n, \ell, j, m_j.$$

with the restriction that j is equal to $\ell + ½$ or $\ell - ½$ (unless $\ell = 0$, in which case $j = ½$, since then the total angular momentum is the same as the spin angular momentum). As Table 24.3 makes clear, the number of distinct states of motion is unchanged by the alternative classification. Either scheme of quantum numbers provides an equally satisfactory basis for understanding the structure of heavy atoms. Since the physical picture underlying our first set (n, ℓ, m_ℓ, m_s) is somewhat easier to visualize, we shall rely on this set in the remaining discussion.

TABLE 24.2 Possible Values of Four Quantum Numbers for the Lowest Two Energy Levels in the Hydrogen Atom

n	ℓ	m_ℓ	m_s
1	0	0	−½
			+½
2	0	0	−½
			+½
	1	−1	−½
			+½
		0	−½
			+½
		+1	−½
			+½

TABLE 24.3 Alternative Classification of Possible States of Motion for the Lowest Two Energy Levels in the Hydrogen Atom. In this scheme **L** and **S** are first combined into a total angular-momentum vector **J**, which is then oriented in space

n	ℓ	j	m_j
1	0	½	−½
			+½
2	0	½	−½
			+½
	1	3/2	−3/2
			−½
			+½
			+3/2
	1	½	−½
			+½

24.10 *The central-field approximation*

150 The theory of atoms containing two or more electrons is based on the central-field approximation. This idea can best be explained by example. Consider a sodium nucleus, containing 11 protons, alone in space, completely separated from its normal complement of 11 electrons. Surrounding the nucleus is an electric field 11 times stronger than the electric field surrounding a proton at the center of a hydrogen atom. Apart from this difference in strength, the two fields are identical. Both are central Coulomb fields varying inversely as the square of the distance from the nucleus. If a single electron approaches the sodium nucleus, it will find an available set of allowed states of motion exactly the same as the allowed quantum states in hydrogen, except for certain changes of scale. Its ground state ($n = 1$, $\ell = 0$)

will be 11 times closer to the sodium nucleus than it would be to the hydrogen nucleus, and its binding energy will be 121 times greater than the binding energy in hydrogen. It is left as an exercise to demonstrate the reason for these scale changes.

Suppose now that the single electron radiates a series of photons and cascades down to the lowest possible energy state about the sodium nucleus. A second electron approaches. It is strongly attracted to the nucleus and weakly repelled by the first electron. The rapidly moving inner electron is sometimes more distant than the nucleus, sometimes closer, sometimes to the left of the nucleus, sometimes to the right. Its *average* position is *at* the nucleus. The average force experienced by the second electron is therefore directed toward the nucleus. The central-field approximation consists of ignoring deviations from the average. The first electron is regarded as a cloud of negative charge surrounding the nucleus. The second electron is then assumed to move in a central field of force created by the nucleus plus the inner electron cloud. The second electron finds an available set of discrete states of motion differing only slightly from those available to the first electron. In fact, the quantum numbers differ not at all. The energies differ somewhat, as do the sizes. Left to itself, the second electron will cascade down to its lowest available state of motion. With the addition of 9 more electrons, the system of sodium nucleus plus 11 electrons becomes a normal sodium atom. According to the central-field approximation, each of the electrons occupies an allowed quantum state in an average central field of force created by the nucleus and the other 10 electrons.

The meaning of the central-field approximation can be clarified by thinking about the solar system. To a good approximation, each planet traces an orbit around the sun as if no other planets were present. Very accurate predictions of planetary orbits require corrections to the central-field approximation arising from the direct forces acting between pairs of planets. In the solar system, the validity of the central-field approximation rests on the great mass of the sun and on the relatively great separation of the planets. Earth is much more strongly attracted to the sun than to Venus, Mars, or Jupiter. A similar pair of factors is at work in atoms. The nuclear charge is greater than the charge on any electron, as much as 100 times greater for the heaviest atoms, making the nucleus the strongest center of force in the atom. Also, for a reason to be clarified below, electrons arrange themselves into "shells" at successively greater distances from the nucleus, with wave amplitudes that are only partially interpenetrating.

The most significant difference between the average central field in a heavy atom and the average central field in the solar system comes about from the opposite sign of the electronic and nuclear charge and the overall neutrality of the atom. In the solar system, the most distant planet, Pluto, experiences the full force of the sun (at that distance), scarcely modified by the presence of eight other planets between it and the sun. In sodium, by contrast, the force exerted by the nucleus on the most distant electron is greatly reduced by the intervening ten electrons. About ten elevenths of the attractive nuclear force is canceled by the repulsive force of the inner electrons. The outermost electron "sees" a net charge of +1 (in units of the proton charge), and accordingly has available a set of quantum states of motion almost exactly the same as the set available to an excited electron in hydrogen. The innermost electron in sodium, on the other hand, "sees" the full nuclear charge of +11. It resides in a small tightly bound state of motion.

24.11 *The periodic table*

Sometimes in science explanation follows hard on the heels of discovery. Sometimes explanation may even precede discovery. Then we call it prediction. At other times, facts or regularities of nature must wait many decades for an explanation. Such was the fate of the set of chemical and physical regularities summarized by the periodic system of the elements. The periodic table, now a fixture in many classrooms and laboratories, was first advanced

TABLE 24.4 Periodic Table of the Elements

Group → Period	I	II											III	IV	V	VI	VII	O
1	1 H 1.00797																	2 He 4.0026
2	3 Li 6.939	4 Be 9.0122											5 B 10.811	6 C 12.01115	7 N 14.0067	8 O 15.9994	9 F 18.9984	10 Ne 20.183
3	11 Na 22.9898	12 Mg 24.312					Transition elements						13 Al 26.9815	14 Si 28.086	15 P 30.9738	16 S 32.064	17 Cl 35.453	18 Ar 39.948
4	19 K 39.102	20 Ca 40.08	21 Sc 44.956	22 Ti 47.90	23 V 50.942	24 Cr 51.996	25 Mn 54.9380	26 Fe 55.847	27 Co 58.9332	28 Ni 58.71	29 Cu 63.54	30 Zn 65.37	31 Ga 69.72	32 Ge 72.59	33 As 74.9216	34 Se 78.96	35 Br 79.909	36 Kr 83.80
5	37 Rb 85.47	38 Sr 87.62	39 Y 88.905	40 Zr 91.22	41 Nb 92.906	42 Mo 95.94	43 Tc (99)	44 Ru 101.07	45 Rh 102.905	46 Pd 106.4	47 Ag 107.870	48 Cd 112.40	49 In 114.82	50 Sn 118.69	51 Sb 121.75	52 Te 127.60	53 I 126.9044	54 Xe 131.30
6	55 Cs 132.905	56 Ba 137.34	57–71 *	72 Hf 178.49	73 Ta 180.948	74 W 183.85	75 Re 186.2	76 Os 190.2	77 Ir 192.2	78 Pt 195.09	79 Au 196.967	80 Hg 200.59	81 Tl 204.37	82 Pb 207.19	83 Bi 208.980	84 Po (210)	85 At (210)	86 Rn (222)
7	87 Fr (223)	88 Ra (227)	89– †															

*Lanthanide rare-earth elements	57 La 138.91	58 Ce 140.12	59 Pr 140.907	60 Nd 144.24	61 Pm (145)	62 Sm 150.35	63 Eu 151.96	64 Gd 157.25	65 Tb 158.924	66 Dy 162.50	67 Ho 164.930	68 Er 167.26	69 Tm 168.934	70 Yb 173.04	71 Lu 174.97
†Actinide rare-earth elements	89 Ac (227)	90 Th 232.038	91 Pa (231)	92 U 238.03	93 Np (237)	94 Pu (242)	95 Am (243)	96 Cm (245)	97 Bk (249)	98 Cf (249)	99 Es (254)	100 Fm (252)	101 Md (256)	102 No (254)	103 Lw

Numbers in parentheses are mass numbers of most stable isotopes of those elements.

in a useful form by Dmitri Mendeleeff in 1869. Certain family groups of elements had been recognized earlier—for instance, the chemically similar alkali metals: sodium, potassium, and rubidium; and the halogen family: fluorine, chlorine, bromine, and iodine. Mendeleeff proposed that an arrangement of all the elements in order of increasing atomic weight would reveal regularly recurring properties, leading to family grouping for all the elements. The 62 elements known to Mendeleeff provided considerable evidence in support of such regular recurrence, but they could not be arranged to show a completely orderly periodicity without gaps. Moved by the faith in simplicity that has inspired so much progress in science, Mendeleeff predicted that yet-to-be-discovered new elements would fill in the gaps. Moreover, because of the expected regularities, he could predict in some detail the chemical and physical properties of these unknown substances. Guided by these predictions, chemists discovered in the period 1874-1885 three new elements: scandium, which fit between calcium and titanium; and germanium and gallium, which filled two gaps between zinc and arsenic. There could be no further doubt about the general correctness of Mendeleeff's periodic system of the elements.

A modern version of the periodic table is given in Table 24.4. No gaps remain in the table of 103 presently known elements. Among the numerous atomic quantities showing periodic behavior are the valence (to be discussed below), which governs the modes of chemical combination of the elements; the size of the atom; the normal phase of the element (solid, liquid, or gas); the properties of its spectrum; and its ionization energy—the energy required to remove a single electron from the atom. Indeed there is scarcely an atomic property that does not vary periodically as atomic number increases, aligning itself with the family groups of the periodic table.

In the absence of a theory of atomic structure, there could be no hope of explaining the periodicities of atomic properties. From 1879 to 1911, Mendeleeff's periodic table was extended, corrected, and reinforced as new elements were discovered (including a whole new family, the rare gases), spectral analysis was refined, and atomic-weight measurements were improved. Yet the underlying reason for the periodic system remained nearly as deep a mystery as ever. Finally Rutherford, by exposing the interior of the atom, cleared the way for an explanation of atomic periodicity. Even then, the explanation was not immediate. In his first paper on atoms in 1913, Bohr made an attempt to account for the structure of atoms heavier than hydrogen in terms of rings of electrons circling the nucleus at various distances. Even earlier J. J. Thomson had proposed an atom model with electrons arranged in successive shells. Although no other aspect of Thomson's model survived, this idea of shell structure persisted, made imperative by the demands of experimental fact. By the early 1920s, the following model of heavy atoms was accepted, and, in the light of later developments, was largely correct. Orbits of various dimensions are available to the electrons. Each orbit, or state of motion, can accommodate only a limited number of electrons. When this limit is exceeded, additional electrons must occupy larger orbits. Periodicity is associated with the capacity of particular orbits. Atoms with the same number of electrons in the outermost shell have similar properties. The alkali metals, for example, have a single electron in their outermost shells. The halogens are one electron shy of a filled shell. Atoms with filled shells are the rare gases, particularly stable and almost inert.

24.12 The exclusion principle and atomic structure

Although successful, the shell model of the atom of the early 1920s was perplexing. Why was the capacity of orbits limited? Why would some orbits accommodate more electrons than others? In order to answer these questions, Wolfgang Pauli postulated in 1925 that electrons obey an "exclusion principle." One way to state the exclusion principle is this: No two electrons can be in identically the same state of motion at the same time. Since the state of motion of an electron in an atom is characterized by a set of four quantum numbers, the exclusion principle may also be stated in this way: No two electrons may have precisely the same set of quantum numbers. This rather strange principle, completely at variance with classical reasoning, found a deeper explanation the following year (1926) in terms of a rather subtle aspect of quantum mechanics. Its basis may be subtle, but its impact on the world could not be plainer. The entire structure of the material world is dictated by the exclusion principle. We know now that electrons, protons, neutrons, and all other particles with half-odd-integral spin $\frac{1}{2}\hbar$, $\frac{3}{2}\hbar$, . . .) obey the exclusion principle, whereas photons and mesons, with integral spin (0, $\hbar$, $2\hbar$, . . .) do not.

151 A good way to understand the impact of the exclusion principle on atomic structure is to do some hypothetical atom building. We start with a stock of bare nuclei, one of each charge—a proton, a helium nucleus, a lithium nucleus, and so on—and a collection of electrons. To each nucleus we add electrons one at a time and think about the states of motion that result. Before starting this Jovian task, we can number the nuclei as well as name them. The atomic number Z is defined as the total charge of the nucleus in units of the proton charge. It is the number of protons in the nucleus and the number of electrons in the neutral atom. If no elements have been overlooked, it also labels the position of the element in the periodic table. Sodium, for instance, with 11 electrons, and 11 protons in its nucleus, is the eleventh element in the periodic table. These facts, simple today, were startling and wonderful insights in 1913. We shall touch again later on the importance of atomic number in helping to establish the quantum theory of atomic structure.

In order to build the world—or at least its constituent atoms—methodically, let us begin with hydrogen ($Z = 1$). An electron brought near to a proton will cascade through successive states of motion to end in the state of lowest energy, the ground state or the normal state of the atom. Here the electron's quantum numbers are

$$n = 1,\ \ell = 0,\ m_\ell = 0,\ m_s = \tfrac{1}{2} \text{ or } -\tfrac{1}{2}\,.$$

Since it has no orbital angular momentum in this state, its orbital orientation quantum number is zero. Its spin may be oriented in either of two directions.

Consider next helium ($Z = 2$). The first electron will fall into the lowest state, with quantum numbers the same as the ground state of hydrogen. According to the exclusion principle, a second electron cannot join the first with identical quantum numbers. However, there is available at the lowest energy level a second state of motion differing from the first in spin orientation. The second electron follows the first to a state with principal quantum number $n = 1$, and orbital angular momentum $\ell = 0$, but with opposite spin orientation. Since these two electrons use up the available two states of motion, this situation is described as a closed shell. Helium is the first closed-shell atom and the first member of the rare-gas family. There are many unique things about helium, all arising from its tightly bound closed-shell structure. To name a few: It is chemically the most inert of all elements; it has the highest ionization energy of any atom; and it has the lowest boiling point of any substance, 4.2°K, because its atoms have so little affinity for each other.

It should be remarked at this point that hydrogen too has a uniqueness apart from being the first element in the periodic table. It is the only atom that has a single electron in a shell and is at the same time one electron shy of a closed shell. Note in Table 24.4 that hydrogen stands alone.

Hydrogen and helium complete the first period. A second period of eight members begins with lithium ($Z = 3$) and ends with the second rare gas, neon ($Z = 10$). In this period electrons fill successively the eight states of motion available at the second principal quantum number, $n = 2$ (see Table 24.2). The first two electrons added about a lithium nucleus can drop into the n = 1 state with oppositely directed spin. The third electron is excluded. The lowest state of motion not already occupied is at the $n = 2$ level. The third electron drops down to this level and can go no further. The power of the exclusion principle is evident. Because of it, lithium has one relatively loosely bound electron. The properties of the atom are in every way vastly different than they would be if the exclusion principle did not act to prevent this electron from joining the first two in the lowest state of motion.

At the $n = 2$ level, the $\ell = 0$ state can accommodate two electrons and the $\ell = 1$ state six more (three orbital orientations, each with two spin orientations). When these eight states of motion are fully occupied, another closed shell results, this one at $Z = 10$, neon. Sodium ($Z = 11$) begins the third period. Its first ten electrons occupy closed shells. Its eleventh electron, excluded from the $n = 1$ and $n = 2$ shells, must stop at the $n = 3$ level. This single eleventh electron can be considered the active agent in sodium. It is responsible for most of the chemical and physical properties of sodium, including its similarity to lithium.

Although it is unnecessary to pursue this atom-by-atom survey further, one important point about atom building remains to be clarified. This is the distortion of the energy-level pattern available to one electron by the other electrons in a heavy atom. Consider again the eleventh electron in sodium. It experiences approximately a central force, but this is not a Coulomb force. At a great distance from the nucleus, attracted by a net charge of $+e$, it experiences a force whose magnitude is

$$F = \frac{e^2}{r^2} \qquad \text{(large } r\text{)} . \tag{24.65}$$

If it penetrates close to the nucleus, it experiences the full force of eleven protons,

$$F = \frac{Ze^2}{r^2} \qquad \text{(small } r\text{)} . \tag{24.66}$$

In between, the form of the law of force varies gradually between these two limits. Because of this variation, states of motion that would be degenerate in hydrogen are separated in energy in heavier atoms. Occupying the $n = 3$ level in sodium, the last electron could have zero, one, or two units of angular momentum. The state of maximum angular momentum corresponds to a circular orbit, the state of minimum angular momentum to a thin ellipse penetrating close to the nucleus. This correspondence is reflected in the wave amplitudes of the different states (see Figure 24.23). The wave of the $\ell = 0$ state penetrates close to the nucleus into the region of stronger attraction. Consequently the three states, all of the same principal quantum number, are spread apart in energy, the $\ell = 0$ state being lowest (tightest binding), the $\ell = 2$ state being highest (loosest binding). This energy distortion is pictured schematically in Figure 24.24 for the first four principal quantum levels. At $n = 3$, the distortion is sufficient to influence the shell structure in an important way. The "circular" state, with $\ell = 2$, is so loosely bound that it is pushed effectively up into the fourth shell. Instead of the eighteen elements that might be expected in the third period, based on counting the total number of states of motion available at $n = 3$, there are only eight elements—from sodium ($Z = 11$) through argon ($Z = 18$). A period of 18 elements shows itself first in the fourth row of the periodic table.

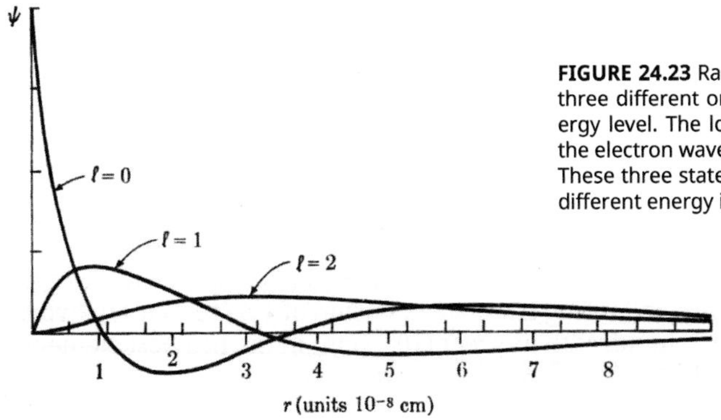

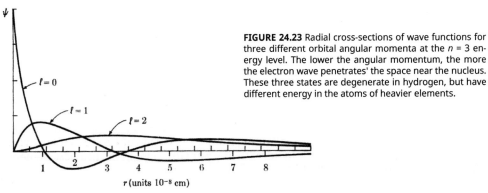

FIGURE 24.23 Radial cross-sections of wave functions for three different orbital angular momenta at the $n = 3$ energy level. The lower the angular momentum, the more the electron wave penetrates' the space near the nucleus. These three states are degenerate in hydrogen, but have different energy in the atoms of heavier elements.

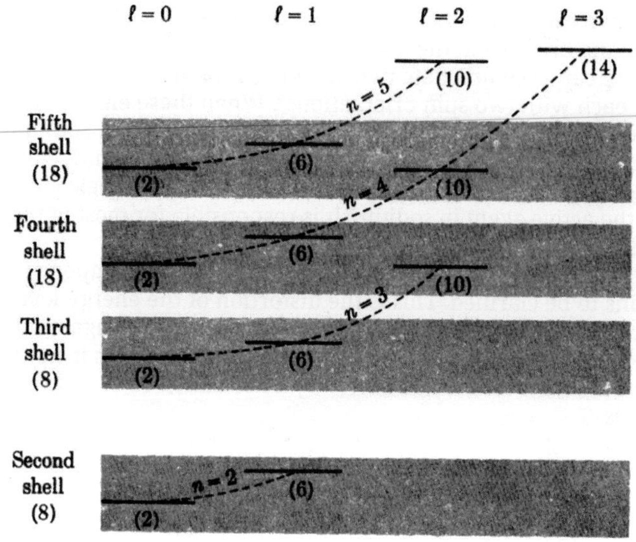

FIGURE 24.24 Distorted energy level diagram produced by stronger attractive force experienced by electrons in states of motion with small angular momentum. The distortion influences the shell structure. An $n = 3$ level, for instance, appears in the fourth shell. (This diagram is schematic only, not to scale.)

In the distorted energy diagram of Figure 24.24, the individual energy levels are labeled by the largest number of electrons they can hold without violating the exclusion principle. These numbers explain at once the observed lengths of the first five periods of elements (see Table 24.5). For the remainder of the periodic table, the energy distortion complicates the pattern further. Nevertheless, a very well defined periodicity persists through all the known elements.

TABLE 24.5 First Five Periods of the Elements

Period	Number of Elements	First Element	Last Element
1	2	H ($Z = 1$)	He ($Z = 2$)
2	8	Li ($Z = 3$)	Ne ($Z = 10$)
3	8	Na ($Z = 11$)	Ar ($Z = 18$)
4	18	K ($Z = 19$)	Kr ($Z = 36$)
5	18	Rb ($Z = 37$)	Xe ($Z = 54$)

Section 24.9 began with emphasis on the importance for atomic structure of two discoveries: electron spin and the exclusion principle. By itself, spin would have been an interesting discovery, since it is a property of an elementary particle. In conjunction with the exclusion principle, it is more than interesting. It is momentous. By doubling the number of states of motion available to the electron, the apparently innocuous fact that the electron spins shapes the periodic table and thereby shapes the world.

24.13 *Inner and outer electrons*

Imagine a set of cards, on each of which a particular atom is represented pictorially by a sketch of the wave amplitudes of its innermost and outermost electrons. If these cards are arranged in order of atomic number and are flipped rapidly to produce the impression of motion, what does one see? Drawn inward by the increasing nuclear charge, the wave of the innermost electron shrinks steadily, collapsing from an extension of 1 A (10^{-8} cm) in hydrogen to an extension only about one hundredth as great (10^{-10} cm) in the heaviest elements. The extension of the outermost electron wave varies in quite a different way. It oscillates slightly inward and outward as the cards are flipped, its dimension never straying far from about 1 A. This outermost electron wave determines the size of the atom. Like most atomic properties, it shows a periodic variation. The periodicity of atomic size was known in Mendeleeff's time, and was used by his contemporary Lothar Meyer as supporting evidence for the periodic system of the elements. Atomic radii are plotted as a function of atomic number in Figure 24.25.

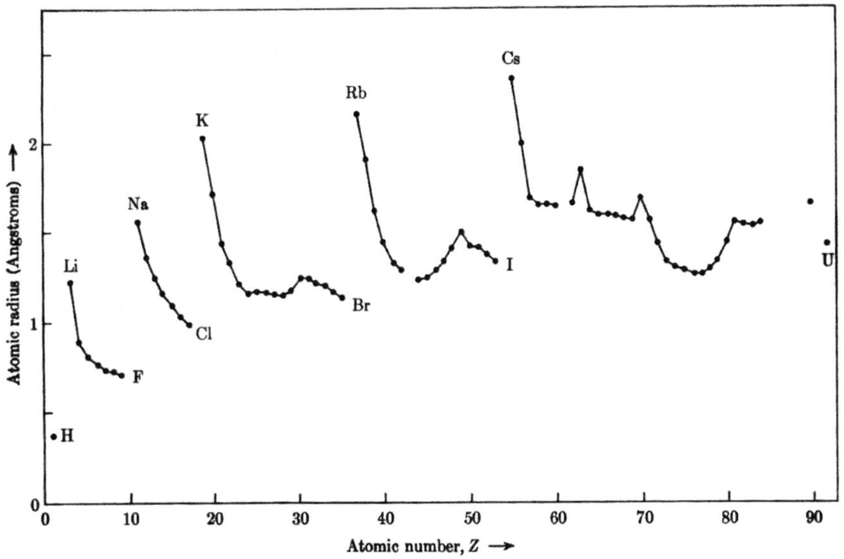

FIGURE 24.25 Atomic radii vs. atomic number. Atomic radii cannot be precisely defined. These values are derived from interatomic spacing in covalent bonding and in the solid form of the element. [Data from M. J. Sienko and R. A. Plane, *Chemistry: Principles and Properties* (New York: McGraw-Hill, 1966), Figure 2.19.]

From our modern point of view, even more remarkable than the periodicity of atomic size is the near constancy of atomic size. In terms of the quantum theory of atomic structure, the fact that atoms of uranium and hydrogen are not very different in size must be regarded as more fortuitous than fundamental. Because the outermost electrons in uranium occupy states of large principal quantum number (n = 5, 6, and 7), they "tend" to spread out to dimensions much greater than the size of the hydrogen atom. Countering this tendency is the imperfect shielding of the nucleus by the inner electrons, which causes the outermost electrons to be drawn inward by a force considerably greater than the force of a single positive charge. These contrary effects nearly cancel. An atom of uranium is not much larger than an atom of hydrogen.

Just as the radii of the innermost and outermost states of motion vary in quite a different way through the periodic table, so do the energies associated with these states of motion. The ionization energy of the atom, which is the energy required to remove the most loosely bound outer electron, never differs greatly from 10 eV from one end of the periodic table to the other. The ionization energy of hydrogen is 13.53 eV; of radium, 5.25 eV. The cesium atom (Z = 55) is most easily ionized, requiring only 3.87 eV to remove its outermost electron. Note in the periodic table that cesium, a member of the alkali metal family, has a single electron outside of closed shells. The largest ionization energy, 24.46 eV, belongs to helium. Figure 24.26 shows the periodicity of ionization energy among the elements.

Just as conditions in a deep cave are perpetually uniform, uninfluenced by the surface fluctuations of wind, weather, and season, so the innermost electron states of motion in an atom exhibit a simple regularity undisturbed by variations in the behavior of the loosely bound electrons up at the "surface" of the atom. In a heavy atom, the outermost electrons can absorb or radiate energy, change their state of motion, or even be set free by an ionization process with scarcely any effect on the energies or states of motion of the innermost electrons. Similarly, the fluctuations in atomic size and ionization energy from one atom to the next have hardly any effect on the innermost electrons, whose properties change smoothly with atomic number.

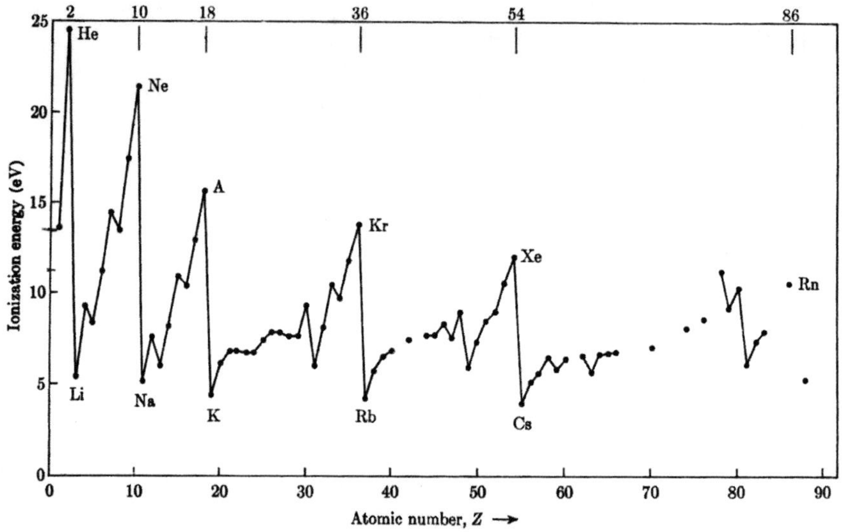

FIGURE 24.26 Ionization energy vs. atomic number.

In the very early history of modern atomic theory, the simplicity associated with the inner tightly bound electrons was used to good advantage in support of the Rutherford-Bohr model of the atom. Normally, energy absorption and emission by an atom involve only its outer electrons. A sufficiently violent disturbance, however, can excite or remove inner electrons. The photons emitted after such a disturbance, as the atom is settling back to normal, provide a way to get information about the innermost electrons. If, by some means, an electron is removed from the most tightly bound state of motion (the state with principal quantum number $n = 1$), the atom finds itself with an available unoccupied state of motion at low energy. One of the electrons perched at a higher energy, previously prevented by the exclusion principle from dropping down, is now free to emit a photon and jump to the lower energy. The photon emitted when it does so is called a K X-ray. The simplest transition to consider is one in which an electron drops from the $n = 2$ level to the $n = 1$ level. This is called the K_α transition. (Following this transition, another electron will drop into the now empty state of motion left at the $n = 2$ level, and so on, but these successive transitions need not concern us.)

In 1913-1914, Henry G. J. Moseley, first in Manchester with Rutherford, then at Oxford, measured with great accuracy the wavelengths of the K_α X-rays for a large number of elements. Generating these X-rays, it should be mentioned, is not very difficult. It requires only that the element under study be bombarded with a stream of electrons energetic enough to eject atomic electrons from the innermost shells. Then X-rays are emitted as the atoms readjust. To measure X-ray wavelengths, Moseley used crystals with known atomic spacing as diffraction gratings. Expressing his results in terms of frequency instead of wavelength, he discovered that his data rather accurately obeyed the following proportionality:

$$f \sim (Z-1)^2. \tag{24.67}$$

Since photon energy is proportional to photon frequency, this means that the energy of the K_α transition is proportional to $(Z-1)^2$, where Z is the atomic number. Moseley realized that this proportionality provided direct support for Bohr's theory. The transition energy from $n = 2$ to $n = 1$ is proportional to, among other things, the square of the nuclear charge. (The factor e^4 on the right side of Equation 24.34 is equal to the square of the electron charge multiplied by the square of the proton charge. For atomic number greater than 1, this factor must be replaced by $(Ze)^2e^2$, or Z^2e^4. Moseley could assume that the electron responsible for the K_α X-ray is close to the nucleus and very little perturbed by the other more distant electrons. Later, the reason for the proportionality to $(Z-1)^2$ rather than Z^2 became clear. The innermost shell in all atoms contains only two electrons, the maximum permitted by the exclusion principle. When one is removed, one remains and shields the nucleus. The net charge attracting the electrons in the $n = 2$ level is then $(Z-1)e$ rather than Ze.

The first line in the Lyman series in hydrogen can be considered a K_α X-ray, although it is not usually so called because its frequency is much lower than the frequency of most X-rays. With ascending atomic number, X-ray frequencies increase rapidly. The K_α X-ray of sodium ($Z = 11$) has a photon energy of about 1,000 eV, or 1 KeV, about 500 times greater than the energy of a typical visible photon. The K_α X-ray of tungsten, sometimes used in medical work, has a photon energy of 58 KeV. Because of its power to ionize atoms and molecules and disrupt their inner structure, this type of radiation is, at least in large doses, a biological hazard. Of course in medical use, its advantages usually outweigh its dangers.

Apart from providing support for Bohr's planetary model of the atom, Moseley's measurements had great practical value. They made it possible to ascertain the nuclear charge of almost every known element, and therefore the proper placement of the elements in the periodic table. Whereas Rutherford had only been able to guess that the atomic number is approximately half the atomic weight, Moseley could state confidently that the atomic number of gold, for instance (atomic weight 197), is exactly 79. Because of the regular variation of X-ray frequency with atomic number, he was able to pinpoint three missing elements, at $Z = 43$, $Z = 61$, and $Z = 75$. All were later discovered. Finally, he confirmed what had earlier been suspected, that the order of increasing atomic weight and the order of increasing atomic number are not always the same. Some years had to pass before the discovery of the neutron and the theory of nuclear structure made clear the reason for this.

Although the inner electrons have an appealing simplicity, and contributed in an important way to the understanding of atoms, it is the outer electrons that, quite literally, give life and color to our world. Among the properties of matter that can be attributed to the outer electrons are these: the union of atoms to form molecules; the temperatures of boiling and freezing; electrical resistance; the magnetic properties of materials; ionization energy; spectra in visible and nonvisible regions; the taste, feel, appearance, and color of substances. Because all of these properties are associated with the outer electrons, all show a periodicity as atomic number is increased.

24.14 Groups of atoms

To deal with groups of atoms in a single section may seem presumptuous, for this is a subject without boundaries. All of chemistry, molecular biology, and the physics of solids, liquids, and plasmas, among other branches of science, are concerned with groups of atoms. We cannot even scratch the surface of these disciplines. In keeping with our emphasis on the fundamental theories of physics, and on phenomena that directly illuminate these theories, we must limit attention to a few of the physical foundations underlying the rich complexity of atomic grouping that make up our world.

Valence

A simple and useful characterization of the outer electrons in an atom is the idea of valence. The valence of an atom is equal to the number of electrons in the outermost shell of the atom (positive valence) or to the number of "holes" in the outermost shell, that is, the number of unoccupied states of motion (negative valence).* This may be called the theoretical definition of valence. There is also an empirical definition of valence: the number of electrons an atom gains, loses, or shares when it combines with other atoms to form molecules. The empirical valence is not always the same as the theoretical valence, since compounds may be formed without the full participation of the outermost shells of all the atoms. A carbon atom, for example, may unite with one atom of oxygen (CO, carbon monoxide) or with two (CO_2, carbon dioxide). Oxygen has a valence of –2. In one compound, carbon exhibits a valence of 2, in the other a valence of 4. With four electrons in its outer shell, carbon has a theoretical valence of 4.

Hydrogen, with one electron, is unique in having valence both +1 and –1. Helium ($Z = 2$), a closed-shell atom, has zero valence. The next four elements in the periodic table—lithium, beryllium, boron, and carbon—containing respectively 1, 2, 3, and 4 electrons in the $n = 2$ shell, have ascending valences of +1, +2, +3, and +4. Lithium, with its single outer electron, is particularly active chemically, combining readily with atoms at the other side of the periodic table that are one electron short of a closed shell. Carbon, in the middle of the second period, is an interesting special case. Since it has four electrons in its outer shell, and is also four electrons short of a closed shell, its valence may be considered to be either +4 or –4. In fact, it never fully gains or loses four electrons in forming a compound, but rather shares electrons with other atoms in a kind of union called covalent bonding. Because of its position in the middle of a period and its tendency to share electrons, carbon can form an enormous number of different compounds. Carbon atoms provide the backbone of every organic molecule in every living thing.

Moving on through the second period, we find atoms of negative valence. Nitrogen ($Z = 7$), whose five outer electrons leave three unoccupied states of motion in the second shell, has valence –3. This is reflected in the molecule of ammonia, NH_3. Oxygen ($Z = 8$), two positions short of the closed shell, has valence –2. (Recall oxygen's most familiar compound, H_2O.) Next comes fluorine ($Z = 9$), with valence –1, then neon ($Z = 10$), a closed-shell atom with zero valence. Sodium ($Z = 11$), beneath lithium in the periodic table, again has valence +1. Magnesium ($Z = 12$), like boron above it, has valence +2, and so on. Evidently valence, like other properties associated with the outer electrons, is periodic.

Molecular Bonding

Valence is a simplifying idea, useful for understanding the formation of compounds. Yet the union of atoms to form molecules is rarely simple. Consider the combination of two hydrogen atoms to form a hydrogen molecule, H_2. When far apart, the two atoms influence each other only very weakly, since both are electrically neutral. Nevertheless, a small force acts, called the Van der Waals force. This weak attractive force arises from the fact that the electron and proton, although equal and opposite in charge, are not coincident in position. In response to an outside electric field, the electron wave is slightly distorted or displaced, a response that causes the atom in turn to create an electric field that reaches out to other atoms. When two hydrogen atoms come very close together, each greatly distorts the electron wave of the other. The resulting force can cause the atoms to scatter, that is, to be de-

* The idea of valence entered chemistry in the midnineteenth century, long before electrons were discovered or atomic structure understood. We shall not explore the historical background of the idea.

flected and move apart again. Or, if the atoms can rid themselves of extra energy, usually by transferring it to another nearby atom or molecule, the force can hold the atoms together as a hydrogen molecule (Figure 24.27). Despite the fact that a hydrogen molecule is indicated symbolically as H_2, it actually bears little resemblance to a pair of hydrogen atoms side by side. It is an entirely new structure, composed of two protons and two electrons. The protons establish themselves about 0.74 A apart, while the two electrons occupy states of motion spreading around and between the two protons. Since the state of motion of an electron in the molecule is quite different than the state of motion of an electron in a single atom, the molecular spectrum differs greatly from the atomic spectrum. Such a simple series as the Balmer series is not to be found in the hydrogen molecular spectrum.

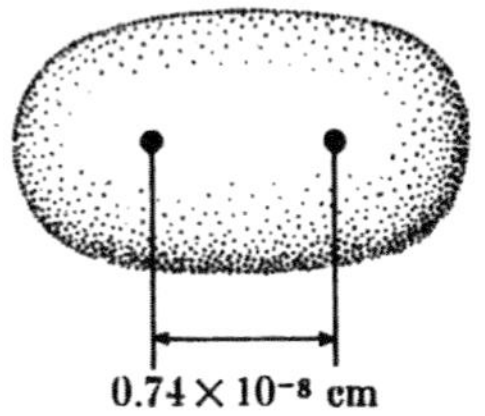

FIGURE 24.27 Schematic representation of hydrogen molecule, H_2. The protons vibrate about a fixed average separation distance, while the two electrons spread themselves over the whole shaded region. The molecule bears little resemblance to a pair of adjacent atoms.

Every atom, no matter how complicated its electronic structure, has two simplifying features. All of its positive charge is concentrated at its center. So is most of its mass. Even the simplest molecules obviously do not possess these features. Because the positive charge is distributed over two or more "widely" separated nuclei, the central-field approximation is not valid in molecules. Because the mass is similarly distributed, new kinds of molecular motion are possible that are not found in atoms. In molecules, as in atoms, electrons can jump from higher to lower states of motion, emitting photons, some of which produce visible light. In addition to this kind of transition, molecules can undergo two other kinds of transition, resulting from the vibration and rotation of the nuclei. In the hydrogen molecule, for example, the protons can vibrate along the line joining them. The energies of this mode of motion are quantized. Transitions from higher to lower vibrational states of motion result in infrared spectra. Even lower energy photons, in the far infrared, result from molecular rotational transitions. Because angular momentum is quantized, the rotational energy too is quantized.

With its completely shared electrons, the hydrogen molecule affords a good example of covalent bonding. Air is composed largely of similarly bound molecules of nitrogen and oxygen, N_2 and O_2. A different kind of binding, ionic binding, occurs when one or more electrons, instead of being shared, are largely transferred from one atom to another. Even in ionic binding, however, the closeness of the atoms causes each partially to lose its atomic individuality. The molecule CsF (see page 411) is a definite entity, quite distinct from separate Cs^+ and F^- ions, just as a hydrogen atom is quite distinct from the separate particles, proton and electron.

Force and Energy

Although atomic and molecular combinations occur in endless variety, the forces acting between atoms and molecules all have certain features in common. Between any pair of atoms or any pair of molecules there acts a weak attractive force if the structures are well separated. As they get quite close together and the electron waves begin to interpenetrate significantly, the force becomes strongly repulsive. This repulsion is created largely by the exclusion principle that prevents more than one electron from occupying any state of motion. Figure 24.28 shows the general character of an interatomic or intermolecular force curve, and the associated potential energy curve. Although the force and energy behave in approximately this way for any pair of atoms, the detailed differences from one pair to another can have enormously important consequences. Atoms of carbon can hold themselves together in graphite up to temperatures of 3,900°K, whereas the force between a pair of helium atoms is so weak that helium does not solidify at all except under high pressure near absolute zero. Hydrogen and oxygen are easily ignited to react explosively. Other atoms combine slowly, even at high temperature. It is important to be aware that the interatomic (or intermolecular) force is quite complicated, depending not only on the electrical forces acting among all of the constituent particles, but on details of the quantum structures of the individual atoms and molecules. It is for this reason that chemistry has developed as a separate discipline of science, with its own techniques and conceptual schemes that have proved useful in classifying, correlating, and controlling the subtle and wonderful variety of atomic and molecular interactions.

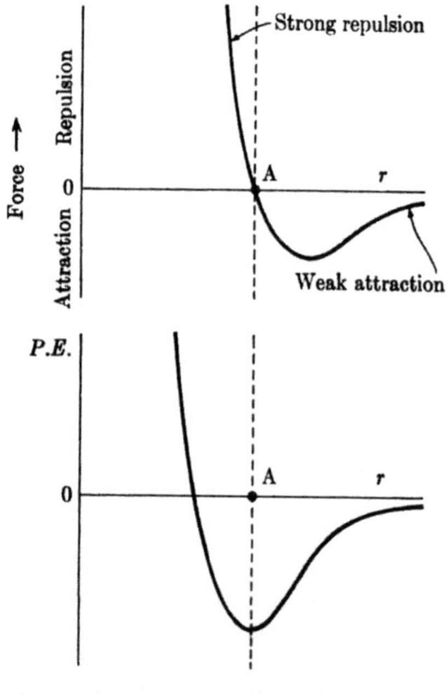

FIGURE 24.28 Approximate form of force and of potential energy between two atoms or molecules. At the point A, where the force changes from attractive to repulsive, the potential energy reaches its lowest value.

For illustrative purposes in the paragraphs above, we have referred primarily to diatomic (two-atom) molecules. Many more complex molecules are of course known, containing three, four, five, six, or a million atoms. Within a man, a DNA molecule containing information about his genetic heritage is composed of billions of atoms. Not surprisingly, the most elaborate known molecules are to be found in living matter. These are long chain molecules strung together like the words in a sentence, the "words" being composed of "letters" which are single atoms or small groups of atoms. Regardless of complexity, one unchanging feature of the interatomic world is the scale of energy. About 4 eV is required to split a hydrogen molecule into two hydrogen atoms. A long-chain protein molecule can be broken in two with about the same energy. This same energy, in turn, is characteristic of the photons of atomic spectra, in the near ultraviolet. Life on earth is possible because the mean thermal energies of molecular motion at our earthly temperatures are much less than the binding energies holding molecules together. Ordinary thermal agitation is unable to disrupt and destroy the molecules of living matter.

Crystalline Solids

Atoms can join together not only in molecules but in the endless regular arrays of crystalline solids. Some solids are merely frozen molecules, that is, molecules immobilized by intermolecular force and close packing. More often, however, a molecule partially or wholly loses its identity in the solid state because of the powerful effect of its neighboring molecules, much as an atom loses its identity when packed close to other atoms in a molecule. As a result, the crystal as a whole must be looked upon as a single giant molecule. Of course its size is no measure of its complexity, since it consists of identical repetitions of a basic pattern.

The types of binding in solid crystals are analogous to the binding in individual molecules. ln a crystal of sodium chloride, there is very little sharing of electrons. Ionic bonding is dominant between Na^+ and Cl^- ions, which are situated at alternate comers of a cubic lattice (Figure 24.29). A metal crystal represents an extreme example of covalent bonding. Electrons are shared not only between adjacent atoms, but among all atoms. In a crystal of copper, for example, a sea of conduction electrons, one contributed by each atom, wander without restriction over the whole latticework of copper ions, Cu^+. Although unrestricted in position within the crystal, the electrons are not wholly unimpeded. Interacting with imperfections in the lattice, an electron can lose some of its kinetic energy, transferring it to vibrational energy of the ions. This is a process that contributes to the electrical resistance of the metal.

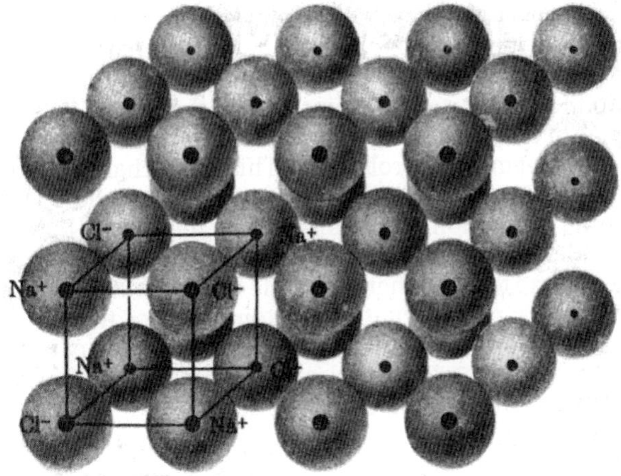

FIGURE 24.29 Ionic binding in sodium chloride crystal. Lines joining nuclei are drawn to show the crystal lattice structure.

Some crystals are held together by another type of interatomic binding, intermediate between ionic and metallic, which has developed special technological significance in recent years. In materials called semiconductors, electrons are not free to wander unrestricted through the crystal, nor are they held in fixed states of motion at single ionic sites. They can jump from site to site, hopping through the crystal under the action of an externally applied electric field (Figure 24.30). In "n-type" semiconductors (n for negative), atoms with almost empty outer shells enable electrons to jump from atom to atom, carrying negative charge through the crystals. In "p-type" semiconductors (p for positive) are atoms with almost filled outer shells. In these semiconductors, it is the "holes" that move through the crystal, effectively carrying positive charge [Figure 24.30(b)]. If an electron jumps to the left to fill a shell, it leaves behind an unoccupied state of motion. This in turn is filled by an electron jumping into it from further to the right, which in turn relocates the unoccupied state further to the right. In this way the "hole" propagates to the right. Transistors, used for control and amplification of electrical signals in radios and many other devices, are composed of combinations of both n-type and p-type semiconductors. Manufacturers can justifiably trumpet the advantages of "solid-state circuitry," for transistors require very little power, and in normal use last indefinitely.

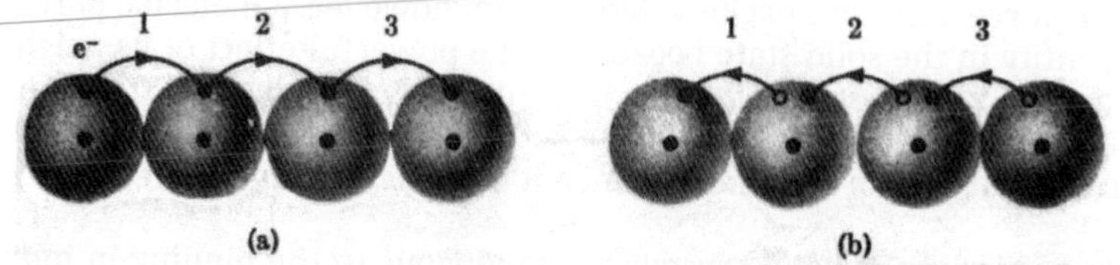

FIGURE 24.30 Schematic representation of semiconductors. **(a)** in an n-type semiconductor, an electron jumps from ion to ion to transport negative charge through the crystal (here to the right). **(b)** In a p-type semiconductor, an electron jumping from one ion to another leaves behind a hole to be filled by another electron, which in turn leaves a hole to be filled by another. Here the leftward jumps effectively transport positive charge (a "hole") to the right.

24.15 Lasers and stimulated emission

152 By postulating quantum jumps, Bohr created new problems to be solved. How does an electron in an excited state of motion "know" when to jump? How does it "decide" which lower state to choose? Under what circumstances can an electron absorb energy and jump to a state of higher energy? These questions were ultimately answered in terms of transition probabilities, calculable in the theory of quantum mechanics. Even before the theory was developed to make possible these probability calculations, Einstein (in 1917) called attention to certain relationships among the transition probabilities. Associated with any given pair of energy levels are three transition probabilities (Figure 24.31). One is called the spontaneous-emission probability. This is the chance per unit time that an atom that finds itself at the higher level will spontaneously emit a photon and jump to the lower level. The second is the absorption probability. This is the chance that an atom at the lower level illuminated by photons of the right frequency and of specified intensity will absorb one of the photons and jump to the higher level. The existence of these two probabilities is not surprising. They account for the excitation of atoms and for the emission of spectra. In addition, there is a third transition probability associated with the pair of levels, which does seem surprising. If an atom at the upper level is illuminated by photons of the right frequency, it may be caused to emit a photon identical to one already present, jumping down to the lower level. The chance that it does so is called the stimulated-emission probability. The downward transition, which would have occurred spontaneously in time anyway, can be triggered to occur sooner by striking the atom with a photon identical to the one it is to emit.

Stimulated emission, it should be remarked, is not exclusively a quantum phenomenon. A classically oscillating particle may either emit or absorb energy if accelerated by an electromagnetic wave. However, it emits and absorbs at the *same* frequency—the frequency of its vibration—and, on the average, absorption is more likely than emission when it is exposed to radiation of that frequency. A quantum system in a particular state of motion, on the other hand, usually absorbs and emits radiation of *different* frequencies. Therefore a photon may be of the right frequency to stimulate emission from a particular state without running any danger of being absorbed by the system in that state. Because of this all-important difference, stimulated emission in the quantum domain has some exceedingly interesting consequences.

Consider what might happen. Suppose a photon of appropriate frequency enters a substance containing numerous excited atoms, each one of them in the same excited state. If it stimulates one of the atoms to emit a photon, the number of photons grows from one to two. If. each of these two stimulates the emission of another photon, the result is four pho-

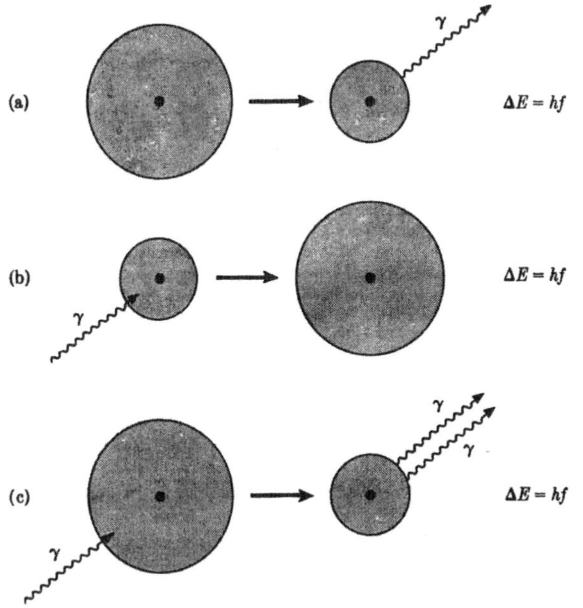

FIGURE 24.31 Types of transition probabilities. **(a)** Spontaneous emission. Atom jumps to lower-energy state of motion (indicated by smaller size) and emits a photon. **(b)** Absorption. Atom absorbs a photon and jumps to higher-energy state of motion. **(c)** Stimulated emission. Incident photon causes excited atom to emit another photon identical in properties to the incident photon. In all three cases, photon energy hf and atomic energy change ΔE *are* related by $\Delta E = hf$.

tons, all identical. The process could continue so that a single photon entering one side of the substance produces a flood of identical photons coming out of the other side. It might be objected that the stimulated emission has only accelerated the emission of many photons that would have been emitted spontaneously anyway, so that it is without special interest. Actually, because of a special feature of stimulated emission that we have so far only hinted at, the stimulated photons are quite different from spontaneously emitted photons. A stimulated photon is identical in *all* respects to the stimulating photon. Not only has it the same energy and frequency, it has the same direction and the same polarization. Whereas a million excited atoms spontaneously emitting photons would send them out in a million different directions, the same million atoms in a cascade of stimulated emission would create an intense beam of directed radiation, one million photons in parallel.

Simple in principle, but difficult in practice, a chain reaction of stimulated emission was first achieved for microwaves in 1955 and for light in 1960. The successful devices are called masers and lasers, their names being acronyms for "Microwave (or Light) Amplification by Stimulated Emission of Radiation." Figure 24.32(a) shows a typical modern laser, a gas-discharge tube with a bright pencil of monochromatic radiation emerging from its end. In order to bring into being this marvel of applied physics, a practical problem had to be overcome: how to keep atoms in an excited state, receptive to stimulated emission. As we have already emphasized, if all atoms are in the same excited state, one photon can stimulate more photons to stimulate still more photons, all without risk of absorption. Two factors work against this state of affairs. First, whenever an atom emits a stimulated photon, it drops down to a lower energy level and becomes a potential absorber of exactly the kind of photon it just emitted. Second, in normal thermal equilibrium, more atoms will be found at the lower than the higher energy level, so that more of them are ready to absorb than to emit. In order to keep the stimulated photons free of absorption, so that they can in turn stimulate more photons and produce the desired chain reaction, the laser must be designed so that atoms at the lower energy level are promptly elevated back to the upper level by means other than direct photon absorption.

It is said that at Disneyland, the average used candy wrapper spends only 20 seconds on the ground before it is picked up. Probably the customer is never more than 30 seconds away from a candy stand able to supply him with a fresh candy wrapper. The aim of the management is to keep more candy wrappers in the hands of the customers than on the ground. This takes some trouble, since in normal equilibrium there would be more on the ground, but it is worth it. Business is stimulated, and a chain reaction of purchase and profit results. There are at least a few points of similarity between Disneyland and a laser. In a helium-neon laser, such as the one pictured in Figure 24.32, neon atoms are the customers, and helium atoms the employees. The neon atoms are encouraged to discard energy in photon emission. As soon as they do so, it is the task of the helium atoms to furnish the neon atoms with more energy to reestablish them at the upper energy level where they can again emit photons. The laser operates when more neon atoms are maintained at the upper than at the lower energy level.

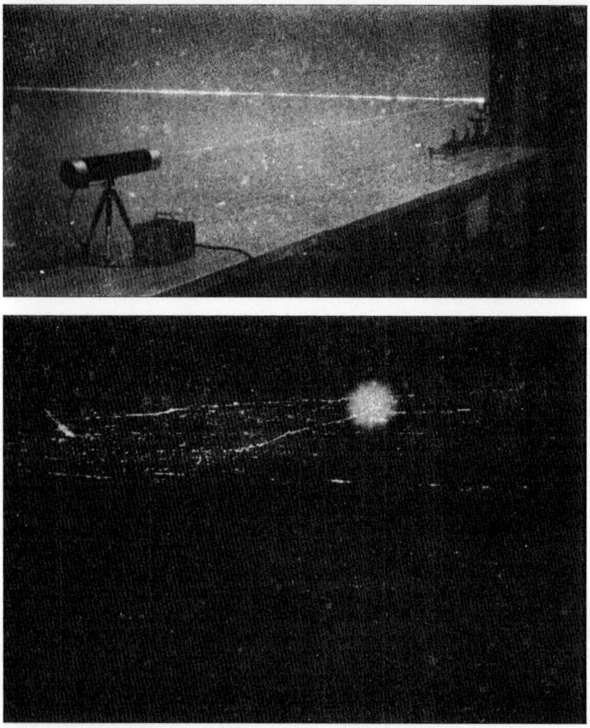

FIGURE 24.32 (a) A modern laser. Its narrow beam is made visible in this photograph by superimposing on a snapshot of the laser and mirror a time exposure of the beam striking vapor clouds in the darkened laboratory room. (Photograph by Thomas E. Stark and the author.) **(b)** In this view from Mt. Hamilton in California, the output power of a laser 25 miles away, although only 0.05 watt, is so narrowly directed that the laser stands out clearly amidst the city lights. (Courtesy of Spectra-Physics, Inc.)

Figure 24.33 shows the energies of the relevant states of motion in helium and neon. Energetic electrons, caused by an external power supply to flow through the gas, excite many helium atoms to a metastable (long-lived) state 20.61 eV above the ground state. An excited helium atom prowls through the gas, ready to give up its store of extra energy to any neon atom it finds without energy. In a collision, the energy can be transferred from the helium atom to the neon atom, leaving the neon atom at an energy level 20.66 eV above its ground state. (The slight discrepancy between 20.61 eV and 20.66 eV is made up by kinetic energy of thermal motion to maintain energy conservation.) By stimulated emission, the neon atom drops down 1.96 eV to a state with 18.70 eV of excitation energy, then dissipates this energy by other means, principally collisions with the wall of the tube. The laser radiation consists of a narrowly directed beam of red light, of wavelength 6,328 A and photon energy 1.96 eV. Mirrors at the ends of the tube cause most of this radiation to traverse the tube many times so that every photon has many opportunities to stimulate the emission of other photons in the same direction. It is only because of these mirrors that the stimulated emission builds up to a high intensity in one particular direction. The useful part of the laser beam is the small fraction, about 1%, that escapes through the mirror at the end.

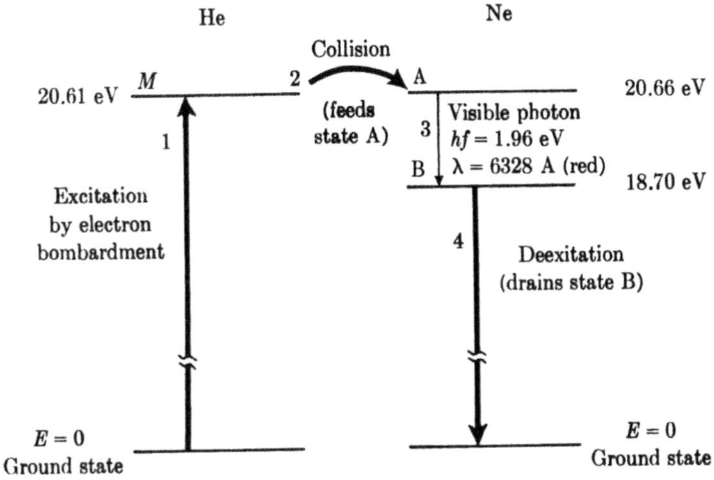

FIGURE 24.33 Partial energy level diagrams for helium and neon atoms relevant to laser action. **1.** Electrons excite helium atoms to metastable state M. **2.** Helium atom transfers its energy to neon, exciting it to state A. **3.** Stimulated emission causes neon atom to jump from state A to state B, emitting photon of red light. **4.** In collision with tube walls, neon dissipates the remainder of its excitation energy. By this indirect means, more neon atoms find themselves in state A than in state B at any one time, so that a photon of 1.96 eV is more likely to stimulate emission than to be absorbed.

It should be emphasized that the laser is in no sense a *source* of energy. It is a converter of energy, taking advantage of the phenomenon of stimulated emission to concentrate a certain fraction of its energy into radiation of a single frequency moving in a single direction. Because of this concentration, a laser beam with less total power than an ordinary light bulb can burn a hole through a metal plate or send a message over hundreds of miles.

EXERCISES

24.1. In what ways does the photon as a particle violate the laws of Newtonian mechanics?

24.2. In experiments on the photoelectric effect what happens if **(a)** the intensity of the incident light is doubled? **(b)** the frequency of the incident light is doubled? How does each of these changes support the photon theory of light?

24.3. For a material particle as well as for a photon, the total energy of the particle and its wave frequency are related by the equation, $E = hf$. **(1)** Use this fact together with the de Broglie equation relating wavelength and momentum to derive an expression for the wave speed of a particle, v_{wave}, in terms of its energy and momentum. **(2)** Show that for a material particle, this wave speed exceeds the speed of light. (This result does not violate the principles of relativity, for the speed at which energy and momentum are transmitted forward is the average velocity, or "group velocity," of a wave packet, which is always less than c. Within a wave packet, wave oscillations can be moving faster.)

24.4. Many phenomena other than the photoelectric effect now support the photon theory of light. Name two such phenomena, either naturally occurring or stimulated by experiment. How does each provide evidence for photons?

24.5. Suggest an experiment that would demonstrate that a *single* photon has wave properties. (HINT: Consider experiments that prove electromagnetic radiation to be a wave phenomenon. Can any of them be performed with photons one at a time?)

24.6. The index of refraction of air for red light is approximately 1.00028. Accordingly, light propagates somewhat more slowly in air than in vacuum, but with no change in frequency. Is the wavelength of the red line of the Balmer series greater or less in air than in vacuum (Equation 24.8)? By approximately how much?

24.7. **(1)** Give a one-sentence definition of the Ritz combination principle. **(2)** Place yourself in the year 1908, armed with all of the facts and theories of that time, but ignorant of all details of atomic structure. If you accepted Einstein's photon and Ritz's combination principle, and believed in the universal validity of the law of energy conservation, what could you conclude about the nature of an atom?

24.8. A spectral series observed in the radiation from atomic hydrogen has inverse wave lengths, or wave numbers, equal to 0.823×10^5 cm^{-1}; 0.975×10^5 cm^{-1}; 1.028×10^5 cm^{-1}; 1.053×10^5 cm^{-1}; 1.066×10^5 cm^{-1}; . . . . **(1)** Write a formula similar to the Balmer series formula that fits these data. **(2)** What is the name of this series? **(3)** Higher members of this series cluster closer and closer to a series limit. What are the wave number and the wavelength of this series limit? In what part of the spectrum does it lie?

24.9. In the light emitted by sodium atoms, a spectral series analogous to the Lyman series in hydrogen is described by the formula,

$$\frac{1}{\lambda} = \mathscr{R}\left(0.378 - \frac{1}{m^2}\right),$$

where $\mathscr{R}$ is the Rydberg constant and m is an integer. (The formula is valid for large m.) **(1)** In terms of the structure of the atom, explain why the second term on the right has the same simple form as in the series of hydrogen. **(2)** From the formula, calculate the ionization energy of sodium. Express the answer in eV.

24.10. Below Equation 24.15 it is stated that the charge-to-mass ratio of the electron is "large." Explain carefully what this means.

24.11. In carrying out his experiments to determine e/m for electrons, Thomson used accelerating potentials of 2,000 to 30,000 volts. Did the effects of relativity, unknown to Thomson, influence the results of any of his experiments? Explain.

24.12. In an electric field, a particle experiences a force proportional to its own charge. Magnetic force is also proportional to charge. Prove that accelerations of electrons produced only by electric and magnetic fields could never be used to determine the charge or the mass of the electron separately, but only the ratio of charge to mass.

24.13. Assume that you have determined the charge-to-mass ratio of the electron by Thomson's method of electric accelerations and magnetic deflection. Show that you can calculate the speed of the electrons without further measurement. Write a mathematical expression for the speed in terms of known quantities. (You may wish to refer back to Section 17.4.)

24.14. Consider a Millikan oil-drop experiment (Figure 24.7) in which the electric field is 900 volts/cm or 3 statvolts/cm. **(1)** If a droplet has a net charge of just one quantum unit and is barely supported by the electric field (electric force = gravitational force), what is the mass of the droplet? **(2)** What is the diameter of the spherical droplet if its density is 0.6 gm/cm^3? Is this "reasonable," i.e., should it be possible to make and to observe droplets this small?

24.15. Give the practical reason why the Balmer series was the first among the various series emitted by hydrogen to be identified and quantified.

24.16. Bohr knew that to provide an explanation for the size of atoms, an atomic length scale was required, and he knew that no combination of the constants e^2 and m could provide such a scale. For this reason he believed that the constant h must play a role in atomic mechanics. Prove to your own satisfaction, either by trial and error, or by a more rigorous method, that no combination of e^2 and m can provide a quantity with the dimension of length. (Recall that the dimension of e^2 is the same as the dimension of energy × length). Optional: Verify that the dimension of the "Bohr radius," $a = \hbar^2/me^2$, is length.

24.17. Equation 24.26 gives an energy formula that accounts for the spectrum of hydrogen. Another formula that works equally well is $E_n = A - (hc\mathcal{R}/n^2)$, where A is any constant. Explain why Bohr assumed $A = 0$.

24.18. The energy of the lowest state of motion of the electron in the hydrogen atom can be written

$$E_1 = -\tfrac{1}{2}mc^2\alpha^2.$$

(1) Express α both **(a)** algebraically as a combination of fundamental constants, and **(b)** numerically. **(2)** Show that α is dimensionless. **(3)** Comment on the significance of the fact that α is much smaller than 1.

24.19. Calculate the ionization energy of hydrogen in eV. If a photon had this energy, in what part of the electromagnetic spectrum would it lie?

24.20. An electron rotates in a circular orbit about a proton—a valid description for states of motion far from the nucleus. Apply laws of classical mechanics to prove that the magnitude of its potential energy is twice the magnitude of its kinetic energy ($P.E. = -2K.E.$), and that its binding energy is therefore equal to its kinetic energy ($W = K.E.$).

24.21. According to classical mechanics, the binding energy W of an electron in a circular orbit about a proton is equal to its kinetic energy (Exercise 24.20). Use this fact as a starting point to derive Equation 24.28, which gives the frequency of the electron's motion as a function of its binding energy.

24.22. **(1)** Verify that the dimension of the Rydberg constant as expressed by Equation 24.33 is inverse length, as required by Equation 24.7. **(2)** The inverse of the Rydberg constant is a distance approximately equal to 912 A. What is the significance of this distance? **(3)** Find a simple algebraic expression for the ratio of this distance to the Bohr radius.

24.23. Astronomers have identified states of motion in hydrogen with principal quantum number n greater than 50. **(1)** Why are such states of motion not identifiable in radiation from hydrogen on earth? **(2)** For a circular orbit with $n = 50$, give **(a)** the speed of the electron; **(b)** the diameter of its orbit; and **(c)** the frequency of its rotation in this orbit; **(3)** What is the frequency of radiation emitted by hydrogen in a transition from $n = 50$ to $n = 49$? In which part of the spectrum does this radiation lie?

24.24. Bohr's early theory provides one way to understand the size of the hydrogen atom. The wave nature of the electron provides another way. A semiquantitative line of reasoning, similar to what was presented in Section 23.5, goes as follows: The electron kinetic energy is about $e^2/2r$, that is (Equation 24.36),

$$\frac{1}{2}mv^2 = \frac{e^2}{2r},$$

where r is the distance from proton to electron. Because the electron is confined within the atom, its half-wavelength is about equal to the atomic diameter:

$$½\lambda = 2r.$$

Finally, the de Broglie relation connects the average electron wavelength and the average electron momentum:

$$\lambda = \frac{h}{p}.$$

(1) Combine these three equations to find a formula for the approximate atomic radius r, and evaluate it numerically. **(2)** Explain why this method of finding the atomic radius, although mathematical, is only approximate, (HINT: Note the marked difference between the wave function graphed in Figure 24.20(a) and a sine wave.)

24.25. **(1)** What is the quantum number n of the state of motion of a particle in a box whose wave function appears in Figure 24.14(b)? **(2)** If the particle is a proton and the walls are separated by 10^{-12} cm (this simulates a nucleus), what is the energy of the particle in this state of motion? Express the answer in ergs and in MeV.

24.26. Examine Equation 24.58 for the energy of a particle in a box. Discuss the dependence of the energy on each of the four quantities, h, n, m, and d, explaining variations in the quantum behavior of the particle as each of these quantities is varied separately. (For purposes of this discussion, pretend that the magnitude of Planck's constant h can be changed.)

24.27. **(1)** Sketch the wave function and the probability distribution for a particle in a box in its lowest state of motion ($n = 1$). **(2)** What are the relative probabilities to find the particle **(a)** at one wall, **(b)** at the midpoint between the walls, and **(c)** one quarter of the way from one wall to the other?

24.28. **(1)** An electron moving in one dimension between rigid walls makes a quantum jump from the second to the first state of motion and emits an ultraviolet photon of energy 6 eV. What is the distance d between the walls? **(2)** Four more electrons are added, and each moves in the lowest state of motion available to it, consistent with the exclusion principle. What states do the five electrons occupy? (Include electron spin in your reasoning.) **(3)** What is the total kinetic energy of this group of five electrons?

24.29. Sketch the wave functions for the first and fourth states of motion of a particle in a box. Compare your sketches with the hydrogenic wave functions graphed in Figures 24.20(a) and (b), and comment on the similarities and differences. (HINT: Every wave function for a particle in a box is a pure sine wave.)

24.30. Consider a "superrelativistic" particle (one whose total energy E is much greater than its rest energy mc^2) moving between rigid walls, as in Figure 24.14. For such a particle Equations 24.51 and 24.57 remain valid, but Equation 24.58 does not. **(1)** Find a formula to replace Equation 24.58 for the discrete energies of this system. **(2)** Evaluate the zero-point energy of an electron confined to a dimension of 10^{-12} cm, comparable to the size of a nucleus. Express the answer in eV. Verify that the calculated energy does indeed make the electron superrelativistic.

24.31. Below are listed quantum numbers and wave functions for three of the lower states in the hydrogen atom:

$$n = 1,\ \ell = 0 \quad \psi = e^{-r/a}\ ;$$

$$n = 2,\ \ell = 1 \quad \psi = re^{-r/2a}\ ;$$

$$n = 3,\ \ell = 2 \quad \psi = r^2e^{-r/3a}\ .$$

(For simplicity, constants of proportionality are omitted.) For the first two of these states, plot ψ vs. r and ψ^2 vs. r—with some care, but without great attention to complete precision. What would you call the radius of the atom in each case? In these formulas, a is the Bohr radius, 0.53 Angstrom. (The third wave function is provided only for information, or for optional attention.)

24.32. Suppose that a magnetic field $\mathscr{B}$ is directed along the z-axis in Figure 24.19. Which of the three orientations of electron motion will have the least energy? Which will have the greatest energy? (Remember to take account of the negative sign of the electron charge. Remember too that the preferred orientation, based on considerations of force and torque, is the orientation of least energy.)

24.33. Extend Table 24.2 to catalogue all of the possible states of motion of the electron at the third principal energy level ($n = 3$) in the hydrogen atom.

24.34. Use the n, ℓ, j, m_j quantum numbers (as in Table 24.3) to catalogue all of the possible states of motion of the electron in the hydrogen atom for $n = 3$. Show that there are 18 such states.

24.35. The diagram shows the $n = 2$ portion of the term diagram of hydrogen, with its "fine structure" of three different energy levels. (The scale is greatly magnified compared with Figure 24.18.) The symbol j is used for total angular momentum, orbital plus spin. **(1)** How many different states of motion are there at each of these levels? What are the quantum numbers n, ℓ, j, m_j of these states? **(2)** Sketch in an approximate way how you would expect the energy level diagram to be changed if the atom is placed in a magnetic field. (HINT: Every state of motion then has a different energy.)

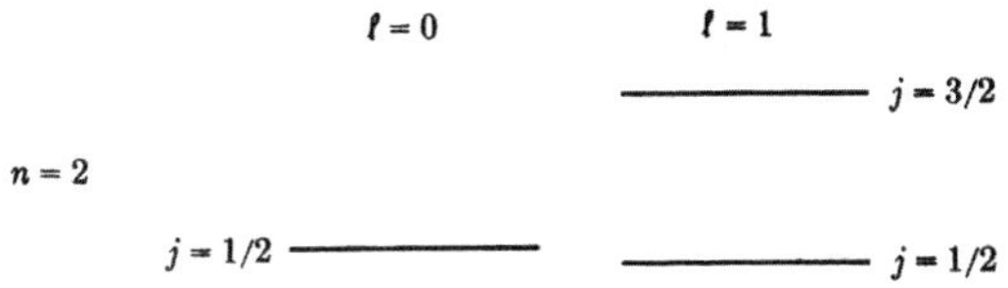

24.36. Consider the Bohr atom theory applied to the motion of a *single* electron in the vicinity of a sodium nucleus, containing 11 protons. Show **(a)** that the Bohr radius in sodium is 11 times smaller than in hydrogen; and **(b)** that the binding energy is 121 times greater in sodium than in hydrogen.

24.37. **(1)** What are the two quantum numbers n and ℓ for the outermost, or valence, electron in lithium? If energy is added to excite this electron to the next higher unoccupied state of motion, what are its quantum numbers n and ℓ (refer to Figure 24.24)? **(2)** Answer the same questions for sodium.

24.38. **(1)** What are the occupied states of motion of the 36 electrons in a krypton atom? **(2)** Why is krypton a particularly stable, chemically inert atom?

24.39. Discuss hypothetical worlds in which **(a)** the exclusion principle operates but electrons are spinless; **(b)** the exclusion principle operates and electrons have spin equal to $\frac{3}{2}\hbar$; and **(c)** the exclusion principle does not operate.

24.40. Compare the approximate speed of an electron in the K shell of a uranium atom with the electron speed in the first Bohr orbit of a hydrogen atom. Is relativity important for either or both of these electrons?

24.41. The K_α X-ray of a certain element has a photon energy of 1.49 KeV. If the effective charge experienced by the electron making this transition is approximately $Z - 1$, what is the element? (Recall that the K_α X-ray results from the $n = 2$ to $n = 1$ transition.)

24.42. Carbon is the backbone of the organic molecules necessary for life. **(1)** In terms of the structure of the carbon atom ($Z = 6$) and the concept of valence, explain why carbon is in a favored position to form many complicated molecules. **(2)** It has sometimes been suggested that silicon ($Z = 14$) might provide a basic element of life on other planets. Why is silicon considered a candidate?

24.43. A "radical" is a group of atoms that act as a single unit in combining with other atoms or other radicals to form molecules. **(1)** The hydroxyl radical, OH, contains one oxygen atom and one hydrogen atom. Explain why its valence is –1. What common compound does it form? **(2)** The ammonium radical, NH_4, contains one nitrogen atom and four hydrogen atoms. Explain why it has valence +1.

24.44. Consider the equally massive nuclei of a diatomic molecule executing rotational motion as shown in the diagram. **(1)** If the mass of each nucleus is M, the speed of each nucleus is v, and their separation is d, what is the angular momentum of the molecule? (Electrons can be neglected because of their small mass.) **(2)** Show that if this rotational angular momentum is quantized according to the rule, $L = \ell\hbar$, where ℓ is an integer, the kinetic energy of rotation is

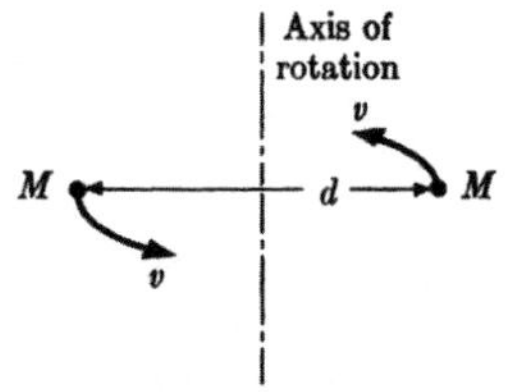

$$\lambda = \frac{h}{p}.$$

(3) Calculate the kinetic energy of the $\ell = 2$ rotational state of the hydrogen molecule ($M = 1.7 \times 10^{-24}$ gm, $d = 0.742 \times 10^{-8}$ cm). Express the answer in eV and compare the magnitude of this energy with the magnitude of a typical energy of electronic motion in a hydrogen atom.

24.45. The energy of thermal agitation at normal temperatures is not sufficient to break the bonds of organic molecules and destroy living matter. Why is life unlikely at even "safer" temperatures much less than the average temperature on earth?

TWENTY-FIVE

Nuclei

Twice in history the atomic nucleus has forced itself upon the attention of man. In 1896, 153
Henri Becquerel in Paris was astonished to discover that a salt containing uranium emitted a new kind of radiation powerful enough to darken a photographic plate through its opaque wrapping. In 1945, all mankind came to know and fear the nucleus when 1 kg of uranium devastated Hiroshima.

These events had something in common. In both, nuclear energy was being released, energy vastly greater *per atom* than the mechanical energy of the most terrifying avalanche, the chemical energy of the most violent combustion, or the electrical energy of the most awesome thunderstorm. The world of science, prepared for the unexpected, could capitalize swiftly on Becquerel's chance discovery. He triggered a world-wide search for the nature of matter, and opened a road of new discovery to Rutherford's nuclear atom and beyond. The world at large was less well prepared for the exploitation of nuclear fission. Man's mastery of nuclear energy on a large scale has produced a state of personal and international tension known as the Atomic Age (more properly, it should be called the Nuclear Age). Of the numerous problems facing the human race in the second half of the twentieth century, none is more urgent than the control of nuclear energy. Numerous other forces in modern society have the potential to alter drastically the nature of human life. Only nuclear energy—in an amount no greater than is already contained in the world's arsenals—has clearly the power to destroy all human life.

As a physical being, man is a complicated collection of atoms held together by binding energies of about 1 eV per atom. He is designed to withstand easily the normal kinetic energy of thermal agitation, about 0.04 eV per molecule—in fact, not only to withstand it, but to require it as well. If this thermal energy is decreased to 0.03 eV per molecule, the dynamic processes sustaining life slow to such an extent that he freezes to death. If the kinetic energy per molecule is increased to 0.05 eV, destruction of molecules begins and he burns to death. A human being is delicately balanced in a narrow energy range. Chemical explosions releasing as much as 1 eV of energy per atom are easily able to destroy life, and have often done so. Yet with one pound of TNT it is difficult to kill more than a few people. The energy scale of chemical explosions and the energy scale of human life are of the same order of magnitude. For this reason, chemical explosives continue to represent a threat to individual life, even to life in whole villages and towns, but not to life on earth. So great is the scale of nuclear energy by comparison that it represents a unique threat, not only greater in magnitude, but different in kind. In ten days in 1943, 40,000 people were killed in Hamburg by 16,000,000 pounds of chemical explosives. In a few moments in August, 1945, 100,000 people in Hiroshima were killed by 2 pounds of uranium. Measured in human lives, the uranium was 20 million times more destructive than the chemical explosives. The energy ratio is similar. On a pound-for-pound basis, the fission of uranium releases 17 million times more energy than the explosion of TNT. Each single fission event releases about 200 MeV of nuclear energy. In addition, the radioactive products of one nuclear-fission event release several MeV more in later disintegration. High-energy particles capable of ionizing atoms and breaking apart molecules may be emitted by the fission products after milliseconds, hours, years, or millennia. The energy of fallout is *nuclear* energy, almost as potent as the original explosive energy of fission, but diffused in space and time.

These facts about energy summarize in capsule form the power and the threat of nuclear explosions. It is impossible to study nuclei without having in the back of one's mind a concern for the grave practical implications of nuclear energy. Yet it would be out of place in this text to pursue the politics, sociology, or even the technology of the Nuclear Age. Our purpose in this chapter is to focus on the physical principles underlying nuclear structure and nuclear behavior, especially those principles related to the quantum-mechanical description of all of submicroscopic nature.

25.1 Facts about nuclei

Although fascinating, the early history of nuclear physics is labyrinthian.* Scientists studied radioactivity for fifteen years before they knew that the atom had a nucleus; twenty more years passed before the interior structure of the nucleus began to show itself. Therefore we shall take at once a modern point of view, postponing to Section 25.8 some remarks on the early development of nuclear physics.

Nuclei are composed of protons and neutrons, particles collectively known as nucleons. The lighter nuclei contain protons and neutrons in about equal number. In heavier nuclei, neutrons outnumber protons (Figure 25.1). The lightest and simplest nucleus is a single proton alone. Heavy nuclei containing more than 100 protons and 150 neutrons have also been identified. The nuclear force binding the nucleons together belongs to the class of strong interactions, and it is experienced by numerous other elementary particles as well, although not by electrons. Electrons feel only the electrical force of the nucleus, and to a lesser extent its magnetic force, but they are immune to the nuclear force that strongly influences nucleons.

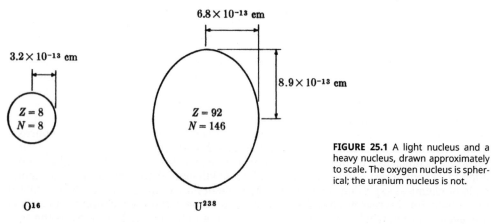

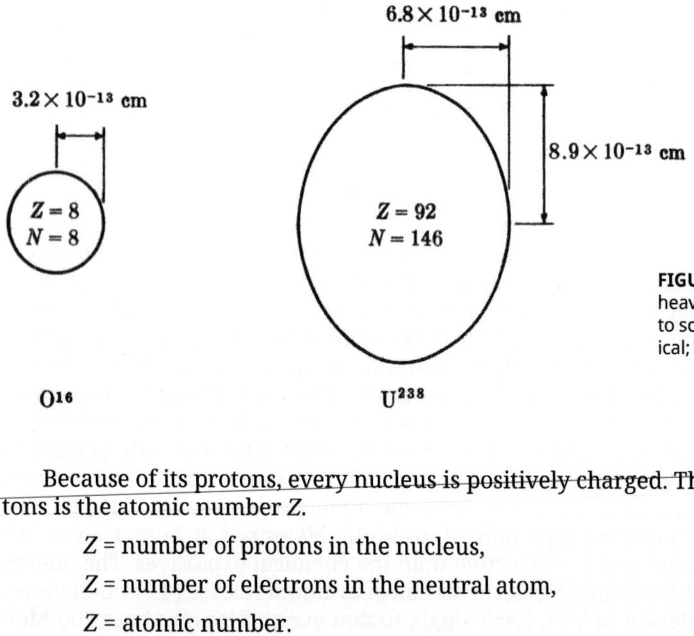

FIGURE 25.1 A light nucleus and a heavy nucleus, drawn approximately to scale. The oxygen nucleus is spherical; the uranium nucleus is not.

Because of its protons, every nucleus is positively charged. The number of nuclear protons is the atomic number Z.

$$Z = \text{number of protons in the nucleus,}$$

$$Z = \text{number of electrons in the neutral atom,}$$

$$Z = \text{atomic number.}$$

The atomic number also identifies the name of the element and locates it in the periodic table. For hydrogen, $Z = 1$; for carbon, $Z = 6$; for tin, $Z = 50$; for uranium, $Z = 92$.

* A good paperback on the early history of nuclear physics is Alfred Romer's *The Restless Atom* (Garden City, New York: Doubleday, 1960, Anchor S12).

Being neutral, neutrons have practically no influence on the atomic electrons. However, they have an important influence on the atomic mass. Neutrons and protons are almost equally massive. Except for small relativistic corrections arising from the binding energy, the nuclear mass is equal to the sum of the masses of its constituent protons and neutrons. Because the electron mass is so much less than the nucleon mass (see Table 2.1), the mass, of the entire atom in turn differs very little from the mass of the central nucleus. In addition to atomic number (or proton number) Z, we define neutron number N, and mass number A:

N = number of neutrons in the nucleus;

$A = Z + N$,

A = number of nucleons in the nucleus,

A = mass number.

At this point, a few more items of notation and definition are in order. All atoms with the same atomic number (the same nuclear charge) belong to the same element. Atoms of the same atomic number but different mass number are called *isotopes* of an element. The nucleus of a particular isotope is identified both by its proton number (atomic number) and by its neutron number. Evidently different isotopes of the same element differ only in the number of neutrons in their nuclei. Typically, an isotope is identified by its atomic number, its name, and its mass number, thus: ${}_6C^{14}$. This designates carbon 14, whose nucleus contains six protons and eight neutrons, altogether fourteen nucleons. The word "isotope" is used somewhat loosely to mean either a particular nucleus, or an atom containing that nucleus, or a collection of identical atoms. The presently adopted atomic mass standard is carbon 12 (${}_6C^{12}$), the most abundant isotope of carbon, whose atomic weight is chosen to be exactly 12.0000. Normalized to this standard, the atomic weight of every other isotope is closely equal to the mass number of the isotope. The biggest discrepancy occurs for the lightest isotope of hydrogen, whose atomic weight of 1.008 differs by 0.8% from its mass number of 1. The atomic weight of an *element* (as opposed to an isotope) is the average atomic weight of the normal mixture of isotopes of that element.*

In studying matter, one of the most important points to keep in mind is the independence of atomic and nuclear properties. The electric charge of the nucleus is the single determining factor of atomic structure. All of the other properties of the nucleus are almost irrelevant to the behavior of the electrons. In their atomic properties, for instance, carbon 12 and carbon 14 are nearly identical, yet their nuclei are totally different in structure. One is stable, the other radioactive. The electrons, in turn, have almost no influence on the nucleus. The situation is much the same as for the earth and its satellites. The satellites stay in orbit because of the gravitational pull of the earth, but are unaffected by a myriad of other interesting properties of the earth. The earth, for its part, does not respond to the presence of the satellites.

Because of this "decoupling" of nuclei and electrons, almost all of the properties of matter can be identified as either primarily atomic (meaning attributable to atomic electrons) or primarily nuclear, but not both. The nucleus is the source of the mass of an atom, its position in the periodic table, and its radioactivity, if any. The electrons account for all of the physical and chemical properties of the elements and for their characteristic spectra.

Now let us ignore the electrons and focus on the nucleus as a mechanical system, a structure in its own right. Most nuclei are prolate spheroids (Figure 25.1), slightly deformed from the spherical shape in the direction of a cigar or football shape. For most purposes, it is sufficient to regard them as spheres, with fairly well-defined boundaries. (The wave nature of matter prohibits sharp boundaries.) Nuclear radii lie between 10^{-13} cm and 10^{-12} cm (one

---

* With rare exceptions, the isotopic mixture of an element is independent of the source of the element. (Lead, whose different isotopes are products of different radioactive decay chains, is an exception.) For elements not found in nature at all, the atomic weight of the "normal mixture" of course has no meaning.

ten thousandth to one hundred thousandth the size of an atom), and follow approximately the formula

$$R = (1.2 \times 10^{-13}\ \text{cm}) \times A^{1/3} \tag{25.1}$$

The proportionality of radius to the cube root of the mass number means that the nuclear volume ($\frac{4}{3}\pi R^3$) is proportional to A, the number of nucleons, almost as if every nucleon represents another brick on the structure (but see Section 25.5). This is quite different from the behavior of atomic volume, which, except for some periodic variations, remains nearly constant, regardless of the number of electrons. Since the nuclear volume is proportional to the number of nucleons contained, the density of matter within the nucleus is about the same for all nuclei. The mass of a nucleus is given by $M = (1.67 \times 10^{-24}\ \text{gm}) \times A$, where the factor multiplying A is the mass of one nucleon. The density of nuclear matter can therefore be calculated:

$$\frac{\text{Mass}}{\text{Volume}} = \frac{1.67 \times 10^{-24}\ \text{gm} \times A}{\frac{4}{3}\pi(1.2 \times 10^{-13}\ \text{cm})^3 A} = 2.3 \times 10^{14}\ \frac{\text{gm}}{\text{cm}^3}. \tag{25.2}$$

A solid lump of nuclear matter one centimeter in diameter would weigh 133 million tons.

Like other quantum systems, a nucleus has a set of stationary states of motion, at discrete energy levels, between which quantum jumps can occur. On the average, nuclear binding energies are about one million times greater than atomic binding energies, and energy changes in nuclear quantum transitions are also typically about one million times greater than the energy changes of the atomic transitions responsible for the emission of visible photons. It is easy to remember that energies associated with outer electrons are measured in electron volts (eV), whereas energies associated with nucleons in the nucleus are measured in millions of electron volts (MeV).

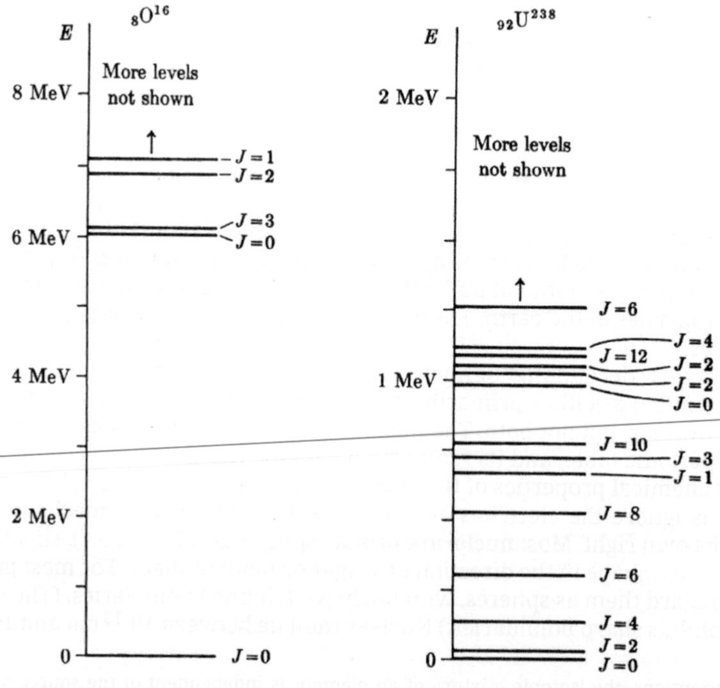

FIGURE 25.2 Partial energy-level diagrams for the same two nuclei pictured in Figure 25.1. Note that the two energy scales are not the same. The symbol J designates the angular momentum of the nuclear state in units of $\hbar$.

Energy-level diagrams can be drawn for nuclear states just as for atomic states (Figure 25.2). Transitions between these states of motion occur in much richer variety for nuclei than for atoms. An isolated atom loses energy spontaneously in only one way, by photon emission. For an isolated nucleus, four quite different modes of energy loss are possible. (1) It may, like an atom, emit a photon. This is called gamma decay. (2) It may emit an electron or a positron, along with an antineutrino or neutrino. In this process, called beta decay, the nucleus increases or decreases its charge by one unit—in short, it becomes a different nucleus, belonging to a different element. The emitted electron or positron is called a beta particle. (3) It may emit an alpha particle, which is the same as a helium nucleus, two neutrons and two protons bound together. In alpha decay, the nucleus literally emits a chunk of itself, losing four units of mass and two units of charge in the process. (4) It may split apart more drastically, breaking into two nearly equal fragments. This is nuclear fission, a process that occurs spontaneously only for the heaviest elements.

Nuclei are normally not completely isolated, for they are surrounded by orbital electrons. Two other modes of nuclear energy loss involve these electrons. (5) A nucleus may transfer its excess energy to an atomic electron, causing the electron to shoot from the atom. This process, called internal conversion, is closely related to gamma decay. (6) A nucleus may capture and annihilate an orbital electron, simultaneous with the creation and emission of a neutrino. This process of orbital capture is brought about by the same interaction that is responsible for beta decay.

A few excited nuclear states have three or more alternative modes of energy loss available. Most have only one or two. Figure 25.3 illustrates an example in which a single excited nuclear state can choose among beta decay, gamma decay, and electron capture, a choice it makes according to quantum laws of probability.

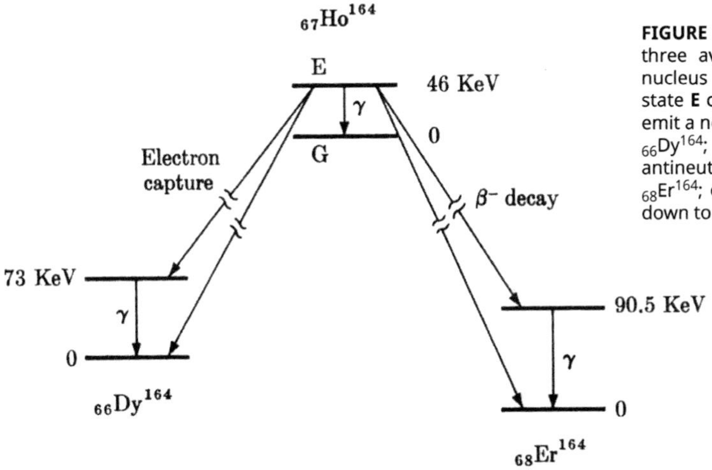

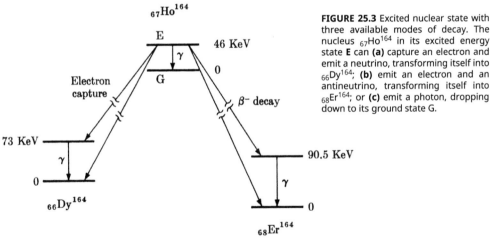

FIGURE 25.3 Excited nuclear state with three available modes of decay. The nucleus $_{67}Ho^{164}$ in its excited energy state **E** can **(a)** capture an electron and emit a neutrino, transforming itself into $_{66}Dy^{164}$; **(b)** emit an electron and an antineutrino, transforming itself into $_{68}Er^{164}$; or **(c)** emit a photon, dropping down to its ground state G.

In terms of the modern picture of stationary states and quantum jumps between them, there is nothing mysterious about radioactivity. Radioactivity means simply spontaneous emission of energy in a nuclear quantum transition. Compared to atomic transitions, nuclear transitions are both more energetic and more diverse. Of particular interest is the fact that transitions can occur between states of motion in two different nuclei. Whereas the ground state of an atom is stable, enduring forever if the atom is undisturbed, the ground state of a nucleus may be unstable. Although it is the lowest energy state of that particular nucleus, alpha decay, beta decay, orbital capture, or spontaneous fission may enable it to make a quantum jump to a still lower energy state in another nucleus. In this way, radioactive decay chains can occur through a series of many nuclei. A nucleus of ${}_{92}U^{238}$, for instance, decays eventually to a nucleus of ${}_{82}Pb^{206}$, an isotope of lead with 10 fewer protons and 22 fewer neutrons than its uranium isotope parent.

Nuclei can interact with one another, and sometimes do, with dramatic consequences—nuclear interaction lights the stars. For the most part, however, each nucleus lives in splendid isolation, shielded from the rest of the world by its own encircling electrons. Even if an atom is stripped of all electrons by ionization, the force of electrical repulsion normally keeps it from coming into contact with other nuclei. Whereas atoms interact with one another incessantly, around us and within us, nuclei rarely interact. One way to produce a nuclear reaction is by bombardment with high-energy particles. A beam of protons, alpha particles, or heavier nuclei, accelerated to high energy (many MeV) and directed at a target, can penetrate the atomic electrons, surmount the electric repulsion, and induce nuclear transformations. Cosmic-ray particles do the same thing. Neutrons, being uncharged, can induce nuclear transformations at lower energy. The fission of uranium 235, for instance, is triggered by slow neutrons. Another way to produce nuclear reactions is with extreme heat. At sufficiently high temperatures—tens of millions of degrees—nuclei of the light elements, stripped of their electrons, acquire enough kinetic energy of thermal motion to strike against one another despite the force of electrical repulsion. Reactions produced in this way are called *thermonuclear reactions*. They account for the energy production in stars and in thermonuclear weapons (H bombs). In the billions of years since the earth was formed, the vast majority of its nuclei have led totally sedentary lives, never once excited or disturbed by outside influences. It is ironic that so rare an event on earth as a nuclear reaction should have such catastrophic implications for mankind.

25.2 Nuclear composition

When the neutron was discovered in 1932, the world of physics was more than ready for it. Within a few months of its discovery, Werner Heisenberg advanced a theory of nuclear composition, according to which every nucleus is composed only of neutrons and protons. He went on to speculate that these particles, of about equal mass and both strongly interacting, might in some sense be two versions of the "same" particle, differing only in charge. Since then, we have found no reason to correct Heisenberg's theory of nuclear composition, except to recognize also the transitory existence of pions and other mesons within the nucleus. Moreover, his suggestion about the sameness of neutron and proton has borne unexpected fruit. Many different particles have been found to come in groups of nearly equal mass, alike except in charge. When we refer now to a nucleon, we mean not only a constituent of a nucleus. We also mean that particular particle type whose two manifestations are the neutron and the proton.

154 Before the discovery of the neutron, if nuclei were to be described in terms of known particles at all, it had to be in terms of protons and electrons. No other building blocks were known. An alpha particle, for example, could be described as a closely bound structure of four protons and two electrons. The four protons contributed the necessary mass; the two electrons canceled the excess charge. Similarly, a nucleus of uranium 238 could be assumed to contain 238 protons and 146 electrons. Quantum mechanics raised two serious objections against this view of nuclear composition. First, to hold an electron within the confines of a nuclear volume requires an energy larger than is actually associated with nuclear binding. If, for instance, an electron is held within the "box" of a helium nucleus, its wave amplitude must undergo at least one half cycle of oscillation in a distance of about 0.5×10^{-12} cm. This means that its wavelength can be at most 10^{-12} cm. Then the de Broglie equation reveals its least possible momentum:

$$p = \frac{h}{\lambda} = \frac{6.63 \times 10^{-27} \text{ gm cm}^2/\text{sec}}{10^{-12} \text{ cm}} = 6.63 \times 10^{-15} \text{ gm cm/sec} . \qquad (25.3)$$

At this point we must proceed with caution. Is this momentum in the domain of relativistic mechanics or of classical mechanics? If we guess it to be classical, we can calculate the speed of the electron to be:

$$v_{\text{classical}} = \frac{p}{m} = \frac{6.63 \times 10^{-15} \text{ gm cm/sec}}{9.11 \times 10^{-28} \text{ cm}} = 7.3 \times 10^{12} \text{ cm/sec} .$$

We have guessed wrong. This speed is 220 times greater than the speed of light, a physical impossibility. Therefore we must turn to a relativistic equation. Of interest is the electron energy, related to momentum by Equation 21.28:

$$E^2 = p^2c^2 + m^2c^4 \quad \text{or} \quad E^2 = (pc)^2 + (mc^2)^2 . \tag{25.4}$$

The quantity mc^2, the rest energy of the electron, is, in electron volt units, $mc^2 = 0.511$ MeV. The quantity pc can be calculated for this example:

$$pc = (6.63 \times 10^{-16} \text{ gm cm/sec}) \times (3 \times 10^{10} \text{ cm/sec}) = 1.99 \times 10^{-4} \text{ erg} = 124 \text{ MeV} .$$

(We have used the conversion factor, 1.6×10^{-12} erg/eV.) Since pc is so much greater than mc^2, the second term in Equation 25.4 contributes very little, and the energy-momentum relation becomes, approximately, $E = pc$ (see the discussion following Equation 21.20). In round numbers, then, an electron held within a helium nucleus would have a kinetic energy in excess of 100 MeV. So energetic an electron should in fact fly out of the nucleus, since the total binding energy holding all the particles together in the helium nucleus is only 28 MeV, or 7 MeV per nucleon. This was one difficulty with the idea of electrons in the nucleus.

Another difficulty concerns nuclear spin. According to the proton-electron model of nuclear composition, a nucleus of $_7N^{14}$ contains 21 particles, 14 protons and 7 electrons. According to the proton-neutron model, the same nucleus contains only 14 particles: 7 protons and 7 neutrons. Since all the particles in question have one-half unit of spin, it makes an important difference whether the nucleus contains an odd or an even number of particles. If the number is odd, the total nuclear spin must be equal to an integer plus one half (in units of $\hbar$). If even, the nuclear spin must be integral. Early evidence that the nitrogen nucleus has integral spin was evidence against electrons in the nucleus. Since then, the spins of several hundred nuclei have been determined. All are consistent with Heisenberg's theory of neutron-proton composition.

Except for very subtle isotopic differences, atomic properties depend on only a single number, the atomic number. Nuclear properties, by contrast, depend in an important way on two different numbers, the proton number and the neutron number. Nuclei of different isotopes of the same element can be in every way as distinct as nuclei of different elements. Because nuclei are built of two basic particles, a chart of the nuclei requires a two-dimensional array. Typically, in such a chart, proton number is plotted vertically, neutron number horizontally. One such chart, prepared by the General Electric Company in 1966, contains entries for 1,500 known nuclei.

Figure 25.4 shows the lower left corner of a chart of the nuclei, beginning with the single proton, $Z = 1$ and $N = 0$, and the single neutron, $Z = 0$, $N = 1$. The shaded boxes indicate stable nuclei. The others are radioactive. The numbers in the shaded boxes show the percentage abundance of that isotope in nature. In the unshaded boxes are shown the half life of the nucleus and the mode of its decay. Many other nuclear properties are known, of course, but no attempt is made to summarize them in this chart.

Beginning with hydrogen, we see three isotopes, H^1, H^2, and H^3, containing, respectively, zero, 1, and 2 neutrons. The nucleus of H^1 is of course the proton. The nucleus of H^2 is known as a deuteron; the nucleus of H^3 is called a triton. Deuterons, making up 0.015% of hydrogen nuclei, are stable. The triton, with a half life of 12.26 years, undergoes beta decay, transforming itself into He^3, a nucleus with two protons and one neutron. For this and other examples of beta decay with emission of an electron, short arrows pointing to the left and upward show the direction of the transformation. Beta decay with emission of a positron is indicated by a short arrow to the right and downward, as from C^{10} to B^{10}. Also indicated by an arrow to the right and downward is the closely related process of orbital electron capture, in which the nucleus captures (and annihilates) one of the innermost atomic electrons simultaneous with the creation and emission of a neutrino. Be^7 transforms itself into Li^7 by means of orbital capture, with a half life of 53 days. Also shown in this small part of the chart of nuclei are two examples of alpha decay. One of them, the decay mode of Be^8, is a very special example, for when Be^8 emits an alpha particle, what is left behind is also an alpha particle. Actually, the alpha decay of Be^8 is equivalent to spontaneous fission, for the nucleus breaks into equal fragments that fly apart with equal speed. The transformation can be written,

$${}_4Be^8 \rightarrow {}_2He^4 + {}_2He^4 .$$

As shown in Figure 25.4, the half life of this decay is only about 3×10^{-16} sec. By contrast, a nucleus containing two more neutrons, Be^{10}, has a half life of 2.7 million years. In between is Be^9, a stable nucleus.

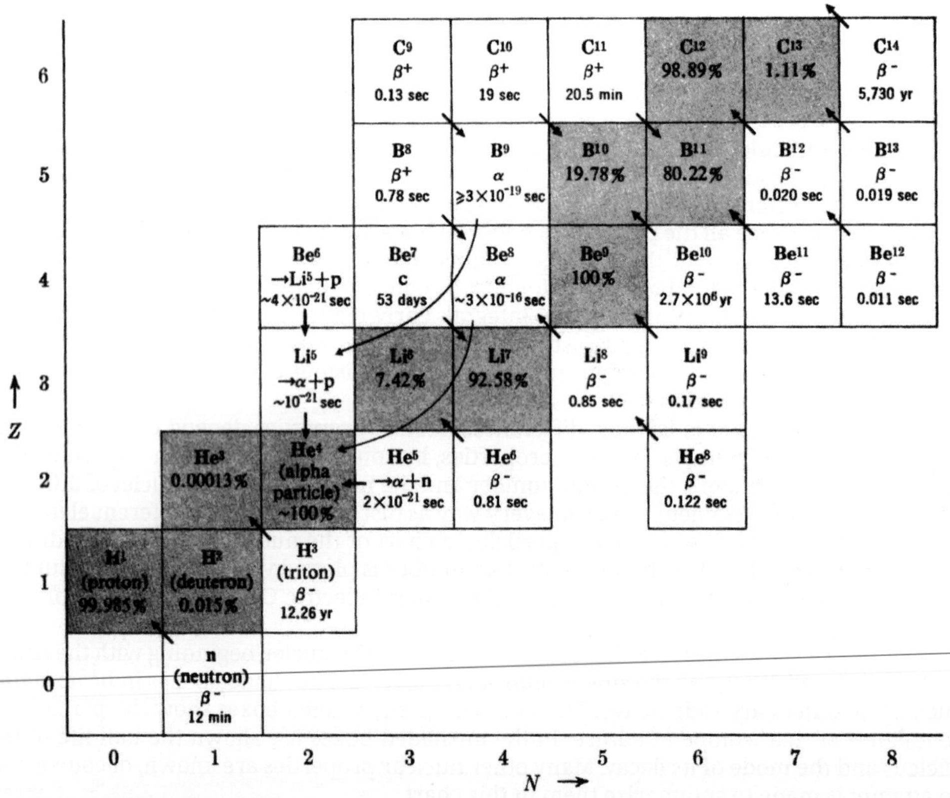

FIGURE 25.4 Small portion of a chart of the nuclei. Proton number (or atomic number) is plotted vertically, neutron number horizontally. The short arrows to the left and upward represent beta decay in which a negative electron and an antineutrino are emitted and a neutron in the nucleus changes into a proton. Then Z increases by 1, N decreases by 1. The opposite arrow direction, downward to the right, represents either beta decay with positron emission or orbital capture, in which proton number decreases by 1, neutron number increases by 1. Numbers in the shaded boxes are per cent abundances of stable isotopes. Numbers in the unshaded boxes are half-lives of unstable nuclei. Alpha decay, beta decay, and orbital capture are indicated by the letters α, β, and c.

All of the unstable nuclei in Figure 25.4 are said to be artificially radioactive. This means that they have been created and studied in the laboratory, but that they are not found in nature, at least not to an appreciable extent.* Naturally radioactive nuclei are those found in nature. The distinction between natural and artificial radioactivity is itself somewhat artificial, especially since man has learned how to cause large-scale nuclear reactions. Many isotopes not found in nature before 1952 are now found in nature.

One naturally radioactive nucleus of special historical importance is ${}_{92}U^{238}$, whose half life is 4.5 billion years. When Becquerel's photographic plate was darkened by uranium, it was because an isotope of uranium, formed billions of years earlier, probably in the interior of a star that later exploded, was still acting out its inevitable fate to follow a law of probability to its ultimate complete decay. Since only half of it disappears each 4.5 billion years, a sizable fraction had endured through the explosion of a star, the formation of our solar system, the creation of the earth, and the billions of years of earth history before man's arrival. Some of what had waited so long chose the year 1896 to decay, when scientists pounced on the event.

It is clear in Figure 25.4 that the stable nuclei of low mass cluster approximately along 155
the line of equal proton and neutron number. As is shown in Figure 25.5, the path traced through the nuclear chart by the stable nuclei starts out in this way, then bends over increasingly in the direction of a neutron excess. For the heaviest nuclei, the line of stability, as this path is called, has a slope corresponding to the addition of about two neutrons for every added proton, instead of the one-to-one ratio that characterizes the light nuclei. Also, as Figure 25.5 makes clear, the line of stability ends. Beyond a certain point, there are no stable nuclei, nor even any long-lived nuclei. Both aspects of the line of stability, its shape and its termination, can be explained rather simply in terms of two effects: the action of the Pauli exclusion principle, and the action of the electrical repulsive force between protons.

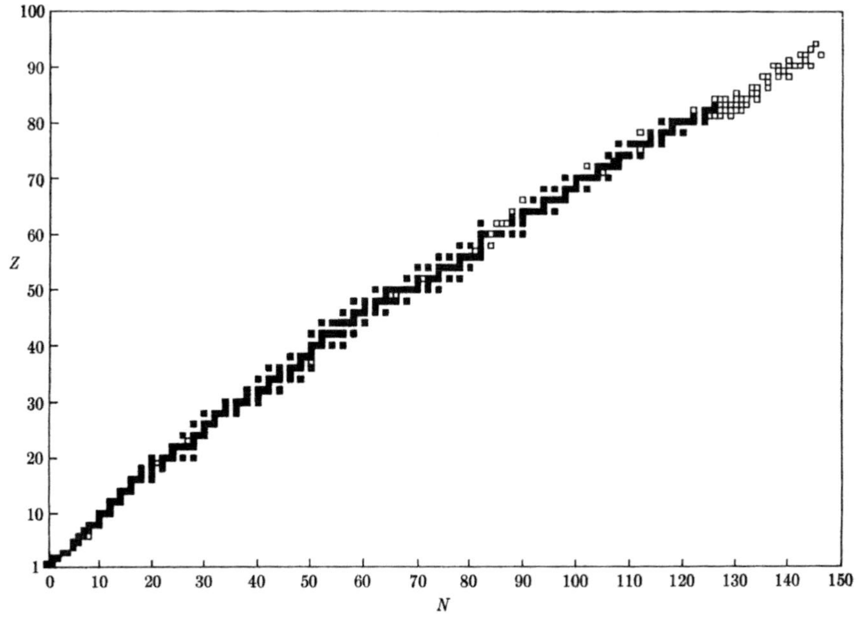

FIGURE 25.5 Nuclear chart. Dark squares indicate stable nuclei; open squares indicate unstable or radioactive nuclei found in nature. Artificially radioactive nuclei are not included in this chart.

* Carbon 14, although very rare in nature, is valuable because of its role in archaeological dating.

Protons and neutrons, like electrons, obey the exclusion principle. Because of this, there is a strong tendency toward equal neutron and proton number. If a proton occupies a given state of motion within the nucleus, an added proton must go into a different state of motion, often with greater energy. An added neutron, on the other hand, can drop into the same state of motion as the proton, since no exclusion acts between neutrons and protons. Moreover, once in identical states of motion, a neutron and a proton are in the most favorable position to experience maximum attractive force, and to contribute as much as possible to the binding energy.

Were it not for electrical forces in the nucleus, the line of stability would follow equal neutron and proton number indefinitely. Instead of about 100 elements we might have thousands. The mutual electrical repulsion of protons both bends and ends the line of stability. Only because of the great strength of the nuclear forces can protons be held together in a nucleus at all. The greater the number of protons, the greater is their tendency to blow the nucleus apart. Thus, despite the action of the exclusion principle, it becomes energetically favorable for heavy nuclei to contain more neutral particles than charged particles. Eventually, beyond 83 protons, there are no stable nuclei at all. Beyond 100 protons, there are no long-lived nuclei. Figure 25.6 shows in detail the present limits of exploration in the upper right end of the chart of nuclei.

Z ↑									
104							260 f 0.3 sec		
103 Lw Lawrencium				Lw 256 45 sec	Lw 257 α 8 sec				
102 No Nobelium			No 254 α ~50 sec	No 255 α ~15 sec	No 256 α, f 8 sec				
101 Md Mendelevium					Md 255 α, c 0.5 hr	Md 256 α, c 1.5 hr	Md 257 α, c, f 3 hr		
100 Fm Fermium	Fm 250 α 30 min	Fm 251 c 7 hr	Fm 252 α 23 hr	Fm 253 α, c ~4.5 day	Fm 254 α, f 3.24 hr	Fm 255 α, f 20 hr	Fm 256 α, f 2.7 hr	Fm 257 α, f 29 day	Fm 258 f (?)
99 Es Einsteinium	Es 249 α, c 2 hr	Es 250 c 8 hr	Es 251 α, c 1.5 day	Es 252 α ~140 day	Es 253 α, f 20 day	Es 254 α, β^-, f 39 hr	Es 255 α, β 40 day	Es 256 β^- (?)	
98 Cf Californium	Cf 248 α, f 350 day	Cf 249 α, f 360 yr	Cf 250 α, f 10 yr	Cf 251 α ~800 yr	Cf 252 α, f 2.55 yr	Cf 253 α, β^- 18 day	Cf 254 f 60 day		
N →	150	151	152	153	154	155	156	157	158

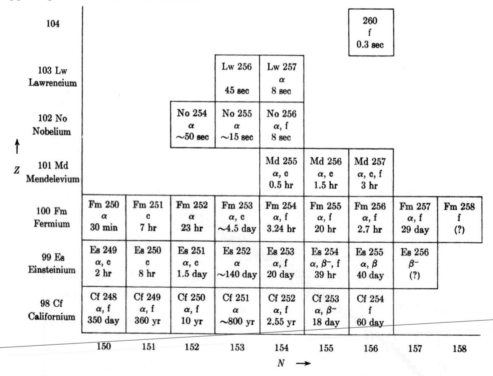

FIGURE 25.6 Portion of chart of the nuclei showing heaviest known nuclei. Numbers in the boxes show half-lives of the nuclei. Modes of decay are indicated by α (alpha), β (beta), c (electron capture), and f (fission).

A useful measure of nuclear stability is the total binding energy divided by the number of nucleons in the nucleus—that is, the binding energy per nucleon. Along the line of stability, as shown in Figure 25.7, the binding energy per nucleon rises quickly in the light elements, has a broad maximum, and declines slowly in the medium-weight and heavy elements. Binding energies are deduced from nuclear masses. The mass of a nucleus is less than the sum of the masses of its individual constituents. The difference multiplied by the square of the speed of light is the binding energy. The most startling fact about the curve of nuclear binding energy per nucleon is its near constancy. The binding energy of almost every nucleus lies between 7 MeV/nucleon and 9 MeV/nucleon. Nevertheless, the differences, small though they appear, are important. A uranium nucleus, when it divides by fission into two lighter nuclei, gains almost 1 MeV of extra binding per nucleon, or about 200 MeV altogether, which is released as kinetic energy and accounts for the explosive energy of fission bombs. For very light nuclei, binding energy is gained by fusion rather than fission. Fusion is the energy source of thermonuclear reactions.

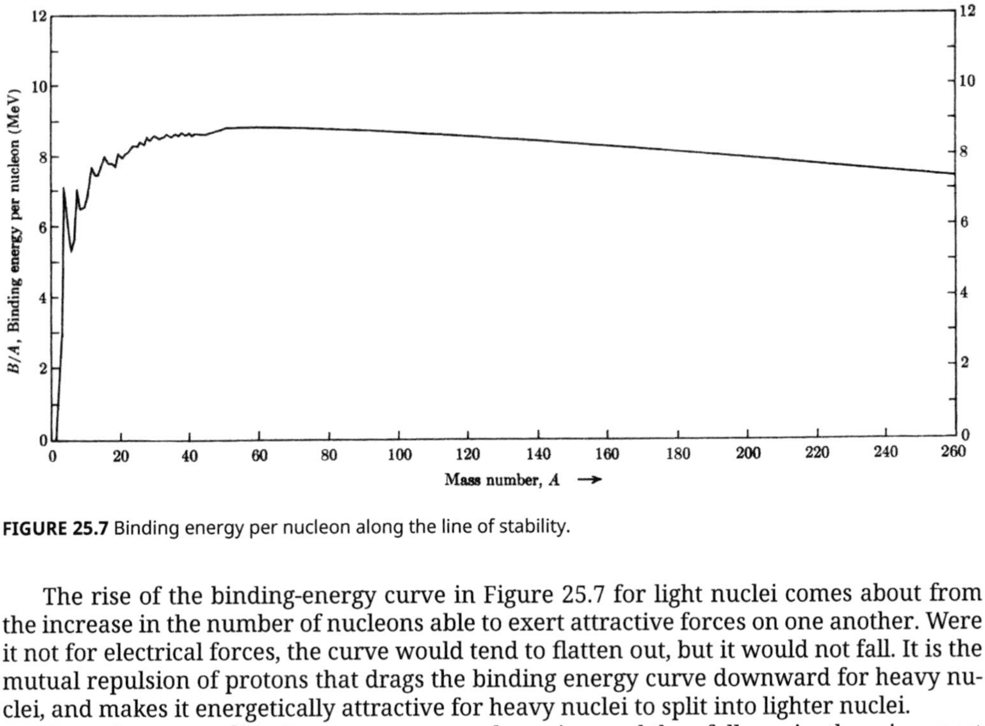

FIGURE 25.7 Binding energy per nucleon along the line of stability.

The rise of the binding-energy curve in Figure 25.7 for light nuclei comes about from the increase in the number of nucleons able to exert attractive forces on one another. Were it not for electrical forces, the curve would tend to flatten out, but it would not fall. It is the mutual repulsion of protons that drags the binding energy curve downward for heavy nuclei, and makes it energetically attractive for heavy nuclei to split into lighter nuclei.

Since the curve of binding energy per nucleon rises and then falls again, there is a most stable nucleus. It is an isotope of nickel, $_{28}Ni^{62}$. The fact that the binding energy curve reaches its maximum near mass number 60 may account for the relatively large abundance of iron and nickel in the universe. Hydrogen and helium seem to make up the bulk of the sun and most other stars. However, it is believed that in the late stages of the life of a star, after its hydrogen has been consumed, fusion reactions proceed through a series of heavier elements until iron and nickel are reached, at which point little fuel remains to feed the thermonuclear flame.*

25.3 Pions and the nuclear force

The discovery of the nuclear force can be said to have coincided with the discovery of the neutron in 1932, for as soon as the view of nuclei as aggregates of nucleons emerged, it was

* Elements heavier than nickel are also formed in stars, but in small abundances. Most stellar energy comes from the fusion of nuclei of elements lighter than nickel.

clear that there must exist a new kind of force—the nuclear force—with two important properties: It must be considerably stronger than the electrical force, for it holds protons together within the nucleus in spite of their electrical repulsion; it also must act only over very short distances, not more than 10^{-12} cm, for nuclei are no larger than this, and their effect upon passing particles extends no farther than this. It was mainly to explain the second fact, the short range of the strong interactions, that Hideki Yukawa in 1935 postulated the existence of the pion, a particle whose exchange among the nucleons was supposed to provide the nuclear glue (see Section 2.3).

The pion has the distinction of being one of the few particles for which the theorist got there first. Most of the particles have been discovered first, and "explained," if at all, later. But Yukawa predicted the existence of the pion more than ten years before its discovery.

156 We must now grapple with a very important idea in contemporary thinking about the submicroscopic world, the idea of a "virtual particle." This idea provides a very beautiful example of the workings of the Heisenberg uncertainty principle at the elementary level, and will provide a clue not only to the nature of the strong interactions, but to the nature of all forces and interactions.

In the form in which it was written down in Chapter Twenty-Three, the Heisenberg uncertainty principle looked like this:

$$\Delta x \Delta p = \hbar \, . \tag{25.5}$$

The uncertainty in the position of a particle (Δx) multiplied by the uncertainty in its momentum (Δp) is equal to the constant $\hbar$. Now, in fact, this fundamental uncertainty in nature arising from the wave nature of particles manifests itself in more ways than in the measurability of position and momentum. Another form of the same basic uncertainty in nature can be written:

$$\Delta t \Delta E = \hbar \, . \tag{25.6}$$

The uncertainty of time (Δt) multiplied by the uncertainty of energy (ΔE) is also equal to the constant $\hbar$. This means a precise measurement of energy (small ΔE) requires a long time (large Δt). Or if an event occurs at a very precisely known time (small Δt), its energy cannot be accurately determined (large ΔE). Never can time and energy both be precisely known at once. In particular, tests of the law of energy conservation require processes extending over some time.

That these two different forms of the uncertainty principle both occur is not surprising when we recall that, according to the theory of relativity, space and time are closely linked, as are energy and momentum. The origin of both forms of the law is the wave nature of matter. Just as a wave cannot be localized in a region of space significantly smaller than its own wavelength, so it cannot be pinpointed in a time interval much shorter than one period of its vibration. The only way to squeeze a wave more in space is to shorten its wavelength; the only way to squeeze it more in time is to shorten its period, that is, to make it vibrate faster. But the higher the rate of vibration, the higher the energy. Turning to music, the only way to get a pure tone, free of harmonic overtones, is to let the note vibrate through many cycles of oscillation. Similarly, the only way to get a "pure" energy, precisely defined without uncertainty, is to let the particle wave vibrate many times, that is, to allow a large span of time, or a large uncertainty of time. The wave nature of matter leads in a simple way to the time-energy uncertainty as it led to the position-momentum uncertainty.

Before applying this new form of the uncertainty principle to the nuclear force, we must consider one fact about the relation of pions to nucleons. A typical process of pion creation in an accelerator may be written symbolically in the form,

$$p + p \rightarrow p + n + \pi^{+}$$

A high-energy proton strikes a proton at rest in a target and from the collision emerge a proton, a neutron, and a positive pion. The simplest interpretation of this event is to say that one of the protons breaks up into a neutron and a pion:

$$p \rightarrow n + \pi^+ .$$

This process obeys all the conservation laws but one—energy. The masses of neutron and pion add up to considerably more than the proton mass, so that a single free proton left to itself would never decay in this way; to do so, it would need to violate energy conservation. But when it is struck by another energetic proton, some of the energy of motion in the collision can be converted to mass energy and the process becomes allowed. The extra needed energy is made available in the collision. The modern description of the event is something like this: A proton "wishes" to convert into a neutron and a positive pion. The strong interaction between nucleons and pions makes such a transformation always a possibility. But the transformation cannot be realized unless some energy is supplied.

Yukawa did not know these facts about pion production, but he surmised them. His further reasoning, reduced to nonmathematical terms, can be explained as follows: Although energy conservation prevents such a transformation as

$$p \rightarrow n + \pi^+$$

from occurring as a definite and irrevocable step, the uncertainty principle introduces into the law of energy conservation a kind of leniency which permits such a transformation to occur as a transitory phenomenon. We might say that the policeman enforcing the law of energy conservation is willing to look the other way if the violation lasts a short enough time. The pion may dart briefly outside the walls of its nucleon prison, and in again, before any enforcement procedures are initiated.

The talk of leniency and enforcement has to do simply with the uncertainty principle: $\Delta t \Delta E = \hbar$. If we wish to "violate" the law of energy conservation, that is, introduce into the energy an uncertainty of amount ΔE, we may do so provided the time of violation, Δt, is no longer than that set by this uncertainty relation. The excess energy needed for a proton to convert into a neutron and a pion is roughly the energy equivalent of the pion mass, or 140 MeV. The allowed time duration of this much energy uncertainty is

$$\Delta t = \hbar/\Delta E . \qquad (25.7)$$

Dividing $\hbar$ (6.6×10^{-22} MeV sec) by ΔE (140 MeV) gives the time uncertainty, $\Delta t = 4.7 \times 10^{-24}$ sec, a very short time indeed! How far could the temporarily freed pion move in this time? Traveling as fast as possible (at nearly the speed of light) it would only cover 1.4×10^{-13} cm.

According to this new view, nucleons are far from quiescent. A proton, even when all alone, is in a constant state of activity. It may eject and then immediately (after less than 10^{-23} sec) recall a positive pion,

$$p \leftrightarrow n + \pi^+$$

The double arrows indicate the two-way nature of the process. Or it may eject and recall a neutral pion,

$$p \leftrightarrow p + \pi^0$$

Because the interaction responsible for this activity is strong, these processes occur repeatedly and the proton must be regarded as a center of continual activity. The pions which come into momentary existence are called "virtual" pions. They are not "real" pions because energy conservation prevents their escape and they can never dart away to leave a track in a bubble chamber or otherwise be observed. Nevertheless, the success of Yukawa's theory is enough to make us believe in this model of the proton. Like a chauffeur on the sidewalk of Fifth Avenue surrounded by a barely controllable group of French poodles on leashes, the proton is surrounded by its cloud of virtual pions, darting this way and that, but leashed by the uncertainty principle to remain within little more than 10^{-13} cm of the nucleon core.

This view of the proton received its strongest direct support from electron-scattering experiments carried out at Stanford University by Robert Hofstadter.* He fired electrons of several hundred MeV energy at targets containing protons. Some of these electrons pass "through" the protons, that is, through the cloud of virtual pions surrounding the proton cores. In doing so, they are deflected, some through a small angle, a few through large angles. Detailed study of the fraction of electrons emerging in each direction reveals something about the size and composition of the pion cloud. The virtual pions do extend to a bit more than 10^{-13} cm, as our calculation above with the uncertainty principle indicated that they should. The average penetration of a pion away from the nucleon core is about eight tenths of a fermi, 0.8×10^{-13} cm. (Actually, the proton cloud contains etas, kaons, and a few other kinds of particles, but it consists mostly of pions.)

The final step in this line of reasoning has to do with the force between two nucleons. All that the uncertainty principle requires is that each virtual pion in the cloud surrounding the proton must vanish almost immediately after it is created, to clear the books of this excess mass energy. If a nucleon stands alone, the pion must be reabsorbed by the same nucleon from which it emerged. But if two are close together, a pion could be emitted by one and absorbed by the other. Suppose, for instance, that a neutron approaches close to a proton. At a particular instant, the proton may have transformed itself momentarily into a neutron and a positive pion. The other neutron can absorb the pion to become, itself, a proton. The net result is that a pion has jumped across from proton to neutron and, in the process, proton and neutron have changed roles. Yukawa realized that this kind of pion exchange could produce a strong attractive force between the two nucleons, a force now called an exchange force (Figure 25.8). Within a nucleus, virtual pions have to be thought of as constantly coming and going, frequently being exchanged back and forth between the nucleons. It is this incessant juggling with pions (and, to a lesser extent, with other mesons) that provides the nuclear glue holding neutrons and protons together.

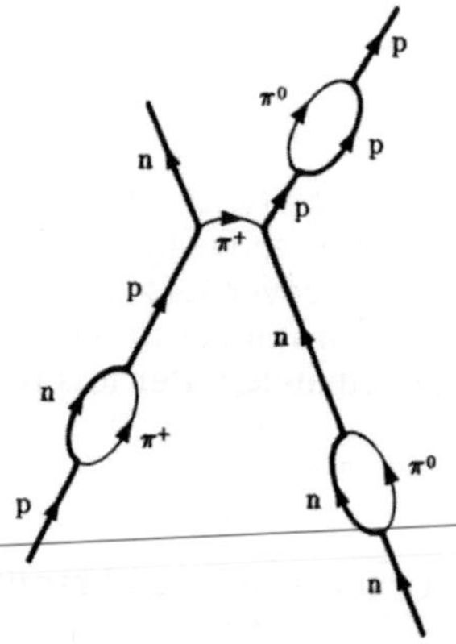

FIGURE 25.8. Schematic representation of exchange force. Well-separated nucleons may emit and reabsorb virtual pions. Close together, the nucleons may exchange virtual pions (or other mesons), a process that creates a force between them. Although this diagram contains the essence of what goes on, it is very highly oversimplified.

Because he knew roughly the range of the nuclear force, Yukawa was able to predict approximately the mass of a pion. The more massive a virtual particle, the more seriously it strains the law of energy conservation; therefore the more briefly it is permitted to ex-

* For his work on the structure of nucleons, Hofstadter received the 1961 Nobel Prize in physics, which he shared with Rudolph Mössbauer, the discoverer of gamma decay without nuclear recoil—see Section 22.4.

ist. Since it can move no faster than light, the shorter-lived virtual particle will penetrate less far from its parent particle and create a tighter, smaller cloud. A second nucleon, in order to feel the exchange force from the first, must come up to the edge of the cloud of virtual particles. The range of the force is therefore about the same as the size of the cloud. Virtual particles heavier than pions would produce an even shorter-ranged force. Virtual particles lighter than pions would produce a longer-ranged force. Since the pion is the lightest strongly interacting particle, it is primarily responsible for determining the range of the nuclear force.

Yukawa was not the first to propose particle exchange as a mechanism of force. In fact, he built on an exchange theory of electromagnetic force developed about five years earlier. Fermi and Dirac had shown that the ordinary electric force between two charged particles could be attributed to the incessant exchange of photons between the particles. In this beautiful application of the new quantum mechanics to the centuries-old problem of electric force, they unified the description of photon emission, photon absorption, and electromagnetic interaction. To put it another way, exactly the same theory explains how atoms gain and lose energy and how they are held together. Since photons are massless, the "range" of the electric force is infinite. Actually, of course, the electric force decreases as the inverse square of the distance; but this is a manifestation of the geometry of three-dimensional space, not a reflection of any inhibition on how far out the exchange photons can reach. It is a familiar fact that electric forces act over macroscopic distances. The truly short-range nuclear force, on the other hand, has no significant effect over a distance greater than 10^{-12} cm.

25.4 Stabilization of the neutron

Another fact about nuclear force needs to be discussed, not because it illustrates any new ideas or points to areas of ignorance, but because of its enormous practical significance. This is the stabilization of the neutron. If a neutron were not stabilized in the presence of one or more protons, the world would have not 92 elements occurring naturally but just one, hydrogen. A neutron alone undergoes the beta decay process, transforming to proton, electron, and antineutrino after an average time of 17 minutes. Joined to a proton, it can acquire an infinite lifetime, making possible the building of all the elements heavier than hydrogen. 157

What stabilizes the neutron is a rather peculiar "chance," that the pion exchange force between a neutron and proton happens to be somewhat stronger than the same force acting between two protons. The nucleus of deuterium, or heavy hydrogen, consists of one proton and one neutron. The mass of this combination (the deuteron) is not simply the mass of a proton plus the mass of a neutron, but somewhat less than this sum. Here are the figures, expressed in atomic mass units* (the subscripts p, n, and d refer to the proton, neutron, and deuteron):

$$\begin{aligned} m_\text{p} &= 1.00728 \text{ a.m.u.} \\ m_\text{n} &= 1.00866 \text{ a.m.u.} \\ \hline m_\text{p} + m_\text{n} &= 2.01594 \text{ a.m.u.} \\ m_\text{d} &= 2.01355 \text{ a.m.u.} \end{aligned}$$

The attractive force that pulled neutron and proton together has released energy, and this lost energy, the binding energy, is reflected in a decreased mass of the deuteron. The mass decrease of 0.00239 a.m.u. is equivalent to about 2.2 MeV. Now, the neutron in the deuteron

* One atomic mass unit (a.m.u.) is defined as one twelfth of the mass of an atom of carbon 12 (its nucleus plus its six electrons). The energy equivalent of 1 a.m.u. is 931 MeV.

has a natural inclination to undergo beta decay, This is normally a "downhill" process, for the neutron transforms to a lighter combination of proton, electron, and antineutrino. As the following numbers make clear, the free neutron is "barely" unstable.

Before	*After*
m_{n} = 1.00866 a.m.u.	m_{p} = 1.00728 a.m.u.
	m_{e} = 0.00055 a.m.u.
	$m_{\bar{\nu}} = 0$
	m_{total} = 1.00783 a.m.u.

—— Allowed Decay →

If the deuteron's neutron decides to follow its inclination for beta decay, the deuteron becomes suddenly a pair of protons. But the nuclear force attracts these protons somewhat less strongly than it attracts a neutron-proton pair; in fact, the force is not quite strong enough to bind the proton pair together. In this hypothetical beta decay process, therefore, the deuteron would transform itself into a pair of free protons. The masses before and after compare as follows:

Before	*After*
m_{d} = 2.01355 a.m.u.	m_{p} = 1.00728 a.m.u.
	m_{p} = 1.00728 a.m.u.
	m_{e} = 0.00055 a.m.u.
	$m_{\bar{\nu}} = 0$
	m_{total} = 2.01511 a.m.u.

—— Forbidden Decay →

The gain in going from heavier neutron to lighter proton is more than offset by the loss of binding energy. The law of energy conservation therefore prevents the neutron decay. The neutron is stabilized by the energy binding it to the proton. (In heavier stable nuclei, the neutron beta decay would merely reduce, not eliminate, the binding energy.) This is all a very delicate balance, the stabilization of the neutron amounting to less than one part in a thousand of the neutron mass. Yet we have reason to be grateful for this rather strange combination of circumstances—that the neutron happens to be so little heavier than the proton, that the pions happen to hold neutron and proton together more tightly than two protons. Viewed in terms of our present knowledge of elementary-particle interactions, it is a remarkable miracle that nature has some 90 atomic building blocks available instead of just one.

25.5 Nuclear structure

Probably the most striking fact about the structure of nuclei is the approximate validity of the central-field approximation. Each nucleon in the nucleus moves throughout the nuclear volume in an average field of force created by the other nucleons. At least to a fairly good approximation, each nucleon occupies its own quantum state of motion independent of the details of the motion of other nucleons. This means that nuclear structure bears many points of similarity to atomic structure, a quite remarkable fact, a fact that physicists recognized and accepted only after the weight of experimental evidence forced them to do so.

A nucleus has no massive center of force, as an atom does. In the nucleus, nucleons all share the same volume, in contrast to the partial separation in space of different electron shells in atoms. Nuclear forces are strong, and each nucleon in the nucleus is always within range of several other nucleons. These are some of the reasons that the approximate independent motion of nucleons within the nucleus seems surprising. In the 1930s, a heavy nucleus was looked upon as a system physically more like a liquid than like a gas. The fact that nuclear volume increases in proportion to the number of nucleons is consistent with this "liquid droplet model" of the nucleus. The fact that a high-energy proton or neutron striking the nucleus is almost sure to disrupt the nucleus rather than pass through it undisturbed is also consistent with the liquid-droplet model. But the most stunning success of the liquid droplet model came in 1939 when Niels Bohr and John Wheeler used it as the basis of the theory of nuclear fission. An important ingredient of their theory (see Section 25.9) was the concept of nuclear surface tension. An ordinary liquid derives its surface tension from the fact that molecules at the surface, being only half surrounded by other molecules, are less tightly bound than molecules in the interior. As a result, a drop of liquid tends to minimize its surface area by assuming a spherical shape. If nucleons in the nucleus behave like molecules in a liquid, the nucleus too should form a spherical droplet. Also, like a liquid droplet, the nucleus should be capable of deformation and fission. According to the theory of Bohr and Wheeler, the Coulomb repulsion between the protons could, under certain circumstances, overcome the force of surface tension and cause the nucleus to split.

Despite the successes of the liquid droplet model of the nucleus, certain bits of evidence 158
began to accumulate in the late 1940s that pointed to a very different view of nuclear structure. *Periodicity* of nuclear properties began to show up, not unlike some of the periodicities of atomic properties, and quite unlike the smooth regularity of properties to be expected of nuclei that behave like liquid droplets. Nuclear spins and nuclear binding energies showed a pattern consistent with independent motion of nucleons in a set of individual quantum states of motion within the nucleus. For certain particular numbers of neutrons and/or protons, nuclei showed unusual stability, much as the rare-gas atoms show unusual stability. These particular numbers, known at first as the "magic numbers," are

$$2, 8, 20, 28, 50, 82, 126.$$

One striking effect of the magic numbers is shown in Figure 25.9, which compares the set of energy levels up to a few MeV of excitation for several light nuclei. In the "doubly-magic" nucleus of $_8O^{16}$ (8 neutrons and 8 protons), considerable energy is required to excite the nucleus. In "singly-magic" $_8O^{17}$ (9 neutrons and 8 protons), less energy separates the ground state and the first excited state. In $_9F^{19}$, containing neither 8 protons nor 8 neutrons, the low-energy levels are clustered even closer together.

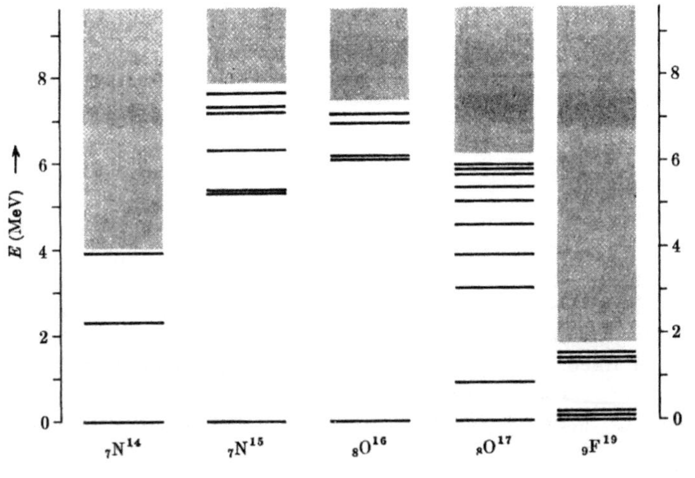

FIGURE 25.9 The first few energy levels in each of several light nuclei. The unusual stability associated with eight neutrons or eight protons is reflected in higher-than-normal excitation energies. In O^{16}, with 8 neutrons and 8 protons, the energy of the first excited state is particularly high. (The shaded regions contain many additional energy levels whose number and position are unknown.)

Apart from the evidence on periodicity associated with certain special numbers, new experiments on the scattering of neutrons by nuclei began to indicate that at low energy the neutron in fact had a good chance to pass right through the nucleus, deflected by the average action of all the nucleons, but not stopped or impeded by single encounters. Finally the evidence was too strong to ignore. Surely nucleons in the nucleus, like electrons in the atom, move approximately independently in an average field of force, successively filling allowed states of motion. In independent papers published at about the same time (1949), Maria Goeppert Mayer in the United States and J. H. D. Jensen (with O. Haxel and H. E. Suess) in Germany summarized the evidence and proposed a nuclear version of the central-field approximation with its resultant shell structure.* The magic numbers ceased to be magic. They simply indicated the numbers of protons or neutrons required to fill successive shells.

In nuclei, as in atoms, the lowest state of particle motion has principal quantum number 1, and orbital angular momentum zero. Because of the two possible spin orientations, this energy level holds two identical particles. A proton occupying a given state of motion excludes another proton from that state. Similarly, neutrons obey the exclusion principle. However, since the exclusion principle applies only to identical particles, it does not act to prevent a neutron and a proton from occupying the same state of motion. In the nuclear shell structure, two overlapping and interpenetrating sets of particle states must be visualized, one set for protons, one for neutrons. The alpha particle, or helium 4 nucleus, with two protons and two neutrons, is the simplest double-closed-shell nucleus (previously called a double magic nucleus). The nucleus of helium 3 has a closed shell of two protons, but an unclosed shell of one neutron.

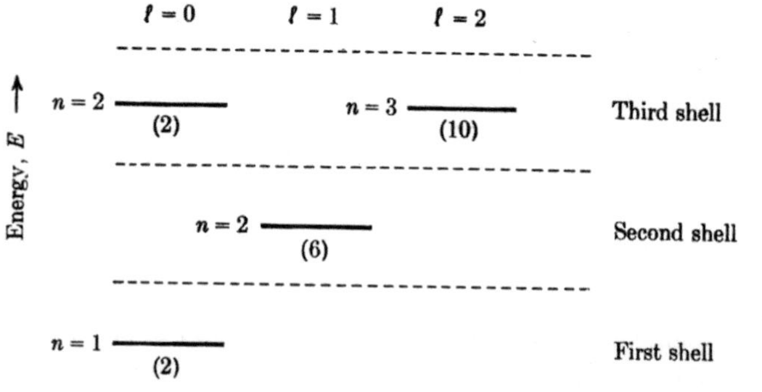

FIGURE 25.10 Arrangement of energies of first few states of motion of nucleons in the nucleus. The first three shells accommodate respectively 2, 6, and 12 nucleons. Compare this energy diagram with the diagram in Figure 24.24.

As shown in Figure 25.10, particle states of motion fill in a somewhat different order in nuclei than in atoms. The difference arises from the quite different shape of the average potential energy curve for the two structures (Figure 25.11). Because of this difference, helium is the only atom whose nucleons and electrons both fill closed shells. The second nuclear shell has quantum numbers

$$n = 2, \qquad \ell = 1 .$$

The three possible orientations of angular momentum multiplied by the two possible orientations of spin gives the number of identical particles that can occupy this energy level. Six protons fill the second proton shell; six neutrons fill the second neutron shell. Thus the next magic number is 8, and ${}_8O^{16}$ is the next doubly-closed-shell nucleus. In the third shell is room for 12 nucleons of each type. At calcium 40 (20 neutrons and 20 protons), both neutron and proton shells are again filled. Beyond calcium, the trend toward a neutron excess

* In 1963, Mayer and Jensen shared the Nobel Prize with Eugene Wigner, another pioneer in modern nuclear theory.

in nuclei prevents the appearance of another double-closed-shell nucleus until near the end of the periodic table. By chance the line of stability passes near $Z = 82$, $N = 126$, both closed-shell numbers. The nucleus $_{82}Pb^{208}$ is a doubly-closed-shell nucleus, a fact clearly indicated by the unusual amount of energy required to excite it (Figure 25.12).

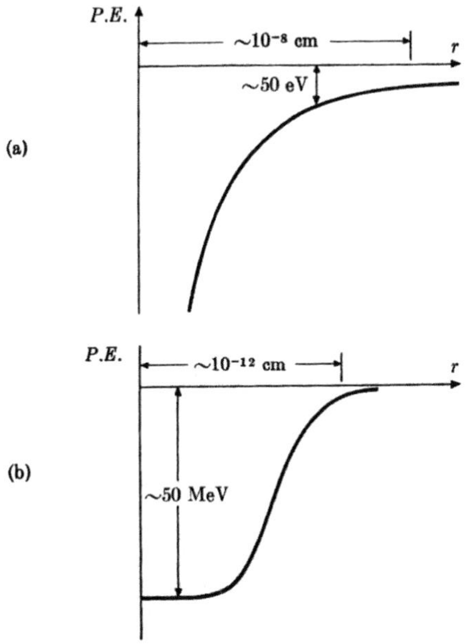

FIGURE 25.11 Comparison of average shapes of potential energy curves for **(a)** the motion of electrons in a heavy atom, and **(b)** the motion of nucleons in a heavy nucleus. Note the difference in length scales and energy scales as well as the differences in shape.

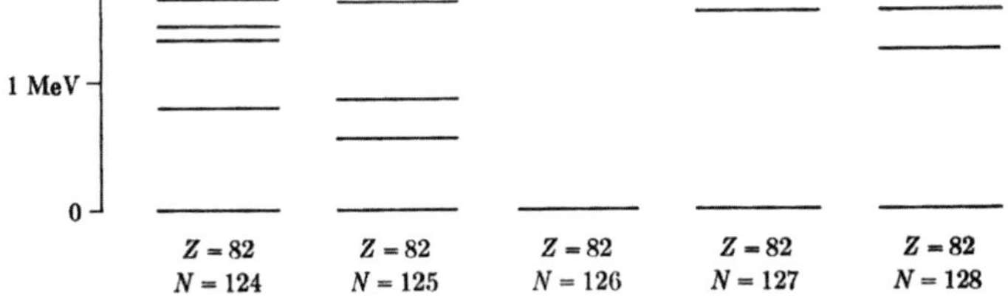

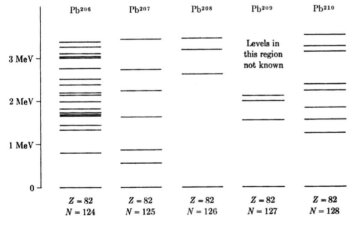

FIGURE 25.12 Partial energy-level diagrams of nuclei of several lead isotopes. The relatively large energy required to excite the nucleus of Pb^{208} is one of the pieces of evidence that nucleon shells close at both 82 and 126 nucleons.

The use of the phrase "shell model" to describe the central-field approximation and its implications, although common, is somewhat misleading. It suggests a spatial separation of particles into layers. In atoms, this picture of separate layers is roughly correct, but far from perfect, since the different shells overlap considerably. In a nucleus, the layer description has no validity whatever. The nuclear shells must be thought of as energy-level shells, not spatial shells. Each nuclear shell strongly overlaps all other nuclear shells.

In recent years, the theory of nuclear structure has reached a state of refinement and detail comparable to the theory of the structure of solids. To pursue the details would be out of place here. The amount of nuclear data collected in recent years is staggering, and the extent of understanding of the data, impressive. Among other aids to understanding is what has come to be called the unified model, best described as a mixture of the best of the liquid droplet model and the nuclear shell model. According to the unified model, nucleons move approximately independently within an average field of force, but this field of force (or region of potential energy) is neither rigid nor static. It can (and usually does) deform into a nonspherical shape. It has a surface tension. It can rotate and vibrate. It can become so elongated that fission results.

To this point in the chapter, we have focused on nuclear energy and nuclear structure. In the remaining sections, the focus shifts to nuclear *change*—transitions and reactions.

25.6 Alpha decay

The characteristic spectra of atoms were identified and classified long before the atomic structure responsible for the radiation was understood. Similarly, nuclei made themselves known first by their radiation—alpha, beta, and gamma rays—before scientists had any inkling of the size and structure of nuclei, or indeed of the existence of nuclei. From the study of nuclear radiation in the first decade of this century came the understanding of atomic structure. Later followed the understanding of nuclear structure.

When atomic line spectra were discovered, some tools of spectroscopy were already at hand for their analysis. With nuclear radiation, however, scientists had no experience and no ready-made detectors. The tools of nuclear physics had to be developed in parallel with the development of understanding. By chance, Becquerel discovered that radioactivity can darken a photographic plate. In the following year (1897), Thomson and Rutherford in Cambridge discovered that the radiation from uranium has the power to ionize a gas, making it a better conductor of electricity. This discovery, although not chance, represented a remarkable confluence of scientific events. X-rays, cathode rays, and the rays of uranium, three very different phenomena, were knitted together in Thomson's laboratory. In the year following the discovery of X-rays, Thomson discovered the ionizing power of X-rays. At the same time he was experimenting with cathode rays, pinning down their properties as small negatively charged particles. This interest, in turn, led him, with his new assistant Rutherford, to offer a theory of ionization as the separation of electrons from atoms. Then, on the heels of the discovery of radioactivity, Thomson and Rutherford set about studying the ionization produced by the rays of uranium. Later both Rutherford and Marie Curie used the conductivity of gas as a way to detect radioactivity. The rate of discharge of an electroscope is roughly proportional to the intensity of the ionizing radiation (Figure 25.13). For the measurement of total radiation dose, this early method of nuclear radiation detection is still in use. Electroscopes about the size of a fountain pen can be carried by persons working near radiation. The earliest method of all, the photographic emulsion, also still finds use, both in research and as a recorder of exposure.

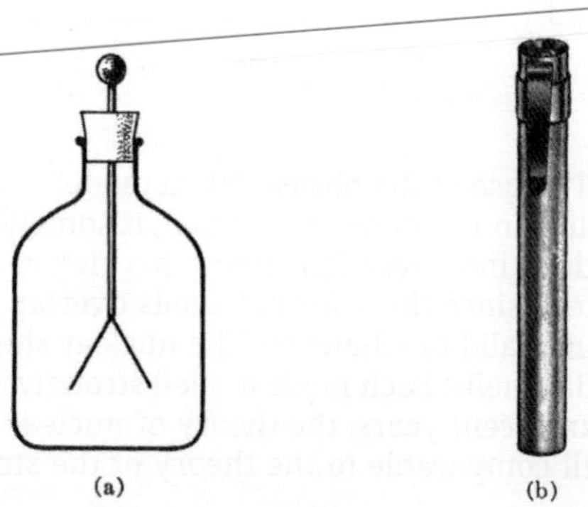

FIGURE 25.13 (a) An electroscope, whose leaves of metal foil repel one another when charged. The rate of discharge of the electroscope provided an early way to measure the intensity of radioactivity. **(b)** A modern electroscope used to monitor exposure to radiation. The tiny electroscope is viewed through a microscope built into the barrel of this fountain-pen sized dosimeter, [(b) After a photograph, courtesy of The Bendix Corporation.]

Within a few years of the discovery of radioactivity, several other methods of detection were devised—the cloud chamber, the Geiger counter, and the fluorescent screen. To this list have been added in more recent years scintillation counters, proportional counters, solid-state counters, bubble chambers, and spark chambers. All of these devices have one thing in common—they respond to the ionizing power of high-energy particles.

Having assisted Thomson in detecting the ionizing power of nuclear radiation, Rutherford set about studying the radiation in greater detail. His first discovery (1897) was an important one. He found that the radiation from uranium consists of at least two distinct kinds. To the radiation of greater ionizing power and less range he gave the name alpha rays. Another radiation, of less ionizing power and greater range, he called beta rays. A modern picture of alpha and beta ray tracks in special photographic emulsion (Figure 25.14) shows this characteristic difference. A thin layer of matter is sufficient to stop alpha particles. In a radium dial watch, for example, no alpha particles penetrate the case or crystal, but beta rays (and gamma rays) can escape. The difference observed by Rutherford is easy to understand in terms of the later understanding of these particles. First, alpha particles have twice the charge of beta particles, and therefore they more easily eject electrons from atoms as they pass through them or close to them. Second, because of its greater mass, an alpha particle of a given kinetic energy moves more slowly than a beta particle of the same energy. Lower speed is also more favorable for ionization (provided, of course, that the speed is not so low that insufficient energy is available). In a typical event of ionization, an alpha particle loses about 30 eV. Thus an alpha particle starting with 3 MeV of kinetic energy can ionize about 100,000 atoms or molecules before it is brought to rest. Its range in air (the distance required for it to dissipate this much energy) is about 150 cm. In aluminum, it is stopped in 0.08 cm.

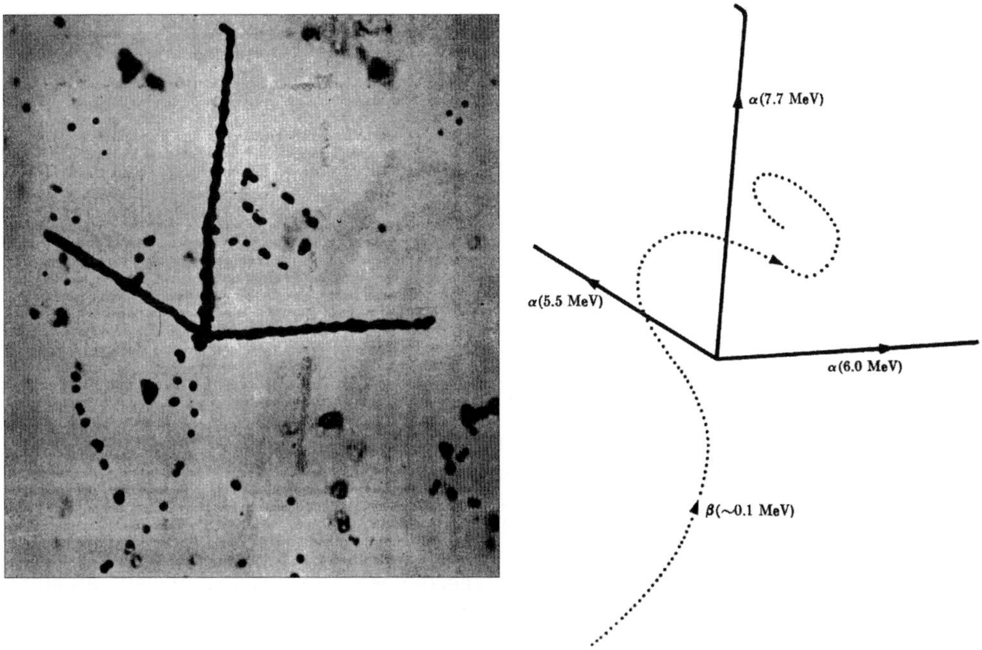

FIGURE 25.14 Tracks of alpha particles and a beta particle (electron) in a modern photographic emulsion. The alpha tracks are much heavier than the electron track, indicating a greater rate of energy loss by the alpha particles. The longest alpha track in the photograph covers about 3.9×10^{-3} cm. The wandering electron covers about 8×10^{-3} cm. Magnification is by a factor of 2,000. (Photograph courtesy of Harry H. Heckman, Lawrence Radiation Laboratory, University of California, Berkeley.)

In the history of alpha particles from their discovery to the present day, three landmarks stand out. First was their identification as helium nuclei. Second was the use of alpha particles as projectiles to study the interior of the atom. Third was the quantum mechanical explanation of alpha decay. The first two advances were made by Rutherford and his associates. The third was the work of George Gamow and (independently) Edward Condon and Ronald Gurney.

In 1902 Rutherford succeeded in deflecting alpha particles with a magnetic field, and showed them to be positively charged particles much heavier than electrons. The clinching evidence on their nature came from spectra. In 1908 Rutherford was able to bring to rest and collect in an evacuated space enough alpha particles to study their spectra. Once stopped, of course, the alpha particles collect electrons and become helium atoms. These atoms, if excited, emit the characteristic line spectrum of helium. Here was clear evidence that radioactivity was indeed a catastrophic event on the atomic scale, bringing into being entirely new atoms. The discovery explains too why helium is found in radioactive mineral deposits. It is a fascinating thought that in the thousands of cubic feet of helium pumped from the earth every day, practically every atom had its origin in the tiny violence of a radioactive disintegration some time within the past few billion years.

In Rutherford's laboratory in Manchester in 1908 began the historic series of experiments on the deflection of alpha particles by matter that was to culminate five years later in Bohr's theory of the hydrogen atom. Carried out by Hans Geiger and Ernest Marsden, the experiments seemed at first routine, unglamorous compared to the recent triumph of the proof that alpha particles are helium ions. But before long they took on central interest in the laboratory, for the results were unexpected. A small fraction of alpha particles directed at a metal foil were deflected through large angles. Even this small fraction was too much to be consistent with the then current view that the positive charge in the atom is spread over the whole atomic volume. Such an atom is too "soft" to deflect an alpha particle through a large angle. The light-weight electrons in the atom cannot do the job; they respond to an alpha particle like a ten-pin to a bowling ball. Nor can a spread-out positive charge do the job. Its electric field is nowhere strong enough to greatly alter the course of the juggernaut alpha particle. Rutherford seized on the results of Geiger and Marsden and interpreted them correctly. In the heart of the atom, he said, must reside an object that is massive enough and forceful enough to turn back an alpha particle that comes close to it. Since electrons, known to be atomic constituents, are light and negative, the central nucleus ought to be heavy and positive.

In 1911, Rutherford went on to calculate the deflection probability for an alpha particle approaching a small positive nucleus. This probability is a probability of ignorance, since for any particular alpha particle, it is impossible to know how it is aimed relative to a nuclear center. Figure 25.15 shows some typical trajectories for fast alpha particles repelled by a nucleus, and also a curve of deflection probability versus deflection angle. Subsequent experiments confirmed this deflection probability curve in detail. By a stroke of good fortune, Rutherford's classical calculation agrees with a quantum calculation carried out many years later. According to the modern view, his well-defined trajectories must be replaced by propagating waves; yet, remarkably, the results based on hyperbolic trajectories of Newtonian mechanics agree with the quantum results.

159 Alpha decay, in contrast to alpha-particle deflection, is a uniquely quantum phenomenon completely at variance with classically expected behavior. Inside a nucleus, an alpha particle feels an attractive nuclear force. This is a region of negative potential energy, as shown in Figure 25.11(b). Outside the nucleus, an alpha particle feels an electrical force of repulsion. This is a region of positive potential energy. As shown in Figure 25.16, the potential-energy diagram for the alpha particle-nucleus system contains a barrier near the nuclear surface. According to classical mechanics, an alpha particle with insufficient energy to surmount the barrier, if fired at the nucleus from outside, is turned back by the bar-

rier (the electric repulsion) and cannot penetrate the nucleus. This is the phenomenon of alpha-particle scattering, as studied by Geiger and Marsden. Classically, an alpha particle of the same energy, if inside the nucleus, is constrained by the same barrier to remain there forever. In short, if classical mechanics were valid in the nuclear domain, the potential-energy barrier would be impenetrable from both sides for any alpha particle with energy less than the peak of the barrier. However, alpha particles of less than this energy do shoot out of nuclei, and quantum mechanics explains how they do it. Because of the wave nature of particles, quantum mechanics puts fuzzy edges on much of the sharpness in a classical description of motion. Classical orbits become distributions of wave amplitude, and the terminus of a trajectory at a potential-energy barrier becomes a region of rapidly changing probability rather than a point of sudden change. The wave amplitude of an alpha particle in the nucleus penetrates into the classically forbidden barrier region. A small part of the wave [Figure 25.16(b)] penetrates through the barrier and emerges on the other side as an oscillatory wave of small amplitude. This means that the alpha particle has a certain chance to pop through the barrier and appear outside the nucleus, where the electrical force accelerates it to a high kinetic energy, typically a few MeV. This is alpha decay. The inverse process, barrier penetration from outside to inside, can also be important in the scattering of alpha particles by nuclei.

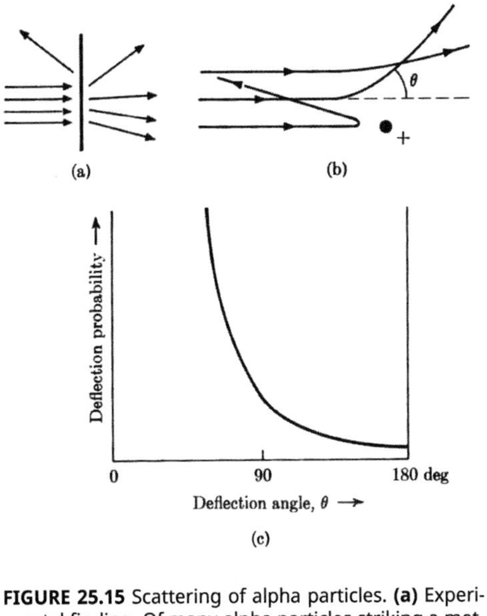

FIGURE 25.15 Scattering of alpha particles. **(a)** Experimental finding. Of many alpha particles striking a metal foil, a few are deflected through a large angle. **(b)** Rutherford's theoretical picture. Influenced by a small central positively charged nucleus, alpha particles follow hyperbolic paths, a few being scattered through large angles. (This diagram is not to scale.) **(c)** Graph of deflection probability vs. deflection angle for alpha particles approaching nuclei randomly.

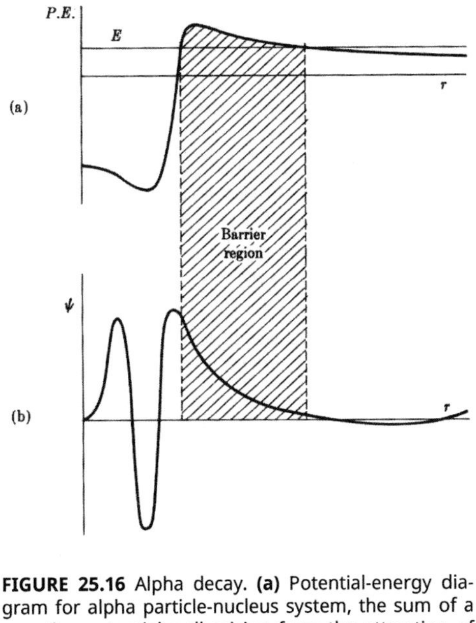

FIGURE 25.16 Alpha decay. **(a)** Potential-energy diagram for alpha particle-nucleus system, the sum of a negative potential well arising from the attraction of nuclear forces and a positive potential hill resulting from electric repulsion. **(b)** Wave function of alpha particle (approximate), showing large probability to be within nucleus and small probability to appear outside.

25.7 *Beta decay*

Beta decay has been a rich source of fundamental information. It led to the prediction of the neutrino and provided the means of discovering this particle. It stimulated the first theory of the annihilation and creation of matter. Through beta decay came the proofs that the neutrino is left-handed (Chapter Twenty-Six) and that parity is not conserved (Chapter Twenty-Seven). As a manifestation of the weak interaction, beta decay has been a source of much detailed knowledge about this fundamental type of interaction. That the beta particles shot from atoms are electrons was established by Becquerel in 1900. Using both electric and magnetic fields, he deflected beta rays, as Thomson had deflected cathode rays a few years earlier. He found that beta rays were negatively charged, and that their charge-to-mass ratio agreed with the charge-to-mass ratio of cathode rays. Since atoms were believed to contain electrons, the idea that radioactive atoms ejected electrons was easy to accept.

After Rutherford established the nuclear atom, it became clear that the nucleus must be the seat of radioactivity—for the vast energy of radioactivity was not to be found in the orbital electrons, and alpha particles, at least, could come from nowhere else than the nucleus. If the electrons of beta decay came from the nucleus, the nucleus must contain electrons—or so it seemed, for the idea of particle creation was years in the future. This idea of electrons in the nucleus was also easy to accept. In fact, it was an attractive idea, for it nicely explained why, for all the elements beyond hydrogen, mass number exceeds atomic number. Electrons in the nucleus would cancel some of the charge of the protons, leaving mass number (number of protons, supposedly) greater than atomic number (net charge, protons minus electrons).

Over the years, more extensive and more accurate measurements on beta decay built up gradually to the crisis of a real puzzle for physics. Apart from the fact that quantum mechanics cast doubt on the presence of electrons in the nucleus, a more serious problem loomed: the apparent nonconservation of energy. In alpha decay and in gamma decay, as in photon emission by atoms, the energy carried away by the ejected particle (its mass energy, if any, plus its kinetic energy) is exactly equal to the difference in energy between the initial and final states of motion. To borrow an expression from optical spectra, we can say that alpha and gamma radioactivity exhibit *line spectra*. Beta decay, by contrast, exhibits a *continuous spectrum*. From a group of identical radioactive atoms undergoing beta decay, electrons of all energies from zero to some maximum are emitted. Apparently (and we now know this indeed to be the case) any particular atom undergoing a particular transition gives to the emitted electron an unpredictable amount of energy.

160 By this point in history (the late 1920s), physicists had been conditioned by the successes of relativity and quantum mechanics to expect the unexpected. Some were willing to abandon the law of energy conservation in the nuclear domain. Wolfgang Pauli in 1930 made a different suggestion that was at once conservative and bold. In order to "save" the law of energy conservation, he postulated an entirely new particle, uncharged and so far unseen, which might exist within nuclei and be emitted along with an electron in beta decay. The letter in which he made this suggestion is a document of considerable interest. It reveals much about the approach of a creative twentieth-century physicist to problems of nature. Pauli's letter* is as follows.

Open letter to the radioactive group at the regional meeting in Tübingen.

Physical Institute of the
Eidgenossischen Technischen Hochschule
Zürich

Zürich
December 4,1930
Gloria St.

Dear Radioactive Ladies and Gentlemen:

I beg you to receive graciously the bearer of this letter who will report to you in detail how I have hit on a desperate way to escape from the problems

* Permission to reprint courtesy of Mrs. Franca Pauli. (Translated from the German.)

of the "wrong" statistics of the N and Li^6 nuclei and of the continuous beta spectrum in order to save the "even-odd" rule of statistics and the law of energy conservation. Namely the possibility that electrically neutral particles, which I would like to call neutrons, might exist inside nuclei; these would have spin ½, would obey the exclusion principle, and would in addition di er from photons through the fact that they would not travel at the speed of light. The mass of the neutron ought to be about the same order of magnitude as the electron mass, and in any case could not be greater than 0.01 proton masses. The continuous beta spectrum would then become understandable by assuming that in beta decay a neutron is always emitted along with the electron, in such a way that the sum of the energies of the neutron and electron is a constant.

Now, the question is, what forces act on the neutron? The most likely model for the neutron seems to me, on wave mechanical grounds, to be the assumption that the motionless neutron is a magnetic dipole with a certain magnetic moment μ (the bearer of this letter can supply details). The experiments demand that the ionizing power of such a neutron cannot exceed that of a gamma ray, and therefore μ probably cannot be greater than $e(10^{-13}$ cm).

At the moment I do not dare to publish anything about this idea, so I rst turn trustingly to you, dear radioactive friends, with the question: how could such a neutron be experimentally identi ed if it possessed about the same penetrating power as a gamma ray or perhaps 10 times greater penetrating power?

I admit that my way out might look rather improbable at rst since if the neutron existed it should have been seen long ago. But nothing ventured, nothing gained. The gravity of the situation with the continuous beta spectrum was illuminated by a remark of my distinguished predecessor in o ice, Mr. Debye, who recently said to me in Brussels, "Oh, that's a problem like the new taxes; one had best not think about it at all." So one ought to discuss seriously any way that may lead to salvation. Well, dear radioactive friends, weigh it and pass sentence! Unfortunately, I cannot appear personally in Tübingen, for I cannot get away from Zürich on account of a ball which is held here on the night of December 6-7. With best regards to you and to Mr. Baek,

Your most obedient servant,

W. Pauli

Despite his smokescreen of banter, Pauli undoubtedly meant this letter to be taken seriously. With unerring instinct, he pinpointed two of the most important puzzles of nuclear physics, the apparent nonconservation of energy in beta decay and the seemingly wrong spin of certain nuclei, and pointed out that a new lightweight neutral particle might resolve *both* puzzles. At the same time, he recognized the speculative nature of his suggestion, and chose this informal method of publicizing it.

We shall not endeavor to explain what Pauli calls the statistics of nuclei in his first paragraph. What he is getting at is the fact that lithium 6 and nitrogen 14 behave as if they contain an even number of spin-one-half particles, whereas, according to the proton-electron theory of nuclear composition they should each contain an odd number of particles. To the 14 protons and 7 electrons that N^{14} was supposed to contain, Pauli wanted to add 7 "neutrons," to make the total number even. If one of these "neutrons" and one of the electrons were emitted together in beta decay, the sharing of energy between the two emitted particles could explain the continuous energy spectrum observed in beta decay.

As happens often at the frontier of science, an important forward step is a mixture of truth and error. Pauli was right about several important things. The nuclei Li^6 and N^{14} *do* contain an even number of particles. A neutral particle with one half unit of spin *is* emitted in beta decay, and energy *is* conserved in the process. He erred in supposing that what is emitted from the nucleus must have existed in the nucleus before the emission process. To suppose otherwise would have been an unjustified break with the accumulated evidence of past experience. What we usually think of as the giant strides of physics are really small steps. Several more steps were required after Pauli's 1930 letter before the creation of matter could be sensibly and seriously proposed. First had to come the discovery of the neutron, the understanding of nuclear composition, and further development of the quantum theory of photon emission.

Eventually it took two neutral particles to fill the roles Pauli had assigned to one, and to solve all the puzzles that beta decay had created. The discovery of the neutron in 1932 took care of the problem of nuclear spin, and also dealt with a puzzle Pauli had not mentioned, the puzzle of electron containment within the nucleus. The neutron banished the electron from the nucleus. But this neutron could not be the lightweight neutral particle emitted in beta decay. If energy was to be conserved, a second neutral particle, much less massive than the neutron, was required. To this new hypothetical particle Enrico Fermi gave the name neutrino (little neutron).

161 At last the stage was set for a quantum-mechanical explanation of beta decay. The successful theory was Fermi's, published in 1934. Borrowing the general ideas of the quantum theory of photon emission to which he had recently contributed, Fermi constructed a theory of beta decay based on particle creation. He supposed that an electron and an antineutrino are created at the instant of the radioactive transformation, just as a photon is created at the instant of an atomic quantum transition. Simultaneous with the creation of the pair of particles (both of which at once escape from the nucleus), the nucleus undergoes a quantum transition to a new state of motion that is not only lower in energy, but is different in composition. The final state contains one more proton and one less neutron than the initial state. Beta decay seems to be a process very different from photon emission. Yet, in what we now regard as the essentials, both are quite similar. One happens to involve the creation of a particle with mass; the other involves only the creation of a particle without mass. In one, the emitting particle (a nucleon) happens to change its charge; in the other, the emitting particle (an electron) does not. One thing that *is* very different is the strength of the interaction responsible for the transition. The electromagnetic interaction that creates photons is many orders of magnitude stronger than the weak interaction that creates electrons and neutrinos.

The Fermi theory of beta decay has stood very well the test of time. Only slightly modified, it successfully accounts for all nuclear beta decay, with both electron and positron emission; for the process of orbital capture, in which the nucleus annihilates an atomic electron instead of emitting a positron; for the beta decay of the free neutron; and for a variety of other weak interactions involving elementary particles, such as the beta decay of a muon, in which a muon transforms itself into an electron, a neutrino, and an antineutrino. So apparent were the successes of the Fermi theory that the neutrino was generally counted as a reliably classified member of the elementary-particle family many years before it was actually observed. The observation of the neutrino, which proved to be much more difficult than Pauli had foreseen, is described in Chapter Twenty-Six.

It is clear that alpha decay and beta decay are processes of very different kinds. Both release comparable energy, since both accompany nuclear transitions. Both happen to span an overlapping range of half lives. Beyond that, they have little in common. The alpha particle, or at least the raw material of an alpha particle, is always present in the nucleus. When alpha decay is energetically possible, the only thing that prevents it happening in a time of 10^{-21} sec or less is a potential-energy barrier. When beta decay is energetically possible, it is delayed by seconds or minutes or centuries by the intrinsic weakness of the fundamental interaction that produces it. Weak though it may be, this interaction, when it does finally act, does so with the same explosive violence on the nuclear scale that characterizes all radioactivity.

25.8 Radioactive decay chains

When radioactivity first forced itself on the attention of man, man had no clear picture of 162
the nature of the atom, much less of its central nucleus. We might say that the nucleus offered to discover itself. But until the genius of Rutherford opened up the atom and revealed its nucleus in 1911, radioactivity was known only as a new and interesting *atomic* phenomenon. In its early history, its significance—which was very great—lay in what it revealed about atoms.

Luckily for the progress of atomic science, radioactivity occurs naturally in the heavy elements. It was the attempt to unravel the mysteries of natural radioactivity that led to the rapid development of atomic understanding around the turn of this century.* Although natural radioactivity contains a store of valuable information, one basic fact about it made particularly difficult the task of decoding its message about submicroscopic nature: It occurs in long chains of successive transformation and transmutation. In a natural sample of uranium are juxtaposed 18 different radioactive isotopes of 10 elements with half lives ranging from 164 microseconds to 4 billion years, all undergoing simultaneously their individual processes of decay. It was a triumph of human intellect, as well as of international cooperation and communication, that only fifteen years elapsed from Becquerel's discovery of radioactivity to Rutherford's discovery of the nucleus.

Among the new insights gained from the study of radioactive decay chains, two stand out as dramatic revolutions in thinking about the structure of matter: (1) Transmutation accompanies radioactivity; one element can transform itself into another. (2) There exist different versions of the same element; isotopes are identical chemically, but differ in atomic weight and in radioactive properties. This pair of discoveries toppled two nineteenth-century axioms of chemistry—that each element is composed of identical atoms of a single kind, and that atoms are immutable. It is ironic that after alchemists had expended so much fruitless effort in trying to transmute elements, transmutation was finally discovered to be a spontaneous process in nature.

The large number of radioactive isotopes among the heavy elements fit into three series, each based on a very long-lived parent isotope. The uranium series springs from ${}_{92}U^{238}$, whose half life is 4.5 billion years. The thorium series springs from ${}_{90}Th^{232}$, whose half life is 14 billion years. The actinium series (named for one of its members, not for its parent) springs from ${}_{92}U^{235}$, whose half life is 713 million years. Because of its shorter half life, most of the uranium 235 that was present when the earth was formed has since decayed away, but enough remains—one atom of U^{235} for every 138 atoms of U^{238}—to give rise to a significant chain of decay. For illustrative purposes, it is sufficient to focus attention on one decay chain. The uranium series is illustrated in Figure 25.17. This diagram is a selected part of the upper right end of a chart of the nuclei, with proton number (atomic number) plotted vertically, neutron number plotted horizontally, as in Figures 25.4, 25.5, and 25.6.

The key discoveries of transmutation and of isotopes, which flowed from the analysis of radioactive decay chains, were closely tied to the discovery of several new elements. When radioactivity was discovered, only two elements heavier than bismuth were known, thorium and uranium. Since the atomic numbers of these elements were unknown, the number

* Often the outcome of scientific inquiry is hard to foresee. In an interesting twist of history, the radioactivity of the heaviest natural element, uranium, launched a series of discoveries that culminated in a successful theory of the lightest element, hydrogen.

of missing elements between bismuth and uranium was also unknown. After radioactivity became available as a new tool of analysis, the missing pieces began to fill in rapidly. In 1898, Marie and Pierre Curie isolated, identified, and named two new elements, polonium and radium. Although they did not know it at the time, they actually isolated particular isotopes of these elements, those produced by the uranium decay chain: ${}_{84}Po^{210}$, whose half life is 138 days, and ${}_{88}Ra^{226}$, whose half life is 1,620 years. (This isotope of radium is now commonly used in the luminous paint of watch dials.) In the following year, André Debierne identified another new element, actinium (now known to have atomic number 89). At about the same time, Rutherford in Canada and Ernst Dorn in Germany were finding evidence for new radioactive gases. Rutherford named his new substance emanation. It came from thorium, had a half life of 1 minute, and, as he and Frederick Soddy later learned, was chemically inert. Dorn's gas, also chemically inert, came from radium and had a half life of 3.8 days. He called it radon. Now both are recognized as isotopes of radon ($Z = 86$), the heaviest of the rare gases. Rutherford's emanation (also known for a time as thoron) is ${}_{86}Rn^{220}$. Dorn's "radon" is ${}_{86}Rn^{222}$. With the discovery of these isotopes began a decade of confusing multiplicity in the apparent number of elements.

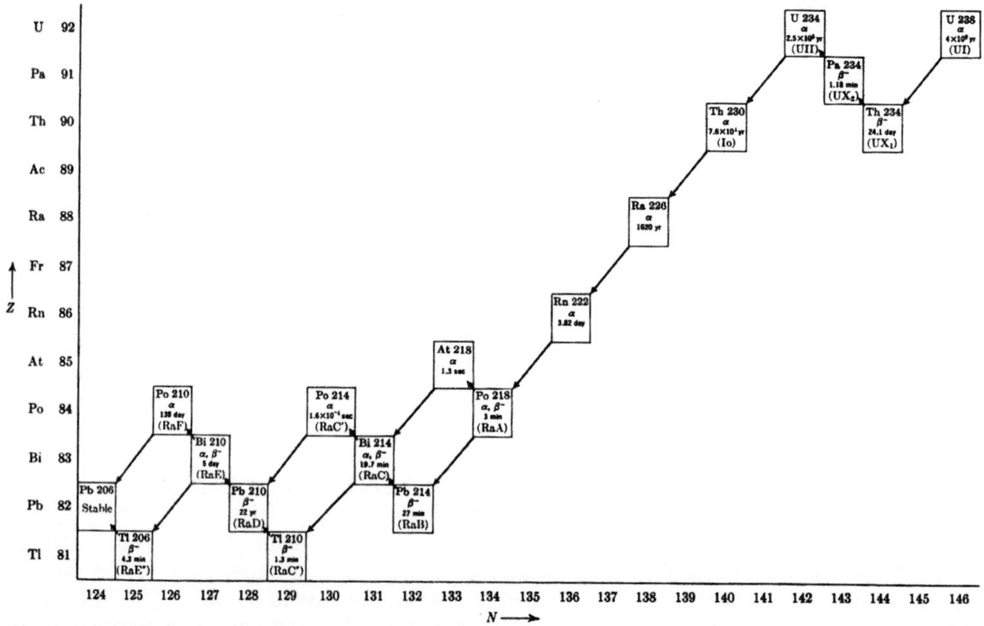

FIGURE 25.17 The uranium series of natural radioactive transformation, beginning with U^{238} and terminating in Pb^{206}, a stable isotope of Pb. Some squares in the chart show the names assigned to particular members of the series before their true isotopic identities were established. Half lives are also shown.

A century of progress in chemistry had seemingly made clear that all atoms of a given element are identical. Therefore scientists engaged in the study of radioactivity quite naturally assumed that if two substances, both identifiable as elements, differed in any way at all—for instance, in half life—they must be different elements. As radioactive decay chains were gradually untangled, new elements seemed to proliferate remarkably. Some of the names assigned to isotopes in the uranium series are shown in Figure 25.17. At one time, six isotopes of thorium bore six different names, as shown in Table 25.1. Before long, the proliferation of "elements" began to frustrate the chemists. In 1907, Herbert McCoy and William Ross in Chicago found thorium and radiothorium to be chemically inseparable. In 1908,

Otto Hahn in Berlin found it impossible to separate ionium and thorium by chemical means. Finally, in 1910, totally frustrated in his efforts to separate these substances, Soddy took the courageous step. He suggested that a single element might exist in two or more forms, different in mass and in radioactive properties, but identical chemically. Later he named these separate forms *isotopes*. Clinching evidence for isotopes came from two quite different experiments in 1913. In that year, Soddy showed the spectra of ionium and thorium to be identical. At about the same time, J. J. Thomson magnetically deflected a beam of neon ions moving in an evacuated space, and discovered that some of the ions were of atomic mass about 20, others of atomic mass 22. He thereby demonstrated that multiple isotopes are not exclusively an attribute of radioactive elements.

TABLE 25.1 Early Nomenclature for Thorium Isotopes

Original Name	Modern Designation	Comments
Radioactinium	Th^{227}	Member of actinium series
Radiothorium	Th^{228}	Member of thorium series
Ionium	Th^{230}	Member of uranium series
Uranium Y	Th^{231}	Member of actinium series
Thorium	Th^{232}	Member of thorium series
Uranium Xi	Th^{234}	Member of uranium series

Working with Rutherford, Soddy had also figured in another courageous conclusion in 1902. Stimulated by a suggestion of Becquerel, Rutherford and Soddy studied a radioactive "impurity" in thorium which they named thorium X.* Finding that thorium and thorium X were undeniably distinct chemically, and that thorium X was a product created by thorium, they concluded that *transmutation* accompanies radioactivity. The Curies, less willing than Rutherford to abandon the solid rock of immutability of atoms, were at first reluctant to accept this revolutionary proposal. But in 1903 (the year Marie Curie received her doctorate and a Nobel Prize), they too accepted the idea of transmutation. Before long it was established beyond question by the combined evidence of chemistry and radioactivity.

Following up their proposal of transmutation, Rutherford and Soddy drew two other related conclusions about the nature of radioactivity that stand as landmarks of discovery in this period. First, they proposed the "one-step" theory of transmutation, that an atom does not gradually evolve into another as it releases radioactive energy, but instead transforms itself instantaneously at the moment of radioactive decay. In this proposal, they were coming very close to the idea of probability working at a fundamental level, since radioactive decay was known to follow rules of probability. But in the absence of forcing evidence, to abandon certainty in science was inconceivable. The hint provided by radioactive decay was ignored for many years. Second, Rutherford and Soddy, analyzing the energy released by radioactivity, concluded that radioactive transmutation must involve at least 100,000 times as much energy per atom as does chemical change. From their work emerged the picture of radioactive decay that is still valid—a violently disruptive explosion of a single atom.

With the advent of accelerators, reactors, and nuclear explosions, radioactivity is no longer a property of a few heavy elements alone.† Unstable nuclei have been identified, studied, and catalogued for every one of the 103 known elements. Twenty-five isotopes of tin are known, fifteen of them radioactive. The Curies' "element," polonium 210, is now only one of twenty-seven known radioactive isotopes of polonium. Altogether, more than 1,200 radioactive isotopes have been studied. Some are of interest only for what they teach about

* Thorium X is now recognized as an isotope of radium, $_{88}Ra^{224}$.

† Apart from the natural radioactivity dating from the origin of the earth, and the man-made radioactivity of very recent origin, there has always been a limited amount of radioactivity produced by cosmic rays. Carbon 14, used for archaeological dating, is—or was—produced primarily by cosmic rays.

nuclear structure; others are of interest as tools of research, others as tools of medicine and technology, still others—some of the many species in fallout—as nuisances and hazards for man.

Apart from half life and mode of decay, a matter of practical concern in dealing with radioactivity is its intensity. Two units are used to measure different aspects of the strength of a sample of radioactive material. First is the curie, a measure of the number of radioactive disintegrations per second in the sample:

$$1 \text{ curie} = 3.7 \times 10^{10} \text{ disintegrations/sec} . \tag{25.8}$$

Example: What is the radioactivity in curies of one gram of radium 226? It is necessary to find out how many atoms of radium are in one gram, then to multiply this number by the probability that any one of the atoms will decay in a second. The mass of one atom of Ra^{226} is approximately 226 times the mass of a nucleon, or (226) (1.67 $\times 10^{-24}$ gm) = 3.78×10^{-22} gm. The number of these atoms in one gram is therefore given by

$$N = \frac{1 \text{ gm}}{3.78 \times 10^{-22} \text{ gm}} = 2.65 \times 10^{21} .$$

To find the chance that an atom decays in one second requires knowledge of another fact, that the decay probability per unit time is the inverse of the *mean life*, not of the half life. Since the half life is 69.4% of the mean life (see Figure 23.1), the mean life of Ra^{226} is (1,620 years/0.694) = 2,340 years = 7.4×10^{10} sec. The inverse of this mean life is the probability per second that an atom will decay:

$$p = \frac{1}{7.4 \times 10^{10}} = 1.36 \times 10^{-11} \text{ sec}^{-1} .$$

This means that in any one second, any single atom of Ra^{226} has about one chance in one hundred billion to decay. This probability multiplied by the number of radioactive atoms in the sample gives the probable number of disintegrations in one second:

$$Np = (2.65 \times 10^{21})(1.36 \times 10^{-11} \text{ sec}^{-1}) = 3.6 \times 10^{10} \text{ disintegrations/sec} .$$

It is no coincidence that this is almost exactly one curie. The curie is defined to be approximately the number of radioactive decay events in one gram of radium 226 in one second.

In assessing the effect of radioactivity on living matter, as well as for some other purposes, it is more important to know the total ionizing power of the emitted radiation over some period of time than to know the nuclear decay rate. Therefore a second unit of measurement of the strength of radioactivity has come into use. This unit is the rad,* a measure of the energy deposited in one gm of matter by alpha, beta, or gamma rays traversing the matter:

$$1 \text{ rad} = 100 \text{ ergs/gm} . \tag{25.9}$$

To raise one gm of water one degree Centigrade requires about 40 million ergs. Therefore a radiation dose of one hundred rad, even if delivered almost instanta-

* A unit closely related to the rad is the roentgen, defined as the amount of electromagnetic energy that dislodges 1 e.s.u. of electrons from the molecules contained in 1 cm^3 of dry air at standard conditions of temperature and pressure. The rep (roentgen physical) is the amount of ionizing radiation that deposits 93 ergs in 1 gm of living tissue. Still another unit is the rem (roentgen equivalent man). The rad, the rem, the rep, and the roentgen do not differ greatly from one another.

neously, would produce no perceptible sensation of warmth. Nevertheless, because of the disruptive effect of high-energy particles on molecules, one hundred rad delivered to all parts of the body is a dangerous dose. It is likely to produce sickness. Several hundred rad delivered to major parts of the body within a few weeks or less may be fatal. Typically, in the United States, a man is exposed to about 0.13 rad per year from cosmic rays and natural radioactivity, or about 10 rad in a lifetime. Workers at nuclear installations are supposed to receive less than 10 rad per year. The *average* level of radioactivity from fallout is less than the natural background. However, the facts that fallout can be localized geographically and that certain isotopes such as strontium 90 are concentrated in the body increase the hazard of fallout.

25.9 Fission

Nuclear fission has played only a minor role in the development of fundamental understanding of submicroscopic nature. Nuclear fusion—the subject of the next section—is of greater scientific importance because it is the energy source of stars. Whatever their importance in the evolution of science, both fission and fusion deserve a place in a modern survey of physics. Each is a basic nuclear phenomenon depending on the principles of quantum mechanics and relativity; together their impact on contemporary society is enormous and unprecedented.

In its essentials, fission is quite similar to alpha decay. In both processes, a piece of a nucleus breaks off, and the two parts fly apart. In both, a potential-energy barrier prevents very rapid decay. Although the energetic possibility of fission has been known for years, its discovery in 1939 came as a surprise—both to its discoverers* and to the rest of the scientific community. This is not to say that fission upset any theories or disagreed with any predictions. It was simply unexpected, since in forty years of work in radioactivity and nuclear transformations, no one had seen it before.

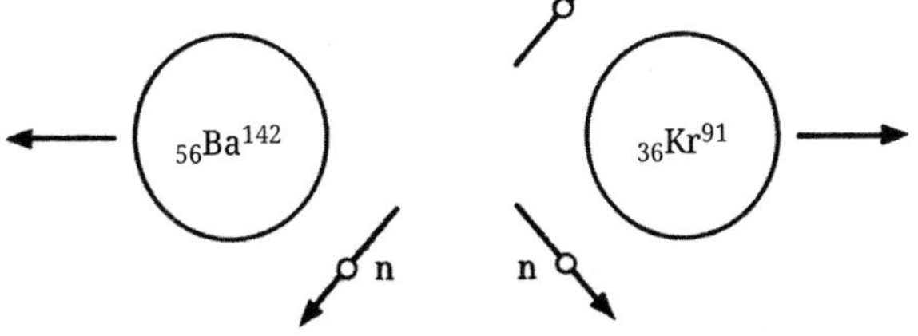

FIGURE 25.18 One of many possible modes of fission. A nucleus of U^{235} after absorbing a neutron to become an excited nucleus of U^{236}, might break apart into fragments which are isotopes of barium and krypton, and at the same time emit several neutrons.

* Fission products were probably first observed (through their radioactivity) by Enrico Fermi and coworkers in 1934, but not recognized as such. Late in 1938, Otto Hahn and Fritz Strassmann in Berlin found definite chemical evidence for the production of barium ($Z = 56$) by uranium bombarded by neutrons, but were not quite able to believe their own results. In the same month, December, 1938, Lise Meitner and Otto Frisch in Sweden, having in hand the same chemical evidence, more bravely proposed that uranium, activated by neutron bombardment, could break apart into fragments of nearly equal size.

Among the naturally occurring elements, spontaneous fission is rare.* However, if a heavy nucleus is excited, by neutron absorption or in some other way, fission may occur swiftly. In a typical fission process, one or more neutrons are emitted, in addition to the major nuclear fragments (Figure 25.18). Because these neutrons can serve as triggers of further fission events, a fission chain reaction is possible.

163 The discovery of fission was a seed dropped on fertile ground. With remarkable speed, Niels Bohr and John Wheeler published a theory of the dynamic process of fission. With equal speed, physicists grasped its practical potentialities. In assessing the prospects of large-scale energy release by means of fission, three basic questions needed to be answered. The first, the easiest, and the least important was: How much energy is released in a fission event? From a binding-energy curve like that in Figure 25.7, the answer could be predicted to be about 1 MeV per nucleon, or about 200 MeV per fission event. This prediction was soon verified by experiment. The reason the precise magnitude of the energy release is not very important is that this energy, quickly transformed to heat, is irrelevant to the nuclear course of events. Although it is obviously of great practical significance, fission energy contributes nothing to the chain-reaction process.

The second question was: Which nuclei are most fissionable? and the third: How many neutrons are emitted in a fission event? Once these questions were answered, it could be decided whether a fission chain reaction could be self-sustaining, and if so, with what isotopes.

Nuclear fission would have been a dramatic discovery at any time. Coming as it did on the eve of a major war and in the midst of a period of persecution of Jews in Germany, its drama was heightened. The persecution drove many of Europe's leading scientists to the United States, and the war drove fission work into secrecy. Half a dozen years after the discovery of fission, the United States emerged as the scientific leader of the world, atomic energy was a household word, and ties forged between science and government set a pattern for the support of large-scale research that has lasted to the present day.

One small act in the total drama of fission was the theoretical work of Bohr and Wheeler in 1939. Using the liquid-droplet model of the nucleus, they provided an explanation of the fission process and predicted which nuclei should be most fissionable (our second question above). Apprised of the ideas of Frisch and Meitner just before leaving Copenhagen early in January, 1939, Bohr pondered the idea of nuclear fission all the way across the Atlantic. By the time he greeted his young colleague John Wheeler on the pier in New York, he had a half-completed theory of fission in his mind (and therefore no inclination to question the validity of the evidence for fission). After five months of effort in Princeton, Bohr and Wheeler submitted for publication a paper that provided both a mathematical theory and a pictorial model of fission.

In essence, the Bohr-Wheeler theory is simple. Two forces are at work in a heavy nucleus: the nuclear force, holding the nucleus together, and the Coulomb force, tending to blow the nucleus apart (see Figure 25.19). For all the nuclei we know, the nuclear force is in control, but for the heaviest known nuclei, it is only barely in control. The problem of the fissionability of a nucleus can be posed this way: If a nucleus is stretched into an elongated shape, what is greater—the repulsive electric force tending to push it into an even more elongated shape, or the attractive nuclear force tending to restore it to a spherical or near spherical shape? If a nucleus like uranium is slightly distorted from its normal shape, the nuclear force wins out, tending to restore it to its original shape. If it is distorted much further, the electric force wins out and it splits in two (or occasionally three). Between these two regions is an energy barrier. In slow spontaneous fission, this barrier can be penetrated, just as a barrier is penetrated in alpha decay. For the rapid fission that occurs in reactors

* For some of the transuranic elements discovered in recent years, spontaneous fission is a principal mode of radioactive decay—see Figure 25.6.

or bombs, the barrier must be surmounted. The magnitude of the energy barrier to be overcome depends sensitively on the relative magnitude of two energies of opposite sign: the Coulomb energy, arising from the mutual repulsion of the protons; and the surface-tension energy, arising from the nuclear forces. From approximate considerations of these energies, we can extract a significant parameter, which measures nuclear fissionability.

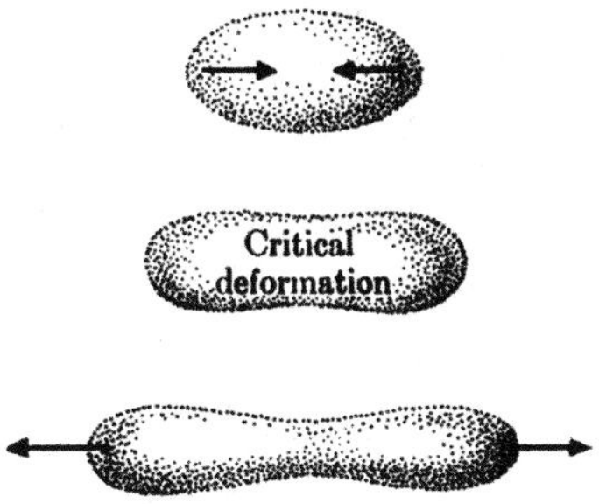

FIGURE 25.19 The mechanism of fission. Attractive nuclear forces create an effective surface tension tending to keep the nucleus in near-spherical form. Electrical repulsion between protons tends to blow the nucleus apart. Below a critical deformation, the surface tension wins. Beyond the critical deformation, the electrical repulsion wins.

Consider first the electric energy. Associated with each pair of protons is an energy $e^2/\bar{r}$, where $\bar{r}$ is an average distance between the two protons. The larger the nucleus, the larger this average separation, in direct proportion to the nuclear radius R. Therefore the Coulomb energy of a pair of protons is *proportional* to $1/R$. The number of distinct proton pairs is equal to $Z(Z-1)$, since each of the Z protons can pair off with $Z-1$ other protons. Therefore, the total Coulomb energy is proportional to $Z(Z-1)/R$, or approximately Z^2/R if Z is large:

$$E_{\text{coulomb}} \sim \frac{Z^2}{R}\,. \qquad (25.10)$$

The other relevant energy is the nuclear surface-tension energy. It is proportional to the surface area of the nucleus, or to the square of the nuclear radius:

$$E_{\text{surface}} \sim R^2\,. \qquad (25.11)$$

The ratio, $E_{\text{coulomb}}/E_{\text{surface}}$, is then proportional to Z^2/R^3. Since the cube of the nuclear radius is proportional to the mass number A (see Equation 25.1), we can write

$$\frac{E_{\text{coulomb}}}{E_{\text{surface}}} \sim \frac{Z^2}{A}\,. \qquad (25.12)$$

Bohr and Wheeler recognized that this ratio, Z^2/A, is an important parameter of fission. The greater its magnitude, the more nearly does the repulsive electric force win out over the attractive nuclear force, and the more easily fissionable is the nucleus. Consider nuclei of the two principal isotopes of uranium, to each of which a neutron is added:

$$U^{235} + n = U^{236} : Z^2/A = 35.9$$

$$U^{238} + n = U^{239} : Z^2/A = 35.4\,.$$

Because of this small difference, the fission energy barrier is slightly lower for U^{236} than for U^{239}. Another and even more important contributor to the distinction between the fissionability of these two isotopes is the somewhat greater excitation energy provided when U^{235} absorbs a neutron than when U^{238} absorbs a neutron. In U^{235}, the absorption of a slow neutron provides enough energy to surmount the fission barrier. In U^{238} it does not. The difference is all-important. Most of the neutrons emitted during fission are of relatively low energy, capable of inducing further fission in U^{235} but not in U^{238}.

Considerations of this kind led to a prediction and a gamble. The transuranic elements, with Z greater than 92, should be still more easily fissionable than uranium. Once a self-sustaining reactor had been achieved, a new prospect appeared. Transuranic isotopes might be "manufactured" in quantity, then used in new reactors or in bombs. The practical isotope to consider was plutonium 239 ($Z = 94$). It can be created from the abundant isotope uranium 238 by the following sequence:

$$n + {}_{92}U^{238} \rightarrow {}_{92}U^{239}$$

$${}_{92}U^{239} \rightarrow {}_{93}Np^{239} + e^- + \overline{\nu_e} \text{ (half life 24 minutes)}$$

$${}_{93}Np^{239} \rightarrow {}_{94}Pu^{239} + e^- + \overline{\nu_e} \text{ (half life 2.35 days).}$$

A nucleus of U^{238} absorbs a neutron to become U^{239}. In a few days, the U^{239} nucleus transforms itself spontaneously via two beta decays into ${}_{94}Pu^{239}$, a highly fissionable isotope.

$$Pu^{239} + n = Pu^{240} : Z^2/A = 36.8\,.$$

The plutonium lives long enough for the immediate purposes of man. It undergoes alpha decay (to U^{235}) with a half life of 24,360 years. It was first identified in trace amount in 1940. The first uranium reactor performed successfully at the end of 1942. Two and a half years later, an enormous and uncertain investment in plutonium-producing reactors had paid off. Transmutation on a large scale was successful. By the summer of 1945, a number of pounds of plutonium had been created, enough for the bomb tested in New Mexico and for the bomb that devastated Nagasaki. (The Hiroshima bomb was an untested model with a core of U^{235}.)

We come now to the third and most crucial question that had to be answered to reach a verdict on the feasibility of large-scale energy release by means of fission: How many neutrons are emitted in a fission event? This was a question that had to be settled by experiment. Nuclear theory was not refined enough to deal with it. This question of neutron multiplicity is of obviously paramount importance for the functioning of a reactor or a fission bomb. The vital requirement for a chain reaction is that every nucleus which undergoes fission must trigger the fission of at least one other nucleus. This means that from among the number of neutrons emitted by the fissioning nucleus, at least one must survive to cause another fission event.

In 1940 Enrico Fermi and his collaborators, working at Columbia University, determined the average neutron multiplicity in the fission of U^{235} to be 2.5. As with most other submicroscopic events, laws of probability govern the fission event. One nucleus of U^{235} might split without emitting any neutrons. Another might emit six. The average number is 2.5. The absorption of even a single neutron can induce fission. Consequently, the average production of 2.5 neutrons for each one absorbed by a fissioning nucleus appears to be quite favorable for a chain reaction. Indeed it is, in a solid chunk of U^{235} or Pu^{239}. These are what lie in the core of an atomic bomb. In them, the chain reaction is not only self-sustaining but is explosively multiplying. The neutron multiplication in each "generation" produces an *exponential* increase of energy release until the reaction is slowed by a combination of depletion of the fissionable material and expansion of the core.

In the early 1940s, neither purified U^{235} nor Pu^{239} was available. Fermi and his coworkers had to work with ordinary uranium, less than one per cent of which is the easily fissionable isotope, U^{235}. In this material, a neutron has available several possible fates other than to cause fission. Chief among these are escape from the material into nonfissionable surroundings, and absorption by U^{238} without inducing fission. To counter escape, the designers of the first reactor—or atomic pile, as it was then called—made it large, about 20 feet across, with its pieces of uranium buried in 400 tons of graphite. The carbon nuclei in graphite happen to have a very low probability to absorb neutrons, and they are light enough so that their recoil drains energy from neutrons in collisions between neutrons and nucleus. The carbon also serves to enhance the probability that a neutron finds its way into a U^{235} nucleus instead of a U^{238} nucleus. The lower the energy of the neutron, the greater is the relative probability of its absorption by U^{235}. The graphite "moderator" helps to degrade the energy of the neutrons. A reactor design that is too "successful" would be not a reactor, but a bomb. A reactor needs control. This is achieved by control rods made of material whose nuclei have a high probability to absorb neutrons. The control rods provide a way to absorb a selected fraction of the neutrons and to maintain a steady rate of energy release. Fully inserted into the reactor, they halt the chain reaction. If fully removed, they could permit the reaction to build up to explosive level or at least to a level sufficient to melt the uranium.

Under the leadership of Enrico Fermi, the first reactor, a latticework of uranium and graphite with cadmium control rods, achieved a self-sustaining chain reaction on December 2, 1942. On that day, Arthur Compton, who had witnessed the event, telephoned James Conant, then president of Harvard University, and said,* "Jim, you'll be interested to know that the Italian navigator has just landed in the new world."

Among other things, nuclear fission has contributed to our society some new phrases, "critical mass" and "critical size." For any particular composition and design, a reactor must be greater than some minimum size and minimum mass in order to function. Otherwise, neutron loss makes it impossible for one fission event to trigger one or more others and propagate the chain reaction. A reactor is subcritical if its rate of energy generation is decaying, supercritical if its rate of energy generation is increasing, and just critical if its rate of energy generation is steady. The same ideas of critical size and critical mass apply to bombs. For a sphere of pure Pu^{239}, the critical size is a few inches in diameter.

The aim of bomb design is to create a supercritical mass and to keep it that way for as long as possible, perhaps a few tenths of a microsecond. A supercritical mass can be created by bringing together two subcritical pieces or by compressing a single subcritical piece. In either case, speed is important. In the uranium bomb dropped on Hiroshima, an explosive charge drove one piece of uranium against another. In the plutonium bomb dropped on Nagasaki, a small sphere of plutonium at the center was compressed by the implosion of a surrounding mass of high explosive. Sitting now in the arsenals of the world's great powers are thousands of bombs, each containing a not quite critical mass of fissionable material and a charge of chemical explosive capable of making it go supercritical. A good many of these atomic bombs in turn stand ready to trigger even more devastating thermonuclear explosions.

25.10 Fusion

The light and heat of the sun are derived principally from the energy released by the fusion of nucleons to form alpha particles. In stars older than our sun, alpha particles probably fuse to form the nuclei of heavier elements. According to the best current theories of stellar evolution, all of the elements (or, more accurately, the nuclei of all of the elements) up to and beyond uranium are created by nucleus-building reactions in the late stages of the life of a star. In a nova explosion, these newly created nuclei are sprayed into the cosmos. As they cool, they gather their protective shells of electrons about them to become docile neutral atoms. It is an interesting thought that the atoms making up the elaborate order of living matter on earth probably originated in the nuclear chaos at the centers of remote and now long dead stars. Only the residual radioactivity of the heavy elements reminds us of the turbulent beginnings of terrestrial matter.

Among elements lighter than iron, most fusion reactions are *exothermic*, or energy producing. This comes about because of the increasing trend of the binding energy per nucleon

---

* See Arthur Compton, *Atomic Quest* (New York: Oxford, 1956).

(Figure 25.7) from hydrogen to iron. The most vigorously exothermic reactions occur among the lightest elements; there the binding-energy curve is rising most rapidly.

164 In the sun, which still consists mostly of protons and electrons, protons unite to form alpha particles according to the following sequence of reactions:

$$(1)\quad {}_1\mathrm{H}^1 + {}_1\mathrm{H}^1 \rightarrow {}_1\mathrm{H}^2 + \mathrm{e}^+ + \nu_\mathrm{e}$$

$$(2)\quad {}_1\mathrm{H}^1 + {}_1\mathrm{H}^2 \rightarrow {}_2\mathrm{He}^3$$

$$(3)\quad {}_2\mathrm{He}^3 + {}_2\mathrm{He}^3 \rightarrow {}_2\mathrm{He}^4 + 2{}_1\mathrm{H}^1 .$$

Although we use here a notation similar to the notation of chemical reactions, it is important to remember that these are reactions of bare nuclei. Enough electrons are present to maintain electrical neutrality, but, because of the high temperature, they are not bound to nuclei, nor do they participate in the reaction. The first of these three solar reactions is especially interesting because it involves strong and weak interactions simultaneously. Normally a neutron undergoes beta decay to become a proton (a negative electron and an antineutrino are emitted). According to the theory of weak interactions, a proton is equally favorably disposed to undergo beta decay, by emitting a positron and a neutrino, transforming itself into a neutron. Energy conservation prevents a free proton from doing so. In many radioactive nuclei, however, the inhibition is overcome by a release of nuclear binding energy. That is what happens in this solar fusion reaction. The binding energy of the deuteron (2.2 MeV) is more than sufficient to provide the neutron-proton mass difference plus the mass of the emitted positron. Once two protons create a deuteron in this way, the deuteron can fuse with another proton to make a helium 3 nucleus. Finally, two helium 3 nuclei react to form an alpha particle and two protons. The net result of two reactions of type 1, two of type 2, and one of type 3 can be summarized by a single reaction equation,

$$4{}_1\mathrm{H}^1 \rightarrow {}_2\mathrm{He}^4 + 2\mathrm{e}^+ + 2\nu_\mathrm{e} ,$$

or, in somewhat simpler notation,

$$4\mathrm{p} \rightarrow \alpha + 2\mathrm{e}^+ + 2\nu_\mathrm{e} . \qquad (25.13)$$

Note, in this net reaction, the conservation of charge, of baryon number, and of electron family number.

In this so-called proton-proton cycle in the sun, 6.2 MeV of energy per proton is released. A few per cent of this energy is given to neutrinos, which escape at once from the center of the sun without impediment. Nearly 10^{11} of these solar neutrinos strike each square centimeter of the earth each second. The rest of the energy goes at first into kinetic energy of charged particles, eventually into electromagnetic energy—photons. In addition, by annihilating with electrons, the positrons add another increment of energy to the total. In each annihilation event,

$$\mathrm{e}^+ + \mathrm{e}^- \rightarrow 2\gamma ,$$

the complete conversion of mass to energy yields about 1 MeV. Although positron annihilation involves neither nuclei nor nuclear force, it is another exothermic contributor to the proton-proton cycle, raising the total energy release from 6.2 to 6.7 MeV per nucleon. Including positron annihilation, the net reaction is

$$2\mathrm{e}^- + 4\mathrm{p} \rightarrow \alpha + 2\nu_\mathrm{e} + k\gamma . \qquad (25.14)$$

The notation $k\gamma$ on the right designates an unknown number of photons. Eventually, for each proton burned in the solar furnace, millions of photons are radiated at the surface.

The net energy release in the sun's proton-proton cycle may be calculated readily from the masses of the participating particles.

Before		
Mass of 4 protons	= (4) (1.007276)	= 4.029104 a.m.u.
Mass of 2 electrons	= (2) (0.000549)	= 0.001098 a.m.u.
Total mass of reacting particles		= 4.03020 a.m.u.
After		
Mass of alpha particle		= 4.00150 a.m.u.
Net loss of mass		= 0.02870 a.m.u.

Net loss of energy = (0.02870 a.m.u.) (931 MeV/a.m.u.) = 26.7 MeV.

In the last line we have used the mass-to-energy conversion factor that is especially useful in the nuclear domain, 931 MeV/a.m.u. Dividing 26.7 MeV by 4, we get 6.7 MeV, the energy release per nucleon quoted above. The fractional conversion of mass to energy in the sun's fusion reactions is 0.7%, or one part in 140. This is about eight times the fractional conversion in uranium fission, and more than 100 million times the fractional conversion in a typical chemical explosion. Each day the sun burns 5.3×10^{16} kg of its hydrogen into alpha-particle ashes, thereby transforming about 3.7×10^{14} kg of its mass into energy, a rate of output equivalent to the explosion of billions of multimegaton weapons each second.

Fusion reactions in the sun and other stars are examples of *thermonuclear* reactions— 165
that is, nuclear reactions brought about by high temperature. Thermonuclear reactions are, quite literally, nuclear burning. They are completely analogous to ordinary chemical combustion. In chemical (or atomic) burning, high temperature initiates the reaction, and energy released by the reaction keeps the temperature high and spreads the fire. The chemical reaction may proceed in several steps; its net result is a combination of atoms into more tightly bound molecules. All of these features characterize nuclear burning as well. More tightly bound nuclei are produced by a reaction or series of reactions initiated and propagated by high temperature. Apart from the nature of the reacting particles, the difference is one of scale. At a temperature of ten million degrees or more, the thermonuclear flame is at least 10^4 times hotter and at least 10^6 times more potent as an energy source than a typical chemical flame. For the design of thermonuclear weapons, one more common feature of nuclear and atomic combustion is important—the harmlessness of cold reactants. The concept of critical mass, important for fissionable material, is not applicable to thermonuclear fuel. Just as there is no limit to the size of a lumber yard or an oil storage depot, there is no theoretical limit to the size of a thermonuclear weapon. Any amount of nuclear fuel can be stored with safety—until it is ignited. Although a match is sufficient to set ablaze the lumber yard or the oil depot, nothing less than a fission bomb can ignite a thermonuclear reaction.

The sun is forced, as it were, to make do with protons as its nuclear fuel. On earth, man has available a variety of other nuclei, some of which react with higher probability and at lower temperatures than do protons. Actually, the sun's proton-proton cycle proceeds at a very leisurely pace. There is no possibility to use the same reaction as an energy source on earth.* More reactive are the heavier isotopes of hydrogen, deuterium and tritium. Especially favorable is the reaction between a deuteron and a triton, known as the DT reaction, which can proceed explosively at a temperature of about 60 million degrees:

$$ {}_1H^1 + {}_1H^3 \rightarrow {}_2He^4 + n \,, $$

* Even in the sun, the proton-proton cycle is a significant energy source only in the hottest and densest central core of the sun. Outside this relatively small core, there is very little thermonuclear energy production.

or, in simpler notation,

$$d + t \rightarrow \alpha + n\,. \tag{25.15}$$

This reaction releases 17.6 MeV, or about 3.5 MeV per nucleon. Most of this energy—14 MeV—appears as kinetic energy of the neutron. The DT reaction is one of the key reactions in thermonuclear weapons.

Except for trace amounts produced by cosmic rays, tritium does not occur in nature. It must be manufactured at great expense in nuclear reactors,* and once manufactured, it decays into He^3 with a half life of 12.26 years. If fusion is to be practical as a *controlled* power source in the future, it must rely on plentiful deuterium for fuel. The DD reaction proceeds with approximately equal probability in two ways:

$$\text{(1)}\ d + d \rightarrow He^3 + n \ \text{(energy release = 3.3 MeV)}\,, \tag{25.16}$$

$$\text{(2)}\ d + d \rightarrow t + p \quad \text{(energy release = 4.0 MeV)}\,. \tag{25.17}$$

In conditions that can be visualized on earth (even in a thermonuclear explosion), the helium 3 produced does not significantly react further. The tritium, however, is consumed by the DT reaction, adding a large increment of energy. The net reaction in the burning of pure deuterium is approximately,

$$5d \rightarrow He^3 + He^4 + 2n + p \quad \text{(energy release = 25 MeV)}. \tag{25.18}$$

The high-temperature requirement for thermonuclear reactions can be understood quite simply in terms of the small size of nuclei and the electrical repulsive force between them. In order for a pair of deuterons, or a deuteron and a triton, to "touch"—that is, to be close enough together so that their wave amplitudes overlap significantly—their centers must be separated by not more than about 10^{-12} cm. At this separation, the electric potential energy associated with their relative position can readily be calculated:

$$P.E. = \frac{e^2}{d} = \frac{(4.80 \times 10^{-10}\ \text{e.s.u.})^2}{10^{-12}\ \text{cm}} = 2.30 \times 10^{-7}\ \text{ergs} = 1.44 \times 10^{5}\ \text{eV} = 144\ \text{KeV}\,. \tag{25.19}$$

Although small compared with energies available in cyclotrons and other accelerators, this energy is enormous compared with ordinary thermal energy. Normally the electric repulsion between hydrogen nuclei, or any other nuclei, very effectively keeps them apart, even if their atomic electrons are stripped away. Even at a temperature of 10,000°K, capable of vaporizing all matter, the mean kinetic energy of thermal motion is only 1.3 eV per particle, more than 100,000 times smaller than the potential barrier of 144 KeV calculated above. What temperature would be required in order that an average pair of deuterons, in a head-on collision, could climb the potential barrier and come within 10^{-12} cm of each other? To answer this question, let us suppose that each deuteron has an energy $\frac{1}{2}kT$ equal to half of the total required, or

$$\tfrac{1}{2}kT = 72\ \text{KeV} = 1.15 \times 10^{-7}\ \text{ergs}\,.$$

Using $k = 1.38 \times 10^{-16}$ ergs/°K, we find

$$T = 560\ \text{million °K}\,. \tag{25.20}$$

In energy units, $kT = 48$ KeV. The actual temperature requirement to ignite deuterium is not quite this extreme, for two reasons. First, some nuclei have considerably more ki-

* The tritium-producing reactor in Savannah River, South Carolina, was also the site of the first identification of the neutrino—see Section 26.4.

netic energy than the average. They can more easily overcome the force of electric repulsion. Second, barrier penetration can be significant. The nuclei need not all surmount the potential-energy barrier in order to react. In practice, thermonuclear explosions proceed at temperatures of about 60 million °K ($kT = 5$ KeV). A fission bomb can provide ignition at this temperature.

Deuterium is a satisfactory thermonuclear fuel in principle. In practice it has a serious disadvantage as a major component of a bomb. In order to be stored compactly at high density, it must be cooled to its liquefaction temperature of 20°K (–253°C). This requires elaborate refrigeration techniques. The fusion fuel actually used in the "H bomb" is a particular isotopic form of lithium hydride, called lithium 6 deuteride (${}_3Li^6\,{}_1H^2$), which is a solid at normal temperature. The lithium plays two important roles as a constituent of the thermonuclear fuel. First, through its chemical combination with deuterium, it holds the deuterons close together, thereby helping to make possible the DD reaction (Formulas 25.16 and 25.17). Second, the lithium itself participates in a key reaction that "manufactures" tritium to stoke the DT reaction.

The significant nuclear reactions in lithium 6 deuteride are these:

$$(1)\quad {}_1H^2 + {}_1H^2 \rightarrow {}_2He^3 + n \qquad \text{(energy release = 3.3 MeV)},$$
$$(2)\quad {}_1H^2 + {}_1H^2 \rightarrow {}_1H^3 + p \qquad \text{(energy release = 4.0 MeV)},$$
$$(3)\quad {}_1H^2 + {}_1H^3 \rightarrow {}_2He^4 + n \qquad \text{(energy release = 17.6 MeV)},$$
$$(4)\quad {}_3Li^6 + n \rightarrow {}_1H^3 + {}_2He^4 \qquad \text{(energy release = 4.9 MeV)}.$$

The first two are the branches of the DD reaction. The third is the DT reaction. The fourth reaction, like the second branch of the DD reaction, is a supplier of tritium for the DT reaction. This neutron-induced reaction is not itself a thermonuclear reaction, since the neutron has no energy barrier to overcome to reach the Li^6 nucleus. It is in fact a fission reaction (although the word "fission" is usually reserved to describe the splitting of heavy nuclei). The Li^6 nucleus absorbs a neutron to become temporarily a Li^7 nucleus that splits into a triton and an alpha particle. Note that in these four interlocked reactions both tritons and neutrons play important roles, although neither is initially present in the fuel. Reactions 1 and 3 produce neutrons to stimulate the Li^6 reaction; reactions 2 and 4 produce tritium to burn in the DT reaction.

One way to increase both the rate of burning and the energy release of a thermonuclear weapon is to encase its lithium 6 deuteride fuel in U^{238}. The 14-MeV neutrons produced by the DT reaction, many of which escape from the combustion zone, are energetic enough to induce fission in U^{238}. From the fission event emerge several lower energy neutrons which can stimulate the tritium-producing reaction in Li^6. Through this sequence of fission-fusion-fission-fusion, the energy of the atomic bomb trigger can be multiplied a thousandfold, using only cheap and readily available materials—lithium 6, deuterium, and uranium 238. In such a uranium-encased weapon, about half of the energy and almost all of the hazardous fallout result from uranium fission. In so-called "clean" bombs, most of the energy comes from fusion.

In summary, most nuclei on earth live out their lives in splendid isolation, protected from outside influences by their surrounding electrons and/or their own electric charge. Occasionally a nucleus is struck by a high-energy particle, man-made or of cosmic origin; or it is infiltrated by an uncharged neutron; or its protective potential is broached by another nucleus in the inferno of a thermonuclear explosion. When any of these things happen, the nucleus reacts in a characteristically violent way, transmuting itself into one or more other nuclei, with associated energy changes per unit mass a million times greater than the everyday energy of chemical change. All of these things happen because of the great strength of the still imperfectly understood nuclear force, a force that governs the behavior of most of the known elementary particles and that quite probably determines the structure and masses of the particles as well.

EXERCISES

25.1. Which of the following properties of a cadmium atom are attributable primarily to its nucleus alone, and which depend also on its electronic structure: **(a)** its mass; **(b)** its ionization energy; **(c)** its valence; **(d)** its spectrum of visible light; **(e)** its spectrum of X-rays; **(f)** its spectrum of gamma rays; **(g)** its ability to absorb neutrons?

25.2. **(1)** Express the density of nuclear matter in units of nucleons per cubic fermi (1 fermi = 10^{-13} cm; 1 cubic fermi = 10^{-39} cm^3). **(2)** Make up and solve a numerical example designed to illustrate the enormity of nuclear density. Consider, for instance, the gravitational effect of a hypothetical macroscopic chunk of nuclear matter.

25.3. **(1)** Calculate the speed of a "slow" neutron, one whose energy is 0.02 eV. **(2)** What is the wavelength of this neutron? How does this compare with the size of a typical nucleus? **(3)** At what energy in eV is the wavelength of a neutron equal to 10^{-12} cm?

25.4. **(1)** Explain in simple terms why there is no nuclear analog of molecule formation—that is, why two or more nuclei do not associate to form multinucleus systems without loss of their individual identities. **(2)** What is the nuclear analog of ionization? What is its consequence ?

25.5. Nuclear binding energies are determined from nuclear masses. **(1)** Do you think atomic binding energies (binding electrons to atoms) can be determined in the same way? Why or why not? **(2)** Suggest a method other than mass measurement that could be used to determine either nuclear or atomic binding energies.

25.6. Before the discovery of the neutron it was supposed that the atomic nucleus contains electrons. In 1934 Enrico Fermi wrote: "Another difficulty for the theory of nuclear electrons is the fact that the current relativistic theory of light-weight particles is unable to explain in a satisfactory way how such particles can be bound in orbits of nuclear dimensions." Explain exactly what Fermi had in mind.

25.7. The only stable isotope of yttrium is ${}_{39}Y^{89}$. An unstable nucleus with the same number of nucleons is ${}_{36}Kr^{89}$. Give two reasons why the mass of krypton 89 is greater than the mass of yttrium 89.

25.8. Two nuclei that differ by a complete interchange of neutrons and protons are called "mirror nuclei." Examples of such mirror nuclei are hydrogen 3 and helium 3; lithium 7 and beryllium 7; boron 11 and carbon 11 (see Figure 25.4). **(1)** Suggest a reason why ${}_2He^3$ is more stable than ${}_1H^3$, despite the repulsive force between the two protons in the helium nucleus. **(2)** Suggest a reason why ${}_5B^{11}$ is more stable than ${}_6C^{11}$.

25.9. The number of radioactive atoms in a sample decreases by a factor of about one million in 20 half lives. **(1)** Verify the correctness of this statement. **(2)** A sample containing 10^{12} atoms with a half life of two hours is prepared on Monday morning. Which of the following phrases best characterizes the probability that one or more radioactive atoms remain in the sample on Saturday afternoon: **(a)** definitely, **(b)** probably, **(c)**

possibly, **(d)** very unlikely, **(e)** impossible? State the reason for your answer.

25.10. The mean lifetime of the first excited state of the hydrogen atom is 1.6×10^{-8} sec. What is the energy uncertainty of the first line of the Lyman series (the transition from the first excited state to the ground state)? Use the uncertainty principle. Express the answer in eV, and note the ratio of this energy uncertainty to the transition energy, which is 10.2 eV.

25.11. Lithium 5 is an extremely unstable nucleus with a mean life of about 10^{-21} sec. Use the time-energy form of the uncertainty principle (Equation 25.6) to estimate the uncertainty in its total energy. Is this uncertainty large enough to have an influence on the result of an experiment designed to measure the mass of the nucleus to an accuracy of 10^{-27} gm ?

25.12. **(1)** Based on the uncertainty principle arguments following Equation 25.7, prove that the range of the nucleon exchange force is approximately

$$d \cong \hbar/mc\,,$$

where m is the mass of a pion. This distance is called the Compton wavelength of the pion. **(2)** Evaluate d numerically.

25.13. **(1)** Suggest two ways in which the energy required to break a deuteron into a neutron and a proton might be supplied. **(2)** If a neutron and a proton coalesce to form a deuteron, a gamma ray is radiated. What is its energy in eV?

25.14. The graph of binding energy per nucleon in Figure 25.7 shows a jagged structure for light nuclei. **(1)** Suggest a reason why it is not smooth. **(2)** In particular, why is there a peak at $A = 4$ (the alpha particle) ?

25.15. If the pattern of nucleon energy levels shown in Figure 25.10 were continued into the next "shell," what would be the next closed shell number, or "magic number" ?

25.16. The following data show the range of alpha particles of several energies in a certain photographic emulsion:

Energy (MeV)	Range (cm)
5	2.0×10^{-3}
10	5.7×10^{-3}
15	11.0×10^{-3}
20	18.0×10^{-3}

(1) Find the average rate of energy loss in MeV/cm in each of four energy ranges: **(a)** 20 to 15 MeV; **(b)** 15 to 10 MeV; **(c)** 10 to 5 MeV; **(d)** 5 to 0 MeV. **(2)** If an alpha particle loses 30 eV per ionization event, how many atoms or molecules does it ionize in the last 0.002 cm of its track? **(3)** By examining the track of an alpha particle in a photographic emulsion, how could you determine the direction in which the particle moved?

25.17. **(1)** Explain carefully why the beta decay of a nucleus is represented by an arrow pointing to the left and upward or by a narrow pointing to the right and downward in Figure 25.4. What distinguishes the two directions? **(2)** What sort of arrow would be appropriate to indicate gamma decay in such a nuclear chart?

25.18. An isotope of argon decays via electron capture as follows:

$$_{18}A^{37} + e^- \rightarrow {}_{17}Cl^{37} + \nu_e \,.$$

(1) Explain why this process is closely related to beta decay. **(2)** What happens to the atomic electrons following this radioactive decay event?

25.19. Fill in the blanks on the right for these three examples of radioactive decay:

(a) $_{27}Co^{56} \rightarrow {}_{26}Fe^{56} +$

(b) $_{27}Co^{60} \rightarrow e^- + \overline{\nu_e} +$

(c) $_{84}Po^{212} \rightarrow {}_{82}Pb^{208} +$

25.20. **(1)** Why do many heavy nuclei spontaneously emit alpha particles but do not emit single neutrons or single protons? **(2)** If neutron emission is energetically possible from a nucleus, the process usually occurs with a half life very much shorter than typical half lives for alpha decay or beta decay. Why is this?

25.21. According to Figure 25.17, the radium 226 used in luminous watch dials gives rise to a dozen product nuclei through successive radioactive transformations. **(1)** For a watch dial a few years old, which of these isotopes will be the most populous, assuming that the original radium was chemically purified? **(2)** After 50 years, which product isotope will be most populous? **(3)** After 1,000 years, how many elements would a chemical analysis of the original radium and its products reveal?

25.22. At present the earth contains 138 atoms of U^{238} for every atom of U^{235}. What was the approximate ratio 4.3 billion years ago? (The half life of U^{238} is 4.5 billion years; the half life of U^{235} is 713 million years.)

25.23. A Geiger counter near a sample of Bi^{212} (a radioactive isotope of bismuth, also known as thorium C) recorded the following number of beta particles.

Time (min)	Counts/min	Time (min)	Counts/min
0	2,500	41	1,634
2	2,517	50	1,339
4	2,433	51	1,325
7	2,438	63	1,246
24	1,966	64	1,212
25	1,983	69	1,165
40	1,609	72	1,197

(1) Draw a graph of the number of counts/min as a function of time. From your graph deduce the half life of Bi^{212}. **(2)** Estimate approximately how accurately you have determined the half life. **(3)** State *two* ways in which these actual experimental data tend to support the idea that fundamental processes in nature are governed by laws of probability.

25.24. The roentgen is defined in the footnote on page 726. Assume that a gamma ray passing through air loses on the average 30 eV for each electron it dislodges from a molecule. If this is so, express the roentgen in the rad unit defined by Equation 25.9. Take the density of air to be 1.23×10^{-3} gm/cm^3.

25.25. The probability per unit time for a radioactive nucleus to decay is equal to the inverse of its mean life. Calculate the number of radioactive decay events to be expected per sec in a sample containing 10^{22} atoms of U^{238}. (The half life of U^{238} is 4.5 billion years. Recall that the half life of a radioactive material is less than its mean life by the factor 0.694.)

25.26. **(1)** If a nucleus of $_{90}Th^{232}$ absorbs a neutron, and the resulting nucleus undergoes two successive beta decays (emitting negative electrons), what nucleus results? **(2)** Should this nucleus be more or less easily fissionable than U^{235}?

25.27. A nuclear explosion occurs under water. Suppose it is a small bomb (by modern standards) in which 1 kg of U^{235} undergoes fission. Each fissioning nucleus releases 200 MeV of energy. Calculate the total energy release in ergs, and calculate the mass of water in kg that could be heated 50°C by this amount of energy. This is largely an exercise in the conversion of units, but it illustrates an important point about nuclear energy. What is the point?

25.28. A common unit of energy applied to nuclear weapons is the ton (or kiloton, or megaton). It is approximately the energy released in the explosion of one ton of TNT. The conversion factor relating it to the cgs energy unit is 4.2×10^{16} ergs/ton. **(1)** If each uranium nucleus releases 200 MeV of energy in fission, what mass of uranium undergoes fission in a 20-kiloton bomb? **(2)** If the fuel in a thermonuclear weapon releases on the average 2 MeV per nucleon, what mass of thermonuclear fuel is required in a 20-megaton bomb?

25.29. **(1)** Explain why water should be a good neutron moderator. **(2)** Can you think of a reason why heavy water (containing deuterium, $_1H^2$, instead of ordinary hydrogen, $_1H^1$) might be preferred to ordinary water as a moderator and neutron reflector in a reactor?

25.30. A neutron starting at one side of a sphere of plutonium has a certain chance to cross the diameter of the sphere and escape out the other side without being absorbed. **(1)** If the sphere is compressed (same mass, smaller radius), does the chance of crossing and escaping increase, decrease, or stay the same? Give a reason for your answer. **(2)** Explain the connection between your answer to part (1) and the fact that compression can make a subcritical mass go supercritical.

25.31. The total energy released in the DT reaction is 17.6 MeV. Show that the neutron which is produced acquires about 80% of this energy, or 14 MeV, if the initial energies of deuteron and triton are relatively small.

25.32. An approximate value of the ratio of nuclear energy to atomic energy can be deduced from the wave nature of particles and a knowledge of nuclear and atomic dimensions. **(1)** Prove that the ratio of nucleon energy to atomic electron energy is given approximately by

$$\frac{E_n}{E_e} \cong \frac{m_e \lambda_e^2}{m_n \lambda_n^2},$$

where m_e and m_n are the masses of electron and nucleon, and λ_e and λ_n are the approximate wavelengths of electron and nucleon. **(2)** Evaluate this expression for $\lambda_e = 4 \times 10^{-8}$ cm, $\lambda_n = 10^{-12}$ cm.

25.33. Find out something about the life of Pauli, Chadwick, Heisenberg, or Fermi. In a single paragraph, summarize what you learned that is of special interest—preferably something about his human qualities and his approach to his work, not just a list of his scientific achievements.

TWENTY-SIX

Particles and interactions

Almost seventy years have gone by since Planck introduced into the description of nature a new quantum constant. More than forty years have passed since quantum mechanics reached near final form. In recent decades knowledge has multiplied and facts have exploded. Yet the search for fundamental understanding, for a new theory with the simplicity and generality to match relativity or quantum mechanics, continues at the slow and difficult pace that has always characterized man's struggle at the frontier of the unknown. Few physicists doubt that such a new theory awaits discovery, for the clues are too numerous, the hints too tantalizing, to allow either complacency about what we know already or discouragement about what we do not yet know. And few doubt that a great step will take place at the submicroscopic frontier, where the profusion of particles and the variety of interactions cry out for unification.

In this chapter and the next, we explore the frontier of the very small, where elementary particles serve admirably to bridge the gap from the known to the unknown. The particles illustrate, often simply and directly, the key ideas of quantum mechanics. At the same time they reveal new facts and new puzzles, pointing the way to future discovery. The theme of this chapter is the intimate connection between the nature of the individual particles and the way in which they interact with each other.

26.1 The classes of interactions

When an electron and a positron meet, they annihilate each other. A positron and a proton approaching each other exert a mutual electrical repulsion and turn aside. A proton and a neutron are drawn together by the powerful nuclear force. A neutron and an electron (to complete the circle) are practically oblivious of one another and can coexist with scarcely any mutual interaction. Not unlike the denizens of the animal world, the members of the elementary-particle family interact with each other in a rich variety of ways, sometimes with attraction, sometimes with repulsion, sometimes strongly, sometimes weakly, sometimes with annihilation in mind, sometimes with a live-and-let-live attitude. About these mutual interactions we have learned a great deal in recent years—their study is the central theme in particle research—but in many important ways they remain baffling.

166 A particle and its interactions are inseparable; the best way to approach understanding of the properties and behavior of the particles is through their interactions. We sometimes speak of the "intrinsic" properties of a particle, those properties which are associated with the particle alone, irrespective of how it behaves in the presence of other particles. Mass, charge, and spin are among those identifying characteristics called intrinsic. Of course, even an intrinsic property can be discovered and measured only via interactions. We know that a particle is charged because it can emit and absorb photons and exert forces on other charged particles. We measure its mass by finding out how quickly it responds to a force (an interaction) that is exerted upon it. We learn its spin only through reactions with other particles. Interactions are the glue joining the parts of the world into a coherent whole—not just in the literal sense of forces holding atoms and molecules and planetary systems together, but also as transmitters of information. Only through the elementary-particle in-

teractions is one part of the world made aware of other parts; in particular, only through these interactions does man gain information about his environment and about the rest of the universe. Directly or indirectly, for instance, photons interacting with charged particles bring to man most of his knowledge as well as most of his energy.

Needless to say, a particle with no interactions at all is a nonentity. Since it has no way to make its presence felt, it is, for practical purposes, nonexistent. Every citizen of a free country may invent as many such particles as he likes, and no physicist will prosecute. Noninteracting particles are entirely harmless, and are quite outside of science. When Einstein banished the ether from science at the beginning of this century, it was because the ether had become totally noninteracting and unobservable, and therefore superfluous.

Before being able to catalogue the particles and their more interesting properties in a useful way, we must catalogue their interactions. By far the most striking fact about the particle interactions known so far is that they all belong to one or another of just four distinct and markedly different classes:

(1) the strong interactions,

(2) the electromagnetic interactions,

(3) the weak interactions,

(4) the gravitational interactions.

The photon is associated with the electromagnetic interactions, the neutrinos with the weak interactions, and the graviton with the (still weaker) gravitational interactions. The pion is a principal carrier of the strong interactions. Since, for the thirty-six different particles in Table 2.1, taken two at a time, there are 630 different pairs to interact, it is a giant step toward simplicity to discover that only four different kinds of interactions govern what goes on between all of these pairs.

Although the strong and the weak interactions are imperfectly understood (as their uninspired names suggest), and though the connections, if any, among the different kinds of interaction are entirely unknown, there are some intriguing facts in hand about the four types of interaction—facts which nature dangles tantalizingly before the theoretical physicist, but which he has not yet been able to use.

First is the quite remarkable disparity of strength among the different interactions. The strong interactions exceed the gravitational interactions in strength by the fantastically large factor of about 10^{41}. The nuclear glue (strong interaction) makes the cosmic glue (gravity) look puny indeed. The electromagnetic interactions fall short of the strong interactions by a factor of "only" about 100, but exceed the weak interactions in strength by the enormous factor of 10^{11}. The weak interactions in turn are about 10^{28} times stronger than the gravitational interactions. In fact, these numbers do not have a very precise meaning, but they indicate clearly that there exist vast gulfs between the strength of one interaction and another.

The second intriguing fact about the four types of interaction is a rule which can be stated roughly: the stronger, the fewer. Of the fourteen kinds of particles listed in Table 2.1 eight (including all of the mesons and baryons) experience strong interactions, eleven (all but the neutrinos and the graviton) experience electromagnetic interactions, thirteen (all but the graviton) experience weak interactions, and all fourteen experience gravitational interactions.* Or, turning the rule around: the weaker the interaction, the larger the number of particles it embraces. Moreover, a particle that experiences any interaction in the hierarchy always experiences all weaker interactions as well; as one goes down the list to weaker interactions, names are added to the list of particles, but none is subtracted. Thus, the eight strongly interacting families of particles also experience electromagnetic, weak,

* Recall that the influence of gravity on massless particles is a consequence of the Principle of Equivalence (Section 22.3).

and gravitational interactions. It is interesting that the heaviest eight families of particles (pions and upward) are the ones that interact strongly. At the next rung down the ladder, the lighter muon, electron, and photon join the list. At each of the last two rungs, massless particles are added. None of these facts is understood.

Finally there is a most interesting connection between strengths of interaction and conservation laws. The seven absolute conservation laws discussed in Chapter Four govern all of the particle interactions. But, in addition to these, there are some partial conservation laws obeyed by some interactions and not by others. The rule is that the stronger the interaction, the more it is hemmed in by additional conservation laws which limit the possible transformations among the particles. The strong interactions are subject to laws of conservation of parity, of charge conjugation, of isotopic spin, and of strangeness. (What these peculiar-sounding conservation laws are all about is explained in Chapter Twenty-Seven. We are here interested only in the number of laws.) The weaker interactions then become lawbreakers. The electromagnetic interactions "violate" the law of isotopic spin conservation (in other words, electromagnetic interactions are not limited by this law). The weak interactions go further and violate all four of these special conservation laws. Since nothing is actually known of the gravitational interaction on the elementary submicroscopic level, it remains an open question whether the gravitational interaction goes even further in lawlessness and breaks one or more of the sacred "absolute" conservation laws. If it does so, the consequences would probably be significant only in the cosmological domain. For example, violation of the law of baryon conservation by the gravitational interaction could lead to the gradual creation of new protons and neutrons, as proposed in the continuous-creation theory of the universe, or to a gradual erosion of these units of matter, undermining the material structure of the world. Fortunately we know experimentally that the latter process, if it occurs at all, is too slow to be of any consequence over hundreds of billions of years. Gravity could also be responsible for giving to space and time a nonuniform structure (as the general theory of relativity predicts that it should do), in which case it would remove the fundamental symmetry supporting the laws of energy, momentum, and angular-momentum conservation. But again, in this case, these conservation laws would be so nearly true that their failure could have significant consequences only over the vast cosmological stretches of space and time where gravity is the ruling interaction.

Among all the facts about elementary particles, two kinds stand out as especially interesting, and probably especially important. In surveying the particles in the remainder of this chapter we pick out only selected excerpts from the catalogue of facts that fall into one or the other of the two categories. Either they illustrate in a particularly simple and beautiful way some fundamental aspect of a law of nature, or they underline clearly some important area of ignorance. The very existence of the electron as a stable particle is a fact belonging to the first group, for it illustrates the simple power of the law of charge conservation and incidentally verifies the accuracy of that law to an extremely high degree of precision. The relationships mentioned above among the strengths of interactions, the number of particles affected, and the number of conservation laws enforced are facts of the second kind. They surely must be significant and will, no doubt, one day rightly be regarded as clues that should have led physicists of today toward an understanding of the links between the different kinds of interactions. Nevertheless, they stand at present only as landmarks of ignorance on the frontier of knowledge.

26.2 Muons and electrons

In the muon and the electron, nature has presented us with a puzzling pair of "identical" twins, exact copies of each other except that one is a giant, the other a dwarf. The muon is about two hundred times more massive than the electron, yet in almost every other way it is indistinguishable from its small brother.

Actually, the *family* of electron together with its neutrino is a twin system to the two members of the muon family. Because of the near identity of the two neutrinos, one can speak of the muon and electron separately as almost twins. Each is a negatively charged particle with a positively charged antiparticle; each has one-half unit of spin; each participates in a simple family-number conservation law; and, most important, each of these two small families seems to have precisely the same interactions with all of the other elementary particles. This is the only such pairing known in the particle world, and it is like a semaphore signal waving before the eyes of the particle physicist. But so far the message has not been read.

There must be a "reason" for the chasm between the masses of these twin particles, a reason that ought to be reflected in some other differences in their properties. So far, however, no other difference has shown up, in spite of the fact that the muon is by far the most accurately studied of the short-lived particles. Nor has any convincing theoretical explanation for the mass difference been offered. The puzzle of the muon-electron mass difference belongs to that class of questions that we have learned enough to ask, but not enough to answer.

In spite of the enormous factor of difference in their masses, the muon and the electron are the two lightest charged particles, and they are the only two massive particles that experience no strong interactions. Although a long way apart, they are at least nearest neighbors in the catalogue of particles. All heavier particles are strongly interacting. The only lighter particles are massless. As the discussion below will make clear, the muon-electron puzzle is primarily an electromagnetic puzzle. Perhaps the electromagnetic interaction—the best understood of all interactions—has still some new surprises to reveal.

Before examining some of the points of identity between the muon-electron twins, we must consider two questions. Why is the big discrepancy in the mass of the twins so worrisome? Why is the vast difference in the lifetimes of these two particles not a problem (that is, why is it not considered a real "difference" at all) ?

We have no deep theoretical understanding of the nature of mass, of why the proton—or any other particle—has just so much energy locked up within it in the form of mass, and not more or less. Yet, through the raw experimental data of particle masses, nature has yielded up a few clues about the origin and the magnitude of masses. It is on the basis of the superficial knowledge built on these clues that the muon-electron mass difference seems paradoxical.

TABLE 26.1 Some Pairs of Particle Masses

First Particle	Mass of First Particle (in energy units of MeV)	Second Particle	Mass of Second Particle (in energy units of MeV)	Mass Difference Between First Particle and Second Particle (MeV)
π^+	139.6	π^-	139.6	0
π^+	139.6	π^0	135.0	4.6
K^+	493.8	π^+	139.6	354.2
K^+	493.8	K^0	497.9	-4.1
p	938.2	n	939.5	-1.3
p	938.2	K^+	493.8	444.4
p	938.2	$\bar{p}$	938.2	0
Σ^+	1189.5	Σ^0	1192.6	-3.1
Σ^+	1189.5	Σ^-	1197.4	-7.9
Σ^+	1189.5	Λ^0	1115.6	73.9
Σ^+	1189.5	$\bar{\Sigma}^+$	1189.5	0
Ξ^-	1321.2	Σ^-	1197.4	123.8

To seek these clues on your own, examine the data of Table 26.1, which displays some masses and mass differences for a number of different particle pairs. In order that your clues may be gathered without reference to muon or electron, these particles are omitted from the table, and the masses are expressed in terms of their equivalent energies given in MeV rather than in units of electron mass. The point is to find clues about mass from among the heavier particles, pretending that we do not know about the electron and muon, then to apply these clues to predict roughly how big the electron and muon masses ought to be.

Inspection of Table 26.1 reveals, first of all, that the mass differences fall into no simple pattern. None is, for example, just twice or three times another. However, the differences are interesting in one way. They fall into three groups in magnitude. Some are exactly zero. Some are "small," 1 to 7 MeV, and some are "large," 74 to 444 MeV. Moreover, closer inspection shows that a particular kind of particle pair is associated with each of these three ranges of magnitude. The three pairs in the table with no mass difference are particle-antiparticle pairs: π^+-π^-, p-$\bar{\text{p}}$, and Σ^+-$\overline{\Sigma^+}$. The equality of mass of particle and antiparticle is a universal rule and is, in fact, understood theoretically.

For the pairs with small mass difference, we find in the table a close kinship. Each such pair consists of particles that differ only electrically. The proton and the neutron, or the positive pion and the neutral pion, or two different Σ particles, are particles with different electric charge and slightly different mass, but otherwise identical. Proton and neutron are both baryons with spin whose strong interactions are exactly the same. Their electromagnetic interactions are obviously different because one is charged and one is not.

The natural conclusion is that the particles must have different mass "because" they have different charge; that, in some way, the electromagnetic interaction is producing the mass difference. Quantum mechanics does tell us that a particle's interactions should contribute to its mass, but it does not reveal how much. The experimental evidence is that the electromagnetic contribution to mass amounts to a few MeV. It is striking that for particles as completely different in mass and in other properties as the pion and the Σ particles, the electromagnetic mass differences are roughly the same. How an "interaction" can contribute to the mass of an isolated particle is an important question to which we return below.

If the electromagnetic interactions contribute only a few MeV of the total mass, where does the rest come from? Table 26.1 provides a hint about the answer to this question. Going on to the large mass differences, we find these between particles that differ more drastically. The real key is that these particle pairs do not have identical strong interactions. The proton and the kaon, although both strongly interacting (as is every particle in the table), do not have precisely the same strong-interaction properties. Perhaps the strong interactions are responsible for these large mass differences. This is an appealing idea and a reasonable one. The strong interactions are about one hundred times stronger than the electromagnetic interactions, and the large mass differences in the table are about one hundred times greater than the small mass differences.

Table 26.1 contains one more vital clue. Looking now at the particle masses themselves, not merely at their differences, we find them to lie in the range 130 to 1,300 MeV. But the large mass differences lie in the same range. A logical guess is that whatever is producing the large mass differences could, in fact, be producing practically all of the mass. This job can fall only on the shoulders of the strong interactions.

In order to appreciate the concept of particle structure to which we are led in this way, think of a builder who has at his disposal granite slabs 25 feet high, some ordinary bricks, a can of paint, and a piece of emery paper. The size of any structure he creates is determined mainly by the number of granite slabs employed. These are his "strong interaction" blocks. With one slab he builds a "pion structure," with two a "kaon structure," and with three a "nucleon structure." (Nature's big blocks are, of course, not all the same size, which complicates our task of trying to deduce the details of construction from examination of the finished product.) For slight variations of height, he can add one or two layers of bricks, his units of "electromagnetic interaction." A single slab alone might represent the neutral pion, a slab topped off with a layer of bricks the charged pion. (Again nature is not quite so simple. The addition of charge may cause a particle's mass to go either up or down slightly.) Now, for still finer control of the size of his structure he might add a few layers of paint, or whisk it once over lightly with the emery paper. His can of paint would be labeled "weak interactions," and stamped on the emery paper would appear "gravitational interactions." A light dusting with emery paper would not be likely to produce any measurable change in the size of the structure, and it is probable that the gravitational interactions contribute nothing significant to particle masses. Even the weak interactions are expected to produce only very tiny contributions to mass.

The concept of particle structure that emerges from a study of the experimental masses (Figure 26.1) mixed with some suggestions from the theory of quantum mechanics, is this. The big blocks of elementary matter are contributed by the strong interactions. Superimposed on this coarse structure is a finer structure (about 100 times finer) of bricks of matter contributed by the electromagnetic interactions. A very much finer structure (the layers of paint) which is normally negligibly small can be contributed by the weak interactions. Mass differences between particles arise from differences in their interactions. If their strong interactions differ, the particles differ greatly in mass. If their strong interactions are the same but the particles have different electrical interaction, they differ in mass by a much smaller amount. If the particles are the same in both strong and electromagnetic interactions, as are particle-antiparticle pairs (except for the sign of the charge, which turns out not to matter), they have no mass difference at all.

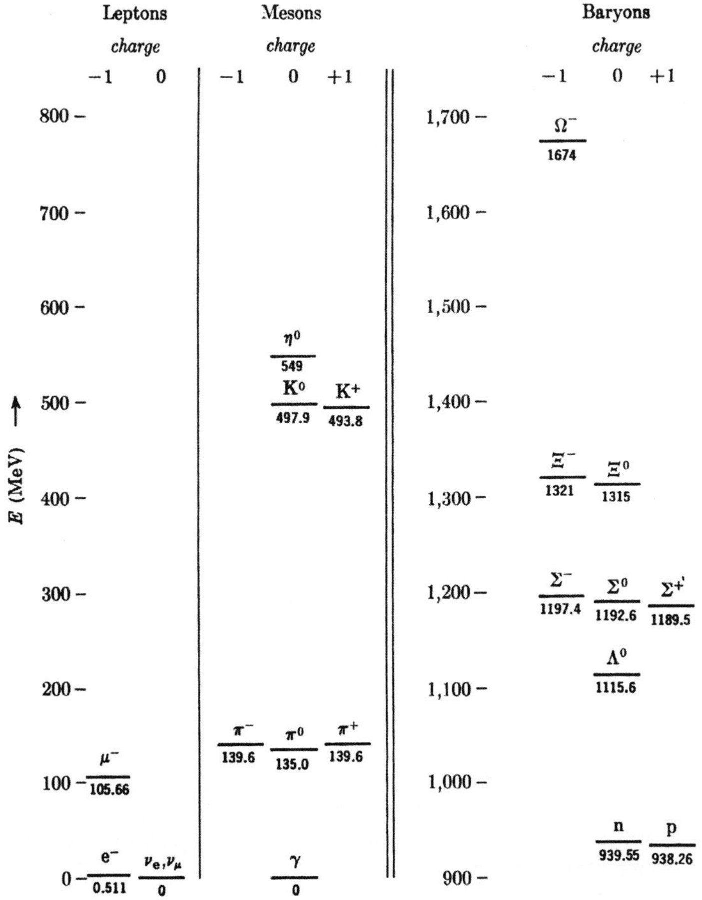

FIGURE 26.1 Energy-level diagram for the elementary particles of Table 2.1 (antiparticles are not included). Plotted vertically is the particle mass expressed in energy units. Plotted horizontally in each of three categories is particle charge. Note the change of energy scale for the baryons. Implicit in such a presentation is the idea that different particles are different energy states of some underlying substratum. Particles differing only in charge have small mass differences. Particles differing in their strong interaction properties have large mass differences.

The theoretical basis of this picture of particle masses is to be found in the idea of transitory creation of virtual particles (Section 25.3). A particle, even if alone, is never quiescent. It is always "interacting," for it is always creating and annihilating its own cloud of virtual particles. This process of self-interaction contributes to the particle's own mass, for the particle obviously represents a localized bundle of energy, which is nothing other than mass. The particles that interact strongly—pions and all heavier particles—presumably have the most intense clouds of virtual particles and correspondingly the most self-energy, or mass. Thus the blocks and bricks and paint of particle masses are representative of something much more evanescent—swarms of virtual particles in a ceaseless dance of annihilation and creation. The mass is no less real on that account.

On the basis of this much understanding of particle masses, what prediction can we make about electron and muon? These are particles with no strong interactions at all. They ought to be constructed with electromagnetic bricks alone without the use of the big blocks of strong interaction. The natural prediction to make is that structures which do not interact strongly at all should have small masses and small mass differences, probably less than about 7 MeV. With the electron and neutrinos, all is well. The neutrinos are massless, and the electron mass is 0.5 MeV, even a bit less than we should happily tolerate. We may even think of the electron as just a charged neutrino. The electron and its neutrino have the same strong interaction (namely, none) and differ only electrically, being in roughly the same relation to each other as the π^0 and π^- mesons. But what of the muon, with a mass of 105 MeV? It blatantly violates our hard-won rules. Its mass, not much less than that of the pion, suggests that it contains a strong-interaction block, but a number of careful tests have revealed no strong interaction whatever for the muon.

By this little excursion into the question of particle masses, we gain some insight into the reason why the muon is "anomalous." It does not belong. First, it violates the rule that only strongly interacting particles have large mass. Second, it violates the rule that particles must have different properties in order to have different masses—for electron and muon appear (so far) to have all the same properties. We can get off the hook of the second paradox just by assuming that muon and electron must have some differences which have not yet been discovered. But there is no escape from the first paradox. If a particle interacts strongly, it interacts strongly, and there is no missing the fact. No amount of careful study of a flea will convert it to an elephant. There can be no doubt; the muon does not interact strongly. Since we cannot escape the paradox, we must obviously have some faulty ideas about mass which need recasting. That is the present state of the dilemma. The muon, being the most bothersome particle in refusing to obey the rules, may turn out to be the particle that teaches us how to change the rules.

The second question to be disposed of was: Why is the lifetime difference not an important difference? We have the audacity to say that muon and electron are alike in every respect except mass; yet the muon lives two millionths of a second while the electron lives forever. The first thing to say about this difference is that two millionths of a second is practically forever. In this length of time, the muon can travel a distance about 10^{18} times its own size. To borrow an analogy from Chapter Three, this is comparable to an automobile traveling more than 10^{15} miles (which, in fact, is farther than the total distance covered by all the automobiles ever built) before falling apart. But the extremely long life of the muon on the microscopic time scale is not the main point. The key fact is that it is only "by chance" that the electron lives forever. The law of charge conservation thwarts the natural inclination of the electron to end its life by transforming into lighter particles, whereas the muon can decay without violating any conservation principle, and therefore it does so.

Just what is the evidence that muon and electron are "identical"? There is a large number of experimental facts which point to this conclusion, and we shall mention three of them here. The first has to do with the way in which the pion decays. A positive pion, for example, may end its life in either of two ways:

$$\pi^+ \rightarrow \mu^+ + \nu_\mu$$
$$\pi^+ \rightarrow e^+ + \nu_e .$$

Out of a large number of positive pions, 99.986% will transform to a positive muon plus a neutrino, and only 0.014% will transform to a positron plus a neutrino. So overwhelming is the pion's preference for the muon mode of decay that, for a long time, it was believed that pions never decayed to electrons. Had this proved to be true, it would have been a serious blow to the "twin" theory of muon and electron. Fortunately, the infrequent electron mode of decay was discovered in 1958 to bolster the idea that electron and muon (together with their respective neutrinos) are really alike except in mass.

According to the modern form of Fermi's theory of weak interactions, the chance that a pion decays into a muon or into an electron should be proportional, among other things, to the difference between the speed of light and the speed of the created muon or electron; in symbols:

$$P \sim c - v ,$$

where c is the speed of light, v is the speed of the muon or electron, and P stands for probability of decay. Now, the muon created in the usual mode of pion decay is rather speedy. It flies away from its point of creation at about one fourth of the speed of light. But if the pion had chosen to decay into an electron instead, this much lighter particle would have acquired a far greater speed, in fact more than 99% of the speed of light. The difference $c - v$ is, therefore, very much smaller for the swift electron than for the heavier more sluggish muon. As a consequence, the chance of pion-to-electron decay is much smaller than the chance of pion-to-muon decay.

It can be regarded as merely a mathematical quirk of the theory of weak interactions that the pion happens to prefer to decay into a heavier slower particle rather than a lighter speedier one. The important point is that this mathematical quirk accounts perfectly for the observed preponderance of pion-to-muon decays, and no difference at all between muon- and electron aside from their mass is needed to account for the big difference in the two ways a pion may decay. Until the tiny, but significant, fraction of pion-to-electron decays was discovered it looked as though muon and electron showed a "real" difference. Moreover, the principle that nature does everything not forbidden to it by a conservation law was saved from an unexplained exception.

Among other functions in nature, electrons build atoms. If muons and electrons are basically the same, then it should be possible to build atoms with muons as well as with electrons. This is quite true, but the job is not so easy. The physicist trying to do some atom-building with muons is like an artist trying to complete a painting using paints that fade into invisibility after one second. Before he could apply a second color, the first would have vanished. Still, if he worked like lightning, he might complete some very rudimentary paintings in a single color. The physicist, working literally like lightning—in times of a millionth of a second or less—can build some simple muon atoms and study them for an instant before the decay of the muon terminates their existence. These muon atoms have provided some valuable data supporting the twin theory of muon and electron.

The problem is getting enough muons in the same place at the same time. If a beam of muons strikes a target material, a particular atom in the target may capture a muon and begin the process of converting itself from an electron atom into a muon atom. But before it can catch a second muon, the first will have vanished. So far, only single-muon atoms have been constructed. However, even these provide useful information.

An important fact about a muon atom is that it is about two hundred times smaller than an electron atom. This comes about because of two aspects of the wave nature of matter discussed in preceding chapters. The first is the de Broglie wavelength relation, which says that more momentum means shorter wavelength. A muon stands in the same relation to an electron as a freight car to an automobile. At the same speed, the heavier object has more momentum. Since muon and electron have about the same speed within an atom, the muon has a much greater momentum and much shorter wavelength. The second fact is simply that the size of an atom is determined by the wavelength of its electrons (or muons). A particle cannot be forced into a region of space shorter than its own wavelength. So when a muon attaches itself to an ordinary atom, it cascades into ever smaller orbits, soon coming within the innermost electron orbit, until finally it falls down to its own lowest state of motion, circling the nucleus in a tiny orbit two hundred times closer than the nearest electron. Mathematically, this difference in radii for electron-atoms and muon-atoms shows itself in Formula 24.41. This formula reveals an inverse proportionality between Bohr radius and particle mass (for fixed charge *e*), so that a 200-fold increase of mass produces a 200-fold decrease in the Bohr radius.

As the muon jumps successively to lower orbits on its journey toward the nucleus, it emits photons whose study has yielded information about the muon— especially the precise value of its mass—as well as some information about the shape and size of the nucleus at the center of the atom.

Once lodged in its lowest orbit (which it reaches in much less than a millionth of a second), the muon may do what it would have done anyway, that is, decay into an electron, a neutrino, and an antineutrino, which fly away at high speed. But it has another course of action open to it, a course it follows more readily the larger the nucleus to which it is attached. The muon may combine with one of the protons in the nucleus to produce a neutron and a neutrino:

$$\mu^- + \mathrm{p} \rightarrow \mathrm{n} + \nu_\mu \,.$$

This reaction conforms to all of the conservation laws and releases considerable energy—about 100 MeV. Some of this energy is carried off by the neutrino; the rest goes into disrupting the nucleus. This process of muon capture is exactly analogous to the process of electron capture exhibited by some radioactive nuclei. Both are direct manifestations of the weak interaction, closely related to beta decay. Comparisons of the muon-capture process and the electron-capture process have provided more evidence that muon and electron differ only in mass. They are otherwise captured in exactly the same way.

By far the most impressive demonstration of the equivalence of electrons and muons has come from the measurements of the intrinsic magnetism of these particles. So far as we know, every spinning charged particle is a tiny electromagnet. (Some neutral particles are also magnetic.) Modern techniques of measurement have made it possible to determine the strengths of these one-particle magnets with phenomenal accuracy.

Specifically, the magnetic moment of the electron, expressed in terms of a unit called the Bohr magneton,* has been measured to be

$$1.00115962\,.$$

The muon's magnetic moment, in terms of its corresponding Bohr magneton, is

$$1.001164\,.$$

Interest attaches to these numbers in part because of the remarkably high accuracy of the measurements, as indicated by the number of figures employed. The special significance of these magnetic moments is that they are among the few measured quantities of the ele-

* The Bohr magneton is $eh/2mc$, where e, h, and c are now-familiar fundamental constants, and m is the mass of the particle in question. This happens to be a convenient unit because it is equal to the magnetic moment a particle would possess if it were a classically rotating object with one unit of spin.

mentary-particle world which can be accurately calculated from theory—that is, which are "understood." Dirac's quantum theory of the electron (1928) has been joined with Maxwell's theory of electromagnetism and the theory of photons to give a theory of the electric and magnetic properties of electrons (and, apparently, of muons as well) which bears the name of quantum electrodynamics. Mathematical difficulties beset this theory for twenty years, and were cleared away only around 1948.* It then became possible to calculate the theoretically expected values of the electron's magnetic moment,

$$1.00115962\,;$$

and, assuming it differs from the electron only in mass, of the muon's magnetic moment,

$$1.001165\,.$$

Within the limits of experimental error these values agree with those which have been measured. Because of the high precision of the measurements, this constitutes an excellent test of the idea that the muon and the electron are twins.

The reason such accurate values of magnetic strength can be calculated for these particles is that muon and electron have no strong interactions, and the weak interactions have no appreciable effect on the magnetism of the particles. Only the electromagnetic interactions play a role, and, of the four classes of interactions, these are the best understood.

Most physicists are tantalized by the muon-electron puzzle. We know so much about their similarities, but so little about the reason for their vastly different masses. Nature is trying to tell us something through these two particles free of all the complexities of strong interactions, but we do not know what. In the words of Abdus Salam,† one of the leading theoretical physicists interested in these problems: "I believe [our present theories] are but stepping-stones to an inner harmony, a deep pervading symmetry. The muon may seem out of place today. When we discover its real nature we shall marvel how neatly it fits into the Great Scheme, how integral a part it is of something deeper, more profound, more transcending. Faith in the inner harmony of nature has paid dividends in the past. I am confident it will continue to do so in the future."

26.3 Strongly interacting particles

Neutrons and protons are strongly interacting particles, and the nuclear force that binds them together is one manifestation of the strong interactions. The pion, whose exchange contributes most to the nuclear force, is another strongly interacting particle. Although these are the particles most important for nuclear structure, they are but a small fraction of the numerous strongly interacting particles now known. Besides the eta, the kaon, the lambda, the sigma, the xi, and the omega, all listed in Table 2.1, dozens of other particles have been discovered in recent years with lifetimes so short that they can leave no measurable track in a bubble chamber, nor can they otherwise be directly detected. All of these supershort-lived particles—or *resonances*, as they have come to be called—are also strongly interacting. Indeed the very profusion of strongly interacting particles is itself a feature of special interest about these particles. It tells us clearly that these particles can surely not all be elementary, but must be various manifestations of some deeper and simpler structure. A second feature of special interest about the strongly interacting particles is the special set of

* Actually, quantum electrodynamics still contains some mathematical puzzles. What happened in 1948 was that enough of the mathematical roadblocks to progress were cleared away to permit accurate calculation of some electron and muon properties. But the theory as a whole still has some troubling features. It is unlikely to survive the next few decades in its present form.

† In *Endeavour*, April 1958. Reprinted by permission.

conservation laws that constrain the strong interactions alone, not the weak or electromagnetic interactions. (A strongly interacting particle may also interact weakly or electromagnetically. When it does, it freely violates these partial conservation laws.)

Strange Particles

The pion, discovered in 1947, was expected. In the same year began a parade which still continues of newly discovered particles, most of them unexpected. The first among the new particles made themselves known through some unexplained V-shaped tracks in the cloud chamber of George Rochester and C. C. Butler of the University of Manchester. A more recent picture of such a V track is shown in Figure 26.2. The track of an incoming kaon terminates in the chamber and, separated from it by a few centimeters, is a V, with its vertex pointing toward the end of the incoming kaon track (point A in the figure). Measurement of momentum of the visible tracks makes it possible to infer that a neutral particle was created at point A and that it decayed at point B into a pair of oppositely charged particles whose visible tracks gave the V.

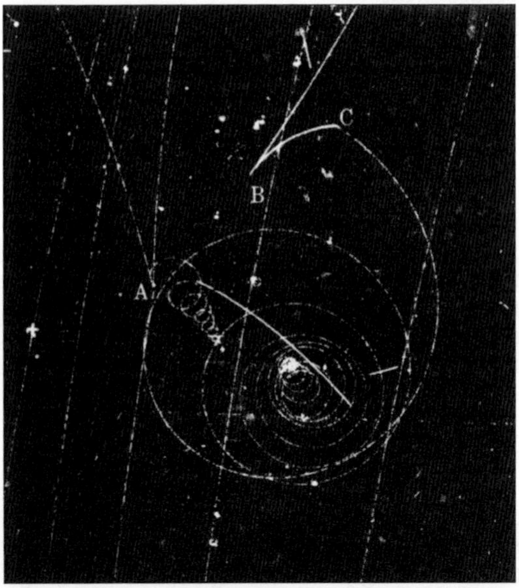

FIGURE 26.2 Example of V-shaped track characteristic of strange particle decay. A lambda particle created at point A decays at point B into a proton, a negative muon, and an unseen antineutrino. The muon in turn decays at point C. The lambda, one of the so-called strange particles, was born at point A in the reaction, $K^- + p \rightarrow \Lambda + \pi^- + \pi^+$. The kaon track enters the chamber from below; the two pions fly upward from point A. (Photograph courtesy of Lawrence Radiation Laboratory, University of California, Berkeley.)

Typically, magnetic fields are applied to bubble chambers and cloud chambers to cause moving charged particles to be deflected and leave curved tracks. In Figure 26.2, the negative particles moving upward bend to the right, the positive particles bend to the left. A high-momentum particle such as the incoming kaon is deviated only slightly from a straight course. A lower momentum particle such as the negative electron from the muon decay is more strongly deflected. By measuring the curvature of the tracks, the experimenter can determine the momentum of each charged particle. Because of the law of momentum conservation, he can then deduce the momentum of unseen neutral particles.

Immediately following the 1947 discovery of V particles at Manchester, similar tracks were noticed by other experimenters, and it was not long before the properties of these new particles were being pinned down by careful measurements in a number of laboratories. We now know that among the first V particles observed were neutral kaons and lambdas, which were decaying according to the schemes:

$$K^0 \rightarrow \pi^+ + \pi^-,$$

$$\Lambda^0 \rightarrow p + \pi^- .$$

Within a few years, their charged brothers had been identified and the sigma and xi particles had been added to the roster. (For the characteristic open-V track of a negative xi, look back to Figure 2.8.) The eta meson was discovered in 1961. By 1964 enough order had appeared in the catalogue of particles to make possible the prediction of the omega baryon before its discovery in a Brookhaven bubble chamber.

The study of the new particles had not proceeded very far before something peculiar about them appeared. In spite of the fact that they had gone unnoticed during many years of cloud-chamber studies, they were, in fact, not particularly rare. In very energetic nuclear collisions, the chance of creating one of the new particles was so great that there was no escape from the conclusion that the new particles must interact strongly, like the pions and nucleons. Particles which experience only electromagnetic and/or weak interactions could not be produced as frequently as the new particles. Expressed in terms of a time, one of the new particles could be produced after only 10^{-22} sec if enough energy were available. Yet, once produced, they lived a million million times longer than that, about 10^{-10} sec. That, said the physicists, is very strange. And the new particles became the "strange particles."

The long life of a pion or a muon is understandable because it decays into particles that interact only weakly. But, in the decay of a lambda, for example,

$$\Lambda^0 \rightarrow p + \pi^- ,$$

the products of the decay—proton and pion—are both strongly interacting particles, and so is the lambda itself, as proved by the ease with which it can be created. It was very hard to see what kept the lambda alive so long, why it did not take advantage of the strong interactions to convert itself immediately into proton and pion.

An answer to the puzzle of the lambda's long life was provided independently in 1953 by two young physicists, Murray Gell-Mann in America (then aged 23) and Kazuhiko Nishijima in Japan (then 26). What makes the electron live forever is the law of charge conservation. What makes the proton live forever is the law of baryon conservation. Suppose, said Gell-Mann and Nishijima, that there is another conservation law, a new one, that makes the lambda live "almost forever." Some new physical quantity, some new "thing," has to be held constant, or conserved. The tongue-in-cheek name given to the new conserved quantity is "strangeness."

According to Gell-Mann and Nishijima, each of the particles has to carry a "strangeness number," just as it carries an electric charge, or a muon-family number, or a baryon number. The pion and the nucleon are not strange; they have zero strangeness number. The lambda and the sigma are assigned strangeness number –1 (and their antiparticles +1). The kaon is given strangeness number +1, and xi, strangeness number –2 (their antiparticles being opposite in sign).

It all sounds rather fanciful, but it works. In the absence of any deeper knowledge, the strangeness assignments are as good a guide as we have to the nature of the strange particles. The conservation law is this: In every strong interaction process, the total strangeness is conserved.

In a pion-production process, such as

$$p + p \rightarrow p + n + \pi^+ ,$$

the new conservation law is satisfied, for the total strangeness number is zero before the collision and zero afterward. But what if a strange particle is produced? This can be managed only if at least two strange particles are produced together, with opposite signs for their strangeness numbers. A typical allowed process is

$$p + p \rightarrow p + \Lambda^0 + K^+ .$$

The strangeness of the two colliding protons is zero. The strangeness numbers –1 and +1 for the lambda and kaon cancel to conserve the total of zero. This phenomenon, christened "associated production," was proposed first by Abraham Pais shortly before the strangeness theory of Gell-Mann and Nishijima. There is now ample evidence that, in the collisions of pions and nucleons, strange particles are always produced two (or more) at once. This was not noticed first because one member of the pair frequently escaped the cloud chamber undetected, leaving only one to be seen. Figure 2.7 provides a good example of associated production, and another example is shown in Figure 26.3.

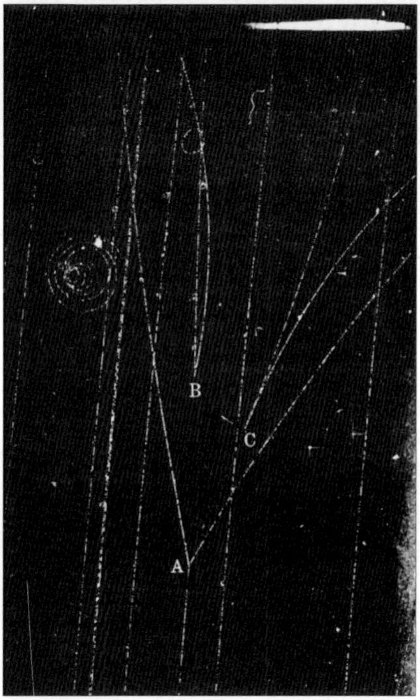

FIGURE 26.3 Associated production. A lambda particle and a kaon are created together at point A in the reaction, $p + p \rightarrow \Lambda^0 + K^0 + p + \pi^+$. These neutral particles decay with characteristic V tracks at points B and C. (Photograph courtesy of Brookhaven National Laboratory.)

If a strange particle strikes a nucleon, again the strangeness-conservation law dictates what may happen. A typical allowed process is

$$\Lambda^0 + p \rightarrow n + p + \overline{K^0}\,.$$

A lambda particle vanishes in collision with a proton, but to preserve the total strangeness number of –1, another strange particle must be formed, in this case an antikaon. In Figure 2.8, the reaction

$$\overline{K^+} + p \rightarrow K^+ + \Xi^-\,.$$

was illustrated. Test this for strangeness conservation (recalling that the antiparticle $\overline{K^+}$ has charge and strangeness opposite to those of the K^+).

As these examples show, the concept of strangeness constitutes more than a feeble attempt at humor. The power of this new conservation law, as of every conservation law, lies in what it forbids. There are a vast number of strong-interaction processes that are forbidden only by the conservation of strangeness; none of them has ever been detected. There can be no doubt that strangeness, whatever its deeper meaning might be, is an important attribute of the particles that strongly limits their possible transformations.

What does strangeness conservation have to say about strange-particle decay? As the simplest example, consider the kaon. It is the lightest strange particle (as the electron is the lightest charged particle, and the proton is the lightest baryon). If strangeness conservation were an absolute law, the kaon could not decay at all, and would join the ranks of the stable particles. But since strangeness conservation governs only the strong interactions, but not the weak, the kaon is inhibited only from undergoing the very speedy decay (after about 10^{-22} sec) which would characterize the strong interactions. Since the weak interactions violate strangeness conservation, they act in their own leisurely fashion and bring about the kaon decay after about 10^{-10} sec.

Strangeness conservation is just one of several new partial conservation laws (others are discussed in Chapter Twenty-Seven) that apply to the strong interactions but not to the weak. Why some conservation laws are absolute and others partial; why strong interactions are hemmed in by more conservation laws than are weak interactions, are questions to which no one knows the answers. They remain important challenges for the future, and it seems likely that, without answers to these, no deep understanding of the particles will be possible.

Resonances

Resonances* belong to the uninhibited generation of elementary particles. Our knowledge of resonances is still very fragmentary, but some things are known about them. It is known that there are quite a few, that some belong to the group of strange particles and some do not, that some are baryons and some are not, and that *all* are strongly interacting.

The main thing the resonances have taught us is that the thirty-five particles listed in Table 2.1 (not counting the graviton) are far from the end of the story. The particles in Table 2.1 are those that are stable and whose decay is mediated by weak or electromagnetic interactions, not by strong interactions. The resonances round out the picture. They are the particles that interact strongly and that are not prevented by any conservation law from following their natural inclination for very rapid decay. Hence, in the short time characteristic of strong interactions, they vanish and give way to lighter particles.

A typical resonance is born and dies all within a space considerably smaller than a single atom. That it ever existed can only be inferred by studies of the longer-lived products of its decay. Suppose, for example, that a proton and antiproton annihilate to create five pions,

$$\bar{p} + p \rightarrow \pi^+ + \pi^+ + \pi^- + \pi^- + \pi^0 .$$

This process is illustrated in Figure 26.4. The bubble-chamber photograph shows the track of the incoming antiproton and the tracks of the four charged pions, apparently all emerging from exactly the same point. Energy and momentum conservation require that one unseen neutral pion also flew from this point. Study of many such events shows that groups of these pions tend to come off in a certain relation to each other, which implies that they must be the decay products of a single particle. This correlation of the final pions shows that what actually happens (some of the time) is a two-stage process in which, first, an eta-one particle is created,

$$\bar{p} + p \rightarrow \eta_1{}^0 + \pi^+ + \pi^- ;$$

* A mechanical system is said to "resonate" when its natural frequency of vibration is stimulated by an external force acting at the same frequency. Quantum-mechanically, the idea of resonance has been extended to include a match between the externally available energy and the energy difference between two states of motion of the system. Since a resonant state is excited by adding energy to a system, it is always an unstable state that can dissipate its energy as the system returns to a lower-energy state of motion. It is because of this connection between resonance and instability that the very short-lived particles are called resonances.

and then the eta-one decays,

$$\eta_1{}^0 \longrightarrow \pi^+ + \pi^- + \pi^0 .$$

There is no doubt that the eta-one existed as an independent entity, even though it had no time to move a measurable distance away from the point of its creation.

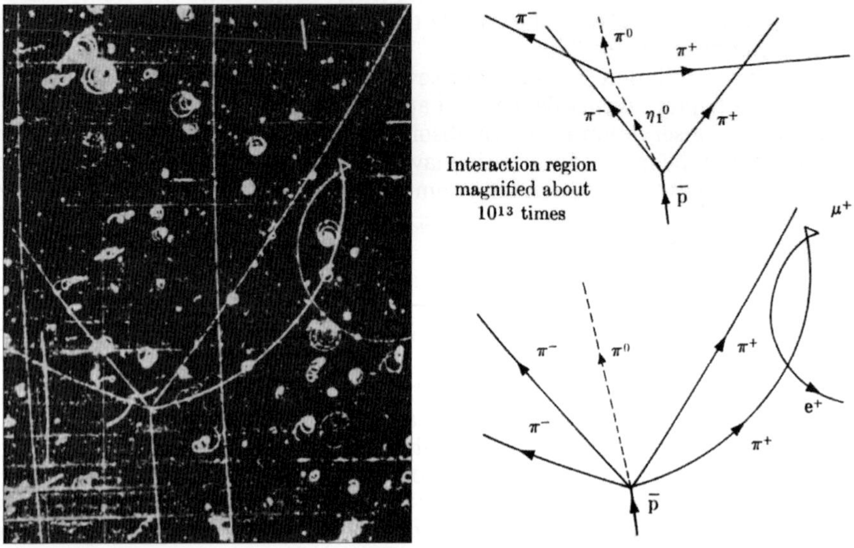

FIGURE 26.4 Production of an eta resonance. The antiproton entering from the bottom annihilates with a proton in the bubble chamber to form two pions and an eta-one particle. After a time of about 10^{-22} sec, the eta-one decays into three more pions. The transitory existence of the eta-one is inferred only from studies of the pion tracks. (Notice that one of the positive pions is also observed to undergo decay into a muon through the reaction, $\pi^+ \longrightarrow \mu^+ + \nu_\mu$, and the muon, in turn, decays into a positron according to $\mu^+ \longrightarrow e^+ + \nu_e + \overline{\nu_\mu}$. The neutrinos are of course unseen.) (Photograph courtesy of Lawrence Radiation Laboratory, University of California, Berkeley.)

Quantum mechanics plays an interesting direct role in the measurement of resonance lifetimes. Because of the uncertainty principle, the shorter the lifetime of a resonance, the easier it is to measure the lifetime. The energy-time form of the uncertainty principle is directly applicable:

$$\Delta E \Delta t = \hbar .$$

The more tightly is the particle squeezed in time (Δt), the greater must be the uncertainty of its energy (ΔE). This energy uncertainty in turn appears as a variability in the apparent mass of the particle. Consider, for example, a resonance whose lifetime is 10^{-22} sec. In the appropriate units for this example, $\hbar$ is equal to 6.58×10^{-22} MeV sec. The energy uncertainty for this short a lifetime is

$$\Delta E = \frac{6.58 \times 10^{-22} \text{ MeV sec}}{10^{-22} \text{ sec}} = 6.6 \text{ MeV} .$$

This is the energy equivalent of about 13 electron masses. No measurement can determine the mass of the resonance to an accuracy greater than this. In a series of many identical measurements, a distribution of values of mass would be obtained, spread over a range of about 13 electron masses. In this way resonances are identified and their lifetimes measured.

The resonances may be thought of as perched precariously on the upper rungs of a set of energy ladders. On the bottom rung of each ladder is a long-lived particle, stabilized or partially stabilized by a conservation law, or by the fact that it has no strong interactions, or by the fact that it is massless and is already as low as it can get. On the bottom rung of one ladder is the pion, the lightest strongly interacting particle. Above it are several resonances, which can tumble rapidly down to the level of the pion. At the foot of another ladder is the proton, the lightest baryon. Occupying rungs above it are a number of short-lived nucleon resonances. On the lowest rung of still another ladder is the lightest strange baryon, the lambda. Up the lambda ladder are various strange resonances. Counting different charges and antiparticles separately, approximately 200 resonances have already been identified.

The big lesson of the resonances, if it was not obvious before, is that there is not a muon problem, and a pion problem, and a kaon problem. There is just one problem, the particle problem. There are far too many particles and far too many intricate interconnections among them to imagine that any one or a few particles will be understood until they are all understood.

Particle Spectroscopy

The proliferation of particles, especially strongly interacting particles, has produced a situation reminiscent of spectroscopy in the nineteenth century. More and more spectral lines were discovered and classified. Finally Balmer, Rydberg, and others brought some of these lines into mathematical order. Through the idea of the photon, Einstein tied spectral frequency to energy. Bohr related the radiated energy quantum to the energy difference of stationary states of atomic motion. Finally, the thousands of known spectral lines were unified by atomic theory which accounts for all of them in terms of the transitions of a single fundamental particle, the electron.

Particle physicists today see themselves following a similar path. Now that so many different particles have been identified, the most intensive effort is being devoted to the search for order—for simple regularities in the masses and other properties of the particles. The particles of course do not correspond exactly to spectral lines. A better analogy is to a set of atoms whose properties are so markedly altered from one stationary state to another that every state of every atom is regarded as a separate particle.

The modern methods of seeking order among the particles are more sophisticated than Balmer's numerology, yet neither so simple nor so deep as Bohr's theory of the atom. Nevertheless, there has been real and encouraging progress. Through a new classification scheme, Gell-Mann and others have delineated family groups of eight and ten particles, a unification that goes beyond two and three particle groupings such as the neutron-proton group or the group of three pions. According to this scheme, for instance, the nucleon, lambda, sigma, and xi particles (eight altogether) are various faces of a single entity, an entity with different states of charge and different states of strangeness.

The difficulties of particle spectroscopy spring primarily from a single source—the large magnitude of energy differences. When a hydrogen atom is excited from its ground state to its first excited state of motion, its mass increases by about one part in 10^8, one millionth of one per cent. When a nucleon is excited to a resonant state, its mass increases by about ten per cent. In the particle world, every change is a drastic change. Yet, as particle data multiply, physicists hope that a combination of their own free imagination plus the hard evidence from more than two hundred particle states will lead to the unifying principle that is still missing.

26.4 Neutrinos

The neutrinos are unique and uniquely interesting. They are all weak. What makes them particularly interesting is what they lack—charge, mass, electromagnetic interactions, and strong interactions. What makes them vital in the structure of modern physics is what they have—energy, momentum, angular momentum, and either electron-family number or muon-family number. Without neutrinos, four fundamental conservation laws would be in trouble.

Because of its heroic role as the savior of conservation laws in nuclear beta decay, the electron's neutrino was for many years an article of faith among physicists, even though it escaped direct observation until 1956. Like the antiproton and the antineutron (which were observed in 1955 and 1956), the neutrino was long a ghostly member of the elementary-particle family, implied by theory but itself unseen.

A simple example that adequately illustrates the role of the neutrino is the beta decay of the free neutron, indicated by

$$n \rightarrow p + e^- + \overline{\nu}_e\,.$$

A proton, an electron, and an antineutrino are created. Figure 26.5 shows a possible result of the decay of a neutron. As a reminder that the neutrino is not actually seen, the "after" diagram is repeated with the path and spin of the neutrino deleted.

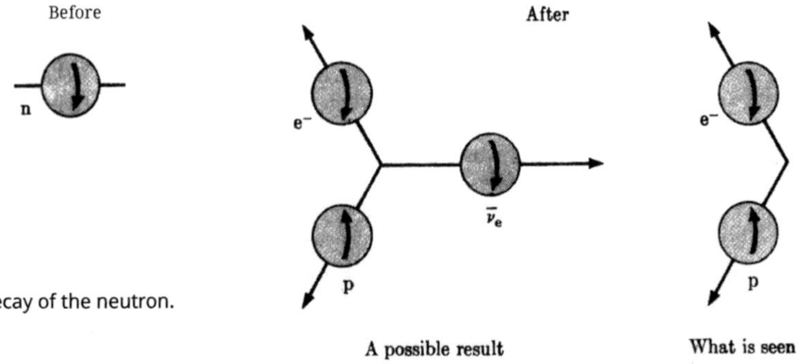

FIGURE 26.5 Beta decay of the neutron.

Considering the observable result of proton and electron only, we can see how, without the neutrino, four fundamental conservation laws would be violated. First, the energies do not balance. Electron energy plus proton energy add up to less than the initial neutron energy (its rest mass). The neutrino was invented to fill this gap. Second, as Figure 26.5 shows, momentum appears not to be conserved. Since the neutron is initially at rest, it has no momentum. To preserve zero momentum, the electron and proton would have to fly off back to back, with equal and opposite momenta canceling to zero. In fact, they do not. But the neutrino, which has the right energy to save the conservation of energy, also has some momentum, and this turns out to be exactly the momentum needed to combine with the momenta of electron and proton to make all three together add up to zero. At no extra charge (with no additional assumptions) the neutrino has saved a second and equally fundamental conservation law. Third, without the neutrino, angular momentum would not be conserved. In Figure 26.5, electron and proton spins are canceling, but the initial neutron had a nonzero spin. If the neutrino is supposed to have one-half unit of spin itself, then it can add to its chores the saving of angular-momentum conservation. The neutrino spin has never been measured directly, but there is ample indirect evidence from successful predictions of beta-decay phenomena that it does indeed have one-half unit of spin.

Finally, the electron's neutrino saves the law of electron-family conservation, for the antineutrino accompanying the electron gives a total electron-family number of zero, the same after the decay as before. Actually, the law of electron-family conservation was suggested only after the neutrino had come to be accepted. Since the underlying symmetry principle upon which this law is based is unknown, we could not claim at the present stage of history to be upset if the law were violated. Nevertheless, the empirical evidence supports this conservation law, and it is a good guess that there exists an undiscovered symmetry principle upon which it is founded. If this is so, then we should be as disturbed ex post facto by a breakdown of the law of electron-family conservation as we now would be by any violation of the laws of energy, momentum, and angular-momentum conservation, which are based upon the uniformity of space and time.

Details of the Fermi theory of beta decay are out of place here, but it is important to be aware that the neutrino in this theory does much more than preserve four conservation laws. It makes possible specific predictions about the fraction of electrons that come out in each direction and with each energy, and predictions about the relative rates of beta decay of different radioactive nuclei, all of which have been verified by experiment. (Actually there were some ambiguities in the Fermi theory that were resolved only in 1957.) In addition, the Fermi theory has been extended with perfect success from the decay of the neutron to the decay of the muon, given by

$$\mu^- \rightarrow \nu_\mu + e^- + \overline{\nu_e}\,.$$

This is also a beta-decay process, for in it an electron is created along with its antineutrino.

The process in nature whereby a stable particle comes into existence is usually quite similar to the process whereby it is absorbed, or annihilated. Thus, a photon is born by the oscillation of electric charge and, upon absorption, sets charge into motion. An electron may be born simultaneously with its antiparticle, the positron; it can die if it meets a positron with which to annihilate. The neutrino comes into existence in processes of beta decay or other decay processes brought about by the weak interactions. It can be absorbed and observed only via further weak interactions. In this lies the key to understanding the difficulty of observing the neutrino. It is observable only when it interacts in some way with matter, but the chance that it interacts is very small indeed.

It will be sufficient here to consider the particular process that was first used to catch the elusive neutrino (actually the antineutrino). If an antineutrino encounters a proton, it may, with a certain very small probability, be annihilated in what is known as the inverse beta-decay process,

$$\overline{\nu_e} + p \rightarrow n + e^+\,.$$

Proton and antineutrino disappear, and a neutron and a positron are created. The capture process involves the same particles as the creation process—except that the laws of charge conservation and electron-family conservation require the antielectron, or positron, to appear in the capture process, whereas the electron appeared in the creation process.

Can we estimate the chance of capturing an antineutrino in this way? It is easy to do and 167
the result is rather startling, providing a clear demonstration that a few minutes is a very long time in the elementary-particle world. We start with the fact that a free neutron, on the average, lives about 17 min, or 1,000 sec, before it spontaneously undergoes beta decay, turning into a proton plus electron and antineutrino. This means that if another antineutrino of comparable energy is in contact with a proton for about 17 min, it will be captured and it will stimulate the process of inverse beta decay. Since an antineutrino travels always at the speed of light, it will not stick close to any proton for 17 min, not even for a small fraction of a second; but as it travels through matter, it will be exposed to many protons, one after the other, each for an instant, and the inverse beta decay will occur on the average after a total exposure of about 17 min. (The precise time depends on the antineutrino energy, and on exactly what nuclei the antineutrino is brushing by on its flight through matter, but the figure 17 min will do for our rough estimate.)

In 17 min an antineutrino moves 3×10^{13} cm, about twice the distance from earth to sun. Now imagine a solid wall 3×10^{13} cm thick. Picture the neutrino as it flies through this wall, a tiny fuzzy ball whose size is about equal to its own wavelength, which we can take to be 4×10^{-11} cm, about one five-hundredth the diameter of an atom. The wave nature of the antineutrino is all-important. If it were truly a point particle, it would have practically no chance at all of striking the nucleus as it passed through an atom. But the wave nature of particles enables the antineutrino to reach out, to make its presence felt out to a distance of its wavelength. This wavelength, however, is short, and the effective size of the antineutrino is still tiny compared with the size of an atom. If we could take a series of snapshots of our fuzzy, extended antineutrino as it passed through matter, we should almost always see it occupying a space between nuclei, not actually touching a nucleus. Only one in many snapshots would show it in contact with a nucleus. So, after spending 17 min traveling through the 200-million-mile solid wall, the antineutrino has been exposed to protons within nuclei a great deal less than 17 min, for most of the time it was not touching a nucleus at all. In fact it spent only about one hundred-millionth of that time in contact with nuclei, so that the thickness of the wall must be extended one hundred million times before the required 17-min total contact between antineutrinos and protons is achieved. The antineutrino would have to penetrate into a solid wall an average distance of 2×10^{16} miles, or 3,500 light-years, before it would be absorbed and induce the inverse beta-decay reaction! This distance, two hundred million times greater than the earth-sun separation, is about one-tenth the size of our galaxy.

It is instructive to carry out a similar calculation for a photon of visible light. To begin with, an atom requires only about 10^{-8} sec to emit such a photon. This photon needs to spend only about 10^{-8} sec in contact with an atom in order to be absorbed. In 10^{-8} sec, a photon proceeds about 300 cm (10 feet). However, our photon is so large that, at any given instant, it encompasses very many atoms. The wavelength of a typical photon of light is about one thousand times greater than the size of an atom. The photon can therefore be pictured as a spread-out fuzzy ball which can encompass about one billion atoms at once ($1{,}000 \times 1{,}000 \times 1{,}000$). As it proceeds through matter—quite unlike the tiny antineutrino—its exposure to the absorbing atoms is enormous. Instead of having to proceed the full ten feet to be absorbed, our photon has to go only one billionth so far, because at each instant a billion different atoms have the opportunity to absorb it. Its penetration distance is, therefore, only about 3×10^{-7} cm, or fifteen layers of atoms, quite a difference from the three thousand light-years the antineutrino travels. Typically, when we look at a solid object, we see only its outside surface because photons of visible light cannot penetrate any significant distance into it.* There are a few exceptions, such as glass and lucite, into which visible photons can penetrate many feet, but we shall not go into the special reasons for this. Even for the exceptions, it is only photons of certain wavelengths that can get very far. Ordinary glass, for instance, stops ultraviolet photons in a very short distance (thereby providing protection against sunburn).

Note in the examples above that two different factors accounted for the vast difference in how far neutrinos and photons could penetrate into solid matter. First was the difference in wavelength of the two particles. Second, and more fundamental, was the difference in exposure times required to bring about absorption. The photon needed an exposure time to atoms of 10^{-8} sec; the neutrino needed an exposure time to nuclei of 10^{3} sec, one hundred billion times longer. The difference in these exposure times is a reflection of the fundamental difference in interaction strengths. The neutrino's absorption is brought about by the weak interactions, the photon's absorption by the electromagnetic interactions.

* Photons of short wavelength usually penetrate matter more deeply than photons of long wavelength because the short-wavelength photons encompass fewer atoms at each instant. Penetrating X-rays have wavelengths considerably shorter than those of visible-light photons.

How was the antineutrino ever detected, if it can penetrate through the whole earth and through any and all experimental apparatus as if nothing were there, leaving no trace? The laws of probability that are part of quantum mechanics come to the rescue of the physicist trying to stop the antineutrino. It is true that, on the average, this elusive particle can penetrate three thousand light-years of solid matter. But some antineutrinos travel even farther than this average before being absorbed, and others are stopped sooner, a few—a very few—even within a distance of a few feet or a few centimeters. It is for this tiny fraction that "by chance" die a premature death, being absorbed in a distance much less than all the rest of their brothers, that the physicists trying to confirm the existence of the antineutrino had to look.

The experiment that definitively demonstrated the existence of the antineutrino was 168
carried out at Savannah River, South Carolina, in 1956, by Clyde Cowan and Frederick Reines, two physicists from the Los Alamos Scientific Laboratory in New Mexico. What Savannah River had to offer them, that Los Alamos did not, was a powerful nuclear reactor, a rich source of antineutrinos. The interior of a reactor is highly radioactive, second only to a nuclear explosion in the intensity of its beta radioactivity. This comes about because of the neutron excess in fission fragments. In one among many possible modes of fission, a nucleus of U^{236} might split into nuclei of molybdenum and tin, plus several extra neutrons:

$$ {}_{92}U^{236} \rightarrow {}_{42}Mo^{104} + {}_{50}Sn^{128} + 4n \, . $$

Both of these nuclei are overly neutron-rich, the heaviest stable isotope of molybdenum being Mo^{100}, and of tin, Sn^{124}. Accordingly both are radioactive. In fact, each undergoes a chain of two beta decays (to ${}_{44}Ru^{104}$ and ${}_{52}Te^{128}$), so that, in this example, a single fission event results finally in four antineutrinos.

The rate of production of antineutrinos in the Savannah River reactor could be calculated* as could the chance that any one of them would be stopped within the experimental apparatus. The infinitesimally small chance of catching one antineutrino multiplied by the enormous number of antineutrinos available gave the number that should be caught in the apparatus. Reines and Cowan reckoned that they could stop one antineutrino every twenty minutes—not many, but enough.†

The heart of the experimental arrangement was a sandwich of tanks, ordinary water in the three-inch-thick "meat" layer, and a special liquid called a scintillator in the two-foot-thick "bread" layers. In order to increase the chances of catching antineutrinos, the experimenters built up a "club sandwich" (Figure 26.6) taller than a man and nearly as broad and as deep as it was tall. Surrounding the scintillator tanks were over a hundred photoelectric cells, or "electric eyes," looking at the darkened interior of the tanks. These were connected, in turn, to an array of electronic circuitry whose function was to analyze and report to the experimenters what the photo cells saw.

Whenever an energetic charged particle moves through the liquid, losing energy as it goes, the liquid gives off a weak pulse of light to signal the passage of the charged particle; that is, the liquid scintillates. Scintillators, both liquid and solid, are frequently used to detect elementary particles. A scintillator can also indirectly react to the passage of a neutral particle through it. A photon, for example, may be absorbed by an electron, and give to the electron enough energy so that it causes a scintillation. A neutron may be absorbed by a nucleus, stimulating the nucleus to emit radiation, which in turn gives energy to electrons which then provoke the scintillation pulse. In a similarly indirect way, the capture of an antineutrino may be recorded by the scintillating liquid.

* The actual production rate of antineutrinos in the Savannah River reactor has never been officially revealed because it is directly related to the total reactor power, which is a secret. However, it is certainly a very large number, probably at least 10^{18} antineutrinos per second.

† In a more recent experiment two miles underground in a gold mine in South Africa, Reines and his collaborators detected neutrinos produced by cosmic rays at the rate of one per month.

The trick is to separate and definitely identify the pulses of energy resulting from the antineutrino capture amid energy pulses arising from a variety of other elementary events continually going on in the tanks. Besides the antineutrinos, other particles—neutrons, gamma rays, and some charged particles, both from the reactor and from the normal cosmic radiation—are entering the tank in a steady stream, and are being stopped there. The first, and easiest, thing to do is to surround the tank with a thick wall of shielding material—earth and lead blocks. This keeps out most of the unwanted particles, but offers no impediment at all to the antineutrinos. The second and more difficult thing to do is to discriminate electronically between antineutrino-capture events and extraneous events known as background. It is this task that required the special ingenuity of the experimenters.

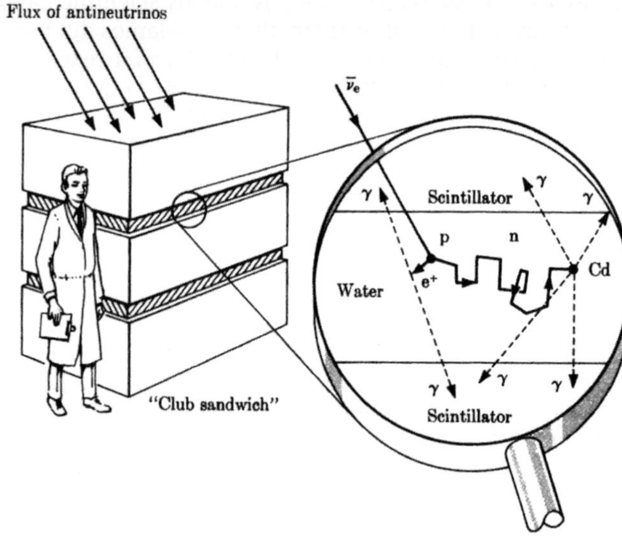

FIGURE 26.6 The antineutrino-detection experiment of Reines arid Cowan. The "meat" in the club sandwich is a three-inch-thick layer of water (rich in protons) seasoned with cadmium chloride, because cadmium nuclei capture neutrons readily. The "bread" is a two-foot-thick tank of liquid scintillator viewed by photoelectric tubes. The chain of events set off by an antineutrino capture is described in the text.

Let us consider now exactly what happens on the submicroscopic scale when an antineutrino is captured. Each molecule of water in the "meat" layers contains two protons (the hydrogen nuclei). It is the job of these protons to capture the antineutrinos through the reaction

$$\overline{\nu_\mu} + \mathrm{p} \rightarrow \mathrm{n} + \mathrm{e}^+ .$$

Figure 26.6 shows a typical sequence of events following such a capture event. Where there was a proton sitting quietly at the center of a hydrogen atom, there is now suddenly a neutron and a positron, each with some kinetic energy. These two new particles separate and fly off in different directions, but then each gradually loses energy in the water and is slowed down. The positron, being charged, experiences an electric interaction with all of the electrons in neighboring atoms and is very rapidly brought to rest (in about 10^{-9} sec, after covering less than one centimeter). Almost immediately, it annihilates with one of the atomic electrons as their mass is converted into energy according to the reaction

$$\mathrm{e}^+ + \mathrm{e}^- \rightarrow \gamma + \gamma .$$

Each of the two photons acquires an energy equivalent to the mass of one electron. Since they fly apart in opposite directions, the photons will usually enter two neighboring scintillation tanks, where they will produce characteristic pulses of light seen by the photo cells.

The events initiated by the positron all occur in a time much less than a millionth of a second. Meanwhile, the uncharged neutron is making its way through the water in more leisurely fashion, caroming off nuclei for several millionths of a second before being brought to rest (actually to a low speed, not exactly zero). Wanting to make sure that the neutron, once slowed down, is promptly captured, the experimenters added to the water some cadmium chloride, since cadmium nuclei absorb slow neutrons with particular enthusiasm. Following the neutron capture, the cadmium nucleus emits one or more gamma rays that fly into the scintillation tank to signal the capture event.

Altogether, then, three pulses of energy result from the capture of an antineutrino by a proton. The first two are simultaneous pulses in each of two neighboring, scintillators as a pair of photons signal the annihilation of a positron. The third is a pulse of one or more photons occurring several millionths of a second later signaling the capture of a neutron. Moreover, each of these three pulses has a characteristic identifying energy which the photo cells can measure—0.5 MeV for each of the annihilation photons, and about 9 MeV total for the neutron-capture photons. In 1956, five years after the first efforts to trap the antineutrino began, Reines and Cowan announced the definite observation of the electron's antineutrino through the identification of this characteristic sequence of scintillation pulses about three times per hour while the reactor was in operation.

The observation of the muon's neutrino in 1962 by a group from Columbia University required the use of a wholly different set of experimental tools. The basic capture reactions,

$$\overline{\nu_\mu} + \mathrm{p} \rightarrow \mathrm{n} + \mu^+ \qquad \text{and} \qquad \nu_\mu + \mathrm{n} \rightarrow \mathrm{p} + \mu^-,$$

can take place only if the neutrinos (and antineutrinos) have sufficient energy—over 100 MeV—to create the mass of a muon. The chance that the capture will occur increases as the energy of the neutrinos increases, so that a prerequisite for success in this experiment was high energy, which dictated the use of the largest available accelerator. The successful experiment was carried out at the Alternating Gradient Synchrotron at Brookhaven (Figure 2.9). Actually, for a reason to be mentioned below, the machine was operated at only 15 GeV for this experiment, about half of its maximum energy.

Via a short chain of intermediate events, the Brookhaven AGS readily provides a copious supply of high-energy neutrinos (see Figure 26.7). Energetic protons within the machine collide with nuclei in a target, producing a spray of secondary particles. Numerous among these are pions. The charged pions, as they fly from the accelerator, undergo spontaneous decay to provide the needed neutrinos, according to the transformations

$$\pi^+ \rightarrow \mu^+ + \nu_\mu \qquad \text{and} \qquad \pi^- \rightarrow \mu^- + \overline{\nu_\mu}.$$

Because the secondary particles must carry away from the collision the same large momentum that the high-energy protons brought into it, these neutrinos and antineutrinos, along with all of the other debris from the collision, fly forward in a narrow cone rather than fanning out in all directions.

In order to get rid of unwanted particles and study the neutrinos alone, the experimenters erected, in the way of this heterogeneous beam, a solid wall of iron 44 feet thick. Behind the wall stood a spark chamber (Figure 26.7). At the chosen machine energy of 15 GeV, the iron battlement was sufficient to stop all of the secondary particles except neutrinos, but to neutrinos it offered no impediment whatever. (At a higher operating energy, 44 feet of iron would not have been quite sufficient to shield the spark chamber from all other particles.)

Coursing through the spark chamber were neutrinos and antineutrinos in large numbers, each having a chance of less than one in a million million to be caught. If the capture reaction indicated by

$$\overline{\nu_\mu} + \mathrm{p} \rightarrow \mathrm{n} + \mu^+$$

occurs, then, once in a while, a positive muon will be created within the chamber and leave a track. But if the muon's neutrino were the same as the electron's neutrino, the creation of positrons should also be possible, just as in the Savannah River experiment; in fact, positron creation should occur equally often. Similarly, negative muons and electrons would be created in equal numbers. In about 300 hours of operation, the Columbia-Brookhaven group observed 29 significant muon tracks in their chamber, and no electron tracks. They thereby positively identified the muon's neutrino and, at the same time, showed it to be distinct from the electron's neutrino.

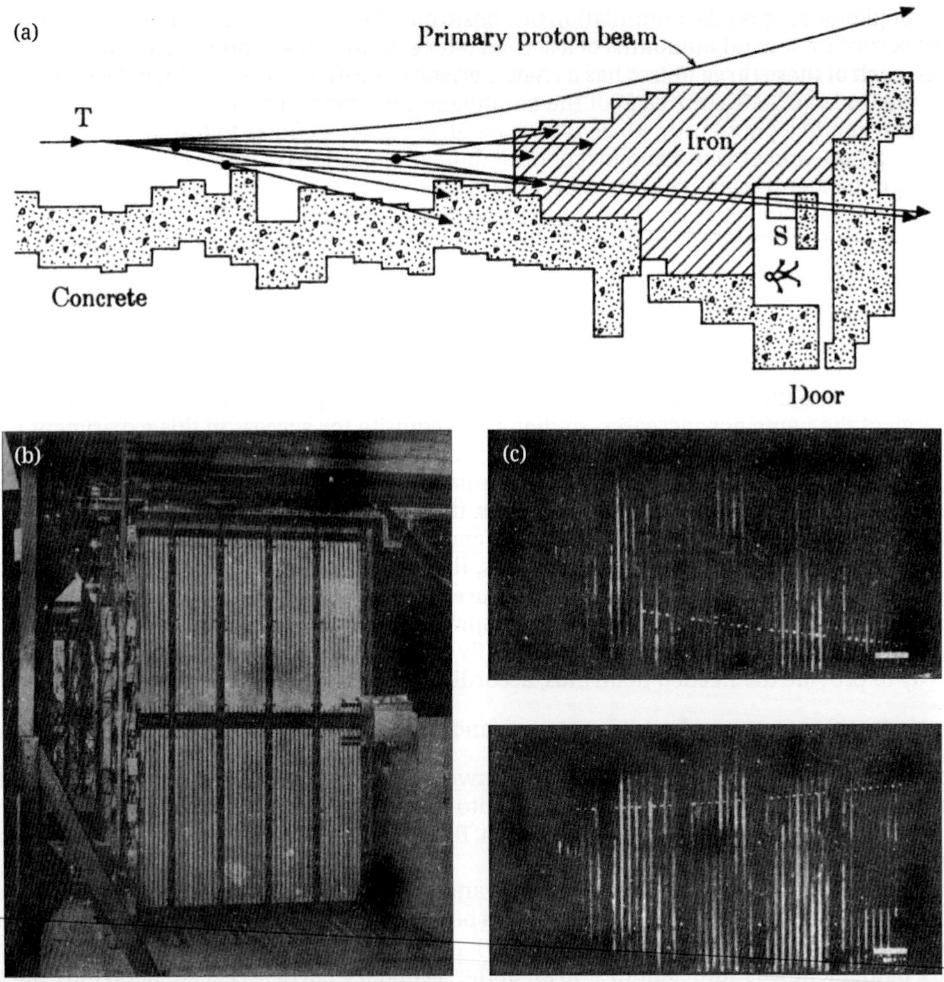

FIGURE 26.7 Discovery of the second neutrino. **(a)** The circulating proton beam in the Brookhaven AGS strikes a target T; secondary particles of all kinds spray to the right. Lines penetrating 44 feet of iron and the spark chamber S represent neutrinos from pion decay. The iron or concrete stops almost all other particles (shorter lines). A reclining experimenter in the spark-chamber room shows the scale. (Magnets guiding the primary proton beam in its circle are not shown.) **(b)** The 10-ton spark chamber containing 90 parallel aluminum plates, each about 4 feet square. Sparks jump between plates to signal the passage of a charged particle. **(c)** Two photographs of spark tracks left by muons created when neutrinos were captured in the chamber. To record 29 significant muon events, about 3×10^{17} protons were accelerated in the AGS and 10^{14} neutrinos passed through the chamber. [(b) Photograph courtesy of Brookhaven National Laboratory; (c) photograph courtesy of Nevis Laboratories, Columbia University.]

If the antineutrino had not been detected in 1956, physicists the world over would probably still be in a state of shock. We could not easily have parted with the conservation laws of energy, momentum, and angular momentum all at one blow. But even as it was giving reassurance to scientists that these sacred laws were not being violated, the neutrino was figuring prominently in the destruction of another law—the law of parity conservation (the meaning of parity is discussed in Chapter Twenty-Seven). Here we wish to mention only an unexpected new property of the neutrino that emerged from the study of parity. The neutrino is left-handed.

Because of the quantum rule governing the orientation of angular momentum (page 673), the same rule that contributes in a vitally important way to the explanation of atomic shell structure, a particle such as a neutrino with one-half unit of spin may have that spin directed in only two possible ways, parallel or antiparallel to a chosen axis of quantization. If this axis is taken to be the line of flight of the particle,* the spin angular momentum may be directed along the flight direction, or opposite to the flight direction. When the spin vector is parallel to the momentum vector, the motion is called right-handed, because then the spin motion and direction of flight are combined in the same way as in the advance of a "normal" or "right-handed" screw or bolt (Figure 26.8).

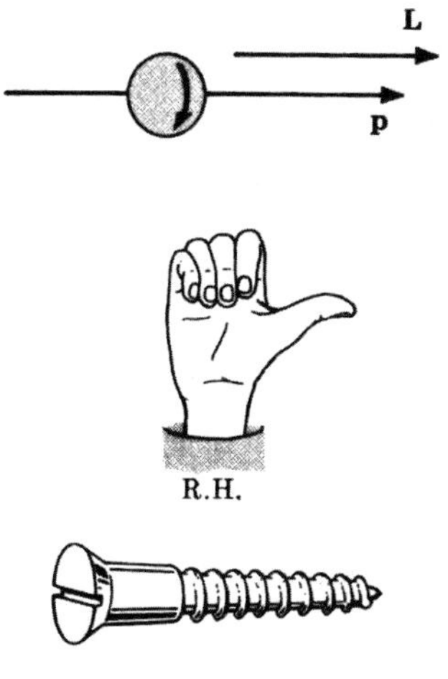

FIGURE 26.8 Right-handed motion of a spinning particle. Its spin angular-momentum vector and its momentum vector are parallel.

An electron or a proton or a neutron—particles with one-half unit of spin—can move in either a right-handed sense or a left-handed sense, that is, with their spin vectors along the flight path or opposite to the flight path. The remarkable fact discovered about the neutrino in 1957 is that it is always left-handed (and the antineutrino is always right-handed).

This left-handed character has been verified for both the electron's neutrino and the muon's neutrino. Here we mention only why the muon's neutrino is known to be left-handed. A positive pion disintegrates into a positive muon and a neutrino, which fly apart energetically (Figure 26.9). Because of the conservation laws, all of the neutrino's properties can be inferred from observation of the muon alone. What is found is that all of the muons

* The reason that spin orientation is quantized regardless of the chosen direction of the axis of quantization is a subtle and difficult point that is closely tied to the fundamental role of probability in quantum mechanics. If the spin of a particle is *definitely* oriented along a certain direction, it has at the same time a certain *probability* to be oriented along a different direction. If it is definitely oriented in the second direction, it still has some probability to be oriented in the first direction. Definiteness follows only from measurement. Whatever the chosen direction for measurement, a particle of spin one-half will always be found to be spinning parallel or antiparallel to that direction.

emerge with a left-handed motion, spin directed backward along the path of flight, as in Figure 26.9. Momentum conservation implies that the neutrino flight path must be just opposite to the muon flight path. And angular-momentum conservation implies that the spin directions of muon and neutrino must also be opposite. Therefore, the neutrino must have come out with left-handed motion also. Finally, the argument is turned around to say that the muon comes out left-handed in this decay because the neutrino is left-handed.

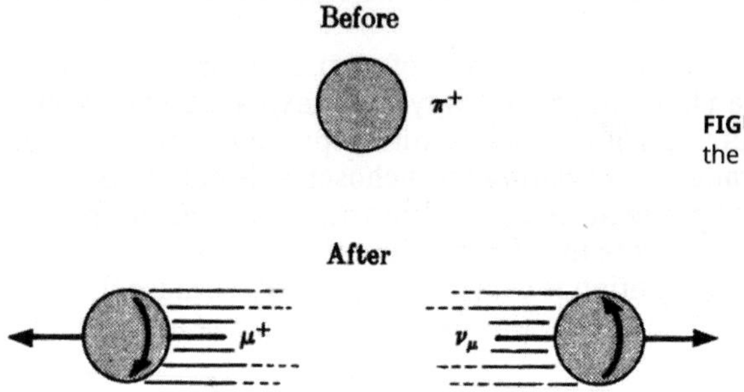

FIGURE 26.9 Creation of a left-handed neutrino in the decay of a positive pion.

There is a subtlety here associated with which of the emerging particles is "really" left-handed. Suppose that we board a rocket ship and overtake the muon on its flight away from the point where the pion decayed. As we come abreast of the muon, we will see its "north pole" pointing toward the rear of our rocket ship, its "south pole" toward the front, just as in Figure 26.9. But if we look out of the window of the rocket ship at the muon as we pass it by, it will seem to be moving to our rear, that is, with its "north pole" leading. Relative to our rocket-ship vantage point, the muon is now right-handed. But we cannot perform the same trick on the neutrino. Because it is moving at the speed of light we cannot overtake it, and no matter how we view it, it will always appear left-handed. It is, therefore, legitimate to speak of the neutrino as a left-handed particle. The muon, on the other hand, may be in a left-handed state of motion or a right-handed state of motion, and is not intrinsically one thing or the other.

We end this section on neutrinos with a discussion of neutrinos and the cosmos—a fascinating subject because the universe is so copiously supplied with neutrinos, yet a frustrating subject because the chance of making any useful observations of these cosmic neutrinos is so slim. What do we know about the comings and goings of neutrinos on the cosmic scale? The interesting thing is that they seem to be coming, but not going. Every star is producing and supplying to the universe a steady stream of neutrinos. But so weak is the interaction of these neutrinos with the rest of the matter in the universe, that practically none of them is ever reabsorbed. Nearly every neutrino ever born in the ten billion years that we call the age of the universe is still alive.

The fact that we can see galaxies billions of light years away shows that photons have a good chance to travel through the universe unstopped. This means, of course, that neutrinos have an even much better chance. In fact, a rough estimate is that, in traveling across the known part of the universe for about ten billion years, a neutrino has only about one chance in 10^{25} of being captured! Our universe is quite tenuous; tiny specks of matter are scattered through the vastness of space. It falls very far short indeed of providing the solid wall three thousand light-years thick that would be required to stop a low-energy neutrino. Over the succeeding billions of years, stars will continue to pour a substantial part of their energy into swarms of neutrinos coursing about the universe in private, as it were, having no further effect on the universe. It seems exceedingly unlikely that man will be able to tap cosmic neutrinos for their energy or even for much information—for they do, in principle, carry interesting information, being the only direct messengers from the interiors of stars.

Just as the earth receives most of its light, photons, from the sun, so it receives most of its neutrinos from our own sun. For every million neutrinos striking the atmosphere, one comes from outer space and the other 999,999 come from the sun. On each square centimeter of the earth in each second, the sun pours about 4×10^{10} neutrinos and about ten million times as many photons. This means that, at any instant, each cubic centimeter of volume in and near the earth contains about one neutrino. One physicist is said to have given a friend as a gift an empty matchbox labeled, "Guaranteed to contain 100 neutrinos." To put the matter on a more human scale, we can say that each of us, in the time it takes to blink an eye, is struck by somewhat over 10^{12} neutrinos.

26.5 *The graviton*

The graviton as a particle has not played much of a role in physics. The reason for this is not so much that it has never been observed—for the neutrino was an important part of the elementary-particle world well before its direct observation—as that it is not known to have any connection with the other particles or with their transformations. The graviton stands alone, being responsible for the transmission of gravitational forces among macroscopic and astronomical bodies, but having nothing to do—so far as we now know—with the elementary events of creation and annihilation occurring in the submicroscopic world. Another way to say this is that the quantum aspects of gravitation have not yet made themselves felt in any way.

It is instructive to compare gravity with electricity. We know various "classical" macroscopic aspects of electricity—currents in wires, bolts of lightning, radio and television waves. These are aspects of electricity in which the quantum aspect, or photon aspect, is entirely unimportant. Yet the quantum or photon aspect of electricity becomes of dominant importance in the atomic or nuclear world, where individual photons have vastly greater energies than, for example, individual radio photons. Moreover, the electrical force is what holds atoms together and is what makes uranium fissionable. Electricity spans the whole spectrum from single elementary-particle events to the large-scale macroscopic world of our senses, from the dominance of discrete-particle aspects to the dominance of continuous-wave aspects.

For gravity, we know only the large-scale end of this spectrum. The reason for this is quite simple. The gravitational force is extraordinarily weak, weaker by far than the "weak" interaction experienced by the neutrino. The gravitational force between an electron and a proton is smaller than the electrical force by the enormous factor of about 10^{39}. Small wonder that it is the electrical force that holds these particles together to form the hydrogen atom, and that when the hydrogen atom changes its energy, it does so by emitting or absorbing photons, not by emitting or absorbing gravitons. In the everyday world, the simple act of picking up a nail with a small magnet illustrates this factor of difference. The magnet pits its upward pulling power against the downward gravitational force of the entire earth, and easily wins the contest.

Yet, in the solar system and the galaxy, gravity, the weakest of all forces, becomes the number one pulling agent. It outstrips the strong interactions and the weak interactions because these are of short range. Although basically stronger than gravity, their effect is negligible beyond about 10^{-13} cm, and they are of no consequence in holding the solar system together. Electricity, like gravity, is a long-range force. But the sun and planets are (almost precisely) electrically neutral, made up of equal numbers of positive and negative particles. The electrical force is, therefore, canceled out and gravity is left to dominate the scene. The observed gravitational forces involve such an incredibly great number of gravitons that it is quite hopeless, in the foreseeable future, to imagine detecting a single graviton whose effect, in isolation, is even weaker than the effect of a neutrino.

A question frequently asked among physicists, but one whose answer no one knows, is this: Will the graviton always remain a particle apart, unconnected with elementary events in the submicroscopic world, or will it be found to "belong," contributing in some way to the deeper structure of the particles? The idealist answers that nature is in the habit of tying its separate parts together into a related whole, that it would be surprising if the graviton were not significant in the world of the very small, and that moreover, the graviton may be just what is needed to help overcome the mathematical roadblocks to progress that have appeared when theorists try to describe the particles without including the graviton in the family. The pragmatist answers that there is at present no evidence whatever that the graviton has anything to do with the transformations or the structure of the other particles, and that it is idle to speculate about that possibility now.

EXERCISES

26.1. Which among the four classes of interactions come into play when you grasp this book and lift it from a desk?

26.2. So far as we know, a conservation law is not subject to laws of probability. It either governs a particular process at all times, or it never does so. Nevertheless we speak of "partial conservation laws." Explain exactly what is meant by a partial conservation law, and contrast it with an absolute conservation law.

26.3. Which of the four classes of interaction influence each of the following particles: **(a)** neutron; **(b)** pion; **(c)** neutrino; **(d)** electron? Call attention to the correlation between number of interactions and mass of particle.

26.4. Instead of an electron, either a negative muon or a negative pion can join a proton to form a hydrogen-like atom. Discuss the size, the energy, the characteristic spectra, and the eventual fate of the muon-atom and the pion-atom. Include some illustrative numerical calculations.

26.5. If a lead nucleus ($Z = 82$) were a point charge, the radius of the lowest Bohr orbit of a muon about the lead nucleus would be approximately equal to $a/17{,}000$, where a is the electron Bohr radius in hydrogen. **(1)** Why is the factor 17,000? **(2)** Evaluate this distance numerically. **(3)** Calculate the radius of the lead nucleus (see Section 25.1) and compare it with this muon Bohr radius. **(4)** The finite size of the nucleus has important effects on the muon. Suggest at least one such effect.

26.6. What is the direction of the magnetic field applied to the bubble chamber pictured in Figure 26.2 ?

26.7. Write two reaction processes among strongly interacting particles that would violate strangeness conservation if they occurred, but are consistent with all of the absolute conservation laws.

26.8. The mass of a resonance is not well defined. Measurements of the mass of the Ξ_1 resonance, for instance, yield values scattered over a range $\Delta m = 15m_e$ about an average value $m_{av} = 2{,}990m_e$, where m_e is the mass of an electron. **(1)** What is the approximate mean life of the Ξ_1? **(2)** At top speed, how far does the Ξ_1 move on the average before it decays ?

26.9. **(1)** The omega particle has strangeness –3. A typical mode of its decay is $\Omega^- \rightarrow \Xi^0 + \pi^-$. Explain why this is a "slow" decay process (mean life about 10^{-10} sec). **(2)** The Ξ_1 resonance referred to in Exercise 26.8 can decay into the same pair of product particles as the Ω^-, but it does so with a very much shorter lifetime than the lifetime of the Ω^-. What is the strangeness of the Ξ_1?

26.10. Cosmic ray protons bombard the earth with about equal intensity from all directions. Most of these create pions near the top of the atmosphere. Many of the pions decay in the atmosphere, creating neutrinos. Explain why, at the surface of the earth, neutrinos of extremely high energy are more likely to be traveling horizontally than vertically, (HINT: Time dilation is relevant.)

26.11. In an inverse beta decay process which experimenters hope to detect, chlorine 37 is transformed to argon 37. Complete the following reaction formula, indicating whether a neutrino or an antineutrino should be added to the left side and indicating what particle should be added to the right side:

$$+ \; {}_{17}\mathrm{Cl}^{37} \rightarrow {}_{18}\mathrm{Ar}^{37} \; +$$

What conservation laws play a role?

26.12. If hydrogen is the thermonuclear fuel in stars, explain why stellar fusion reactions produce neutrinos, not antineutrinos, (HINT: One or more conservation laws is at work.)

26.13. **(1)** Imagine creatures on another planet who see neutrinos of a few MeV instead of photons of a few eV. Describe their view of the world. Would the sun be brighter than the stars? Would there be a difference between day and night? Between up and down? **(2)** Why is the existence of such creatures "impossible," that is, inconsistent with what we know about nature ?

TWENTY-SEVEN

The submicroscopic frontier

At the frontier of the very large are the stars, the galaxies, and the universe itself. At the frontier of the very complex are solids, liquids, plasmas, and organic molecules.* At the frontier of the very small are the elementary particles. In this final chapter, we shall explore some aspects of the contemporary scene at the third frontier—the frontier of the very small. Here, in the submicroscopic area of particle interactions, are to be found both insights and puzzles as profound as any in science.

On every frontier of physics, quantum mechanics is relevant, helping to explain what is known, and helping man to push into the unknown. Nowhere does quantum mechanics reveal itself more primitively than on the submicroscopic frontier, where the accoutrements of macroscopic nature have been shed and the smoothness of classical behavior abandoned. In the world of particles, *granularity* shows itself not only in mass, energy, and other physical variables, but in every event of particle action. Governed by laws of *probability*, these elementary events produce action and interaction via the *annihilation and creation* of particles. *Wave and particle* ideas merge most fully in the description of particles, and the *uncertainty principle* accounts for the chaotic swarm of virtual particles that populate the small-scale world.

In contemporary thinking about the smallest domains of nature, two main ideas are dominant: the idea of the quantum field, the basic stuff of the universe underlying particle structure and particle action; and the idea of symmetry and invariance in nature, the principle of order constraining an otherwise chaotic world. The bulk of this chapter is given to explanation and illustration of these two ideas. Invariance principles and associated conservation laws have been encountered repeatedly in earlier chapters. In this chapter we shall examine some new principles of invariance which have come to light only in the submicroscopic world. Some of them provide partial constraints on particle behavior; one is, perhaps, absolute.

27.1 Quantum fields: granular action

Electric and magnetic fields were described in Chapter Fifteen as key concepts of the electromagnetic theory. In this section we shall review some aspects of these classical fields and follow the evolution of the field idea to the modern quantum version.

169 When the field entered the house of science over a hundred years ago, it came in by the back door, for to nineteenth-century physicists the field stood not for an independent entity, but for a disturbed state of something else—the ether. According to this view of fields, an electric charge isolated in an ether-filled space is like a fish in the depths of a quiet ocean. Although surrounded by water, the fish is uninfluenced by the water (or would be, if he could glide without friction through it). When another charge is introduced near the first one, the ether is "strained"; the particle feels the "field" established in its own neighborhood and is pushed or pulled, much as the fish might be affected by a depth-bomb explosion not

* We refer here to the frontiers of physics. In biology and psychology the frontiers push further into complexity—to cells, nervous systems, and the mind of man—with tools of analysis and understanding that shift from quantitative to qualitative as the complexity increases.

far away. He would think he was pushed by something quite real—which he might decide to call a field—even though it is only the water that was there all along that is doing the pushing. However, Maxwell's electromagnetic field was more than a name. Even though the mental image made the ether basic and the field auxiliary, in fact, the field could be defined mathematically and used in the equations of electromagnetic theory, for it was only the field, the nonuniformity in the ether, that was measurable, not the underlying ether itself.

When Einstein rejected the ether altogether as unobservable, and therefore meaningless, there was no upset in electromagnetic field theory. The mental image of the field underwent a revolution, but the equations of the theory remained absolutely unchanged. The revolution of the image, however, was all-important in establishing the psychological climate for the future progress of field theory. The field became a real physical entity, a thing existing in otherwise truly empty space.

As a rough analogy, we can say that the idea of the field as a disturbance in a fluid was replaced by the idea of the field as itself a fluid—but not an all-pervasive one, rather one scattered about here and there, some places intense, some places thin, and frequently traveling through space in bundles of wave motion. Space became comparable to a dry creek bed rather than to the depths of a calm ocean; an electromagnetic wave, like a sudden spring freshet flowing down the creek bed, instead of like a wave of pressure traveling through the sea. An electric charge absorbing electromagnetic radiation was to be thought of as an idle prairie dog in the creek bed suddenly struck by a wall of water instead of as a fish in the sea struck by a pulse of pressure.

The recognition of fields as real physical entities was the first of two big revolutions in the physicists' understanding of the field. The second was the change from the classical view of a field, as a fluid-like substance distributed smoothly through a region of space, to the quantum view of a field, as something slightly particle-like which could be created and destroyed in lumps.

Einstein's relativity had something to do with the first of these revolutions. His photon theory of the photoelectric effect gave strong support to the second. The photon concept required that the electromagnetic field must be created, not like a stream of water pouring from a hose, but like the rattle of bullets from a machine gun. This particle-like behavior is emphasized only during acts of emission and absorption. In between, the field has to be pictured more like a stream of water than like the bullets, a fluid-like substance, not concentrated at points, but spread out over a region of space and propagating as a wave from one point to another. Moreover—and this is all-important to the apparent smoothness and continuity of our large-scale world—a wavelike behavior is also induced even in the acts of emission and absorption when a sufficient multitude of bullets act in consort. We are already quite accustomed to the idea that the apparent smoothness and divisibility of ordinary matter is illusory, coming about only because of the minute size of atoms and the vast number of them in any sample of material that can be seen by the eye, or even with a microscope. To this idea of indivisible units of *matter* must now be added the idea of indivisible units of *action*. Just as apparently smooth matter is really granular at the submicroscopic level, so an apparently smooth flow of events is really an uneven, jerky succession of tiny explosions. Almost without exception, all of the smoothness and continuity we observe, in happenings as well as in material substances, is the result of the superposition of a vast array of elementary units.

The idea of the quantum field and the wave-particle duality began with the electromagnetic field and the photon in 1905. But not until the late 1920s and the early 1930s did our present more general view of fields as the basic stuff of the universe emerge. As late as 1926, when quantum mechanics was developed, particles and fields were regarded as two different things. The dropping of the last barrier between particles and fields was forced on the physicists rather unexpectedly when they tried to merge the theories of relativity and quantum mechanics. It just turned out—for mathematical reasons that no one had foreseen—that this merger could be effected only if *all* particles, material as well as immaterial (electrons as well as photons), were regarded as the quantum lumps of an underlying field. Thus it became suddenly necessary to add to the electromagnetic field an electron field and a proton field and, as more and more particles were discovered, more and more distinct fields. The essential feature of this field theory of particles is that all particles, and not just photons, must be capable of being created and annihilated. Of course, we are now up against an embarrassing richness of fields which no one really believes in. It seems certain that each of the known particles cannot be the quantum manifestation of a distinct underlying field, but that, in some way not yet understood, all of the particles must arise from one or very few basic fields. (This view is at the moment an article of faith, not of evidence.)

To summarize the modern view of fields and particles: There is a nebulous physical substance called a field that can propagate as a wave through space and that can carry energy and momentum and mass and charge and other measurable quantities. Whenever any part of a field comes into or goes out of existence, it does so with catastrophic suddenness at a particular point of space and time. A peculiarity of every field is that it has a definite mass associated with it. A lump of electron field can be created with any kinetic energy or any momentum, but it has always the same invariable mass, which is precisely the mass of the particle we call an electron. The two biggest mysteries of field theory today are the apparent multiplicity of different fields, and the origin of the mass carried by the field. For reasons no one knows, some fields lock up a great deal of energy in the form of mass, others very little, and others none at all.

We shall be concerned in the remainder of this chapter with the creation and annihilation of fields, that is, with the particle-like aspects of the fields. The interaction of one field with another—the source of all the action and events in the world—emphasizes the particle properties of fields and is, therefore, easy to visualize. It will be possible to speak of, and think of, a particle being created at one point, flying somewhere else, and then being annihilated. Yet in the back of our minds should be the more complex picture of a lump of field being created, propagating elsewhere as a wave, and then being again absorbed all at once.

27.2 Feynman diagrams

170 In picturing the interactions of particles it is important to think of the when as well as the where, hence to think of paths through time as well as paths in space. Such space-time paths, or world lines, were discussed in Section 20.7. A full space-time map would require four dimensions, but often two suffice, one for a particular direction in space, one for time. As a reminder of the appearance of world-line diagrams, Figure 27.1 shows a simplified space-time map for an airplane trip from Flagstaff to Albuquerque (due east). The axes are labeled x (distance to the east) and t (time). Recall that there is no such thing as standing still in a space-time map. Something that remains at rest—the city of Flagstaff— still moves through time. The world line of Flagstaff is shown by a vertical dashed line in Figure 27.1, and another vertical dashed line farther east is the world line of Albuquerque. Now what about the airplane? While parked on the field of Flagstaff, it also moves only through time, not through space, and also traces out a vertical world line (the segment AB in the diagram). Then the plane takes off and flies east, moving both in space and time, and tracing out the world line BC. Having landed at Albuquerque, it once again moves only in time, and its world line continues upward.

We have attached arrows to the world line of the airplane. They may seem redundant, since there is, after all, only one possible direction to move through time—forward. But they do no harm and will serve a useful purpose in the elementary-particle world.

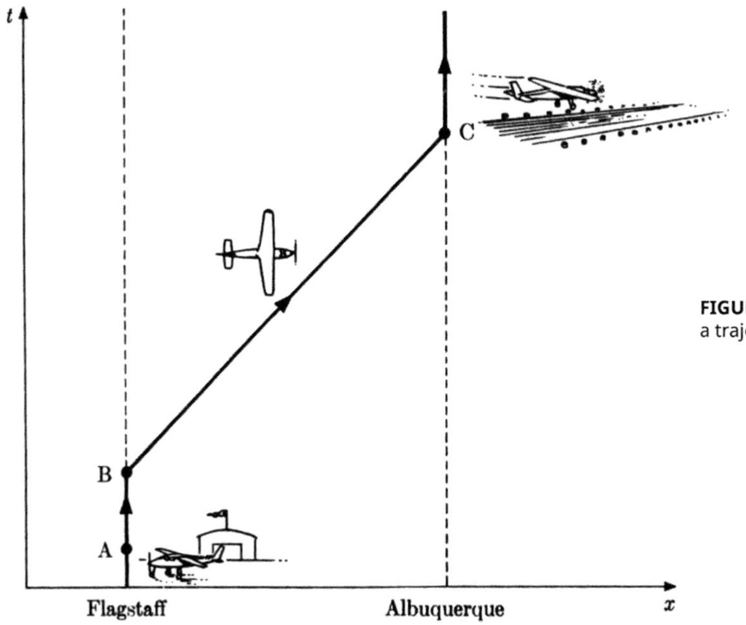

FIGURE 27.1 World-line diagram: a trajectory in a space-time map.

This simple space-time map describes motion only along a single straight line, but that will be sufficient for understanding particle interactions. We may imagine more complicated maps with two or even three space dimensions and one time dimension; but it is better to restrict the diagrams to the form illustrated here.

Turning now to the particle world, we sketch the world lines of a few simple occurrences (Figure 27.2). The first one shows the process of photon emission by an atom. Initially (starting at the bottom of the diagram) an atom is at rest and, like the city of Flagstaff, traces out a straight vertical world line. It then emits a photon that flies off to the right, and the atom itself, having fired the photon, recoils and moves off more slowly to the left. Note that the more slowly a particle moves, the more nearly vertical is its world line—no motion at all is exactly vertical. Conversely, the faster the particle moves, the more its world line is tipped over toward the horizontal. But it can never become exactly horizontal; for that, the particle would have to travel from one place to another in no time at all. The photon line is tipped the maximum amount; the inverse of its slope is the speed of light.

The next diagram, Figure 27.2(b), illustrates pion decay:

$$\pi^- \rightarrow \mu^- + \overline{\nu}_\mu .$$

At some point, indicated by the black dot, the negative pion ceases to exist. It is annihilated and its world line terminates. But at that same place and same time (that is, at that same point in space-time) the negative muon and the antineutrino are born and fly apart, the antineutrino line being tipped at the angle corresponding to the speed of light. The black dot locates an event, an occurrence at a single point in space and time. In each of the other diagrams, there is also at least one important event. In the world of particles, every significant event is marked by the creation and/or annihilation of particles.

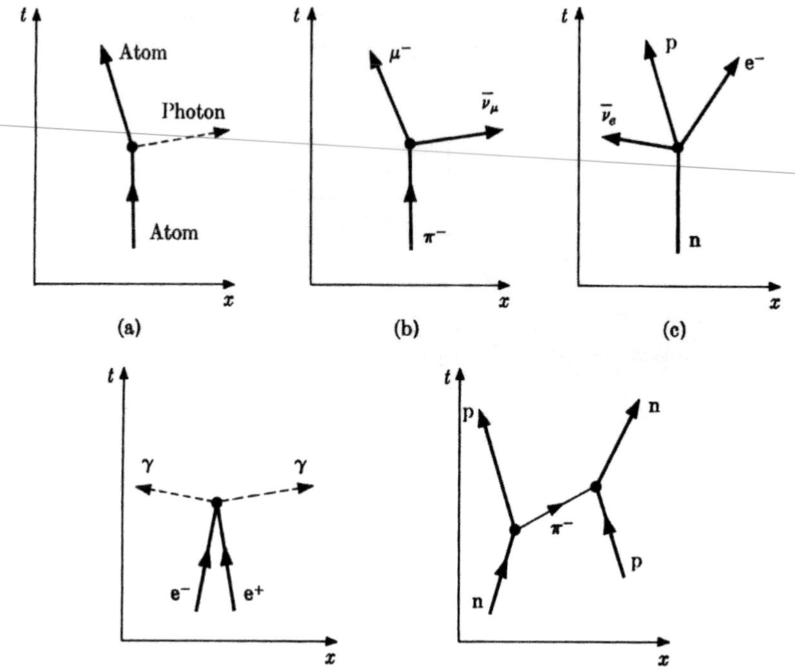

FIGURE 27.2 World lines of various occurrences in the particle world. **(a)** Emission of photon by an atom. **(b)** Pion decay. **(c)** Beta decay of a neutron. **(d)** Positron-electron annihilation. **(e)** Pion exchange process.

In diagram 27.2(c) is illustrated the beta decay of the neutron:

$$n \rightarrow p + e^- + \overline{\nu_\mu} \, .$$

This time, the crucial event in space-time involves the destruction of one particle and the creation of three others. The next diagram shows the annihilation of an electron and a positron to create two photons (gamma rays),

$$e^- + e^+ \rightarrow \gamma + \gamma \, .$$

Finally, diagram 27.2(e) illustrates a pion-exchange process that contributes to the force between a neutron and a proton. Initially (bottom of diagram) a neutron and a proton are present. They exchange a pion, and exchange roles, emerging with different speeds. According to the Yukawa theory, the force between two nucleons arises from this exchange process, together with various more complicated exchanges that can also take place.

Before we go on to the question of how these pictures are related to what is "really" going on at the submicroscopic level, one flag of warning must be raised. It is possible, indeed probable, that some of the single "events" indicated by the black dots are really complex sequences of events which all happen within such a tiny domain of space and a short span of time that they seem to be events at only one single space-time point. It is already known, for example, that the phenomenon of electron-positron annihilation does not proceed exactly as pictured in Figure 27.2(d). Rather, the two photons actually emerge from slightly different points, as in Figure 27.3.

We must be prepared for the possibility that the future will reveal an inner structure to other apparently simple events, even for the possibility that the apparently catastrophic acts of sudden annihilation and creation are really the result of a smooth, continuous flow of events over tinier regions of space and time than it has been possible to investigate so far. This is pure speculation. Down to the shortest distances (10^{-14} cm) and shortest times (10^{-24} sec) to which man has probed, the elementary events of the particle world still seem to be the catastrophic events of sudden creation and destruction of the packets of field energy we call particles.

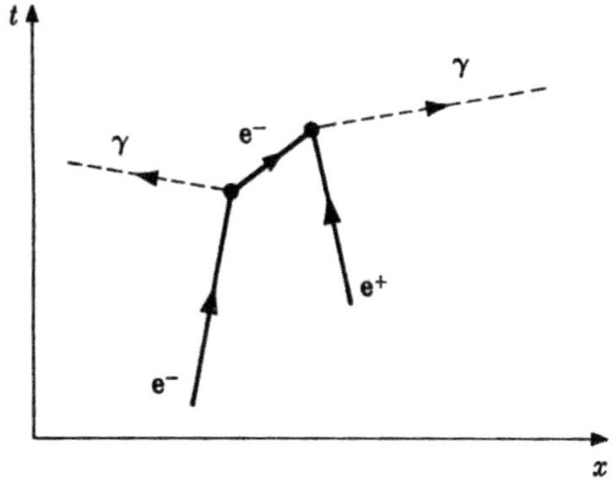

FIGURE 27.3 What "really" happens in a positron-electron annihilation process, a corrected version of Figure 27.2 (d).

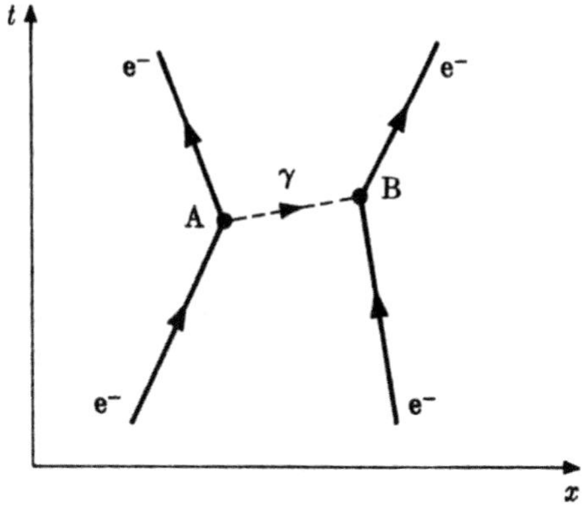

FIGURE 27.4 World-line diagram for the mutual interaction and deflection of two electrons.

The accumulated evidence from experiments probing the world of the very small, together with the confirming support of the quantum theory of fields, leads to the following very important general conclusion: All of the interactions in nature arise from acts of annihilation and creation of particles at definite points in space and time. There are two important ideas here. First, all interactions involve the creation and annihilation of particles; second, these creations and annihilations do not take place over a region of space or over a span of time, but are instantaneous and localized at points. By an "interaction" is meant simply the influence of anything on anything else. Thus, all ordinary forces—pushes or pulls of one thing on another—are interactions. Also, the decay of an unstable particle is the manifestation of an interaction. The final particles are "influenced" by the initial particle—they come into existence only because the initial particle was there.

Contrast this new view of interactions with the classical view. The sun and the earth "interact," for the earth is pulled by the sun. Nothing is apparently being created or destroyed, and nothing seems to be happening instantaneously at isolated points of space or time. But, according to the new view, gravitons are constantly being emitted and absorbed, both by the sun and by the earth. Each act of emission or absorption occurs at an instant of time and a point in space. The "force" the earth experiences is nothing but the accumulated effect of all of these graviton interactions.

To turn to an example in the particle world that is more surely understood, consider the scattering of two electrons. According to the old view, the electrons approach each other, feel a mutual repulsive force, and are deflected. The new view goes deeper; it explains "why" the electrons exert a force on each other. In Figure 27.4 we see two electrons approaching one another. At point A, the electron on the left emits a photon and changes *its* speed. At point B, the electron on the right absorbs the photon and changes its speed. The two electrons have interacted, or exerted a force on each other, since their motion was altered. It was the exchange of a photon that brought about this apparent interaction. Strictly speaking, the basic interaction was not between the two electrons at all, but between each electron and a photon. The second electron is only indirectly aware of the presence of the first. The old idea of action at a distance, of a force "reaching out" from one to the other, is completely abandoned. It is replaced by the idea of a "local interaction," of each electron interacting locally—that is, at its own location—with a photon.

Of course, the particular diagram illustrated here is only one of many; the others involve more complicated exchanges between the electrons. The net effect of all the possible exchanges is the deflection of the electrons according to the ordinary repulsive electric force, but with motion that is a series of hops rather than a smooth flow.

According to the present theory of the interactions of electrons and photons, Figure 27.4 is a picture of what "really" happens on the submicroscopic scale. Such a diagram is called a Feynman diagram after Richard Feynman, who showed in 1949 that such pictures have an exact correspondence to mathematical expressions in the field theory of electrons and photons. These diagrams, therefore, portray what is "really" happening, and provide a convenient way to catalogue the various possible processes of creation, annihilation, and exchange.

The key points of a Feynman diagram are the "vertices," representing those points at which (in this example) photons are created or absorbed. All processes involving photons, and therefore *all* of the interactions associated with electromagnetism, arise from elementary events of photon creation or photon annihilation. These fundamental interaction events can be pictured by a single kind of vertex that looks like either of the diagrams in Figure 27.5. The solid lines represent charged particles and the dashed line represents a photon. The points A and B in Figure 27.4 are vertices of this kind.

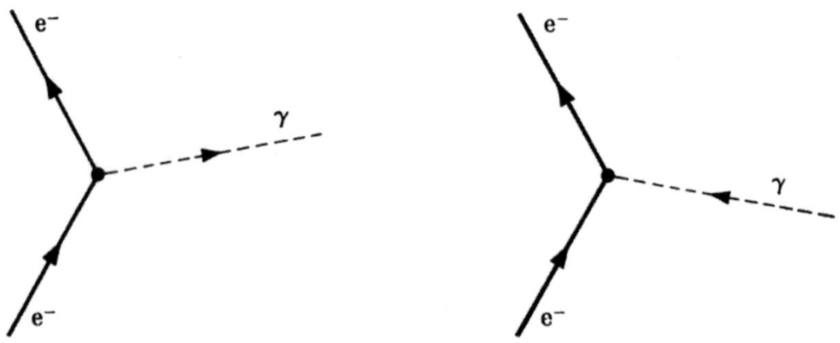

FIGURE 27.5 Basic electron-photon interaction vertices.

If the solid lines represent the world lines of electrons, for example, it appears that the fundamental interaction event may be regarded as an event in which a photon is created or absorbed, and an electron simultaneously changes its state of motion. There is, however, a more general and more fruitful interpretation. The vertex may be taken to represent a point where the world line of one electron terminates and the world line of another begins. According to this view, the vertex represents a truly catastrophic event. Nothing survives it. Instead of thinking of a single electron being changed at the vertex, one can think of one electron being destroyed and another electron being created. Since all electrons are indistinguishable, it has no real meaning to say that the outgoing electron is the same as or different from the incoming electron. To think of the outgoing electron as a new and different electron, however, corresponds more closely to the mathematical theory of the fundamental interaction. The creation-destruction interpretation also leads to a simple, unified description of particle events and antiparticle events.

The vertex on the right in Figure 27.3 seems to differ from those in Figures 27.4 and 27.5. Instead of being a point where one electron world line ends and another begins, this vertex is a point where both an electron world line and a positron world line end. There is a simple artistic trick by which we can change the picture significantly. Suppose we turn around the arrow on the positron line. The arrowhead was, after all, redundant, since all particles move forward in time. We can use it instead as a label to distinguish particles from antiparticles. An arrowhead pointing in the "right" direction will indicate a particle (for example, an electron); an arrowhead pointing in the "wrong" direction will indicate an antiparticle (a positron). Using this revised notation, we show in Figure 27.6(a) an electron-positron annihilation vertex, and in Figure 27.6(b) an electron-positron creation vertex. Now these vertex diagrams involving positrons look like twisted versions of the fundamental vertex diagrams in Figure 27.5. The generalized conclusion is that the fundamental electron-photon vertex with its limbs twisted in all possible directions in space-time represents all the possible basic interactions among electrons, positrons, and photons. This provides a magnificently simple and general view of the underlying basis of all electromagnetic phenomena.

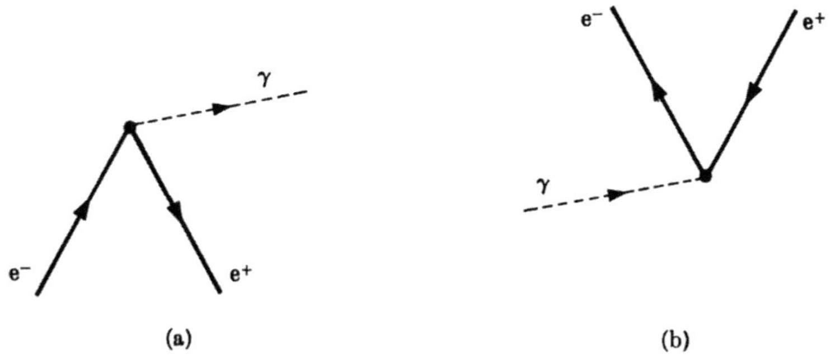

FIGURE 27.6 Additional fundamental electron-photon interaction vertices which include antiparticles (positrons).

What Feynman showed when he discussed the connection between such world-line diagrams and the structure of the mathematical theory of the electron-positron-photon interaction was that the device of the reversed arrowhead is much more than an artistic trick. According to the field theory of electrons, the creation of a positron is "equivalent" to the annihilation of an electron (they are not identical processes, but the theory says that whenever one can happen, the other must also be able to happen). Moreover, the mathematical description of a positron field propagating forward in time is identical with the description of an electron field propagating backward in time. It is perfectly possible and consistent to think of particles moving backward in time as well as forward.

This circumstance need not lead us to deep philosophical conclusions, although philosophical implications are hard to escape. The positron *may* be described as an electron moving backward in time, but it does not *have* to be so described. An alternative description is equally possible in which the positron is a normal particle moving forward in time. Nevertheless, this picture of backward motion in time is a tantalizing one that simplifies the view of elementary interactions and provides a "natural explanation" for the existence of antimatter. Consider the Feynman diagram in Figure 27.7, for example. According to the normal view of time unrolling in one direction, we start at the bottom of the diagram and read up. First an electron and a photon are approaching each other. At vertex A, the photon creates an electron-positron pair. The new electron flies away, while the positron collides with the first electron at vertex B. There they undergo mutual annihilation and a new photon is born.* The alternative view, which Feynman showed to be also consistent, is

* This process is electron-photon scattering, known as the Compton effect. When it was discovered by Arthur Compton in 1923, it added strong support to the photon concept of light.

to picture the first electron proceeding to point B, where it emits a photon and reverses its path through time. It "then" travels to point A, where it absorbs the incoming photon and once again reverses its course through time, flying off in the "right" direction. Either view is permissible and logically consistent.

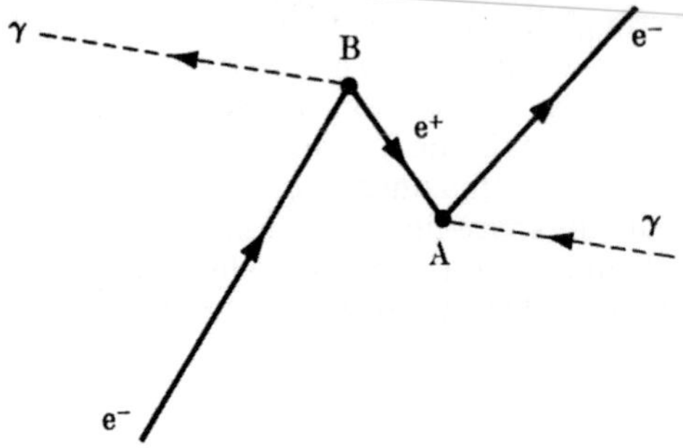

FIGURE 27.7 Feynman diagram for the process of photon-electron scattering.

171 In thinking of the philosophical consequences of this remarkable view of motion in both directions through time, we must ask: What about man? Why do we move only forward through time, not enjoying the same freedom the particles have? The answer is that we *do* enjoy the same freedom, but fortunately are prevented from exercising it by a chance. Man is composed of particles, not antiparticles; particles always move forward in time. By chance (?), our comer of the universe is constructed almost exclusively from particles and contains very few antiparticles. (Whether there exist other parts of the universe where antiparticles dominate is unknown.) Therefore we cannot find the antimatter with which to annihilate and start our path backward in time. Of course, an occasional positron enters every man and destroys one of his electrons, but we can always spare a few electrons.

This answer may seem unsatisfying, for the question could be phrased somewhat differently. One might ask, must the antimatter already be present? Could man not emit a barrage of photons and reverse his course in time as the electron did? The answer is that if a man were going to do this in the future, he would know it already. We are aware of what is around us *now*, and a path that later on reversed itself in time would again pass through the present *now*. In the Feynman diagram of Figure 27.8 the present moment, "now," is indicated by a straight horizontal line. If the world line of a man were to reverse, then the man and the "later" antiman would both be present side by side at this moment of time. Even if there is a science-fiction ring about it, this argument is perfectly sound and really does not differ from the argument of the preceding paragraph. Because we see no significant quantity of antimatter around us now, we can be sure that we are safe from future annihilation or time reversal.

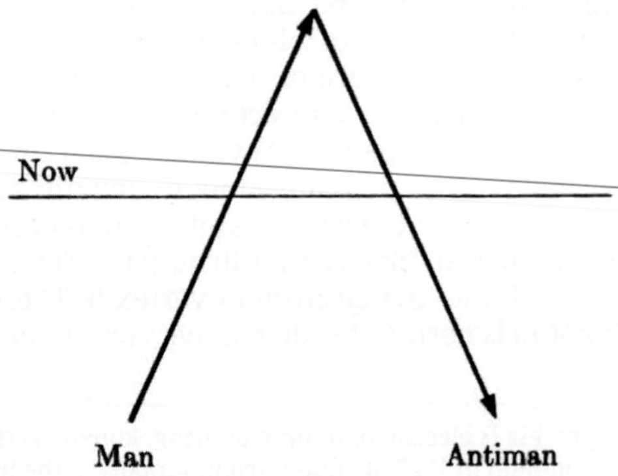

FIGURE 27.8 "Feynman diagram" for a man who reversed his course through time.

According to the theory of relativity, there is nothing special about one direction of time. It is therefore satisfying to know that particles may have the same freedom to move forward and backward in time that they have to move left and right or up and down in space. This view of the world restores the symmetry of space and time and attributes the apparent one-way flow of time to the fact that we happen to live in a world with an enormous disparity between the number of particles and the number of antiparticles.

Unfortunately, no one knows how to test the theory of time-reversed trajectories experimentally. It must be accepted (if at all) for the symmetry it introduces into our picture of the world and for the simplicity it introduces into the description of antiparticles. To see that the direction of motion through time cannot be measured, imagine yourself a submicroscopic observer of the scene depicted in Figure 27.7, If you put a ruler horizontally across the bottom of the diagram and then push it slowly upward across the diagram, the intersections of the world line with the moving ruler edge give a rough history of your observations. (Actually the ruler should be slightly tilted to take account of the fact that information conveyed by light from other points takes some time to arrive, but this is an unimportant detail for this discussion.) The point is that it is the ruler edge of your observation that is moving in time, not the world lines themselves. The particle world lines may be regarded as perfectly static, simply there, painted in space-time like lines on a map. To a creature capable of comprehending the whole span of time as we comprehend the span of space, the activity of annihilation and creation represented in this diagram is not activity at all. It is a stationary display, a picture painted in space and time. It is the fact that man, the observer, can comprehend only one instant of time at each moment that converts the stationary display into motion and activity. At one time, the human observer "sees" an electron in one place. At a later time, he sees it at a different place. It is natural for him to believe that the electron, like the man, had moved forward in time to get from the first place to the second place. But there is no essential reason to believe this. We know only that the electron world line has traced out a certain path in space-time, but we have no possible way of knowing in what order the points of that path were traced out, or even whether it makes sense to speak of an order or a direction of tracing the world line.

These considerations lead to philosophical problems outside the scope of this book. According to the theory of relativity, forward and backward motion in time are equally acceptable, and the best way to imagine world lines may be as static lines in a four-dimensional geography. Quantum field theory has added the insight that the simplest description of antiparticles is in terms of particles moving backward through time (an idea that can also be blended with the idea of a static display). We are still faced with the fact that man certainly moves through time in only one direction. We remember the past and do not remember the future. It is human memory *and nothing else* that tells us in which direction we move through time. Man experiences an asymmetry of time he does not share with the elementary particles. Can man's asymmetry of time be reconciled with the particles' symmetry of time if man is nothing but a collection of particles? Can man mold the course of future history if world lines are really fixed features of a four-dimensional geography? It is this author's opinion that the answer to both of these questions (which are merely twentieth-century versions of much older questions about determinism and free will) is *yes*, and that the resolution of the apparent paradoxes lies in the fantastic complexity and high degree of organization of the human being, complexity and organization so extraordinarily far beyond any power of prediction from the basic laws of the submicroscopic world as to constitute an independent spirit.

27.3 The TCP theorem

The connection between time-reversed trajectories and antiparticles, a connection that reflects a basic space-time symmetry in the elementary-particle world, has found expression recently in the "TCP theorem," whose three letters stand for three hypothetical operations: T, time reversal; C, charge conjugation, the technical name for interchanging particles and antiparticles; P, space reversal, which is approximately equivalent to taking a mirror image of space. This is sometimes called the parity operation, hence the P. The TCP theorem is actually a conservation law of a special kind, possibly an absolute conservation law governing all the interactions in nature. (It remains to be accurately tested for the weak interactions.) It says that if the three operations of T, C, and P are all applied to any physical process, the result of this mangling of what actually happened is another physical process that could have happened. This is not as complicated as it sounds; you may easily perform the TCP operations on any of the Feynman diagrams illustrated in these pages. All the equipment needed is one wall mirror and a little imagination.

First, turn to any page containing a "proper" Feynman diagram—one with its antiparticle arrowheads pointing downward. Figure 27.6 or any later one will do. Now, to perform the operation C, simply imagine all arrowheads turned around; this interchanges all particles and antiparticles. (A backward-pointing arrow on a photon is all right, for a photon is its own antiparticle, a property it shares with the neutral pion.) For example, the operation C converts the negative pion decay,

$$\pi^- \rightarrow \mu^- + \overline{\nu_\mu},$$

into the positive pion decay,

$$\pi^+ \rightarrow \mu^+ + \nu_\mu,$$

since the positive pion is the antiparticle of the negative pion.

To perform the operation P, space reversal, turn the page of the book away from you and look at the diagram reflected in the mirror. This interchanges left and right. (There are fuller implications of P than this, concerned with particle spin; they are considered in Section 27.5.)

Finally, turn the book upside down and look at the inverted diagram. This has obviously turned time around—the operation T—but it has also turned all the arrowheads around (operation C) and has interchanged left and right (operation P). The upside-down diagram therefore represents the application of all three transformations, T, C, and P, to the original physical process. The upside down picture is another Feynman diagram that illustrates an actual, physically allowed process (*if* the TCP theorem is valid). In general it will not be the process you started with; it may even be quite different. But, according to the TCP theorem, since the original diagram represented a real physical process, the triply inverted diagram does also.

To see the effect of time reversal alone, we must undo the arrowhead reversal and the left-right inversion. Turn the book upside down, then look at the upside-down diagram in the mirror *and* imagine the arrowheads reversed. The diagram resulting from this manipulation illustrates the purely time-reversed process. Other combinations may be tried. The mirror view with arrowheads reversed is the result of the double operation PC, and so on.

The fact that the TCP theorem is a conservation law is not obvious, for one wonders just what quantity remains constant. Although we can define the "TCP quantity," study of it is not particularly instructive. It is more helpful to note that the TCP theorem, like all conservation laws, is a law of prohibition. Only those events can occur, it says, whose TCP inversions are also possible real physical events. If the process represented by the triply inverted diagram were not allowed, then the original process would be also forbidden.

27.4 *Time-reversal invariance*

In Section 14.9, the time-reversal invariance that seems to govern nature in the small was contrasted with the "arrow of time," the unique sense of time direction that is evident in everyday affairs. For complex systems, one direction of the unfolding of events (the entropy-increasing direction) is so much more probable than the opposite direction (the entropy-decreasing direction) that we call one direction in time possible, the other direction impossible.

To see nature's basic symmetry of time, we must examine simple processes involving very few objects. Suppose a space traveler bound for another galaxy takes along a moving picture of our solar system to show his hosts. If the picture had been taken from a point a few hundred billion miles distant in the general direction of the North Star, it would show the planets as tiny dots tracing out their elliptical orbits in a counterclockwise direction about a central sun. The creatures of the other galaxy, being well versed in the laws of mechanics, would watch the picture with interest and conclude that there were no tricks, that it did indeed depict an actual physical chain of events. But had the film been run off backward, they would have been equally convinced. The backward view of planetary motion, although "untrue," since, in fact, our planets do not move that way, is nevertheless "possible," for it is consistent with the same laws of mechanics.

Now allow our space traveler to proceed to another galaxy whose residents are intelligent and mathematically adept, but scientifically primitive. One audience of these creatures is shown the planetary film in the forward direction; another audience is shown the same film run backward. Both audiences are asked to deduce the law of gravitational attraction and the law of mechanical motion from what they have seen. If they are collectively as clever as Newton, they will put their heads together and both groups will arrive at the correct, and identical, laws.

This is the real significance of time-reversal invariance. Under a hypothetical reversal of the direction of time, the laws of nature are unchanged. This is the statement of the principle that emphasizes *invariance*. To emphasize the *constraint* imposed by the law, we must phrase it somewhat differently. Only those things can happen that could also happen in the opposite order. Or, still more negatively: if a time-reversed process is impossible, then the process itself must be impossible.

Time-reversal invariance finds its simplest application in the world of particles, where it appears to govern the strong and electromagnetic interactions, and possibly also the weak.* Figure 27.3, for example, illustrates electron-positron annihilation. The time-reversed process is the creation of an electron-positron pair by the collision of two photons. Recall the directions given in Section 27.3 for time-inverting a Feynman diagram. Turn the page upside down, view it in a wall mirror, and mentally turn the arrows around. You then see two photons coming together from the bottom, and an electron and a positron flying apart at the top. According to time-reversal invariance, this reverse process is not only possible, but can occur in every detail as the inverted sequence of fundamental interaction events. Since the "strength" of interaction at each vertex is unchanged, there is a definite numerical ratio between the probability of pair creation and the probability of pair annihilation which is implied by the law.

The role of probability in time-reversed events, so painfully obvious for a secretary untyping or a barber unclipping, makes itself felt in the particle world as well. A simple

* In 1965 a group at Princeton University learned by studying kaon decay that the product of space inversion and charge conjugation (PC) is not a completely valid invariance principle for the weak interactions. This discovery has a possible implication for time-reversal invariance. If the TCP theorem is valid, T-invariance must also be violated to a small degree in order that the product of all three operations may provide an absolute invariance. On the other hand, time-reversal invariance may survive further tests. If it does, the TCP theorem will fail as a completely general principle, although it will surely retain an approximate validity.

process of pion creation, for example, is impossible to reverse in practice. Two protons may collide to yield proton, neutron, and positive pion, as shown in Figure 27.9. The time-reversed process requires the nearly simultaneous collision of three particles, too unlikely to achieve in practice. But the requirement that *each* fundamental event is time reversible is a constraining condition that influences the possible form of the pion-nucleon interaction and therefore has an important effect on the forward-in-time process, whether or not the time-reversed process is likely, or experimentally possible. This is an exceedingly important aspect of invariance principles related to their increasingly central role in physics. The laws of mechanics to which the planets are subjected are partially determined by the condition of time-reversal invariance. Therefore, the particular way in which the planets arc through the sky is dictated in part by time-reversal invariance, even though it is entirely out of the question to stop the planets and turn them back in their tracks. Similarly, time-reversal invariance may be tested in particle interactions even if it is impractical actually to reverse the particular process being studied.

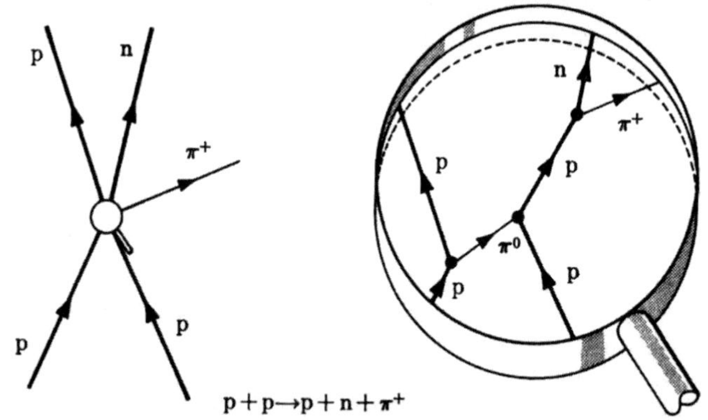

FIGURE 27.9 Creation of a pion in a proton-proton collision. The time-reversed process is possible but far less likely than the direct process.

27.5 Parity

The parity principle, or the principle of space-inversion invariance, says that there is a symmetry between the world and its mirror image. To state this in terms closer to those we used for time-reversal invariance: The mirror image of any physical process depicts a possible physical process, one that is governed by the same laws as the process itself. Most of us perform parity transformations every day, whenever we look in a mirror. There is nothing "strange" about a mirror view of most things or most events. A mirror view is not "right," but it appears quite reasonable and possible. The mirror view of a person may not be an exact replica of any person, but we are quite prepared to believe that a real person *could* appear just that way.

The mirror view of a printed page looks "wrong." It is obviously quite different from the direct view of the page. But there is nothing impossible about it. A printer could design inverted type and produce a page which, viewed directly, would be identical to the mirror view of the normal page. Most children put this symmetry to work by learning "mirror writing," which appears normal only when viewed in a mirror (Figure 27.10).

The situation with the space-inverted view of the normal world differs completely from that with the time-inverted, view. The mirror view of the world looks quite normal, on the whole, and prepares us to believe in parity conservation, or space-inversion invariance. The time-reversed view of the world, on the other hand, looks ridiculous and impossible and prepares us to resist the idea of time-reversal invariance. The particles have fooled us on both counts. Time-reversal invariance proves to be an absolute conservation law, so far as we know now, and space-inversion invariance turns out to be only a partial conservation law, one that is completely violated by the weak interactions. This means that the mirror view of an actual weak interaction process such as beta decay shows something that *cannot* happen. Even scientists had discarded their normal caution, and come to regard parity conservation as an absolute law. When, following a suggestion by Tsung Dao Lee and Chen Ning Yang, it was verified in 1956 that the weak interactions do *not* have mirror symmetry,* it came as rather a shock to the scientific community, a useful reminder that an untested theory is a house built on sand.

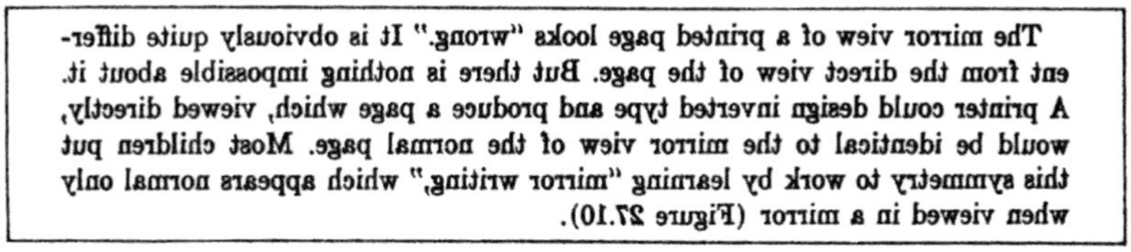

FIGURE 27.10 Parity transformation applied to a printed page.

A slide projector or a movie projector is a better device than a mirror for illustrating the idea of parity conservation in the everyday world. Most of us are so accustomed to mirrors that we mentally compensate for the space inversion in the mirror view. But, with a projector, an element of mystery can be introduced that makes the invariance more obvious. If you were shown an unfamiliar landscape on the screen, you would be quite unable to tell whether the slide had been inserted correctly or the wrong way around. It would look reasonable and possible in either case. Even for a moving picture with the film inverted left for right, it might be difficult to decide whether you were seeing the true view or the space-inverted view. Of course, if nine out of ten of the actors seemed to be left-handed, or if cars in an American city were being driven on the left, or if a sign-post showed backward writing, the secret would be out. But even with these clues to give the show away, nothing in the inverted view would violate common sense or be obviously impossible. This is no mere coincidence. The fact that mirror views of the ordinary world look perfectly reasonable is directly connected with the fact that all of the laws of nature that govern the large-scale world obey the principle of space-inversion invariance. If they did not—if, for example, the weak interactions had any direct effect on the world of our sense experiences—we should be made obviously aware that a mirror view of an actual sequence of events was all wrong, a physical impossibility.

The radioactive nucleus of cobalt 60 ($_{27}Co^{60}$), the same nucleus that could threaten hu- 172
man life if nations made and used "cobalt bombs,"† was responsible for bringing the first enlightenment on the violation of parity conservation in weak interactions. Reduced to its essentials, the experiment of Chien-Shiung Wu was exceedingly simple. The cobalt nuclei were lined up so that their intrinsic rotational motion, viewed from the top, was counterclockwise. See the "direct view" in the upper left of Figure 27.11, which shows a Co^{60} nucleus with its "north pole" on top, its "south pole" on the bottom. In the experiment, a large number of Co^{60} nuclei were oriented in exactly the same way. It was then observed that, as the nuclei, one by one, underwent their explosive process of beta decay, the ejected electrons almost all flew off in the downward direction. The solid arrows in the diagram represent

* For this prediction and related work, Lee and Yang shared the Nobel Prize in physics in 1957. At the time they challenged the law of parity conservation, Lee was 29 years old and Yang was 33.

† A "cobalt bomb" is a thermonuclear weapon encased in Co^{59}. Neutron capture in the case would create in quantity the isotope Co^{60}, whose half life is 5.26 years.

the preferred flight direction of the electrons. Now, the mirror view of this process shows a Co^{60} nucleus apparently rotating in the opposite direction, but with the electrons still coming out mostly downward. (It is to be borne in mind, of course, that a single Co^{60} nucleus ejects only one electron.) If, on the other hand, the whole experimental apparatus, including all the nuclei, were turned upside down, the direction of rotation of the nuclei would be changed *and* the direction of emission of the electrons would also be changed. The inescapable conclusion is that the upside-down view and the mirror view of the original process are inconsistent with each other. One or the other of them (or both!) must be impossible.

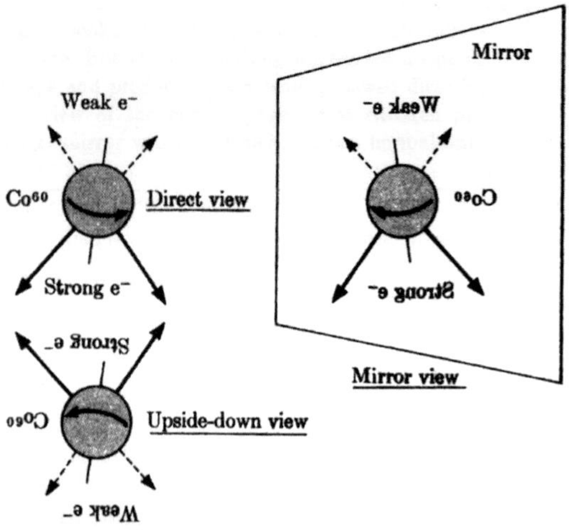

FIGURE 27.11 Decay of an oriented Co^{60} nucleus and other views of the process. The turned-around labels are without significance. Only the physically measurable things have meaning in the various views.

Our first inclination is to say that the upside-down view is obviously possible. There is nothing to prevent the experimenter from turning his apparatus upside down or, more simply, from turning himself upside down, to get a new view of the process. This is quite correct, although it is important to realize that this obviousness rests upon an invariance principle, the isotropy of space, which underlies the law of angular-momentum conservation. That an upside-down experiment must yield the same results as the original experiment is "obvious" only to the extent that our everyday experience conditions us to accept the invariance of laws of nature under rotations as a self-evident truth. In fact, we have strong reasons to believe in angular-momentum conservation and the uniformity of space as absolute laws. Therefore the upside-down view of the Co^{60} beta-decay process is indeed a picture of a physically possible process.

To someone who believes with equal conviction in the invariance of laws of nature under space inversion, it would be equally "obvious" that the mirror view must also represent a physically possible process. Most physicists were nearly in-that position. All the classical laws of physics possessed mirror invariance, and in the quantum world of particles it was already known that the strong interactions and the electromagnetic interactions possessed mirror invariance. Parity conservation was in danger of becoming a self-evident truth. Yet the mirror view of the Co^{60} experiment certainly represented an impossible process. Mirror invariance for the weak interactions had to be abandoned. It must be admitted that, even for physicists, the common sense derived from everyday experience is still a much more powerful force than the common .sense derived from mathematical insight. The failure of space-rotation invariance would have caused an even more violent stir in the world of science than did the failure of space-inversion invariance.

Imagine that there exist tiny creatures whose everyday life is strongly affected by the weak interactions. They would probably find the mirror view of the Co^{60} decay quite diverting. They would recognize at once that it showed an absurdly impossible event and would perhaps find it as amusing as we find the backward-run film of an automobile race.

The value of every conservation law or invariance principle is that it imposes constraints, restricting nature's freedom. For the Co^{60} experiment, mirror invariance would require that just as many electrons fly up as down, in fact, that the pattern of emerging electrons is exactly symmetric between up and down, for only then would the upside-down view and the mirror view agree. If the law of parity conservation were suspended, this restriction would be removed and the electrons would be permitted to fly out in any way at all (consistent with *other* conservation laws). As it happens, the weak interactions violate parity conservation to the maximum possible extent.

Why did more than fifty years elapse between the discovery of beta decay and the discovery that the beta-decay process lacks mirror invariance? There is a simple thermodynamic reason for this. The Co^{60} experiment was actually a good deal more complicated than our discussion has indicated. Lining up nuclei all to rotate in the same direction is a difficult process, for the energy of thermal agitation at normal temperatures far exceeds the energy of alignment that comes from the interaction of the nuclear magnetic moment with an externally applied magnetic field. Even if a strong magnetic field is applied at room temperature to get a "handle" on the nuclei, thermal agitation prevents alignment. It is clear that the Co^{60} experiment would have been useless had half the nuclei been spinning one way and half the other way. In that case, as many electrons would have come out upward as downward, and nothing would have been learned about mirror invariance. To line up the nuclei and keep them that way, Madame Wu enlisted the aid of a group at the National Bureau of Standards who were expert at achieving very low temperatures. At a temperature of less than one-tenth of a degree above absolute zero, the thermal agitation was reduced sufficiently that the cobalt nuclei could be maintained with oriented spins, and the desired experiment could be carried out.*

Among the many further proofs of the failure of mirror invariance, none is simpler or more convincing than the discovery that the neutrino is left-handed. The mirror view of a left-handed neutrino (Figure 27.12) is a right-handed neutrino, but there is no such thing as a right-handed neutrino; therefore the mirror view is a view of the impossible.

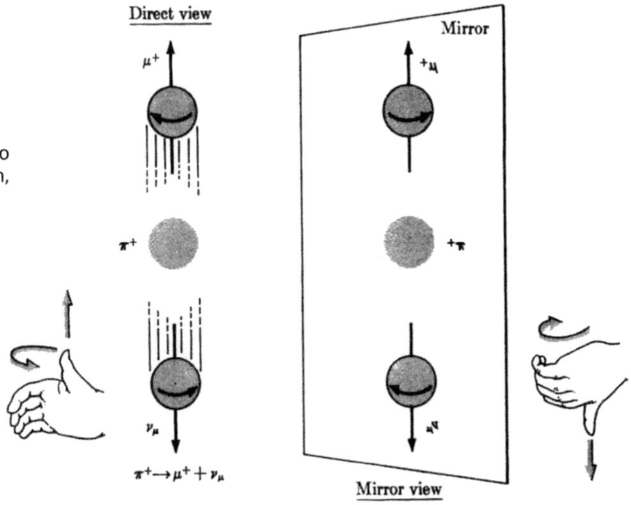

FIGURE 27.12 The left-handed neutrino created in the decay of a positive pion, and its right-handed mirror image.

* Actually, temperatures much less than one degree cannot be maintained for very long. The cobalt was cooled to about one-hundredth of a degree and the measurement carried out during the ten minutes required for the well-insulated material to warm up one degree.

So far we have emphasized the *failure* of parity conservation. But in all but the weak interactions it is a powerful and valid law limiting what can happen. For example, had the Co^{60} nuclei undergone gamma decay instead of beta decay, as many photons would have emerged upward as downward. In that case, only the electromagnetic interaction would have been at work, not the weak interaction, and mirror invariance would have required the symmetric result.

Similarly, the decay of the neutral pion into two photons,

$$\pi^0 \rightarrow \gamma + \gamma ,$$

which is the result of the combined action of strong and electromagnetic interactions, is quite different from the weak decay of the charged pion. Suppose, for a particular decay, that the photons come out left-handed, as did the muon and neutrino in Figure 26.9. The mirror view of this process (Figure 27.13) shows right-handed photons, so mirror invariance requires that decay into right-handed photons is also possible. In fact, it requires that decay into right-handed photons is exactly equally probable, for all the laws of nature (governing *this* process) are the same in the mirror world as in the real world. If decay into left-handed photons were preferred, by however small a margin, in the real world, then right-handed photons would be preferred by the same margin in the mirror world, and this would be a violation of space-inversion invariance.

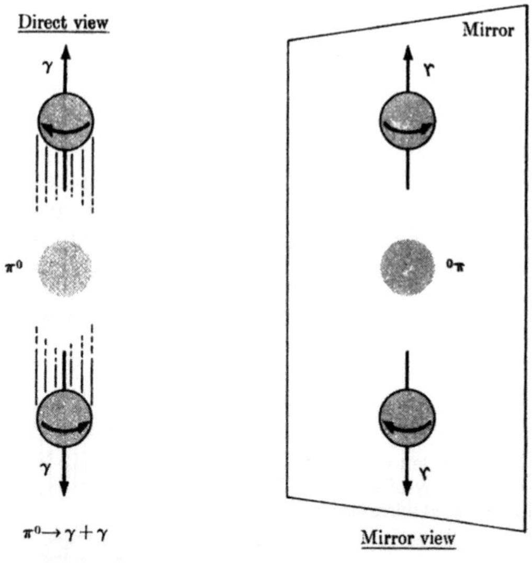

FIGURE 27.13 Decay of the neutral pion, and its mirror image. Because the interactions governing this decay possess mirror invariance, the mirror view pictures a possible mode of neutral pion decay. The photons may be either left-handed or right-handed.

In prequantum physics, parity conservation was recognized but thought to be unimportant. In quantum physics its power was recognized because of the restrictions it imposed on the flow of events governed by laws of probability. Now it has become doubly important through the discovery that some interactions obey the law and some do not. A good reason why the weak interactions are parity violators has yet to be found.

27.6 Charge conjugation

The third member of the T, P, C triumvirate is charge conjugation, the interchange of particles and antiparticles. As emphasized in Section 27.2, the fact that the particle-antiparticle inversion has anything to do with time inversion and space inversion comes about because antiparticles may be described as particles moving backward in time. Hand in hand with the overthrow of parity went the less heralded overthrow of charge-conjugation invariance. The present situation is this: P invariance and C invariance are grossly violated by the weak interactions, but in such a way that combined PC invariance is nearly valid. T invariance is at least approximately, and perhaps absolutely, valid for the weak interactions (see footnote on page 779). For the strong and electromagnetic interactions, so far as we know now, all three inversion invariances, T, C, and P, are separately valid laws.

The left-handed neutrino, which violates space-inversion invariance, also violates charge-conjugation invariance. Consider the decay of the positive pion, written symbolically:

$$\pi^+ \rightarrow \mu_L{}^+ + \nu_{\mu L},$$

The subscripts L indicate that neutrino and positive muon fly apart with left-handed spin. We know that the action of C on this process is to interchange particles and antiparticles, while the action of P is to convert left-handed motion to right-handed motion (subscripts R). Thus we get the following transformed processes:

C:	$\pi^- \rightarrow \mu_L{}^- + \overline{\nu_{\mu L}},$	No.
P:	$\pi^+ \rightarrow \mu_R{}^+ + \nu_{\mu R},$	No.
PC:	$\pi^- \rightarrow \mu_R{}^- + \overline{\nu_{\mu R}},$	Yes.

The C transformation leads to an "impossible" process (one that has never been seen), since it converts a left-handed neutrino to a left-handed antineutrino, but antineutrinos are right-handed. Therefore this weak interaction decay violates C invariance. The P transformation applied to the original process of positive pion decay converts the left-handed neutrino into a right-handed neutrino, giving again a process that has never been seen. But P and C together change the left-handed neutrino to a right-handed antineutrino. The last line in our list of symbolic decays represents what is actually seen in the decay of the negative pion. The PC transformation applied to a physically allowed process yields a process that is also physically allowed. Through this and a number of other examples, it has been verified that even the undisciplined weak interactions do not grossly violate the combined PC invariance. Unfortunately for the view of simplicity in nature, for which the physicist constantly struggles, the weak decay of the kaon has shown combined PC invariance to be only approximately, not absolutely, valid.

To complete the TCP picture, we list the results of two more kinds of inversion on the process of positive pion decay:

T:	$\nu_{\mu L} + \mu_L{}^+ \rightarrow \pi^+,$	Yes.
TCP:	$\overline{\nu_{\mu R}} + \mu_R{}^- \rightarrow \pi^-,$	Yes.

Time inversion alone changes the order of events in the original process. All three inversions together change left to right, particles to antiparticles, and before to after. Each of these transformed processes is almost certainly physically possible, but there is no hope of testing either of them experimentally.

Although the left-handed neutrino is a parity violator and a charge-conjugation violator, it is probably not the answer to the question: *Why* do the weak interactions violate P invariance and C invariance? Some processes that do not involve neutrinos nevertheless violate these conservation laws. It is known, for instance, that, in the decay of the lambda particle,

$$\Lambda^0 \rightarrow p + \pi^-,$$

the proton tends to come off in the direction in which the "north pole" of the lambda was pointing, and the pion in the opposite, or "south pole" direction. This process is indicated in the upper left-hand diagram of Figure 27.14. This asymmetry of decay violates P invariance, for, in the mirror image, the proton is seen to emerge in the south-pole direction, contrary to what would happen if the Λ^0 particle had simply been turned upside down.

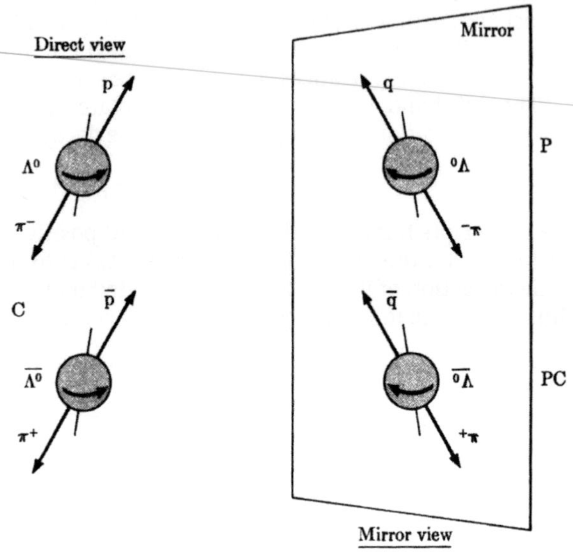

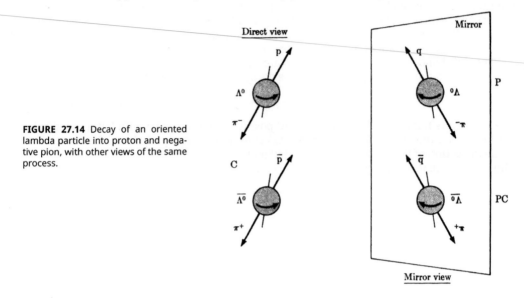

FIGURE 27.14 Decay of an oriented lambda particle into proton and negative pion, with other views of the same process.

The lower left-hand diagram pictures the hypothetical process resulting from particle-antiparticle exchange:

$$\overline{\Lambda^0} \longrightarrow \overline{\mathrm{p}} + \pi^+ .$$

Although the preferred direction of emission of the pion has not yet been observed in the decay of the antilambda, other less direct evidence makes it certain that the pictured process will *not* happen.

Finally, the lower right-hand diagram pictures the mirror image of the charge-conjugation process, representing the combined action of P and C. This is again a physically allowed process, with the antiproton emerging in the direction of the antilambda's south pole.

27.7 *Isotopic spin*

Every fact of nature finds its origin in a conservation law. This appealing, but still unproved, assertion receives some support from the history of our final two conservation laws—strangeness and isotopic spin.

The strange facts of the rapid production and slow decay of the new particles, the existence of some processes and the absence of others, found a simple explanation in terms of the conservation of some new property of matter which was christened *strangeness*. Its underlying invariance principle remains unknown. All we know is that if, to every particle, we assign a strangeness number—in addition to its charge and spin and baryon or electron or muon family number—then the conservation of strangeness in strong interactions serves to account for a large array of experimental facts, in particular, for the absence of a number of processes that should otherwise be observed.

Isotopic spin, with an even more peculiar name than strangeness, is a property of the same type. It is some extra "thing" carried by the strongly interacting particles, and the total amount of this "thing" remains unchanged during all strong interaction processes. Unfortunately, isotopic spin is not just a number that can be assigned to each particle. It behaves like a vector. To make matters worse, the isotopic-spin vector does not point in any direction in ordinary space, but exists in an entirely different space, outside the range of man's perception, something called isotopic-spin space, or simply I space.

It should be obvious that not even the boldest and most imaginative physicist would have dared to suggest seriously concepts like this unless driven to do so by the accumulated weight of experimental evidence. At every step along the road from the common-sense picture of the large-scale world around us to the remarkably different and very noncommonsense view of the world of the very small, man has embraced strange new concepts only when they have afforded the simplest possible explanation of some collection of experimental facts. Sometimes, new experimental facts have suggested directly a new concept, as when the photoelectric effect suggested the photon. Sometimes the new way of looking at nature evolved slowly, and with the help of mathematics. Quantum mechanics, for example, was a complete success as a mathematical description of atomic phenomena well before its full conceptual content was realized. The revolutionary view of the world implied by quantum mechanics was developed over a period of years as more and more of the mathematical consequences of the theory were worked out.

In order to see how the concept of the invisible and not directly knowable I space made its way into physics, we must move back to 1932 and the discovery of the neutron. The previously known particles—the electron, the photon, and the proton—were very different from one another. But the new particle, the neutron, very obviously belonged with the proton. Werner Heisenberg, immediately impressed by the kinship of neutron and proton, seized on the similarities and in the same year showed that these two particles could be regarded as two states of a *single* particle, the nucleon. His mathematical stratagem, described in words, will hardly seem worth the effort. It required the invention of the new I space, in which the nucleon is a vector that may point "up" (which has nothing to do with the "up" of ordinary space) or "down." If it points up, we see it as a proton. If it points down, we see it as a neutron. If two or more nucleons exist together, as in a nucleus, their separate I vectors may add up to a total I vector that may point up, down, or in other possible directions.

According to the Heisenberg theory, the nucleon is a doublet, that is, it can exist in just two possible states, as neutron or as proton. But this was not the first such doublet behavior noticed in nature. A particle with one half unit of spin, such as the proton or the electron, can exist with its spin pointing either up or down (in *ordinary* space). The quantization of orientation restricts its allowed spin directions to two. Heisenberg's method of describing proton and neutron as two states of the same particle was mathematically equivalent to the description of spin, so the new property of the nucleon was also named a kind of spin —a poor name, in fact, since it has nothing whatever to do with ordinary spin. Since the change of a neutron to a proton changes one nucleus to another, and since individual nuclei are called isotopes, the new kind of "spin" was labeled isotopic spin. For a group of nucleons, a rotation of the total isotopic-spin vector in I space corresponds to changing from one nucleus to another without changing the total number of nucleons. I spin, like ordinary spin, is quantized both in magnitude and in orientation.

Several years had to elapse before it could be decided whether Heisenberg's I space was mere mathematical construction or something with real physical content. The decision in favor of its real significance rested on two developments. First of all, as new particles were discovered, they were found to come in closely linked groups, like proton and neutron. There were three pions which formed a triplet and could be described as three states of a single pion with one unit of isotopic spin. The sigma particles also formed a triplet, and the kaons, like the nucleons, a doublet. Thus the kaon was assigned one-half unit of isotopic spin. The lambda particle stands alone as a singlet; it has zero isotopic spin.

Second, and more important, an isotopic-spin conservation law is found to be obeyed by the strong interactions. Technically this means that the probability of any process, or the strength of any interaction, is unchanged if the total isotopic-spin vector is rotated in I space. One consequence of this law is that the interaction between a proton and a positive pion must be exactly the same as that between a neutron and a negative pion, for the change from ($p + \pi^+$) to ($n + \pi^-$) is equivalent to turning over both the nucleon and pion isotopic spins, from "up" to "down."

There are a wealth of consequences of isotopic-spin conservation, but most of them are difficult to comprehend. Roughly, however, the significance of the law can be understood simply as independence of charge. In the strong interactions, nature does not care how much electric charge a particle carries. Proton and neutron are on an equal footing; positive, negative, and neutral pions all interact in the same way. It is as if different charges come in different colors and the strong interactions can see only black and white.

A photon, on the other hand, can see the colors very clearly. It interacts with charged particles, but not readily with neutral particles.* Therefore the electromagnetic interactions violate the law of isotopic-spin conservation. As far as we know now, this is the *only* conservation law not common to the strong interactions and the electromagnetic interactions. The reason? No one knows.

It should not be surprising to learn that the weak interactions also flaunt the law of isotopic-spin conservation. In the decay of the lambda, for example,

$$\Lambda^0 \rightarrow p + \pi^- \qquad \text{or} \qquad \Lambda^0 \rightarrow n + \pi^0 ,$$

the initial isotopic spin is zero. But the final nucleon has one-half unit of isotopic spin and the final pion one unit. These two can combine to a total of either one half or three halves, in either case different from zero. According to present evidence, the lambda chooses the less flagrant violation, going to the final combination with one-half unit.

27.8 Submicroscopic chaos

173 Most of the Feynman diagrams presented earlier in this chapter represent, more or less, what is actually observed. But a few—those involving transitory intermediate particles—do not. Diagram (e) of Figure 27.2, for example, shows an intermediate or "virtual" pion exchanged between two nucleons to produce the force between them. The experimenter sees only the nucleons and must infer from their behavior that in a time of about 10^{-23} sec a pion went from one to the other. You may discover other virtual particles in Figures 27.4 and 27.7.

As emphasized in Section 25.3, virtual particles play a particularly important and interesting role in what is called self-interaction. We want to reexamine this phenomenon now pictorially with the help of Feynman diagrams. A free particle's tendency to interact with itself displays most clearly the new view of the submicroscopic world—a view of continual chaotic activity, from which no particle can be isolated.

Our first guess about the world line of a single free particle (a proton, for instance), sitting motionless and alone in free space, might be the uninteresting vertical line of diagram (a) in Figure 27.15. As far as macroscopic observation goes, that is the whole story—an unchanging, unmoving particle tracing out its straight course through time. Since we know that nucleons and pions interact, we might inquire about the possibility of diagram (b). The proton emits a positive pion and converts itself into a neutron, the fundamental interaction event of the Yukawa theory, as illustrated, for example, by the vertices in Figure 27.2(e).

* If a neutral particle possesses a magnetic moment (as the neutron does), it too may interact with photons, but usually more weakly than a charged particle interacts.

However, it is easy to see that energy conservation prohibits this process for a proton all alone. Since the proton is not moving, its total energy is just its mass energy. But neutron mass plus pion mass add up to considerably more than the proton mass. There is simply not enough energy available to create the neutron and the pion. So, as we knew already, the proton does not disintegrate into other particles.

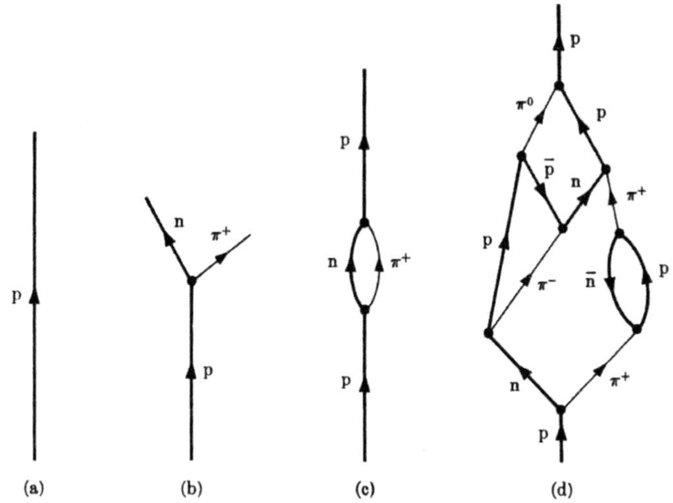

FIGURE 27.15 Feynman diagrams associated with a single isolated proton.

Having rejected diagram (b) and others of its type because of the prohibitions of conservation laws, we are apparently forced back to diagram (a). But as we learned from studying the consequences of the Heisenberg uncertainty principle, nature is willing to overlook a violation of the law of energy conservation, provided the violation lasts a short enough time. The more flagrant the violation, the briefer its duration must be. This casts a new light on diagram (b). Suppose the violation of the law of energy conservation perpetrated by the event of diagram (b) could be caused to last only a short time. It can. If the neutron reabsorbs the pion and becomes again a proton, as in diagram (c), the energy violation has been confined in time. The pion, instead of being allowed to fly off as a free particle, is never quite released. It remains a virtual particle, reabsorbed after about 10^{-23} sec, the longest time the uncertainty principle can allow this violation of energy conservation to endure.

So we must conclude that our isolated proton, if it were possible to view it through a sufficiently high-powered microscope, would be seen to be in a state of some agitation, emitting and reabsorbing pions continually, and existing part of the time as a neutron. This is the state of affairs described in Chapter Twenty-Five without the help of Feynman diagrams. It leads to the image of the proton surrounded by its swarm of virtual pions. The "high-powered microscope" needed to see this pion cloud has actually been supplied in the form of Hofstadter's high-energy electrons.

Given the possibility of transitory violation of the law of energy conservation, all sorts of complex virtual states of motion can arise. Diagram (d) in Figure 27.15 pictures one such sequence of events, quite horrendous looking, but perfectly real. Every proton occasionally goes through exactly this dance of creation and destruction, to emerge unscathed at the other end—as well as through every other tortuous chain that is in consonance with the other conservation laws and with the uncertainty principle. As far as we know, violations of charge conservation and of the three family-number conservation laws are not permitted, even for an instant of time. Therefore, at every vertex in diagram (d), these laws are satisfied. Each vertex involves an "incoming" baryon line, an "outgoing" baryon line, and a pion line. Altogether, diagram (d) includes protons, antiprotons, neutrons, antineutrons, and positive, negative, and neutral pions.

Since even a single particle alone is in such a continual state of agitation, we might ask about the still simpler situation of plain empty space. Field theory provides the answer that empty space, far from being truly void, is a rather lively place. Transitory violations of energy conservation permit particles to be formed out of nothing and vanish again. The "vacuum diagrams" illustrated in Figure 27.16 show some of the things that can (and do) transpire in empty space. The name "physical vacuum" has been given to space filled continually with all of these momentary comings and goings, to distinguish it from the unreal "bare vacuum." In a similar vein, the hypothetical, purely inert particle of diagram (a) in Figure 27.15 is termed a "bare particle" to distinguish it from the real "physical particle" or "dressed particle" that exists part of the time in states of activity such as are pictured in diagrams (c) and (d).

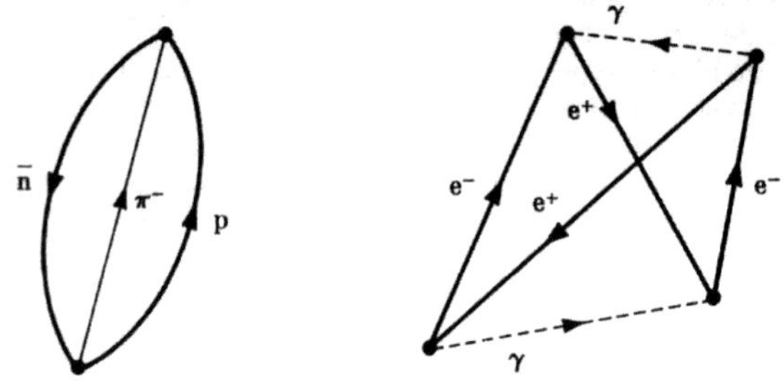

FIGURE 27.16 Vacuum diagrams illustrating the transitory existence of particles in empty space.

By its very complexity, diagram (d) carries a special message about the submicroscopic world, a message of chaos: the *chaos* provoked by the fundamental events of annihilation and creation underlying an *order* imposed by the conservation laws. This theme of order and chaos, already repeatedly touched upon, illustrates as clearly as anything can the complete revolution in our view of the world that has been brought about by the achievements of physical science in this century.

Briefly stated, the new view is a view of chaos *beneath* order—or, what is the same thing, of order imposed upon a deeper and more fundamental chaos. This is in startling contrast to the view developed and solidified in the three centuries from Kepler to Einstein, a view of order beneath chaos. In spite of the haphazard and unpredictable nature of the world around us, ran the old argument, nature's basic laws are fundamentally simple and orderly, and therefore the behavior of nature at the submicroscopic level is fundamentally simple and orderly too. The building blocks of the universe are elementary objects, colorless, unemotional, identical, comprehensible, and predictable, moving in calculable paths, interacting in a known way with other elementary objects.

A modern computing machine illustrates fairly well this classical view of elementary simplicity and orderliness. The basic units of the machine, such as its transistors, are simple objects, each capable of performing only a very elementary function in a predictable and easily controllable way. The electrical engineer might be excused for waxing rhapsodic about the magnificent beauty and simplicity of a transistor and of the laws governing its action. The layman might be excused for regarding it as, on the whole, a rather dull object. But both must agree that when a few million simple units are connected together in the right way, a complicated organism comes into existence, with a rich and rewarding variety of functions and behavior patterns. A few million is a small enough number that the engineer can still undertake to predict just what the machine will do under all circumstances, but even he will be astonished by the complexity engendered by mere size.

Such was the view of the physical world until recent times. Simple objects and simple laws—a basic, underlying orderliness—lay beneath the rich chaos and complexity of the world of our senses. To lay bare the orderliness and probe to ever deeper and simpler layers of reality was the task of science, a task carried on with unprecedented speed and success in the most recent few centuries of man's existence.

But, in this century, the theories of relativity and quantum mechanics and the experimental evidence from the world of elementary particles have combined to reveal a deeper-lying and more fundamental chaos. Particles are found to have a transitory existence; empty space is a beehive of disordered activity; laws of probability have replaced laws of certainty; an isolated particle is engaged in a constant frenzied dance whose steps are random and unpredictable; a principle of uncertainty prevents too close scrutiny or precisely accurate measurements in the world of the very small.

This is not to say that the older idea of simplicity in the small and complexity in the large has been wholly abandoned. An electron is still a simple object, even if far from inert; nature's most fundamental laws still appear to govern the submicroscopic world; and the complexity of size and organization is still very real. The revolution has appeared primarily in shifting the source of order from the elementary interactions and *activity* of particles to the overriding *constraints* of conservation laws. The emerging picture of the world is that of a nearly limitless chaos governed only by a set of constraining laws, a world in which apparently everything that *can* happen, subject only to the straitening effect of these conservation laws, *does* happen. The fields and particles of the submicroscopic world must be regarded as an unruly lot who carry on in every conceivable way that is not absolutely forbidden by the overriding restriction of a conservation principle.

Is this fundamental chaos of nature a temporary phenomenon in science that will be replaced by a deeper order in the future? Perhaps. There is no evidence at all upon which to base an answer to this question, but two main possibilities need to be cited. On the one hand, an elementary event of creation and annihilation which now appears to occur catastrophically at a single point in space-time may, upon closer inspection, prove to be a swift but smooth, and more orderly, unrolling of a chain of events. The probability of quantum mechanics may prove to rest only upon the great complexity of the things we now regard as simple. On the other hand, our view of the world of the very small could easily become more chaotic, not less so. Lying dormant thus far in our view of the world is space-time itself. While fields and particles come and go, space and time lie inert, providing the stage upon which the actors play their roles. There is some reason to believe that the future theory of particles may involve space and time as actors, not merely as stage. If so, weird convolutions of space and time and/or the quantization of space-time may contribute more to the chaos in our view of the world.

In whatever direction the future theory of particles proceeds (and speculation on this score is rather an idle pastime, unlikely to bear fruit), it must be emphasized that present theory is much more likely to be supplemented than to be rejected. Just as Newtonian mechanics is still entirely adequate for describing the motion of planets, present theories of particles are likely to remain adequate for describing all those features of the particle world that have been understood quantitatively so far. Nevertheless, it is the deepest theory that most strongly affects our *image* of the world, and this image may be drastically altered in the future.

In the seventeenth century, man looked upward and outward into the universe and was humbled as his earth took its diminutive place as a speck of matter in a corner of the cosmos. In this century, we look downward and inward and find new reasons for humility. Where we might have expected to find some firm lumps of matter as the building blocks of man and his world, we find a chaos of annihilation and creation, a swarm of transitory bits of matter, and the tenuous substance of wave fields. Where we might have expected to find laws of certainty, we find laws of probability, and seem to see the hand of chance working at every turn—chance that any particles are stable, chance that the neutron can live forever within a nucleus, chance that we are free of the threat of annihilation by antiparticles. Above the chaos and the probability stand the conservation laws, imposing their order upon the undisciplined energy of the universe to make possible the marvelously intricate, incredibly organized structures of the world around us.

EXERCISES

27.1. Sketch world-line diagrams for the following sequences of events in the macroscopic world. **(1)** A vertically fired shell explodes into three fragments which fly off to the left and right. Include only horizontal components of motion in the diagram. **(2)** A baseball player intending to bunt holds his bat stationary. Ball and bat remain briefly in contact. The ball then reverses the direction of its velocity and the bat recoils. **(3)** A cargo spacecraft drifts toward an orbiting laboratory. A mail bag is tossed from one to the other and they drift apart. For each of these three situations, name an elementary particle phenomenon characterized by a very similar world-line diagram.

27.2. In each of the family number conservation laws, particles are assigned positive family number and antiparticles are assigned negative family number. Discuss this fact in terms of the vertices of Feynman diagrams, explaining how this sign convention is simply and logically connected to the behavior of lines coming into and going out of the fundamental interaction vertices.

27.3. Feynman diagrams illustrate many of the conservation laws in a clear pictorial way. **(1)** Explain how the form of the basic electron-photon vertices in Figures 27.5 and 27.6 shows the conservation of charge and of electron family number. **(2)** Find Feynman diagrams in this chapter that illustrate two other conservation laws.

27.4. **(1)** What physical process is represented by this Feynman diagram? Write a reaction formula for it. **(2)** What physical process is represented if the charge conjugation operation C is applied to the diagram? Is it possible? Is it likely to be observed in practice? **(3)** What physical process is represented if the time reversal operation T is applied to the diagram? Is it possible? Is it likely to be observed in practice?

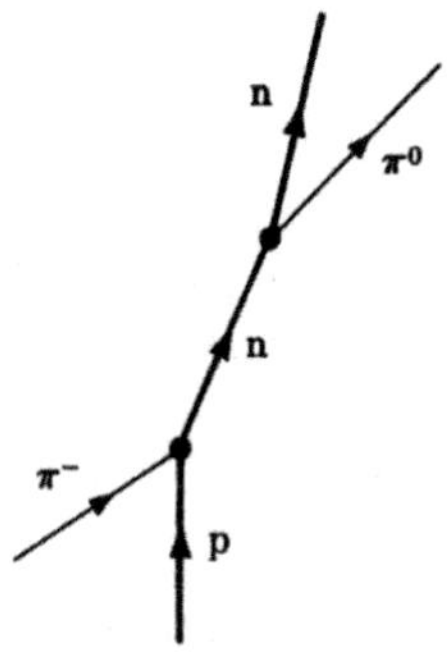

27.5. The Feynman diagram in Exercise 27.4 represents a physically allowed process. What is the implication of the TCP theorem as applied to this process?

27.6. Planetary motion is an example of a process in the macroscopic world that possesses time reversal invariance. **(1)** Name one other macroscopic process that exhibits, at least approximately, time reversal invariance. **(2)** Name two macroscopic processes that violate time reversal invariance.

27.7. Name **(a)** a specific object in the macroscopic world whose mirror image depicts something that does exist in the world; and **(b)** a specific object in the macroscopic world whose mirror image depicts something that does not exist in the world. Explain why the existence of the latter object does not violate the law of space-inversion invariance.

27.8. **(1)** Explain how satellite orbits possess mirror invariance. **(2)** Invent a rule for satellite orbits that violates mirror invariance, (HINT: To violate mirror invariance, the satellite must move in three dimensions, not two.)

27.9. If an antineutron could avoid annihilation long enough to undergo spontaneous decay, **(1)** what would be the products of its decay? **(2)** What would be its lifetime?

27.10. A negative kaon decays sometimes into a negative muon and one other particle. **(1)** What is the other particle? **(2)** Is the spin of the muon constrained to point in a certain direction? If so, what direction? **(3)** This decay must involve the weak interaction. Why?

27.11. An antineutrino strikes a proton and stimulates the following inverse beta decay reaction:

$$\overline{\nu_{eR}} + p \rightarrow n + e_R^+ .$$

The subscript R indicates right-handed motion. **(1)** What is the charge conjugate process? **(2)** Is it possible in principle? Why or why not? **(3)** Is it possible in practice? Why or why not?

27.12. What is the isotopic spin of **(a)** the xi particle; **(b)** the omega particle? Give a reason for each answer.

27.13. Discuss the reason why the strong interaction between a Σ^+ and a K^+ is identical to the strong interaction between a Σ^- and a K^0, whereas the electromagnetic interactions between these two pairs are not the same.

27.14. An electron-positron pair materializes out of nothing in vacuum. **(1)** About how long a time may elapse before this pair annihilates itself? **(2)** Over how great a distance in space can this transitory phenomenon spread itself? Compare this distance with the size of an atom and with the size of a nucleus.

22.8. (1) Explain [illegible] (2) [illegible] three dimensions [illegible]

22.9. [illegible] (1) [illegible] (2) [illegible]?

22.10. A [illegible] and one other particle. (1) What is the [illegible] particle? (2) In the space of the [illegible] constrained to [illegible] in a [illegible] (3) [illegible] involve the weak interaction? Why?

22.11. An antineutrino strikes a proton and stimulates the following inverse beta decay reaction:

Epilogue

BERTRAND RUSSELL
(1872 – 1970)

"We know very little, and yet it is astonishing that we know so much, and still more astonishing that so little knowledge can give us so much power."

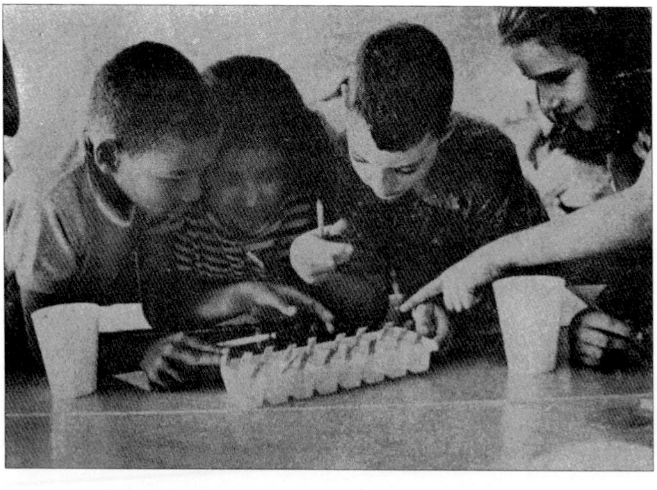

TWENTY-EIGHT

The progress of science

Thrilling insights and awesome power. These are the fruits of science. In this final chapter, we shall examine the nature and progress of man's most successful enterprise, scientific inquiry, and hazard some guesses about its future. 174

To science we owe most of our comforts, our leisure, our health and longevity, our ability to mold the environment, to communicate instantly, and to move swiftly over the earth. To science we also owe our ability to wipe out populations with devices of mass destruction and to numb populations with devices of mass communication. What are the features that set science apart from other human activity? What are the reasons for its remarkable growth? What does the future hold for scientific progress? We can give at best partial answers to these questions, but one thing seems clear: Future scientific understanding can be no better legislated and controlled than can any other creative activity of man.

What might be called the defining feature of science is its quantitative character. By "quantitative" we mean not merely numerical, but mathematical in a general sense—subject to rules of logic and order, and reproducible. A heckler could find some examples of nonquantitative science and of quantitative nonscience, but in the main it is this quantitative aspect that distinguishes science from other avenues of search for understanding. Because of this feature, science gives to man the power to predict and to control, not merely the power to describe.

Paradoxically, it is the modest goals of science that are most responsible for its mushrooming growth. It is commonplace to refer to the vast scope and generality of natural laws, which are indeed magnificent and impressive. Yet it has to be borne in mind—and the scientist is perhaps particularly conscious of this fact— that of all the important questions man has posed to himself in recorded history, science has so far provided answers only to the easiest ones. Science encompasses a limited range of human experience, yet with no other range of experience has human endeavor been so successful.

Besides the self-imposed limitation to "easy" questions, another key to understanding the progress of science is the generality of application that springs from an economically few fundamentals. This "amplification factor" (or "leverage" as it is called in the business world) could as well be called a quality of nature as a property of science. Man has found himself able not only to describe nature quantitatively, but to do so with relatively few basic ideas and relationships. From each of the few important theories and laws in physics flows a wealth of application. The blossoming of a whole area of technology in a few years' time can spring from the discovery of a single fundamental fact. What is just one significant advance in science may appear as an incredible rapid-fire series of advances because of the latent power of each forward step.

Has the growth of science actually been "explosive"? It is commonly said to be so. In the long view of human history it certainly seems explosive. In this century both the accumulation of scientific facts and the growth of technology merit the adjective "explosive." But the pace of scientific progress can be overemphasized. It is important to distinguish between merely *rapid* growth and *accelerating* growth. The expansion of the world's population is an example of accelerating growth. Every decade the increase is greater than in the previous decade. There is indeed a population "explosion." Without question, the most evident aspects of science—its catalogue of facts and its application for practical goals—have also shown accelerated growth. So have the number of scientific workers and the amount of money spent on science. Nevertheless, there is little evidence in recent history for an accelerated growth of fundamental understanding of nature. It is more accurate to think of science as having undergone a metamorphosis in the seventeenth century which has led to its continual steady (and rapid) growth since then, rather than to think of an ever accelerating pace of progress.

The true landmarks of scientific progress are the occasional revolutions in our view of the workings of nature and the enlargements of our horizons in the physical world. Such key points of progress in the twentieth century have been the new view of space, time, and gravitation initiated by the theory of relativity; the elucidation of atomic structure through the theory of quantum mechanics; the discovery of the subatomic world of transitory particles; the elevation of principles of invariance to a primary position among the laws of nature; and the discovery of the molecular structure of the basic units of living matter. This last advance, representing the union of physical science and biological science at the submicroscopic level, may prove to be as important as any in this century. Looking back to the nineteenth century, one finds that revolutions of scientific thought occurred nearly as frequently as in the present century. In the nineteenth century the theory of electromagnetism revealed the nature of light and the unity of electricity and magnetism; the kinetic theory of matter, besides explaining the nature of heat and temperature, revealed the turmoil of random molecular motion underlying our apparently solid material world; the theory of thermodynamics brought scientific precision to the concepts of order and disorder and provided a basis for understanding why we experience only a one-way flow of time; the analysis of all matter into a small number of elements and the orderly arrangement of these elements into the periodic table brought a new simplicity to the view of the physical world; the universal scope of the law of conservation of energy brought hope (and some overoptimism) for the construction of a single overriding theory of natural phenomena; the theory of natural selection revolutionized man's view of human history; the discovery of laws of genetics showed the existence of quantitative simplicity in the living world as well as in the inanimate world.

It is, of course, impossible to measure exactly the relative importance of different advances in science or to arrive at any unique list of "truly fundamental" advances. The purpose of the brief catalogue of scientific progress above is to emphasize two facts about modern science. First, the technological marvels of the present day and the ever greater human resources poured into scientific research should not be confused with true progress in understanding the design of nature. The list is notably deficient in inventions (radio) or technical feats (space flight) or in reference to mere accumulation of data. Second, the rate of generation of fundamental ideas in science has shown no marked trend of acceleration, at least over the past 150 years. Progress in both the nineteenth and the twentieth centuries has been rich, but scarcely any richer at the present time than 100 years ago. It seems that any important new idea requires some time to be thoroughly assimilated and appreciated before it can serve as the basis of further advance. Perhaps the instant communication and high-speed travel of the present era serve to stimulate scientific progress through cross-fertilization of ideas. However, it is clear—and it is important to keep in mind—that the pace of fundamental scientific progress is by no means comparable either to the rate of technological development or to the input of human effort.

Can history teach us the best way to push forward the frontiers of understanding in science? Probably very little. In discussing the future of science, scientists agree only that the future is unpredictable. Past progress has revealed no single scientific method. Idealized versions of scientific induction or of the sequence of experimental and theoretical steps in the evolution of a theory have borne little resemblance to actual progress so far. But two of the aspects of past scientific progress stand out as features likely to persist. (1) Theory and experiment have both been necessary. On the one hand, the mere accumulation of data has been barren without the insight provided by hypothesis and theory. On the other hand, flights of fancy untempered by experiment have not been fruitful. (2) All of the great advances have been but small forward steps in the overall view, resting heavily on what has gone before. In discussing an important advance in science, there is a natural human tendency to emphasize its novelty or the creative genius which it reflects. The less heralded developments which preceded and made possible the giant stride tend to be forgotten.

The modern world abounds in scientific cranks, men who ignore one or the other of these paramount aspects of science, seeking radical new departures in science without reference to experiment, or without reference to the past stream of scientific thought. Notwithstanding the oft-repeated charge by the cranks that the "high priests of science" have closed minds, most scientists try to keep an open mind about the possible direction from which future progress might come. Anyone with an idea in science can get a hearing. However, past history makes it seem likely that the fruitful ideas will be generated by men who build upon their own deep comprehension of past achievement, not by men who attempt the great leap from no firm foundation.

It is easy to predict one near-certainty about the future, that scientific progress will continue. This sounds like an obvious statement, yet there have been, times in the past when both mathematicians and physicists thought that their fields were drying up. In contemporary science, complacency about the state of our understanding is completely absent. Every practicing scientist is painfully aware of his areas of ignorance at the frontiers of science. Although a future unrolling of scientific progress accompanying ever deeper understanding of nature is easy enough to foresee, it is more fascinating to consider how future science might differ from past science. Is science getting more difficult to comprehend, and may progress therefore be slowed or stopped? Will the scope of science increase to make it encompass a larger fraction of man's experience?

Fundamental science is undoubtedly becoming more difficult. The new concepts are further removed from everyday experience and are less easy to visualize. The world of man's senses changes very little. The world of the fundamental theories of nature has been changing rapidly. Albert Einstein and Leopold Infeld* expressed this trend in these words: "The simpler and more fundamental our assumptions become, the more intricate is our mathematical tool of reasoning; the way from theory to observation becomes longer, more subtle, and more complicated." A warning is sounded for the future. Will the progress of science be slowed by the ever longer route which man must follow from the world of his senses to the elementary world of nature's design? So far, we humans have shown ourselves adept at learning to think in terms foreign to our direct sense experience. We believe in the annihilation and creation of matter, the wave nature of particles, the relativity of time, the quantum of energy, and the curvature of space-time. How far this adaptability will carry us no one can predict.

About the future *scope* of science there can be little doubt. The range of experience encompassed by science has increased gradually, and a further enlargement of what comprises science is almost certain. Although the expansion of biology and the incorporation into science of parts of psychology will have the greatest impact on man, growth has occurred and will continue within physical science itself. A few centuries ago, the question, Why does the sun give out light? had no answer outside of religion. A hundred years ago, the question, Why is energy conserved? belonged to philosophy. Both questions are now answered within the framework of physics. Other scientific-sounding questions are easy to pose which in fact lie outside of science today but may be incorporated in the future: Is the universe boundless? What was going on fifty billion years ago? What is the true nature of time? How did the matter in the universe come into existence?

In the realm of biology, consider disease, which has moved from anger of the gods to a very scientific matter of germs and viruses. What the future undoubtedly holds is scientific

* *The Evolution of Physics* (New York: Simon and Schuster, 1950).

understanding of some patterns of human behavior. The past problems raised by science have been mainly technological—industrial automation, and implements of war. The future problems are likely to be more directly concerned with man's behavior and are sure to be even more vexing. It is probably wiser to prepare for an expanding scope of scientific understanding—which means the power to predict and control—than to cry out for the bliss of past ignorance. Fortunately the human being is a mechanism of fantastic complexity. It is unlikely that science will ever impinge on the pleasurable uncertainties that characterize some human relations.

Upon the principles of physics rests the whole of science—biological as well as physical—whose growth provides ever deepening understanding and therefore ever broadening responsibility, and ever increasing opportunity for betterment of the human condition.

EXERCISES

28.1. Which of the following are quantitative statements: **(1)** $F = ma$. **(2)** Faulkner is a greater novelist than Hemingway. **(3)** Sound travels at about 300 meters per second through air. **(4)** Dogs can hear higher frequencies than can people. **(5)** Philosophers of the eighteenth century were one hundred per cent more effective than philosophers of the nineteenth century.

28.2. It was stated in Chapter One that fundamental simplicity does not imply ease of comprehension. Illustrate this point with several specific facts or theories of physics.

28.3. Einstein has characterized a great theory as one possessed of "inner perfection" and verified by "external confirmation." Explain carefully what these phrases mean. To what extent does any one of the fundamental theories of physics fit Einstein's criteria?

28.4. **(1)** Name two "hard questions" that lie outside the scope of science. **(2)** Give one question, the answer to which now lies outside of science, but might reasonably be expected later to lie within science.

28.5. Name **(a)** something *not* in science concerned with inanimate matter; **(b)** something *not* in science concerned with cataloging of facts; **(c)** something *in* science concerned with human life; **(d)** something *in* science concerned with philosophy.

28.6. Science is the discovery of the reality of nature. It is also the creation of a simple description of nature. Based on specific things you have learned about physics, write briefly in support of the thesis that these two views of science are consistent and both correct.

28.7. Kepler discovered his first two laws of planetary motion at about the same time that Shakespeare wrote *King Lear*. Support the proposition that the lives of modern students are more influenced by Kepler than by Shakespeare, even though many students have read Shakespeare and few have studied Kepler.

28.8. Following are two brief definitions of science: **(1)** Science is what scientists do. **(2)** Science is the quantitative part of man's creative activity. Discuss the strengths and weaknesses of either one of these definitions. *Optional:* Make up an equally succinct definition of science of your own, and discuss it.

28.9. Past scientific advance has resulted from a constant interplay between theory and experiment. Do you think that either theory or experiment will play a more dominant role in the future progress of physics? Which one, and why?

Appendices

APPENDIX ONE

Greek alphabet

Α	α	alpha
Β	β	beta
Γ	γ	gamma
Δ	δ	delta
Ε	ε	epsilon
Ζ	ζ	zeta
Η	η	eta
Θ	θ	theta
Ι	ι	iota
Κ	κ	kappa
Λ	λ	lambda
Μ	μ	mu
Ν	ν	nu
Ξ	ξ	xi
Ο	ο	omicron
Π	π	pi
Ρ	ρ	rho
Σ	σ	sigma
Τ	τ	tau
Υ	υ	upsilon
Φ	φ	phi
Χ	χ	chi
Ψ	ψ	psi
Ω	ω	omega

APPENDIX TWO

Numerical data

A. *Constants of Nature*

Constant	Symbol	Value
Gravitational constant	G	6.67×10^{-8} dyne cm^2/gm^2 (or cm^3/gm sec^2)
Avogadro's number	N_0	6.0225×10^{23} (particles/mole or a.m.u./gm)
Boltzmann's constant (microscopic gas constant)	k	1.3805×10^{-16} erg/°K
Macroscopic gas constant	$R = (N_0k)$	8.3143×10^{7} erg /°K
Quantum unit of charge	e	4.8030×10^{-10} e.s.u. or 1.6021×10^{-19} coulomb
Faraday constant (one mole of electricity)	$F = (N_0e)$	2.8926×10^{14} e.s.u. or 9.6487×10^{4} coulomb
Speed of light	c	2.99793×10^{10} cm/sec
Planck's constant	h	6.6256×10^{-27} erg sec or 4.1356×10^{-21} MeV sec
	$\hbar = (h/2\pi)$	1.0545×10^{-27} erg sec or 6.582×10^{-22} MeV sec
Mass of electron	m_e	9.109×10^{-28} gm or 5.4860×10^{-4} a.m.u.
Mass of proton	m_p	$1.67\,25 \times 10^{-24}$ gm or 1.0072766 a.m.u.
Mass of neutron	m_n	1.6748×10^{-24} gm or 1.008665 a.m.u.
Intrinsic energy of electron	m_ec^2	0.5110 MeV
Intrinsic energy of protron	m_pc^2	938.26 MeV
Rydberg constant for infinitely massive nucleus	$\mathscr{R}_\infty = \left(\frac{m_e e^4}{4\pi\hbar^3 c}\right)$	1.097373×10^{5} cm^{-1}
Rydberg constant for hydrogen	$\mathscr{R}_H$	1.096776×10^{5} cm^{-1}
Fine structure constant	$\alpha = \left(\frac{e^2}{\hbar c}\right)$	7.2972×10^{-3} or 1/137.039

B. *Terrestrial Data*

Acceleration of gravity near the earth's surface (g)	980 cm/sec^2 (varies from 978 cm/sec^2 at equator to 983 cm/sec^2 at poles)
Mass of earth ($M_\oplus$)	5.98×10^{27} gm
Radius of earth ($R_\oplus$)	6.37 × 10s cm or 3960 miles (varies from 6378 km at equator to 6357 km at poles)
Equatorial circumference of earth	4.008×10^{9} cm or 24,900 miles
Standard air pressure at sea level	1.013×10^{6} dyne/cm^2
Standard dry air density at sea level and 0°C	1.293×10^{-3} gm/cm^3
Speed of sound in standard air at sea level and 0°C	3.31×10^{4} cm/sec or 740 miles/hr

C. *Astronomical Data*

Distance from center of earth to center of moon	3.844×10^{10} cm or 2.389×10^{5} miles
Period of moon	27.32 days or 2.360×10^{6} sec
Mass of moon	7.34×10^{25} gm
Radius of moon	1.738×10^{8} cm
Acceleration of gravity at the surface of the moon	167 cm/sec^2
Distance from center of earth to center of sun (1 A.U.)	1.496×10^{13} cm or 9.30×10^{7} miles
Mass of sun ($M_\odot$)	1.987×10^{33} gm
Radius of sun ($R_\odot$)	6.96×10^{10} cm
Period of earth	365.26 days or 3.156×10^{7} sec
Average orbital speed of earth	2.98×10^{6} cm/sec
Average orbital acceleration of earth	0.593 cm/sec^2

APPENDIX THREE

Conversion of units

For convenience in units arithmetic, this appendix lists conversion factors directly (such as 2.54 cm/in.) rather than equations (such as 1 inch = 2.54 cm). Any quantity can be multiplied or divided by appropriate conversion factors, since each conversion factor is equivalent to unity.

1. Length

2.54 cm/in.

30.48 cm/ft

10^5 cm/km

1.609×10^5 cm/mile

9.46×10^{17} cm/light-year

10^{-8} cm/Angstrom

10^{-13} cm/fermi

2. Time

3600 sec/hour

3.156×10^7 sec/year

3. Speed

44.7 (cm/sec)/(miles/hr)

4. Angle

57.30 (= $180/\pi$) deg/radian

60 sec/min

60 min/deg

6.283 (= 2π) radian/revolution

5. Mass

453.6 gm/lb

1.6604×10^{-24} gm/a.m.u.

6. Force

10^5 dyne/newton

7. Energy

10^7 erg/joule

4.187×10^7 erg/calorie

3.60×10^{13} erg/kilowatt hour

1.602×10^{-12} erg/eV

1.602×10^{-6} erg/MeV

4.2×10^{19} erg/kiloton

0.0434 (eV/molecule) / (kilocarie/mole)

8. Power

10^7 (erg/sec)/watt

746 watt/horsepower

9. Temperature

1.80 °F/°C

T(°K) = T(°C) + 273.15

10. Electrical quantities

3.00×10^9 e.s.u./coulomb

3.00×10^9 (e.s.u./sec)/ampere

300 volts/statvolt

INDEX

For names and topics that appear often in the book, only the more important page references are given. Italic letters following page numbers are used with the following meanings: *e*, exercise; *f*, figure; *n*, footnote; and *t*, table. An exercise is referenced only if it contains factual information that does not occur elsewhere in the book.

END OF ORIGINAL BOOK

BASIC PHYSICS

KENNETH W. FORD

FEATURES INDEX